Fading and Shadowing in Wirel

P. Mohana Shankar

Fading and Shadowing in Wireless Systems

Second Edition

P. Mohana Shankar
Electrical and Computer Engineering
Drexel University
Philadelphia, PA
USA

ISBN 978-3-319-85087-0 ISBN 978-3-319-53198-4 (eBook)
DOI 10.1007/978-3-319-53198-4

© Springer International Publishing AG 2012, 2017
Softcover reprint of the hardcover 2nd edition 2017
This work is subject to copyright. All rights are reserved by the Publisher, whether the whole or part of the material is concerned, specifically the rights of translation, reprinting, reuse of illustrations, recitation, broadcasting, reproduction on microfilms or in any other physical way, and transmission or information storage and retrieval, electronic adaptation, computer software, or by similar or dissimilar methodology now known or hereafter developed.
The use of general descriptive names, registered names, trademarks, service marks, etc. in this publication does not imply, even in the absence of a specific statement, that such names are exempt from the relevant protective laws and regulations and therefore free for general use.
The publisher, the authors and the editors are safe to assume that the advice and information in this book are believed to be true and accurate at the date of publication. Neither the publisher nor the authors or the editors give a warranty, express or implied, with respect to the material contained herein or for any errors or omissions that may have been made. The publisher remains neutral with regard to jurisdictional claims in published maps and institutional affiliations.

Printed on acid-free paper

This Springer imprint is published by Springer Nature
The registered company is Springer International Publishing AG
The registered company address is: Gewerbestrasse 11, 6330 Cham, Switzerland

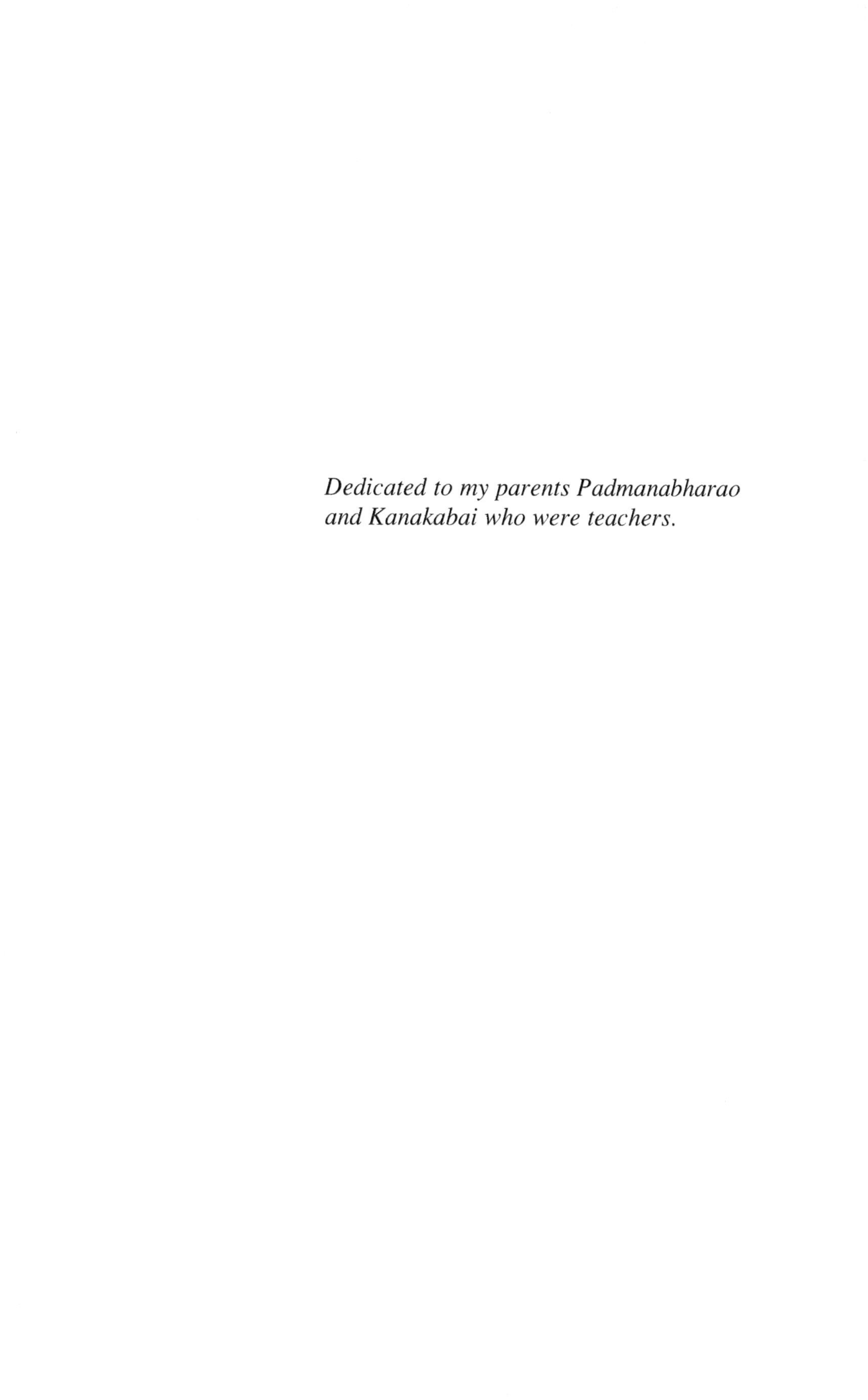

Dedicated to my parents Padmanabharao and Kanakabai who were teachers.

Preface to the Second Edition

While maintaining all the contents from the first edition, the updated edition provides an expanded view of the subject matter in modeling of fading, shadowing, and shadowed fading as well as diversity techniques. In order to provide a didactic component, more background has been provided to expand the concept of random variables (Chap. 2), newer models in fading and shadowing (Chap. 4), and analysis of diversity (Chap. 5). All these new sections now also provide appropriate Matlab scripts necessary to follow and interpret the results. A new chapter on cognitive radio (Chap. 7) has been added to complete the description of fading, shadowing, and shadowed fading in wireless systems.

It would not have been possible to complete this book without the support and assistance provided by my wife Raja and our daughter Raji. Their diligence and active participation were vital for the successful and timely completion of the project.

I thank my editor Ms. Mary James for all her support and prompt actions to keep moving the project along. The support provided by Springer, particularly Mr. Brian Halm and his colleagues in USA and Mr. T. Murugesan, R. Santhamurthy and their team in India, is also gratefully acknowledged.

I extend my sincere appreciation to Drexel University for providing the sabbatical leave during the current academic year.

Philadelphia, PA, USA P. Mohana Shankar

Preface to the First Edition

During the past two decades, there has been a substantial growth in research in wireless communications. The number of journals published from various parts of the world catering to the research community has grown exponentially. Despite such a growth, the engineering community still needs more information so as to thoroughly comprehend wireless channel characteristics. What specifically must be understood are the effects of channel degradation brought on by statistical fluctuations in the channel. This must be grasped fully and comprehensively before successful attempts can be made to mitigate the channel impairments. Such statistical fluctuations do manifest as variations in signal powers, which are observed in the channel generally modeled using a variety of probability distributions, both in straight forms and in compound forms. While the former might explain some of the effects, the latter, namely, the compound models, which incorporate both short-term and long-term power fluctuations in the channel, help explain the much more complex nature of the signals in these channels. Often, we see newer and more powerful models being proposed, presented, and tested to see how they fit the observed power fluctuations.

This book addresses the needs of graduate students and instructors who are exploring the various aspects of power fluctuations, which are generally expressed in terms of fading, shadowing, and shadowed fading channels. This work grew out of my studies and explorations during a two-quarter sabbatical (September 2009–March 2010) granted by Drexel University. The chapters are based on both my teaching and research into the statistical aspects of wireless channels. Specifically, this work focuses on the analysis and study of several models currently available in the literature of wireless communications (books, archival journals, and conference proceedings). This expansive research describes the statistical characteristics of the signals that account for the fading, shadowing, and shadowed fading seen in these channels. The book is not intended as a catalog or encyclopedia of fading and shadowing. Indeed, the thrust of the book is a pedagogical approach to the topics of fading and shadowing. It provides insight into the modeling and analysis of fading and shadowing. Starting with statistical background and digital communications, the chapters are formulated to follow the details of modeling of the statistical fluctuations of signals in these channels. The degradations in the channels arising from the statistical fluctuations are quantitatively described in terms of various measures. This is followed by the discussion of diversity and associated signal processing algorithms that mitigate the effects of statistical fluctuations in the channel and the quantitative measures of improvements brought on by diversity. The book also examines the effects of interference from other channels. It is my expectation that this work will provide an in-depth and unique coverage of topics for graduate-level study in wireless communications.

The book would not have been possible without the full support of my wife Raja and our daughter Raji. Besides reading the early drafts of the chapters, they provided insights into the chapter organization and pointed out the need for further explanation. Their efforts made it possible to complete the project in a reasonable period; I am proud to say that the book has been a family project.

I am very grateful to our friend Ms. Maura Curran who graciously agreed to proofread the chapters on very short notice, despite having a full workload as a compositor and editor.

I thank my editor Mr. Brett Kurzman. Without his enthusiastic support, this work could not have been undertaken. The support provided by Springer, particularly Mr. Brian Halm and his colleagues in New York and Mr. D. Raja and his team at SPi Global, is acknowledged. I also extend my grateful appreciation to Drexel University for their support and cooperation.

Philadelphia, PA, USA P. Mohana Shankar

Contents

Chapter 1
Overview

1.1 Outline

Wireless communications are ubiquitous, covering all aspects of everyday life from simple voice communication to vital transmission of critical information in battle fields and healthcare complexes. This expansive use of wireless channels is an important tool for fast and efficient means of data transmission. It requires a careful study of the characteristics of the channel so that the communication can be maintained with high fidelity and reliability. In this context, the study of the signal transmission and deterioration in the signal quality and characteristics as they pass through these channels assumes great significance. In this book, the signal strengths in these channels are described using appropriate statistical models. Such modeling and study of the models present a challenge to the instructors, graduate students, and researchers who are in the forefront of developing techniques to improve signal quality and to enhance overall reliability of the communication link between the sender and the recipient. This book is an effort to address this challenge. It provides a thorough discussion of the models used to describe signal strength fluctuations. It also examines the diversity techniques which are developed to mitigate the effects of these fluctuations.

The fundamental issues described in this work can be summarized in a few simple equations that relate the transmitted and received signals. The received signal in an ideal channel can be expressed as

$$r = s + n. \tag{1.1}$$

In (1.1), r is the received signal, s is the transmitted signal, and n is the noise. Noise n is typically Gaussian distributed with zero mean. We expect the received signal strength r to have a mean value of s as per (1.1). However, measurements in wireless channels suggest that the observed fluctuations do not follow such a simple pattern and that the representation in (1.1) is inadequate. Taking into account the

© Springer International Publishing AG 2017

P.M. Shankar, *Fading and Shadowing in Wireless Systems*,
DOI 10.1007/978-3-319-53198-4_1

random nature of the fluctuations observed, a more appropriate way to represent the received signal in a wireless channel is

$$r = As + n. \tag{1.2}$$

In (1.2), A represents the fluctuations in the channel. In terms of accurately describing the characteristics of the channel, it is necessary to have the best fit of the statistical distribution of A to the measured or observed data. Such an exploration has led to the development of a number of different (statistical) models to describe the fluctuations in the signal strengths observed, with each new model providing better and better matches to the experimental observations. While the model described through (1.2) provides reasonable fits to the observed data in some of the geographical regions, it was shown to be insufficient to comprehensively describe the signal fluctuations in all the geographical regions. To address such anomalies, (1.2) can be modified as

$$r = ABs + n. \tag{1.3}$$

In (1.3), a new parameter B has been introduced with its own characteristic statistical behavior. Thus, by having two different statistical distributions representing A and B in (1.3), it might be possible to increase the accuracy of modeling the signal fluctuations taking place in wireless channels. In terms of standard terminology in wireless communications, (1.1) represents an ideal channel, (1.2) represents a fading channel, and (1.3) represents a shadowed fading channel. As in the case of A, it is possible to explore various statistical distributions for B to provide the best possible fit for the observations and make the representation in (1.3) reasonably versatile. The product concept presented in (1.3) can be further extended to a cascaded one indicating the product of three or more parameters as in

$$r = \left[\prod_{k=1}^{N} A_k\right] s + n. \tag{1.4}$$

In (1.4), N is the number of cascaded components. One can view (1.3) as a special case of (1.4) with $N = 2$ and (1.2) as a special case with $N = 1$. Note that (1.1) is another special case of (1.2) when the density function of A is a delta function.

As illustrated through the equations above, this work provides descriptions of the models to fit the observed signal strength fluctuations along with the rationales for their appropriateness. Their inadequacies necessitate the need for newer models and appropriate measures to evaluate the performance of the signal transmission through the wireless channels. Such a detailed view is essential so that strategies might be developed to mitigate the adverse effects of shadowed fading channels. Mitigation is achieved through diversity implementation. The book explores the characteristics of signals in detail with each of the models available to describe fading and shadowing before and after the implementation of diversity.

Chapter 2 provides an advanced review of random variables with specific emphasis on probability densities and cumulative distribution functions (CDFs) (which are of particular interest in wireless communications). A number of densities and distributions are presented and their associations with the study of fading channels are clearly enunciated to illustrate the relationships among various densities and their special forms.

This chapter also provides several examples of interest in wireless communications including functions of multiple random variables, derivation of their densities, and distribution functions. Special attention has been paid to the functions of random variables generated from variables of gamma type. This includes cascaded ones since several models of fading and shadowing make use of gamma-related random variables. Order statistics is also discussed in detail. Its relationship to the selection combining as well as GSC algorithms (discussed later in the text) is shown. Several bivariate random variables of interest in wireless communications are introduced. Statistical decision theory is presented for the case of signal detection in Gaussian as well as non-Gaussian noise. This chapter also includes a presentation on Chebyshev inequality and Chernoff bounds.

An overview of the digital communications is given in Chap. 3. Starting with the basics of pulses, pulse shapes, and their spectra, the various modems which form the core of the wireless communications are presented. These modems include both linear and nonlinear forms of modulation as well as binary and M-ary modulation schemes. The specific waveforms of the modulated outputs are illustrated so as to emphasize the differences and similarities among the modems. The calculation of error rates, the effects of phase mismatches, problems with timing, and the overall impact of these factors on the error rates are explored. Chapter 3 contains detailed discussion of the spectra associated with the various modems. Orthogonal frequency division multiplexing (OFDM) techniques which allow greater flexibility in achieving higher data rates while providing some fading mitigation are presented. All the special functions of interest such as complimentary error functions, Q functions, Marcum's Q functions, and generalized Marcum's Q functions are included, along with such topics of interest as noise and eye patterns.

Chapter 4 provides a complete description of all the models currently available for the statistical characterization of the signal strengths in wireless channels. Starting from the basic concepts of multipath propagation phenomenon, the chapter first presents the simple Rayleigh model for the envelope of the received signal. The importance of the phase of the received signal is demonstrated in connection with the description of the multipath channels when a direct path between the transmitter and receiver also exists. While these early and simple models of the short-term fading are depicted along with the Nakagami model, other simple models based on gamma, generalized gamma, and Weibull are presented. Establishing the differences between short-term fading and long-term fading, the simplest model for the long-term fading or shadowing is introduced. The relationships among the simple lognormal density for the shadowing and equivalent models for shadowing using the gamma and Gaussian inverse densities are illustrated. This is followed by the discussion of the so-called compound models for shadowed fading channels to characterize the signal when both short-term fading and shadowing are present in the channel. The Suzuki and Nakagami-lognormals are introduced initially before presenting equivalent models using the generalized K distribution, the Nakagami inverse Gaussian distributions, and the generalized gamma distribution. The notion of multiple scattering to explain short-term fading is advanced through the demonstration of the so-called double Rayleigh, double Nakagami, and other mixture distributions. From there, the concept of cascaded channels as a complete means of describing the short-term fading is explained. Then we extend the notion of cascading to encompass shadowing leading to the Nakagami-N-gamma channels to model shadowed fading channels. The model is compared with the Nakagami-lognormal for shadowed fading channels.

Beginning with the simplest Rayleigh model, each model is characterized in terms of such quantitative measures as the amount of fading, signal-to-noise ratio (SNR), ergodic capacity, outage probability, and error rate. In this way the degradation in channels from fading and shadowing with each of these models can be understood and compared. The reader will understand the justification for more complex models to describe the statistical characteristic of the wireless channels. Chapter 4 introduces such complex models for fading as the $\eta - \mu$, the $\kappa - \mu$, and so forth. To complete the picture of fading and shadowing this chapter concludes with a discussion of some second-order statistical properties. Two examples of these are the level crossing rates and average fade duration.

Chapter 5 explores the concepts of diversity. It begins with the three basic forms of diversity combining algorithms: the selection combining (SC), equal gain combining (EGC), and maximal ratio combining (MRC). It also introduces the variation of the selection combining, specifically the switch and stay combining (SSC). While Chap. 4 delineates all the models for fading and shadowing, Chap. 5 emphasizes fading and shadowing using either the Nakagami (or gamma) distribution or densities of products of gamma variables. The density functions and distribution functions of the SNR following diversity are derived in each case. The densities and distribution functions are obtained using random number simulations in cases where analytical expressions for the densities and distributions do not exist. The effects of diversity are analyzed in terms of SNR enhancement, reduction in the amount of fading (AF), shifts in the peaks of probability density functions of the SNR, changes in the slopes of the CDF, reduction in error rates, and lowering of the outage probabilities. In addition to the combining algorithms mentioned above, we also examine the GSC algorithm to mitigate short-term fading. The use of hybrid combining with MRC to mitigate short-term fading and SC to mitigate shadowing is examined so as to gauge the overall improvements in performance in shadowed fading channels. The view of diversity in terms of where and how it is physically implemented is discussed by presenting the notion of "microdiversity" and "macrodiversity."

Chapter 6 covers the effect of cochannel interference in wireless systems. Once again, the densities and distribution functions of the SNR incorporating the cochannel interference are obtained, with the emphasis on Nakagami-related fading and shadowing channels. To examine the effect of cochannel interference, both the pure short-term faded channels and shadowed fading channels are studied. The outage probabilities and error rates in the presence of CCI are estimated in each case so that the impact of CCI can be fully understood. The improvements gained through diversity are evaluated using MRC in Nakagami channels in the presence of CCI.

Each chapter contains several plots of densities, distribution functions, error rates, outages, spectra, and so forth. Those results are obtained through two main software packages, *Matlab* (Version R2009b and earlier ones) and *Maple* (Version 10 and later ones). Even though there was no preference for either one, *Maple* (Version 10) offered a means to check and verify some of the analytical expressions involving hypergeometric and Meijer G-functions. All the plots were generated in *Matlab* even when the data sets were generated using *Maple*. The *Symbolic Toolbox* in Matlab with *MuPad* was necessary to handle functions such as the Meijer G. *Matlab* was used extensively to undertake numerical integration using *quadl* and associated routines performing single, double, and triple integrals. Occasionally trapezoidal integration using *trapz* in *Matlab* was used for single and double integration.

For most of the analytical manipulations, three main sources were used. Two of them were books/monographs (Mathai 1993; Gradshteyn and Ryzhik 2007) while the third one was the library of functions provided by *Mathematica* (Wolfram, Inc.). These sources are listed in the bibliography at the end of the chapter. *Maple* also provided verification of some of the analytical solutions mentioned earlier.

As previously suggested, this work is intended for graduate students in wireless communications. Even though it is expected that the student cohort has reasonable exposure to concepts of random variables and digital communications, Chaps. 2 and 3 are written to provide the tools necessary to undertake the study of fading and shadowing. The chapters are self-contained so that minimal cross-referencing of equations and figures is necessary. Consequently, a few equations from Chaps. 2 and 3 are repeated in subsequent chapters to provide a continuity of the discussion and to avoid referring to previous chapters or pages. Even though separate exercise sets are not provided, students can be expected to work out some or most of the derivations and demonstrations of the results.

The bibliography below provides a list of useful and relevant books, and information on the software packages used in this work.

Second edition of the book provides an expanded and updated coverage of the topics in fading, shadowing, and diversity. It also includes a new chapter on cognitive radio. As a pedagogic tool, most of the Matlab and Maple scripts used in the updated edition are provided alongside the results. Here is a summary of what is new in Edition 2.

Chapter 2 has been updated to provide additional background necessary to handle probability densities and their applications in wireless. Specifically, parameter estimation techniques are presented within the context of modeling the signal strength fluctuations. The two main techniques for parameter estimation, method of moments and maximum likelihood, are described with Matlab simulations in every case. This is followed by the study of statistical testing for the validation of models. In light of recent interest in cognitive radio, receiver operating characteristics analysis is discussed. A detailed exposition of Laplace and Mellin transforms for application in wireless channel is presented with examples (along with Matlab and Maple scripts).

While no changes are made to Chap. 3, some new models for fading and shadowing are described in Chap. 4. Initially, the McKay model for fading is presented and analyzed in detail. The effect of shadowing is also incorporated into the McKay model to provide a complete description of the signal strength fluctuations in wireless channels. The McKay model is followed by introducing a complementary approach based on mixture densities for modeling signal strength fluctuations in wireless channels. Using the concepts developed in updated Chap. 2, parameter estimation and hypothesis testing are now covered for these new models in addition to evaluation of wireless channel performance.

Diversity analysis has been undertaken in the updated version of Chap. 5. This analysis covers the models introduced in the updated version of Chap. 4.

While there are no changes to Chap. 6, Chap. 7 is new for the Second Edition. It is devoted to cognitive radio and is presented with a detailed exposition of performance evaluation in an ideal channel as well as when the channel undergoes fading and shadowing.

One of the new features of the Second Edition is the inclusion of Matlab scripts for the results obtained. The scripts are detailed and annotated for the benefit of the readers. They have been included on the basis of the some of the requests received from the readers of the first edition. The Matlab scripts cover updated sections of Chaps. 2, 4, and 5 and all of Chap. 7. In a couple of cases where Maple provides simpler options such as in the case of Laplace and Mellin transforms, Maple scripts are also made available. Each of the updated chapters and the new Chap. 7 also contain sections devoted to random number simulations relevant to the topics presented.

The Matlab version used is 2016a. The scripts rely on the following toolboxes:

Communications Systems Toolbox

Image Processing Toolbox

Optimization Toolbox

Signal Processing Toolbox

Statistics and Machine Learning Toolbox

Symbolic Math Toolbox

The Maple version used is Maple 16.

While every effort has been made to ensure that the Matlab and Maple scripts are correct, the author and the publishers are not responsible for any errors or omissions in the Matlab and Maple scripts.

References

Gradshteyn, I. S., & Ryzhik, I. M. (2007). *Table of integrals, series and products.* Oxford: Academic.
Mathai, A. M. (1993). *A handbook of generalized special functions for statistical and physical sciences.* Oxford: Oxford University Press.

Further Readings

Abramowitz, M., & Segun, I. A., (Eds.). (1972). Handbook of mathematical functions with formulas, graphs, and mathematical tables. New York: Dover.
Evans, M., Hastings N. A. J., et al. (2000). Statistical distributions. New York, Wiley.
Gagliardi, R. M. (1988). Introduction to communications engineering. New York, Wiley.
Goldsmith, A. (2005). Wireless communications. New York, Cambridge University Press.
Helstrom, C. W. (1968). Statistical theory of signal detection. Oxford, New York, Pergamon Press.
Jakes, W. C. (1994). Microwave mobile communications. Piscataway, NJ, IEEE Press.
Mathai, A. M., & Haubold H. J. (2008). Special functions for applied scientists. New York, Springer.
Middleton, D. (1996). An introduction to statistical communications theory. Piscataway, NJ: IEEE Press.
Molisch, A. F. (2005). Wireless communications. Chichester: Wiley.
Papoulis, A., & Pillai S. U. (2002). Probability, random variables, and stochastic processes. Boston, McGraw-Hill.
Proakis, J. G. (2001). Digital communications. Boston, McGraw-Hill.
Rohatgi, V. K., & Saleh A. K. M. E. (2001). An introduction to probability and statistics. New York, Wiley.
Schwartz, M. (1980). Information transmission, modulation, and noise: A unified approach to communication systems. New York, McGraw-Hill.
Schwartz, M. et al. (1996). Communication systems and techniques. Piscataway, NJ, IEEE Press.
Simon, M. et al. (1995). Digital communication techniques: Signal design and detection. Englewood Cliffs, NJ: Prentice Hall PTR.
Simon, M. K., & Alouini M.-S. (2005). Digital communication over fading channels. Hoboken, NJ: Wiley-Interscience.
Sklar, B. (2001). Digital communications: Fundamentals and applications. Upper Saddle River, NJ, Prentice-Hall PTR.
Steele, R., & Hanzó L. (1999). Mobile radio communications: Second and third generation cellular and WATM systems. New York, Wiley
Van Trees, H. L. (1968). Detection, estimation, and modulation theory: Part I. New York: Wiley.

Information on Software

Maple www.maplesoft.com (Waterloo, ON, Canada, N2 V 1 K8).
Matlab http://www.mathworks.com (Natick, MA 01760–2098).
Wolfram http://functions.wolfram.com (100 Trade Center Drive, Champaign, IL 61820, USA).

Chapter 2
Concepts of Probability and Statistics

2.1 Introduction

The analysis of communication systems involves the study of the effects of noise on the ability to transmit information faithfully from one point to the other using either wires or wireless systems (in free space), including microwave signals (Lindsey and Simon 1973; Schwartz 1980; Schwartz et al. 1996; Middleton 1996). Thus, it is important to understand the properties of noise, which requires the study of random variables and random processes. Additionally, the presence of fading and shadowing in the channel (mainly the wireless ones which are more susceptible to uncertainty) demands an understanding of the properties of random variables, distribution, density functions, and so on (Helstrom 1968; Rohatgi and Saleh 2001; Papoulis and Pillai 2002).

In this chapter, we will review, among others, the concepts of random variables, the properties of the distribution and density functions, estimation of parameters of the densities, and moments. We will also look at characterizing two or more random variables together so that we can exploit their properties when we examine techniques such as the diversity combining algorithms employed to mitigate fading and shadowing seen in wireless systems (Brennan 1959; Suzuki 1977; Hansen and Meno 1977; Shepherd 1977; Jakes 1994; Sklar 1997a, b; Vatalaro 1995; Winters 1998). In very broad terms, fading and shadowing represent the random fluctuations of signal power observed in wireless channels. Their properties are characterized in terms of their statistical nature, mainly through the density functions (Saleh and Valenzuela 1987; Vatalaro 1995; Yacoub 2000, 2007a, b; Cotton and Scanlon 2007; Nadarajah and Kotz 2007; da Costa and Yacoub 2008; Karadimas and Kotsopoulos 2008, 2010; Shankar 2010; Papazafeiropoulos and Kotsopoulos 2011) and we will review all the pertinent statistical aspects to facilitate a better understanding of signal degradation and mitigation techniques to improve the signal quality.

© Springer International Publishing AG 2017

P.M. Shankar, *Fading and Shadowing in Wireless Systems*,
DOI 10.1007/978-3-319-53198-4_2

In the updated edition, parameter estimation techniques are presented within the context of modeling the statistical fluctuations (Davenport et al. 1988; Iskander et al. 1999). The two main techniques based on method of moments and maximum likelihood are detailed with Matlab simulations in every case (Cheng and Beaulieu 2001; Tepedelenlioglu and Gao 2005). This is followed by the study of statistical testing for the validation of models (Mann and Wald 1942; Papoulis and Pillai 2002). In light of recent interest in cognitive radio, receiver operating characteristics are discussed (van Trees 1968; Helstrom 1968; Hanley and McNeil 1982; van Erkel and Pattynama 1998). A detailed exposition of Laplace and Mellin transforms for application in wireless channel is presented with examples (Epstein 1948; Erdelyi 1953; Kilicman and Ariffin 2002; Rossberg 2008). In addition to Matlab scripts, Maple scripts are also provided in connection with Laplace and Mellin transforms.

2.2 Random Variables, Probability Density Functions, and Cumulative Distribution Functions

A random variable is defined as a function that maps a set of outcomes in an experiment to a set of values (Rohatgi and Saleh 2001; Papoulis and Pillai 2002). For example, if one tosses a coin resulting in "heads" or "tails," a random variable can be created to map "heads" and "tails" into a set of numbers which will be discrete. Similarly, temperature measurements taken to provide a continuous set of outcomes can be mapped into a continuous random variable. Since the outcomes (coin toss, roll of a die, temperature measurements, signal strength measurements etc.) are random, we can characterize the random variable which maps these outcomes to a set of numbers in terms of it taking a specific value or taking values less than or greater than a specified value, and so forth. If we define X as the random variable, $\{X \leq x\}$ is an event. Note that x is a real number ranging from $-\infty$ to $+\infty$. The probability associated with this event is the distribution function, more commonly identified as the cumulative distribution function (CDF) of the random variable. The CDF, $FX(x)$, is

$$F_X(x) = \text{Prob}\{X \leq x\}. \tag{2.1}$$

The probability density function (pdf) is defined as the derivative of the CDF as

$$f_X(x) = \frac{d[F_X(x)]}{dx}. \tag{2.2}$$

From the definition of the CDF in (2.1), it becomes obvious that CDF is a measure of the probability and, therefore, the CDF has the following properties:

$$
\begin{aligned}
&F_X(-\infty) = 0, \\
&F_X(\infty) = 1, \\
&0 \leq F_X(x) \leq 1, \\
&F_X(x_1) \leq F_X(x_2), \quad x_1 \leq x_2.
\end{aligned} \tag{2.3}
$$

Based on (2.2) and (2.3), the probability density function has the following properties:

$$
\begin{aligned}
&0 \leq f_X(x) \\
&\int_{-\infty}^{\infty} f_X(x)dx = 1, \\
&F_X(x_1) = \int_{-\infty}^{x} f_X(\alpha)d\alpha, \\
&\text{Prob}\{x_1 \leq X \leq x_2\} = F_X(x_2) - F_X(x_1) = \int_{x_1}^{x_2} f_X(\alpha)d\alpha.
\end{aligned} \tag{2.4}
$$

A few comments regarding discrete random variables and the associated CDFs and pdfs are in order. For the discrete case, the random variable X takes discrete values $(1, 2, 3, \ldots, n)$. The CDF can be expressed as (Helstrom 1991; Rohatgi and Saleh 2001; Papoulis and Pillai 2002)

$$
F_X(x) = \sum_{m=1}^{n} \text{Prob}\{X = m\} U(x - m). \tag{2.5}
$$

In (2.5), $U(.)$ is the unit step function. The pdf of the discrete random variable becomes

$$
f_X(x) = \sum_{m=1}^{n} \text{Prob}\{X = m\} \delta(x - m). \tag{2.6}
$$

In (2.6), $\delta(.)$ is the delta function (Abramowitz and Segun 1972).

It can be easily seen that for a continuous random variable there is no statistical difference between the probabilities of outcomes where the random variable takes a value less than ($<$) or less than or equal to ($\leq$) a specified outcome because the probability that a continuous random variable takes a specified value is zero. However, for the case of a discrete random variable, the probabilities of the two cases would be different. We will now examine a few properties of the random variables and density functions.

The moments of a random variable are defined as

$$
\mu_k = \langle X^k \rangle = E(X^k) = \int_{-\infty}^{\infty} x^k f(x)\, dx. \tag{2.7}
$$

The first moment ($k = 1$) is the mean or the expected value of the random variable and $<\cdot>$ as well as $E(.)$ represents the statistical average. The variance (var) of the random variable is related to the second moment ($k = 2$) and it is defined as

$$\text{var}(x) = \sigma^2 = \langle X^2 \rangle - \langle X \rangle^2. \tag{2.8}$$

The quantity σ is the standard deviation. Variance is a measure of the degree of uncertainty and

$$\sigma^2 \geq 0. \tag{2.9}$$

The equality in (2.9) means that the degree of uncertainty or randomness is absent; we have a variable that is deterministic. Existence of higher values of variance suggests higher level of randomness. For the case of a discrete random variable, the moments are expressed as

$$\langle X^k \rangle = \sum_{m=1}^{n} m^k \text{Prob}\{X = m\}. \tag{2.10}$$

The survival function $S(x)$ of a random variable is defined as the probability that $\{X > x\}$. Thus,

$$S(x) = \int_x^{\infty} f_X(\alpha) d\alpha = 1 - F_X(x). \tag{2.11}$$

The coefficient of skewness is defined as (Evans et al. 2000; Papoulis and Pillai 2002)

$$\eta_3 = \frac{\mu_3}{\sigma^3}. \tag{2.12}$$

The coefficient of kurtosis is defined as

$$\eta_4 = \frac{\mu_4}{\sigma^4}. \tag{2.13}$$

The entropy or information content of a random variable is defined as

$$I = -\int_{-\infty}^{\infty} f(x) \log_2 [f(x)] dx. \tag{2.14}$$

The mode of a pdf $f(x)$ is defined as the value x where the pdf has the maximum. It is possible that a pdf can have multiple modes. Some of the density functions will not have any modes. The median x_m of a random variable is defined as the point where

$$\int_{-\infty}^{x_m} f(x)dx = \int_{x_m}^{\infty} f(x)dx. \tag{2.15}$$

2.3 Characteristic Functions, Moment Generating Functions and Laplace Transforms

The characteristic function (CHF) of a random variable is a very valuable tool because of its use in the performance analysis of wireless communication systems (Nuttall 1969, 1970; Tellambura and Annamalai 1999; Papoulis and Pillai 2002; Tellambura et al. 2003; Goldsmith 2005; Withers and Nadarajah 2008). This will be demonstrated later in this chapter when we discuss the properties of the CHFs. The CHF, $\psi(x)$, of a random variable having a pdf $f(x)$ is given by

$$\psi_X(\omega) = \langle \exp(j\omega X)\rangle = \int_{-\infty}^{\infty} f(x)\exp(j\omega x)dx. \tag{2.16}$$

Equation (2.16) shows that CHF is defined as the statistical average of exp $(j\omega x)$. Furthermore, it is also the Fourier transform of the pdf of the random variable. Rewriting (2.16) by replacing $(j\omega)$ by s, we get the moment generating function (MGF) of the random variable as (Alouini and Simon 2000; Papoulis and Pillai 2002)

$$\phi_X(s) = \langle \exp(sX)\rangle = \int_{-\infty}^{\infty} f(x)\exp(sx)dx. \tag{2.17}$$

Defining (2.16) slightly differently, we can obtain the expression for the Laplace transform of pdf as (Beaulieu 1990; Tellambura and Annamalai 1999, 2000; Papoulis and Pillai 2002)

$$L_X(s) = \int_{-\infty}^{\infty} f(x)\exp(-sx)dx. \tag{2.18}$$

Consequently, we have the relationship between the bilateral Laplace transform and the MGF of a random variable:

$$L_X(s) = \phi_X(-s). \tag{2.19}$$

Going back to (2.16), we can define the inverse Fourier transform of the CHF to obtain the pdf as

$$f_X(x) = \frac{1}{2\pi}\int_{-\infty}^{\infty} \psi(\omega)\exp(-j\omega x)d\omega. \tag{2.20}$$

By virtue of the property (2.20), CHF will uniquely determine the probability density function. We will explore the use of CHFs in diversity analysis later in this chapter and we will use the Laplace transform, CHF, and MGF to estimate the error rates and outage probabilities to be seen in later chapters (Tellambura and Annamalai 1999).

2.4 Some Commonly Used Probability Density Functions

We will now look at several random variables which are commonly encountered in wireless communication systems analysis. We will examine their properties in terms of the pdfs, CDFs and CHFs. Within the context of wireless communications, we will explore the relationships among some of these random variables.

2.4.1 *Beta Distribution*

The beta distribution is not commonly seen in wireless communications. However, it arises in wireless system when we are studying issues related to signal-to-interference ratio (Jakes 1994; Winters 1984, 1987). The beta random variable is generated when the ratio of certain random variables is considered (Papoulis and Pillai 2002). The beta density function is

$$f(x) = \begin{cases} \dfrac{x^{a-1}(1-x)^{b-1}}{\beta(a,b)} & 0 < x < 1, \\ 0 & \text{elsewhere.} \end{cases} \tag{2.21}$$

In (2.21), $\beta(a,b)$ is the beta function given in integral form as

$$\beta(a,b) = \int_0^1 x^{a-1}(1-x)^{b-1}dx. \tag{2.22}$$

Equation (2.22) can also be written in terms of gamma functions (Abramowitz and Segun 1972):

$$\beta(a,b) = \frac{\Gamma(a)\Gamma(b)}{\Gamma(a+b)}, \tag{2.23}$$

where Γ(.) is the gamma function. The CDF of the beta random variable is

$$F(x) = \begin{cases} 0, & x \leq 0, \\ \int_0^x \frac{y^{a-1}(1-y)^{b-1}}{\beta(a,b)} dy, & 0 < x < 1, \\ 1, & x \geq 1. \end{cases} \tag{2.24}$$

The moments of the beta random variable are

$$\langle X^k \rangle = \frac{\Gamma(k+a)\Gamma(k+b)}{\Gamma(k+a+b)\Gamma(k)}. \tag{2.25}$$

The mean and variance of the beta variable are

$$\langle X \rangle = \frac{a}{a+b}, \tag{2.26}$$

$$\mathrm{var}\langle X \rangle = \frac{ab}{(a+b+1)(a+b)^2}. \tag{2.27}$$

The beta density function is shown in Fig. 2.1 for three sets of values of (a,b).

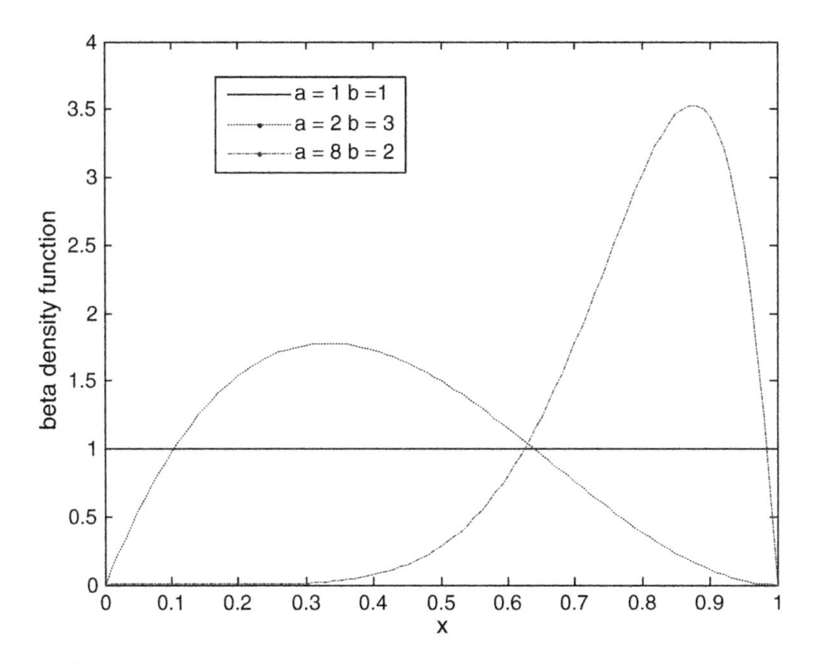

Fig. 2.1 The density function of the beta random variable

It can be seen that for the case of $a = b = 1$, the beta density becomes the uniform density function as seen later in this chapter. Note that X is a beta variate with parameters a and b; then $Y = 1 - X$ is also a beta variate with parameters b and a.

The beta density function allows significant flexibility over the standard uniform distribution in the range $\{0, 1\}$ as one can see from Fig. 2.1. As mentioned in the beginning if X_1 and X_2 are two gamma variables with parameters (a, A) and (b, A) respectively, the ratio $X_1/(X_1 + X_2)$ will be a beta variate with parameters (a,b).This aspect will be shown later in this chapter when we examine the density function of linear and nonlinear combination of two or more random variables.

2.4.2 *Binomial Distribution*

The binomial distribution is sometimes used in wireless systems when estimating the strength of the interfering signals at the receiver, with the number of interferers contributing to the interfering signal being modeled using a binomial random variable (Abu-Dayya and Beaulieu 1994; Shankar 2005). For example, if there are n interfering channels, it is possible that all of them might not be contributing to the interference. The number of actual interferers contributing to the interfering signal can be statistically described using the binomial distribution. This distribution is characterized in terms of two parameters, with the parameter n representing the number of Bernoulli trials and the parameter p representing the successes from the n trials. (A Bernoulli trial is an experiment with only two possible outcomes that have probabilities p and q such that $(p + q) = 1$). While n is an integer, the quantity p is bounded as $0 < p < 1$. The binomial probability is given by (Papoulis and Pillai 2002)

$$\mathrm{Prob}\{X = k\} = \binom{n}{k} p^k (1 - p)^{n-k}, \quad k = 0, 1, \ldots. \tag{2.28}$$

In (2.28),

$$\binom{n}{k} = C_k^n = \frac{n!}{(n - k)!k!}. \tag{2.29}$$

The mean of the variable is given by (np) and the variance is given by (npq) where $q = (1 - p)$. The binomial probabilities are shown in Fig. 2.2.

The binomial variate can be approximated by a Poisson variate (discussed later in this section) if $p \ll 1$ and $n < 10$. Later we will describe how the binomial distribution approaches the normal distribution, when $npq > 5$ and $0.1 < p < 0.9$. The transition toward the normal behavior is seen in Fig. 2.2.

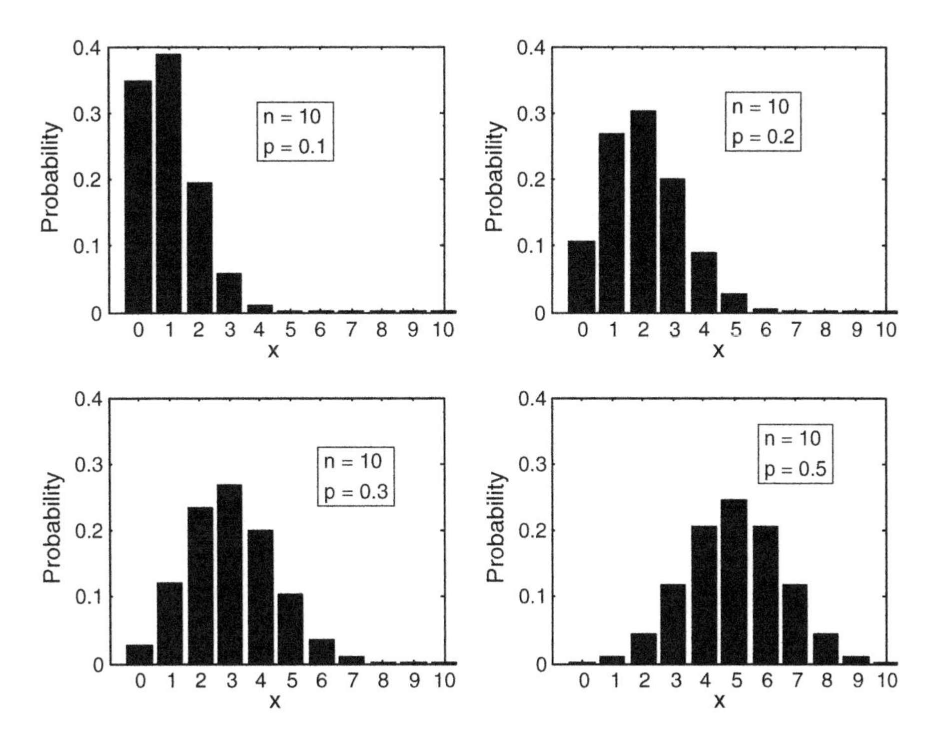

Fig. 2.2 Binomial probabilities

We alluded to Bernoulli trials which had only two outcomes. We can extend such experiments to have more than two outcomes resulting in the generalized Bernoulli trials. Instead of two outcomes, let us consider the case of r outcomes (mutually exclusive) such that the total probability

$$p_1 + p_2 + \cdots + p_r = 1. \tag{2.30}$$

Now, we repeat the experiment n times. Our interest is in finding out the probability that outcome #1 occurs k_1 times, #2 occurs k_2 times, and so on. In other words, we are interested in the Prob{#1 occurs k_1 times, #2 occurs k_2 times,. . ., #r occurs k_r times}. Noting that the number of ways in which these events can occur is $n! = (k_1!k_2!.\ldots k_r!)$, the required probability becomes

$$p(k_1, k_2, \ldots., k_r) = \frac{n!}{k_1!k_2!\ldots k_r!} p_1^{k_1} p_2^{k_2} \cdots p_r^{k_r} \quad \sum_{j=1}^{r} k_r = n. \tag{2.31}$$

We will use some of these results when we examine the order statistics later in this chapter.

2.4.3 Cauchy Distribution

The Cauchy distribution arises in wireless systems when we examine the pdfs of random variables which result from the ratio of two Gaussian random variables (Papoulis and Pillai 2002). What is unique in terms of its properties is that its moments do not exist. The Cauchy pdf is expressed as

$$f(x) = \frac{1}{\pi\beta[1 + ((x - \alpha)/\beta)]}, \quad -\infty < x < \infty. \tag{2.32}$$

The associated CDF is

$$F_X(x) = \frac{1}{2} + \frac{1}{\pi}\tan^{-1}\left(\frac{x - a}{\beta}\right). \tag{2.33}$$

The CHF is

$$\psi(\omega) = \exp(j\alpha\omega - |\omega|\beta). \tag{2.34}$$

As mentioned, its moments do not exist and its mode and median are (each) equal to α. Note that Cauchy distribution might appear similar to the normal (Gaussian) distribution. While both the Cauchy and Gaussian distributions are unimodal (only a single mode exists) and are symmetric (Gaussian around the mean and Cauchy around α), the Cauchy distribution has much heavier tails than the Gaussian pdf. The Cauchy pdf is shown in Fig. 2.3 for the case of $\alpha = 0$. The heavier tails are seen as β goes up.

2.4.4 Chi-Squared Distribution

This distribution arises in statistics (and in wireless systems) when we examine the pdf of the sum of the squares of several Gaussian random variables. It is also used in hypothesis testing such as the χ^2 goodness of fit test (Papoulis and Pillai 2002).

The chi square (or chi squared) random variable has a pdf given by

$$f(x) = \frac{x^{(n-2)/2}\exp(-(x/2))}{2^{n/2}\Gamma(n/2)}, \quad 0 < x < \infty. \tag{2.35}$$

The shape parameter n is designated as the degrees of freedom associated with the distribution. This density function is also related to the Erlang distribution used in the analysis of modeling of the grade of service (GOS) in wireless systems and the gamma distribution used to model fading and shadowing seen in wireless systems. The gamma and Erlang densities are described later in this section. The

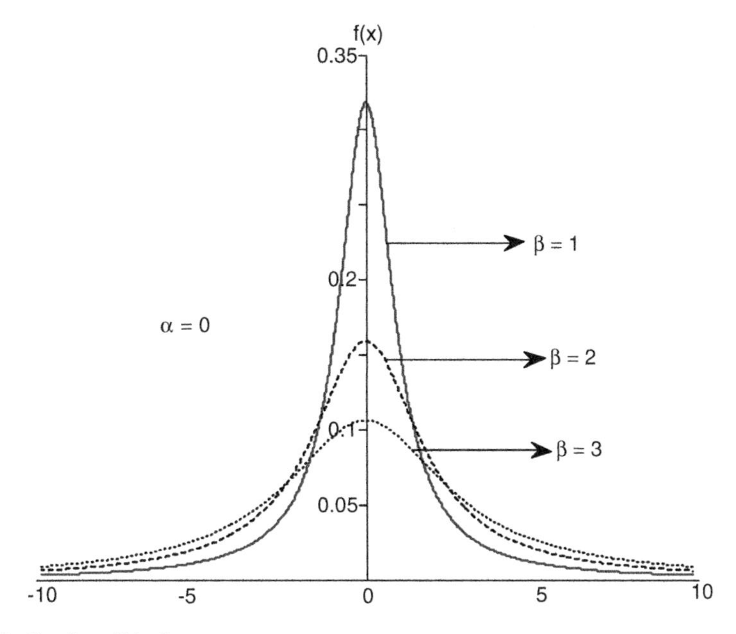

Fig. 2.3 Cauchy pdf is shown

distribution is often identified as $\chi^2(n)$. The density function also becomes the exponential pdf in the limiting case ($n = 2$).

There is no simple analytical expression for the CDF, but, it can be expressed in terms of incomplete gamma function as

$$F_X(x) = 1 - \frac{\Gamma\left(\frac{n}{2}, \frac{x}{2}\right)}{\Gamma\left(\frac{n}{2}\right)} = \frac{\gamma\left(\frac{n}{2}, \frac{x}{2}\right)}{\Gamma\left(\frac{n}{2}\right)}, \tag{2.36}$$

where

$$\Gamma(a,b) = \int_b^{\infty} x^{a-1} \exp(-x) dx, \quad \gamma(a,b) = \int_0^b x^{a-1} \exp(-x) dx. \tag{2.37}$$

$\Gamma(.,.)$ is the (upper) incomplete gamma function, $\gamma(.,.)$ the (lower) incomplete gamma function and $\Gamma(.)$ is the gamma function (Abramowitz and Segun 1972; Gradshteyn and Ryzhik 2007). Note that the pdf in (2.35) is a form of gamma pdf of parameters (n/2) and 2 (as we will see later). The CHF is given by

$$\psi(\omega) = (1 - 2j\omega)^{-(n/2)}. \tag{2.38}$$

The mean of the random variable is n and the variance is $2n$. The mode is (n – 2), $n > 2$ and the median is (n – 2/3). The density function is plotted in Fig. 2.4. It can

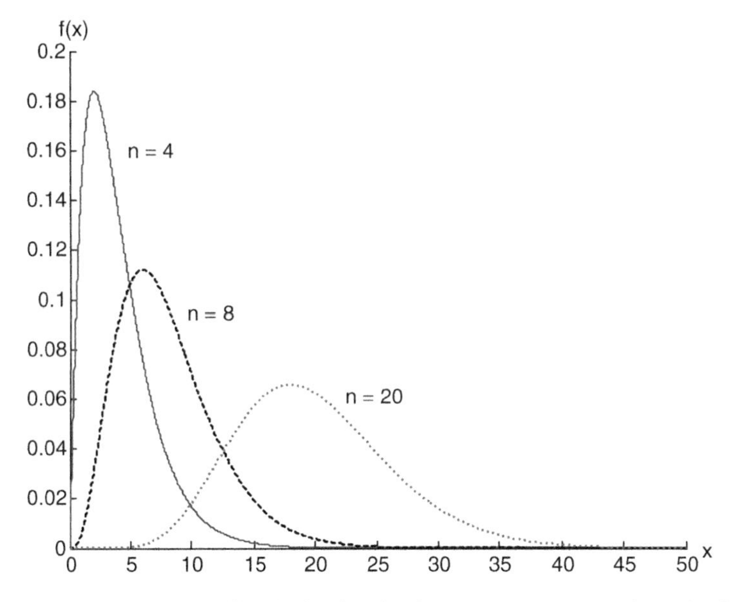

Fig. 2.4 The Chi squared pdf. The density function becomes more symmetric as the degree of freedom n goes up

be seen that as the degrees of freedom (n) increases, the symmetry of the pdf increases and it will approach the Gaussian pdf.

There is a related distribution called the non-central chi squared distribution which is described later in this chapter.

An associated distribution is the *chi* distribution. The *chi* variable is the positive square root of a *chi* squared variable. The chi pdf is given by

$$f(x) = \frac{x^{n-1}\exp(-(x^2/2))}{2^{(n/2)-1}\Gamma(n/2)}, \quad 0 \le x \le \infty. \tag{2.39}$$

The cumulative distribution becomes

$$F_X(x) = 1 - \frac{\Gamma\left(\frac{n}{2}, \frac{x^2}{2}\right)}{\Gamma\left(\frac{n}{2}\right)} = \frac{\gamma\left(\frac{n}{2}, \frac{x^2}{2}\right)}{\Gamma\left(\frac{n}{2}\right)} \tag{2.40}$$

When $n = 2$, the *chi* pdf becomes the Rayleigh pdf, described later in this section.

2.4.5 Erlang Distribution

When we examine the density of a sum of exponential random variables, we get the so-called Erlang pdf. The Erlang pdf is given by

$$f(x) = \frac{(x/\beta)^{c-1}\exp(-(x/\beta))}{\beta(c-1)!}, \quad 0 \le x \le \infty. \tag{2.41}$$

Note that the shape parameter c is an integer. Equation (2.41) becomes the gamma pdf if c is a non-integer. The CHF is given by

$$\psi(\omega) = (1 - j\beta\omega)^{-c}. \tag{2.42}$$

The mean of the random variable is (βc) and the variance is $(\beta^2 c)$. The mode is $\beta(c-1)$, $c \ge 1$. The CDF can be expressed as

$$F_X(x) = 1 - \left[\exp\left(-\frac{x}{\beta}\right)\right]\left(\sum_{k=0}^{c-1}\frac{(x/\beta)^k}{k!}\right). \tag{2.43}$$

Note that the Erlang density function and chi-squared pdf in (2.35) have similar functional form with $c = n/2$ and $\beta = 2$.

2.4.6 Exponential Distribution

The exponential distribution (also known as the negative exponential distribution) arises in communication theory in the modeling of the time interval between events when the number of events in any time interval has a Poisson distribution. It also arises in the modeling of the signal-to-noise ratio (SNR) in wireless systems (Nakagami 1960; Saleh and Valenzuela 1987; Simon and Alouini 2005). Furthermore, the exponential pdf is a special case of the Erlang pdf when $c = 1$ and a special case of chi-squared pdf when $n = 2$. The exponential pdf is given by

$$f(x) = \frac{1}{\beta}\exp\left(-\frac{x}{\beta}\right), \quad 0 \le x \le \infty. \tag{2.44}$$

The associated CDF is given by

$$F(x) = 1 - \exp\left(-\frac{x}{\beta}\right). \tag{2.45}$$

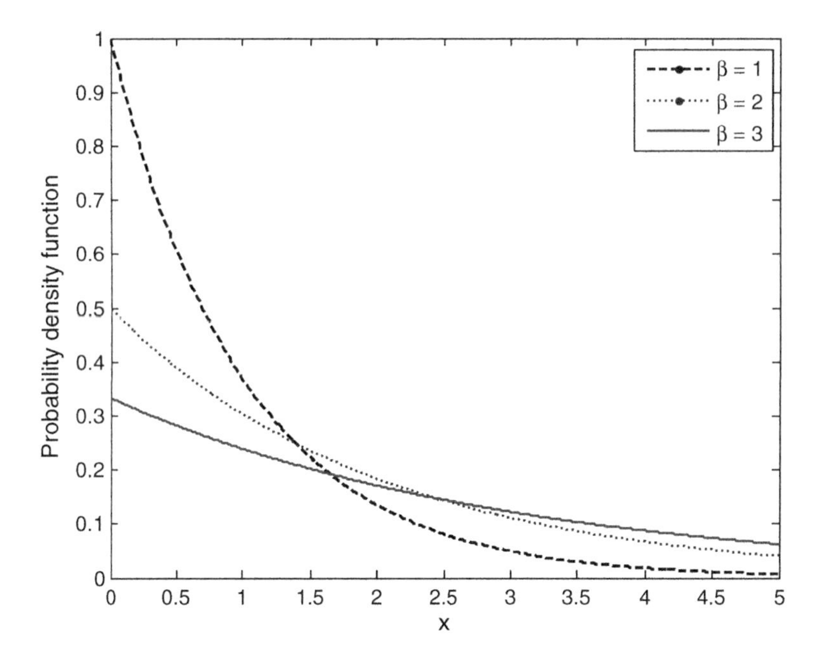

Fig. 2.5 The exponential densities are plotted for three values of the mean

The CHF is given by

$$\psi(\omega) = \frac{1}{1 - j\omega\beta}. \tag{2.46}$$

The exponential has no mode. The mean is β and the variance is β^2 and, thus, the exponential pdf is uniquely characterized in terms of the fact that the ratio of its mean to its standard deviation is unity. The exponential pdf is shown in Fig. 2.5 for three values of β.

The exponential CDF is a measure of the outage probability (probability that the signal power goes below a threshold) is shown in Fig. 2.6.

2.4.7 F (*Fisher-Snedecor) Distribution*

The SNR in fading is often modeled using gamma distributions (Simon and Alouini 2005; Shankar 2004). The ratio of two such variables is of interest when examining the effects of interfering signals in wireless communications (Winters 1984). The *F* distribution arises when we examine the density function of the ratio of two chi-squared random variables of even degrees of freedom or two Erlang variables or two gamma variables of integer orders (Lee and Holland 1979; Nadarajah 2005;

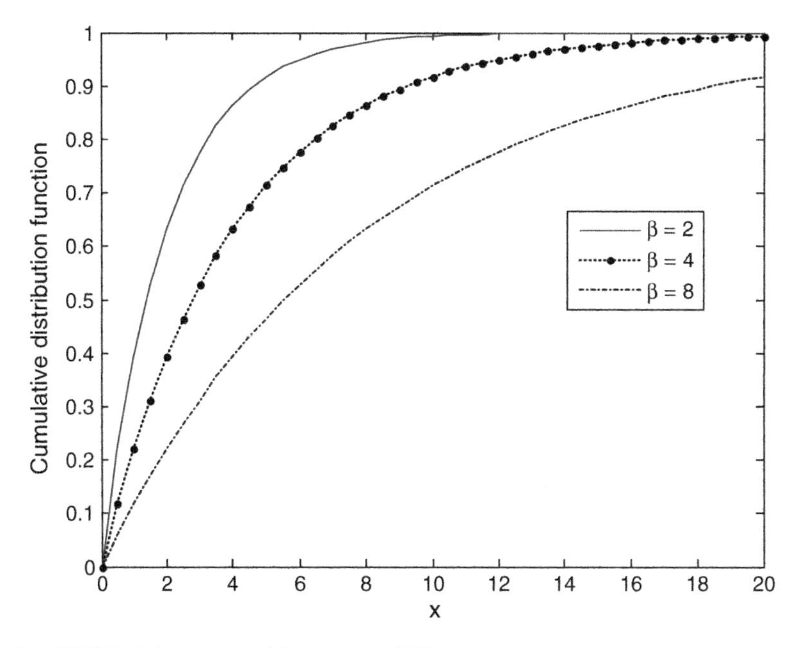

Fig. 2.6 (CDFs) of the exponential random variables are shown

Nadarajah and Gupta 2005; Nadarajah and Kotz 2006a, b). The F density can be written as

$$f(x) = \frac{\Gamma((m+n)/2)m^{m/2}n^{n/2}}{\Gamma(m/2)\Gamma(n/2)} \frac{x^{m/2-1}}{(n+mx)^{[(m+n)/2]}} U(x). \tag{2.47}$$

In (2.47), m and n are integers. The mean is given by

$$\langle X \rangle = \frac{n}{n-2}, \quad n > 2. \tag{2.48}$$

The variance is

$$\mathrm{var}(x) = 2n^2 \frac{(n+m-2)}{m(n-4)(n-2)^2}, \quad n > 4. \tag{2.49}$$

The density function is identified as $F(m,n)$, with (m,n) degrees of freedom. The F distribution is shown in Fig. 2.7. The density function is sometimes referred to as Fisher's variance ratio distribution.

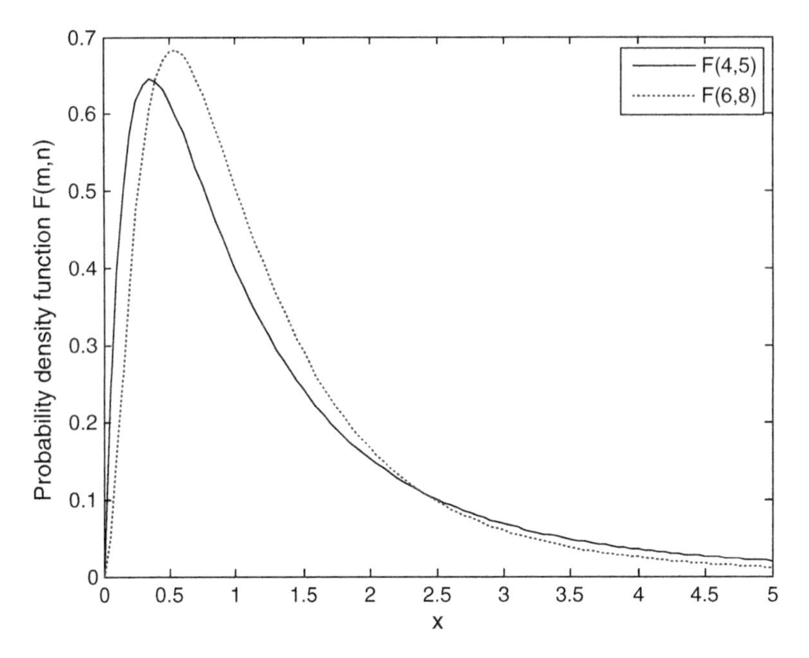

Fig. 2.7 The F distribution is plotted for two sets of values

2.4.8 Gamma Distribution

The gamma distribution is used extensively in wireless communications to model the power in fading channels (Nakagami 1960; Abdi and Kaveh 1999; Atapattu et al. 2010a, b). As mentioned earlier, the gamma pdf is a special case of the Erlang distribution in (2.41). The pdf is given by

$$f(x) = \frac{(x/\beta)^{c-1}\exp(-(x/\beta))}{\beta\Gamma(c)}, \quad 0 < x < \infty. \tag{2.50}$$

Comparing (2.41) and (2.50), we see that for the gamma pdf c can be any positive number while for the Erlang pdf, c must be an integer. The CDF can be expressed in terms of the incomplete gamma function as

$$F_X(x) = \frac{\gamma(c,(x/\beta))}{\Gamma(c)} = 1 - \frac{\Gamma(c,(x/\beta))}{\Gamma(c)}. \tag{2.51}$$

The moments are given by

$$E(X^k) = b^k \frac{\Gamma(c+k)}{\Gamma(c)}. \tag{2.52}$$

The moments are identical to that of the Erlang distribution. The exponential distribution is a special case of the gamma pdf with $c = 1$. The received SNR or power is modeled using the gamma pdf. The CHF of the gamma distribution is

$$\psi(\omega) = (1 - j\beta\omega)^{-c}. \tag{2.53}$$

The gamma pdf in (2.50) becomes the chi-squared pdf in (2.35) when $n = 2c$ and $\beta = 2$. The relationship of the gamma pdf to other distributions is given in Table 2.1.

The gamma pdf is plotted in Fig. 2.8 for three values of c, all having identical mean of unity. One can see that as the order of the gamma pdf increases, the peaks

Table 2.1 Relationship of gamma pdf to other distributions

c	β	Name of the distribution	Probability density function
1	–	Exponential	$\frac{1}{\beta}\exp\left(-\frac{x}{\beta}\right)$
Integer	–	Erlang	$\frac{(x/\beta)^{c-1}\exp(-(x/\beta))}{\beta(c-1)!}$
$c = 2n$	2	Chi-squared	$\frac{x^{(n-2)/2}\exp(-(x/2))}{2^{n/2}\Gamma(n/2)}$

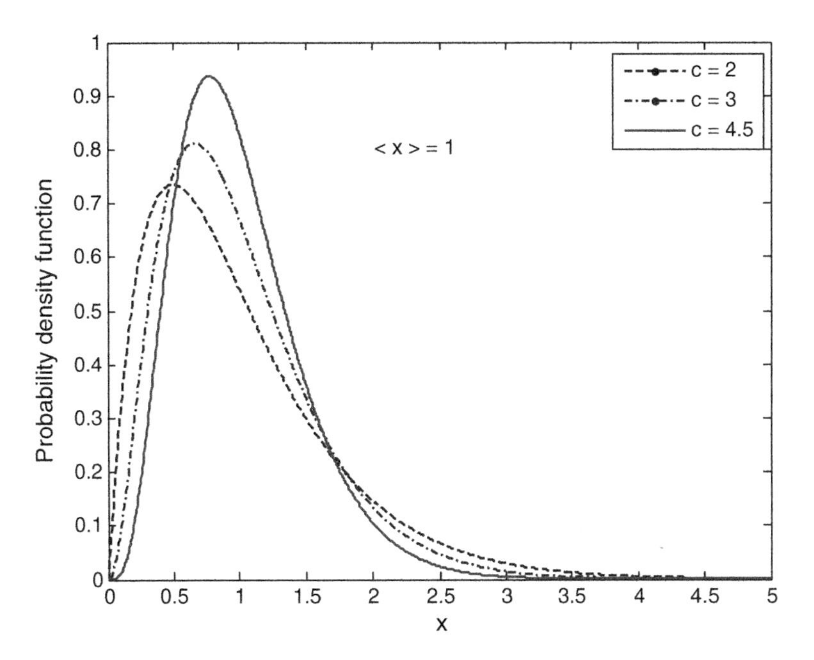

Fig. 2.8 The density functions of the gamma random variables for three values of the order. All have the identical means (unity)

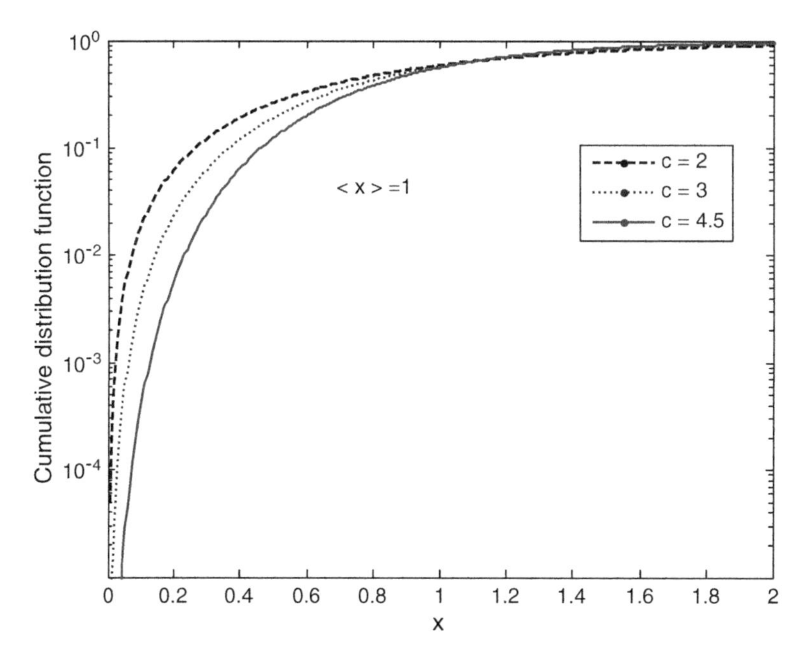

Fig. 2.9 The CDFs of the gamma random variables for three values of the order. All have the identical means (unity)

of the densities move farther to the right. This aspect will be seen later in Chap. 5 when we examine the impact of diversity in fading channels. The corresponding CDFs are shown in Fig. 2.9.

2.4.9 Generalized Gamma Distribution

Instead of the two parameter distribution in (2.50), the three parameter gamma distribution known as the generalized gamma distribution can be expressed as (Stacy 1962; Stacy and Mihram 1965; Lienhard and Meyer 1967; Griffiths and McGeehan 1982; Coulson et al. 1998a, b; Gupta and Kundu 1999; Bithas et al. 2006)

$$f(x) = \frac{\lambda x^{\lambda c-1}}{\beta^{\lambda c}\Gamma(c)} \exp\left[-\left(\frac{x}{\beta}\right)^{\lambda}\right], \quad 0 < x < \infty. \tag{2.54}$$

There is no simple expression for the CHF of the generalized gamma density function. The generalized gamma random variable can be obtained from a gamma random variable by scaling the random variable by $(1/\lambda)$. This distribution is also known as the Stacy distribution (Stacy 1962), which is expressed in a slightly different form as

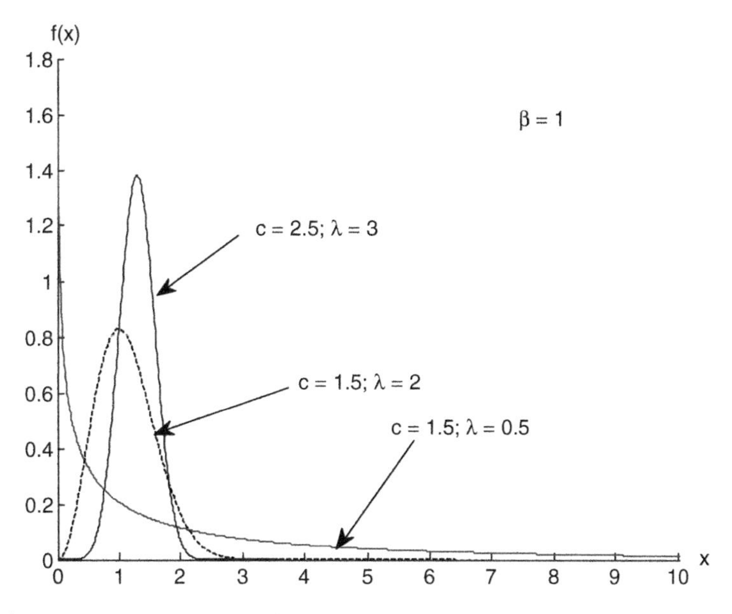

Fig. 2.10 The generalized gamma pdf in (2.56)

$$f(x) = \left(\frac{p}{a^d}\right)x^{d-1}\frac{\exp\left(-\left(\frac{x}{a}\right)^p\right)}{\Gamma\left(\frac{d}{p}\right)}, \quad x \geq 0, a > 0, p > 0, d > 0. \tag{2.55}$$

Note that the generalized gamma pdf can also be expressed in yet another form as

$$f(x) = \frac{\lambda x^{\lambda c-1}}{\beta^c \Gamma(c)} \exp\left[-\frac{x^\lambda}{\beta}\right], \quad 0 < x < \infty. \tag{2.56}$$

The GG pdf in (2.56) is plotted in Fig. 2.10 for the case of $\beta = 1$. It can be seen that as λ increases the density function moves to the right, indicating that the density function will approach Gaussian as is possible with the case of Erlang pdf. The same effect will be present when c increases as well.

The generalized gamma (GG) distribution can be used to model power (or SNR) as well as the magnitude (or envelope) values in wireless communications (Coulson et al. 1998a, b). This can be accomplished by varying c and λ. Also, the GG distribution can morph into other distributions in limiting cases (Griffiths and McGeehan 1982). The CDF can once again be represented in terms of the incomplete gamma functions in (2.37) and (2.52). Using the representation in (2.56), the expression for the CDF becomes

$$F_X(x) = 1 - \frac{\Gamma(c, (x^{\lambda}/\beta))}{\Gamma(c)}. \tag{2.57}$$

If we express the generalized gamma pdf in (2.54) as

$$f(x) = Ax^n \exp\left(-\frac{x^m}{b^m}\right), \quad 0 \le x \le \infty, \tag{2.58}$$

where b is a scaling factor and A is a normalization factor such that

$$\int_0^{\infty} f(x)dx = 1, \tag{2.59}$$

we can relate the pdf in (2.58) to several density functions as described in Table 2.2 (Griffiths and McGeehan 1982).

There are also other forms of gamma and generalized gamma distribution. Even though they are not generally used in wireless communications, we will still provide them so as to complete the information on the class of gamma densities. One such gamma pdf is (Evans et al. 2000)

$$f(x) = \frac{(x-\gamma)^{c-1} \exp - (x-\gamma)/\beta}{\beta^c \Gamma(c)}, \quad x > \gamma > 0, c > 0, \beta > 0. \tag{2.60}$$

Note that when $\gamma = 0$, (2.60) becomes the standard two parameter gamma density defined in (2.50), another form of generalized gamma density function has four parameters and it is of the form

Table 2.2 The generalized gamma distribution and its special cases

m	n	b	Name of the distribution	Probability density function
1	–	–	Gamma	$Ax^n \exp\left(-\frac{x}{b}\right)$
1	>0 (integer)	–	Erlang	$Ax^n \exp\left(-\frac{x}{b}\right)$
1	>1 (integer)	2	Chi-squared	$Ax^n \exp\left(-\frac{x}{b}\right)$
2	–	–	Nakagami	$Ax^n \exp\left(-\frac{x^2}{b^2}\right)$
–	–	–	Stacy	$Ax^n \exp\left(-\frac{x^m}{b^m}\right)$
–	$n = m - 1$	–	Weibull	$Ax^{m-1} \exp\left(-\frac{x^m}{b^m}\right)$
1	0	–	Exponential	$A\exp\left(-\frac{x}{b}\right)$
2	0	–	One sided Gaussian	$A\exp\left(-\frac{x^2}{b^2}\right)$
	0	–	Generalized exponential	$A\exp\left(-\frac{x^m}{b^m}\right)$
2	1	–	Rayleigh	$Ax\exp\left(-\frac{x^2}{b^2}\right)$
–	1	–	Generalized Rayleigh	$Ax\exp\left(-\frac{x^m}{b^m}\right)$

In all the expressions in the last column, A is a normalization factor

$$f(x) = \frac{\lambda(x-\gamma)^{\lambda c-1}}{\beta^c \Gamma(c)} \exp\left[-\left(\frac{x-\gamma}{\beta}\right)^{\lambda}\right], \quad x > \gamma > 0, \lambda > 0, c > 0, \beta > 0. \quad (2.61)$$

Note that (2.61) becomes the generalized gamma distribution (Stacy's pdf) in (2.54) when $\gamma = 0$.

While the gamma pdf and the generalized gamma densities mentioned so far are defined for positive values (single-sided), there also exists a double-sided generalized gamma density function which has been used to model noise in certain cases (Shin et al. 2005). The two-sided generalized gamma pdf is

$$f(x) = \frac{\lambda |x|^{c\lambda-1}}{2\beta^c \Gamma(c)} \exp\left(-\frac{|x|^{\lambda}}{\beta}\right), \quad -\infty < x < \infty. \quad (2.62)$$

The two-sided generalized gamma pdf is flexible enough that it can become Gaussian, Laplace, generalized gamma or gamma.

2.4.10 Inverse Gaussian (Wald) Distribution

This pdf is sometimes used to model shadowing in wireless systems because of the closeness of its shape to the lognormal density function (Karmeshu and Agrawal 2007; Laourine et al. 2009). The pdf is expressed as

$$f(x) = \left(\frac{\lambda}{2\pi x^3}\right)^{1/2} \exp\left[-\frac{\lambda(x-\mu)^2}{2\mu^2 x}\right], \quad 0 < x < \infty. \quad (2.63)$$

Note that both λ and μ are positive numbers. There is no simple analytical expression for the CDF. The CHF is given by

$$\psi(\omega) = \exp\left\{\frac{\lambda}{\mu}\left[1 - \left(1 - \frac{2j\mu^2\omega}{\lambda}\right)^{1/2}\right]\right\}. \quad (2.64)$$

The mean is given by μ and the variance is given by μ^3/λ. The density function is shown in Fig. 2.11 for a few values of the parameters.

2.4.11 Laplace Distribution

This distribution is generally not used in wireless communication systems even though research exists into its use in communication systems (Sijing and Beaulieu 2010).

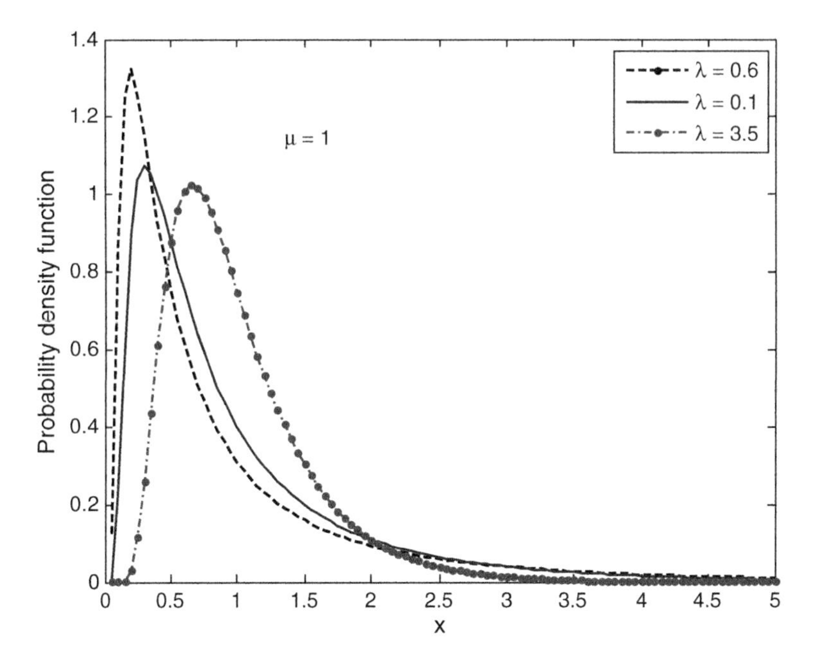

Fig. 2.11 Inverse Gaussian distribution

It is related to the exponential distribution in that it extends the range of the variable down to $-\infty$. The pdf is expressed as

$$f(x) = \frac{1}{2\beta}\exp\left(-\frac{|x-\alpha|}{\beta}\right), \qquad -\infty < x < \infty. \tag{2.65}$$

The associated CDF is given by

$$F(x) = \begin{cases} \frac{1}{2}\exp\left(-\frac{\alpha - x}{\beta}\right), & x < \alpha, \\ 1 - \frac{1}{2}\exp\left(-\frac{x-\alpha}{\beta}\right), & x \geq \alpha. \end{cases} \tag{2.66}$$

The CHF is given by

$$\psi(\omega) = \frac{\exp(j\omega\alpha)}{1+\beta^2\omega^2}. \tag{2.67}$$

The mean, mode, and median are all equal to α, and the variance is equal to $2\beta^2$. The Laplace density function is shown in Fig. 2.12.

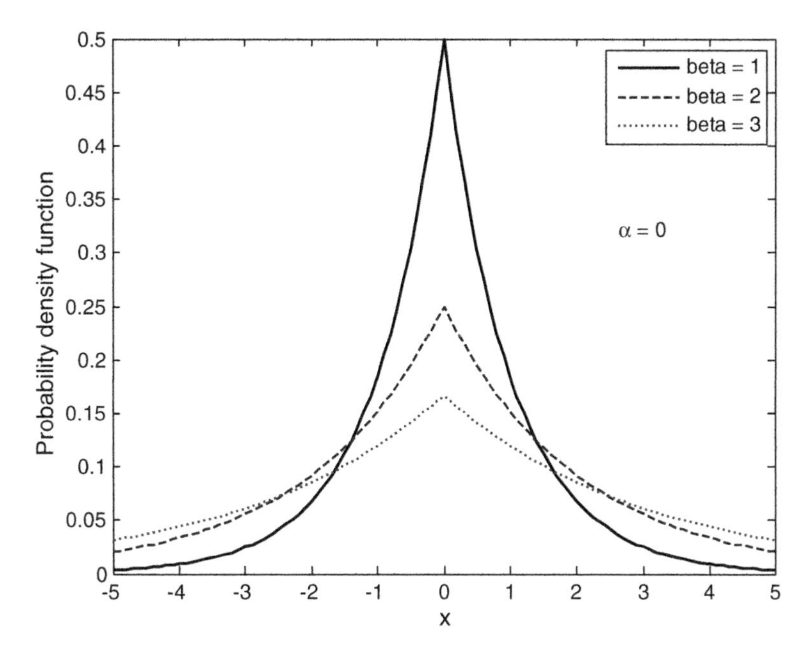

Fig. 2.12 The Laplace density functions

2.4.12 Lognormal Distribution

The lognormal pdf arises when central limit theorem for products of random variables is applied (Papoulis and Pillai 2002). It is used to model long-term fading or shadowing seen in wireless systems (Hudson 1996; Tjhung and Chai 1999; Coulson et al. 1998a, b; Patzold 2002; Kostic 2005; Stuber 2000; Cotton and Scanlon 2007). In certain cases, it finds applications in modeling short-term fading as well (Cotton and Scanlon 2007). The pdf of the lognormal random variable is given by

$$f(x) = \frac{1}{\sqrt{2\pi\sigma^2}x}\exp\left[-\frac{(\log_e(x)-\mu)^2}{2\sigma^2}\right], \quad 0 < x < \infty. \tag{2.68}$$

The CDF and CHF are not readily available in analytical form. The mean and variance can be expressed as

$$E(X) = \exp\left(\mu + \frac{1}{2}\sigma^2\right), \tag{2.69}$$

$$\text{var}(X) = \left[\exp\left(\sigma^2 - 1\right)\right]\exp\left(2\mu + \sigma^2\right). \tag{2.70}$$

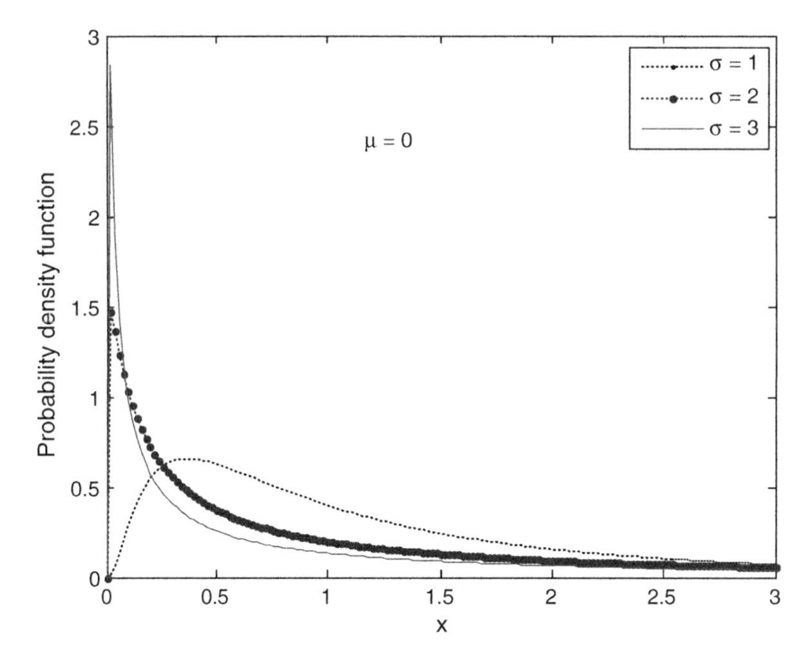

Fig. 2.13 Lognormal density function in (2.68) is shown

Equation (2.68) can also be expressed in a slightly different form using the logarithm with the base 10 as

$$f(x) = \frac{[10/\log_e(10)]}{\sqrt{2\pi\sigma^2 x^2}} \exp\left[-\frac{(10\log_{10}(x) - \mu)^2}{2\sigma^2}\right], \quad 0 < x < \infty. \tag{2.71}$$

In (2.71), μ and σ are in decibel units. Note that if the lognormal variable is converted to decibel units, the pdf of the variable in dB will be Gaussian. This density function (in terms of decibel units) is discussed in detail in Chap. 4.

Figure 2.13 shows the plots of the lognormal pdf in (2.68).

2.4.13 *Nakagami Distribution*

Even though there are several forms of the Nakagami distribution, the most commonly known one is the "Nakagami-m distribution" with a pdf given by (Nakagami 1960; Simon and Alouini 2005)

$$f(x) = 2\left(\frac{m}{\Omega}\right)^m \frac{x^{2m-1}}{\Gamma(m)} \exp\left(-m\frac{x^2}{\Omega}\right) U(x), \quad m \geq \frac{1}{2}. \tag{2.72}$$

In (2.72), m is the Nakagami parameter, limited to values greater than or equal to ½. The moments of the Nakagami distribution can be expressed as

$$\langle X^k \rangle = \frac{\Gamma(m + (k/2))}{\Gamma(m)} \left(\frac{\Omega}{m}\right)^{1/2}. \tag{2.73}$$

The mean is

$$E(X) = \frac{\Gamma\left(m + \frac{1}{2}\right)}{\Gamma(m)} \sqrt{\frac{\Omega}{m}}. \tag{2.74}$$

The variance is

$$\mathrm{var}(X) = \Omega\left[1 - \frac{1}{m}\left(\frac{\Gamma(m + (1/2))}{\Gamma(m)}\right)^2\right]. \tag{2.75}$$

Note that the Nakagami pdf becomes the Rayleigh density function when m is equal to unity. The square of the Nakagami random variable will have a gamma pdf. Under certain limiting conditions, the Nakagami density function can approximate the lognormal distribution (Nakagami 1960; Coulson et al. 1998a, b). The Nakagami pdf is plotted in Fig. 2.14.

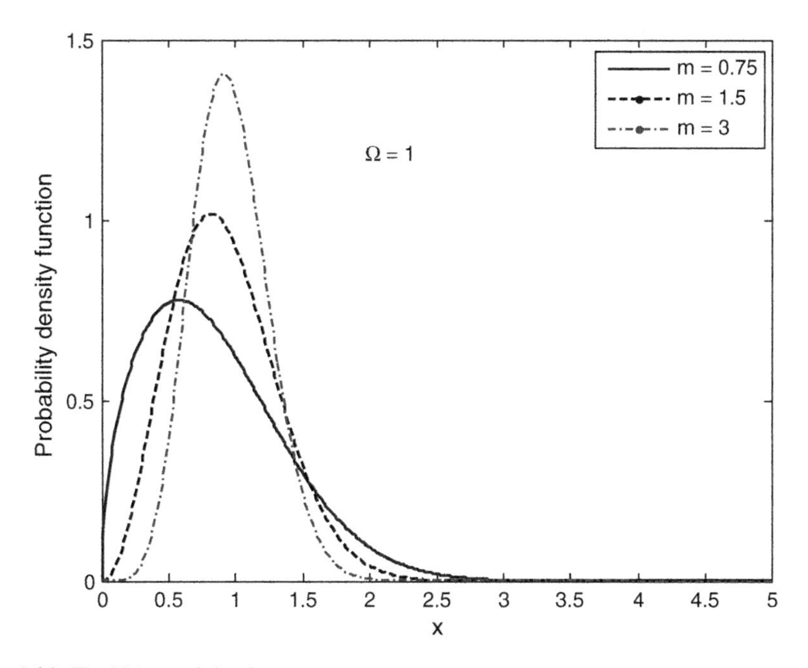

Fig. 2.14 The Nakagami density

The CDF associated with the Nakagami-m pdf can be expressed in terms of incomplete gamma functions. The Nakagami-m CDF is

$$F_X(x) = 1 - \frac{\Gamma(m, (m/\Omega)x^2)}{\Gamma(m)}. \tag{2.76}$$

There is no simple analytical expression for the CHF. The Nakagami pdf is used to model the magnitude or envelope of the signals in communication systems. The density function of the square of the Nakagami random variable,

$$Y = X^2 \tag{2.77}$$

with X having the Nakagami pdf in (2.72) is

$$f_Y(y) = 2\left(\frac{m}{\Omega}\right)^m \frac{y^{m-1}}{\Gamma(m)} \exp\left(-\frac{m}{\Omega}y\right), \quad y \geq 0, m \geq \frac{1}{2}. \tag{2.78}$$

It can be seen that there is no difference between the pdf of the power (square of the magnitude) expressed in (2.78) and the gamma pdf in (2.50) except that the order of the gamma pdf m must be larger than ½ for it be associated with the Nakagami pdf for the magnitude.

The CDF associated with the Nakagami-m distribution of power is

$$F_Y(y) = 1 - \frac{\Gamma(m, (m/\Omega)y)}{\Gamma(m)}. \tag{2.79}$$

In (2.79),

$$\Gamma\left(m, \frac{m}{\Omega}y\right) = \int_{(m/\Omega)y}^{\infty} x^{m-1} \exp(-x) dx \tag{2.80}$$

is the incomplete gamma function (Gradshteyn and Ryzhik 2007). The other forms of the Nakagami distribution such as the Nakagami-Hoyt and Nakagami-Rice will be discussed in Chap. 4 where specific statistical models for fading will be presented (Okui 1981; Korn and Foneska 2001; Subadar and Sahu 2009).

It is also possible to define a generalized form of the Nakagami-m pdf called the generalized Nakagami density function. This density function is obtained by the exponential scaling of the Nakagami variable X as (Coulson et al. 1998a, b; Shankar 2002a, b)

$$Y = X^{(1/\lambda)}, \quad \lambda > 0. \tag{2.81}$$

The density function of Y is the generalized density function given as

$$f(y) = \frac{2\lambda m^m y^{2m\lambda-1}}{\Gamma(m)\Omega^m} \exp\left(-\frac{m}{\Omega} y^{2\lambda}\right) U(y). \tag{2.82}$$

The generalized Nakagami pdf in (2.82) becomes the Rayleigh density function in (2.94) when m and λ are each equal to unity. It can be easily observed that the generalized gamma random variable is obtained by squaring the generalized Nakagami random variable.

2.4.14 Non-Central Chi-Squared Distribution

While we looked at chi-squared distribution which arises from the sum of the squares of zero mean identical random variables, the non-central chi-squared distribution arises when the Gaussian variables have non-zero means. The density function can be expressed in several forms as (Evans et al. 2000; Papoulis and Pillai 2002; Wolfram 2011)

$$f(x) = \frac{\sqrt{\lambda}}{2(\lambda x)^{r/4}} \exp\left(-\frac{x+\lambda}{2}\right) x^{(r-1)/2} I_{(r/2)-1}\left(\sqrt{\lambda x}\right), \quad r > 0, \tag{2.83}$$

$$f(x) = \frac{1}{2^{-(r/2)}} \exp\left(-\frac{x+\lambda}{2}\right) x^{(r-1)/2} \sum_{k=0}^{\infty} \frac{(\lambda x)^k}{2^{2k} k! \Gamma(k + (r/2))}, \quad r > 0, \tag{2.84}$$

$$f(x) = 2^{-(r/2)} \exp\left(-\frac{x+\lambda}{2}\right) x^{(r-2)-1} {}_0F_1\left([], \left[\frac{r}{2}\right], \frac{\lambda x}{4}\right), \quad r > 0. \tag{2.85}$$

When λ is zero, the non-central chi-squared distribution becomes the chi squared distribution. In (2.83), $I(.)$ is the modified Bessel function of the first kind and in (2.85) $F(.)$ is the hypergeometric function (Gradshteyn and Ryzhik 2007). The mean and variance of the pdf in (2.83) are

$$\langle X \rangle = \lambda + r, \tag{2.86}$$

$$\mathrm{var}(x) = 2(2\lambda + r). \tag{2.87}$$

This distribution in its limiting form becomes the Rician distribution (Rice 1974; Papoulis and Pillai 2002), which is discussed later. This density function also occurs in clustering-based modeling of short-term fading in wireless systems. The CDF can be expressed in terms of Marcum Q functions (Helstrom 1968; Nuttall 1975; Helstrom 1992, 1998; Chiani 1999; Simon 2002; Gaur and Annamalai 2003; Simon and Alouini 2003).

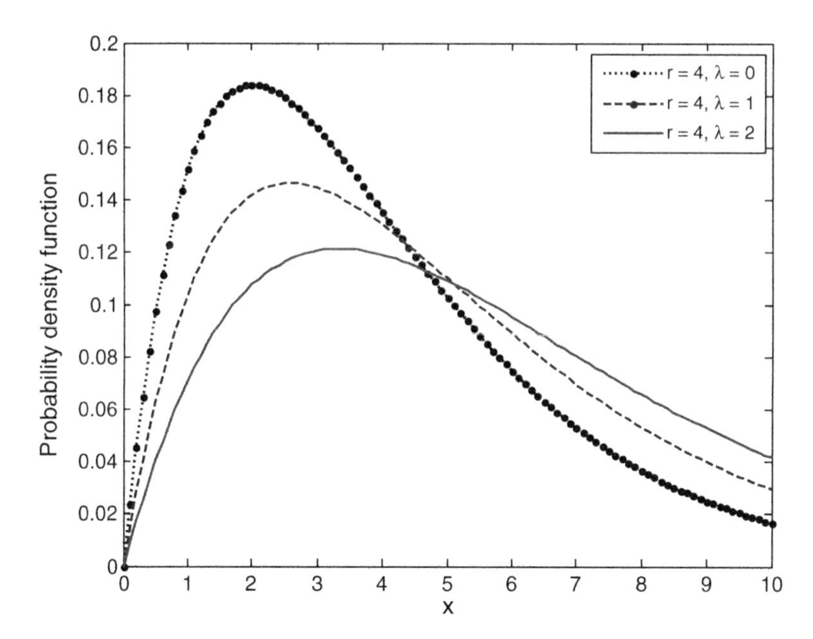

Fig. 2.15 Non-central chi-squared density function

The non-central chi-squared distribution is plotted in Fig. 2.15 for a few values of r and λ.

2.4.15 Normal (Gaussian) Distribution

The normal pdf is expressed as

$$f(x)=\frac{1}{\sqrt{2\pi\sigma^2}}\exp\left[-\frac{(x-\mu)^2}{2\sigma^2}\right],\quad -\infty<x<\infty. \tag{2.88}$$

The CDF can be expressed in terms of error functions or Q functions (Haykin 2001; Proakis 2001; Sklar 2001; Papoulis and Pillai 2002; Simon and Alouini 2005). The CDF is

$$F(x)=1-Q\left(\frac{x-\mu}{\sigma}\right). \tag{2.89}$$

In (2.89), the Q function is given by

$$Q(\alpha)=\int_{\alpha}^{\infty}\frac{1}{\sqrt{2\pi}}\exp\left(-\frac{w^2}{2}\right)dw. \tag{2.90}$$

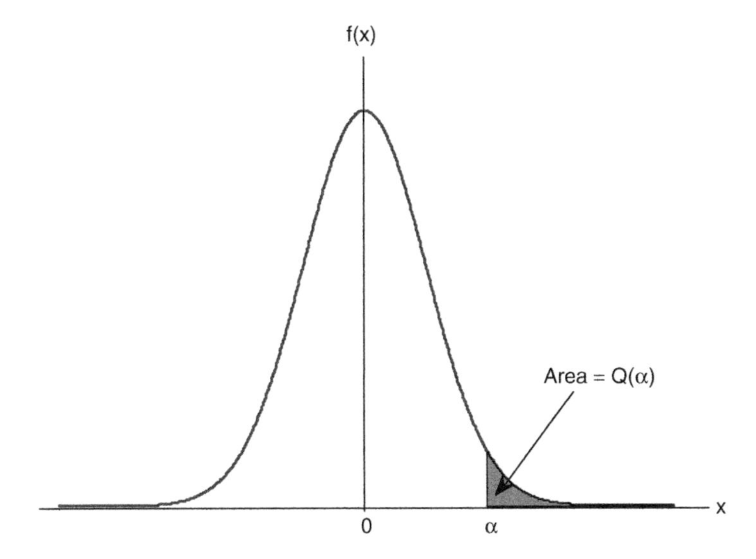

Fig. 2.16 The CDF of the Gaussian variable

The shaded area in Fig. 2.16 represents the value of the *Q*(.). The function and its properties are discussed in detail in Chap. 3.

The CHF of the normal random variable is

$$\psi(\omega) = \exp\left(j\mu\omega - \frac{1}{2}\sigma^2\omega^2\right). \tag{2.91}$$

The mean is μ and the standard deviation is σ. The mode and median are also equal to μ. Note that the normal pdf is used in the modeling of white noise (zero mean) in communication systems. The pdf of the sum of the squares of two independent identically distributed (zero mean) normal random variables leads to an exponentially distributed random variable with a pdf in (2.44). The sum of the squares of several Gaussian random variables (zero mean) also leads to a chi-squared distribution.

The normal distribution in (2.88) is identified in literature as $N(\mu,\sigma)$.

2.4.16 Poisson Distribution

The Poisson distribution is of the discrete type and is commonly used in communications to model the frequency of telephone calls being made (Stuber 2000; Papoulis and Pillai 2002; Molisch 2005; Gallager 2008). Since the random variable is of the discrete type, we need the probability that the number of outcomes equals a specific non-negative integer. This can be expressed as

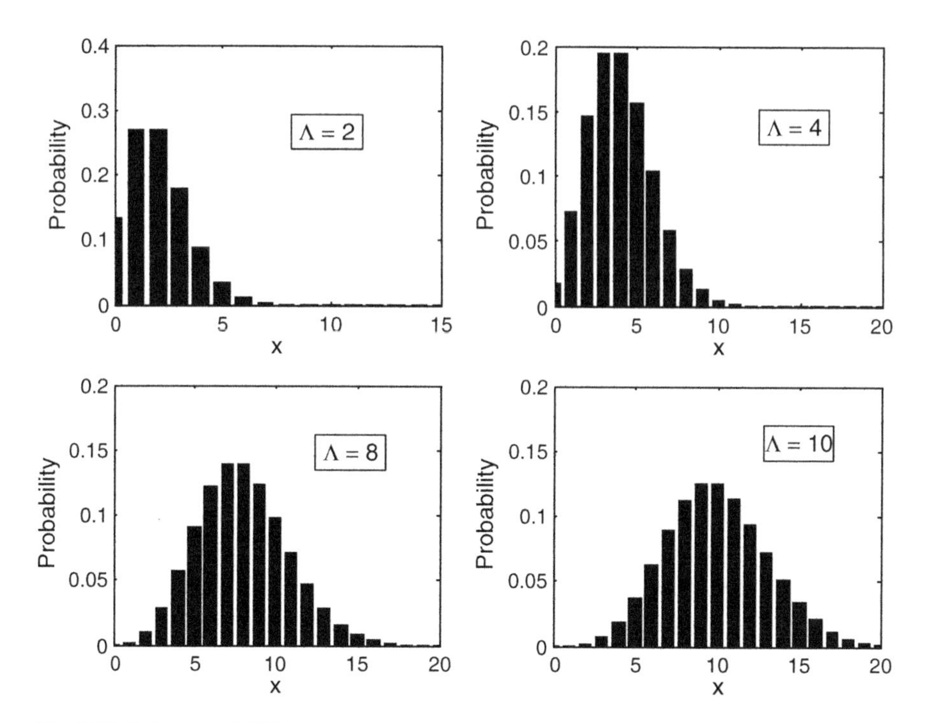

Fig. 2.17 Poisson probabilities

$$\text{Prob}\{X = k\} = \frac{\Lambda^k}{k!}\exp(-\Lambda), \quad k = 0, 1, 2, \ldots. \tag{2.92}$$

In (2.92), Λ is the average of the Poisson variable. It is also equal to the variance. When Λ increases, the density function approaches the normal or Gaussian pdf (Papoulis and Pillai 2002). This is illustrated in Fig. 2.17.

2.4.17 Rayleigh Distribution

The short-term fading observed in wireless channels is modeled by treating the magnitude of the signal as having the Rayleigh pdf (Jakes 1994; Schwartz et al. 1996; Sklar 1997a, b; Steele and Hanzó 1999; Patzold 2002; Rappaport 2002; Shankar 2002a, b). The density function results from the square root of the sum of two independent and identically distributed zero mean Gaussian random variables. In other words, if X_1 and X_2 are independent and identically distributed zero mean Gaussian variables, the Rayleigh variable X will be

$$X = \sqrt{X_1^2 + X_2^2}. \tag{2.93}$$

The Rayleigh pdf can be expressed as

$$f(x) = \frac{x}{\beta}\exp\left(-\frac{x^2}{2\beta}\right), \quad 0 < x < \infty. \tag{2.94}$$

The CDF can be expressed as

$$F(x) = 1 - \exp\left(-\frac{x^2}{2\beta}\right). \tag{2.95}$$

There is no simple expression for the CHF. The mean and variance can be expressed as

$$E(X) = \sqrt{\frac{\beta\pi}{2}}, \tag{2.96}$$

$$\mathrm{var}(X) = \left(2 - \frac{\pi}{2}\right)\beta. \tag{2.97}$$

Note that the Rayleigh random variable and exponential random variable are related, with the square of a Rayleigh random variable having an exponential distribution. The Rayleigh density function is also the special case of the Nakagami pdf in (2.72) when the Nakagami parameter $m = 1$. The Rayleigh pdf is shown in Fig. 2.18. It must be noted that the Rayleigh pdf is characterized by the fact that the

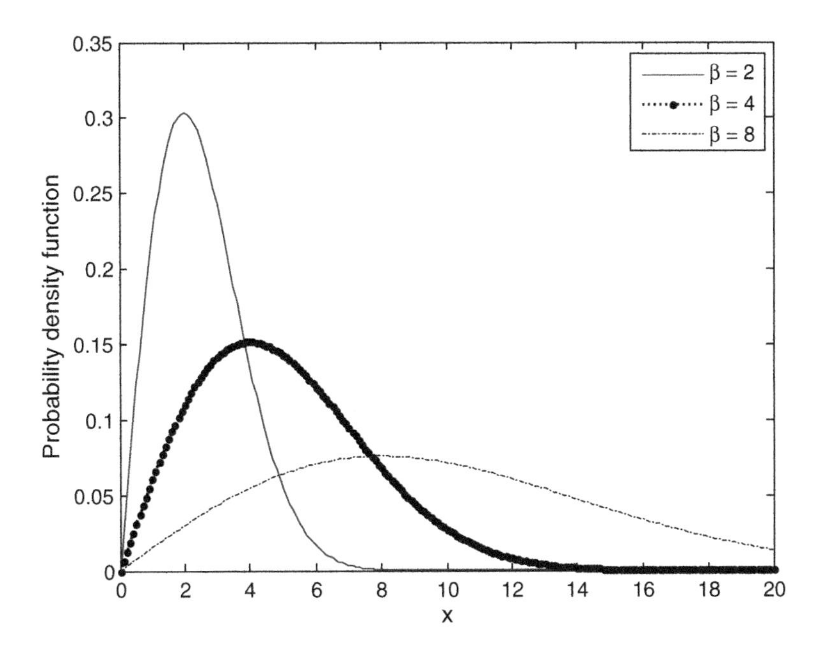

Fig. 2.18 Plot of Rayleigh pdf for three values of β

ratio of its mean to its standard deviation is fixed at 1.91 and has no dependence on the parameter β. Thus, regardless of the fact that the peak of the density function moves to the right as β increases, such a shift would have no impact on the level of fading in Rayleigh fading channels (as we will see in Chap. 4).

There is also another density function closely related to the Rayleigh pdf. This is the so called Generalized Rayleigh distribution with a density function (Blumenson and Miller 1963; Kundu and Raqab 2005; Voda 2009) of

$$f(x) = 2\alpha\beta x\exp\left(-\beta x^2\right)\left[1 - \exp\left(-\beta x^2\right)\right]^{\alpha-1}, \quad 0 < x < \infty. \tag{2.98}$$

Note that (2.98) becomes the conventional Rayleigh density function when $\alpha = 1$. Another form of generalized Rayleigh is identified by the pdf

$$f(x) = \frac{m}{b^2\Gamma(2/m)} x\exp\left(-\frac{x^m}{b^m}\right). \tag{2.99}$$

Equation (2.99) is a special case of the generalized Gamma distribution in (2.54). It becomes the simple Rayleigh density when $m = 2$. There is no analytical expression for the CDF associated with the density function in (2.99). The density function in (2.99) is shown in Fig. 2.19.

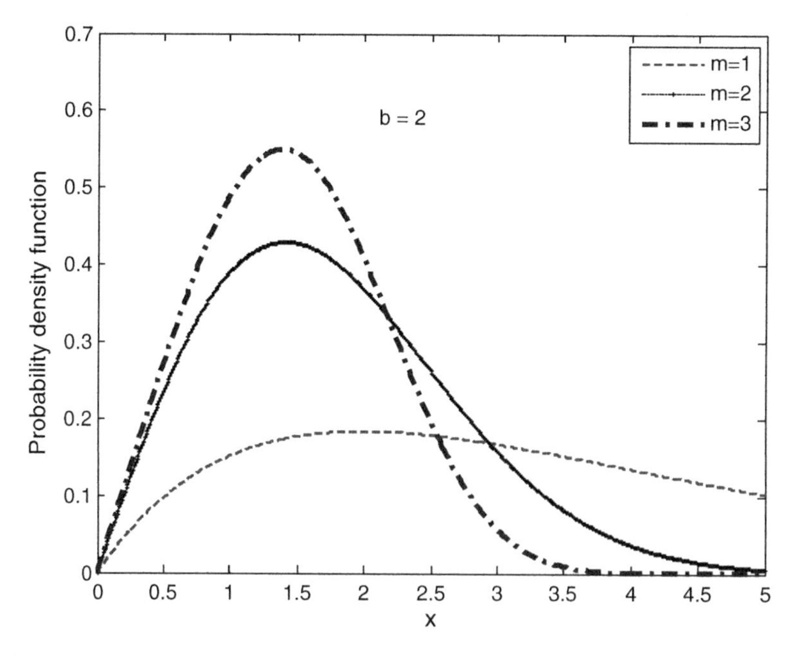

Fig. 2.19 The plot of the generalized Rayleigh pdf in (2.99) for three values of m and $b = 2$. The case of $m = 2$ corresponds to the Rayleigh pdf in (2.94)

There is another form of generalized Rayleigh distribution (Blumenson and Miller 1963; Kundu and Raqab 2005; Voda 2009) as follows:

$$f(x) = \frac{2\alpha^{k+1}}{\Gamma(k+1)} x^{2k+1} \exp\left(-\alpha x^2\right), \quad x \geq 0, \alpha > 0, k \geq 0. \tag{2.100}$$

Equation (2.100) becomes the Rayleigh density function when $k = 0$. Once again, there is no simple analytical expression for the CDF associated with the pdf in (2.100). It must also be noted that the generalized Rayleigh distribution in (2.100) is a more general form of the chi distribution in (2.39). Other aspects of the Rayleigh density function are discussed in Chap. 4.

2.4.18 Rectangular or Uniform Distribution

Uniform distribution is widely used in communication systems to model the statistics of phase (Stuber 2000; Papoulis and Pillai 2002; Shankar 2002a, b; Vaughn and Anderson 2003). If there are two independent normal random variables (X and Y) with zero means and identical variances, the pdf of the random variable

$$Z = \tan^{-}\left(\frac{Y}{X}\right) \tag{2.101}$$

will have a uniform distribution. The density function $f(z)$ can be expressed as

$$f(z) = \frac{1}{\beta - \alpha}, \quad \alpha < z < \beta. \tag{2.102}$$

The CDF is given by

$$f(z) = \frac{z - a}{\beta - \alpha}. \tag{2.103}$$

The CHF is given by

$$\psi(\omega) = \frac{[\exp(j\beta\omega) - \exp(j\alpha\omega)]}{j\omega(\beta - \alpha)}. \tag{2.104}$$

The mean and variance are given by

$$E(Z) = \frac{\alpha + \beta}{2}, \tag{2.105}$$

$$\mathrm{var}(Z) = \frac{(\beta - \alpha)^2}{12}. \tag{2.106}$$

In multipath fading, the phase is uniformly distributed in the range $[0,2\pi]$. The phase statistics can be displayed in two different ways. One is the conventional way of sketching the pdf as shown in Fig. 2.20. The other one is using polar plots, in which case uniformly distributed phase will appear as a circle. This is illustrated in Fig. 2.21 which was obtained by taking the histogram of several random numbers with the uniform pdf in the range $[0,2\pi]$. The histogram is seen as a circle.

The latter representation using the polar plot is convenient in understanding and interpreting the fading seen in Rician fading channels, as we will see in Chap. 4.

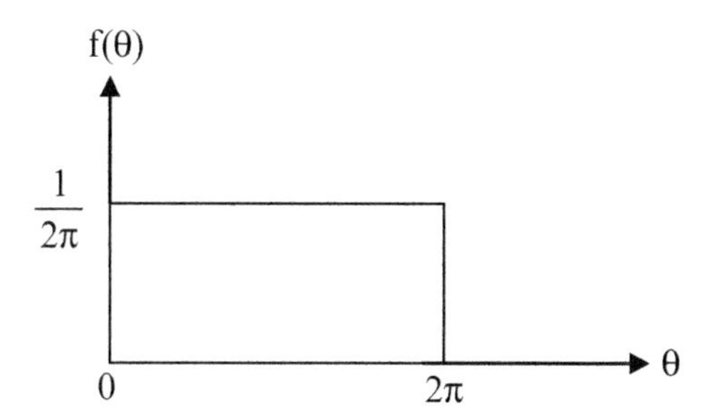

Fig. 2.20 Rectangular or uniform pdf

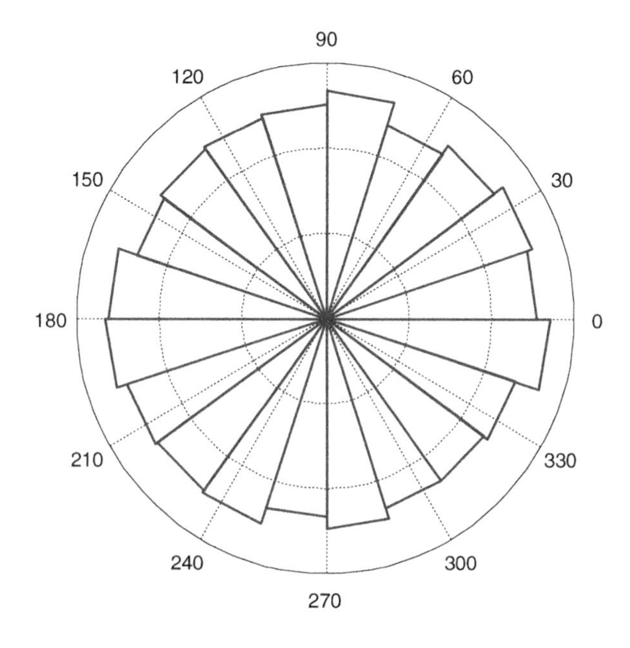

Fig. 2.21 Polar representation of the uniform random variable $[0, 2\pi]$

2.4.19 Student's t Distribution

The student t distribution in its shape looks very similar to the Gaussian and Cauchy distributions, and in the limiting cases it approaches either the Gaussian or the Cauchy distribution. Even though this density function is not directly used in wireless communications it can arise when we examine the ratio of a normal random variable to a normalized chi variable. This will be shown later in this chapter when we explore the random variables generated by mixing two or more random variables. This density function is also extensively used for testing to validate the statistical fits to densities.

A random variable z has a Student t distribution with n degrees of freedom if the pdf of Z is

$$fz(z) = \frac{\Gamma((n+1)/2)}{\sqrt{n\pi}\Gamma(n/2)} \frac{1}{\sqrt{(1+(z^2/n))^{n+1}}}, \qquad -\infty < x < \infty. \tag{2.107}$$

When $n = 1$, the Student t pdf in (2.107) becomes the Cauchy pdf in (2.32). The Student t density is often identified as $t(n)$.

As n becomes large, the density function in (2.107) approaches a Gaussian pdf since

$$\left(1+\frac{z^2}{n}\right)^{-((n+1)/2)} \rightarrow \exp\left(-\frac{z^2}{2}\right) \quad \text{as } n \rightarrow \infty. \tag{2.108}$$

A simple expression for the CDF is not readily available. The moments of the Student t distribution become

$$\langle Z^k \rangle = \begin{cases} 0, k\,\text{odd}, \\ \dfrac{1.3.5...(k-1)n^{k/2}}{(n-2)(n-4)...(n-k)}, & k\,\text{even}, \\ k, k\,\text{even}, n > k. \end{cases} \tag{2.109}$$

Thus, this random variable has a mean of zero and variance of $n = (n-2)$; $n > 2$.

The student t distribution is shown in Fig. 2.22 for three values of n. The heavier "tails" of the pdf are clearly seen at the lower values of n, indicating that the Student t distribution belongs to a class of density functions with "heavy tails" or heavy tailed distributions (Bryson 1974).

The transition to the Gaussian distribution is demonstrated in Fig. 2.23 where the CDFs of the Student t variable and the normal variable with identical variance are plotted. As n goes from 4 to 10, the CDFs almost match, indicating that with increasing values of n, the Student t distribution approaches the Gaussian distribution.

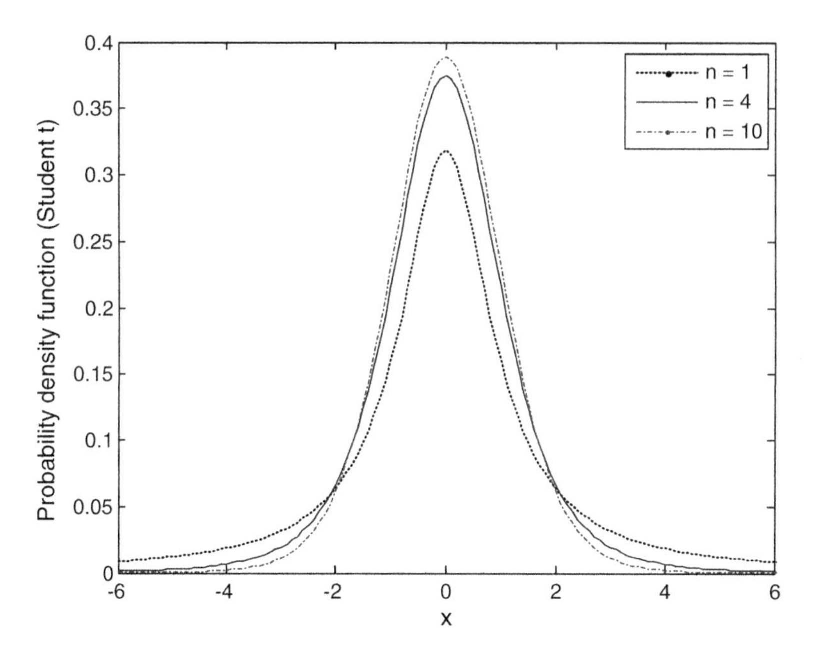

Fig. 2.22 Student *t* distribution

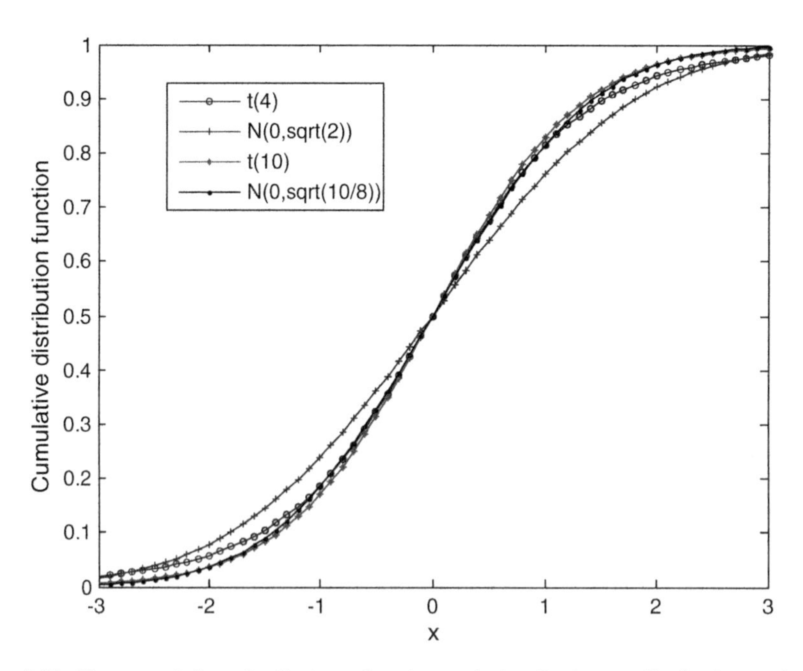

Fig. 2.23 The cumulative distributions functions of the Student *t* distribution and the corresponding Gaussian CDF with identical variances are shown

2.4.20 Weibull Distribution

In wireless communication systems, the Weibull pdf is also used to model the SNR in short-term fading (Shepherd 1977; Cheng et al. 2004; Sahu and Chaturvedi 2005; Alouini and Simon 2006; Ismail and Matalgah 2006). The Weibull density function can be expressed as

$$f(x) = \frac{\eta x^{\eta-1}}{\beta^{\eta}} \exp\left[-\left(\frac{x}{\beta}\right)^{\eta}\right], \quad 0 < x < \infty. \tag{2.110}$$

The Weibull pdf can also be written in a much simpler form as

$$f(x) = \alpha x^{\alpha-1} \exp(-x^{\alpha}), \quad x \geq 0, \alpha > 0. \tag{2.111}$$

The pdf in (2.111) is shown in Fig. 2.24. The CDF associated with the pdf in (2.110) is given by

$$F(x) = 1 - \exp\left[-\left(\frac{x}{\beta}\right)^{\eta}\right]. \tag{2.112}$$

There is no simple analytical expression for the CHF. The mean and variance can be expressed as

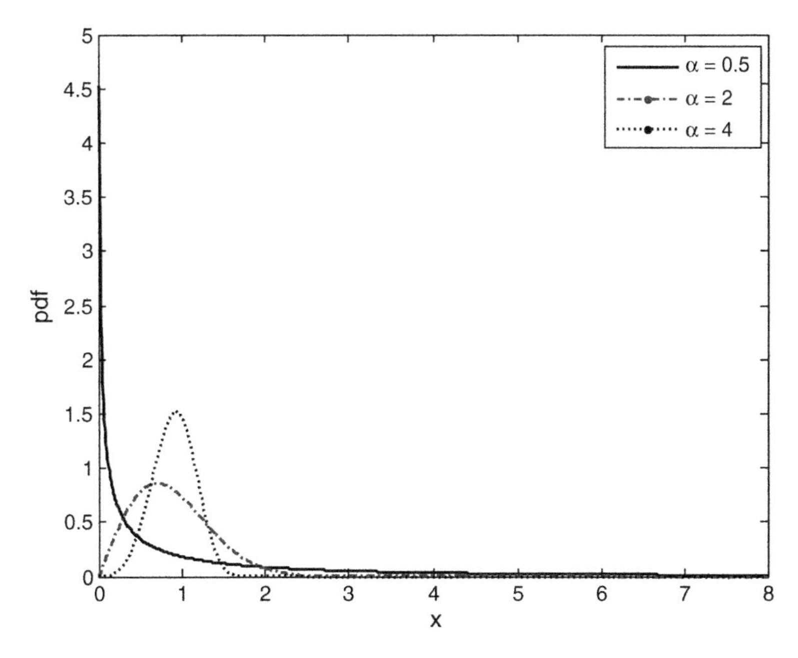

Fig. 2.24 The Weibull densities

$$E(X) = \beta\Gamma\left(1 + \frac{1}{\eta}\right), \tag{2.113}$$

$$\mathrm{var}(X) = \beta^2\left[\Gamma\left(1 + \frac{2}{\eta}\right) - \Gamma^2\left(1 + \frac{1}{\eta}\right)\right]. \tag{2.114}$$

Note that the Weibull and generalized gamma (GG) random variables are related. If we put $c = 1$ and $\eta = \lambda$ in (2.56), the GG variable becomes a Weibull random variable.

2.5 Joint, Marginal and Conditional Densities

If $X_1,\ldots, X_k$ are a set of random variables, the joint CDF is defined as (Rohatgi and Saleh 2001; Papoulis and Pillai 2002)

$$F(x_1, \ldots, x_k) = P(X_1 \leq x_1, \ldots, X_k \leq x_k). \tag{2.115}$$

The joint pdf is obtained by differentiating (2.115) with respect to $x_1,\ldots, x_k$,. We have

$$f(x_1, \ldots, x_k) = \frac{\partial^k}{\partial x_1 \ldots \partial x_k}[F(x_1, \ldots, x_k)]. \tag{2.116}$$

Marginal densities can be obtained from the joint density. The marginal density of x_1 is given as

$$f(x_1) = \int_{x_2=-\infty}^{\infty}\int_{x_3=-\infty}^{\infty}\cdots\int_{x_k=-\infty}^{\infty} f(x_1, \ldots, x_k)dx_2dx_3\ldots dx_k. \tag{2.117}$$

Similarly,

$$f(x_1, x_2) = \int_{x_3=-\infty}^{\infty}\int_{x_4=-\infty}^{\infty}\cdots\int_{x_k=-\infty}^{\infty} f(x_1, \ldots, x_k)dx_3\ldots dx_k. \tag{2.118}$$

If the random variables are independent, we have

$$f(x_1, \ldots, x_k) = f(x_1)f(x_2)\ldots f(x_k), \tag{2.119}$$

$$F(x_1, \ldots, x_k) = F(x_1)F(x_2)\ldots F(x_k). \tag{2.120}$$

We can also define conditional densities. If we have a joint pdf of two random variables $f(x,y)$, the conditional density function of Y, conditioned on $X = x$ is defined as

$$f(y|X = x) = f(y|x) = \frac{f(x,y)}{f(x)}. \tag{2.121}$$

It is obvious from (2.121), that if X and Y are independent, the conditional density of Y is unaffected by the presence of X. It is given by $f(y)$ itself. The conditioning expressed in (2.121) can also be extended to multiple random variables. On the other hand, if X and Y are dependent, the marginal density function of Y can be obtained as

$$f(y) = \int_{-\infty}^{\infty} f(y|x)f(x)dx = \int_{-\infty}^{\infty} f(x,y)dx. \tag{2.122}$$

Equation (2.122) can be interpreted as the Bayes Theorem for continuous random variables (Papoulis and Pillai 2002). Extending Eq. (2.121) to multiple random variables, we can express

$$f(x_n,\ldots,x_{k+1}|x_k,\ldots,x_1) = \frac{f(x_1,x_2,\ldots,x_k,\ldots x_n)}{f(x_1,\ldots,x_k)}. \tag{2.123}$$

2.6 Expectation, Covariance, Correlation, Independence, and Orthogonality

The expected value of a function of random variables, $g(x_1,\ldots, x_k)$ is defined as (Papoulis and Pillai 2002)

$$\begin{aligned} E[g(x_1,\ldots,x_k)] &= \langle g(x_1,\ldots,x_k)\rangle \\ &= \int_{-\infty}^{\infty}\cdots\int_{-\infty}^{\infty} g(x_1,\ldots,x_k)f(x_1,\ldots,x_k)dx_1\ldots dx_k. \end{aligned} \tag{2.124}$$

Let us now look at the case of two random variables, X and Y. The joint expected value of the product of the two random variables,

$$E(XY) = \int_{-\infty}^{\infty}\int_{-\infty}^{\infty} xyf(x,y)\,dx\,dy. \tag{2.125}$$

The *covariance* C_{xy} of two random variables is defined as

$$C_{xy} = E[(X-\eta_x)(Y-\eta_y)]. \tag{2.126}$$

In (2.126),

$$\eta_x = \int\int x f(x, y)\, dx\, dy \tag{2.127}$$

and

$$\eta_y = \int\int y f(x, y)\, dx\, dy. \tag{2.128}$$

Expanding (2.126), we have

$$C_{xy} = E(XY) - \eta_x \eta_y. \tag{2.129}$$

Note that if X and Y are independent random variables, using (2.119), we have

$$C_{xy} = 0 \tag{2.130}$$

The *correlation coefficient* of two random variables, ρ_{xy} is defined as

$$\rho_{xy} = \frac{C_{xy}}{\sigma_x \sigma_y}. \tag{2.131}$$

It can be shown that

$$\left|\rho_{xy}\right| \le 1. \tag{2.132}$$

Two random variables X and Y are said to be *uncorrelated* if

$$\rho_{xy} = 0 \quad \text{or} \quad C_{xy} = 0 \quad \text{or} \quad E(XY) = E(X)E(Y). \tag{2.133}$$

It can be easily seen that if the two random variables are independent, they will be uncorrelated. The converse is true only for Gaussian random variables. Two random variables are called *orthogonal* if

$$E(XY) = 0 \tag{2.134}$$

2.7 Central Limit Theorem

If we have n independent random variables X_i, $i = 1,2,\ldots,n$, the pdf $f(y)$ of their sum Y

$$Y = \sum_{i=1}^{n} X_i \tag{2.135}$$

approaches a normal distribution,

$$f_Y(y) \simeq \frac{1}{\sqrt{2\pi\sigma^2}} \exp\left[-\frac{(x-\eta)^2}{2\sigma^2}\right]. \tag{2.136}$$

In (2.136), η and σ are the mean and standard deviation of Y. If the random variables are identical, a lower value of n would be adequate for the pdf to become almost Gaussian while a higher value of n will be required if the random variables are not identical. In fact, for the case of independent identically distributed random variables with marginal pdfs that are smooth, a value of n of 5 or 6 would be enough to make the pdf of the sum approach the Gaussian distribution. Note that one of the requirements for the CLT to hold true is that none of the random variables can have variances of infinity. This would means that CLT is not applicable if the random variables have Cauchy densities. It can also be concluded that the chi-squared pdf in (2.35) will also approximate the Gaussian pdf when the number of degrees of freedom n is large.

Instead of the sum of n independent random variables, we can consider the product of n independent random variables, i.e.,

$$Z = \prod_{i=1}^{n} X_i. \tag{2.137}$$

Then the pdf of Z is approximately lognormal.

$$f_Z(z) \simeq \frac{1}{\sqrt{2\pi z\sigma^2}} \exp\left[-\frac{1}{2\sigma^2}(\ln y-\eta)^2\right]. \tag{2.138}$$

In (2.138),

$$\eta = \sum_{i=1}^{n} E(\ln X_i), \tag{2.139}$$

$$\sigma^2 = \sum_{i=1}^{n} \mathrm{var}(\ln X_i). \tag{2.140}$$

Note that in the equations above, var is the variance and ln is the natural logarithm. The central limit theorem for products can be restated by defining Y as

$$Y = \ln Z \sum_{i=1}^{n} \ln (X_i). \tag{2.141}$$

Now, if we use the CLT for the sum, the pdf of Y will be Gaussian. One of the justifications for the lognormal pdf for modeling the shadowing component in wireless channels arises from the fact that the shadowing process resulting in the lognormal pdf is a consequence of multiple scattering/reflections and consequent applicability of the central limit theorem for products. This aspect will be covered in Chap. 4 when we discuss the various models for describing, fading and shadowing. It will be possible to determine the lower limit on the value of n such that the central limit theorem for the products would hold.

2.8 Transformation of Random Variables

Although we examined the properties of random variables, often in wireless communications, the signals pass through filters. We are interested in obtaining the statistical properties of the outputs of the filters. The input to the filters may be single variable or multiple variables. For example, in systems operating in the diversity mode, the output might be the sum of the inputs or the strongest of the inputs (Brennan 1959). The output might also be a scaled version of the input. The output of interest might be the ratio of two random variables such as in the case of the detection of desired signals in the presence of cochannel interference.

We will now look at techniques to obtain the density functions of the outputs, knowing the density functions of the input random variables.

2.8.1 *Derivation of the pdf and CDF of* Y = g(X)

As mentioned above, in wireless communications it is necessary to determine the statistics of the signal when it has passed through filters, linear and nonlinear. For example, if a signal passes through a square law device or an inverse law device, we need to find out the density function of the output from the density function of the input random variable. As a general case, we are interested in obtaining the pdf of Y which is the output of a filter as shown in Fig. 2.25.

If we consider the transformation that is monotonic, i.e., $dy = dx$ is either positive or negative, it is easy to determine the pdf of Y knowing the pdf of X. An example of monotonic transformation is shown in Fig. 2.26. Starting with the definition of the random variable and CDF, the probability that the variable Y lies between y and $y + \Delta y$ is

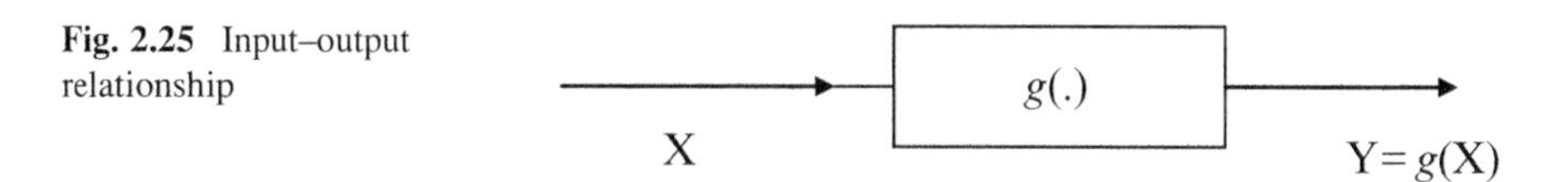

Fig. 2.25 Input–output relationship

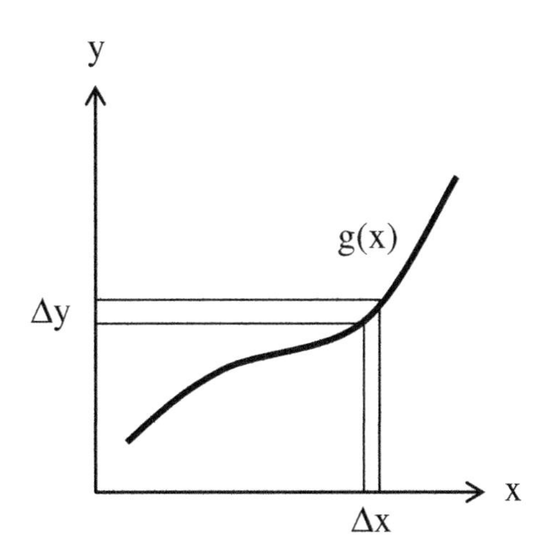

Fig. 2.26 Monotonic transformation of the random variable

$$P\{y < Y < y + \Delta y\} = P\{x < X < x + \Delta x\}. \tag{2.142}$$

Equation (2.142) is possible because of the monotonic nature of the transformation. Once again, using the definition of the CDF we can rewrite (2.142) as

$$f_Y(y)\Delta y = f_X(x)\Delta x. \tag{2.143}$$

By letting Δx and, hence, $\Delta y \to 0$, (2.143) becomes

$$f_Y(y) = \left.\frac{f_X(x)}{|dy/dx|}\right|_{x=g^{-1}(y)}. \tag{2.144}$$

The absolute sign in (2.144) merely reflects inclusion of both the monotonically increasing and decreasing nature of Y and shows that the pdf is always positive. The CDF can be found either from (2.144) or directly from the definition of the CDF as

$$F_Y(y) = P\{Y < y\} = P\{g(x) < y\} = F_X[g(x)]. \tag{2.145}$$

We can now consider the case of a non-monotonic transformation shown in Fig. 2.27.

If the transformation from X to Y is not monotonic as shown in Fig. 2.7, then, (2.144) can be modified to

$$f_Y(y) = \left.\frac{f_X(x)}{|dy/dx|}\right|_{x_1=g^{-1}(y)} + \left.\frac{f_X(x)}{|dy/dx|}\right|_{x_2=g^{-1}(y)} + \ldots \left.\frac{f_X(x)}{|dy/dx|}\right|_{x_n=g^{-1}(y)}. \tag{2.146}$$

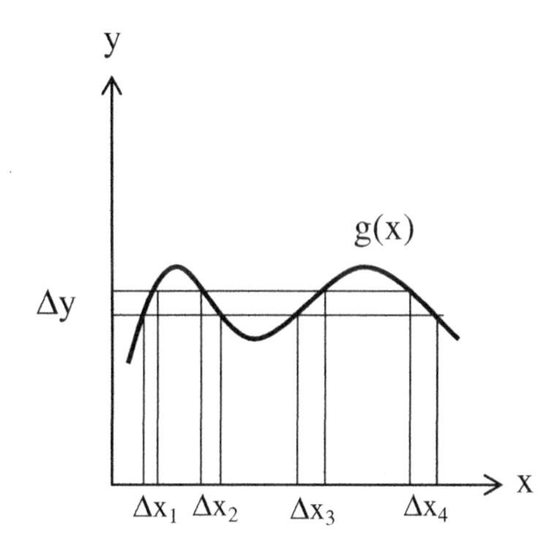

Fig. 2.27 Non-monotonic transformation. Multiple solutions are seen

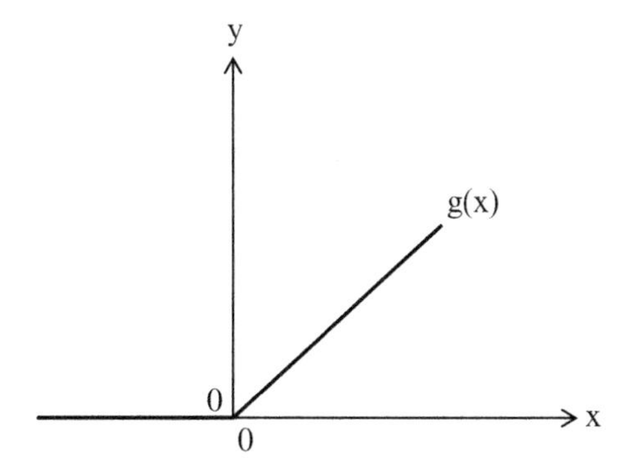

Fig. 2.28 Many-to-one transformation. For all negative values of X, there is only a single solution ($Y = 0$)

In (2.146), x_1, x_2,…, x_n are the n roots of the non-monotonic transformation between X and Y.

It is also possible that there may be instances where non-monotonic transformation might have infinite roots. Consider the case where X is a Gaussian random variable with a zero mean and standard deviation of s. For example, let us consider the case of a half wave rectifier, i.e.,

$$Y = \begin{cases} X, & X \geq 0, \\ 0, & X < 0. \end{cases} \tag{2.147}$$

This is shown in Fig. 2.28.

From (2.147), we have

$$F_Y(y) = \begin{cases} \frac{1}{2}, & y = 0, \\ F_X(y), & y > 0. \end{cases} \tag{2.148}$$

The pdf is obtained by differentiating keeping in mind that $Y = 0$ is an event with a probability of 0.5, resulting in

$$f_Y(y) = \frac{1}{2}\delta(y) + f_X(y)U(y). \quad (2.149)$$

2.8.2 Probability Density Function of Z = X + Y

Let us find out the density function of the sum of two random variables

$$Z = X + Y. \quad (2.150)$$

Using the fundamental definition of the probability, we have

$$F_Z(z) = P\{Z < z\} = P\{X + Y < z\} = \iint_{x+y<z} f(x, y)dx\,dy. \quad (2.151)$$

The region defined by (2.151) is shown in Fig. 2.29.
Rewriting (2.151),

$$F_Z(z) = \int_{y=-\infty}^{\infty} \int_{x=-\infty}^{\infty} f(x, y)dxdy. \quad (2.152)$$

The pdf $f(z)$ is obtained by differentiating (2.152) with respect to z. Using the Leibnitz rule (Gradshteyn and Ryzhik 2007), we get

$$f_Z(z) = \int_{-\infty}^{\infty} f(z - y, y)\,dy. \quad (2.153)$$

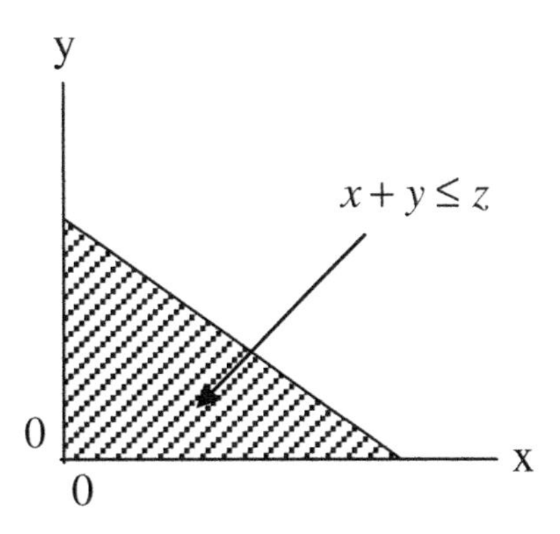

Fig. 2.29 The region defined by $x + y < z$

If X and Y are independent, (2.153) becomes

$$f_Z(z) = \int_{-\infty}^{\infty} f_X(z-y) f_Y(y)\, dy. \tag{2.154}$$

Equation (2.154) shows that the pdf of the sum of two independent random variables is the convolution of the marginal density functions. If X and Y exist only for positive values, (2.154) becomes

$$f_Z(z) = \int_0^z f_X(z-y) f_Y(y)\, dy. \tag{2.155}$$

2.8.3 *Joint pdf of Functions of Two or More Random Variables*

We will now look at ways of obtaining the joint pdf of two variables which are functions of two random variables. For example, if U and V are two random variables given expressed as

$$\begin{aligned} U &= g(X,Y), \\ V &= h(X,Y). \end{aligned} \tag{2.156}$$

We are interested in obtaining $f(u,v)$ given the joint pdf of X and Y, namely $f(x,y)$. Extending the concept used in the derivation of (2.144) and (2.146), the joint pdf of U and V can be expressed as

$$f_{U,V(u,v)} = \left. \frac{f_{X,Y}(x,y)}{|J(x,y)|} \right|_{x=[u,v]^{-1},\, y=[u,v]^{-1}}. \tag{2.157}$$

In (2.157), $J(x,y)$ is the Jacobian given by

$$J(x,y) = \begin{vmatrix} \dfrac{\partial u}{\partial x} & \dfrac{\partial u}{\partial y} \\ \dfrac{\partial v}{\partial x} & \dfrac{\partial v}{\partial y} \end{vmatrix}. \tag{2.158}$$

If one is interested in the pdf of U or V, it can easily be obtained as

$$f_U(u) = \int f(u,v) dv. \tag{2.159}$$

We can also use (1.16) to obtain the pdf of the sum of two random variables. Defining

$$\left.\begin{array}{l} Z = X + Y \\ V = Y \end{array}\right\}. \tag{2.160}$$

The Jacobian will be

$$J(x, y) = 1. \tag{2.161}$$

We have

$$f(z, v) = f_{x,y}(z - v, v). \tag{2.162}$$

Using (2.159), we have

$$f_Z(z) = \int_{-\infty}^{\infty} f_{z,v}(z - v, v)\, dv. \tag{2.163}$$

Equation (2.163) is identical to (2.153) obtained earlier directly. We will now find the density function of the sum of two-scaled random variables such as

$$W = aX + bY. \tag{2.164}$$

In (2.164), a and b are real-valued scalars. Defining an auxiliary variable

$$V = Y, \tag{2.165}$$

we have the Jacobian of the transformation as

$$J(x, y) = a. \tag{2.166}$$

Using (2.157), we have the joint pdf

$$f(w, v) = \left.\frac{f_{X,Y}(x, y)}{|J(x, y)|}\right|_{y=v;x=\frac{w-by}{a}} = \frac{1}{|a|} f_{X,Y}\left(\frac{w - by}{a}, y\right). \tag{2.167}$$

The density function of the random variable in (2.164) now becomes

$$f(w) = \frac{1}{|a|} \int_{-\infty}^{\infty} f_{X,Y}\left(\frac{w - by}{a}, y\right) dy. \tag{2.168}$$

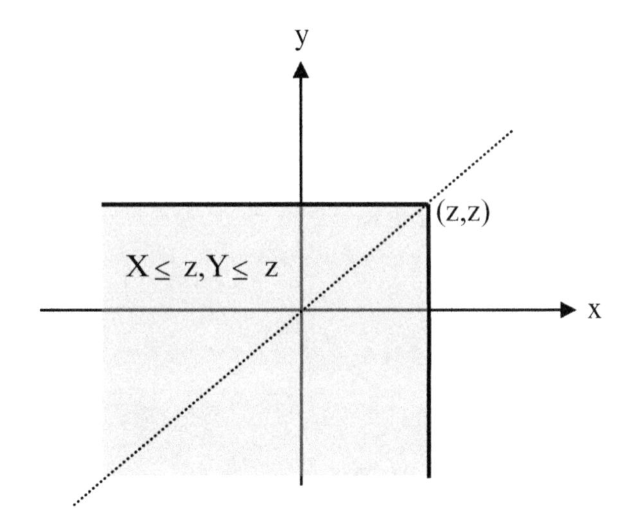

Fig. 2.30 The region of interest to obtain the pdf of the maximum of two random variables

We will look at a few more cases of interest in wireless systems. For example, one of the diversity combining algorithm uses the strongest signal from several branches. Let us determine the density function of

$$Z = \text{Max}\{X, Y\}. \tag{2.169}$$

The CDF of Z becomes

$$F_Z(z) = P\{\text{Max}(X, Y) < z\} = P\{X < z, Y < z\} = F_{X,Y}(z, z). \tag{2.170}$$

The CDF is the volume contained in the shaded area in Fig. 2.30.

If X and Y are independent, (2.170) becomes

$$F_Z(z) = F_X(z)F_Y(z). \tag{2.171}$$

The pdf of the maximum of two independent random variables is obtained by differentiating (2.171) w.r.t. z as

$$f_Z(z) = f_X(z)F_Y(z) + f_Y(z)F_X(z). \tag{2.172}$$

Furthermore, if X and Y are identical, (2.172) becomes

$$f_Z(z) = 2f_X(z)F_X(z) = 2f_Y(z)F_Y(z). \tag{2.173}$$

We can easily find out the density function of the minimum of two random variables. Let us define W as

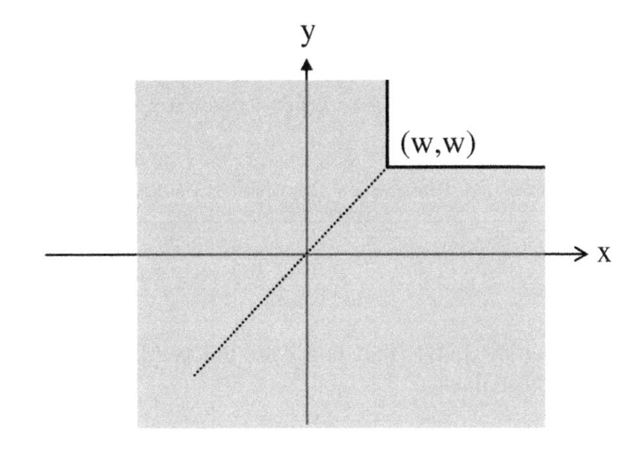

Fig. 2.31 The region of interest to obtain the CDF of the minimum of two variables

$$W = \mathrm{Min}\{X, Y\} = \begin{cases} Y, & X > Y, \\ X, & X \le Y. \end{cases} \tag{2.174}$$

The CDF of W will be

$$F_W(w) = P\{\mathrm{Min}(X,Y) < w\} = P\{Y < w, X > Y\} + P\{X < w, X \le Y\}. \tag{2.175}$$

The volume contained in the shaded area in Fig. 2.31 corresponds to this CDF. Assuming X and Y to be independent, (2.175) can result in

$$\begin{aligned} F_W(w) &= 1 - P\{W > w\} = 1 - P\{X > w, Y > w\} \\ &= 1 - P\{X > w\}P\{Y > w\}. \end{aligned} \tag{2.176}$$

Equation (2.176) simplifies to

$$F_W(w) = F_X(w) + F_Y(w) - F_X(w)F_Y(w). \tag{2.177}$$

We get the pdf by differentiating (2.177) w.r.t w, and we have

$$f(w) = f_X(w) + f_Y(w) - f_X(w)F_Y(w) - f_Y(w)F_X(w). \tag{2.178}$$

Another random parameter of interest is the product of two random variables. If U is the product of the two random variables

$$U = XY. \tag{2.179}$$

The density function of U might be obtained in a couple of different ways. First, let us define a dummy variable V as X,

$$V = Y. \tag{2.180}$$

Thus, the joint pdf of U and V can be written using (2.157) as

$$f(u,v) = \frac{f_{X,Y}((u/v),v)}{|v|}. \tag{2.181}$$

The density function of the product of two random variables is obtained as

$$f_U(u) = \int_{-\infty}^{\infty} \frac{1}{|v|} f_{X,Y}\left(\frac{u}{v}, v\right) dv. \tag{2.182}$$

Assuming that both X and Y are independent and exist only for positive values, (2.182) becomes

$$f_U(u) = \int_{0}^{\infty} \frac{1}{|y|} f_{X,Y}\left(\frac{u}{y}, y\right) dy. \tag{2.183}$$

One can also obtain the pdf of U from the fundamental definition of the CDF as well. The CDF of U can be expressed as

$$F_U(u) = P\{XY < u\} = P\left\{Z < \frac{u}{y}\right\} = \int_0^{\infty} \int_0^{u/y} f(X,Y)\,dx\,dy. \tag{2.184}$$

The region with horizontal lines in Fig. 2.32 shows the region of interest for the calculation of the probability volume in (2.184).

The pdf is obtained by differentiating (2.184) w.r.t u and using the Leibniz's rule. Therefore, we have

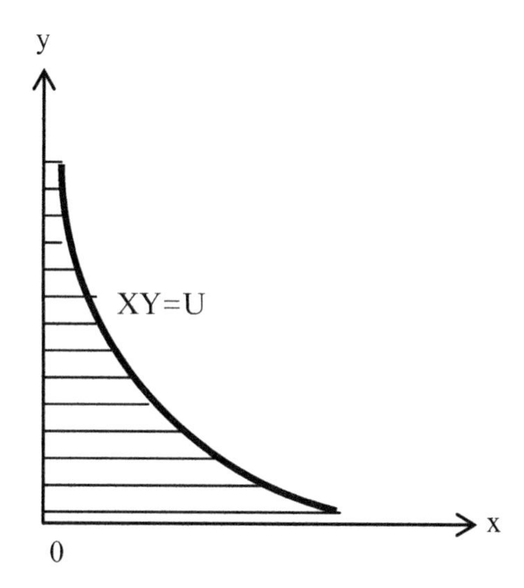

Fig. 2.32 The probability (volume) contained in the shaded region corresponds to the CDF

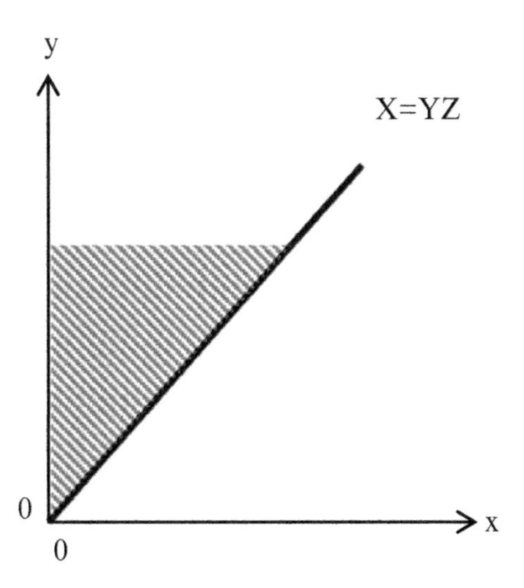

Fig. 2.33 The probability (volume) in the shaded region corresponds to the CDF $X = YZ$

$$f_U(u) = \int_{-\infty}^{\infty} \frac{1}{y} f_{X,Y}\left(\frac{u}{y}, y\right) dy. \tag{2.185}$$

Working similarly, if

$$Z = \frac{X}{Y}. \tag{2.186}$$

The pdf of the ratio of two random variables ($X,Y > 0$) becomes,

$$f_Z(z) = \int_0^{\infty} y f_{X,Y}(zy, y)\, dy. \tag{2.187}$$

It is also possible to obtain the CDF first. This was done for the case of the product of random variables in (2.184). Differentiating the CDF is given by the shaded region ($x > 0$, $y > 0$) in Fig. 2.33, we can also find the density function of

$$W = X^2 + Y^2. \tag{2.188}$$

Using the fundamental definition of CDF, we can write the CDF of the variable W as

$$F_W(w) = \text{Prob}\{X^2 + Y^2 < w\}. \tag{2.189}$$

Since $X^2 + Y^2 \le w$ represents a circle of radius $\sqrt{w}$, the CDF becomes

$$F_W(w) = \int_{y=-\sqrt{z}}^{\sqrt{z}} \int_{-\sqrt{z-y^2}}^{\sqrt{z-y^2}} f(x,y)\,dx\,dy. \tag{2.190}$$

The pdf is obtained by differentiating the CDF in (2.190) resulting in

$$f(w) = \int_{-\sqrt{w}}^{\sqrt{w}} \frac{1}{2\sqrt{w-y^2}} \left[f_{x,y}\left(\sqrt{w-y^2}, y\right) + f_{x,y}\left(-\sqrt{w-y^2}, y\right) \right] dy. \tag{2.191}$$

2.8.4 Use of CHF to Obtain pdf of Sum of Random Variables

One of the main uses of the CHF or the MGF is in diversity combining (Beaulieu 1990; Tellambura and Annamalai 2003; Annamalai et al. 2005). If the output of a diversity combining algorithm is given by the sum of several outputs such as

$$Y = X_1 + X_2 + \cdots + X_M, \tag{2.192}$$

and if we assume that the random variables $X_1, \ldots, X_M$ are independent, the pdf Y will be obtained by the M-fold convolution of the pdfs of those random variables. If the CHFs of X's are available, the CHF of Y can instead be expressed as

$$\psi_Y(\omega) = \langle \exp(j\omega X_1 + j\omega X_2 + \cdots j\omega X_M) \rangle. \tag{2.193}$$

Using the Fourier transform properties, we have

$$\psi_Y(\omega) = \prod_{k=1}^{M} \psi_{X_k}(\omega). \tag{2.194}$$

If the random variables X's are identical, (2.194) becomes

$$\psi_Y(\omega) = [\psi_X(\omega)]^M. \tag{2.195}$$

The pdf of Y can now be obtained using the inverse Fourier transform property as

$$f_Y(y) = \frac{1}{2\pi} \int_{-\infty}^{\infty} [\psi_X(\omega)]^M \exp(-j\omega y)\,d\omega. \tag{2.196}$$

Note that (2.196) is a single integral which replaces the M-fold convolution required if one were to use the marginal pdfs directly.

2.8.5 Some Transformations of Interest in Wireless Communications

We will now look at a few examples of transformations of random variables such as the sum or difference of two random variables, the sum of the squares of random variables, and the products and ratios of random variables.

Example #1 Let X and Y be two independent identically distributed random variables each with zero mean. The joint pdf is

$$f(x,y) = f(x)f(y) = \frac{1}{2\pi\sigma^2}\exp\left(-\frac{x^2+y^2}{2\sigma^2}\right). \tag{2.197}$$

We will find the joint pdf of

$$R = \sqrt{X^2+Y^2}, \tag{2.198}$$

$$\Theta = \tan^{-1}\left(\frac{Y}{X}\right). \tag{2.199}$$

Note that R is the envelope (or magnitude) and Θ is the phase. The Jacobian $J(x,y)$ defined in (2.158) for the set of these two variables

$$J(x,y) = \begin{vmatrix} \dfrac{\partial r}{\partial x} & \dfrac{\partial r}{\partial y} \\ \dfrac{\partial \theta}{\partial x} & \dfrac{\partial \theta}{\partial y} \end{vmatrix} = \begin{vmatrix} \dfrac{x}{\sqrt{x^2+y^2}} & \dfrac{y}{\sqrt{x^2+y^2}} \\ \dfrac{-y}{x^2+y^2} & \dfrac{x}{x^2+y^2} \end{vmatrix} = \frac{1}{r}. \tag{2.200}$$

The joint pdf now becomes

$$f(r,\theta) = \frac{f(x,y)}{|J(x,y)|} = \frac{r}{2\pi\sigma^2}\exp\left(-\frac{r^2}{2\sigma^2}\right), \quad 0 \le r \le \infty, 0 \le \theta \le 2\pi. \tag{2.201}$$

The marginal density function of R is

$$f(r) = \int_{-\pi}^{\pi} f(r,\theta)d\theta = \frac{r}{\sigma^2}\exp\left(-\frac{r^2}{2\sigma^2}\right), \quad 0 \le r \le \infty. \tag{2.202}$$

The marginal density function of the phase is

$$f(\theta) = \int_{0}^{\infty} f(r,\theta)dr = \frac{1}{2\pi}, \quad 0 \le \theta \le 2\pi. \tag{2.203}$$

Note that R and Θ are independent with R having a Rayleigh distribution and Θ having a uniform distribution. Often, the range of the phase is also expressed as $-\pi < \theta < \pi$. A discussion on the phase statistics appears in connection with the next example.

Example #2 Another related case of interest arises when one of the Gaussian random variables in Example #1 has a non-zero mean. Let

$$f(x,y) = \frac{1}{2\pi\sigma^2}\exp\left[-\frac{(x-A)^2}{2\sigma^2}\right]\exp\left(-\frac{y^2}{2\sigma^2}\right). \tag{2.204}$$

Our interest is still the joint and marginal pdfs of R and Θ in (2.198) and (2.199). The Jacobian for the transformation will be unaffected by the existence of the mean A of the random variable X. Thus, the joint pdf becomes

$$f(r,\theta) = \frac{r}{2\pi\sigma^2}\exp\left(-\frac{r^2+A^2}{2\sigma^2}\right)\exp\left[\frac{rA\cos(\theta)}{\sigma^2}\right], \quad 0 \le r \le \infty, 0 \le \theta \le 2\pi. \tag{2.205}$$

The pdf of the magnitude R is

$$f(r) = \frac{r}{\sigma^2}\exp\left(-\frac{r^2+A^2}{2\sigma^2}\right)\int_0^{2\pi}\frac{1}{2\pi}\exp\left[\frac{rA\cos(\theta)}{\sigma^2}\right]d\theta. \tag{2.206}$$

Noting the relationship between the integral in (2.206) and the modified Bessel function of the first kind $I_0(.)$ (Abramowitz and Segun 1972; Gradshteyn and Ryzhik 2007)

$$I_0(w) = \frac{1}{2\pi}\int_0^{2\pi}\exp[w\cos(\theta)]\,d\theta. \tag{2.207}$$

We can write (2.206) as

$$f(r) = \frac{r}{\sigma^2}\exp\left[-\frac{r^2-A^2}{2\sigma^2}\right]I_0\left(\frac{rA}{\sigma^2}\right), \quad 0 \le r \le \infty. \tag{2.208}$$

Equation (2.208) is known as the Rician distribution of the magnitude or envelope. This pdf arises in wireless systems when a direct path exists between the transmitter and receiver in addition to the multiple diffuse paths (Nakagami 1960; Rice 1974; Polydorou et al. 1999). When A $\to$ 0, the Rician pdf in (2.208) becomes the Rayleigh pdf in (2.202). Another interesting observation on the Rician pdf is its characteristics when A becomes large. If

$$\frac{A}{\sigma^2} \gg 1 \tag{2.209}$$

we can use the approximation (Abramowitz and Segun 1972)

$$I_0(x) = \frac{\exp(x)}{\sqrt{2\pi x}}. \tag{2.210}$$

The Rician pdf now becomes (strong direct path and weak multipath component),

$$f(r) = \frac{1}{\sqrt{2\pi\sigma^2}}\left(\frac{r}{A}\right)^{1/2}\exp\left[-\frac{1}{2\sigma^2}(r-A)^2\right]. \tag{2.211}$$

Equation (2.211) has an approximate Gaussian shape but for the factor of $\sqrt{r/A}$. This situation also arises in electrical communication systems when we examine the sum of a strong sine wave signal and a weak narrow band additive Gaussian noise.

We can now look at the density function of the power Z,

$$Z = R^2. \tag{2.212}$$

The pdf can be obtained using the properties of the transformation of variables as

$$f(z) = \frac{f(r)}{|dz/dr|}. \tag{2.213}$$

Defining the Rician factor K_0 as

$$K_0 = \frac{A^2}{2\sigma^2} \tag{2.214}$$

and

$$Z_R = 2\sigma^2 + A^2 \tag{2.215}$$

(2.213) becomes

$$f(z) = \frac{(1+K_0)}{Z_R}\exp\left[-K_0 - (1+K_0)\frac{z}{Z_R}\right] I_0\left[2\sqrt{K_0(1+K_0)\frac{z}{Z_R}}\right], \quad 0 \le z \le \infty. \tag{2.216}$$

Equation (2.216) is the Rician distribution of the power or SNR. Another point to note is that the density function of the phase Θ will not be uniform (Goodman 1985;

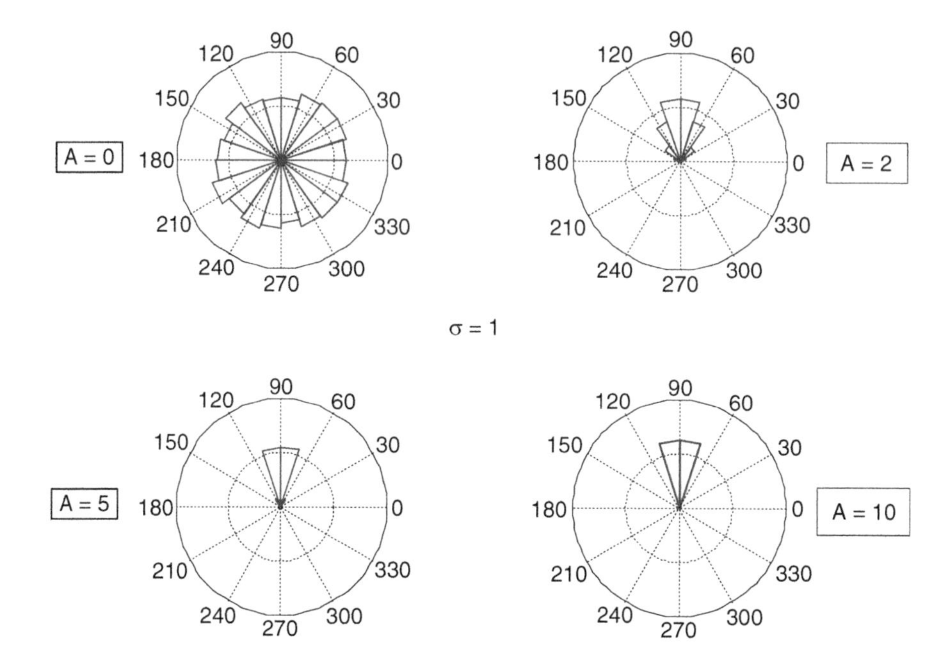

Fig. 2.34 The histogram of the phase associated with the Rician pdf

Papoulis and Pillai 2002). By observing the joint pdf of the envelope and phase in (2.205), we can also note that the variables R and Θ are not independent.

The density functions of the phase and the changes in the densities as the Rician factor changes can be observed in Fig. 2.34.

Example #3 Let X and Y be independent identically distributed exponential random variables. The joint density function becomes

$$f(x,y) = \frac{1}{a^2}\exp\left(-\frac{x+y}{a}\right), U(x)U(y). \tag{2.217}$$

We can find the pdf of

$$Z = X + Y. \tag{2.218}$$

Using (2.155), the pdf of Z becomes

$$f(z) = \int_0^z \frac{1}{a^2}\exp\left(-\frac{x}{a}\right)\exp\left[-\left(\frac{z-x}{a}\right)\right]dx = \frac{z}{a^2}\exp\left(-\frac{z}{a}\right)U(z). \tag{2.219}$$

Note that the pdf of Z in (2.219) is the Erlang distribution in (2.41) for the case of $c = 2$. If we now write

$$W = \sum_{k=1}^{n} X_k \tag{2.220}$$

where each of the random variable on the right hand side is exponentially distributed with the parameter a, proceeding similarly we can get the pdf of W to be

$$f(w) = \frac{1}{a^N} \frac{w^{N-1}}{\Gamma(N)} \exp\left(-\frac{w}{a}\right) U(w). \tag{2.221}$$

Equation (2.221) is the Erlang pdf in (2.41).

Example #4 We will repeat Example #2 when the two random variables are gamma distributed with a joint pdf

$$f(x,y) = \frac{(x/\beta)^{c-1} \exp(-(x/\beta))}{\beta\Gamma(c)} \frac{(y/\beta)^{c-1} \exp(-(y/\beta))}{\beta\Gamma(c)}, \quad 0 < x < \infty, 0 < y < \infty. \tag{2.222}$$

Let

$$R = X + Y. \tag{2.223}$$

Using (2.155), the pdf of R becomes

$$f(r) = \int_0^r \frac{(x/\beta)^{c-1} \exp(-(x/\beta))}{\beta\Gamma(c)} \frac{((r-x)/\beta)^{c-1} \exp(-(r-x)/\beta)}{\beta\Gamma(c)} dx. \tag{2.224}$$

Using the table of integrals (Gradshteyn and Ryzhik 2007), the pdf of the sum of the two gamma random variable becomes

$$f(r) = \frac{r^{2c-1}}{\beta^{2c}\Gamma(2c)} \exp\left(-\frac{r}{\beta}\right), \quad 0 \le r \le \infty. \tag{2.225}$$

In arriving at (2.225), we have made use of the following identity of doubling formula for gamma functions (Gradshteyn and Ryzhik 2007; Wolfram 2011)

$$\Gamma(2c) = \frac{2^{2c-1}}{\sqrt{\pi}} \Gamma(c)\Gamma\left(c + \frac{1}{2}\right). \tag{2.226}$$

A comparison of (2.225) and (2.50) suggests that R is also a gamma random variable with order $2c$ and mean of $2\beta c$. In other words, R is a gamma random variable of parameters $2c$ and β.

Example #5 Let us continue with the case of two independent and identically distributed exponential random variables with a joint pdf in (2.217). We will now find out the pdf of the product of the two variables,

$$U = XY. \tag{2.227}$$

Using the approach given in (2.182), we have

$$f_U(u) = \int_0^{\infty} \frac{1}{x} f_{X,Y}\left(\frac{u}{x}, x\right) dx = \int_0^{\infty} \frac{1}{a^2 x} \exp\left(-\frac{u}{ax}\right) \exp\left(-\frac{x}{a}\right) dx. \tag{2.228}$$

Using tables of integrals (Gradshteyn and Ryzhik 2007), Eq. (2.228) becomes

$$f(u) = \frac{2}{a^2} K_0\left(\frac{2}{a}\sqrt{u}\right), \quad 0 \le u \le \infty. \tag{2.229}$$

In (2.229), $K_0()$ is the modified Bessel function of the second kind of order zero (Gradshteyn and Ryzhik 2007). Note that the pdf in (2.229) arises when we examine the SNR in shadowed fading channels or cascaded channels (Shankar 2004; Bithas et al. 2006; Andersen 2002; Nadarajah and Gupta 2005; Salo et al. 2006; Nadarajah and Kotz 2006a, b).

If we define a new random variable as

$$W = \sqrt{U}, \tag{2.230}$$

we can identify the pdf of W as the double Rayleigh pdf (Salo et al. 2006). Using the property of the monotonic transformation of random variables in (2.144), we can write

$$f(w) = \frac{f(u)}{|du/dw|} = \left(2\sqrt{u}\right) \frac{2}{a^2} K_0\left(\frac{2}{a}\sqrt{u}\right) = \frac{4w}{a^2} K_0\left(\frac{2}{a} w\right), \quad 0 \le w \le \infty. \tag{2.231}$$

Example #6 We will repeat Example #5 for the case of two gamma random variables, independent and identically distributed with a joint pdf in (2.222).

Using (2.182) the pdf of the product becomes

$$f(u) = \int_0^{\infty} \frac{1}{x} \frac{(x/\beta)^{c-1} \exp(-(x/\beta))}{\beta \Gamma(c)} \frac{(u/\beta x)^{c-1} \exp(-(u/\beta x))}{\beta \Gamma(c)} dx. \tag{2.232}$$

Using the table of integrals (Gradshteyn and Ryzhik 2007), we have

$$f(u) = \frac{2}{\beta^{2c}\Gamma(c)} u^{c-1} K_0\left(\frac{2}{\beta}\sqrt{u}\right), \quad 0 \leq u \leq \infty. \tag{2.233}$$

Note that (2.233) becomes (2.229) when $c = 1$.

If we limit the value of $c > 1/2$, and the pdf of W in (2.230) is compared with the pdf of U in (2.233) the latter is identified as the double Nakagami pdf (Wongtrairat and Supnithi 2009; Shankar and Gentile 2010). In general, (2.233) is also known as the double gamma pdf and we will study its properties in Chap. 4 when we examine cascaded fading channels.

Using the procedure adopted in connection with (2.231), we get

$$\begin{aligned}(w) = \frac{f(u)}{\left|du/dw\right|} &= \left(2\sqrt{u}\right) \frac{2}{\beta^{2c}\Gamma^2(c)} u^{c-1} K_0\left(\frac{2}{\beta}\sqrt{u}\right) \\ &= \frac{4w^{2c-1}}{\beta^{2c}\Gamma^2(c)} \times K_0\left(\frac{2}{\beta}w\right), \quad 0 \leq w \leq \infty.\end{aligned} \tag{2.234}$$

Equation (2.234) is the pdf associated with the product of two Nakagami variables.

Example #7 Another interesting case in wireless systems is variable created from the product of two nonidentical gamma random variables. Let X and Y be two gamma distributed variables with pdfs

$$f(x) = \frac{(x/\alpha)^{c-1} \exp(-(x/\alpha))}{\alpha\Gamma(c)}, \tag{2.235}$$

$$f(y) = \frac{(y/\beta)^{m-1} \exp(-(y/\beta))}{\beta\Gamma(m)}. \tag{2.236}$$

Let

$$S = XY. \tag{2.237}$$

Once again, pdf of S can be written using (2.183) as

$$f(s) = \int_0^\infty \frac{1}{x} \frac{(x/\alpha)^{c-1} \exp(-(x/\alpha))}{\alpha\Gamma(c)} \frac{(s/\beta x)^{m-1} \exp(-(s/\beta x))}{\beta\Gamma(m)} dx. \tag{2.238}$$

Using the table of integrals (Gradshteyn and Ryzhik 2007), the pdf of S becomes

$$f(s) = \frac{2}{\left(\sqrt{\alpha\beta}\right)^{m+c}\Gamma(m)\Gamma(c)} s^{((m+c)/2)-1} \left(2\sqrt{\frac{s}{\alpha\beta}}\right), 0 \leq s \leq \infty. \tag{2.239}$$

The pdf in (2.239) is the gamma–gamma pdf or the generalized K distribution (Lewinsky 1983; McDaniel 1990; Abdi and Kaveh 1998; Anastassopoulos et al. 1999; Shankar 2004). Note that if $\alpha = \beta = b$ and $m = c$, (2.239) becomes (2.233). Furthermore, if $c = 1$, (2.233) is the so called K distribution or the K pdf (Jakeman and Tough 1987; Abdi and Kaveh 1998; Iskander et al. 1999).

Example #8 We will now extend the Example #7 to two generalized gamma random variables. Let

$$f(x) = \frac{\lambda x^{\lambda m-1}\exp\left[-(x/\alpha)^{\lambda}\right]}{\alpha^{\lambda m}\Gamma(m)}, \quad 0 \le x \le \infty, \tag{2.240}$$

$$f(y) = \frac{\lambda y^{\lambda n-1}\exp\left[-(y/\beta)^{\lambda}\right]}{\beta^{\lambda m}\Gamma(n)}, \quad 0 \le y \le \infty. \tag{2.241}$$

The pdf of $S = XY$ can be obtained using the integral in (2.183) as

$$f(s) = \frac{2\lambda s^{[(\lambda/2)(m+n)]-1}}{(\alpha\beta)^{(\lambda/2)(m+n)}\Gamma(m)\Gamma(n)} K_{m-n}\left[2\left(\frac{s}{\alpha\beta}\right)^{(\lambda/2)}\right], \quad 0 \le s \le \infty. \tag{2.242}$$

Expressing

$$\chi = \alpha\beta, \tag{2.243}$$

(2.242) becomes a four parameter pdf given as

$$f(s) = \frac{2\lambda s^{[(\lambda/2)(m+n)]-1}}{(\chi)^{(\lambda/2)(m+n)}\Gamma(m)\Gamma(n)} K_{m-n}\left[2\left(\frac{s}{\chi}\right)^{(\lambda/2)}\right], \quad 0 \le s \le \infty. \tag{2.244}$$

The pdf in (2.242) is known as the generalized Bessel K distribution (GBK) which becomes the GK distribution in (2.239) when $\lambda = 1$ and becomes the K pdf when $\lambda = 1$ and $n = 1$ (Iskander and Zoubir 1996; Anastassopoulos et al. 1999; Frery et al. 2002). Note that the GBK pdf is a five parameter distribution as in (2.242) with shape parameters m, n, and λ and scaling factors α and β or a four parameter distribution in (2.244) with shape parameters m, n, and λ and scaling factor χ. Several density functions can be obtained from the GBK pdf by varying some of these shape parameters and following the details in Table 2.3.

Example #9 Let us look at another case of interest in wireless communications where the new random variable is the sum of several gamma random variables (Moschopoulos 1985; Kotz and Adams 1964; Provost 1989; Alouini et al. 2001; Karagiannidis et al. 2006). Let

Table 2.3 The relationship of GBK distribution to other pdfs

GBK pdf $f(x;m,n,\chi,\lambda)=\frac{2\lambda x^{[(\lambda/2)(m+n)]-1}}{(\chi)^{(\lambda/2)(m+n)}\Gamma(m)\Gamma(n)}K_{m-n}\left[2\left(\frac{x}{\chi}\right)^{(\lambda/2)}\right]$	Probability density functions (special case)
$f(x;\,1,\infty,\chi,\,1)$	Exponential
$f(x;\,1,\infty,\chi,\,2)$	Rayleigh
$f(x;\,m,n,\chi,\,1)$	GK distribution
$f(x;\,m,\,1,\chi,\,1)$	*K* distribution
$f(x;\,m,\infty,\chi,\,1)$	Gamma distribution
$f(x;\,m,\infty,\chi,\lambda)$	Generalized gamma distribution
$f\left(x;\frac{1}{2},\infty,\chi,2\right)$	Half Gaussian

$$Z=\sum_{k=1}^{N}X_k. \tag{2.245}$$

In (2.245), there are N independent and identically distributed random variables X's, each with a pdf of the form in (2.50). We will use the relationship between density functions and CHFs to obtain the pdf in this case. Since the pdf of Z will be a convolution of the pdfs of N identical density functions, the CHF of Z can be written as the product of N identical CHFs, each of them of the form given in (2.53). The CHF of Z is

$$\psi_z(\omega)=[\psi_X(\omega)]^N=(1-j\beta\omega)^{-cN}. \tag{2.246}$$

The pdf of Z is now obtained from the Fourier relationship between CHF and pdf.

$$f_Z(z)=\frac{1}{2\pi}\int_{-\infty}^{\infty}\frac{1}{(1-j\beta\omega)^{cN}}\exp(-j\omega z)\,d\omega. \tag{2.247}$$

Using the Fourier integral tables (Gradshteyn and Ryzhik 2007), Eq. (2.247) becomes

$$f_Z(z)=\frac{z^{cN-1}}{\beta^{cN}\Gamma(cN)}\exp\left(-\frac{z}{\beta}\right),\quad z\geq 0. \tag{2.248}$$

From (2.248), it is seen that the sum of N identically distributed gamma random variables with parameters c and β is another gamma random variable with parameters cN and β.

Example #10 Shadowing in wireless systems is modeled using the lognormal pdf. In diversity systems, it might be necessary to obtain the density function of the sum of several lognormal random variables, each having a pdf of the form in (2.71). If

$$Z = \sum_{k=1}^{N} X_k \tag{2.249}$$

7a simple analytical expression for the density function of Z is not readily available. Several researchers have proposed approximate forms for the density function (Beaulieu et al. 1995; Slimane 2001; Beaulieu and Xie 2004; Cardieri and Yacoub 2005; Lam and Le-Ngoc 2006; Lam and Tho 2007; Liu et al. 2008). One such approximation results in the pdf of N independent identically distributed lognormal variables is expressed as a shifted gamma pdf (Lam and Le-Ngoc 2006)

$$f_Z(z) = \begin{cases} \dfrac{[10\log_{10}(z) - \delta]^{\alpha-1}}{[10/\log_e(10)\beta^{\alpha}\Gamma(\alpha)z]}\exp\left[-\dfrac{10\log_{10}(z) - \delta}{\beta}\right], & z > 10^{\delta/10} \\ 0, \quad z \leq 10^{\delta/10} \end{cases} \tag{2.250}$$

The three parameters, namely α, β, δ (all in decibel units), can be obtained by matching the first three moments of the variable in (2.249) and the moments of the pdf in (2.250).

Example #11 We will explore another interesting case in wireless communications involving the product of several random variables (Karagiannidis et al. 2007; Shankar 2010). Let Z be the product of N random variables which forms a set of independent random variables with the same density functions but with different parameters.

$$Z = \prod_{k=1}^{N} X_k \tag{2.251}$$

Let the density of X_k be given by a gamma pdf as

$$f(x_k) = \frac{x_k^{mk-1}}{b_k^{mk}\Gamma(m_k)}\exp\left(-\frac{x_k}{b_k}\right), \quad k = 1, 2, \ldots, N. \tag{2.252}$$

Using MGFs and Laplace transforms, the density function of Z can be obtained as (Kabe 1958; Stuart 1962; Podolski 1972; Mathai and Saxena 1973; Carter and Springer 1977; Abu-Salih 1983; Nadarajah and Kotz 2006a, b; Karagiannidis et al. 2007; Mathai and Haubold 2008):

$$f(z) = \frac{1}{z\Pi_{k=1}^{N}\Gamma(mk)} G_{0,N}^{N,0}\left(\frac{z}{\Pi_{k=1}^{N}\Gamma(b_k)}\middle| \begin{matrix} - \\ m_1, m_2, \ldots, m_N \end{matrix}\right) U(z). \tag{2.253}$$

In (2.253), G () is the Meijer's G-function. The CDF can be obtained using the differential and integral properties of the Meijer's G-function as (Springer and Thompson 1966, 1970; Mathai and Saxena 1973; Mathai 1993; Adamchik 1995):

$$F(z) = \frac{1}{\Pi_{k=1}^{N}\Gamma(mk)} G_{1,N+1}^{N,1}\left(\frac{z}{\Pi_{k=1}^{N}\Gamma(b_k)}\middle| \begin{matrix} 1 \\ m_1, m_2, \ldots, m_N, 0 \end{matrix}\right)U(z). \tag{2.254}$$

If X's are identical and $m_k = m = 1$ and $b_k = b$, we have the pdf of the products of exponential random variables,

$$f(z) = \frac{1}{z} G_{0,N}^{N,0}\left(\frac{z}{b^N}\middle| \begin{matrix} - \\ \underbrace{1, 1, \ldots, 1}_{N\text{-terms}} \end{matrix}\right)U(z). \tag{2.255}$$

Using the relationship between the Meijer's G-function and modified Bessel functions (Mathai and Saxena 1973; Wolfram 2011)

$$G_{0,1}^{1,0}\left(\frac{z}{b}\middle| \begin{matrix} - \\ m \end{matrix}\right) = \left(\frac{z}{b}\right)^m \exp\left(-\frac{z}{b}\right) \tag{2.256}$$

(2.253) becomes the gamma pdf for $N = 1$ and

$$G_{0,2}^{2,0}\left(\frac{z}{b^2}\middle| \begin{matrix} - \\ m, n \end{matrix}\right) = 2\left(\frac{z}{b^2}\right)^{(1/2)(m+n)} K_{m-n}\left(\frac{2}{b}\sqrt{z}\right) \tag{2.257}$$

(2.253) becomes the GK pdf which is obtained for the pdf of the product of two gamma random variables (Shankar 2005; Laourine et al. 2008).

It is also interesting to find out the pdf of the cascaded channels when lognormal fading conditions exist in the channel. In this case, we take

$$f(x_k) = \frac{[10/\log_e(10)]}{\sqrt{2\pi\sigma_k^2 x_k^2}} \exp\left[-\frac{(10\log_{10}(x_k) - \mu_k)^2}{2\sigma_k^2}\right], \quad 0 < x < \infty, \ k = 1, 2, \ldots, N. \tag{2.258}$$

The cascaded output in (2.251) can be expressed in decibel form as

$$W = 10\log_{10}(Z) = \sum_{k=1}^{N} 10\log_{10}(X_k). \tag{2.259}$$

Since X's are lognormal and each term in the summation in (2.259) is therefore a Gaussian random variable, the density function of the random variable W will be Gaussian,

$$f(w) = \frac{1}{\sqrt{2\pi \sum_k^N \sigma_k^2}} \exp\left[-\frac{\left(w - \sum_{k=1}^N \mu_k\right)^2}{2\sum_k^N \sigma_k^2} \right]. \tag{2.260}$$

Converting back, the density function of Z will be lognormal given by

$$f(z) = \frac{[10/\log_e(10)]}{\sqrt{2\pi\left(\sum_{k=1}^N \sigma_k^2\right) z^2}} \exp\left[-\frac{\left(10\log_{10}(z) - \sum_{k=1}^N \mu_k\right)^2}{2\sum_{k=1}^N \sigma_k^2} \right], \quad 0 < z < \infty. \tag{2.261}$$

Example #12 There is also interest in wireless communications to determine the density function of the ratio of two random variables. For example, the received signal generally has to be compared against an interference term which is also random (Winters 1984; Cardieri and Rappaport 2001; Shah et al. 2000). Thus, if X represents the signal (power) and Y represents the interference (power), we are interested in finding out the pdf of

$$Z = \frac{X}{Y}. \tag{2.262}$$

In practical applications, we would want Z to be a few dB so that the signal power will be stronger than the interference. Since we can assume that the signal and interference are independent, the pdf of Z can be written from (2.187). If both X and Y are gamma distributed (originating from Nakagami-m distributed envelope values), we have

$$f(x) = \left(\frac{m}{\alpha}\right)^m \frac{x^{m-1}}{\Gamma(m)} \exp\left(-m\frac{x}{\alpha}\right), \quad 0 < x < \infty, \tag{2.263}$$

$$f(y) = \left(\frac{n}{\beta}\right)^m \frac{x^{n-1}}{\Gamma(n)} \exp\left(-n\frac{y}{\beta}\right), \quad 0 < y < \infty, \tag{2.264}$$

The density function of Z is written using (2.187) as

$$f(z) = \int_0^\infty y\left(\frac{m}{\alpha}\right)\frac{(yz)^{m-1}}{\Gamma(m)} \exp\left(-m\frac{yz}{\alpha}\right)\left(\frac{n}{\beta}\right)^n \frac{y^{n-1}}{\Gamma(n)} \exp\left(-n\frac{y}{\beta}\right) dy. \tag{2.265}$$

Equation (2.265) can be solved easily. We have the pdf for the ratio of two gamma distributed powers as

$$f(z) = \frac{\Gamma(m+n)}{\Gamma(m)\Gamma(n)} (\alpha n)^n (\beta m)^m \frac{z^{m-1}}{(mz\beta + n\alpha)^{m+n}}, \quad 0 < z < \infty. \tag{2.266}$$

For the special case of $m = n$ (corresponds to exponential distribution of power), we have the pdf of the ratio of the powers as

$$f(z) = \frac{\alpha\beta}{(\alpha + \beta z)^2}, \quad 0 < z < \infty. \tag{2.267}$$

Another interesting case in wireless communications arises when both the signal power and the interfering component power have lognormal distributions (Sowerby and Williamson 1987; Ligeti 2000). In this case, the density function of the ratio can be determined in a straight forward fashion. From (2.71) we have

$$f(x) = \frac{[10/\log_e(10)]}{\sqrt{2\pi\sigma_x^2 x^2}} \exp\left[-\frac{(10\log_{10}(x) - \mu_x)^2}{2\sigma_x^2}\right], \quad 0 < x < \infty. \tag{2.268}$$

$$f(x) = \frac{[10/\log_e(10)]}{\sqrt{2\pi\sigma_y^2 y^2}} \exp\left[-\frac{(10\log_{10}(y) - \mu_y)^2}{2\sigma_y^2}\right], \quad 0 < y < \infty. \tag{2.269}$$

Taking the logrithm and converting into decibel units, (2.262) becomes

$$10\log_{10}(Z) = 10\log_{10}(X) - 10\log_{10}(Y). \tag{2.270}$$

Since X and Y are lognormal random variables, the variables on the right-hand side of (2.270) will be Gaussian. Thus, the density function of

$$W = 10\log_{10}(Z) \tag{2.271}$$

will be Gaussian with a mean equal to the difference of the means of the two variables on the right-hand side of (2.270) and variance equal to the sum of the variances of the two variables on the right-hand side of (2.270). The pdf of W can now be expressed as

$$f(w) = \frac{1}{\sqrt{2\pi\left(\sigma_x^2 + \sigma_y^2\right)}} \exp\left\{-\frac{\left[w - \left(\mu_x - \mu_y\right)\right]^2}{2\left(\sigma_x^2 + \sigma_y^2\right)}\right\}. \tag{2.272}$$

Converting back from W to Z we can see that the density function of the ratio will also be lognormal. It can be expressed as

$$f(z) = \frac{[10/\log_e(10)]}{\sqrt{2\pi\sigma^2 z^2}} \exp\left[-\frac{(10\log_{10}(z) - \mu)^2}{2\sigma^2}\right], \quad 0 < z < \infty \tag{2.273}$$

with

$$\begin{aligned} \mu &= \mu_x - \mu_y, \\ \sigma^2 &= \sigma_x^2 + \sigma_y^2. \end{aligned} \tag{2.274}$$

Example #13 In wireless communications we must also address the situation that occurs when the receiver gets both the signal of interest (desired signal) and unwanted cochannels in addition to the noise (Winters 1984; Shah et al. 2000; Yacoub 2000; Aalo and Zhang 1999). The SNR at the receiver can be expressed as

$$Z = \frac{S_i}{N + S_c}. \tag{2.275}$$

In (2.275), S_i is the signal power and S_c is the cochannel power. Both are random, N is the noise power. Rewriting (2.275), we have

$$Z = \frac{X}{1 + Y}. \tag{2.276}$$

In (2.276), X is the SNR of the desired signal and Y is the SNR of the cochannel. Defining an auxiliary variable W as

$$W = Y. \tag{2.277}$$

The Jacobian for the transformation involving Z and W becomes

$$J(x, y) = \begin{vmatrix} \dfrac{1}{1+y} & 0 \\ -\dfrac{x}{(1+y)^2} & 1 \end{vmatrix} = \frac{1}{1+y} = \frac{1}{1+w}. \tag{2.278}$$

The joint pdf of W and Y now becomes

$$f(w, z) = |1 + w| f_{X,Y}(z(1 + w), w). \tag{2.279}$$

The pdf of Z now becomes

$$f(z) = \int_{-\infty}^{\infty} |1 + w| f_{X,Y}(z(1 + w), w) dw. \tag{2.280}$$

Since X and Y represent random variables which only take nonzero values, (2.280) can be rewritten as

$$f(z) = \int_{0}^{\infty} (1+w) f_{X,Y}(z(1+w), w) dw. \tag{2.281}$$

Example #14 Instead of the SNR defined in (2.276), we might see cases where the random variable might be of the form

$$U = \frac{X}{X+Y}. \tag{2.282}$$

Note that the random variable U can be rewritten as

$$U = \frac{(X/Y)}{X/Y+1} = \frac{V}{1+V}. \tag{2.283}$$

Let us examine a specific case of interest in wireless systems where both X and Y are gamma distributed, as in (2.235) and (2.236) with the special case of $\alpha = \beta$. The CDF of the variable in (2.283) can be expressed as

$$F_U(u) = \text{Prob}\left[\frac{V}{1+V} \leq u\right] = \text{Prob}\left[V \leq \frac{u}{1-u}\right]. \tag{2.284}$$

Rewriting, we have

$$F_U(u) = F_V\left(\frac{u}{1-u}\right). \tag{2.285}$$

The density function becomes

$$f_U(u) = \frac{1}{(1-u)^2} f_V\left(\frac{u}{1-u}\right). \tag{2.286}$$

Using (2.266) for the density of the ratio of two gamma variables, we have

$$f(u) = \frac{\Gamma(m+n)}{\Gamma(m)\Gamma(n)} u^{m-1}(1-u)^{n-1}, \quad 0 < u < 1. \tag{2.287}$$

Note that (2.287) is the beta pdf described earlier.

Example #15 We will now establish the relationship among the Gaussian, Chi-squared and Student t distributions. Let

$$Z = \frac{X}{\sqrt{Y/n}}. \tag{2.288}$$

In (2.288), X is a Gaussian variable having a zero mean and unit standard deviation identified as $N(0,1)$ and Y is the chi squared variable identified by $\chi^2(n)$. Using the concept of an auxiliary or "dummy" variable, let us define

$$W = Y. \tag{2.289}$$

The Jacobian of the transformation becomes

$$J(x,y) = \sqrt{\frac{n}{w}}. \tag{2.290}$$

The joint density function of Z and W now becomes

$$\begin{aligned} f_{Z,W}(z,w) &= \sqrt{\frac{w}{n}} f(x,y) \\ &= \sqrt{\frac{w}{n}} \frac{1}{\sqrt{2\pi}} \exp\left(-\frac{w}{2n} z^2\right) \frac{w^{(n/2)-1}}{2^{(n/2)}\Gamma(n/2)} \exp\left(-\frac{w}{2}\right). \end{aligned} \tag{2.291}$$

The density function of Z is obtained as

$$f_Z(z) = \int_0^\infty f(z,w)\ dw. \tag{2.292}$$

The limits of integration reflect the fact that $\chi^2(n)$ density function in (2.35) exists only in the range 0 to ∞. Carrying out the integration, we have

$$f(z) = \frac{\Gamma((n+1)/2)}{\sqrt{n\pi}\Gamma(n/2)} \frac{1}{\sqrt{(1+(z^2/n))^{n+1}}}, \qquad -\infty < x < \infty. \tag{2.293}$$

Note that (2.293) is identical to the Student t-distribution seen in (2.107).

2.9 Some Bivariate Correlated Distributions of Interest in Wireless Communications

We will now look at a few joint distributions that are used in the analysis of wireless communication systems. While the joint pdf is the product of marginal pdfs, when the variables are independent (this property was used in some or most of the examples given above), their functional forms are unique when correlation exists between the two variables.

2.9.1 Bivariate Normal pdf

$$f(x,y) = A\exp\left\{-\frac{1}{2(1-\rho^2)}\left[\frac{(x-\eta_1)^2}{\sigma_1^2} - \frac{2\rho(x-\eta_1)(y-\eta_2)}{\sigma_1\sigma_2} + \frac{(x-\eta_2)^2}{\sigma_2^2}\right]\right\} \tag{2.294}$$

In (2.294),

$$A = \frac{1}{2\pi\sigma_1\sigma_2\sqrt{1-\rho^2}}, \quad |\rho| \le 1 \tag{2.295}$$

X and Y are Gaussian with means of η_1 and η_2 and standard deviations of σ_1 and σ_2 respectively, and ρ is the correlation coefficient defined earlier in (2.131). Note that when ρ is zero, (2.294) becomes the product of the marginal density functions of X and Y. Note that for the bivariate Gaussian, uncorrelatedness also implies independence. The joint Gaussian pdf is plotted in Figs. 2.35, 2.36, and 2.37 for two zero mean variables each with unit variance for three values of the correlation coefficient ($\rho = 0, 0.8$ and -0.9). One can see that the joint pdf which is symmetric for the independent case ($\rho = 0$) takes on a ridge-like shape as the correlation increases (Papoulis and Pillai 2002).

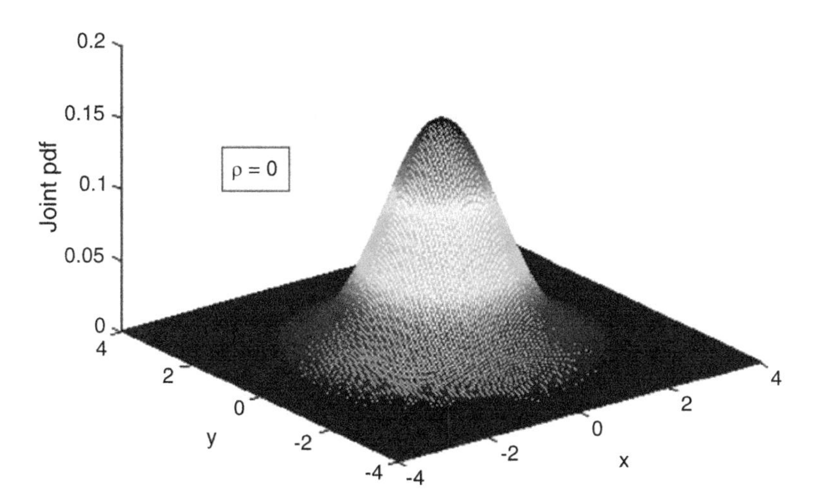

Fig. 2.35 Joint pdf of two Gaussian variables ($\rho = 0$)

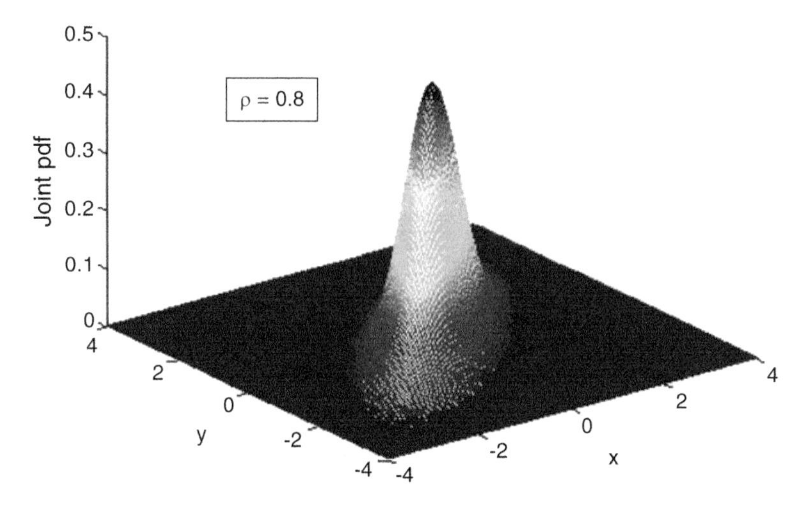

Fig. 2.36 Joint pdf of two Gaussian variables ($\rho = 0.8$)

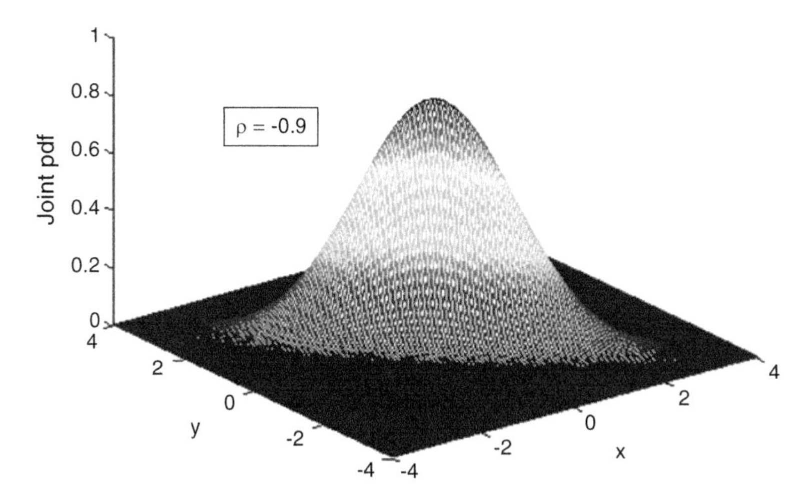

Fig. 2.37 Joint pdf of two Gaussian variables ($\rho = -0.9$)

2.9.2 *Bivariate Nakagami pdf*

The bivariate Nakagami pdf can be written as (Nakagami 1960; Tan and Beaulieu 1997; Karagiannidis et al. 2003a, b)

$$f(x,y) = B\exp\left[-\frac{m}{(1-\rho)}\left(\frac{x^2}{\Omega_x}+\frac{y^2}{\Omega_y}\right)\right]I_{m-1}\left[\frac{2mxy\sqrt{\rho}}{(1-\rho)\sqrt{\Omega_x\Omega_y}}\right] \tag{2.296}$$

with

$$B = \frac{4m^{m+1}(xy)^m}{\Gamma(m)\Omega_x\Omega_y(1-\rho)\left(\sqrt{\rho\Omega_x\Omega_y}\right)^{m-1}}. \tag{2.297}$$

In (2.296),

$$\Omega_x = E(X^2), \quad \Omega_y = E(Y^2). \tag{2.298}$$

The parameter ρ is the power correlation coefficient given by

$$\rho = \frac{\mathrm{cov}(X^2, Y^2)}{\sqrt{\mathrm{var}(X^2)\mathrm{var}(Y^2)}}. \tag{2.299}$$

In (2.299), cov(,) is the covariance defined in (2.126) and var is the variance. Note that m is the Nakagami parameter which has been considered to be identical for the two variables, X and Y. $I_{m-1}()$ is the modified Bessel function of the first kind of order $(m-1)$.

The bivariate Rayleigh pdf is obtained from (2.299) by putting $m = 1$. We have

$$f(x,y) = \frac{4xy}{(1-\rho)\Omega_x\Omega_y}\exp\left[-\frac{1}{(1-\rho)}\left(\frac{x^2}{\Omega_x}+\frac{y^2}{\Omega_y}\right)\right]I_0\left[\frac{2xy\sqrt{p}}{(1-\rho)\sqrt{\Omega_x\Omega_y}}\right]. \tag{2.300}$$

2.9.3 *Bivariate Gamma pdf*

There are several forms of the bivariate gamma pdf. One of these representations is (Kotz and Adams 1964; Mathai and Moschopoulos 1991; Tan and Beaulieu 1997; Yue et al. 2001; Holm and Alouini 2004; Nadarajah and Gupta 2006; Nadarajah and Kotz 2006a, b)

$$f(x,y) = \frac{(xy/\rho)^{\left(\frac{m-1}{2}\right)}}{\alpha^{m+1}\Gamma(m)(1-\rho)}\exp\left[-\frac{x+y}{\alpha(1-\rho)}\right]I_{m-1}\left[\frac{2\sqrt{\rho xy}}{\alpha(1-\rho)}\right]. \tag{2.301}$$

Note that ρ is the correlation coefficient of X and Y (of identical order m and parameter α), and the density function in (2.301) is identified as the Kibble's bivariate gamma distribution. It must be noted that a pdf similar to (2.301) can also be obtained from (2.296) by converting to power values and replacing Ω/m by α. The plots of the bivariate correlated gamma density functions are shown in Figs. 2.38, 2.39, and 2.40 for three values of the correlation $(m = 1)$.

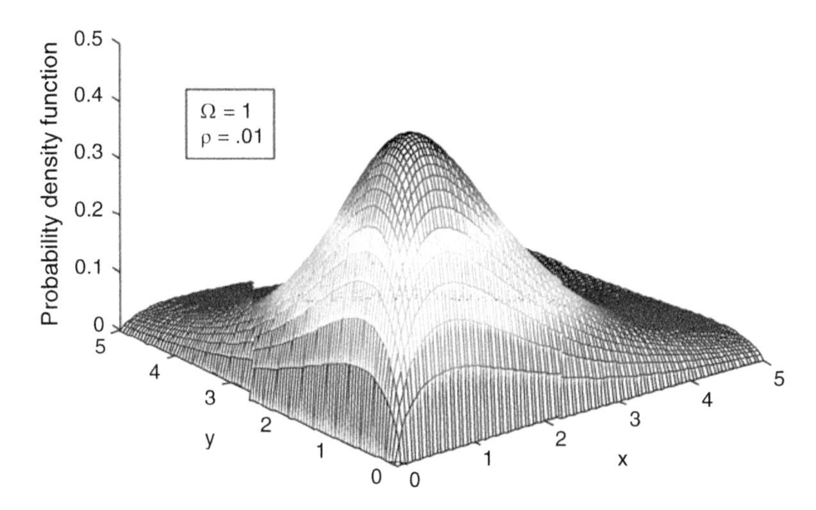

Fig. 2.38 Bivariate gamma pdf ($\rho = 0.01$ and $\alpha = \Omega = 1$)

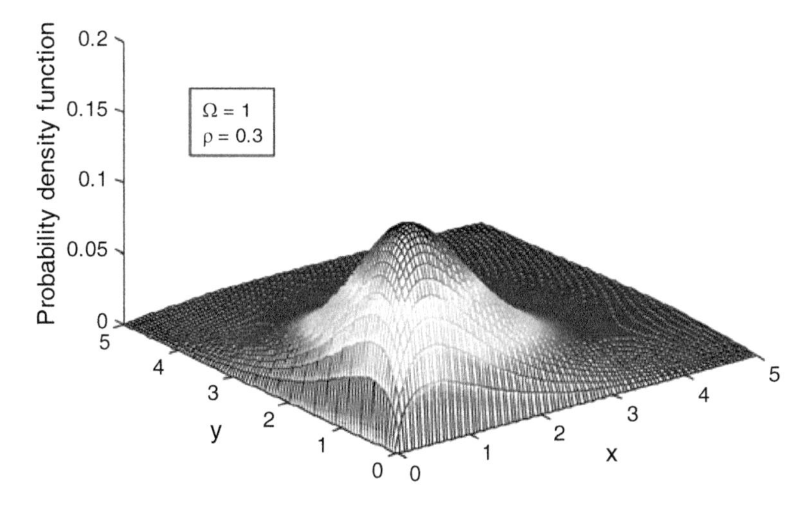

Fig. 2.39 Bivariate gamma pdf ($\rho = 0.3$ and $\alpha = \Omega = 1$)

Another form of a bivariate gamma distribution is known as McKay's bivariate gamma distribution (Nadarajah and Gupta 2006) and the density function is given by

$$f(x, y) = \frac{a^{p+q} x^{p-1}}{\Gamma(p)\Gamma(q)} (y - x)^{q-1} \exp(-ay), \\ y > x > 0, \; a > 0, \; p > 0, \; q > 0. \tag{2.302}$$

An examination of the density function in (2.302) clearly shows that the two random variables X and Y are not independent.

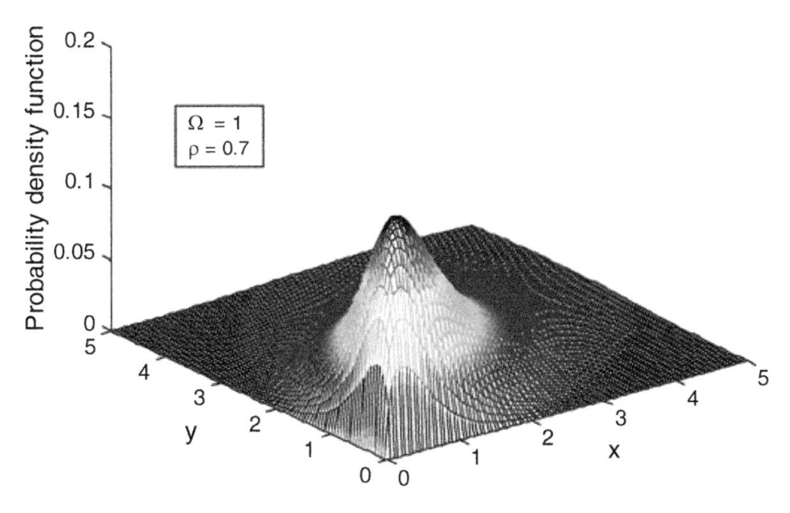

Fig. 2.40 Bivariate gamma pdf ($\rho = 0.7$ and $\alpha = \Omega = 1$)

A different form of bivariate gamma pdf is known as the Arnold and Strauss's bivariate gamma distribution. It has the joint pdf expressed as (Nadarajah 2005)

$$\begin{aligned} & f(x,y) = Kx^{\alpha-1}y^{\beta-1}\exp[-(ax+by+cxy)], \\ & x>0,\ \ y>0,\ \ a>0,\ \ b>0,\ \ c>0,\ \ \alpha>0,\ \ \beta>0. \end{aligned} \tag{2.303}$$

The parameter K is the normalization factor and it can be seen that the two random variables X and Y are not independent.

2.9.4 *Bivariate Generalized Gamma pdf*

The bivariate generalized gamma pdf can be obtained from the bivariate gamma pdf using the transformation of variables (Aalo and Piboongungon 2005). We have

$$\begin{aligned} f(z,w) = & \frac{\lambda^2 m^{m+1}(zw)^{((\lambda(m+1))/2)-1}\rho^{((1-m)/2)}}{(\Omega_z\Omega_w)^{((1+m)/2)}(1-\rho)\Gamma(m)} \\ & \times\exp\left[-\frac{m}{(1-\rho)}\left(\frac{z^\lambda}{\Omega_z}+\frac{w^\lambda}{\Omega_w}\right)\right]I_{m-1}\left[\frac{2m}{1-\rho}\sqrt{\frac{\rho(zw)^\lambda}{\Omega_z\Omega_w}}\right]. \end{aligned} \tag{2.304}$$

In (2.304),

$$\Omega_z = E\left(Z^{\lambda}\right), \quad \Omega_w = E\left(W^{\lambda}\right) \tag{2.305}$$

and ρ is the correlation coefficient between Z^2 and W^2. Note that by putting $m = 1$ in (2.305), we can obtain an expression for the bivariate Weibull pdf.

2.9.5 Bivariate Weibull pdf

The bivariate Weibull pdf can be obtained from the bivariate generalized gamma pdf in (2.304) by putting $m = 1$ as

$$f(z,w) = \frac{\lambda^2 (zw)^{\lambda-1}}{(\Omega_z\Omega_w)(1-\rho)} \exp\left[-\frac{1}{(1-\rho)}\left(\frac{z^{\lambda}}{\Omega_z} + \frac{w^{\lambda}}{\Omega_w}\right)\right] I_{m-1}\left[\frac{2}{1-\rho}\sqrt{\frac{\rho(zw)^{\lambda}}{\Omega_z\Omega_w}}\right] \tag{2.306}$$

with

$$\Omega_z = E\left(Z^{\lambda}\right), \quad \Omega_w = E\left(W^{\lambda}\right). \tag{2.307}$$

2.9.6 Bivariate Rician Distribution

The bivariate Rician distribution of the SNR values Z and W can be expressed as (Zogas and Karagiannidis 2005; Bithas et al. 2007; Panajotovic et al. 2009)

$$f(z,w) = \frac{(1+K_0)^2}{2\pi Z_R^2(1-\rho^2)} \exp\left[-\frac{2K_0}{1+\rho} - \frac{(1+K_0)(z+w)}{2Z_R(1-\rho^2)}\right] R. \tag{2.308}$$

In (2.308),

$$R = \int_0^{2\pi} \exp\left[\frac{2\rho(1+K_0)\sqrt{zw}\cos(\theta)}{(1-\rho^2)Z_R}\right] I_0\left[\sqrt{\frac{4K_0(1+K_0)(z+w+2\sqrt{zw}\cos(\theta))}{Z_R(1+\rho)^2}}\right] d\theta. \tag{2.309}$$

The parameter ρ is correlation coefficient between the two envelope values A_1 and A_2 corresponding to the SNR values of Z and W respectively as

$$Z = A_1^2, \quad W = A_2^2. \tag{2.310}$$

The two variables (Z and W) are considered to be identical with equal average SNR values equal to Z_R. The Rician factor is given by K_0 and was defined in (2.214).

2.10 Order Statistics

Another important and interesting statistical entity of interest in wireless systems is the order statistics (Rohatgi and Saleh 2001; Papoulis and Pillai 2002). For example, in selection combining (SC) algorithm, we are interested in finding the largest of a set of outputs and in generalized selection combining (GSC), we are interested in picking the M largest of a total of L outputs (Alouini and Simon 1999; Ma and Chai 2000; Alouini and Simon 2006; Annamalai et al. 2006). Both of these outcomes can be analyzed in terms of the order statistics. Let $X_1, X_2, \ldots, X_L$ correspond to the a set of L random variables. We will assume that the random variables are independent and identical. Our interest is in finding out the joint pdf of the largest M variables, i.e., in finding the joint pdf of

$$\{X_1, X_2, \ldots, X_M\}, X_1 > X_2 > X_3 > \cdots > X_M, \quad M \leq L. \tag{2.311}$$

Even though the random variables are the same, to make matters simple and eliminate confusion because the outputs and inputs of the selection process will contain X's, the output set will be identified as $\{Y_1, Y_2, \ldots, Y_M\}$. The joint CDF of $\{Y_1, Y_2, \ldots, Y_M\}$can be written as

$$F(y_1, y_2, \ldots, y_M) = \sum_{i=1}^{L} F_{X_i}(y_1) \sum_{j=1, j \neq i}^{L} F_{X_j}(y_2) \ldots \prod_{n=1, n \neq i, j, \ldots}^{L} F_{Xn}(y_M), \quad y_1 > y_2 > y_3 > \cdots > y_M. \tag{2.312}$$

Since the random variables have been considered to be identical, (2.312) becomes

$$F(y_1, y_2, \ldots, y_M) = LF(y_1)(L-1)F(y_2) \ldots (L-M)F(y_{M-1})[F(y_M)]^{L-M+1}. \tag{2.313}$$

Equation (2.313) can be explained as follows. There are L different ways of choosing the largest. Thus, this probability will be $LF(y_1)$. There are $(L - 1)$ ways of picking the second largest, and the probability of doing so will be $(L - 1)F(y_2)$ and so on. There are $(L - M)$ ways of picking the next to the last in that order, and the probability of doing so will be

$(L - M)F(y_{M-1})$. Since the rest of the variables $(L - M + 1)$ will be either equal to or smaller than y_M, the probability of this event will be the last term in (2.313). The joint pdf can be obtained by differentiating (2.313) with respect to $y_1, y_2, \ldots, y_M$, and we have

$$f(y_1, y_2, \ldots, y_M) = Lf(y_1)(L-1)f(y_2)\ldots(L-M+1)f(y_m)[F(y_m)]^{L-M}. \tag{2.314}$$

Equation (2.314) can be easily expressed as

$$f(y_1, y_2, \ldots, y_M) = \Gamma(M+1)\binom{L}{M}[F(y_m)]^{L-M}\prod_{k=1}^{M} f(y_k), \quad y_1 > y_2 > y_3 > \cdots > y_M. \tag{2.315}$$

In (2.315),

$$\binom{L}{M} = \frac{L!}{M!(L-M)!}. \tag{2.316}$$

If we are choosing the largest output, $M = 1$, and we have the expression for the pdf of the maximum of L independent and identically distributed variables as

$$f(y_1) = L[F(y_1)]^{L-1}f(y_1). \tag{2.317}$$

If one puts $L = 2$ in (2.317), we have the pdf of the largest of the two random variables, obtained earlier in (2.173).

We can now obtain the pdf of the kth largest variable. Let Y_k be the kth largest variable. If $f_M(y)$ is the pdf of the variable Y_k, we can write

$$f_k(y)dy = \text{Prob}\{y \le Y_k \le y + dy\}. \tag{2.318}$$

The event $\{y \le Y_k \le y + dy\}$ occurs *iff* exactly $(k - 1)$ variables less than y, one in the interval $\{y, y + dy\}$, and $(L - k)$ variables greater than y. If we identify these events as I and II and III respectively,

$$I = (x \le y), \quad II = (y \le x \le y + dy), \quad III = (x > y + dy). \tag{2.319}$$

The probabilities of these events are

$$\text{Prob}(I) = F_X(y), \quad \text{Prob}(I) = f_X(y)dy, \quad \text{Prob}(III) = 1 - F_X(y). \tag{2.320}$$

Note that event I occurs $(k - 1)$ times, event II occurs once, and event III occurs $L - k$ times. Using the concept of generalized Bernoulli trial and (2.31), (2.318) becomes

$$f_k(y)\ \mathrm{dy} = \frac{L!}{(k-1)!1!(L-k)!}[F_X(y)]^{k-1} f_X(y)\ dy[1 - F_X(y)]^{L-k}. \tag{2.321}$$

Equation (2.321) simplifies to

$$f_k(y) = \frac{L!}{(k-1)!(L-k)!}[F_X(y)]^{k-1}[1 - F_X(y)]^{L-k} f_X(y). \tag{2.322}$$

When k equals L, we get the pdf of the largest variable (maximum), and (2.322) becomes (2.317). If $k = 1$, we get the pdf of the smallest of the random variable (minimum) as described below.

We can also obtain the density function of the minimum of L random variables. Let us define

$$Z = \min\{x_1, x_2, \ldots, x_L\}. \tag{2.323}$$

Noting the probability that at least one of the variables is less than Z, we can express the CDF as

$$F_Z(z) = 1 - \mathrm{Prob}\{x_1 > z, x_2 > z, \ldots, x_L > z\} = 1 - [1 - F_X(z)]^L. \tag{2.324}$$

In (2.324), $F_X(.)$ and $f_X(.)$ are the marginal CDF and pdf of the X's which are treated to be identical and independent. Differentiating the CDF, we have the density function of the minimum of a set of random variables as

$$F_Z(z) = L[1 - F_X(z)]^{L-1} f_X(z). \tag{2.325}$$

Note that (2.322) can also be used to obtain the pdf of the minimum of the random variables by setting $k = 1$ and (2.322) reduces to (2.325).

Before we look at the specific cases of interest in wireless communications, let us look at the special case of the exponential pdf. If X's are independent and identically distributed with exponential pdf,

$$f(x_i) = \frac{1}{\alpha}\exp\left(-\frac{x_i}{\alpha}\right), \quad i = 1, 2, \ldots, L \tag{2.326}$$

the density function of the minimum of the set can be written using (2.325) as

$$f(z) = \frac{L}{\alpha}\exp\left(-\frac{L}{\alpha}z\right). \tag{2.327}$$

In other words, the minimum of a set of independent and identically distributed exponential random variables is another exponentially distributed random variable with a mean of (α/L), i.e., with a reduced mean.

2.10.1 A Few Special Cases of Order Statistics in Wireless Communications

We will now look at a few special cases of interest in wireless systems, specifically, the cases of bivariate Nakagami (or gamma) and bivariate lognormal distributions. We will start with the bivariate Nakagami distribution. The bivariate Nakagami pdf given in (2.296) is

$$\begin{aligned} f(a_1,a_2) &= \frac{4m^{m+1}(a_1a_2)^m}{\Gamma(m)P_{01}P_{02}(1-\rho)(\sqrt{P_{01}P_{02}\rho})^{m-1}} \\ &\times \exp\left[-\frac{m}{(1-\rho)}\left(\frac{a_1^2}{P_{01}}+\frac{a_2^2}{P_{02}}\right)\right] I_{m-1}\left(2m\frac{a_1a_2\sqrt{\rho}}{(1-\rho)\sqrt{P_{01}P_{02}}}\right). \end{aligned} \tag{2.328}$$

In (2.328), m is the Nakagami parameter considered to be identical for the two variables, a_1 and a_2. The other parameters are

$$P_{01} = \langle a_1^2 \rangle, \quad P_{02} = \langle a_2^2 \rangle, \quad \rho = \frac{\mathrm{cov}\left(a_1^2, a_2^2\right)}{\mathrm{var}\left(a_1^2\right)\mathrm{var}\left(a_2^2\right)}. \tag{2.329}$$

In (2.329), cov is the covariance and var is the variance. Note that by putting $m = 1$ in (2.328), we get the bivariate Rayleigh pdf

$$f(a_1,a_2) = \frac{4(a_1a_2)}{P_{01}P_{02}(1-\rho)}\exp\left[-\frac{1}{(1-\rho)}\left(\frac{a_1^2}{P_{01}}+\frac{a_2^2}{P_{02}}\right)\right] I_0\left(2\frac{a_1a_2\sqrt{\rho}}{(1-\rho)\sqrt{P_{01}P_{02}}}\right). \tag{2.330}$$

One of the interesting uses of the bivariate correlated pdf is in diversity combining. We will obtain the pdf of the selection diversity combiner in which the output of the diversity algorithm is the stronger of the two input signals. We will also simplify the analysis by assuming that the signals have identical powers, i.e., $P_{01} = P_{02} = P_0$. Since the comparison is made on the basis of power or SNR, we will rewrite (2.328) in terms of the powers

$$X = a_1^2, \quad Y = a_2^2 \tag{2.331}$$

as

$$f(x,y) = \frac{m^{m+1}(xy)^{((m-1)/2)}\rho^{((1-m)/2)}}{\Gamma(m)P_0^{m+1}(1-\rho)\left(\sqrt{\rho}\right)^{m-1}}\exp\left[-\frac{m(x+y)}{P_0(1-\rho)}\right] I_{m-1}\left(2m\frac{\sqrt{xy\rho}}{P_0(1-\rho)}\right). \tag{2.332}$$

Note that (2.332) is identical to the Kibble's bivariate gamma pdf given in (2.301) with an appropriately scaled average power. If we define

$$Z = \max(X, Y). \tag{2.333}$$

The pdf of the selection combining can be expressed as

$$f(z) = \frac{d}{dz}[\mathrm{Prob}(X < z, Y < z)]. \tag{2.334}$$

The expression for the pdf becomes

$$f(z) = 2\frac{m^m}{\Gamma(m)P_0}\left(\frac{z}{P_0}\right)^{m-1}\exp\left(-m\frac{z}{P_0}\right)\left[1 - Q_m\left(\sqrt{\frac{2mp}{1-\rho}\left(\frac{z}{P_0}\right)}, \sqrt{\frac{2m}{1-\rho}\left(\frac{z}{P_0}\right)}\right)\right]. \tag{2.335}$$

In (2.335), $Q_m(,)$ is the generalized Marcum Q function given by (Simon 2002)

$$Q_m(\alpha, \beta) = \frac{1}{\alpha^{m-1}}\int_{\beta}^{\infty} w^m \exp\left(-\frac{w^2 + \alpha^2}{2}\right) I_{m-1}(\alpha w)\ dw. \tag{2.336}$$

We will also obtain the pdf of the maximum of two correlated lognormal random variables. As we will discuss in Chap. 3, lognormal density function is often used to model shadowing. It has been shown that short-term fading in some of the indoor wireless channels can be modeled using the lognormal density functions. Consequently, one can find the use of bivariate lognormal densities in the analysis of diversity algorithms (Ligeti 2000; Alouini and Simon 2003; Piboongungon and Aalo 2004).

Using the relationship between normal pdf and lognormal pdf, it is possible to write the expression for the joint pdf of two correlated lognormal random variables X and Y. The pdf is given by

$$f(x,y) = \frac{D}{xy}\exp\left\{-\frac{1}{2(1-\rho^2)}\left[\left(\frac{10\log_{10}(x) - \mu_x}{\sigma_x}\right)^2 + \left(\frac{10\log_{10}(y) - \mu_y}{\sigma_y}\right)^2 - 2\rho g(x,y)\right]\right\}. \tag{2.337}$$

In (2.337),

$$D = \frac{A_0^2}{2\pi\sigma_x\sigma_y\sqrt{1-\rho^2}}, \tag{2.338}$$

$$g(x,y)=\left(\frac{10\log_{10}(x)-\mu_x}{\sigma_x}\right)\left(\frac{10\log_{10}(y)-\mu_y}{\sigma_y}\right). \quad (2.339)$$

Note that μ's are the means and σ's the standard deviations of the corresponding Gaussian random variables in decibel units. A_0 was defined earlier and it is given by

$$A_0=\frac{10}{\log_e(10)}. \quad (2.340)$$

As in the case of the Nakagami random variables, ρ is the correlation coefficient (power). If Z represents the maximum of the two random variables, the pdf of the maximum can be written either as in (2.334) or as

$$f(z)=\int_0^z f_{X,Y}(z,y)dy+\int_0^z f_{X,Y}(x,z)dx. \quad (2.341)$$

In (2.341), $f(x,y)$ is the joint pdf of the dual lognormal variables expressed in (2.337). The integrations in (2.341) can be performed by transforming the variables to Gaussian forms by defining

$$\alpha=\frac{10\log_{10}(x)-\mu_x}{\sigma_x}, \quad (2.342)$$

$$\beta=\frac{10\log_{10}(y)-\mu_y}{\sigma_y}, \quad (2.343)$$

$$z_{\mathrm{dB}}=10\log_{10}(z). \quad (2.344)$$

We get

$$f(z)=\exp\left[-\frac{1}{2}\left(\frac{z_{\mathrm{dB}}-\mu_x}{\sigma_x}\right)^2\right]g_1+\exp\left[-\frac{1}{2}\left(\frac{z_{\mathrm{dB}}-\mu_y}{\sigma_y}\right)^2\right]g_2. \quad (2.345)$$

In (2.345), g_1 and g_2 are given by

$$g_1=Q\left[\frac{(z_{\mathrm{dB}}-\mu_y)/\sigma_y-\rho((z_{\mathrm{dB}}-\mu_x)/\sigma_x)}{\sqrt{1-\rho^2}}\right], \quad (2.346)$$

$$g_2=Q\left[\frac{(z_{\mathrm{dB}}-\mu_x)/\sigma_x-\rho((z_{\mathrm{dB}}-\mu_y)/\sigma_y)}{\sqrt{1-\rho^2}}\right], \quad (2.347)$$

In (2.346) and (2.347),

$$Q(\lambda) = \int_{\lambda}^{\infty} \frac{1}{\sqrt{2\pi}} \exp\left(-\frac{\phi^2}{2}\right) d\phi. \tag{2.348}$$

The CDF of the maximum of two lognormal random variables can now be expressed as

$$F(z) = 1 - Q\left(\frac{\alpha - \mu_x}{\sigma_x}\right) - Q\left(\frac{\beta - \mu_y}{\sigma_y}\right) + Q\left(\frac{\alpha - \mu_x}{\sigma_x}, \frac{\beta - \mu_y}{\sigma_y}; \rho\right). \tag{2.349}$$

In (2.349), the last function is given by (Simon and Alouini 2005)

$$Q(u, v; \rho) = \frac{1}{2\pi\sqrt{1 - \rho^2}} \int_{u}^{\infty} \int_{v}^{\infty} \exp\left[-\frac{x^2 + y^2 - 2\rho xy}{2(1 - \rho^2)}\right] dx\ dy. \tag{2.350}$$

2.11 Decision Theory and Error Rates

Digital communication mainly involves the transmission of signals with discrete values (0's and 1's or any other set of M values $M > 1$) and reception of those signals (Van Trees 1968; Helstrom 1968; Middleton 1996). The channel adds noise.

Noise as stated earlier is typically modeled as a Gaussian random variable with a means of zero and a standard deviation. But, noise could also be non-Gaussian. We will examine the two cases separately.

2.11.1 Gaussian Case

As mentioned above, the received signal consists of the transmitted signal plus noise. Thus, the problem of identifying the received discrete values becomes one of hypothesis testing (Fig. 2.41). This problem is shown in Fig. 2.37. We will examine the case of a binary system where 0's and 1's are transmitted. Note that these two bits are represented by the values a_0 and a_1. (Benedetto and Biglieri 1999; Simon et al. 1995; Proakis 2001; Haykin 2001; Couch 2007.)

Let us first examine what happens when the signals move through the channel. Because of the noise, "0" might be detected as a "1" or "0." Similarly, "1" might be detected as a "0" or "1." This scenario points to two distinct ways of making an error at the output. These two are the detection of "0" as a "1" and detection of "1" as a "0." This is shown in Fig. 2.42. We can invoke Bayes theorem in probability theory to estimate this error (Papoulis and Pillai 2002). If we have two events, A and B, each with probabilities of $P(A)$ and $P(B)$, respectively, the conditional probability can be expressed in terms of Bayes theorem as

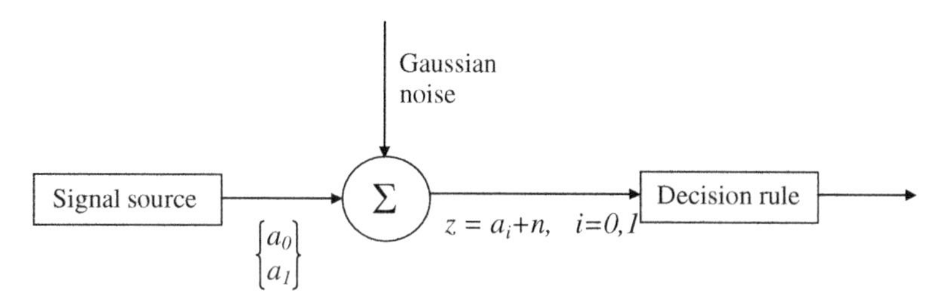

Fig. 2.41 Hypothesis testing problem

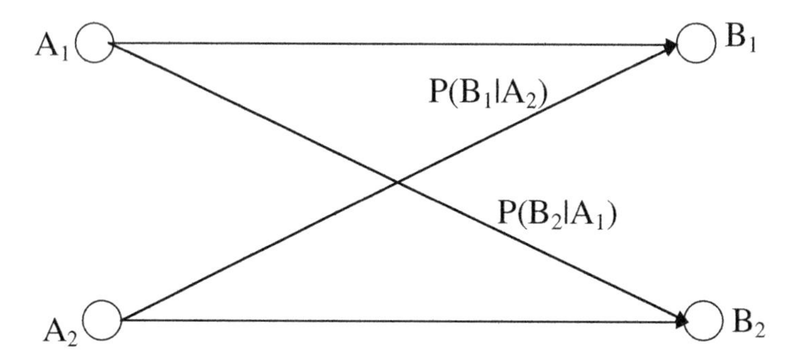

Fig. 2.42 The binary channel

$$P(A|B) = \frac{P(B|A)P(A)}{P(B)}. \tag{2.351}$$

Extending this notion to multiple events $A = \{A_1, A_2, \ldots, A_M\}$ and $B = \{B_1, B_2, \ldots, B_M\}$, the Bayes rule becomes

$$P(A_i|B_j) = \left.\frac{P(B_j|A_i)P(A_i)}{P(B_j)}\right\} \quad \begin{matrix} i = 1, 2, \ldots, M \\ j = 1, 2, \ldots, M \end{matrix}, \tag{2.352}$$

where

$$P(B_j) = \sum_{i=1}^{M} P(B_J|A_i)P(A_i). \tag{2.353}$$

In our description, A's represent the input and B's represent the received signals. We also identify $P(A_i)$ as the a priori probability, $P(B_j|A_i)$ as the conditional probability and $P(A_i|B_j)$ as the a posteriori probability. Equation (2.353) shows that each output will have contributions from the complete input set. We can now introduce the notion of errors by looking at a binary case ($M = 2$). In the case of binary signal transmission, we have two inputs $\{A_1, A_2\}$ and two corresponding outputs $\{B_1,B_2\}$. Since there is noise in the channel, B_1 could be read as A_2 and B_2 could be read as A_1. Thus, we have two ways in which error can occur. If the error is represented by e, the probability of error can be written using the Bayes rule as

$$p(e) = P(B_2|A_1)P(A_1) + P(B_1|A_2)P(A_2). \tag{2.354}$$

We will now expand this notion to the transmission of digital signals through a channel corrupted by noise. The received signal can be written as

$$z = a_i + n, \quad i = 1, 2. \tag{2.355}$$

Since the noise is Gaussian, we can represent the density function of noise as (Shanmugam 1979; Schwartz 1980; Schwartz et al. 1996; Sklar 2001)

$$f(n) = \frac{1}{\sqrt{2\pi\sigma^2}} \exp\left(-\frac{n^2}{2\sigma^2}\right). \tag{2.356}$$

Note that even though a's only takes discrete values, z will be continuous as the noise is not discrete. The density function of the received signal can be represented as conditional density functions,

$$f(z|a_1) = \frac{1}{\sqrt{2\pi\sigma^2}} \exp\left[-\frac{(z-a_1)^2}{2\sigma^2}\right], \tag{2.357}$$

$$f(z|a_2) = \frac{1}{\sqrt{2\pi\sigma^2}} \exp\left[-\frac{(z-a_2)^2}{2\sigma^2}\right]. \tag{2.358}$$

We will now identify the two hypotheses we have as H_0 and H_1, the former corresponding to the transmission "0" (signal strength a_0) and the latter corresponding to the transmission of "1" (signal strength a_1). We will rewrite (2.357) and (2.358) as

$$f(z|a_0) \equiv f(z|H_0), \tag{2.359}$$

$$f(z|a_1) \equiv f(z|H_1). \tag{2.360}$$

These two density functions are plotted in Fig. 2.43.

Using simple logic, it would seem that we can decide on whether the received signal belongs to hypothesis H_0 or H_1 as (Van Trees 1968)

$$f(H_1|z) \underset{H_0}{\overset{H_1}{\gtrless}} f(H_0|z). \tag{2.361}$$

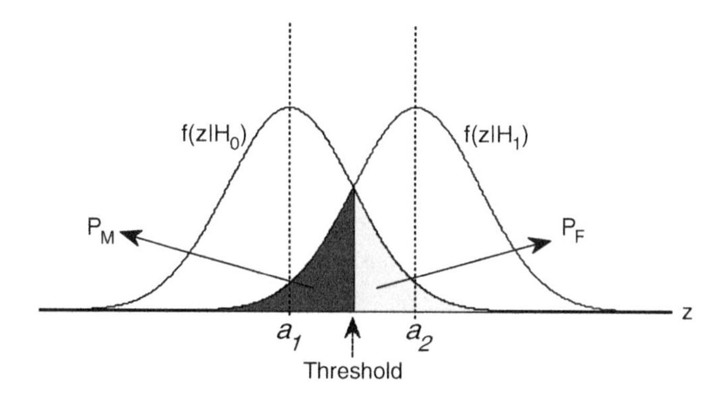

Fig. 2.43 Concept of hypothesis testing

Using the Bayes rule in (2.351)

$$f(z|H_1)P(H_1) \underset{H_0}{\overset{H_1}{\gtrless}} f(z|H_0)P(H_0) \tag{2.362}$$

or

$$\left[\frac{f(z|H_1)}{f(z|H_0)}\right] \underset{H_0}{\overset{H_1}{\gtrless}} \left[\frac{P(H_0)}{P(H_1)}\right]. \tag{2.363}$$

The left-hand side of (2.363) is known as the likelihood ratio; the entire equation is identified as the likelihood ratio test. If both hypotheses are equally likely, the right-hand side of (2.363) will be unity and take logarithms. We arrive at the log likelihood ratio:

$$\log_e\left[\frac{f(z|H_1)}{f(z|H_0)}\right] \underset{H_0}{\overset{H_1}{\gtrless}} 0. \tag{2.364}$$

Substituting (2.357) and (2.358), we have the threshold for the decision as

$$z = z_{\text{thr}} = \frac{(a_1 + a_2)}{2} \tag{2.365}$$

Thus, (2.365) demonstrates that the threshold for making decision on whether H_1 is accepted for H_0 is accepted is the midpoint of the two amplitude values. The errors can occur in two ways. First, even when a "0" was transmitted, because of noise, it can be read/detected as a "1." This is called the probability of false alarm (P_F) and is given by the probability indicated by the shaded area in Fig. 2.12.

$$P_F = \int_{z_{\text{thr}}}^{\infty} f(z|H_0)\ dz. \tag{2.366}$$

Second, when "1" was sent, it can be read/detected as "0." This is called the probability of miss (P_M) and it is given by the shaded area corresponding to the probability

$$P_M = \int_{-\infty}^{z_{\text{thr}}} f(z|H_1)\ dz. \tag{2.367}$$

The average probability of error is given by

$$p(e) = p(e|H_1)P(H_1) + p(e|H_0)P(H_0) \tag{2.368}$$

or

$$p(e) = P_F p(H_1) + P_M P(H_0). \tag{2.369}$$

If both hypotheses are equally likely (0's and 1's being transmitted with equal probability), the average probability of error will be

$$p_e = P_F \quad or \quad P_M. \tag{2.370}$$

It can be seen that the average probability of error can be expressed in terms of the CDF of the Gaussian random variable as well as the complimentary error functions (Sklar 2001).

2.11.2 Non-Gaussian Case

While the detection of signals in additive white Gaussian noise is generally encountered in communications, there are also instances when the receiver makes decisions based on the envelope of the signal. Examples of these include envelope detection of amplitude shift keying and frequency shift keying. It also includes distinguishing the signals at the wireless receiver from the signal of interest and the interference, both of which could be Rayleigh distributed when short-term fading is present. As discussed earlier the statistics of the envelope might also be Rician distribution. There is a need to minimize the error rate by determining the optimum value of the threshold.

We would like to examine three separate cases, the first where both hypotheses lead to Rayleigh densities, the second where one of the hypothesis leads to Rayleigh while the other hypothesis results in Rician, and the third where both hypothesis result in Rician densities. Let the two density functions be

$$f(z|H_0) = \frac{z}{\sigma_0^2}\exp\left[-\frac{z^2}{2\sigma_0^2}\right], \quad z > 0, \tag{2.371}$$

$$f(z|H_1) = \frac{z}{\sigma_1^2}\exp\left[-\frac{z^2}{2\sigma_1^2}\right], \quad z > 0 \tag{2.372}$$

with

$$\sigma_1 > \sigma_0. \tag{2.373}$$

From (2.364) the log likelihood ratio test leads to a threshold of

$$z_{\mathrm{thr}} = \sigma_0\sigma_1\sqrt{\frac{\log_e(\sigma_1/\sigma_0)}{2(\sigma_1^2 - \sigma_0^2)}}. \tag{2.374}$$

The decision region and the two regions for the computation of error are shown in Fig. 2.44.

The values of σ_0 and σ_1 for the Fig. 2.44 are 2 and 5, respectively, and the threshold is at 4.177.

Next we consider the case where the density function of the envelope under the hypothesis H_0 is Rayleigh distributed, as previously given in (2.371), while the pdf of the envelope is Rician distributed for the second hypothesis, namely H_1. The pdf for the envelope under the hypothesis H_1 is as in (2.208)

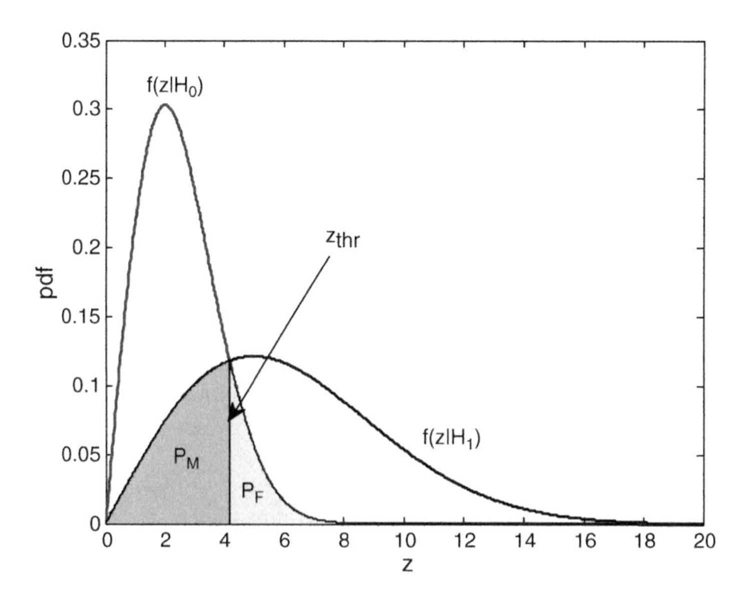

Fig. 2.44 Hypothesis testing (non-Gaussian case)

$$f(z|H_1) = \frac{z}{\sigma_0^2}\exp\left[-\frac{z^2+A^2}{2\sigma_0^2}\right] I_0\left(\frac{zA}{\sigma_0^2}\right), \quad z>0. \tag{2.375}$$

Note that both the density functions in (2.371) and (2.375) come from Gaussian random variables of identical standard deviation σ_0 and A is >0. A direct solution of (2.364) is going to be difficult because of the existence of the Bessel function in (2.375). We will use an indirect approach for solving the threshold. Let us assume that η is the threshold. The probability of false alarm P_F will be given by (Van Trees 1968; Cooper and McGillem 1986)

$$P_F = \int_\eta^\infty f(z|H_0)\ dz = \exp\left(-\frac{\eta^2}{2\sigma_0^2}\right) \tag{2.376}$$

The probability of miss P_M will be

$$P_M = \int_0^\eta f(z|H_1)\ dz = \int_0^\eta \frac{z}{\sigma_0^2}\exp\left[-\frac{z^2+A^2}{2\sigma_0^2}\right] I_0\left(\frac{zA}{\sigma_0^2}\right)\ dz. \tag{2.377}$$

Since there is no analytical solution to (2.377), we will examine what happens if we assume that the direct component $A \gg \sigma_0$. Invoking the approximation to the modified Bessel function of the first kind mentioned in (2.210), and rewriting (2.304) as

$$P_M = 1 - \int_\eta^\infty \frac{z}{\sigma_0^2}\exp\left[-\frac{z^2+A^2}{2\sigma_0^2}\right] I_0\left(\frac{zA}{\sigma_0^2}\right)\ dz \tag{2.378}$$

we get

$$P_M = 1 - \int_\eta^\infty \frac{\sqrt{z/A}}{\sqrt{2\pi\sigma_0^2}}\exp\left[-\frac{(z-A)^2}{2\sigma_0^2}\right]\ dz. \tag{2.379}$$

The integrand in (2.379) is sharply peaked at $z = A$, and the slowly varying factor $\sqrt{z/A}$ may be replaced with its value at the peak, i.e., unity, leading to a simpler expression for the probability of miss as

$$P_M = 1 - \int_\eta^\infty \frac{1}{\sqrt{2\pi\sigma_0^2}}\exp\left[-\frac{(z-A)^2}{2\sigma_0^2}\right]\ dz. \tag{2.380}$$

Using the Q function defined in (2.90), the probability of miss becomes

$$P_M = 1 - Q\left[\frac{\eta - A}{\sigma_0}\right]. \tag{2.381}$$

The total error will be

$$p(e) = \frac{1}{2}[P_F + P_M]. \tag{2.382}$$

Taking the derivative of (2.382) and setting it equal to zero, we have

$$\frac{d[P_F + P_M]}{d_\eta} = 0. \tag{2.383}$$

Using the derivative of the Q function as

$$\frac{d[Q(x)]}{dx} = -\frac{1}{\sqrt{2\pi}}\exp\left(-\frac{x^2}{2}\right) \tag{2.384}$$

(2.384) becomes

$$-\frac{\eta}{\sigma_0^2}\exp\left(-\frac{\eta^2}{2\sigma_0^2}\right) + \frac{1}{\sqrt{2\pi\sigma_0^2}}\exp\left[-\frac{(\eta - A)^2}{2\sigma_0^2}\right] = 0. \tag{2.385}$$

Once again, a solution to (2.385) is not straightforward. Instead, we import the concept from the Gaussian pdfs in (2.365). At high values of the SNR, as seen in (2.380), the envelope follows a Gaussian under the hypothesis H_1. We can argue that the threshold should be the midway point between the peaks of the two density functions. This is shown in Fig. 2.45. This would mean that the optimum threshold is

$$z_{\text{thr}} = \frac{1}{2}[\sigma_0 + A]. \tag{2.386}$$

Note that in (2.386), A is the mode of the pdf of the envelope under hypothesis H_1 and σ_0 the mode of the pdf under the hypothesis H_0. Since $A \gg \sigma_0$, the optimum threshold is

$$z_{\text{thr}} \approx \frac{A}{2}. \tag{2.387}$$

We can now look at the last case where the envelopes follow Rician pdf under both hypotheses, with direct components of A_0 and A_1 respectively. Using the Gaussian approximation to Rician, the optimum threshold will be

$$z_{\text{thr}} = \frac{1}{2}[A_1 + A_0]. \tag{2.388}$$

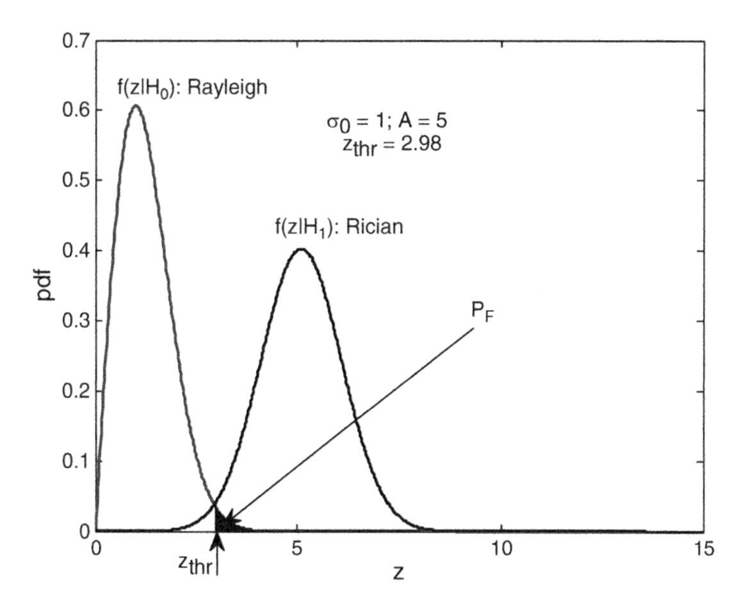

Fig. 2.45 Hypothesis testing (Rayleigh vs. Rician)

2.12 Upper Bounds on the Tail Probability

In wireless communications, it is often necessary to evaluate the probability that the SNR exceeds a certain value or that SNR fails to reach a certain threshold value. In Sect. 2.13, we saw two measures of error probabilities, a probability of false alarm, i.e., the area under the upper tail of the pdf curve beyond a certain value, and a probability of miss, i.e., the area under the lower tail below a certain value. Often, it might not be possible to evaluate such areas. It might be necessary to get an estimate of the upper bounds of such probabilities which correspond to areas under the tails of the pdf curves. We will look at two bounds, one based on the Chebyshev inequality and the other one called the Chernoff bound (Schwartz et al. 1996; Haykin 2001). The former one is a loose bound while the latter is much tighter.

2.12.1 Chebyshev Inequality

Let $f(x)$ be the pdf of a random variable X with a mean of m and standard deviation of σ. If δ is any positive number, the Chebyshev inequality is given by (Papoulis and Pillai 2002)

$$P(|X-m|\geq\delta)\leq\frac{\sigma^2}{\delta^2}. \tag{2.389}$$

In (2.389), $P(.)$ is the probability. Eq. (2.389) can be established from the definition of variance as

$$\sigma^2 = \int_{-\infty}^{\infty} (x - m)^2 f(x)\ dx. \tag{2.390}$$

Changing the lower limit in (2.390), we have

$$\sigma^2 \geq \int_{|x-m|\geq\delta}^{\infty} (x - m)^2 f(x)\ dx = \delta^2 \int_{|x-m|\geq\delta}^{\infty} f(x)\ dx. \tag{2.391}$$

Equation (2.391) simplifies to

$$\sigma^2 \geq \delta^2 P(|X - m| \geq \delta) \tag{2.392}$$

which is the Chebyshev inequality in (2.389). The Chebyshev inequality can also be obtained in a slightly different way that would provide more insight into its meaning. If we assume that the random variable has a mean of zero (i.e., define a new random variable $Y = X - m$), we can define a new function

$$g(Y) = \begin{cases} 1, & |Y| \geq \delta, \\ 0, & |Y| \geq \delta. \end{cases} \tag{2.393}$$

The left-hand side of the Chebyshev inequality in (2.389) is related to the mean of the new variable in (2.393)

$$\langle g(Y) \rangle = P(|Y| \geq \delta) = P(|X - m| \geq \delta). \tag{2.394}$$

The concept of this approach to Chebyshev inequality is shown in Fig. 2.46. From the figure, it is seen that the $g(Y)$ is bounded (upper) by the quadratic $(Y/\delta)^2$,

$$g(Y) \leq \left(\frac{Y}{\delta}\right)^2. \tag{2.395}$$

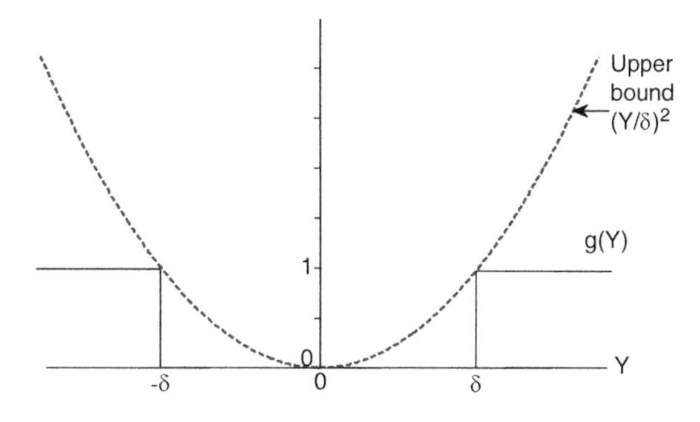

Fig. 2.46 Concept of Chebyshev inequality

Note that Y is a zero mean random variable and hence

$$\langle Y^2 \rangle = \left\langle (X-m)^2 \right\rangle = \sigma^2. \tag{2.396}$$

Thus,

$$\langle g(Y) \rangle \leq \left\langle \frac{Y^2}{\delta^2} \right\rangle = \frac{\langle Y^2 \rangle}{\delta^2} = \frac{\sigma^2}{\delta^2}. \tag{2.397}$$

Since (2.394) is the tail probability, (2.397) provides the Chebyshev inequality in (2.389).

2.12.2 *Chernoff Bound*

In some of the cases we are interested in the area only under one tail. For most of the wireless communications, SNR takes positive values only. We are often interested in the probability that the SNR exceeds a certain value, i.e., the area under the tail from δ to ∞. Since only a single tail is involved, we could use an exponential function instead of the quadratic function in the Chebyshev inequality (Proakis 2001; Haykin 2001). Let $g(Y)$ be such that

$$g(Y) \leq \exp[v(Y-\delta)] \tag{2.398}$$

with

$$g(Y) = \begin{cases} 1, & Y \geq \delta, \\ 0, & Y < \delta. \end{cases} \tag{2.399}$$

In (2.398), the parameter v needs to be found and optimized. The exponential upper bound is shown in Fig. 2.47.

The expected value of $g(Y)$ is

$$\langle g(Y) \rangle = P(Y \geq \delta) \leq \langle \exp[v(Y-\delta)] \rangle. \tag{2.400}$$

Note that v must be positive. Its value can be obtained by minimizing the expected value of the exponential term in (2.400). A minimum occurs when

$$\frac{d}{dv} \langle \exp[v(Y-\delta)] \rangle = 0. \tag{2.401}$$

Changing the order of integration and differentiation in (2.401), we have

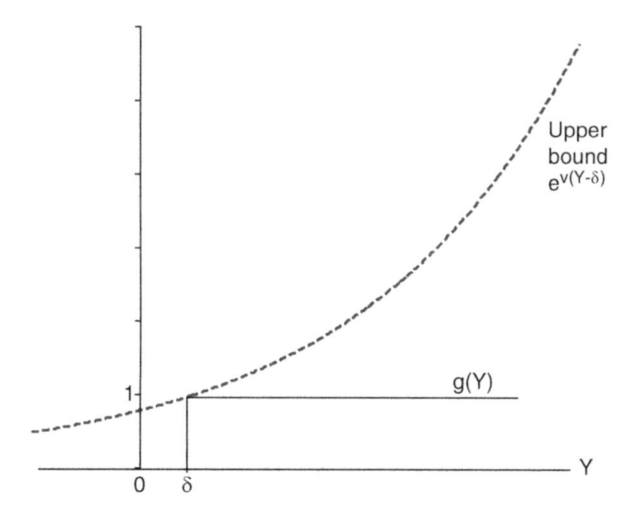

Fig. 2.47 Concept of the Chernoff bound

$$\exp(-v\delta)[\langle Y\exp(vY)\rangle - \delta\langle\exp(vY)\rangle] = 0. \tag{2.402}$$

The tightest bound is obtained from

$$\langle Y\exp(vY)\rangle - \delta\langle\exp(vY)\rangle = 0. \tag{2.403}$$

If v is the solution to (2.403), the upper bound on the one sided tail probability from (2.400) is

$$P(Y \geq \delta) \leq \exp(-\widehat{v}\delta)\langle\exp(\widehat{v}\delta)\rangle. \tag{2.404}$$

As it can be observed, the Chernoff bound is tighter than the tail probability obtained from the Chebyshev inequality because of the minimization in (2.401).

2.13 Stochastic Processes

We have so far examined the randomness of observed quantities, such as the signal amplitude or power and noise amplitude or power at a certain fixed time instant. While the signal and noise are spread over a long period of time, the random variable we studied constituted only a sample taken from the time domain function associated with the signal or noise (Taub and Schilling 1986; Gagliardi 1988; Papoulis and Pillai 2002). In other words, we need to characterize the temporal statistical behavior of the signals. This is accomplished through the concept of random processes. Consider a simple experiment. Take, for example, the measurement of a current or voltage through a resistor. Let us now set up several such

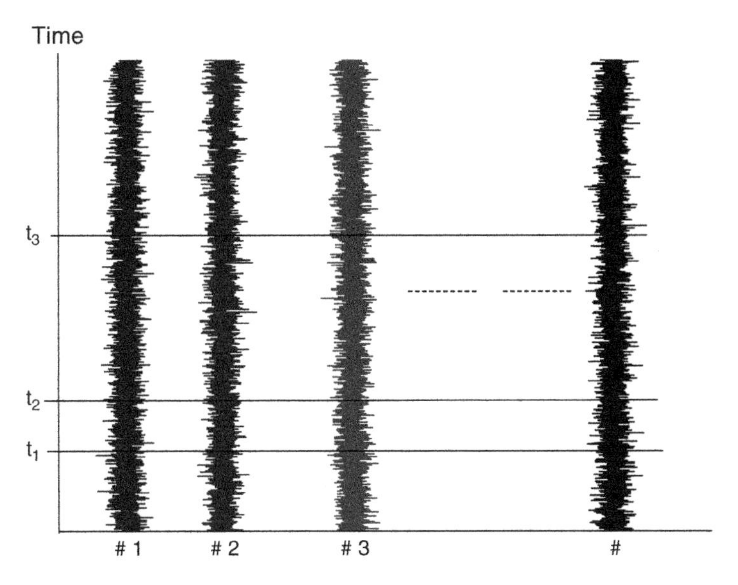

Fig. 2.48 Concept of a random process

experiments, all of them with identical characteristics. In each of these experiments, we measure the signal over a certain time period and we expect to see several time functions as illustrated in Fig. 2.48.

The voltages are plotted along the X-axis for each experiment while the temporal information is plotted along the Y-axis. For a fixed time (t_1), if we take the samples along the time line, we get different sample values which are random. In other words, if each of the temporal functions are represented by X, we can express it as a function of two variables, namely the "ensembles" ξ and "time" t. All such time functions together constitute the random process which describes the noise voltage. In other words, $X(\xi,t)$ represents the random process and for a fixed time instant t_k, $X(\xi,t_k)$ represents a random variable, simply represented by $X(t_k)$. If the sample value is fixed, i.e., for a fixed value of ξ, say ξ_n, we have a pure time function as $X(\xi_n,t)$. In other words, by fixing the time, we create a random variable from a process, and the random process becomes a pure time function when we fix the ensemble value. Thus, the random process provides both temporal and ensemble information on the noise or any other noise phenomenon. For example, we can create a random process by making the phase of a cosine wave random as

$$X(t) = \cos(2\pi f_0 t + \Theta). \tag{2.405}$$

In (2.405), Θ is a random variable. Since the random process contains both time and ensemble information, we can obtain two different types averages, autocorrelation values, and so on. Let us first define the density functions. For any value of t, $X(t)$ is a random variable, and hence, the first-order CDF is expressed as

$$F(x,t) = P[x(t) < x]. \tag{2.406}$$

The first order density function $f(x,t)$ is obtained by differentiating (2.406) w.r.t x as

$$f(x,t) = \frac{\partial F(x,t)}{\partial x}. \tag{2.407}$$

First, if we have time at two instants, t_1 and t_2, we can similarly obtain the second density function

$$f(x_1, x_2; t_1, t_2) = \frac{\partial^2 F(x_1, x_2; t_1, t_2)}{\partial x_1 \partial x_2}. \tag{2.408}$$

Similarly we can create higher-order density functions. Once we have the first-order density function, we have the so called first-order statistics and we get the nth order statistics of the random process by defining the density function for ensembles taken at the n unique time instants, $t_1, t_2, \ldots, t_n$.

We define the ensemble average or statistical average $\eta(t)$ as

$$\eta(t) = \langle X(t) \rangle = \int x f(x,t)\ dx. \tag{2.409}$$

The autocorrelation of $X(t)$ is the joint moment of $X(t_1)X(t_2)$ defined as

$$R(t_1, t_2) = \int\int x_1 x_2 f(x_1, x_2; t_1, t_2)\ dx_1 dx_2. \tag{2.410}$$

We can define the autocovariance as

$$C(t_1, t_2) = R(t_1, t_2) - \eta(t_1)\eta(t_2). \tag{2.411}$$

Equation (2.411) provides the variance of the random process when $t_1 - t_2 = 0$. The average values defined so far have been based on the use of density functions. Since the random process is a function of time as well, we can also define the averages in the time domain. Let

$$s = \int_a^b X(t)\ dt \tag{2.412}$$

and

$$s^2 = \int_a^b \int_a^b X(t_1)X(t_2)\ dt_1 dt_2. \tag{2.413}$$

The averages in (2.412) and (2.413) are random variables. To remove the randomness and to get the true mean, we need to take a further expectation (statistical) of the quantities.

$$\eta(s) = \langle s \rangle = \int_a^b \langle X(t) \rangle \; dt = \int_a^b \eta(t) \; dt, \tag{2.414}$$

$$\langle s^2 \rangle = \int_a^b \int_a^b \langle X(t_1)X(t_2) \rangle \; dt_1 dt_2 = \int_a^b \int_a^b R(t_1, t_2) \; dt_1 dt_2. \tag{2.415}$$

In communications, it is easy to perform temporal averaging by observing the signal in the time domain as in (2.412) or (2.413). If the temporal averages and statistical averages are equal, it would be necessary for the integrals in (2.414) and (2.415) to be equal. Such properties are associated with stationary processes. A stochastic process $X(t)$ is called strict sense stationary (SSS) if its statistical properties are invariant to a shift in the origin. Thus, a process is first-order stationary if

$$f(x,t) = f(x,t+c) = f(x), \quad c > 0. \tag{2.416}$$

This would mean that the temporal average and statistical average are identical. A process is second order stationary if the joint pdf depends only on the time difference and not on the actual values of t_1 and t_2 as

$$f(x_1, x_2; t_1, t_2) = f(x_1, x_2; \tau), \tag{2.417}$$

$$\tau = t_1 - t_2. \tag{2.418}$$

A process is called wide stationary sense (WSS) if

$$\langle X(t) \rangle = \eta, \tag{2.419}$$

$$\langle X(t_1)X(t_2) \rangle = R(\tau) \tag{2.420}$$

Since τ is midway point between t and $t + \tau$, we can write

$$R(\tau) = \left\langle X\left(t - \frac{\tau}{2}\right) X^* \left(t + \frac{\tau}{2}\right) \right\rangle. \tag{2.421}$$

In (2.421), we have treated the process as complex in the most general sense. Note that a complex process

$$X(t) = X_r(t) + jX_i(t) \tag{2.422}$$

is specified in terms of the joint statistics of the real processes $X_r(t)$ and $X_i(t)$. A process is called ergodic if the temporal averages are equal to the ensemble averages. For a random process to be ergodic, it must be SSS. Using this concept of ergodicity, we can see that a wide sense of stationary process is ergodic in the mean and ergodic in the autocorrelation. These two properties are sufficient for most of the analysis of communication systems.

The power spectral density (PSD) $S(f)$ of a wide sense stationary process $X(t)$ is the Fourier transform of its autocorrelation, $R(\tau)$. As defined in (2.421) the process may be real or complex. We have

$$S(f) = \int_{-\infty}^{\infty} R(\tau)\exp(-j2\pi f_0\tau)\ d\tau. \tag{2.423}$$

Note that $S(f)$ is real since $R(-\tau) = R^*(\tau)$. From the Fourier inversion property,

$$R(\tau) = \int_{-\infty}^{\infty} S(f)\exp(j2\pi f_0\tau)\ df. \tag{2.424}$$

Furthermore, if $X(t)$ is real and $R(\tau)$ is real and even, then

$$S(f) = \int_{-\infty}^{\infty} R(\tau)\cos(j2\pi f_0\tau)\ d\tau = 2\int_{0}^{\infty} R(\tau)\cos(j2\pi f_0\tau)\ d\tau, \tag{2.425}$$

$$R(\tau) = 2\int_{0}^{\infty} S(f)\cos(j2\pi f_0\tau)\ df. \tag{2.426}$$

As discussed earlier, the noise in communication systems is modeled as a random process. The primary characteristic of the noise in communication systems, referred to as thermal noise, is that it is zero mean Gaussian and that it is white (Taub and Schilling 1986). This means that the thermal noise at any given time instant has a Gaussian distribution with zero mean. The whiteness refers to the fact that its spectral density is constant (shown in Fig. 2.49).

The spectral density $G_n(f)$ of the noise $n(t)$ has a constant value of ($N_0/2$) over all the frequencies. This means that its autocorrelation is a delta function as shown

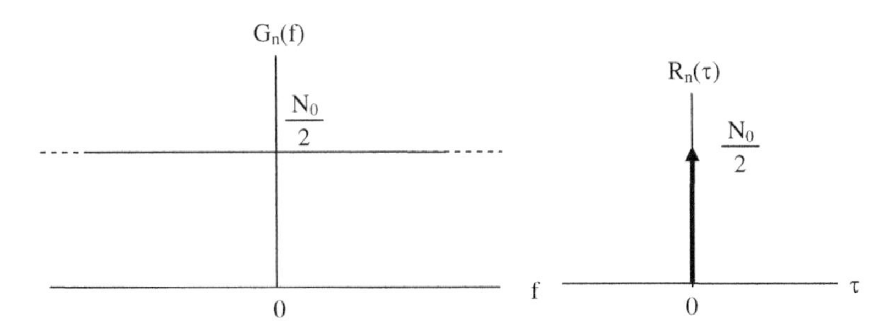

Fig. 2.49 Spectral density and the autocorrelation of noise

$$R(\tau) = \frac{N_0}{2}\delta(\tau). \tag{2.427}$$

Equation (2.427) suggests that any two samples of the noise taken at any separation (however small this separation may be) will be uncorrelated. When the noise passes through a low pass filter of bandwidth B, the noise energy will be $((N_0/2)2B)$ or N_0B.

2.14 Parameter Estimation and Testing

In statistical modeling such as the description of signal strength fluctuations seen in wireless channels, it is necessary to validate the underlying model. While the models that have been in existence for a long time such as the Rayleigh, Nakagami (or gamma), Nakagami-lognormal (or gamma-lognormal), and Rician have been tested against the data collected, some of the newer family of models such as the generalized K distribution, K distribution, $\eta - \mu$ distribution, $\kappa - \mu$ distribution, McKay distribution, Meijer G family of functions, and mixture densities require testing to validate their match to the actual data. Since all these densities contain a number of parameters, methods to estimate these parameters need to be examined, and the estimates have to be tested to ensure that the density functions actually fit the data.

The parameter estimation methods primarily fall in two categories, moment based estimates or the method of moments (MoM) and maximum likelihood estimates (MLE). While the method of moments is simple and direct, maximum likelihood estimation requires a more rigorous theoretical exercise and it is more likely to result in reliable estimates (Papoulis and Pillai 2002). Along with parameter estimation, it is also necessary to test whether the model and its parameters fit the measured data. This step requires hypothesis testing.

The methods of parameter estimation will be explored first before examining the hypothesis testing approaches.

2.14.1 *Method of Moments*

A number of probability densities commonly seen in wireless are studied to illustrate the concepts of parameter estimation. A detailed picture of these densities and their properties appear in Chap. 4.

Consider the case of a gamma random variable with a density (Papoulis and Pillai 2002)

$$f_Z(z) = \frac{1}{\Gamma(\alpha)\beta^{\alpha}} z^{\alpha-1} e^{-\frac{z}{\beta}} U(z). \tag{2.428}$$

Note that the gamma density associated with the Nakagami fading channel is obtained when

$$\begin{aligned}\alpha &= m\\ \beta &= \frac{\Omega}{m}.\end{aligned} \tag{2.429}$$

In Eq. (2.429), m is the Nakagami fading parameter and Ω is the average SNR. The moments of the gamma density are

$$E(Z^k) = \frac{\Gamma(k+\alpha)}{\Gamma(\alpha)}\beta^k = \frac{\Gamma(k+\alpha)}{\Gamma(\alpha)}\left(\frac{\Omega}{m}\right)^k, \quad k = 1, 2, \ldots \tag{2.430}$$

The first two moments are

$$\begin{aligned}m_1 &= E(Z) = \alpha\beta = \Omega\\ m_2 &= E(Z^2) = \alpha^2\beta^2 + \alpha\beta^2 = \frac{(m+1)\,\Omega^2}{m}\end{aligned} \tag{2.431}$$

The parameters of the gamma density function can be obtained from the first and second moments of the data using Eqs. (2.429) and (2.431).

The Rician density (SNR) is expressed as (Simon and Alouini 2005)

$$f(z) = \frac{K_0+1}{\Omega}e^{-K}e^{-\frac{K_0+1}{\Omega}z}I_0\left(2\sqrt{\frac{K_0(K_0+1)}{\Omega}z}\right)U(z) \tag{2.432}$$

Note that K_0 is the Rician factor given by the ratio of the power of the direct path (line-of-sight or LOS) to the power of the multipath (excluding the LOS) components. When $K_0 = 0$, the Rician density becomes the exponential density. The moments of the Rician density are

$$E(Z^k) = \frac{\Gamma(k+1)}{(K_0+1)^k}\,{}_1F^1 1(-k, 1; -K_0)\Omega^k, \quad k = 1, 2, \ldots \tag{2.433}$$

In Eq. (2.433), ${}_1F_1(-k, 1; -K_0)$ is the hypergeometric function (Gradshteyn and Ryzhik 2007). The first two moments become

$$\begin{aligned}m_1 &= E(Z) = \alpha\beta = \Omega\\ m_2 &= E(Z^2) = \frac{(4K_0 + K_0^2 + 2)\,\Omega^2}{(K_0+1)^2}\end{aligned} \tag{2.434}$$

The two parameters, Ω and K_0, can be obtained by solving the set of equations given in Eq. (2.434).

Now, consider the case where the SNR in a fading channel follows a generalized gamma distribution (Aalo and Piboongungon 2005)

$$f(z) = \frac{c}{b^{a\,c}\Gamma(a)} z^{a\,c-1} e^{-\left(\frac{z}{b}\right)^{c}} U(z). \tag{2.435}$$

The moments are

$$E\left(Z^k\right) = \frac{\Gamma\left(\frac{k+ac}{c}\right)}{\Gamma(a)} b^k, \quad k = 1, 2, \ldots \tag{2.436}$$

The first three moments are

$$\begin{aligned} m_1 &= E(Z) = \frac{\Gamma\left(\frac{1+ac}{c}\right)}{\Gamma(a)} b \\ m_2 &= E\left(Z^2\right) = \frac{\Gamma\left(\frac{2+ac}{c}\right)}{\Gamma(a)} b^2 \\ m_3 &= E\left(Z^3\right) = \frac{\Gamma\left(\frac{3+ac}{c}\right)}{\Gamma(a)} b^2. \end{aligned} \tag{2.437}$$

The three parameters of the generalized gamma density can be obtained by solving the set of equations in Eq. (2.437).

The three densities discussed above clearly indicate that the number of parameters to be determined match the number of moments needed to solve for them. Thus, as the number of parameters increases, the order of moments also goes up and in some cases such as in the case of the generalized gamma density in Eq. (2.435), solving the three equations to get the parameters a, b, and c would require graphical or numerical approaches since these equations are of the transcendental type. As the number of parameters of the density function increase, higher and higher order moments are needed to estimate the parameters and the reliability of the moments becomes questionable as the order increases. Thus, MoM is less dependable when accurate parameter estimation is necessary and other methods of parameter estimation need to be explored.

2.14.2 Maximum Likelihood Estimation

The other method, maximum likelihood estimate (MLE), available for estimation of parameter is based on the likelihood function (LF) of a random variable (Redner and Walker 1984; Iskander et al. 1999; Papoulis and Pillai 2002; Bowman and Shenton 2006). Consider a random variable with a single parameter (for example, an exponential density function with parameter θ). The density function $f(z)$ can be expressed as

$$f_Z(z) = f(z;\theta). \tag{2.438}$$

If a number of observations of the variable are undertaken resulting in a vector of observables (i.e., data) $Z_1, Z_2, \ldots, Z_n$, the joint density becomes

$$L(\theta; z_1, z_2, \cdots, z_n) = f(z_1;\theta) f(z_2;\theta) \cdots f(z_n;\theta) = \prod_{i=1}^{n} f(z_i;\theta). \tag{2.439}$$

In Eq. (2.439) the number of observations is n. The observations are treated as independent. The quantity or function on the left-hand side of Eq. (2.439), $L(\theta; z_1, z_2, \cdots, z_n)$ is called the likelihood function associated with the density of Z in Eq. (2.438). The principle of MLE consists of choosing an estimator that will maximize the likelihood function.

Since the likelihood function appears in product form, the convenient option for getting the estimator is to use the logarithm of the likelihood function which allows the analysis to be undertaken using summation instead of the product. Note that logarithm is a monotonic operator preserving all the properties of the likelihood function intact. As it will be shown using a number of sample densities, θ need not be a scalar. Indeed $\boldsymbol{\theta}$ can be a vector and multi-parameter densities can be handled using the same approach allowing the estimation of the parameters of the gamma, Rician, generalized gamma, or other densities.

The maximization is achieved through the equation

$$\frac{\partial}{\partial\theta} \log[L(\theta; z_1, z_2, \cdots, z_n)] = 0 \tag{2.440}$$

If multiple parameters exist, the maximization is achieved through the set of equations formed from

$$\frac{\partial}{\partial\theta_j} \log\left[L\left(\vec{\theta}; z_1, z_2, \cdots, z_n\right)\right] = 0, j = 1, 2, \cdots, J. \tag{2.441}$$

Equation (2.441) consists of a set of J equations, one for each parameter expressed in vector form as

$$\vec{\theta} = \begin{bmatrix} \theta_1 \\ \theta_2 \\ \cdot \\ \cdot \\ \theta_J \end{bmatrix}. \tag{2.442}$$

A non-trivial solution of the set of equations in Eq. (2.441) provides the maximum likelihood estimates of the parameters contained in Eq. (2.442). A few points are necessary to clarify the strength and weakness of the MLE approach. While MLE generally leads to more reliable results, MLE can also lead to problems if the non-trivial solutions are not unique. Additionally, there is no assurance that the derivative of the log likelihood function exists. It is also possible that even if a derivative exists, there is no simple analytical expression for the derivative, making it necessary to invoke numerical techniques for solving the set of equations.

The strengths and weaknesses of the MLE will now be illustrated using the same density functions presented earlier in connection with MoM along with additional density functions.

Consider the example of an exponential density,

$$f(z) = f(z;\theta) = \frac{1}{\theta} e^{-\frac{z}{\theta}} U(z). \tag{2.443}$$

The likelihood function is

$$L(\theta; z_1, z_2, \cdots, z_n) = \left(\frac{1}{\theta}\right)^n \prod_{i=}^{n} e^{-\frac{z_i}{\theta}} \tag{2.444}$$

The log likelihood function is

$$\log[L(\theta; z_1, z_2, \cdots, z_n)] = -n\log(\theta) - \frac{1}{\theta}\sum_{i=1}^{n} z_i \tag{2.445}$$

Taking the logarithm and setting it equal to zero,

$$\frac{\partial}{\partial\theta}\log[L(\theta; z_1, z_2, \cdots, z_n)] = 0 = -\frac{n}{\theta} + \frac{1}{\theta^2}\sum_{i=1}^{n} z_i. \tag{2.446}$$

Simplifying Eq. (2.446),

$$\frac{n}{\theta} = \frac{1}{\theta^2}\sum_{i=1}^{n} z_i \Rightarrow \theta_{\mathrm{MLE}} = \frac{1}{n}\sum_{i=1}^{n} z_i. \tag{2.447}$$

Equation (2.447) clearly shows the expected result. The estimate of the parameter θ is the sample mean itself. It is of interest to note that the MLE of the parameter and moment based estimate match in this case.

Now, consider the case of a gamma density with two parameters (Cheng and Beaulieu 2001)

$$f(z) = f(z; m, \Omega) = \left(\frac{m}{\Omega}\right)^m \frac{z^{m-1}}{\Gamma(m)} e^{-\frac{m}{\Omega}z} U(z) \tag{2.448}$$

The likelihood function will be

$$L(m, \Omega; z) = \left[\left(\frac{m}{\Omega}\right)^m \frac{1}{\Gamma(m)}\right]^n \prod_{i=1}^{n} z_i^{m-1} e^{-\frac{m}{\Omega}z_i} \tag{2.449}$$

The log likelihood function will be

$$\log[L(m, \Omega; z)] = n\left[m\log\left(\frac{m}{\Omega}\right) - \log(\Gamma(m))\right] + (m-1)\sum_{i=1}^{n} \log(z_i) - \left(\frac{m}{\Omega}\right)\sum_{i=1}^{n} z_i. \tag{2.450}$$

Simplifying,

$$\log[L(m,\Omega;z)] = m\log\left(\frac{m}{\Omega}\right) - \log(\Gamma(m)) + (m-1)\left(\frac{1}{n}\sum_{i=1}^{n} \log(z_i)\right) - \left(\frac{m}{\Omega}\right)\left(\frac{1}{n}\sum_{i=1}^{n} z_i\right) \tag{2.451}$$

Since there are two parameters, the two equations for maximization result in

$$\frac{\partial}{\partial \Omega}[\log[L(m, \Omega; z)]] = 0. \tag{2.452}$$

$$\frac{\partial}{\partial m}[\log[L(m, \Omega; z)]] = 0. \tag{2.453}$$

Equation (2.452) results in

$$-\left(\frac{m}{\Omega}\right) + \left(\frac{m}{\Omega^2}\right)\left(\frac{1}{n}\sum_{i=1}^{n} z_i\right) = 0 \tag{2.454}$$

Equation (2.454) simplifies to

$$\Omega_{mle} = \frac{1}{n}\sum_{i=1}^{n} z_i. \tag{2.455}$$

Equation (2.455) is the expected result showing that the parameter Ω is the sample mean. Thus, the MLE of Ω matches the parameter obtained from the method of moments. Eq. (2.453) results in

$$\frac{\partial}{\partial m}\left(m\log\left(\frac{m}{\Omega_{\text{mle}}}\right)-\log(\Gamma(m))+(m-1)\frac{1}{n}\sum_{i=1}^{n}\log(z_i)-\left(\frac{m}{\Omega_{\text{mle}}}\right)\Omega_{\text{mle}}\right)=0. \tag{2.456}$$

In Eq. (2.456) Ω has been replaced with Ω_{mle} obtained from Eq. (2.455). Equation (2.456) leads to

$$\log(m)-\psi(m)=\log(\Omega_{\text{mle}})-\langle\log(z_i)\rangle. \tag{2.457}$$

In Eq. (2.457), $\psi(.)$ is the digamma function defined as

$$\psi(x)=\frac{d}{dx}[\log\Gamma(x)] \tag{2.458}$$

One can see the advantage and disadvantage of the MLE based estimate of the Nakagami parameter. The advantage is that the estimation depends only on the first moment and the first moment of the logarithm of the data. The disadvantage is that there is no direct way to solve Eq. (2.457) to get the value of m and numerical techniques are required to get the solution for m.

2.14.3 Comparison of MoM and MLE

To compare the approaches, a simple Matlab code was written to establish the relationship between the actual value of m and estimates obtained using the MoM and MLE. A number of examples on the use of MLE and MOM are presented next.

2.14.3.1 Gamma Distribution

The gamma distribution is of interest in wireless communication since the Nakagami model is the most commonly used to describe the statistics of the amplitude in fading channels. When the amplitude is Nakagami distributed, the power or the signal-to-noise ratio is gamma distributed. Even though Nakagami samples can be generated in Matlab, gamma random variables were generated for the simulation. It should be noted that the parameter can also be estimated directly in Matlab using *gamfit*(.). The simple code and the results are given below.

```
% estimation_nakagami1
% p m shankar, June 2016
% compare MoM and MLE
clear;clc;close all
N=1e5; % number of samples
Z0=15; % mean
mn=0.5:.1:1.5; % values of m for simulation
for k=1:length(mn)
    mm=mn(k);
x=gamrnd(mm,Z0/mm,1,N); % gamma random variable with mean Z0
X1=mean(x);              VX=var(x);              XL=mean(log(x));
m_mom(k)=X1^2/VX; % m from MoM
m=0.4:.005:2;
y=log(m)-psi(m)-log(X1)+XL; %MLE expression for m
ind=find(abs(y)<0.006); % obtain the solution examining when the EQN crosses 0
m_mle(k)=median(m(ind)); % if there is more than one value, take the median
end;
plot(mn,m_mom,'*r',mn,m_mle,'b^')
legend('MoM','MLE','location','southeast'),ylim([0.5,2])
xlabel('m parameter')
ylabel('estimate of m')
text(0.55,1.94,'     m          m_{MOM}          m_{MLE}','color','r','fontweight','bold')
text(0.6, 1.4,num2str([mn;m_mom;m_mle]'))
```

Figure 2.50 shows the results of the Nakagami parameter estimation using both approaches. It clearly points to the fact that the parameter estimates obtained from both methods are very close.

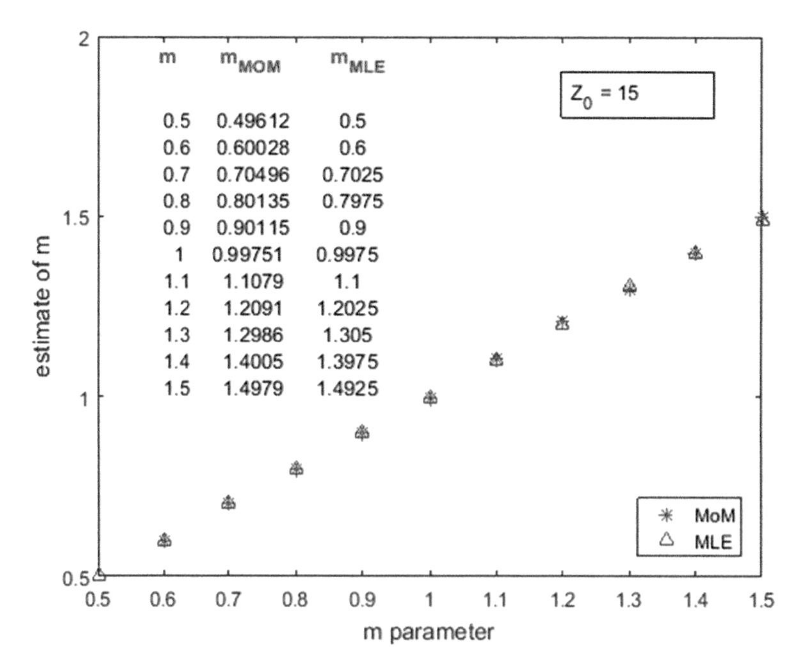

Fig. 2.50 Comparison of MoM and MLE: Nakagami parameter estimation

2.14.3.2 Generalized Gamma Density

Another uniquely interesting case is the generalized gamma density,

$$f(z) = f(z; a, b, c) = \frac{c}{b^a \Gamma a} z^{ac-1} e^{-\frac{z^c}{b}} U(z). \tag{2.459}$$

There are several forms of the generalized gamma density functions and Eq. (2.459) is one of these. The generalized gamma density becomes a gamma density when $c = 1$ and Weibull density when $a = 1$. The likelihood function associated with the density in Eq. (2.459) becomes (Papoulis and Pillai 2002; Kumar and Shukla 2010)

$$L(a, b, c; z) = \left[\frac{c}{b^a \Gamma a}\right]^n \prod_{i=1}^{n} z_i^{ac-1} e^{-\frac{z_i^c}{b}} \tag{2.460}$$

Samples of generalized gamma random variables can be generated from samples of gamma variables using the concept of transformation of random variables. To create a generalized gamma random variable sample, first consider a gamma random variable Y with the density

$$f(y) = f(y; a, b) = \frac{1}{\Gamma(a) b^a} z^{a-1} e^{-\frac{z}{b}} U(z). \tag{2.461}$$

The generalized gamma density in Eq. (2.459) is obtained by transforming Y to Z as

$$Z = Y^{\frac{1}{c}}, c > 0. \tag{2.462}$$

Equation (2.462) thus provides a simple means to generate random numbers which follow the density in Eq. (2.459). The log likelihood function of the generalized gamma density in Eq. (2.460) becomes

$$\log(L(a, b, c; z)) = n\log\left[\frac{c}{b^a \Gamma a}\right] + (ac - 1)\sum_{i=1}^{n} \log(z_i) - \frac{1}{b}\sum_{i=1}^{n} z_i^c \tag{2.463}$$

The log likelihood function in Eq. (2.463) can be rewritten as

$$\log(L(a, b, c; z)) = n\log c - na\log b - n\log(\Gamma(a)) + (ac - 1)\sum_{i=1}^{n} \log(z_i) - \frac{1}{b}\sum_{i=1}^{n} z_i^c. \tag{2.464}$$

The three equations for maximization become

$$\frac{\partial}{\partial a}[\log(L(a,b,c;z))] = -n\log b - n\psi(a) + c\sum_{i=1}^{n}\log(z_i) = 0 \tag{2.465}$$

$$\frac{\partial}{\partial b}[\log(L(a,b,c;z))] = -\frac{na}{b} + \frac{1}{b^2}\sum_{i=1}^{n} z_i^c = 0 \tag{2.466}$$

$$\frac{\partial}{\partial c}[\log(L(a,b,c;z))] = \frac{n}{c} + a\sum_{i=1}^{n}\log(z_i) - \frac{1}{b}\sum_{i=1}^{n} z_i^c\log(z_i) = 0 \tag{2.467}$$

To obtain the estimates of the three parameters, there is a need to simultaneously solve three transcendental equations that can be obtained from Eqs. (2.465), (2.466), and (2.467). These are

$$\begin{aligned} &-n\log b - n\psi(a) + c\sum_{i=1}^{n}\log(z_i) = 0 \\ &-\frac{na}{b} + \frac{1}{b^2}\sum_{i=1}^{n} z_i^c = 0 \\ &\frac{n}{c} + a\sum_{i=1}^{n}\log(z_i) - \frac{1}{b}\sum_{i=1}^{n} z_i^c\log(z_i) = 0 \end{aligned} \quad . \tag{2.468}$$

One might also notice from Eq. (2.436) that three moments needed to solve for a, b, and c are expressed in terms of gamma functions

$$E\left[Z^k\right] = b^{\frac{k}{c}}\frac{\Gamma\left(\frac{ac+k}{c}\right)}{\Gamma(a)}. \tag{2.469}$$

Equation (2.469) indicates that even MoM requires the use of numerical techniques. The estimation is demonstrated using the "mle" function in Matlab. It should be noted that MLE equation set in Eq. (2.468) is not required for the parameter estimation in Matlab directly using the "mle" command. The Matlab script appears below.

```
function  generalized_gamma_estimation_demo
close all
% estimate the parameters of the generalized gamma density
% the expression is created using symbolic toolbox and converted to the
% in-line function that is the input to mle(.)
% done in three steps:
%
% step # 1 b and c fixed, a varying: estimate a, b (mean), c (mean)
% step # 2 a and c fixed, b varying: estimate a (mean), b, c (mean)
% step # 3 a and b fixed, c varying: estimate a (mean), b (mean), c
%
% p m shankar, October 2016
syms a b c z f_Z(z)
ff=c*(z^(a*c-1))*exp(-(z^c)/b)/(gamma(a)*b^a);
pdf=[f_Z(z)==ff]; % create the pdf expression for plot display
N=1e5;
A=[0.5:.1:1.2]; % values of a for random number generation
L=length(A);
for k=1:L
    aa=A(k);
    cc=0.6;% value of b (fixed)
    bb=5; % value of c (fixed)
  [pp] = gg_estimatef(aa,bb,cc,N);
  AB(k)=pp(1);
  C(k)=pp(3);
  B(k)=pp(2);
end;
figure,plot(A,AB,'*')
xlabel('a'),ylabel('a_{est}')
title(['$' latex(pdf) '$'],'interpreter','latex','fontsize',14)
data={['b_{input}=',num2str(bb),'; ','c_{input}=',num2str(cc)];...
    ['mean b_{est}=',num2str(mean(B)),', ','mean c_{est}=',num2str(mean(C))]};
text(0.55,1,data,'color','r','fontweight','bold')
text(.9,.45,'[b and c fixed input]','backgroundcolor','y')
clear A
B=[2:12]; % values of b
L=length(B);
for k=1:L
    aa=0.7; % value of a fixed
    cc=0.6; % value of c fixed
    bb=B(k);
  [ pp] = gg_estimatef(aa,bb,cc,N);
  A(k)=pp(1);
  C(k)=pp(3);
  BB(k)=pp(2);
end;
figure,plot(B,BB,'*')
xlabel('b'),ylabel('b_{est}')
title(['$' latex(pdf) '$'],'interpreter','latex','fontsize',14)
data={['a_{input}=',num2str(aa),'; ','c_{input}=',num2str(cc)];...
    ['mean a_{est}=',num2str(mean(A)),', ','mean c_{est}=',num2str(mean(C))]};
text(2.55,10,data,'color','r','fontweight','bold')
```

```
text(8,2,'[a and c fixed input]','backgroundcolor','y')
clear B
C=[0.5:.1:1.2]; % values of c
L=length(C);
for k=1:L
    aa=.8; % value of a fixed
    cc=C(k);
    bb=8; % value of b fixed
  [ pp] = gg_estimatef(aa,bb,cc,N);
  A(k)=pp(1);
  CK(k)=pp(3);
  B(k)=pp(2);
end;
figure,plot(C,CK,'*')
xlabel('c'),ylabel('c_{est}')
title(['$' latex(pdf) '$'],'interpreter','latex','fontsize',14)
data={['a_{input}=',num2str(aa),'; ','b_{input}=',num2str(bb)];...
    ['mean a_{est}=',num2str(mean(A)),', ','mean b_{est}=',num2str(mean(B))]};
text(0.55,1.1,data,'color','r','fontweight','bold')
text(.9,.55,'[a and b fixed input]','backgroundcolor','y')


end


function [ pp] = gg_estimatef(aa,bb,cc,N)
X=gamrnd(aa,bb,1,N);
x=X.^(1/cc);% generalized gamma variable
syms XX w y z % x, a, b, c
myp=z*(XX^(w*z-1))*exp(-(XX^z)/y)/(gamma(w)*y^w);
mydensity=matlabFunction(myp);
clear XX w y z a b c
opt = statset('mlecustom');
opt = statset(opt,'FunValCheck','off','MaxIter',5000,'TolX',1e-5,'TolFun',1e-5,'TolBnd',1e-
4','MaxFunEvals', 1000);
a=0.45; % initial guess
b=4;  % initial guess
c=0.25;  % initial guess
pp = mle(x,'pdf',mydensity,'start',[a b c], 'options',opt); % fit mypdf to the data Y
% pp is [a,b,c]

end
```

Results from the generalized gamma parameter estimation are displayed in Figs. 2.51, 2.52, and 2.53. They clearly show that the MLE approach provides simultaneous solutions to all the three parameters and the estimates are very close to the original input set.

The results displayed are obtained by keeping two parameters fixed and varying the third parameter. Such a step allows the determination of the average values of the estimates.

2.14.3.3 Generalized *K* Distribution

Another example is the generalized *K* distribution proposed as an approximate model for describing the SNR the Nakagami-lognormal fading channel (see Chap. 4). The generalized *K* distribution is the density of the product of two independent non-identical gamma random variables (Lomnicki 1967; Springer and Thompson 1970; Shankar 2005)

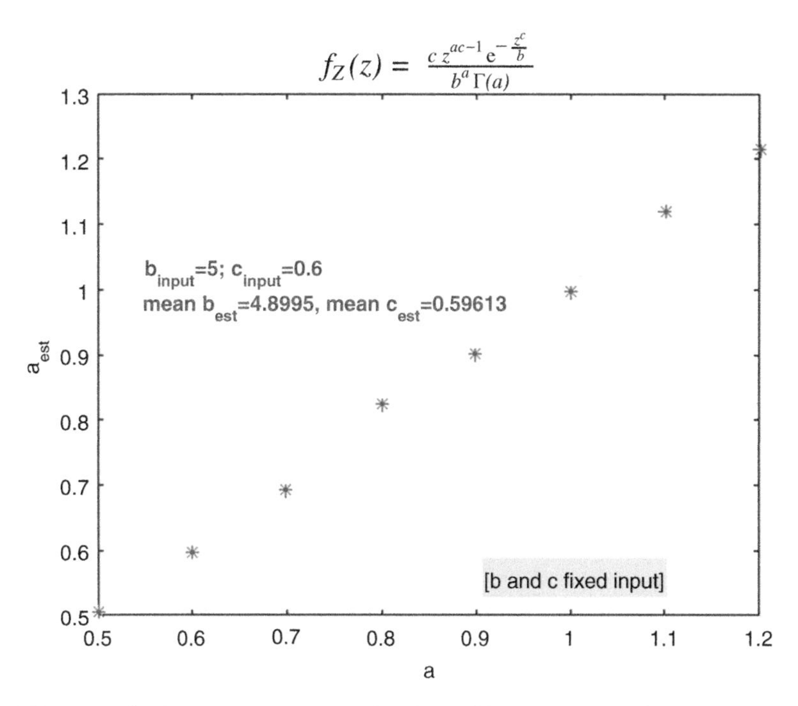

Fig. 2.51 Generalized gamma parameter estimation (MLE): fixed value of b and c. The mean estimates of b and c are also displayed

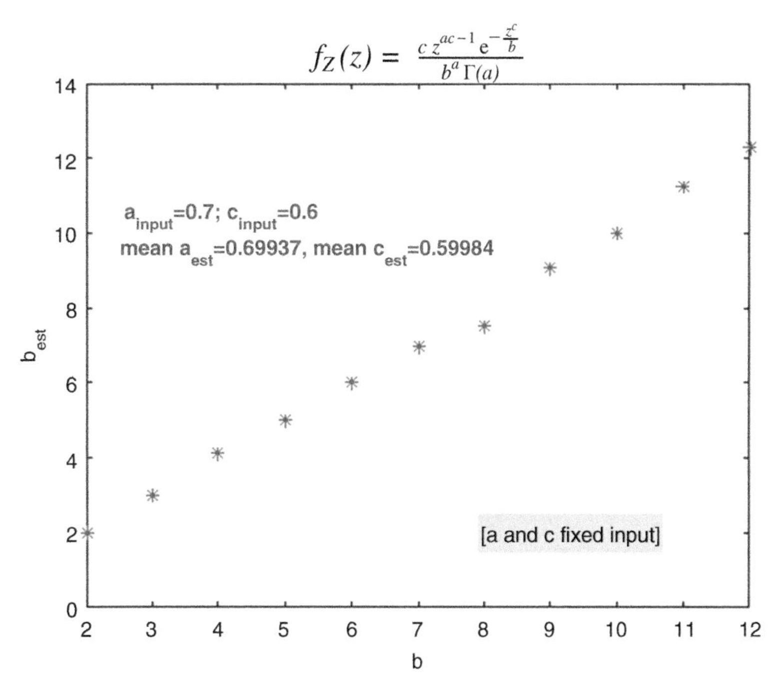

Fig. 2.52 Generalized gamma parameter estimation (MLE): fixed value of a and c. The mean estimates of a and c are also displayed

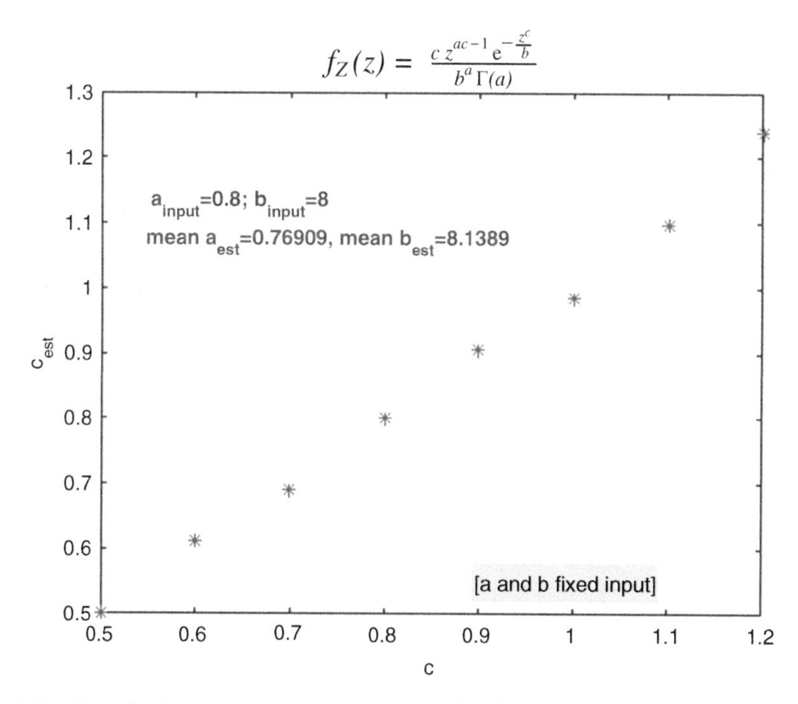

Fig. 2.53 Generalized gamma parameter estimation (MLE): fixed value of a and b. The mean estimates of a and b are also displayed

$$Z = XY \tag{2.470}$$

$$f(x) = \frac{\left(\frac{x}{a}\right)^{c-1} \exp\left(-\frac{x}{a}\right)}{a\Gamma(c)} \tag{2.471}$$

$$f(y) = \frac{\left(\frac{y}{b}\right)^{m-1} \exp\left(-\frac{y}{b}\right)}{b\Gamma(m)} \tag{2.472}$$

$$f(z) = \frac{2}{\left(\sqrt{ab}\right)^{m+c} \Gamma(m)\Gamma(c)} z^{\frac{m+c}{2}-1} K_{m-c}\left(2\sqrt{\frac{z}{ab}}\right) U(z). \tag{2.473}$$

The moments of the generalized K distribution are (Jouchin et al. 1993; Dogandzic and Jin 2004)

$$E\left(Z^k\right) = \frac{\Gamma(m+k)\Gamma(c+k)}{\Gamma(m)\Gamma(c)} (ab)^k. \tag{2.474}$$

It is obvious that the generalized K distribution in Eq. (2.473) is a 3-parameter distribution of m, c, and ab requiring simultaneous solution of three equations for obtaining the estimates. The MoM based method would require the use of numerical techniques and MLE using Matlab offers a simple and easy alternative to the MoM approach. One would notice that a and b are scaling factors and one of these can be taken to be unity while the other one is varied (for the simulation). Also, once the parameters m and c are obtained, the product term ab can be obtained from the first moment, completing the three parameters needed for the generalized K-density.

Results of the simulation are shown in Figs. 2.54 and 2.55. The advantage of the "mle" command in Matlab directly is seen again from the fact that there is no need to derive the expressions for the parameters. The only requirement is the availability of the density function. The Matlab script is also capable of keeping track of the time required for computation so that the two methods could be compared on the basis of computational time.

```
function  generalized_K_estimation_mle
close all
% estimate the parameters of the K distribution
% the expression is created using symbolic toolbox and converted to the
% in-line function that is the input to mle(.)
% done in three steps:
%
% step # 1 ab and c fixed, m varying: estimate m, c (mean), ab(mean)
% step # 2 ab and m fixed, c varying: estimate  m (mean), c, ab (mean)

%
% p m shankar, June 2016
syms m c a b z f_Z(z)
tic
ff=2*(z^((m+c)/2-1))*besselk(m-c,2*sqrt(z/(a*b)))/(gamma(m)*gamma(c)*(sqrt(a*b))^(m+c));
pdf=[f_Z(z)==ff]; % create the pdf expression for plot display
clear m c a b z f_Z(z)
N=1e5;
mm=[0.52:.1:1.2]; % values of a for random number generation
L=length(mm);
for k=1:L
    m1=mm(k);
    cc=4.5;% value of c (fixed)
    ab=5; % value of ab or alpha*beta (fixed)
  [pp] = gk_estimatef(m1,ab,cc,N);
  mk(k)=pp(1);
  AB(k)=pp(2); % a*b
  C(k)=pp(3);
end;
 figure,plot(mm,mk,'*')
 ylim([0.4,1.4])
 xlim([0.4,1.4])
 xlabel('m'),ylabel('m_{est}')
 title(['$' latex(pdf) '$'],'interpreter','latex','fontsize',14)
 data={['a*b_{input}=',num2str(ab),';  ','c_{input}=',num2str(cc)];...
    ['mean ab_{est}=',num2str(mean(AB)),',  ','mean c_{est}=',num2str(mean(C))]};
 text(0.55,1.25,data,'color','r','fontweight','bold')
 text(.9,.45,'[a*b and c fixed input]','backgroundcolor','y')
% clear A
CD=[3:9]; % values of a for random number generation
L=length(CD);
for k=1:L
    m1=0.7;% value of m (fixed)
    cc=CD(k);%
    ab=8; % value of ab or alpha*beta (fixed)
  [pp] = gk_estimatef(m1,ab,cc,N);
  mk(k)=pp(1);
  AB(k)=pp(2); % a*b
  C(k)=pp(3);
```

```
end;
 figure,plot(CD,C,'*')
 ylim([2,10])
 xlim([2,10])
 xlabel('m'),ylabel('m_{est}')
 title(['$' latex(pdf) '$'],'interpreter','latex','fontsize',14)
 data={['a*b_{input}=',num2str(ab),'; ','m_{input}=',num2str(m1)];...
    ['mean ab_{est}=',num2str(mean(AB)),', ','mean m_{est}=',num2str(mean(mk))]};
 text(3,9,data,'color','r','fontweight','bold')
 text(7,3,'[a*b and m fixed input]','backgroundcolor','y')
toc
end


function [ pp] = gk_estimatef(m,ab,cc,N)
X=gamrnd(m,ab,1,N);
Y=gamrnd(cc,1,1,N);
x=X.*Y; %GK variable
syms XX w y z % z, m,ab,c
myp=2*(XX^((w+z)/2-1))*besselk(w-z,2*sqrt(XX/(y)))/(gamma(w)*gamma(z)*(sqrt(y))^(w+z));
mydensity=matlabFunction(myp);
clear XX w y z a b c
opt = statset('mlecustom');
opt = statset(opt,'FunValCheck','off','MaxIter',5000,'TolX',1e-5,'TolFun',1e-5,'TolBnd',1e-
4','MaxFunEvals', 1000);
a=0.5; % initial guess
b=6;  % initial guess
c=2.2;  % initial guess
pp = mle(x,'pdf',mydensity,'start',[a b c], 'options',opt); % fit mypdf to the data Y
% pp is [a,b,c]

end
```

```
Elapsed time is 207.199643 seconds.
```

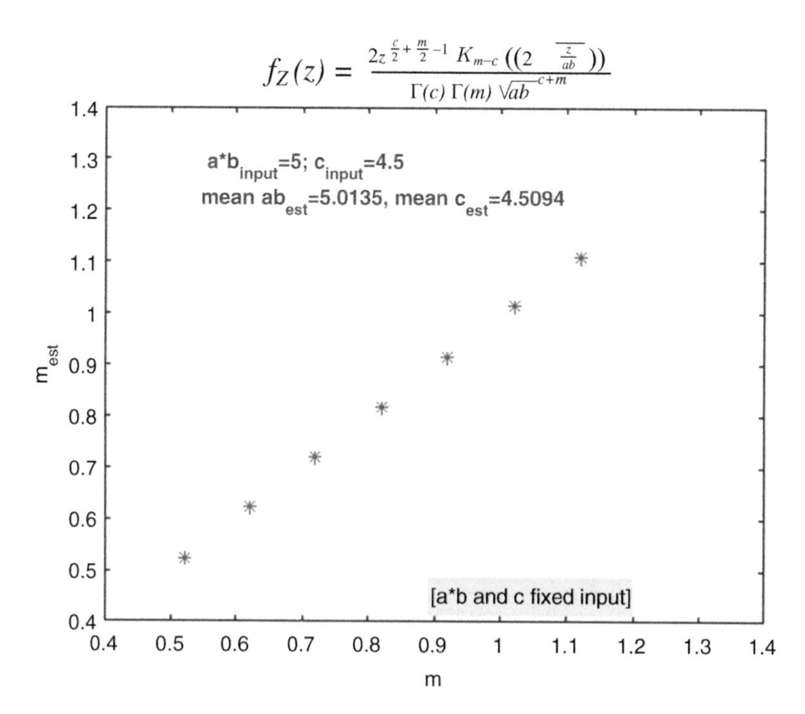

Fig. 2.54 Generalized K parameter estimation (MLE): fixed value of a, b, and c

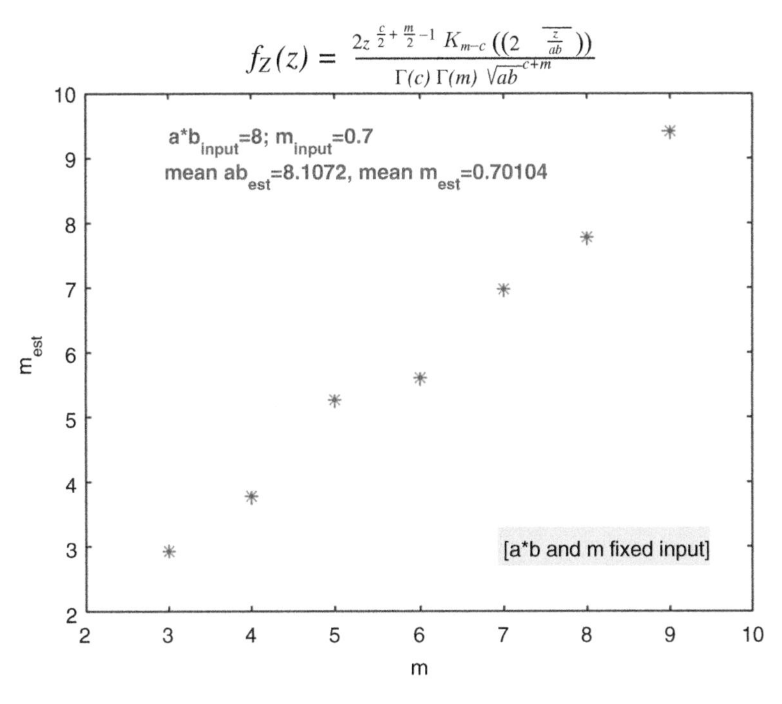

Fig. 2.55 Generalized K parameter estimation (MLE): fixed value of a, b, and m

An examination of the elapsed time shows that the time taken to estimate parameter based on MLE is a little bit on the high end. It should be noted that the computational time can be reduced by operating at a lower accuracy by setting the tolerances to appropriate values ("mle" options).

The method of moments can be utilized to see whether it offers any quicker and simpler way to estimate the parameters. While MoM might not always provide a simple means of solution, the generalized K distribution offers a unique case where it is possible to get parameter estimates more easily. The first two moments of the generalized K distribution are (Jouchin et al. 1993; Dogandzic and Jin 2004)

$$E(Z) = mcab \tag{2.475}$$

$$E\left(Z^2\right) = m(m+1)c(c+1)(ab)^2 \tag{2.476}$$

Instead of using the third moment and risking errors associated with higher order moments, one can obtain the half order moment from Eq. (2.474) as

$$E\left(Z^{\frac{1}{2}}\right) = \frac{\Gamma\left(m+\frac{1}{2}\right)\Gamma\left(c+\frac{1}{2}\right)}{\Gamma(m)\Gamma(c)}(ab)^{\frac{1}{2}}. \tag{2.477}$$

The scaling factors of the gamma densities $\alpha\beta$ can be eliminated using the normalized half and normalized second moments as

$$M_{\frac{1}{2}} = \frac{E\left(Z^{\frac{1}{2}}\right)}{\sqrt{E(Z)}} = \frac{\Gamma\left(m+\frac{1}{2}\right)\Gamma\left(c+\frac{1}{2}\right)}{\Gamma(m)\Gamma(c)\sqrt{mc}} \tag{2.478}$$

$$M_2 = \frac{E\left(Z^2\right)}{E^2(Z)} = \left(1+\frac{1}{m}\right)\left(1+\frac{1}{c}\right) \tag{2.479}$$

The availability of these moments of the data offers a simple way to solve for m and c first and then, estimate $\alpha\beta$ from Eq. (2.475). The solution can be carried out numerically in Matlab. The results shown below clearly illustrate the level of ease in obtaining the solution using MoM.

The results are displayed in Figs. 2.56 and 2.57. They clearly demonstrate that MoM based methods can lead to reliable values.

```
function generalized_K_estimation_MoM
close all
% estimate the parameters of the K distribution using MoM
% the expression is created using symbolic toolbox and converted to the
% in-line function
% the external function is used to create the two equations necessary for
% obtaining the two parameters. The third parameter can be eliminated since
% the moments are simpler.
% step # 1 ab and c fixed, m varying: estimate m, c (mean), ab(mean)
% step # 2 ab and m fixed, c varying: estimate  m (mean), c, ab (mean)

%
% p m shankar, June 2016
syms m c a b z f_Z(z)
tic
ff=2*(z^((m+c)/2-1))*besselk(m-c,2*sqrt(z/(a*b)))/(gamma(m)*gamma(c)*(sqrt(a*b))^(m+c));
pdf=[f_Z(z)==ff]; % create the pdf expression for plot display
clear m c a b z f_Z(z)
N=1e5;
mm=[0.52:.1:1.2]; % values of a for random number generation
L=length(mm);
global MH M2
for k=1:L
    m1=mm(k);
    cc=4.5;% value of c (fixed)
    ab=5; % value of ab or alpha*beta (fixed)
    X=gamrnd(m1,ab,1,N);
    X=X.*gamrnd(cc,1,1,N);
    M0=mean(X);
    MH=mean(sqrt(X))/sqrt(M0); % normalized half moment
    M2=mean(X.^2)/M0^2; % normalized second moment
options = optimset('Display','off','MaxFunEvals',5000,'TolFun',1e-4,'TolX',1e-5);%fsolve is
```

```
repeated to prevent solutions not being found
 x0=[0.7;4.15]; % initial guess
[x]=fsolve(@momf,x0,options);
mk(k)=min(x); % m
C(k)=max(x); % c
AB(k)=M0/(mk(k)*C(k)); % ab
end;
  figure,plot(mm,mk,'*')
  ylim([0.4,1.4])
  xlim([0.4,1.4])
  xlabel('m'),ylabel('m_{est}')
  title(['$' latex(pdf) '$'],'interpreter','latex','fontsize',14)
  data={['a*b_{input}=',num2str(ab),'; ','c_{input}=',num2str(cc)];...
     ['mean ab_{est}=',num2str(mean(AB)),', ','mean c_{est}=',num2str(mean(C))]};
  text(0.55,1.25,data,'color','r','fontweight','bold')
  text(.9,.45,'[a*b and c fixed input]','backgroundcolor','y')
 clear A
 CD=[3:9]; % values of c for random number generation
 L=length(CD);
 for k=1:L
     m1=0.7;% value of m (fixed)
     cc=CD(k);%
   ab=8; % value of ab or alpha*beta (fixed)
 X=gamrnd(m1,ab,1,N);
    X=X.*gamrnd(cc,1,1,N);
    M0=mean(X);
    MH=mean(sqrt(X))/sqrt(M0);
    M2=mean(X.^2)/M0^2;
options = optimset('Display','off','MaxFunEvals',5000,'TolFun',1e-4,'TolX',1e-5);%fsolve is
repeated to prevent solutions not being found
 x0=[0.7;4.15];
[x]=fsolve(@momf,x0,options);
mk(k)=min(x);
C(k)=max(x);
AB(k)=M0/(mk(k)*C(k));
 end;
 figure,plot(CD,C,'*')
 ylim([2,10]), xlim([2,10]), xlabel('m'),ylabel('m_{est}')
 title(['$' latex(pdf) '$'],'interpreter','latex','fontsize',14)
  data={['a*b_{input}=',num2str(ab),'; ','m_{input}=',num2str(m1)];...
     ['mean ab_{est}=',num2str(mean(AB)),', ','mean m_{est}=',num2str(mean(mk))]};
 text(3,9,data,'color','r','fontweight','bold')
  text(7,3,'[a*b and m fixed input]','backgroundcolor','y')
  toc
end


function y = momf(x)
global MH M2
xx=x(1)*x(2); % m *c
xx1=sqrt(xx); % sqrt(mc)
f1=MH-gamma(x(1)+.50)*gamma(x(2)+.50)/((gamma(x(1))*gamma(x(2)))*xx1);
f2=M2-(1+1/x(1))*(1+1/x(2));
y=[f1;f2];
end
```

```
Elapsed time is 1.571311 seconds.
```

Examining the elapsed times, it can be seen that the time taken with MoM is only a few seconds compared to more than 3 min with MLE. This suggests that there is a need to make a judicious choice of the approach to parameter estimation and each

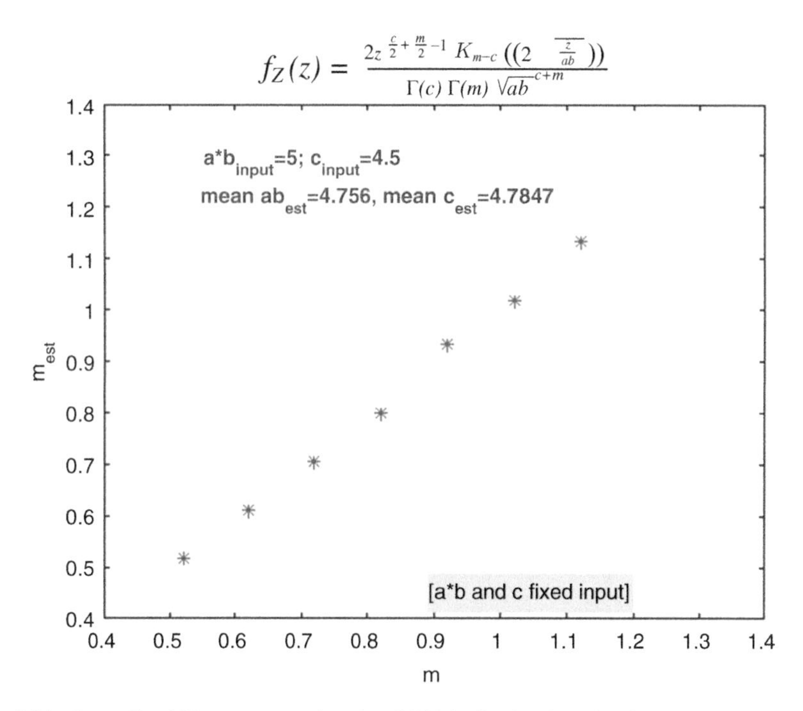

Fig. 2.56 Generalized K parameter estimation (MOM): fixed value of a, b, and c

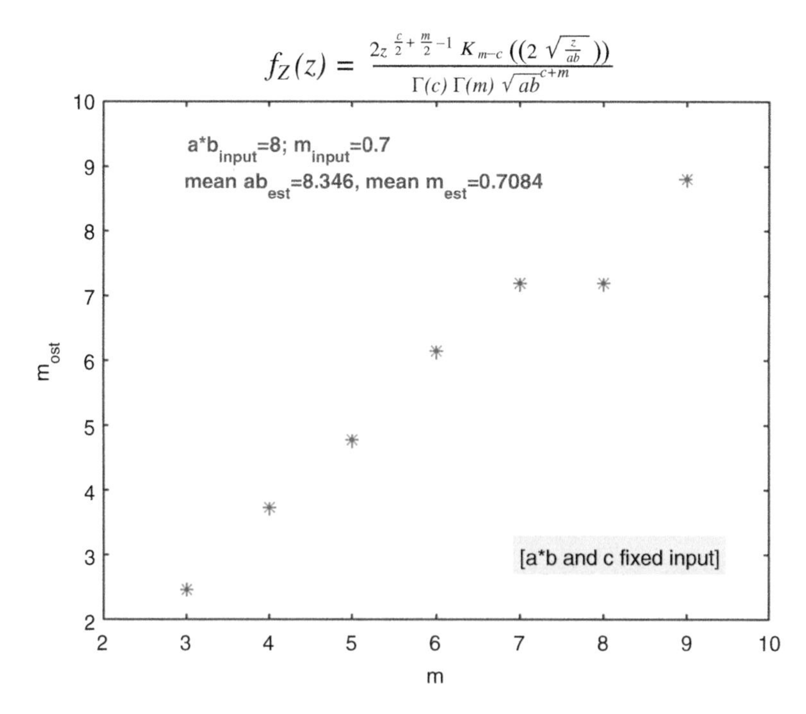

Fig. 2.57 Generalized K parameter estimation (MOM): fixed value of a, b, and m

density function might require a different approach to get efficient estimates of the parameters. The time taken and the accuracy of the estimates also depend on the actual values of the parameters being estimated.

2.14.3.4 Gamma-Lognormal (Shadowed Fading Channel)

Cases also exist in the analysis of wireless channels where the pdf of the SNR is not in analytical form. In such instances, MLE is not the best option for parameter estimation and the method that could provide parameters within a reasonable amount of time is based on the method of moments. An example of this exists in the modeling of shadowed fading channels. The signal strength fluctuations in a shadowed fading channel are modeled in terms of the gamma-lognormal density. The gamma density represents the fading component and lognormal density represents the shadowing component. Using the method of transformation of random variables where one random variable is gamma distributed and the other has a lognormal density, the density of the SNR or power in a shadowed fading channel is given in integral form as (Abdi and Kaveh 1998; Simon and Alouini 2005; Shankar 2010)

$$f(z) = \int_0^\infty \left(\frac{m}{w}\right)^m e^{-\frac{m}{w}z} \frac{K}{\sqrt{2\pi\sigma^2 w^2}} e^{-\frac{\left(10\log_{10}(w)-\mu\right)^2}{2\sigma^2}} dw. \tag{2.480}$$

In Eq. (2.480), μ is the average SNR in dB (or average power in dB_m) and σ is the shadowing level in dB. The Nakagami parameter is m and K is the logarithmic conversion factor

$$K = \frac{10}{\log_e(10)}. \tag{2.481}$$

The moments of the density in Eq. (2.480) are given as (Simon and Alouini 2005)

$$E(Z^k) = \frac{\Gamma(m+k)}{\Gamma(m)} e^{\frac{k}{K}\mu + \frac{1}{2}\left(\frac{k}{K}\right)^2 \sigma^2} \tag{2.482}$$

The normalized second and third moments, respectively, are

$$M_2 = \frac{E(Z^2)}{E^2(Z)} = \frac{(m+1)}{m} e^{\frac{\sigma^2}{K^2}} \tag{2.483}$$

$$M_3 = \frac{E(Z^3)}{E^3(Z)} = \frac{(m+1)(m+2)}{m^2} e^{\frac{3\sigma^2}{K^2}}. \tag{2.484}$$

Equations (2.483) and (2.484) can be solved for m and σ. The average signal level μ in dB is obtained from the relationship between the levels in dB

$$\mu_{\text{est}} = 10\log_{10}[E(Z)] - \frac{\sigma^2}{2K}. \tag{2.485}$$

The Matlab script is given below.

```
function  gamma_lognormal_MoM
close all
% estimate the parameters of the gamma-lognormal density
% p m shankar, June 2016
eqn='$$ f(z)=\int_0^{\infty}\frac{m}{w}e^{-\frac{m}{w}z}\frac{K}{\sqrt{2\pi \sigma^2w^2}}e^{-
\frac{(10log_{10}w-\mu)^2}{2\sigma^2}} dw $$';
N=5e7;
mm=[0.52 0.7 0.95 1.1 1.3 2.5]; % values of m for random number generation
L=length(mm);
global M2 M3   % normalized moments
K=10/log(10);
for k=1:L
    m1=mm(k);
    mu=9;% dB
    sig=4; % dB
    Xg=gamrnd(m1,1/m1,1,N);
    X2=normrnd(mu,sig,1,N);
    X3=10.^(X2/10);
    X=Xg.*X3; %gamma_lognormal random number
    M0=mean(X);
    M2=mean(X.^2)/M0^2; % normalized second moment
    M3=mean(X.^3)/M0^3; % normalized second moment
options = optimset('Display','off','MaxFunEvals',6000,'TolFun',1e-5,'TolX',1e-5);
 x0=[0.65;12.8]; % initial guess
[x]=fsolve(@momf,x0,options);
mk(k)=x(1); % m
C(k)=x(2) ;% sigma sq
 mmu(k)=10*log10(M0)-x(2)/(2*K);
end;
 figure,plot(mm,mk,'*')
   ylim([0.4,1.4])
  xlim([0.4,1.4])
  xlabel('m'),ylabel('m_{est}')
  title(eqn,'interpreter','latex','fontsize',14)
  data={['\mu_{input}=',num2str(mu),' dB_m; ','\sigma_{input}=',num2str(sig),' dB'];...
      ['mean mu_{est}=',num2str(mean(mmu)),' dB_m, ','mean
\sigma_{est}=',num2str(mean(sqrt(C))),' dB']};
  text(0.42,1.25,data,'color','r','fontweight','bold')
   text(.9,.5,'[\mu and \sigma fixed input]')
end

function y = momf(x)
global  M2 M3
% m is x(1) and x(2) is sigmasq
K=10/log(10);
K2=K^2;
x1=x(1);
x2=x(2);
f1=M2-(x1+1)*exp(x2/K2)/x1;
f2=M3-(x1+1)*(x1+2)*exp(3*x2/K2)/x1^2;
y=[f1;f2];
end
```

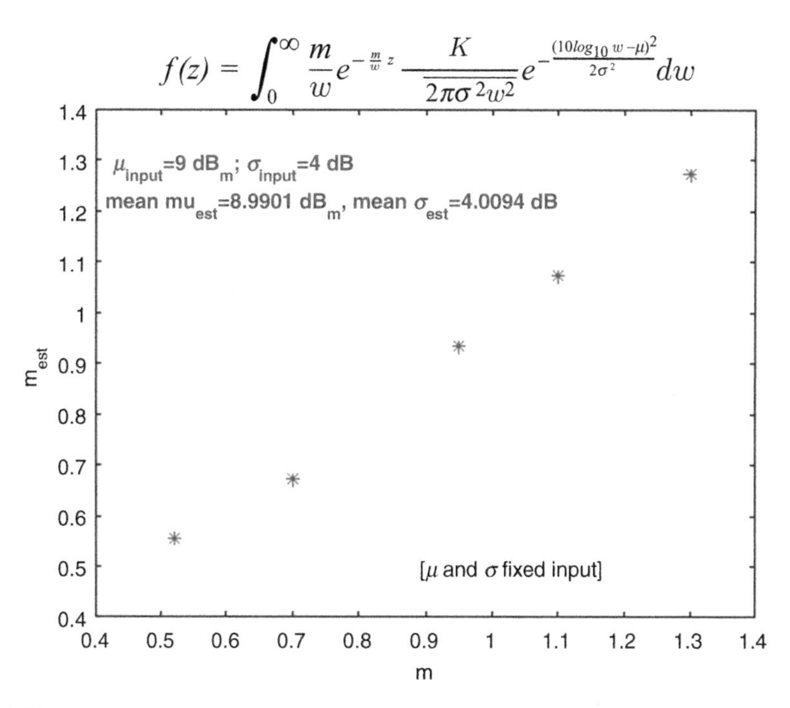

Fig. 2.58 Parameter estimation: gamma-lognormal channel (MoM): fixed value of μ and σ

Results of the parameter estimation of the Nakagami-lognormal (gamma-lognormal for the SNR) are shown in Fig. 2.58. Note that the procedure requires the solution of two simultaneous equations and the optimization toolbox (*fslove*) in Matlab is required.

It is also possible to obtain the parameters using a single equation by eliminating the shadowing level σ using Eqs. (2.483) and (2.484) resulting in a single equation for the Nakagami parameter m as

$$\frac{m}{m+1} M_2 = \left[M_3 \frac{m^2}{(m+1)(m+2)} \right]^{\frac{1}{3}}. \tag{2.486}$$

The shadowing level σ can be obtained from the normalized second moment and the average level μ is once again obtained from Eq. (2.485).

The Matlab script and results are displayed below. To get reasonably stable estimates, a large sample set is required (in this example a sample size of $N = 4e7$ was used). The script here does not require the use of optimization toolbox. Instead, it uses the symbolic toolbox (*solve*) to find the solution.

```
function  gamma_lognormal_MoM_single
close all
% estimate the parameters of the gamma-lognormal density
% uses a single equation to get m first (uses symbolic solve)
% sigma square is then calcluated. Similarly mu is calculated.
% p m shankar, June 2016
eqn='$$ f(z)=\int_0^{\infty}\frac{m}{w}e^{-\frac{m}{w}z}\frac{K}{\sqrt{2\pi \sigma^2w^2}}e^{-
\frac{(10log_{10}w-\mu)^2}{2\sigma^2}} dw $$';
N=4e7;
mm=[0.52 0.62 0.7 0.8 0.94 1.2]; % values of m for random number generation
L=length(mm);
K=10/log(10);
for k=1:L
    m1=mm(k);
    mu=9;% dB
    sig=4; % dB
    Xg=gamrnd(m1,1/m1,1,N);
    X2=normrnd(mu,sig,1,N);
    X3=10.^(X2/10);
    X=Xg.*X3; %gamma_lognormal random number
    M0=mean(X);
    M2=mean(X.^2)/M0^2; % normalized second moment
    M3=mean(X.^3)/M0^3; % normalized second moment
syms xm yy
yy=sym(M2)*xm/(xm+1)-(sym(M3)*xm^2/((xm+1)*(xm+2)))^(1/3);
mx=double(solve(yy==0,xm));
mt(k)=max(mx); % three solutions exisit: choose the largest one
sq2(k)=K*K*log(M2*mt(k)/(mt(k)+1)); %sigma sq
% mean dB; using the relationship between the dB of the power and mu
 mmu(k)=10*log10(M0)-sq2(k)/(2*K);% mu
 clear Xg X2 X3 X
end;
  figure,plot(mm,mt,'*')
    ylim([0.4,1.4])
  xlim([0.4,1.4])
  xlabel('m'),ylabel('m_{est}')
  title(eqn,'interpreter','latex','fontsize',14)
 data={['\mu_{input}=',num2str(mu),' dB_m; ','\sigma_{input}=',num2str(sig),' dB'];...
      ['mean mu_{est}=',num2str(mean(mmu)),' dB_m, ','mean
\sigma_{est}=',num2str(mean(sqrt(sq2))),' dB']};
   text(0.42,1.25,data,'color','r','fontweight','bold')
    text(.9,.5,'[\mu and \sigma fixed input]')
 end
```

Matlab results are displayed in Fig. 2.59.

From all these examples it is clear that the choice of the method for estimating parameters is a critical one and each one of these methods has its own limitations both in terms of computational complexities and accuracy of the estimates. The availability of a single equation (including transcendental type) is a little bit more convenient since a single equation offers multiple options for obtaining the solution either using a simple *find*(.) command in Matlab, symbolic solution or the traditional approaches using numerical methods based on *fsolve*(.) and *fzero*(.) in Matlab.

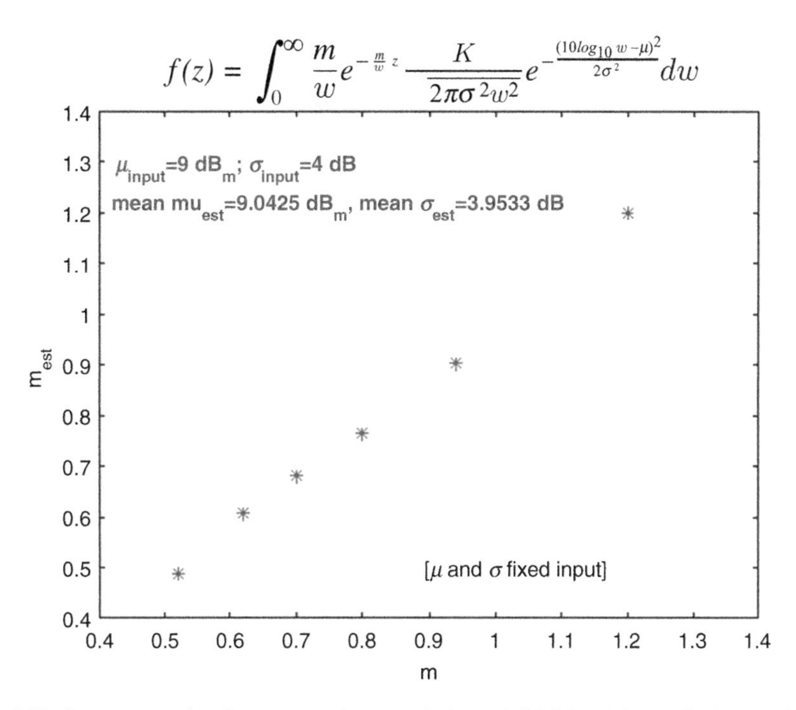

Fig. 2.59 Parameter estimation: gamma-lognormal channel (MoM, solving *a* single equation): fixed value of μ and σ

Parameter estimation is only the first part of the development of a model for appropriately describing the signal strength fluctuations observed in wireless channel. When multiple models exist, it is also necessary to determine the best (model) fit to the data collected. Methods of finding the best fit lead to the field of hypothesis testing. One specific form of hypothesis testing is the chi-square test. This is described next.

2.15 Chi-Square Tests

Once the parameter estimates are obtained, it is necessary to verify that the samples fit the presumed probability density and quantify the closeness of the fit. Chi-square testing (sometimes also called chi-squared goodness-of-fit test) allows this task to be undertaken (Mann and Wald 1942; Papoulis and Pillai 2002). The test can be applied to any univariate distribution. It requires access to the cumulative distribution in analytical form. If analytical expression is unavailable, access to numerical values of the CDF is sufficient to perform the test. The test relies on partitioning the samples. The data samples are "binned" or "grouped" into a number of bins, groups, or classes and the frequency of the bins is compared to expected number of observations based on the theoretical model (pdf). A test statistic is obtained from this comparison first. This test statistic is then compared to a threshold and decision is made to accept or reject the hypothesis that the samples follow the expected density function depending on whether the test statistic falls below the threshold or above the threshold. The various steps in the chi-square testing procedure involved are described first. This is followed by a few examples of the test and a discussion of the limitations of the chi-square test.

Step #1 The data set is sorted and values are arranged in ascending order. This procedure allows an easy way to create the binned data and estimate the relative frequencies.

Step #2 Estimate the parameters of the underlying assumed distribution from the data. This would require the estimation of one, two, three, or more parameters as discussed in the previous sections. While it is not necessary to use MLE, it is expected that reliable estimates can be obtained.

Note that steps #1 and # 2 are interchangeable.

Step #3 The sorted data (Step # 1) is divided into k bins.

Even though the choice of the number of bins is rather arbitrary, the number must be large enough to provide a reasonable distribution of the data in the bins. The number of bins should not be made arbitrarily large since it invariably leads to excessive computation. Higher number of bins might also lead to empty bins. Thus, it is expected that the number of bins must be such that each bin is populated and no bin exists with a frequency of 0. Generally, a number between 5 and 10 is sufficient and a number below 5 is too low and a number above 20 might be too high. The choice of the bin size is also a factor of how sparse the data set is and this aspect will be discussed when the limitations of chi-square tests are described. Let the observed frequency be denoted by O_i. It is expected that if the total number of samples is N, O_i is an integer lying between 1 and N, and

$$\sum_{i=1}^{k} O_i = N. \tag{2.487}$$

Step #4 Use the estimated parameters and recreate the theoretical (model) cumulative distribution function and estimate the probabilities for the random variable to stay within the corresponding bins.

If E_i is the number of observations expected using the theoretical model,

$$E_i = Np_i. \tag{2.488}$$

In Eq. (2.488), p_i is the probability that the random variable lies in the ith bin and is given by

$$p_i = F_X(x_{i+1}) - F_X(x_i), \quad i = 1, 2, \cdots, k. \tag{2.489}$$

Note that Eq. (2.489) simply implies that if k bins exist, the random variable will have $(k + 1)$ discrete values. The test statistic χ^2 is formed as

$$\chi^2 = \sum_{i=1}^{k} \frac{(O_i - E_i)^2}{E_i}. \tag{2.490}$$

Step #5 The test statistic χ^2 follows a chi-squared distribution. It is necessary to determine the number of degrees of freedom of the chi-squared distribution. The number of degrees of freedom is $(k - c)$, with c to be defined.

If no parameters are estimated from the data, $c = 1$. In this case, χ^2 follows a chi-squared distribution with $(k - 1)$ degrees of freedom.

If the number of parameters estimated from the data is q (note that q is an integer), the test statistic χ^2 follows a chi-squared distribution with $(k - q - 1)$ degrees of freedom or $c = q + 1$. Thus, larger number of parameters estimated from the data leads to a lower number of degrees of freedom of the chi-squared distribution associated with the test.

The hypothesis that the data set follows a specified density which resulted in Eq. (2.489) is rejected if the test statistic exceeds a threshold chosen on the basis of a significance level α. The threshold value is

$$\chi_{\text{thr}} = \chi^2_{1-\alpha,k-c}. \tag{2.491}$$

The chi-squared density with ν degrees of freedom is (Papoulis and Pillai 2002)

$$f(\lambda;\nu) = \frac{1}{2^{\frac{\nu}{2}}\Gamma\left(\frac{\nu}{2}\right)} \lambda^{\frac{\nu-2}{2}} e^{-\frac{\lambda}{2}} U(\lambda). \tag{2.492}$$

It can be easily seen that for the case of a Rayleigh random variable with a single parameter, $c = 2$. For a gamma distribution with two parameters $c = 3$ and for the case of a generalized gamma distribution with 3 parameters $c = 4$ etc. In most cases, a value of $\alpha = 0.05$ is sufficient.

While Matlab provides a means to perform chi-square tests for known densities, it is worthwhile to undertake the steps involved in chi-square testing using examples which show its use for densities that are custom made such as the generalized K distribution, K distribution, and McKay density, arising from the various models presented in Chap. 4.

2.15.1 Exponential Density

The first example is the case of an exponential random variable with the density

$$f(x) = \frac{1}{b} e^{-\frac{x}{b}} U(x). \tag{2.493}$$

The parameter b of the exponential density in Eq. (2.493) is the mean of the variable. Instead of the built in Matlab chi-square tests, a script is written to demonstrate the chi-square testing procedure. With Rayleigh and exponential random variable sets as inputs, the input data sets are tested to see if they follow the exponential density. It also tests to confirm that the density of the sum of the squares two independent and identically distributed zero mean Gaussian random variables is also exponential. In other words, the data sets of Rayleigh, exponential, and sums of squares of identical zero mean Gaussian random variables are tested to explore the fit to an exponential density.

```
function chisquare_test_exponential_example
% shows the way to create chi square test. In this case, the exponential
% density is used.
% generate a set of Rayleigh random variables. The test should reject the
% hypothesis that the set follows an EXPONENTIAL pdf.
% also tests with exponential set.
% another set is created from two i.i.d zero mean Gaussian random variables
% and the data set is the sum of the squares of the Gaussian variables.
% Must follow the exponential density
% results display the test statistic, the degrees of freedom the threshold,
% and the thrshodl value based on alpha=0.05 and the number of degrees of
% freedom and finally statement on whether the hypothesis should be
% rejected or NOT. The test statistic and the threshold values have been
% rounded to the nearest integer.
%
% p m shankar, June 2016

x=raylrnd(20,1,1000);
mb=10;
disp('----------------------------------------------')
disp('input is Rayleigh distributed')
disp(' ')
display_results(x,mb)
disp('----------------------------------------------')
disp(' ')
clear x mb
x=raylrnd(20,1,1000);
mb=15;
disp('----------------------------------------------')
disp('input is Rayleigh distributed')
disp(' ')
display_results(x,mb)
disp('----------------------------------------------')
disp(' ')
clear x mb
xx=exprnd(4,1,1000); % generate exponential random variable
mb=10;
disp('----------------------------------------------')
disp('input is Exponential distributed')
disp(' ')
display_results(xx,mb)
disp('----------------------------------------------')
disp(' ')
clear xx mb
x1=normrnd(0,5,1,2000); % Gaussian variable of mean 0, sigma =5
```

```
y1=normrnd(0,5,1,2000); % Gaussian variable of mean 0,sigma =5
y=x1.^2+y1.^2; % will be exponential
mb=10;
disp('----------------------------------------------')
disp('input is Exponetial distributed: sum of squares of zero mean Gaussian i.i.d')
disp(' ')
display_results(y,mb)
disp('----------------------------------------------')
disp(' ')
clear x1 y1 y mb
end

function [qq,QC,mq,deg ] = exponential_test(y,m)
%Tests for exponential distribution
y=sort(y); % random samples
N=length(y);%size of the sample
my=mean(y);%mean assuming exponential density
range=max(y)-min(y); %interval for chi-square test
intst=range/m; %size of sub-interval
intvl=[min(y):intst:max(y)]; %samples at the sub-intervals: should be (m+1)
k=[zeros(1,m)]; %number of samples in each sub-interval intialization
% get the frequencies Oi
for i=1:N
   for j=1:m
      if y(i)<=intvl(j+1)
         k(j)=k(j)+1;
         break;
      end;
   end;
end;
% get the CDF
p=zeros(1,m);
for i=1:m
   p(i)=expcdf(intvl(i+1),my)-expcdf(intvl(i),my); %probabilities at the sub-interval samples
end;
np=p.*N; % this is Ei
q=0;
for i=1:m
   q = q + ((k(i)-np(i))^2)/np(i); %test statistic
end;
%QC the critical value based on the number of bins (mm) and the two estimated are Z and m (2)
qq=round(q);
deg=m-1-1;
QC=chi2inv(.95, m-1-1);% 1 is the number of parameters calculated from the DATA
QC=round(QC);
if q<=QC
    mq=1;
else
    mq=0;
end;

end
```

```
function display_results(x,mb)
[qq,QC,mq,deg] = exponential_test(x,mb);
disp(['The number of bins = ',num2str(mb)])
disp(['Test statistic =',num2str(qq)])
disp(['Degrees of freedom =',num2str(deg),'; Threshold value (95% confidence) =',num2str(QC)])
if mq==1
  disp('PASS: DO NOt REJECT the hypothesis that the data set follows exponential density')
else
    disp( 'FAIL: Reject the hypothesis that the data set follows exponential density')
end;
end
```

```
--------------------------------------------
input is Rayleigh distributed

The number of bins = 10
Test statistic =404
Degrees of freedom =8; Threshold value (95% confidence) =16
FAIL: Reject the hypothesis that the data set follows exponential density
--------------------------------------------

--------------------------------------------
input is Rayleigh distributed

The number of bins = 15
Test statistic =441
Degrees of freedom =13; Threshold value (95% confidence) =22
FAIL: Reject the hypothesis that the data set follows exponential density
--------------------------------------------

--------------------------------------------
input is Exponential distributed

The number of bins = 10
Test statistic =10
Degrees of freedom =8; Threshold value (95% confidence) =16
PASS: DO NOt REJECT the hypothesis that the data set follows exponential density
--------------------------------------------

--------------------------------------------
input is Exponetial distributed: sum of squares of zero mean Gaussian i.i.d

The number of bins = 10
Test statistic =6
Degrees of freedom =8; Threshold value (95% confidence) =16
PASS: DO NOt REJECT the hypothesis that the data set follows exponential density
--------------------------------------------
```

The results above demonstrate the method of chi-square testing and way of interpreting the results. It should be noted that the chi-square testing for the exponential density can be undertaken directly in Matlab using *chi2gof* (Statistics and Machine Learning toolbox). However, the detailed outputs shown above are not available in the built in chi-square testing command in Matlab.

As discussed earlier, parameter estimation is crucial for undertaking the chi-square tests for the validation of the models. In the exponential density example above, the parameter estimation was straightforward, primarily due to the fact that

CDF calculation was straightforward. In many cases, MoM involving more than one moment or MLE needs to be used to obtain parameter estimates. The next example illustrates this. It also shows that if the CDF of the random variable is not analytically available, numerical integration needs to be carried out to get the probabilities.

2.15.2 K *Distribution*

Consider the case of wireless channel with fading modeled in terms of a double gamma process (Chap. 4). This means that the SNR in the channel is the product of two identical independent gamma random variables. The density can be obtained from Eq. (2.473) by assuming that $m = c$ and $a = b = X_0$. The density is given by (Shankar and Gentile 2010)

$$f(z) = \frac{2}{X_0^{2m}\Gamma^2(m)} z^{m-1} K_0\left(\frac{2}{X_0}\sqrt{z}\right) U(z). \tag{2.494}$$

Note that Eq. (2.494) implies the expressions for the two marginal densities and the product of two random variables is Z,

$$\begin{aligned} f(x) &= \frac{x^{m-1}}{X_0^m\Gamma(m)}\exp\left(-\frac{x}{X_0}\right) \\ f(y) &= \frac{y^{m-1}}{X_0^m\Gamma(m)}\exp\left(-\frac{y}{X_0}\right) \\ Z &= XY. \end{aligned} \tag{2.495}$$

The moments of the double gamma variable having the density in Eq. (2.494) are

$$E\left(Z^k\right) = \frac{\Gamma^2(k+m)}{\Gamma^2(m)} X_0^{2k}. \tag{2.496}$$

The half order moment is

$$E\left(Z^{\frac{1}{2}}\right) = \frac{\Gamma^2\left(\frac{1}{2}+m\right)}{\Gamma^2(m)} X_0. \tag{2.497}$$

The first order moment or the mean is

$$E(Z) = m^2 X_0^2 \tag{2.498}$$

The normalized half moment is

$$M_h = \frac{E(\sqrt{Z})}{\sqrt{E(Z)}} = \frac{\Gamma^2\left(m+\frac{1}{2}\right)}{m\Gamma^2(m)}. \tag{2.499}$$

Equation (2.499) allows a simple way to obtain the estimate of m by solving a single equation. Once the estimate of m is found, the parameter X_0 can be estimated from Eq. (2.497) or (2.498). Even though the CDF of the density is analytically available either as a modified Bessel function or as a Meijer G-function (see the Appendix in Chap. 4), the chi-square test will be carried out assuming that no analytical solution is available. This would be the case with many of the custom density functions where closed form expressions of the pdf and CDF might not be available.

```
function chisquare_test_dblgamma_example
% shows the way to create chi square test. In this case, the case of a
% double gamma density is used
% The parameters are estimated and the probabilities evaulated numerically
% assuming that CDF is not available. The parameters are estimated using
% the Method of Moments. Symbolic toolbox is used to create the in-line
% function for the solution of the parameter estimates. Symbolic toolbox is
% also used to obtain the in-line function for the integrand for the
% numerical integration using integral(.) command
%
% results display the test statistic, the degrees of freedom the threshold,
% and the thrshodl value based on alpha=0.05 and the number of degrees of
% freedom and finally statement on whether the hypothesis should be
% rejected or NOT. The test statistic and the threshold values have been
% rounded to the nearest integer.
%
% p m shankar, June 2016

x=exprnd(20,1,5000);
mb=10;
disp('-----------------------------------------------')
disp('input is exponentially distributed')
disp(' ')
display_results(x,mb)
disp('-----------------------------------------------')
disp(' ')
clear x mb
x1=gamrnd(2.5,4,2,5000); % generate 2 sets of gamma random variables
x=prod(x1);% one set of double gamma variable
 mb=15;
 disp('-----------------------------------------------')
 disp('input is double gamma distributed')
 disp(' ')
 display_results(x,mb)
disp('-----------------------------------------------')
 disp(' ')
 clear x1 x mb
x1=gamrnd(1.5,10,2,5000); % generate 2 sets of gamma random variables
x=prod(x1);% one set of double gamma variable
 mb=10;
 disp('-----------------------------------------------')
 disp('input is double gamma distributed')
 disp(' ')
```

```
 display_results(x,mb)
disp('---------------------------------------------')
 disp(' ')
  clear x1 x mb
  % % this is a mixture of gamma and exponential variables
x=[gamrnd(1.5,10,1,5000) exprnd(1,1,5000)];
 mb=10;
 disp('---------------------------------------------')
 disp('input is gamma_exponential mixture')
 disp(' ')
 display_results(x,mb)
disp('---------------------------------------------')
 disp(' ')
end

function [qq,QC,mq,deg ] = double_gamma_test(y,m)
%Tests for exponential distribution
zh=mean(sqrt(y));
zm=mean(y);
zmh=zh/sqrt(zm);
% create the equation for solving for the parameter using symbolic toolbox
syms mq positive
yq=(gamma(mq+1/2))^2/(mq*gamma(mq)*gamma(mq));
% conver the equation to an in-line form
yx=matlabFunction(yq); % convert to an inline function
clear mq yq
% now solve for
mm=0.4:.02:4; % input m values to find the estimate of m
yy=yx(mm);
idx=find(abs(yy-zmh)<.005); % find where the function is less than abs(.005)
mest=median(mm(idx)); % if multiple values exist, take the median
X=sqrt(zm)/mest; % get the second variable of the double gamma density
clear mm
% now we have the estimate of m and X
y=sort(y); % random samples
N=length(y);%size of the sample
range=max(y)-min(y); %interval for chi-square test
intst=range/m; %size of sub-interval
intvl=[min(y):intst:max(y)]; %samples at the sub-intervals: should be (m+1)
k=[zeros(1,m)]; %number of samples in each sub-interval intialization
% get the frequencies Oi
for i=1:N
   for j=1:m
      if y(i)<=intvl(j+1)
         k(j)=k(j)+1;
         break;
      end;
   end;
end;
% create the pdf for computing the CDF
% create a symbolic expression for the pdf
syms x
mdf= 2*besselk(0,2*sqrt(x)/X)*x^(mest-1)/(gamma(mest)*gamma(mest)*X^(2*mest));
% convert the symbolic expression to an in-line form
myfunc=matlabFunction(mdf); % this is the integrand for the computation of the
% probabilities. This is nothing but the pdf
p=zeros(1,m);
for i=1:m
    CDFm=integral(myfunc,intvl(i),intvl(i+1)); % CDF is evaluated by integration
    % defined by the two bins
   p(i)=CDFm; %probabilities at the sub-interval samples
end;
```

```
np=p.*N; % this is Ei
q=0;
for i=1:m
   q = q + ((k(i)-np(i))^2)/np(i); %test statistic
end;
%QC the critical value based on the number of bins (mm) and the two estimated are Z and m (2)
qq=round(q);
deg=m-1-2;% 2 is the number of parameters calculated from the DATA
QC=chi2inv(.95, deg);%
QC=round(QC);
if q<=QC
    mq=1;
else
    mq=0;
end;

end


function display_results(x,mb)
[qq,QC,mq,deg] = double_gamma_test(x,mb);
disp(['The number of bins = ',num2str(mb)])
disp(['Test statistic =',num2str(qq)])
disp(['Degrees of freedom =',num2str(deg),'; Threshold value (95% confidence) =',num2str(QC)])
if mq==1
  disp('PASS: DO NOt REJECT the hypothesis that the data set follows double gamma density')
else
    disp( 'FAIL: Reject the hypothesis that the data set follows double gamma density')
end;
end
```

```
------------------------------------------------
input is exponentially distributed

The number of bins = 10
Test statistic =36
Degrees of freedom =7; Threshold value (95% confidence) =14
FAIL: Reject the hypothesis that the data set follows double gamma density
------------------------------------------------

------------------------------------------------
input is double gamma distributed

The number of bins = 15
Test statistic =16
Degrees of freedom =12; Threshold value (95% confidence) =21
PASS: DO NOt REJECT the hypothesis that the data set follows double gamma density
------------------------------------------------

------------------------------------------------
input is double gamma distributed

The number of bins = 10
Test statistic =11
Degrees of freedom =7; Threshold value (95% confidence) =14
PASS: DO NOt REJECT the hypothesis that the data set follows double gamma density
------------------------------------------------

------------------------------------------------
input is gamma_exponential mixture

The number of bins = 10
Test statistic =367
Degrees of freedom =7; Threshold value (95% confidence) =14
FAIL: Reject the hypothesis that the data set follows double gamma density
------------------------------------------------
```

Before the discussion on chi-square testing is concluded, the example of the double gamma density is now carried out using the custom CDF available in place of the numerical integration that had to be carried out. For the marginal density given in Eq. (2.495), the density function of Z can also be expressed in terms of Meijer G-functions as (Shankar 2013)

$$f(z) = \frac{1}{z\Gamma^2(m)} G_{0,2}^{2,0}\left(\frac{m^2 z}{Z_0}\middle| \begin{matrix} - \\ m, m \end{matrix}\right). \tag{2.500}$$

In Eq. (2.500), Z_0 is the average SNR and it is given by Eq. (2.498) as

$$Z_0 = m^2 X^2 \tag{2.501}$$

The cumulative distribution function (CDF) is

$$F(z) = \frac{1}{\Gamma^2(m)} G_{1,3}^{2,1}\left(\frac{m^2 z}{Z_0}\middle| \begin{matrix} - \\ m, m, 0 \end{matrix}\right) \tag{2.502}$$

```
function chisquare_test_dblgamma_example_Meijer G
% the pdf and CDF are analytically expressed in terms of the Meijer G
% functions.
% shows the way to create chi square test. In this case, the case of a
% double gamma density is used
% The parameters are estimated and the probabilities evaulated numerically
% assuming that CDF is not available. The parameters are estimated using
% the Method of Moments. Symbolic toolbox is used to create the in-line
% function for the solution of the parameter estimates.
% CDF is obtained directly using Meijer G functions
% results display the test statistic, the degrees of freedom the threshold,
% and the thrshodl value based on alpha=0.05 and the number of degrees of
% freedom and finally statement on whether the hypothesis should be
% rejected or NOT. The test statistic and the threshold values have been
% rounded to the nearest integer.
%
% p m shankar, June 2016

x=exprnd(20,1,5000);
mb=10;
disp('----------------------------------------------')
disp('input is exponentially distributed')
disp(' ')
display_results(x,mb)
```

```
disp('---------------------------------------------')
disp(' ')
clear x mb
x1=gamrnd(2.5,4,2,5000); % generate 2 sets of gamma random variables
x=prod(x1);% one set of double gamma variable
 mb=15;
 disp('---------------------------------------------')
 disp('input is double gamma distributed')
 disp(' ')
 display_results(x,mb)
disp('---------------------------------------------')
 disp(' ')
 clear x1 x mb
x1=gamrnd(1.5,10,2,5000); % generate 2 sets of gamma random variables
x=prod(x1);% one set of double gamma variable
 mb=10;
 disp('---------------------------------------------')
 disp('input is double gamma distributed')
 disp(' ')
 display_results(x,mb)
disp('---------------------------------------------')
 disp(' ')
  clear x1 x mb
  % % this is a mixture of gamma and exponential variables
x=[gamrnd(1.5,10,1,5000) exprnd(1,1,5000)];
 mb=10;
 disp('---------------------------------------------')
 disp('input is gamma_exponential mixture')
 disp(' ')
 display_results(x,mb)
disp('---------------------------------------------')
 disp(' ')
end

function [qq,QC,mq,deg ] = double_gamma_test(y,m)
%Tests for exponential distribution
zh=mean(sqrt(y));
zm=mean(y);
zmh=zh/sqrt(zm);
% create the equation for solving for the parameter using symbolic toolbox
syms mq positive
yq=(gamma(mq+1/2))^2/(mq*gamma(mq)*gamma(mq));
% conver the equation to an in-line form
yx=matlabFunction(yq); % convert to an inline function
clear mq yq
% now solve for
mm=0.4:.02:4; % input m values to find the estimate of m
yy=yx(mm);
idx=find(abs(yy-zmh)<.005); % find where the function is less than abs(.005)
mest=median(mm(idx)); % if multiple values exist, take the median
% the second parameter in this case is zm; the mean
clear mm
% now we have the estimate of m and mean needed for the Meijer G function
```

```
y=sort(y); % random samples
N=length(y);%size of the sample
range=max(y)-min(y); %interval for chi-square test
intst=range/m; %size of sub-interval
intvl=[min(y):intst:max(y)]; %samples at the sub-intervals: should be (m+1)
k=[zeros(1,m)]; %number of samples in each sub-interval intialization
% get the frequencies Oi
for i=1:N
   for j=1:m
      if y(i)<=intvl(j+1)
         k(j)=k(j)+1;
         break;
      end;
   end;
end;
p=zeros(1,m);
gm=1/gamma(mest)^2;
for i=1:m
     CDFUpper=gm*double(evalin(symengine,sprintf('Meijer G([[1], []], [[%e,%e], [0]],
%e)',mest,mest,(mest^2)*intvl(i+1)/zm)));
    CDFLower=gm*double(evalin(symengine,sprintf('Meijer G([[1], []], [[%e,%e], [0]],
%e)',mest,mest,(mest^2)*intvl(i)/zm)));
   p(i)=CDFUpper-CDFLower; %probabilities at the sub-interval samples
end;
np=p.*N; % this is Ei
q=0;
for i=1:m
   q = q + ((k(i)-np(i))^2)/np(i); %test statistic
end;
%QC the critical value based on the number of bins (mm) and the two estimated are Z and m (2)
qq=round(q);
deg=m-1-2;% 2 is the number of parameters calculated from the DATA
QC=chi2inv(.95, deg);%
QC=round(QC);
if q<=QC
    mq=1;
else
    mq=0;
end;

end


function display_results(x,mb)
[qq,QC,mq,deg] = double_gamma_test(x,mb);
disp(['The number of bins = ',num2str(mb)])
disp(['Test statistic =',num2str(qq)])
disp(['Degrees of freedom =',num2str(deg),'; Threshold value (95% confidence) =',num2str(QC)])
if mq==1
  disp('PASS: DO NOt REJECT the hypothesis that the data set follows double gamma density')
else
    disp( 'FAIL: Reject the hypothesis that the data set follows double gamma density')
```

```
end;
end
```

```
--------------------------------------------
input is exponentially distributed

The number of bins = 10
Test statistic =51
Degrees of freedom =7; Threshold value (95% confidence) =14
FAIL: Reject the hypothesis that the data set follows double gamma density
--------------------------------------------

--------------------------------------------
input is double gamma distributed

The number of bins = 15
Test statistic =21
Degrees of freedom =12; Threshold value (95% confidence) =21
PASS: DO NOt REJECT the hypothesis that the data set follows double gamma density
--------------------------------------------

--------------------------------------------
input is double gamma distributed

The number of bins = 10
Test statistic =11
Degrees of freedom =7; Threshold value (95% confidence) =14
PASS: DO NOt REJECT the hypothesis that the data set follows double gamma density
--------------------------------------------

--------------------------------------------
input is gamma_exponential mixture

The number of bins = 10
Test statistic =361
Degrees of freedom =7; Threshold value (95% confidence) =14
FAIL: Reject the hypothesis that the data set follows double gamma density
--------------------------------------------
```

2.16 MSE and Chi-Square Test

While the chi-square tests provide a means to achieve the goal of verifying whether the data set fits the statistical model, one of the short comings of the chi-square testing is its binary nature. The test provides a threshold for rejection of the hypothesis leaving the user with some ambiguity regarding the validity of the hypothesis. It is also possible to suggest that the value of the test statistic itself can be used as a means to compare two or more hypotheses, with the hypothesis resulting in a lower value of the test statistic appearing to be a better fit than the ones having higher values of the test statistic. Such qualitative descriptions can also be quantitatively described using mean square error (MSE) values. If $f_{\mathrm{th}}(x_k)$, $k = 1,2,\ldots,n$ are the samples of the probability density estimated assuming a specific hypothesis and $f_d(x_k)$ are the samples of the density estimated from the histogram of the data,

$$\mathrm{MSE} = \sum_{k=1}^{n} [f_{\mathrm{th}}(x_k) - f_d(x_k)]^2. \tag{2.503}$$

The theoretical density with the lowest value of MSE is the better fit.

The following example illustrates a comparison of multiple hypotheses using chi-square tests and MSE values. A set of data is tested to determine which one of the three hypotheses is the best fit. The three hypotheses are the K distribution, double gamma distribution, and the Weibull distribution, all three densities being two parameter densities making the comparison of the test statistic fair and meaningful. The K distribution is obtained from the generalized K distribution in Eq. (2.473) by putting $m = 1$ resulting in

$$f_K(z) = \frac{2}{\lambda^{1+c}\Gamma(c)} z^{\frac{1+c}{2}-1} K_{c-1}\left(\frac{2}{\lambda}\sqrt{z}\right) U(z). \tag{2.504}$$

The moments are

$$E\left(Z^k\right) = \frac{\Gamma(1+k)\Gamma(c+k)}{\Gamma(c)} \lambda^{2k} \tag{2.505}$$

$$E\left(Z^{\frac{1}{2}}\right) = \frac{\Gamma\left(\frac{3}{2}\right)\Gamma\left(c+\frac{1}{2}\right)}{\Gamma(c)} \lambda \tag{2.506}$$

$$E(Z) = c\lambda^2 \tag{2.507}$$

The parameters c and λ may be evaluated using the half order and first order moments. The double gamma density was given in Eq. (2.494) and it is

$$f_G(z) = \frac{2}{X_0^{2m}\Gamma^2(m)} z^{m-1} K_0\left(\frac{2}{X_0}\sqrt{z}\right) U(z). \tag{2.508}$$

The parameters m and X can be determined from the half order and first order moments given in Eqs. (2.497) and (2.498).

The Weibull density is (Papoulis and Pillai 2002)

$$f_L(z) = \left[\frac{b}{a}\right]\left(\frac{z}{a}\right)^{b-1} e^{-\left(\frac{z}{a}\right)^b} U(z). \tag{2.509}$$

The moments of the Weibull density are

$$E\left(Z^k\right) = a^k \Gamma\left(\frac{k+b}{b}\right). \tag{2.510}$$

The two parameters can be estimated from the first two moments. It is also possible to estimate the parameters directly in Matlab. The results are shown and discussion follows the displays of the script and figures.

```
function hypotheses_comparison_example
% Three hypotheses are compared. The chi-square tests are carried out and
% the MSE values estimated. The three densities are double gamma, K, and
% Weibull. All the parameters are estimated. Since Weibull is available in
% Matlab, the parameters of Weibull are estimated from Matlab.
%
% results display the test statistic, the degrees of freedom the threshold,
% and the threshold value based on alpha=0.05 and the number of degrees of
% freedom. The estimated pdfs and the data histogram (pdf) are plotted and
% MSE values are estimated. The maximum value of the difference in each case
% is also estimated.

% P M Shankar, July 2016

% use a bin size of 10
close all
global nam % identification for display
mbin=10;
x1=gamrnd(1.2,1,2,1e6); %generate 2 sets of identical gamma variables
x1=prod(x1);%take the product
nam='double gamma input';
display_results(x1,mbin)
clear x1
%generate 2 sets of non-identical gamma variables and take the product
x1=gamrnd(.7,2,1,1e6).*gamrnd(1,1,1,1e6);
nam='gamma product input';
display_results(x1,mbin)
clear x1
x1=gamrnd(0.4,1,1,1e6); % a single set of single gamma variable
nam='gamma input';
 display_results(x1,mbin)
end

function [X,mest,qq] = double_gamma_test(y,m)
%Tests for double gamma distribution
zh=mean(sqrt(y));
zm=mean(y);
zmh=zh/sqrt(zm);
% create the equation for solving for the parameter using symbolic toolbox
syms mq positive
yq=(gamma(mq+1/2))^2/(mq*gamma(mq)*gamma(mq));
% conver the equation to an in-line form
yx=matlabFunction(yq); % convert to an inline function
clear mq yq
% now solve for
mm=0.4:.02:4; % input m values to find the estimate of m
yy=yx(mm);
idx=find(abs(yy-zmh)<.005); % find where the function is less than abs(.005)
mest=median(mm(idx)); % if multiple values exist, take the median
X=sqrt(zm)/mest; % get the second variable of the double gamma density
clear mm
% now we have the estimate of m and X
y=sort(y); % random samples
N=length(y);%size of the sample
range=max(y)-min(y); %interval for chi-square test
intst=range/m; %size of sub-interval
intvl=[min(y):intst:max(y)]; %samples at the sub-intervals: should be (m+1)
k=[zeros(1,m)]; %number of samples in each sub-interval intialization
% get the frequencies Oi
```

```
for i=1:N
    for j=1:m
        if y(i)<=intvl(j+1)
            k(j)=k(j)+1;
            break;
        end;
    end;
end;
% create the pdf for computing the CDF
% create a symbolic expression for the pdf
syms x
mdf= 2*besselk(0,2*sqrt(x)/X)*x^(mest-1)/(gamma(mest)*gamma(mest)*X^(2*mest));
% convert the symbolic expression to an in-line form
myfunc=matlabFunction(mdf); % this is the integrand for the computation of the
% probabilities. This is nothing but the pdf
p=zeros(1,m);
for i=1:m
     CDFm=integral(myfunc,intvl(i),intvl(i+1)); % CDF is evaluated by integration
     % defined by the two bins
    p(i)=CDFm; %probabilities at the sub-interval samples
end;
np=p.*N; % this is Ei
q=0;
for i=1:m
    q = q + ((k(i)-np(i))^2)/np(i); %test statistic
end;
qq=round(q);

end


function [lam,Cest,qq] = K_test(y,m)
%Tests for K distribution distribution
zh=mean(sqrt(y));
zm=mean(y);
zmh=zh/sqrt(zm);
% create the equation for solving for the parameter using symbolic toolbox
syms cq positive
yq=gamma(sym(3/2))*gamma(cq+1/2)/(gamma(cq)*sqrt(cq));
% conver the equation to an in-line form
yx=matlabFunction(yq); % convert to an inline function
clear mq yq
% now solve for
cm=0.4:.02:4; % input m values to find the estimate of m
yy=yx(cm);
idx=find(abs(yy-zmh)<.005); % find where the function is less than abs(.005)
Cest=median(cm(idx)); % if multiple values exist, take the median
lam=sqrt(zm)/sqrt(Cest); % get the second variable of the double gamma density
clear cm
% now we have the estimate of m and X
y=sort(y); % random samples
N=length(y);%size of the sample
range=max(y)-min(y); %interval for chi-square test
intst=range/m; %size of sub-interval
intvl=[min(y):intst:max(y)]; %samples at the sub-intervals: should be (m+1)
k=[zeros(1,m)]; %number of samples in each sub-interval intialization
% get the frequencies Oi
for i=1:N
    for j=1:m
        if y(i)<=intvl(j+1)
            k(j)=k(j)+1;
            break;
```

```
        end;
    end;
end;
% create the pdf for computing the CDF
% create a symbolic expression for the pdf
syms x
mdf= 2*besselk(Cest-1,2*sqrt(x)/lam)*x^(Cest/2-1/2)/(gamma(Cest)*lam^(1+Cest));
% convert the symbolic expression to an in-line form
myfunc=matlabFunction(mdf); % this is the integrand for the computation of the
% probabilities. This is nothing but the pdf
p=zeros(1,m);
for i=1:m
     CDFm=integral(myfunc,intvl(i),intvl(i+1)); % CDF is evaluated by integration
     % defined by the two bins
   p(i)=CDFm; %probabilities at the sub-interval samples
end;
np=p.*N; % this is Ei
q=0;
for i=1:m
   q = q + ((k(i)-np(i))^2)/np(i); %test statistic
end;
qq=round(q);
end

function [aest,best,qq] = weibull_test(y,m)
%Tests for weibull
[par]=wblfit(y);
aest=par(1);
best=par(2);
y=sort(y); % random samples
N=length(y);%size of the sample
range=max(y)-min(y); %interval for chi-square test
intst=range/m; %size of sub-interval
intvl=[min(y):intst:max(y)]; %samples at the sub-intervals: should be (m+1)
k=[zeros(1,m)]; %number of samples in each sub-interval intialization
% get the frequencies Oi
for i=1:N
   for j=1:m
      if y(i)<=intvl(j+1)
         k(j)=k(j)+1;
         break;
      end;
   end;
end;

p=zeros(1,m);
for i=1:m
     CDFm=wblcdf(intvl(i+1),aest,best)-wblcdf(intvl(i),aest,best);
     % defined by the two bins
   p(i)=CDFm; %probabilities at the sub-interval samples
end;
np=p.*N; % this is Ei
q=0;
```

```
for i=1:m
    q = q + ((k(i)-np(i))^2)/np(i); %test statistic
end;
qq=round(q);
end

function display_results(x1,mbin)
global nam
 [X,m,qq_D] = double_gamma_test(x1,mbin) ;
[LAM,c,qq_K] = K_test(x1,mbin);
[aest,best,qq_W] = weibull_test(x1,mbin);
% get the histogram or the data pdf
[fi,xr]=ksdensity(x1);
% ksdensity might give negative values of xr because of the fitting
% technique used by Matlab; eliminate them for getting nice plots
idr=find(xr<0);
z=xr(idr+1:end);
fr=fi(idr+1:end);
m2=2*m;
fG=2*(z.^(m-1)).*besselk(0,2*sqrt(z)/X)/(X^(m2)*gamma(m)^2);
fG=real(fG);% Bessel function might give very small imaginary values
fK=2*(z.^(c/2-1/2)).*besselk(c-1,2*sqrt(z)/LAM)/(gamma(m)*LAM^(c+1));
fK=real(fK);
fW=wblpdf(z,aest,best);
figure,plot(z,fr,'r-',z,real(fG),'k*',z,fK,'bo',z,fW,'g^')
text(0.03*median(z),0.9*max(fG),'[PDF      MSE         \chi^2 test Statistic]','color','b')
text(0.03*median(z),0.975*max(fG),nam,'backgroundcolor','y')
legend('data-histogram','Dbl-gamma','K','Weibull')
xlabel('SNR z'),ylabel('estimated pdf')
xlim([0,0.75*median(z)])
disp('---------------------------------------')
disp('Parameters of the double gamma density fit')
disp(['m = ',num2str(m),'; X =',num2str(X)])
disp('---------------------------------------')
disp('Parameters of the K-density fit')
disp(['c = ',num2str(c),'; lambda =',num2str(LAM)])
disp('---------------------------------------')
disp('Parameters of the Weibull density fit')
disp(['a = ',num2str(aest),'; lambda =',num2str(best)])
disp('---------------------------------------')
frG=fr-fG;
fgM=max(abs(frG));
frK=fr-fK;
fkM=max(abs(frK));
frW=fr-fW;
fwM=max(abs(frW));
disp('Absolute Maximum of the difference between pdf(theory) and pdf (fit)')
disp(['Double gamma -->',num2str(fgM),'; K-density -->',num2str(fkM),'; Weibull-density --
>',num2str(fwM)])
MSED=sum(frG.^2)/length(fr);
MSEK=sum(frK.^2)/length(fr);
MSEW=sum(frW.^2)/length(fr);
text(0.03*median(z),0.75*max(fG),{'DG';'K ';'W '},'color','b','fontweight','bold')
```

```
text(0.1*median(z),0.75*max(fG),{num2str(MSED);num2str(MSEK);num2str(MSEW)},'color','b','fontweig
ht','bold')
text(0.25*median(z),0.75*max(fG),{num2str(qq_D);num2str(qq_K);num2str(qq_W)},'color','b','fontwei
ght','bold')
% text(0.06*median(z),0.75*max(fG),{['DG      ',num2str(MSED),'              ',num2str(qq_D)];...
%    ['K      ',num2str(MSEK),'              ',num2str(qq_K)]; ['W      ',num2str(MSEW),'
',num2str(qq_W)]})
deg=mbin-1-2;% 2 is the number of parameters calculated from the DATA
QC=chi2inv(.95, deg);%
QC=round(QC);
title(['Threshold \chi^2 statistic: ',num2str(mbin), ' bins = ',num2str(QC)])

end
```

```
--------------------------------------
Parameters of the double gamma density fit
m = 1.21; X =0.99088
--------------------------------------
Parameters of the K-density fit
c = 1.51; lambda =0.97571
--------------------------------------
Parameters of the Weibull density fit
a = 1.1819; lambda =0.74346
--------------------------------------
Absolute Maximum of the difference between pdf(theory) and pdf (fit)
Double gamma -->0.0045621; K-density -->0.024841; Weibull-density -->0.043244
--------------------------------------
Parameters of the double gamma density fit
m = 0.83; X =1.4264
--------------------------------------
Parameters of the K-density fit
c = 0.7; lambda =1.415
--------------------------------------
Parameters of the Weibull density fit
a = 0.87629; lambda =0.58199
--------------------------------------
Absolute Maximum of the difference between pdf(theory) and pdf (fit)
Double gamma -->0.00073196; K-density -->0.033058; Weibull-density -->0.0034453
--------------------------------------
Parameters of the double gamma density fit
m = 0.88; X =0.71927
--------------------------------------
Parameters of the K-density fit
c = 0.79; lambda =0.71213
--------------------------------------
Parameters of the Weibull density fit
a = 0.24403; lambda =0.53493
--------------------------------------
Absolute Maximum of the difference between pdf(theory) and pdf (fit)
Double gamma -->0.34346; K-density -->0.50335; Weibull-density -->0.12894
```

The results are shown in Figs. 2.60, 2.61, and 2.62, respectively, for double gamma, gamma product (non-identical), and gamma random variables. Three data sets were tested to determine the fits to three different fading models described earlier. The results demonstrate a means to compare multiple hypotheses to test whether a specific model for fading in wireless is better than the others. While chi-square tests provide a binary marker, the actual chi-square test statistic value presents a quantitative measure to judge the proximity of the density to the actual density. Since the test statistic value goes up when the actual density and fit move farther and farther, a lower value of the chi-square test statistic is a clear indicator of

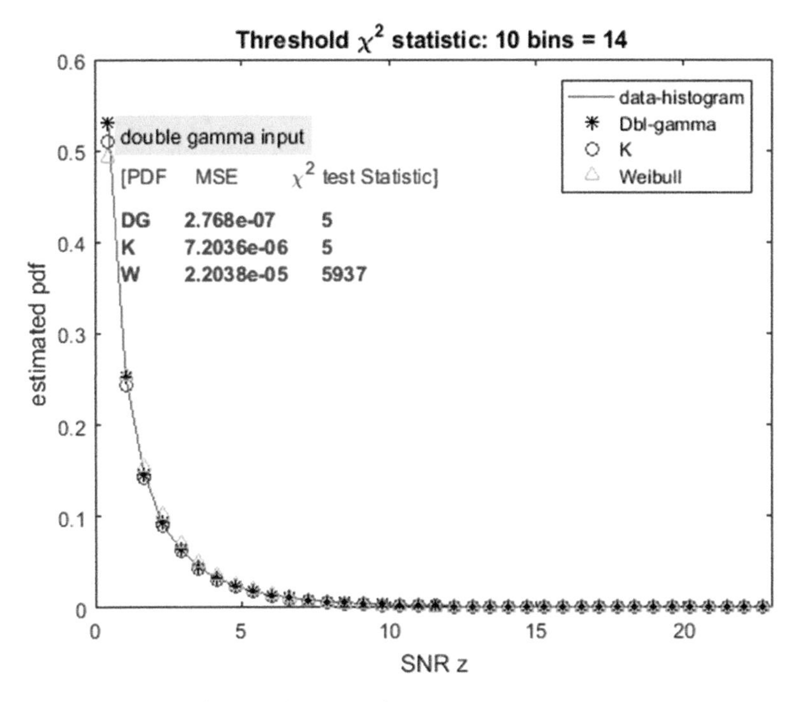

Fig. 2.60 Density validation (double gamma input)

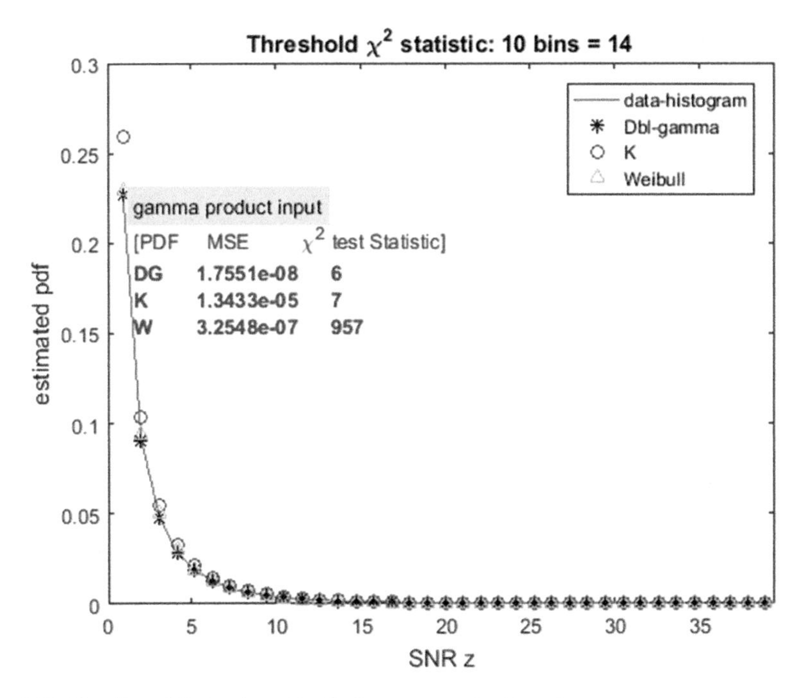

Fig. 2.61 Density validation (non-identical gamma product input)

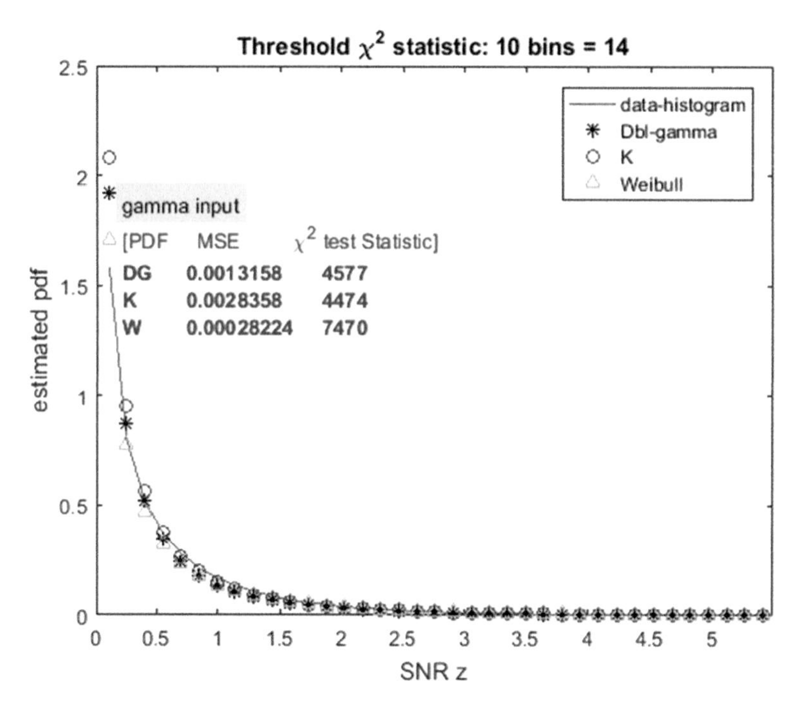

Fig. 2.62 Density validation (gamma input)

the best fit of the multiple models being tested. While the Weibull model clearly results in the highest value of the test statistic in all three sample data sets, in the last example, all the values of the test statistic are above the threshold value. However, the *K* distribution (and the double gamma) has the lowest value of the test statistic, so in the absence of any other models, the *K*-density is a reasonable way to model the data. One can also see that there is a high degree of correlation between MSE values and chi-square test statistic, in most cases, the values are very small and MSE only provides a simple way of seeing how the densities match graphically.

2.17 Mixture Densities in Wireless Channel Modeling

The various densities normally used to describe the statistical fluctuations of signal strengths or SNR in wireless channels can be termed as "pure" or "standard" densities. These densities are formed either from direct examinations of the fundamental physical processes resulting in the observed statistical fluctuations or through secondary modeling process such as those that generate the Nakagami-lognormal or Rician-lognormal densities that might also explain the observed statistical fluctuations. Thus, the Rayleigh, Rician, Nakagami, Hoyt, lognormal densities can be categorized as "pure" or "standard" densities. Improved modeling of the statistical fluctuations leads to Suzuki, Nakagami-lognormal, $\eta - \mu$, McKay, Meijer G, etc., densities which are also in standard forms. Standard forms of densities also arise when we examine diversity combining algorithms employed to mitigate the effects of fading or shadowing. On the other hand, observations can be used to model the statistical fluctuations by examining whether multiple classes of target regions exist in wireless channels and explore efficient ways of modeling such situations. These approaches are extensively used in pattern recognition, image analysis, boundary detection, etc., where the signal from any region of interest (ROI) is treated as having created from two or more distinct "species" which allows the combination of densities to provide an overall model of the statistical fluctuations. Such approaches are also used in modeling the lifetimes of chemical and biochemical phenomena. A simple description of this modeling effort is to express the probability density as the sum of weighted densities (Papoulis and Pillai 2002; Liu et al. 2007; Atappatu et al. 2011; Jung et al. 2014; Büyükçorak et al. 2015; Selim et al. 2015). If $f(z)$ is the density of the SNR,

$$f(z) = p_1 f_1(z) + p_2 f_2(z) + \cdots + p_n f_n(z). \tag{2.511}$$

Total probability theorem requires that

$$p_1 + p_2 + \cdots + p_n = 1, \quad 0 < p_i < 1, \quad i = 1, 2, \cdots, n. \tag{2.512}$$

The number of component densities in the mixture is n and each $f(.)$ on the right-hand side of Eq. (2.511) is a valid density function,

$$1 = \int_0^\infty f_i(z) dz, \quad i = 1, 2, \cdots, n \tag{2.513}$$

The lower limit of 0 in Eq. (2.513) merely is an acceptance of the fact that all the densities are used for modeling the SNR and therefore, the lowest value of the variable is 0. It should be noted that results are also available in literature where the lower limit of integration is $-\infty$. An example of this case would be a mixture of Gaussian densities. The mixture model in Eq. (2.511) also is flexible such that one does not require that $f_i(z)$ be of identical type (such as gamma, exponential, and Rician). It can easily be seen that if $n = 1$, the mixture model reverts to the pure or standard model of fading. While specific mixture models for fading, shadowing, and shadowed fading are described in later chapters, statistical aspects of the mixture models can be understood from basic principles of probability and random variables. We will make use of the method of moments (MoM), maximum likelihood estimation (MLE), chi-square testing, etc., presented in previous sections for this purpose (Redner and Walker 1984; Davenport et al. 1988; Bowman and Shenton 2006).

A simple example of mixture density is described next. A mixture of exponential densities (with two component densities) is considered. The density is expressed as

$$f(z)=(1-p)\frac{1}{a}e^{-\frac{z}{a}}+p\frac{1}{b}e^{-\frac{z}{b}},\quad 0<p<1;\quad a,b,\ z>0. \tag{2.514}$$

The mixture requires the estimation of three parameters, namely a, b, and p with the limitation on p being $0 < p < 1$. The parameters can be estimated using the method of moments (MoM) or MLE as discussed in previous sections. Since both approaches require the use of numerical techniques, the MoM is described first by examining the moments while the MLE is implemented in Matlab. The moments of the exponential mixture are

$$E(Z^k)=\left[p\left(b^k-a^k\right)+a^k\right]\Gamma(1+k). \tag{2.515}$$

Since there are three unknown, three moments are required for the parameter estimation. Instead of using the first three integer moments, the third moment can be avoided by using the half order moment and the first and second moments, providing the three equations to solve for a, b, and p. These moments are

$$E_h=E\left(\sqrt{Z}\right)=\left[p\left(b^{\frac{1}{2}}-a^{\frac{1}{2}}\right)+a^{\frac{1}{2}}\right]\Gamma\left(1+\frac{1}{2}\right)=\frac{\sqrt{\pi}}{2}\left[p\left(b^{\frac{1}{2}}-a^{\frac{1}{2}}\right)+a^{\frac{1}{2}}\right] \tag{2.516}$$

$$E_1=E(Z)=p(b-a)+a \tag{2.517}$$

$$E_2=E(Z^2)=2\left[p\left(b^2-a^2\right)+a^2\right] \tag{2.518}$$

Using these moments,

$$p=\frac{E_1-a}{b-a}=\frac{E_2-2a^2}{2b^2-2a^2} \tag{2.519}$$

$$p=\frac{E_1-a}{b-a}=\frac{E_h-\frac{\sqrt{\pi}}{2}\sqrt{a}}{\frac{\sqrt{\pi}}{2}\sqrt{b}-\frac{\sqrt{\pi}}{2}\sqrt{a}} \tag{2.520}$$

Using Eqs. (2.519) and (2.520), an equation connecting a and b can be written as

$$\frac{E_1-a}{\sqrt{b}+\sqrt{a}}=\frac{E_h-\frac{\sqrt{\pi}}{2}\sqrt{a}}{\frac{\sqrt{\pi}}{2}}. \tag{2.521}$$

Using the first and second moments,

$$b=\frac{\frac{E_2}{2}-E_1a}{E_1-a}. \tag{2.522}$$

Combining Eqs. (2.521) and (2.522), one can get an equation in a single variable a that can be solved easily. The MLE can be implemented directly in Matlab as shown earlier.

The validity of the parameters and the mixture model were examined using a chi-square test and MSE. Starting with a mixture of exponential densities, the process of the mixture density modeling is given in the Matlab script containing the complete set of steps including MoM, MLE, chi-square testing, and plots of the data histogram and pdf estimates and MSE.

The Matlab script and the results are given below. The discussion on the results is provided following the Matlab work.

```
function exponential_mixture_example
% Mixture density demonstration using a mixture two exponential densities.
% Two sets of exp random variables are created with different means, each
% set containing different number of samples to mimic the weights. If N1
% and N2 are the respective number of samples, then the weights or
% probabilities are N1/(N1+N2) and N2/(N1+N2). First the MoM is used to
% estimate the three parameters.  The equations are reduced to a simngle
% variable one in 'a' and then, values of 'b' and 'a' are obtained. These
% values are then used to test the hypothesis that the mixture is indeed
% one comprised of two exponentials through the chi-square test.
%
% next, MLE is used to estimate the parameters and the chi square test is
% repeated. Ksdensity is used to get the data histogram and estimated
% densities are plotted alongside the actual data histogram in each case.
%
% p m shankar, July 2016
close all
global best aest pest
x=[exprnd(3,1,45000) exprnd(5,1,25000)]; % create two sets of exp random numbers
% with different number of samples in each set.
%
N=length(x);
xx=x';
clust=kmeans(xx,2); % get an initial estimate of the two populations by clustering
Ns=hist(clust,2);%
% Ns(1) and Ns(2) are the number of samples in two clusters and Ns/N will
% be the approximate probabilities if an iterative procedure is used. NOT
% used here
[fx,xr]=ksdensity(x);
idr=find(xr<=0);% for removing any -ve values of the xr that arise from ksdensity
xrr=xr(idr+1:end);
fr=fx(idr+1:end);
[a1,b1,p1] =parameter_est_MOM(x);
fsim1=(1-p1)*exppdf(xrr,a1)+p1*exppdf(xrr,b1); % pdf from MoM
pest=p1;aest=a1;best=b1;
mb1=10; % number of bis for the chi square test
[qq1] =chiexp_demo(x,mb1);
QC=chi2inv(0.95,10-1-2);
state1={['Chi square test statistic (MoM) = ',num2str(round(qq1))];...
    ['Chi-Square Threshold (number of bins=10) =',num2str(round(QC))]};
disp(state1)
figure,plot(xrr,fr,'ro',xrr,fsim1,'*')
xlim([.01,0.8*max(xrr)])
xlabel('SNR z'),ylabel('pdf estimate')
legend('data histogram','mixture pdf estimate (MoM)')
title(state1,'color','b')
MSE2=sum((fr-fsim1).^2)/length(xrr);
text(0.1*max(xrr),0.8*max(fsim1),['MSE_{MoM} = ',num2str(MSE2)])
vals1={['a = ',num2str(round(a1*100)/100)];['b = ',num2str(round(b1*100)/100)];...
    ['p = ',num2str(round(p1*100)/100)]};
text(0.5*max(xrr),0.35*max(fsim1),vals1)
% now get the parameter estimates using MLE
```

```
[ a2,b2,p2 ] =parameter_est_MLE(x);
pest=p2;aest=a2;best=b2;
[qq2] =chiexp_demo(x,mb1);
state2={['Chi square test statistic (MLE) = ',num2str(round(qq2))];...
    ['Chi-Square Threshold (number of bins=10) =',num2str(round(QC))]};
disp(state2)
fsim2=(1-p2)*exppdf(xrr,a2)+p2*exppdf(xrr,b2); % pdf from MLE
figure,plot(xrr,fr,'ro',xrr,fsim2,'*')
xlim([.01,0.8*max(xrr)])
xlabel('SNR z'),ylabel('pdf estimate')
legend('data histogram','mixture pdf estimate (MLE)')
title(state2,'color','b')
MSE2=sum((fr-fsim2).^2)/length(xrr);
text(0.1*max(xrr),0.8*max(fsim2),['MSE_{MLE} = ',num2str(MSE2)])
vals2={['a = ',num2str(round(a2*100)/100)];['b = ',num2str(round(b2*100)/100)];...
    ['p = ',num2str(round(p2*100)/100)]};
text(0.5*max(xrr),0.35*max(fsim2),vals2)
end

function [ a1,b1,p1 ] =parameter_est_MOM(x)
% parameter estimation using method of Moments
Mh=mean(sqrt(x)); % half order mean
M1=mean(x);% mean
M2=mean(x.^2); %second moment
syms a b  positive
y=(sym(Mh)-sym(sqrt(pi))*sqrt(a)/2)/(sqrt(sym(pi))/2);
yy=(sym(M1)-a)/(sqrt(b)+sqrt(a));
b1=(0.5*sym(M2)-sym(M1)*a)/(sym(M1)-a);
y3=subs(yy,b,b1);
ys=y-y3; % get a single equation in a that can be easily solved
ab=0.1*M1:.1:M1; % initial value of a
yys=matlabFunction(ys); % convert the equation to  inline form
idx=find(abs(yys(ab))<.002);% find the index where values are less than .002
a1=median(ab(idx));% estimate of a; there might be more than values of a
b1=(0.5*M2-M1*a1)/(M1-a1);% estimate of b
p1=(M1-a1)/(b1-a1);% estimate of p

end


function [ q] =chiexp_demo(yy,m)
global aest best pest
y=sort(yy); % sort samples
N=length(y);
inter=max(y)-min(y); %interval for chi-square test
intstep=inter/m; %size of sub-interval
intv=[min(y):intstep:max(y)]; %samples at the sub-intervals
k=[zeros(1,m)]; %number of samples in each sub-interval intialization
for i=1:N
   for j=1:m
      if y(i)<=intv(j+1)
         k(j)=k(j)+1;
         break;
      end;
   end;
end;
p=zeros(1,m);
for i=1:m
    CDF1=(1-pest)*expcdf(intv(i+1),aest)+pest*expcdf(intv(i+1),best);
     CDF2=(1-pest)*expcdf(intv(i),aest)+pest*expcdf(intv(i),best);
   p(i)=CDF1-CDF2; %probabilities at the sub-interval samples
end;
```

```
np=p.*N;
q=0;
for i=1:m
    q = q + ((k(i)-np(i))^2)/np(i); %chi-square test
end;

end

function [a1,b1,p1 ] =parameter_est_MLE(x)
% parameter estimation using MLE
M1=mean(x);
syms a b p xx
myfun=(1-p)*(1/a)*exp(-xx/a)+p*(1/b)*exp(-xx/b);
mydensity=matlabFunction(myfun);
clear xx a b p
aa=0.4*M1;
bb=0.3*M1;
pp=0.2;
 opt = statset('mlecustom');
opt = statset(opt,'FunValCheck','off','MaxIter',3000,'TolX',1e-4,'TolFun',1e-5,'TolBnd',1e-
5','MaxFunEvals', 1000);
pq = mle(x,'pdf',mydensity,'start',[aa bb pp], 'options',opt); % fit mydensity to the data
a1=pq(1); b1=pq(2); p1=pq(3);

end
```

```
'Chi square test statistic (MoM) = 9'
'Chi-Square Threshold (number of bins=10) =14'

'Chi square test statistic (MLE) = 8'
'Chi-Square Threshold (number of bins=10) =14'
```

Results are displayed in Figs. 2.63 (MoM) and Fig. 2.64 (MLE).

It can be seen that both approaches lead to acceptable levels of statistical match based on chi-square testing. Note that the identification of a and b is arbitrary and what matters is the nature of the fit. The MoM becomes less convenient and more time consuming when the number of components in the mixture goes up and the mixture requires larger and larger number of parameters. The MoM also is difficult when non-identical types of density functions constitute the mixture. Consider the case of a pdf mixture with two different types of densities,

$$f(z) = (1-p)\left(\frac{1}{a}\right)e^{-\frac{z}{a}} + p\frac{z^{m-1}}{b^m}e^{-\frac{z}{b}}. \tag{2.523}$$

The mixture consists of exponential and gamma densities and it requires the estimation of four parameters a, m, b, and p making MoM inconvenient and error prone since higher order moments would be necessary. In the example, MLE was used. The Matlab script appears below. Figure 2.65 shows the results on the mixture of exponential and gamma variables.

```
function general_mixture_example
% Mixture density demonstration using a mixture of exponential and gamma
% densities. MLE is used to estimate the parameters
%
% p m shankar, July 2016
close all
global best aest pest mest
x=[gamrnd(1.5,3,1,5000) gamrnd(1,2,1,2000)];
% create two sets of gamma  random numbers: second one is an exponential
% random variable
% with different number of samples in each set. The ratio of the number of
% samples to the total number of samples gives the weights or probabilities
N=length(x);
[fx,xr]=ksdensity(x);
idr=find(xr<=0);% for removing any -ve values of the xr that arise from ksdensity
xrr=xr(idr+1:end);
fr=fx(idr+1:end);
mb1=10; % number of bis for the chi square test
% now get the parameter estimates using MLE
 pq =parameter_mix_MLE(x);
 pest=pq(1);aest=pq(2);mest=pq(3);best=pq(4);
 [qq] =chimix_demo(x,mb1); % get the results of the chi square test
 QC=chi2inv(0.95,mb1-1-4);
 state2={['Chi square test statistic (MLE) = ',num2str(round(qq))];...
     ['Chi-Square Threshold (number of bins=10) =',num2str(round(QC))]};
 disp(state2)
 fsim=(1-pest)*exppdf(xrr,aest)+pest*gampdf(xrr,mest,best); % pdf from MLE
 figure,plot(xrr,fr,'ro',xrr,fsim,'*')
 xlim([.01,0.8*max(xrr)])
 xlabel('SNR z'),ylabel('pdf estimate')
 legend('data histogram','mixture pdf estimate (MLE)')
 title(state2,'color','b')
 MSE=sum((fr-fsim).^2)/length(xrr);
 text(0.1*max(xrr),0.8*max(fsim),['MSE_{MLE} = ',num2str(MSE)])
 vals={['a = ',num2str(round(aest*100)/100)];['m = ',num2str(round(mest*100)/100)];...
    [' b= ',num2str(round(best*100)/100)];['p = ',num2str(round(pest*100)/100)]};
 text(0.5*max(xrr),0.35*max(fsim),vals)
 % create the equation for display
 syms a b m p f(z)
 eqn=(1-p)*(1/a)*exp(-z/a)+p*z^(m-1)*exp(-z/b)/(b^m*gamma(m));
 ff=[f(z)==eqn];
 text(0.3*max(xrr),0.65*max(fsim),['$' latex(ff) '$'],...
    'interpreter','latex','fontsize',14)
end



function [ q] =chimix_demo(yy,m)
global aest pest mest best
y=sort(yy); % sort samples
N=length(y);
inter=max(y)-min(y); %interval for chi-square test
intstep=inter/m; %size of sub-interval
intv=[min(y):intstep:max(y)]; %samples at the sub-intervals
k=[zeros(1,m)]; %number of samples in each sub-interval intialization
for i=1:N
   for j=1:m
      if y(i)<=intv(j+1)
         k(j)=k(j)+1;
         break;
      end;
   end;
end;
```

```
p=zeros(1,m);
for i=1:m
    % estimate the CDF using the mixture density
    CDF1=(1-pest)*expcdf(intv(i+1),aest)+pest*gamcdf(intv(i+1),mest,best);
     CDF2=(1-pest)*expcdf(intv(i),aest)+pest*gamcdf(intv(i),mest,best);
   p(i)=CDF1-CDF2; %probabilities at the sub-interval samples
end;
np=p.*N;
q=0;
for i=1:m
   q = q + ((k(i)-np(i))^2)/np(i); %chi-square test
end;
end

function pq =parameter_mix_MLE(x)
% parameter estimation using MLE
N=length(x);
[Ns]=hist(x,2);
pinit=Ns(1)/N; % initial guess of the weight
[pha]=gamfit(x); % initial estimate of the parameters of the gamma dist
M1=mean(x); % initial estimate of the parameter of the exp density
mypdf = @(x,pq1,aq,mq,bq) (1-pq1)*exppdf(x,aq) + pq1*gampdf(x,mq,bq);
PQ=[pinit,0.8*M1,pha(1),pha(2)];
opt = statset('mlecustom');
opt = statset(opt,'FunValCheck','off','MaxIter',6000,'TolX',1e-4,'TolFun',1e-3,'TolBnd',1e-
3','MaxFunEvals', 4000);
pq = mle(x,'pdf',mypdf,'start',PQ, 'options',opt); % fit mypdf to the data
% pq=[p1,a1,m1,b1];

end
```

```
'Chi square test statistic (MLE) = 2'
'Chi-Square Threshold (number of bins=10) =11'
```

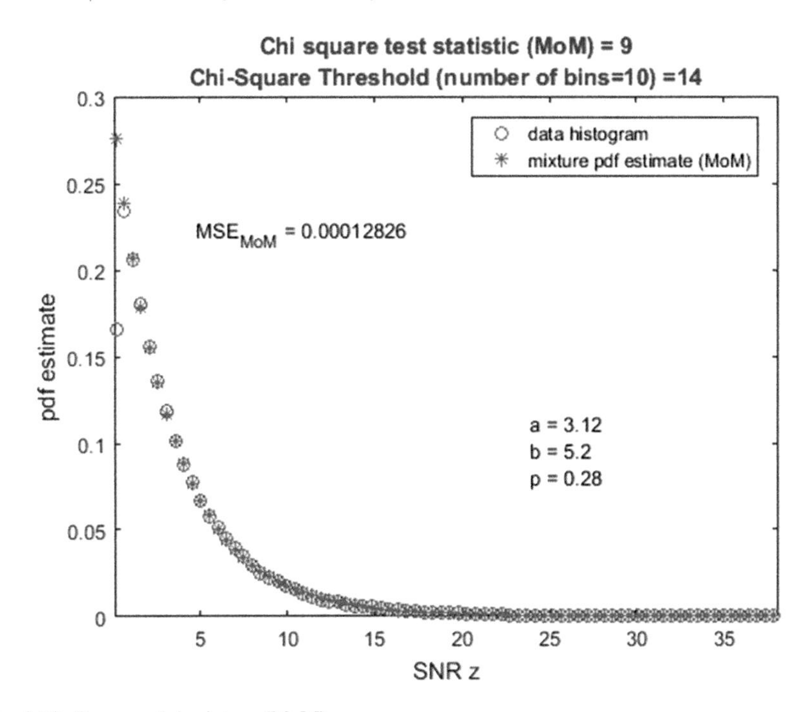

Fig. 2.63 Exponential mixture (MoM)

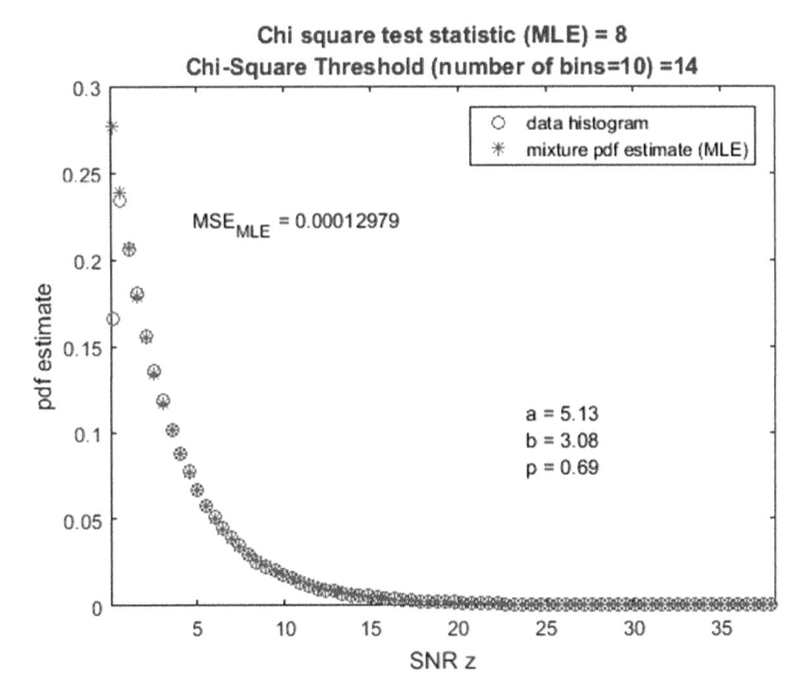

Fig. 2.64 Exponential mixture (MLE)

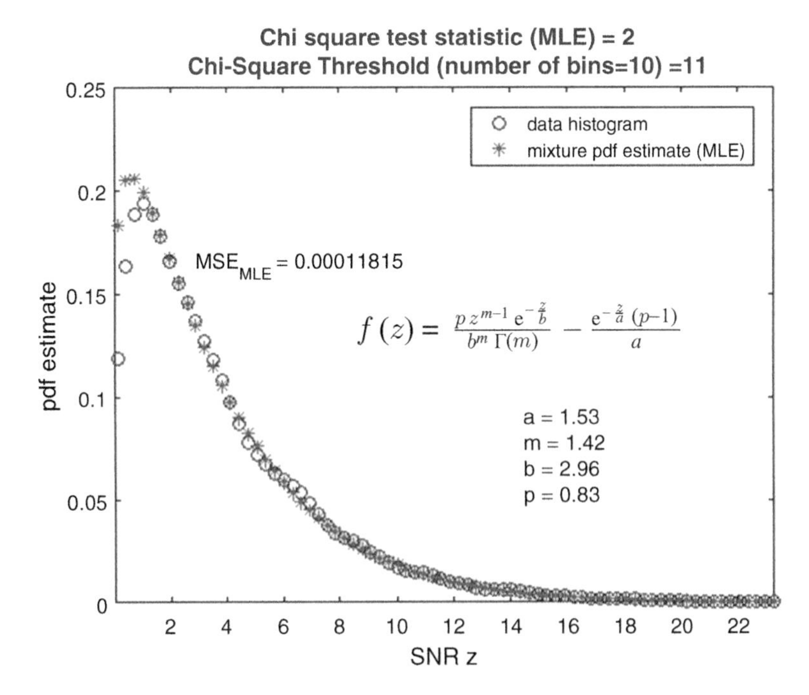

Fig. 2.65 Mixture of exponential and gamma (MoM)

It should be noted while MoM might give unacceptable results because of the uncertainty in higher order moments, MLE might also provide unreliable parameter estimates since MLE depends heavily on the initial guesses chosen for the parameters. It is therefore recommended that appropriate choice be made on some "a priori" information on the parameter as it was done in the case of the exponential gamma mixture above. The initial guesses were chosen on the basis of putting the data in two categories (for the initial weight) and the parameter guesses were made on the assumption of the particular density. There is yet another method, namely the expectation maximization (EM), also available for the estimation of parameters of density mixtures (Redner and Walker 1984; da Silva and Yongacoglu 2015).

Additional analysis of the mixture densities will be provided when specific models for fading and shadowing using density mixtures are discussed (Chaps. 4 and 5). In general, one also needs to determine the best fit to the data by varying the number of mixture of components (n). As n goes up, chi-square testing is of limited use and one has to explore other means of finding the best fits based on Bayes Information Criterion (Kass and Raftery 1995; Neath and Cavanaugh 2012). Details on BIC will be provided in Chap. 5.

2.18 Receiver Operating Characteristics

Receiver operating characteristics (ROC), originally being used in radar detection theory, is now extensively being used in medical and behavioral statistics (Van Trees 1968; Helstrom 1968; Hanley and McNeil 1982). Recent research in cognitive radio and the need to quantify the energy detection schemes in wireless fading channels have spawned a new interest in undertaking the ROC analysis in fading channels (Atapattu et al. 2010a, b; Shankar 2016). The performances of the various energy detection schemes can be compared in terms of quantitative measures obtained through the ROC studies.

In simple terms, ROC analysis examines a system with binary outputs using some form of a detector. Consider the case of a radar detector or the case of a specific cancer detection algorithm which makes a positive identification of a target or of the presence of cancer on the basis of some observed or measurable quantity. In the case of radar, it might be the amplitude of the backscattered signal and in the case of clinical diagnostics, it might be a numeric value assigned to a particular set of data based on some criteria used by the clinicians. These steps can be broken down into two simple hypotheses and one of the hypotheses will be accepted based on the numerical values assigned. For the case of radar, the two hypotheses are the following:

Hypothesis H_0 The detected voltage or signal arose out of noise only.

Hypothesis H_1 The detected voltage is the result of a target existing in the region along with noise.

Often, the first hypothesis H_0 is also called the null hypothesis. In order to decide which hypothesis will be accepted, the observed quantity is compared against a threshold. If the observed quantity is below the threshold the null hypothesis H_0 is accepted. Otherwise, it is rejected and hypothesis H_1 is accepted. If the observed quantity is represented by Z, the two hypotheses correspond to two density functions

$$\begin{aligned} f_Z(z|H_0) &: \quad \text{Hypothesis } H_0 \\ f_Z(z|H_1) &: \quad \text{Hypothesis } H_1 \end{aligned}. \tag{2.524}$$

The major issues in hypothesis testing are the development and access to the probability densities required for modeling the observed quantity and an optimum way to design a threshold so that the user will have the most favorable detection performance. The "most favorable detection performance" needs to be quantified for practical applications. There are a number of ways to achieve quantification of the performance of the detectors and depending on the nature of its use, these quantification metrics could be different. These include the probabilities of detection and false alarm, probability of error, area under the ROC, etc. The probabilities of false alarm and detection were discussed earlier in Sect. 2.11. These will be discussed briefly first.

Assuming that additive white Gaussian noise is present in the system, the probability densities associated with the hypotheses can be expressed as

$$f(z|H_0) = \frac{1}{\sqrt{2\pi\sigma^2}} \exp\left(-\frac{z^2}{2\sigma^2}\right) \tag{2.525}$$

$$f(z|H_1) = \frac{1}{\sqrt{2\pi\sigma^2}} \exp\left(-\frac{(z-\mu)^2}{2\sigma^2}\right). \tag{2.526}$$

In Eqs. (2.525) and (2.526),μ is the mean voltage and σ^2 is the variance of the noise. If a threshold is chosen at Z_T, the probabilities of false alarm (P_F) and miss (P_M) are given as

$$P_F = \int_{Z_T}^{\infty} f(z|H_0)dz = \int_{Z_T}^{\infty} \frac{1}{\sqrt{2\pi\sigma^2}} \exp\left(-\frac{z^2}{2\sigma^2}\right) dz \tag{2.527}$$

$$P_M = \int_{-\infty}^{Z_T} f(z|H_1)dz = \int_{-\infty}^{Z_T} \frac{1}{\sqrt{2\pi\sigma^2}} \exp\left(-\frac{(z-\mu)^2}{2\sigma^2}\right) dz. \tag{2.528}$$

The probability of error $p(e)$ is

$$p(e) = P_F p(H_0) + P_M p(H_1). \tag{2.529}$$

As described in Sect. 2.11, $p(H_0)$ and $p(H_1)$ are the probabilities associated with the respective hypotheses. Methods to choose the threshold were also described there.

For applications in communication theory, the probability of error is accepted as sufficient metric regarding the performance of the detector. It should also be noted that in communication theory related applications, the hypotheses are considered to be equiprobable. In radar as well as in clinical applications, the hypotheses do not exist with equal probability and the metric of the error rate or probability of error is not a sufficient indicator of the performance. In most of the non-communication theory applications, the interest is in finding a critical threshold that will provide a specific probability of false alarm or probability of detection. This will also be the case with energy detectors in cognitive radio (Chap. 7).

The goal of finding such a threshold is accomplished through plots known as the receiver operating characteristic (ROC) curves (Hanley and McNeil 1982; Metz 2006). These are created by plotting the probabilities of false alarm against probabilities of detection as the threshold is varied. Consider the case when $\mu = 0, 2, 4, \ldots$, and $\sigma^2 = 4$, providing plots for each set of values of $\mu =$ and σ^2. To generate these plots, the probabilities of detection are varied and the corresponding threshold values are calculated. The probabilities of false alarm are then calculated for these thresholds. For the two hypotheses in Eqs. (2.527) and (2.528), the probabilities of detection (P_D) and probabilities of false alarm are

$$P_D = \int_{Z_T}^{\infty} \frac{1}{\sqrt{2\pi\sigma^2}} \exp\left(-\frac{(z-\mu)^2}{2\sigma^2}\right) dz \tag{2.530}$$

$$P_F = \int_{Z_T}^{\infty} \frac{1}{\sqrt{2\pi\sigma^2}} \exp\left(-\frac{z^2}{2\sigma^2}\right) dz \tag{2.531}$$

Note that both PF and PD can be expressed in terms of erfc(.) or Gaussian Q(.) functions. For the plots, the SNR is defined as

$$Z_0 = 20\log_{10}\left(\frac{\mu}{\sigma}\right) \ dB \tag{2.532}$$

A Matlab script written to demonstrate several aspects associated with ROC is given below.

```
function ROC_analysis_f

% ROC_Analysis
% ROC_analysis for the Gaussian channel and an energy detector in Rayleigh
% channel. For the Gaussian channel, three figures are displayed starting
% from the simple ROC plot to the one in logarithmic scale (for X-axis) and
% the ROC curve with areas calculated. The values of Az are indicated along
% with the SNR values. Note that for the Gaussian channel the lowest value
% of -INF dB while for the Rayleigh channel it will be 0 dB. For the
% Rayleigh channel, only the plots with the areas computed is shown.
%
% P M Shankar, October 2016
clear;clc;close all
PD=0:.005:1; % probability of detection values chosen to get threshold
sig=2;% sigma
mu=0:5;
SNR=round(20*log10(mu/sig));% SNR
PF=zeros(length(mu),length(PD));
for k=1:6
```

```
% norminv(.) is the CDF while PD is 1-CDF.
% PD=1-normcdf(thr,mu,sig)-->[1-PD]=normcdf(thr,mu,sig)-->
% thr=norminv(1-PD,mu,sig)
ZT=norminv(1-PD,mu(k),sig); % calculate threshold based on the detection probability
PF(k,:)=1-normcdf(ZT,0,sig); % probability of false alarm
end;
figure
for k=1:length(mu)
   plot(PF(k,:),PD,'linewidth',1.5)
   mu1=mu(k);
   hold on
 ZT1=norminv(0.8,0,sig);% obtain threshold for PF=0.2 for indicating plots
 PDT=1-normcdf(ZT1,mu1,sig);% this is PD for PF=.2 for indicating the SNR
 if mu1==0
 text(0.15,PDT,'Z_0 = -\infty dB','fontweight','bold')
 else
  text(0.15,PDT,['Z_0 = ',num2str(SNR(k)),' dB'],'fontweight','bold')
 end;
 hold on
Az1(k)=0.5+polyarea(PF(k,:),PD);
end;
xlabel('Probability of False Alarm')
ylabel('Probability of Detection')
clear PF

% % change mu and plot the data with PF in logarithmic scale
mu=[5,7,9,11,4];
SNR=round(20*log10(mu/sig));% SNR
PF=zeros(length(mu),length(PD));
for k=1:length(mu)
ZT=norminv(1-PD,mu(k),sig); % calculate threshold based on the detection probability
PF(k,:)=1-normcdf(ZT,0,sig); % probability of false alarm
end;
figure
for k=1:length(mu)
   semilogx(PF(k,:),PD,'linewidth',1.5)
   mu1=mu(k);
   SN=20*log(mu1/sig);
   hold on
   PFT=1e-3;
 ZT1=norminv(1-1e-4,0,sig);% obtain threshold for PF=1e-4 for indicating plots
 PDT=1-normcdf(ZT1,mu1,sig);% this is PD for PF=1e-4 for indicating the SNR
  text(.5e-4,PDT,[num2str(SNR(k)),' dB'],'fontweight','bold')
 hold on
Az1(k)=0.5+polyarea(PF(k,:),PD);
end;
xlim([1e-15,1])
xlabel('Probability of False Alarm')
ylabel('Probability of Detection')
%

mu=[0,2,4,6];
SNR=round(20*log10(mu/sig));% SNR
```

```
PF=zeros(length(mu),length(PD));
for k=1:length(mu)
ZT=norminv(1-PD,mu(k),sig); % calculate threshold based on the detection probability
PF(k,:)=1-normcdf(ZT,0,sig); % probability of false alarm
end;
% change mu and show the areas
figure
for k=1:length(mu)
   subplot(2,2,k),plot(PF(k,:),PD,'linewidth',1.5)
   xlabel('P_F')
ylabel('P_D')
   mu1=mu(k);
    SN=round(20*log(mu1/sig));
Az1(k)=0.5+polyarea(PF(k,:),PD);
hold on
PFF=PF(k,:);
% to fill the areas properly supply the missing values
PFF=[PFF,ones(1,10)];
PDF=[PD,ones(1,10)];
fill(PFF,PDF,'b') % this only fills the upper region below the plots
fill([0,.5,1,1,1,.5,0],[0,0,0,.5,1,.5,0],'b')
aqz=round(Az1(k)*1000)/1000;% just keep only three decimal places
text(.5,.6,['A_z = ',num2str(aqz)],'backgroundcolor','w')
if mu1==0
  text(.5,0.3,['Z_0 = -\infty dB'],'backgroundcolor','y')
else
text(.5,0.3,['Z_0 = ',num2str(SN),' dB'],'backgroundcolor','y')
end;
hold off

end;

% Rayleigh channel
BA=[1,2,4,6]; % this B/A
SN=round(10*log10(BA));% SNR
PD=0:.005:1; % probability of detection values chosen to get threshold
PF=zeros(length(BA),length(PD));
for k=1:length(BA)
ZT=BA(k)*(-log(PD));% calculate threshold based on the detection probability
PF(k,:)=exp(-ZT*BA(k)); % probability of false alarm
end;
figure
for k=1:length(BA)
   subplot(2,2,k),plot(PF(k,:),PD,'linewidth',1.5)
   xlabel('P_F')
ylabel('P_D')
   AB1=BA(k);
Az1(k)=0.5+polyarea(PF(k,:),PD);
hold on
PFF=PF(k,:);
% to fill the areas properly supply the missing values
PFF=[PFF,ones(1,10)];
PDF=[PD,ones(1,10)];
```

```
fill(PFF,PDF,'b') % this only fills the upper region below the plots
fill([0,.5,1,1,1,.5,0],[0,0,0,.5,1,.5,0],'b')
aqz=round(Az1(k)*1000)/1000;% just keep only three decimal places
text(.5,.6,['A_z = ',num2str(aqz)],'backgroundcolor','w')
 if AB1==0
   text(.5,0.3,['Z_0 = -\infty dB'],'backgroundcolor','y')
 else
 text(.5,0.3,['Z_0 = ',num2str(SN(k)),' dB'],'backgroundcolor','y')
 end;
hold off
end;



end
```

Results are displayed in Figs. 2.66, 2.67, 2.68, and 2.69.

For the case of the mean equal to the standard deviation ($Z_0 = -\infty$ dB), the diagonal line clearly informs that the chance of detection is 50%. As the ratio of the mean to standard deviation increases, it is clear that the probabilities of detection for a fixed value of probability of false alarm increase clearly demonstrating the improving detection performance with the increase in mean μ. The information in Fig. 2.66 might still be insufficient to pick an appropriate threshold since the interest is to have probabilities of false alarm $<1e{-}5$. Changing the X-axis to logarithmic units, it is possible to get a better grasp of the P_F vs P_D as seen in Fig. 2.67.

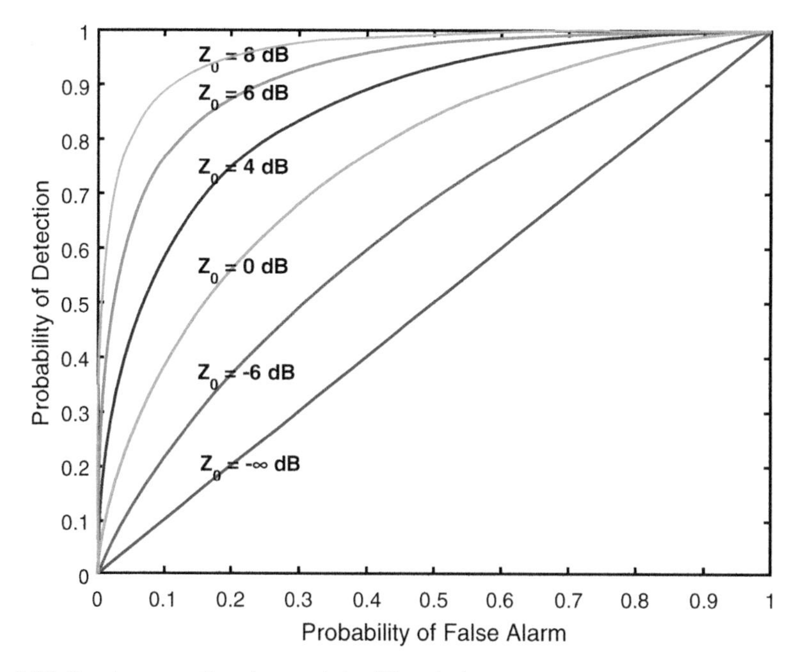

Fig. 2.66 Receiver operating characteristics (Gaussian)

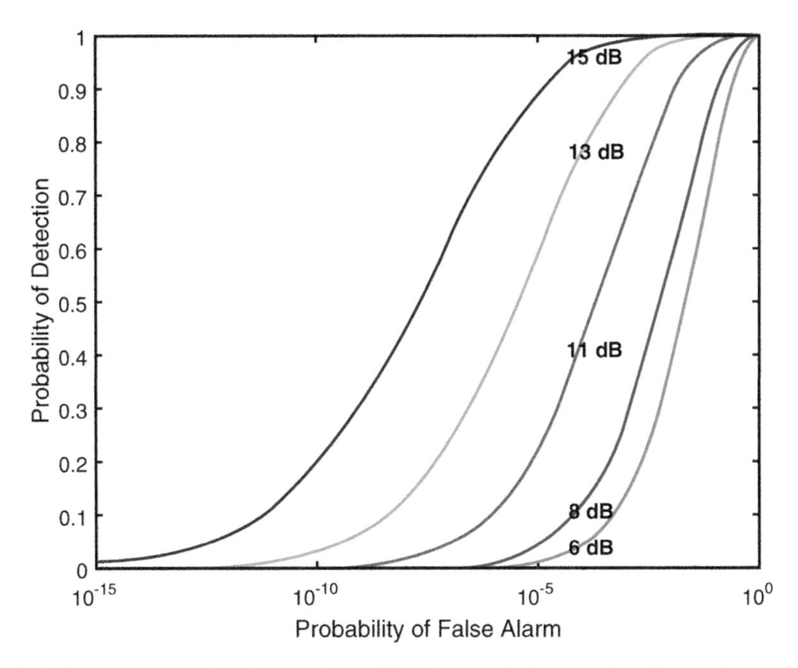

Fig. 2.67 Receiver operating characteristics (Gaussian) with *X*-axis in logarithmic format

While the plot of ROC with logarithmic scale is an improvement, quantification of the performance of the detector is still not fully accomplished. The quantification of the detector performance is carried out by calculating the area enclosed by the plots keeping in mind that the best detector will have an area of unity. From a plot of P_F vs. P_D, the area under the ROC curve, typically represented by A_z (called the A_z value), is (Hanley and McNeil 1982; McClish 1989; Obuchowski 2003)

$$A_z = \int_0^1 P_D(y)\, d(P_F(y)). \tag{2.533}$$

By examining the probabilities associated with the two hypotheses, Eq. (2.533) can be expressed as

$$A_z = 1 - \int_{-\infty}^{\infty} f(y|H_0) F(y|H_1) dy. \tag{2.534}$$

In Eq. (2.534), $F(.)$ is the CDF. It is therefore obvious that in the Gaussian model described in connection with the densities in Eqs. (2.525) and (2.526), when the $\mu = \sigma$, $Z_0 = 0$ dB, the ROC plot is the diagonal line and the area under the ROC will be 0.5. This means that the success or failure is equally likely, a very unacceptable outcome. As the value of SNR increases, the A_z values increase, providing a clearer means of quantification of the performance of the detector. Note that the area under

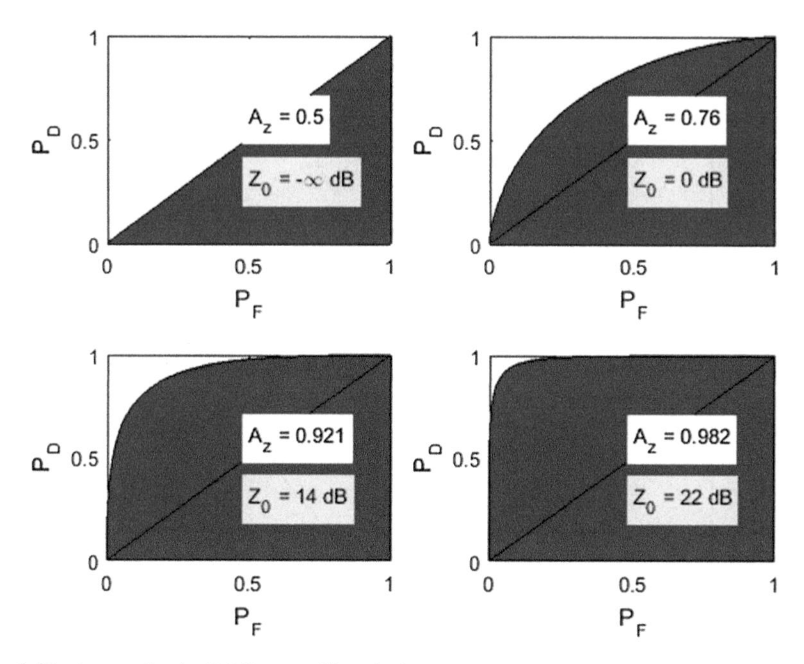

Fig. 2.68 Area under the ROC curve (Gaussian)

the ROC curve can be estimated in Matlab using the command *polyarea*(.). The results are displayed in Fig. 2.68. The value of the area under the ROC curve (also expressed as AUC) continues to increase with increasing value of the ratio of the mean to standard deviation, providing a quantitative measure of the performance compared to the qualitative measure seen in ROC plots.

While the previous example dealt with the detection performance based on Gaussian densities, the energy detection schemes discussed in Chap. 7 would have densities that only take positive values. A typical example will be the case of an energy detector which attempts to make a decision to determine whether the hypothesis H_0 or H_1 can be accepted in Rayleigh fading channel conditions. In a Rayleigh channel,

$$f(z|H_0) = \frac{1}{A}\exp\left(-\frac{z}{A}\right) \tag{2.535}$$

$$f(z|H_1) = \frac{1}{B}\exp\left(-\frac{z}{B}\right), B \geq A \tag{2.536}$$

In this case,

$$Z_0 = 10\log_{10}\left(\frac{B}{A}\right)\ dB. \tag{2.537}$$

The results on the area under the ROC curve are shown in Fig. 2.69. Notice that the plots are not symmetric for the Rayleigh channel while they were symmetric for the Gaussian channel.

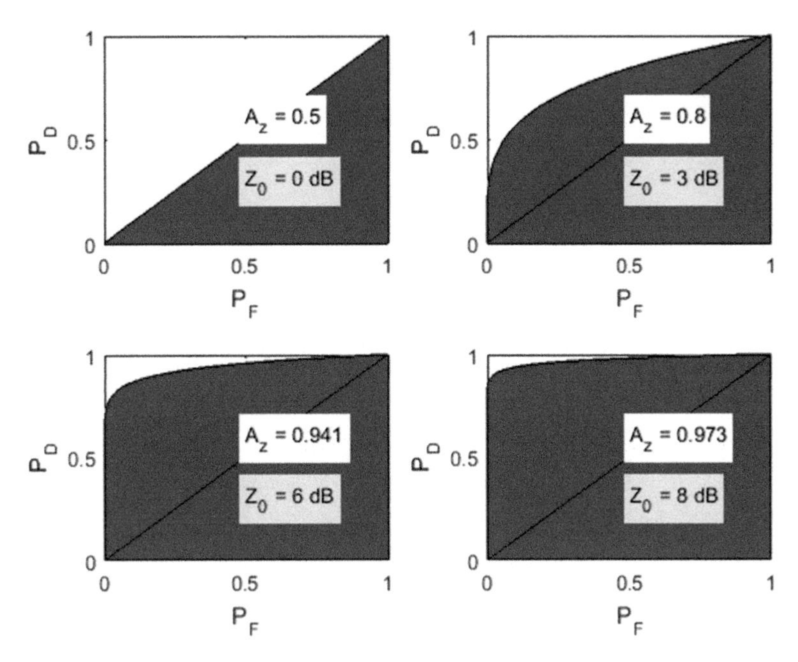

Fig. 2.69 Area under the ROC curve (Rayleigh)

2.19 Laplace and Mellin Transforms

While almost all books in probability and random variables describe characteristic functions of random variables and its relationship to Fourier transforms, there is little discussion of Laplace transforms and even less discussion of Mellin transforms (Epstein 1948; Erdelyi 1953; Rossberg 2008). For wireless systems operating in fading and shadowed fading channels, the fading models evolve out of summation of random variables as well as products of random variables (Wells et al. 1962; Lomnicki 1967; Springer and Thompson 1970; Block and Savits 1980; Shankar 2013). The summation arises when diversity techniques are employed to mitigate the effects of fading. There are also a few instances of summation of random variables in the modeling of fading such as those leading to $\eta-\mu$, $\kappa-\mu$, and McKay densities. The products of random variables arise in double, triple, quadruple, etc., types of scattering channels resulting in densities that are expressed in terms of Meijer G-functions. The shadowed fading channels are also modeled in terms of SNR evolving from the product of two random variables. Since almost all these cases arise from random variables which only take positive values, Laplace and Mellin transforms are likely to offer some interesting possibilities for the statistical analysis undertaken in wireless communications (Wells et al. 1962; Subrahmanaim 1970; Block and Savits 1980; Park 1986; Abate 1995; Tellambura et al. 2003; Ahmed et al. 2011) The Laplace transform was defined earlier in Sect. 2.8 of a random variable X with a pdf $f_X(x)$ is

$$L[f(x)] = L_X(s) = \int_0^\infty f(x)e^{-xs}dx = E(e^{-sX}). \tag{2.538}$$

The Mellin transform of the random variable X is

$$M[f(x)] = M_X(s) = \int_0^\infty f(x)x^{s-1}dx. \tag{2.539}$$

The inverse Laplace transform and the inverse Mellin transforms generate the density $f(x)$ as

$$f(x) = L^{-1}[L_X(s)] = \frac{1}{2\pi j}\int_{\lambda-j\infty}^{\lambda+j\infty} L_X(s)e^{xs}ds \tag{2.540}$$

$$f(x) = M^{-1}[M_X(s)] = \frac{1}{2\pi j}\int_{\lambda-j\infty}^{\lambda+j\infty} M_X(s)x^{-s}ds \tag{2.541}$$

In general Eqs. (2.540) and (2.541) may be useful in gaining considerable insight into derivation of density functions of sums or products of random variables. One rarely takes the inverse transforms by performing integrations and they are usually obtained from the tables of Laplace and Mellin transforms. Even though Matlab can provide both Laplace and inverse Laplace transforms, Matlab does not provide Mellin transforms presently. In general, Mellin transforms can be generated using the integral in Eq. (2.539). The inverse Mellin transforms are available in Maple and Mathematica.

Without getting into the details of deriving the convolution relationships, it can be stated that the Laplace transform of the convolution of two densities $f_X(x)$ and $f_Y(y)$ is the product of Laplace transforms $L_X(s)$ and $L_Y(s)$,

$$L_X(s)L_Y(s) = L[f(x)*f(y)]. \tag{2.542}$$

Equation (2.542) shows that if X and Y are independent random variables, the density of

$$Z = X + Y \tag{2.543}$$

is

$$f_Z(z) = \int_0^z f_X(x)f_Y(z-x)dx = f_X(x)*f_Y(y). \tag{2.544}$$

The density of the sum of X and Y can be obtained from the inverse Laplace transform of the product of the two marginal Laplace transforms,

$$f_Z(z) = L^{-1}[L_X(s)L_Y(s)]. \tag{2.545}$$

The concept the product of Laplace transforms can be extended to obtain the probability density of the sum of a number of independent random variables (positive).

The convolution property of the Mellin transforms takes a different form (Epstein 1948; Kilicman and Ariffin 2002). While the convolution integral in Eq. (2.544) has an integrand of the form $f_X(x)f_Y(z-x)$ with the limit of integration of 0 to z, the "convolution integral" in the Mellin transform domain is

$$f(w) = \int_0^{\infty} f_X(x)\frac{1}{x}f_Y\left(\frac{w}{x}\right)dx. \tag{2.546}$$

The convolution theorem in the Mellin domain is

$$M_W(s) = M_W[f(w)] = M_W\left[\int_0^{\infty} f_X(x)\frac{1}{x}f_Y\left(\frac{w}{x}\right)dx\right] = M_X(s)M_Y(s) \tag{2.547}$$

Note that Eq. (2.546) gives the density of the product of two independent random variables (each taking only positive values)

$$W = XY. \tag{2.548}$$

In other words, the pdf of W is obtained by taking the inverse Mellin transform of the product of the Mellin transforms of X and Y,

$$f_W(w) = M^{-1}[M_X(s)M_Y(s)]. \tag{2.549}$$

Mellin transforms are also useful in obtaining the density of the ratio of two independent random variables U,

$$U = \frac{X}{Y}. \tag{2.550}$$

Using the concept of transformation of random variables,

$$f_U(u) = \int_0^{\infty} yf_X(uy)f_Y(y)dy. \tag{2.551}$$

The Mellin convolution result now takes a different form

$$M_U(s) = M_U[f(u)] = M_W\left[\int_0^{\infty} y f_X(uy) f_Y(y) dy\right] = M_X(s) M_Y(2-s). \quad (2.552)$$

Thus, the density is obtained by taking the inverse Mellin transform of $M_U(s)$ as

$$f_U(u) = M^{-1}[M_X(s) M_Y(2-s)]. \quad (2.553)$$

The specific applications of Laplace and Mellin transforms to problems in wireless channels will now be examined starting with the Laplace and inverse Laplace transforms.

2.19.1 Laplace Transforms and Sums of Random Variables

Some of the results on the probability density of sum of independent variables were presented in Sect. 2.8 earlier. These densities will be revisited now through the application of the transforms. Before specific problems are studied, it is necessary to have access to the properties of Laplace transforms and Laplace transforms of known functions as well as Laplace transforms of specific probability densities. The table of Laplace transforms is given in Table 2.4

Table 2.4 Laplace transform

$y(t)$	$Y(s) = \int_0^{\infty} f(t)dt$
1	$\frac{1}{s}$
e^{at}	$-\frac{1}{a-s}$
t^n	$\frac{\Gamma(n+1)}{s^{n+1}}$
$\sin(at)$	$\frac{a}{a^2+s^2}$
$\cos(at)$	$\frac{s}{a^2+s^2}$
$e^{bt}\cos(at)$	$-\frac{b-s}{a^2+(b-s)^2}$
$e^{bt}\sin(at)$	$\frac{a}{a^2+(b-s)^2}$
$t^n e^{at}$	$\frac{\Gamma(n+1)}{(s-a)^{n+1}}$
$t^2\cos(at)$	$\frac{8s^3}{(a^2+s^2)^3} - \frac{6s}{(a^2+s^2)^2}$

$s > 0; \quad a, b, c > s; \quad n = 0, 1, 2, 3, \ldots$

Table 2.5 Laplace transforms (properties)

$g(t)$	$G(s) = \int_0^\infty f(t)dt$
$bh(t) + ay(t)$	$bH(s) + aY(s)$
$y'(t)$	$sY(s) - y(0)$
$y''(t)$	$s^2Y(s) - sy(0) - y'(0)$
$y'''(t)$	$s^3Y(s) - sy'(0) - s^2y(0) - y''(0)$
$\int_0^t h(\tau)y(t-\tau)dt$	$H(s)Y(s)$

Table 2.6 Laplace transforms (densities)

Probability density function	Laplace transform
$f_X(x) = x^{m-1} \frac{e^{-\frac{mx}{a}} \left(\frac{m}{a}\right)^m}{\Gamma(m)}$	$\frac{m^m}{(m + as)^m}$
$f_Y(y) = x^{m-1} \frac{e^{-\frac{mx}{b}} \left(\frac{m}{b}\right)^m}{\Gamma(m)}$	$\frac{m^m}{(m + bs)^m}$
$\int_0^z f_X(x) f_Y(z-x)dx$	$\frac{m^{2m}}{(m + bs)^m (m + as)^m}$
$f_W(w) = w^{n-1} \frac{e^{-\frac{nw}{b}} \left(\frac{n}{b}\right)^n}{\Gamma(n)}$	$\frac{n^n}{(n + bs)^n}$
$\int_0^z f_X(x) f_W(z-x)dx$	$\frac{m^m n^n}{(n + bs)^n (m + as)^m}$
$f_X(x) = \frac{xe^{-\frac{x^2}{2b^2}}}{b^2}$	$1 - \frac{\sqrt{2}bs\sqrt{\pi}e^{\frac{b^2s^2}{2}}\text{erfc}\left(\frac{\sqrt{2}bs}{2}\right)}{2}$

A few additional properties pertaining to derivatives and integrals are listed in Table 2.5. The Laplace transforms of some of the common density functions are given in Table 2.6.

The first example considered is the sum of two exponential variables. This case represents the density of the maximal ratio combining algorithm (See Chap. 5) in a Rayleigh channel. Treating the two variables X and to be independent and identically distributed, the pdf of $Z = X + Y$ is

$$f(z) = f_X(x) * f_Y(y). \tag{2.554}$$

The density function of X (or Y) is

$$f_X(x) = \frac{1}{a} e^{-\frac{x}{a}}. \tag{2.555}$$

Using the principle of convolution, the Laplace transform of $f(z)$ will be the square of the Laplace transform of X,

$$F_X(s) = \frac{1}{as + 1} \tag{2.556}$$

Equation (2.556) is obtained by putting $m = 1$ in the first equation in Table 2.6. The pdf of Z will be

$$f(z) = L^{-1}\left[\frac{1}{(as+1)^2}\right] = \frac{z}{a^2} e^{-\frac{z}{a}} U(z) \tag{2.557}$$

Consider the case of two independent and identically distributed gamma variables with densities of the form for X (or Y)

$$f_X(x) = \frac{1}{a^m \Gamma(m)} x^{m-1} e^{-\frac{x}{a}}. \tag{2.558}$$

The Laplace transform of X (or Y) will be

$$L_X(s) = L_Y(s) = \frac{1}{(as+1)^m} \tag{2.559}$$

The pdf of $Z = X + Y$ will be

$$f(z) = L^{-1}\left[\frac{1}{(as+1)^{2m}}\right] = \frac{z^{2m-1}}{a^2 \Gamma(2m)} e^{-\frac{z}{a}} U(z). \tag{2.560}$$

Consider the case of X and Y being independent gamma variables with different orders, and different scaling factors or averages such as

$$f_X(x) = \frac{1}{a^m \Gamma(m)} x^{m-1} e^{-\frac{x}{a}} \tag{2.561}$$

$$f_Y(y) = \frac{1}{a^n \Gamma(n)} x^{n-1} e^{-\frac{y}{a}} \tag{2.562}$$

The Laplace transforms of these densities are

$$L_X(s) = \frac{1}{(as+1)^m} \tag{2.563}$$

$$L_Y(s) = \frac{1}{(as+1)^n} \tag{2.564}$$

The density of Z is

$$f(z) = L^{-1}\left[\frac{1}{(as+1)^{m+n}}\right] = \frac{z^{m+n-1}}{a^{m+n} \Gamma(m+n)} e^{-\frac{z}{a}} U(z) \tag{2.565}$$

The Matlab work relating to these transformations is included below (following the next example). It can be seen that the Laplace based approach is simple to implement. Matlab might not provide a solution using the traditional integral approach of convolution as in the case of two gamma random variables with identical orders with non-identical scaling parameters

$$f_X(x) = \frac{1}{a^m \Gamma(m)} x^{m-1} e^{-\frac{x}{a}} \tag{2.566}$$

$$f_Y(y) = \frac{1}{b^m \Gamma(m)} y^{m-1} e^{-\frac{y}{b}} \tag{2.567}$$

The expression for density of the sum using the Laplace transform is

$$f(z) = L^{-1}\left[\frac{1}{(as+a)^m (bs+1)^m}\right]. \tag{2.568}$$

For reasons unknown, Matlab does not provide a solution for the density as seen from the Matlab script below. On the other hand, Maple does provide a solution and the density of z is

$$f(z) = \frac{1}{\Gamma(m)} \sqrt{\frac{\pi}{ab}} \left(\frac{z}{|a-b|}\right)^{m-\frac{1}{2}} e^{-\frac{z}{2}\left(\frac{1}{a}+\frac{1}{b}\right)} I_{m-\frac{1}{2}}\left(\frac{1}{2}\frac{|a-b|}{ab} z\right). \tag{2.569}$$

Note that the density in Eq. (2.569) is known as the McKay density and this density and its properties are discussed in Chap. 5.

```
% laplace examples % July 2016 P M Shankar
% four cases; two IID exponential variables, two IID gamma variables, two
% independent gamma variables of different orders and identical scaling
% factors, two independent gamma variables of identical orders and
% different scaling factors
```

2.19.1.1 Two Exponential Variables

```
clear;clc
syms a x y z s positive
fx=(1/a)*exp(-x/a);
fy=(1/a)*exp(-y/a);
fxL=laplace(fx,x,s);
fyL=laplace(fy,y,s);
disp(['Laplace transform of Z=X+Y-->',char(fxL*fyL)])
fzL=ilaplace(fxL*fyL,s,z);
% verify using convolution
fy=subs(fx,x,y); % create fy using substitution
fyz=subs(fy,y,z-x);% created the shifted density
fzdirect=int(fx*fyz,x,0,z);
fzdirect=simplify(fzdirect,'steps',100);
% verify that the pdf of Z is correct
pr=simplify(int(fzL,z,0,inf));
disp([' Verification: Integral of f(z)-->',num2str(double(pr))])
disp(['density of Z=X+Y directly-->',char(fzdirect)])
disp(['density of Z=X+Y using Laplace-->',char(fzL)])
```

```
Laplace transform of Z=X+Y-->1/(a^2*(s + 1/a)^2)
 Verification: Integral of f(z)-->1
density of Z=X+Y directly-->(z*exp(-z/a))/a^2
density of Z=X+Y using Laplace-->(z*exp(-z/a))/a^2
```

2.19.1.2 Two Identical Gamma Variable: *G*(*m*,*a*) and *G*(*m*,*a*)

```
syms a x y m z s positive
clear fx fy
fx=(1/a)^m*exp(-x/a)*x^(m-1)/gamma(m);
fy=(1/a)^m*exp(-y/a)*y^(m-1)/gamma(m);
fxL=laplace(fx,x,s);
fyL=laplace(fy,y,s);
disp(['Laplace transform of Z=X+Y-->',char(fxL*fyL)])
fz=ilaplace(fxL*fyL,s,z); %pdf using Laplace
disp(['density of Z=X+Y using Laplace-->',char(fz)])
disp('verify that the density obtained using Laplace transform is valid')
disp(int(fz,z,0,inf))
```

```
Laplace transform of Z=X+Y-->1/(a^(2*m)*(s + 1/a)^(2*m))
density of Z=X+Y using Laplace-->(z^(2*m - 1)*exp(-z/a))/(a^(2*m)*gamma(2*m))
verify that the density obtained using Laplace transform is valid
1
```

2.19.1.3 Two Non-Identical Gamma Variables (Different Orders): $G(m,a)$ and $G(n,a)$.

```
syms a b x y m n z s positive
clear fx fy
fx=(1/a)^m*exp(-x/a)*x^(m-1)/gamma(m);
fy=(1/a)^n*exp(-y/a)*y^(n-1)/gamma(n);
fxL=laplace(fx,x,s);
fyL=laplace(fy,y,s);
disp(['Laplace transform of Z=X+Y-->',char(fxL*fyL)])
fz=ilaplace(fxL*fyL,s,z); %pdf using Laplace
disp(['density of Z=X+Y using Laplace-->',char(fz)])
disp('verify that the density is valid')
disp(int(fz,z,0,inf))
```

```
Laplace transform of Z=X+Y-->1/(a^m*a^n*(s + 1/a)^m*(s + 1/a)^n)
density of Z=X+Y using Laplace-->(z^(m + n - 1)*exp(-z/a))/(a^m*a^n*gamma(m + n))
verify that the density is valid
1
```

2.19.1.4 Two Non-Identical Gamma Variables (Identical Order): $G(m,a)$ and $G(m,b)$.

```
syms a b x y m  z s positive
clear fx fy
fx=(1/a)^m*exp(-x/a)*x^(m-1)/gamma(m);
fy=(1/b)^m*exp(-y/b)*y^(m-1)/gamma(m);
fxL=laplace(fx,x,s);
fyL=laplace(fy,y,s);
disp(['Laplace transform of Z=X+Y-->',char(fxL*fyL)])
fz=ilaplace(fxL*fyL,s,z); %pdf using Laplace
disp(['density of Z=X+Y using Laplace-->',char(fz)])
disp('verify that the density is valid')
disp(int(fz,z,0,inf))
```

```
Laplace transform of Z=X+Y-->1/(a^m*b^m*(s + 1/a)^m*(s + 1/b)^m)
density of Z=X+Y using Laplace-->ilaplace(1/((s + 1/a)*(s + 1/b))^m, s, z)/(a^m*b^m)
verify that the density is valid
int(ilaplace(1/((s + 1/a)*(s + 1/b))^m, s, z)/(a^m*b^m), z, 0, Inf)
```

While Matlab had some difficulties in providing solutions to a few of these cases, Maple provides solutions to all of these problems including matching results with Laplace/inverse Laplace and convolution.

The following segment (Maple 16) shows the results of convolution in Eq. (2.565).

- #pdf of $Z = X + Y$, $X = G(m, a)$ and $X = G(n, a)$, X and Y independent
- $fx := \dfrac{\left(\frac{1}{a}\right)^m \times x^{m-1} \times \exp\left(-\frac{1}{a}x\right)}{\Gamma(m)}$; #pdf of X

$$fx := \frac{\left(\frac{1}{a}\right)^m x^{m-1} e^{-\frac{x}{a}}}{\Gamma(m)}$$

- $fy := \dfrac{\left(\frac{1}{a}\right)^n \times y^{n-1} \times \exp\left(-\frac{1}{a}y\right)}{\Gamma(n)}$; #pdf of Y

$$fy := \frac{\left(\frac{1}{a}\right)^n y^{n-1} e^{-\frac{y}{a}}}{\Gamma(n)}$$

- $fyy :=$ subs($y = z - x, fy$); # shifted pdf of Y

$$fyy := \frac{\left(\frac{1}{a}\right)^n (z-x)^{n-1} e^{-\frac{z-x}{a}}}{\Gamma(n)}$$

- $fz := \int_0^z fx \cdot fyy\, dx$; #convolution to get pdf of $Z = X + Y$

$$fz := \int_0^z \frac{\left(\frac{1}{a}\right)^m x^{m-1} e^{-\frac{x}{a}} \left(\frac{1}{a}\right)^n (z-x)^{n-1} e^{-\frac{z-x}{a}}}{\Gamma(m)\Gamma(n)}\, dx$$

- $fzz :=$ simplify(fz, assume = positive);

$$fzz := \frac{z^{n+m-1} a^{-m-n} e^{-\frac{z}{a}}}{\Gamma(m+n)}$$

- with(inttrans):
- $fL :=$ laplace(fzz, z, s); # find the laplace transform of the pdf

$$fL := \frac{a^{-m-n}\mathrm{laplace}\left(z^{n+m-1} e^{-\frac{z}{a}}, z, s\right)}{\Gamma(m+n)}$$

- $fLs :=$ simplify(fL, assume = positive); # Laplace transform of the pdf $X + Y$

$$fLs := (as+1)^{-m-n}$$

- $fIX := \text{simplify}(\text{laplace}(fx, x, s), \text{assume} = \text{positive});$

$$fIX := (as + 1)^{-m}$$

- $fIY := \text{simplify}(\text{laplace}(fy, y, s), \text{assume} = \text{positive});$

$$fIY := (as + 1)^{-n}$$

The segment below shows the results of the convolution to obtain the density of Z in Eq. (2.568).

- #pdf of $Z = X + Y$, $X = G(m, a)$ and $X = G(m, b)$, X and Y independent
- $fx := \dfrac{\left(\frac{1}{a}\right)^m \times x^{m-1} \times \exp\left(-\frac{1}{a}x\right)}{\Gamma(m)}$; #pdf of X

$$fx := \frac{\left(\frac{1}{a}\right)^m x^{m-1} e^{-\frac{x}{a}}}{\Gamma(m)}$$

- $fy := \dfrac{\left(\frac{1}{b}\right)^m \times y^{m-1} \times \exp\left(-\frac{1}{b}y\right)}{\Gamma(m)}$; #pdf of Y

$$fy := \frac{\left(\frac{1}{b}\right)^m y^{m-1} e^{-\frac{y}{b}}}{\Gamma(m)}$$

- $fyy := \text{subs}(y = z - x, fy)$; # shifted pdf of Y

$$fyy := \frac{\left(\frac{1}{b}\right)^m (z - x)^{m-1} e^{-\frac{z-x}{b}}}{\Gamma(m)}$$

- $fz := \int_0^z fx \times fyy\, dx$; #convolution to get pdf of $Z = X + Y$

$$fz := \frac{1}{\Gamma(m)}\left(\left(\frac{1}{a}\right)^m \left(\frac{1}{b}\right)^m z^{2m-1} e^{-\frac{1}{2}\frac{z(a+b)}{ba}} \sqrt{\pi}\left(\frac{(a-b)z}{ba}\right)^{-m+\frac{1}{2}} \text{BesselI}\left(m - \frac{1}{2}, \frac{1}{2}\frac{(a-b)z}{ba}\right)\right)$$

- with(inttrans):
- $fL := \text{laplace}(fz, z, s)$; # find the laplace transform of the pdf

$$fL := \frac{1}{\Gamma(m)}\left(\left(\frac{1}{a}\right)^m \left(\frac{1}{b}\right)^m \sqrt{\pi}\left(\frac{a-b}{ba}\right)^{-m+\frac{1}{2}} \text{laplace}\left(e^{-\frac{1}{2}\left(\frac{1}{b}+\frac{1}{a}\right)z} \text{BesselI}\left(m - \frac{1}{2}, \frac{1}{2}\left(\frac{1}{b} - \frac{1}{a}\right)z\right) z^{m-\frac{1}{2}}, z, s\right)\right)$$

- $fLz := \text{simplify}(fL, \text{assume} = \text{positive});$ # Laplace transform of $X + Y$;

$$fLz := \left(abs^2 + as + bs + 1\right)^{-m}$$

- $fLx := \text{simplify}(\text{laplace}(fx, x, s), \text{assume} = \text{positive});$ # Laplace transform of X

$$fLx := (as + 1)^{-m}$$

- $fLy := \text{simplify}(\text{laplace}(fy, y, s), \text{assume} = \text{positive});$ # Laplace transform of Y

$$fLy := (bs + 1)^{-m}$$

The segment below shows the use of inverse Laplace transform to confirm the results of the pdf (Maple 16).

- restart;
- assume($a > 0$);
- assume($m > 0$);
- assume($n > 0$);
- assume($b > 0$);
- $flz1 := (a\,s + 1)^{-2\,m};$ # Laplace transform of $Z = X + Y$ with X, Y are IID

$$flz1 := (a \sim s + 1)^{-2m\sim}$$

- with(MTM);

[*ElementwiseAnd, ElementwiseNot, ElementwiseOr, Map, Minus, Mod, Zip*, abs, *acos, acosh, acot, acoth, acsc, acsch, array_dims, asec, asech, asin, asinh, atan, atanh, besseli, besselij, besselk, bessely, ccode*, ceil, *char, coeffs, collect, colspace, compose, conj*, cos, cosh, *cosint*, cot, coth, csc, csch, *ctranspose, det, diag, diff, digits, dirac, disp, double, dsolve, eig, end, eq*, erf, exp, *expand, expm, ezcontour, ezcontourf, ezmesh, ezmeshc, ezplot, ezplot3, ezpolar, ezsurf, ezsurfc, factor, findsym, finverse, fix*, floor, *fortran, fourier*, frac, γ, *gcd, ge, gt, heaviside, horner, horzcat*, hypergeom, *ifourier, ilaplace, imag, int, int16, int32, int64, int8, inv, isreal, iztrans, jacobian, jordan, lambertw, laplace, latex, lcm, ldivide, le, limit*, log, log10, *log2, lt, mfun, mldivide, mpower, mrdivide, mtimes, ne, null, numden, numel, plus, poly, poly2sym, power, pretty, procread, prod, quorem, rank, rdivide, real*, round, *rref*, sec, sech, *simple, simplify*, sin, *single*, sinh, *sinint, size, solve, sort*, sqrt, *struct, subs, subsagn, subsref, sum, svd, sym2poly, symsum*, tan, tanh, *taylor, times, transpose, tril, triu, uint16, unit32, unit64, unit8, vertcat, vpa*, ζ, *ztrans*]

- $fz1 = \text{ilaplace}(flz1, s, z);$ # pdf of $Z = X + Y$, $G(m, a)$, $Y : G(m, a)$

$$fz1 = \frac{e^{-\frac{z}{a\sim}} a\sim^{-2m\sim} z^{2m\sim-1}}{\Gamma(2m\sim)}$$

- $flz2 := (a\,s+1)^{-m-n}$; # Laplace transform of $Z=X+Y$ with $X: G(m,a), Y: G(n,a)$

$$flz2 := (a \sim s+1)^{-m\sim -n\sim}$$

- $fz2 = \text{ilaplace}(flz2, s, z)$; # Laplace pdf of $Z=X+Y$ with $X: G(m,a), Y: G(n,a)$

$$fz2 = \frac{\mathrm{e}^{-\frac{z}{a\sim}} a\sim^{-m\sim -n\sim} z^{m\sim +n\sim -1}}{\Gamma(m\sim +n\sim)}$$

- $flz3 := (a\,b\,s^2 + a\,s + b\,s + 1)^{-m}$; # Laplace transform of $Z=X+Y$ with $X: G(m,a), Y: G(m,b)$

$$flz3 := \left(a \sim b \sim s^2 + a \sim s + b \sim s + 1\right)^{-m\sim}$$

- $fz3 = \text{ilaplace}(flz3, s, z)$; # *pdf* of $Z=X+Y$ with $X: G(m,a), Y: G(m,b)$

$$fz3 = \frac{1}{\Gamma(m\sim)}\left(\sqrt{\frac{\pi}{a\sim b\sim}}\left(\frac{z}{|a\sim -b\sim|}\right)^{m\sim -\frac{1}{2}} \mathrm{e}^{-\frac{1}{2}\left(\frac{1}{b\sim}+\frac{1}{a\sim}\right)z} \right.$$
$$\left. \text{BesselI}\left(m\sim -\frac{1}{2}, \frac{1}{2}\text{signum}(a\sim -b\sim)\left(\frac{1}{b\sim}-\frac{1}{a\sim}\right)z\right)\right)$$

These results are summarized in Tables 2.7, 2.8, 2.9, and 2.10.

Even though the results are shown only for the case of sum of two gamma random variables, the concept of Laplace transforms can be extended to a number of random variables (Karagiannidis et al. 2006). These results show that the sum of a number of independent and identically distributed random variables is another gamma random variable as seen in Eq. (2.560). The same property is applicable to a

Table 2.7 Densities of sums of gamma variables (gamma density from Nakagami)

$f_X(x) = x^{m-1}\dfrac{e^{-\frac{mx}{q}}\left(\frac{m}{q}\right)^m}{\Gamma(m)}$	pdf of X
$f_Y(y) = y^{m-1}\dfrac{e^{-\frac{mx}{q}}\left(\frac{m}{q}\right)^m}{\Gamma(m)}$	pdf of Y
$L_X(s) = \dfrac{m^m}{(m+qs)^m}$	Laplace transform of $f_X(X)$
$L_Y(s) = \dfrac{m^m}{(m+qs)^m}$	Laplace transform of $f_Y(Y)$
$L_Z(s) = \dfrac{m^{2m}}{(m+qs)^{2m}}$	Laplace transform of $f_Z(z) = L_X(s) \times L_Y(s)$
$f_Z(z) = m^{2m}z^{2m-1}\dfrac{e^{-\frac{mz}{q}}}{q^{2m}\Gamma(2m)}$	pdf of $Z = X + Y \Rightarrow G\left[2m, \frac{2q}{2m}\right]$

Table 2.8 Densities of sums of gamma variables (gamma density)

$f_X(x) = x^{m-1}\dfrac{e^{-\frac{x}{b}}}{b^m\Gamma(m)}$	pdf of X
$f_Y(y) = y^{m-1}\dfrac{e^{-\frac{y}{b}}}{b^m\Gamma(m)}$	pdf of Y
$L_X(s) = (bs+1)^{-m}$	Laplace transform of $f_X(x)$
$L_Y(s) = (bs+1)^{-m}$	Laplace transform of $f_Y(y)$
$L_Z(s) = (bs+1)^{-2m}$	Laplace transform of $f_Z(z) = L_X(s) \times L_Y(s)$
$f_Z(z)z^{2m-1}\dfrac{e^{-\frac{z}{b}}}{b^{2m}\Gamma(2m)}$	pdf of $Z = X + Y \Rightarrow G[2m, b]$

Table 2.9 Densities of sums of gamma variables (non-identical orders)

$f_X(x) = x^{m-1}\dfrac{e^{-\frac{x}{a}}}{a^m\Gamma(m)}$	pdf of X
$f_Y(y) = y^{n-1}\dfrac{e^{-\frac{y}{a}}}{a^n\Gamma(n)}$	pdf of Y
$L_X(s) = (as+1)^{-m}$	Laplace transform of $f_X(x)$
$L_Y(s) = (as+1)^{-n}$	Laplace transform of $f_Y(y)$
$L_Z(s) = (as+1)^{-m-n}$	Laplace transform of $f_Z(z) = L_X(s) \times L_Y(s)$
$f_Z(z) = z^{m+n-1}\dfrac{e^{-\frac{z}{a}}}{a^{m+n}\Gamma(m+n)}$	pdf of $Z = X + Y \Rightarrow G[a + c, b]$

Table 2.10 Densities of sums of gamma variables (identical orders)

$f_X(x) = x^{m-1}\dfrac{e^{-\frac{x}{a}}}{a^m\Gamma(m)}$	pdf of X
$f_Y(y) = y^{m-1}\dfrac{e^{-\frac{y}{b}}}{b^m\Gamma(m)}$	pdf of Y
$L_X(s) = (a\,s+1)^{-m}$	Laplace transform of $f_X(x)$
$L_Y(s) = (b\,s+1)^{-m}$	Laplace transform of $f_Y(y)$
$L_Z(s) = ((a\,s+1)(b\,s+1))^{-m}$	Laplace transform of $f_Z(z) = L_X(s) \times L_Y(s)$
$f_Z(z) = z^{m-\frac{1}{2}}\sqrt{\pi}e^{-\frac{z\left(\frac{a}{2}+\frac{b}{2}\right)}{ab}}\lvert a-b\rvert^{\frac{1}{2}-m}\dfrac{I_{m-\frac{1}{2}}\left(\frac{(z\lvert a-b\rvert)}{(2ab)}\right)}{\sqrt{a}\sqrt{b}\Gamma(m)}$	

set of gamma variables of non-identical orders having identical scaling factors as seen in Eq. (2.565).

The example of the sum of two random variables in Eq. (2.568) with identical orders having different means cannot be easily extended to multiple variables. While Laplace transforms play a crucial role in obtaining the density of sum of independent random variables, Mellin transforms are useful in obtaining the density of the products of random variables.

2.19.2 Mellin Transforms and Products of Random Variables

As mentioned earlier, Mellin transforms allow the derivation of the density of products or ratios of Rayleigh, gamma, gamma variables, or any other random variable which exists for positive values (Epstein 1948; Lomnicki 1967; Subrahmanaim 1970; Park 1986; Abate 1995; Ahmed et al. 2011). While the conceptual approach is relatively simple, the practical implementation of Mellin transforms for obtaining the densities is less straightforward because of the limited availability and access to information on Mellin transforms in books. Another limiting factor is the limited options in Matlab even though (as stated earlier), Mellin transforms can be found by using the integral directly. Finding the inverse Mellin is still a daunting task. Before examining examples of transformation, it is appropriate to look at the properties of Mellin transforms and the Table of Mellin transforms (Epstein 1948; Erdelyi 1953).

The properties of Mellon transforms are shown in Table 2.11.

Mellin transforms of common densities are given in Tables 2.12 and 2.13. One of the interesting aspects of Mellin transform is the fact that the transform integral in

Table 2.11 Mellin transforms

$f(x)$	$M_f(s)=\int_0^{\infty}x^{s-1}f(x)dx$
$x^v f(x)$	$M_f(s+v)$
$f(bx)$	$b^{-s}M_f(s),\quad b>0$
$f(x^{\rho})$	$\frac{1}{\rho}M_f\left(\frac{1}{\rho}\right),\quad \rho>0$
$\frac{d}{dx}f(x)$	$-(s-1)M_f(s-1)$
$\int_0^x f(w)dw$	$\frac{-1}{s}M_f(s+1)$
$Af(\alpha x)+Bh(\beta x)$	$A\alpha^{-s}M_f(s)+B\beta^{-s}M_h(s)$

Table 2.12 Mellin transforms of probability densities

Probability density function	Mellin transform
$x^{c-1}\frac{e^{-\frac{x}{b}}}{b^c\Gamma(c)}$	$\frac{b^{s-1}\Gamma(c+s-1)}{\Gamma(c)}$
$x^{m-1}\frac{e^{-\frac{mx}{b}}\left(\frac{m}{b}\right)^m}{\Gamma(m)}$	$\frac{b^{s-1}m^{1-s}\Gamma(m+s-1)}{\Gamma(m)}$
$\frac{e^{-\frac{x}{b}}}{b}$	$b^{s-1}\Gamma(s)$
e^{-x}	$\Gamma(s)$
$\frac{ax^{ac-1}e^{-\left(\frac{x}{b}\right)^a}}{b^{ac}\Gamma(c)}$	$\frac{b^{s-1}\Gamma\left(\frac{s+ac-1}{a}\right)}{\Gamma(c)}$
$\frac{xe^{-\frac{x^2}{2a^2}}}{a^2}$	$2^{\frac{s}{2}-\frac{1}{2}}a^{s-1}\Gamma\left(\frac{s}{2}+\frac{1}{2}\right)$

Table 2.13 Mellin transforms of probability densities

Probability density function	Mellin transform
$\frac{2x^{c-1}K_0\left(\frac{(2\sqrt{x})}{b}\right)}{b^{2c}\Gamma(c)^2}$	$\frac{b^{2s-2}\Gamma(c+s-1)^2}{\Gamma(c)^2}$
$2x^{\frac{a}{2}+\frac{b}{2}-1}\frac{K_{a-b}\left(\frac{(2\sqrt{x})}{\sqrt{AB}}\right)}{\Gamma(a)\Gamma(b)(AB)^{\frac{a}{2}+\frac{b}{2}}}$	$\frac{A^{s-1}B^{s-1}\Gamma(a+s-1)\Gamma(b+s-1)}{\Gamma(a)\Gamma(b)}$
$\frac{zK_0\left(\left(\sqrt{\frac{1}{a^2}}\sqrt{\frac{z^2}{b^2}}\right)\right)}{a^2b^2}$	$2^{s-1}a^{s-1}b^{s-1}\Gamma\left(\frac{s}{2}+\frac{1}{2}\right)^2$
$\frac{2K_0\left(\left(2\sqrt{\frac{z}{ab}}\right)\right)}{ab}$	$a^{s-1}b^{s-1}\Gamma(s)^2$

Table 2.14 Mellin transforms of probability densities expressed in terms of Meijer G-functions

Probability density function	Mellin transform
$\frac{1}{x\Gamma(m)}G_{0,1}^{1,0}\left(\frac{mx}{Z_0}\middle\vert\begin{matrix}-\\ m\end{matrix}\right)$	$\frac{m^{1-s}Z_0^{s-1}\Gamma(m-1+s)}{\Gamma(m)}$
$\frac{1}{x\Gamma^2(m)}G_{0,2}^{2,0}\left(\frac{m^2x}{Z_0}\middle\vert\begin{matrix}-\\ m,m\end{matrix}\right)$	$\frac{m^{2-2s}Z_0^{s-1}\Gamma^2(m-1+s)}{\Gamma^2(m)}$
$\frac{1}{x\Gamma^3(m)}G_{0,3}^{3,0}\left(\frac{m^3x}{Z_0}\middle\vert\begin{matrix}-\\ m,m,m\end{matrix}\right)$	$\left[\frac{m^{1-s}Z_0^{s-1}\Gamma(m-1+s)}{\Gamma(m)}\right]^3$
$\frac{1}{x\Gamma^N(m)}G_{0,N}^{N,0}\left(\frac{m^Nx}{Z_0}\middle\vert\begin{matrix}-\\ \underbrace{m,m,\ldots,m}_{N-terms}\end{matrix}\right)$	$\left[\frac{m^{1-s}Z_0^{s-1}\Gamma(m-1+s)}{\Gamma(m)}\right]^N$
$\frac{1}{x\Gamma(m)\Gamma(n)}G_{0,2}^{2,0}\left(\frac{mnx}{ab}\middle\vert\begin{matrix}-\\ m,n\end{matrix}\right)$	$\frac{(mn)^{1-s}(ab)^{s-1}\Gamma(m-1+s)\Gamma(n-1+s)}{\Gamma(m)\Gamma(n)}$
$\frac{1}{\sqrt{2\pi}\ ab}G_{0,3}^{3,0}\left(\frac{x^2}{8a^2b^2}\middle\vert\begin{matrix}-\\ 0,\frac{1}{2},\frac{1}{2}\end{matrix}\right)$	$\sqrt{2^{s-1}}(ab)^{s-1}\Gamma(s)\Gamma\left(\frac{s+1}{2}\right)$

Eq. (2.539) appears to be similar to the integral representation of a gamma function if $f(x)$ is exponential. In that case, it can be easily concluded that the Mellin transforms are likely to be expressed in terms of gamma functions. Since this section is devoted to a discussion of probability densities and transformations, it is necessary to look at the transforms of some of the common and not-so-common density functions that are likely to be observed in the study of fading and shadowing in wireless channels. Three separate tables are given below. Table 2.12 contains the Mellin transforms of some of the simple densities while Table 2.13 contains the transforms of densities that are observed in doubly scattering fading channels. Table 2.14 lists the Mellin transforms of densities expressed in terms of Meijer G-functions.

The procedure used to create these tables is described next. One of the most important properties utilized pertaining to the convolution in the Mellin domain in

Eqs. (2.547) and (2.553) was established using combinations of Rayleigh, exponential, and gamma densities. The Maple script (with annotation) is reproduced below. The Mellin transforms were obtained using the integral form in Eq. (2.539) and compared to the result obtained directly using the *mellin*(.) command in Maple. The density functions of the products and ratios were obtained using the concepts of transformation of random variables with the appropriate integrals evaluated in Maple, with the probability densities verified every time by checking that the total probability is unity.

Example # 1 Product of two independent and identically distributed gamma variables

- #pdf of $Z = XY$, X and Y are IID gamma distributed
- $fx := \frac{x^{c-1}\exp\left(-\frac{x}{b}\right)}{b^c \times \Gamma(c)}$; #input Gamma density fx

$$fx := \frac{x^{c-1}e^{-\frac{x}{b}}}{b^c\Gamma(c)}$$

- $fyz := \mathrm{subs}\left(x = \frac{z}{x}, fx\right)$; #input density fy following substitution

$$fyz := \frac{\left(\frac{z}{x}\right)^{c-1}e^{-\frac{z}{xb}}}{b^c\Gamma(c)}$$

- $\mathrm{int}G := \left(\frac{1}{x}\right) \times fx \times fyz$; #Create integrand for the Transformation

$$\mathrm{int}G := x^{c-1}e^{-\frac{x}{b}}\left(\frac{z}{x}\right)^{c-1}\frac{e^{-\frac{z}{xb}}}{x(b^c)^2\Gamma(c)}$$

- $fz := \mathrm{simplify}\left(\int_0^\infty \mathrm{int}G\, dx, \mathrm{assume} = \mathrm{positive}\right)$; #pdf of $Z = XY$

$$fz := \frac{2b^{-2c}z^{c-1}\mathrm{BesselK}\left(0, \frac{2\sqrt{z}}{b}\right)}{\Gamma(c)^2}$$

- $\mathrm{simplify}\left(\int_0^\infty fz\, dz, \mathrm{assume} = \mathrm{positive}\right)$; #verify the pdf

$$1$$

- with(inttrans):

- $Mxs := \text{simplify}(\text{mellin}(fx, x, s), \text{assume} = \text{positive});$
  #Mellin transform of fx using builtin command

$$Mxs := \frac{b^{s-1}\Gamma(s+c-1)}{\Gamma(c)}$$

- $Mxs1 := \text{simplify}\left(\int_0^\infty fx \times x^{s-1}dx, \text{assume} = \text{positive}\right);$
  #Mellin transform of fx using the integral representation

$$Mxs1 := \frac{b^{s-1}\Gamma(s+c-1)}{\Gamma(c)}$$

- $Mxz := \text{simplify}(\text{mellin}(fz, z, s), \text{assume} = \text{positive});$
  #Mellin transform of fz using the builtin command

$$Mxz := \frac{b^{2s-2}\Gamma(s+c-1)^2}{\Gamma(c)^2}$$

- $Mxz1 := \text{simplify}\left(\int_0^\infty fz \times z^{s-1}dz, \text{assume} = \text{positive}\right);$
  #Mellin transform of fz using the integral representation

$$Mxz1 := \frac{b^{2s-2}\Gamma(s+c-1)^2}{\Gamma(c)^2}$$

- #Mellin transform of $Z = XY$ is the product of Mellin transforms of X
  and Y

The results show that the Mellin transform of the density of the product of two independent and identically distributed gamma variables is the square of the Mellin transform of the marginal density as seen in Eq. (2.547).

Example # 2 Product of two independent non-identical gamma variables

- #pdf of $Z = XY$, X and Y are non - identical gamma distributions
- $fx := \dfrac{x^{c-1}\exp\left(-\frac{x}{b}\right)}{b^c \times \Gamma(c)}$; #input Gamma density fx

$$fx := \frac{x^{c-1}e^{-\frac{x}{b}}}{b^c\Gamma(c)}$$

- $fy := \dfrac{y^{m-1}\exp\left(-\frac{y}{q}\right)}{q^m \times \Gamma(m)}$; #input Gamma density *fy*

$$fy := \frac{y^{m-1}e^{-\frac{y}{q}}}{q^m\Gamma(m)}$$

- $fyz := \text{subs}\left(y = \frac{z}{x}, fy\right)$; #input density *fy* following substitution

$$fyz := \frac{\left(\frac{z}{x}\right)^{m-1}e^{-\frac{z}{xq}}}{q^m\Gamma(m)}$$

- $\text{int}G := \left(\frac{1}{x}\right) \times fx \times fyz$; #Create integrand for the Transformation

$$\text{int}G := x^{c-1}\frac{e^{-\frac{x}{b}}\left(\frac{z}{x}\right)^{m-1}e^{-\frac{z}{xq}}}{b^c\Gamma(c)q^m\Gamma(m)x}$$

- $fz := \text{simplify}\left(\int_0^\infty \text{int}G\,dx, \text{assume} = \text{positive}\right)$; #pdf of $Z = XY$

$$fz := \frac{1}{\Gamma(c)\Gamma(m)}\left(2z^{\frac{1}{2}c-1+\frac{1}{2}m}q^{-\frac{1}{2}c-\frac{1}{2}m}b^{-\frac{1}{2}c-\frac{1}{2}m}\,\text{Bessel}K\left(c-m, \frac{2\sqrt{z}}{\sqrt{b}\sqrt{q}}\right)\right)$$

- $\text{simplify}\left(\int_0^\infty fz\,dz, \text{assume} = \text{positive}\right)$; #verify the pdf

$$1$$

- with(inttrans):
- $Mxs := \text{simplify}(\text{mellin}(fx, x, s), \text{assume} = \text{positive})$;
  #Mellin transform of *fx* using the builtin command

$$Mxs := \frac{b^{s-1}\Gamma(s+c-1)}{\Gamma(c)}$$

- $Mxs := \text{simplify}(\text{mellin}(fy, y, s), \text{assume} = \text{positive})$;
  #Mellin transform of *fy* using builtin command

$$Mxs := \frac{q^{s-1}\Gamma(s+m-1)}{\Gamma(m)}$$

- $Mxs1 := \text{simplify}\left(\int_0^\infty fx \times x^{s-1}dx, \text{assume} = \text{positive}\right);$
  #Mellin transform of fx using the integral representation

$$Mxs1 := \frac{b^{s-1}\Gamma(s+c-1)}{\Gamma(c)}$$

- $Mys1 := \text{simplify}\left(\int_0^\infty fy \times y^{s-1}dy, \text{assume} = \text{positive}\right);$
  #Mellin transform of fy using the integral representation

$$Mys1 := \frac{q^{s-1}\Gamma(s+m-1)}{\Gamma(m)}$$

- $Mxz := \text{simplify}(\text{mellin}(fz, z, s), \text{assume} = \text{positive});$
  #Mellin transform of fz using builtin command

$$Mxz := \frac{q^{s-1}b^{s-1}\Gamma(s+c-1)\Gamma(s+m-1)}{\Gamma(c)\Gamma(m)}$$

- $Mzs1 := \text{simplify}\left(\int_0^\infty fz \times z^{s-1}\text{d}z, \text{assume} = \text{positive}\right);$
  #Mellin transform of fz using the integral representation

$$Mzs1 := \frac{q^{s-1}b^{s-1}\Gamma(s+c-1)\Gamma(s+m-1)}{\Gamma(c)\Gamma(m)}$$

- #Mellin transform of $Z = XY$ is the product of Mellin transforms of X and Y

The results show that the Mellin transform of the density of the product of two independent distributed gamma variables is the product of the Mellin transform of the marginal densities as seen in Eq. (2.547).

The next example shows the extension to the case of the product of three identically distributed (and independent) gamma variables. The pdf of the product of two is obtained from first and then the product of the three.

Example # 3 Product of three independent identical gamma variables

- #pdf of $W = X1 \,.\, X2 \,.\, X3 \ldots$. Products of IID gamma
- $fx1 := \frac{x1^{m-1} \times \exp\left(-\frac{x1}{b}\right)}{b^m \times \Gamma(m)}$; #input Gamma density $fx1$

$$fx1 := \frac{x1^{m-1}e^{-\frac{x1}{b}}}{b^m\Gamma(m)}$$

- #$Z = X1 \,.\, X2$ # products of two

- $fx2z := subs\left(x1 = \frac{z}{x1}, fx1\right)$; #input density $fx2$ following substitution

$$fx2z := \left(\frac{z}{x1}\right)^{m-1} \frac{e^{-\frac{z}{x1b}}}{b^m\Gamma(m)}$$

- $\text{int}G1 := \left(\frac{1}{x1}\right) \times fx1 \times fx2z$; #Create integrand for the Transformation

$$\text{int}G1 := x1^{m-1}e^{-\frac{x1}{b}}\left(\frac{z}{x1}\right)^{m-1} \frac{e^{-\frac{z}{x1b}}}{x1(b^m)^2\Gamma(m)^2}$$

- $fz := \text{simplify}\left(\int_0^\infty \text{int}G1 dx1, \text{assume} = \text{positive}\right)$; #pdf of $Z = X1 \times X2$

$$fz := \frac{2b^{-2m}z^{m-1}\text{Bessel}K\left(0, \frac{2\sqrt{z}}{b}\right)}{\Gamma(m)^2}$$

- $\text{simplify}\left(\int_0^\infty fz\, dz, \text{assume} = \text{positive}\right)$; #verify the pdf

$$1$$

- $\#W = X1 \,.\, X2 \,.\, X3 = Z \times X3$
- $fx3z := \text{subs}\left(x1 = \frac{w}{z}, fx1\right)$; #density of $X3$ following substitution

$$fx3z := \frac{\left(\frac{w}{z}\right)^{m-1} e^{-\frac{w}{zb}}}{b^m\Gamma(m)}$$

- $\text{int}G2 := \left(\frac{1}{z}\right) \times fz \times fx3z$; #Create integrand for the Transformation

$$\text{int}G2 : \frac{2b^{-2\,m}z^{m-1}\text{Bessel}K\left(0, \frac{2\sqrt{z}}{b}\right)\left(\frac{w}{z}\right)^{m-1} e^{-\frac{w}{zb}}}{z\Gamma(m)^3 b^m}$$

- $fw := \text{simplify}\left(\int_0^\infty \text{int}G2\, dz, \text{assume} = \text{positive}\right)$; #pdf of $W = X1 \times X2 \times X3$

$$fw : \frac{b^{3-3\,m}w^{-2+m}\text{MeijerG}\left([[], []], [[1,1,1], []], \frac{w}{b^3}\right)}{\Gamma(m)^3}$$

- $\text{simplify}\left(\int_0^\infty fw\, dw, \text{assume} = \text{positive}\right)$; #verify the pdf

$$1$$

- with(inttrans):

- $Mxs := \text{simplify}(\text{mellin}(fx1, x1, s), \text{assume} = \text{positive});$
  #Mellin transform of $fx1$ using builtin command

$$Mxs := \frac{b^{s-1}\Gamma(s+m-1)}{\Gamma(m)}$$

- $Mxs1 := \text{simplify}\left(\int_0^\infty fx1 \times x1^{s-1}dx1, \text{assume} = \text{positive}\right);$
  #Mellin transform of fx using the integral representation

$$Mxs1 := \frac{b^{s-1}\Gamma(s+m-1)}{\Gamma(m)}$$

- $Mws := \text{simplify}(\text{mellin}(fw, w, s), \text{assume} = \text{positive});$
  #Mellin transform of fz using builtin command

$$Mws := \frac{b^{-3+3s}\Gamma(s+m-1)^3}{\Gamma(m)^3}$$

- #Mellin transform of $W = X1.X2.X3$ is the product of Mellin transform of individual ones.In this case power of 3
- #In general Mellin transform of the pdf of the product of N IID variables is the Mellin transform of a single variable raised to the power of N

The results show that the Mellin transform of the density of the product independent gamma variables is the product of the Mellin transform of the marginal densities as seen in Eq. (2.547). One can see the benefit of the use of Mellin transforms in obtaining the densities of products of random variables that exist only for positive values provided the Mellin transform is available. This is shown in the next example which displays the relationship between the Mellin transform of the densities expressed in Meijer G-functions and the Mellin transforms of the marginal densities.

Example # 4 Mellin transforms of densities expressed as Meijer G-functions

- #Mellin transform of densities expressed as Meijer$G$ functions:
- #each marginal density is gamma arising from the amplitude being in the Nakagami formal
- $fx4 := \dfrac{\text{MeijerG}\left([[],[]],[[m,m,m,m],[]],\dfrac{m^4x}{Z}\right)}{x \times (\Gamma(m))^4};$
  #density of quadruple gamma prod

$$fx4 := \frac{\text{MeijerG}\left([[],[]],[[m,m,m,m],[]],\frac{m^4 x}{Z}\right)}{x\Gamma(m)^4}$$

- $\text{simplify}\left(\int_0^{\infty} fx4\, dx, \text{assume} = \text{positive}\right);$

$$1$$

- $Mz4 := \text{simplify}\left(\int_0^{\infty} fx4 \times x^{s-1} dx, \text{assume} = \text{positive}\right);$
  #Mellin transform

$$Mz4 := \frac{m^{4-4s} Z^{s-1} \Gamma(m-1+s)^4}{\Gamma(m)^4}$$

- with(inttrans):
- $Mz41 := \text{simplify}(\text{mellin}(fx4, x, s), \text{assume} = \text{positive});$
  #Mellin transform of *fx*4 using builtin command

$$Mz41 := \frac{m^{4-4s} Z^{s-1} \Gamma(m-1+s)^4}{\Gamma(m)^4}$$

- $fx3 := \frac{\text{MeijerG}\left([[],[]],[[m,m,m],[]],\frac{m^3 x}{Z}\right)}{x \times (\Gamma(m))^3};$
  #density of triple gamma prod

$$fx3 := \frac{\text{MeijerG}\left([[],[]],[[m,m,m],[]],\frac{m^3 x}{Z}\right)}{x\Gamma(m)^3}$$

- $\text{simplify}\left(\int_0^{\infty} fx3\, dx, \text{assume} = \text{positive}\right);$

$$1$$

- $Mz3 := \text{simplify}\left(\int_0^{\infty} fx3 \times x^{s-1} dx, \text{assume} = \text{positive}\right);$

$$Mz3 := \frac{Z^{s-1} m^{3-3s} \Gamma(m-1+s)^3}{\Gamma(m)^3}$$

- $Mz31 := \text{simplify}(\text{mellin}(fx3, x, s), \text{assume} = \text{positive});$
  #Mellin transform of *fx*3 using builtin command

$$Mz31 := \frac{Z^{s-1} m^{3-3s} \Gamma(m-1+s)^3}{\Gamma(m)^3}$$

- $fx2 := \dfrac{\text{MeijerG}\left([[],[]],[[m,m],[]],\dfrac{m^2 x}{Z}\right)}{x \times (\Gamma(m))^2};$
  #density of double gamma prod

$$fx2 := \frac{\text{MeijerG}\left([[],[]],[[m,m],[]],\frac{m^2 x}{Z}\right)}{x\Gamma(m)^2}$$

- $\text{simplify}\left(\int_0^\infty fx2\, dx, \text{assume} = \text{positive}\right);$

$$1$$

- $Mz2 := \text{simplify}\left(\int_0^\infty fx2 \times x^{s-1}\, dx, \text{assume} = \text{positive}\right);$

$$Mz2 := \frac{m^{2-2s} Z^{s-1} \Gamma(m-1+s)^2}{\Gamma(m)^2}$$

- $Mz21 := \text{simplify}(\text{mellin}(fx2, x, s), \text{assume} = \text{positive});$
  #Mellin transform of $fx2$ using builtin command

$$Mz21 := \frac{m^{2-2s} Z^{s-1} \Gamma(m-1+s)^2}{\Gamma(m)^2}$$

- $fx1 := \dfrac{\text{MeijerG}\left([[],[]],[[m],[]],\dfrac{m \times x}{Z}\right)}{x \times (\Gamma(m))};$
  #density of single gamma

$$fx1 := \frac{\text{MeijerG}\left([[],[]],[[m],[]],\frac{mx}{Z}\right)}{x\Gamma(m)}$$

- $\text{simplify}\left(\int_0^\infty fx1\, dx, \text{assume} = \text{positive}\right);$

$$1$$

- $Mz1 := \text{simplify}\left(\int_0^\infty fx1 \times x^{s-1}\, dx, \text{assume} = \text{positive}\right);$

$$Mz1 := \frac{m^{1-s} Z^{s-1} \Gamma(m-1+s)}{\Gamma(m)}$$

- $Mz11 := \text{simplify}(\text{mellin}(fx1, x, s), \text{assume} = \text{positive});$
  #Mellin transform of $fx1$ using builtin command

$$Mz11 := \frac{m^{1-s}Z^{s-1}\Gamma(m-1+s)}{\Gamma(m)}$$

- #Mellin transform of the pdf of the product of gamma random variables is the product of individual Mellin transforms (identical)
- #now examine the pdf of the product of non - identical gamma variables
- $fz := \frac{\text{MeijerG}\left([[],[]],[[m,n],[]],\frac{m \times n \times z}{a \times b}\right)}{z \times (\Gamma(m)) \times \Gamma(n)};$
  #Z is the product of two independent non-identical gamma variables

$$fz := \frac{\text{MeijerG}\left([[],[]],[[n,m],[]],\frac{mnz}{ab}\right)}{z\Gamma(m)\Gamma(n)}$$

- $\text{simplify}\left(\int_0^\infty fz\,dz, \text{assume} = \text{positive}\right);$ #verify the pdf

$$1$$

- $MZ := \text{simplify}\left(\int_0^\infty fz \times z^{s-1}\,dz, \text{assume} = \text{positive}\right);$
  #Mellin transform of fz using integral

$$fnxy := \frac{m^{1-s}a^{s-1}b^{s-1}n^{1-s}\Gamma(m-1+s)\Gamma(n-1+s)}{\Gamma(m)\Gamma(n)}$$

- $MZ1 := \text{simplify}(\text{mellin}(fz, z, s), \text{assume} = \text{positive});$
  #Mellin transform of fz using builtin command

$$MZ1 := \frac{m^{1-s}a^{s-1}b^{s-1}n^{1-s}\Gamma(m-1+s)\Gamma(n-1+s)}{\Gamma(m)\Gamma(n)}$$

- $fw := \frac{\text{MeijerG}\left([[],[]],[[m,n,q],[]],\frac{m \times n \times q \times w}{a \times b \times c}\right)}{w \times (\Gamma(m)) \times \Gamma(n) \times \Gamma(q)};$
  #W is the product of three independent non-identical gamma variables

$$fw := \frac{\text{MeijerG}\left([[],[]],[[q,n,m],[]],\frac{mnqw}{abc}\right)}{w\Gamma(m)\Gamma(n)\Gamma(q)}$$

- simplify$\left(\int_0^\infty fw\,dw, \text{assume} = \text{positive}\right)$; #verify the pdf

$$1$$

- $MW := \text{simplify}\left(\int_0^\infty fw \times w^{s-1}\,dw, \text{assume} = \text{positive}\right)$;
  #Mellin transform of *fw* using integral

$$MW := \frac{1}{\Gamma(m)\Gamma(n)\Gamma(q)}\left(m^{1-s}n^{1-s}q^{1-s}a^{s-1}b^{s-1}c^{s-1}\Gamma(m-1+s)\Gamma(n-1+s)\Gamma(q-1+s)\right)$$

- $MW1 := \text{simplify}(\text{mellin}(fw, w, s), \text{assume} = \text{positive})$;
  #Mellin transform of *fw* using builtin command

$$MW1 := \frac{1}{\Gamma(m)\Gamma(n)\Gamma(q)}\left(m^{1-s}n^{1-s}q^{1-s}a^{s-1}b^{s-1}c^{s-1}\Gamma(m-1+s)\Gamma(n-1+s)\Gamma(q-1+s)\right)$$

The next two examples extend the results to products of identical Rayleigh variables and product of Rayleigh and exponential variables expanding the possibilities.

Example # 5 Mellin transform of the density of the product of two Rayleigh variables

- #pdf of $Z = XY$, X and Y are non - identical Rayleigh
- $fx := \dfrac{x \times \exp\left(-\frac{x^2}{2\times a^2}\right)}{a^2}$; #input Rayleigh density of X

$$fx := \frac{xe^{-\frac{1x^2}{2a^2}}}{a^2}$$

- $fy := \dfrac{y \times \exp\left(\frac{-y^2}{2\times b^2}\right)}{b^2}$; #input Rayleigh density of Y

$$fy := \frac{ye^{-\frac{1y^2}{2b^2}}}{b^2}$$

- $fyz := subs\left(y = \frac{z}{x}, fy\right)$; #input density *fy* following substitution

$$fyz := \frac{ze^{-\frac{1z^2}{2b^2}}}{xb^2}$$

- $\text{int}G := \left(\frac{1}{x}\right) \times fx \times fyz$, #Create integrand for the Transformation

$$\text{int}G := e^{-\frac{1}{2}\frac{x^2}{a^2}} \frac{ze^{-\frac{1}{2}\frac{z^2}{x^2 b^2}}}{xa^2b^2}$$

- $fz := \text{simplify}\left(\int_0^\infty \text{int}G\, dx, \text{assume} = \text{positive}\right)$; #pdf of $Z = XY$

$$fz := \frac{z\text{BesslK}\left(0, \frac{z}{ab}\right)}{b^2a^2}$$

- $\text{simplify}\left(\int_0^\infty fz\, dz, \text{assume} = \text{positive}\right)$; #verifyf the pdf

$$1$$

- with(inttrans):
- $Mxs := \text{simplify}(\text{mellin}(fx, x, s), \text{assume} = \text{positive})$;
  #Mellin transform of fx using builtin command

$$Mxs := 2^{\frac{1}{2}s-\frac{1}{2}} a^{s-1} \Gamma\left(\tfrac{1}{2}s+\tfrac{1}{2}\right)$$

- $Mxs := \text{simplify}(\text{mellin}(fy, y, s), \text{assume} = \text{positive})$;
  #Mellin transform of fy using builtin command

$$Mxs := 2^{\frac{1}{2}s-\frac{1}{2}} b^{s-1} \Gamma\left(\tfrac{1}{2}s+\tfrac{1}{2}\right)$$

- $Mxs1 := \text{simplify}\left(\int_0^\infty fx \times x^{s-1} dx, \text{assume} = \text{positive}\right)$;
  #Mellin transform of fx using the integral representation

$$Mxs1 := 2^{\frac{1}{2}s-\frac{1}{2}} a^{s-1} \Gamma\left(\tfrac{1}{2}s+\tfrac{1}{2}\right)$$

- $Mxs1 := \text{simplify}\left(\int_0^\infty fy \times y^{s-1} dy, \text{assume} = \text{positive}\right)$;
  #Mellin transform of fy using the integral representation

$$Mxs1 := 2^{\frac{1}{2}s-\frac{1}{2}} b^{s-1} \Gamma\left(\tfrac{1}{2}s+\tfrac{1}{2}\right)$$

- $Mxz := \text{simplify}(\text{mellin}(fz, z, s), \text{assume} = \text{positive})$;
  #Mellin transform of fz using builtin command

$$Mxz := a^{s-1} b^{s-1} 2^{s-1} \Gamma\left(\frac{1}{2}s + \frac{1}{2}\right)^2$$

- $Mzs1 := \text{simplify}\left(\int_0^\infty fz \times z^{s-1} dz, \text{assume} = \text{positive}\right)$;
  #Mellin transform of fz using the integral representation

$$Mxs1 := a^{s-1}b^{s-1}2^{s-1}\Gamma\left(\frac{1}{2}s+\frac{1}{2}\right)^2$$

- #Mellin transform of $Z = XY$ is the product of Mellin transforms of X and Y

Example # 6 Mellin transform of the density of the product of a Rayleigh and an exponential variables

- #pdf of $Z = XY$, X and Y are non - identical gamma Rayleigh different
- $fx := \frac{x \times \exp\left(-\frac{x^2}{2\times a^2}\right)}{a^2}$; #input Rayleigh desity of X

$$fx := \frac{xe^{-\frac{1x^2}{2a^2}}}{a^2}$$

- $fy := \frac{1}{b} \times \exp\left(-\frac{y}{b}\right)$; #input exponential density of Y

$$fy := \frac{e^{-\frac{y}{b}}}{b}$$

- $fyz := \text{subs}\left(y = \frac{z}{x}, fy\right)$; #input density fy following subsititution

$$fyz := \frac{e^{-\frac{z}{xb}}}{b}$$

- $\text{int}G := \left(\frac{1}{x}\right) \times fx \times fyz$; #Create integrand for the Transformation

$$\text{int}G := e^{-\frac{1x^2}{2a^2}}\frac{e^{-\frac{z}{xb}}}{a^2b}$$

- $fz := \text{simplify}\left(\int_0^\infty \text{int}G dx, \text{assume} = \text{positive}\right)$; #pdf of $Z = XY$

$$fz := \frac{1}{2}\frac{\sqrt{2}\text{MeijerG}\left([[\,], [\,]]\left[\left[\frac{1}{2}, \frac{1}{2}, 0\right], [\,]\right], \frac{1}{8}\frac{z^2}{a^2b^2}\right)}{b\,a\sqrt{\pi}}$$

- $\text{simplify}\left(\int_0^\infty fz\,dz, \text{assume} = \text{positive}\right)$; #verify the pdf

$$1$$

- with(inttrans):
- $Mxs := \text{simplify}\left(\text{mellin}(fx, x, s), \text{assume} = \text{positive}\right)$;
  #Mellin trasnform of fx using builtin command

$$Mxs := 2^{\frac{1}{2}s-\frac{1}{2}}a^{s-1}\Gamma\left(\tfrac{1}{2}s+\tfrac{1}{2}\right)$$

- $Mxs := \text{simplify}\left(\text{mellin}(fy, y, s), \text{assume} = \text{positive}\right);$ #Mellin trasnform of *fy* using builtin command

$$Mxs := b^{s-1}\Gamma(s)$$

- $Mxs1 := \text{simplify}\left(\int_0^\infty fx \times x^{s-1}dx, \text{assume} = \text{positive}\right);$ #Mellin transform of *fx* using the integral representation

$$Mxz := 2^{\frac{1}{2}s-\frac{1}{2}}b^{s-1}a^{s-1}\Gamma(s)\Gamma\left(\tfrac{1}{2}s+\tfrac{1}{2}\right)$$

- $Mys1 := \text{simplify}\left(\int_0^\infty fy \times y^{s-1}dy, \text{assume} = \text{positive}\right);$ #Mellin transform of *fy* using the integral representation

$$Mys1 := b^{s-1}\Gamma(s)$$

- $Mxz := \text{simplify}\left(\text{mellin}(fz, z, s), \text{assume} = \text{positive}\right);$ #Mellin transform of *fz* using builtin command

$$Mxz := 2^{\frac{1}{2}s-\frac{1}{2}}b^{s-1}a^{s-1}\Gamma(s)\Gamma\left(\tfrac{1}{2}s+\tfrac{1}{2}\right)$$

- $Mzs1 := \text{simplify}\left(\int_0^\infty fz \times z^{s-1}dz, \text{assume} = \text{positive}\right);$ #Mellin transform of *fz* using the integral representation

$$Mzs1 := 2^{\frac{1}{2}s-\frac{1}{2}}b^{s-1}a^{s-1}\Gamma(s)\Gamma\left(\tfrac{1}{2}s+\tfrac{1}{2}\right)$$

- #Mellin transform of $Z = XY$ is the product of Mellin transforms of X and Y

In the next example, the concept of Mellin convolution is demonstrated to obtain the density of the ratio of two gamma variables.

Example # 7 The Mellin transform of the density of the ratio of two independent and identically distributed gamma variables

- #pdf of $Z = X$ by Y, X and Y are IID gamma
- $fy = \dfrac{y^{c-1}\exp\left(-\frac{y}{b}\right)}{b^c \times \Gamma(c)}$; #input Gamma density *fy*

$$fy := y^{c-1}\frac{e^{-\frac{y}{b}}}{b^c\Gamma(c)}$$

- $fx := \text{subs}(y = x, fy)$; #input density fx

$$fx := x^{c-1}\frac{e^{-\frac{x}{b}}}{b^c\Gamma(c)}$$

- $fxz := \text{subs}(x = y \times z, fx)$; #create the pdf with z

$$fxz := (yz)^{c-1}\frac{e^{-\frac{yz}{b}}}{b^c\Gamma(c)}$$

- $\text{int}G := y \times fy \times fxz$; #Create integrand for the Transformation

$$\text{int}G := yy^{c-1}\frac{e^{-\frac{y}{b}}(yz)^{c-1}e^{-\frac{yz}{b}}}{(b^c)^2\Gamma(c)^2}$$

- $fz := \text{simplify}\left(\int_0^{\text{Ë}} \text{int}G\, dy, \text{assume} = \text{positive}\right)$;
  #density of Z equal to X by Y

$$fz := \frac{1}{2}\frac{4^c z^{c-1}(1 + 2z + z^2)^{-c}\Gamma\left(c + \frac{1}{2}\right)}{\sqrt{\pi}\Gamma(c)}$$

- $\text{simplify}\left(\int_0^\infty fz\, dz, \text{assume} = \text{positive}\right)$;
  #veritf total probability equal to one

$$1$$

- $Mz := \text{simplify}\left(\int_0^\infty fz \times z^{s-1}\, dz, \text{assume} = \text{positive}\right)$;
  #Mellin transform of fz the integral

$$Mz := \frac{\Gamma(1 + c - s)\Gamma(c - 1 + s)}{\Gamma(c)^2}$$

- with (inttrans)
- $Mz1 := \text{simplify}(\text{mellin}(fz, z, s), \text{assume} = \text{positive})$;
  #Mellin transform of fz using builtin command

$$Mz1 := \frac{\Gamma(1 + c - 1 + s)\Gamma(c - 1 + s)}{\Gamma(c)^2}$$

- $Mx := \text{simplify}(\text{mellin}(fx, x, s), \text{assume} = \text{positive})$;
  #Mellin transform of fx using builtin command

$$Mx := \frac{b^{s-1}\Gamma(c - 1 + s)}{\Gamma(c)}$$

- $Mxy := \text{subs}(s = w, Mx);$
  #first step to create the shifted Mellin transform to get MY(2-s)

$$Mxy := \frac{b^{w-1}\Gamma(c-1+w)}{\Gamma(c)}$$

- #Mellin transform of Y with s replaced by (2-s) to obtain the Mellin convolution of ratio
- $Mq = \text{subs}(w = 2 - s, Mxy)$

$$Mq := \frac{b^{-s+1}\Gamma(1+c-s)}{\Gamma(c)}$$

- ##The Mellin transform MZ of the pdf of X by Y is the product of the $M_X(s)$ and $M_Y(2-s)$

The results demonstrate the validity of Eq. (2.553) which show that the Mellin transform of the ratio is the product of the Mellin transform of the density of the variable in the numerator and the shifted version of the Mellin transform of the density of the variable in the denominator.

Example # 8 The case of ratio of two non-identical gamma variables

#pdf of $Z = X$ by Y, X and Y are IID gamma

- $fy := \frac{y^{c-1}\exp\left(-\frac{y}{b}\right)}{b^c \times \Gamma(c)}$; #input Gamma density fy

$$fy := \frac{y^{c-1}e^{-\frac{y}{b}}}{b^c\Gamma(c)}$$

- $fx := \frac{x^{m-1}\exp\left(-\frac{x}{a}\right)}{a^m \times \Gamma(m)}$; #input gamma density of x, fx

$$fx := \frac{x^{m-1}e^{-\frac{x}{a}}}{a^m\Gamma(m)}$$

- $fxz: = \text{subs}(x = z, fx)$; # create the pdf with z

$$fxz := \frac{(yz)^{m-1}e^{-\frac{yz}{a}}}{a^m\Gamma(m)}$$

- $\text{int}G: = y \times fy \times fxz$; # Create integrand for the Transformation

$$\text{int}G := yy^{c-1}\frac{e^{-\frac{y}{b}}(yz)^{m-1}e^{-\frac{yz}{a}}}{b^c\Gamma(c)a^m\Gamma(m)}$$

- $fz := \text{simplify}\left(\int_0^\infty \text{int}G\, dy, \text{assume} = \text{positive}\right);$
  #denity of Z equal to X by Y

$$fz := \frac{z^{m-1} a^c (a+zb)^{-c-m} b^m \Gamma(c+m)}{\Gamma(c)\Gamma(m)}$$

- $\text{simplify}\left(\int_0^\infty fz\, dz, \text{assume} = \text{positive}\right);$
  #verify total probability equal to one

$$1$$

- $Mz := \text{simplify}\left(\int_0^\infty fz \times z^{s-1}\, dz, \text{assume} = \text{positive}\right);$
  #Mellin transform of fz using the integrat

$$Mz := \frac{b^{-s+1} a^{s-1} \Gamma(1+c-s)\Gamma(s-1+m)}{\Gamma(c)\Gamma(m)}$$

- with(inttrans):
- $Mz1 := \text{simplify}(\text{mellin}(fz, z, s), \text{assume} = \text{positive});$
  #Mellin transform of fz unsing builtin command

$$Mz1 := \frac{a^{s-1} b^{-s-1} \Gamma(c-m) B(s-1+m, 1+c-s)}{\Gamma(c)\Gamma(m)}$$

- $My := \text{simplify}(\text{mellin}(fy, y, s), \text{assume} = \text{positive});$
  #Mellin transform of fx unsing builtin command

$$My := \frac{b^{s-1}\Gamma(s+c-1)}{\Gamma(c)}$$

- $Mx := \text{simplify}(\text{mellin}(fx, x, s), \text{assume} = \text{positive});$
  #Mellin transform of fx unsing builtin command

$$Mx := \frac{a^{s-1}\Gamma(s-1+m)}{\Gamma(m)}$$

- $My1 := \text{subs}(s = w, My);$
  #first step to create the shifted Mellin transform to get $MY(2-s)$

$$Mx1 := \frac{b^{w-1}\Gamma(w+c-1)}{\Gamma(c)}$$

- Mellin transform of Y with s replaced by $(2-s)$ to obtain the Mellin convolution of ratio
- $My2 = \text{subs}(w = 2 - s, My1);$

$$My2 := \frac{b^{-s+1}\Gamma(1+c-s)}{\Gamma(c)}$$

- The Mellin transform MZ of the pdf of X by Y is the product of the $M_X(s)$ and $M_Y(2-s)$

2.20 Properties of Densities Revisited

Several densities seen earlier are now presented in a concise format (Fig. 2.70).

CDF,pdf, moments and Characteristic Function

CDF → $F_X(x) = P(X \le x), \quad F_X(-\infty) = 0, \quad F_X(\infty) = 1$

pdf → $f_X(x) = \dfrac{d\mathrm{F}x(x)}{dx}, \quad 0 \le f_X(x) \le \infty$

CDF → $F_X(x) = \displaystyle\int_{-\infty}^{x} f(y)dy, \quad 0 \le F_X(x) \le 1$

Probability → $P[x_1 \le X \le x_2] = \displaystyle\int_{x_1}^{x_2} f(x)dx = F_X(x_2) - F_X(x_1)$

n^{th} moment → $E[X^n] = \displaystyle\int_{-\infty}^{\infty} x^n f(x)dx$

Characteristic Function → $\phi_X(w) = E[e^{iXw}] = \displaystyle\int_{-\infty}^{\infty} e^{ixw} f(x)dx$

Laplace Transform → $f_X(x)$, x>0 $\quad \mathrm{L}_X(s) = E[e^{-Xs}] = = \phi_X(w)|_{(-jw=s)} = \displaystyle\int_{0}^{\infty} e^{-xs} f(x)dx$

$P[X \le x] \equiv P[X < x] = \displaystyle\int_{-\infty}^{x} f(\alpha)d\alpha$

$P[X = x] = 0, \quad \text{Continuous density}$

$P[X \ge x] \equiv P[X > x] = 1 - F_X(x) = \displaystyle\int_{x}^{\infty} f(x)dx$

Mean → $E[X] = \displaystyle\int_{-\infty}^{\infty} xf(x)dx$

Variance → $var[X] = E[X^2] - (E[X])^2$

Fig. 2.70 CDF, pdf, and characteristic functions

pdf and CDF

pdf	→	$f_X(x) = \frac{e^{-\frac{x}{a}}}{a}$	$0 \le x \le \infty$
CDF	→	$F_X(x) = 1 - e^{-\frac{x}{a}}$	
Mean	→	$E_X(x) = a$	
Variance	→	$\mathrm{var}_X(x) = a^2$	
Char. Fn	→	$G(w) = -\frac{1}{-1 + a\,w\,i}$	
Laplace Transform	→	$L(s) = \frac{1}{a\,s+1}$	

Exponential pdf

pdf and CDF

pdf	→	$f_X(x) = \frac{x^{a-1}\,e^{-\frac{x}{b}}}{b^a\,\Gamma(a)}$	$0 \le x \le \infty$
CDF	→	$F_X(x) = 1 - \frac{\Gamma\left(a, \frac{x}{b}\right)}{\Gamma(a)}$	
Mean	→	$E_X(x) = a\,b$	
Variance	→	$\mathrm{var}_X(x) = a\,b^2$	
Char. Fn	→	$G(w) = (1 - b\,w\,i)^{-a}$	
Laplace Transform	→	$L(s) = (b\,s + 1)^{-a}$	

Gamma pdf

pdf and CDF

pdf → $f_X(x) = \frac{\sqrt{2}\,e^{-\frac{(m-\log(x))^2}{2s^2}}}{2\,s\,x\,\sqrt{\pi}}$ $0 \le x \le \infty$

CDF → $F_X(x) = \frac{1}{2} - \frac{\operatorname{erf}\left(\frac{\sqrt{2}\,m - \sqrt{2}\log(x)}{2s}\right)}{2}$

Mean → $E_X(x) = m\,e^{\frac{s^2}{2}}$

Variance → $\operatorname{var}_X(x) = m^2\,e^{s^2}\left(e^{s^2} - 1\right)$

Char. Fn → NO Analytical expression

Laplace Transform → No Analytical Expression

Lognormal pdf

pdf and CDF

pdf → $f_X(x) = \frac{\sqrt{2}\,e^{-\frac{(m-x)^2}{2s^2}}}{2\,s\,\sqrt{\pi}}$ $-\infty \le x \le \infty$

CDF → $F_X(x) = \frac{1}{2} - \frac{\operatorname{erf}\left(\frac{\sqrt{2}\,(m-x)}{2s}\right)}{2}$

Mean → $E_X(x) = m$

Variance → $\operatorname{var}_X(x) = s^2$

Char. Fn → $G(w) = e^{-\frac{s^2 w^2}{2} + m\,w\,i}$

Laplace Transform → Undefined $(-\infty < x < \infty)$

Normal pdf

pdf and CDF

pdf → $f_X(x) = \frac{x\, e^{-\frac{x^2}{2b^2}}}{b^2}$ $0 \le x \le \infty$

CDF → $F_X(x) = 1 - e^{-\frac{x^2}{2b^2}}$

Mean → $E_X(x) = \frac{\sqrt{2}\, b \sqrt{\pi}}{2}$

Variance → $\mathrm{var}_X(x) = -\frac{b^2 (\pi - 4)}{2}$

Char. Fn → $G(w) = \frac{b\, w\, e^{-\frac{b^2 w^2}{2}} \sqrt{-2\,\pi}}{2} - \frac{b\, w\, e^{-\frac{b^2 w^2}{2}}\, \mathrm{erfi}\left(\frac{\sqrt{2}\, b\, w}{2}\right) \sqrt{2\,\pi}}{2}$.

Laplace Transform → $L(s) = 1 - \frac{\sqrt{2}\, b\, s \sqrt{\pi}\, e^{\frac{b^2 s^2}{2}}\, \mathrm{erfc}\left(\frac{\sqrt{2}\, b\, s}{2}\right)}{2}$

Rayleigh pdf

pdf and CDF

pdf → $f_X(x) = \frac{b\, x^{b-1}\, e^{-\left(\frac{x}{a}\right)^b}}{a^b}$ $0 \le x \le \infty$

CDF → $F_X(x) = 1 - e^{-\left(\frac{x}{a}\right)^b}$

Mean → $E_X(x) = \frac{a\, \Gamma\left(\frac{1}{b}\right)}{b}$

Variance → $\mathrm{var}_X(x) = a^2\, \Gamma\left(\frac{2}{b} + 1\right) - \frac{a^2\, \Gamma\left(\frac{1}{b}\right)^2}{b^2}$

Char. Fn → NO Analytical expression

Laplace Transform → No Analytical Expression

Weibull pdf

pdf and CDF

pdf → $f_X(x) = -\frac{1}{a-b}$ $a \leq x \leq b$

CDF → $F_X(x) = \frac{a-x}{a-b}$

Mean → $E_X(x) = \frac{a}{2} + \frac{b}{2}$

Variance → $\mathrm{var}_X(x) = \frac{(a-b)^2}{12}$

Char. Fn → $G(w) = -\frac{\left(e^{awi} - e^{bwi}\right) i}{w(a-b)}$

Laplace Transform → Undefined $(-\infty < x < \infty)$

Uniform pdf

CDF, pdf, nth moment & Characteristic Function

Probability Mass Function → $P_X(k) = P(X = k)$

pdf → $f_X(x) = \sum_k P_X(k)\delta(x-k)$

CDF → $F_X(x) = \sum_k P_X(k)U(x-k)$

nth moment → $E[X^n] = \sum_k k^n P_X(k)$

Characteristic Function → $\phi_X(w) = E\left[e^{ixw}\right] = \sum_k e^{iwk} P_X(k)$

2.21 Summary

In this chapter, we examined some theoretical aspects of probability density functions and distributions encountered in the study of fading and shadowing in wireless channels. We started with the basic definition of probability, then discussed density functions and properties relating to the analysis of fading and shadowing. We also examined the transformations of random variables in conjunction with relationships of different types of random variables. This is important in the study of diversity and modeling of specific statistical behavior of the wireless channels. The density functions of some of the functions of two or more random variables were derived. We examined order statistics, placing emphasis on the density functions of interest in diversity analysis. Concepts of stochastic processes and their properties were outlined. Similarly, the characteristics of noise were delineated within the context of signal detection. In addition, this study included an exploration of ways of expressing some of the densities in more compact forms using the hypergeometric functions and Meijer's G-function.

In the updated sections, readers are exposed to parameter estimation and statistical testing methods that are necessary in understanding the statistical models of fading and shadowing. Two methods, one based on maximum likelihood and the other one based on moments for parameter estimation are detailed and compared using a number of examples. The statistical approaches to appreciate the mixture densities for modeling signal strength fluctuations are provided. The receiver operating characteristics, essential for the understanding of cognitive radio, are introduced. To facilitate a better understanding of the statistics of sums and products of random variables, detailed discussion of Laplace and Mellin transforms is provided along with several examples. In addition, an overview of properties of all the key random variables is once again given. The pedagogy of each of these topics has been enhanced through detailed exposition of Matlab scripts. In specific instances where Matlab is deficient, Maple scripts are provided to support the study of Mellin transforms.

References

Aalo, V., & Zhang, J. (1999). On the effect of cochannel interference on average error rates in Nakagami-fading channels. *IEEE Communications Letters, 3*(5), 136–138.

Aalo, V. A., & Piboongungon T. (2005). On the multivariate generalized gamma distribution with exponential correlation. In *Global Telecommunications Conference, 2005. GLOBECOM '05* (pp. 3–5). IEEE.

Abate, J. (1995). Numerical inversion of Laplace transforms of probability distributions. *ORSA Journal on Computing, 7*, 36–43.

Abdi, A., & Kaveh, M. (1998). K distribution: An appropriate substitute for Rayleigh-lognormal distribution in fading-shadowing wireless channels. *Electronics Letters, 34*(9), 851–852.

Abdi, A., & Kaveh M. (1999). On the utility of gamma PDF in modeling shadow fading (slow fading). In *1999 I.E. 49th Vehicular Technology Conference* (pp. 2308–2312).

Abramowitz, M., & Segun, I. A. (Eds.). (1972). *Handbook of mathematical functions with formulas, graphs, and mathematical tables*. New York: Dover Publications.

Abu-Dayya, A. A., & Beaulieu, N. C. (1994). Outage probabilities in the presence of correlated lognormal interferers. *IEEE Transactions on Vehicular Technology, 43*(1), 164–173.

Abu-Salih, M. (1983). Distributions of the product and the quotient of power-function random variables. *Arabian Journal of Mathematics, 4*, 1–2.

Adamchik, V. (1995). The evaluation of integrals of Bessel functions via G-function identities. *Journal of Computational and Applied Mathematics, 64*(3), 283–290.

Ahmed, S., Yang, L.-L., & Hanzo, L. (2011). Probability distributions of products of Rayleigh and Nakagami-m variables using Mellin transform. In *2011 I.E. International Conference on Communications (ICC)* (pp. 1–5).
Alouini, M., & Simon, M. (2003). Dual diversity over correlated log-normal fading channels. *IEEE Transactions on Communications, 50*(12), 1946–1959.
Alouini, M. S., & Simon, M. K. (1999). Performance of coherent receivers with hybrid SC/MRC over Nakagami-m fading channels. *IEEE Transactions on Vehicular Technology, 48*(4), 1155–1164.
Alouini, M. S., & Simon, M. K. (2000). An MGF-based performance analysis of generalized selection combining over Rayleigh fading channels. *IEEE Transactions on Communications, 48*(3), 401–415.
Alouini, M. S., & Simon, M. K. (2006). Performance of generalized selection combining over Weibull fading channels. *Wireless Communications and Mobile Computing, 6*(8), 1077–1084.
Alouini, M. S., Abdi, A., & Kaveh, M. (2001). Sum of gamma variates and performance of wireless communication systems over Nakagami-fading channels. *IEEE Transactions on Vehicular Technology, 50*(6), 1471–1480.
Anastassopoulos, V., Lampropoulos, G. A., Drosopoulos, A., & Rey, N. (1999). High resolution radar clutter statistics. *IEEE Transactions on Aerospace and Electronic Systems, 35*(1), 43–60.
Andersen, J. B. (2002). *Statistical distributions in mobile communications using multiple scattering*. Presented at the 27th URSI General Assembly, Maastricht, Netherlands.
Annamalai, A., Tellambura, C., & Bhargava, V. K. (2005). A general method for calculating error probabilities over fading channels. *IEEE Transactions on Communications, 53*(5), 841–852.
Annamalai, A., Deora, G., & Tellambura, C. (2006). Analysis of generalized selection diversity systems in wireless channels. *IEEE Transactions on Vehicular Technology, 55*(6), 1765–1775.
Atapattu, S., Tellambura, C., & Jiang, H. (2010). Representation of composite fading and shadowing distributions by using mixtures of gamma distributions. In *Wireless Communications and Networking Conference (WCNC) in Sydney*, 18–21 April 2010 (pp. 1–5). IEEE.
Atapattu, S., Tellambura, C., & Jiang, H. (2010b). Analysis of area under the ROC curve of energy detection. *IEEE Transactions on Wireless Communications, 9*, 1216–1225.
Atapattu, S., Tellambura, C., & Jiang, H. (2011). A mixture gamma distribution to model the SNR of wireless channels. *IEEE Transactions on Wireless Communications, 10*(12), 4193–4203.
Beaulieu, N., & Xie, Q. (2004). An optimal lognormal approximation to lognormal sum distributions. *IEEE Transactions on Vehicular Technology, 53*(2), 479–489.
Beaulieu, N. C. (1990). An infinite series for the computation of the complementary probability distribution function of a sum of independent random variables and its application to the sum of Rayleigh random variables. *IEEE Transactions on Communications, 38*(9), 1463–1474.
Beaulieu, N. C., Abu-Dayya, A. A., & McLane, P. J. (1995). Estimating the distribution of a sum of independent lognormal random variables. *IEEE Transactions on Communications, 43* (12), 2869.
Benedetto, S., & Biglieri, E. (1999). *Principles of digital transmission: With wireless applications*. New York: Kluwer Academic/Plenum Press.
Bithas, P. S., Sagias, N. C., & Mathiopoulos, P. T. (2007). Dual diversity over correlated Ricean fading channels. *Journal of Communications and Networks, 9*(1), 67–74.
Bithas, P. S., Sagias, N. C., Mathiopoulos, P. T., Karagiannidis, G. K., & Rontogiannis, A. A. (2006). On the performance analysis of digital communications over generalized-K fading channels. *IEEE Communications Letters, 10*(5), 353–355.
Block, H. W., & Savits, T. H. (1980). Laplace transforms for classes of life distributions. *The Annals of Probability, 8*(3), 465–474.
Blumenson, L. E., & Miller, K. S. (1963). Properties of generalized Rayleigh distributions. *The Annals of Mathematical Statistics, 34*(3), 903–910.
Bowman, K. O., & Shenton, L. R. (2006). Maximum likelihood estimators for normal and gamma mixtures. *Far East Journal of Theoretical Statistics, 20*, 217–240.
Brennan, D. G. (1959). Linear diversity combining techniques. *Proceedings of the IRE, 47*(6), 1075–1102.

Bryson, B. C. (1974). Heavy tailed distributions: Properties and tests. *Technometrics, 16*(1), 61–68.
Büyükçorak, S., Vural, M., & Kurt, G. K. (2015). Lognormal mixture shadowing. *IEEE Transactions on Vehicular Technology, 64*(10), 4386–4398.
Cardieri, P., & Rappaport, T. (2001). Statistical analysis of co-channel interference in wireless communications systems. *Wireless Communications and Mobile Computing, 1*(1), 111–121.
Cardieri, P., & Yacoub, M. D. (2005). Simple accurate lognormal approximation to lognormal sums. *Electronics Letters, 41*(18), 1016–1017.
Carter, B., & Springer, M. (1977). The distribution of products, quotients and powers of independent H-function variates. *SIAM Journal on Applied Mathematics, 33*(4), 542–558.
Cheng, J., & Beaulieu, N. C. (2001). Maximum likelihood based estimation of the Nakagami m-parameter. *IEEE Communications Letters, 5*, 101–103.
Cheng, J., Tellambura, C., & Beaulieu, N. C. (2004). Performance of digital linear modulations on Weibull slow-fading channels. *IEEE Transactions on Communications, 52*(8), 1265–1268.
Chiani, M. (1999). Integral representation and bounds for Marcum *Q*-function. *Electronics Letters, 35*(6), 445–446.
Cooper, G. R., & McGillem, C. D. (1986). *Modern communications and spread spectrum*. New York: McGraw-Hill.
Cotton, S. L., & Scanlon, W. G. (2007). Higher order statistics for lognormal small-scale fading in mobile radio channels. *IEEE Antennas and Wireless Propagation Letters, 6*, 540–543.
Couch, L. W. (2007). *Digital and analog communication systems*. Upper Saddle River, NJ: Pearson/Prentice Hall.
Coulson, A. J., Williamson, A. G., & Vaughan, R. G. (1998a). Improved fading distribution for mobile radio. *IEE Proceedings-Communications, 145*(3), 197–202.
Coulson, A. J., Williamson, A. G., & Vaughan, R. G. (1998b). A statistical basis for lognormal shadowing effects in multipath fading channels. *IEEE Transactions on Communications, 46*(4), 494–502.
da Costa, B. D., & Yacoub, M. D. (2008). Moment generating functions of generalized fading distributions and applications. *IEEE Communications Letters, 12*(2), 112–114.
da Silva, V. R., & Yongacoglu, A. (2015, November). EM algorithm on the approximation of arbitrary PDFs by Gaussian, gamma and lognormal mixture distributions. In *2015 7th IEEE Latin-American Conference on Communications (LATINCOM)* (pp. 1–6). IEEE.
Davenport, J. W., Bezdek, J. C., & Hathaway, R. J. (1988). Parameter estimation for finite mixture distributions. *Computers & Mathematcs with Applications, 15*(10), 819–828.
Dogandzic, A., & Jin, J. (2004). Maximum likelihood estimation of statistical properties of composite gamma-lognormal fading channels. *IEEE Transactions on Signal Processing, 52*, 2940–2945.
Epstein, B. (1948). Some applications of Mellin transform in statistics. *Annals of Mathematical Statistics, 19*(3), 370–379.
Erdelyi, A. (1953). *Table of integral transforms*. New York: McGraw-Hill.
Evans, M., Hastings, N. A. J., & Peacock, J. B. (2000). *Statistical distributions*. New York: Wiley.
Frery, A. C., Muller, H. J., Yanasse, C. D. C. F., & Sant'Anna, S. J. S. (1997). A model for extremely heterogeneous clutter. *IEEE Transactions on Geoscience and Remote Sensing, 35*(3), 648–659.
Gagliardi, R. M. (1988). *Introduction to communications engineering*. New York: Wiley.
Gallager, R. G. (2008). *Principles of digital communication*. Cambridge; New York: Cambridge University Press.
Gaur, S., & Annamalai, A. (2003). Some integrals involving the Qm(a√x, b√x) with application to error probability analysis of diversity receivers. *IEEE Transactions on Vehicular Technology, 52*(6), 1568–1575.
Goldsmith, A. (2005). *Wireless communications*. New York: Cambridge University Press.
Goodman, J. W. (1985). *Statistical optics*. New York: Wiley.

Gradshteyn, I. S., & Ryzhik, I. M. (2007). *Table of integrals, series and products*. Oxford: Academic.

Griffiths, J., & McGeehan, J. P. (1982). Inter relationship between some statistical distributions used in radio-wave propagation. *IEE Proceedings F Communications, Radar and Signal Processing, 129*(6), 411–417.

Gupta, R. D., & Kundu, D. (1999). Generalized exponential distributions. *Australian & New Zealand Journal of Statistics, 41*(2), 173–188.

Hanley, J. A., & McNeil, B. (1982). The meaning and use of the area under a receiver operating characteristic (ROC) curve. *Radiology, 143*, 29–36.

Hansen, F., & Meno, F. I. (1977). Mobile fading; Rayleigh and lognormal superimposed. *IEEE Transactions on Vehicular Technology, 26*(4), 332–335.

Haykin, S. S. (2001). *Digital communications*. New York: Wiley.

Helstrom, C. W. (1968). *Statistical theory of signal detection*. Oxford; New York: Pergamon Press.

Helstrom, C. W. (1991). *Probability and stochastic processes for engineers*. New York: Macmillan.

Helstrom, C. W. (1992). Computing the generalized Marcum *Q* function. *IEEE Transactions on Information Theory, 38*(4), 1422–1428.

Helstrom, C. W. (1998). Approximate inversion of Marcum's Q-function. *IEEE Transactions on Aerospace and Electronic Systems, 34*(1), 317–319.

Holm, H., & Alouini, M. S. (2004). Sum and difference of two squared correlated Nakagami variates in connection with the McKay distribution. *IEEE Transactions on Communications, 52*(8), 1367–1376.

Hudson, J. E. (1996). A lognormal fading model and cellular radio performance. In *Global Telecommunications Conference, 1996. GLOBECOM '96. Communications: The Key to Global Prosperity* (Vol. 2, pp. 1187–1191).

Iskander, D. R., Zoubir, A. M., & Boashash, B. (1999). A method for estimating the parameters of the K distribution. *IEEE Transactions on Signal Processing, 47*(4), 1147–1151.

Iskander, R. D., & Zoubir, A. M. (1996). On coherent modeling of non-Gaussian radar clutter. In *Proceedings of the 8th IEEE Signal Processing Workshop on statistical Signal and Array Processing* (pp. 226–229).

Ismail, M. H., & Matalgah, M. M. (2006). On the use of Pade approximation for performance evaluation of maximal ratio combining diversity over Weibull fading channels. *EURASIP Journal on Wireless Communications and Networking, 6*, 62–66.

Jakeman, E., & Tough, R. (1987). Generalized K distribution: A statistical model for weak scattering. *Journal of the Optical Society of America A, 4*(9), 1764–1772.

Jakes, W. C. (1994). *Microwave mobile communications*. Piscataway, NJ: IEEE Press.

Joughin, I. R., Percival, D. B., & Winebrenner, D. P. (1993). Maximum likelihood estimation of K distribution parameters for the SAR data. *IEEE Transactions on Geoscience and Remote Sensing, 51*, 989–999.

Jung, J., Lee, S. R., Park, H., Lee, S., & Lee, I. (2014). Capacity and error probability analysis of diversity reception schemes over generalized-fading channels using a mixture gamma distribution. *IEEE Transactions on Wireless Communications, 13*(9), 4721–4730.

Kabe, D. G. (1958). Some applications of Meijer-G functions to distribution problems in statistics. *Biometrika, 45*(3/4), 578–580.

Karadimas, P., & Kotsopoulos, S. A. (2008). A generalized modified Suzuki model with sectored and inhomogeneous diffuse scattering component. *Wireless Personal Communications, 47*(4), 449–469.

Karadimas, P., & Kotsopoulos, S. A. (2010). A modified loo model with partially blocked and three dimensional multipath scattering: Analysis, simulation and validation. *Wireless Personal Communications, 53*(4), 503–528.

Karagiannidis, G. K., Zogas, D. A., & Kotsopoulos, S. A. (2003a). Performance analysis of triple selection diversity over exponentially correlated Nakagami-m fading channels. *IEEE Transactions on Communications, 51*(8), 1245–1248.

Karagiannidis, G. K., Zogas, D. A., & Kotsopoulos, S. A. (2003b). On the multivariate Nakagami-m distribution with exponential correlation. *IEEE Transactions on Communications, 51*(8), 1240–1244.

Karagiannidis, G. K., Sagias, N. C., & Mathiopoulos, P. T. (2007). N*Nakagami: A novel statistical model for cascaded fading channels. *IEEE Transactions on Communications, 55* (8), 1453–1458.

Karagiannidis, G. K., Sagias, N. C., & Tsiftsis, T. A. (2006). Closed-form statistics for the sum of squared Nakagami-m variates and its applications. *IEEE Transactions on Communications, 54* (8), 1353–1359.

Karmeshu, J., & Agrawal, R. (2007). On efficacy of Rayleigh-inverse Gaussian distribution over K-distribution for wireless fading channels. *Wireless Communications and Mobile Computing, 7*(1), 1–7.

Kass, R. E., & Raftery, A. E. (1995). Bayes factors. *Journal of the American Statistical Association, 90*, 773–795.

Kilicman, A., & Ariffin, M. R. K. (2002). A note on the convolution in the Mellin sense with generalized functions. *Bulletin of the Malaysian Mathematical Sciences Society (Second Series), 25*, 93–100.

Korn, I., & Foneska, J. P. (2001). M-CPM with MRC diversity in Rician-, Hoyt-, and Nakagami-fading channels. *IEEE Transactions on Vehicular Technology, 50*(4), 1182–1118.

Kostic, I. M. (2005). Analytical approach to performance analysis for channel subject to shadowing and fading. *IEE Proceedings Communications, 152*(6), 821–827.

Kotz, S., & Adams, J. (1964). Distribution of sum of identically distributed exponentially correlated gamma-variables. *The Annals of Mathematical Statistics, 35*(1), 277–283.

Kumar, V., & Shukla, G. (2010). Maximum likelihood estimation in generalized gamma type model. *Journal of Reliability and Statistical Studies, 3*(1), 43–51.

Kundu, K., & Raqab, M. Z. (2005). Generalized Rayleigh distribution: Different methods of estimations. *Computational Statistics and Data Analysis, 49*, 187–200.

Lam, C. L. J., & Le-Ngoc, T. (2006). Estimation of typical sum of lognormal random variables using log shifted gamma approximation. *IEEE Communications Letters, 10*(4), 234–223.

Lam, C. L. J., & Tho, L.-N. (2007). Log-shifted gamma approximation to lognormal sum distributions. *IEEE Transactions on Vehicular Technology, 56*(4), 2121–2129.

Laourine, A., Alouini, M. S., Affes, S., & Stephenne, A. (2008). On the capacity of generalized-K fading channels. *IEEE Transactions on Wireless Communications, 7*(7), 2441–2445.

Laourine, A., Alouini, M. S., Affes, S., & Stéphenne, A. (2009). On the performance analysis of composite multipath/shadowing channels using the G-distribution. *IEEE Transactions on Communications, 57*(4), 1162–1170.

Lee, R., & Holland, B. (1979). Distribution of a ratio of correlated gamma random variables. *SIAM Journal on Applied Mathematics, 36*(2), 304–320.

Lewinsky, D. J. (1983). Nonstationary probabilistic target and cluttering scattering models. *IEEE Transactions on Aerospace and Electronic Systems, 31*, 490–449.

Lienhard, J. H., & Meyer, P. L. (1967). A physical basis for the generalized gamma distribution. *Quarterly of Applied Mathematics, 25*(3), 330–334.

Ligeti, A. (2000). Outage probability in the presence of correlated lognormal useful and interfering components. *IEEE Communications Letters, 4*(1), 15–17.

Lindsey, W. C., & Simon, M. K. (1973). *Telecommunication systems engineering*. Englewood Cliffs, NJ: Prentice-Hall.

Liu, Z., Almhana, J., Wang, F., & McGorman, R. (2007). Mixture lognormal approximations to lognormal sum distributions. *IEEE Communications Letters, 11*(9), 711–713.

Liu, Z., Almhana, J., & McGorman, R. (2008). Approximating lognormal sum distributions with power lognormal distributions. *IEEE Transactions on Vehicular Technology, 57*(4), 2611–2617.

Lomnicki, Z. A. (1967). On the distribution of products of random variables. *Journal of the Royal Statistical Society: Series B Methodological, 29*(3), 513–524.

Ma, Y., & Chai, C. C. (2000). Unified error probability analysis for generalized selection combining in Nakagami fading channels. *IEEE Journal on Selected Areas in Communications, 18*(11), 2198–2210.

Mann, H. B., & Wald, A. (1942). On the choice of the number of class intervals in the application of the chi square test. *The Annals of Mathematical Statistics, 13*, 306–317.

Mathai, A., & Moschopoulos, P. (1991). On a multivariate gamma. *Journal of Multivariate Analysis, 39*(1), 135–153.

Mathai, A. M. (1993). *A handbook of generalized special functions for statistical and physical sciences*. Oxford: Oxford University Press.

Mathai, A. M., & Haubold, H. J. (2008). *Special functions for applied scientists*. New York: Springer.

Mathai, A. M., & Saxena, R. K. (1973). *Generalized hypergeometric functions with applications in statistics and physical sciences*. Berlin; New York: Springer.

McClish, D. K. (1989). Analyzing a portion of the ROC curve. *Medical Decision Making, 9*, 190–195.

McDaniel, S. T. (1990). Seafloor reverberation fluctuations. *The Journal of the Acoustical Society of America, 88*(3), 1530–1535.

Metz, C. E. (2006). Receiver operating characteristic analysis: A tool for the quantitative evaluation of observer performance and imaging systems. *Journal of the American College of Radiology, 3*(6), 413–422.

Middleton, D. (1996). *An introduction to statistical communications theory*. Piscataway, NJ: IEEE Press.

Molisch, A. F. (2005). *Wireless communications*. Chichester: Wiley.

Moschopoulos, P. (1985). The distribution of the sum of independent gamma random variables. *Annals of the Institute of Statistical Mathematics, 37*(1), 541–544.

Nadarajah, S. (2005). Products, and ratios for a bivariate gamma distribution. *Applied Mathematics and Computation, 171*(1), 581–595.

Nadarajah, S., & Gupta, A. (2005). On the product and ratio of Bessel random variables. *International Journal of Mathematics and Mathematical Sciences, 2005*(18), 2977–2989.

Nadarajah, S., & Gupta, A. (2006). Some bivariate gamma distributions. *Applied Mathematics Letters, 19*(8), 767–774.

Nadarajah, S., & Kotz, S. (2006a). On the product and ratio of gamma and Weibull random variables. *Econometric Theory, 22*(02), 338–344.

Nadarajah, S., & Kotz, S. (2006b). Bivariate gamma distributions, sums and ratios. *Bulletin of the Brazilian Mathematical Society, 37*(2), 241–274.

Nadarajah, S., & Kotz, S. (2007). A class of generalized models for shadowed fading channels. *Wireless Personal Communications, 43*(4), 1113–1120.

Nakagami, M. (1960). The m-distribution—A general formula of intensity distribution of rapid fading. In W. C. Hoffman (Ed.), *Statistical methods in radio wave propagation*. Elmsford, NY: Pergamon.

Neath, A. A., & Cavanaugh, J. E. (2012). The Bayesian information criterion: Background, derivation, and applications. *Wiley Interdisciplinary Reviews: Computational Statistics, 4*, 199–203.

Nuttall, A. (1975). Some integrals involving the Q_M function (Corresp.) *IEEE Transactions on Information Theory, 21*(1), 95–96.

Nuttall, A. H. (1969). Numerical evaluation of cumulative probability distribution functions directly from characteristic functions. *Proceedings of the IEEE, 57*(11), 2071–2072.

Nuttall, A. H. (1970). Alternate forms for numerical evaluation of cumulative probability distributions directly from characteristic functions. *Proceedings of the IEEE, 58*(11), 1872–1873.

Obuchowski, N. A. (2003). Receiver operating characteristic curves and their use in radiology. *Radiology, 229*, 3–8.

Okui, S. (1981). Probability distributions for ratios of fading signal envelopes and their generalization. *Electronics and Communications in Japan, 64-B*(3), 72–80.

Panajotović, A. S., Stefanović, M. Č., & Drača, D. L. (2009). Effect of microdiversity and macrodiversity on average bit error probability in shadowed fading channels in the presence of interference. *ETRI Journal, 31*(5), 500–505.

Papazafeiropoulos, A. K., & Kotsopoulos, S. A. (2011). The a-l-m and a-m-m small scale general fading distributions: A unified approach. *Wireless Personal Communications, 57*, 735–751.

Papoulis, A., & Pillai, S. U. (2002). *Probability, random variables, and stochastic processes.* Boston: McGraw-Hill.

Park, C. S. (1986). The Mellin transform in probabilistic cash flow modeling. *The Engineering Economist, 32*(2), 115–134.

Patzold, M. (2002). *Mobile fading channels.* Chichester: Wiley.

Piboongungon, T., & Aalo, V. A. (2004). Outage probability of L-branch selection combining in correlated lognormal fading channels. *Electronics Letters, 40*(14), 886–888.

Podolski, H. (1972). The distribution of a product of n independent random variables with generalized gamma distribution. *Demonstratio Mathematica, 4*(2), 119–123.

Polydorou, D. S., Babalis, P. G., & Capsalis, C. N. (1999). Statistical characterization of fading in LOS wireless channels with a finite number of dominant paths. Application in millimeter frequencies. *International Journal of Infrared and Millimeter Waves, 20*(3), 461–472.

Proakis, J. G. (2001). *Digital communications.* Boston: McGraw-Hill.

Provost, S. (1989). On sums of independent gamma random variables. *Statistics, 20*(4), 583–591.

Rappaport, T. S. (2002). *Wireless communications: Principles and practice.* Upper Saddle River, NJ: Prentice Hall PTR.

Redner, R., & Walker, H. F. (1984). Maximum likelihood estimation and the EM algorithm. *SIAM Review, 26*, 195–239.

Rice, S. (1974). Probability distributions for noise plus several sine waves–the problem of computation. *IEEE Transactions on Communications, 22*(6), 851–853.

Rohatgi, V. K., & Saleh, A. K. M. E. (2001). *An introduction to probability and statistics.* New York: Wiley.

Rossberg, A. G. (2008). Laplace transforms of probability distributions and their inversions are easy on logarithmic scales. *Journal of Applied Probability, 45*(2), 531–541.

Sahu, P. R., & Chaturvedi, A. K. (2005). Performance analysis of predetection EGC receiver in Weibull fading channel. *Electronics Letters, 41*(2), 85–86.

Saleh, A., & Valenzuela, R. A. (1987). A statistical model for indoor multipath propagation. *IEEE Journal on Selected Areas in Communications, 5*, 128–137.

Salo, J., El-Sallabi, H. M., & Vainikainen, P. (2006). The distribution of the product of independent Rayleigh random variables. *IEEE Transactions on Antennas and Propagation, 54*(2), 639–643.

Schwartz, M. (1980). *Information transmission, modulation, and noise: A unified approach to communication systems.* New York: McGraw-Hill.

Schwartz, M., Bennett, W. R., & Stein, S. (1996). *Communication systems and techniques.* Piscataway, NJ: IEEE Press.

Selim, B., Alhussein, O., Muhaidat, S., Karagiannidis, G. K., & Liang, J. (2015). Modeling and analysis of wireless channels via the mixture of Gaussian distribution. *IEEE Transactions on Vehicular Technology, 65*(10), 8309–8321.

Shah, A., Haimovich, A. M., Simon, M. K., & Alouini, M. S. (2000). Exact bit-error probability for optimum combining with a Rayleigh fading Gaussian cochannel interferer. *IEEE Transactions on Communications, 48*(6), 908–912.

Shankar, P. (2002a). *Introduction to wireless systems.* New York: Wiley.

Shankar, P. (2002b). Ultrasonic tissue characterization using a generalized Nakagami model. *IEEE Transactions on Ultrasonics Ferroelectrics and Frequency Control, 48*(6), 1716–1720.

Shankar, P. M. (2004). Error rates in generalized shadowed fading channels. *Wireless Personal Communications, 28*(3), 233–238.

Shankar, P. M. (2005). Outage probabilities in shadowed fading channels using a compound statistical model. *IEE Proceedings on Communications, 152*(6), 828–832.

Shankar, P. M. (2010). Statistical models for fading and shadowed fading channels in wireless systems: A pedagogical perspective. *Wireless Personal Communications, 60*(2), 191–213. doi:10.1007/s11277-010-9938-2.

Shankar, P. M. (2013). Diversity in cascaded N*Nakagami channels. *Annals of Telecommunications, 68*(7), 477–483.

Shankar, P. M. (2016). Performance of cognitive radio in N*Nakagami cascaded channels. *WIRE, 88*(3), 657–667.

Shankar, P. M., & Gentile, C. (2010). Statistical analysis of short term fading and shadowing in ultra-wideband systems. In *IEEE International Conference on Communications (ICC), 2010* (pp. 1–6).

Shanmugam, K. S. (1979). *Digital and analog communication systems*. New York: Wiley.

Shepherd, N. H. (1977). Radio wave loss deviation and shadow loss at 900 MHZ. *IEEE Transactions on Vehicular Technology, 26*(4), 309–313.

Shin, J. W., Chang, J. H., & Kim, N. S. (2005). Statistical modeling of speech signals based on generalized gamma distribution. *IEEE Signal Processing Letters, 12*(3), 258–261.

Sijing, J., & Beaulieu N. C. (2010). BER of antipodal signaling in Laplace noise. In *2010 25th Biennial Symposium on Communications (QBSC)*.

Simon, M. (2002). The Nuttall Q function-its relation to the Marcum *Q* function and its application in digital communication performance evaluation. *IEEE Transactions on Communications, 50* (11), 1712–1715.

Simon, M. K., & Alouini, M. S. (2003). Some new results for integrals involving the generalized Marcum *Q* function and their application to performance evaluation over fading channels. *IEEE Transactions on Wireless Communications, 2*(4), 611–615.

Simon, M. K., & Alouini, M.-S. (2005). *Digital communication over fading channels*. Hoboken, NJ: Wiley-Interscience.

Simon, M. K., Hinedi, S. M., & Lindsey, W. C. (1995). *Digital communication techniques: Signal design and detection*. Englewood Cliffs, NJ: Prentice Hall PTR.

Sklar, B. (1997a). Rayleigh fading channels in mobile digital communication systems. I. Characterization. *IEEE Communications Magazine, 35*(7), 90–100.

Sklar, B. (1997b). Rayleigh fading channels in mobile digital communication systems. II. Mitigation. *IEEE Communications Magazine, 35*(7), 102–109.

Sklar, B. (2001). *Digital communications: Fundamentals and applications*. Upper Saddle River, NJ: Prentice-Hall PTR.

Slimane, B. (2001). Bounds on the distribution of a sum of independent lognormal random variables. *IEEE Transactions on Communications, 49*(6), 975–978.

Sowerby, K. W., & Williamson, A. G. (1987). Outage probability calculations for a mobile radio system having two log-normal interferers. *Electronics Letters, 23*(25), 1345–1346.

Springer, M., & Thompson, W. (1966). The distribution of products of independent random variables. *SIAM Journal on Applied Mathematics, 14*(3), 511–526.

Springer, M., & Thompson, W. (1970). The distribution of products of beta, gamma and Gaussian random variables. *SIAM Journal on Applied Mathematics, 18*(4), 721–737.

Stacy, E. (1962). A generalization of the gamma distribution. *The Annals of Mathematical Statistics, 33*(3), 1187–1192.

Stacy, E., & Mihram, G. (1965). Parameter estimation for a generalized gamma distribution. *Technometrics, 7*(3), 349–358.

Steele, R., & Hanzó, L. (1999). *Mobile radio communications: Second and third generation cellular and WATM systems*. Chichester; New York: Wiley.

Stuart, A. (1962). Gamma-distributed products of independent random variables. *Biometrika, 49* (3–4), 564.

Stuber, G. L. (2000). *Principles of mobile communication*. New York: Kluwer Academic.

Subadar, R., & Sahu, P. R. (2009). Performance analysis of dual MRC receiver in correlated Hoyt fading channels. *IEEE Communications Letters, 13*(6), 405–407.

Subrahmanaim, K. (1970). On some applications of mellin transforms to statistics: Dependent random variables. *SIAM Journal on Applied Mathematics, 19*(4), 658–662.

Suzuki, H. (1977). A statistical model for urban radio propagation. *IEEE Transactions on Communications, 25*, 673–680.

Tan, C. C., & Beaulieu, N. C. (1997). Infinite series representations of the bivariate Rayleigh and Nakagami-distributions. *IEEE Transactions on Communications, 45*(10), 1159–1161.

Taub, H., & Schilling, D. L. (1986). *Principles of communication systems*. New York: McGraw-Hill.

Tellambura, C., & Annamalai, A. (1999). An unified numerical approach for computing the outage probability for mobile radio systems. *IEEE Communications Letters, 3*(4), 97–99.

Tellambura, C., & Annamalai, A. (2000). Efficient computation of erfc(x) for large arguments. *IEEE Transactions on Communications, 48*(4), 529–532.

Tellambura, C., & Annamalai A. (2003). Unified performance bounds for generalized selection diversity combining in fading channels. In *Wireless Communications and Networking (WCNC), 2003 IEEE*.

Tellambura, C., et al. (2003). Closed form and infinite series solutions for the MGF of a dual-diversity selection combiner output in bivariate Nakagami fading. *IEEE Transactions on Communications, 51*(4), 539–542.

Tepedelenlioglu, C., & Gao, P. (2005). Estimators of the Nakagami-m parameters and performance analysis. *IEEE Transactions on Wireless Communications, 4*, 519–527.

Tjhung, T. T., & Chai, C. C. (1999). Fade statistics in Nakagami-lognormal channels. *IEEE Transactions on Communications, 47*(12), 1769–1772.

van Erkel, A. R., & Pattynama, P. M. T. (1998). Receiver operating characteristic (ROC) analysis: Basic principles and applications in radiology. *European Journal of Radiology, 27*, 88–94.

Van Trees, H. L. (1968). *Detection, estimation, and modulation theory, part I*. New York: Wiley.

Vatalaro, F. (1995). Generalised Rice-lognormal channel model for wireless communications. *Electronics Letters, 31*(22), 1899–1900.

Vaughn, R. A., & Anderson, J. B. (2003). *Channels, propagation and antennas for mobile communications*. Herts: IEE.

Voda, V. G. (2009). A method constructing density functions: The case for a generalized Rayleigh variable. *Applications of Mathematics, 54*(5), 417–443.

Wells, W. T., Anderson, R. L., & Cell, J. W. (1962). The distribution of the product of two central or non-central Chi-Square variates. *The Annals of Mathematical Statistics, 33*(3), 1016–1020.

Winters, J. (1987). Optimum combining for indoor radio systems with multiple users. *IEEE Transactions on Communications, 35*(11), 1222–1230.

Winters, J. H. (1984). Optimum combining in digital mobile radio with cochannel interference. *IEEE Transactions on Vehicular Technology, 33*(3), 144–155.

Winters, J. H. (1998). The diversity gain of transmit diversity in wireless systems with Rayleigh fading. *IEEE Transactions on Vehicular Technology, 47*(1), 119–123.

Withers, C., & Nadarajah, S. (2008). MGFs for Rayleigh random variables. *Wireless Personal Communications, 46*(4), 463–468.

Wolfram. (2011). http://functions.wolfram.com/, Wolfram Research.

Wongtrairat, W., & Supnithi, P. (2009). Performance of digital modulation in double Nakagami-m fading channels with MRC diversity. *IEICE Transactions on Communications, E92b*(2), 559–566.

Yacoub, M. D. (2000). Fading distributions and co-channel interference in wireless systems. *IEEE Antennas and Propagation Magazine, 42*(1), 150–160.

Yacoub, M. D. (2007a). The a-m distribution: A physical fading model for the Stacy distribution. *IEEE Transactions on Vehicular Technology, 56*(1), 27–34.

Yacoub, M. D. (2007b). The m-m distribution and the Z-m distribution. *IEEE Antennas and Propagation Magazine, 49*(1), 68–81.

Yue, S., et al. (2001). A review of bivariate gamma distributions for hydrological applications. *Journal of Hydrology, 246*, 1–18.

Zogas, D., & Karagiannidis, G. K. (2005). Infinite-series representations associated with the bivariate Rician distribution and their applications. *IEEE Transactions on Communications, 53*(11), 1790–1179.

Chapter 3
Modems for Wireless Communications

3.1 Introduction

Digital modulation and demodulation techniques together (modem) form the fundamental building block of data transmission in communication systems in general and wireless communication systems in particular (Lucky et al. 1968; Oetting 1979; Amoroso 1980; Feher 1995; Benedetto and Biglieri 1999; Proakis 2001; Simon and Alouini 2005; Schwartz 2005; Couch 2007). The modems can be classified in a multitude of ways (Simon et al. 1995; Sklar 1993, 2001; Haykin 2001). They can be identified in terms of the signal property that is modulated such as amplitude, phase, or frequency. They can also be classified in terms of the number of levels of values (binary, quaternary, or in general M-ary) attributable to the property. Detection methods such as coherent or noncoherent ones can also be used for the classification. We can, in addition, use terms such as "linear" and "nonlinear" modulation to classify the modulation types (Sundberg 1986; Anderson et al. 1986; Gagliardi 1988; Gallager 2008). The output of the frequency modulated system has a constant envelope while the output of the amplitude or phase modulated system has a constant frequency with time varying amplitudes. This makes the amplitude and phase modulation a form of "linear" modulation and the frequency modulation a form of "nonlinear" modulation. Even for a specific modulation type such as phase modulation (phase shift keying, for example), it is possible to have a coherent or a noncoherent detector or receiver. Modems cover a wide range of possibilities with some common themes such as the initial building blocks of the "signal space." The concepts of signal space make it possible to analyze modems, design them so that they meet certain criteria, provide a uniform framework, and make it possible to compare and contrast the different modems.

We also look at efficient ways of managing the most precious commodity, the spectrum, through appropriate schemes of pulse shaping (Helstrom 1960, 1968; Amoroso 1980; Sklar 1993; Aghvami 1993; Sundberg 1986; Anderson 2005).

© Springer International Publishing AG 2017

P.M. Shankar, *Fading and Shadowing in Wireless Systems*,
DOI 10.1007/978-3-319-53198-4_3

We compare the efficiencies of these modems with appropriate trade-off characteristics so that an optimum choice of a modem can be made for a specific application.

3.2 Optimum Receiver, Pulse Shaping, and Nyquist's Criteria

Let us look at a basic model of the digital communication system depicted in Fig. 3.1. Note that most of the information is in analog form, we start with an A/D converter. This is followed by transforming the digitized information into bits (0's and 1's) and combining a few of them into a sequence of bits, and so on. The digital source puts out the bits which are encoded into symbols.

The encoding is generally done using gray coding (discussed later), which reduces the errors during the detection process (Schwartz 1980; Simon et al. 1995; Schwartz et al. 1996). If we combine "k" bits into a sequence, we get a "symbol." Thus, a symbol size of M corresponds to 2^k bits. For $k = 1$, we have a binary system and for $k = 2$, we have a quaternary system, and for $k = 3$ we have an 8-ary system and, in general, we have an M-ary system. If Tb represents the bit duration, the symbol duration Ts will be kTb, with the symbol and bit duration being identical for the binary case.

Once the bit values have been picked, such a step needs to be followed by the choice of appropriate functions to represent these bits. This means that the shapes of the functions, such as rectangular, Gaussian, raised cosine, and others, also become part of the fabric and language of the digital communication (Haykin 2001; Proakis 2001). Between the transmitter and receiver, noise gets into the system and the received signal is corrupted by this noise. The channel through which the signal moves can also impact the shape of these functions as the signal passes through the wireless channel, distorting them, making it necessary to reconstruct and reshape the pulse before it is detected. This detection will be affected by the noise in the system. Before we expand the discussion to the concept of signal space and M-ary

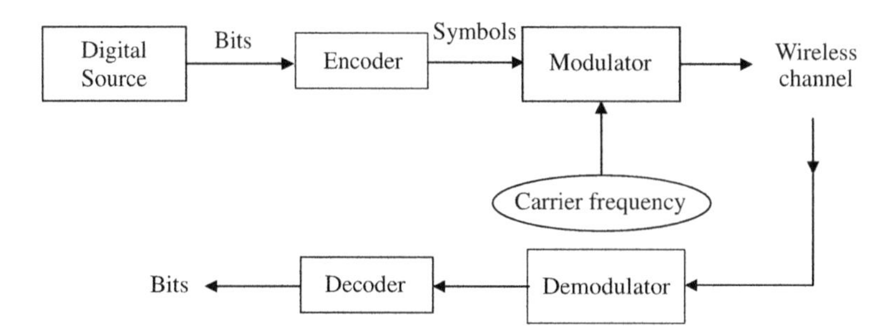

Fig. 3.1 Concept of digital data transmission

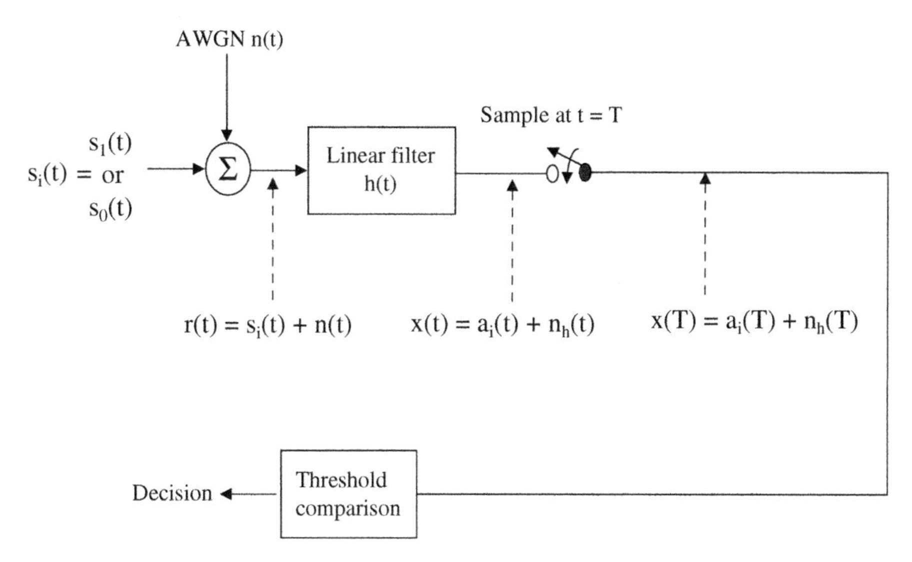

Fig. 3.2 Generic block diagram of a receiver

signaling (Taub and Schilling 1986; Sklar 1993, 2001), we discuss the detection of binary signals.

A generic block diagram of digital transmission is shown in Fig. 3.2. No carrier wave is present. Consider the transmission of two "bits," a "1" represented by a waveform $s_1(t)$ and a zero represented by a waveform $s_0(t)$. We shall examine the appropriate waveshapes later. Over a symbol period ($Ts = Tb = T$), the transmitted signal $s(t)$ is given by

$$s(t) = \begin{cases} s_1, & 0 \le t \le T \quad '1', \\ s_0, & 0 \le t \le T \quad '0'. \end{cases} \tag{3.1}$$

The channel, wired or wireless, is modeled as a linear filter of impulse response $h(t)$.

As shown in Fig. 3.2, the channel adds noise $n(t)$ to the signal and the received signal $r(t)$ can be expressed as

$$r(t) = s_i(t) + n(t), \quad i = 0, 1, \;\; 0 \le t \le T. \tag{3.2}$$

We will treat the noise as a zero mean Gaussian white noise (Papoulis and Pillai 2002). We will also assume that pulse has not been distorted or modified by the channel. For us to determine whether the transmitted bit is a "1" or "0," we need to have a single value at the receiver so as to compare it with a threshold value to determine whether the bit is a 1 or 0. This transformation of the information in (3.2) to a single value (known as the test statistic in literature) can take place if we have a filter of impulse response $h(t)$ followed by sampling at the end of the bit duration T.

If this sampled value is represented by $x(T)$, we can write

$$x(T) = a_i(T) + n_h(T) \quad i = 0, 1. \tag{3.3}$$

In (3.3), the first term on the right-hand side is the signal component and the second term is the noise component, with both components arising out of the filtering and sampling indicated in Fig. 3.2. Since we have assumed that the noise is Gaussian, the probability density function (pdf) of the noise component can be written as

$$f(n_h) = \frac{1}{\sqrt{2\pi\sigma^2}} \exp\left(-\frac{n_h^2}{2\sigma^2}\right). \tag{3.4}$$

In (3.4), s^2 is the noise variance. The density function of the sampled output can now be written as two conditional pdfs as

$$f(x|s_1) = \frac{1}{\sqrt{2\pi\sigma^2}} \exp\left[-\frac{(x - a_1)^2}{2\sigma^2}\right]. \tag{3.5}$$

$$f(x|s_0) = \frac{1}{\sqrt{2\pi\sigma^2}} \exp\left[-\frac{(x - a_0)^2}{2\sigma^2}\right]. \tag{3.6}$$

If we assume that 0's and 1's are equally likely, it can be easily shown (Chap. 2) that the optimal threshold X_0 that minimizes the error in the channel is given by (Taub and Schilling 1986)

$$X_0 = \frac{1}{2}(a_1 + a_0). \tag{3.7}$$

The average error $p(e)$ can be expressed as

$$p(e) = p(e|s_0)p(s_0) + p(e|s_1)p(s_1). \tag{3.8}$$

Since the bits are equally likely, (3.8) becomes

$$p(e) = \frac{1}{2}[p(e|s_0) + p(e|s_1)]. \tag{3.9}$$

Furthermore, because of symmetry, we have (Helstrom 1968; Haykin 2001)

$$p(e) = p(e|s_0) = p(e|s_1) \int_{X_0}^{\infty} f(x|s_0)\mathrm{d}x = \int_{-\infty}^{X_0} f(x|s_1)\mathrm{d}x = Q\left[\frac{a_1 - a_2}{2\sigma}\right]. \tag{3.10}$$

In (3.10), $Q(.)$ is the Gaussian Q function (Chap. 2) defined as (Borjesson and Sundberg 1979; Sklar 2001; Simon and Alouini 2005)

$$Q(z) = \frac{1}{\sqrt{2\pi}} \int_{z}^{\infty} \exp\left(-\frac{u^2}{2}\right) \mathrm{d}u. \tag{3.11}$$

To obtain the error expressed in (3.10), we need the characteristics of the appropriate filter, indicated in Fig. 3.2. If $H(f)$ is the transfer function of the filter, we can write the expression for the signal-to-noise ratio of the sampled output as

$$\left(\frac{S}{N}\right) = \frac{\left|\int_{-\infty}^{\infty} H(f)S(f)\exp(j2\pi fT)\mathrm{d}f\right|^2}{(N_0/2)\int_{-\infty}^{\infty} |H(f)|^2 \mathrm{d}f}. \tag{3.12}$$

In (3.12), the two-sided power spectral density of the noise is given by $(N_0/2)$ Watts/Hz. The remaining issue is the determination of the transfer function of the filter which maximizes the signal-to-noise ratio. We invoke the Schwarz inequality given by (Van Trees 1968; Taub and Schilling 1986; Gagliardi 1988; Haykin 2001)

$$\left|\int_{-\infty}^{\infty} g(x)h(x)\mathrm{d}x\right|^2 \leq \int_{-\infty}^{\infty} |g(x)|^2 \mathrm{d}x \int_{-\infty}^{\infty} |h(x)|^2 \mathrm{d}x \tag{3.13}$$

with the equality existing when

$$g(x) = k_0 h^*(x). \tag{3.14}$$

Note that in (3.14), * represents the complex conjugate and k_0 is a constant. Using (3.13) and (3.14), (3.12) for the maximum SNR becomes

$$\left(\frac{S}{N}\right)_{max} = \frac{2E}{N_0}, \tag{3.15}$$

where

$$E = \int_{-\infty}^{\infty} |S(f)|^2 \mathrm{d}f. \tag{3.16}$$

Equation (3.15) shows that the maximum SNR depends only on the signal energy; the input pulse shape is immaterial. Equations (3.12) and (3.14) lead to an expression for the optimum transfer function as

$$H(f) = k_0 h^*(f)\exp(-j2\pi fT). \tag{3.17}$$

The corresponding impulse response is

$$h(t) = \begin{cases} k_0 s(T-t), & 0 \leq t \leq T \\ 0 & \text{elsewhere} \end{cases}. \tag{3.18}$$

Fig. 3.3 Matched filter

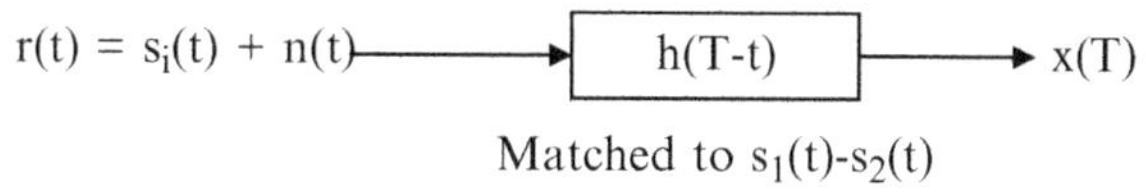

r(t) = s_i(t) + n(t) → h(T-t) → x(T)

Matched to s_1(t)-s_2(t)

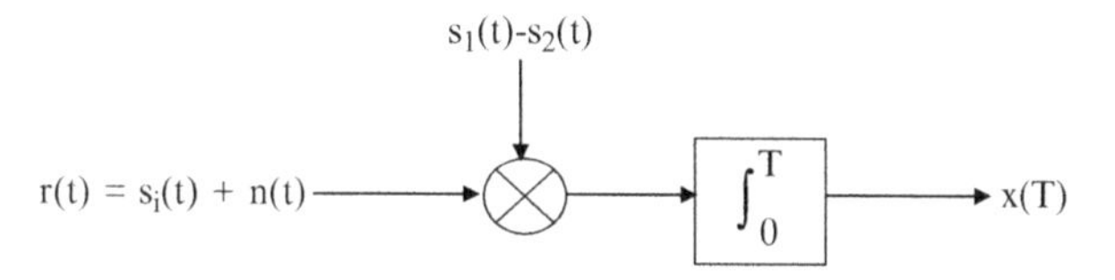

Fig. 3.4 Correlator

Equation (3.18) means that the impulse response of the optimum filter is a mirror image of the message signal delayed by the bit or symbol duration. The optimum filter is also known as the matched filter (Taub and Schilling 1986; Haykin 2001). If we now have a filter matched to the difference signal as shown in Fig. 3.3 with

$$h(t) = s_1(T - t) - s_0(T - t), \tag{3.19}$$

the maximum signal-to-noise ratio now becomes

$$\left(\frac{S}{N}\right)_{\text{max}} = \frac{(a_1 - a_0)^2}{\sigma^2} = \frac{2E}{N_0} \tag{3.20}$$

with

$$E_0 = \int_0^T [s_1(t) - s_0(t)]^2 \mathrm{d}t. \tag{3.21}$$

Using (3.21), the error in (3.10) becomes

$$p(e) = Q\left(\sqrt{\frac{E_0}{2N_0}}\right). \tag{3.22}$$

The matched filter is often replaced by the so-called correlator or a product integrator. All these signal processing elements behave the same way (Taub and Schilling 1986; Sklar 2001). The correlator or product integrator is shown in Fig. 3.4. Besides the incoming signal, the input to the correlator is the difference between the two signals, $s_1(t) - s_2(t)$.

We have so far used two signals $s_1(t)$ and $s_0(t)$ without specifying the shape of the pulses that represent the signals. The most common pulse shape is rectangular as shown in Fig. 3.5. But such a pulse when transmitted through any finite bandwidth channel will be broadened and possibly distorted. Broadened pulses lead to intersymbol interference (ISI) from the overlap of adjoining symbols which can lead to problems in correctly reconstructing the bits. Therefore, it is necessary to

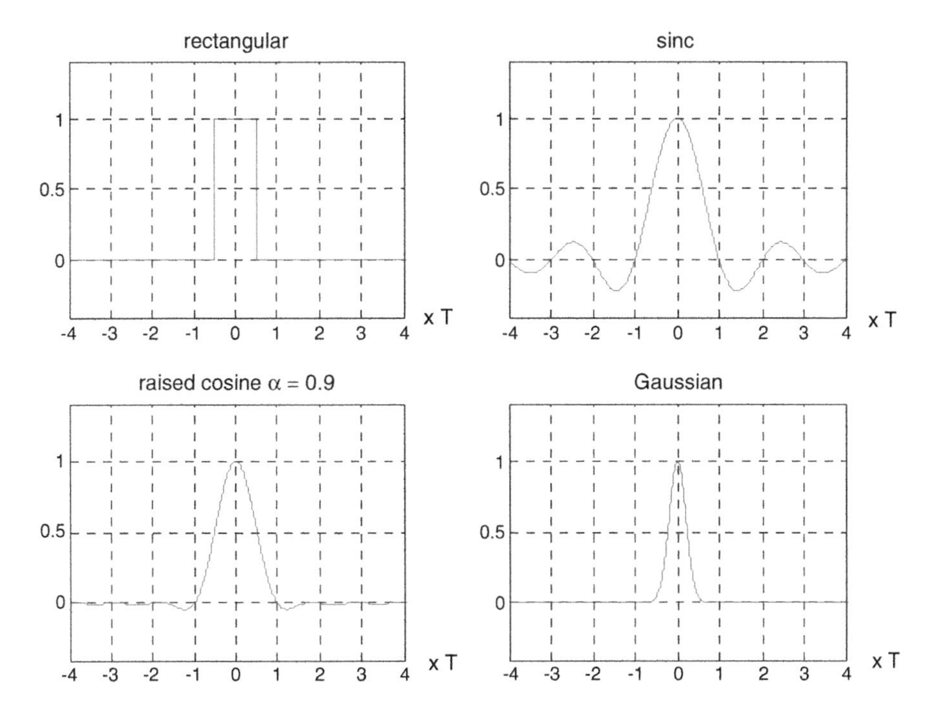

Fig. 3.5 Four common pulse shapes (rectangular, sin *c*, raised cosine, and Gaussian)

look at other pulse shapes along with the rectangular one to determine the optimum pulse shape with the least amount of distortion. Such a choice can be made on the basis of the so-called Nyquist's criteria (Lucky et al. 1968; Franks 1968; Lindsey and Simon 1973; Nyquist 2002; Anderson 2005). We will consider four pulse shapes: rectangular, sinc, raised cosine, and Gaussian. These are shown in Fig. 3.5. The shapes can be expressed in mathematical form as

$$s(t) = \begin{cases} \dfrac{1}{2T}\ 0 \le t \le T, & \text{rectangular,} \\ \dfrac{\sin((\pi/T)t)}{(\pi/T)t}, & \sin c, \\ \dfrac{\sin((\pi/T)t)}{(\pi/T)t}\dfrac{\cos((\pi/T)\alpha t)}{1-[(4\alpha/2T)t]^2} < 0 < \alpha < 1, & \text{raised cosine,} \\ \dfrac{\sqrt{\pi}}{\beta}\exp\left(-\dfrac{\pi^2 t^2}{\beta^2}\right),\ \beta = \dfrac{0.5887}{B}, & \text{Gaussian.} \end{cases} \tag{3.23}$$

The corresponding spectra are shown in Fig. 3.6. For the Gaussian pulse, the parameter B is the 3 dB bandwidth of the baseband Gaussian shaping filter.

A pulse that satisfies Nyquist's first criterion must pass through zero $t = kT$, $k = \pm 1, \pm 2, \ldots$ except at $t = 0$. By this requirement, the Gaussian pulse shape does

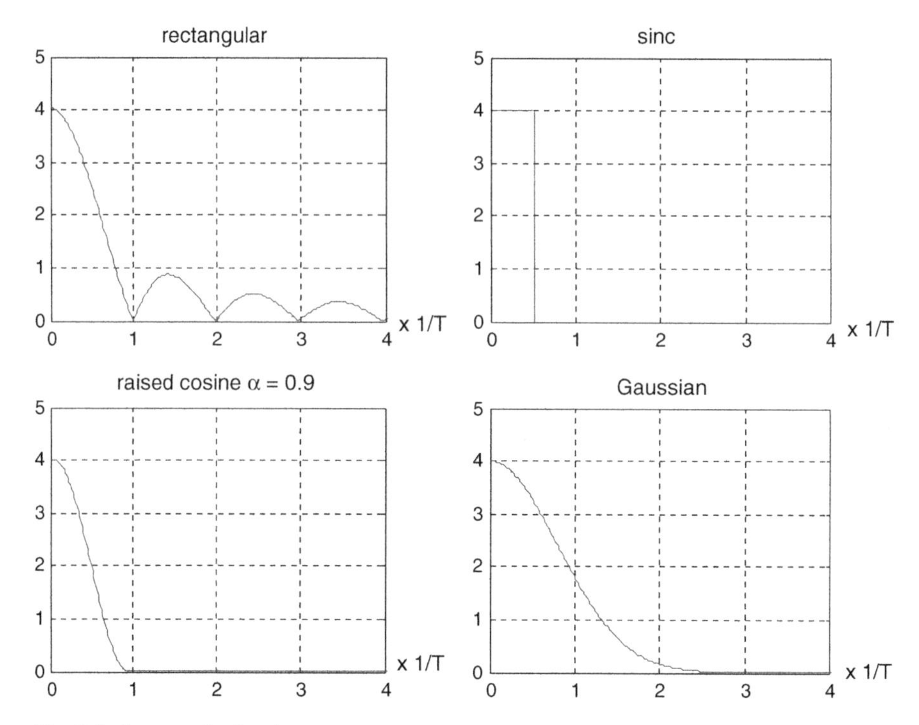

Fig. 3.6 Spectra of pulse shapes

not satisfy this criterion and the other three pulse shapes do. This does not mean that Gaussian pulses shapes are not used in wireless communication. Indeed, they are used in constant envelope modulation schemes, as we will see later. Of the three pulses that meet this criterion, the rectangular pulse has no crossover to the next bit interval. In fact, the ISI would vanish if the overall impulse response of the system was to match these pulse shapes. Note that the sin *c* pulse and raised cosine pulses become identical when the roll-off factor α is zero.

If one examines the pulse shapes using the frequency domain properties, one can see that the rectangular pulse has sidelobes that do not decay at a sufficiently high rate and leave a significant amount of energy in the sideband. This poses a problem. The bandwidth of the channel will be limited and the sidelobes will reappear since the filters used to limit the bandwidth do not have sharp cut-offs leading to interchannel cross stalk from the sidelobes of the next user occupying the next bit. Thus, the lack of tightness of the spectrum of the rectangular pulse makes it a less-than-ideal pulse. The sin *c* pulse, on the other hand, has the most limited bandwidth making it an ideal candidate on the basis of the "tight" spectrum. However, its sidelobes in the time domain decay very slowly and any synchronization problems will lead to increased ISI levels in the presence of timing jitter (i.e., if the sampling instants are not exactly at $t = \pm kT$). Thus, the raised cosine pulse is the most preferred pulse shape since it is possible to control both the temporal behavior of the pulse and the spectral characteristics using the roll-off factor α, even though use of $\alpha > 0$ will cause an increase in the bandwidth required to transmit the pulse (Kisel 1999; Nyquist 2002).

Nyquist's second criterion refers to the bandwidth associated with the pulse.

The Nyquist's pulse must have a bandwidth of at least $1/2T$ Hz. This statement is also known as the Nyquist's sampling theorem. From Fig. 3.4 it is seen that the sin c pulse occupies the lowest bandwidth of all the pulses and still satisfies the Nyquist's bandwidth criterion. Since the data rate R = $1/T$, the Nyquist's bandwidth corresponds to R/2 Hz, the lowest possible bandwidth that will be required for transmission (Sayar and Pasupathy 1986; Nyquist 2002; Beaulieu and Damen 2004).

These two Nyquist's criteria might not be sufficient to improve the performance of digital receivers. When multiple symbols are sent, correlation type receivers are used and, in this case, it is essential that the pulse shapes also satisfy the criterion that they are orthogonal under a shift of T's, i.e.,

$$\int_{-\infty}^{\infty} s(t)s(t-kT)\mathrm{d}t = 0, \quad k = \pm 1, \ \pm 2, \ldots. \tag{3.24}$$

From (3.24) it is clear that if the pulses satisfy Nyquist's criteria, including the orthogonality criterion, there will be less ISI (zero values from other pulses at sampling instants) and minimal or no effect of overlap integral (orthogonality).

3.3 Efficiency of Digital Modulation Techniques

Before we look at signal representation and modulation/demodulation techniques, let us also examine ways of quantifying the modems so that it will be possible to compare among them (Cahn 1959; Helstrom 1960, 1968; Salz 1970; Prabhu 1976a, b; Oetting 1979; Amoroso 1980; Sklar 1993, 2001). The two most important quantities (or qualities) are the power efficiency and bandwidth or spectral efficiency. Power efficiency is a measure of a modem to achieve a certain quality or fidelity in terms of a minimum value of probability of error. A particular modem which requires a lower level of SNR to maintain a fixed error rate has a better power efficiency than another modem which requires a higher level of SNR. The bandwidth or spectral efficiency is the ability of a modem to transmit more data in bits/s at a given bandwidth. Thus, a particular modem capable of transmitting data at a rate or R bps (bit/se) and B is the bandwidth occupied by the signal; the BW efficiency is defined as R/B. It will not be possible to have high spectral efficiency and high power efficiency at the same time; trade-offs will have to be made in practical situations. Furthermore, use of nonlinear amplifiers in communication systems may also force us to examine the out-of-band power since inadequate filtering followed by nonlinear amplifiers results in spectral regrowth, that can cause problems in linear modulation schemes such as phase shift keying systems.

All these considerations—the power efficiency, spectral efficiency, and out of band power—must be taken into account for the proper choice of the modem.

3.4 Geometric Representation of Signals and Orthonormal Functions

As discussed in the introduction, when we use an M-ary transmission system, we must have appropriate functional forms for the waveforms representing the signals. It will be convenient to study the behavior of these M-ary modulation formats and the effects of noise on these and compare the performance of the multiple M-ary level as well as different types of modulation formats. One of the ways in which this can be accomplished is through the use of signal space concepts (Proakis 2001; Sklar 2001; Haykin and Moher 2005; Shankar 2002). Let us define an N-dimensional orthogonal space as consisting of N linearly independent functions, called basis functions represented by the set {ci(t)} which satisfy the orthogonality conditions

$$\int_0^T \psi_i(t)\psi_k^*(t)\mathrm{d}t = K_i\delta_{ik}. \tag{3.25}$$

In (3.25), K is a constant and d is the Kroenecker delta function (Abramowitz and Segun 1965). If K is unity, the space becomes orthonormal, and it can be easily seen that K is related to the energy Ek since

$$E_k = \int_0^T \psi_k^2(t)\mathrm{d}t. \tag{3.26}$$

The waveforms of interest used in the previous analysis (for example, $s_0(t)$ and $s_1(t)$) can be expressed as linear combination of these orthogonal (orthonormal) set of functions:

$$s_i(t) = \sum_{k=1}^{N} a_{ik}\psi_k(t), \quad k = 1, 2, \ldots, M, N < M. \tag{3.27}$$

Using the orthogonality property, we have (Davenport and Root 1958; Cooper and McGillem 1986; Papoulis and Pillai 2002)

$$a_{ik} = \frac{1}{K_k}\int_0^T s_i(t)\psi_{ik}(t)\mathrm{d}t. \tag{3.28}$$

Let us look at the earlier case pertaining to $s_0(t)$ and $s_1(t)$. We now have only two waveforms. Looking at the two simplest examples of orthogonal functions, namely cos() and sine() functions, we have

$$\psi(t) = \begin{cases} \sqrt{\frac{2}{T}}\cos(2\pi f_0 t), & 0 \le t \le T \text{ or} \\ \sqrt{\frac{2}{T}}\sin(2\pi f_0 t), & 0 \le t \le T. \end{cases} \tag{3.29}$$

In (3.29), f_0 is the carrier frequency. Using (3.29), we have

$$s_1(t) = \begin{cases} \sqrt{\frac{2E}{T}}\cos(2\pi f_0 t), & 0 \le t \le T \text{ or} \\ \sqrt{\frac{2E}{T}}\sin(2\pi f_0 t), & 0 \le t \le T \end{cases} \tag{3.30}$$

$$s_2(t) = \begin{cases} -\sqrt{\frac{2E}{T}}\cos(2\pi f_0 t), & 0 \le t \le T \text{ or} \\ -\sqrt{\frac{2E}{T}}\sin(2\pi f_0 t), & 0 \le t \le T. \end{cases} \tag{3.31}$$

In (3.30) and (3.31), E is the energy. We can now go back to (3.22) and express the error in terms of the energy obtained through the orthogonal representations. Consider the case where we treat the two waveforms as orthogonal. In this case, we have

$$s_1(t) = \sqrt{\frac{2E}{T}}\cos(2\pi f_0 t), \quad 0 \le t \le T \tag{3.32}$$

$$s_2(t) = \sqrt{\frac{2E}{T}}\sin(2\pi f_0 t), \quad 0 \le t \le T. \tag{3.33}$$

Using (3.21), we have

$$E_0 = \int_0^T s_1^2(t)\mathrm{d}t + \int_0^T s_2^2(t)\mathrm{d}t - 2\int_0^T s_1(t)s_2(t)\mathrm{d}t = 2E. \tag{3.34}$$

The third integral in (3.34) is zero (by virtue of the orthogonality). The probability of error expressed in (3.22) now becomes

$$p(e) = Q\left(\sqrt{\frac{E}{N_0}}\right). \tag{3.35}$$

If we now take the case where $s_1(t)$ and $s_0(t)$ are not orthogonal and treat them in bipolar form, we have

$$s_1(t) = \sqrt{\frac{2E}{T}}\cos(2\pi f_0 t), \quad 0 \le t \le T \tag{3.36}$$

$$s_2(t) - s_1(t) = \sqrt{\frac{2E}{T}} \cos(2\pi f_0 t), \quad 0 \le t \le T. \tag{3.37}$$

In this case, using (3.21), we have

$$E_0 = \int_0^T s_1^2(t)\ \mathrm{d}t + \int_0^T s_1^2(t)\mathrm{d}t + 2\int_0^T s_1^2(t)\mathrm{d}t = 4E. \tag{3.38}$$

The probability of error now becomes

$$p(e) = Q\left(\sqrt{\frac{2E}{N_0}}\right). \tag{3.39}$$

We can now express (3.35) and (3.39) together in one equation using the Euclidian distance between the waveforms as (Proakis 2001; Haykin 2001)

$$p(e) = Q\left(\sqrt{\frac{d_{\mathrm{min}}^2}{2N_0}}\right). \tag{3.40}$$

In (3.40), *d*min is the Euclidian distance between the two waveforms (shown in Fig. 3.7). When we have a set of orthogonal waveforms, we have the example of binary orthogonal signaling; when we have the case of two identical waveforms of opposite signs, we have the case of bipolar signaling. If we had taken one of the two

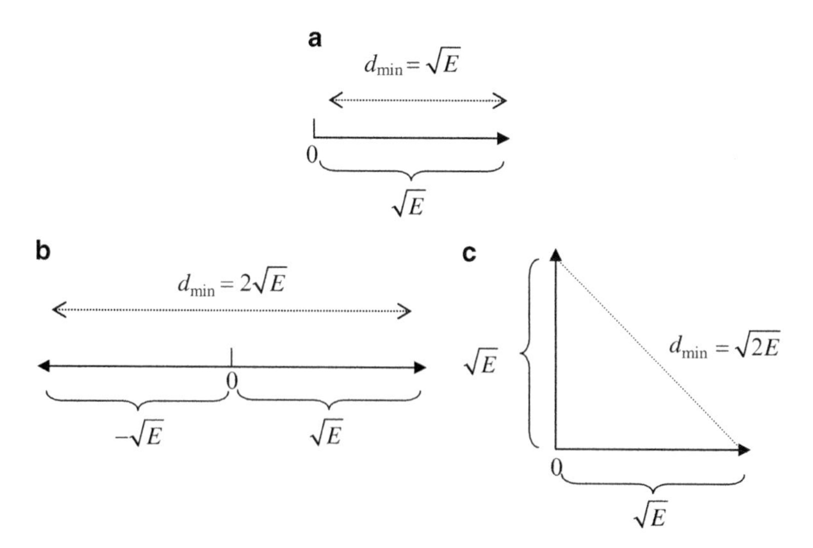

Fig. 3.7 Euclidian distance

signals to be zero, we would have the case of a unipolar signaling and the probability of error would be

$$p(e) = Q\left(\sqrt{\frac{E}{2N_0}}\right). \tag{3.41}$$

Thus, we see that the signal space representation makes it possible to represent the waveforms as well as estimate the error probabilities using the Euclidian distance between the signals. We will later extend this to M-ary signaling schemes.

3.5 Modulation Techniques

We will examine several digital modulation schemes (Lindsey and Simon 1973; Oetting 1979; Shanmugam 1979; Akaiwa and Nagata 1987; Akaiwa 1997; Sklar 2001; Proakis 2001; Stuber 2002). As mentioned in the introduction, the schemes include both linear and nonlinear modulations. They use amplitude, phase, amplitude and phase, and frequency. The demodulation techniques include those based on coherent techniques which require frequency and phase matching at the receiver as well as noncoherent techniques which constitute a wide variety of approaches. We will start with the linear ones first.

3.5.1 Amplitude Shift Keying

As the name suggests, this modulation technique involves the transmission of varying amplitudes of the carrier wave. The M-ary ASK signal can be expressed as

$$s(t) = \sqrt{\frac{2E_i}{T}}\cos(2\pi f_0 t)m \quad 0 \le t \le T, i = 1, 2, \ldots, M. \tag{3.42}$$

If $M = 2$, we have a binary ASK, and if $E_2 = 0$, we have an on–off-keying (OOK). In this case, when one uses a matched filter, the Euclidian distance will be $\sqrt{E}$ and the error at the output of the receiver will be given by

$$p(e) = Q\left(\sqrt{\frac{E}{2N_0}}\right). \tag{3.43}$$

If, however, we have a binary ASK and the two amplitudes are equal and take positive and negative values, $+\sqrt{E}$or $-\sqrt{E}$, the Euclidian distance is $2\sqrt{E}$ and we have a bipolar ASK and the error becomes

$$p(e) = Q\left(\sqrt{\frac{2E}{N_0}}\right). \tag{3.44}$$

As the value of M goes above 2, the error rate increases sharply and M-ary ASK is not used in wireless communications. Note that the error rate in (3.43) for OOK (unipolar ASK) is obtained under the assumption that the matched filter function is cos $(2\pi f_0 t)$ as required by (3.19). If instead, the matched filter function is 2 cos $(2\pi f_0 t)$, the error rate becomes

$$p(e) = Q\left(\sqrt{\frac{E}{N_0}}\right). \tag{3.45}$$

Use of this doubling ensures that the error rate comparison to binary phase shift keying (BPSK) can be made with identical powers of the local oscillator since BPSK needs a signal of 2 cos $(2\pi f_0 t)$ as required in (3.19). This is discussed in the next section on binary phase shift keying. One can also argue that in the OOK scheme, since one of the signals is zero, the average energy is $E\text{av} = E/2$, therefore replacing E by the average energy. Equation (3.43) becomes

$$p(e) = Q\left(\sqrt{\frac{E_{\text{av}}}{N_0}}\right). \tag{3.46}$$

3.5.2 *Phase Shift Keying*

This modulation technique involves the transmission of a carrier wave with varying values of the phase (Prabhu 1969, 1973, 1976a, b; Prabhu and Salz 1981). The M-ary PSK (MPSK) waveform can be expressed as

$$s(t) = \sqrt{\frac{2E}{T}} \cos\left[2\pi f_0 t + \phi_i(t)\right], \quad 0 \le t \le T, \; i = 1, 2, \ldots, M. \tag{3.47}$$

In (3.47), the phase term $\phi_i(t)$ will have M discrete values given by

$$\phi_i(t) = \frac{2\pi}{M} i, \quad i = 1, 2, \ldots, M. \tag{3.48}$$

Note that T is the symbol duration and E is the symbol energy. For the case of $M = 2$, we have

$$\begin{aligned} s(t) &= \sqrt{\frac{2E}{T}} \cos\left[2\pi f_0 t\right], \quad 0 \le t \le T, \text{“1”}. \\ s(t) &= \sqrt{\frac{2E}{T}} \cos\left[2\pi f_0 t + \pi\right], \quad 0 \le t \le T, \text{“0} \end{aligned} \tag{3.49}$$

As we can see, the bit "0" actually corresponds to "−1". Because of this, in PSK, "0" and "−1" are interchangeably used. One can also notice that the binary PSK is identical to bipolar ASK. Furthermore, we can use sin() instead of the cosine() term for BPSK and get

$$\begin{aligned} s(t) &= \sqrt{\frac{2E}{T}}\sin(2\pi f_0 t), \quad 0 \le t \le T; \quad \text{“1”}. \\ s(t) &= -\sqrt{\frac{2E}{T}}\sin(2\pi f_0 t), \quad 0 \le t \le T, \quad \text{“0”}. \end{aligned} \tag{3.50}$$

Thus, the BPSK involves the transmission of two phases, either 2π or π representing the two bits. This transmission is also referred to as antipodal because the two vectors representing the bits are in the opposite direction, regardless of whether one uses sine or cosine carrier waves. The generic forms of a BPSK transmitter and coherent BPSK receiver are shown in Fig. 3.8.

The binary data is encoded with either one of the two phase values in the transmitter. The receiver shown in Fig. 3.8 is of a coherent type and requires phase and frequency match. Assuming that the matches with the phase and frequency of the incoming BPSK signal are perfect, the multiplier followed by the integrator acts as the coherent detector. While a bandpass filter at the transmitter prevents any spectral overlap with signals from adjoining channels, if any, the low pass filter at the receiver eliminates any high frequencies that might come through.

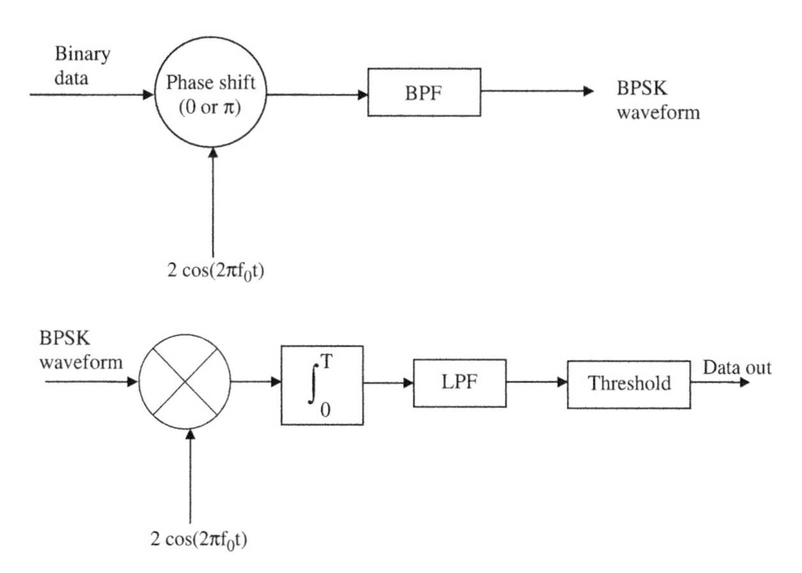

Fig. 3.8 BPSK modulator (*top*), BPSK demodulator (*bottom*)

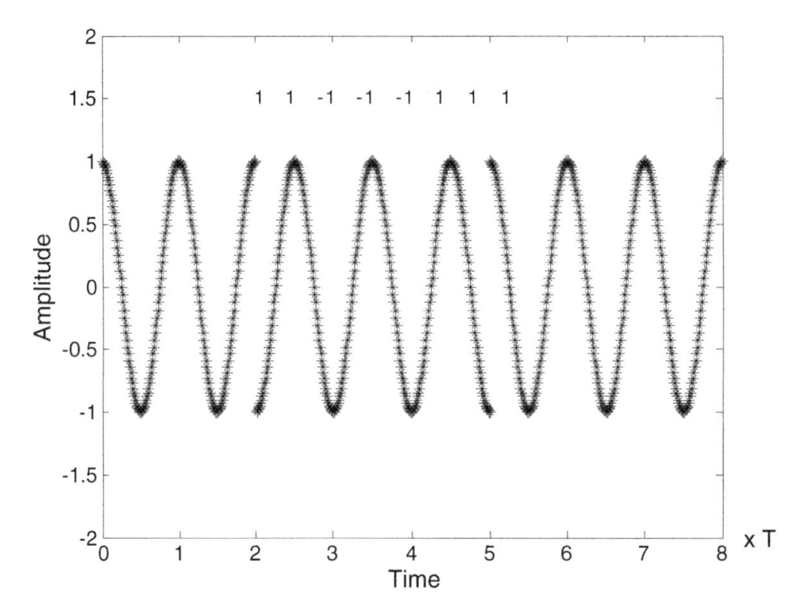

Fig. 3.9 A typical BPSK waveform

Since the two waveforms are antipodal or bipolar, the error rate becomes

$$p(e) = Q\left(\sqrt{\frac{2E}{N_0}}\right). \tag{3.51}$$

A typical BPSK waveform is shown in Fig. 3.9. It shows the abrupt changes in phase that can take place. As we shall see later when we discuss M-ary PSK, BPSK suffers from poor spectral efficiency and, therefore, it is not well suited for data transmission systems when there is premium on the bandwidth. Because of this, while BPSK offers a simple modulation format, it is generally not used in wireless communications (Feher 1991; Rappaport 2002; Molisch 2005; Goldsmith 2005; Schwartz 2005).

3.5.3 Frequency Shift Keying

The general expression for the FSK waveform can be expressed as

$$s(t) = \sqrt{\frac{2E}{T}}\cos\left[2\pi f_i t + \phi(t)\right], \quad 0 \le t \le T,\ i = 1, 2, \ldots, M. \tag{3.52}$$

The frequency, ϕ_i, takes on M discrete values and ϕ is an arbitrary constant value of the phase (Aulin and Sundberg 1981; Sklar 2001; Proakis 2001). The binary FSK

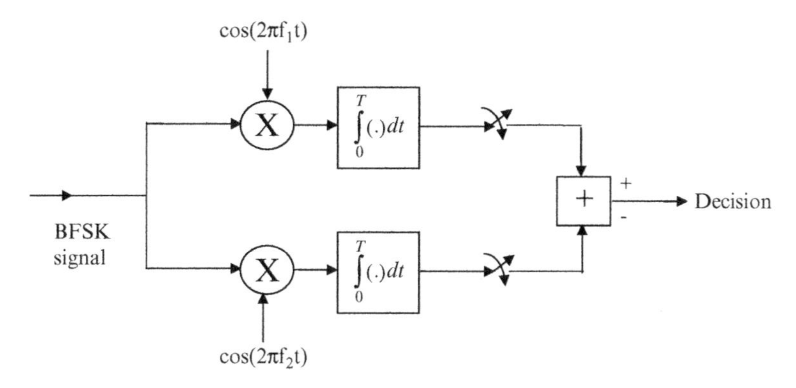

Fig. 3.10 Block diagram of a BFSK modulator

waveform is sketched in Fig. 3.10. One can see that there is an abrupt change in the frequency when bits change. If the frequencies are chosen so that the M signals are mutually orthogonal, we have the case of an orthogonal FSK. The requirement on the frequency separation to create FSK is discussed later in the description of M-ary FSK where it is shown that the minimum separation is $(1/2T)$.

For an orthogonal BFSK, the error rate in coherent reception can be written from the results in (3.35) as

$$p(e) = Q\left(\sqrt{\frac{E}{N_0}}\right). \tag{3.53}$$

The coherent receiver can be constructed similarly to the one for the BPSK, with two different modulators in parallel, each tuned to the appropriate carrier frequency.

It is seen that BFSK will have a poorer error rate performance than BPSK. But, the error rate is identical to that of OOK in (3.45). Since we are using two frequencies, intuitively we can see that the bandwidth efficiency will be poorer than BPSK. We will compare the efficiencies of all the modems later.

An example of orthogonal BFSK is shown in Fig. 3.11. One can see that BFSK can also result in undesirable discontinuities in the signal waveform. This binary form of FSK is also known as discontinuous FSK. This existence of discontinuities can be eliminated using a different form of FSK known as the continuous phase FSK (CPFSK) (discussed later).

The bit error rates for BPSK and BFSK are plotted in Fig. 3.12. For a bit error rate of 1e–6, the BPSK requires an SNR of about 10.5 dB while BFSK requires about 13.5 dB. This shows that the power efficiency of BFSK is 3 dB worse than that of BPSK.

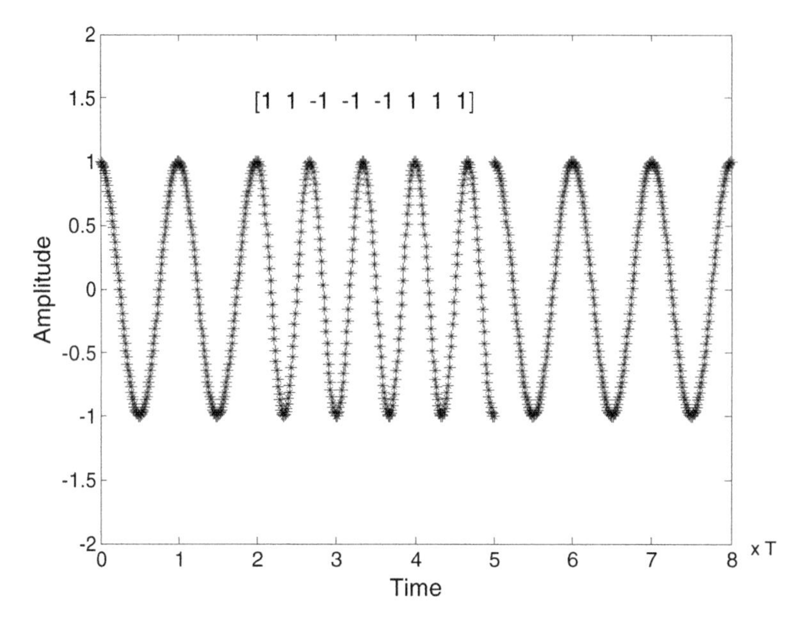

Fig. 3.11 An orthogonal FSK waveform

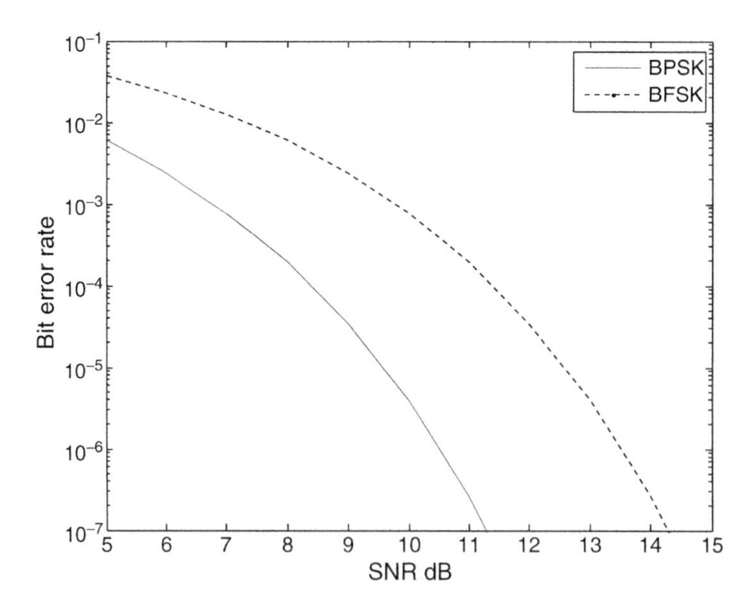

Fig. 3.12 Error rates for BPSK and BFSK

3.5.4 Amplitude and Phase Shift Keying

There is another form of digital modulation which combines amplitude and phase shift keying schemes for values of $M > 2$. This modulation format is called the M-ary Quadrature Amplitude Modulation (MQAM) and we will examine this modulation scheme along with rest of the M-ary modulation schemes (Campopiano and Glazer 1962; Cahn 1960; Thomas et al. 1974; Prabhu 1980; Sklar 1983a, b; Webb 1992; Proakis 2001).

3.5.5 Limitations of BPSK and Justification for MPSK

While the BPSK scheme is rather simple to implement (modulation and detection), its bandwidth efficiency is poor. Since there is a premium on available bandwidth, wireless systems require schemes that have high bandwidth efficiencies. The bandwidth efficiency can be understood by plotting the spectrum of BPSK (Fig. 3.13). The spectra are shifted so as to be centered around the zero frequency. They are all normalized to the value at $\phi = 0$. On the basis of a null-to-null criterion, BPSK requires a bandwidth of B Hz for a data rate of B bits/s. This high bandwidth requirement of BPSK can be compared with the bandwidth required for a 4-level PSK (QPSK), which has two bits/symbol. Based on the null-to-null criterion, QPSK requires a bandwidth of (B/2) Hz, thus doubling the

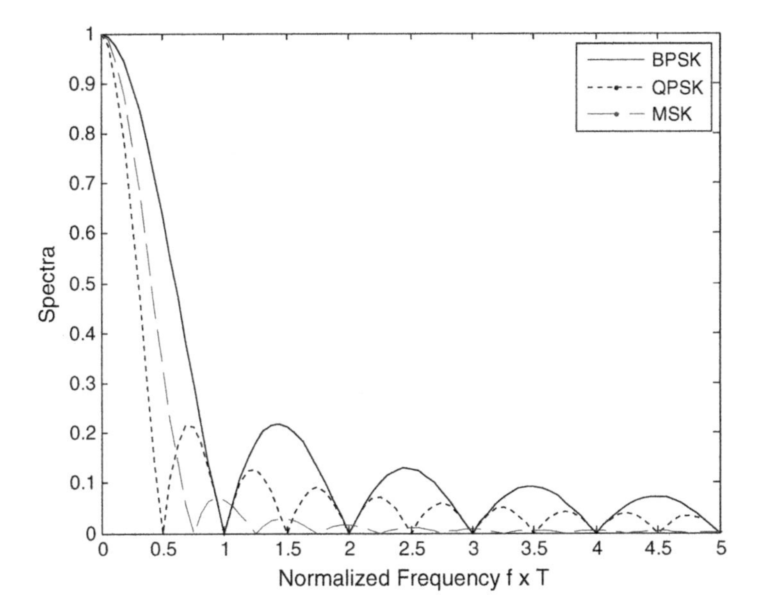

Fig. 3.13 Spectra of digital signals

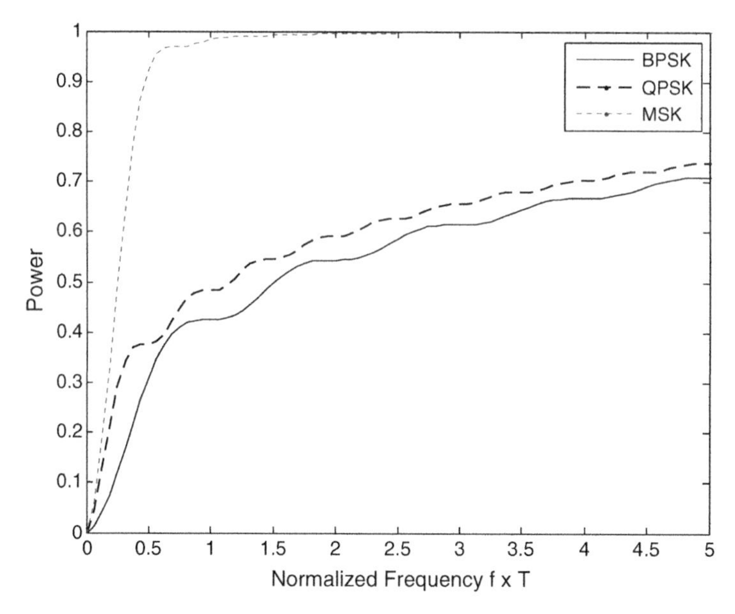

Fig. 3.14 Integrated power of digitally modulated signals

spectral efficiency (Helstrom 1960; Amoroso 1980; Forney et al. 1984; Kim and Polydoros 1988; Feher 1991). The QPSK modulation scheme (and possibly other higher level modulation schemes) offers a more bandwidth efficient scheme in wireless systems. The spectrum of minimum shift keying (MSK) modem is also shown. The MSK modem will be discussed later in this section and we will return to this Fig. 3.13 later.

The power in the sidebands is also a crucial factor in the choice of the modems. The integrated power up to a certain frequency band is shown in Fig. 3.14.

The considerable amount of power in the sidebands for the BPSK and QPSK is clearly seen. On the other hand, for MSK, significant amount of power resides within a narrow band of frequencies and point out the spectral characteristic typical of linear (PSK) and nonlinear (MSK) modulation schemes. We will revisit this issue later when we compare the modems.

3.5.5.1 M-ary Signals

One of the reasons for using M-ary signaling schemes ($M > 2$) is the spectral efficiencies that can be gained over the binary schemes (specifically BPSK). This was seen in Fig. 3.13. One can see that as M goes up from 2 to 4 (BPSK to QPSK), the bandwidth requirement will proportionately decrease in the case of PSK. But bandwidth efficiency and power efficiency cannot improve simultaneously and we need to take a close look at the symbol error rates in M-ary schemes to understand the trade-off between these two important characteristics of digital modems.

In a typical M-ary signaling, sequence of bits are grouped into k bits at a time such that we have $M = 2k$ waveforms, with the binary case occurring when $k = 1$. For $M = 4$ in PSK, the receiver is supposed to detect correctly one of the four phases, 0, $\pi/2$, π, $3\pi/2$ which differ only by $\pi/2$ (Proakis 2001; Gagliardi 1988; Haykin 2001). Thus, even under the best case scenario, the detection experiments will be less reliable compared with a BPSK case where the two phase values differ by double the difference in QPSK, namely π. Intuitively, this suggests that the symbol error in QPSK is likely to be higher than that in BPSK. If we extend this notion of declining difference between adjoining symbols as M increases, we can see that the power efficiencies decline as M increases.

If we extend the arguments for MPSK to MFSK, we will come to a different conclusion. When M increases, the bandwidth required to transmit M waveforms goes up since we expect them to be orthogonal. Therefore, a minimum spectral separation between adjoining signals needs to be established. Thus, for MFSK, the bandwidth efficiency goes down as k ($M = 2k$) increases and, consequently, we expect that the power efficiency will improve as k increases. Thus, M-ary signaling permits us to transmit data efficiently with a trade-off between power and spectral efficiencies. If bandwidth is a premium, PSK is likely to be a preferred choice. If power efficiency is a must, then FSK is the choice. We will now concentrate on MPSK and explore such a trade-off in greater detail.

3.5.5.2 Signal Space Picture of MPSK

We had seen earlier that the signal space view of unipolar, bipolar, and orthogonal waveforms can represent binary ASK, PSK, and FSK. We had also seen that the optimum threshold is the midway point between the two energy levels. Keeping that in mind, Fig. 3.15 shows the signal space for MPSK with $k = 1$, 2, and 3.

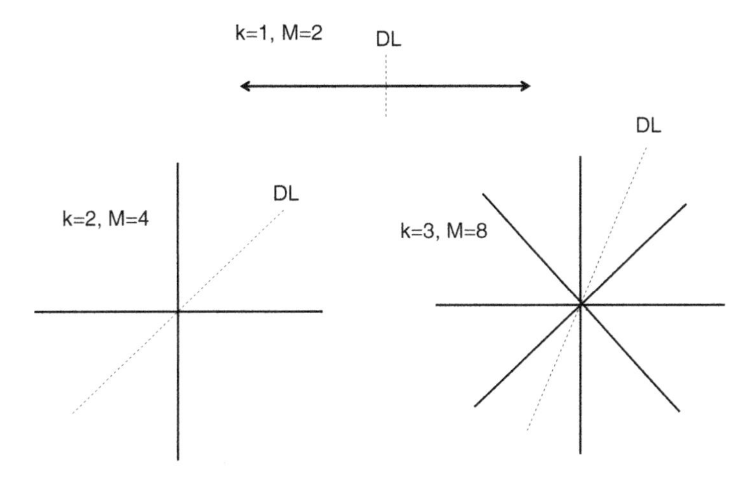

Fig. 3.15 Signal space aspects of BPSK, QPSK, and 8-PSK

The length of the vectors (the arrows from the origin) is all identical. The decision line (DL) is shown as the midway point between the vectors (only a single DL is shown for $k > 1$). Note that each symbol consists of k bits. The dotted lines show the boundary of the decision line (mid point between adjoining vectors).

It can easily be seen that as k increases, we have more and more signals vectors occupying the limited space, decreasing the separation between the adjoining ones, and leading to increased errors in detection. While this is happening, each symbol carries more bits with increasing k allowing a reduction in bandwidth and, hence, a gain in spectral efficiency. The only means to reduce the error while maintaining the gain spectral efficiency is to increase the length of the vectors. This will lead to an increase in the energy, point to a decline in power efficiencies, and place an upper limit on k based on the maximum power that can be transmitted. Let us now look at the case of a 4-level PSK or QPSK.

Since for MPSK schemes where $k > 1$, we have more than one bit/symbol, we resort to gray encoding (Schwartz 1980; Simon et al. 1995). The pairing of bits based on gray encoding is also shown in Table 3.1.

If b's represent the binary code values and g's represent the corresponding gray code values, we can obtain the gray encoding as

$$g_1 = b_1 \tag{3.54}$$

$$g_m = b_m \oplus b_{m-1}, \quad m > 1. \tag{3.55}$$

In (3.55), $\oplus$ represents modulo-2 addition of binary numbers. As one can see, gray encoding ensures that adjacent symbols differ only by a single bit out of the

Table 3.1 Gray encoding ($k = 2, 3, 4$)

	Binary code				Gray code			
Digit	b1	b2	b3	b4	g1	g2	g3	g4
0	0	0	0	0	0	0	0	0
1	0	0	0	1	0	0	0	1
2	0	0	1	0	0	0	1	1
3	0	0	1	1	0	0	1	0
4	0	1	0	0	0	1	1	0
5	0	1	0	1	0	1	1	1
6	0	1	1	0	0	1	0	1
7	0	1	1	1	0	1	0	0
8	1	0	0	0	1	1	0	0
9	1	0	0	1	1	1	0	1
10	1	0	1	0	1	1	1	1
11	1	0	1	1	1	1	1	0
12	1	1	0	0	1	0	1	0
13	1	1	0	1	1	0	1	1
14	1	1	1	0	1	0	0	1
15	1	1	1	1	1	0	0	0

$\log_2 M$ bits. This ensures that if noise causes errors, the error in the adjoining symbol will be accompanied by one and only one bit error.

3.5.5.3 Quaternary Phase Shift Keying

The waveforms for QPSK can be expressed as

$$s_i(t) = \sqrt{\frac{2E}{T_s}} \cos\left[2\pi f_0 t + \varphi_i\right], \quad i = 1, 2, 3, 4. \tag{3.56}$$

In (3.56), *Ts*, the symbol duration is equal to twice the bit duration (*T*b) and the phase ϕ_i will take one of the four values. Rewriting (3.56), we have the familiar representation of QPSK as

$$s_i(t) = \sqrt{\frac{2E}{T_s}} \cos\left[2\pi f_0 t\right] \cos\left(\phi_i\right) - \sqrt{\frac{2E}{T_s}} \sin\left[2\pi f_0 t\right] \sin\left(\phi_i\right). \tag{3.57}$$

The phase constellation corresponding to this waveform is shown in Fig. 3.16. It is easy to see that a different constellation can be obtained by shifting the phase of the carrier by $\pi/4$ as

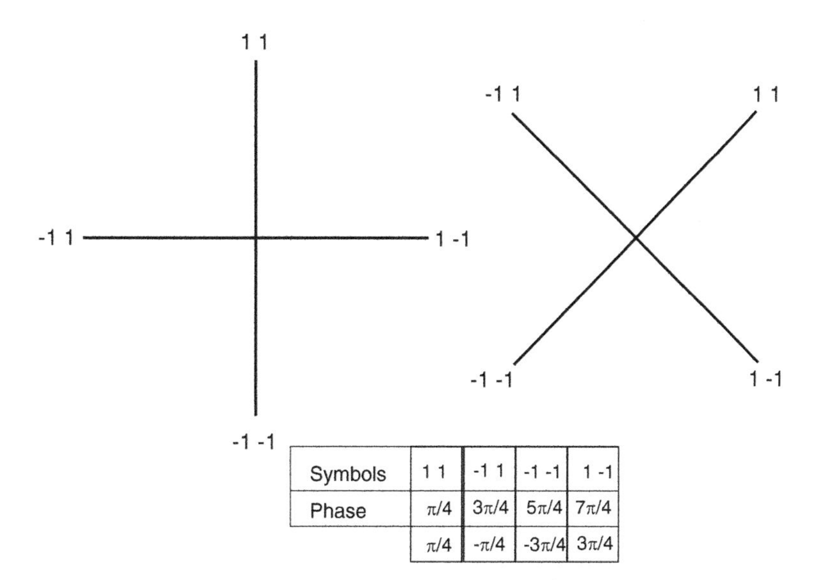

Symbols	1 1	-1 1	-1 -1	1 -1
Phase	π/4	3π/4	5π/4	7π/4
	π/4	-π/4	-3π/4	3π/4

Fig. 3.16 The QPSK constellations, symbols, and corresponding phases

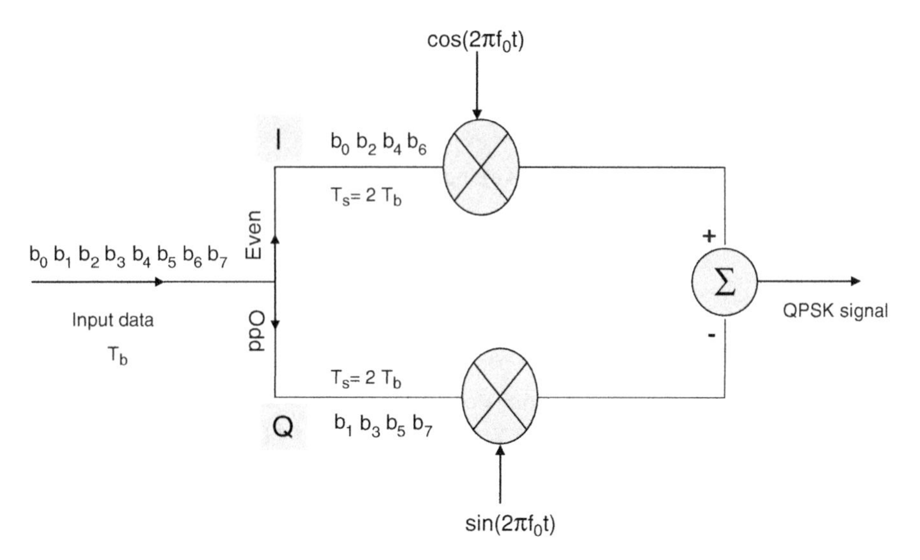

Fig. 3.17 QPSK modulator

$$
\begin{aligned}
s_i(t) = & \sqrt{\frac{2E}{T_s}} \cos\left[2\pi f_0 t + \pi/4\right] \cos\left(\phi_i\right) \\
& - \sqrt{\frac{2E}{T_s}} \sin\left[2\pi f_0 t + \pi/4\right] \sin\left(\phi_i\right).
\end{aligned}
\tag{3.58}
$$

One can choose either one of the two sets of phases given in the two rows. We can also observe that QPSK is a combination of two BPSK signals in quadrature, with the first being identified as the inphase term and the second one as the quadrature term. The phase encoding associated with the two constellations is given in Table 3.1.

A block diagram of the QPSK modulator is shown in Fig. 3.17.

It shows that the input data is split into the I stream (even) and Q stream (odd) and the duration changing from Tb to Ts (2Tb). The carrier frequency components could be either $(2\pi f_0 t)$ or $(2\pi f_0 t + \pi/4)$.

Typical time domain waveforms based on both constellations are shown in Fig. 3.18. While the linear aspect of the modulation is seen, one can also observe that the waveforms have discontinuities with phase jumps of 0, $\pm\pi/2$, $\pm\pi$. This discontinuity can lead to problems of regrowth of sidelobes after filtering; it will be ideal if these phase jumps can be reduced. Note that the significant amount of spectral power in the sidebands was observed in Fig. 3.14 for the case of QPSK.

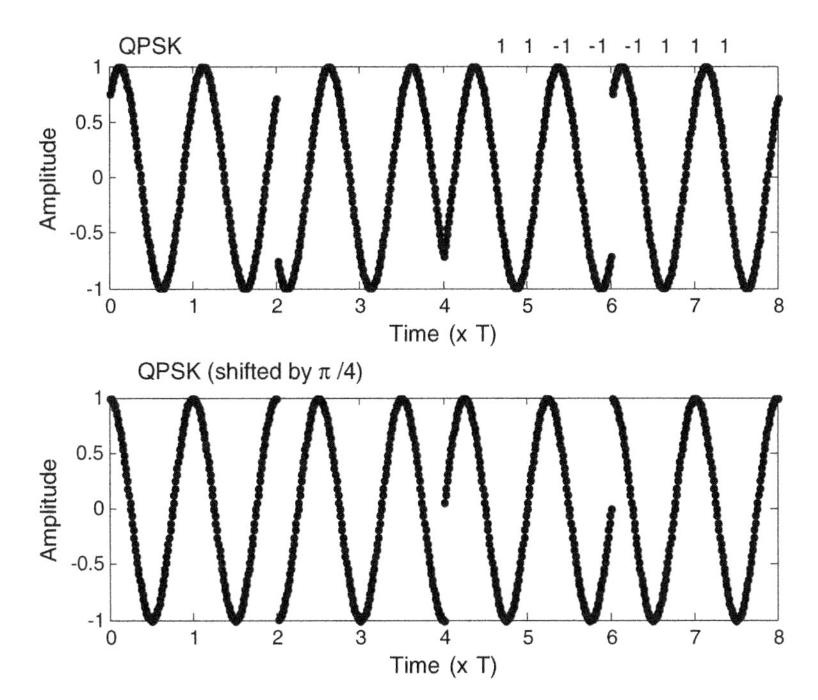

Fig. 3.18 QPSK waveforms

3.5.5.4 Offset QPSK

One of the ways in which the issue of phase discontinuity can be reduced is through the use of an offset-QPSK scheme (OQPSK). In this case, the I and Q streams are delayed/offset by one bit period (Gronemeyer and McBride 1976; Sampei 1997; Sklar 2001; Proakis 2001). A block diagram to implement OQPSK is shown in Fig. 3.19 and a typical waveform associated with OQPSK corresponding to the same bit stream in QPSK is shown in Fig. 3.20. It can be seen that the phase jumps are now limited to 0, $\pm\pi/2$ even though the phase jumps happen more often than in QPSK.

3.5.5.5 $\pi/4$-QPSK

These phase discontinuities are still unacceptable and efforts can be made to reduce them by choosing the scheme known as the $\pi/4$-QPSK which utilizes both the constellations of QPSK, one with and without the phase shift of $\pi/4$ in (3.57) and (3.58). The constellations alternate between the two for the consecutive symbols and this can be expressed as

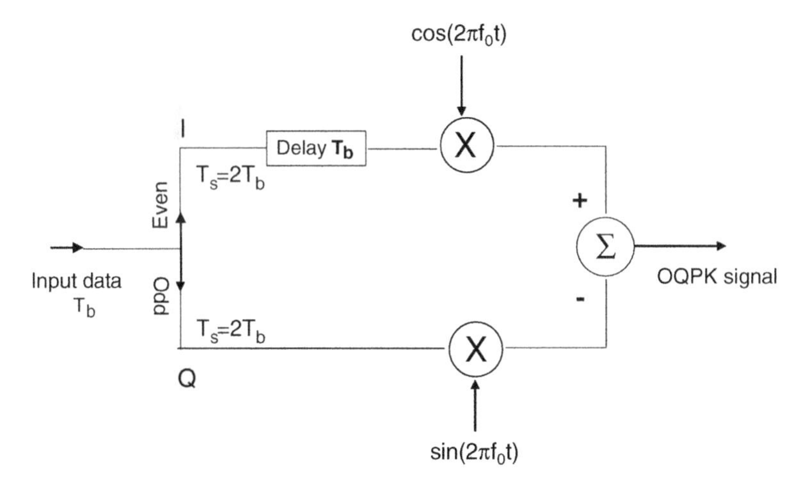

Fig. 3.19 OQPSK modulator

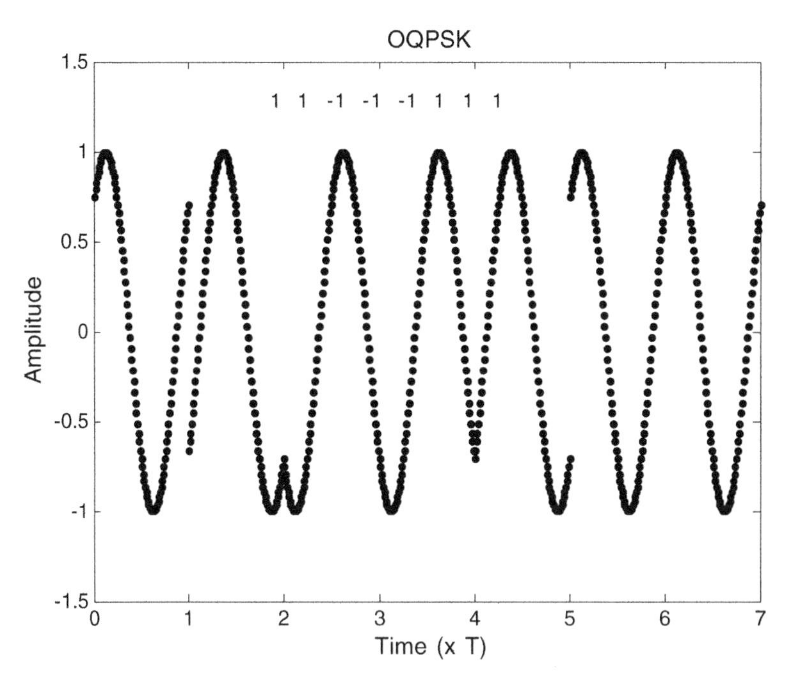

Fig. 3.20 OQPSK waveform

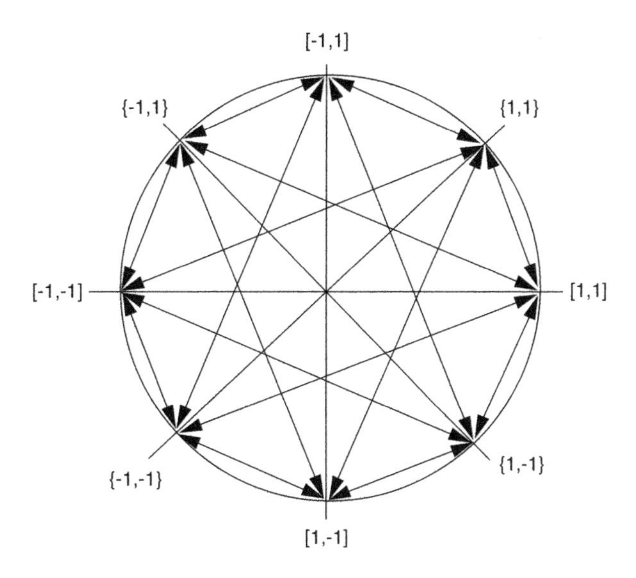

Fig. 3.21 $\pi/4$ QPSK constellation and transition diagram

$$s_2(t) = \begin{cases} \cos(2\pi f_0 t)\cos(\phi_n) \pm \sin(2\pi f_0 t)\sin(\phi_n) \rightarrow n\text{th symbol} \\ \cos\left(2\pi f_0 t + \frac{\pi}{4}\right)\cos(\phi_n) \pm \sin\left(2\pi f_0 t + \frac{\pi}{4}\right)\sin(\phi_n) \rightarrow (n+1)\text{th symbol.} \end{cases} \tag{3.59}$$

Note that in (3.59), ϕ_n is the phase value for the nth symbol in Fig. 3.16. The complete phase constellation associated with $\pi/4$-QPSK is shown in Fig. 3.21.

Since the waveforms alternate between the two, there will be no phase jumps of 0, π, and $\pi/2$ in the $\pi/4$ QPSK waveform reducing the sharp jumps seen with QPSK and OQPSK waveforms. A waveform of the $\pi/4$ QPSK signal is shown in Fig. 3.22. It is seen from the waveform that the phase difference between successive symbols is now limited to an odd multiple of $\pi/4$, and the jumps of $\pi/2$ and π have been eliminated. A slightly modified version of $\pi/4$-QPSK is called $\pi/4$-DQPSK (Differential QPSK) which uses the accumulated phases so that the waveform can be written as

$$s(t) = \cos\left(2\pi f_0 t + \frac{\pi}{4}\right)\cos(\theta_n) \pm \sin\left(2\pi f_0 t + \frac{\pi}{4}\right)\sin(\theta_n), \tag{3.60}$$

where

$$\theta_n = \theta_{n-1} + \phi_n. \tag{3.61}$$

The waveform corresponding to $\pi/4$-DQPSK is shown in Fig. 3.23 and identical to the one for $\pi/4$-QPSK. The advantage of $\pi/4$-DQPSK is the fact that differential encoding (accumulated phases) is used, permitting an easy differential detection as well.

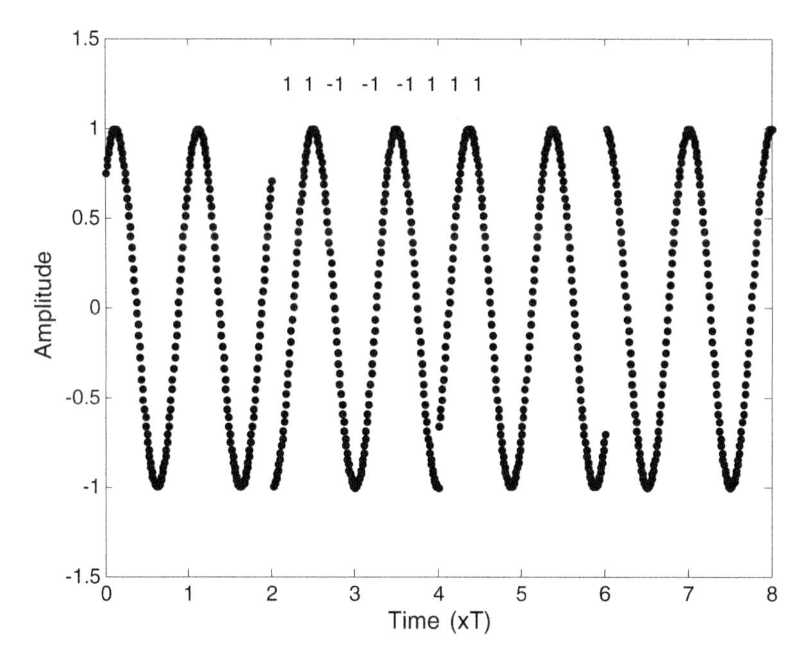

Fig. 3.22 $\pi/4$ QPSK waveform

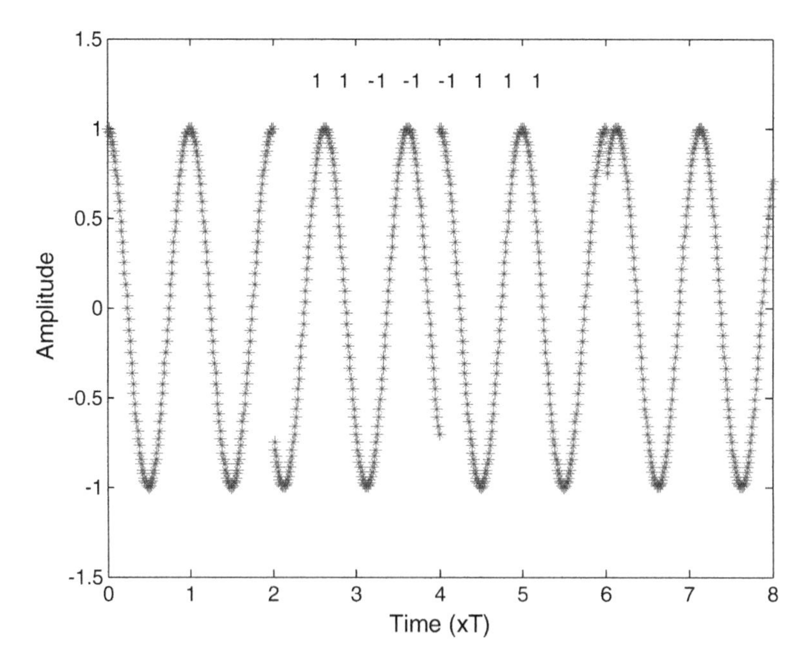

Fig. 3.23 $\pi/4$DQPSK waveform

3.5.5.6 Symbol Error in M-ary PSK

When a sequence of bits (i.e.,) symbols is transmitted as in the case with M-ary systems, the calculation of error rates is not as straightforward as in the case of a binary case which only has one bit/symbol. Since it is possible to encode or map the k bits into M symbols in many ways, we need to ensure that mapping is done to minimize errors. For example, if the mapping is done such that adjacent symbols differ only by a single bit, then, if and when an error occurs, that symbol will be misidentified as the one above it or below it. This form of mapping is known as the gray encoding shown in Table 3.1, and in this case only a single bit error in a sequence exists (Cahn 1959; Prabhu 1969; Schwartz 1980; Gagliardi 1988; Simon et al. 1995; Haykin 2001; Anderson 2005).

For the M-ary PSK, we will assume that we have a gray level encoding. A constellation of an M-ary PSK is shown in Fig. 3.24.

We assume that M is large and just as in the case of the BPSK and QPSK, the length of the vector is $\sqrt{2E}$, which is the radius of the circle. Since the angular separation between the adjacent vectors $2\pi/M$ rad, we can calculate the separation between the adjoining symbols vectors, $d$min, as it was defined for the case of binary modulation schemes. Using the law of cosines, we have (Gagliardi 1988; Taub and Schilling 1986)

$$d_{\min}^2 = 4E\sin^2\left(\frac{\pi}{M}\right). \tag{3.62}$$

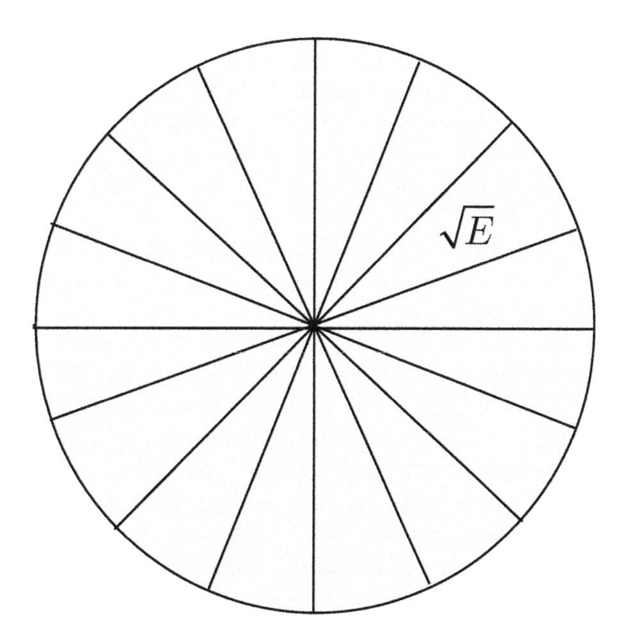

Fig. 3.24 A generic MPSK constellation

We can now obtain an approximate expression for the symbol error rate without undertaking a rigorous analysis. If we have gray encoding, the error results from misidentifying the symbol either as the one above it or below it. Since the separations are equal, these two errors will also be equal. Thus, the symbol error rate will be twice the error in misidentifying the symbol. The expression for the symbol error can now be written using (3.40) as

$$p_s(e) \approx 2Q\left(\sqrt{\frac{d_{\min}^2}{2N_0}}\right) = 2Q\left[\sqrt{\frac{2E}{2N_0}}\sin\left(\frac{\pi}{M}\right)\right]. \tag{3.63}$$

Specifying E as the symbol energy and identifying it as Es and the bit energy as Eb, (3.63) becomes

$$p_s(e) \approx 2Q\left[\sqrt{\frac{2E_s}{2N_0}}\sin\left(\frac{\pi}{M}\right)\right] = 2Q\left[\sqrt{\frac{2kE_b}{N_0}}\sin\left(\frac{\pi}{M}\right)\right]. \tag{3.64}$$

The equivalent bit error rate for M-ary PSK can be obtained from (3.64). Consider the case of QPSK. A symbol error occurs when either the first bit or the second bit, or both bits are in error. Therefore,

$$p_s(e) = p_b(e) + p_b(e) - p_b(e)p_b(e) \approx 2p_b(e). \tag{3.65}$$

Extending this argument, we can write that in general for MPSK

$$p_s(e) \approx kp_b(e). \tag{3.66}$$

The symbol error rates for coherent MPSK are plotted in Fig. 3.25. One can see that the error rates increase as M increases, and for $M = 2$ and 4 there is very little difference between the values of the error rates. This suggests the advantage of using QPSK over BPSK without incurring the need for additional SNR to maintain a required error rate, yet taking advantage of the lower bandwidth requirement of QPSK. From (3.64) and (3.66), it is clear that the bit error rates of coherent BPSK and QPSK are identical (Sklar 2001; Proakis 2001). Since QPSK, OQPSK, and $\pi/4$ QPSK are simply different versions of the 4-level PSK schemes, they all have the same bit error rates.

We can now revisit the earlier discussion on spectral efficiencies. We have seen that QPSK requires half as much bandwidth as BPSK (Fig. 3.13). As M increases, the symbol duration $T = T$s increases. Since Tb is identified as the bit duration, we have Ts $= kT$b. Thus, as M increases, the null-to-null bandwidth is reduced by k. This means that for the M-ary PSK system, the spectral efficiency goes up with M. On the other hand, we see that the symbol error continues to increase with k and, therefore, the power efficiency of M-ary PSK starts going down as M increases. Beyond a certain point, the increase in the minimum SNR required to maintain an error rate overwhelms any advantage gained through an

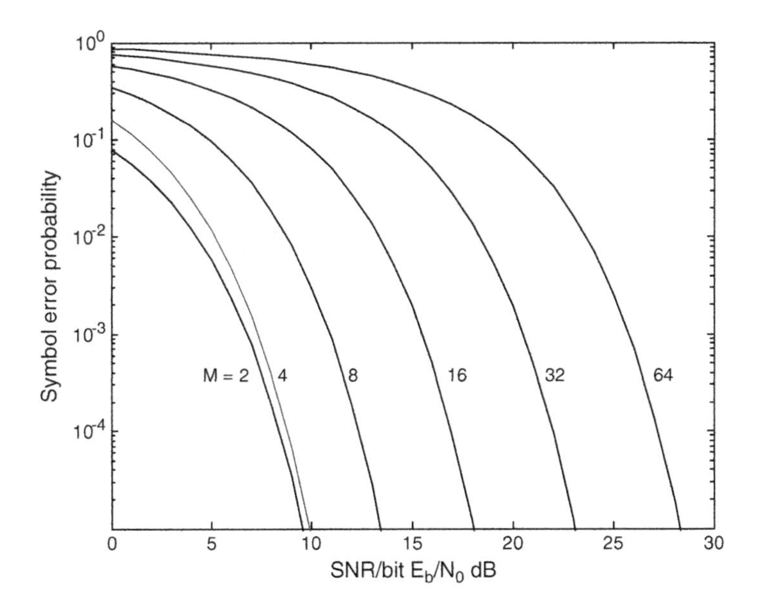

Fig. 3.25 Symbol error rates for coherent MPSK

Table 3.2 Spectral efficiencies of MPSK and the SNR required to maintain a BER of 1e–6

M	E_b/N_o dB	η bits/s/Hz
2	10.5	0.5
4	10.5	1
8	14	1.5
16	18.5	2
32	23.4	2.5
64	28.5	3

improvement in bandwidth efficiency; M-ary PSK ceases to be a viable option. This is given in Table 3.2 which shows the spectral efficiencies and the minimum SNR required to maintain a specified BER.

Both 8-level PSK and $\pi/4$ DQPSK are used in traditional wireless systems. The former is the modem used in the third generation GSM systems (EDGE) while the latter was the modem for the American Digital Cellular and Japanese Digital Cellular systems, before the providers decided to switch to CDMA based systems which use QPSK (Winters 2000; Olivier et al. 2003).

Prior to looking at nonlinear modulation schemes, we will look at a hybrid amplitude phase shift keying scheme, mentioned earlier (Prabhu 1980; Sklar 1983a, b; Webb 1992; Proakis 2001).

Even though this scheme is not used in wireless communications, modems based on this scheme offer a few advantages over MPSK. This scheme is more commonly known as M-ary Quadrature Amplitude Modulation (MQAM). The details of this scheme appear next.

3.5.5.7 M-ary QAM

A quadrature modulation scheme is also used in analog modulation. It is a means by which two channel amplitude modulation can be achieved requiring only the bandwidth required to transmit one channel, thus providing a significant bandwidth saving over transmitting two messages using frequency division multiplexing two carrier frequencies. The modulated signal in an analog QAM can be written as

$$s(t) = m_1(t)\cos(2\pi f_0 t) - m_2(t)\sin(2\pi f_0 t). \tag{3.67}$$

In (3.67), $m_1(t)$ and $m_2(t)$ are two analog baseband signals. It can be seen that the QPSK waveform in (3.57) and the QAM signal in (3.67) are same. Thus, it can be seen that the inphase and quadrature representation in (3.67) also represent a digital signal. If $m_1(t)$ and $m_2(t)$ are allowed to take only ± 1, we have the 4-level QAM. The difference between MQAM and MPSK is that in the former the information bits are encoded in both amplitude and phase while in the latter, the information bits are encoded in phase only (Thomas et al. 1974; Webb 1992; Proakis 2001). Thus, if we replace (3.57) as given below,

$$s(t) = A_i\cos(\phi_i)\cos(2\pi f_0 t) - A_i\sin(\phi_i)\sin(2\pi f_0 t). \tag{3.68}$$

We get the MQAM waveform. Note that for a 4-QAM, $A_i = \sqrt{2E_s/T_s}$, we see that 4-QAM and 4-PSK are exactly identical. Both of them have identical vector lengths. However, for $M > 4$, the vectors can take different lengths. The case of $M = 8$ is shown in Fig. 3.26 with four possible ways of generating 8-QAM.

All the constellations shown are "rectangular" indicating that there are unique phases and vector lengths. It is clear that the symbol error rates will be a function of the particular signal constellations used. As M increases, the multiplicity of constellations makes it difficult to derive a single expression for the symbol error. However, it can be easily seen from Fig. 3.26 that while for MPSK each symbol only has phase encoding, for MQAM there are two pieces of information for each symbol, the length of the vector and the phase. Thus, for $M > 4$, we intuitively expect that in a given average symbol energy (or SNR), the symbol error rates for MQAM will be less than the error rates for MPSK. In other words, while the spectral efficiency of MQAM and MPSK are identical, MQAM provides a higher power efficiency. A quantitative measure of this improvement b has been shown to be (Gagliardi 1988; Simon et al. 1995; Proakis 2001; Sklar 2001)

$$\beta = \frac{1}{2\sin^2(\pi/M)}\left(\frac{3}{M-1}\right). \tag{3.69}$$

As mentioned earlier, for $M = 4$, the power efficiency of PSK and QAM are identical. For $M = 4$, a few values of β are given in Table 3.3

Comparing QPSK and 4-QAM, for the trade-off with a slightly complex receiver, the power efficiency can be increased with QAM.

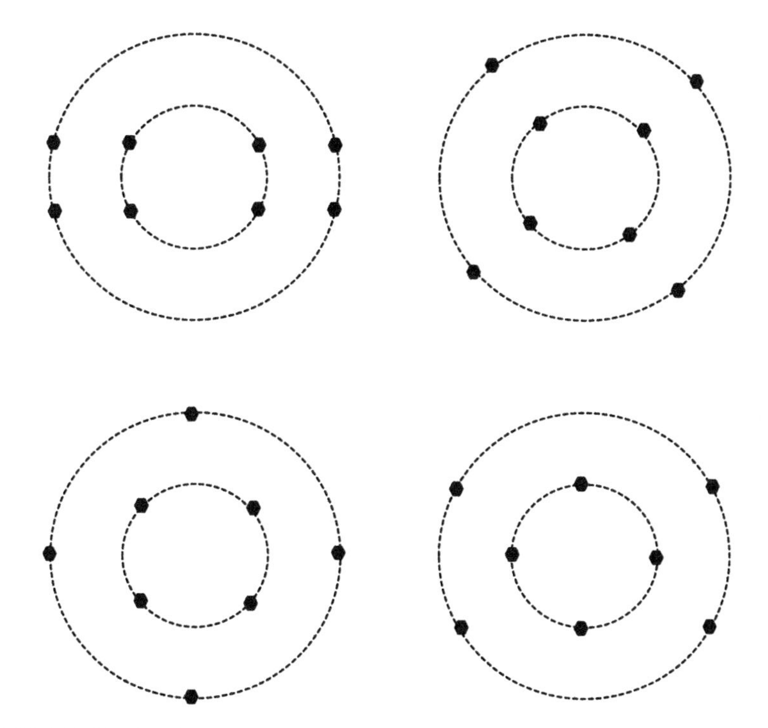

Fig. 3.26 MQAM signal space

Table 3.3 Improvement in power efficiency of MQAM over MPSK

M	β (dB)
4	1
8	1.65
16	4.2
32	7.02
64	9.95

3.6 General Nonlinear Modulation Schemes

One of the problems with linear modulation schemes is the presence of phase discontinuities and the consequences arising from it such as the regeneration of the sidelobes resulting in cross talk (Amoroso 1980; Prabhu 1981; Proakis 2001; Anderson 2005). This brings up the examination of the nonlinear modulation schemes which result in constant envelopes. One such modulation scheme discussed earlier was the frequency shift keying scheme with orthogonal functions or orthogonal FSK. There are variations of this theme of constant envelope modulations through the use of different pulse shapes, which can lead to modulation schemes different from FSK, e.g., the minimum shift keying (MSK) or the Gaussian minimum shift keying (GMSK). Before we look at those schemes, let us determine the minimum frequency separation required to make the FSK orthogonal.

3.6.1 Frequency Shift Keying

We take the example of a binary FSK, we have the two signals

$$\begin{aligned} s_1(t) &= \cos(2\pi f_1 t + \theta) \\ s_2(t) &= \cos(2\pi f_2 t). \end{aligned} \tag{3.70}$$

In (3.70), f_1 and f_2 are the two carrier frequencies and y is a constant phase. Use of this constant phase of arbitrary angle will allow us to separate the coherent FSK and noncoherent FSK. For the two signals to be orthogonal, we must have

$$\int_0^T s_1(t)s_2(t)\mathrm{d}t = \int_0^T \cos(2\pi f_1 t + \theta)\sin(2\pi f_2 t)\mathrm{d}t = 0. \tag{3.71}$$

Equation (3.71) simplifies to

$$\cos(\theta)\sin(2\pi\Delta f T) + \sin(\theta)[\cos(2\pi\Delta f T) - 1] = 0, \tag{3.72}$$

where

$$\Delta f = f_1 - f_2. \tag{3.73}$$

For a coherent FSK, there will be no arbitrary phase and hence for orthogonal FSK,

$$\sin(2\pi\Delta f T) = 0. \tag{3.74}$$

This leads to

$$\Delta f = \frac{1}{2T} \tag{3.75}$$

On the other hand, for noncoherent FSK, the arbitrary phase $\theta \neq 0$ and, therefore, the only solution to (3.72), exists when simultaneously

$$\begin{aligned} &\sin(2\pi\Delta f T) = 0 \\ &[\cos(2\pi\Delta f T) - 1] = 0. \end{aligned} \tag{3.76}$$

This happens only when

$$2\pi\Delta f T = 2\pi. \tag{3.77}$$

This leads to

$$\Delta f = \frac{1}{T}. \tag{3.78}$$

Comparing (3.75) and (3.78) we see that the minimum separation of frequencies for coherent orthogonal FSK is only half as much as the value for noncoherent FSK, suggesting that coherent FSK is more bandwidth efficient than noncoherent FSK. We had already argued that the error rate in binary orthogonal FSK can be obtained from the earlier results.

We can now look at a form of FSK which does not lead to discontinuities in the waveform.

3.6.2 Digital Frequency Modulation and Minimum Shift Keying

But, the FSK still does not have a smooth waveform since there are sudden jumps in frequency as bits alter between 1 and 0. This can be overcome using the technique of continuous phase frequency shift keying (CPSFK) (Nuttall and Amoroso 1965; Miyagaki et al. 1978; Sundberg 1986). This method is also sometimes referred to as Digital Frequency Modulation (DFM) because the approach is similar to an analog frequency modulation, shown in Fig. 3.27 (Aulin and Sundberg 1982; Noguchi et al. 1986; Proakis 2001; Shankar 2002).

Instead of applying an analog signal to a frequency modulator (FM), the bit stream is the input. If b's represent the bit stream, the input to a frequency modulator is expressed as

$$v(t) = \sum_{m=-\infty}^{\infty} b_n g(t - nT). \tag{3.79}$$

In (3.79), b's take values of $\pm$ and $g(t)$ is a rectangular pulse with unit area of 1/2

$$\begin{aligned} g(t) &= \frac{1}{2T}, \quad 0 \le t \le T, \\ &= 0 \quad \text{otherwise.} \end{aligned} \tag{3.80}$$

The pulse shape $g(t)$ is shown in Fig. 3.28.

The output of the FM is given by

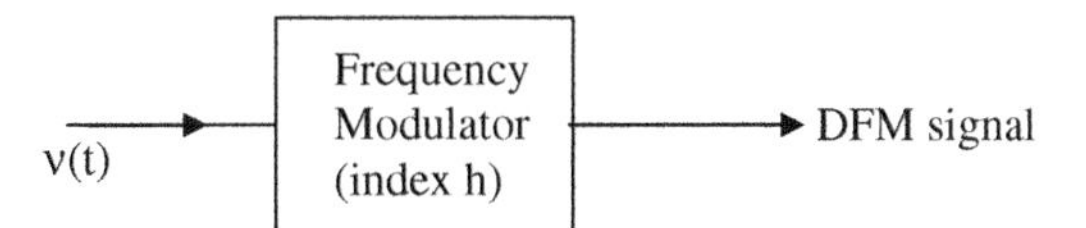

Fig. 3.27 Block diagram of a digital frequency modulator (DFM)

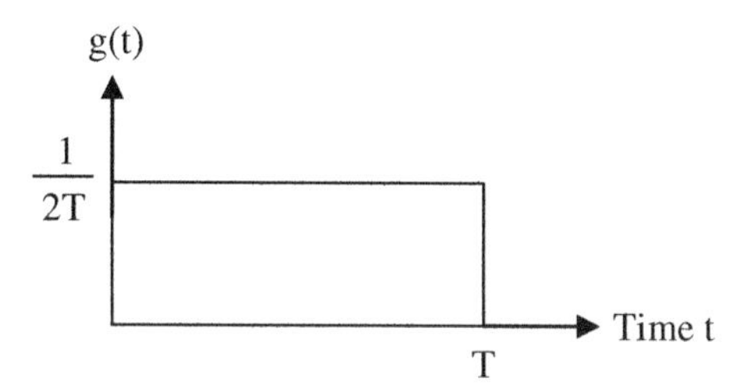

Fig. 3.28 DFM pulse shape

$$s(t) = A_0 \cos\left[2\pi f_0 t + 2\pi h \int_{-\infty}^{t} v(t)\mathrm{d}\tau + \phi_0\right]. \tag{3.81}$$

In (3.81), ϕ_0 a constant phase (can be zero) and h is the FM index given by

$$h = 2f_\mathrm{d}T, \tag{3.82}$$

where *fd* is the frequency deviation. The DFM signal can be expressed as

$$s(t) = A_0 \cos\left[2\pi f_0 t + \theta(t)\right], \tag{3.83}$$

where

$$\theta(t) = 2\pi h \int_{-\infty}^{t} v(\tau)\mathrm{d}\tau. \tag{3.84}$$

The continuous nature of the phase term in (3.84) makes this modulation format CPFSK. Defining $q(t)$ as

$$q(t) = \int_{0}^{t} v(\tau)\mathrm{d}\tau, \quad 0 \le t \le T \tag{3.85}$$

the DFM signal in (3.81) becomes

$$s(t) = A_0 \cos\left[2\pi f_0 t + \pi h \sum_{k=-\infty}^{n-1} b_k + b_n h q(t')\right], \quad (n-1)T \le t \le nT. \tag{3.86}$$

With $t' = t - (n-1)T$. If $h = 0.5$, (3.86) becomes

$$s(t) = A_0 \cos\left[2\pi f_0 t + b_n \frac{\pi t'}{2T} + \frac{\pi}{2} \sum_{k=-\infty}^{n-1} b_k\right], \quad (n-1)T \le t \le nT. \tag{3.87}$$

For the two bits represented by ± 1, the two frequencies corresponding to the two bits become $(f_0 - (1/4T))$ and $(f_0 + (1/4T))$, which makes the difference equal to $1/2T$ giving the case of orthogonal FSK. For the value of $h = 0.5$, (3.87) is also a

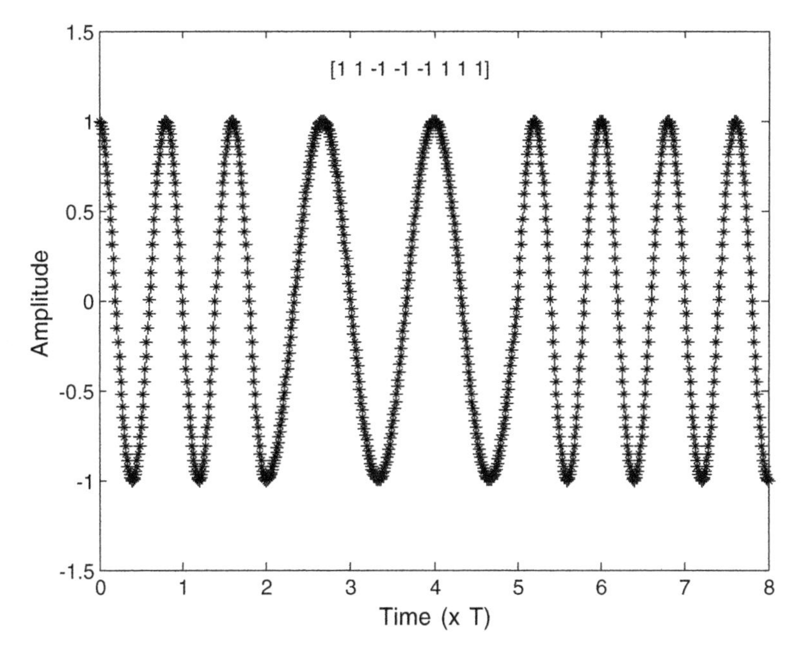

Fig. 3.29 MSK waveform

form of minimum shift keying (MSK). Thus, one way to create MSK is to use CPFSK with a modulation index $h = 0.5$ as shown in Fig. 3.27.

A typical MSK waveform is shown in Fig. 3.29. One can see that there is a smooth transition between the symbols/bits which provides a constant envelope.

A more common means of generating MSK is to modify the arrangement used for OQPSK (Pasupathy 1979). The diagram of the MSK modulator is shown in Fig. 3.30. An additional multiplication with a second carrier is introduced in the OQPSK modulator to realize the MSK. This second carrier has a frequency of $1/4T$ which produces the same effect as that seen in CPFSK, where T is once again the bit duration to generate the orthogonal modulation format (Simon 1976; Prabhu 1981; Ziemer and Ryan 1983; Svensson and Sundberg 1985; Sadr and Omura 1988; Klymyshyn et al. 1999).

The MSK waveform generated is shown in Fig. 3.31 which appears to be different from the one obtained using the DFM. Both forms are MSK waveforms. They differ because of the direct nature of generation from CPFSK in one case and staggered nature from OQPSK in the other case (Amoroso and Kivett 1977; Akaiwa 1997; Pasupathy 1979; Hambley and Tanaka 1984). The power spectrum of MSK is shown together with that of BPSK and QPSK. One can see the disadvantage of MSK. It has a higher bandwidth based on the first zero crossing. However, MSK has a much lower bandwidth if we measure 35 dB bandwidth, suggesting that, compared with QPSK or BPSK (Amoroso 1980), MSK will result in negligible problems from power amplifier nonlinearities.

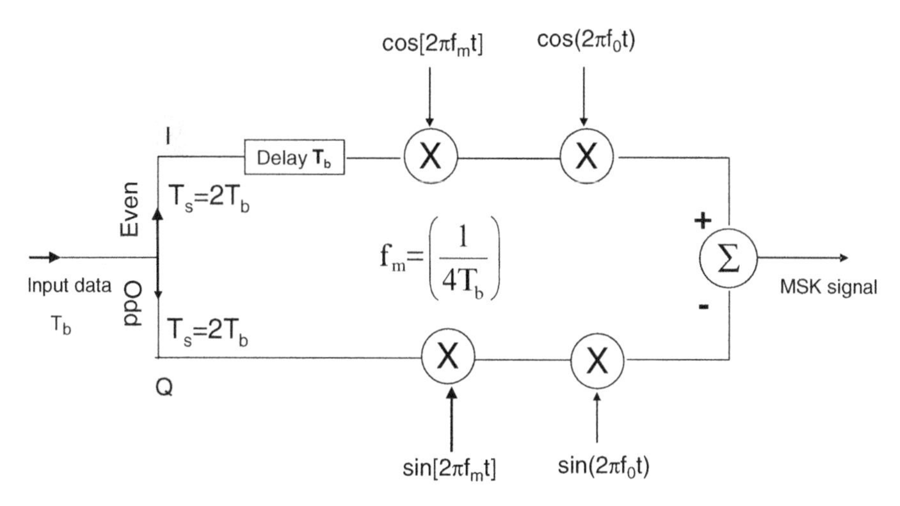

Fig. 3.30 MSK modulator using the OQPSK modulator approach

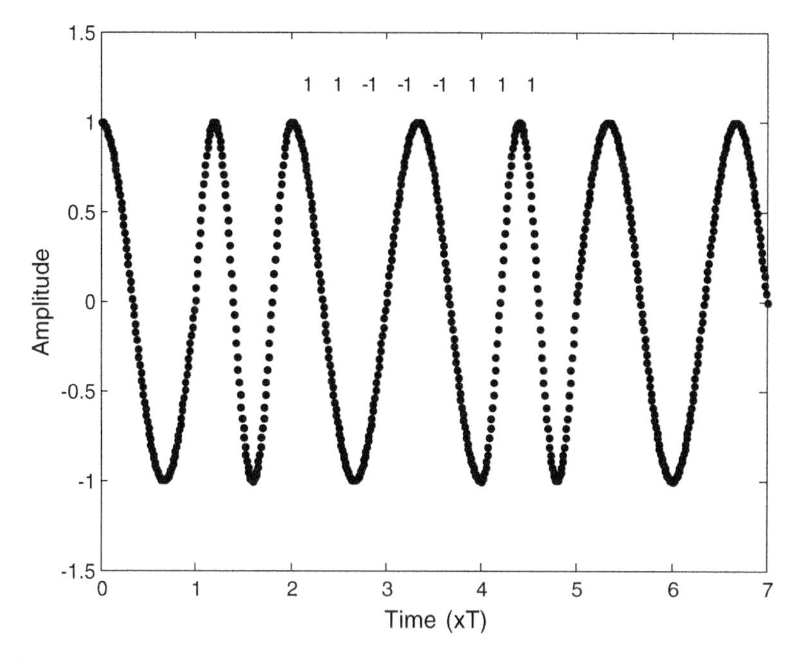

Fig. 3.31 MSK waveform generated using the OQPSK type modulator

While MSK is interpreted as a form of FSK using a half sinusoidal pulse shape, other pulse shapes can also be used. One of the pulse shapes that was discussed earlier was the Gaussian, which did not meet the Nyquist's criterion of the required zero crossing in the time domain. But the Gaussian pulse shape does offer an easy

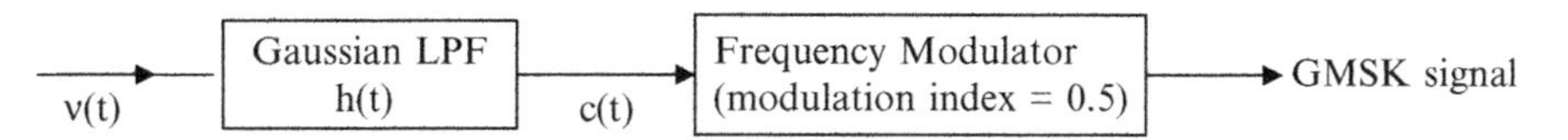

Fig. 3.32 A typical GMSK modulator using the concepts of DFM

means of controlling the bandwidth requirement for transmission. Combined with DFM it becomes an ideal means of generating modulated signals in wireless.

This leads to the so-called Gaussian Minimum Shift Keying (GMSK) which is the modulation format in the second generation GSM systems (Amoroso 1976, 1979; Linz and Hendrickson 1996).

3.6.3 Gaussian Minimum Shift Keying

A GMSK signal can be created by introducing a Gaussian low pass filter prior to the frequency modulator in the DFM setup shown in Fig. 3.32.

If $h(t)$ is given by

$$h(t) = \frac{\sqrt{\pi}}{\alpha}\exp\left(-\frac{\pi^2 t^2}{\alpha^2}\right) \tag{3.88}$$

the input to the modulator, $v(t)$, is given by

$$v(t) = \sum_{k=-\infty}^{\infty} b_n h(t) * g(t - nT) = \sum_{k=-\infty}^{\infty} b_n c(t - nT). \tag{3.89}$$

where $c(t)$ is $h(t) * g(t)$. The parameter α is related to the 3 dB bandwidth (B) of the low pass filter and expressed earlier in (3.23) as

$$\alpha = \frac{0.5887}{B} \tag{3.90}$$

and the GMSK modulation is characterized by the parameter BT. The GMSK pulse shapes are shown in Fig. 3.33. The corresponding spectra are shown in Fig. 3.34.

After some algebra, the pulse shape at the input to the FM modulator can be written as

$$c(t) = \frac{1}{4T}\left[\operatorname{erf}\left(\frac{nT - t}{\sqrt{2}\sigma}\right) - \operatorname{erf}\left(\frac{(n-1)T - t}{\sqrt{2}\sigma}\right)\right]. \tag{3.91}$$

A typical GMSK signal (BT = 0.3) is plotted in Fig. 3.35. The GMSK spectrum is plotted in Fig. 3.36 (Rowe and Prabhu 1975; Murota and Hirade 1981; Kuchi and Prabhu 1999; Murota 2006; Elnoubi 2006).

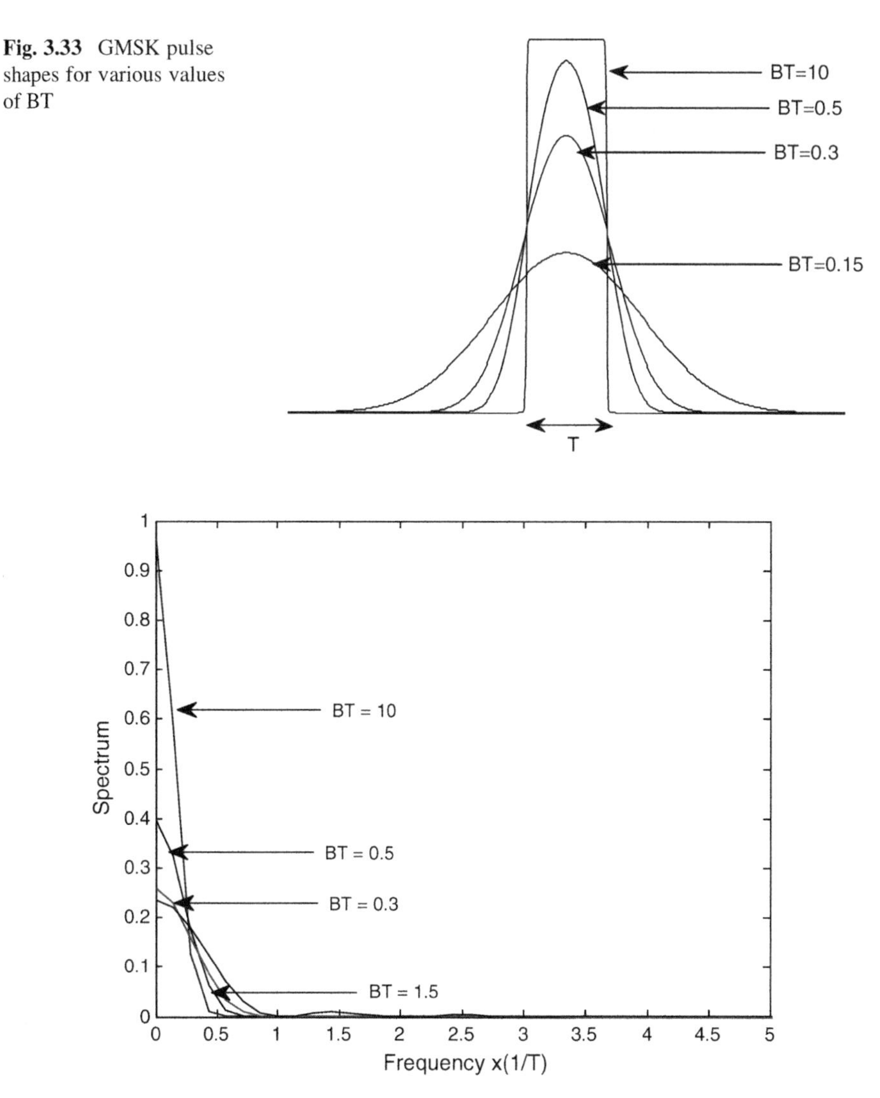

Fig. 3.33 GMSK pulse shapes for various values of BT

Fig. 3.34 Spectra of the GMSK pulse shapes for various values of BT

3.6.4 *Orthogonal M-ary FSK*

We can obtain the symbol error rates in MFSK modems using the approach that was used in connection with MPSK. Typical BFSK and 3-level FSK constellations are shown in Fig. 3.37.

Since all the M vectors are orthogonal, the separation between any one symbol and any other remaining symbol will be the same as $d$min given by $\sqrt{2E_s}$ (Simon et al. 1995; Sklar 2001). Also, when a symbol is received, an error can occur such

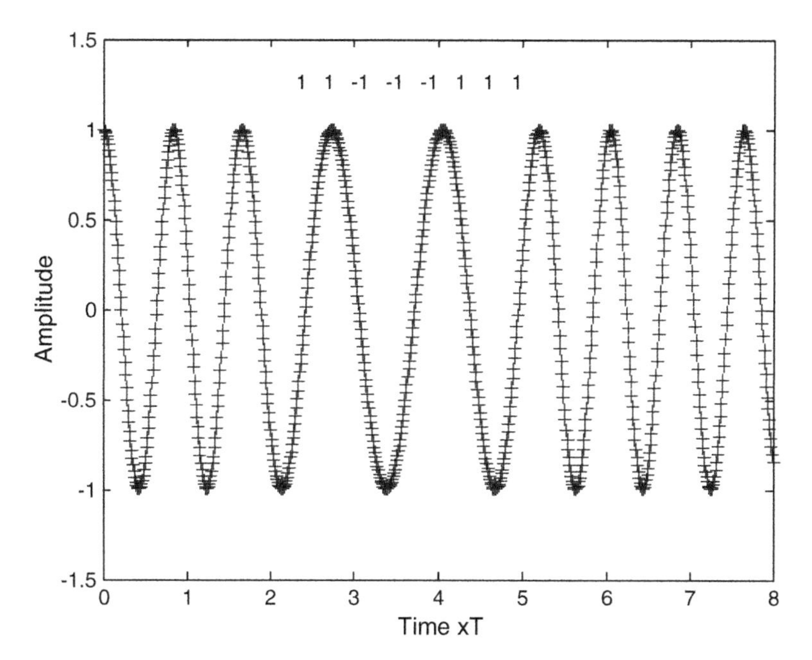

Fig. 3.35 GMSK signal for BT = 0.3

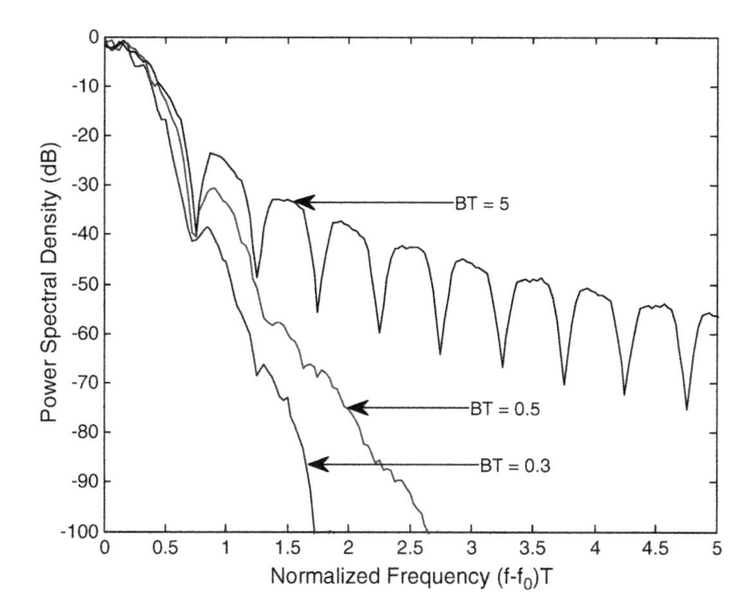

Fig. 3.36 GMSK spectra

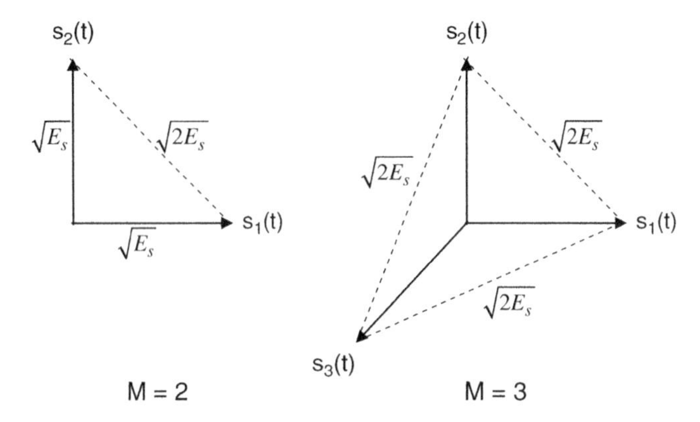

Fig. 3.37 Orthogonality of the signals results in identical separation between the signals regardless of the value of M, namely $\sqrt{2E_s}$.

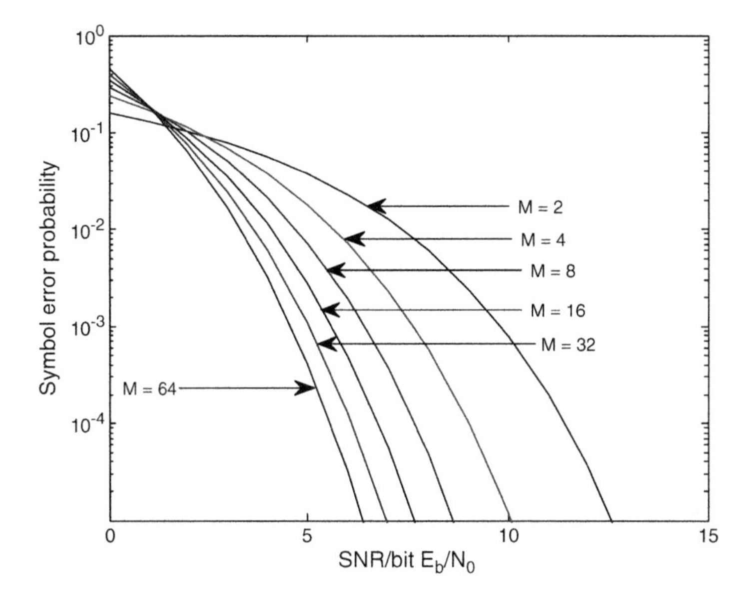

Fig. 3.38 Symbol error rates for MFSK

that it can be identified as any one of the remaining $(M - 1)$ symbols. Since all of them are separated equally, the symbol error rate will be $(M - 1)$ times the error in detecting one. In terms of (3.40), the symbol error in coherent orthogonal MFSK becomes

$$p_s(e) \approx (M - 1)Q\left(\sqrt{\frac{E_s}{N_0}}\right). \tag{3.92}$$

The symbol error rate for MFSK is plotted in Fig. 3.38.

We can also determine the relationship between the symbol error and bit error in MFSK. Consider the case of $M = 8$. In this case, an error in detecting any one symbol will lead to an incorrect decision of any one of the remaining $(M - 1)$ or 7 signals. Thus, the error might occur with equal probability in any one of the $(2^k - 1)$ symbols. If we examine the first bit of M symbols, there will be equal number of 0's and 1's. Thus, if there was a symbol error (one of the bits incorrectly identified), it could happen in $(M/2 = 2^{k-1})$ ways since there are $(M/2)$ bits of the other type.

Therefore, the ratio of the bit error to symbol error can be expressed as

$$\frac{p_{\mathrm{b}}(e)}{p_{\mathrm{s}}(e)} = \frac{(M/2)}{M - 1} = \frac{2^{k-1}}{2^k - 1}. \tag{3.93}$$

When the number of bits in a sequence is large (large M and large k), the ratio of the error rates becomes

$$\frac{p_{\mathrm{b}}(e)}{p_{\mathrm{s}}(e)} \approx \frac{1}{2}. \tag{3.94}$$

Combining the MFSK and MPSK schemes, we can write that

$$\frac{p_{\mathrm{s}}(e)}{k} \leq p_{\mathrm{b}}(e) \leq \frac{2^{k-1}}{2^k - 1} p_{\mathrm{s}}(e). \tag{3.95}$$

The relationship between the two error rates can also be expressed in terms of M as

$$\frac{p_{\mathrm{s}}(e)}{\log_2 M} \leq p_{\mathrm{b}}(e) \leq \frac{(M/2)}{(M - 1)} p_{\mathrm{s}}(e). \tag{3.96}$$

Note that the quantity on the left is the bit error rate (best case) for MPSK.

3.6.5 Error Rates for MSK, OQPSK, π/4 QPSK, and GMSK

As mentioned earlier, OQPSK and $\pi/4$ QPSK are forms of QPSK. The error rates for those modulation schemes are all identical. Since MSK is a form of OQPSK with a different form of pulse shaping, the error rates for MSK are identical to those of QPSK when a coherent receiver is used. Since GMSK and MSK are similar except for the pulse shape (of limited bandwidth), the bit error rate for GMSK can be written as (Rappaport 2002; Shankar 2002)

$$p(e) = Q\left(\sqrt{\varepsilon\frac{2E}{N_0}}\right). \tag{3.97}$$

The parameter ε accounts for the limited bandwidth and is

$$\varepsilon = \begin{cases} 0.68 & \mathrm{BT} = 0.3, \\ 0.85 & \mathrm{BT} \to \infty. \end{cases} \tag{3.98}$$

3.7 Error Rates for Differentially Encoded Signals

So far we have discussed coherent detection schemes. It is also possible to detect the signals noncoherently. There exist differential encoding schemes such as differentially encoded BPSK as well as differentially encoded QPSK, $\pi/4$ QPSK, and so on. Such differentially encoded signals can be detected using a coherent receiver. But often, these differentially encoded signals are also detected using a variety of noncoherent receivers and, hence, the error rates will vary depending on the specific detection scheme used (Edbauer 1992; Simon and Divsalar 1997; Miller and Lee 1998; Gagliardi 1988; Proakis 2001; Haykin 2001). We will look at the example of noncoherent receiver for orthogonal FSK and extend the results to noncoherent BPSK. We will start with the signals for orthogonal FSK as

$$s_i(t) = \sqrt{\frac{2E}{T}}\cos\left(2\pi f_i t + \phi\right), \quad 0 \le t \le T, i = 1, 2. \tag{3.99}$$

Note that the two frequencies are chosen so that the two signals form an orthogonal set. The receiver structure is shown in Fig. 3.39.

While the outputs of the coherent detectors were Gaussian variables, the presence of the envelope detectors leads to non-Gaussian statistics for the outputs.

If ri is the received signal, we have

$$r_i(t) = s_i(t) + n(t). \tag{3.100}$$

Note that $n(t)$ is the same additive white Gaussian noise considered earlier.

If x_1 and x_2 are the sampled outputs, error probability is given by

$$\begin{aligned} p(e) = \; & \frac{1}{2}\text{Prob}[\text{the correct channel}(1)\text{sample} < \text{the incorrect channel}(2)\text{sample}] \\ & \frac{1}{2}\text{Prob}[\text{the correct channel}(2)\text{sample} < \text{the incorrect channel}(1)\text{sample}]. \end{aligned} \tag{3.101}$$

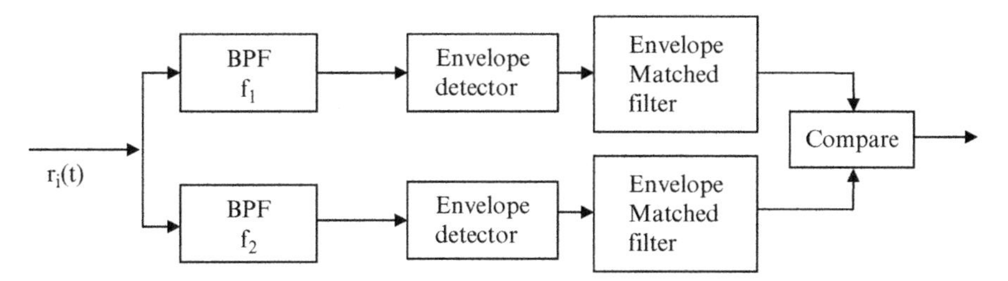

Fig. 3.39 Noncoherent detection of orthogonal BFSK. The bandpass filters have a bandwidth of ($Wf = 1/T$). The outputs are sampled every T s

Since the two channels are identical, we have

$$p(e) = \text{Prob}[\text{the correct channel sample} < \text{the incorrect channel sample}]. \tag{3.102}$$

Thus,

$$p(e) = \int_0^\infty \int_{x_1}^\infty f(x_1, x_2)\mathrm{d}x_1\,\mathrm{d}x_2. \tag{3.103}$$

Since the two signals are uncorrelated,

$$p(e) = \int_0^\infty f(x_1)\left[\int_{x_1}^\infty f(x_2)\mathrm{d}x_1\right]\mathrm{d}x_2. \tag{3.104}$$

Because of the BPF and matched filter, there will only be noise in the incorrect channel; in the correct channel there will be signal and noise. The noise is modeled in inphase and quadrature form, with each components having zero mean Gaussian statistics, the density function of x_2 (incorrect channel) will be Rayleigh distributed as

$$f(x_2) = \frac{x_2}{\sigma_c^2}\exp\left(-\frac{x_2^2}{2\sigma_c^2}\right)U(x_2). \tag{3.105}$$

The noise variance is

$$\sigma_c^2 = 2\left(\frac{N_0}{2}\right)W_f, \tag{3.106}$$

where N_0 was defined earlier in connection with (3.12). In the channel, the presence of the signal makes the envelope arising out of two Gaussian random variables with one of them nonzero because of the presence of the signal. This leads to Rician statistics for the envelope (Davenport and Root 1958; Gagliardi 1988; Cooper and McGillem 1986; Haykin 2001)

$$f(x_1) = \frac{x_1}{\sigma^2}\exp\left(-\frac{x_1^2 + A^2}{2\sigma_c^2}\right) I_0\left(\frac{x_1 A}{\sigma_c^2}\right) U(x_1). \tag{3.107}$$

In (3.107), $I_0(.)$ is the modified Bessel function of the first kind and A is the signal amplitude (envelope). The error rate becomes

$$p(e) = \int_0^\infty \frac{x_1}{\sigma^2}\exp\left(-\frac{x_1^2 + A^2}{2\sigma_c^2}\right) I_0\left(\frac{x_1 A}{\sigma_c^2}\right) \int_{x_1}^\infty \frac{x_2}{\sigma^2}\exp\left(-\frac{x_2^2}{2\sigma_c^2}\right) \mathrm{d}x_2 \ \mathrm{d}x_1. \tag{3.108}$$

Solving (3.108), we have

$$p(e) = \frac{1}{2}\exp\left(-\frac{A^2}{4\sigma_c^2}\right) = \frac{1}{2}\exp\left(-\frac{A^2}{4N_0 W_\mathrm{f}}\right) = \frac{1}{2}\exp\left(-\frac{A^2 T}{4N_0}\right). \tag{3.109}$$

The envelope of the signal is related to energy/bit as

$$E = A^2 \frac{T}{2}. \tag{3.110}$$

The error rate now becomes

$$p(e) = \frac{1}{2}\exp\left(-\frac{E}{2N_0}\right). \tag{3.111}$$

The noncoherent receiver for BFSK is easier to implement since it does not require the use of two separate local oscillators with each matched to the respective frequencies of the BFSK signals.

We will now look at noncoherent reception of BPSK, specifically what is known as differential phase shift keying (DPSK). Let the two signals be (Gagliardi 1988; Haykin 2001)

$$s_1(t) = \sqrt{\frac{2E}{T}}\cos\left(2\pi f_0 t + \phi\right), \quad 0 \le t \le T, \tag{3.112}$$

$$s_2(t) = \sqrt{\frac{2E}{T}}\cos\left(2\pi f_0 t + \phi + \pi\right), \quad 0 \le t \le T. \tag{3.113}$$

There is a unique difference between the noncoherent receiver for BPSK and the one for BFSK. While there are two unique envelopes for BFSK, for BPSK the envelopes are the same. Thus, there is no fixed decision region. The decision will be based on the successively received signal. For the case of DPSK, we can view the transmitted signals or received signals as if each bit transmitted as a binary signal pair as

$$\begin{aligned} x_1(t) &= (s_1, s_1) \quad \text{or} \quad (s_2, s_2) \quad 0 \le t \le 2T, \\ x_2(t) &= (s_1, s_2) \quad \text{or} \quad (s_2, s_1) \quad 0 \le t \le 2T. \end{aligned} \tag{3.114}$$

In (3.114), $(s_i; s_j)(i; j = 1; 2)$ denotes that $s_i(t)$ is followed by $s_j(t)$. Note that the first T s of the waveform of one of the signals in (3.114) will be the last T s of the previous waveform, and so on. Also, from (3.114), it is clear that x_1 and x_2 will be orthogonal (over a period of $2T$ s) as

$$z(2T) = \int_0^{2T} x_1(t)x_2(t)\mathrm{d}t = \int_0^{T} [s_1(t)]^2\mathrm{d}t - \int_0^{T} [s_2(t)]^2\mathrm{d}t = 0. \tag{3.115}$$

Now that we have seen that we can create orthogonal signals, we can use (3.111) to obtain an expression of the error rate of the noncoherent binary PSK receiver as

$$p(e) = \frac{1}{2}\exp\left(-\frac{E}{N_0}\right). \tag{3.116}$$

In (3.116), the energy is double because of the duration of the orthogonal signal "$2T$." A receiver structure for detection is shown in Fig. 3.40

An optimal noncoherent receiver has been shown to be one that uses a local oscillator without the need for a phase match (Fig. 3.41).

The differentially encoded BPSK can be easily generated from the binary data. Instead of transmitting a phase of "0" and "π," differentially encoded phases are transmitted by comparing the data with an arbitrary bit (for example, a "1").

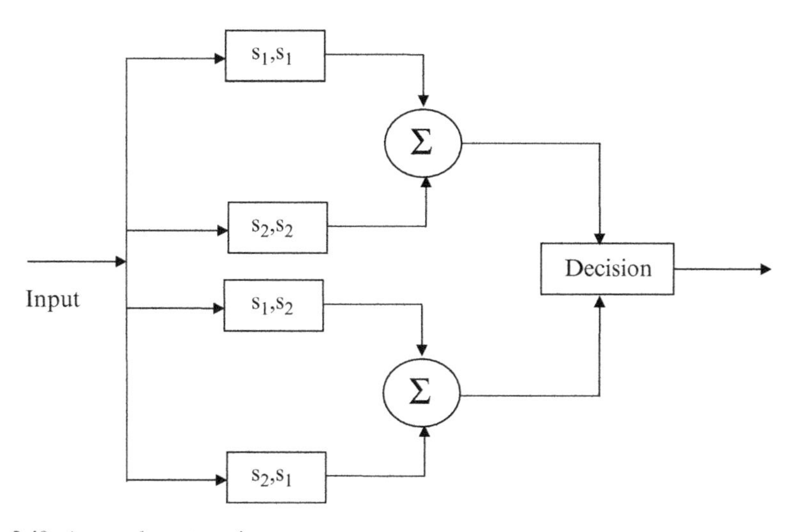

Fig. 3.40 A noncoherent receiver

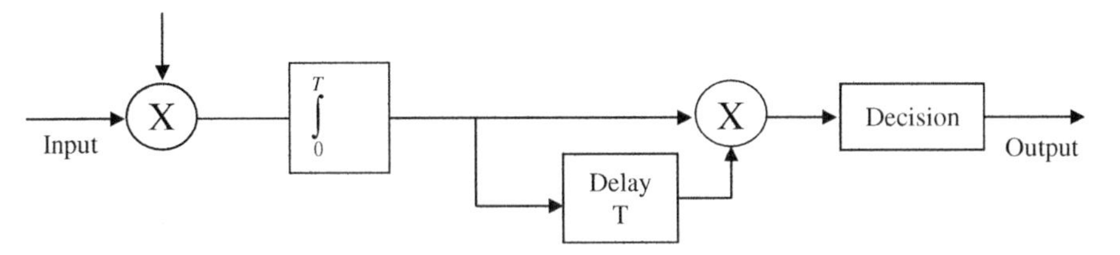

Fig. 3.41 Optimal noncoherent receiver

Table 3.4 Phase encoding for DPSK

k	0	1	2	3	4	5	6	7
b_k		1	−1	−1	1	1	−1	1
c_k	1	1	−1	1	1	1	−1	−1
ϕ_k	0	0	π	0	0	0	π	π

Table 3.4 Phase encoding for DPSK

Let b_k denote the raw data and c_k denote the encoded data. An example of the data encoding is given in Table 3.4. The relationship between the raw data and encoded data is expressed as

$$c_k = b_k \oplus c_{k-1} \tag{3.117}$$

or

$$c_k = \overline{b_k \oplus c_{k-1}}. \tag{3.118}$$

In (3.117) and (3.118), $\oplus$ represents the modulo-2 addition and the overbar represents the complement. Equation (3.118) is used for encoding data in Table 3.4.

The transmitted bits (symbols) are

$$s_k(t) = \sqrt{\frac{2E}{T}} \cos(2\pi f_0 t + \phi_k), \quad k = 0, 1, 2, \ldots. \tag{3.119}$$

The error rates for binary phase and frequency shift keying modems are compared for the case of coherent and noncoherent receivers in Fig. 3.42.

We will also examine the 8-level PSK scheme used in EDGE offered by GSM. The simple approach, very similar to the differential encoding employed with BPSK, can be extended (with some modifications) to differential 8PSK. First, we can write the 8PSK signal as

$$s_k(t) = \sqrt{2\frac{E_s}{T_s}} \cos(2\pi f_0 t + \phi_k), \quad (k-1)T_s \le t \le kT_s. \tag{3.120}$$

In (3.120), *Es* is the symbol energy and *Ts* is the symbol duration, equal to three times the bit duration. The phase *fk* will be a multiple of ($2\pi/8$) chosen from the

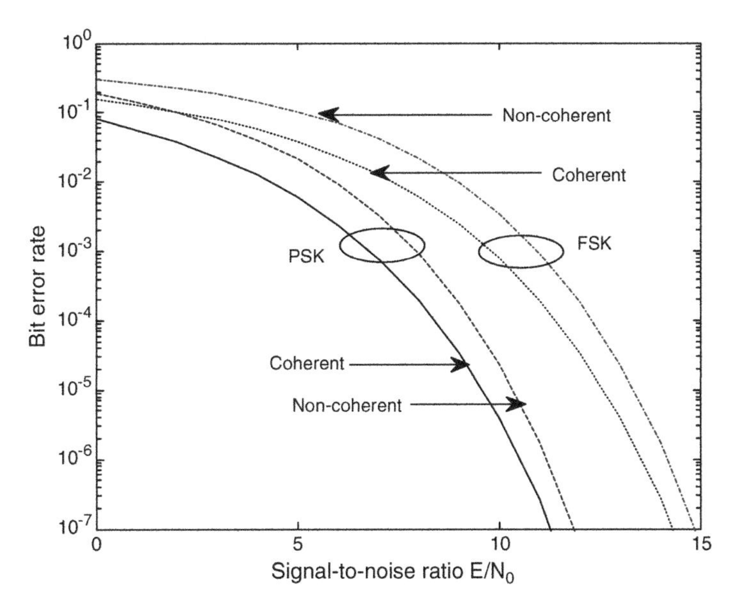

Fig. 3.42 Comparison of error rates for coherent and noncoherent modems

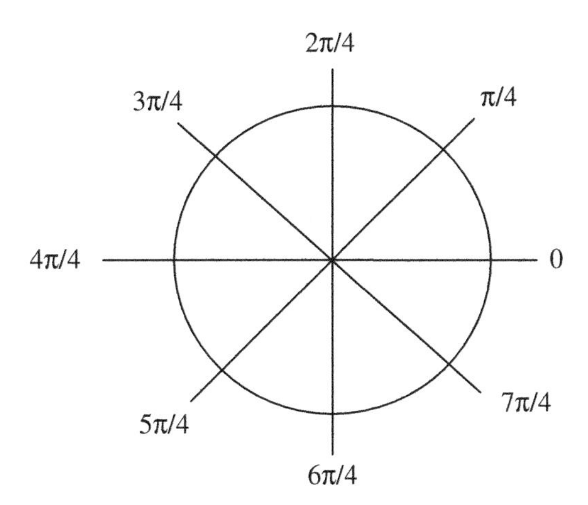

Fig. 3.43 Phase constellation for 8-PSK

constellation shown in Fig. 3.43. The constellation is based on the gray level encoding discussed earlier.

The complete gray code is given in Table 3.1 and the phase encoding for the 8-PSK is given in Table 3.5. Note that for each symbol (*k*th), we have three bits, x_k, y_k, and z_k.

Instead of transmitting the *k*th symbol with the phases listed in table, we will use the differential encoding so that the transmitted phase of the *k*th symbol is

Table 3.5 Phase encoding for 8-PSK

x_k	y_k	z_k	ϕ_k
0	0	0	0
0	0	1	$\pi/4$
0	1	1	$2\pi/4$
0	1	0	$3\pi/4$
1	1	0	$4\pi/4$
1	1	1	$5\pi/4$
1	0	1	$6\pi/4$
1	0	0	$7\pi/4$

$$\theta_k = \phi_k + \theta_{k-1}. \tag{3.121}$$

For the first symbol,

$$\theta_k = \phi_k. \tag{3.122}$$

The transmitted signal can now be written as

$$\begin{aligned} s_k(t) &= \sqrt{\frac{2E_s}{T_s}} \cos\left(2\pi f_0 t + \phi_k\right) \\ &= \sqrt{\frac{2E_s}{T_s}} \left[\cos\left(2\pi f_0 t\right) \cos\left(\theta_k\right) - \sin\left(2\pi f_0 t\right) \sin\left(\theta_k\right)\right]. \end{aligned} \tag{3.123}$$

Rewriting (3.123) in quadrature form, we have

$$s_k(t) = \sqrt{\frac{2E_s}{T_s}} \left[p_{kx} \cos\left(2\pi f_0 t\right) - p_{ky} \sin\left(2\pi f_0 t\right)\right], \tag{3.124}$$

where the first term in (3.124) is the inphase term and the second one is the quadrature term. Thus, an 8-PSK, differential, or otherwise can be generated using a quadrature modulator geometry as shown in Fig. 3.44.

Thus, QPSK as well as other higher order phase modulation schemes can be implemented using two BPSK modulators in quadrature. A simple demodulator operating in quadrature can estimate p_{kx} and p_{ky}, namely $\widehat{p_{kx}}$ and $\widehat{p_{ky}}$ (as shown in Fig. 3.45), and decode the symbols.

A typical 8-level PSK signal is shown in Fig. 3.46.

There are other ways of detecting differentially modulated signals. These include differential detection, intermediate frequency (IF), differential detection, as well as FM discriminator (Taub and Schilling 1986; Makrakis and Feher 1990; Haykin 2001). These approaches result in slightly different error rates. As in the case of the BPSK, the power efficiencies will be slightly lower than the coherent schemes. But noncoherent approaches offer an advantage in terms of not requiring perfect phase match.

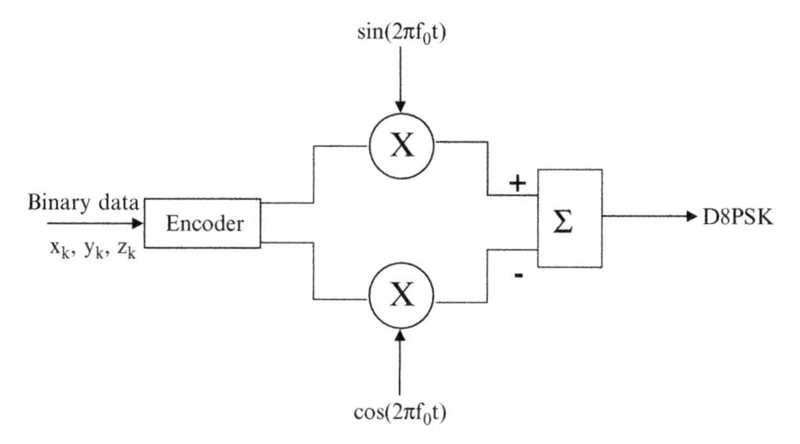

Fig. 3.44 Modulator for 8-PSK

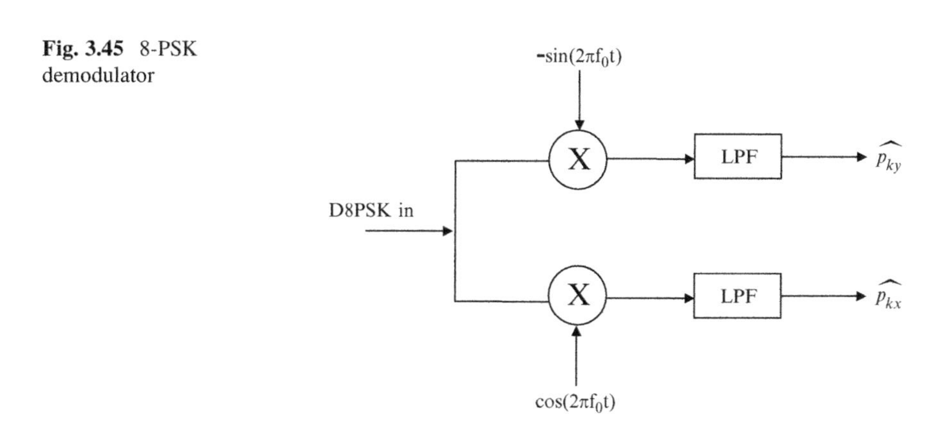

Fig. 3.45 8-PSK demodulator

We will now examine one of the multiplexing techniques being used to transmit data at very high rates, namely orthogonal frequency division multiplexing.

3.8 Orthogonal Frequency Division Multiplexing

As discussed in Chap. 2, the chances of frequency selective fading go up with increased data rates. One can develop an alternate strategy to transmit a large volume of data by dividing the data into smaller segments and transmitting them in parallel by modulating individual carriers with these data segments. By keeping the bandwidth of these data segments low enough, we can alleviate the problems of frequency selective fading. While in a conventional frequency division multiplexing, each channel occupies a large bandwidth and adjoining channels do not overlap. Typically, these bandwidths are also high. Use of nonoverlapping

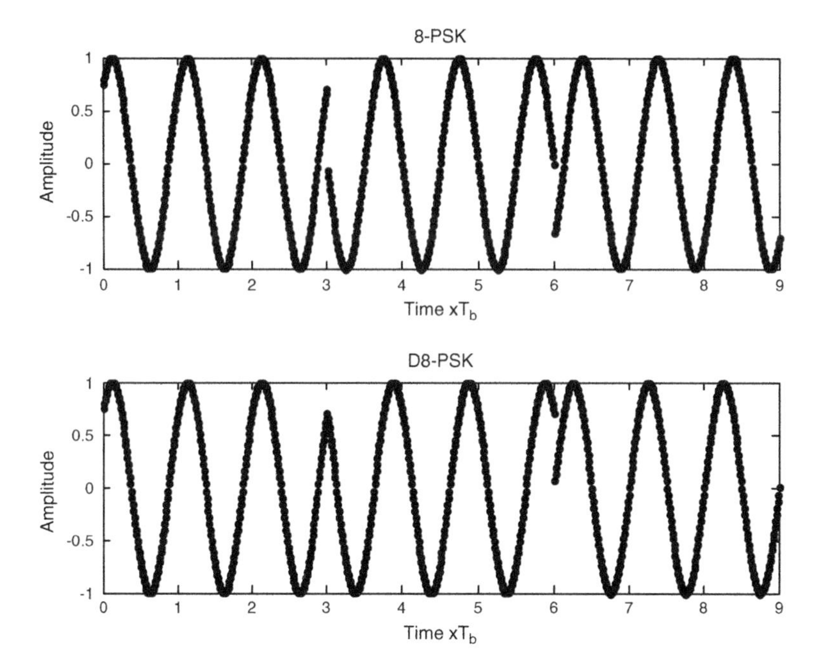

Fig. 3.46 Waveform of 8-PSK and D8-PSK

channels is very inefficient in terms of frequency use. Instead, if we allow the channels to overlap and still maintain orthogonality of adjoining channels, the overall transmission capability could be enhanced (Bingham 1990, 2000). In this case, the individual carriers form an orthogonal set and, hence, the scheme is identified as orthogonal frequency division multiplexing (OFDM). The concepts of a conventional frequency division multiplexing (FDM) and OFDM are illustrated in Fig. 3.47. In the FDM schemes, guard bands are inserted between the multiplexed channels to prevent any spillover among adjacent channels. This results in a waste of spectrum (Cimini 1985; Prasad 2004; Goldsmith 2005; Molisch 2005; Gao et al. 2006). In OFDM, we see that the spectra do overlap, and this overlapping of the spectra is made possible by choosing an appropriate spectrum for the individual channels so that inter channel interference (in this case, inter carrier interference) does not occur.

Block diagram in Fig. 3.48 shows the elementary principle of the OFDM system. The input serial data stream is broken into smaller streams or symbols. Let us assume that there are N serial data elements and each of these is used to modulate N subcarrier frequencies. Each subcarrier frequency is

$$f_k = f_0 + k\Delta f, \tag{3.125}$$

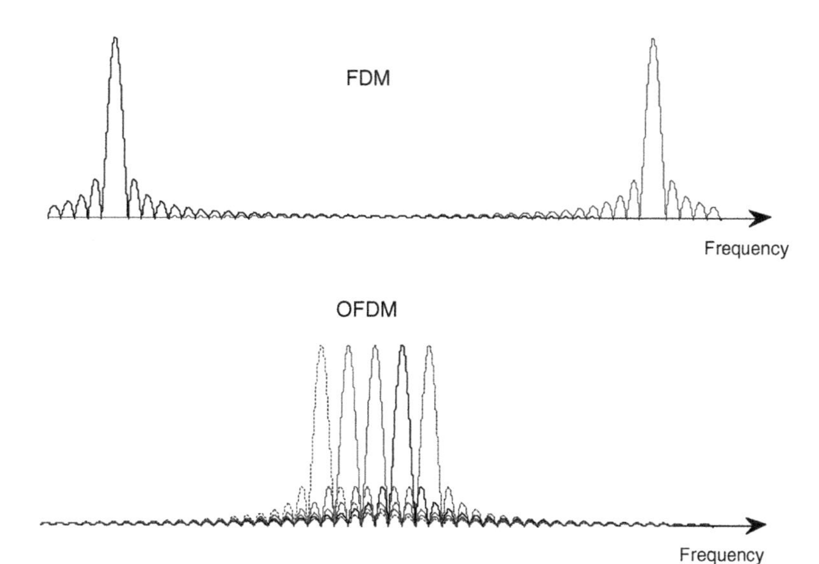

Fig. 3.47 Frequency domain representations of FDM and OFDM

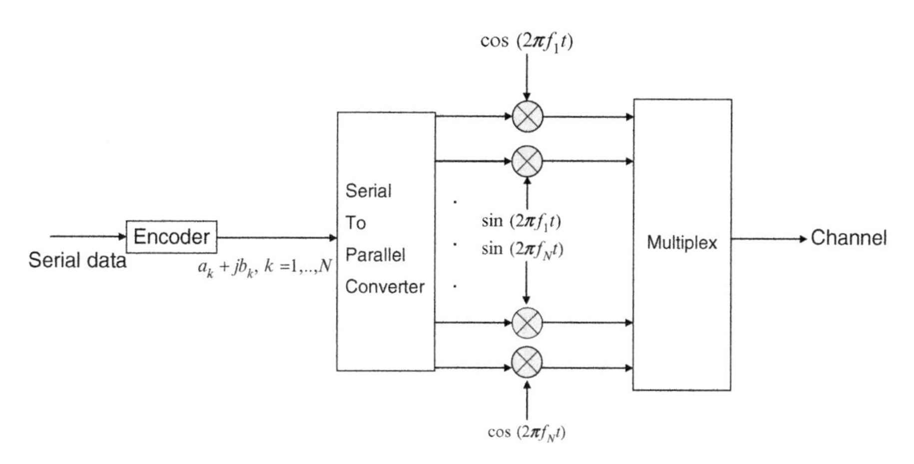

Fig. 3.48 Concept of an OFDM transmitter

$$\Delta f = \frac{1}{N\Delta f}. \tag{3.126}$$

In (3.126), Δt is the inverse of symbol rate. One can view the arrangement as akin to the transition from BPSK to QPSK. The signaling interval now has gone up N times the signaling interval of the input. The subcarrier frequencies are also now spaced $1/N\Delta t$ apart from each other with little or no signal distortion. Because of this spacing and consequent orthogonality, interchannel interference does not

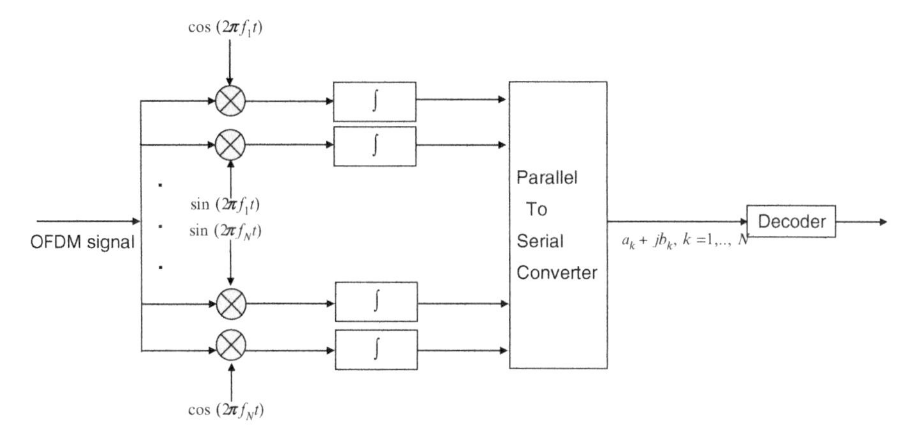

Fig. 3.49 Concept of an OFDM receiver

occur. This assumes ideal carrier regeneration as shown in Fig. 3.49. Later we will examine the departure from this ideal situation and the subsequent problem.

If one examines the transmitter and receiver for OFDM, it is clear that one requires the necessary equipment to generate multiple carrier frequencies, making the practical implementation of the OFDM system difficult. One also would require appropriate filters in addition to the multiple carrier frequencies. This complexity can be solved using the principle of digital Fourier transforms. Let us rewrite the transmitted OFDM signal as seen in Fig. 3.48 as

$$S(t) = \mathrm{Re}\left\{\sum_{k=0}^{N-1}[a_k + jb_k]\exp(-j2\pi f_k t)\right\}. \tag{3.127}$$

If express $t = m\Delta t$, (3.127) can be expressed in terms of discrete set as

$$S(m) = \mathrm{DFT}[S(t)] = \left\{\sum_{k=0}^{N-1}[a_k + jb_k]\exp\left(-j\frac{2\pi}{N}mk\right)\right\}. \tag{3.128}$$

Comparison of (3.127) and (3.128) shows that we can avoid the use of multiple carrier frequencies through the use of discrete Fourier transform. We can therefore replace the transmitter shown in Fig. 3.48 by a DFT based one shown in Fig. 3.50.

Note that in Fig. 3.50 the DFT is accomplished through Fast Fourier Transforms (FFT) algorithm. Note that the simplicity gained through the use of a DFT is coming at the cost of truncating the signal in the computer to be in $(0, N\Delta t)$ which would lead to interchannel interference. It is possible to use the Inverse Fast Fourier Transform (IFFT) approach at the receiver to recover the data (Sari et al. 1994; Wu and Zou 1995; Cimini 1985; Liu and Tureli 1998).

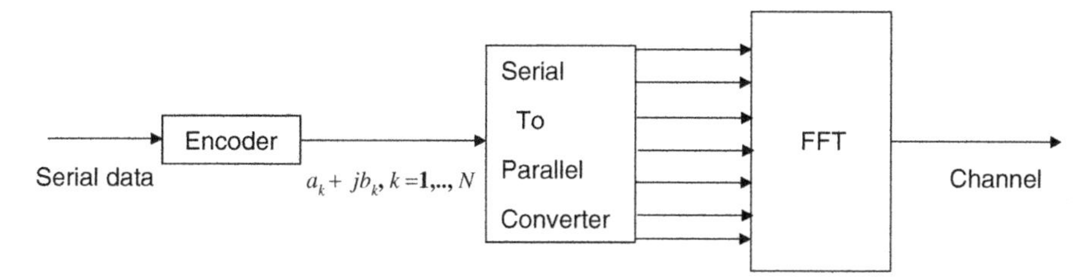

Fig. 3.50 A DFT (FFT) based OFDM transmitter

Output of the coherent receiver for the kth symbol can be expressed as

$$y_k = x_k d_0 + \sum_{l=0,\, l \neq k}^{N-1} d_{l-k} x_l + n, \quad k = 0, 1, \ldots, N-1. \tag{3.129}$$

In (3.129), n is the additive white noise, and the first term is the contribution from the kth symbol with x's representing the symbols values and d_0 representing the energy. The term in the middle represents the interchannel interference (ICI), if any. Thus, in the absence of any, (3.129) is no different from the output of the coherent receiver in a typical signal detection problem in additive white Gaussian noise. We will examine the effects of ICI later in this chapter (Russell and Stuber 1995; Armada 2001; Sathananthan and Tellambura 2001).

3.9 Summary

A review of various modems of interest in wireless communications is presented. We begin with the basics of pulses and pulse shapes typically used in different digital modulation schemes. Their properties and the differences in terms of spectral characteristics are introduced. Particular modulation formats are emphasized. We discuss in detail the different linear and nonlinear digital modulation schemes. We obtain the error rates in ideal channels, and then compare the power efficiencies of the modems. In addition, we provide details on several special functions generally encountered in the analysis of digital communications.

Appendix

We will now explore a few topics of interest that were mentioned in passing earlier in this chapter. These include the difference between signal-to-noise ratio and energy-to-noise ratio, bandwidth concepts as they pertain to digital signals, synchronization, intersymbol interference, and so on. We will examine the effects of phase mismatch in coherence detection and also problems with timing.

Noise, Signal-to-Noise Ratio, Symbol Energy, Bit Energy

We had introduced the notion of noise at the beginning of this chapter by stating that the noise in wireless systems is assumed to be additive white Gaussian noise (Taub and Schilling 1986; Sklar 1993, 2001; Proakis 2001). Typically signal-to-noise ratio (SNR) is defined as the ratio of the signal power (Pr) to the noise power (Nr)

$$\mathrm{SNR} = \frac{P_r}{N_r}. \tag{3.130}$$

The noise power is given by the product of $2B$ and the spectral density of noise is given by ($N_0/2$) where B is the message bandwidth. Since we are dealing with signals of finite energy (bit or symbol), the SNR can be written in terms of the energy as

$$\mathrm{SNR} = \frac{E_\mathrm{s}}{N_0 B T_\mathrm{s}} = \frac{E_\mathrm{b}}{N_0 B T_\mathrm{b}}. \tag{3.131}$$

However, in most of the analysis of wireless systems, the SNR is expressed as either (Es/N_0) or (Eb/N_0) for the SNR per symbol or SNR per bit. The performance is generally measured using the average bit error rate. Thus, if one uses MPSK or MFSK, the symbol error rate must be converted to bit error rate for the purpose of comparison.

Bandwidth of Digital Signals

While the notion of bandwidth is reasonably clear in analog systems, the bandwidth of digital signals has several definitions, often leading to problems in understanding the spectral content (Amoroso 1980; Aghvami 1993; Sklar 2001; Haykin 2001). There are several ways of defining the bandwidth, each with different meanings. Primarily this is because the basic shape of the message signal is a time limited one and, thus, ideally the bandwidth is infinite. We will now look at the multiple definitions of bandwidth and understand the relationships among them. We will choose the example of a binary ASK or PSK. The modulated output can be represented as

$$p(t) = p_T(t)\cos(2\pi f_0 t), \quad 0 \le t \le T. \tag{3.132}$$

In (3.132), the pulse train is represented by

$$p_T(t) = \sum_{n=-\infty}^{\infty} m(t - nT) \tag{3.133}$$

with

$$m(t) = \pm 1, \ \ NT \le t \le (N + T)T. \tag{3.134}$$

The data rate $R = (1/T)$. The spectrum of $p(t)$ is given as

$$P(f) = \frac{T}{4}\left[\frac{\sin \pi T(f - f_0)}{\pi T(f - f_0)}\right]^2 + \frac{T}{4}\left[\frac{\sin \pi T(f + f_0)}{\pi T(f + f_0)}\right]^2. \tag{3.135}$$

The spectrum of the modulated signal is shown along with the spectrum of the basic pulse, as seen in Fig. 3.51.

We will use Fig. 3.51 to define and compare the multiple definitions.

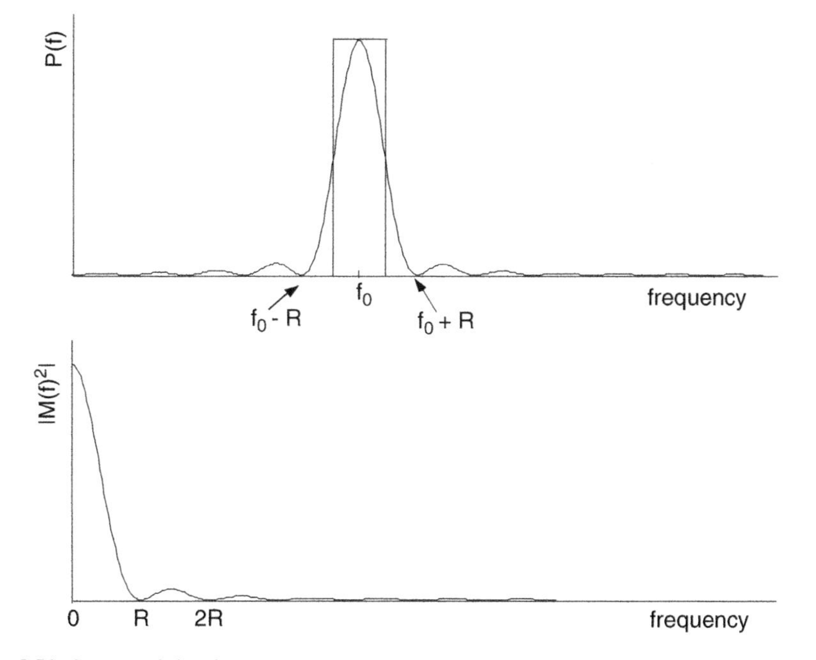

Fig. 3.51 Spectra of signals

Absolute Bandwidth

Absolute bandwidth of a signal is defined as the frequency range outside (positive frequencies) of which the power is zero. This is infinite both for the bandpass signal $p(t)$ and baseband signal $m(t)$. Thus, this absolute bandwidth is not useful in comparing digital signals in terms of bandwidth since both baseband and bandpass signals have infinite bandwidth.

3 dB Bandwidth

The 3 dB bandwidth or half power bandwidth is the frequency range (positive frequencies, f_1, f_2 with $f_2 > f_1$), where the power drops to 50% of the peak value.

The bandwidth is given by $(f_2 - f_1)$. It will have a different value for baseband and bandpass signals.

Equivalent Bandwidth

This is the extent of the positive frequencies occupied by a rectangular window such that it has a height equal to the peak value of the positive spectrum and a power equal to the power contained within the positive frequencies. This is shown by the rectangular window in Fig. 3.51 (top), and has a value of R.

Null-to-Null Bandwidth

This corresponds to the frequency band between two "nulls" on either side of the peak. For the bandpass spectrum, this corresponds to $2R$.

Bounded Spectrum Bandwidth

This is the value of $(f_2 - f_1)$ such that outside the band $f_1 < f < f_2$, the power spectrum is down by at least a significant amount (about 50 dB) below the peak value.

Table 3.6 Definition of bandwidths of BPSK

Definition	Bandwidth
Absolute bandwidth	∞
3 dB bandwidth	0.88*R*
Equivalent bandwidth	1.0*R*
Null-to-null bandwidth	2.0*R*
Bounded spectrum bandwidth	201.5*R*
Power (99%) bandwidth	20.5*R*

Table 3.7 Bandwidths comparison

Modulation technique	99% bandwidth	
BPSK	20.5*R*	
QPSK	10.3*R*	
MSK	1.2*R*	
Modulation technique	Null-to-null bandwidth	
BPSK	2P	
QPSK	*R*	
MSK	1.5*R*	
Modulation technique	35 dB bandwidth	50 dB bandwidth
BPSK	35.12*R*	201*R*
QPSK	17.56*R*	100.5*R*
MSK	3.24*R*	8.18*R*
Modulation technique	Noise bandwidth	Half power bandwidth
BPSK	1.0*R*	0.88*R*
QPSK	0.5*R*	0.44*R*
MSK	0.62*R*	0.59*R*

Power (99%) Bandwidth

This is the range of frequencies $(f_2 - f_1)$ such that 99% of the power resides in that frequency band.

For the BPSK signal, these values are given in Tables 3.6 and 3.7.

Carrier Regeneration and Synchronization

For the matched filter for both BPSK and QPSK, we require a carrier wave of matching frequency and phase at the receiver. There are two ways of accomplishing this (Gagliardi 1988; Haykin 2001; Anderson 2005). One is to transmit a pilot tone along with the modulated signal. This leads to the waste of transmit power since the pilot tone also carries some power. The second method involves the regeneration of the carrier wave from the BPSK or QPSK signal. Since the carrier recovery

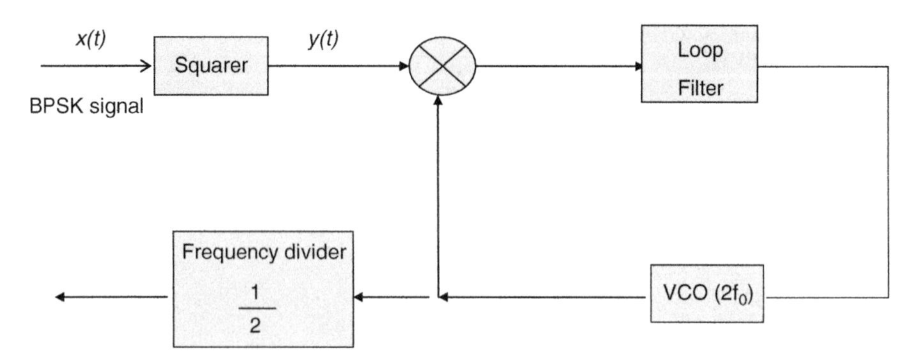

Fig. 3.52 Block diagram of carrier recovery

approaches are different for BPSK and QPSK (as we will see), we will start with the carrier recovery system for BPSK first.

A block diagram of the carrier recovery system for BPSK is shown in Fig. 3.52. It is based on the squaring operation on the received signal. If $x(t)$ represents the BPSK signal we have

$$x(t) = Am(t)\cos\left[2\pi f_0 t + \phi(t)\right] + n(t). \tag{3.136}$$

In (3.136), $m(t)$ represents the bipolar amplitude values of ± 1. The noise is $n(t)$ and the phase of the signal which must also be tracked along with the carrier frequency of f_0. The output of the squarer is

$$y(t) = x^2(t) = \frac{A^2}{2} + \frac{A^2}{2}\sin\left[2\pi(2f_0)t + 2\phi(t)\right] + n^2(t) + \cdots. \tag{3.137}$$

Note that the second term does not depend on $m(t)$ at all and contains the double frequency term. A phase locked loop with a voltage controlled oscillator (VCO) at $2f_0$ tracks this double frequency term. The VCO output will give a signal at a frequency of $2f_0$ and phase of $2f(t)$. This is fed through a frequency divider providing a signal at frequency of f_0 and an approximate phase of $f(t)$, the required signal. We will later examine the problems arising out of phase mismatching which can pose serious problems.

For the case of QPSK modulation, a squarer will not suffice since there are four symbols and, therefore, four possible combinations of the bit pairs. Thus, instead of the squarer, one must use an mth law device where $m = 4$, i.e., one needs to take the fourth power of the incoming QPSK signal and use a frequency division by 4 to get the matching phase and frequency as shown in Fig. 3.53. Note that even in this case, the phase mismatch will create problems. We will explore those problems in the next section.

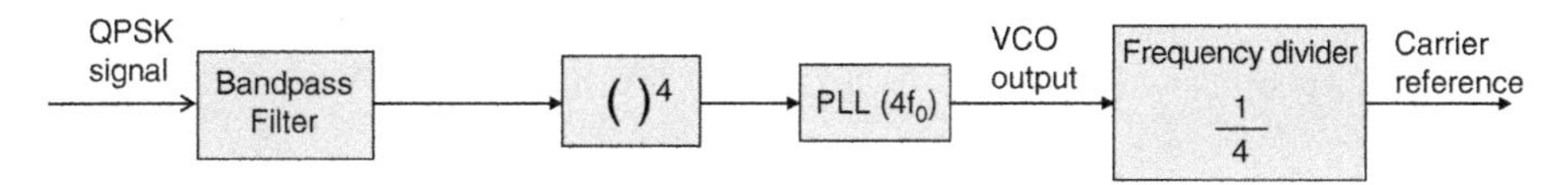

Fig. 3.53 QPSK carrier reference generator

For higher order phase shift keying, the concept of the QPSK carrier phase reference generation can be extended by replacing the mth law device with the Mth law device and using a $(1/M)$ frequency divider.

The same approach needs to be used with the binary and M-ary FSK with the phase and carrier frequency circuits for each of the carrier frequencies of the incoming modulated signal at the receiver.

Digital reception also requires timing information so that the symbols and bit can be correctly recovered at the receiver. The output is sampled at regular intervals of T plus a small time delay that can account for the traverse of the data from the transmitter to the receiver. This requires the need for a clock signal at the receiver. The procedure of generating this clock signal is called symbol synchronization or timing recovery.

Symbol synchronization can be accomplished in several ways. Often, the transmitter and receiver clocks are synchronized to a master clock which takes care of the clock recovery issues. The receiver can estimate the transit time delay and generate the complete information necessary. The other option is to transmit a clock frequency $(1/T)$. The receiver can use a very narrow bandpass signal and receive this information. However, such a step will apportion a part of the transmit power for the clock signal. This signal will also occupy a small fraction of a bandwidth as well. Thus, other techniques using the received data must be used for the timing recovery. One such technique uses the symmetry properties of the signal at the output of the matched filter. This gives a peak at $t = T$ for a rectangular pulse. Similar characteristics can be used with other pulse shapes as well. As in the case of carrier phase recovery, there can be timing errors.

There are also techniques that allow the simultaneous estimation of carrier phase and symbol timing. But the purpose of this brief discussion is to indicate that while the carrier phase and symbol timing are critical in demodulation of digitally modulated signals, errors in the estimation of these two will impact the overall performance of the wireless system since such errors can lead to increased error rates. We will explore such effects in the following section.

Problems of Phase Mismatch: Deterministic and Random, Timing Error, etc.

In reference to phase and frequency, the coherent detector requires a perfect match of the local oscillator. A common problem is the phase mismatch (Prabhu 1969,

1976b; Gagliardi 1988; Shankar 2002). If the phase mismatch is ψ_{m}, the error rate for BPSK becomes

$$p(e;\psi_{\mathrm{m}}) = Q\left[\sqrt{\frac{2E}{N_0}}\cos(\psi_{\mathrm{m}})\right]. \tag{3.138}$$

Often, because of the instabilities in the local oscillator, the phase mismatch becomes a random variable. If the pdf of the phase mismatch is Gaussian,

$$f(\psi) = \frac{1}{\sqrt{2\pi\sigma_{\psi}^{2}}}\exp\left[-\frac{(\psi-\psi_0)^2}{2\sigma_{\psi}^{2}}\right]. \tag{3.139}$$

The error rate becomes

$$p(e) = \int_{-\infty}^{\infty} Q\left[\sqrt{\frac{2E}{N_0}}\cos(\psi_{\mathrm{m}})\right]\frac{1}{\sqrt{2\pi\sigma_{\psi}^{2}}}\exp\left[-\frac{(\psi-\psi_0)^2}{2\sigma_{\psi}^{2}}\right]\mathrm{d}\psi. \tag{3.140}$$

The error rate in (3.140) is plotted in Fig. 3.54 for two values of fixed phase mismatch.

Figure 3.55 shows the effects of angular mismatch for a fixed SNR.

The error rate in the presence of random phase mismatch is shown in Fig. 3.56 for a few of the average phase mismatch ($\psi_0 = 0$) and standard deviation (σ_{ψ}).

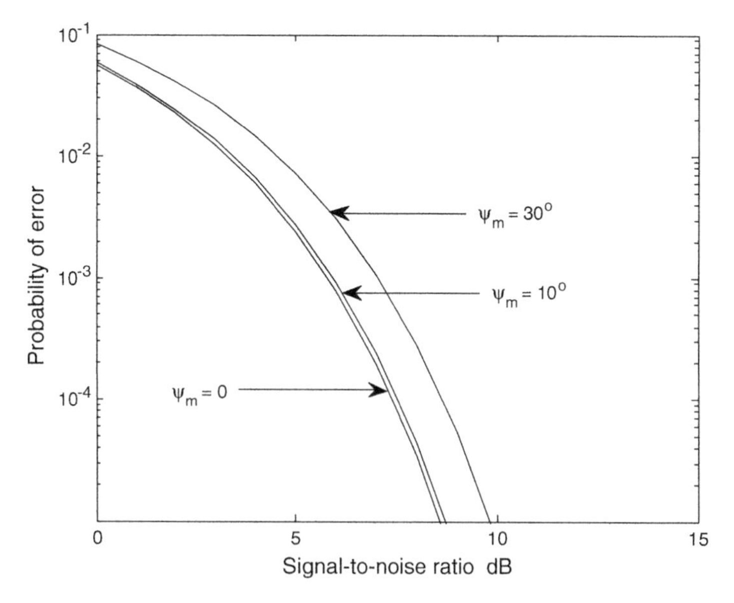

Fig. 3.54 Error rate in BPSK as a function of phase mismatch

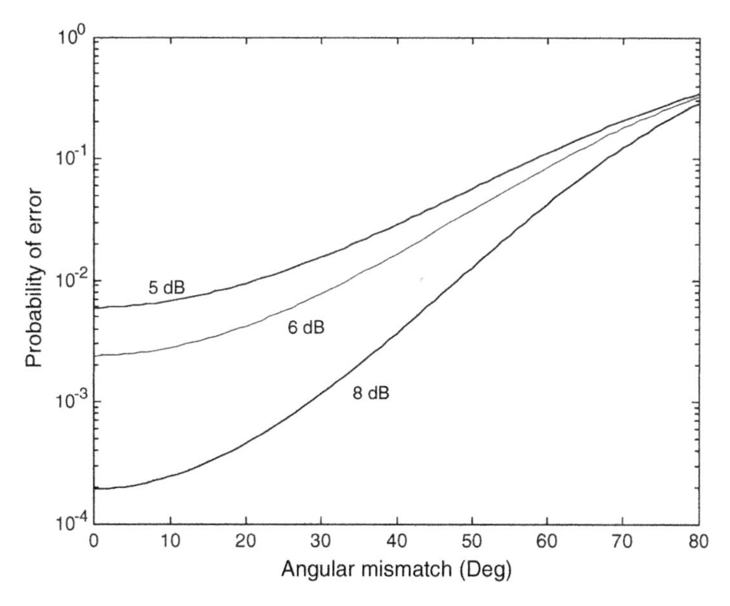

Fig. 3.55 Error rates as a function of angular mismatch for three values of the SNR

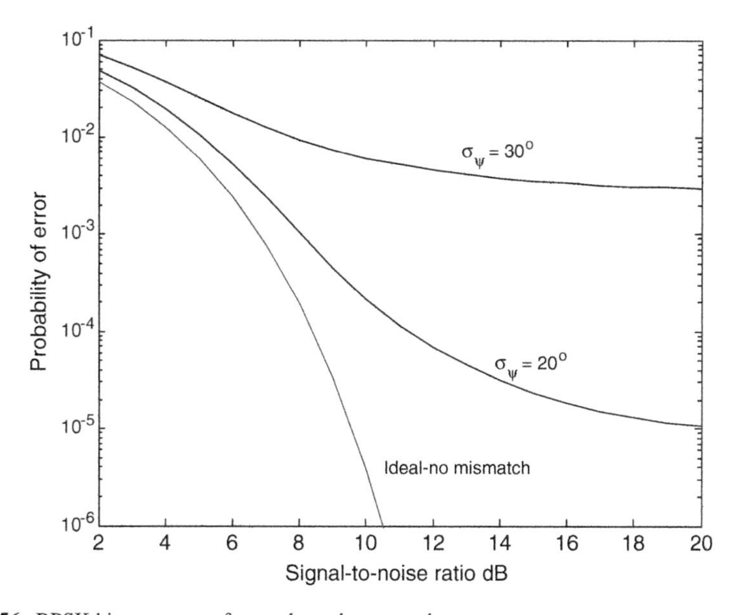

Fig. 3.56 BPSK bit error rates for random phase match

Note that as the standard deviation of the mismatch increases, the error curves flatten out and one sees that beyond a certain SNR, the error rate does not come down, signifying the existence of an error floor. It is also possible to see that even

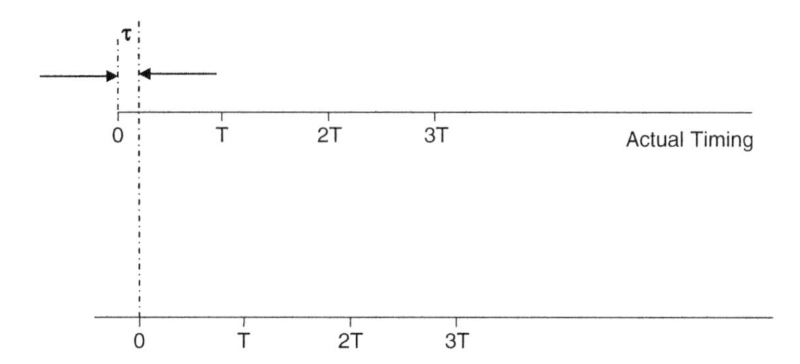

Fig. 3.57 Timing offset

when the mismatch is deterministic, as the mismatch approaches $\pi/2$, the argument of the Q function in (3.138) approaches zero, suggesting a significant increase in the error rate.

The importance of maintaining phase matching is seen from these results. The performance of a BPSK system is ultimately determined by the integrity of the local oscillator and its ability to provide a perfect match.

Another problem in BPSK is assumed from the assumption of perfect timing and absence of any timing "jitter" (Cooper and McGillem 1986; Gagliardi 1988). This means that there is some imprecision in the location of the timing instant. The signal correlation which produces the output of the matched filter does not result in the ideal value because the deviation from the ideal timing instant is t (timing offset) (Fig. 3.57). We can quantify the effect of timing errors on the error probability by considering the receiver structure of the matched filter discussed earlier.

The output of the matched filter is

$$x(T,\tau) = \pm\int_0^{T-\tau} s(t+\tau)s(t)\mathrm{d}t \pm \int_{T-\tau}^{T} s(t+\tau-T)s(t)\mathrm{d}t. \tag{3.141}$$

In (3.141), $s(t)$ is the signal. The first integral arises from the symbol/bit being considered (0,T), and the next integral comes from the adjoining symbol/bit (T, 2T).

The $\pm$ accounts for the bipolar nature, with the first $\pm$ depending on the bit in (0,T) and the second $\pm$ depending on the adjoining window. In an ideal case of no timing errors, $t=0$ and the second term vanishes and the first term provides the maximum overlap and will be a scaled version of the energy in the bit. The integrals are forms of correlation and the correlation function can be written as

$$\rho(\alpha) = \frac{1}{E}\int_0^{T} s(t+\alpha)s(t)\mathrm{d}t. \tag{3.142}$$

For the BPSK case

$$s(t) = \pm\sqrt{\frac{2E}{T}}\cos(2\pi f_0 t). \tag{3.143}$$

Using (3.143), (3.142) becomes

$$\rho(\alpha) = \cos(2\pi f_0 \alpha). \tag{3.144}$$

Thus, if the bits are same, (3.141) yields

$$x(T,\tau) = \pm E\rho(\tau) = \pm E\cos(2\pi f_0 \tau). \tag{3.145}$$

If the bits are different, we have

$$x(T,\tau) = \pm\left\{E\cos(2\pi f_0\tau) - 2\left|\int_{T-\tau}^{T} s(t+\tau-T)s(t)\mathrm{d}t\right|\right\}. \tag{3.146}$$

Simplifying,

$$x(T,\tau) = \pm\left\{E\cos(2\pi f_0\tau) - 2E\left|\frac{\tau}{T}\cos(2\pi f_0\tau)\right|\right\}. \tag{3.147}$$

Note that when t is equal to zero, both (3.145) and (3.147) yield the same value, namely E. The error rate in the presence of the timing error now becomes

$$p(e;\tau) = \frac{1}{2}Q\left(\sqrt{\frac{2E}{N_0}}\right) + \frac{1}{2}Q\left[\sqrt{\frac{2E}{N_0}}(1-2|\varepsilon|)\right]. \tag{3.148}$$

In (3.148), the normalized timing offset is

$$\varepsilon = \frac{\tau}{T}. \tag{3.149}$$

It has been assumed that the carrier frequency is high and that

$$\cos(2\pi f_0\tau) \approx 1. \tag{3.150}$$

Just as in the case of the phase mismatch, if the timing offset is also random, the error rate will increase further. It is essential that timing offset be reduced to a small fraction of the bit duration. If there are problems of phase mismatch and timing offset simultaneously, the error rate can be written as

$$p(e;\tau,\psi) = \frac{1}{2}Q\left[\sqrt{\frac{2E}{N_0}}\cos(\psi)\right] + \frac{1}{2}Q\left[\sqrt{\frac{2E}{N_0}}\cos(\psi)\left(1-\frac{2|\tau|}{T}\right)\right]. \tag{3.151}$$

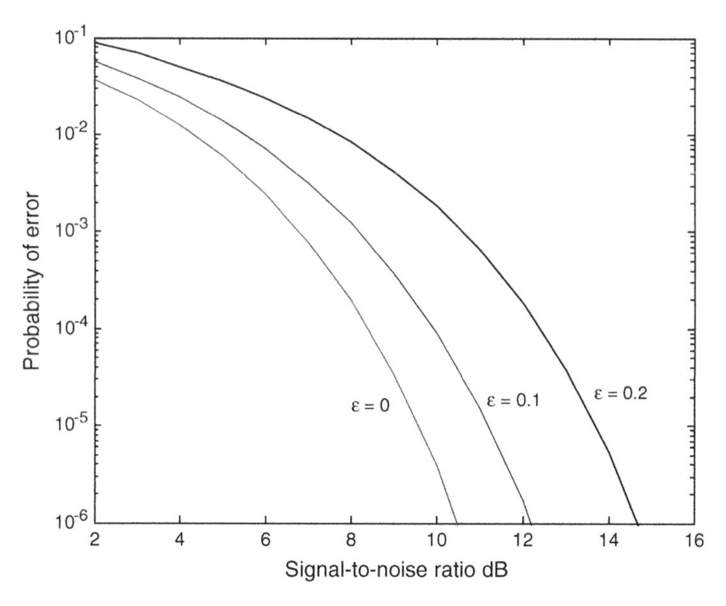

Fig. 3.58 BER plotted for the normalized timing offset for BPSK

In (3.148) and (3.151), E is the bit energy. The error rates in the presence of timing mismatch are plotted in Fig. 3.58.

A similar analysis for phase mismatch in coherent QPSK can be carried out (Gagliardi 1988). Noting that the bit error rate for QPSK is half of the symbol error rate, the bit error rate with no phase mismatch from (3.63) is

$$p_{\mathrm{b}}(e) = Q\left[\sqrt{\frac{2E}{N_0}}\sin\left(\frac{\pi}{4}\right)\right] = Q\left[\sqrt{\frac{E}{N_0}}\right]. \tag{3.152}$$

The bit error rate for coherent QPSK when phase mismatch exists becomes

$$p_{\mathrm{b}}(e) = \frac{1}{2}Q\left\{\sqrt{\frac{E}{N_0}}[\cos(\psi) + \sin(\psi)]\right\} + \frac{1}{2}Q\left\{\sqrt{\frac{E}{N_0}}[\cos(\psi) - \sin(\psi)]\right\}. \tag{3.153}$$

Note that E is the symbol energy equal to twice the bit energy in (3.152) and (3.153).

For the case of OQPSK, a slightly different result is obtained as

$$p_{\mathrm{b}}(e) = \frac{1}{4}Q\left\{\sqrt{\frac{E}{N_0}}[\cos(\psi) + \sin(\psi)]\right\} + \frac{1}{4}Q\left\{\sqrt{\frac{E}{N_0}}[\cos(\psi) - \sin(\psi)]\right\},$$
$$+\frac{1}{2}Q\left\{\sqrt{\frac{E}{N_0}}[\cos(\psi)]\right\} \tag{3.154}$$

Comparison of Digital Modems

We explored several linear and nonlinear modulation schemes, all with coherent detection. While MPSK and MQAM are examples of linear modulation formats, orthogonal MFSK is an example of a nonlinear modulation scheme. The other modulation schemes discussed such as OQPSK and $\pi/4$-QPSK are similar to QPSK while GMSK, a nonlinear scheme, uses a different pulse shape. MSK can be treated as a derived form of OQPSK, albeit with a different pulse shape and, therefore, the error rates and bandwidth capabilities of MSK and GMSK will depend on the pulse shape. Hence, we will exclude them from this discussion and devote our attention to MPSK, MQAM, and MFSK, all using the same pulse shape, e.g., a rectangular one. Let T be the pulse duration. Because of the rectangular shape of the pulse, the null-to-null bandwidth required for transmission will be $2(1/T)$. Since MPSK and MQAM are identical in their pulse shape/symbol shape characteristics, with MPSK having identical vector lengths and MQAM having non-identical vector lengths, the bandwidth for these two schemes can be written as

$$W = 2\left(\frac{1}{T_{\mathrm{b}}}\right)\frac{1}{\log_2 M} = \frac{2R_b}{\log_2 M}. \tag{3.155}$$

In (3.155) Rb is the data (bit) rate equal to ($1/T$b). If we now define the spectral efficiency r as

$$\rho = \frac{R_{\mathrm{b}}}{W}, \tag{3.156}$$

using (3.155), the spectral efficiency becomes

$$\rho = \frac{\log_2 M}{2} = \frac{k}{2}. \tag{3.157}$$

Thus, we see that the spectral or bandwidth efficiency of the linear modulation schemes such as MPSK and MQAM go up with M and since $M = 2^k$, spectral efficiency is directly proportional to $k/2$.

The bandwidth requirement of MFSK is different from MPSK since the M signals are created from carrier frequencies that are orthogonal. This requires that they be separated by $1/2T$. Thus, the bandwidth required for the transmission of M orthogonal FSK signals is

$$W = M\left(\frac{1}{2T}\right) = \frac{M}{2\left(\frac{k}{R_b}\right)} = \frac{M}{2\log_2 M} R_b. \quad (3.158)$$

The spectral efficiency for MSFK is

$$\frac{R_b}{W} = \frac{2\log_2 M}{M} = \frac{2k}{2^k} = \frac{k}{2^{k-1}}. \quad (3.159)$$

We see from (3.159) that the spectral efficiency decreases as M or k increases, clearly demonstrating that MFSK is inferior to MFSK and MQAM in terms of bandwidth utilization.

While spectral efficiency allows us to compare the modems, we also need to examine the power or energy efficiency of the modems so that the minimum SNR required to maintain a certain bit error rate can be used for comparison. We have discussed the fact that the minimum SNR required to have a certain error rate increases with $M > 4$ for MPSK and MQAM. However, since MQAM has dual encoding (amplitude and phase), the SNR required for MQAM will be less than that for MPSK. On the other hand, the SNR required to achieve an acceptable error rate in MFSK decreases as M increases. The values are tabulated in Table 3.8. Note that the spectral efficiency is redefined in terms of the two-sided spectra such that the newly defined efficiency $\eta = \rho/2$ where r is defined in (3.156).

A more general way of showing the power and spectral efficiencies of the modems is through the use of Shannon's channel capacity theorem. The channel capacity C can be written as

$$C = W \log_2\left(1 + \frac{P_{\text{av}}}{WN_0}\right). \quad (3.160)$$

In (3.160), Pav is the average power. Converting to energy units, (3.160) becomes

Table 3.8 Comparison of spectral (η) and power efficiencies

MPSK			MQAM		MFSK	
M	η (bits/s/Hz)	E_b/N_o (dB)	η bits/s/Hz	E_b/N_o dB	η bits/s/Hz	E_b/N_o dB
2	0.5	10.5	0.5	10.5	1	13.5
4	1	10.5	1	10.5	1	10.8
8	1.5	14	1.5	12.35	0.75	9.3
16	2	18.5	2	14.3	0.5	8.2
32	2.5	23.4	2.5	16.38	0.3125	7.5
64	3	28.5	3	18.55	0.1875	6.9

$$C = W\log_2\left(1+\frac{RE_b}{WN_0}\right). \tag{3.161}$$

In an ideal system, $C = R$ and (3.161) becomes

$$\frac{R}{W} = \log_2\left(1+\frac{RE_b}{WN_0}\right). \tag{3.162}$$

Rewriting, we have

$$\frac{E_b}{N_0} = \frac{2^{R/W}-1}{(R/W)}. \tag{3.163}$$

The plot of the spectral efficiency vs. the average energy-to-noise ratio is shown in Fig. 3.59. The values for MPSK, MQAM, and MFSK are shown as individual points on the plot. These values are indicated in terms of the improvement over the respective binary modulation schemes (BPSK and BFSK). We can clearly see the trade-off between power and spectral efficiencies of the linear and nonlinear modulation schemes. The region above the X-axis is the bandwidth limited region; the region below the X-axis is the power limited region. If bandwidth is available and power is at a premium, one would use MFSK. If power constraints are absent and bandwidth is limited, one would use the linear modulation scheme such as

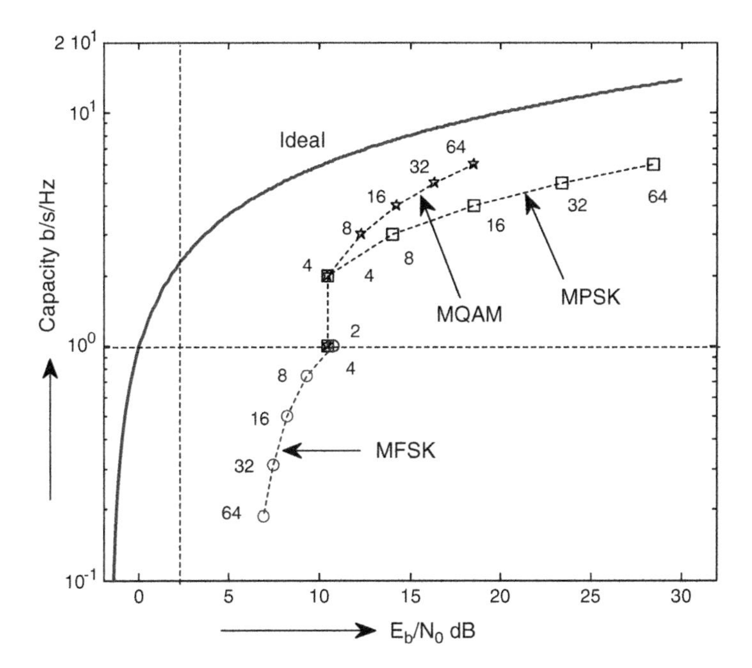

Fig. 3.59 Channel capacity comparison

MPSK and MQAM. Furthermore, in the bandwidth limited region, we can trade-off between the SNR complexity by choosing MQAM, which saves power but uses a complicated transmitter/receiver.

A few additional comments are in order. Even though we had discussed several modems, the error rates were not compared as a group. For the case of noncoherent differential phase shift keying (DPSK), the error rate is given by

$$p_e = \frac{1}{2}\exp\left(-\frac{E_\mathrm{b}}{N_0}\right). \tag{3.164}$$

For noncoherent binary FSK, the error rate is

$$p_e = \frac{1}{2}\exp\left(-\frac{1E_\mathrm{b}}{2N_0}\right) \tag{3.165}$$

For noncoherent MFSK, the (symbol) error rate (upper limit) is

$$p(s)_e < \frac{M-1}{2}\exp\left(-\frac{1E_s}{2N_0}\right). \tag{3.166}$$

Q *Function, Complementary Error Function, and Gamma Function*

We had seen that the error rate is conveniently expressed in terms of Q functions defined as (Helstrom 1968; Sklar 2001; Papoulis and Pillai 2002; Gradshteyn and Ryzhik 2007)

$$Q(x) = \frac{1}{\sqrt{2\pi}}\int_x^\infty \exp\left(-\frac{z^2}{2}\right)\mathrm{d}z. \tag{3.167}$$

The error function erf(.) is defined as

$$\mathrm{erf}(x) = \frac{2}{\sqrt{\pi}}\int_0^x \exp(-z^2)\mathrm{d}z. \tag{3.168}$$

The complementary error erfc(.) function is defined as

$$\mathrm{erfc}(x) = \frac{2}{\sqrt{\pi}}\int_x^\infty \exp(-z^2)\mathrm{d}z = 1 - \mathrm{erf}(x). \tag{3.169}$$

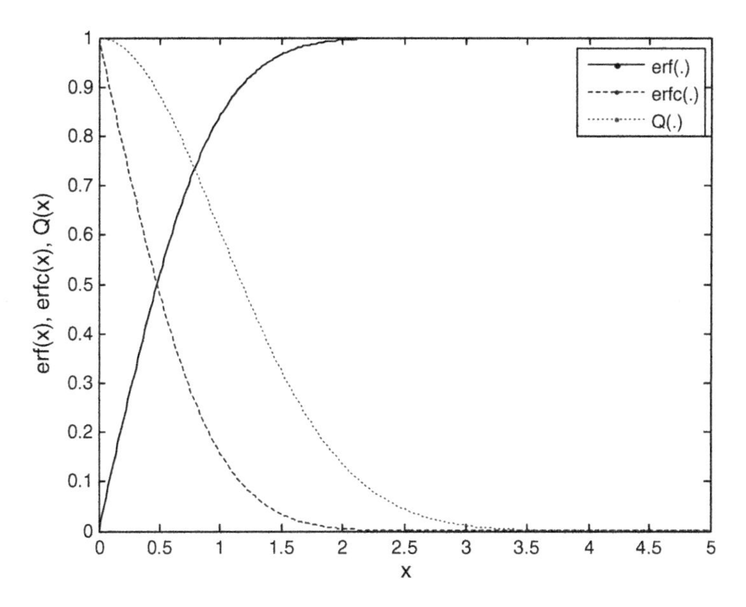

Fig. 3.60 Erf(.), Erfc(.), and Q(.) function

Thus, there is also a relationship between the Q function and the complementary error function

$$\mathrm{erfc}(x) = 2Q\left(x\sqrt{2}\right). \tag{3.170}$$

Inversely,

$$Q(x) = \frac{1}{2}\mathrm{erfc}\left(\frac{x}{\sqrt{2}}\right). \tag{3.171}$$

The three functions, erf(.), erfc(.), and Q(.), are plotted in Fig. 3.60.

The error rate can also be expressed in terms of gamma functions (Wojnar 1986). We had seen that the error rates are functions of the SNR (or energy-to-noise ratio) E/N_0. If we represent the error rate that can be expressed as

$$z = \frac{E}{N_0}, \tag{3.172}$$

where

$$p_{\mathrm{e}}(z) = \frac{\Gamma(b, az)}{2\Gamma(b)}, \tag{3.173}$$

$$\Gamma(b, az) = \int_{az}^{\infty} t^{b-1} \exp(-t) \mathrm{d}t \tag{3.174}$$

$$\Gamma(b) = \int_{0}^{\infty} t^{b} \exp(-t) \mathrm{d}t \tag{3.175}$$

$$b = \begin{cases} \frac{1}{2}, & \text{coherent detection} \\ 1, & \text{noncoherent/differential detection} \end{cases} \tag{3.176}$$

$$a = \begin{cases} \frac{1}{2}, & \text{Orthoganal FSK} \\ 1, & \text{bipolar/bipodal PSK} \end{cases}. \tag{3.177}$$

Since

$$\Gamma\left(\frac{1}{2}\right) = \sqrt{\pi} \tag{3.178}$$

$$\Gamma\left(\frac{1}{2}, x\right) = \sqrt{\pi}\ \mathrm{erfc}\left(\sqrt{x}\right), \tag{3.179}$$

we have for coherent BPSK,

$$p_e(z) = \frac{\Gamma((1/2), z)}{2\Gamma(1/2)} = \frac{1}{2}\mathrm{erfc}\left(\sqrt{z}\right) = Q\left(\sqrt{2z}\right). \tag{3.180}$$

The error probability in (3.173) is plotted in Fig. 3.61 for the four sets of values of (a,b).

In some of the analysis involving bit error rates in fading channels, we would also require the use of the derivative of the complimentary error function. We have

$$\frac{\partial}{\partial z}[erfc(z)] = -\frac{2}{\sqrt{\pi}}\exp\left(-z^2\right) \tag{3.181}$$

$$\frac{\partial}{\partial z}[Q(z)] = -\frac{1}{\sqrt{2\pi}}\exp\left(-\frac{z^2}{2}\right). \tag{3.182}$$

Specifically,

$$\frac{\partial}{\partial z}[p_e(z)] = -\frac{1}{2\sqrt{\pi z}}\exp(-z). \tag{3.183}$$

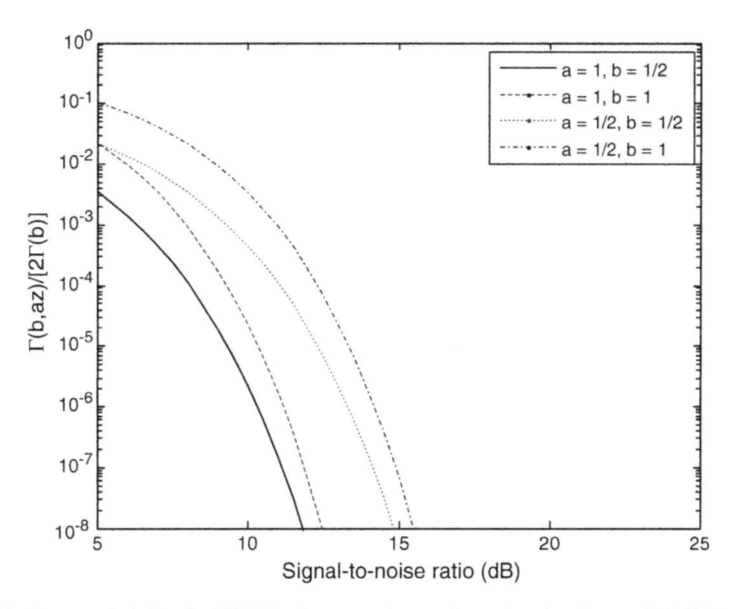

Fig. 3.61 Error probability for BPSK (coherent and noncoherent) and orthogonal BFSK (coherent and noncoherent)

There are several approximations to the Q function (Chiani et al. 2003; Karagiannidis and Lioumpas 2007; Yunfei and Beaulieu 2007; Isukapalli and Rao 2008). The two bounds of the Q functions are

$$\left(1-\frac{1}{\alpha^2}\right)\frac{1}{\sqrt{2\pi\alpha^2}}\exp\left(-\frac{\alpha^2}{2}\right)\leq Q(\alpha)\leq\frac{1}{\sqrt{2\pi\alpha^2}}\exp\left(-\frac{\alpha^2}{2}\right). \tag{3.184}$$

An upper bound for large values of the argument for the Q function is

$$Q(\alpha)\leq\frac{1}{2}\exp\left(-\frac{\alpha^2}{2}\right). \tag{3.185}$$

Marcum Q *Function*

While the Q function defined in (3.167) is seen in error rate calculations involving additive white noise, another form of Q function is seen in equations involving integrals such as the ones associated with the Rician density function mentioned in Chap. 2. One such form is the Marcum Q function defined as (Nuttall 1975; Simon 1998, 2002; Helstrom 1998; Corazza and Ferrari 2002; Simon and Alouini 2003, 2005)

$$Q(a,b) = \int_b^{\infty} x \ \exp\left(-\frac{x^2 + a^2}{2}\right) I_0(ax)\mathrm{d}x. \tag{3.186}$$

Equation (3.186) can also be expressed as an infinite summation,

$$Q(a,b) = \exp\left(-\frac{x^2 + a^2}{2}\right) \sum_{k=0}^{\infty} \left(\frac{a}{b}\right)^k I_k(ab). \tag{3.187}$$

The limiting cases of the Marcum Q function lead to

$$Q(a,a) = \frac{1 + \exp(-a^2)I_0(a^2)}{2} \tag{3.188}$$

$$Q(0,b) = \exp\left(-\frac{b^2}{2}\right) \tag{3.189}$$

$$Q(a,0) = 1. \tag{3.190}$$

The Marcum Q function is shown in Fig. 3.62 for three values of b. The generalized Marcum Q function is defined as

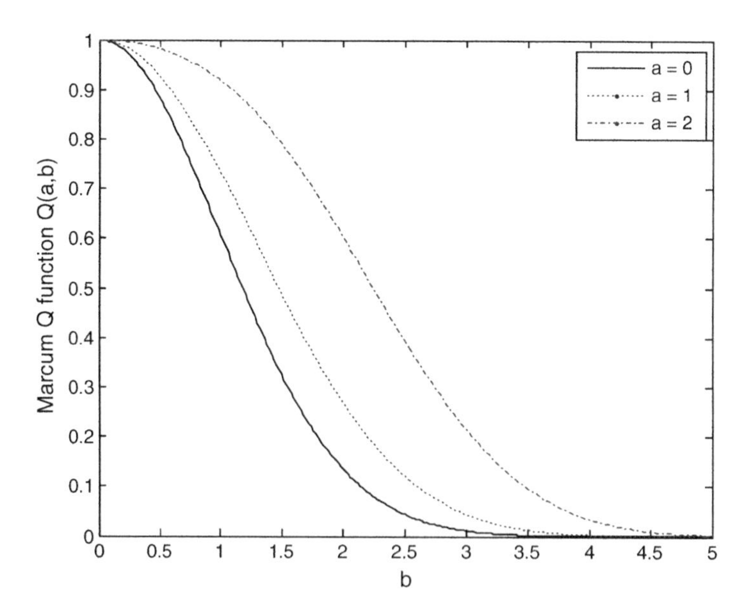

Fig. 3.62 Marcum Q function

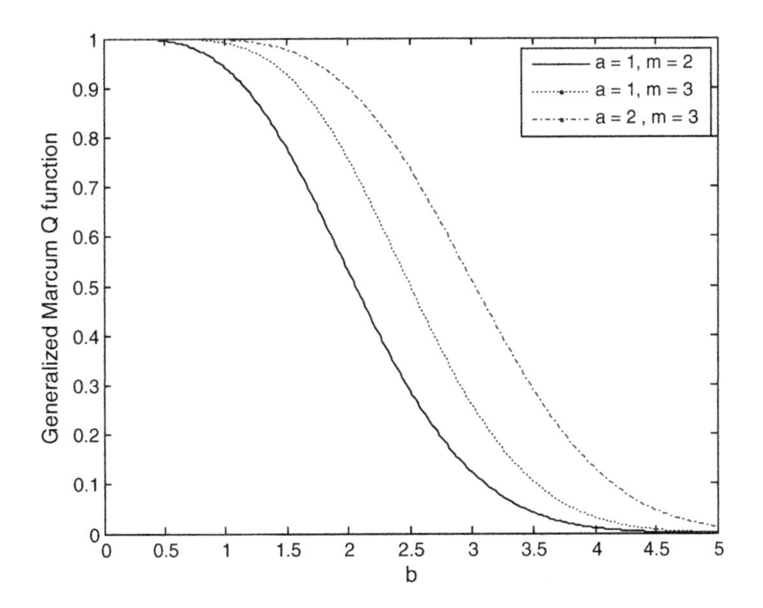

Fig. 3.63 The generalized Marcum Q function

$$Q_m(a,b) = \int_b^{\infty} x^m \exp\left(-\frac{x^2+a^2}{2}\right) I_{m-1}(ax)\mathrm{d}x. \tag{3.191}$$

It is obvious that when $m = 1$, the generalized Marcum function reduces to the Q function in (3.186) as

$$Q_1(a,b) = Q(a,b). \tag{3.192}$$

The generalized Marcum Q function is shown in Fig. 3.63.

The recursion relation for the generalized Marcum Q function is

$$Q_m(a,b) = \left(\frac{b}{a}\right) \exp\left(-\frac{a^2+b^2}{2}\right) I_{m-1}(ab) + Q_{m-1}(a,b). \tag{3.193}$$

For the integer values of m the generalized Marcum Q function can be expressed as

$$Q_m(a,b) = \exp\left(-\frac{a^2+b^2}{2}\right) \sum_{k=1-m}^{\infty} \left(\frac{a}{b}\right)^k I_k(ab). \tag{3.194}$$

For integer values of m, (3.194) becomes

$$Q_m(a,a) = \frac{1}{2} + \exp(-a)\left[\frac{I_0(a^2)}{2} + \sum_{k=1}^{m-1} I_k(a^2)\right]. \tag{3.195}$$

One limiting case of the generalized Marcum Q function is

$$Q_m(0,b) = \frac{\Gamma(m, (b^2/2))}{\Gamma(m)}. \tag{3.196}$$

Equation (3.196) can be simplified for the case of integer values of m as

$$Q_m(0,b) = \sum_{k=0}^{m-1} \frac{1}{k!}\left(\frac{b^2}{2}\right)^k \exp\left(-\frac{b^2}{2}\right). \tag{3.197}$$

A further generalization of the Marcum Q function (Nuttall 1975), which will be useful in integrals involving fading distributions such as the '$T - m$ or $k - m$, is

$$Q_{m,n}(a,b) = \int_b^{\infty} x^m \exp\left(-\frac{x^2 + a^2}{2}\right) I_{n-1}(ax)\mathrm{d}x. \tag{3.198}$$

Note that m and n can take any positive values including zero, and a and b lie between 0 and 1. For the special case of $m = n + 1$, (3.198) can be expressed in terms of the generalized Marcum Q function of (3.191) as

$$Q_{n+1,n}(a,b) = a^n Q_{n+1}(a,b), \tag{3.199}$$

with a further special case of

$$Q_{1,0}(a,b) = Q_1(a,b) = Q(a,b). \tag{3.200}$$

Intersymbol Interference

The issue of intersymbol interference was brought up earlier in connection with the choice of the appropriate pulse shapes (Sklar 2001; Haykin 2001). We will now briefly review the concepts behind intersymbol interference and the way to characterize problems arising from it. Consider an ideal low pass filter with a transfer function as shown in Fig. 3.64. It has a bandwidth of $(1/2T)$, and if one looks at the corresponding time function in Fig. 3.65 we see that the impulse response is a sin c function.

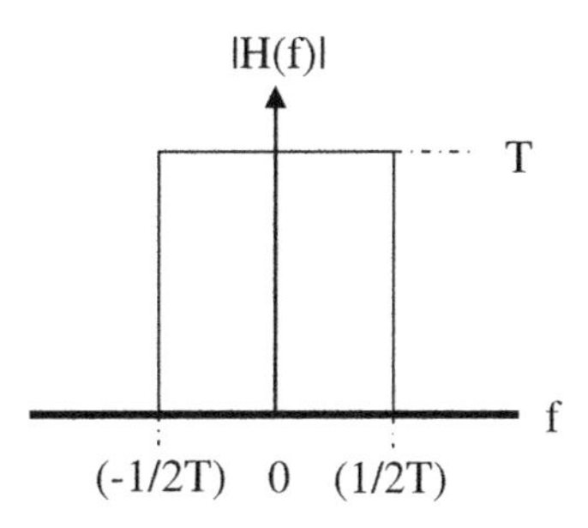

Fig. 3.64 Ideal LPF transfer function

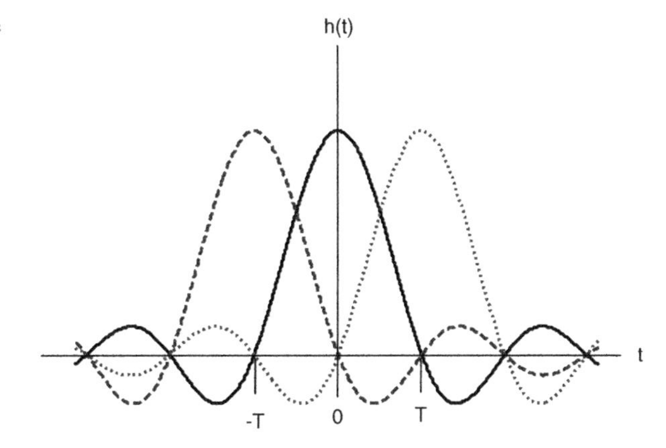

Fig. 3.65 Impulse response (s) of ideal LPF

Two other sin c functions are also plotted, one delayed by $+T$ and the other delayed by $-T$. We can see that the pulse shapes are such that at every T s interval, the maximum of any given pulse has zero contributions from the other pulses since the sin c function goes through zero. In other words, for transmission of data occupying a duration of Ts, there will be no contribution from adjoining bits (or symbols) if we can choose the sin c pulse. Since the bandwidth of such a pulse is $(1/2T)$, use of such pulses will lead to transmission with the best spectral efficiency because this BW is only equal to $(R/2)$ where R is the data rate given by $(1/T)$.

However, such a condition is not ideal for two major reasons. First, the filter response is far from ideal and real filters would have a gradual fall off beyond $f = (1/2T)$. This would mean that the pulses after passage through filters will be distorted leading to overlaps at sampling instants resulting in intersymbol interference (ISI). Second, ideal sampling at exact multiples of T is also not practical since timing errors would be present leading to contributions from adjoining symbols and resulting in ISI which leads to increase in error rates.

One of the ways to mitigate ISI is to use a raised cosine pulse shape as discussed earlier. The raised cosine pulse shape can be expressed as

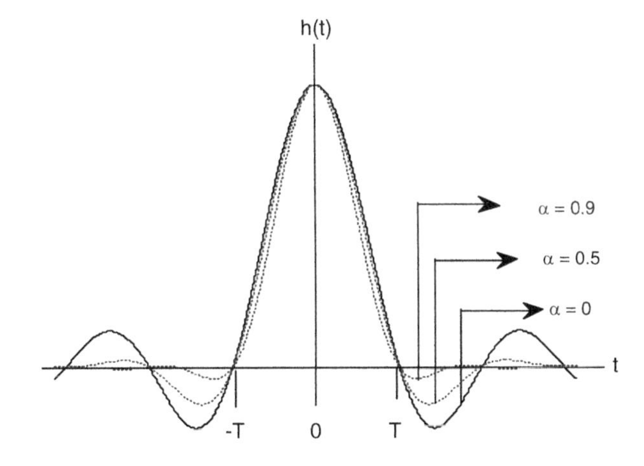

Fig. 3.66 Raised cosine pulses

$$h(t) = \sin c(2Wt)\left[\frac{\cos(2\pi\alpha Wt)}{1 - 16\alpha^2 W^2 t^2}\right]. \tag{3.201}$$

In (3.201), α is the roll-off factor and W is the bandwidth of the ideal low pass filter shown in Fig. 3.66 and corresponding to the ideal pulse for the highest data rate transmission, i.e.,

$$W = \frac{1}{2T}. \tag{3.202}$$

Pulse shapes corresponding to three values of $\alpha = 0$, 0.5, and 0.9 are shown in Fig. 3.66. The corresponding spectra are shown in Fig. 3.67. As can be seen, $\alpha = 0$ corresponds to the case of a sin c pulse and an increase in values demonstrates the shrinking of the sidelobes which can reduce the chances of ISI. The value of $\alpha = 1$ corresponds to the full roll-off raised cosine pulse. One can see that the bandwidths of the pulses increase with α. The BW of the *WR* raised cosine pulse can be written in terms of W as

$$W_R = W(1 + \alpha). \tag{3.203}$$

Thus, for a full raised cosine pulse with $\alpha = 1$ requires twice the BW of a sin c pulse. Defining

$$f_1 = W(1 - \alpha), \tag{3.204}$$

the spectrum of the raised cosine pulse shapes are

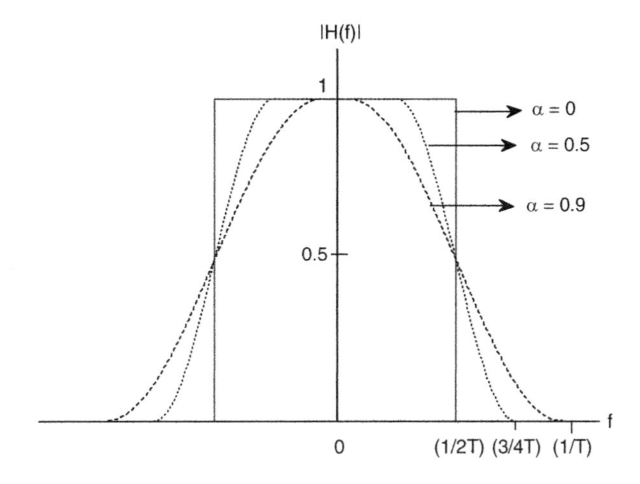

Fig. 3.67 Spectra of the raised cosine pulses

$$H(f) = \begin{cases} \frac{1}{2W}, & 0 \leq |f| < f_1 \\ \frac{1}{4W}\left\{1 - \sin\left[\frac{\pi(|f| - W)}{2W - 2f_1}\right]\right\}, & f_1 \leq |f| 2W - f_1 \\ 0, & |f| \geq 2W - f_1 \end{cases}. \tag{3.205}$$

For α = 1, we have

$$\begin{cases} \frac{1}{4W}\left\{1 - \sin\left[\frac{\pi(|f| - W)}{2W - 2f_1}\right]\right\}, & 0 \leq |f| < 2W \\ 0, & |f| \geq 2W \end{cases} \tag{3.206}$$

and, the corresponding pulse shape becomes

$$h(t) = \frac{\sin c(4Wt)}{1 - 16W^2 - t^2}. \tag{3.207}$$

At the receiver, pulse shaping is employed to make the pulse shapes to be of a raised cosine type so that ISI will be minimum. We will now look at eye patterns which show the effects of ISI.

Eye patterns allow us to study the effect of dispersion (pulse broadening) and noise in digital transmission. An eye pattern is a synchronized superposition of all possible realizations of the signal by the various bit patterns. We can easily create them using Matlab by considering the transmission of rectangular pulse (width T) passing through a low pass filter of bandwidth B. For a given pulse width, increasing values of BT means less and less distortion. Two eye patterns are shown in Figs. 3.68 and 3.69. The one for a low value of BT appears in Fig. 3.68. In both

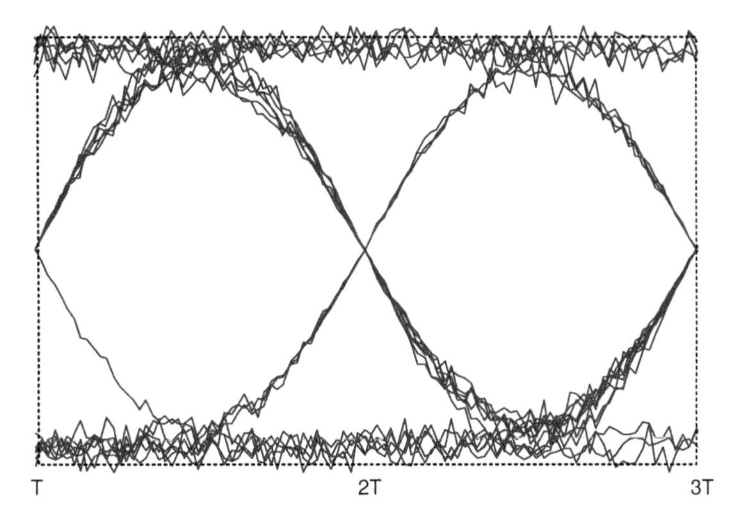

Fig. 3.68 Eye pattern (low value of BT)

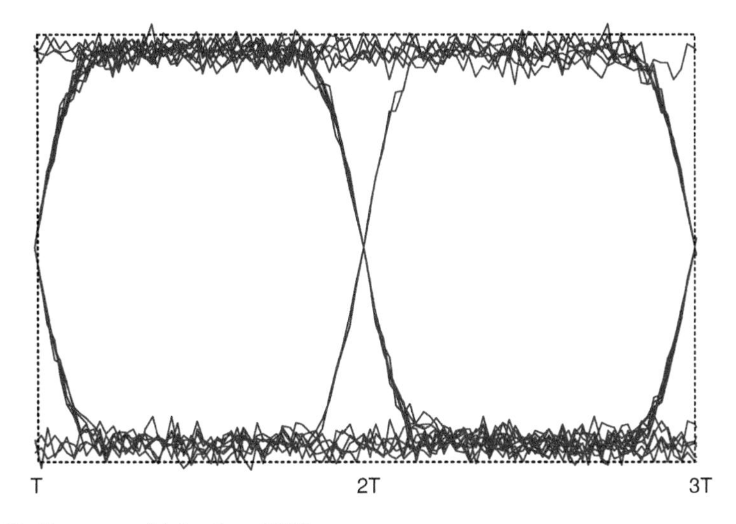

Fig. 3.69 Eye pattern (high value of BT)

cases, additive white noise has also been added. The one with a higher value of BT appears in Fig. 3.69.

The width of the eye opening provides information on the time interval over which the received signal can be sampled without being impacted by ISI. One can see that a higher value of BT results in a wider eye, while a lower value of BT results in a narrow eye.

References

Abramowitz, M., & Segun, I. (1965). *Handbook of mathematical functions: With formulas, graphs, and mathematical tables*. New York, NY: Dover Publications.
Aghvami, A. (1993). Digital modulation techniques for mobile and personal communication systems. *Electronics & Communication Engineering Journal, 5*(3), 125–132.
Akaiwa, Y. (1997). *Introduction to digital mobile communication*. New York, NY: Wiley.
Akaiwa, Y., & Nagata, Y. (1987). Highly efficient digital mobile communications with a linear modulation method. *IEEE Journal on Selected Areas in Communications, 5*(5), 890–895.
Amoroso, F. (1976). Pulse and Spectrum Manipulation in the Minimum (Frequency) Shift Keying (MSK) Format. *IEEE Transactions on Communications, 24*(3), 381–384.
Amoroso, F. (1979). The use of quasi-bandlimited pulses in MSK transmission. *IEEE Transactions on Communications, 27*(10), 1616–1624.
Amoroso, F. (1980). Bandwidth of digital data signals. *IEEE Communications Magazine, 18*(6), 13–24.
Amoroso, F., & Kivett, J. (1977). Simplified MSK signaling technique. *IEEE Transactions on Communications, 25*(4), 433–441.
Anderson, J. B. (2005). *Digital transmission engineering*. Hoboken, NJ: IEEE.
Anderson, J., Aulin, T., et al. (1986). *Digital phase modulation*. New York, NY: Plenum Publishing Corporation.
Armada, A. G. (2001). Understanding the effects of phase noise in orthogonal frequency division multiplexing (OFDM). *Transactions of IEEE on Broadcasting, 47*(2), 153–159.
Aulin, T., & Sundberg, C. (1981). Continuous phase modulation–Part I: Full response signaling. *IEEE Transactions on Communications, 29*(3), 196–209.
Aulin, T., & Sundberg, C. E. (1982). Exact asymptotic behavior of digital FM spectra. *IEEE Transactions on Communications, 30*(11), 2438–2449.
Beaulieu, N. C., & Damen, M. O. (2004). Parametric construction of Nyquist-I pulses. *IEEE Transactions on Communications, 52*(12), 2134–2142.
Benedetto, S., & Biglieri, E. (1999). *Principles of digital transmission: With wireless applications*. New York, NY: Kluwer Academic.
Bingham, J. (1990). Multicarrier modulation for data transmission: An idea whose time has come. *IEEE Communications Magazine, 28*(5), 5–14.
Bingham, J. A. C. (2000). *ADSL, VDSL, and multicarrier modulation*. New York, NY: Wiley.
Borjesson, P., & Sundberg, C. E. (1979). Simple approximations of the error function $Q(x)$ for communications applications. *IEEE Transactions on Communications, 27*(3), 639–643.
Cahn, C. (1959). Performance of digital phase-modulation communication systems. *IRE Transactions on Communication Systems, 7*(1), 3–6.
Cahn, C. (1960). Combined digital phase and amplitude modulation communication systems. *IRE Transactions on Communication Systems, 8*(3), 150–155.
Campopiano, C., & Glazer, B. (1962). A coherent digital amplitude and phase modulation scheme. *IRE Transactions on Communication Systems, 10*(1), 90–95.
Chiani, M., et al. (2003). New exponential bounds and approximations to the computation of error probability in fading channels. *IEEE Transactions on Wireless Communications, 2*(4), 840–845.
Cimini Jr., L. (1985). Analysis and simulation of a digital mobile channel using orthogonal frequency division multiplexing. *IEEE Transactions on Communications, 33*(7), 665–675.
Cooper, G. R., & McGillem, C. D. (1986). *Modern communications and spread spectrum*. New York, NY: McGraw-Hill.
Corazza, G. E., & Ferrari, G. (2002). New bounds for the Marcum Q-function. *IEEE Transactions on Information Theory, 48*(11), 3003–3008.
Couch, L. W. (2007). *Digital and analog communication systems*. Upper Saddle River, NJ: Pearson/Prentice Hall.
Davenport, W., & Root, W. (1958). *Random signals and noise*. New York, NY: McGraw-Hill.

Edbauer, F. (1992). Bit error rate of binary and quaternary DPSK signals with multiple differential feedback detection. *IEEE Transactions on Communications, 40*(3), 457–460.

Elnoubi, S. (2006). Analysis of GMSK with discriminator detection in mobile radio channels. *IEEE Transactions on Vehicular Technology, 35*(2), 71–76.

Feher, K. (1991). MODEMs for emerging digital cellular-mobile radio system. *IEEE Transactions on Vehicular Technology, 40*(2), 355–365.

Feher, K. (1995). *Wireless digital communications: Modulation and spread spectrum applications*. Upper Saddle River, NJ: Prentice-Hall.

Forney Jr., G., Gallager, R., et al. (1984). Efficient modulation for band-limited channels. *IEEE Journal on Selected Areas in Communications, 2*(5), 632–647.

Franks, L. (1968). Further results on Nyquist's problem in pulse transmission. *IEEE Transactions on Communication Technology, 16*(2), 337–340.

Gagliardi, R. M. (1988). *Introduction to communications engineering*. New York, NY: Wiley.

Gallager, R. G. (2008). *Principles of digital communication*. Cambridge, NY: Cambridge University Press.

Gao, X. Q., You, X. H., et al. (2006). An efficient digital implementation of multicarrier CDMA system based on generalized DFT filter banks. *IEEE Journal on Selected Areas in Communications, 24*(6), 1189–1198.

Goldsmith, A. (2005). Wireless communications. Cambridge, NY, Cambridge University Press.

Gradshteyn, I. S., & Ryzhik, I. M. (2007). *Table of integrals, series and products*. Oxford: Academic.

Gronemeyer, S., & McBride, A. (1976). MSK and offset QPSK modulation. *IEEE Transactions on Communications, 24*(8), 809–820.

Hambley, A., & Tanaka, O. (1984). Generalized serial MSK modulation. *IEEE Transactions on Communications, 32*(3), 305–308.

Haykin, S. M. (2001). *Digital communications*. New York, NY: Wiley.

Haykin, S. M., & Moher, M. (2005). *Modern wireless communications*. Upper Saddle River, NJ: Prentice-Hall.

Helstrom, C. (1960). The comparison of digital communication systems. *IRE Transactions on Communications Systems, 8*(3), 141–150.

Helstrom, C. W. (1968). *Statistical theory of signal detection*. Oxford, NY: Pergamon Press.

Helstrom, C. W. (1998). Approximate inversion of Marcum's *Q*-function. *IEEE Transactions on Aerospace and Electronic Systems, 34*(1), 317–319.

Isukapalli, Y., & Rao, B. (2008). An analytically tractable approximation for the Gaussian *Q*-function. *IEEE Communications Letters, 12*(9), 669–671.

Karagiannidis, G. K., & Lioumpas, A. S. (2007). An improved approximation for the Gaussian *Q*-function. *IEEE Communications Letters, 11*(8), 644–646.

Kim, K. & Polydoros, A. (1988). Digital modulation classification: The BPSK versus the QPSK case. In *Proceedings of MILCOM '88* (pp. 24.4.1–24.4.6). San Diego, CA, October 26–29, 1988.

Kisel, A. V. (1999). An extension of pulse shaping filter theory. *IEEE Transactions on Communications, 47*(5), 645–647.

Klymyshyn, D., Kumar, S., et al. (1999). Direct GMSK modulation with a phase-locked power oscillator. *IEEE Transactions on Vehicular Technology, 48*(5), 1616–1625.

Kuchi, K. & V. Prabhu (1999). Power spectral density of GMSK modulation using matrix methods. In *Military Communications Conference Proceedings, 1999. (MILCOM 1999), Vol. 1* (pp. 45–50). IEEE.

Lindsey, W. C., & Simon, M. K. (1973). *Telecommunication systems engineering*. Englewood Cliffs, NJ: Prentice-Hall.

Linz, A., & Hendrickson, A. (1996). Efficient implementation of an IQ GMSK modulator. Circuits and Systems II. *IEEE Transactions on Analog and Digital Signal Processing, 43*(1), 14–23.

Liu, H., & Tureli, U. (1998). A high-efficiency carrier estimator for OFDM communications. *IEEE Communications Letters, 2*(4), 104–106.

Lucky, R. W., Salz, J., & Weldon Jr., E. J. (1968). *Principles of data communication*. New York, NY: McGraw-Hill.

Makrakis, D., & Feher, K. (1990). Optimal noncoherent detection of PSK signals. *Electronics Letters, 26*(6), 398–400.

Miller, L. E., & Lee, J. S. (1998). BER expressions for differentially detected pi/4 DQPSK modulation. *IEEE Transactions on Communications, 46*(1), 71–81.

Miyagaki, Y., Morinaga, N., et al. (1978). Error probability characteristics for CPSK signal through m-distributed fading channel. *IEEE Transactions on Communications, 26*(1), 88–100.

Molisch, A. F. (2005). *Wireless communications*. Chichester: Wiley.

Murota, K. (2006). Spectrum efficiency of GMSK land mobile radio. *IEEE Transactions on Vehicular Technology, 34*(2), 69–75.

Murota, K., & Hirade, K. (1981). GMSK modulation for digital mobile radio telephony. *IEEE Transactions on Communications, 29*, 1044–1050.

Noguchi, T., Daido, Y., et al. (1986). Modulation techniques for microwave digital radio. *IEEE Communications Magazine, 24*, 21–30.

Nuttall, A. (1975). Some integrals involving the *Q_M* > function (Corresp.) *IEEE Transactions on Information Theory, 21*(1), 95–96.

Nuttall, A., & Amoroso, F. (1965). Minimum Gabor bandwidth of M-ary orthogonal signals. *IEEE Transactions on Information Theory, 11*(3), 440–444.

Nyquist, H. (2002). Certain topics in telegraph transmission theory. *Proceedings of the IEEE, 90* (2), 280–305.

Oetting, J. (1979). A comparison of modulation techniques for digital radio. *IEEE Transactions on Communications, 27*(12), 1752–1762.

Olivier, J. C., Sang-Yick, L., et al. (2003). Efficient equalization and symbol detection for 8-PSK EDGE cellular system. *IEEE Transactions on Vehicular Technology, 52*(3), 525–529.

Papoulis, A., & Pillai, S. U. (2002). *Probability, random variables, and stochastic processes*. Boston: McGraw-Hill.

Pasupathy, S. (1979). Minimum shift keying: A spectrally efficient modulation. *IEEE Communications Magazine, 17*(4), 14–22.

Prabhu, V. (1969). Error-rate considerations for digital phase-modulation systems. *IEEE Transactions on Communication Technology, 17*(1), 33–42.

Prabhu, V. (1973). Error probability performance of M-ary CPSK systems with intersymbol interference. *IEEE Transactions on Communications, 21*(2), 97–109.

Prabhu, V. (1976a). Bandwidth occupancy in PSK systems. *IEEE Transactions on Communications, 24*(4), 456–462.

Prabhu, V. K. (1976b). PSK performance with imperfect carrier phase recovery. *IEEE Transactions on Aerospace and Electronic Systems, 12*(2), 275–286.

Prabhu, V. (1980). The detection efficiency of 16-ary QAM'. *Bell System Technical Journal, 59* (1), 639–656.

Prabhu, V. K. (1981). MSK and offset QPSK modulation with bandlimiting filters. *IEEE Transactions on Aerospace and Electronic Systems, AES-17*(1), 2–8.

Prabhu, V., & Salz, J. (1981). On the performance of phase-shift-keying systems. *AT T Technical Journal, 60*, 2307–2343.

Prasad, R. (2004). *OFDM for wireless communications systems*. Boston, MA: Artech House.

Proakis, J. G. (2001). *Digital communications*. Boston, MA: McGraw-Hill.

Rappaport, T. S. (2002). *Wireless communications: principles and practice*. Upper Saddle River, NJ: Prentice Hall PTR.

Rowe, H., & Prabhu, V. (1975). Power spectrum of a digital, frequency-modulation signal. *AT T Technical Journal, 54*, 1095–1125.

Russell, M., & Stuber, G. L. (1995). Interchannel interference analysis of OFDM in a mobile environment. *IEEE Vehicular Technology Conference, 2*, 820–824.

Sadr, R., & Omura, J. K. (1988). Generalized minimum shift-keying modulation techniques. *IEEE Transactions on Communications, 36*(1), 32–40.

Salz, J. (1970). Communications efficiency of certain digital modulation systems. *IEEE Transactions on Communication Technology, 18*(2), 97–102.
Sampei, S. (1997). *Applications of digital wireless technologies to global wireless telecommunications*. Upper Saddle River, NJ: Prentice-Hall.
Sari, H., Karam, G., et al. (1994). An analysis of orthogonal frequency-division multiplexing for mobile radio applications. In *Vehicular Technology Conference, 1994 I.E. 44th* (pp. 1653–1639).
Sathananthan, K., & Tellambura, C. (2001). Probability of error calculation of OFDM systems with frequency offset. *IEEE Transactions on Communications, 49*(11), 1884–1888.
Sayar, B., & Pasupathy, S. (1986). Further results on Nyquist's third criterion. *Proceedings of the IEEE, 74*(10), 1460–1462.
Schwartz, M. (1980). *Information transmission, modulation, and noise: a unified approach to communication systems*. New York, NY: McGraw-Hill.
Schwartz, M. (2005). *Mobile wireless communications*. Cambridge: Cambridge University Press.
Schwartz, *M*. et al. (1996). Communication systems and techniques. Piscataway, NJ., IEEE Press
Shankar, P. (2002). *Introduction to wireless systems*. New York, NY: Wiley.
Shanmugam, K. S. (1979). *Digital and analog communication systems*. New York, NY: Wiley.
Simon, M. (1976). A generalization of minimum-shift-keying (MSK)-type signaling based upon input data symbol pulse shaping. *IEEE Transactions on Communications, 24*(8), 845–856.
Simon, M. (1998). A new twist on the Marcum *Q*-function and its application. *IEEE Communications Letters, 2*(2), 39–41.
Simon, M. K. (2002). The Nuttall *Q* function–its relation to the Marcum *Q* function and its application in digital communication performance evaluation. *IEEE Transactions on Communications, 50*(11), 1712–1715.
Simon, M. K., & Alouini, M. S. (2003). Some new results for integrals involving the generalized Marcum *Q* function and their application to performance evaluation over fading channels. *IEEE Transactions on Wireless Communications, 2*(4), 611–615.
Simon, M. K., & Alouini, M.-S. (2005). *Digital communication over fading channels*. Hoboken, NJ: Wiley.
Simon, M. K., & Divsalar, D. (1997). On the optimality of classical coherent receivers of differentially encoded M-PSK. *IEEE Communications Letters, 1*(3), 67–70.
Simon, M., Hinedi, S., et al. (1995). *Digital communication techniques: Signal design and detection*. Englewood Cliffs, NJ: Prentice Hall PTR.
Sklar, B. (1983a). A structured overview of digital communications—A tutorial review–Part I. *IEEE Communications Magazine, 21*(5), 4–17.
Sklar, B. (1983b). A structured overview of digital communications—A tutorial review–Part II. *IEEE Communications Magazine, 21*(7), 6–21.
Sklar, B. (1993). Defining, designing, and evaluating digital communication systems. *IEEE Communications Magazine, 31*(11), 91–101.
Sklar, B. (2001). *Digital communications: Fundamentals and applications*. Upper Saddle River, NJ: Prentice-Hall PTR.
Stuber, G. L. (2002). *Principles of mobile communication*. New York, NY: Kluwer Academic.
Sundberg, C. E. (1986). Continuous phase modulation. *IEEE Communications Magazine, 24*(4), 25–38.
Svensson, A., & Sundberg, C. (1985). Serial MSK-type detection of partial response continuous phase modulation. *IEEE Transactions on Communications, 33*(1), 44–52.
Taub, H., & Schilling, D. L. (1986). *Principles of communication systems*. New York, NY: McGraw–Hill.
Thomas, C., Weidner, M., et al. (1974). Digital amplitude-phase keying with M-ary alphabets. *IEEE Transactions on Communications, 22*(2), 168–180.
Van Trees, H. L. (1968). *Detection, estimation and modulation theory*. New York, NY: Wiley.
Webb, W. (1992). QAM: the modulation scheme for future mobile radio communications? *Electronics & Communication Engineering Journal, 4*(4), 167–176.

Winters, J. H. (2000). Smart antennas for the EDGE wireless TDMA system. In *Adaptive Systems for Signal Processing, Communications, and Control Symposium 2000. AS-SPCC*. The IEEE.

Wojnar, A. H. (1986). Unknown bounds on performance in Nakagami channels. *IEEE Transactions on Communications, 34*(1), 22–24.

Wu, Y., & Zou, W. Y. (1995). Orthogonal frequency division multiplexing: a multi-carrier modulation scheme. *IEEE Transactions on Consumer Electronics, 41*(3), 392–399.

Yunfei, C., & Beaulieu, N. C. (2007). Solutions to infinite integrals of Gaussian *Q*-function products and some applications. *IEEE Communications Letters, 11*(11), 853–855.

Ziemer, R., & Ryan, C. (1983). Minimum-shift keyed modem implementations for high data rates. *IEEE Communications Magazine, 21*(7), 28–37.

Chapter 4
Modeling of Fading and Shadowing

4.1 Introduction

In this chapter, we will review the statistics of signal degradation in wireless channels. This degradation arises from fading and shadowing individually as well the simultaneous existence of fading and shadowing. Several models are available in literature for the description of the random variations in signal power (Clark 1968; Bullington 1977; Aulin 1979; Loo 1985; Jakes 1994). Simple fading models will be reviewed first and then the models for shadowing. This will be followed by models which describe cases where fading and shadowing are present simultaneously (Hansen and Meno 1977; Suzuki 1977; Shankar 2004, 2010a, b). Most of the simple models fail to account for the signal fluctuations fully and accurately, and modifications to these models, or new ones, are necessary to achieve a better description of the randomness in signal strength. A different generation of models is also necessary in light of the fact that developments on wireless data communications have taken us from the traditional communications involving a single transmitter site and a receiver site to a more complex one involving relay stations, multiple input and multiple output arrangements, and so on (Andersen 2002a; Salo et al. 2006a, b; Karagiannidis et al. 2006a, 2007; Shankar 2011a, b). This makes it necessary to include cascaded approaches to modeling of the signal strengths. Toward the end of the chapter, we will study special models which encompass most if not all of the simple models described earlier. We will use unified approaches which will permit us to describe the newer generation of wireless systems (Yacoub 2007a, b; Karadimas and Kotsopoulos 2008, 2010; Papazafeiropoulos and Kotsopoulos 2011a).

While models are important in expanding the understanding of wireless channels, they must also be characterized in terms of known quantities. This allows us to compare several of them to see which one offers the best fit for a specific wireless environment. This means examining both the first order and second order statistics of the density functions of the power and comparing the different quantitative

© Springer International Publishing AG 2017

P.M. Shankar, *Fading and Shadowing in Wireless Systems*,
DOI 10.1007/978-3-319-53198-4_4

measures available to contrast these models. Furthermore, we also need to establish the association and relationships among the various models making it possible to choose a simpler analytical model over a complex one. The quantitative measures of the statistical characteristics of these models include the amount of fading, average bit error rates, outage probabilities, and entropies (Simon and Alouini 2005).

Three new sections have been added in the updated edition. These cover descriptions and analysis of some of the newer models for fading and shadowing. For the benefit of the readers, one of the new sections is devoted to random number simulations pertaining to all the models described in this chapter. This covers simulation of densities, distributions, error rates, and outage probabilities.

We begin with a description of the McKay model for fading (McKay 1932; Holm 2002; Holm and Alouini 2004; Shankar 2015). The effect of shadowing is also incorporated to provide a complete description of the signal strength fluctuations in wireless channels. A complementary approach based on mixture densities for modeling signal strength fluctuations in wireless channel is also presented (Atapattu et al. 2011). For each analysis undertaken, Matlab scripts are presented with full annotations to illustrate the analytical concepts described in this chapter. These Matlab scripts only cover the new sections.

4.2 Background

In wireless communications, the transmitted signals often do not reach the receiver directly. As the power is lost due to attenuation and absorption by the intervening medium, the signals reach the receiver after undergoing scattering, diffraction, reflection, etc., from the buildings, trees, and other structures in the medium (channel) between the transmitter and the receiver (Gilbert 1965; Clark 1968; Stein 1987; Braun and Dersch 1991; Jakes 1994). Thus, there exist multiple paths for the signal to reach the receiver and the signals arriving through these paths add inphase. Since the amplitude and phase of the signal from each of these paths can be treated as random variables, the received power will also be random. This random fluctuation of power is identified as "fading" in wireless systems (Hansen and Meno 1977; Loo 1985; Shankar 2002a, b; Simon and Alouini 2005). These power fluctuations have a very short period and hence, the fading is referred to as "short-term fading." The geometry of the transmission and reception of signals is sketched in Fig. 4.1.

It shows four different paths (the actual number of paths may be higher) between the transmitter and the receiver. These paths are independent and do not encounter other objects or obstructions. Instead of the simple scenario drawn in Fig. 4.1, signals might be reaching the receiver after multiple scattering in the channel: the signal in a path encounters more than one object in its path. This is shown in Fig. 4.2.

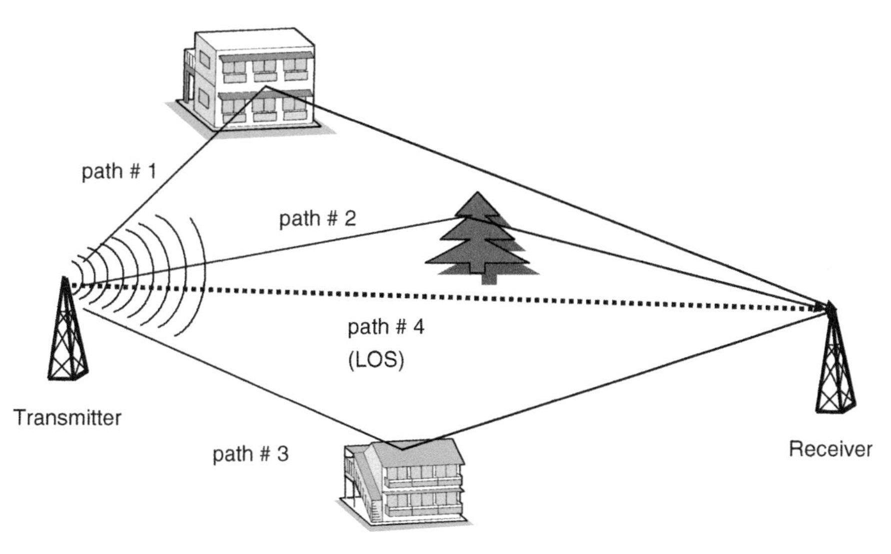

Fig. 4.1 The concept of multipath transmission. While path # 4 is the line-of-sight (LOS) path, the other paths take either reflection, refraction, scattering, or diffraction as the mechanism to reach the receiver

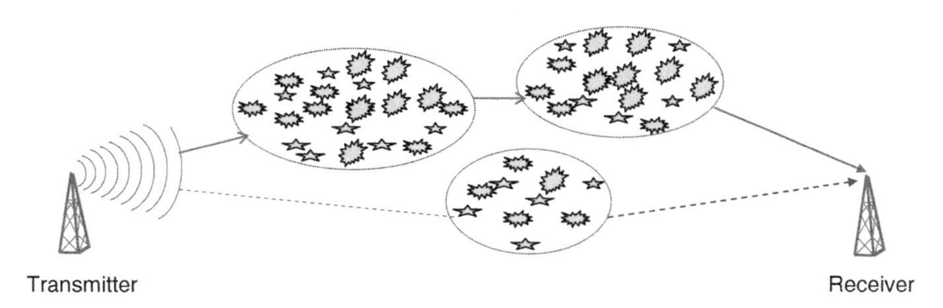

Fig. 4.2 The concept of multiple scattering. Each of the multiple paths depicted in Fig. 4.1 could take multiple bounces to reach the receiver (path shown with a continuous line on the *top*)

This phenomenon causes fluctuations in the received signal that have a period longer than those associated with short-term fading. These fluctuations are identified as "long-term fading" or "shadowing" (Suzuki 1977; Hansen and Meno 1977; Loo 1985; Coulson et al. 1998a; Kim and Li 1999; Fraile et al. 2007). Figure 4.3 shows the typical received signal profiles. The thick line represents the attenuated signal reaching the receiver. The heavy dotted line with a slow variation is the shadowing are long-term fading and the "short-term fluctuations" seen riding on the heavy dotted line represent the short-term fading. Thus, the realistic case of the received signal consists of the transmitted signal that reaches the receiver as it undergoes attenuation and passes through the "shadowed fading channel" resulting in randomly varying and attenuated signals.

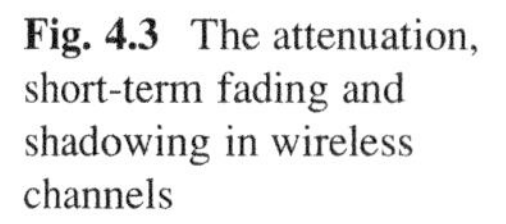
Fig. 4.3 The attenuation, short-term fading and shadowing in wireless channels

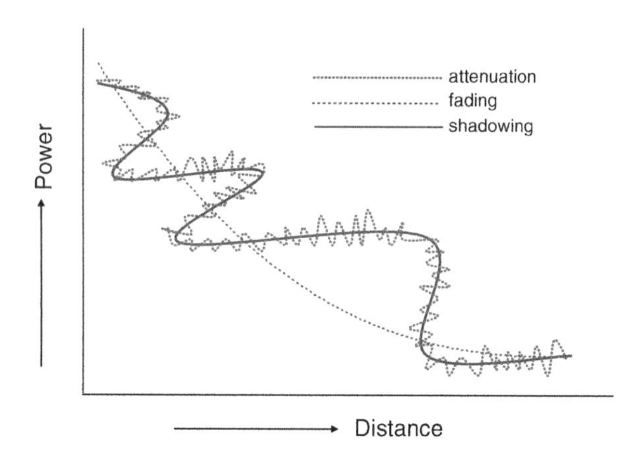

It must be noted that the short-term fading might be accompanied by frequency dependent effects, which limit the bandwidth capability of the channel. In this work, our discussion will not address such "frequency selective fading" channels; we will only examine channels that are considered as "flat," implying that the short-term fading does not alter the frequency characteristics or bandwidth capability of the channel (Jakes 1994; Steele and Hanzó 1999; Simon and Alouini 2005). We also assume that neither the transmitter nor the receiver is in motion; we exclude the effects of any frequency modulation caused by the Doppler effect, commonly labeled as "Doppler fading." In other words, we will only be considering channels that are considered as "slow" (relative speed of the transmitter/receiver negligible) instead of the "fast" (relative speed of the transmitter/receiver high) channels. Thus, the analysis below deals with wireless channels for which fading can be described as "slow" and "flat." We will briefly discuss the effects of motion of the transmitter/receiver when we examine the second order statistics of a few selected cases toward the end of this chapter.

We will now explore different ways to describe the statistical fluctuations of the received signal arising from fading and shadowing occurring separately as well as concurrently. We will also examine the effects of fading on the transmission of information through the channel using several quantitative measures so that we can compare the different models in terms of their ability to take all the possible channel conditions.

4.3 Models for Short-Term Fading

Short-term fading in wireless channels has been described using several models. These include the simple models such as Rayleigh as well as complex models such as the $\kappa -$ or $\eta - \mu$ ones and the cascaded ones (Jakes 1994; Andersen 2002b; Uysal 2005; Karagiannidis et al. 2007; Sagias and Tombras 2007; Papazafeiropoulos and

Kotsopoulos 2011b). The primary difference among these models is the flexibility in modeling the statistical characteristics of the channels that can accommodate mild, moderate, and severe fading conditions existing in the channel. These models also differ on the number of parameters necessary to completely define them: the Rayleigh model with a single parameter contrasted to the complex models with 2, 3, and more parameters. These models also come with varying degrees of analytical complexities, making the application of these models to the study of wireless channels often difficult.

We will now look at all these models starting with the Rayleigh fading model and followed by the other ones.

4.3.1 Rayleigh Fading

To understand fading, we have to examine the manner in which the signals from the transmitter reach the receiver. The simplest way to visualize this situation is through the use of the multipath phenomenon (Jakes 1994; Pahlavan and Levesque 1995, 2005; Steele and Hanzó 1999; Shankar 2002a, b; Prabhu and Shankar 2002). A typical multipath scenario is shown in Fig. 4.1 where the transmitter sends a simple sinusoidal signal at a carrier frequency of f_0. Use of a sinusoidal signal is a reasonable approach since we are only dealing with a "flat" channel that does not introduce any frequency dependent changes. The received signal $er(t)$ arising from the propagation of the signal via multiple paths in the channel can be expressed as

$$e_r(t) \sum_{i=1}^{N} a_i \cos(2\pi f_0 t + \phi_i). \tag{4.1}$$

The number of multiple paths is N, which can be treated as equivalent to the number of scattering/reflecting/diffracting centers or objects in the channel. The ith multipath signal component has an amplitude a_i and a phase ϕ_i. Equation (4.1) can be rewritten in terms of inphase and quadrature notation as

$$e_r(t) \cos(2\pi f_0 t) \sum_{i=1}^{N} a_i \cos(\phi_i) - \sin(2\pi f_0 t) \sum_{i=1}^{N} a_i \sin(\phi_i) \tag{4.2}$$

where the first summation (associated with the cosine term) is identified as the inphase term and the second summation (associated with the sine term) is identified as the quadrature term. If the locations of the structures are completely random, one can safely assume that the phase ϕ's will be uniformly distributed in the range $\{0,2\pi\}$. The amplitude of the received signal can then be expressed as

$$e_r(t) = X \cos(2\pi f_0 t) - Y \sin(2\pi f_0 t), \tag{4.3}$$

where

$$X = \sum_{i=1}^{N} a_i \cos(\phi_i), \quad Y = \sum_{i=1}^{N} a_i \sin(\phi_i). \tag{4.4}$$

Under conditions of large N, X, and Y will be independent identically distributed (i.i.d) Gaussian random variables of zero mean by virtue of the central limit theorem (Papoulis and Pillai 2002a, b). This Gaussianity of X and Y also leads to the envelope of the received signal,

$$A = \sqrt{X^2 + Y^2} \tag{4.5}$$

to be Rayleigh distributed (Papoulis and Pillai 2002a, b). The probability density function (pdf) of the received signal envelope, $f_R(a)$, will be given by

$$f_R(a) = \frac{a}{\sigma^2} \exp\left(-\frac{a^2}{2\sigma^2}\right) U(a). \tag{4.6}$$

In (4.6), σ^2 is the variance of the random variables X (or Y) and U(.) is the unit step function. The subscript (R) of the pdf in (4.6) and subscripts in all the other pdfs later in this chapter merely indicate the nature of the statistics associated with fading, i.e., in this case, Rayleigh. Note that if the envelope of the signal is Rayleigh distributed, the power, $P = A^2$, will have an exponential pdf, given by

$$f_{\mathrm{R}}(p) = \frac{1}{P_0} \exp\left(-\frac{P}{P_0}\right) U(p). \tag{4.7}$$

Once again, the subscript R relates to the nature of the statistics, which in this case is classified as Rayleigh. In (4.7), $2\sigma^2$ has been replaced by the average power P_0 of the received signal.

The phase y of the received signal is also random and one can obtain the pdf of the phase as well. The phase y is given by

$$\Theta = \tan^{-1}\left(\frac{Y}{X}\right) \tag{4.8}$$

and the pdf of the phase can easily be obtained from the fact that X and Y are zero mean and i.i.d Gaussian random variables. It can be expressed as

$$f_{\mathrm{R}}(\theta) = \frac{1}{2\pi}, \quad 0 \leq \theta \leq 2\pi. \tag{4.9}$$

In other words, the phase is uniformly distributed in the range $\{0,2\pi\}$ and it can also be seen that the phase and the envelope are independent, i.e.,

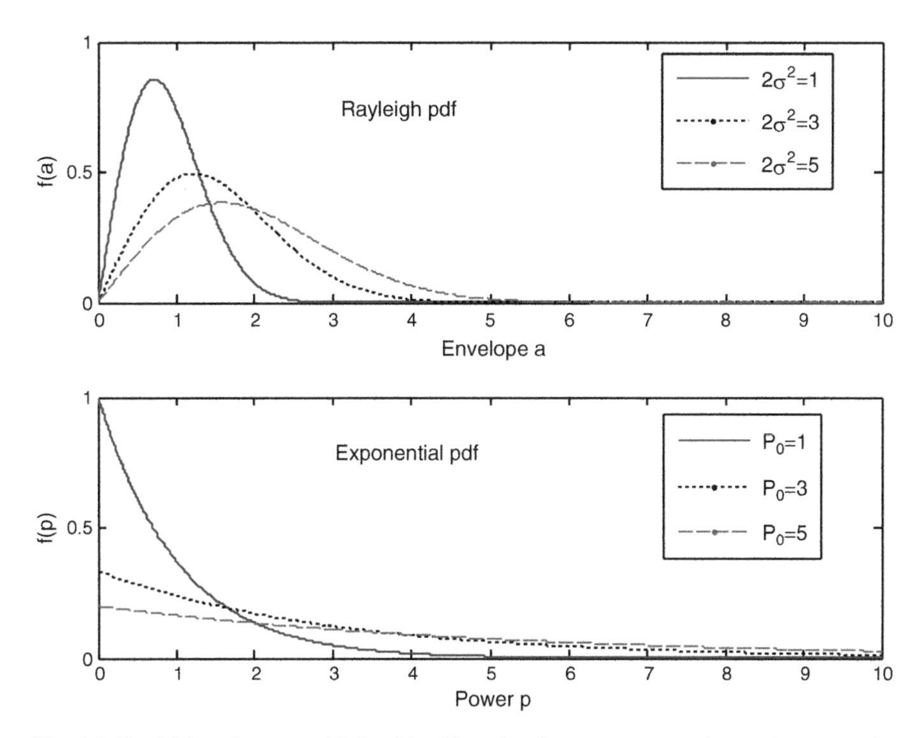

Fig. 4.4 Rayleigh and exponential densities. Note that the average power has no impact on the level of fading

$$f_{\mathrm{R}}(a,\theta) = f_{\mathrm{R}}(a) f_{\mathrm{R}}(\theta) \tag{4.10}$$

Among other characteristics pertaining to fluctuations in signal power, the unique feature of the Rayleigh fading is the independence of the phase and envelope statistics as expressed in (4.10). The phase becomes much more important when we examine the fading conditions which depart from Rayleigh, a special case that will be discussed in Sect. 4.2. Figure 4.4 shows the Rayleigh pdf for several values of the average power. The corresponding exponential densities are also shown.

A typical faded signal generated through simulation ($N = 10$) is shown in Fig. 4.5.

The histogram of the envelope and the Rayleigh fit corresponding to the data used in Fig. 4.5 are shown in Fig. 4.6.

Rayleigh density function is not the only pdf that can be used to model the statistics of short-term fading. It has limited application in a broader context because of its inability to model fading conditions that result in significant degradation in performance of wireless systems. It also cannot model fading conditions where the level of fading is not as severe as it is in a Rayleigh channel (Nakagami 1960; Coulson et al. 1998b). From (4.7) it is seen that the pdf of the power is a single parameter distribution and that the parameter is the average power, limiting

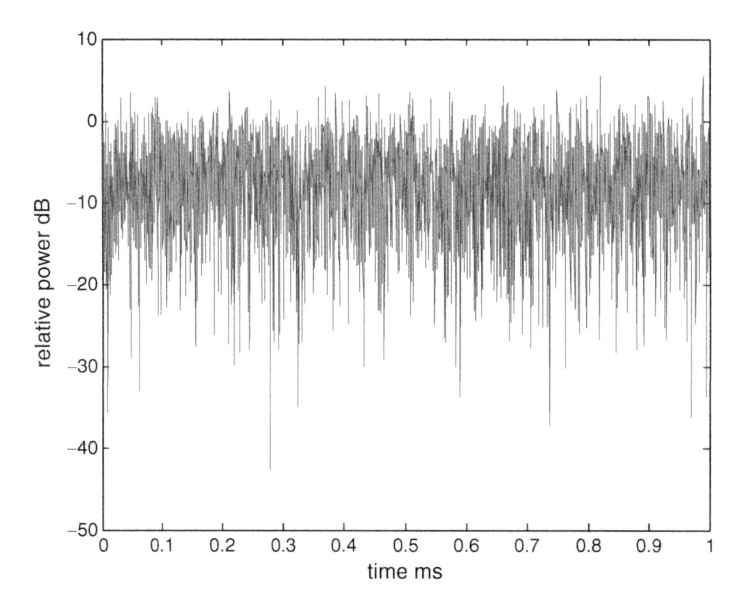

Fig. 4.5 The short-term fading signal (power) for ten multiple paths. The simulation was undertaken for a carrier frequency of 900 MHz

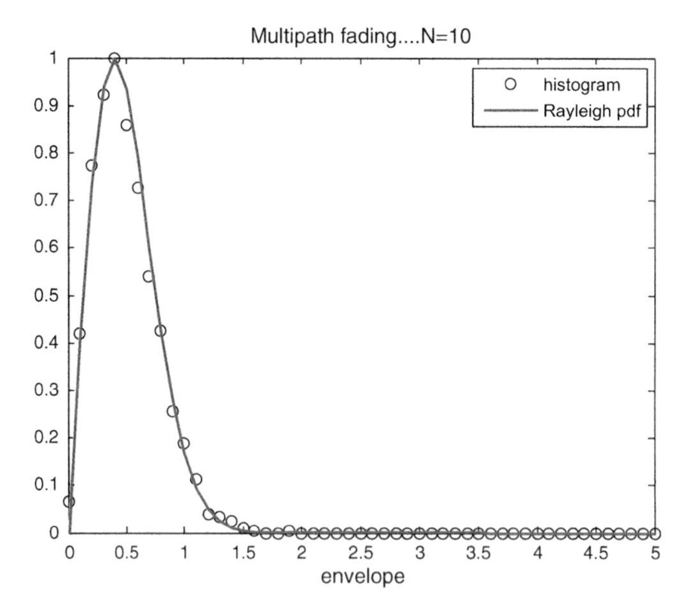

Fig. 4.6 The histogram of the envelope and the Rayleigh fit are shown for the data used in Fig. 4.5 for the short Rayleigh fading

the flexibility of the model. In other words, the Rayleigh model does not allow modeling of channels having the same average power, but exhibits different fading characteristics which can lead to variation in error rates and outage rates (the received power falling below a fixed threshold power, etc.). To understand these issues, it is necessary to quantify the level of fading. The parameter used to measure levels of fading is the amount of fading (AF) defined as (Nakagami 1960; Simon and Alouini 2005)

$$\mathrm{AF} = \frac{\langle P^2 \rangle - \langle P \rangle^2}{\langle P \rangle^2}. \tag{4.11}$$

Note that in (4.11), $\langle . \rangle$ represents the statistical average. Making use of the kth moment of the exponential pdf in (4.7), given by

$$\langle P^k \rangle_{\mathrm{R}} = P_0^{\mathrm{m}} \frac{\Gamma(k+1)}{\Gamma(k)} \tag{4.12}$$

where $\Gamma(.)$is the gamma function, (4.11) becomes

$$\mathrm{AF}_R = 1. \tag{4.13}$$

Thus, the amount of fading in a channel that has Rayleigh pdf for the envelope is equal to unity. If the amount of fading is larger than unity we have severe fading conditions. We will classify that channel as a pre-Rayleigh, and if the fading conditions are better than Rayleigh, AF will be less than one, and we identify such a channel as post-Rayleigh. Thus, it is obvious that Rayleigh pdf is inadequate to model all fading conditions that exist in wireless channels, and other models need to be explored. One such model is based on the Nakagami-*m* pdf, but before we look at the Nakagami-*m* pdf, let us go back to the multipath model described earlier and make minor modifications to it by considering a direct path or a line-of-sight (LOS) between the transmitter and receiver. Such a multipath scenario results in the Rician fading channel as described below (Nakagami 1960; Jakes 1994).

4.3.2 Rician Fading

By including a direct path between the transmitter and receiver as shown in Fig. 4.1, represented by $a_0 \cos(2\pi f_0 t)$, where a_0 is a constant, (4.3) becomes

$$\begin{aligned} e_{\mathrm{Rice}}(t) &= X \cos(2\pi f_0 t) - Y \sin(2\pi f_0 t) + a_0 \cos(2\pi f_0 t) \\ &= (X + a_0)\cos(2\pi f_0 t) - Y \sin(2\pi f_0 t) \end{aligned} \tag{4.14}$$

The received power will now be given by

$$P = (X + a_0)^2 + Y^2 = X'^2 + Y^2 \tag{4.15}$$

where X' is a Gaussian random variable with a nonzero mean equal to a_0. The pdf of the power will be given by

$$f_{\mathrm{Ri}}(p) = \frac{1}{2\sigma^2}\exp\left(-\frac{p + a_0^2}{2\sigma^2}\right) I_0\left(\frac{a_0}{2\sigma^2}\sqrt{p}\right) U(p), \tag{4.16}$$

where $I_0(.)$ is the modified Bessel function of the first kind (Abramowitz and Segun 1972; Gradshteyn and Ryzhik 2007). The corresponding density function for the envelope was given in Chap. 2. Equation (4.16) is the pdf of the received signal power in a Rician fading channel, which differs from the Rayleigh channel because of the existence of an LOS path in addition to multiple indirect paths. In light of the presence of this direct path, the amount of fading will be less than what is observed in Rayleigh fading as will be indicated later.

The mean and the second moment associated with the pdf in (4.16) is

$$\begin{aligned} \langle P \rangle &= 2_0^2 + a_0^2, \\ \langle P \rangle &= 8\sigma_0^4 + a_0^4 + 8a_0^2 + \sigma_0^2. \end{aligned} \tag{4.17}$$

The Rician factor K_0 is defined as

$$K_0 = \frac{a_0^2}{2\sigma^2}. \tag{4.18}$$

The quantity K_0 is a measure of the strength of the LOS component, and when $K_0 \to 0$, we have Rayleigh fading. As K_0 increases, the fading in the channel declines (Simon and Alouini 2005). If the average received power is PRi, it can be expressed as

$$P_{\mathrm{Ri}} = \langle P \rangle_{\mathrm{Ri}} = 2\sigma^2 + a_0^2. \tag{4.19}$$

We now have, and

$$2\sigma^2 \frac{1}{K_0 + 1} P_{\mathrm{Ri}} \tag{4.20}$$

and

$$a_0^2 = \frac{K_0}{K_0 + 1} P_{\mathrm{Ri}}. \tag{4.21}$$

Using (4.19)–(4.21), the pdf of the received power in Rician fading becomes

$$f_{\mathrm{Ri}}(p)=\frac{K_0+1}{P_{\mathrm{Ri}}}\exp\left[-K_0(K_0+1)\frac{p}{P_{\mathrm{Ri}}}\right]I_0\left(2\sqrt{\frac{K_0(K_0+1)}{P_{\mathrm{Ri}}}p}\right)U(p). \tag{4.22}$$

The density function of the amplitude or envelope is

$$f_{\mathrm{Ri}}(a)=\left(\frac{K_0+1}{P_{\mathrm{Ri}}}\right)\exp\left[-K_0(K_0+1)\frac{a^2}{P_{\mathrm{Ri}}}\right]I_0\left(2a\sqrt{\frac{K_0(K_0+1)}{P_{\mathrm{Ri}}}p}\right)U(a). \tag{4.23}$$

The mean and the second moments of a Rician envelope are

$$\langle A\rangle=\frac{\sqrt{\pi P_{\mathrm{Ri}}}}{2\sqrt{K_0+1}}\left[K_0I_0\left(\frac{K_0}{2}\right)+I_0\left(\frac{K_0}{2}\right)+K_0I_1\left(\frac{K_0}{2}\right)\right]\exp\left(-\frac{1}{2}K_0\right). \tag{4.24}$$

$$\langle A^2\rangle=P_{\mathrm{Ri}} \tag{4.25}$$

Note that when $K_0=0$, (4.22) becomes (4.7) and (4.23) becomes the Rayleigh pdf in (4.6), the CDF associated with the Rician distributed envelope can be expressed as

$$F_{\mathrm{Ri}}(a)=\int_0^a 2x\left(\frac{K_0+1}{P_{\mathrm{Ri}}}\right)\exp\left[-K_0-(K_0+1)\frac{x^2}{P_{\mathrm{Ri}}}\right]I_0\left(2x\sqrt{\frac{K_0(K_0+1)}{P_{\mathrm{Ri}}}}\right)\mathrm{d}x. \tag{4.26}$$

Simplifying, we have

$$F_{\mathrm{Ri}}(a)=\int_0^{\sqrt{2((K_0+1)/P_{\mathrm{Ri}})^a}} x\exp\left[-K_0-\frac{x^2}{2}\right]I_0\left(x\sqrt{2K_0}\right)\mathrm{d}x. \tag{4.27}$$

Equation (4.27) can be simplified using the Marcum Q function $Q(\alpha,\beta)$

$$Q(\alpha,\beta)=\int_\beta^\infty x\exp\left(-\frac{x^2+\alpha^2}{2}\right)I_0(\alpha x)dx. \tag{4.28}$$

And the CDF in (4.27) becomes

$$F_{\mathrm{Ri}}(a)=1-Q\left[\sqrt{2K_0},\sqrt{2\left(\frac{K_0+1}{P_{\mathrm{Ri}}}\right)}a\right]. \tag{4.29}$$

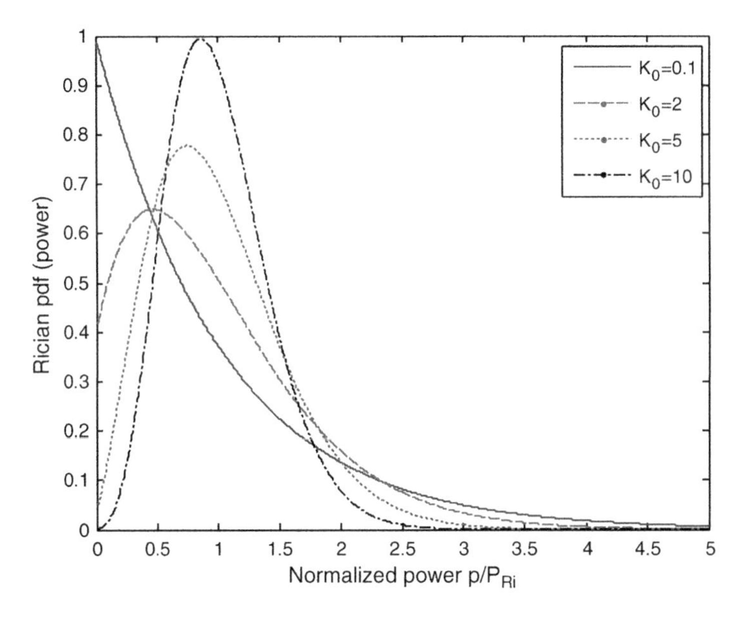

Fig. 4.7 Rician pdf. The average power P_{Ri} has been taken to be unity

It can be easily seen that as K_0 increases, the amount of fading decreases and, as $K_0 \rightarrow 1$, the amount of fading becomes zero. In other words, the existence of the direct path can reduce the levels of fading in wireless channels. When K_0 becomes zero, (4.30) becomes unity, the amount of fading in a Rayleigh channel. The Rician pdf in (4.22) is plotted in Fig. 4.7 for several values of K_0.

The amount of fading in a Rician channel can be obtained from the moments of the pdf in (4.22) and can be expressed as

$$\mathrm{AF}_{\mathrm{Ri}} = \frac{1 + 2K_0}{(1 + K_0)^2}. \tag{4.30}$$

The amount of fading in a Rician channel is plotted in Fig. 4.8, clearly demonstrating the reduction in the level of fading as the Rician factor increases, thus justifying its classification as an example of a post-Rayleigh channel.

Another aspect of the Rician pdf is that as the value of K_0 increases, the density function becomes more symmetric. Indeed, it can be shown that the Rician pdf for the envelope approaches the Gaussian statistics. This aspect has practical implications since the received signal amplitude R in a wireless communication system can be written as

$$r(t) = As(t) + n. \tag{4.31}$$

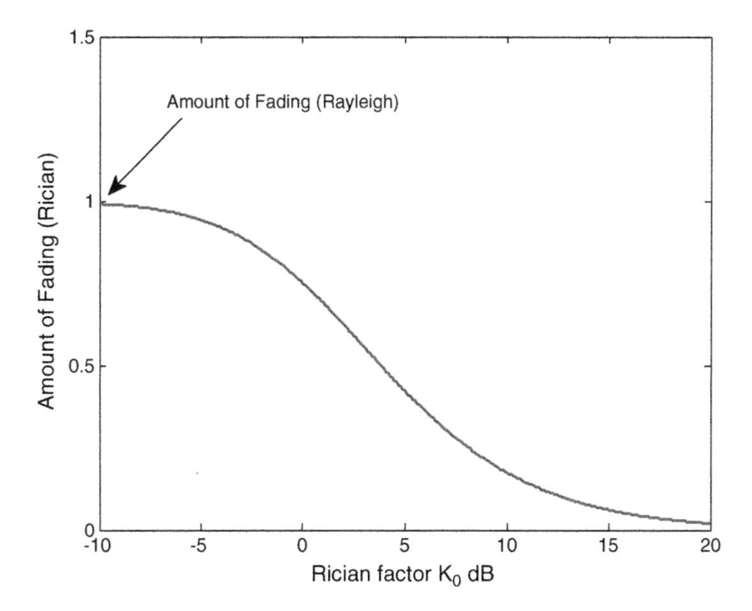

Fig. 4.8 Amount of fading in a Rician channel

In (4.31), s is the deterministic transmitted signal amplitude and n is the additive white Gaussian noise. The parameter A represents the fading as that modeled as having a Rayleigh, Nakagami, or Rician pdf for the envelope. In (4.31) its usefulness in the analysis of wireless systems will be discussed in greater detail in Chap. 4. As K_0 increases, the density function of A approaches Gaussian and the signal detection problem in (4.31) then becomes a simple matter of estimating the mean of a Gaussian variable. This is shown in Fig. 4.9. The Gaussian pdf has a mean and variance equal to the mean and variance of the Rician envelope, respectively.

At low values of the Rician factor K_0, the two density functions are different. But, as the value of K_0 increases, the curves start overlapping more and more. For K_0 values of 5 dB and beyond, it is impossible to see the difference between the two densities.

There is another interesting aspect of the Rician fading channel. While it was noted that the phase of the received signal is uniform in the range $\{0{,}2\pi\}$ in a Rayleigh channel, the phase in a Rician channel will be neither independent of the envelope nor uniformly distributed in the range $\{0{,}2\pi\}$ (as mentioned in Chap. 2).

The pdf of the phase (Goodman 1985)

$$f_{\mathrm{Ri}}(\theta) = \frac{\exp(-K_0)}{2\pi} + \sqrt{\frac{K_0}{\pi}}\cos(\theta)\exp\left[-K_0\sin^2(\theta)\right]\left[\frac{1}{2}+\frac{1}{2}\mathrm{erf}\left(\sqrt{K_0}\cos(\theta)\right)\right],$$
$$\times 0 \leq \theta \leq 2\pi. \tag{4.32}$$

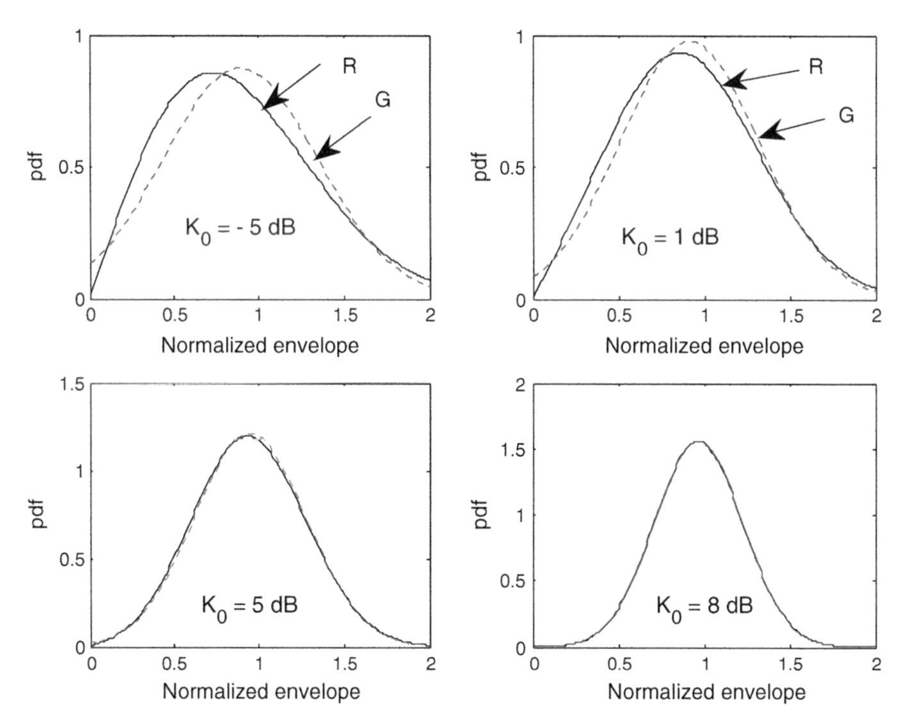

Fig. 4.9 Rician (R) compared to Gaussian (G). As the Rician parameter K_0 increases, the pdf becomes Gaussian

In (4.32), erf() is the error function. When K_0 goes to zero, (4.32) reverts back to a uniform distribution as in (4.9). The density function of the phase is plotted in Fig. 4.10, which shows that the pdf becomes narrower as K_0 increases, and it will cease to be random when $K_0 \rightarrow$ a. When K_0 is zero, the phase has a uniform density.

We will now look at the most commonly used model to describe short-term fading in wireless channels, namely the Nakagami model (Nakagami 1960).

4.3.3 Nakagami Fading

Based on the original work by Nakagami, the pdf of the received signal envelope in short-term fading can be expressed as (Nakagami 1960)

$$f_N(a) = \frac{2m^m a^{2m-1}}{P_0^m \Gamma(m)} \exp\left(-m\frac{a^2}{P_0}\right) U(a), \quad m \geq \frac{1}{2}, \tag{4.33}$$

where m is called the Nakagami parameter and Γ(.) is the gamma function. The average power is P_0. The moments of the Nakagami density in (4.33) are

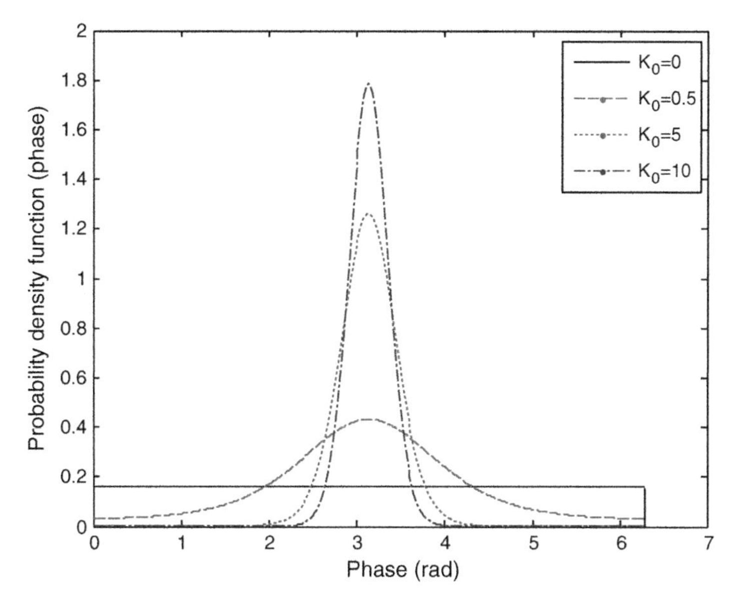

Fig. 4.10 The density function of the phase associated with the Rician channel. The pdf of the phase becomes narrower as K_0 increases and becomes a delta function as $K_0 \to 1$

$$\langle A^k \rangle = \frac{\Gamma(m + (k/2))}{\Gamma(m)} \left(\frac{P_0}{m} \right)^{k/2}. \tag{4.34}$$

The corresponding density function of the power is

$$f_N(p) = \frac{m^m p^{m-1}}{P_0^m \Gamma(m)} \exp\left(-m \frac{p^2}{P_0} \right) U(p), \quad m \geq \frac{1}{2}. \tag{4.35}$$

Equation (4.33) is commonly identified as the Nakagami-m type pdf and often (4.35) is identified as the Nakagami-m pdf of the power or signal-to-noise ration in a short-term fading channel. There are other types of Nakagami density functions used in the analysis of fading channels and we will examine them later. The Nakagami-m pdf for the envelope in (4.33) is plotted in Fig. 4.11 for a few values of m, all having the same average power. Comparing the Nakagami-m pdf to Rayleigh, one can see that for the same average power, the peak of the pdf moves farther and farther to the right as m increases, allowing more flexibility to model short-term fading since the dependence on the average power has been removed and (4.33) only depends on m.

The cumulative distribution of the power associated with the Nakagami-m model for short-term fading is obtained as

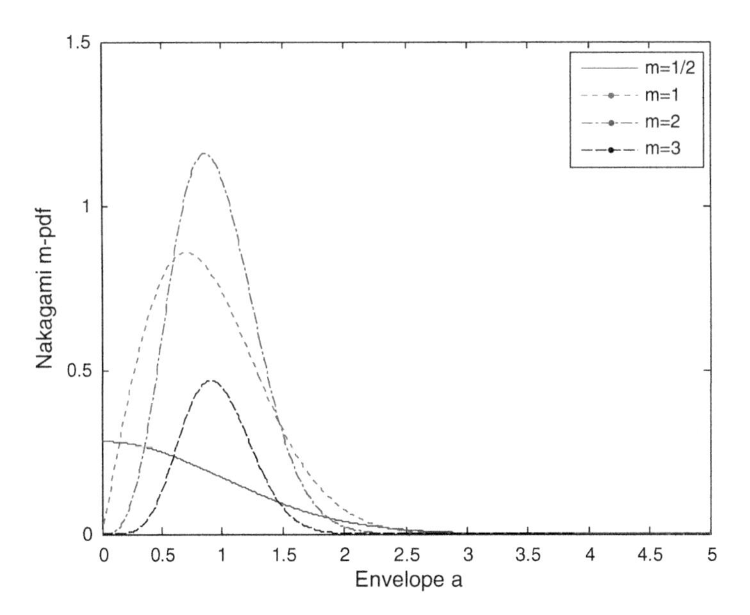

Fig. 4.11 Nakagami-m envelope densities with equal average powers

$$
\begin{aligned}
F_N(p) &= \int_0^p f_N(\alpha)\mathrm{d}\alpha = \int_0^p \frac{m^m \alpha^{m-1}}{P_0^m \Gamma(m)} \exp\left(-m\frac{\alpha}{P_0}\right) \mathrm{d}\alpha \\
&= \gamma\left(m, \frac{mp}{P_0}\right) [T(m)]^{-1}.
\end{aligned} \tag{4.36}
$$

In (4.36), γ(,) is the incomplete gamma function defined in Chap. 2. We will come back to the use of the cumulative distribution function (CDF) when we examine the performance of wireless channels in terms of outage probability.

Using the moments of the Nakagami-m pdf in (4.35) given as

$$
\langle P^k \rangle_N = \frac{\Gamma(m+k)}{\Gamma(m)m^k} P_0^k, \tag{4.37}
$$

the amount of fading in a Nakagami channel becomes,

$$
\mathrm{AF}_N = \frac{1}{m}. \tag{4.38}
$$

Equation (4.38) also provides a rationale for calling "m" the fading parameter since the amount of fading depends entirely on it.

Even though the Nakagami-m pdf has shown to be a good fit based on experimental observation, we can also provide a simple semi-analytical means to justify its use. The approach is based on the concept of "clustering" or "bunching" of scatterers. Consider the case of a wireless channel which contains a large number of scattering centers (buildings, trees, and so forth, which reflect, scatter, diffract, refract) as described earlier. However, instead of all of them being located in a purely random way in the channel, they are now clustered together. Let us assume that there are n such clusters of scatterers within the channel, with each cluster having sufficiently large number of scatterers. We will also assume that these clusters are located randomly within the confines of the channel, with the received signal coming from these clusters or groups. Since we assumed that each of these groups has sufficiently large number of scatterers, without any loss of generality, the inphase X and quadrature Y components from them would be Gaussian distributed (Yacoub 2007c, d). Let us define Z as

$$Z = \sum_{i=1}^{n} X_1^2, \tag{4.39}$$

where n is an even number ≥ 2. This means that for $n = 2, I = 1$ represents X and $i = 2$ represents Y, and no information is lost in expressing Z in terms of X alone.

Note also that Z represents the power. We will now extend (4.39) to include the cases of n being an odd number as well since (4.39) is simply the sum of the squares of Gaussian random numbers. The pdf of Z will be a chi-square density function given by

$$f(z) = \frac{z^{(n/2)-1}}{z_0^{(n/2)}\Gamma(n/2)} \exp\left(-\frac{z}{z_0}\right) U(z), \tag{4.40}$$

where Z_0 is a constant. If we replace $(n/2)$ by m, and z by p, the density function will be identical to the Nakagami-m pdf for the power in (4.35) with the requirement that $m > 1/2$ since the smallest value of $n = 1$ (except for scaling factors). Rayleigh fading (exponential pdf for the power) is still possible with $m = 1$, so that $n = 2$ and Z will be equal to

$$Z = \sum_{i=1}^{2} X_i^2 = X_1^2 + X_2^2 = X_1^2 + Y_1^2 \tag{4.41}$$

since $X1$ and $Y1$ are identically distributed. The pdf can now be generalized to have m take non-integer values in which case (4.40) will be identical to (4.35). Please note that treating the collection of scatterers as individual ones or treating them as "clusters" is not in conflict since we can have a case of a single cluster with a large number of scatterers giving rise to the Rayleigh case described in connection with (4.7).

Two important aspects of the Nakagami-m pdf can be easily seen. First, when $m = 1$, (4.35) becomes identical to the exponential pdf in (4.7). Second, choice of different values of m permits the amount of fading to vary from zero ($m = 1$) to 2 ($m = 1/2$). This means that at very high values of m ($m \to 1$), the fading vanishes

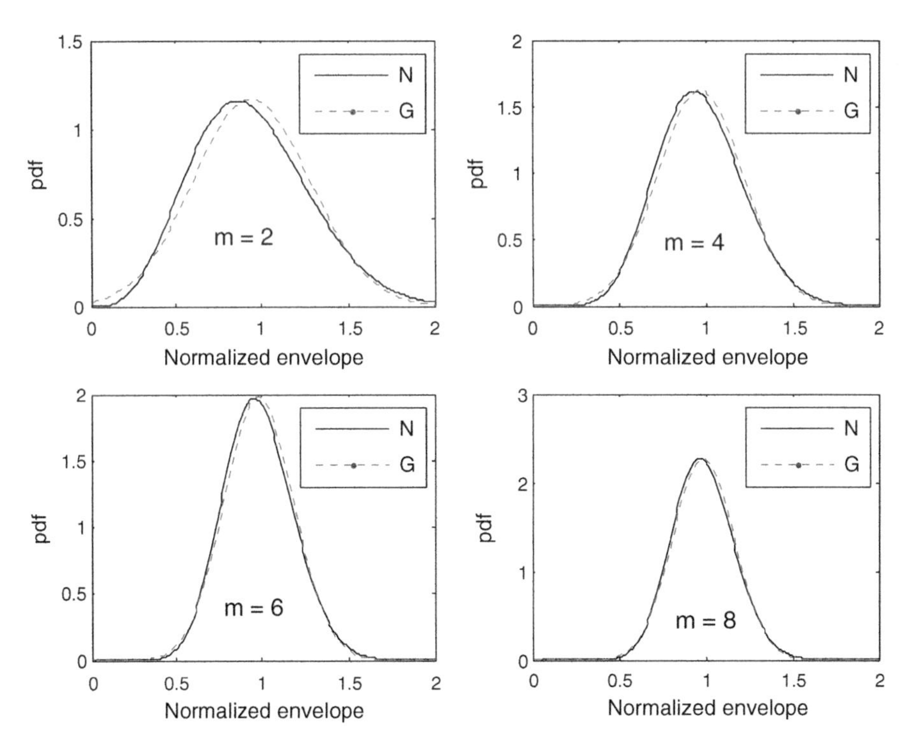

Fig. 4.12 Nakagami (N) and Gaussian (G) densities. As the value of the Nakagami parameter m increases, the Nakagami density starts matching the Gaussian pdf

and the channel becomes a pure Gaussian channel suggesting that the channel suffers only from additive white Gaussian noise (and possibly other forms of noise such as impulsive noise if such exist in the channel). This comparison between Gaussian (G) and Nakagami-m densities is shown in Fig. 4.12. As m goes from 2 to 8, the densities overlap very well.

It is also possible to relate the Nakagami and Rician distributions (Nakagami 1960). As mentioned, Nakagami fading channel is classified as a post-Rayleigh fading channel when m is greater than unity. This suggests that we can establish a relationship between the Rician density and Nakagami-m density when $m > 1$ since Rician is also a form of post-Rayleigh statistics. We can obtain such a relationship by equating the expressions for the amount of fading. From (4.30), we can write

$$\frac{1}{m} = \frac{1 + 2K_0}{(1 + K_0)^2}, \quad m \geq 1. \tag{4.42}$$

Rewriting, we have

$$m = 1 + \frac{K_0^2}{1 + 2K_0}. \tag{4.43}$$

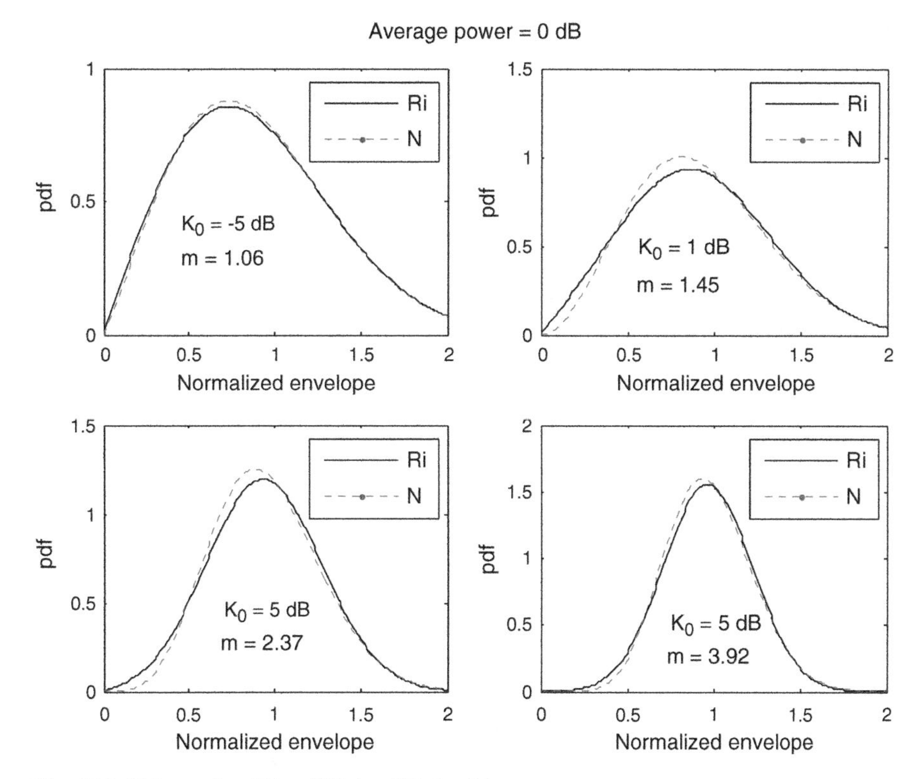

Fig. 4.13 Nakagami-m (N) and Rician (Ri) densities

Figure 4.13 shows the Nakagami-m (N) and Rician (Ri) envelope densities. The remarkable match between the two is clearly seen suggesting that in complicated analysis involving Rician densities, it might be easier to work with the equivalent Nakagami-m densities and get the results that are reasonably close.

Even though the Nakagami-m pdf allows modeling of fading conditions better than Rayleigh fading conditions, the Nakagami-m pdf in (4.35) cannot account for fading conditions that might lead to values of AF > 2. Note that it is possible to know the amount of fading present in the channel (if so required) by conducting measurements over a long period and evaluating AF in (4.11) from the observed signal powers and estimating the moments. Thus, it is necessary to have models which allow descriptions of fading channels which exhibit fading levels far more severe than what are observed in the description using the Nakagami-m pdf.

A wide range of fading conditions can be modeled using other pdfs. We will now examine the use of gamma and Weibull distributions in this context.

4.3.4 Gamma, Generalized Gamma, and Weibull Fading

There are several pdfs that have been used to model such channels where fading is worse than in a Nakagami channel. The first and simplest approach to undertake such a modeling is to rename (19). Note that (4.35) is a gamma pdf. The only difference between the Nakagami-m pdf in (4.35) and the gamma pdf is that for the latter values of m less than 1/2 are allowed. So, we can use the gamma pdf straightforward for the power and the pdf of the power as per the gamma model for short-term fading can be expressed as (Aalo et al. 2005; Yacoub 2007a, b; Shankar 2011a, b)

$$f_{\mathrm{G}}(p) = \frac{p^{m-1}}{P_g^m(m)} \exp\left(-\frac{p}{p_g}\right) U(p), \quad m > 0, \tag{4.44}$$

where P_g is related to the average power. Equations (4.35) and (4.44) have identical forms except for two factors, namely m is allowed to take any positive values and the average powers corresponding to the pdf in (4.35) and (4.44) are different. Thus, use of (4.44) permits us to include conditions where the fading could be worse than what is observed in a Nakagami channel. The average power can be obtained from the moments of the gamma pdf as

$$\langle P \rangle_{\mathrm{G}} = mP_g. \tag{4.45}$$

The amount of fading AFG in a gamma fading channel is identical to that in a Nakagami-m channel,

$$\mathrm{AF}_{\mathrm{G}} = \frac{1}{m}. \tag{4.46}$$

Even removing the restriction on m and consequently changing the Nakagami fading to gamma fading might be insufficient to model the fading observed in wireless channels. To accommodate such channels, a generalized Nakagami or generalized gamma channel (Coulson et al. 1998; Shankar 2002; Aalo et al. 2005; Shankar 2010) can be defined by scaling the power by $(1/s)$, where s is a positive number. Let us define P_{gg} as

$$P_{gg} = P^{1/s}, \quad s > 0. \tag{4.47}$$

The pdf of the power in (4.47) can be obtained and is expressed as

$$f\left(p_{gg}\right) = \frac{s p_{gg}^{ms-1}}{\Gamma(m) P_g^m} \exp\left(-\frac{p_{gg}^s}{p_g}\right) U\left(p_{gg}\right), \quad 0 < s < \infty \tag{4.48}$$

Without any loss of generality, we can replace Pgg by p and we have the expression for the pdf of the received power in a generalized gamma fading channel as

$$f_{\mathrm{GG}}(p) = \frac{sp^{ms-1}}{\Gamma(m)P_g^m}\exp\left(-\frac{p^s}{p_g}\right)U(p), \quad 0 < s < \infty. \tag{4.49}$$

Equation (4.49) will be pdf of the power associated with the generalized Nakagami pdf if m is restricted to values larger than 1/2. The moments of the pdf in (4.49) are given by

$$\langle P^k \rangle_{\mathrm{GG}} = \frac{\Gamma(m + (k/s))}{\Gamma(m)} P_g^{(k/s)} \tag{4.50}$$

resulting in

$$\langle P \rangle_{\mathrm{GG}} = \left[\frac{\Gamma(m + (1/s))}{\Gamma(m)}\right] P_g^{(1/s)}. \tag{4.51}$$

Equation (4.51) reduces to (4.45) when $s = 1$. Thus, the parameter s permits an additional level of flexibility in modeling fading channels. Using the moments of the generalized gamma pdf given in (4.50), the amount of fading in a generalized gamma fading channel becomes

$$\mathrm{AF}_{\mathrm{GG}} = \frac{\Gamma(m + (2/s))\Gamma(m)}{[\Gamma(m + (1/s))]^2} - 1 \tag{4.52}$$

Note that when s becomes unity,

$$\mathrm{AF}_{\mathrm{GG}} = \mathrm{AF}_{\mathrm{G}} = \mathrm{AF}_{N}. \tag{4.53}$$

We can obtain the expression for the CDF of the power or SNR associated with the generalized gamma fading channel. The CDF becomes

$$F_{\mathrm{GG}}(p) = \int_0^p \frac{s\alpha^{ms-1}}{\Gamma(m)P_g^m}\exp\left(-\frac{\alpha^s}{P_g}\right) \mathrm{d}\alpha = \gamma\left(m, \frac{p^s}{P_g}\right)[\Gamma(m)]^{-1}. \tag{4.54}$$

Note that the average power hPi in a GG channel is related to the parameter Pg through (4.51) as

$$P_g = \left[\langle P \rangle \frac{\Gamma(m)}{\Gamma(m + (1/s))}\right]^s. \tag{4.55}$$

One can easily see that when $s = 1$, the CDF in (4.54) reverts to the CDF of the power in the Nakagami-m channel expressed in (4.36).

It is also possible to redefine the generalized gamma distribution so that instead of the dependence of fading on two parameters, namely m and s, a new fading parameter m_{w} can be used. This leads to the Weibull fading model (Shepherd 1977; Alouini and Simon 2006). The simplest way to generate the Weibull fading conditions is to start from the Rayleigh fading or the associated exponential pdf of the power in (4.7). Let us define a new variable W

$$W = P^{1/m_{\mathrm{w}}}, \tag{4.56}$$

where P is exponentially distributed with a pdf of the form in (4.7). Equation (4.56) suggests that the power observed in this fading channel is best described in terms of a scaled version of the power received in a typical Rayleigh channel. Using the concept of transformation of the variables, the pdf of W can be obtained as [5]

$$f(w) = \frac{m_{\mathrm{w}}}{P_0} w^{m_{\mathrm{w}}-1} \exp\left(-\frac{w^{m_{\mathrm{w}}}}{P_0}\right) U(w). \tag{4.57}$$

Without any loss of generality, we can replace w with the variable p and, hence, the pdf of the power in a Weibull fading channel becomes

$$f_{\mathrm{W}}(p) = \frac{m_{\mathrm{w}}}{P_0} p^{m_{\mathrm{w}}-1} \exp\left(-\frac{p^{m_{\mathrm{w}}}}{P_0}\right) U(p). \tag{4.58}$$

Certainly, (4.58) is much simpler than (4.49) for the generalized gamma fading channel and more complicated than the Nakagami-m (or gamma) channels. The fading parameter is identified as m_{w}. It must be noted that (4.58) can also be obtained as a special case of the generalized gamma pdf by putting $m = 1$ and $s = m_{\mathrm{w}}$, pointing to Weibull fading being simpler than the generalized gamma fading and justifying the scaling employed in (4.56).

The moments of the pdf in (4.58) can be expressed as

$$\langle P^k \rangle = \Gamma\left(1 + \frac{k}{m_{\mathrm{w}}}\right) P_0^{(k/m_{\mathrm{w}})}. \tag{4.59}$$

The average power in a Weibull channel becomes

$$\langle P \rangle = \Gamma\left(1 + \frac{k}{m_{\mathrm{w}}}\right) P_0^{(1/m_{\mathrm{w}})}. \tag{4.60}$$

The amount of fading in a Weibull channel, AFW, now becomes

$$\mathrm{AF_W} = \frac{\Gamma(1+(2/m_\mathrm{w}))}{[\Gamma(1+(1/m_\mathrm{w}))]^2} - 1. \tag{4.61}$$

One can now compare the amount of fading existing in the channels by comparing (4.7), (4.44), and (4.49). Note that m in (4.35) and (4.44) is identical. The CDF of the power in a Weibull fading channel can be expressed as

$$\mathrm{F_W}(p) = \int_0^p \frac{m_\mathrm{W}}{P_0} \alpha^{m_\mathrm{w}-1} \exp\left(-\frac{\alpha^{m_\mathrm{w}}}{P_0}\right) \mathrm{d}\alpha = 1 - \exp(-p^{m_\mathrm{w}}/P_0). \tag{4.62}$$

The average power in a Weibull channel with a CDF in (4.62) is related to the quantity P_0 from (4.60) as

$$P_0 = \left[\frac{\langle P \rangle}{T(1+(1/m_\mathrm{w}))}\right]^{m_\mathrm{w}}. \tag{4.63}$$

The Weibull, gamma, and generalized gamma channels are compared in terms of the amount of fading. The results are shown in Fig. 4.14.

It is clear that the generalized gamma channel offers a means to model fading channels with widely varying levels of fading. For values of $s > 1$, the generalized fading channel has less fading than the gamma channel, and for values of $s < 1$ the generalized fading channel has higher levels of fading. Comparing the gamma channel and Weibull channel, it is clear that each of them offers a different way

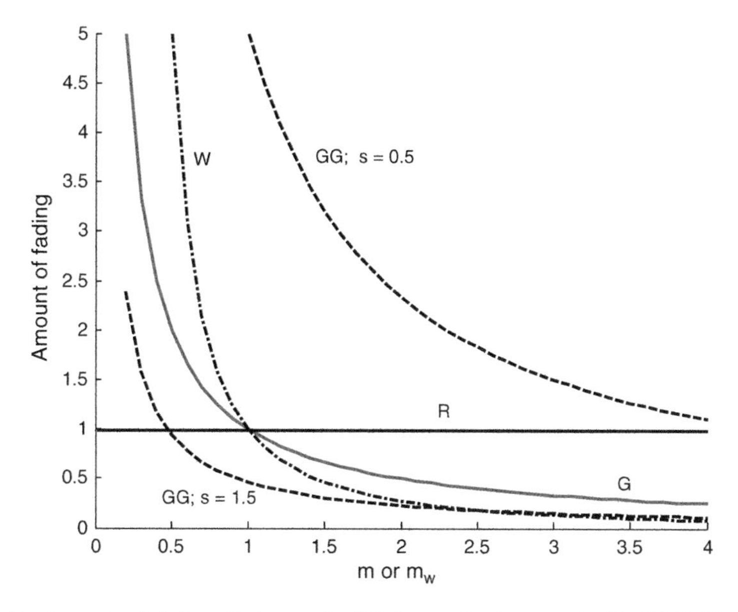

Fig. 4.14 Amount of fading G (gamma), GG (generalized gamma), W (Weibull), and R (Rayleigh)

of looking at the fading taking place in the channel. Note that simplicity is offered by the Nakagami pdf with m taking values equal to and beyond 1/2 (or gamma pdf with m taking values less than 1/2).

Now that we have looked at a few fading models, we will now examine the models to describe shadowing in wireless systems. We will then come back to other complex fading models such as the cascading models, $\kappa - \mu$ and $\eta - \mu$ models.

4.4 Models for Shadowing

As mentioned earlier, in wireless systems it is often observed that local average power varies randomly from location to location within a given geographical region (Fig. 4.2). This has been attributed to the existence of shadowing by terrain, buildings, structures, and so on. Measurements have suggested that the density function of the average power can be modeled in terms of a lognormal pdf or a Gaussian pdf if the power is expressed in decibel (dB) units. The simplest way to argue for the case of a Gaussian pdf for shadowing (expressed in dB) is to invoke the central limit theorem for products mentioned in Chap. 2 (Coulson et al. 1998a, b; Jakes 1994; Rappaport 2002; Laourine et al. 2007).

Shadowing can be described in terms of multiple scattering and the received signal power can be expressed as the product of powers. The received power Z can be expressed as

$$Z = \prod_{i=1}^{J} P_i, \tag{4.64}$$

where J is the number of multiple scattering elements and P_i is the fraction of the power scattered at each instance. Converting (4.64) into decibels (dB), we have

$$10\log_{10}(Z) = Z_{\mathrm{dBm}} = \sum_{i=1}^{J} 10\log_{10}(P_i). \tag{4.65}$$

If J is sufficiently large, the pdf of the power on the left-hand side of equation will be Gaussian and it can be written as

$$f(z_{\mathrm{dBm}}) = \frac{1}{\sqrt{2\pi\sigma_{\mathrm{dB}}^2}} \exp\left[-\frac{(z_{\mathrm{dBm}} - \mu)^2}{2\pi\sigma_{\mathrm{dB}}^2}\right], \tag{4.66}$$

where m is the average power in dBm and σ_{dB} is the standard deviation of shadowing. Converting back to power units in Watts or milliWatts, the pdf of shadowing becomes

$$f_{\mathrm{L}}(z) = \frac{A_0}{\sqrt{2\pi\sigma_{\mathrm{dB}}^2 z^2}} \exp\left[-\frac{(10\log_{10}z - \mu)^2}{2\sigma_{\mathrm{dB}}^2}\right] U(z), \tag{4.67}$$

where

$$A_0 = \frac{10}{\log_e(10)}. \tag{4.68}$$

Equation (4.67) is the well-known lognormal distribution. The notion of the amount of fading defined in (4.11) is still valid when shadowing is present by itself or concurrently with short-term fading. Using the moments of the pdf of (4.67) given as

$$\langle Z \rangle_{\mathrm{L}} = \exp\left[\frac{k}{A_0}\mu + \frac{1}{2}\left(\frac{k}{A_0}\right)^2 \sigma_{\mathrm{dB}}^2\right]. \tag{4.69}$$

the amount of fading in a shadowing channel can be obtained as

$$\mathrm{AF}_{\mathrm{L}} = \exp\left(\frac{\sigma_{\mathrm{dB}}^2}{A_0^2}\right) - 1. \tag{4.70}$$

Note that z and $\langle Z \rangle$ in (4.67) and (4.69) are in watts or milliwatts. The lognormal density functions are plotted for a few values of the shadowing parameter σ_{dB} in Fig. 4.15. A value of $m = 0$ dB has been assumed for these plots.

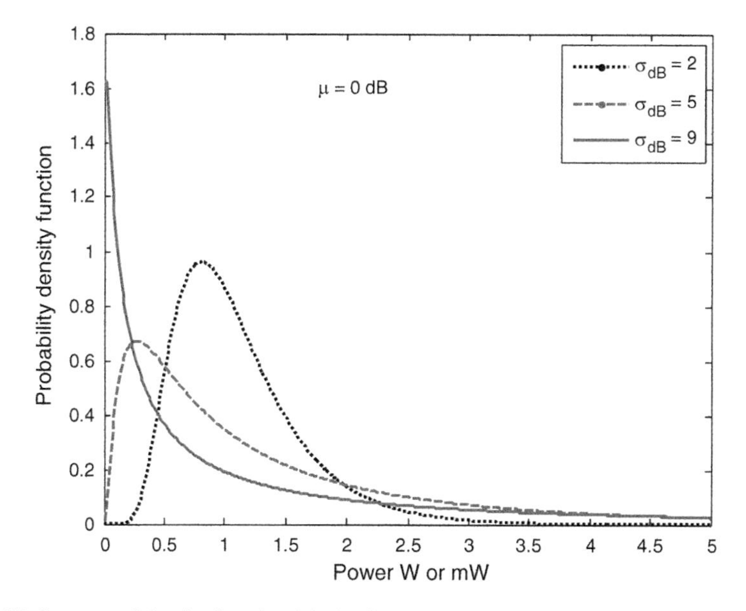

Fig. 4.15 Lognormal density function (shadowing)

Lognormal pdf in (4.67) is not the only pdf that has been proposed for modeling the shadowing seen in wireless systems. Based on the analysis of terrestrial data, it was argued that a simple gamma distribution can also be used as model shadowing (Abdi and Kaveh 1998, 2000; Shankar 2004). The pdf of the shadowing power Z can be expressed as

$$f_{\mathrm{G}}(z) = \frac{z^{c-1}}{y_0^c \Gamma(c)} \exp\left(-\frac{z}{y_0}\right) U(z), \quad c > 0. \tag{4.71}$$

Since the severity of shadowing is expressed in terms of the standard deviation of shadowing σ_{dB}, it is necessary to establish the relationship between (m, σ_{dB}) in (4.66) and (c and y_0) in (4.71). This can be done by comparing the moments of the lognormal pdf in (4.66) and the moments of the pdf in (4.71) after conversion into decibel units. These parameters are related as (Abdi and Kaveh 1998; Shankar 2005)

$$\sigma_{\mathrm{dB}}^2 = A_0^2 \psi'(c), \tag{4.72}$$

$$\mu = A_0[\log(y_0) + (c)], \tag{4.73}$$

where $\psi(.)$ and $\psi'(.)$ are the digamma and trigamma functions (Gradshteyn and Ryzhik 2000, 2007). A plot of σ_{dB} vs. c is plotted in Fig. 4.16. An inverse relationship exists between the shadowing parameter σ_{dB} and the gamma parameter c. A shadowing level of σ_{dB} of 2 corresponds to a value of $c = 5$ (approximately) and σ_{dB} 8 corresponds to a value of $c = 0.62$ (approximately).

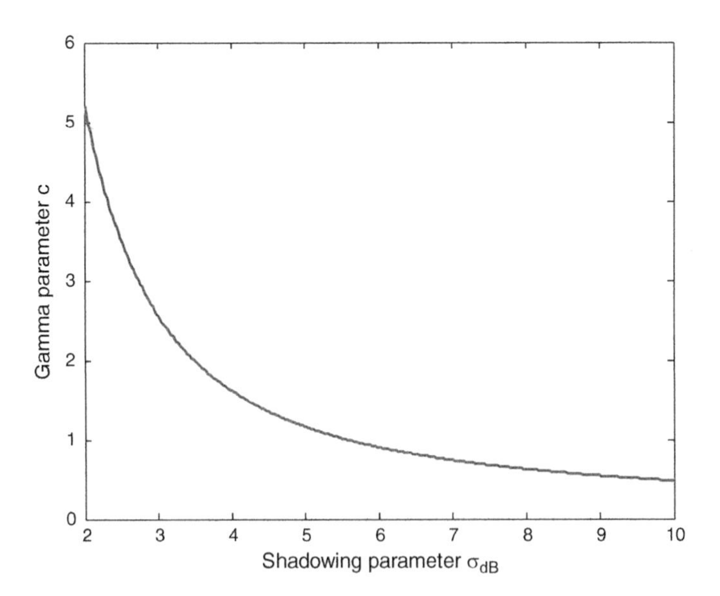

Fig. 4.16 Plot the shadowing parameter σ_{dB} vs. the gamma parameter c

Even though short-term fading and shadowing are two different effects, often, the wireless signal is subject to both at the same time (as seen in Fig. 4.2). We will now look at ways of modeling the statistical characteristics of signals in such shadowed fading channels.

4.5 Models for Shadowed Fading Channels

When evaluating the performance of wireless systems, it is necessary to consider the simultaneous effect of fading and shadowing on the received signal. Such co-existence of fading and shadowing also points to higher values of the amount of fading that is a simple measure of fluctuations in the channel.

4.5.1 *Nakagami-Lognormal Models*

As mentioned earlier, the consequence of shadowing is the loss of the deterministic nature of the mean power of the short-term faded signal. Indeed, the average power becomes random, and, for the case of a Nakagami faded signal, (4.35) needs to be rewritten as

$$f(p|z) = \frac{m^m p^{m-1}}{z^m \Gamma(m)} \exp\left(-m\frac{p}{z}\right) U(p). \tag{4.74}$$

The average power P_0 in (4.35) has been replaced by a random variable z. The pdf in (4.74) is conditioned on $Z = z$ (Suzuki 1977; Hansen and Meno 1977; Simon and Alouini 2005). Taking fading and shadowing simultaneously, the pdf of the received signal power can now be expressed as

$$f(p) = \int_0^\infty f(p|z) f(z)\, \mathrm{d}z, \tag{4.75}$$

where $f(z)$ is the pdf of the mean power. If we treat $f(z)$ to be lognormal as discussed earlier in connection with shadowing, the Nakagami-lognormal pdf for the received power becomes

$$f_{\mathrm{NLN}}(p) = \int_0^\infty \frac{m^m p^{m-1}}{z^m \Gamma(m)} \exp\left(-m\frac{p}{z}\right) \frac{A_0}{\sqrt{2\pi\sigma_{\mathrm{dB}}^2 z^2}} \exp\left[-\frac{\left(10\log_{10} z - \mu\right)^2}{2\sigma_{\mathrm{dB}}^2}\right] \mathrm{d}z. \tag{4.76}$$

Equation (4.76) is plotted in Figs. 4.17 and 4.18. For comparison, the pdf of the power under pure Nakagami-m fading conditions are also shown. At low values of the shadowing parameter, the density functions of the power under pure short-term

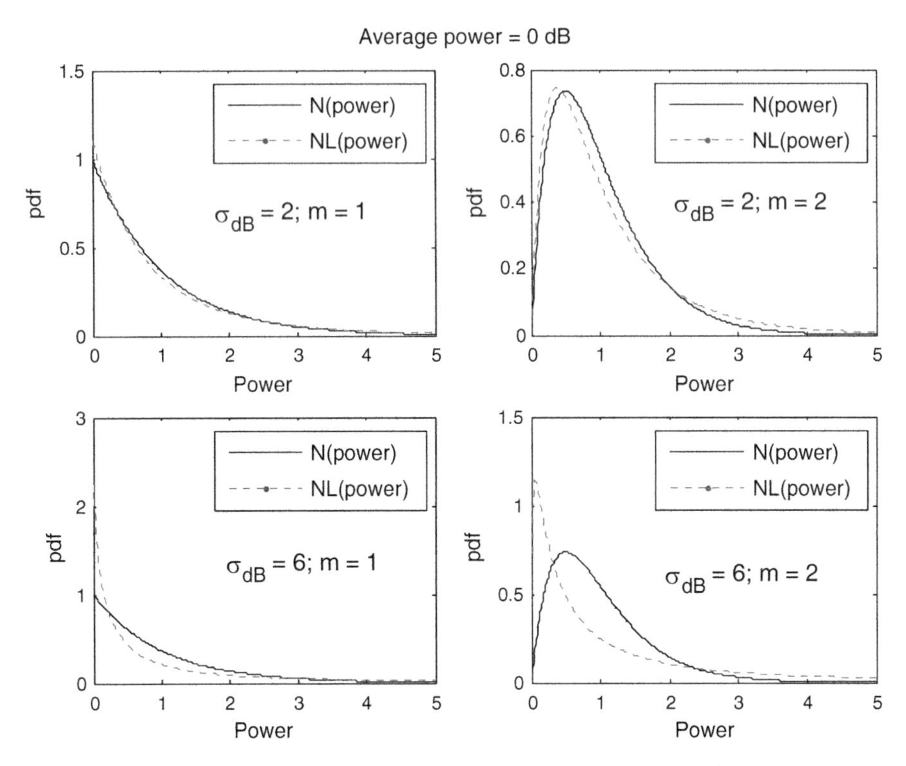

Fig. 4.17 Short-term fading (Nakagami-m power) compared to shadowed fading (Nakagami-lognormal power)

fading and Nakagami-lognormal shadowed fading conditions are very close. However, as the severity of shadowing increases, the density functions of the power in shadowed fading channels move to the left, indicating an increase in randomness in the channel and potential for increased error rates. In Fig. 4.18, one can see the worsening conditions (for a fixed value of $m = 1.5$) with increasing value of σ_{dB}.

4.5.2 *Nakagami-Gamma or Generalized* K *Models*

The Nakagami-lognormal pdf in (4.76) is in integral form; no closed solution exists for this pdf. Therefore, the evaluation of performance of wireless systems in shadowed fading channels (shadowing and fading concurrently present) using (4.76) is very cumbersome. Since it was argued that a gamma shadowing is an excellent match to the lognormal shadowing seen in wireless systems, (4.76) can be rewritten using the gamma pdf in (4.71) for z as

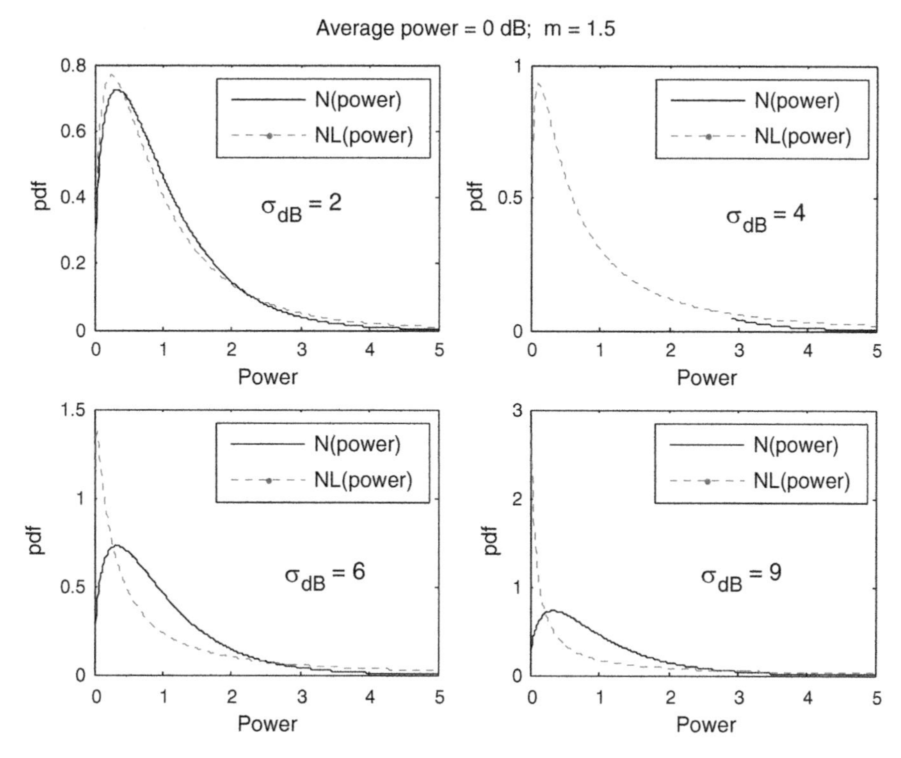

Fig. 4.18 Short-term fading (Nakagami-m power) compared to shadowed fading (Nakagami-lognormal power) for a fixed value of $m = 1.5$

$$f_{\mathrm{NG}}(p) = \int_0^{\infty} \frac{m^m p^{m-1}}{z^m \Gamma(m)} \exp\left(-m\frac{p}{z}\right) \frac{z^{c-1}}{y_0^c \Gamma(c)} \exp\left(-\frac{z}{y_0}\right) \mathrm{d}z. \tag{4.77}$$

The subscript (NG) on the left-hand side identifies the pdf as Nakagami-gamma. An analytical solution exists for (4.77) and the resulting pdf is known as the generalized K distribution given as (Shankar 2004; Bithas et al. 2006)

$$f_{\mathrm{GK}}(p) = \frac{2}{\Gamma(m)\Gamma(c)} \left(\frac{b}{2}\right)^{c+m} p^{((c+m)/2)-1} K_{c-m}\left(b\sqrt{p}\right) \mathrm{U}(p), \tag{4.78}$$

where b is a parameter related to the average power and $Kc-m(.)$ is the modified Bessel function of the second kind of order $(c - m)$. Using the moments of the pdf in (4.78) (Gradshteyn and Ryzhik 2007)

$$\left\langle p^k \right\rangle_{\mathrm{GK}} = \left(\frac{2}{b}\right)^{2k} \frac{\Gamma(m+k)\Gamma(c+k)}{\Gamma(m)\Gamma(c)} \tag{4.79}$$

we have

$$\langle p \rangle_{\mathrm{GK}} = \langle P|Z \rangle_Z = c y_0 = mc\left(\frac{2}{b}\right)^2. \tag{4.80}$$

Note that if $m = 1$, (4.76) is the pdf for the Rayleigh-lognormal channel and (4.78) becomes the K distribution (or K-fading) as (Abdi and Kaveh 1998)

$$f_K(p) = \frac{2}{\Gamma(c)}\left(\frac{b}{2}\right)^{c+1} p^{((c+1)/2)-1} K_{c-1}\left(b\sqrt{p}\right) U(p), \tag{4.81}$$

where the average power is given by

$$\langle P \rangle_K = c\left(\frac{2}{b}\right)^2. \tag{4.82}$$

The generalized K pdf is shown in Fig. 4.19a for the case of $m = 1$ and in Fig. 4.19b for the case of $m = 2$. Four levels of shadowing are shown ranging from heavy shadowing (9 dB) to weak shadowing (2 dB).

The effect of shadowed fading channels is clear from the plots of the density functions. As the shadowing levels increase, the peaks of the density function move toward lower values of the power, hinting that the amount of fading will be higher.

4.5.3 Nakagami-Inverse-Gaussian Model

Another model for shadowed fading channels makes use of the similarity between the lognormal pdf and the inverse Gaussian pdf (Karmeshu and Agrawal 2007; Laourine et al. 2009). The density function of the average power is considered to be given by the inverse Gaussian pdf,

$$f_{\mathrm{IG}}(z) = \sqrt{\frac{\lambda}{2\pi z^3}} \exp\left(\frac{\lambda(z-\theta)^2}{2\theta^2 z}\right) U(z), \tag{4.83}$$

where the two parameters θ and λ can be related to μ and σ_{dB}. Using the first and second moments of the pdf in (4.83), we have

$$\theta = \langle Z \rangle_{\mathrm{IG}} = \exp\left(\frac{\mu}{A_0} + \frac{1}{2}\frac{\sigma_{\mathrm{dB}}^2}{A_0^2}\right), \tag{4.84}$$

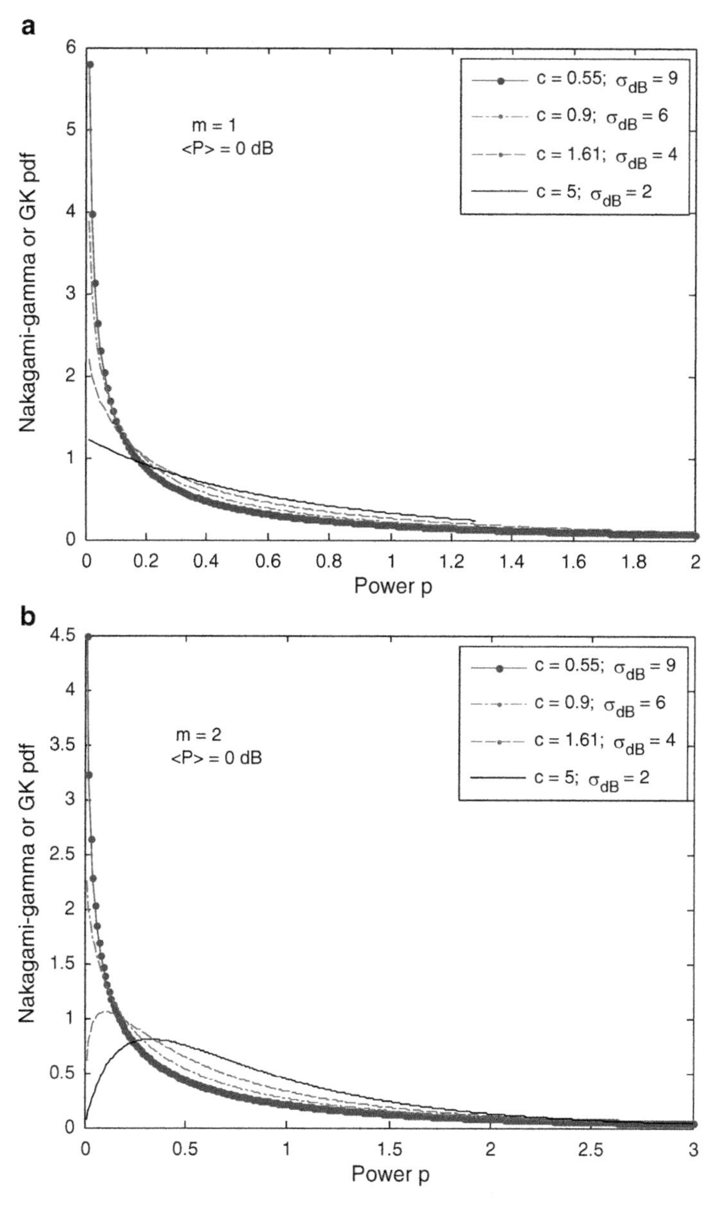

Fig. 4.19 The generalized K distribution or the Nakagami-gamma distribution (**a**) $m = 1$ (**b**) $m = 2$ for a range values of shadowing levels from weak shadowing (2 dB) to strong shadowing (9 dB). (**b**) The generalized K distribution or the Nakagami-gamma distribution for $m = 2$

$$\theta^2\left(\frac{\theta}{\lambda}+1\right)=\exp\left(\frac{2\mu}{A_0}+\frac{2\sigma_{\mathrm{dB}}^2}{A_0^2}\right). \tag{4.85}$$

The pdf of the received signal power when the shadowing is treated in terms of an inverse Gaussian distribution is obtained by putting (4.83) into (4.75) resulting in

$$f_{\mathrm{NIG}}(p)=\frac{4m^m\sqrt{\lambda/2\pi}\left(\lambda/\theta^2\right)^{m+(1/2)}}{\Gamma(m)\left[\sqrt{g(p)}\right]^{m+(1/2)}}\exp\left(\frac{\lambda}{\theta}\right)p^{2m-1}K_{m+(1/2)}\left(\sqrt{g(p)}\right)U(p), \tag{4.86}$$

where

$$g(p)=\frac{2\lambda}{\theta^2}\left(mp^2+\frac{\lambda}{2}\right). \tag{4.87}$$

4.5.4 Generalized Gamma Model

Yet another model for shadowed fading channels can be created by taking a different look at the generalized gamma pdf for short-term fading. It was suggested that the generalized gamma pdf can also model the received power in shadowed fading channels (Coulson et al. 1998a). Let us go back to the pdf in (4.49). This was arrived on the basis of a simple scaling of the power so that the scaled power v is given by

$$v=p^{1/s},\quad s>0. \tag{4.88}$$

It is possible to treat the scaling in (4.88) as akin to an exponential multiplication, similar to what was described in the section on lognormal fading. Thus, scaling should produce a shadowed fading channel. In other words, (4.88) is a case where a short-term faded signal is scaled to produce a shadowed fading case. While s could take any positive value in the absence of shadowing, we will now see that treating (4.49) as the case of a shadowed fading channel will lead to limits on the value of s. Note that v in (4.88) is a dummy variable and we can go back and replace v with p so that the pdf of the signal power under the shadowed fading channel using the generalized gamma model becomes

$$f_{\mathrm{GG}}(p)=\frac{sp^{ms-1}}{\Gamma(m)P_{\mathrm{g}}^m}\exp\left(-\frac{p^s}{P_{\mathrm{g}}}\right)U(p),\quad 0<s<?. \tag{4.89}$$

Equation (4.89) is identical to (4.49) except for the change in the condition on s indicating that the upper limit is yet to be determined as shown by the question mark(?) in (4.89). To obtain the relationship between the parameters of the generalized gamma pdf for shadowed fading channels and those of the Nakagami-lognormal pdf, we can proceed as we had done in the case of the GK distribution. Comparing the moments of the generalized gamma pdf and Nakagami-lognormal, we get the relationship among the parameters of the Nakagami-lognormal and the generalized gamma as (Shankar 2011a, b)

$$A_0^2\psi'(m) + \sigma_{\mathrm{dB}}^2 = \frac{A_0^2}{s^2}\psi'(m). \tag{4.90}$$

In (4.90), $\psi'()$ is the trigamma function (Abramowitz and Segun 1972). Thus, the scaling factor s is related to both the Nakagami parameter m and the standard deviation of shadowing σ_{dB}. Examining (4.90), it is clear that if $\sigma_{\mathrm{dB}} = 0$ (no shadowing), the scaling parameter s is equal to unity and the GG pdf becomes the Nakagami pdf. If σ_{dB} goes to 1 (extreme shadowing), s approaches zero. Thus, the scaling parameter s must be in the range 0–1 in shadowed fading channels for the GG pdf. This is illustrated in Fig. 4.20.

When used to describe the shadowed fading channels, the generalized gamma distribution in (4.89) takes the form

$$f_{\mathrm{GG}}(p) = \frac{sp^{ms-1}}{\Gamma(m)P_{\mathrm{g}}^{m}}\exp\left(-\frac{p^s}{P_{\mathrm{g}}}\right)U(p), \quad 0 < s < 1. \tag{4.91}$$

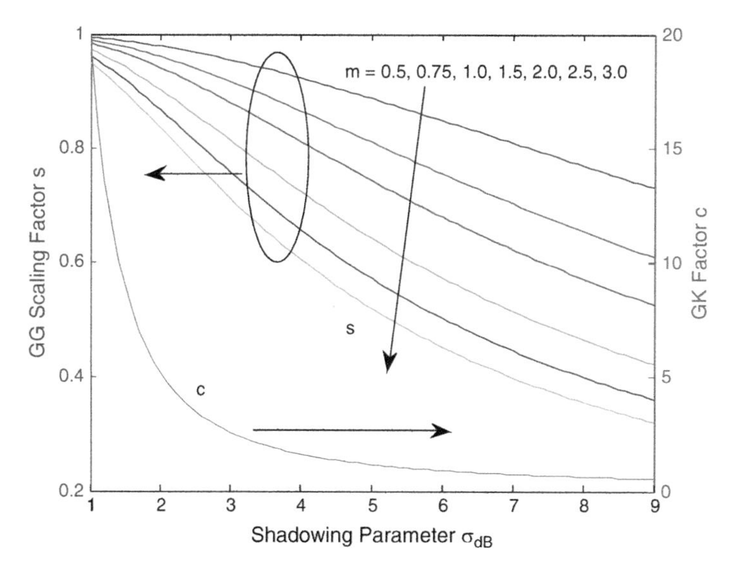

Fig. 4.20 Relationships among the GK parameter (c), GG scaling factor (s), and the shadowing level σ_{dB}

4.5.5 Amount of Fading in Shadowed Fading Channels

Now that we have a few pdfs for the received signal power in shadowed fading channels, we can now use the quantitative measure of AF to compare the power fluctuations that would be observed in those channels. Using the moments of the pdf in (4.76) for the Nakagami-lognormal shadowing, the amount of fading is

$$\mathrm{AF_{NLN}} = \left(\frac{m+1}{m}\right)\exp\left(\frac{\sigma_{\mathrm{dB}}^2}{A_0^2}\right) - 1. \tag{4.92}$$

Using the moments of the GK distribution given in (4.79), the amount of fading is

$$\mathrm{AF_{GK}} = \frac{1}{m} + \frac{1}{c} + \frac{1}{mc}. \tag{4.93}$$

The amount of fading in a GK channel is shown in Fig. 4.21.

Using the moments of the Nakagami-inverse Gaussian distribution in (4.86), the amount of fading becomes

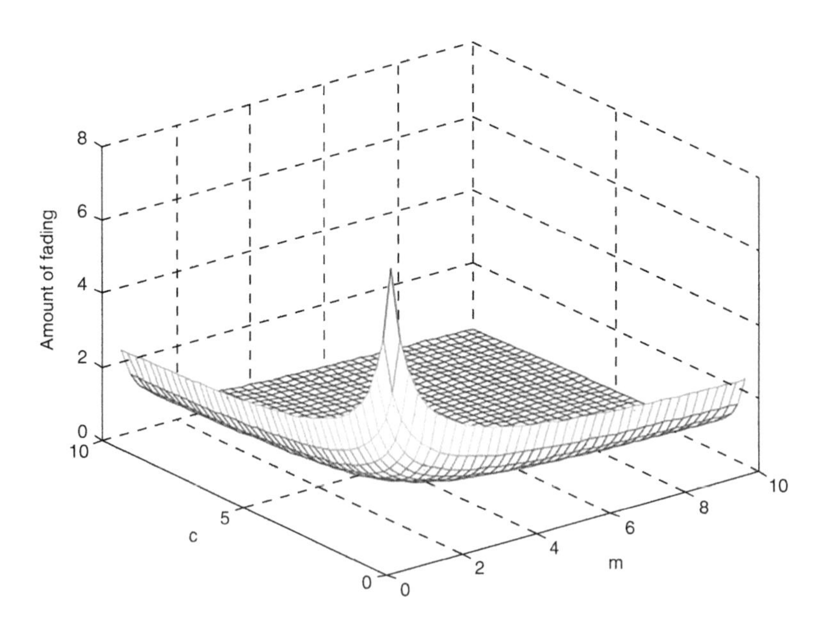

Fig. 4.21 Amount of fading in a GK channel

$$\mathrm{AF}_{\mathrm{NIG}} = \left[\frac{m+1}{m}\right]\left[\frac{\theta+\lambda}{\lambda}\right] - 1$$
$$= \left[\frac{m+1}{m}\right]\left[2\sin h\left(\frac{\sigma_{\mathrm{dB}}^2}{2A_0^2}\right)\exp\left(\frac{\sigma_{\mathrm{dB}}^2}{2A_0^2}\right)\right] - 1. \tag{4.94}$$

The amount of fading in shadowed fading channels having the generalized gamma pdf for the power is still given by (4.52) with $s < 1$. The upper limit of 1 of s is also intuitively obvious if we examine the plot of AF given in Fig. 4.14. Since the amount of fading in shadowed fading channels is worse than in channels with fading alone, the amount of fading in shadowed fading channels with the generalized gamma pdf will be above the curve for the gamma fading alone. This limits the value of s to lie in the range of 0–1.

Note that while exponential (Rayleigh envelope), gamma (Nakagami envelope), Weibull (which can be obtained from gamma pdf), and lognormal densities for the received power are supported by theoretical and experimental observations, the other distributions such as the inverse Gaussian are based on pure empirical matching. This also suggests that the Nakagami-lognormal and the generalized K distribution can also be justified on the basis of experimental and theoretical observations. A strong case of GK distribution is based on a very general model which can describe short-term fading, long-term fading, and concurrent instances of both (shadowed fading channels). This is discussed in the next section where we go back to the clustering model used earlier to justify the Nakagami distribution for fading and take it a few steps in a different direction.

4.6 Composite Model for Fading, Shadowing, and Shadowed Fading

Let us go back to the case of a wireless channel which is modeled so as to consist of a number of clusters of scattering (or reflecting, diffracting, etc.) centers. The concept is illustrated in Fig. 4.2 which shows three clusters, with each cluster having a number of scattering/reflecting/diffracting centers consisting of buildings, trees, people, vehicles, and other such artifacts. In the Nakagami or Rayleigh pdf, we assumed that the clusters were separated so that the signal from each of these clusters arrived at the receiver independently and made up the total signal at the receiver (Andersen 2002a, b; Salo et al. 2006a, b). Now, let us make the channel a bit more closely packed so that there is a likelihood that the signals from the clusters could only reach the receiver after multiple scattering among them instead of arriving independently. The transmitted signal is shown to reach the receiver after passing through two clusters (solid line in Fig. 4.2). Let the signal power from each of the cluster be C_i, $i = 1, 2, \ldots$. The received signal power P can now be expressed as

$$P = C_0 + \alpha_1 C_1 + \alpha_{12} C_1 C_2 + \alpha_{123} C_1 C_2 C_3 + \cdots. \tag{4.95}$$

Equation (4.95) needs some explanation. If there is a possibility that a direct path can exist between the transmitter and receiver, the power contributed to that component is C_0. If there is only a single cluster and, hence, no chance of multiple scattering, the received power will come from the second term, $\alpha_1 C_1$ (dotted line in Fig. 4.2). If there are at least two clusters and the chance of multiple scattering exists, the received signal will come from the third term, $\alpha_{12} C_1 C_2$, and so on. Note that α's are scaling factors and can be made equal to unity. Equation (4.95) further assumes that each of those processes (i.e., power from each term) is independent of the other.

Let us first look at the second term, $\alpha_1 C_1$. If there is no multiple scattering and there is only a single cluster, we will now treat the scatterers within that single cluster acting as miniclusters. This case is similar to the cluster model used to explain the Nakagami or Rayleigh fading in connection with the pdf in (4.40). Therefore, the second term in (4.95) will lead to pure short-term fading with a Nakagami or Rayleigh pdf.

Now consider the third term $\alpha_{12} C_1 C_2$. There are two clusters and the received power is expressed as the product of the powers from the two clusters. If each cluster power can be described in terms of a gamma pdf (Nakagami pdf for the envelope), the received power becomes (treating α_{12} to be unity)

$$P = C_1 \cdot C_2, \tag{4.96}$$

where C_1 and C_2 are gamma distributed. We will now look at a few special cases of (4.96).

If we assume that both clusters result in Rayleigh pdf for the envelopes, pdf of the received power in (4.96) can be achieved using transformation techniques for obtaining density functions of products. This leads to

$$f_{\mathrm{DR}}(p) = 2\left(\frac{u}{2}\right)^2 K_0\left(u\sqrt{p}\right)U(p). \tag{4.97}$$

The parameter u is related to the average power through

$$\langle p \rangle = \left(\frac{2}{u}\right)^2. \tag{4.98}$$

In terms of the average power, the double Rayleigh pdf for the power in (4.97) becomes

$$f_{\mathrm{DR}}(p) = \frac{2}{\langle p \rangle} K_0\left(2\sqrt{\frac{p}{\langle p \rangle}}\right)U(p). \tag{4.99}$$

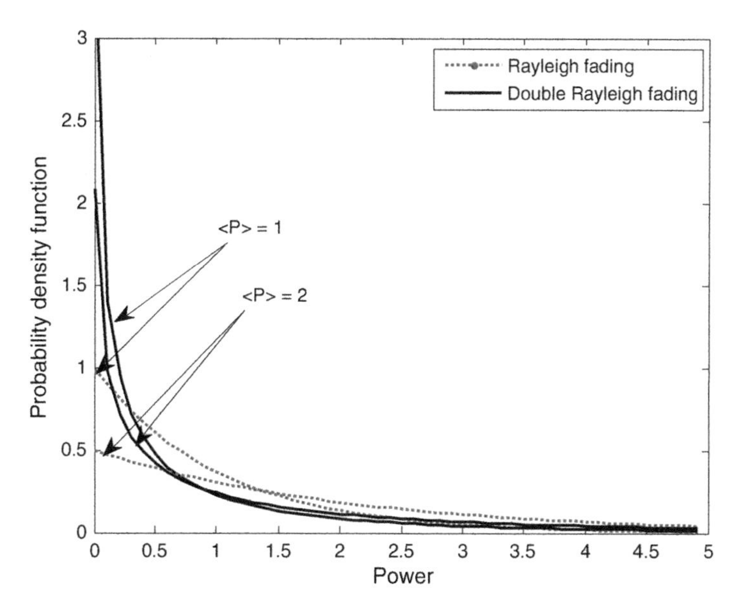

Fig. 4.22 Double Rayleigh and Rayleigh densities for the power are plotted for two values of the average power

This model for shadowed fading channels is often referred to as the double Rayleigh model and is identified by the subscript (DR). Estimating the moments of the pdf one can see that the amount of fading (AF) is equal to 3, a fading level three times worse than what is seen in a Rayleigh faded channel (Salo et al. 2006a, b; Shankar and Gentile 2010). The density function of the power is shown in Fig. 4.22.

The plot of the density functions in Fig. 4.22 clearly illustrates the problem of increased fading level in a double Rayleigh channel. The double Rayleigh pdf is closer to the low power values than to the Rayleigh pdf, indicating that statistically one has a higher probability of seeing lower powers in the former than in the latter.

If we assume that both clusters result in identical Nakagami-m pdf for the envelopes, pdf of the received power in (4.96) can once again be obtained using transformation techniques. This leads to

$$f_{\mathrm{DN}}(p) = \frac{2}{[\Gamma(m)]^2}\left(\frac{v}{2}\right)^{2m} p^{m-1} K_0\left(c\sqrt{p}\right)U(p), \tag{4.100}$$

where

$$\langle p\rangle = \left(\frac{2m}{v}\right)^2. \tag{4.101}$$

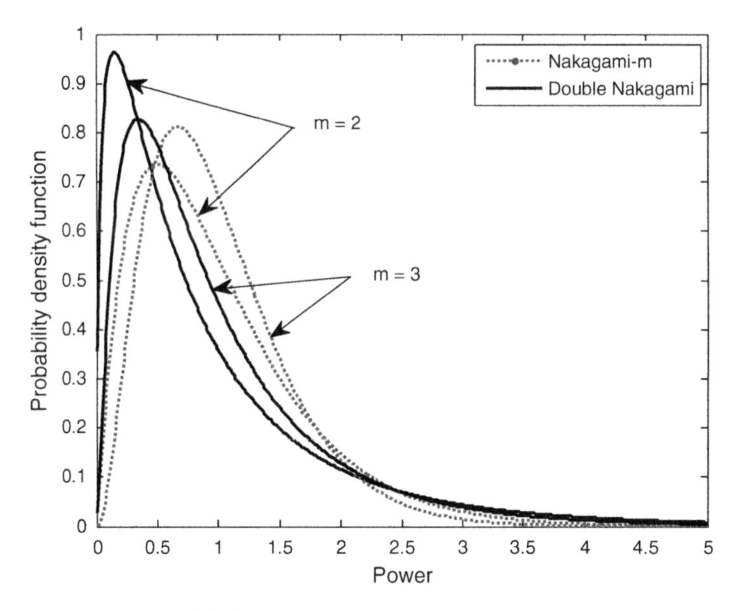

Fig. 4.23 Double Nakagami fading and Nakagami-m fading

In terms of the average power P_0, (4.100) becomes

$$f_{\mathrm{DM}}(p) = \frac{2}{[\Gamma(m)]^2}\left(\frac{m}{\sqrt{\langle p\rangle}}\right)^{2m} p^{m-1} K_0\left(2m\sqrt{\frac{p}{\langle p\rangle}}\right)U(p). \tag{4.102}$$

The pdf in (4.102) is called the double Nakagami (DN) or double gamma distribution. The double Nakagami pdf is compared with the Nakagami-m pdf for power in Fig. 4.23. As it was in the case of the plot of the double Rayleigh pdf and Rayleigh pdf in Fig. 4.22, we see that the peaks of the double Nakagami densities are closer to the lower end of the power in contrast to the Nakagami densities.

We now consider the most general case where C_1 and C_2 are Nakagami-m pdfs for envelopes with non-identical values of the Nakagami-m parameters, let us say m_1 and m_2. We can use (4.78) if we replace m by m_1 and c by m_2. We get,

$$f_{\mathrm{NdN}}(p) = \frac{2}{\Gamma(m_1)\Gamma(m_2)}\left(\frac{w}{2}\right)^{(m_1+m_2)} p^{((m_1+m_2)/2)-1} K_{m_1-m_2}\left(w\sqrt{p}\right)U(p), \tag{4.103}$$

where

$$\langle P\rangle = m_1 m_2\left(\frac{2}{w}\right)^2. \tag{4.104}$$

Equation (4.103) can now be expressed in terms of the average power as

$$f_{\mathrm{NdN}}(p)=\frac{2}{\Gamma(m_1)\Gamma(m_2)}\left(\sqrt{\frac{m_1 m_2}{\langle P\rangle}}\right)^{(m_1+m_2)}p^{((m_1+m_2)/2)-1}K_{m_1-m_2}\left(2\sqrt{\frac{m_1 m_2}{\langle P\rangle}p}\right)U(p). \tag{4.105}$$

In (4.103) and (4.105) the pdf has a subscript (NdN) indicating that the two Nakagami pdfs have identical powers but different levels of fading through m_1 and m_2. It can be seen that (4.103) is identical to the GK pdf in (4.78). The K distribution of (4.81) is a special case of (4.103) when C_1 comes from an exponential pdf and C_2 comes from a gamma pdf.

Note that shadowing occurs generally in conjunction with short-term fading and, therefore, there is no need to look for modeling shadowing as a stand-alone process. It is possible to create and analyze more complex shadowed fading channels by considering the received power as rising out of the product of three or more clusters. We can assume that the condition under which only a direct path exists is not realistic and C_0 can be put equal to zero. This raises the issue of the case of Rician fading channels. To understand how Rician conditions can be included in this model we can go back to the case of a single cluster which gave rise to a Rayleigh or Nakagami fading channel where we assumed that the single cluster can be considered to be made up of several miniclusters. Once that premise is accepted, a single cluster can give rise to Rayleigh, Nakagami, or Rician when one of the "miniclusters" is treated as contributing to the direct path. Another way to visualize pure Rayleigh or Nakagami fading is to reclassify the third term C_1C_2 in (4.96) by arguing that C_2 is a deterministic scalar quantity and C_1 corresponds to the case of an exponential or gamma pdf. Thus, even multiple scattering can result in Rayleigh or Nakagami channels if all but one of the multiple scattering components is deterministic and thus only providing a scaling factor. This notion can now be extended to the other terms in (4.95) so that regardless of the number of product terms, we can always get the case of Rayleigh, Nakagami, double Rayleigh, double Nakagami, or K or generalized K channels.

The representation of the channel in terms of (4.96) permits us to create any number of different short-term fading or shadowed fading channels such as those based on the generalized gamma pdf and Weibull pdf. Each of these separately gives rise to other pdfs described as gamma–Weibull or Weibull–Weibull channels for modeling shadowed fading channels. The representation of fading in this manner also makes it unnecessary even to consider the lognormal shadowing since the random variations in the channel can be described in terms of a single cluster (Rayleigh or Nakagami fading) or two clusters (double Rayleigh, double Nakagami, or generalized K fading). Since the evidence of Rayleigh and Nakagami fading in wireless channels is overwhelming, use of double Rayleigh, double Nakagami, generalized K fading, and K fading is well justified to characterize the statistical fluctuations observed in wireless channels.

But, before examining additional models (general models) for fading, shadowing, and shadowed fading channels, a few comments regarding the signal power and signal-to-noise ratio (SNR) are in order. In communications systems, the primary contribution to the uncertainty in the received signal comes from the additive white Gaussian noise in the channel even though it is possible that other forms of noise such as impulse noise may also be present. Since the noise is primarily additive and independent of the signal, we can define signal-to-noise ratio Z as

$$Z = \frac{P}{N\mathrm{p}}. \tag{4.106}$$

In (4.106), P is the signal power and Np is the noise power. Since noise power is fixed, the density function of Z will be of the same form as the pdf of P, with appropriate scaling from the scaling factor N_p. In other words, we can say that in a Rayleigh channel, the SNR is exponentially distributed, and in a gamma channel the SNR is gamma distributed, and so on. Because of this, we can safely interchange the pdf of the power and pdf of the SNR without losing the statistical meaning and interpretation.

4.7 General Cascaded Models

So far, we have assumed that a simple Nakagami-m pdf is a reasonable model to describe the short-term fading in wireless systems. We also examined the case of a double Nakagami and double Rayleigh channel to expand the range of fading values that we can accommodate. But, for the most part, such models assume that there is little or no multiple scattering in the channel (except when shadowing was specifically considered). Often that is not the case; the signal leaving the transmitter reaches the receiver after multiple scattering (Andersen 2002a, b; Chizhik et al. 2002; Shin and Lee 2004; Salo et al. 2006a, b; Karagiannidis et al. 2007). However, it has been shown by several researchers that a more suitable way to model realistic fading conditions when multiple scattering exists is through the use of a cascaded approach. In this model, the received signal is treated as the product of a number of scattering components, each with its own statistical characteristics (Salo et al. 2006a, b; Karagiannidis et al. 2007; Sagias and Tombras 2007; Yilmaz and Alouini 2009). This approach is also consistent with the notion of amplify-and-forward relay wireless systems where there are multiple terminals between the transmitter and receiver, offering the advantage of using several low power transmitters spread throughout a region instead of a single transmitter operating at a high power (Karagiannidis et al. 2006b; Trigui et al. 2009). Such an approach is characteristic of the cooperative diversity where different users cooperate with one another to use the notion of spatial diversity. It has also been suggested that a cascaded channel approach is well suited to model the wireless channel propagation taking place through several keyholes (Chizhik et al. 2002; Uysal 2005; Shin and Lee 2004). We will examine the modeling of such cascaded channels with a view toward characterizing short-term fading in wireless.

4.7.1 *Statistical Background of Cascaded Fading Channels*

When the transmission between two stations (transmitter and receiver) involves multiple interactions among the "obstructions" in the channel, the received signal can generally be expressed in several ways. If there is no multiple scattering in the channel the received signal-to-noise ratio (SNR) Z will be gamma distributed, considering the channel to be a Nakagami-m type. The pdf of Z can be expressed as (Karagiannidis et al. 2006a, b; Shankar 2011a, b)

$$f_1(z) = \frac{z^{m-1}}{b^m \Gamma(m)} \exp\left[-\frac{z}{b}\right], \quad m > \frac{1}{2}. \tag{4.107}$$

In (4.107), m is the Nakagami parameter and $\Gamma(.)$ is the gamma function. Note that (4.107) is expressed in a slightly different form from (4.35) and therefore, the average SNR in the Nakagami-m channel equal to Z_{01} is given by

$$Z_{01} = mb. \tag{4.108}$$

The subscript of the pdf in (4.107) identifies the absence of any multiple scattering in the channel. If the overall fading in the channel is the result of multiple scattering components (N cascades), the received SNR Z of the cascaded channel can be expressed as the product of N gamma distributed variables

$$Z = \prod_{k=1}^{N} Z_k. \tag{4.109}$$

We will assume that these random variables are independent and identically distributed, each with a density function of the form given in (4.107). The pdf of the SNR expressed in (4.109) a cascaded Nakagami-m channel has been derived by several researchers. It is obtained from the concept of the pdf of a product of multiple random variables. This pdf can be expressed in terms of Meijer G-function, shown in Chap. 2 as (Galambos and Simonelli 2004; Nadarajah and Kotz 2006b; Mathai and Haubold 2008)

$$f_N(z) = \frac{1}{\Gamma^N(m)} b^{-N} G_{0,N}^{N,0}\left(\frac{z}{b^N}\middle| \underbrace{m-1, m-1, \ldots, m-1}_{N-\text{terms}}.\right), \quad N = 1, 2, \ldots. \tag{4.110}$$

In (4.110), the subscript N identifies the pdf as one arising in an N*cascaded channel and the pdf itself is expressed in terms of the Meijer G-function. Equation (4.110) can also be rewritten in a slightly different form using the transformational relationship among Meijer G-functions expressed as (Springer and Thompson 1966, 1970; Gradshteyn and Ryzhik 2007)

$$w^k G_{pq}^{mn}\left(z\middle|{}_{b_q}^{a_p}\right) = G_{pq}^{mn}\left(w\middle|{}_{b_q+k}^{a_p+k}\right). \tag{4.111}$$

Using (4.111), the pdf in (4.110) becomes

$$f_N(z) = \frac{1}{z\Gamma^N(m)} G_{0,N}^{N,0}\left(\frac{z}{b^N}\middle|\underbrace{m, m, \ldots, m}_{N-\text{terms}}.\right), \quad N = 1, 2, \ldots. \tag{4.112}$$

The density function of the SNR in a cascaded Nakagami channel or a cascaded gamma channel often appears in the form expressed in (4.112). Since we have considered the gamma random variables to be independent, the average SNR Z_{0N} in the cascaded channel can easily be written as the product of the averages of the individual gamma variables as,

$$Z_{0N} = Z_0 = (mb)^N. \tag{4.113}$$

In writing down the expression for the average SNR in (4.113), we have dropped the second subscript N and the average SNR is Z_0.

Equation (4.112) can now be rewritten in terms of the average SNR as

$$f_N(z) = \frac{1}{z\Gamma^N(m)} G_{0,N}^{N,0}\left(\frac{m^N}{Z_0}\middle|\underbrace{m, m, \ldots, m}_{N-\text{terms}}.\right), \quad N = 1, 2, \ldots. \tag{4.114}$$

Using the moments of the gamma random variable, the amount of fading (AF) in a cascaded Nakagami channel (or a cascaded gamma channel) becomes

$$\text{AF} = \left[\left(\frac{m+1}{m}\right)^N - 1\right]. \tag{4.115}$$

The density function in (4.114) is plotted in Figs. 4.24 and 4.25. The severe fading resulting from cascading is seen in these figures. As N increases, the density functions are more skewed and move toward the Y-axis or toward lower values of the SNR.

The amount fading is plotted in Fig. 4.26 and also shows the rapidly increasing levels of fading as N increases.

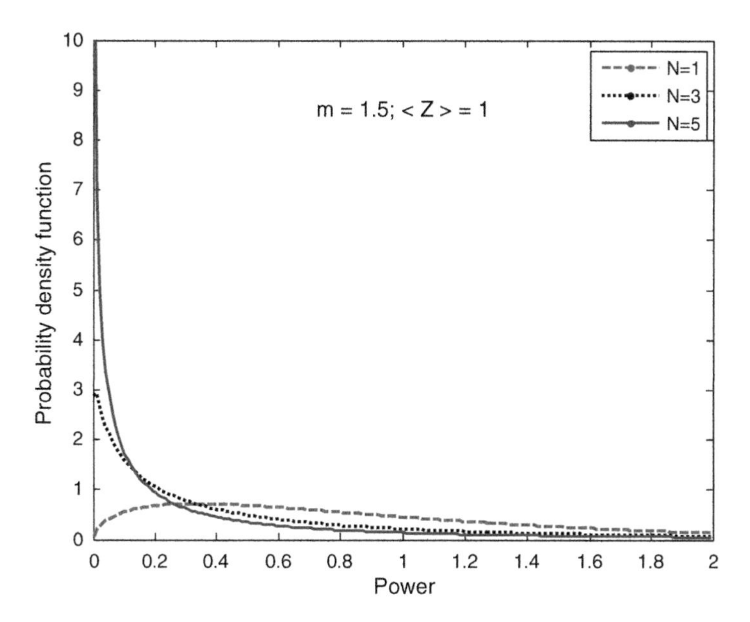

Fig. 4.24 Plot of the density function of the SNR in a cascading channel for $m = 1.5$ for $N = 1$ (no cascading) and $N = 3, 5$

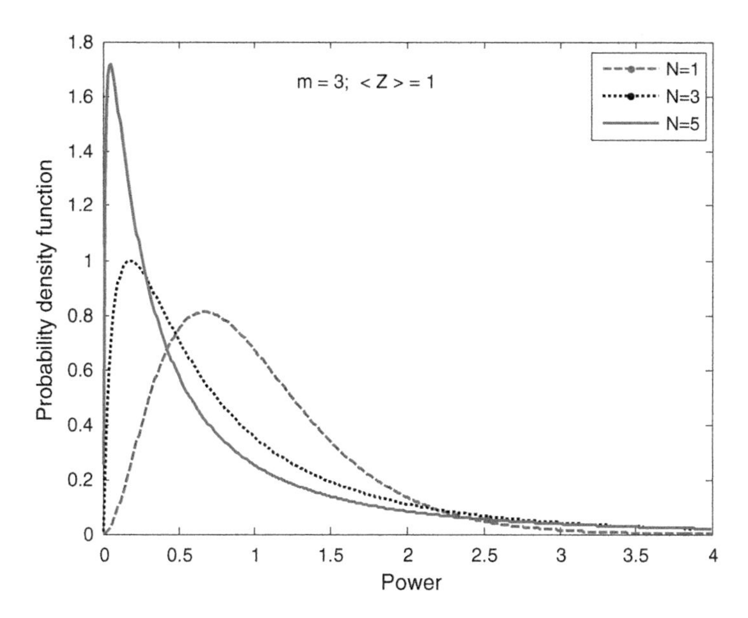

Fig. 4.25 Plot of the density function of the power in a cascading channel for $m = 3$ for $N = 1$ (no cascading) and $N = 3, 5$

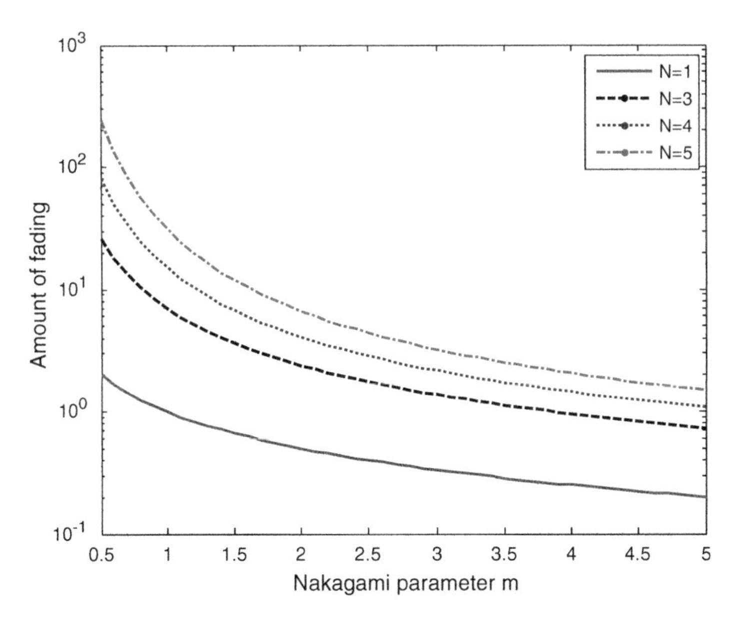

Fig. 4.26 Amount of fading in a cascaded channel

We can clearly see that the cascading approach allows us to model conditions that cover wide range fading scenarios that would not have been possible with a simple Nakagami-m model or a double Nakagami model (Nakagami 1960; Yuan et al. 2008; Wongtrairat and Supnithi 2009). Still, the cascading model includes those two as well and retains the ability to model all types of fading. The cascaded approach also makes it possible to support the hypothesis of the Nakagami-gamma model for the shadowed fading channels, whereby, we can treat the case of a shadowed fading channel to be a form of short-term fading with a much higher value of amount of fading over a traditional short-term faded channel modeled using the Nakagami-m pdf (Shankar 2004; Bithas et al. 2006).

We will now look at the cumulative distributions of the SNR in an N*Nakagami channel. The CDF can be obtained using the differential and integral properties of the Meijer G-functions as (Gradshteyn and Ryzhik 2007)

$$F_N(z) = \int_0^z f_N(\xi)\,\mathrm{d}\xi = \frac{1}{\Gamma^N(m)} G_{1,N+1}^{N,1}\left(\frac{m^N}{Z_0}\middle| \underbrace{m, m, \overset{1}{\ldots}, m, 0}_{N-\text{terms}\,\cdot}\right), \quad N = 1, 2, \ldots. \tag{4.116}$$

The CDF is shown for $m = 0.5$ in Fig. 4.27.

The CDF for $m = 1.5$ is plotted in Fig. 4.28.

The problems associated with the cascading channels seen with the help of the density functions are also supported by the CDF plots. As N increases, the CDF

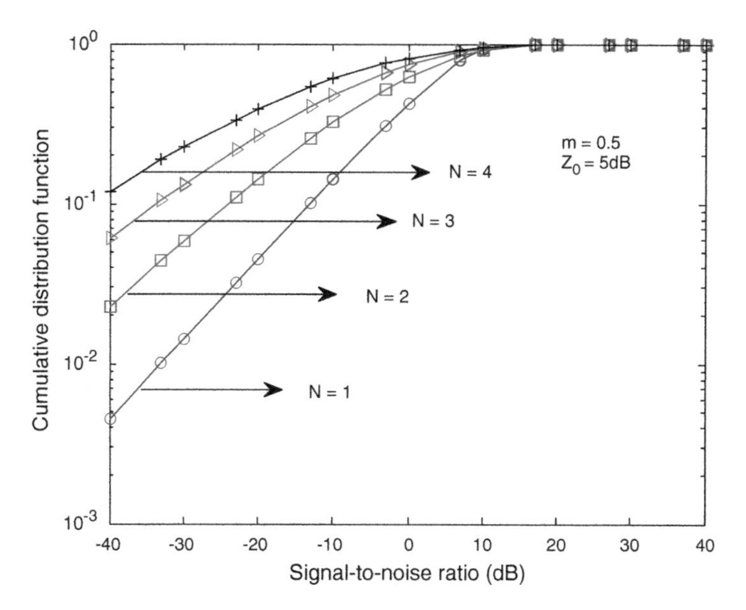

Fig. 4.27 The CDF for $N = 1,2,3,4$ for $m = 0.5$

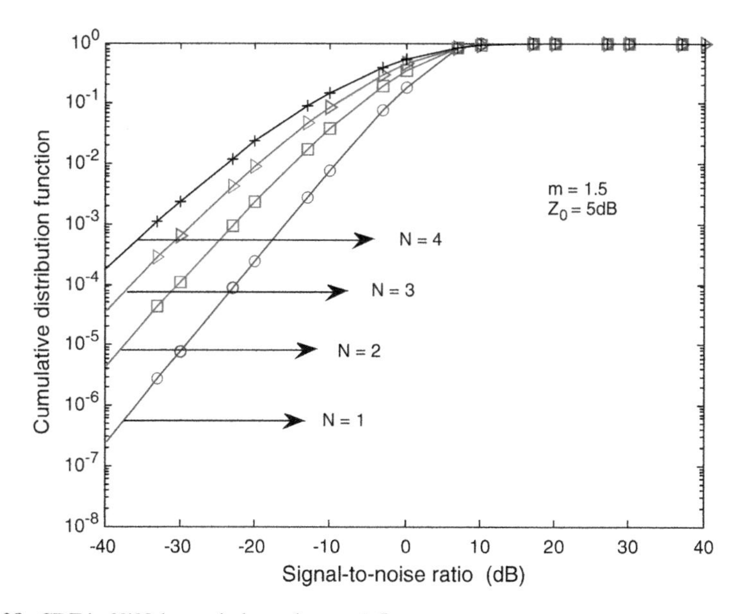

Fig. 4.28 CDF in N*Nakagami channels $m = 1.5$

curves move up indicating that the probability of the SNR staying below a certain threshold goes up with increasing values of N. This is a sign that cascading leads to fading conditions that are far more severe than what is seen in the absence of cascading, an observation that was made earlier on the basis of the shift of the peaks to the lower SNR values for the density functions as N increases.

The description of the cascading so far examined its use in modeling the short-term fading seen in wireless channels. However, as we discussed, short-term fading and shadowing coexist and we will now explore the cascading as a means to model the shadowing in wireless channels. For this, we will treat the short-term fading as a simple Nakagami-m type. This approach is described in the next section.

4.7.2 Cascaded Approach to Shadowed Fading Channels

It was argued earlier that the Nakagami-lognormal density function for shadowed fading channels can be replaced by the Nakagami-gamma pdf in (4.78). This approach produced the so-called GK distribution for the SNR in shadowed fading channels (Abdi and Kaveh 1998; Shankar 2004; Bithas et al. 2006). Another approach to overcome the analytical short coming of the Nakagami-lognormal model was suggested through the use of the inverse Gaussian distribution for the lognormal pdf (Karmeshu and Agrawal 2007; Laourine et al. 2009). While both the Nakagami-gamma (Generalized K) and Nakagami-inverse Gaussian (NiG) approaches provided a simpler means to model shadowed fading, neither of them provide the match over the whole range of fading and shadowing values. Neither of those approximation takes into account the origin of the lognormal pdf as explained in connection with (4.66), namely the product nature of the shadowing which results in the lognormal pdf from the central limit theorem for products (Rohatgi and Saleh 2001; Papoulis and Pillai 2002a, b; Andersen 2002a, b).

Shadowing can be treated as a form of cascading as suggested by some researchers (Coulson et al. 1998a, b; Salo et al. 2006a, b; Andersen 2002a, b). It can be treated as arising out of the product of N independent and identically distributed gamma variables, with N equal to unity resulting in the Nakagami-gamma (GK distribution) distribution for the shadowed fading channels (Shankar 2004; Bithas et al. 2006; Laourine et al. 2008). We can extend this notion of the Nakagami-gamma distribution to a case of Nakagami-N-gamma distribution whereby we replace the lognormal pdf, not with a single gamma pdf, but with a cascaded gamma density with N cascades (Shankar 2010b).

In a shadowed fading channel, the pdf $f(z)$ of the signal-to-noise ratio Z was rewritten earlier as

$$f(z) = \int_0^\infty f(z|y)f(y)\,\mathrm{d}y. \tag{4.117}$$

In (4.117), the conditionality demonstrates the effect of shadowing and for the Nakagami short-term faded channel, we can express the conditional density in (4.117) as

$$f(z|y) = \left(\frac{m}{y}\right)^{m} \frac{z^{m-1}}{\Gamma(m)} \exp\left(-\frac{mz}{y}\right), \quad z > 0, y > 0. \tag{4.118}$$

Modeling shadowing as a cascading process, the average SNR in a short-term faded signal can be expressed as the product of several components as

$$Y = \prod_{k=1}^{N} W_i. \tag{4.119}$$

In (4.119), N is the number of cascades and W_i is the ith element. Since it was shown that Nakagami-gamma was a reasonable approximation to the Nakagami-lognormal, we will treat W's to be gamma distributed so that when $N = 1$, we obtain the Nakagami-gamma pdf in (4.117) as discussed in (4.78). When N becomes large, applying the central limit theorem for products, the density function of log10(Y) in Eq. (4.119) will be Gaussian, and we have the lognormal pdf in (4.67). Thus, by varying N we expect that the pdf of N-Gamma product will tend toward a lognormal pdf and the density function of the SNR in the shadowed fading channel will move toward the Nakagami-lognormal density function (Tjhung and Chai 1999; Vaughn and Andersen 2003; Simon and Alouini 2005; Cotton and Scanlon 2007).

We make the assumption that all W's are independent and identically distributed gamma random variables, each with parameters c and y_0 such that

$$f(w_i) = \left(\frac{w_i}{y_0}\right)^{c} \frac{1}{w_i \Gamma(c)} \exp\left(\frac{w_i}{y_0}\right), \quad w_i > 0. \tag{4.120}$$

The density function of Y in (4.119) can be obtained from the results on products of random variables, namely those of gamma type as shown earlier in (4.112)

$$f_N(y) = \frac{1}{y\Gamma^N(c)} G_{0,N}^{N,0}\left(\frac{y}{y_0^N}\middle| \underbrace{c, c, \ldots, c}_{N-\text{terms}} .^{1}\right), \quad y = 0. \tag{4.121}$$

In (4.121), $G()$ is the Meijer G-function and the relationships of c and y_0 in (4.121) to the parameters m and σ_{dB} of the lognormal pdf are yet to be established (Mathai and Saxena 1973; Gradshteyn and Ryzhik 2007).

As mentioned earlier, (4.121) reduces to the gamma pdf when N is unity. We can now establish the relationship between the parameters of the lognormal pdf and the pdf in (4.121) by taking the logarithm in (4.119) and comparing the moments (Nakagami 1960; Clark and Karp 1965; Ohta and Koizumi 1969). We get,

$$\mu = NA_0[\mathrm{Psi}(c_N) + \log_e(y_{0N})], \tag{4.122}$$

$$\sigma_{\mathrm{dB}}^2 = NA_0^2\mathrm{Psi}(1, c_N). \tag{4.123}$$

The parameter A_0 was defined earlier in Eq. (4.68). Psi(.) is the digamma function $\psi(.)$ and Psi(1,.) is the trigamma function $c'(.)$ (Abramowitz and Segun 1972; Gradshteyn and Ryzhik 2007). The subscript N associated with the two parameters of the gamma pdf merely reflects the fact that as N varies, these two parameters take on different values. Inverting (4.122) and (4.123), we can write

$$c_N = \mathrm{InvPsi}\left(1, \frac{\sigma^2}{NA_0^2}\right), \tag{4.124}$$

$$y_0N = \exp\left[\frac{\mu}{NA_0} - \mathrm{Psi}(c_N)\right], \tag{4.125}$$

In (4.124), InvPsi(1,.) is the inverse of the Psi(1,.) function. Thus, from the values of shadowing parameters (m, σ_{dB}), it is possible to estimate the corresponding values of the gamma parameters of the pdf in (4.121) for each value of N. The relationship between the values of c (for different number of cascaded gamma densities) and the shadowing level sigma is shown in Fig. 4.29 $N = 1,2,3,4$. Since the shadowing has been shown to be in the range of 3–9 dB for most wireless systems, we will limit ourselves to that range. Equation (4.124) was used to get the values of the order of the gamma pdf (c_N). The order of the gamma

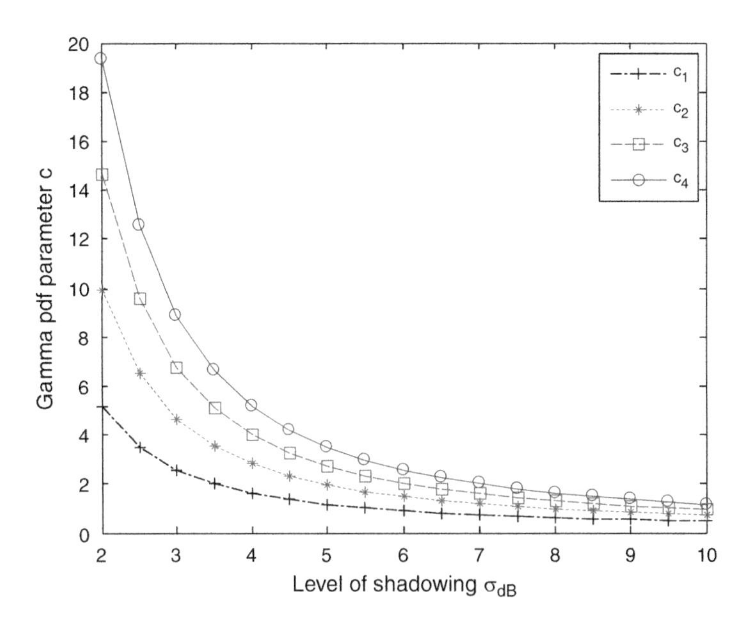

Fig. 4.29 Gamma parameter c as a function of shadowing level

pdf, c_N, increases as the shadowing levels decrease. The values of c_N are also higher with increasing values of N.

The next step is the derivation of the pdf of the SNR in the shadowed fading channel. Inserting (4.121) in (4.117), we have the pdf of the SNR in a shadowed fading channel as

$$f(x) = \int_0^{\infty} \left(\frac{m}{y}\right)^m \frac{x^{m-1}}{\Gamma(m)} \exp\left(-\frac{mx}{y}\right) \frac{1}{y\Gamma^N(c_N)} G_{0,N}^{N,0}\left(\frac{y}{(y_{0N})^N}\middle|\underbrace{c_N, c_N, \ldots, c_N}_{N-\text{terms}}.\right) dy. \tag{4.126}$$

In (4.126), the values of the gamma parameters are identified for each N separately as c_N and y_{0N} (mentioned earlier). Noting that the pdf in (4.126) is the pdf of the product of (N + 1) gamma random variables consisting of N identical gamma variables (parameters c_N and y_{0N}) and one other gamma variable (parameters m and $1/m$), the pdf becomes (Podolski 1972; Adamchik 1995; Gradshteyn and Ryzhik 2007)

$$f(x) = \frac{1}{z\Gamma(m)\Gamma^N(c_N)} G_{0,N+1}^{N+1,0}\left(\frac{mz}{(y_{0N})^N}\middle|\underbrace{m, c_N, c_N, \ldots, c_N}_{N-\text{terms}}.\right), \quad z > 0. \tag{4.127}$$

In terms of the average SNR Z_0, (4.127) becomes

$$f(z) = \frac{1}{z\Gamma(m)\Gamma^N(c_N)} G_{0,N+1}^{N+1,0}\left(\frac{m(c_N)^N z}{(Z_0)^N}\middle|\underbrace{m, c_N, c_N, \ldots, c_N}_{N+1-\text{terms}}.\right), \quad z > 0. \tag{4.128}$$

For $N = 1$, (4.127) becomes the familiar GK pdf based on the relationship between the Meijer G-function and Bessel functions. The density function in (4.127) is plotted in Fig. 4.30 for $m = 0.5$ and very low levels of shadowing. It shows that at low levels of shadowing, a value of $N = 2$ is sufficient to produce a match with the Nakagami-lognormal pdf. Note that the notation of NNG (Nakagami-N-gamma with varying integer values of N) is used to indicate the number of cascades N used. The average SNR has been assumed to be unity in all these plots.

Figure 4.31 examines the match at a moderate level of shadowing (5 dB). One can see that the match with the Nakagami-lognormal improves as N increases. The match at a higher level of shadowing ($\sigma_{\text{dB}} = 8$) is examined in Fig. 4.32. These figures show that at low values of the shadowing, a single gamma variable is sufficient to model the lognormal which results in the Nakagami-gamma pdf for shadowing or the GK pdf for shadowing. As the value of shadowing increases, more and more gamma products are necessary to provide a better match with the lognormal. Thus, the use of the N-gamma pdf or a cascaded gamma model for

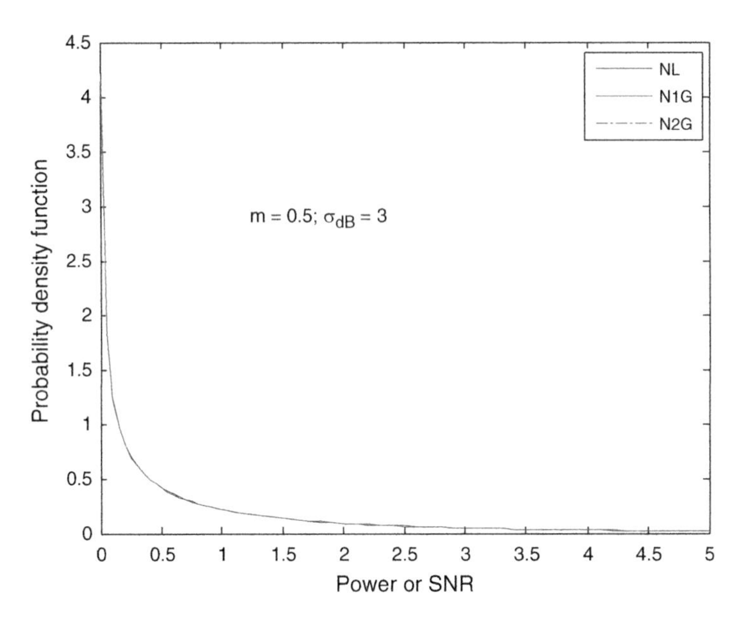

Fig. 4.30 Density functions $m = 0.5$ and $\sigma_{\text{dB}} = 3$ only N1G and N2G are needed for match with the Nakagami-lognormal

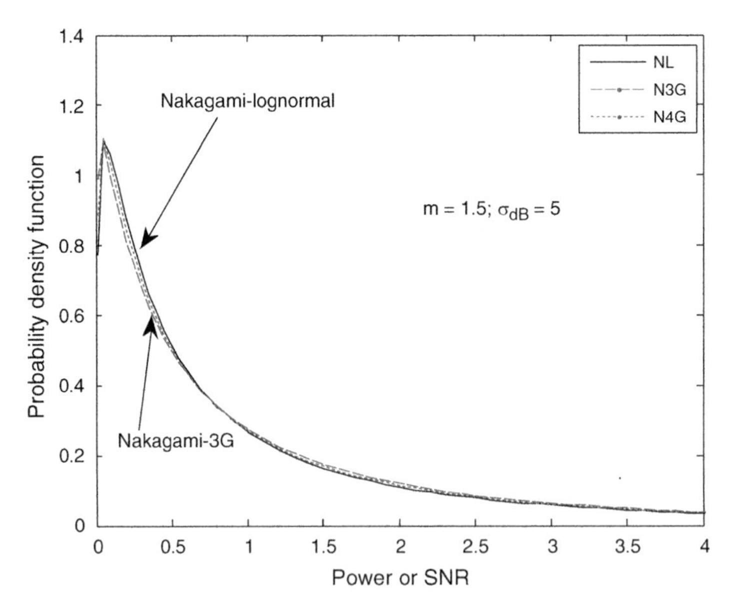

Fig. 4.31 Density function of the pdf. NL—lognormal, N3G—Nakagami-triple gamma, and N4G is the Nakagami-quadruple gamma for $m = 1.5$ and $\sigma_{\text{dB}} = 5$

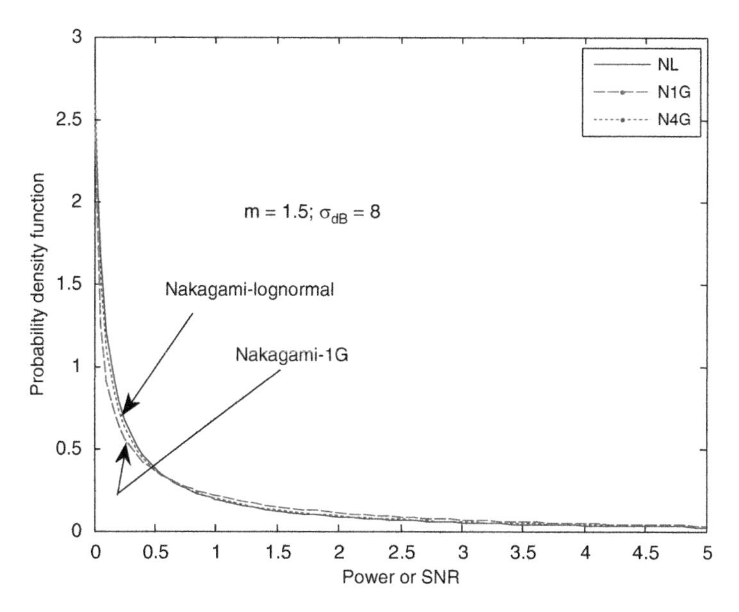

Fig. 4.32 Match at a higher level of shadowing ($\sigma_{\text{dB}} = 8$)

shadowing provides a means to match the lognormal at any level of shadowing. The nature of the match and mismatch will be clearer in later sections when we examine the outage probabilities and error rates in shadowed fading channels.

Taking the moments, the amount of fading in the shadowed fading channel becomes

$$\text{AF} = \left(\frac{m+1}{m}\right)\left(\frac{c_N+1}{c_N}\right)^N - 1. \tag{4.129}$$

For $N = 1$, the amount of fading in (4.129) becomes equal to the AF in a generalized K fading channel. The amount of fading in a shadowed fading channel with $N = 2$ is plotted in Figs. 4.33, 4.34, and 4.35.

The CDF of the SNR in a shadowed fading channel can be obtained using the differential/integral properties of the Meijer G-function as (Gradshteyn and Ryzhik 2007)

$$F(z) = \frac{1}{\Gamma(m)\Gamma^N(c_N)} G_{1,N+2}^{N+1,1}\left(\frac{m(c_N)^N z}{Z_0}\left[\underbrace{m, c_N, c_N, \ldots, c_N, 0}_{N\text{terms}}\right]\right). \tag{4.130}$$

Equations (4.127) and (4.130) clearly show that the use of cascaded approach to shadowing results in closed form solutions to the pdf and CDF of the SNR in a

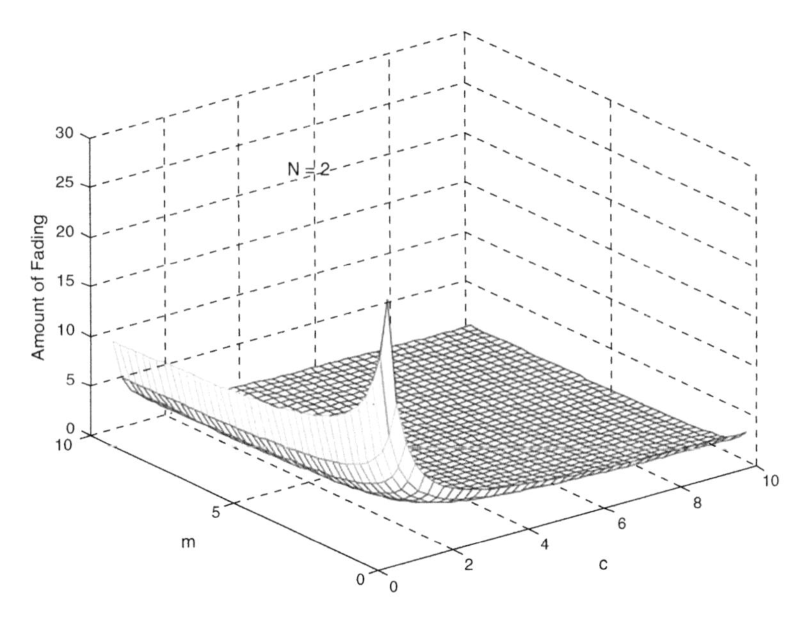

Fig. 4.33 Amount of fading in Nakagami-N-gamma channel; $N = 2$

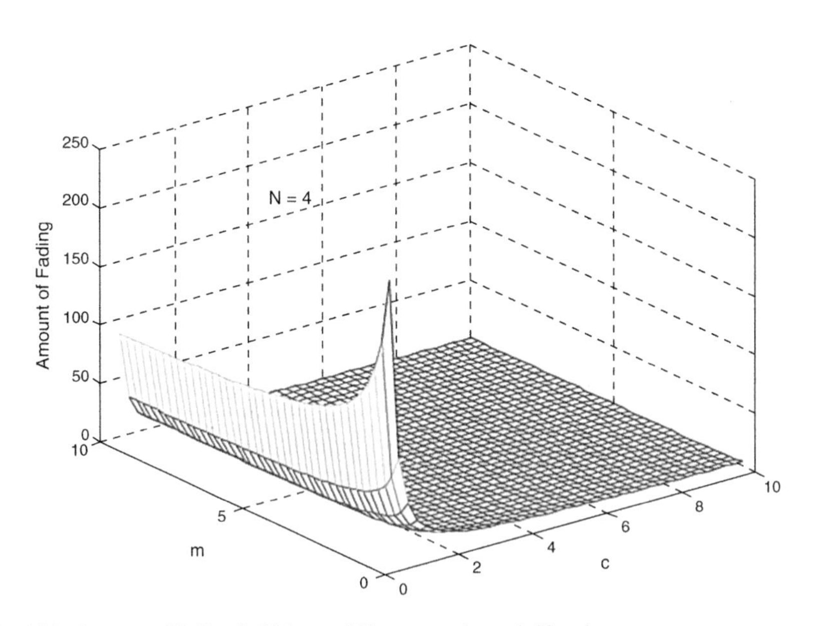

Fig. 4.34 Amount of fading in Nakagami-N-gamma channel; $N = 4$

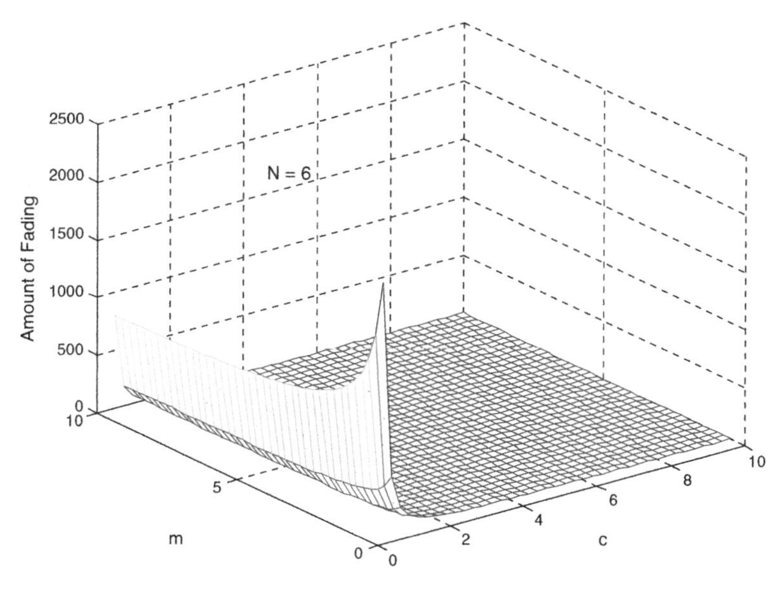

Fig. 4.35 Amount of fading in Nakagami-N-gamma channel; $N = 6$

shadowed fading channel. Note that for $N = 1$, (4.130) can be expressed in terms of hypergeometric functions which one obtains as the CDF associated with the GK pdf (Mathai and Saxena 1973; Mathai and Haubold 2008).

The CDFs of the signal-to-noise ratio in shadowed fading channels were evaluated next. The NL pdf in (4.67) can be used to obtain the corresponding CDF. The CDF associated with the Nakagami-N-gamma model is obtained directly using the expression in (4.130). For the average SNR of the NL channel, the corresponding value of y_0 is calculated from Eq. (4.125) using the values of c_N obtained earlier in connection with Fig. 4.29. CDF was estimated assuming an average SNR of 5 dB. Figure 4.36 shows the case of $m = 1$. The match between the Nakagami-lognormal CDF and the Nakagami-N-gamma CDF is very good even with $N = 1$ for moderate ($\sigma_{\text{dB}} = 5$) and high levels ($\sigma_{\text{dB}} = 8$) of shadowing, with an excellent match with $N = 4$. Figure 4.37 shows the case for $m = 1.5$. As the value of m goes to 1.5 in Fig. 4.37, the mismatch becomes a little bit more pronounced for high levels of shadowing ($\sigma_{\text{dB}} = 8$). Still, the mismatch between the Nakagami-lognormal CDF and the Nakagami-N-gamma CDF exists only at the very low values of signal-to-noise ratio and such a mismatch will have very little impact in practical situations. However, the CDF plots do not show the levels of matching at higher values of the SNR where wireless systems operate. To observe how closely the two models match, we need to look at the outage probabilities and average bit error rates (Shankar 2010b).

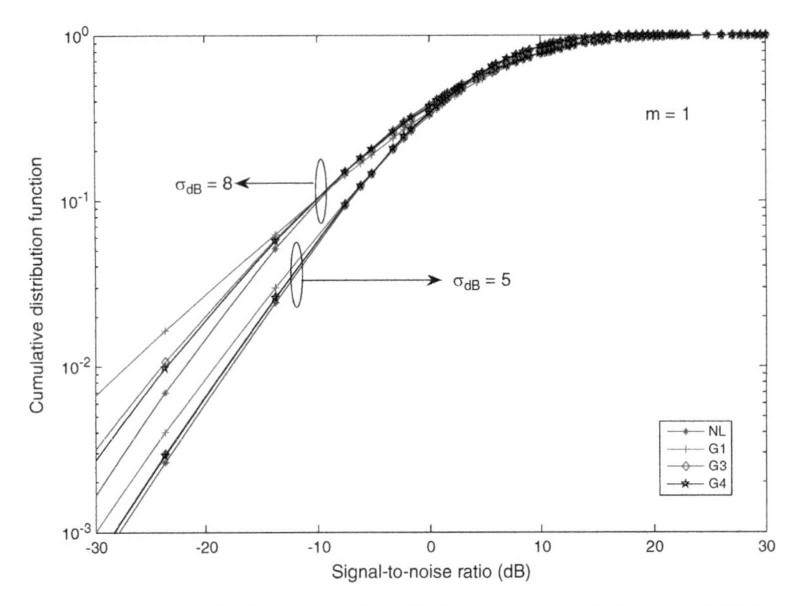

Fig. 4.36 The cumulative distribution function (CDF) of the Nakagami-N-gamma CDF for $m = 1$

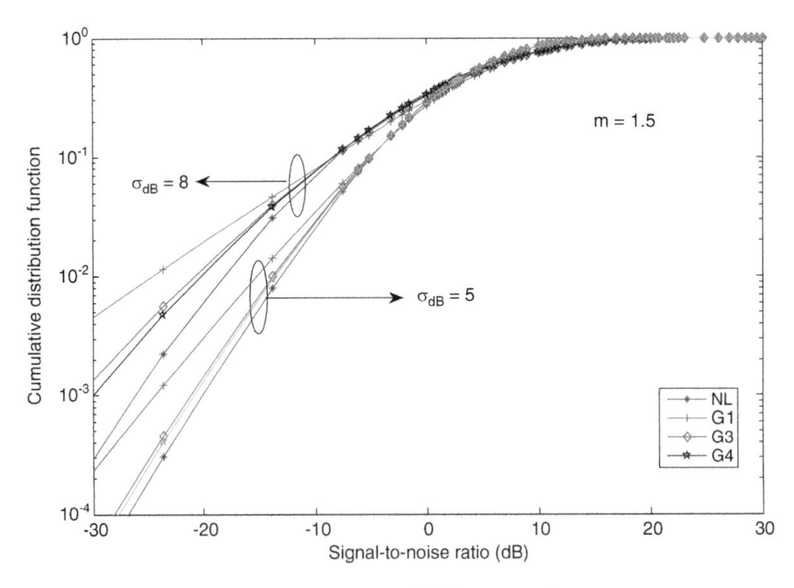

Fig. 4.37 The CDF of the Nakagami-N-gamma CDF for $m = 1.5$

4.7.3 N*Weibull Channels

While N^*gamma models are reasonably good descriptors of fading, it is possible to reexamine the concepts that led to the lognormal distribution to model long-term fading. It has been suggested that the notion of the cascaded channels can be extended to include those where each channel is best described using a Weibull pdf and that the cascaded Weibull channel can approximate the lognormal channel associated with long-term fading conditions (Sagias and Tombras 2007).

4.7.4 Double Rician Channels

Performance of wireless networks can be enhanced through amplify-and-forward relay systems. In a simple example of such a system, a relay amplifies the signal received and retransmits it to the destination receiver (Patzold 2002; Zogas and Karagiannidis 2005; Talha and Patzold 2007; Wongtrairat et al. 2008; Mendes and Yacoub 2007). A simple diagram of the concept is shown in Fig. 4.38. Note that this principle can be extended to multiple relaying stations and we have the equivalent of multihop systems described earlier. If we look at the specific case of amplify-and-forward relay, the signal at the receiver can be expressed as the product of the two channel signals resulting in a double Rician channel if LOS conditions exist in both channels.

Results are also available where double Hoyt channels have been explored to model the behavior of amplify-and-forward relay fading channels (Hajri et al. 2009; de Souza and Yacoub 2009).

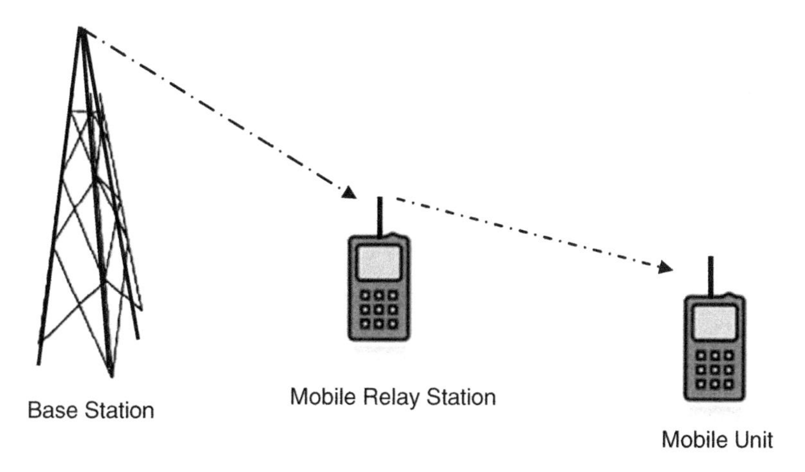

Fig. 4.38 Concept of a relay station possibly resulting in double Rician fading

4.8 Comparison of Different Models

A broad overview of the pdfs in the study of wireless channels was presented. The list of the pdfs for fading is summarized in Table 4.1. It also contains the expressions for the amount of fading (AF). Table 4.2 lists the expressions for pdfs in shadowed fading channels and the amount of fading (AF).

As mentioned earlier, the amount of fading provides a quantitative measure of fading present in wireless channels. Note that the definition of the amount of fading depends only on two moments, namely the first and the second moments of the power. Thus, it is conceivable to get the same value of AF from several other pdfs as well. In other words, the quantification of the level of fading in (4.11) does not completely provide the characteristics of the channel or the impairments caused by the existence of fading or shadowing in the channel. To provide a better quality measure of the degradation of the channel in terms of its ability to carry information and deliver it to the receiver with high confidence, one needs to look at the error rates in fading channels and outage probabilities.

Table 4.1 The probability density functions (pdfs) of the received power in fading channels and the amount of fading

Probability density function ($p \geq 0$)	Amount of fading (AF)	Additional information
$f_{\mathrm{R}}(x)\frac{1}{P_0}\exp\left(-\frac{p}{P_0}\right)$	1	
$f_{\mathrm{N}}(p)=\frac{m^m p^{m-1}}{P_0^m\Gamma(m)}\exp\left(-m\frac{p}{P_0}\right)$	$\frac{1}{m}$	$m\geq\frac{1}{2}$
$f_{\mathrm{G}}(p)=\frac{p^{m-1}}{P_{\mathrm{g}}^m\Gamma(m)}\exp\left(-\frac{p}{P_{\mathrm{g}}}\right)$	$\frac{1}{m}$	$m>0, \quad \langle P\rangle_{\mathrm{G}}=mP\mathrm{g}$
$f_{\mathrm{GG}}(p)=\frac{sp^{ms-1}}{\Gamma(m)P_{\mathrm{g}}^m}\exp\left(-\frac{p^s}{P_{\mathrm{g}}}\right)$	$\frac{\Gamma(m+(2/s))\Gamma(m)}{[\Gamma(m+(1/s))]^2}-1$	$m>0, \quad 0<s<\infty$ $\langle P\rangle_{\mathrm{GG}}=\frac{\Gamma(m+(1/s))}{\Gamma(m)}P_{\mathrm{g}}^{(1/2)}$
$f_{\mathrm{W}}(p)=\frac{m_{\mathrm{w}}}{P_{\mathrm{g}}^m}p^{m_{\mathrm{w}}-1}\exp\left(-\frac{p^{m_{\mathrm{w}}}}{P_{\mathrm{g}}}\right)$	$\frac{\Gamma(1+(2/m_{\mathrm{w}}))\Gamma(m)}{[\Gamma(m+(1/m_{\mathrm{w}}))]^2}-1$	$\langle P\rangle_{\mathrm{w}}=\Gamma(1+(1/m_{\mathrm{w}}))P_{\mathrm{g}}^{(1/m_{\mathrm{w}})}$

Special conditions on the parameters are also provided. Note that all the pdfs exist only for $p\geq 0$ The subscripts with the pdfs and the moments indicate the names associated with the pdfs: Rayleigh (R), Nakagami (N), gamma (G), generalized gamma (GG), Weibull (W)

Table 4.2 The pdf of the received power in shadowing channels and the amount of fading

Probability density function ($p \geq 0$)	Amount of fading (AF)	Additional information
$f_{\mathrm{L}}(p)=\frac{A_0}{\sqrt{2\pi\sigma_{\mathrm{dB}}^2p^2}}\exp\left[-\frac{(10\log_{10}p-\mu)^2}{2\sigma_{\mathrm{dB}}^2}\right]$	$\exp\left(-\frac{\sigma_{\mathrm{dB}}^2}{A_0^2}\right)-1$	$A_0=\frac{10}{\log_e 10}$
$f_{\mathrm{G}}(p)=\frac{p^{c-1}}{y_0^c(c)}\exp\left(-\frac{p}{y_0}\right)$	$\frac{1}{c}$	$\sigma_{\mathrm{dB}}^2=A_0^2\psi'(c)$

Note that all the pdfs exist only for $p\geq 0$. The subscripts with the pdfs and the moments indicate the names associated with the pdfs: Lognormal (L), gamma (G). The trigamma function is represented by $c_0(.)$

4.8.1 Average Probability of Error

Comparison of the error rates in an ideal channel and a channel undergoing fading or shadowing will allow the "cost" of departing from an ideal channel. We will explain the notion of such a cost or "power penalty" by examining the error rates in an ideal (Gaussian) channel and a Rayleigh channel. As suggested in (4.106), one of the consequences of fading or shadowing is that the signal-to-noise ratio in the channel becomes random and acquires the density function of the SNR. We will consider the example of a coherent binary shift keying scheme discussed in Chap. 3. The error rate in an ideal Gaussian channel is a function of the signal-to-noise ratio Z_0 and can be expressed as (Proakis 2001; Simon and Alouini 2005)

$$p_{\mathrm{e}}(Z_0) = Q\left(\sqrt{2Z_0}\right). \tag{4.131}$$

Note that the error rate can also be written in terms of complementary error functions as

$$p_{\mathrm{e}}(Z_0) = \frac{1}{2}\operatorname{erfc}\left(\sqrt{Z_0}\right). \tag{4.132}$$

However, when short-term fading is present, the SNR is a random variable and the error rates need to be averaged as (Alouini and Simon 1998; Proakis 2001)

$$p_{\mathrm{e}}(Z_0) = \int_0^{\infty} p_{\mathrm{e}}(z) f(z)\, \mathrm{d}z. \tag{4.133}$$

In (4.133), Z_0 is the average signal-to-noise ratio and $f(z)$ is the density function of the SNR in a fading channel. In Rayleigh fading, the error rate in a fading channel becomes

$$p_{\mathrm{e}}(Z_0) = \int_0^{\infty} Q\left(\sqrt{2z}\right)\left(\frac{1}{Z_0}\right)\exp\left(-\frac{z}{Z_0}\right) \mathrm{d}z = \frac{1}{2}\left[1 - \sqrt{\frac{Z_0}{1+Z_0}}\right]. \tag{4.134}$$

The two error rates are plotted in Fig. 4.39.

The plot shows that while the error rates drop fast for an ideal channel, the rate of decline in error rates as a function of the average SNR is much slower in a Rayleigh faded channel. For example, if we pick an error rate of 1e-4, one can see that the fading channel requires a higher SNR to achieve the error rate of 1e-4 compared with the ideal channel. This difference in SNR needed to maintain a certain error rate is called "excess SNR" required or "the power penalty." This quantity in fading channels typically expressed in decibels. Going back to our discussion on power or energy efficiencies of digital modems, it is clear that one important consequence of the existence of fading is the reduction in the energy efficiency of the modem. One can also see from Fig. 4.39 that the penalty values are not fixed and that they vary

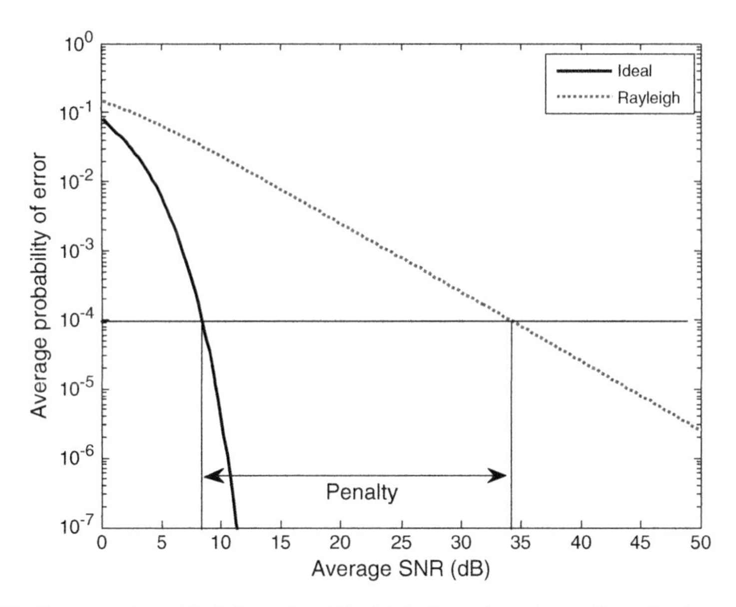

Fig. 4.39 Error rates in an ideal channel and Rayleigh channel are shown. For a fixed error rate of 1e-4, the difference in SNR corresponds to the penalty

with the error rate chosen or required. This is shown in Fig. 4.40 which is obtained by inverting (4.131) and (4.134). We have the expressions for the signal-to-noise ratio values as a function of the error rates as

$$Z_{0\text{ideal}} = \left[\text{erfcinv}(2p_{\text{emin}})\right]^2, \tag{4.135}$$

$$Z_{0\text{fad}} = \frac{(1-2p_{\text{emin}})^2}{4p_{\text{emin}}(1+p_{\text{emin}})}. \tag{4.136}$$

In (4.135) and (4.136), p_{emin} is the minimum acceptable error rate to have a satisfactory performance and erfcinv the inverse of the complementary error function. The excess SNR required is

$$z_{\text{ex}} = Z_{0\text{fad}} - Z_{0\text{ideal}}. \tag{4.137}$$

As the error rate goes down, the excess SNR required to mitigate the presence of fading in the channel increases. This is an important adverse consequence of fading. The quantity, excess SNR, provides more quantitative information on fading than what is available with the amount of fading (AF) defined earlier.

We will now compare the error rate performance in Nakagami channels and examine the effect of increasing values of m (Annamalai and Tellambura 2001; Simon and Alouini 2005). The average probability of error in a Nakagami-m faded channel becomes

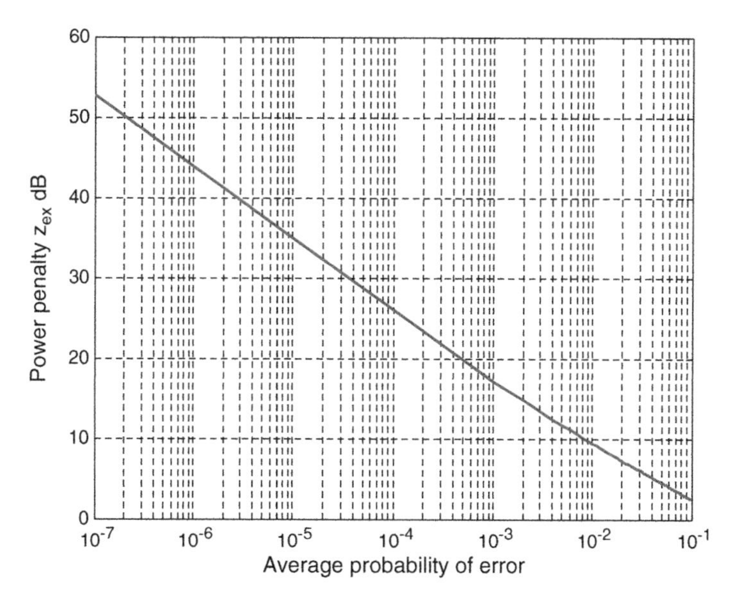

Fig. 4.40 Power penalty or excess SNR required to mitigate Rayleigh fading

$$
\begin{aligned}
p_{\mathrm{e}}(Z_0) &= \int_0^{\infty} Q\left(\sqrt{2z}\right)\left(\frac{m}{Z_0}\right)^{m}\frac{z^{m-1}}{\Gamma(m)}\exp\left(-\frac{m}{Z_0}\right)\mathrm{d}z \\
&= \frac{1}{2}-\frac{\Gamma(m+(1/2))}{\Gamma(m\sqrt{\pi})}\sqrt{\frac{Z_0}{m^2}}{}_2F_1\left(\left[\frac{1}{2},m+\frac{1}{2}\right],\left[\frac{3}{2}\right],-\frac{Z_0}{m}\right).
\end{aligned}
\tag{4.138}
$$

In (4.138), 2F1 is the hypergeometric function (Abramowitz and Segun 1972; Gradshteyn and Ryzhik 2007). The result in (4.138) can be obtained using the relationships among hypergeometric functions, Meijer G-functions, and complementary error functions (Wolfram 2011) (http://functions.wolfram.com/HypergeometricFunctions/). The results are shown in Fig. 4.41.

One can see that as the Nakagami parameter m increases, the error rates are falling steeper and approach the ideal channel case. This was also evident in Fig. 4.11, shown earlier, where the movement of the Nakagami-m pdf toward the Gaussian pdf with increasing values of m was discussed.

It is also possible to see the consequence of fading by examining the SNR required to maintain a fixed value of the probability of error. For example, for a probability of error of 1/1000, in an ideal channel one only requires a signal-to-noise ratio of about 6 dB. To achieve the same error rate, one would require more than 30 dB of SNR for a Nakagami channel with an m value of 0.5, and about 14 dB for the case of a Nakagami channel with an m value of 2.0. This decline in the sensitivity (minimum power needed to maintain an acceptable probability of error) is a major consequence of fading as suggested earlier. It is easy to see that for an

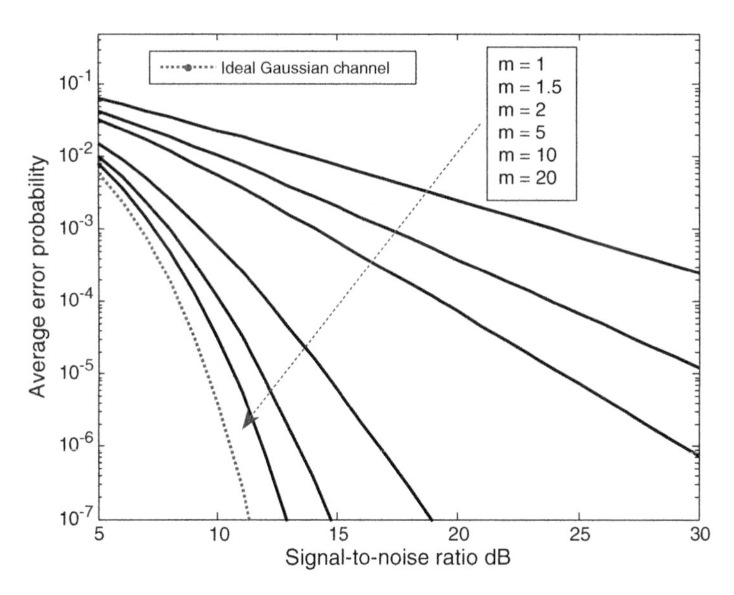

Fig. 4.41 Error rates in Nakagami channels for several values of m

error rate of 1/1000, a user will pay a power penalty of about 8 dB in a Nakagami channel with $m = 2$ and more than 24 dB in a Nakagami channel with an m value of 0.5.

Figure 4.41 supports the notion that increasing values of the Nakagami parameter reduces the fading in the channel. One can obtain the values of the excess SNR or power penalty required to mitigate the effects of fading. It is obvious that with increasing values of m, the excess SNR goes down.

The error rates in a Rician channel mirror the rates in the Nakagami channels for $m > 1$. As the Rician factor K_0 increases, the error rates drop faster with SNR and the error rates will approach those of an ideal Gaussian channel.

We will now look at the case of error rates in channels degraded by lognormal shadowing (Stuber 2002). It will also be noted that lognormal pdf has been suggested as a means to model short-term fading. There also exists a relationship between the Nakagami pdf and the lognormal pdf as discussed by the original work by Nakagami. It is thus reasonable to estimate the error rates in a lognormal channel along with other ones in short-term fading channels. A few points must be made here as we compare the error rates in an ideal white noise limited Gaussian channel to that in a lognormal channel. Note that the quantity m in decibels in (4.67) is the average of the power or SNR measured in decibels, i.e.,

$$\mu = \langle 10\log_{10}(Z) \rangle. \tag{4.139}$$

However, in the absence of shadowing, the power is measured and then averaged. Thus, when we discuss the average SNR in the absence of shadowing, what we have is

$$Z_0 = (\mathrm{dB}) = 10\log_{10}\langle Z\rangle. \tag{4.140}$$

Note that (4.139) and (4.140) lead to different values. Thus, if one is comparing any channel performances where in one case (in a Gaussian channel) the power or SNR is measured in terms of (4.140) and results compared to the case of a lognormal shadowing case, then the values need to be the same. The relationship between the two averages can be derived as follows: For the lognormal case, taking the mean of the lognormal random variable, we have from (4.69)

$$\langle Z\rangle - \exp\left(\frac{\mu}{A_0} + \frac{\sigma_{\mathrm{dB}}^2}{2A_0^2}\right). \tag{4.141}$$

Converting (4.141) into decibels, we have

$$Z_0(\mathrm{dB}) = 10\log_{10}(\langle Z\rangle) = \frac{10\log_{10}(\langle Z\rangle)}{\log_{10}(10)} = A_0\left(\left[\frac{\mu}{A_0} + \frac{\sigma_{\mathrm{dB}}^2}{2A_0^2}\right]\right) = \mu + \frac{\sigma_{\mathrm{dB}}^2}{2A_0^2}. \tag{4.142}$$

Thus,

$$\mu_{\mathrm{dB}} = Z_0(\mathrm{dB}) - \frac{\sigma_{\mathrm{dB}}^2}{2A_0^2}. \tag{4.143}$$

It can be seen that in the absence of shadowing ($\sigma_{\mathrm{dB}} = 0$), both averages are equal. If $\sigma_{\mathrm{dB}} > 0$, a correction factor, m_0, needs to be applied to the average SNR in shadowing such that

$$\mu_{\mathrm{dB}} = Z_0(\mathrm{dB}) - \frac{\sigma_{\mathrm{dB}}^2}{2A_0^2} = Z_0(\mathrm{dB}) - \mu'. \tag{4.144}$$

A plot of the correction factor is shown in Fig. 4.42.

As in the case of Nakagami fading one can see the need to have an excess SNR to mitigate the effects of shadowing. As the shadowing parameter decreases, the error rate curves move closer to the ideal, i.e., Gaussian channel. The error rates in a lognormal channel are shown in Fig. 4.43.

We can also examine the error rates in a generalized gamma channel. The error rates in a generalized gamma fading channel become (Aalo et al. 2005; Nadarajah and Kotz 2006a, c; Malhotra et al. 2009)

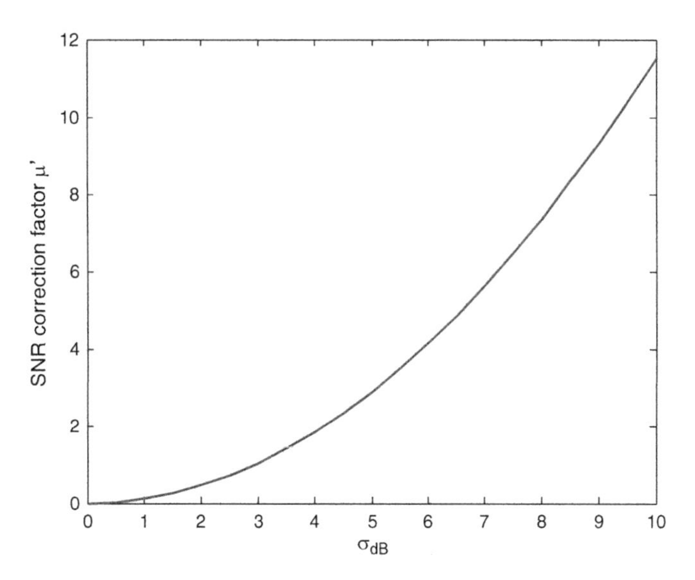

Fig. 4.42 SNR correction factor for shadowing

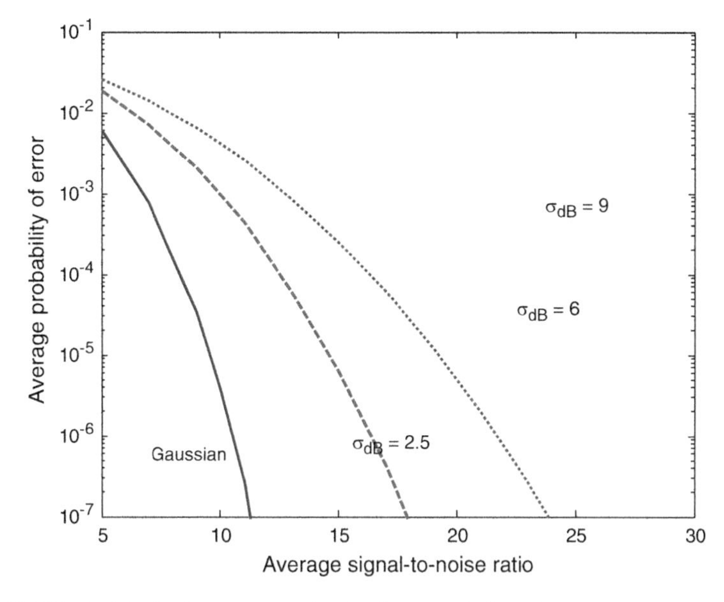

Fig. 4.43 Effect of shadowing on error rates in a lognormal channel

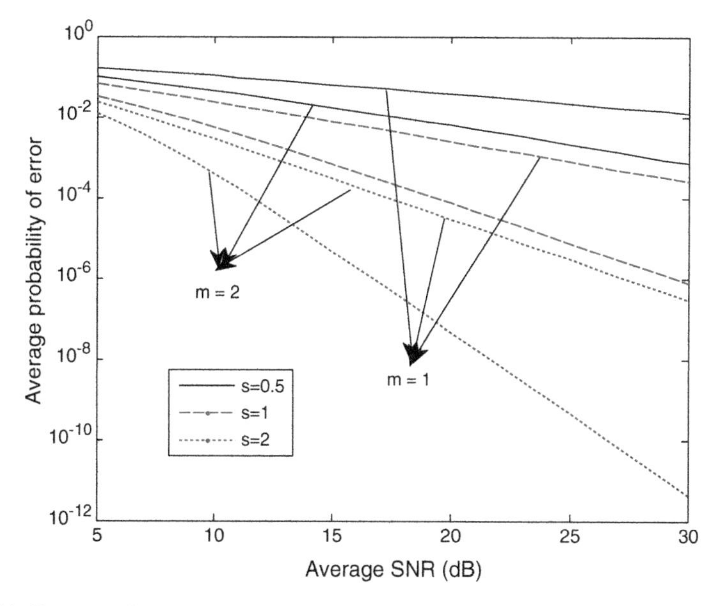

Fig. 4.44 Error rates in a generalized gamma channel

$$p_{\mathrm{e}}(Z_0) = \int_0^{\infty} \frac{1}{2}\operatorname{erfc}\left(\sqrt{z}\right)\frac{sz^{ms-1}}{\Gamma(m)Z_{\mathrm{g}}^{m}}\exp\left(-\frac{z^{s}}{Z_{\mathrm{g}}}\right)\,\mathrm{d}z, \tag{4.145}$$

where

$$Z_{\mathrm{g}} = \left[Z_0\frac{\Gamma(m)}{\Gamma(m+(1/s))}\right]^{s}. \tag{4.146}$$

The results are shown in Fig. 4.44. As expected, higher values of s along with higher values of m result in lower error rates.

Error rates in a Weibull channel are shown in Fig. 4.45. The trends in error rates in Weibull channels follow the patterns observed with the Nakagami and generalized gamma fading channels.

4.8.1.1 Error Rates in Cascaded Channels

As discussed earlier, short-term fading is also modeled using cascaded channels. The average error probability in a cascaded channel can be expressed using (4.133). The average error probability becomes

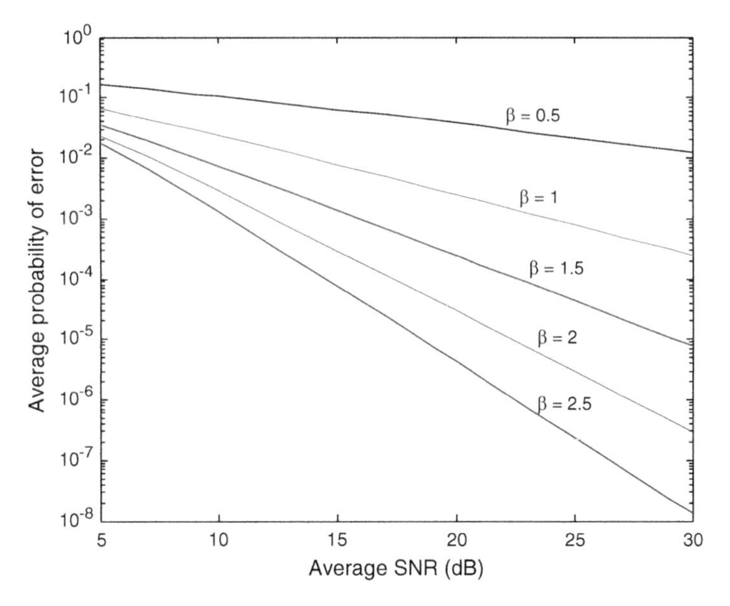

Fig. 4.45 Average error rates in Weibull channels

$$p_{\mathrm{e}}(Z_0) = \int_0^{\infty} Q\left(\sqrt{2z}\right)\frac{1}{z\Gamma^N(m)} G_{0,N}^{N,0}\left(\frac{m^N}{Z_{0N}}z \middle| \underbrace{m, m, \ldots, m}_{N-\text{terms}}.\right) \mathrm{d}z. \tag{4.147}$$

Using the table of integrals (Gradshteyn and Ryzhik 2007; Wolfram 2011), (4.147) becomes

$$p_{\mathrm{e}}(Z_0) = \frac{1}{2} - \frac{1}{2\Gamma^N(m)\sqrt{\pi}} G_{2,N+1}^{N+1,1}\left(\frac{m^N}{Z_{0N}} \middle| \begin{matrix} \frac{1}{2}, 1 \\ 0, \underbrace{m, m, \ldots, m}_{N-\text{terms}} \end{matrix}\right). \tag{4.148}$$

The error rates in cascaded channels are plotted in Figs. 4.46, 4.47, and 4.48. In Fig. 4.46, we see the effect of increasing values of the cascading components N for $m = 1$.

The error rates go up as N increases, leading to higher and higher values of excess SNR needed to mitigate the effects of short-term fading. Slight improvement is seen in Fig. 4.47 for $m = 2.5$ which is expected since the value of the Nakagami parameter m has gone up. The effect of increasing values of m on error rates is depicted in Fig. 4.48 for the case of $N = 4$. The error rates move lower and lower as m increases.

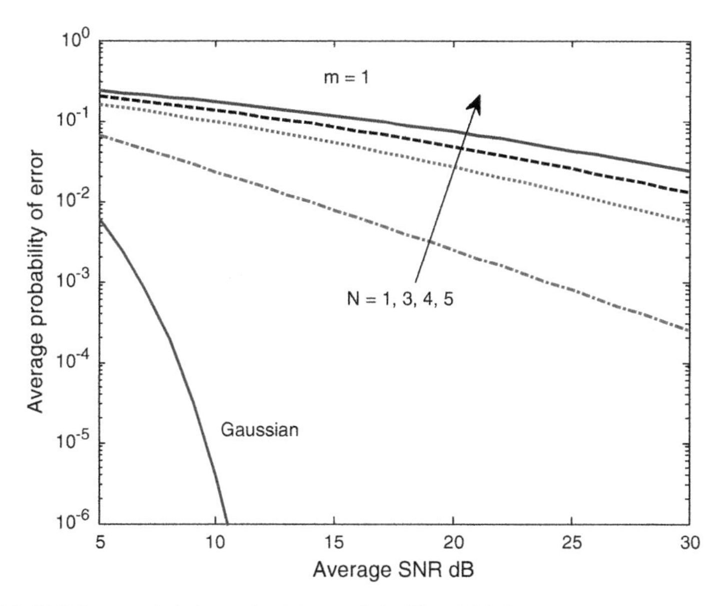

Fig. 4.46 BER in cascaded channels with $m = 1$ for $N = 1,3,4,5$

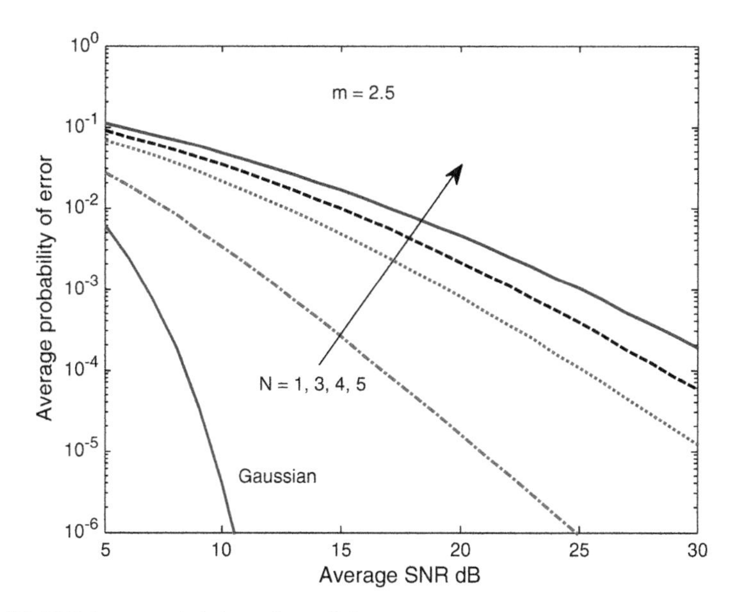

Fig. 4.47 BER in a cascaded channel $m = 2.5$

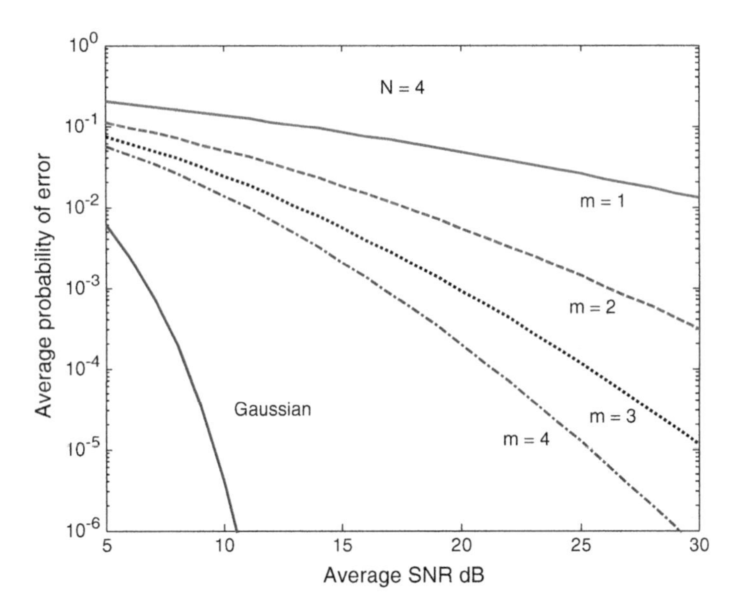

Fig. 4.48 BER in a cascaded channel $N = 4$ for multiple values of m

4.8.1.2 Error Rates in Shadowed Fading Channels

Note that shadowing never occurs alone; it occurs concurrently with fading. We need to examine the effects of shadowed fading channels (Patzold 2002; Kostic 2005), and will now undertake this task by looking at the shadowed fading channels modeled using the cascaded approach. We can now look at the estimation of the average bit error rate in shadowed fading channels. First, we will examine the error rates in shadowed fading channels using the exact models. From (4.76), we can write the expression for the error rate as

$$\begin{aligned} p_{\mathrm{e}}(\mu) = \quad & 0 \int_0^{\infty} \int_0^{\infty} \frac{m^m z^{m-1}}{x^m \Gamma(m)} \exp\left(-m\frac{z}{x}\right) \frac{A_0}{\sqrt{2\pi 2\sigma_{\mathrm{dB}}^2 x^2}} \exp\left[-\frac{\left(10\log_{10} x - \mu\right)^2}{2\sigma_{\mathrm{dB}}^2}\right] \\ & \times \frac{1}{2} \operatorname{erfc}\left(\sqrt{z}\right) \, \mathrm{d}x \, \mathrm{d}z. \end{aligned} \tag{4.149}$$

The average error rate in (4.149) must be evaluated using numerical integration. The bit error rates for BPSK in shadowed fading channels are shown in Figs. 4.49, 4.50, and 4.52.

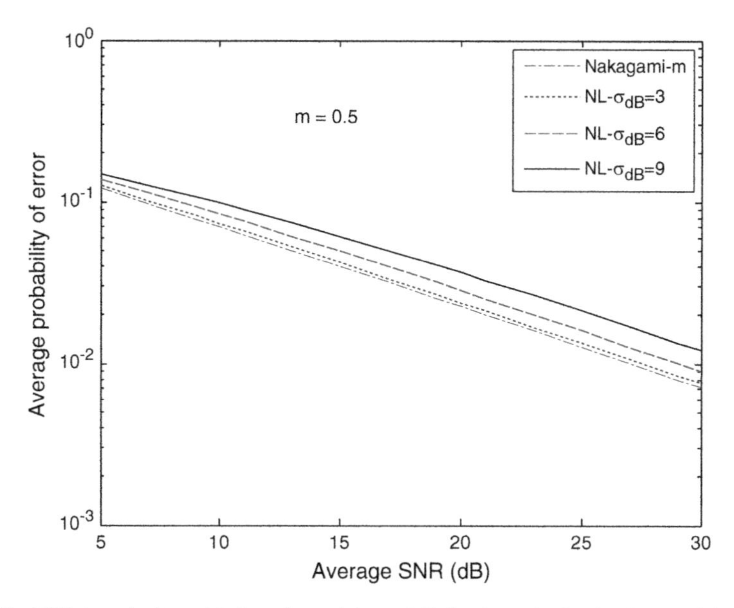

Fig. 4.49 BER in a shadowed fading channel ($m = 0.5$) for three shadowing levels, light (3 dB), moderate (6 dB), and high (9 dB)

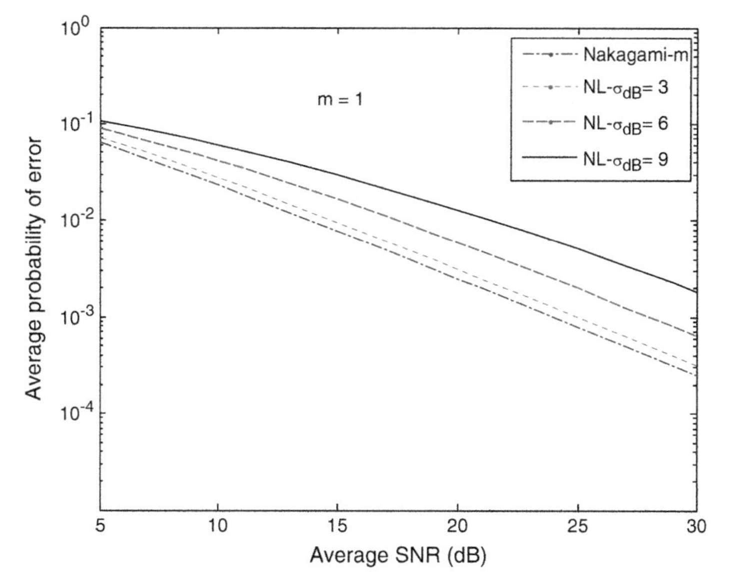

Fig. 4.50 BER in a shadowed fading channel ($m = 1$) for three shadowing levels, light (3 dB), moderate (6 dB), and high (9 dB)

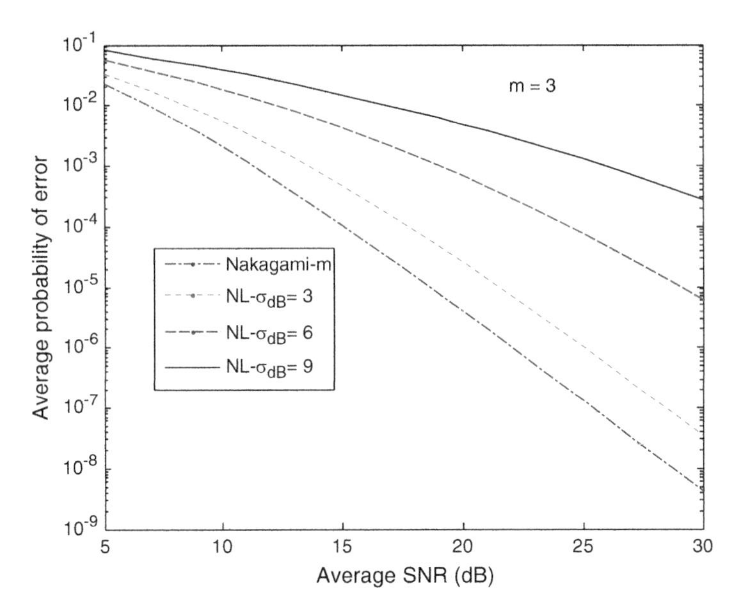

Fig. 4.51 BER in a shadowed fading channel ($m = 3$) for three shadowing levels, light (3 dB), moderate (6 dB), and high (9 dB)

The deleterious effects of the presence of shadowing are seen in all these figures. The error rates stay at unusually high values. The error rates appear high enough rendering transmission of signals difficult since the excess SNR required to mitigate the effects of both fading and shadowing will be high. As the shadowing levels become weak, the error rates curves approach those of the pure short-term fading channels. However, as seen in Fig. 4.51 when the value of m is large ($m = 3$), even weak shadowing shows up at substantially higher error rates compared to those in a pure short-term faded channels. A composite picture of the error rates is shown in Fig. 4.52 where the error rates are plotted for a fixed value of the average SNR as a function of the shadowing levels. The error rates stay at reasonably high values which points to the need for fading and shadowing mitigation techniques such as diversity combining (discussed in Chap. 5).

We can also examine the error rates in shadowed fading channels using the GK model or the Nakagami-N-gamma model discussed earlier in this chapter.

We had showed that we can replace the lognormal pdf with the gamma pdf. In a more general approach, it was shown that a cascaded gamma pdf allows a better match with the Nakagami-lognormal model for the shadowed fading channel. We will now use the Nakagami-N-gamma model to estimate the error rates in shadowed fading channels and examine the fit to the Nakagami-lognormal, varying the number of cascaded components from N to 1 through higher integer values.

As discussed earlier, the average BER in a shadowed fading channel can be expressed either using the pdf or CDF. We will reproduce the error rate estimation in fading channel as

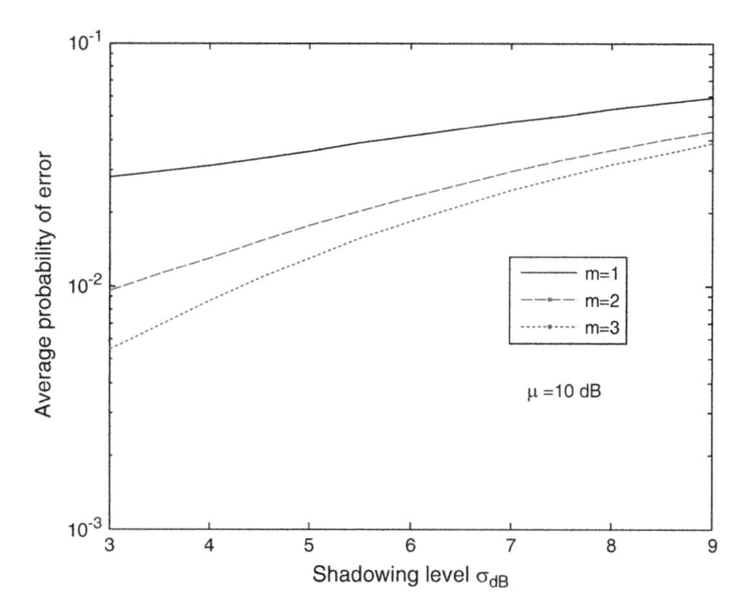

Fig. 4.52 BER in a Nakagami-lognormal for $m = 10$ dB as a function of shadowing levels for three levels of fading

$$p_\mathrm{e}(Z_0) = \int_0^\infty f(x) p_\mathrm{e}(x)\,\mathrm{d}x = -\int_0^\infty F(x)\frac{d[p_\mathrm{e}(x)]}{\mathrm{d}x}\mathrm{d}x. \tag{4.150}$$

Using the CDF in Eq. (4.130) and the derivative of erfc(), (4.150) becomes

$$p_\mathrm{e}(Z_0) = \int_0^\infty \frac{1}{\Gamma(m)\Gamma^N(c_N)} G_{1,N+2}^{N+1,1}\left(\frac{c_N^N mz}{Z_0}\left[\underbrace{m, c_N, c_N, \ldots, c_N, 0}_{N\text{ terms}}^{\;1}\right]\right)\frac{\exp(-z)}{2\sqrt{\pi z}}\mathrm{d}z. \tag{4.151}$$

As mentioned earlier, using the table of integrals (Gradshteyn and Ryzhik 2007), a closed form solution to (4.151) can be obtained and the average error probability becomes

$$p_\mathrm{e}(Z_0) = \frac{1}{2\sqrt{\pi}}\frac{1}{\Gamma\left(m\Gamma^N(c_N)\right)} G_{2,N+2}^{N+1,2}\left(\frac{mc_N^N}{Z_0}\left|\begin{matrix}\frac{1}{2},1\\ \underbrace{c_N, c_N, \ldots, c_N, m, 0}_{N\text{ terms}}\end{matrix}\right.\right). \tag{4.152}$$

For the case of $N = 1$, the average bit error rate can be expressed in terms of hypergeometric functions as (Gradshteyn and Ryzhik 2007)

$$
\begin{aligned}
P_{\mathrm{e}}(Z_0) = & \frac{1}{2} - \frac{1}{2}\frac{\pi^2 \csc(\pi m)\csc(\pi c1)}{\Gamma(1-m)\Gamma(1-c_1)\Gamma(m)\Gamma(c_1)} \\
& + \frac{1}{2\sqrt{\pi}} \frac{\Gamma(m-c_1)\Gamma(c_1+1/2)\left(\frac{mc_1}{Z_0}\right)c_1}{\Gamma(m)\Gamma(c_1+1)} \\
& \times {}_2F_2\left(\left[c_1, c_1+\frac{1}{2}\right], [1+c_1, 1-m+c_1], \frac{mc_1}{Z_0}\right) \\
& + \frac{1}{2\sqrt{\pi}} \frac{\Gamma(c_1-m)\Gamma(m+1/2)(mc_1/Z_0)^m}{\Gamma(m)\Gamma(c_1)} \\
& \times {}_2F_2\left(\left[m, m+\frac{1}{2}\right], [1+m, 1-c_1+m], \frac{mc_1}{Z_0}\right).
\end{aligned}
\tag{4.153}
$$

In (4.153), $F(.)$ is the hypergeometric function and csc(.) is the trigonometric function cosecant(.). Note also that $N = 1$ corresponds to the Nakagami-gamma model which results in the generalized K distribution for the shadowed fading channels. The average bit error rates are plotted in Fig. 4.53 for the case of moderate level of shadowing (5 dB). For low values of m (0.5 and 1), the error rates estimated using the Nakagami-lognormal match very well with the Nakagami-N-gamma model for all values of N; there is no need to use higher values of N. The error rates are plotted in Fig. 4.54 for the case of a higher level of shadowing (8 dB). In this case, for the low value of m, there is very little difference between the error

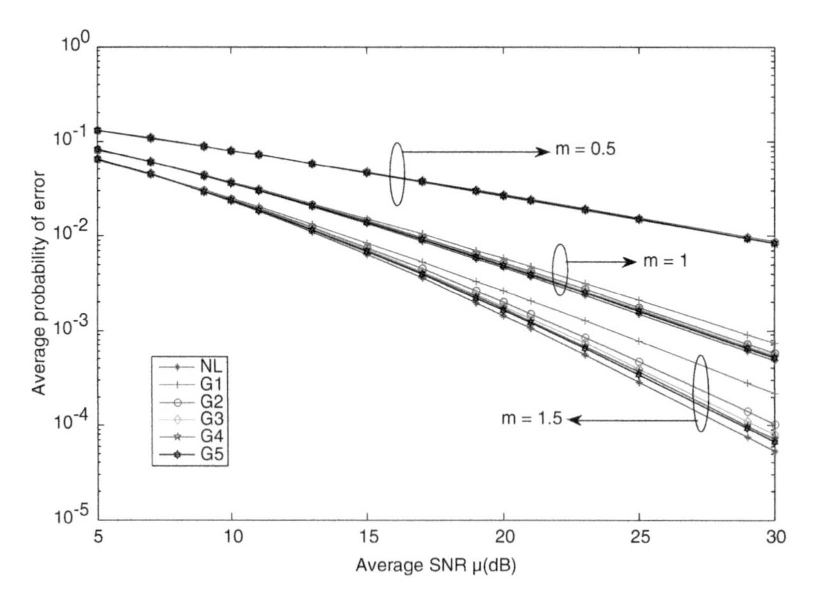

Fig. 4.53 Bit error rates in shadowed fading channels. For $m = 0.5$, 1 and 1.5 for the case of a moderately shadowed channel ($\sigma_{\mathrm{dB}} = 5$)

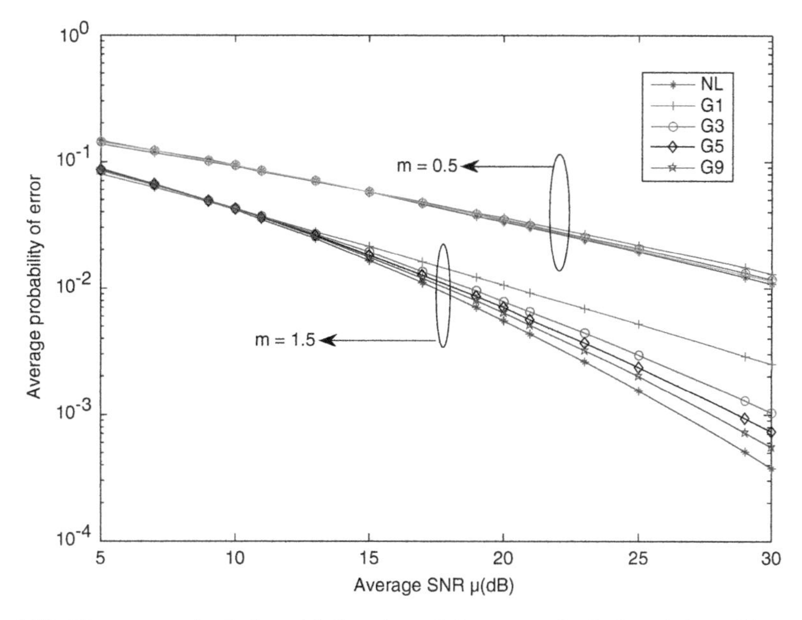

Fig. 4.54 Bit error rates in shadowed fading channels for a severely shadowed channel ($\sigma_{\mathrm{dB}} = 8$)

rates. On the other hand, as the value of m goes to 1.5, the case of $N = 1$ is a weaker match and the match with the Nakagami-lognormal error rates improves as N goes up to 9.

One can also observe the match of the Nakagami-lognormal pdf and the Nakagami-N-gamma pdf based error rates as one looks at higher values of the Nakagami parameter. The error rates for $m = 2$ and $m = 3$ are shown in Fig. 4.55 for the case of moderate shadowing ($\sigma_{\mathrm{dB}} = 5$). It can be seen that the fit of the Nakagami-gamma pdf ($N = 1$) is very poor.

Error probabilities offer one quantitative measure of the performance in wireless fading channels including shadowed fading channels. Even though only the case of a coherent BPSK modem was used, the study could be extended to other modems as well. This would simply mean replacing the error rate in the ideal Gaussian channel with the appropriate equation for the error rate for a particular modem. Even though individual performance would depend on the modem, the general trends seen with coherent BPSK here will be exhibited by all types of modems such as the increase in error rates with increasing levels of fading. The existence of shadowing leads to further erosion in the performance in the wireless channel.

4.8.2 Outage Probability

While error rate performance provides us with a means to compare the models in terms of the excess SNR required to maintain a specific bit error rate, there is yet

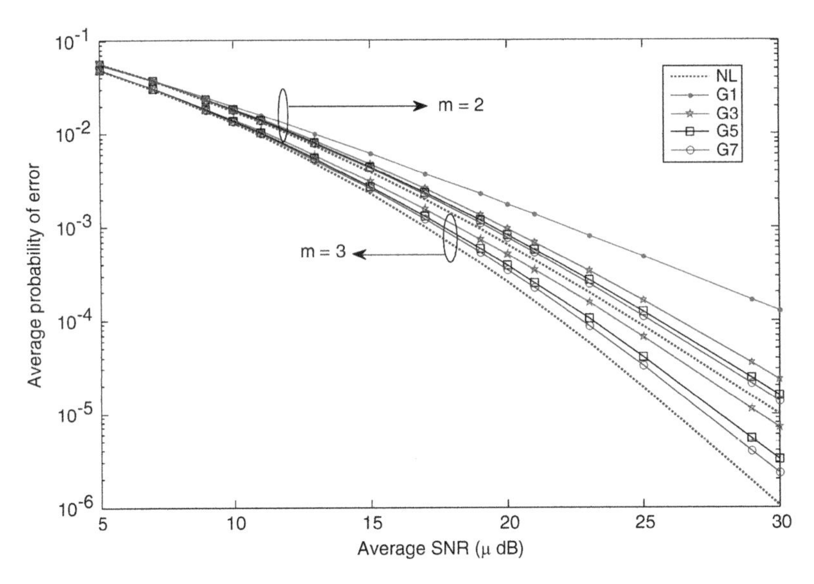

Fig. 4.55 Bit error rates in shadowed fading channels for $m = 2$ and 3 in the presence of moderate shadowing ($\sigma_{\mathrm{dB}} = 5$)

another means to quantify the performance of the wireless communication systems in different channels as mentioned earlier (Simon and Alouini 2005; Tellambura and Annamalai 1999). The models can be compared in terms of the outage probability. This is the probability that the signal-to-noise ratio fails to reach a certain threshold to maintain a specific error rate. One can see the difference in measures obtained through error rates and outage probabilities. The former provides a means to see how much additional SNR is required to maintain the specific error rate as the fading conditions change; the latter provides a measure of the ability to keep the error rate at a specified value (Abu-Dayya and Beaulieu 1994; Annamalai et al. 2001, 2005; Shankar 2005).

Whenever the signal power goes below a set threshold, which depends on the data rate, coding, modulation, demodulation, and so forth, the channel goes into outage. The outage probability associated with fading, shadowing, or shadowed fading can be expressed as (Simon and Alouini 2005; Shankar 2005)

$$P_{\mathrm{out}} = \int_0^{Z_{\mathrm{T}}} f(z)\,\mathrm{d}z = F(Z_{\mathrm{T}}), \tag{4.154}$$

where *ZT* is the threshold SNR. The density function of the SNR is $f(z)$ and $F(.)$ is the CDF of the SNR evaluated at $z =$ *ZT*. We will use coherent BPSK as an example and we will assume a required performance level of error rates no larger than 1e-4. The threshold SNR now becomes

$$10^{-4} = \frac{1}{2}\text{erfc}\left(\sqrt{Z_T}\right). \quad (4.155)$$

Taking the inverse, we have

$$Z_{\text{T}} = \left[\text{erfc}\,\text{inv}\left(2.10^{-4}\right)\right]^2 = 6.9155. \quad (4.156)$$

In this chapter, we will use (4.156) as our standard to compare the outage probabilities even though we could use any other modem as an example to determine the threshold SNR for acceptable operation.

We first look at the outage probabilities in a Nakagami-*m* faded channel. The outage probability can be expressed using the CDF obtained in a Chap. 2 equation as

$$P_{\text{out}} = 1 - \frac{\Gamma(m, (m/Z_0)Z_{\text{T}})}{T(m)}. \quad (4.157)$$

The outage probabilities are plotted against the average SNR in Fig. 4.56. One can see the unacceptably high values of the outage probabilities at low values of the average SNR, a consequence of the fading, low values resulting in the highest values of the outage probabilities. Figure 4.57 shows the outage probabilities as a function of the Nakagami parameter for three values of the average signal-to-noise ratio.

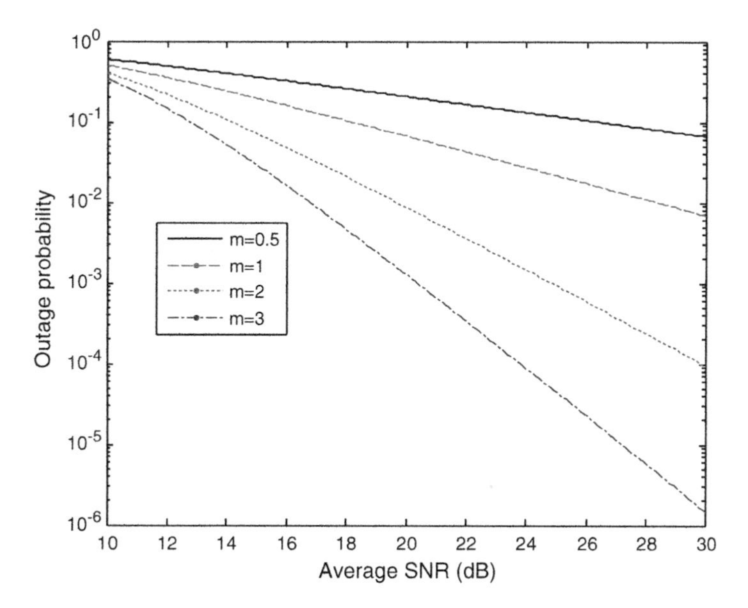

Fig. 4.56 Outage probabilities in a Nakagami-*m* faded channel

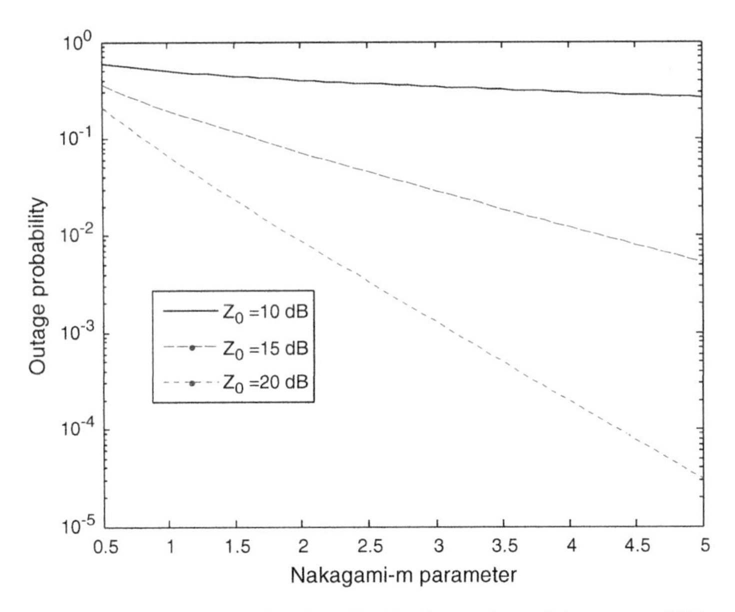

Fig. 4.57 Outage probabilities as a function of m for three values of the average SNR

The outage probabilities in a generalized gamma channel can be obtained from the CDF of the SNR given in (4.54). The outage probabilities corresponding to an error rate of 1e-4 in a coherent BPSK modem are shown in Fig. 4.58. As the value of s increases, the effect of fading declines as seen from declining values of the outage probabilities.

The outage probabilities in a generalized Gaussian channel are shown as a function of the shape parameter s in Fig. 4.59. It can be seen that low values of s correspond to high values of outage probabilities; the outage probabilities come down with increasing values of the shape parameter.

The outage probabilities in a Weibull channel are obtained from the CDF of the Weibull variable in (4.62). Figure 4.60 shows the outage probabilities in a Weibull channel. The outage probabilities as a function of the Weibull parameter are shown in Fig. 4.61 for an average SNR of 20 dB.

The outage probabilities in a channel that undergo lognormal shadowing can be expressed in terms of the Gaussian CDF. They will not be discussed here. The properties of the Gaussian CDF are very well understood; the outage probabilities will increase with the shadowing levels and decrease with increasing values of the mean SNR.

Outage probabilities in channels modeled using a cascaded approach can be estimated from the CDF of the SNR in a cascaded channel. The CDF given in (4.116) and the outage probability becomes

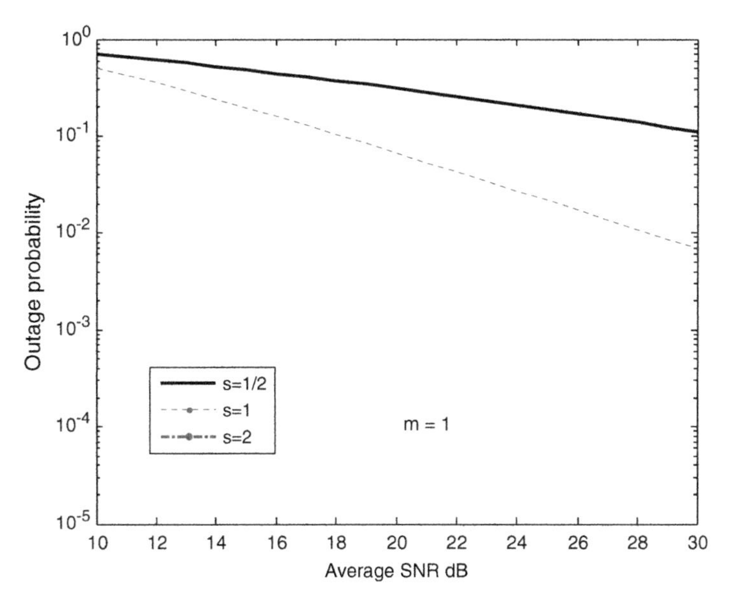

Fig. 4.58 Outage probabilities in a generalized gamma fading channel

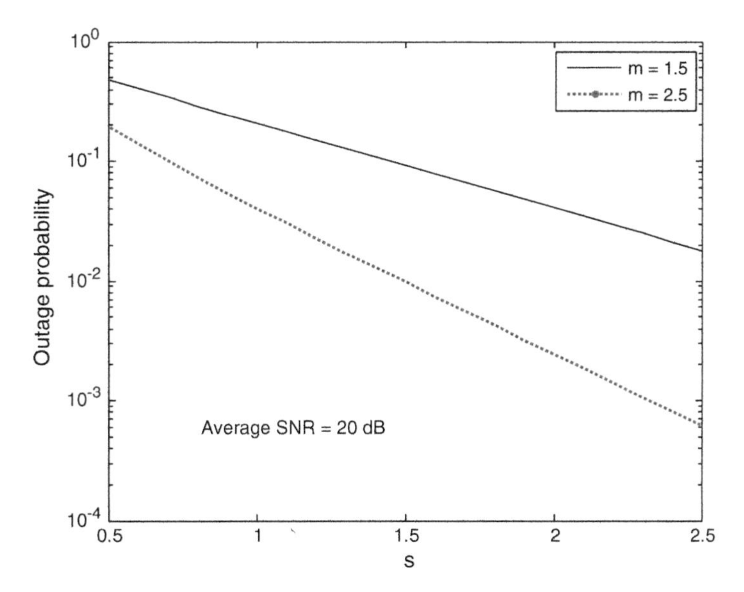

Fig. 4.59 Outage probabilities in a generalized gamma channel: variation with the shape parameter s for an average SNR of 20 dB

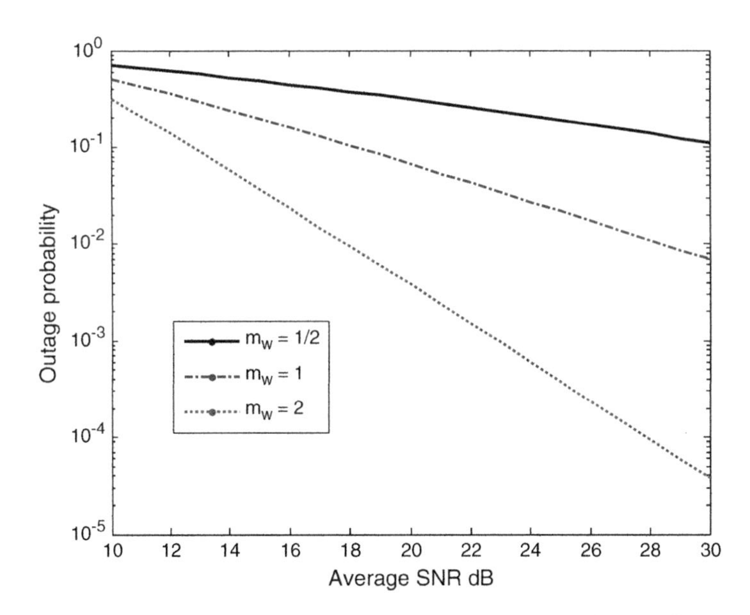

Fig. 4.60 Outage probabilities in Weibull channels for three values of the Weibull parameter m_w

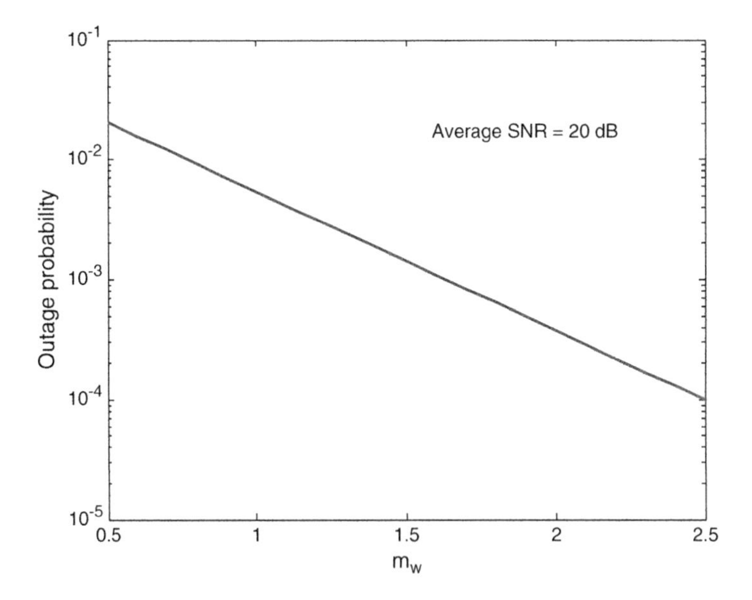

Fig. 4.61 Outage probabilities in Weibull channels (average SNR of 20 dB) as a function of parameter m_w

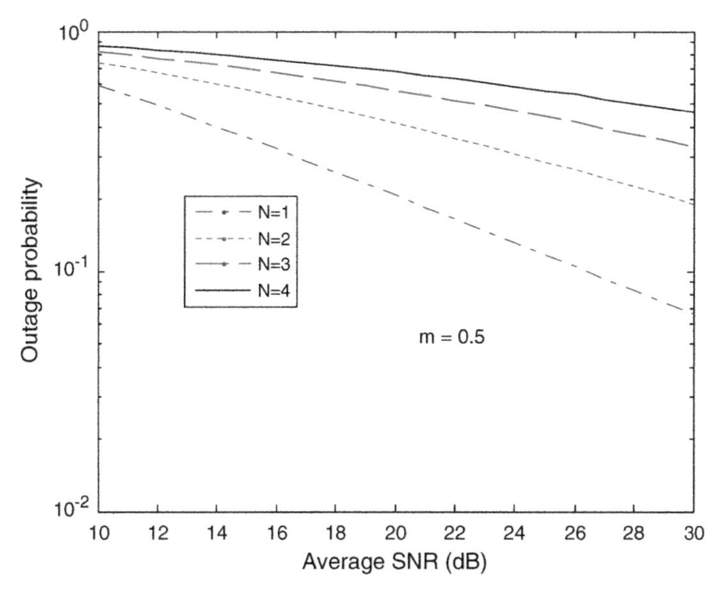

Fig. 4.62 Outage probability in cascaded N*Nakagami channels for $m = 0.5$

$$P_{\text{out}} = \frac{1}{\Gamma^N(m)} G_{1,N+1}^{N,1}\left(\frac{m^N Z_{\text{T}}}{Z_0}\left[\underbrace{m, m, \overset{1}{\ldots}, m, 0}_{N\ \text{trrms}}\right]\right). \tag{4.158}$$

The outage probabilities have been evaluated for a few values of N and are shown in Figs. 4.62, 4.63, and 4.64, respectively, for $m = 0.5$, 1, and 2.

One can see that the outage probabilities rise with increasing values of N and decline with increasing values of m.

4.8.2.1 Outage Probabilities in Shadowed Fading Channels

The presence of both fading and shadowing simultaneously (i.e., existence of the shadowed fading channels) will have serious adverse consequences in communications. We will first examine the outage probabilities in a Nakagami-lognormal channel before we examine the outage probabilities. We will use approximate models of the shadowed fading channels which employ the cascaded approach (with $N = 1$ giving rise to the GK channels). The outage probability in a Nakagami-lognormal channel is

$$P_{\text{out}} = \int_{z=0}^{Z_{\text{T}}} \int_0^{\infty} \frac{m^m z^{m-1}}{x^m \Gamma(m)} \exp\left(-m\frac{z}{x}\right) \frac{A_0}{\sqrt{2\pi\sigma_{\text{dB}}^2 x^2}} \exp\left[-\frac{\left(10\log_{10}x - \mu\right)^2}{2\sigma_{\text{dB}}^2}\right] \mathrm{d}x\, \mathrm{d}z. \tag{4.159}$$

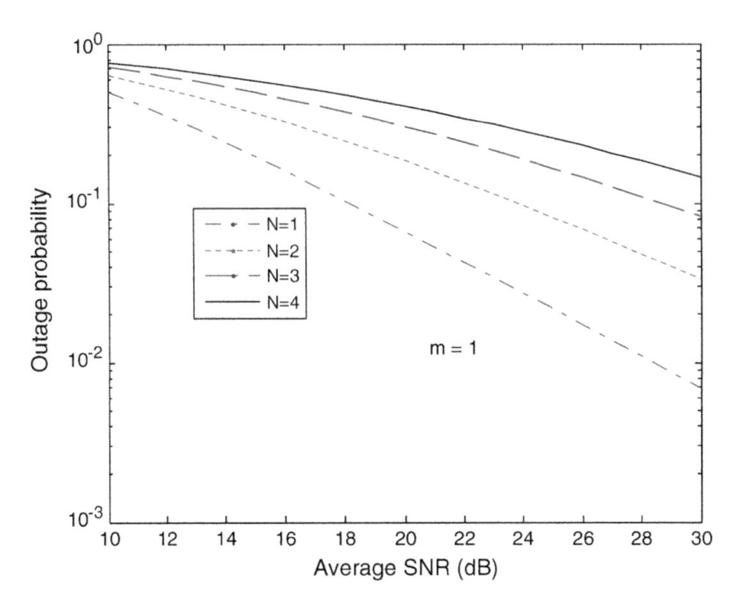

Fig. 4.63 Outage probability in cascaded N*Nakagami channels for $m = 1$

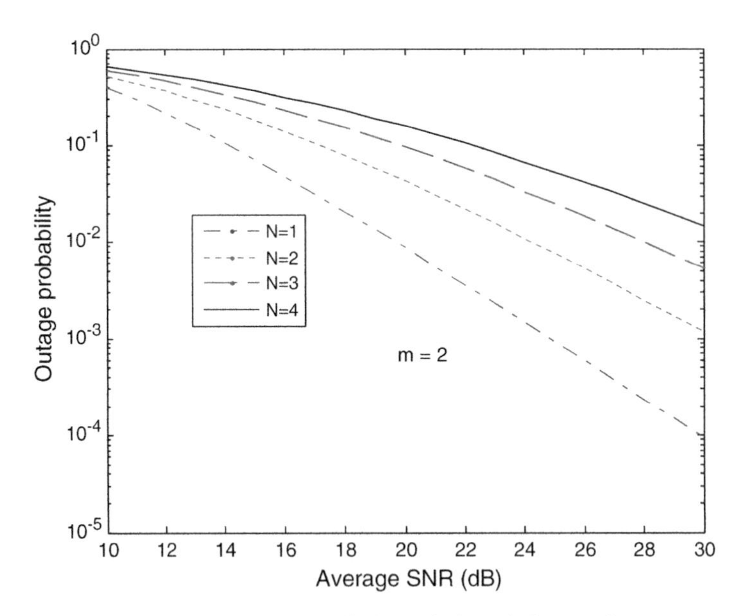

Fig. 4.64 Outage probability in cascaded N*Nakagami channels for $m = 2$

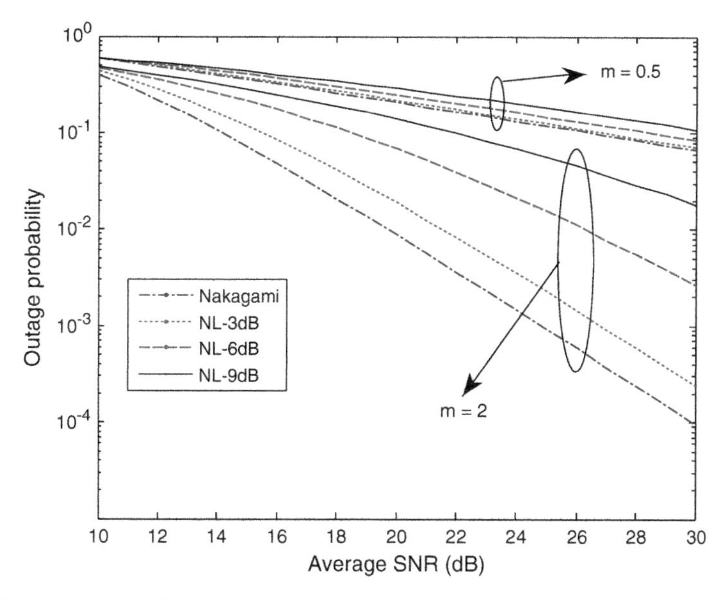

Fig. 4.65 Outage probabilities as a function of the average SNR for two values of the Nakagami parameter and three values of the shadowing levels

This integral needs to be evaluated numerically. The results are shown in Figs. 4.65 and 4.66. In Fig. 4.65, the outage probabilities are plotted as a function of the average SNR; in Fig. 4.66, the outage probabilities are plotted as a function of the shadowing levels. In both figures, one can easily observe the deleterious effects of shadowing leading to increased values of outage, therefore, making it necessary to explore diversity implementation.

4.8.2.2 Outage Probabilities in Shadowed Fading Channels Using the Cascaded Model

We can now examine the outage probabilities in shadowed fading channels modeled using the cascaded approach. The outage probability can be obtained from the CDF in a Nakagami cascaded gamma channel as (Shankar 2011a, b)

$$P_{\text{out}} = \frac{1}{\Gamma(m)\Gamma^{N}(c_N)} G_{1,N+2}^{N+1,1}\left(\frac{m_N^N z}{Z_0} \left[\underbrace{m, c_N, c_N, \ldots, c_N, 0}_{N \text{ trrms}}^{1} \right] \right). \tag{4.160}$$

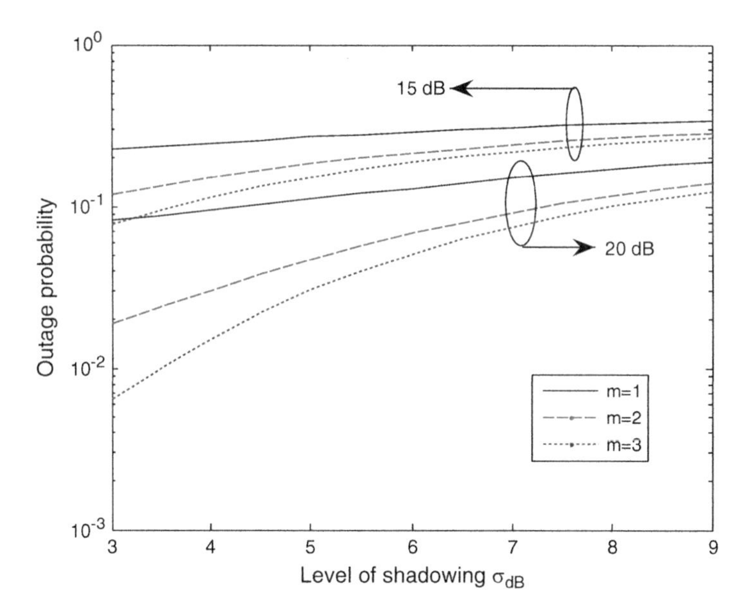

Fig. 4.66 Outage probabilities in Nakagami-lognormal channels as a function of sigma for two values of the average SNR

The outage probability is

$$P_{\text{out}} = \frac{1}{\Gamma(m)\Gamma^N(c_N)} G_{1,N+2}^{N+1,1}\left(\frac{m c_N^N Z_{\text{T}}}{Z_0} \left[\underbrace{m, c_N, c_N, \overset{1}{\ldots}, c_N, 0}_{N \text{ trms}} \right] \right). \tag{4.161}$$

Some of the results are shown in Figs. 4.67 and 4.68. Figure 4.67 shows the outage probabilities at a moderate level of fading (5 dB) at three values of the Nakagami parameter. As the value of the Nakagami parameter m increases, higher values of N are required to match the Nakagami-lognormal results to those of the Nakagami-N-gamma model. Similar observations can be made by observing the results in Fig. 4.68 for a higher level of shadowing.

Regardless of whether one uses the exact Nakagami-lognormal or the approximate Nakagami-N-gamma model for the shadowed fading channel, it is clear that the effect of shadowing is to increase the outage probabilities making it necessary that mitigation approaches to overcome both fading and shadowing need to be implemented to facilitate efficient wireless data transmission.

Table 4.1 provides earlier summaries of the most commonly used density functions for modeling short-term fading while Table 4.2 listed earlier provides the list of density functions used for modeling shadowing. Table 4.3 lists the density function used for modeling shadowed fading channels. Table 4.4 provides a list of the CDF associated with these models (only when analytical expression is available) and the expressions for the outage probabilities for the different density

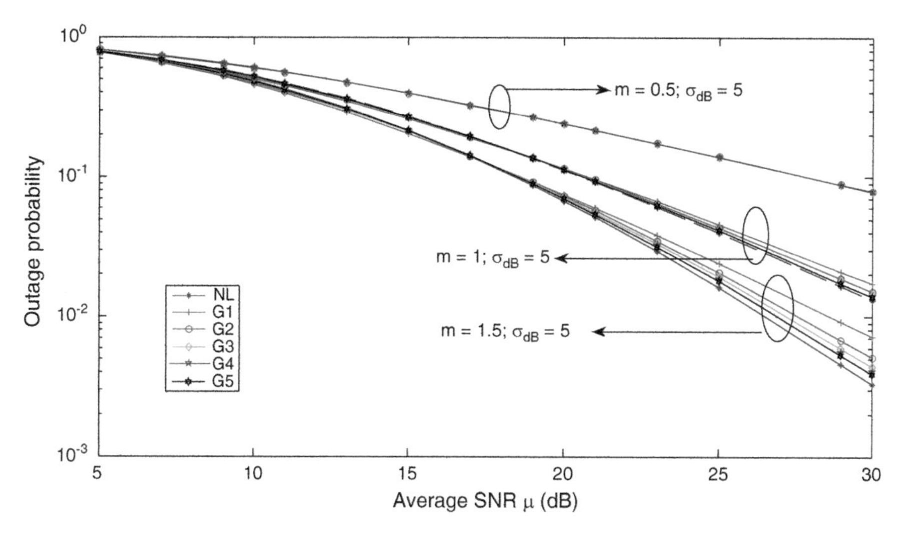

Fig. 4.67 Outage probabilities in a Nakagami-lognormal channel where the lognormal shadowing is modeled using the cascaded approach (moderate level of shadowing)

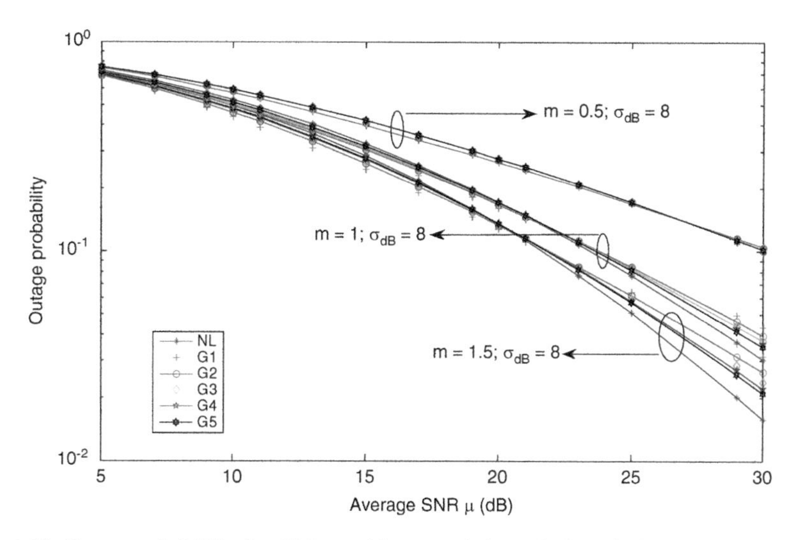

Fig. 4.68 Outage probabilities in a Nakagami-lognormal channel where the lognormal shadowing is modeled using the cascaded approach (higher level of shadowing)

functions which describe the fading and shadowed fading channels. It is seen that neither the Nakagami-lognormal nor the Nakagami-inverse Gaussian distribution leads to a closed form expression for the outage probability while all the other density functions possess an analytical expressions for the outage probabilities. The outage probability in a Nakagami-inverse Gaussian channel can be expressed as an

Table 4.3 The pdf of the received power in shadowed fading channels and the amount of fading (AF) are tabulated

Probability density function ($p \geq 0$)	Amount of fading (AF)	Additional information
$f_{\mathrm{NLM}}(p) = \int_0^\infty \frac{m^m p^{m-1}}{z^m \Gamma(m)} \exp\left(= \frac{mp}{z}\right) \frac{A_0}{\sqrt{2\pi\sigma_{\mathrm{dB}}^2 z^2}} \exp\left[-\frac{(10\log_{10} z - \mu)^2}{2\sigma_{\mathrm{dB}}^2}\right] \mathrm{d}z$	$\left(\frac{m+1}{m}\right)\exp\left(\frac{\sigma_{\mathrm{dB}}^2}{A_0^2}\right) - 1$	$A_0 = \frac{10}{\log_e(10)}$
$f_{\mathrm{GK}}(p) = \frac{2}{\Gamma(m)\Gamma(c)}\left(\frac{b}{2}\right)^{c+m} p^{((c+m)/2)-1} K_{c-m}\left(b\sqrt{p}\right)U(p)$	$\frac{1}{m} + \frac{1}{c} + \frac{1}{mc}$	$\sigma_{\mathrm{dB}}^2 = A_0^2 \psi'(c), \quad \langle P \rangle_{\mathrm{GK}} = mc\left(\frac{2}{b}\right)^2$
$f_{\mathrm{K}}(p) = \frac{2}{\Gamma(c)}\left(\frac{a}{2}\right)^{c+1} p^{((c+1)/2)-1} K_{c-1}\left(a\sqrt{p}\right)U(p)$	$1 + \frac{1}{c}$	GK pdf with $m = 1$ $\sigma_{\mathrm{dB}}^2 = A_0^2 \psi'(c), \quad \langle P \rangle_{\mathrm{GK}} = mc\left(\frac{2}{b}\right)^2$
$f_{\mathrm{DR}}(p) = 2\left(\frac{u}{2}\right)^2 K_0\left(u\sqrt{p}\right)U(p)$	3	GK pdf with $m = c = 1$ $\langle P \rangle_{\mathrm{DR}} = \left(\frac{2}{u}\right)^2$
$f_{\mathrm{DN}}(p) = \frac{2}{[\Gamma(m)]^2}\left(\frac{v}{2}\right)^{2m} p^{m-1} K_0\left(v\sqrt{p}\right)U(p)$	$\frac{2}{m} + \frac{1}{m^2}$	GK pdf with $c = m$ $\langle P \rangle_{\mathrm{DN}} = \left(\frac{2m}{v}\right)^2$
$f_{\mathrm{NdN}}(p) = \frac{2}{\Gamma(m_1)\Gamma(m_2)}\left(\frac{w}{2}\right)^{(m_1+m_2)/2} p^{((m_1+m_2)/2)-1} K_{m_1-m_2}\left(w\sqrt{p}\right)U(p)$	$\frac{1}{m_1} + \frac{1}{m_2} + \frac{1}{m_1 m^2}$	GK pdf with $m = m_1; c = m_2$ $\langle P \rangle_{\mathrm{NdN}} = m_1 m_2 \left(\frac{2}{w}\right)^2$
$f_{\mathrm{GG}}(p) = \frac{s p^{ms-1}}{\Gamma(m) P_g^m} \exp\left(-\frac{p^s}{P_g}\right)$	$\frac{\Gamma(m+(2/s))\Gamma(m)}{[\Gamma(m+(1/s))]^2} - 1$	$0 < s < 1 \quad A_0^2 \psi'(m) + \sigma_{\mathrm{dB}}^2 = \frac{A_0^2}{s^2}\psi'(m)$
$f_{\mathrm{NIG}}(p) = \left(\frac{1}{HP_0}\right)^{m+(1+2)} \sqrt{\frac{P_0}{2\pi H} \frac{4m^m \exp(1/H)}{\Gamma(m)\left[\sqrt{g(p)}\right]^{m+(1/2)}} p^{2m-1} K_{m+(1/2)}\left(\sqrt{g(p)}\right)}$	$\left(\frac{m+1}{m}\right)H - 1$	$g(p) = \frac{2}{HP_0}\left(mp^2 \frac{P_0}{2H}\right)$ $H = 2\sin h\left(\frac{\sigma_{\mathrm{dB}}^2}{2A_0^2}\right)\exp\left(\frac{\sigma_{\mathrm{dB}}^2}{2A_0^2}\right)$

Nakagami-lognormal distribution (NLN), generalized K distribution (GK), K distribution (K), double Rayleigh distribution (DR), double Nakagami distribution (DN), non-identical double Nakagami distribution (NdN), generalized gamma distribution (GG), and Nakagami-inverse Gaussian distribution (NIG)

Table 4.4 The outage probabilities in fading and shadowed fading channels are tabulated

Probability density function	Outage probability (P_T is the threshold)	Additional information
Fading channels		
Rayleigh	$1 - \exp.(-P_T = P)$	
Nakagami	$\gamma\left(m, \frac{mP_T}{P_0}\right)[\Gamma(m)]^{-1}, \quad m \geq \frac{1}{2}$	$\gamma(.,.)$ is the incomplete gamma function
Gamma	$\gamma\left(m, \frac{mP_T}{P_g}\right)[\Gamma(m)]^{-1}, \quad m > 0$	$\gamma(.,.)$ is the incomplete gamma function
Generalized gamma	$\gamma\left(m, \frac{P_T^S}{P_g}\right)[\Gamma(m)]^{-1}$	$\gamma(.,.)$ is the incomplete gamma function
Weibull	$1 - \exp\left(-P_T^{m_w}/P_g\right), \quad m_w > 0$	
Shadowed fading channels		
Nakagami-lognormal	...	No analytical expression
Generalized K	$\frac{\Gamma(m-c)}{\Gamma(m)\Gamma(c+1)} \times {}_1F_2\left(c, [1-m+c, 1+c], \frac{P_T b^2}{4}\right)\left(\frac{P_T b^2}{4}\right)^c + \frac{\Gamma(m-c)}{\Gamma(m)\Gamma(c+1)} \times {}_1F_2\left(m, [1-c+m, 1+m], \frac{P_T b^2}{4}\right)\left(\frac{P_T b^2}{4}\right)^m$	${}_1F_2(., [.,.],.)$ is the hypergeometric function
Generalized gamma	$\gamma\left(m, \frac{P_T^S}{P_g}\right)[\Gamma(m)]^{-1}$	
Double Rayleigh	$\left[1 - 2u\sqrt{\frac{P_T}{P_0}} K_1\left(u\sqrt{\frac{P_T}{P_0}}\right)\right]$	$K_1()$ is the modified Bessel function of order 1
Double Nakagami	$[\Gamma(m)]^{-2} G_{1,3}^{2,1}\left(\frac{1}{4}P_T v^2 \middle\| \begin{matrix} 1 \\ m, m, 0 \end{matrix}\right)$	G is the Meijer G-function
K distribution	$1 - \frac{2}{\Gamma(c)}\left(a\sqrt{\frac{P_T}{2}}\right)^c K_c\left(a\sqrt{P_T}\right)$	$K_c()$ is the modified Bessel function of order c

Additional information on the special functions is also provided

infinite sum. For the case of the pdfs containing the modified Bessel function $K()$, the outage probabilities are in terms of the modified Bessel, Meijer G, and hypergeometric functions.

The models described in the previous sections do not constitute the whole ensemble of models available in literature to describe the statistical fluctuations in signals in fading or shadowed fading channels. Such models include $\alpha - \lambda - \mu$ and $\alpha - \eta - \mu$ distributions for fading. They provide yet another way to look at the characteristics of the signals in wireless channels (Yacoub 2000, 2007a, b; Filho and Yacoub 2005a; Papazafeiropoulos and Kotsopoulos 2011a, b).

4.9 Other General Fading Models

We have examined some of the most commonly used statistical descriptions of fading as described in the previous sections. However, there are still a few other models that researchers have proposed which encompass almost all the common models for small scale fading or short-term fading. The so-called unified model for small scale fading includes the $\alpha - \lambda - \mu$ and $\alpha - \eta - \mu$ distributions or the $\kappa - \mu$ and $\eta - \mu$ distributions (Yacoub 2007a, b; Papazafeiropoulos and Kotsopoulos 2011a, b). The fundamental basis for these models is a different interpretation of the phenomenon of fading itself. While traditional Rayleigh and Rician and other such models assume that the received radio frequency signal can be expressed as a vector sum of scattered/reflected/refracted components coming from individual obstacles, the unified models assume that the received signal power arises from the obstacles which are modeled as scattering clusters. Thus in the former case, a direct application of the central limit theorem is possible to get the inphase and quadrature components to be independent and identically distributed Gaussian random variables if sufficient numbers of individual obstacles exist in the channel (between the transmitter and receiver resulting in the multipath scenario). The latter case of clustered scattering can lead to more complicated and diverse statistical description of the short-term fading in the channel. In other words, a homogenous and diffuse scattering from randomly distributed point-like scatterers is essential for the successful modeling leading to Rayleigh and other related fading channels. In reality, the channel is more likely to be heterogeneous with finite size scatterers (buildings). The number of such obstacles is likely to be much smaller in number than what is required for the central limit theorem to be met. It is also possible that each of these non-point like obstacles themselves is made up of several randomly located scatterers. Thus, we can consider a scenario where central limit theorem might be applicable within each cluster. Because of the distribution of these non-point targets, the received signal power can be written as the sum of the powers of the components coming from each of these objects or clusters. The addition of powers is a reasonable assumption considering the fact that the differential delays among these clusters might be large enough to make them completely uncorrelated even though the differential delays within a cluster among the scatters within a cluster might be very small and vectorial addition is likely within each cluster. In other words, a form of nonlinear behavior exists along with the heterogeneity in the wireless channel. Since the traditional models of short-term fading were explored in detail, we will now look at this different paradigm of clustered scattering and the ensuing statistical models for fading. As we will demonstrate, the new approach will provide the Rayleigh, Rician, and Nakagami channels and more as special cases. Thus, the clustered scattering based modeling can expand the descriptions of scattering, allowing a very wide range of fading conditions to be accurately modeled (Yacoub 2007a, b; Karadimas and Kotsopoulos 2010; Papazafeiropoulos and Kotsopoulos 2011a, b).

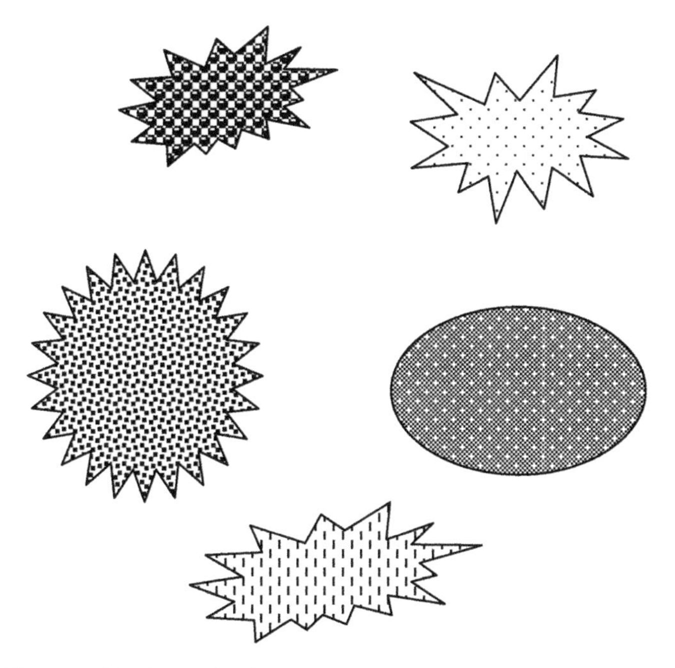

Fig. 4.69 Cluster model of scattering for a generalized model

The cluster based scattering model is shown in Fig. 4.69 (Yacoub 2007a, b). A few clusters from which the scattering/reflection, refraction, and so forth can take place are shown.

It is assumed that each cluster has enough scatterers so that the inphase and quadrature component of the RF signal can be treated as Gaussian random variables X and Y. We also assume that they are independent. If A is the envelope of the signal, the power or SNR at the receiver is Z, we can write

$$Z = A^2 = \sum_{k=0}^{u} X_k^2 + \sum_{k=0}^{u} Y_k^2. \tag{4.162}$$

In (4.162), u is the number of scattering clusters between the transmitter and receiver. From this stage onward, there are a couple of different ways of proceeding. We will take two separate steps and then provide a much more general way of looking at (4.162). First, we assume that X and Y are zero mean variables but have unequal variances, with σ_x and σ_y as the respective standard deviations. Defining the ratio of variances of the inphase and quadrature components as

$$\eta = \frac{\sigma_x^2}{\sigma_y^2}, \tag{4.163}$$

the pdf of the power at the receiver can be expressed as

$$f(z) = \frac{2\sqrt{\pi}}{\Gamma(\mu)\vartheta^{\mu-(1/2)}}\mu^{\mu-(1/2)}\upsilon^{\mu}z^{\mu-(1/2)}\exp(-2\mu\upsilon z)I_{\mu-(1/2)}(2\mu\vartheta z)U(z). \quad (4.164)$$

In (4.164),

$$\upsilon = \frac{1}{4}\left(2+\eta+\frac{1}{\eta}\right) \quad (4.165)$$

$$\vartheta = \frac{1}{4}\left(\frac{1}{\eta}-\eta\right) \quad (4.166)$$

and

$$\mu = \frac{\langle Z\rangle^2}{2\left[\langle Z^2\rangle - \langle Z\rangle^2\right]}\left[1+\left(\frac{\vartheta}{\upsilon}\right)^2\right]. \quad (4.167)$$

Note that $0 < \eta < \infty$. The pdf in (4.164) is often identified as the $\eta - \mu$ density function (Yacoub 2007a, b). This density function was also obtained by Nakagami. It is often referred to as the Nakagami–Hoyt distribution when $m = 1/2$. We will return to this definition later. The $\eta - \mu$ densities are plotted in Figs. 4.70 and 4.71.

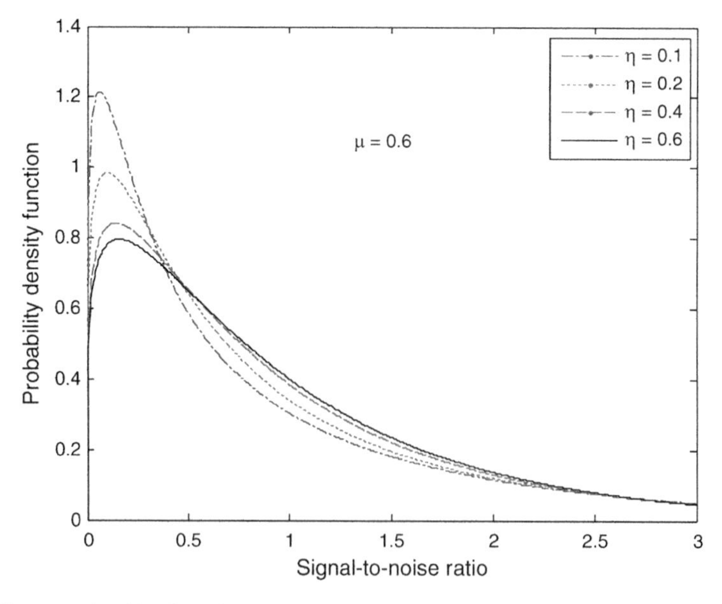

Fig. 4.70 $\eta - \mu$ densities for a fixed value of μ

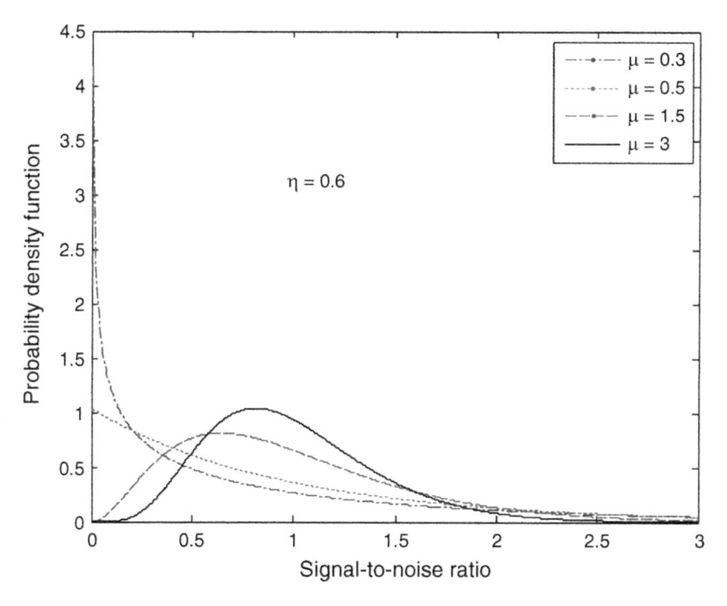

Fig. 4.71 $\eta - \mu$ densities for a fixed value of μ

It is clearly seen that by varying values of η and μ, we can cover the whole range of density functions that are available for modeling fading.

Let us now consider a case where the inphase and quadrature components are identically distributed, suggesting that $\eta = 1$. This would mean that

$$\mu = \frac{\langle Z \rangle^2}{2\left[\langle Z^2 \rangle - \langle Z \rangle^2\right]}. \tag{4.168}$$

Noting that the Nakagami parameter m in fading channels is defined as

$$m = \frac{\langle Z \rangle^2}{\langle Z^2 \rangle - \langle Z \rangle^2}, \tag{4.169}$$

we can obtain a relationship between m and μ as

$$\mu = \frac{m}{2}, \tag{4.170}$$

Making use of the approximation that the modified Bessel function $I_v\,(x)$ for $x \ll 1$ can be approximated to (Abramowitz and Segun 1972)

$$I_v(x) \approx \frac{(x/2)^v}{\Gamma(v)} \tag{4.171}$$

(4.164) becomes

$$f(z) = \frac{(2\mu)^{2\mu}}{\Gamma(2\mu)} z^{2\mu-1} \exp(-2\mu) U(z). \tag{4.172}$$

Note that (4.172) is the pdf of the SNR in Nakagami-m fading with an average SNR of unity. If Z_0 is the average SNR, (4.172) becomes

$$f(z) = \frac{(2\mu)^{2\mu}}{Z_0} \frac{z^{2\mu-1}}{\Gamma(2\mu)} \exp\left(-2\mu \frac{z}{Z_0}\right) U(z). \tag{4.173}$$

With the relationship between m and μ in (4.170), (4.173) matches exactly to the pdf of the SNR in Nakagami-m fading channels defined in (4.35). For different values of $m = 2\mu$, (4.173) can also provide matching to Rayleigh ($m = 1$), single [_ sided Gaussian ($m = 1/2$), and other fading conditions derived directly from the Nakagami-m fading conditions.

To understand and appreciate the flexibility offered by the pdf in (4.164), let us go back to (4.167). It can be rewritten using (4.169) as

$$\mu = \frac{m}{2} \left[1 + \left(\frac{\vartheta}{\upsilon} \right)^2 \right]. \tag{4.174}$$

Since r $J > 1$, (4.174) at high values of η can be rewritten as

$$\mu = \frac{m}{2} [1 + 1] = m \tag{4.175}$$

since

$$\left(\frac{\vartheta}{\upsilon} \right)^2 \Big|_{\eta \to \infty} = 0. \tag{4.176}$$

Thus, in terms of the Nakagami parameter m, the values of μ will be limited to

$$\frac{m}{2} < \mu < m. \tag{4.177}$$

It is interesting to note that for a fixed value of m, (4.174) and (4.177) show that μ can take a range of values determined by the values of the ratio of variances of the inphase and quadrature parts with

$$\left(\frac{\vartheta}{\upsilon} \right)^2 = 2\frac{\mu}{m} - 1. \tag{4.178}$$

Note that m can take multiple values indicating that the pdf in (4.173) offers the flexibility of modeling a wide range of fading conditions. Going back to (4.164) and (4.177), we obtain the Hoyt distribution by choosing a value of $m = 0.5$. The pdf now becomes,

$$f(z) = \frac{1+q^2}{2qZ_0}\exp\left[-\frac{(1+q^2)^2 z}{4q^2 Z_0^2}\right] I_0\left[\frac{(1-q^4)z}{4q^2 Z_0}\right]. \tag{4.179}$$

In (4.179), q has replaced m and the density function in Eq. (4.179) is identified as the Nakagami-q density function or the Hoyt pdf of the SNR in short-term fading channels (Simon and Alouini 2005; Paris 2009a, b). Taking note of (4.177), the value of q will be limited to a positive quantity less than unity. Thus, the Nakagami-q distribution models fading conditions far more severe than Rayleigh ($q = m = 1$) or Nakagami-m channels ($m > 1/2$).

The CDF of the SNR in a Hoyt fading channel has been obtained in an analytical form (Paris 2009a, b) and it is given by

$$F(z) = Q\left[A(q)\sqrt{\frac{z}{4Z_0}}, B(q)\sqrt{\frac{z}{4Z_0}}\right] - Q\left[B(q)\sqrt{\frac{z}{4Z_0}}, A(q)\sqrt{\frac{z}{4Z_0}}\right], \tag{4.180}$$

where

$$A(q)\sqrt{\frac{1+q}{1-q}}\sqrt{\frac{1+q^4}{2q}} \tag{4.181}$$

$$B(q)\sqrt{\frac{1+q}{1-q}}\sqrt{\frac{1+q^4}{2q}} \tag{4.182}$$

and Q is the Marcum's Q function defined in Chap. 3.

Note that we had obtained (4.164) and all the other density functions from (4.164) by assuming that the inphase and quadrature components are independent with identical means of zero but with different variances. It is also possible to arrive at (4.164) and the other density functions by treating the inphase and quadrature components to be identical and of zero mean, but, treat them as two correlated random variables with a correlation coefficient η such that $-1 < \eta < 1$. Thus, whether one looks at the inphase and quadrature components as independent with non-identical variances or correlated with identical variances, the Hoyt pdf will result. We will now look at (4.162) differently and rewrite it as

$$Z = A^2 = \sum_{k=0}^{u}(X_k + \mu_x)^2 + \sum_{k=0}^{u}(Y_k + \mu_y)^2. \tag{4.183}$$

In (4.183), X's and Y's are still zero mean, independent, and identically distributed Gaussian random variables. However, now the inphase and quadrature components each has a constant mean of mx and my, respectively. This suggests that within each cluster, the possibility exists that there is power in the dominant path of some strength over and beyond the power in the diffuse components. Expressing (4.184) in terms of the sum of independent identical variables, we have

$$Z = \sum_{k=1}^{u} Z_k = \sum_{k=1}^{u} A_k^2. \tag{4.184}$$

Note that the pdf of Zk is the well-known Rice distribution given by

$$f(z_k) = \frac{1}{2\sigma^2}\exp\left(-\frac{z_k + d_k^2}{2\sigma^2}\right) I_0\left(\frac{d_k\sqrt{z_k}}{\sigma^2}\right). \tag{4.185}$$

In (4.185), σ^2 is the variance of the inphase and quadrature components and

$$d_k = \sqrt{\mu_x^2 + \mu_y^2}. \tag{4.186}$$

The pdf of Z can now be obtained using the concept of the pdf of sum of independent random variables as

$$f(z) = \frac{1}{\sigma^2}\left(\frac{z}{d^2}\right)^{(u-1)/2} \exp\left(-\frac{z_k + d^2}{2\sigma^2}\right) I_{u-1}\left(\frac{d\sqrt{z}}{\sigma^2}\right) \tag{4.187}$$

with

$$d^2 = \sum_{k=0}^{u} d_k^2. \tag{4.188}$$

Defining the ratio of the total power in the dominant paths to diffuse power (or SNR) as

$$\kappa = \frac{d^2}{2u\sigma^2} \tag{4.189}$$

we have

$$\frac{\langle Z \rangle^2}{\langle Z^2 \rangle - \langle Z \rangle^2} = u\frac{(1+\kappa)^2}{(1+2\kappa)}. \tag{4.190}$$

Equations (4.189) and (4.190) suggest to us that depending on the values of κ and the moments, the quantity u might not always be integer. Recognizing this, we will replace u with μ taking a value larger than or equal to 1. Equation (4.187) can now be rewritten in terms of κ and μ as

$$f(z) = \frac{\mu}{\kappa^{((\mu-1)/2)}\exp(\mu\kappa)}(1+\kappa)^{((\mu-1)/2)}z^{((\mu-1)/2)} \times\exp[-\mu(1+\kappa)z]I_{\mu-1}\left[2\mu\sqrt{\kappa(1+\kappa)z}\right]. \quad (4.191)$$

The pdf in (4.191) is often identified as the $\kappa-\mu$ distribution. For $m=1$, (4.191) is the pdf of SNR in the traditional Rician fading channels if $m=1$ and $\kappa=0$, Eq. (4.191) becomes the exponential distribution for the SNR associated with the Rayleigh fading channels. The $\kappa-\mu$ density is plotted in Fig. 4.72 for a fixed value of m and in Fig. 4.73 for a fixed value of κ.

Redefining Eq. (4.191) for the case of $m=1$, we have

$$\kappa = \frac{d^2}{2\sigma^2} = n^2. \quad (4.192)$$

In Eq. (4.192) n is a positive number and we can rewrite Eq. (4.191) as

$$f(z) = (1+n^2)\exp(n^2)\exp[-\mu(1+n^2)z]I_0\left[2n\sqrt{(1+n^2)z}\right]. \quad (4.193)$$

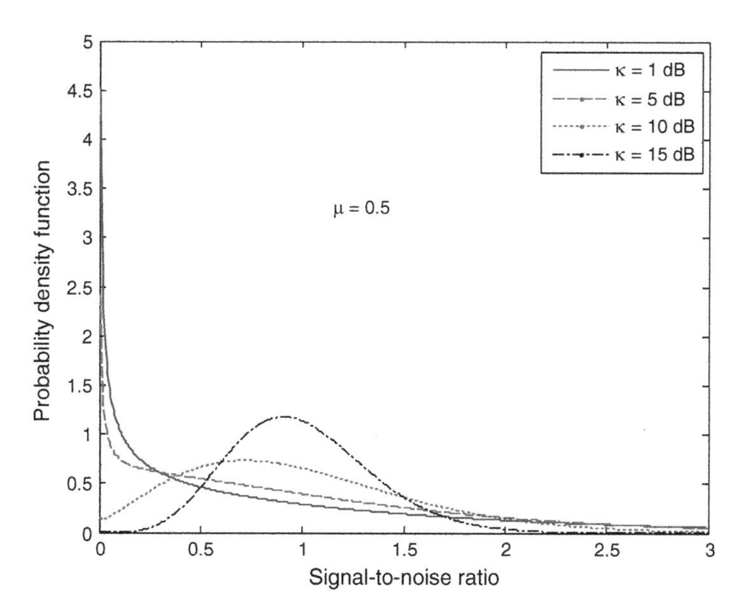

Fig. 4.72 $\kappa-\mu$ density for a fixed value of m

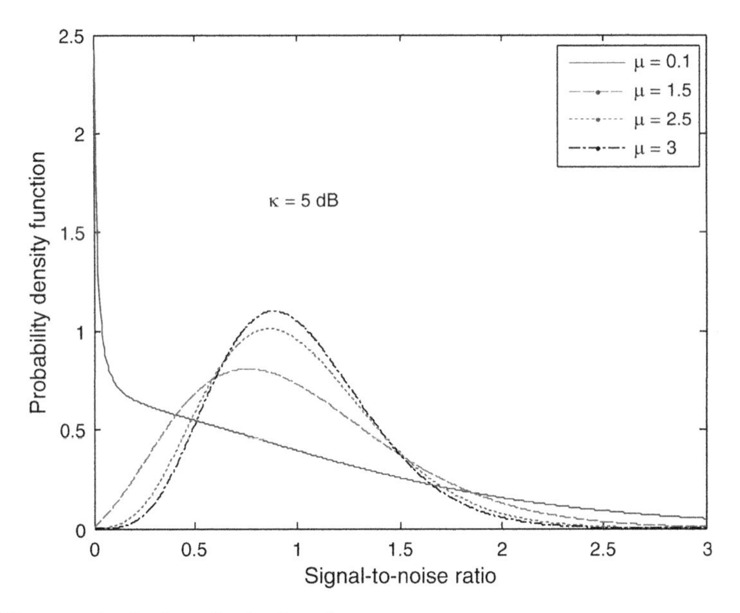

Fig. 4.73 $\kappa - \mu$ density for a fixed value of κ

Scaling by the average SNR Z_0, we get

$$f(z) = \frac{(1+n^2)}{Z_0}\exp(n^2)\exp\left[-\mu\frac{(1+n^2)}{Z_0}z\right]I_0\left[2n\sqrt{(1+n^2)\frac{z}{Z_0}}\right]. \tag{4.194}$$

Equation (4.194) is the well-known Nakagami–Rice or *Nakagami-n* distribution for the SNR where n is the Nakagami-n fading parameter. Note that the parameter κ in (4.192) is identical to the Rician factor K_0 in (4.18) expressed earlier.

We can also look at the relationships among the three Nakagami distributions, namely Nakagami-m, Nakagami-q, and Nakagami-n. Comparing the moments of the pdf, the relationships become

$$m = \frac{(1+q^2)^2}{2(1+q^4)},\quad m \le 1 \tag{4.195}$$

and

$$m = \frac{(1+n^2)^2}{1+2n^2},\quad m \le 1, n \ge 0. \tag{4.196}$$

It is now possible to look at yet another generalization of the fading models described above. Instead of the $\eta - \mu$ and $\kappa - \mu$ distributions, it is possible to introduce another parameter α leading to the so-called $\alpha - \eta - \mu$ and $\alpha - \kappa - \mu$

distributions (Papazafeiropoulos and Kotsopoulos 2011a). Let us go back to (4.162) and rewrite it by scaling the power as

$$Z^{\alpha} = \left[\sum_{k=0}^{u} \left(X_k^2 + Y_k^2\right)\right]. \tag{4.197}$$

The density functions of the appropriate $\alpha - \eta - \mu$ and $\alpha - \kappa - \mu$ variables can be obtained using the transformation of variables discussed in Chap. 2. It is also worth mentioning that (4.197) can lead to the generalized Nakagami pdf or the generalized gamma distribution. Thus, use of an exponential scaling makes it possible to take the $\eta - \mu$ and $\kappa - \mu$ distributions, and with the additional parameter α, the generalized gamma pdf could be incorporated into the mix of the densities that are encompassed in a single density function (Papazafeiropoulos and Kotsopoulos 2011b).

The performance of wireless channels undergoing $\eta - \mu$ and $\kappa - \mu$ fading can be studied in the same manner as was done with the other fading channels. Since the Rayleigh, Rician, Nakagami, generalized gamma, and Weibull channels have been studied, we will look at the remaining one from the class of $\eta - \mu$ and $\kappa - \mu$ distributions, namely the Hoyt channels or the Nakagami–Hoyt channels. The Nakagami–Hoyt pdf is shown in Fig. 4.74.

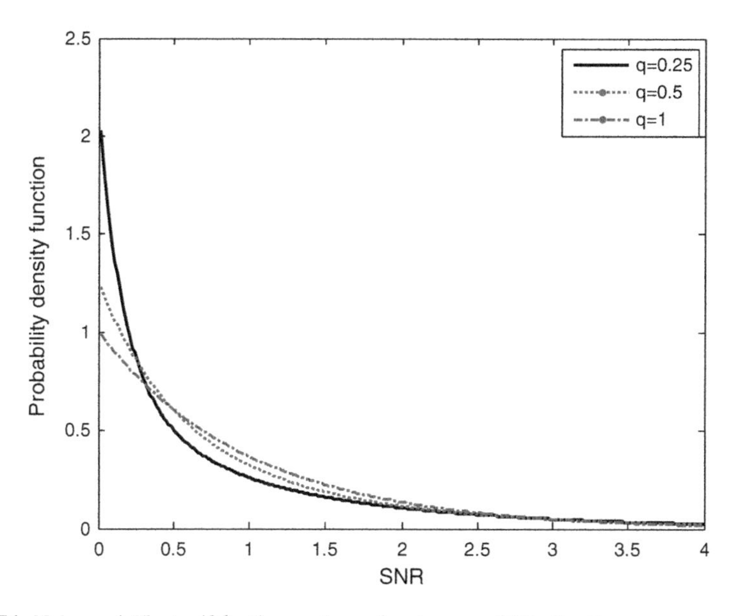

Fig. 4.74 Nakagami–Hoyt pdf for three values of q. Average SNR of unity

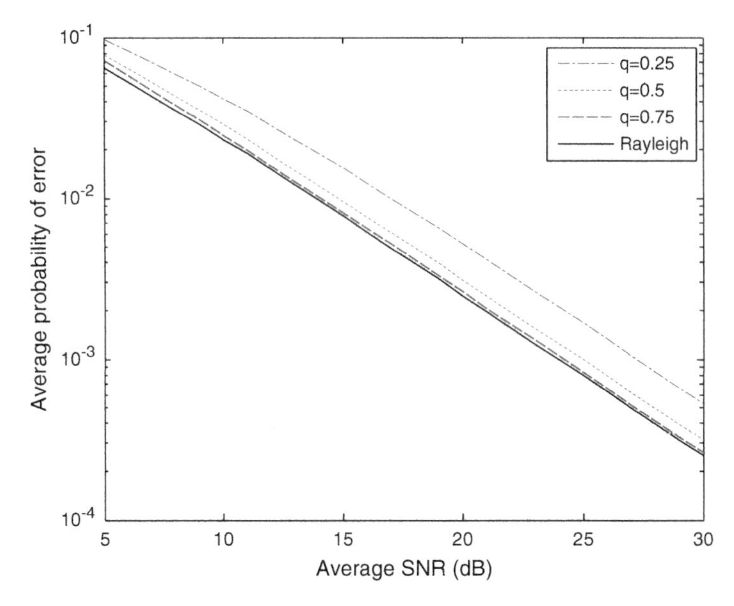

Fig. 4.75 Error rates in Nakagami–Hoyt channels

The moments of the Hoyt density in (4.179) are

$$\left\langle Z^k \right\rangle = \Gamma(1+k)_2F_1\left(-\frac{k-1}{2}, -\frac{k}{2}; 1; \left(\frac{1-q^2}{1-q^2}\right)^2\right) Z_0^k. \tag{4.198}$$

In Eq. (4.198), ${}_2F_1$ is the hypergeometric function (Gradshteyn and Ryzhik 2007). The amount of fading becomes

$$\mathrm{AF} = 2\frac{(1+q^4)}{(1+q^2)^2}, \quad 0 \le q \le 1. \tag{4.199}$$

Thus, the amount of fading ranges from 2 at the high end (same value corresponding to $m = 1/2$ for the Nakagami-m case) to 1 at the low end, the same as the value in Rayleigh channels. The error rates in Hoyt channels can now be estimated using (4.133). The results are shown in Fig. 4.75.

The outage probabilities in Hoyt channels are plotted in Fig. 4.76.

4.10 A Few Additional Quantitative Measures of Fading and Shadowing

We will also discuss two additional quantitative measures. (We will not undertake a detailed analysis which was already carried out with the other measures above.) First of these is the Ergodic Channel capacity which can be estimated using the

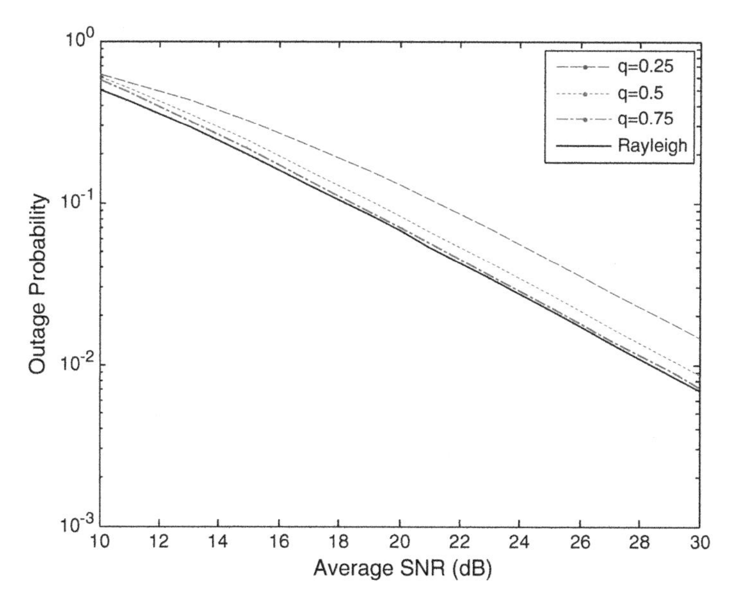

Fig. 4.76 Outage probabilities in Nakagami–Hoyt channels

density functions obtained earlier (Lee 1990; Sagias et al. 2004, 2005; Simon and Alouini 2005; Laourine et al. 2007; Di Renzo et al. 2010). The second measure is the average fade duration which requires the second order statistics of the envelope and takes into account the relative motion of the transmitter/receiver (Jakes 1994).

4.10.1 Ergodic Channel Capacity

Channel capacity provides a measure of the amount of information that can be reliably transmitted over a communications channel. In an ideal channel corrupted by additive white Gaussian noise, the channel capacity C is given by (Haykin 2001; Sklar 2001)

$$C = (\mathrm{BW})\log_2(1 + Z)\mathrm{b/s}. \tag{4.200}$$

In (4.200), BW is the channel bandwidth (Hz) and Z is the signal-to-noise ratio (SNR). When fading, shadowing, or shadowed and fading are present, the SNR Z is a random variable (described earlier) and (4.200) needs to be rewritten to obtain the mean channel capacity as (Lee 1990)

$$\langle C \rangle = (\mathrm{BW}) \int_0^{\infty} \log_2(1 + z) f(z) \mathrm{d}z \ \mathrm{b/s}, \tag{4.201}$$

where $f(z)$ is the pdf of the SNR in the channel undergoing random fluctuations. Normalized average channel capacity in b/s/Hz can be written as

$$\langle C\rangle_n = \frac{\langle C\rangle}{(\mathrm{BW})}\int_0^{\infty} \log_2(1+z)f(z)\mathrm{d}z \ \mathrm{b/s/Hz}. \tag{4.202}$$

We will first look at the case of a pure short-term faded channel undergoing Nakagami fading. In a Nakagami faded channel, (4.202) becomes

$$\langle C\rangle_n = \int_0^{\infty} \log_2(1+z)\left(\frac{m}{Z_0}\right)^m \frac{z^{m-1}}{\Gamma(m)}\exp\left(-m\frac{z}{Z_0}\right)\mathrm{d}z. \tag{4.203}$$

For $m = 1$, (4.203), provides the normalized average capacity in a Rayleigh channel and Z_0 is the average SNR. It is possible to solve (4.203) by representing its different factors in terms of other functions. Using the relationship between Meijer G-function and simple functions, we have (Gradshteyn and Ryzhik 2007)

$$\log(1+z) = G_{2,2}^{1,2}\left[z\middle|_{1,0}^{1,1}\right] \tag{4.204}$$

$$\exp\left(-m\frac{z}{Z_0}\right) = G_{0,1}^{1,0}\left[m\frac{z}{Z_0}\middle|_{0}^{-}\right]. \tag{4.205}$$

Equation (4.203) now becomes

$$\langle C\rangle_n = \frac{1}{\log(2)\Gamma(m)}\left(\frac{m}{Z_0}\right)^m \int_0^{\infty} G_{2,2}^{1,2}\left[z\middle|_{1,0}^{1,1}\right] G_{0,1}^{1,0}\left[m\frac{z}{Z_0}\middle|_{0}^{-}\right] z^{m-1}\,\mathrm{d}z. \tag{4.206}$$

Equation (4.206) can be solved using the integral properties of Meijer G-functions. The normalized average channel capacity becomes

$$\langle C\rangle_n = \frac{1}{\log(2)\Gamma(m)}\left(\frac{m}{Z_0}\right)^m \int_0^{\infty} G_{2,3}^{3,1}\left[\frac{m}{Z_0}\middle|_{0,-m,-m}^{-m,1-m}\right]. \tag{4.207}$$

For the case of Rayleigh fading, the normalized average channel capacity becomes (by putting $m = 1$ in (4.207))

$$\langle C\rangle_n = \frac{1}{\log(2)}\left(\frac{1}{Z_0}\right) G_{2,3}^{3,1}\left[\frac{1}{Z_0}\middle|\begin{matrix}-1,0\\0,-1,-1\end{matrix}\right]. \tag{4.208}$$

The channel capacity is plotted in Fig. 4.77 for Nakagami channels.

Proceeding in a similar way, we can obtain the normalized average channel capacity in a Rician faded channel as

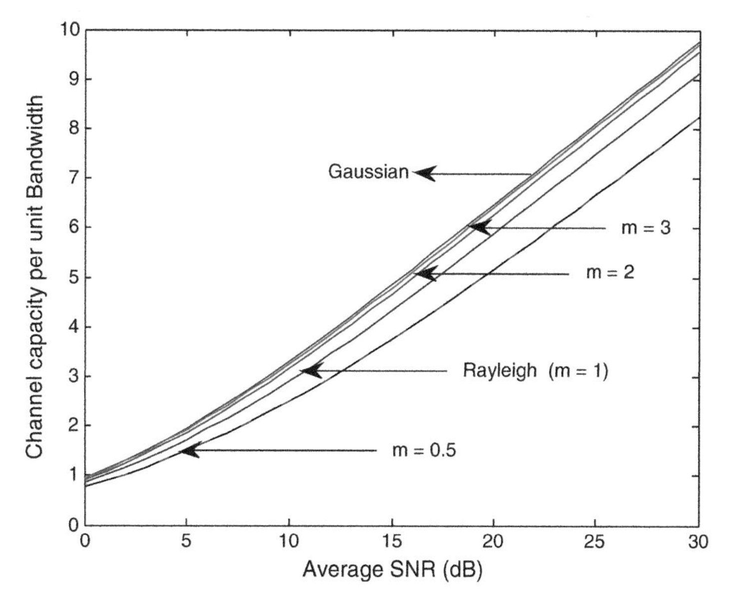

Fig. 4.77 Channel capacities in Nakagami channels

$$\langle C\rangle_n = \int_0^{\infty} \log(1+z)\frac{1}{Z_0}(1+K_0)\exp\left[-K_0-\frac{z}{Z_0}(1+K_0)\right]I_0\left(\sqrt{4K_0(1+K_0)\frac{z}{Z_0}}\right)\mathrm{d}z. \tag{4.209}$$

Note that Z_0 is the average SNR in the Rician channel which includes the contribution from the LOS component. The Rician factor is given by K_0. The modified Bessel function of the first kind I_0 can be expanded in an infinite series. Using the relationship between Meijer G-function and elementary functions expressed in Eqs. (4.204) and (4.205), the channel capacity in (4.209) in the Rician channel can be simplified to

$$\begin{aligned}\langle C\rangle_n = &\frac{(1+K_0)\exp(-K_0)}{\log(2)Z_0}\sum_{n=0}^{\infty}\left(\frac{1}{n!}\right)^2\left[\frac{(1+K_0)K_0}{Z_0}\right]^n G_{2,3}^{3,1}\\ &\times\left[\left(\frac{1+K_0}{Z_0}\right)\Bigg|\begin{matrix}-1-n,\,-n\\ 0,\,-1-n,\,-1-n\end{matrix}\right].\end{aligned} \tag{4.210}$$

The channel capacities in Rician channels are plotted in Fig. 4.78 for a few values of the Rician factor or parameter in dB.

The channel capacity in Nakagami–Hoyt channels can be estimated in a similar fashion. The capacities are plotted in Fig. 4.79.

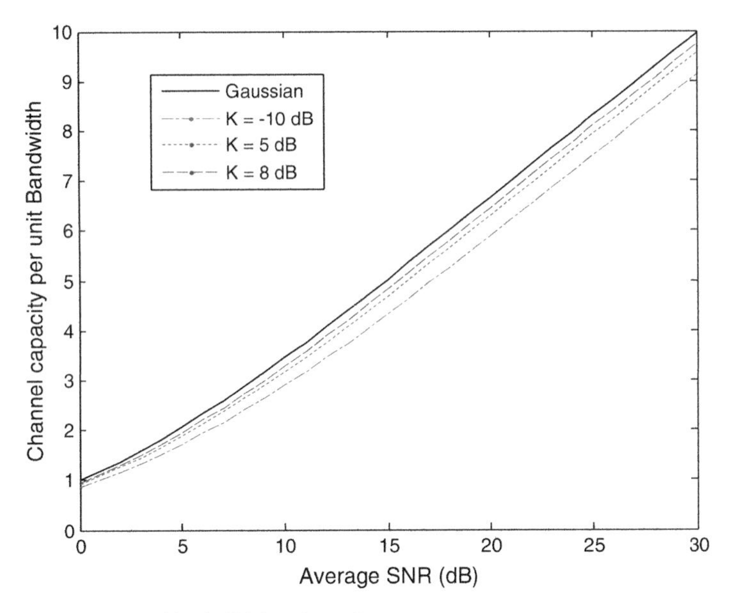

Fig. 4.78 Channel capacities in Rician channels

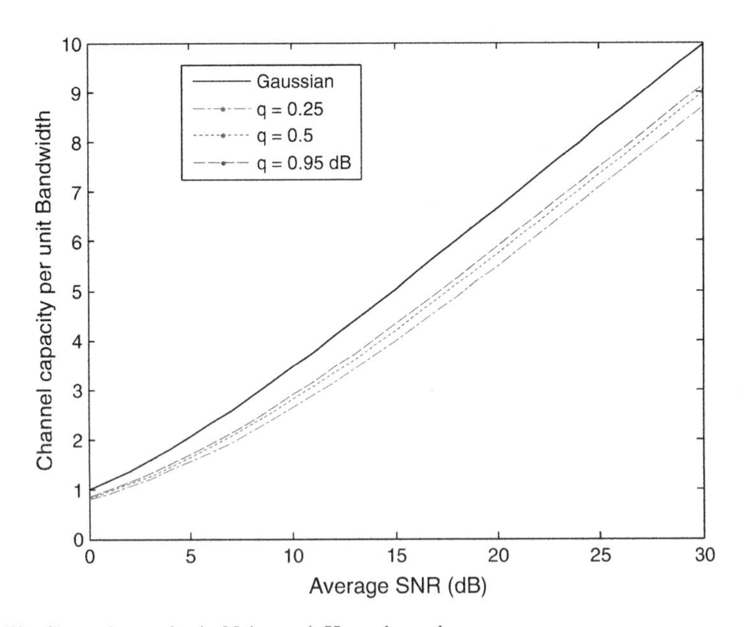

Fig. 4.79 Channel capacity in Nakagami–Hoyt channels

The channel capacity in cascaded short-term fading channels can be obtained as

$$\langle C\rangle = \frac{1}{\log(2)}\int_0^{\infty} G_{2,2}^{1,2}\left[z\middle|_{1,0}^{1,1}\right]\frac{1}{z\Gamma^N(m)}G_{0,N}^{N,0}\left(\frac{m^N}{Z_0}z\middle|_{m,m,\ldots,m}\right)\,\mathrm{d}z. \tag{4.211}$$

In writing (4.211), we have expressed the relationship between log(1 + x) and Meijer G-function in (4.204). The density function of the SNR in a cascaded channel is described in (4.114). Using the table of integrals, (4.211) can be simplified to

$$\langle C\rangle = \frac{1}{\log(2)m^N\Gamma^N(m)}G_{2,2}^{1,2}\left(\frac{m^N}{Z_0}z\middle|\begin{matrix}1,2\\1,1,m+1,m+1,\ldots,m+1\end{matrix}\right). \tag{4.212}$$

Channel capacities are plotted in Figs. 4.80, 4.81, and 4.82.

We will now look at the case of shadowed fading channels. Given that the Nakagami-lognormal channel leads to the unavailability of an analytical expression for the pdf, we will use the Nakagami-gamma model which results in the GK distribution for the SNR in a shadowed fading channel. Taking note of the fact that a doubly cascaded channel with gamma parameters of m and c is the same as the GK distribution, we will use the results of the cascaded channel to estimate the average channel capacity. The density function of the SNR in a shadowed fading channel modeled using the generalized K distribution can be written as

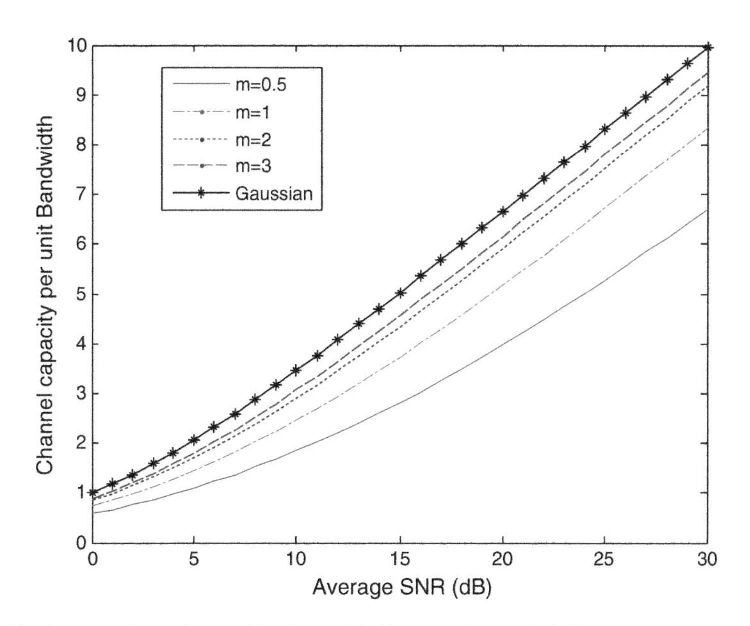

Fig. 4.80 Average channel capacities in double Nakagami cascaded channels

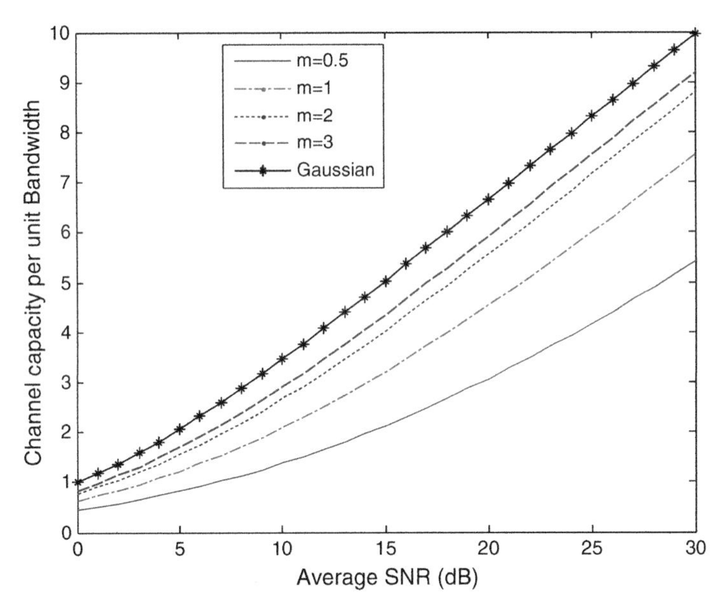

Fig. 4.81 Average channel capacities in triple cascaded channels

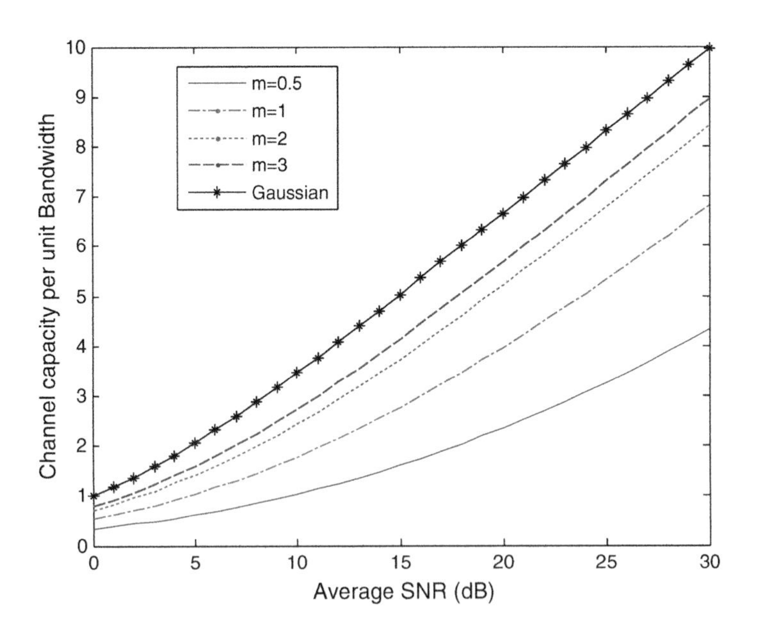

Fig. 4.82 Average channel capacities in quadruple cascaded channels

$$f(z) = \frac{1}{z\Gamma(m)\Gamma(c)} G_{0,2}^{2,0}\left(\frac{mc}{Z_0}z \middle| \begin{matrix} - \\ m, c \end{matrix}\right). \tag{4.213}$$

Note that in (4.213), c is related to the shadowing level measured in terms of σ_{dB} through (4.72). Equation (4.212) can now be used to obtain average channel capacities in a shadowed fading channel as

$$\langle C \rangle = \frac{Z_0}{\log(2) mc\Gamma(m)\Gamma(c)} G_{2,4}^{4,1}\left(\frac{mc}{Z_0} \middle| \begin{matrix} 1, 2 \\ 1, 1, m+1, c+1 \end{matrix}\right). \tag{4.214}$$

In this case, the average channel capacities can now be estimated and plotted as a function of the Nakagami parameter m and shadowing parameter c with high values of c corresponding to low levels of shadowing and low values of c corresponding to high levels of shadowing. The channel capacities are plotted for $m = 1$, 2, and 3 for a few levels of shadowing in Figs. 4.83, 4.84, and 4.85, respectively.

4.10.2 *Second Order Statistics of Fading, Shadowing, and Shadowed Fading Channels*

As mentioned in the introduction, we have concentrated on fading channels that are slow and flat, thereby ignoring the effects of relative motion of the transmitter and

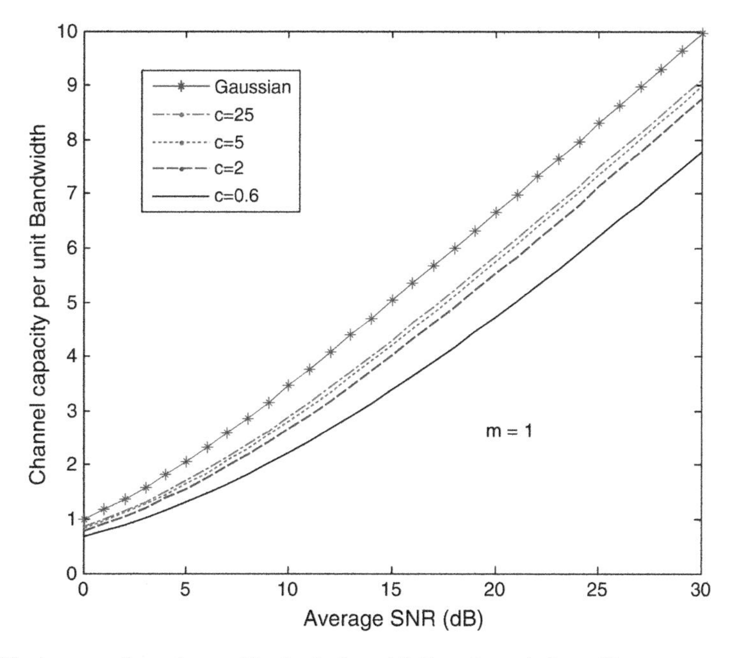

Fig. 4.83 Average channel capacities in shadowed fading channels ($m = 1$)

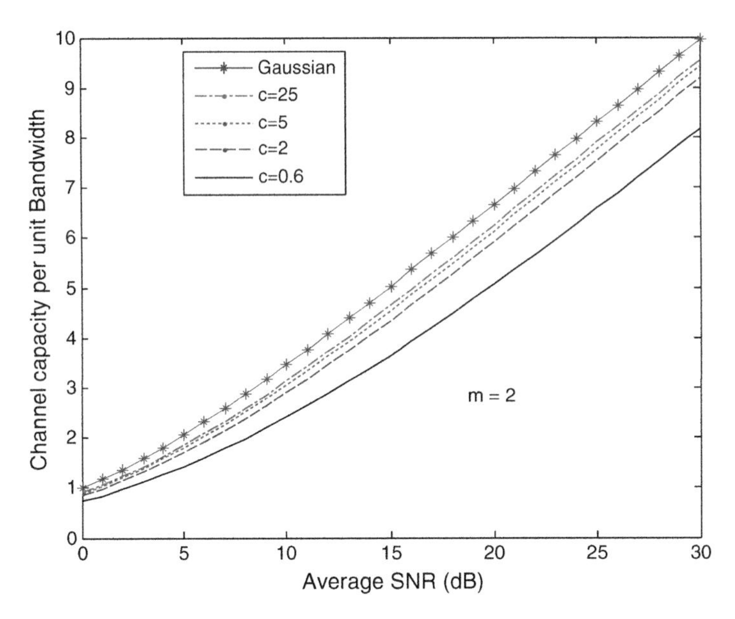

Fig. 4.84 Average channel capacities in shadowed fading channels ($m = 2$)

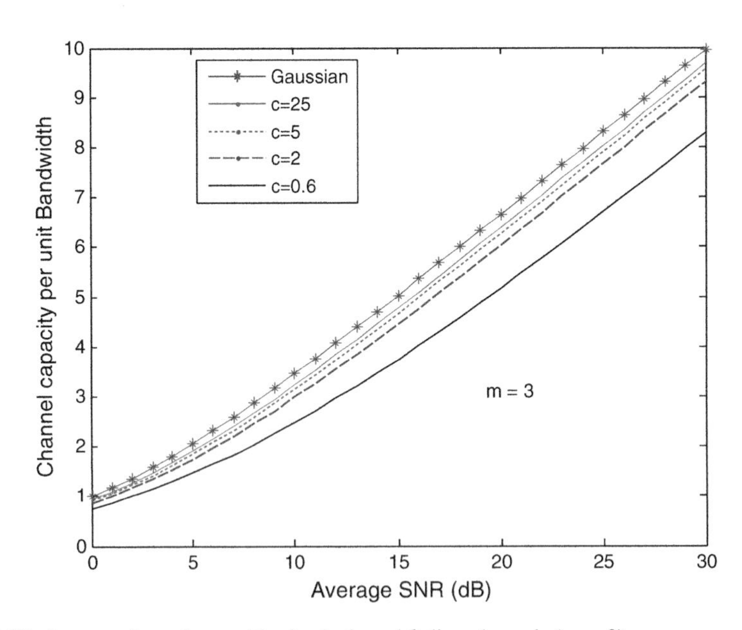

Fig. 4.85 Average channel capacities in shadowed fading channels ($m = 3$)

receiver. Still, some properties of the wireless channel in terms of the temporal characteristics are necessary to fully comprehend the problems in wireless channels that undergo fading, shadowing, as well as shadowing and fading simultaneously. Thus, we will complete the overview of the fading channels by examining such dynamic channels characteristics as the rate at which a channel goes into outage and how long it stays, and so on. Such issues are crucial to the development of a successful strategy which takes into account the fact that often the mobile unit is not stationary and such motion of the mobile unit would also effect the ability to transmit data.

The measures we reviewed so far, namely error probability and outage probability, are quantitative measures of the fading channel based on first order statistics since the only information required is the pdf of the SNR. However, the second order statistics, namely the joint pdf of the SNR, will provide additional information on the fading channel such as the number of times the envelope of the received signal will stay below or above a required threshold value, whether it goes below the threshold, and how long it will stay there. More importantly, dynamic quantitative measures provide means to manage handoff algorithms, estimate package error rates (burst errors), and give an overall estimate of the state of the channel. Two such measures are level crossing rates (LCR) and the average fade duration (AFD). The former is defined as the number of times/unit duration that the envelope crosses the threshold in the negative direction; the latter is defined as the average duration of time the envelope stays below the threshold once it goes below. These are generally accepted as two markers to quantify the second order statistics of the channel. LCR and AFD can be estimated for the fading channels before and after the implementation of diversity (Jakes 1994). Figure 4.86 illustrates the concepts of fade duration and level crossing rates.

We will start the discussion of the second order statistical measures by looking at the simplest case of a Rayleigh faded channel prior to looking at other models for fading, shadowing, and shadowed fading channels before and after diversity. The pdf in (4.6) of the envelope A in a Rayleigh faded channel can be expressed as

$$f(a) = \frac{2a}{P_0}\exp\left(-\frac{a^2}{P_0}\right). \tag{4.215}$$

In (4.215), P_0 is the average power. The level crossing rate (LCR), $N_A(\alpha)$ is defined as the expected rate at which the envelope crosses a specified signal level A:

$$N_A(A) = \int_0^\infty \dot{a} f(A, \dot{a}) \mathrm{d}\dot{a}\,. \tag{4.216}$$

In Eq. (4.216), the upper period above a indicates the derivative with respect to time and $f(A, \dot{a})$ is the joint pdf of a and $\dot{a}$ at $a = A$. The joint pdf of the envelope (at the fixed value of $a = A$) and its derivative can be written in terms of conditional pdfs as

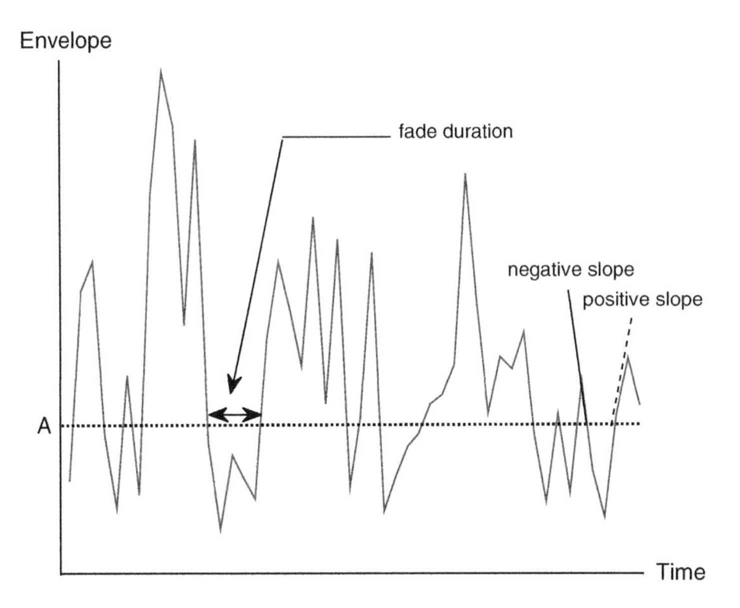

Fig. 4.86 Concepts of fade duration and level crossing rates. A is the threshold

$$f(\dot{a}|a=A)=f(\dot{a}). \tag{4.217}$$

These pdfs can be obtained from the original work of Rice. For Rayleigh, Rician, and Nakagami distributions, the pdf of the derivative of the magnitude is independent of the pdf of the magnitude (Jakes 1994). This pdf has been shown to be

$$f(\dot{a})=\frac{1}{\sqrt{2\pi\sigma^2}}\exp\left(-\frac{\dot{a}^2}{2\sigma^2}\right). \tag{4.218}$$

Furthermore, it was shown that

$$f(\dot{a}|a=A)=f(\dot{a}). \tag{4.219}$$

Equation (4.219) leads to (4.216) becoming

$$N_A(A)=f(A)\int_0^\infty \dot{a}f(\dot{a})\,\mathrm{d}\dot{a}\,. \tag{4.220}$$

In Eq. (4.218),

$$\sigma^2=\pi^2\Omega f_d^2. \tag{4.221}$$

In (4.221), *fd* is the maximum Doppler frequency shift and Ω is given by (Jakes 1994; Yacoub et al. 1998, 1999, 2001; Dong and Beaulieu 2001)

$$\Omega \begin{cases} P_0 \text{ Rayleigh} \\ \dfrac{P_0}{m} \text{ Nakagami} \\ \dfrac{P_0}{K_0+1} \text{ Rician} \end{cases}. \tag{4.222}$$

The average fade duration (AFD) is defined as

$$T(a) = \frac{F(A)}{N_A(A)}. \tag{4.223}$$

In (4.223), $F(A)$ is the probability that a is less than the fixed value A. Note that the numerator in (4.223) is nothing but the CDF which is also the outage probability. Thus the average fade duration provides a measure of the time that the wireless system remains in outage.

We will now look at the Nakagami fading channel. Using (4.218) and (4.221), the level crossing rate in (4.216) becomes

$$N(\rho) = \frac{\sqrt{2\pi f_{\mathrm{d}}^2}}{\Gamma(m)} \left(m\rho^2\right)^{m-(1/2)} \exp\left(-m\rho^2\right). \tag{4.224}$$

The average fade duration in (4.223) becomes

$$T(\rho) = \frac{\Gamma(m, m\rho^2)\exp(m\rho^2)}{\sqrt{2\pi f_{\mathrm{d}}^2}\left(m\rho^2\right)^{m-(1/2)}}. \tag{4.225}$$

In Eqs. (4.224) and (4.225), ρ is the normalized magnitude given by

$$\rho = \frac{A}{\sqrt{\Omega}}. \tag{4.226}$$

and $\Gamma(.,.)$ is the incomplete gamma function defined in (2.37) in Chap. 2.

The LCR and AFD in a Rayleigh cannel can easily be obtained from (4.224) and (4.225) by putting $m = 1$ as

$$N(\rho) = \rho\sqrt{2\pi f_{\mathrm{d}}^2}\exp\left(_{-\rho^2}\right) \tag{4.227}$$

and

$$T(\rho) = \frac{\left[\exp(\rho^2) - 1\right]}{\rho\sqrt{2\pi f_{\mathrm{d}}^2}}. \tag{4.228}$$

Using the Rician pdf in (4.23), the LCR in a Rician channel becomes (Abdi et al. 2000; Abdi and Kaveh 2002; Chen 2007)

$$T(\rho) = \sqrt{2\pi(K_0+1)}f_m\rho\exp\left[-K_0-(K_0+1)\rho^2\right]I_0\left(2\rho\sqrt{K_0-(K_0+1)}\right). \tag{4.229}$$

The average fade duration in a Rician fading channel can be expressed in terms of the CDF obtained in (4.29), and it is given by

$$T(\rho) = \frac{1-Q\left(\sqrt{2K_0},\sqrt{2(K_0+1)\rho^2}\right)}{\sqrt{2\pi(K_0+1)}f_m\rho\exp[-K_0-(K_0+1)\rho^2]I_0(2\rho\sqrt{K_0-(K_0+1)})}. \tag{4.230}$$

In (4.229) and (4.230), $I_0(.)$ is the 0th modified Bessel function of the first kind and $Q(.,.)$ is the Marcum Q function defined in Eq. (3.186) in Chap. 3.

The level crossing rate in Weibull fading can also be obtained similarly (Sagias et al. 2004). The Weibull pdf of the envelope can be expressed as

$$f(a) = \frac{m_w}{P_0}a^{m_w-1}\exp\left(-\frac{a^{m_w-1}}{P_0}\right). \tag{4.231}$$

Note that the Weibull pdf for the power and envelope looks the same and for $m_w = 2$ (4.231) becomes the Rayleigh pdf. Note that

$$E(a^{m_w}) = P_0 = \Omega. \tag{4.232}$$

The CDF of the envelope in Weibull fading (evaluated at $a = A$) is

$$F(A) = 1-\exp\left(-\frac{Am_w}{P_0}\right). \tag{4.233}$$

Noting that the pdf of the derivative of the envelope is given in (4.218), we can obtain the expression for the LCR as

$$N_A(\rho_w) = \sqrt{2\pi f_d^2}\left(\frac{\rho_w}{\sqrt{\beta}}\right)^{\frac{m_w}{2}}\exp\left[-\left(\frac{\rho_w}{\sqrt{\beta}}\right)^{m_w}\right]. \tag{4.234}$$

In (4.234),

$$\beta = \frac{1}{\Gamma(1+(2/m_w))} \tag{4.235}$$

$$\rho_{\mathrm{w}}=\frac{A\sqrt{\beta}}{\Omega^{1/m_{\mathrm{w}}}}. \tag{4.236}$$

The average fade duration becomes

$$T(\rho_{\mathrm{w}})=\frac{1-\exp\left[-(\rho_{w}/\sqrt{\beta})^{m_{\mathrm{w}}}\right]}{\sqrt{2\pi f_{\mathrm{d}}^{2}(\rho_{\mathrm{w}}/\sqrt{\beta})}^{-m_{\mathrm{w}}/2}\exp\left[-(\rho_{w}/\sqrt{\beta})^{m_{\mathrm{w}}}\right]}. \tag{4.237}$$

When $m_{\mathrm{w}}=2$, (4.234) and (4.237) become the corresponding values in Rayleigh channels given in (4.227) and (4.228), respectively.

As mentioned earlier, in some of the indoor propagation channels, short-term fading has been seen to be best described using a lognormal pdf (Cotton and Scanlon 2007). This means that the pdf of the envelope expressed in decibel units

$$a_{\mathrm{dB}}=20\log_{10}(a) \tag{4.238}$$

can be expressed as

$$f(a_{\mathrm{dB}})=\frac{1}{\sqrt{2\pi\sigma_{\mathrm{dB}}^{2}}}\exp\left[-\frac{(a_{\mathrm{dB}}-\mu a_{\mathrm{dB}})^{2}}{2\sigma_{\mathrm{dB}}^{2}}\right]. \tag{4.239}$$

Note that in (4.239), mdB and σ_{dB}, respectively, are the mean and standard deviation of the envelope in decibel units (Loo 1985; Tjhung and Chai 1999; Laourine et al. 2007). Note that the pdf of the derivative of the envelope (in dB) is independent of the envelope and Gaussian distributed as in the previous models of fading, the expression for the LCR becomes

$$N(A_{\mathrm{dB}})=f(A_{\mathrm{dB}})\int_{0}^{\infty}\dot{a}_{\mathrm{dB}}f(\dot{a}_{\mathrm{dB}})d\dot{a}_{\mathrm{dB}}. \tag{4.240}$$

In (4.240),

$$f(\dot{a}_{\mathrm{dB}})=\frac{1}{\sqrt{2\pi\dot{\sigma}^{2}}}\exp\left(-\frac{\dot{a}_{\mathrm{dB}}^{2}}{2\dot{\sigma}^{2}}\right) \tag{4.241}$$

with

Equation (4.240) now becomes

$$\dot{\sigma}=2\pi\sigma_{\mathrm{dB}}f_{\mathrm{d}}. \tag{4.242}$$

$$N(A_{\mathrm{dB}})=f_{\mathrm{d}}\exp\left[-\frac{(A_{\mathrm{dB}}-\mu_{\mathrm{dB}})^{2}}{2\sigma_{\mathrm{dB}}^{2}}\right]. \tag{4.243}$$

The average fade duration now becomes

$$T(A_{\mathrm{dB}}) = \frac{1}{2N(A_{\mathrm{dB}})}\left[1 + \mathrm{erf}\frac{A_{\mathrm{dB}} - \mu a_{\mathrm{dB}}}{\sqrt{2\sigma_{\mathrm{dB}}^{2}}}\right]. \tag{4.244}$$

In (4.244), erf() is the error function defined in Chap. 3.

Now that we have looked at several fading models, we can examine the second order statistics of shadowed fading channels. Shadowed fading channels are typically modeled as the result of a product of a short-term faded component (for example, Nakagami distributed) and a long-term faded or shadowing component (typically lognormal) as (Tjhung and Chai 1999)

$$A(t) = X(t)Y(t). \tag{4.245}$$

In Eq. (4.245), $X(t)$ and $Y(t)$, respectively, are the short-term and long faded components and $A(t)$ is the resulting shadowed fading component (envelope). In a typical Nakagami-lognormal shadowed fading channel, X will be Nakagami distributed while Y will be lognormal. Instead of that approach, we will follow the simplified approach of using the shadowed fading channel modeled using the GK distribution. This allows us to write the right-hand side of Eq. (4.245) as the product of two Nakagami random variables (note that Nakagami distributed envelope results in gamma distributed power (Zlatanov et al. 2008). Note that the double Nakagami process also leads to the GK distribution as discussed earlier in Sect. 4.6. This means that density function of the short-term fading component given as

$$f(x) = 2\left(\frac{m}{P_x}\right)^{m} x^{2m-1}\exp\left(-m\frac{x^2}{P_x}\right). \tag{4.246}$$

The pdf of the shadowing process is given as (Shankar 2004; Zlatanov et al. 2008)

$$f(y) = 2\left(\frac{m}{P_y}\right)^{c} y^{2c-1}\exp\left(-c\frac{y^2}{P_y}\right). \tag{4.247}$$

In (4.246) and (4.247) m and c are the Nakagami parameters and P_x and P_y, respectively, are the average powers. It must be noted that c can take any positive value if we identify the pdf in Eq. (4.247) as one resulting from a gamma density function.

The densities of the derivatives of X and Y will be Gaussian with variances of the form given in (4.221). The corresponding variances becomes

$$\sigma_{\dot{X}}^{2} = \pi^2 f_{\mathrm{d}}^{2}\left(\frac{P_x}{m}\right) \quad \sigma_{\dot{Y}}^{2} = \pi^2 f_{\mathrm{d}}^{2}\left(\frac{P_y}{c}\right). \tag{4.248}$$

The LCR in a showed fading channel becomes

$$N_A(A) = \int_0^\infty \dot{a} f(A, \dot{a})\, \mathrm{d}\dot{a}\,. \tag{4.249}$$

Equation (4.249) can be written in terms of the densities of X and Y as

$$N_A(A) = \int_0^\infty \left(\int_0^\infty \dot{a} f_{\dot{A}|AX}(\dot{a}|A, x)\, \mathrm{d}\dot{a} \right) f_{A|X}(A|x) f_X(x)\, \mathrm{d}x. \tag{4.250}$$

In (4.250), $f_{\dot{A}|AX}(.)$ is the conditional pdf of $\dot{A}$ conditioned on A and X and it can be determined from the derivative of A in (4.245) as

$$\dot{A} = Y\dot{X} + X\dot{Y} = \frac{Z}{X}\dot{X} + X\dot{Y}\,. \tag{4.251}$$

The expression for LCR now becomes (Zlatanov et al. 2008)

$$N_A(A) = \frac{1}{\sqrt{2\pi}} \frac{4z^{2c-1}}{\Gamma(m)\Gamma(c)} \left(\frac{m}{P_x}\right)^m \left(\frac{c}{P_y}\right)^c \tag{4.252}$$

The integral in Eq. (4.252) can be evaluated numerically. The average fade duration can be expressed as the ratio of the CDF of A evaluated at $a = A$ and the LCR in (4.252). The AFD becomes

$$T(A) = \frac{\frac{1}{T(m)T(c)} G_{1,3}^{2,1}\left(A^2 \frac{mc}{P_x P_y} \middle| \begin{matrix} 1 \\ m, c, 0 \end{matrix} \right)}{N(A)}\,. \tag{4.253}$$

In (4.253), $G(.)$ is the Meijer G-function defined earlier in connection with cascaded channels and the numerator is the CDF associated with the double Nakagami pdf.

4.11 Sum, Product, and Mixture Models of Fading

Fading models can also be classified and identified in terms of their mathematical origins. The product model of fading leads to densities described in terms of the Meijer G-functions (Shankar 2013, 2015). In product models, the signal-to-noise ratio is modeled in terms of products of several random variables. The Nakagami-lognormal, Rayleigh-lognormal, and generalized K densities for shadowed fading channels originate from the product approach. The genesis of the lognormal density for shadowing can be attributed to the product model as explained earlier in this chapter where the SNR is best described as the product of several random variables. The sum model, on the other hand, results in SNR values expressed as the sum of two or more random variables. The sum model may be generalized to include the weighted sum of random variables. One example of the sum model discussed in the previous section is the so-called $\eta - \mu$ and $\kappa - \mu$ densities (Andersen 2002b; Asplund et al. 2002; Yacoub 2007a, b; Shankar 2015). Thus, the key characteristic of the sum and product models is the representation of the signal-to-noise ratio as the sum or product of a number of component SNRs. The density functions of the SNR in sum and product models are obtained using the approaches based on the concept of transformation of random variables described in Chap. 2. In some cases, these transformations do not lead to closed form solutions for the densities as in the case of Nakagami-lognormal or Rayleigh-lognormal densities for the amplitudes of the signal strengths (Simon and Alouini 2005). In a complementary approach to modeling the statistical fluctuations in wireless channels, it is possible to invoke the concept of statistical mixtures or mixtures of densities (Chap. 2). While sum and product models might not lead to closed form solutions for densities, the densities in mixture models are created by taking the weighted sum of component densities resulting in analytical forms for the densities of the SNR in these models. Thus, the mixture densities differ from the densities in the sum and product models in terms of the availability of a closed form solutions to the densities of the mixtures. The number of components in the mixtures, their weights, and the parameters of the component densities are to be determined and this aspect introduces computational complexities depending on the number of components in the mixture and basic density used in the mixture.

Another sum model leading to the McKay density is described first. The product model based on Meijer G-functions for the shadowed fading channels is presented next followed by the mixture models. The McKay density has been used to describe the statistics of the density of the SNR following diversity and the density itself was originally proposed and derived in 1932. It will also be shown that the McKay density and $\eta - \mu$ densities are similar in terms of their properties. Combining the cascaded gamma models for fading with cascaded models for shadowing, the product model of shadowed fading can be extended to take a more general form and described in terms of Meijer G-functions.

Once the sum model leading to the McKay density is presented, the mixture models proposed as an alternate form of models will be described.

4.11.1 Sum Model and McKay Density

A model constructed from the sum of random variables can be used to describe the signal strength fluctuations seen in short-term fading in wireless channels. Such a model is expressed in terms of the McKay distribution which was originally proposed in 1932 in connection with the statistics of gamma random variables

(McKay 1932). It treats the signal strength fluctuations in wireless channels as the result of scatters in the channel being grouped into two non-identical clusters. This description of the model differs from the one used in connection with the Nakagami model which assumes that there is a single cluster or there exist multiple clusters of identical scattering properties.

Consider the case of a wireless channel with no dominant scatterers. This suggests that the channel does not support any direct path between the transmitter and receiver (Rician fading conditions do not exist). The scatterers are grouped together into clusters (a model that was presented in connection with the Nakagami channel). The received signal strength in terms of the power or signal-to-noise ratio can then be written as the sum of the powers of the signals from each of these clusters. If there are n clusters, the SNR (or power) of the received signal is (Shankar 2015)

$$Z = \sum_{i=1}^{n} X_i^2. \tag{4.254}$$

In Eq. (4.254), each term of the right-hand side is the result of the sum of the powers of the inphase and quadrature components. Each cluster is assumed to have sufficient number of scatterers so that the criteria of the central limit theorem (Chap. 2) are met. This means that the inphase and quadrature components are treated as zero mean Gaussian random variables. Under these conditions (sufficient number of scatterers in each cluster and independent and identical clusters), the probability density of the SNR Z becomes

$$f(z) = \frac{z^{\frac{n}{2}-1}}{\Gamma\left(\frac{n}{2}\right) Z_0^{\frac{n}{2}}} \exp\left(-\frac{z}{Z_0}\right) U(z). \tag{4.255}$$

The average SNR is denoted by Z_0 in Eq. (4.255). It should be noted (as stated earlier in connection with the discussion of Nakagami fading) that Eq. (4.254) is valid even when there is only one cluster leading to the SNR in a Nakagami fading channel with the Nakagami parameter m being equal to ½ as

$$f(z) = \left(\frac{m}{Z_0}\right) \frac{z^{m-1}}{\Gamma(m)} \exp\left(-\frac{m}{Z_0} z\right) U(z), \quad m \geq \frac{1}{2}. \tag{4.256}$$

Under the Nakagami model, the minimum value of m is ½ and as it will be shown, it is possible for m to take any positive value if we treat the density in Eq. (4.256) as the standard gamma density. In the remaining discussion, m is considered to be a positive number.

One of the limitations of the model described above (and in connection with the Nakagami channel) is that it treats all the clusters to be identical. The fading can be considered to be more general if that condition is relaxed by treating the clusters to be non-identical. A simple case of two clusters will be considered first before considering additional clusters. For the case of two clusters, the SNR becomes

$$Z = X_1^2 + X_2^2. \tag{4.257}$$

If the clusters have sufficient number of scatterers to satisfy the central limit theorem, the densities of X's will be Gaussian. But, since the clusters are not identical, the average powers of the two clusters will be unequal. If the ratio of the average powers from the two clusters is defined as q^2 and σ_1 and σ_0 are the standard deviations of the Gaussian variables,

$$q^2 = \begin{cases} \frac{\sigma_1^2}{\sigma_2^2}, \sigma_1^2 < \sigma_2^2, 0 < q < 1 \\ \frac{\sigma_2^2}{\sigma_1^2}, \sigma_1^2 > \sigma_2^2, \quad 0 < q < 1 \end{cases}. \tag{4.258}$$

Under these conditions, the density function of the variable Z is given by (Nakagami 1960)

$$f(z) = \frac{(1+q^2)}{2qZ_0} \exp\left[-\frac{(1+q^2)^2}{4q^2 Z_0} z\right] I_0\left[\frac{(1-q^4)}{4q^2 Z_0} z\right]. \tag{4.259}$$

In Eq. (4.259), $I_0(.)$ is the modified Bessel function of the first kind of order 0 and Z_0 is the average SNR. The expression in Eq. (4.259) is identified as the Nakagami–Hoyt density.

While the Nakagami–Hoyt density and the Rician density (also referred to as the Nakagami–Rice density) contain $I_0(.)$, the density functions differ significantly in terms of the characteristics of the fading channel. While the Rician fading channel has the amount of fading less than unity (note that the Rician density arises from two independent Gaussian variables of identical variances but differing means), the amount of fading in a Nakagami–Hoyt channel (also called the Hoyt channel) becomes

$$AF = \frac{\langle Z^2 \rangle}{\langle Z \rangle^2} - 1 = 2\frac{(1+q^4)}{(1+q^2)^2}, 0 < q < 1. \tag{4.260}$$

Equation (4.260) leads to

$$1 \le AF_{Hoyt} \le 2. \tag{4.261}$$

This means that the amount of fading in a Hoyt channel is worse than what one observes in a Rayleigh channel. When q approaches 1, the density function in Eq. (4.259) becomes the exponential density

$$f(z) = \left\{ \frac{(1+q^2)}{2qZ_0} \exp\left[-\frac{(1+q^2)^2}{4q^2Z_0} z \right] I_0 \left[\frac{(1-q^4)}{4q^2Z_0} z \right] \right\}_{q \to 1} = \frac{1}{Z_0} e^{\frac{z}{Z_0}} U(z). \quad (4.262)$$

In other words, the Hoyt channel becomes the Rayleigh channels when the average powers are equal. Another way to interpret the results in Eqs. (4.259) and (4.262) is to redefine Eq. (4.257) as

$$Z = Z_1 + Z_2. \quad (4.263)$$

In Eq. (4.263), Z_1 and Z_2 are gamma random variables of order ½. If Z_1 and Z_2 are identical with mean values of $Z_0/2$, it can be shown that the density function of Z will be a gamma random variable with an order equal to (1/2 + 1/2) and mean equal to ($Z_0/2 + Z_0/2$). The density of Z_i can be expressed as

$$f(z_i) = \left(\frac{1\ 2}{2Z_0} \right)^{\frac{1}{2}} z_i^{\frac{-\frac{1}{2}}{\Gamma\left(\frac{1}{2}\right)}} e^{-\left(\frac{1}{2}\right)\frac{2z_i}{Z_0}} U(z), \quad i = 1, 2. \quad (4.264)$$

It can easily be seen that the pdf in Eq. (4.264) is the density of the SNR in a Nakagami channel with $m = 1/2$. In the case of two non-identical gamma variables Z_1 and Z_2 are considered with

$$\langle Z_1 \rangle = \frac{1}{1+q^2} Z_0. \quad (4.265)$$

$$\langle Z_2 \rangle = \frac{q^2}{1+q^2} Z_0. \quad (4.266)$$

The probability density functions are

$$f(z_1) = \left[\frac{1}{2} \frac{(1+q^2)}{Z_0} \right]^{\frac{1}{2}} z_1^{\frac{-\frac{1}{2}}{\Gamma\left(\frac{1}{2}\right)}} e^{-\left(\frac{1}{2}\right)\frac{(1+q^2)z_1}{Z_0}} U(z_1) \quad (4.267)$$

$$f(z_2) = \left[\frac{1}{2} \frac{(1+q^2)}{q^2 Z_0} \right]^{\frac{1}{2}} \frac{z_2^{-\frac{1}{2}}}{\Gamma\left(\frac{1}{2}\right)} e^{-\left(\frac{1}{2}\right)\frac{(1+q^2)z_2}{q^2 Z_0}} U(z_2). \quad (4.268)$$

The density function of the sum becomes

$$f(z) = \int_0^z f_{z_1}(z_1) f_{z_2}(z - z_1) dz_1. \quad (4.269)$$

The integral in Eq. (4.269) leads to the same pdf as in Eq. (4.259). In other words, the density function of two independent gamma variables of order half and unequal means lead to the Nakagami–Hoyt density (Nakagami 1960). Before going back to the presentation and discussion of the McKay model, one must also explore another way to examine the case of two clusters. This involves modeling the powers from the two clusters to be identically distributed with the correlation ρ existing between them. Going back to Eq. (4.263), treating Z_1 and Z_2 to be identically distributed gamma random variables of order ½ and mean power $Z_0/2$, the joint density of the two can be expressed as (Nakagami 1960; Okui et al. 1981; Nadaraja and Kotz 2006; Karagiannidis et al. 2006a, b; Chatelain et al. 2008; Papazafeiropoulos and Kotsopoulos 2011a, b)

$$f(z_1, z_2) = \left(\frac{1}{Z_0}\right)^{\frac{3}{2}} \frac{\left(\frac{z_1 z_2}{\rho}\right)^{-\frac{1}{4}}}{\Gamma\left(\frac{1}{2}\right)(1-\rho)} \exp\left[-\frac{1}{Z_0}\left(\frac{z_1+z_2}{1-\rho}\right)\right] I_{-\frac{1}{2}}\left[\frac{2\sqrt{z_1 z_2 \rho}}{Z_0(1-\rho)}\right]. \tag{4.270}$$

The density of the sum now becomes

$$f(z) = \int_0^z f_{z_1, z_2}(z - z_2, z_2) dz_2. \tag{4.271}$$

Performing the integration, the density in Eq. (4.271) becomes

$$f(z) = \frac{1}{Z_0\sqrt{1-\rho}} \exp\left[-\frac{1}{Z_0(1-\rho)} z\right] I_0\left[\frac{\sqrt{\rho}}{Z_0(1-\rho)} z\right]. \tag{4.272}$$

It can be seen from Eqs. (4.259) and (4.272) that the density functions of the sum of the powers from two clusters treated as independent with identical order but with different means or the clusters being treated as identical but correlated lead to similar forms. It is also possible to relate the parameters q^2 and ρ as

$$\rho = \frac{(1-q^2)^2}{(1+q^2)^2}. \tag{4.273}$$

These multiple forms of the density of the SNR from the two clusters will make it easy to understand some of the properties of the McKay fading channel.

Going back to the discussion of two clusters resulting in the Nakagami–Hoyt distribution in Eq. (4.259), that notion will now be extended to a number of such cluster groups. If n such groups exist in the channel, one can use the results from the work of Nakagami and obtain the density of the resultant SNR (once again represented by Z) as (Radaydeh and Matalgah 2008; Shankar 2013; Shankar 2015)

$$f(z)=\frac{\sqrt{\pi}}{\Gamma\left(\frac{n}{2}\right)\left[\frac{2qZ_{0n}}{n(1+q^2)}\right]^n\left(\frac{(1-q^4)n}{2q^2Z_{0n}}\right)^{\frac{(n-1)}{2}}}\exp\left[-\frac{n(1+q^2)^2}{4q^2Z_{0n}}z\right]z^{\frac{(n-1)}{2}}I_{\frac{(n-1)}{2}}\left[\frac{n(1-q^4)}{4q^2Z_{0n}}z\right]. \tag{4.274}$$

Equation (4.274) contains the average SNR expressed as Z_{0n} and it is given by

$$Z_{0n}=nZ_0. \tag{4.275}$$

In the original work by Nakagami, Eq. (4.274) is identified as the generalized Nakagami–Hoyt density (Nakagami 1960). The interesting part of this analysis is that there is another density that appears very similar to the one in Eq. (4.274) which results from the sum of two correlated gamma variables of identical order and equal means expressed as (Nakagami 1960; Holm and Alouini 2004)

$$f(z)=\frac{\sqrt{\pi}}{\left(2\frac{2m\sqrt{\rho}}{X(1-\rho)}\right)^{m-\frac{1}{2}}\Gamma(m)\left[\left(\frac{X}{2m}\right)^2(1-\rho)\right]^m}\exp\left[-\frac{2m}{X(1-\rho)}z\right]z^{m-\frac{1}{2}}I_{m-\frac{1}{2}}\left(\frac{2m\sqrt{\rho}}{X(1-\rho)}z\right). \tag{4.276}$$

In Eq. (4.276) X is the mean and ρ is the correlation coefficient making

$$X=\langle Z\rangle. \tag{4.277}$$

The two correlated random variables have identical mean of $X/2$ and each is of order m. It can easily be seen that Eq. (4.272) can be obtained from Eq. (4.276) by putting $m=1/2$ and noting that

$$\Gamma\left(\frac{1}{2}\right)=\sqrt{\pi}. \tag{4.278}$$

Comparing Eqs. (4.274) and (4.276) is seen that

$$n=2m. \tag{4.279}$$

Even though the lower limit of m is 1/2 in Nakagami's original work (Nakagami 1960), the density function of the SNR in a Nakagami model is a form of a gamma density and therefore, m can take any positive values. This suggests that one can entertain the possibility that (1) n does not have to be an integer and (2) the minimum value of n can therefore be equal to 0. To remove any symbolism associated with n being an integer, n is replaced by α and Eq. (4.274) is rewritten as

$$f(z)=\frac{\sqrt{\pi}}{\Gamma\left(\frac{\alpha}{2}\right)\left[\frac{2qZ_{0\alpha}}{\alpha(1+q^2)}\right]^{\alpha}\left(\frac{(1-q^4)\alpha}{2q^2Z_{0\alpha}}\right)^{\frac{(\alpha-1)}{2}}}\exp\left[-\frac{\alpha(1+q^2)^2}{4q^2Z_{0\alpha}}z\right]z^{\frac{(\alpha-1)}{2}}I_{\frac{(\alpha-1)}{2}}\left[\frac{\alpha(1-q^4)}{4q^2Z_{0\alpha}}z\right]. \tag{4.280}$$

Noting that $m=\alpha/2$, Eq. (4.276) becomes

$$f(z)=\frac{\sqrt{\pi}\quad z^{\frac{\alpha-1}{2}}}{\left(2\frac{\alpha\sqrt{\rho}}{X(1-\rho)}\right)^{\frac{\alpha-1}{2}}\Gamma\left(\frac{\alpha}{2}\right)\left[\left(\frac{X}{\alpha}\right)^2(1-\rho)\right]^{\frac{\alpha}{2}}}\exp\left[-\frac{\alpha}{X(1-\rho)}z\right]I_{\frac{\alpha-1}{2}}\left(\frac{\alpha\sqrt{\rho}}{X(1-\rho)}z\right). \tag{4.281}$$

Comparing Eqs. (4.280) and (4.281), it is seen that the values of ρ and q^2 are related as indicated by the Eq. (4.273). Furthermore, the mean of Z is

$$\langle Z\rangle = X = Z_{0n}. \tag{4.282}$$

It is clear that the relationship existing between ρ and q is such that when $\rho=0$, $q=1$ and when $\rho=1$, $q=0$. Stated in another way, the density function in Eq. (4.281) can be considered as arising from two independent non-identical gamma variables of identical order or arising out of the sum of two identical correlated gamma variables.

At this point, one can note that there is another density similar to the one in Eq. (4.281) originally proposed by McKay in 1932 (McKay 1932). The McKay density is given by

$$f(z)=E_0\ z^a\exp\left(-\frac{c}{b}z\right)I_a\left(\frac{z}{b}\right),\quad z\geq 0,\ b>0,\quad c>1,\quad a>-\frac{1}{2}. \tag{4.283}$$

It can be seen that

$$a=\frac{(\alpha-1)}{2}>-\frac{1}{2}. \tag{4.284}$$

$$b=\frac{X(1-\rho)}{\alpha\sqrt{\rho}}>0. \tag{4.285}$$

$$c=\frac{1}{\sqrt{\rho}}=\frac{(1+q^2)}{(1-q^2)}>1. \tag{4.286}$$

It must be noted that the density in Eq. (4.281) can also be attributed to the pdf of the SNR in maximal ratio combining with two correlated gamma branches. Thus, whether one treats the fading channel as arising out of a number of Nakagami–Hoyt signals, two correlated gamma clusters or two non-identical independent clusters, the density in Eq. (4.281) represents the density of the SNR in a short-term fading channel. This channel is identified as the McKay fading channel. Since Eqs. (4.280) and (4.281) represent the McKay fading channel, Eq. (4.281) will be used to represent the SNR of the McKay fading channel for the remainder of this analysis. It should be noted that the McKay density is similar to $\eta - \mu$ density except for the simple and single definition of ρ in the McKay density (Yacoub 2007a, b; Shankar 2013, 2015).

4.11.1.1 Moments and Laplace Transform of McKay Density

The moments of the McKay density in Eq. (4.283) are (Holm and Alouini 2004; Gradshteyn and Ryzhik 2007; Shankar 2005)

$$\langle Z^k \rangle = \frac{\Gamma(\alpha+k)}{\Gamma(\alpha)}\left(\frac{c}{c^2-1}\right)^k {}_2F^21\left(\left[-\frac{k-1}{2},-\frac{k}{2}\right];\left[\frac{(\alpha+1)}{2}\right];\frac{1}{c^2}\right)b^k. \tag{4.287}$$

In Eq. (4.287), ${}_2F_1(.)$ is the hypergeometric function (Gradshteyn and Ryzhik 2007). Comparing the original McKay density in Eq. (4.283) and the pdf of the SNR in a Mackay fading channel given in Eq. (4.281), the moments of SNR becomes

$$\langle Z^k \rangle = \frac{\Gamma(\alpha+k)}{\Gamma(\alpha)}{}_2F^21\left(\left[-\frac{k-1}{2},-\frac{k}{2}\right];\left[\frac{(\alpha+1)}{2}\right];\rho\right)\left(\frac{X}{\alpha}\right)^k. \tag{4.288}$$

The amount of fading in a McKay channel is

$$\mathrm{AF} = \frac{\langle Z^2 \rangle}{\langle Z \rangle^2} - 1 = \frac{1+\rho}{\alpha}. \tag{4.289}$$

Equation (4.289) provides the range of the amount fading existing in a McKay fading channel. It is possible to compare AF in Eq. (4.289) to AF in a Nakagami, Rayleigh, or a Nakagami–Hoyt channel. Since $0 \le \rho \le 1$ and $\alpha > 0$, the amount of fading in a McKay channel ranges from 0 to ∞, allowing the modeling of channels that are far worse than the Nakagami or Rayleigh channels.

As seen above, the McKay density can easily be interpreted as the pdf of the sum of two identically distributed gamma variables each having order of $\alpha/2$ and mean of $X/2$ with a correlation of ρ (positive). At this point, it is possible to summarize the relationship between the density of the two correlated gamma variables with identical orders and means resulting in the McKay distribution in Eq. (4.280) and the density of sum of two independent gamma variables of mean Z_{01} and Z_{02} of identical order $\alpha/2$. The marginal densities of these two independent gamma variables of non-identical means are

$$f_{Z_1}(z_1) = \left(\frac{\alpha}{2Z_{01}}\right)^{\frac{\alpha}{2}} z_1^{\frac{\alpha}{2}-1} e^{-\frac{\alpha}{2Z_{01}}z_1} U(z_1) \tag{4.290}$$

$$f_{Z_2}(z_2) = \left(\frac{\alpha}{2Z_{02}}\right)^{\frac{\alpha}{2}} z_2^{\frac{\alpha}{2}-1} e^{-\frac{\alpha}{2Z_{02}}z_2} U(z_2) \tag{4.291}$$

The density of the sum of these two independent and non-identical gamma variables can be obtained using transformation of variables as

$$f(z) = \sqrt{\pi}\left(\frac{\alpha}{2}\right)^{\frac{\alpha}{2}+\frac{1}{2}} \frac{|Z_{01}-Z_{02}|^{\frac{1}{2}-\frac{\alpha}{2}}}{\sqrt{Z_{01}Z_{02}}} z^{\frac{\alpha}{2}-\frac{1}{2}} e^{-\frac{\alpha}{4}\left(\frac{Z_{01}+Z_{02}}{Z_{01}Z_{02}}\right)z} I_{\frac{\alpha}{2}-\frac{1}{2}}\left[\frac{\alpha}{4Z_{01}Z_{02}}|Z_{01}-Z_{02}|z\right] U(z). \tag{4.292}$$

The relationships between the parameters of the McKay distribution in Eq. (4.292) and the McKay distribution in Eq. (4.280) can be expressed as

$$\begin{aligned} Z_{01} &= \frac{X}{2}(1+\sqrt{\rho}). \\ Z_{02} &= \frac{X}{2}(1-\sqrt{\rho}) \end{aligned} \tag{4.293}$$

Some of the simple short-term models of fading can be obtained from the McKay density in Eq. (4.281). If we treat the two clusters that resulted in Eq. (4.281) to be identical and independent, $q = 1$ and $\rho = 0$. To obtain the simple form, the density in Eq. (4.281) can be expressed in series form by expanding the modified Bessel function as (Gradshteyn and Ryzhik 2007)

$$I_\nu(x) = \sum_{k=0}^{\infty} \frac{1}{k!\Gamma(k+\nu+1)} \left(\frac{x}{2}\right)^{\nu+2k}. \tag{4.294}$$

$$f(z) = \frac{\sqrt{\pi} z^{\frac{(\alpha-1)}{2}}}{\left(\frac{2\alpha\sqrt{\rho}}{X(1-\rho)}\right)^{\frac{(\alpha-1)}{2}} \Gamma\left(\frac{\alpha}{2}\right)\left[\left(\frac{X}{\alpha}\right)^2(1-\rho)\right]^{\frac{\alpha}{2}}} \exp\left[-\frac{\alpha}{X(1-\rho)}z\right] \sum_{k=0}^{\infty} \frac{\left(\frac{\alpha\sqrt{\rho}}{2X(1-\rho)}z\right)^{\frac{(\alpha-1)}{2}+2k}}{k!\Gamma\left(k+\frac{1}{2}+\frac{\alpha}{2}\right)}. \tag{4.295}$$

Equation (4.295) can be simplified to

$$\begin{aligned} f(z) = {} & \frac{1}{\Gamma\left(\frac{\alpha}{2}\right)} \left(\frac{\alpha}{X\sqrt{1-\rho}}\right)^{\alpha} \sum_{k=0}^{\infty} \frac{\Gamma\left(k+\frac{\alpha}{2}\right)}{k!\Gamma(2k+\alpha)} \left(\frac{\alpha\sqrt{\rho}}{X(1-\rho)}\right)^{2k} \\ & \exp\left[-\frac{\alpha}{X(1-\rho)}z\right] \times z^{\alpha+2k-1}. \end{aligned} \tag{4.296}$$

Note that when $\rho \to 0$, the only term remaining in the summation will correspond to $\kappa = 0$ and Eq. (4.296) now becomes

$$f(z) = \frac{\sqrt{\pi}}{\Gamma(\frac{\alpha}{2})\Gamma(\frac{1}{2}+\frac{\alpha}{2})2^{\alpha-1}} \exp\left[-\frac{\alpha}{X}z\right]\left(\frac{\alpha}{X}\right)^{\alpha} z^{\alpha-1}. \tag{4.297}$$

Equation (4.297) is simplified using the doubling formula for the gamma functions given as (Gradshteyn and Ryzhik 2007)

$$\Gamma(2x) = \Gamma(x)\Gamma\left(x+\frac{1}{2}\right)\frac{2^{2x-1}}{\sqrt{\pi}}. \tag{4.298}$$

Using Eq. (4.298), the density function in Eq. (4.297) of two identical and independent gamma variables of parameter $\alpha/2$ and mean $X/2$ becomes

$$f(z) = \left(\frac{\alpha}{X}\right)^{\alpha} \frac{z^{\alpha-1}}{\Gamma(\alpha)} \exp\left[-\frac{\alpha}{X}z\right]. \tag{4.299}$$

Equation (4.299) is the pdf of the SNR in a gamma fading channel ($\alpha > 0$) or in a Nakagami fading channel ($\alpha \geq 1/2$) as expected. By putting $\alpha = 1$, the channel becomes a Rayleigh channel.

When $\alpha = 1$ and $\rho > 0$, the density in Eq. (4.281) becomes the Nakagami–Hoyt density

$$f(z) = \frac{1}{X\sqrt{1-\rho}} \exp\left[-\frac{z}{X(1-\rho)}\right] I_0\left[\frac{z\sqrt{\rho}}{X(1-\rho)}\right]. \tag{4.300}$$

Several short-term fading models are related to the McKay fading channel. For reasonable values of $m > 1$, the Nakagami model matches the Rician model and therefore, the McKay fading channel encompasses most of the fading scenarios observed in wireless channels. Once the generalized McKay fading density is derived in Sects. 4.11.1.4, the relationship of the McKay to other fading models will be shown.

It is possible to obtain an expression for the Laplace transform of the McKay density. The Laplace transform can be obtained using Maple directly. The Maple script is given below.

> *restart;# Laplace transform of Meijer G pdf*

$$> pdf := \sqrt{\pi}\cdot z^{\frac{(\alpha-1)}{2}} \cdot \frac{\exp\left(-\frac{\alpha}{X\cdot(1-\rho)}\cdot z\right)\cdot \text{BesselI}\left(\frac{(\alpha-1)}{2}, \frac{\alpha\cdot\sqrt{\rho}}{X\cdot(1-\rho)}\cdot z\right)}{\Gamma\left(\frac{\alpha}{2}\right)\cdot\left(\left(\frac{X}{\alpha}\right)^2\cdot(1-\rho)\right)^{\frac{\alpha}{2}}\cdot\left(\frac{2\cdot\alpha\cdot\sqrt{\rho}}{X\cdot(1-\rho)}\right)^{\frac{(\alpha-1)}{2}}};$$

$$pdf := \frac{\sqrt{\pi}\, z^{\frac{1}{2}\alpha-\frac{1}{2}}\, e^{-\frac{\alpha z}{X(1-\rho)}}\, \text{BesselI}\left(\frac{1}{2}\alpha-\frac{1}{2}, \frac{\alpha\sqrt{\rho}\, z}{X(1-\rho)}\right)}{\Gamma\left(\frac{1}{2}\alpha\right)\left(\frac{X^2(1-\rho)}{\alpha^2}\right)^{\frac{1}{2}\alpha}\left(\frac{2\alpha\sqrt{\rho}}{X(1-\rho)}\right)^{\frac{1}{2}\alpha-\frac{1}{2}}}$$

> *with(inttrans) :*

> *fL := simplify(laplace(pdf, z, s), assume = positive);*

$$fL := \left(\frac{s^2X^2\rho - 2sX\alpha - s^2X^2 - \alpha^2}{-1+\rho}\right)^{-\frac{1}{2}\alpha}\alpha^{\alpha}(1-\rho)^{-\frac{1}{2}\alpha}$$

> *# the above expression can be simplified manually to the following*

$$> McKayLaplace := \frac{\alpha^{\alpha}}{\left(\alpha^2 + s^2X^2 + 2sX\alpha - s^2X^2\rho\right)^{\frac{\alpha}{2}}};$$

$$McKayLaplace := \frac{\alpha^{\alpha}}{\left(\alpha^2 + s^2X^2 + 2sX\alpha - s^2X^2\rho\right)^{\frac{1}{2}\alpha}}$$

The Laplace transform of the McKay density is (Filho and Yacoub 2005b; Ermolova 2008, 2009)

$$L_Z(s) = \frac{\alpha^{\alpha}}{\left[\alpha^2 + 2s\alpha X + X^2s^2(1-\rho)\right]^{\frac{\alpha}{2}}}. \tag{4.301}$$

Note that Eq. (4.301) may also be obtained from the table of Laplace transforms given in Chap. 2 for the case of two non-identical gamma variables along with using the relationship between the correlated and uncorrelated variables in Eq. (4.293).

For the remainder of the analysis, Eq. (4.281) will be identified the density of the SNR in a McKay fading channel.

4.11.1.2 Plots of McKay CDF and PDF

Using Eq. (4.296), the cumulative distribution function (CDF) of the SNR in a McKay fading channel becomes (Shankar 2013, 2015)

$$F_Z(z)=\int_0^z f_Z(w)dw=\frac{\sqrt{\pi}\left(\sqrt{1-\rho}\right)^{\alpha}}{\Gamma\left(\frac{\alpha}{2}\right)}\sum_{k=0}^{\infty}\frac{\rho^k}{k!\Gamma\left(k+\frac{1}{2}+\frac{\alpha}{2}\right)2^{\alpha+2k-1}}\gamma\left(\alpha+2k,\frac{\alpha}{X(1-\rho)}z\right). \tag{4.302}$$

In obtaining the CDF in Eq. (4.302), the density function in the form of an infinite series in Eq. (4.296) is used. In Eq. (4.302) $\gamma(.,.)$ is the lower incomplete gamma function (Gradshteyn and Ryzhik 2007)

$$\gamma(q,x)=\int_0^x \exp(-y)y^{q-1}dy. \tag{4.303}$$

It is thus clear that the McKay distribution, originally proposed in 1932, is an ideal candidate to model the signal strength fluctuations in short-term fading channels. The general nature of the McKay density is evident by examining (4.280) for different values of α and ρ. When $\rho=0$ and $\alpha=1$, we have the Rayleigh channel, $\rho=0$ and $\alpha\geq$ ½, we have a Nakagami channel, $\rho=0$ and $\alpha>0$, we have a gamma channel, and $0<\rho<1$ and $\alpha=1$, we have the Hoyt channel and if $\alpha\geq 1$ and $0<\rho<1$, we have the α–μ channel. The McKay distribution allows the amount of fading to span the widest range (0–∞), making it possible to include channels exhibiting severe fading as well. The additional versatility of the McKay distribution is evident from the initial steps in the derivation, illustrating the fact that it also represents the density of the sum of a number of independent Nakagami–Hoyt variables as well as the sum of two correlated gamma random variables with direct applications in the study of maximal ratio combining (MRC) diversity. It can be seen that the McKay density is also identical to the $\eta-\mu$ density proposed to model short-term fading in wireless channels.

Several plots of the McKay density and cumulative distribution functions are shown in Figs. 4.87, 4.88, 4.89, 4.90, 4.91, and 4.92 for a few values of α, ρ, and X. All the results were generated in Matlab primarily with the aid of the symbolic toolbox. The Matlab script used is provided. It can be seen that as the value of α increases, the peaks of the density plots move to the right as expected. A similar trend is expected when ρ is smaller for a set of values of α and X. The script also performs numerical integration to verify that the pdf integrates to unity as expected.

```
mcKay_displays_pdf_CDF_shankarF(0.15,5)
```

```
pdf integrates to 1:  valid pdf
pdf integrates to 1:  valid pdf
pdf integrates to 1:  valid pdf
```

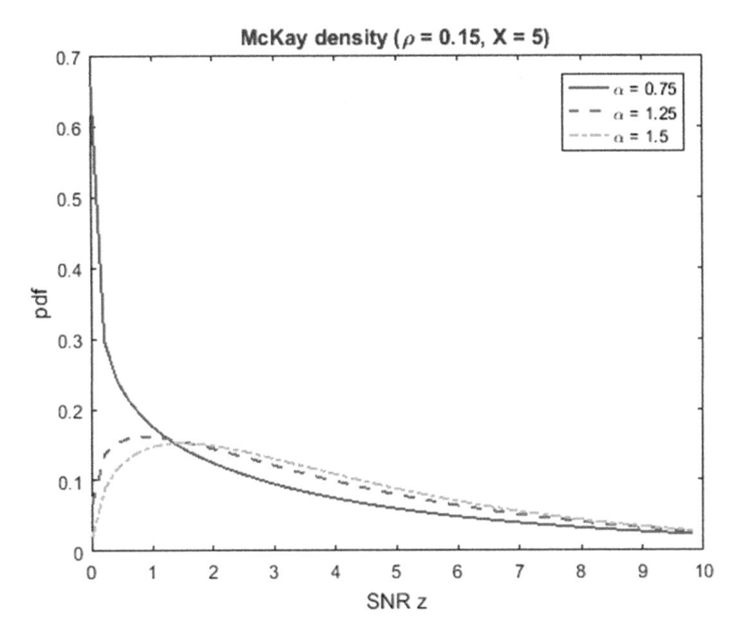

Fig. 4.87 Plots of the McKay density, $\rho = 0.15$

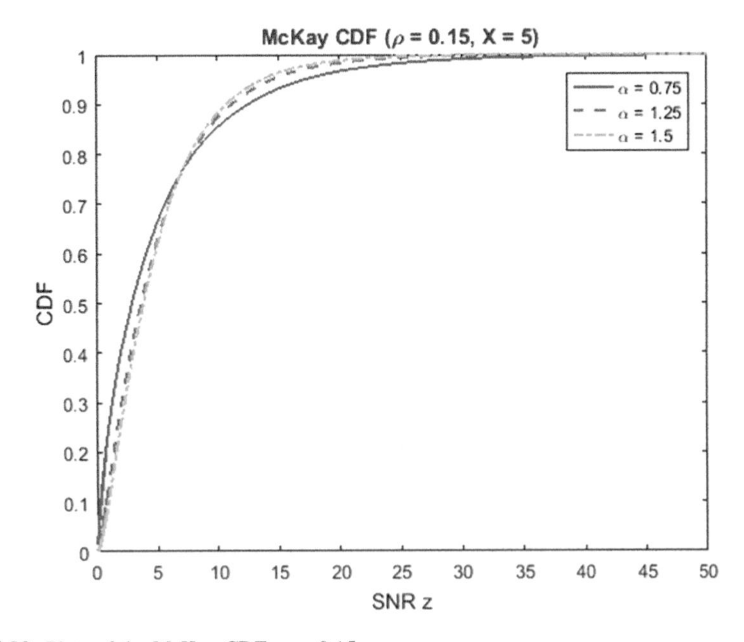

Fig. 4.88 Plots of the McKay CDF, $\rho = 0.15$

```
mcKay_displays_pdf_CDF_shankarF(0.35,5)
```

pdf integrates to 1: valid pdf
pdf integrates to 1: valid pdf
pdf integrates to 1: valid pdf

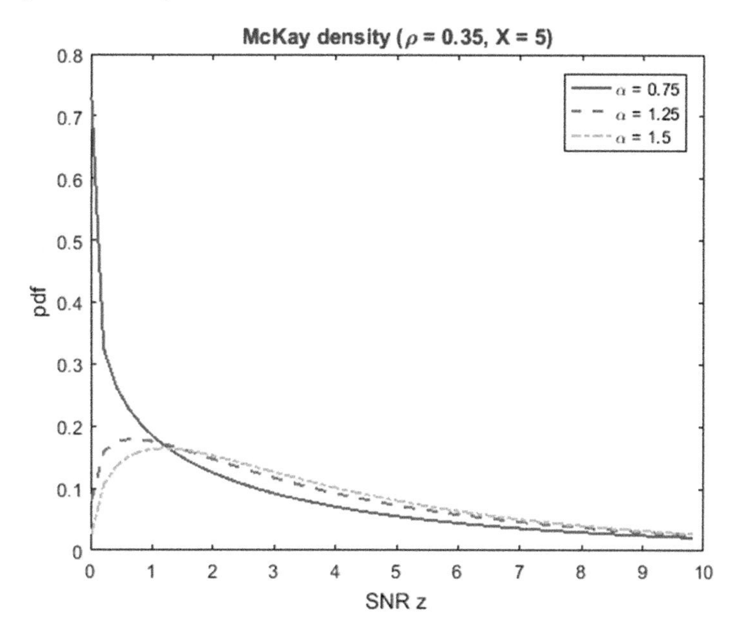

Fig. 4.89 Plots of the McKay density, $\rho = 0.35$

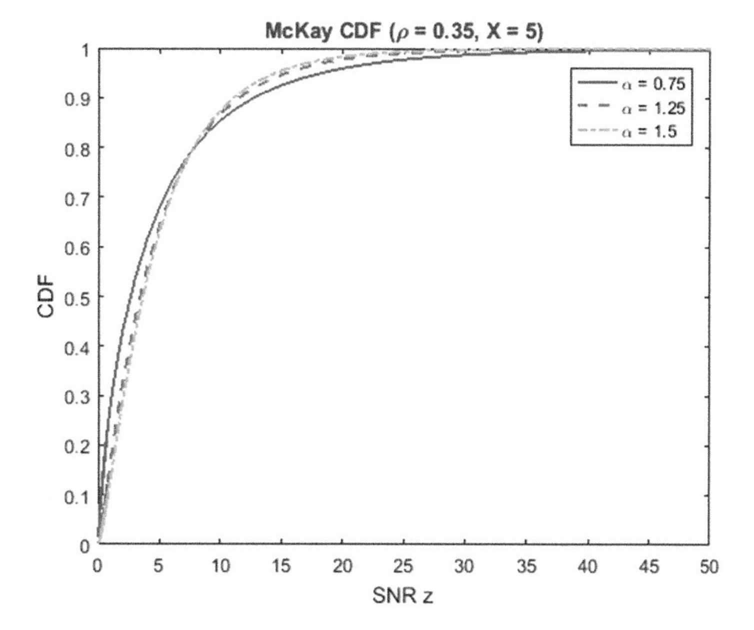

Fig. 4.90 Plots of the McKay CDF, $\rho = 0.35$

```
mcKay_displays_pdf_CDF_shankarF(0.75,10)
```

```
pdf integrates to 1: valid pdf
pdf integrates to 1: valid pdf
pdf integrates to 1: valid pdf
```

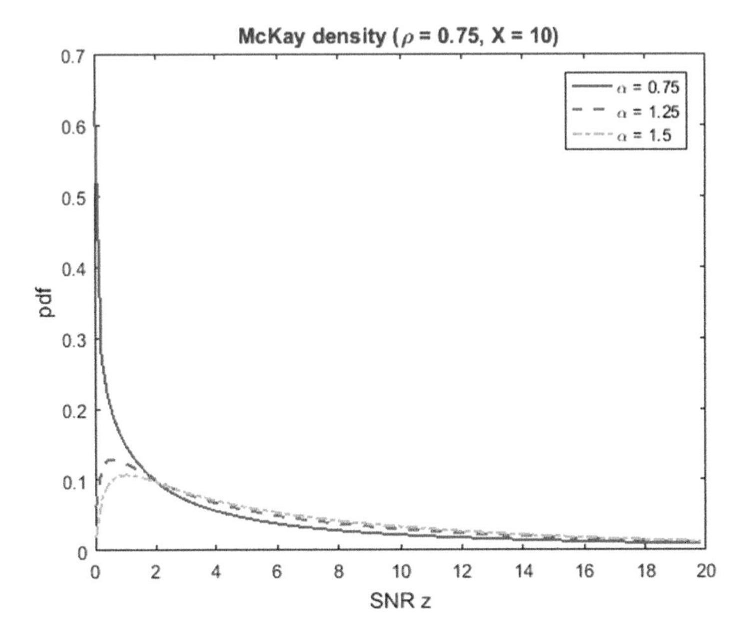

Fig. 4.91 Plots of the McKay pdf, $\rho = 0.75$

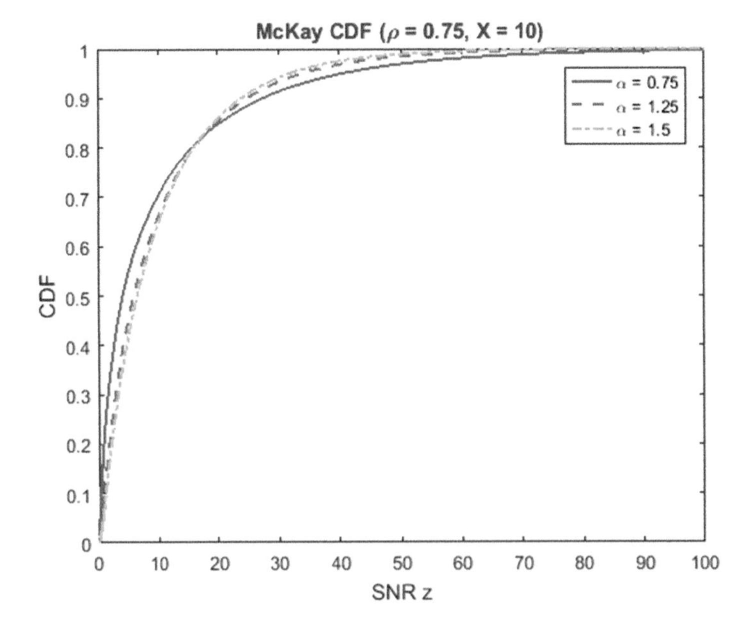

Fig. 4.92 Plots of the McKay CDF, $\rho = 0.75$

```
function mcKay_displays_pdf_CDF_shankarF(rho,XM)
% p m shankar, July 2016
%
% symbolic toolbox is needed
%
% display the McKay density, McKay CDF
% displays verification of pdf (total probability =1) in  command window
% The pdf is generated as an external function. CDF is also generated as an
% external function. In each case, the expression for the pdf is generated
% using the symbolic toolbox and converted to an in-line function. CDF is
% obtained using numerical integration. Integration is carried out by
% transforming the integral to one in tan(.) so that the upper limit will
% be pi/2 instead of INF. Care is taken to prevent the overflow by limiting
% the upper limit to a value slightly less than pi/2. In each case, the
% validity of the pdf is verified by checking to see if total probability
% is unity.
% the input required is rho and mean X. Three values of alpha are already
% included. One can go into the script and choose other values of alpha
% typically lying betweem 0 and 2 or 3 or more. Higher values might create
% issues with numerical integration and the numerical integration has not
% been optimized for higher values of alpha or mean.
close all
global funmckay X
syms z
X=XM;
% set # 1
alpha1=0.75;alpha2=1.25;  alpha3=1.5;
funmckay=mckaypdf1(alpha1,rho,X);
pdf1=matlabFunction(funmckay); % this is the pdf
[zzf,CDF1] = mckayCDF; % this is the CDF
[valueP] = mckaypdftestintegral; % verification of the pdf
disp(['pdf integrates to ',num2str(valueP),':  valid pdf'])
% set # 2
funmckay=mckaypdf1(alpha2,rho,X);
pdf2=matlabFunction(funmckay);
[zzf,CDF2] = mckayCDF;
[valueP] = mckaypdftestintegral;
disp(['pdf integrates to ',num2str(valueP),':  valid pdf'])
% set # 3
funmckay=mckaypdf1(alpha3,rho,X);
pdf3=matlabFunction(funmckay);
[zzf,CDF3] = mckayCDF;
[valueP] = mckaypdftestintegral;
disp(['pdf integrates to ',num2str(valueP),':  valid pdf'])
% plot the pdf
z1=0.01:.2:2*X;
plot(z1,pdf1(z1),'-',z1,pdf2(z1),'--',z1,pdf3(z1),'-.','linewidth',1.5)
legend(['\alpha = ',num2str(alpha1)],['\alpha = ',num2str(alpha2)],...
    ['\alpha = ',num2str(alpha3)])
title(['McKay density (\rho = ',num2str(rho),', X = ',num2str(X),')'],...
    'color','b')
xlabel('SNR z'),ylabel('pdf')
% plot the CDF
figure,plot(zzf,CDF1,'-',zzf,CDF2,'--',zzf,CDF3,'-.','linewidth',1.5)
ylim([0,1])
legend(['\alpha = ',num2str(alpha1)],['\alpha = ',num2str(alpha2)],...
    ['\alpha = ',num2str(alpha3)])
title(['McKay CDF (\rho = ',num2str(rho),', X = ',num2str(X),')'],...
    'color','b')
xlabel('SNR z'),ylabel('CDF')
end
```

```
% create the function to generate the pdf

function [funmckay] = mckaypdf1(alpha,rho,ME)
% create the McKay density
syms z a r X  positive
% get the expression for density directly from the notes
Nr=sqrt(sym(pi))*z^((1/2)*a-1/2)*exp(-a*z/(X*(1-r)))...
    *besseli((1/2)*a-1/2, a*sqrt(r)*z/(X*(1-r)));
Dr=gamma((1/2)*a)*(X^2*(1-r)/a^2)^((1/2)*a)...
    *(2*a*sqrt(r)/(X*(1-r)))^((1/2)*a-1/2);
fun=Nr/Dr; % pdf McKay depends on X, alpha (a) and rho(r)
funmckay=subs(fun,[a X r],[alpha,ME,rho]); % put values [alpha, X and rho]
end

% verification of the mckay density validity.. numerical integration

function [valueP] = mckaypdftestintegral
% verification of the integral
global funmckay
mckaydensityf=matlabFunction(funmckay);
syms z y
fun1=subs(mckaydensityf,z,tan(y));
fun2=fun1*(1+tan(y)*tan(y));
mckayf=matlabFunction(fun2);
valueP=integral(mckayf,0,0.9995*pi/2); % should be equal to unity
% the upper limit needs to be made close to pi/2, but less than it so that
% the result does not become infinite
end
% create the function to generate the CDF
function [zzf,CDF] = mckayCDF
% Now obtain the CDF
global funmckay X
mckaydensityf=matlabFunction(funmckay);
xx1=[1:40]*.02;
xx2=[max(xx1):20]*.1;
xx3=[round(max(xx2)):10*X];
zz=[xx1,xx2,xx3]; % create the X-axis values; uneven
KL=length(zz);
for k=1:KL;
CDF(k)=integral(mckaydensityf,0,zz(k)); % this is the CDF
end;
zzf=zz;
end
```

The next step was to examine the representation of the CDF as a sum and the number of terms required to get sufficient accuracy. The results are shown in Figs. 4.93, 4.94, 4.95, and 4.96. They show the accuracy of the summation and the effect of the number of terms in Eq. (4.296) on the pdf and CDF. It can be seen that the exact pdf and the pdf obtained from the summation match even when the number of terms is about 20. The result of the integration of the pdf is shown separately from what appears in the command window.

```
mcKay_displays_pdfsum_CDFsum_shankarF(.8,.6,10)
```

pdf integrates to 0.98904: valid pdf
pdf integrates to 0.99917: valid pdf
pdf integrates to 0.99951: valid pdf
pdf integrates to 0.99951: valid pdf

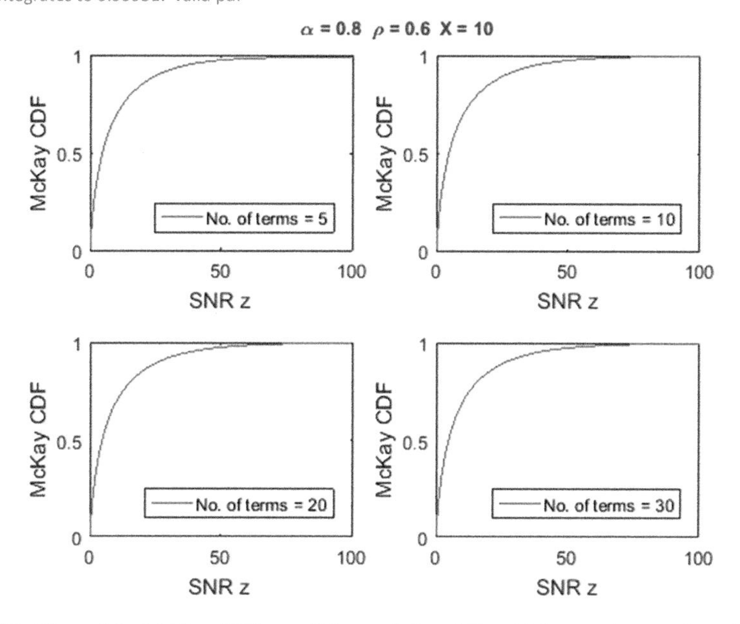

Fig. 4.93 Plots of the McKay CDF, $\alpha = 0.8$, $\rho = 0.6$, and $X = 10$ for varying number of terms

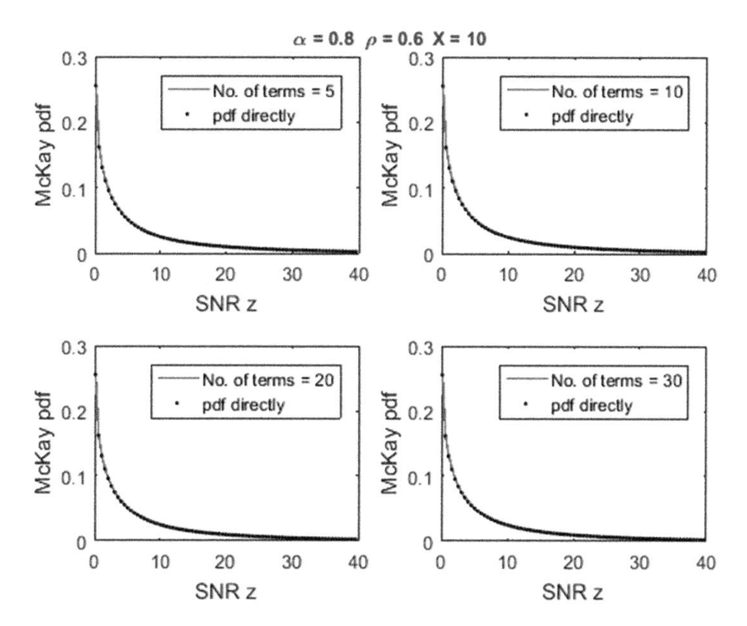

Fig. 4.94 Plots of the McKay pdf, $\alpha = 0.8$, $\rho = 0.6$, and $X = 10$ for varying number of terms

```
mcKay_displays_pdfsum_CDFsum_shankarF(1.4,.4,6)
```

pdf integrates to 0.99794: valid pdf
pdf integrates to 0.99998: valid pdf
pdf integrates to 1: valid pdf
pdf integrates to 1: valid pdf

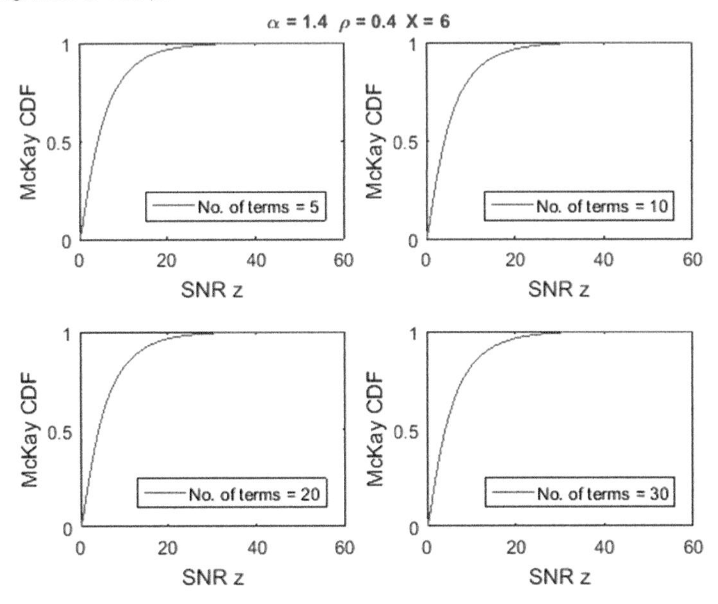

Fig. 4.95 Plots of the McKay CDF, $\rho = 0.4$, $\alpha = 0.1.4$, and $X = 6$ for varying number of terms

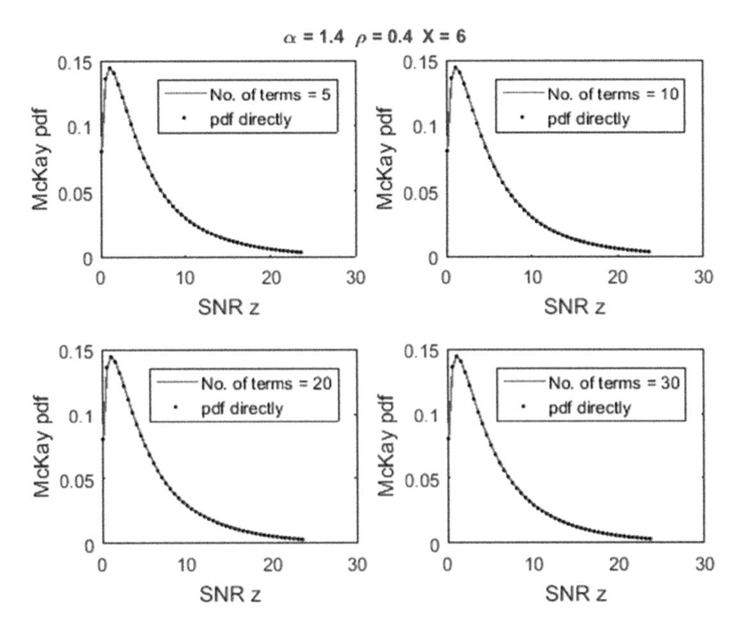

Fig. 4.96 Plots of the McKay pdf, $\rho = 0.4$, $\alpha = 0.1.4$, and $X = 6$ for varying number of terms

```
function mcKay_displays_pdfsum_CDFsum_shankarF(alpha,rho,XM)
%
% requires symbolic toolbox

% display the McKay CDF using the summation for BesselI and verify that
% integral (total probability) is unity. Four values for number of terms
% in the sum is already considered within the code. The CDF expression is
% generated symbolically and converted into an in line function.
% pdf is obtained using the summation and then compared to the direct one.
% P M Shankar July 26
close all
global funmckaysum X
X=XM;
syms z
MM=[5,10,20,30]; % number of terms in the summation
for km=1:length(MM)
    M=MM(km);
    [funmckaysum] = mckaypdfsum(alpha,rho,X,M);
    % step # 1 verify that pdf is VALID; integrate the pdf from 0 to INF
    [valueP] = mckaypdftestintegral;
    disp(['pdf integrates to ',num2str(valueP),':  valid pdf'])
    % Step # 2 Obtain the CDF
    [zzf,FF] = mckaycdff;
    subplot(2,2,km),plot(zzf,FF,'r-','linewidth',1.1)
    if km==1
        text(80,1.15,['\alpha = ',num2str(alpha),'  \rho = ',num2str(rho),...
            '  X = ',num2str(X)]);
    else
    end;
    xlabel('SNR z'),ylabel('McKay CDF')
    legend(['No. of terms = ',num2str(M)],'location','southeast')

end;
figure
for km=1:length(MM)
    M=MM(km);
    [funmckaysum] = mckaypdfsum(alpha,rho,X,M);
    z1=0.1:.5:4*X;
    mckaydens=matlabFunction(funmckaysum);
    mckaypdfd=mckaypdfdirect(alpha,rho,X);
    pdfq=matlabFunction(mckaypdfd);
    subplot(2,2,km)
    plot(z1,mckaydens(z1),'r-',z1,pdfq(z1),'k.','linewidth',1.1)
    if km==1
        text(80,1.15,['\alpha = ',num2str(alpha),...
            '  \rho = ',num2str(rho),'  X = ',num2str(X)]);
    else
    end;
    xlabel('SNR z'),ylabel('McKay pdf')
    legend(['No. of terms = ',num2str(M)],'pdf directly')
end;

end

% verification of the mckay density validity.. numerical integration

function [valueP] = mckaypdftestintegral
% verification of the integral
global funmckaysum
mcaydensityf=matlabFunction(funmckaysum);
syms z y
```

```
fun1=subs(mcaydensityf,z,tan(y));
fun2=fun1*(1+tan(y)*tan(y));
mckayf=matlabFunction(fun2);
valueP=integral(mckayf,0,0.995*pi/2); % should be equal to unity
end
% create a function to check the density obtained using summation
% influence of the summation terms
function [funmckaysum] = mckaypdfsum(alpha,rho,ME,MM)
% create the McKay density as a SUM
syms z a r X  positive
syms k M
a2=(a-1)/2;
r1=1-r;
rq=sqrt(r);
x1=a/(X*r1);
x2=x1*rq;
x22=2*x2;
x3=gamma(a/2)*(r1*X^2/a^2)^(a/2);
ex1=exp(-x1*z);
XX2=x2*z/2;
XX4=XX2^2;
eI=(XX2^a2)*symsum(((XX4)^k)...
    /(factorial(k)*gamma(k+1+a2)),k,0,M); % M is the number of terms SUM
ez=z^a2;
con=sqrt(pi)/(x3*x22^a2);
fun2=con*ex1*eI*ez;
funmckaysum=subs(fun2,[a X r M],[alpha,ME,rho,MM]); % alpha, X, rho, M
end


function [zzf,FF] = mckaycdff
% create the CDF
global funmckaysum X
mcaydensityf2=matlabFunction(funmckaysum);
xx11=[1:40]*.02;
xx22=[max(xx11):20]*.1;
xx33=[max(xx22):10*X];
zzf=[xx11,xx22,xx33]; % create the X-axis values; uneven
KL1=length(zzf);
for kk=1:KL1;
    FF(kk)=integral(mcaydensityf2,0,zzf(kk));
end;
end

function [funmckay] = mckaypdfdirect(alpha,rho,ME)
% create the McKay density direct expression instead of the summation
syms z a r X  positive
% get the expression for density directly from the notes
Nr=sqrt(sym(pi))*z^((1/2)*a-1/2)*exp(-a*z/(X*(1-r)))...
    *besseli((1/2)*a-1/2, a*sqrt(r)*z/(X*(1-r)));
Dr=gamma((1/2)*a)*(X^2*(1-r)/a^2)^((1/2)*a)...
    *(2*a*sqrt(r)/(X*(1-r)))^((1/2)*a-1/2);
fun=Nr/Dr; % pdf McKay depends on X, alpha (a) and rho(r)
funmckay=subs(fun,[a X r],[alpha,ME,rho]); % put values [alpha, X and rho]
end
```

4.11.1.3 McKay Parameter Estimation

The discussion of the McKay density for modeling short-term fading channels will be incomplete without an examination of the approaches for estimation of its parameters. As discussed in Chap. 2, given a set of data collected from a wireless channel, it is essential to estimate the parameters of the McKay distribution, test whether McKay density is an acceptable statistical fit to the data before predicting the performance based on error rates and outage probabilities.

Two methods available for the estimation of parameters, method of moments (MoM) and maximum likelihood estimation (MLE) were presented in Chap. 2. Since three parameters exist, at least three moments are needed for the solution of the parameters if MOM is used. A discussion of MLE is presented first before offering an assessment of the MOM for the estimation of parameters of the McKay density (Chatelain et al. 2008; Shankar 2013).

The choice of the initial guesses of the parameters is critical to the success of MLE. These initial estimates of three parameters were made on the basis of treating the data to fit a gamma density by testing whether the data set passes a chi-square test for gamma. If it passes the gamma test, the initial guess of ρ is taken to be an extremely low value ($1e^{-5}$). If it fails the gamma test, the initial guess for ρ is taken to be 0.15. This makes the computation with three parameters relatively simple.

Once the parameters are estimated, a chi-square testing is undertaken to determine whether McKay density is the appropriate fit. In addition, mean square estimate of the error is also undertaken. Four examples are shown below. The Matlab script (fully annotated) is also given. It must be noted that the data sets were created within the script and therefore, the scripts need to be modified for use for a different set of input data. (In the absence of real data, samples of McKay variables are used as inputs for the study. The script can be easily changed slightly by bringing in data set as the input.) In each case, the values of α, ρ, and X could be extracted as outputs. While the chi-square test statistic appears in the command window, the other results are displayed on a plot. The Matlab script appears only with example # 1. The results are displayed in Figs. 4.97, 4.98, 4.99, 4.100, and 4.101.

Example # 1

```
function mckayparameter_mle_symbf
% estimate parameters of mcKay density;
% October 2016
% create the pdf symbolically first
% July 10: A segment added to check if data is gamma distributed. In this
% case, correlation coefficient will be either -ve or complex. Based on the
% gamma test, the initial guesses for the three parameters are made
% a chi square test for McKay density is also undertaken. The histogram
% (ksdensity) is matched to the theoretical pdf of the mcKay density using
% the parameters of the estimate. MSE value is also estimated.
close all
syms XX aL rh av positive
 % alpha (aL), mean (av) and rho (rh) XX is z, the SNR
 % get the pdf expression from the notes
 Nr=sqrt(sym(pi))*XX^((1/2)*aL-1/2)*exp(-aL*XX/(av*(1-rh)))...
    *besseli((1/2)*aL-1/2, aL*sqrt(rh)*XX/(av*(1-rh)));
Dr=gamma((1/2)*aL)*(av^2*(1-rh)/aL^2)^((1/2)*aL)...
    *(2*aL*sqrt(rh)/(av*(1-rh)))^((1/2)*aL-1/2);
fun=Nr/Dr; % pdf McKay depends on X, alpha (a) and rho(r)
mckaydensityf=matlabFunction(fun); % order the variables is important
% it should be XX (for z) followed by alpha (aL), mean (av) and rho (rh)
% choose symbolic variables such that the function is created such that the
% order of the variables pdf(x,[parameters]).... note that XX represents x
% in MLE
clear XX aL rh av
% modified July 2016; replaced the average values
% CREATE THE DATA
m=1.15; % this is alpha
m2=m/2;
XX=20; % this is X
XX2=XX/2;
rr=0.6; % this is rho
X1=XX2*(1+sqrt(rr));
x1=gamrnd(m2,X1/m2,1,15000); % gamma random variable
X2=XX2*(1-sqrt(rr));
x2=gamrnd(m2,X2/m2,1,15000); % gamma random variable
x=x1+x2; % McKay variable
% both have equal orders (alpha/2) and the ratio of the average powers is h
% h=min(X1,X2)/max(X1,X2);% ratio of the two means
%  rh_th=((1-h)/(1+h))^2 % rho from the ratio of the means
% the above segment is not used since the original data input will consist
% of a data set and there is no apriori information on the ratio of the
% average powers
% check whether the data fits gamma or Nakagami in amplitude
QC=chi2inv(.95, 7);% number of bins 10-1-2; 2 is the number of parameters
phat=gamfit(x);
% get gamma fit and these will the initial guesses for alpha and X
qq = chigammaf(x);
if qq<=QC
    disp('data set is gamma distributed; correletion Coeff negligible')
% it appears that data set follows gamma; Corr. coeff must be very small
    cc=1e-5;
else
    % data set is not gamma distributed
    cc=0.15; % starting estimate of corr coeff
end;
aa=phat(1); % starting estimate of m or alpha
bb=phat(1)*phat(2); % starting estimate of X  or the mean
[my,ny]=size(x);
L=my*ny; % length of the data
x=reshape(x,[L,1]);
```

```
% use the MLE to estimate the parameters
opt = statset('mlecustom');
opt = statset(opt,'FunValCheck','off','MaxIter',3000,'TolX',1e-4,'TolFun',1e-5,'TolBnd',1e-
5','MaxFunEvals', 1000);
pp = mle(x,'pdf',mckaydensityf,'start',[aa bb cc], 'options',opt); % fit mypdf to the data Y
a=pp(1);% alpha
X=pp(2); % X
r=pp(3);% rho
if isreal(r)==0||r<0 % if the corr Coeff estimate is -ve
    r=1e-5;%
    disp('the correlation coefficient is complex or negative')
    disp('Data set probably matches a gamma distribution')
    disp('Corr Coeff. is taken to be a very low value, r=1e-5')
else
end;
% conduct a chi square test for McKay based on these parameters estimated
qq1=chimckay(x,a,X,r);
QC=chi2inv(.95, 6);% number of bins 10-1-3; 3 is the number of parameters
disp(['Chi square test statistic threshold (3 parameters; 10 bins) =',...
    num2str(round(QC))])
disp(['Chi square test statistic (McKay) =',num2str(round(qq1))])
% test to see match of the densities
[f,xr]=ksdensity(x);
NK=[]; % eliminate negative values of xr
NK=find(xr<0);
if isempty(NK)==0
    NK1=max(NK)+1;% find the largest index and add 1 to start the non-zero values
    xr=xr(NK1:end);
    f=f(NK1:end);
else
end;
L=length(xr);
fs=mckaydensityf(xr,a,X,r);
figure,plot(xr,f,'r-',xr,fs,'*')
xlabel('SNR z'),ylabel('probability density function')
legend('histogram','fit')
 tit={ ['\alpha_{input} = ',num2str(m),'; ','X_{input} = ',...
     num2str(XX),'; ','\rho_{input} = ',num2str(rr)]
     ['\alpha_{est} = ',num2str(a),'; ','X_{est} = ',...
    num2str(X),'; ','\rho_{est} = ',num2str(r)]};
 title(tit)
MSE=sum((f-fs).^2)/L;
text(0.7*max(xr),0.2*max(max(f,fs)),['MSE = ',num2str(MSE)])


end
% chi square test for gamma
function qq = chigammaf(yy)
m=10;% number of bins
y=sort(yy); % random samples
N=length(yy);%size of the sample
phat=gamfit(y);
inter=max(y)-min(y); %interval for chi-square test
intstep=inter/m; %size of sub-interval
intval=[min(y):intstep:max(y)]; %samples at the sub-intervals
k=[zeros(1,m)]; %number of samples in each sub-interval intialization
for i=1:N
   for j=1:m
      if y(i)<=intval(j+1)
         k(j)=k(j)+1;
         break;
      end;
```

```
    end;
end;
p=zeros(1,m);
for i=1:m
    p(i)=gamcdf(intval(i+1),phat(1),phat(2))-gamcdf(intval(i),phat(1),phat(2)); %probabilities at
the sub-interval samples
end;
np=p.*N;
q=0;
for i=1:m
    q = q + ((k(i)-np(i))^2)/np(i); %chi-square test
end;
 qq=q;

end

% chi square test for mcKay
function qq = chimckay(yy,a,X,r)
m=10;% number of bins
y=sort(yy); % random samples
N=length(yy);%size of the sample
inter=max(y)-min(y); %interval for chi-square test
intstep=inter/m; %size of sub-interval
intval=[min(y):intstep:max(y)]; %samples at the sub-intervals
k=[zeros(1,m)]; %number of samples in each sub-interval intialization
for i=1:N
    for j=1:m
        if y(i)<=intval(j+1)
            k(j)=k(j)+1;
            break;
        end;
    end;
end;
% create the CDF
syms x aL av rh
Nr=sqrt(sym(pi))*x^((1/2)*aL-1/2)*exp(-aL*x/(av*(1-rh)))...
    *besseli((1/2)*aL-1/2, aL*sqrt(rh)*x/(av*(1-rh)));
Dr=gamma((1/2)*aL)*(av^2*(1-rh)/aL^2)^((1/2)*aL)...
    *(2*aL*sqrt(rh)/(av*(1-rh)))^((1/2)*aL-1/2);
fun=Nr/Dr; %
funx=subs(fun,[aL,rh,av],[a,r,X]);% function of a single variable
mckaydensityf=matlabFunction(funx);

p=zeros(1,m);
for i=1:m
     prob=integral(mckaydensityf,intval(i),intval(i+1)); %CDF for the range
    p(i)=prob; %probabilities at the sub-interval samples
end;
np=p.*N;
q=0;
for i=1:m
    q = q + ((k(i)-np(i))^2)/np(i); %chi-square test
end;
 qq=q;
end
```

```
Chi square test statistic threshold (3 parameters; 10 bins) =13
Chi square test statistic (McKay) =6
```

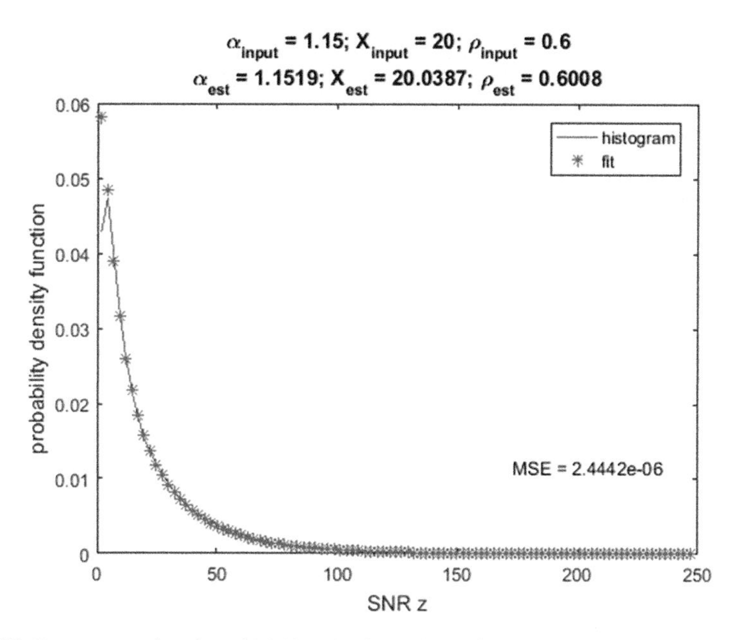

Fig. 4.97 Parameter estimation of McKay density (MLE): Example # 1

Example # 2

```
% CREATE THE DATA
m=0.85; % this is alpha
m2=m/2;
XX=10; % this is X
XX2=XX/2;
rr=0.2; % this is rho
```

```
Chi square test statistic threshold (3 parameters; 10 bins) =13
Chi square test statistic (McKay) =11
```

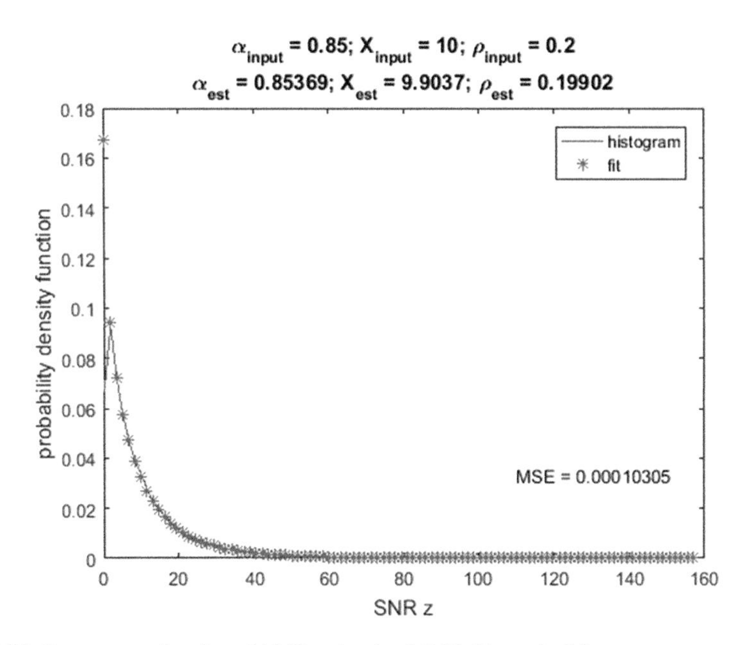

Fig. 4.98 Parameter estimation of McKay density (MLE): Example # 2

Example # 3

```
% CREATE THE DATA
m=2.2; % this is alpha
m2=m/2;
XX=10; % this is X
XX2=XX/2;
rr=0.6; % this is rho
```

```
Chi square test statistic threshold (3 parameters; 10 bins) =13
Chi square test statistic (McKay) =7
```

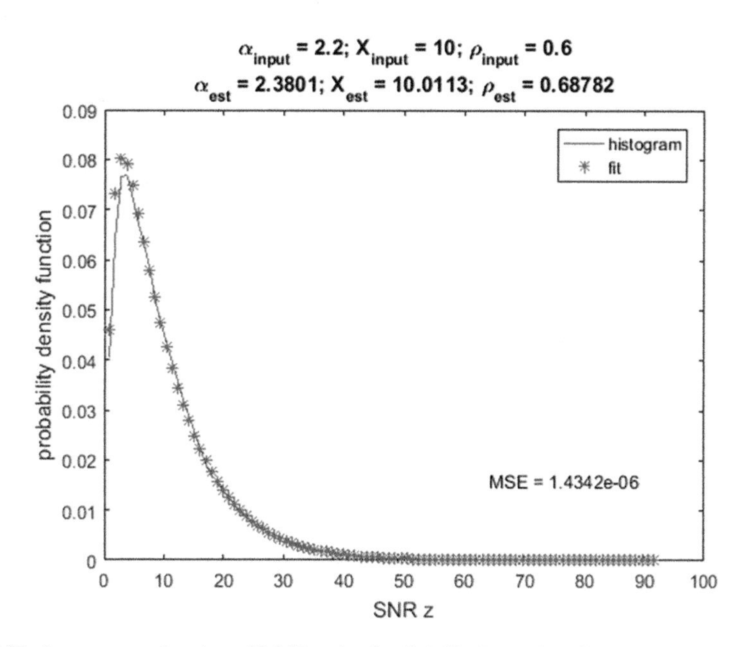

Fig. 4.99 Parameter estimation of McKay density (MLE): Example # 3

Example # 4

```
% CREATE THE DATA
m=.6; % this is alpha
m2=m/2;
XX=5; % this is X
XX2=XX/2;
rr=0.001; % this is rho
data set is gamma distributed; correletion Coeff negligible
Chi square test statistic threshold (3 parameters; 10 bins) =13
Chi square test statistic (McKay) =9
```

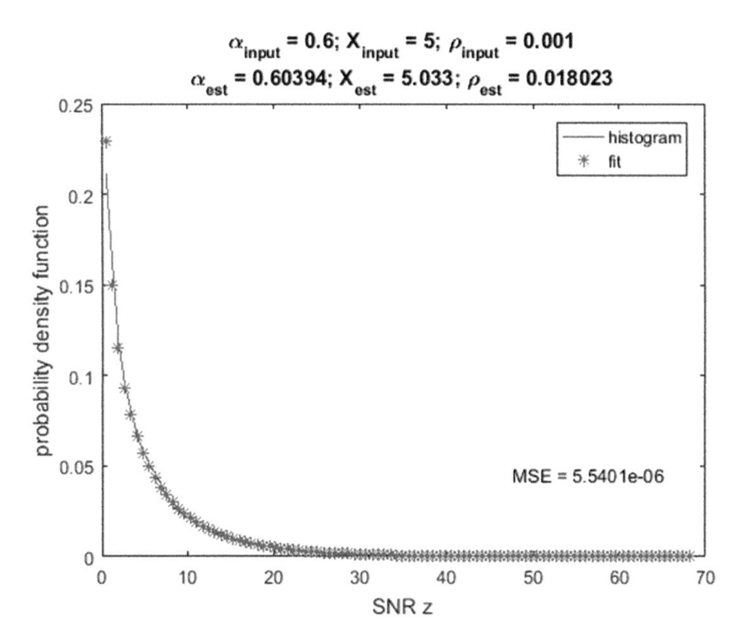

Fig. 4.100 Parameter estimation of McKay density (MLE): Example # 4

Example # 5 In this example, the input is gamma distributed. The Matlab script is coded to display the statement indicating that the correlation coefficient is too low for accurate computation suggesting that the input is probably gamma distributed

```
% CREATE THE DATA
x=gamrnd(1.1,8,1,20000);% a single set % McKay variable
```

```
data set is gamma distributed; correletion Coeff negligible
the correlation coefficient is complex or negative
Data set probably matches a gamma distribution
Corr Coeff. is taken to be a very low value, r=1e-5
Chi square test statistic threshold (3 parameters; 10 bins) =13
Chi square test statistic (McKay) =3
```

.

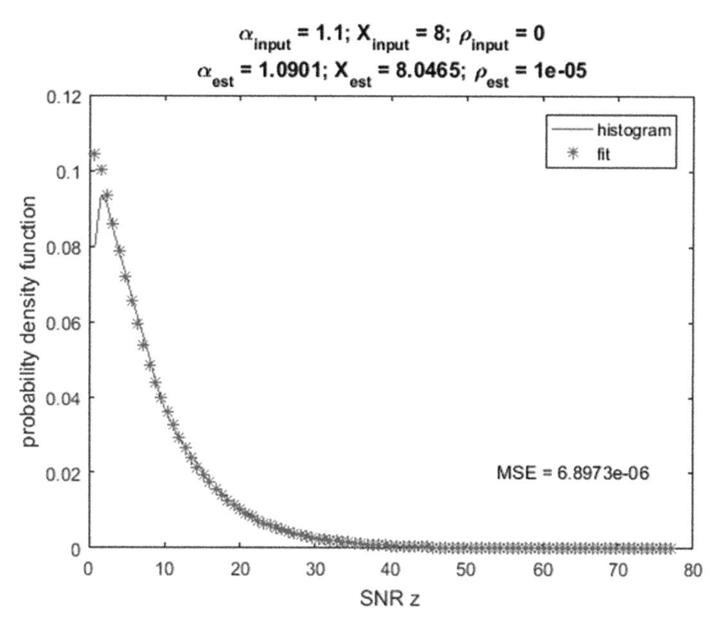

Fig. 4.101 Parameter estimation of McKay density (MLE): Example # 5

While MLE methods can be implemented in Matlab taking advantage of the "mle" command, it is also possible to use the method of moments (MoM). A simple way to implement MoM is to use a half order, first and second order moments to solve for the parameters. These moments are obtained from Eq. (4.288)

$$\left\langle Z^{\frac{1}{2}}\right\rangle = \frac{\Gamma\left(\alpha+\frac{1}{2}\right)}{\Gamma(\alpha)}\,{}_2F_1\left(\left[-\frac{1}{4},\frac{1}{4}\right];\left[\frac{(\alpha+1)}{2}\right];\rho\right)\left(\frac{X}{\alpha}\right)^{\frac{1}{2}}. \quad (4.304)$$

$$\langle Z\rangle = X. \quad (4.305)$$

$$\left\langle Z^{2}\right\rangle = \frac{\alpha+1+\rho}{\alpha}X^{2}. \quad (4.306)$$

Using moments normalized with respect to the first moment, a single equation in α can be created and solved. Only the solution of α was undertaken for the examples mentioned above. It can be seen that MLE provides a better estimate of α and therefore, it is likely to be more reliable means of parameter estimation. The Matlab script and the results on the MoM based estimation are given below.

```
% mcKay_parameter_moments
% estimation of alpha using MoM: The half order, first and second
% order meoments are used.
% P M Shankar, July 2016
clear;clc;close all
for k=1:5
    if k==1
m2=1.15; % this is alpha/2
X1=7;  x1=gamrnd(m2,X1/m2,1,15000); % gamma random variable
X2=16;  x2=gamrnd(m2,X2/m2,1,15000); % gamma random variable
x=x1+x2; % McKay variable
    elseif k==2
 m2=0.63; % this is alpha/2
X1=4;  x1=gamrnd(m2,X1/m2,1,15000); % gamma random variable
X2=18;  x2=gamrnd(m2,X2/m2,1,15000); % gamma random variable
x=x1+x2; % McKay variable
    elseif k==3
m2=0.55; % this is alpha/2
X1=2;  x1=gamrnd(m2,X1/m2,1,15000); % gamma random variable
X2=18;  x2=gamrnd(m2,X2/m2,1,15000); % gamma random variable
x=x1+x2; % McKay variable
    elseif k==4
m2=0.35; % this is alpha/2
X1=17;  x1=gamrnd(m2,X1/m2,1,15000); % gamma random variable
X2=18;  x2=gamrnd(m2,X2/m2,1,15000); % gamma random variable
x=x1+x2; % McKay variable
    else
x=gamrnd(1.1,8,1,20000);% a single set % McKay variable
    end;
x=x/mean(x);% normalize w.r.t mean
M2=mean(x.^2); % normalized second moment
Mh=mean(sqrt(x)); % normalized half moment
syms r a
r=M2*a-(1+a);
y=Mh-gamma(a+1/2)*hypergeom([-1/4, 1/4], [1/2+(1/2)*a], r)/(sqrt(a)*gamma(a));
yf=matlabFunction(y);
aa=0.5:.02:2.5;
idx=find(abs(yf(aa))<.002); % find where the equation changes sign
alpha_est=median(aa(idx)); % if there are multiple points, use the median
disp(['Example # ',num2str(k)])
disp(['alpha_estimate = ',num2str(alpha_est)])
clear r a idx

end;
```

```
Example # 1
alpha_estimate = 2.07
Example # 2
alpha_estimate = 0.99
Example # 3
alpha_estimate = 0.76
Example # 4
alpha_estimate = 0.7
Example # 5
alpha_estimate = 1.11
```

4.11.1.4 Performance in McKay Channels

Following the characterization of the McKay density, we can examine the error rates and outage probabilities. In all the analysis presented below, detailed Matlab scripts have been provided that were used in the creation of the plots.

Error Rates

The case of a coherent BPSK modem is considered as an example. The error rate in the McKay fading channel becomes

$$p_e(X) = \int_0^\infty \frac{\sqrt{\pi}\quad z^{\frac{\alpha-1}{2}} \exp\left[-\frac{\alpha}{X(1-\rho)} z\right]}{\left(2\frac{\alpha\sqrt{\rho}}{X(1-\rho)}\right)^{\frac{\alpha-1}{2}} \Gamma\left(\frac{\alpha}{2}\right) \left[\left(\frac{X}{\alpha}\right)^2 (1-\rho)\right]^{\frac{\alpha}{2}}} I_{\frac{\alpha-1}{2}}\left(\frac{\alpha\sqrt{\rho}}{X(1-\rho)} z\right) \frac{1}{2} erfc\left(\sqrt{z}\right) dz. \tag{4.307}$$

In Eq. (4.307), $\frac{1}{2}erfc(.)$ is the error rate in an ideal Gaussian channel and X is the average SNR. The error rate as a function of the average SNR needs to be evaluated numerically.

If the McKay density in summation form in Eq. (4.296) is used, it is possible to obtain a simple expression for the error rate (as a sum),

$$p_e(X) = \sum_{k=0}^{\infty} \frac{\left(\frac{\alpha}{2X}\right)^{\alpha+2k} \rho^k}{k!\Gamma\left(\frac{\alpha}{2}\right)\Gamma\left(\frac{\alpha_k+1}{2}\right)} \frac{\Gamma\left(\alpha_k+\frac{1}{2}\right)}{\alpha_k(1-\rho)^{2k+\frac{\alpha}{2}} {}_2F^2 1} \left(\left[\alpha_k, \alpha_k+\frac{1}{2}\right], [1+\alpha_k], -X_{\alpha\rho}\right). \tag{4.308}$$

In Eq. (4.308),

$$\alpha_k = \alpha + 2k. \tag{4.309}$$

$$X_{\alpha\rho} = \frac{\alpha}{X(1-\rho)}. \tag{4.310}$$

Note that ${}_2F_1(\cdot)$ is the hypergeometric function (Gradshteyn and Ryzhik 2007).

Since an analytical expression for the Laplace transform of the McKay density is available, it is possible to use Laplace transforms to evaluate the bit error rates. Note that the Laplace transform, $L_X(s)$, and moment generating function $M_X(s)$ of the density of a random variable X are expressed as (da Costa and Yacoub 2008; Peppas et al. 2009; Yilmaz and Alouini 2012)

$$M_X(s) = L_X(-s). \tag{4.311}$$

The error rate in terms of the moment generating function for BPSK is (Simon and Alouini 2005)

$$p_e(X) = \frac{1}{\pi} \int_0^{\frac{\pi}{2}} M_X\left(-\frac{1}{\sin^2(\theta)}\right) d\theta. \tag{4.312}$$

Thus, the error rate in a McKay fading channel becomes

$$p_e(X) = \frac{1}{\pi} \int_0^{\frac{\pi}{2}} L_X\left(\frac{1}{\sin^2(\theta)}\right) d\theta. \tag{4.313}$$

In terms of the Laplace transform of the McKay density given in Eq. (4.301), the error rate becomes

$$p_e(X) = \frac{1}{\pi} \int_0^{\frac{\pi}{2}} \alpha^{\alpha} \left[\alpha^2 + 2\alpha \frac{X}{\sin^2(\theta)} + \frac{X^4}{\sin^4(\theta)}(1-\rho)\right]^{-\frac{\alpha}{2}} d\theta. \tag{4.314}$$

All three approaches presented above in Eqs. (4.307), (4.308), and (4.314) require numerical methods for error rate estimation. The Matlab script used to obtain the error rates and the results are presented next. The script only appears with the first set of results for $\alpha = 1.35$. For the other values of α, only plots are shown. For the estimation of error rates on the basis of summation in Eq. (4.308), ten terms are used. For the case of the lowest correlation, results are compared to the case of a Nakagami channel. They are displayed in Figs. 4.102, 4.103, 4.104, 4.105, 4.106, and 4.107.

```
function bercalculation_mckay
% P M Shankar, October 2016
% bit error calculations for BPSK based on three approachs
% direct integration with the pdf. converted into trigonometric form
%
% using the density expressed as a sum
%
% using the Laplace transforms.
%
% In each case, the integrands & the summation are created symbolically and
% transformed into on-line functions for evaluation. alpha needs to
% be entered manually even though the function can be modified to have
% alpha as the input upon prompt. The SNR is allowed to vary from 5 to 40
% dB, correlation coefficient [0.0001,0.1,0.4,0.6,0.8].
% Summation is created as a separate function since it takes up longer.
% The number of terms in the summation (M) has been kept at 10 and can be
% varied if necessary.
%
% for rho =0.0001, results are compared to the case of pure Nakagami
% channel
close all
```

```
global Z alpha rh
Z0=5:3:40;
Z0=[Z0,40];
Z=10.^(Z0/10);
LK=length(Z);
alpha=0.95;
rh=[0.0001,0.1,0.4,0.6,0.8]; % rho values
ber1= bernakagami; % error rates in a Nakagami channel
LR=length(rh);
syms z a r X  positive
% get the expression for density directly from the notes
Nr=sqrt(sym(pi))*z^((1/2)*a-1/2)*exp(-a*z/(X*(1-r)))...
    *besseli((1/2)*a-1/2, a*sqrt(r)*z/(X*(1-r)));
Dr=gamma((1/2)*a)*(X^2*(1-r)/a^2)^((1/2)*a)...
    *(2*a*sqrt(r)/(X*(1-r)))^((1/2)*a-1/2);
fun=Nr/Dr; % pdf McKay depends on X, alpha (a) and rho(r)
funer=0.5*erfc(sqrt(z))*fun;% error rate integrand
% Laplace transform
syms s
fLzer=a^a/(a^2+s^2*X^2+2*s*X*a-s^2*X^2*r)^((1/2)*a);
peD=zeros(LK,LR); % pe directly
peL=zeros(LK,LR);% pe using Laplace transforms or MGF
for kr=1:LR
    rho=rh(kr);
    for k=1:LK
         % put values [alpha, X and rho]
        funmckay=subs(funer,[a X r],[alpha,Z(k),rho]);
        % convert from dz to dy with z=tan(y)
        syms y
        fun1m=subs(funmckay,z,tan(y));
        fun2m=fun1m*(1+tan(y)*tan(y));
        mckayf1=matlabFunction(fun2m); % converted integral:limit:0,pi/2
        if rho>0.5 % adjust the upper limit to prevent INF
            peD(k,kr)=integral(mckayf1,0,0.9975*pi/2); %
        else
            peD(k,kr)=integral(mckayf1,0,0.9995*pi/2); %
        end;
        fun1L=subs(fLzer,[a, X, r],[alpha,Z(k),rho]);
        fun2L=subs(fun1L,s,1/(sin(y))^2);
         mckayf2=matlabFunction(fun2L);
         peL(k,kr)=(1/pi)*integral(mckayf2,0,0.999975*pi/2);
    end;
end;
peS=berMcKay_using_sum;
% for rho=0.0001, also plot the theoretical error rate in a Nakagami
% channel
figure,semilogy(Z0,peD(:,1),'r*',Z0,peL(:,1),'kd',Z0,peS(:,1),'bo',Z0,ber1,'-m')
legend('direct','Laplace transform','sum','Nakagami')
xlabel('average SNR (dB)')
ylabel('Average probability of error')
title(['\alpha = ',num2str(alpha),'  \rho = ',num2str(rh(1))])
xlim([5,40])
figure,semilogy(Z0,peD(:,2),'r*',Z0,peL(:,2),'kd',Z0,peS(:,2),'bo')
legend('direct','Laplace transform','sum')
xlabel('average SNR (dB)')
ylabel('Average probability of error')
title(['\alpha = ',num2str(alpha),'  \rho = ',num2str(rh(2))])
xlim([5,40])
figure,semilogy(Z0,peD(:,3),'r*',Z0,peL(:,3),'kd',Z0,peS(:,3),'bo')
legend('direct','Laplace transform','sum')
xlabel('average SNR (dB)')
ylabel('Average probability of error')
title(['\alpha = ',num2str(alpha),'  \rho = ',num2str(rh(3))])
xlim([5,40])
```

```
figure,semilogy(Z0,peD(:,4),'r*',Z0,peL(:,4),'kd',Z0,peS(:,4),'bo')
legend('direct','Laplace transform','sum')
xlabel('average SNR (dB)')
ylabel('Average probability of error')
title(['\alpha = ',num2str(alpha),'  \rho = ',num2str(rh(4))])
xlim([5,40])

figure,semilogy(Z0,peD(:,5),'r*',Z0,peL(:,5),'kd',Z0,peS(:,5),'bo')
legend('direct','Laplace transform','sum')
xlabel('average SNR (dB)')
ylabel('Average probability of error')
title(['\alpha = ',num2str(alpha),'  \rho = ',num2str(rh(5))])
xlim([5,40])

figure,semilogy(Z0,peD(:,1),'r*',Z0,peD(:,3),'bo',...
    Z0,peD(:,4),'-k^',Z0,peD(:,5),'b-->')
legend(['  \rho = ',num2str(rh(1))],['  \rho = ',num2str(rh(3))],...
    ['  \rho = ',num2str(rh(4))],['  \rho = ',num2str(rh(5))])
xlabel('average SNR (dB)')
ylabel('Average probability of error')
title(['\alpha = ',num2str(alpha)])
xlim([5,40])
text(6,2e-6,'numerical integral results','backgroundcolor','y')
end

function ber1= bernakagami
% error rates in a Nakagami channel
global  Z alpha
syms a X
F=((a/X)^a)*gamma(a+1/2)/(2*sqrt(sym(pi))*a*gamma(a));
er=F*hypergeom([a,a+1/2],[a+1],(-a/X));
for k=1:length(Z)
    ber1(k)=double(subs(er,[a,X],[alpha,Z(k)]));
end;
end

function [ peS ] =berMcKay_using_sum
% p m shankar, BER calculation using the summation
global Z alpha rh
syms a X k M r
ak=a+2*k;
Xa=a/(X*(1-r));
X2=2*X;
Nr1=(a/X2)^ak*r^k*gamma(ak+1/2);
Dr1=factorial(k)*gamma(1/2+ak/2)*gamma(a/2)*ak*(1-r)^(2*k+a/2);
NDR=Nr1/Dr1;
pes=symsum(NDR*hypergeom([ak,ak+1/2],[ak+1],(-Xa)),k,0,M);
LK=length(Z);
LR=length(rh);
peS=zeros(LK,LR);% pe using summation
for kr=1:LR
    rho=rh(kr);
for kk=1:LK
    MM=10;
    pesk=subs(pes,[a,X,r,M],[alpha,Z(kk),rho,MM]);
   peS(kk,kr)=double(pesk);
end;
end;

end
```

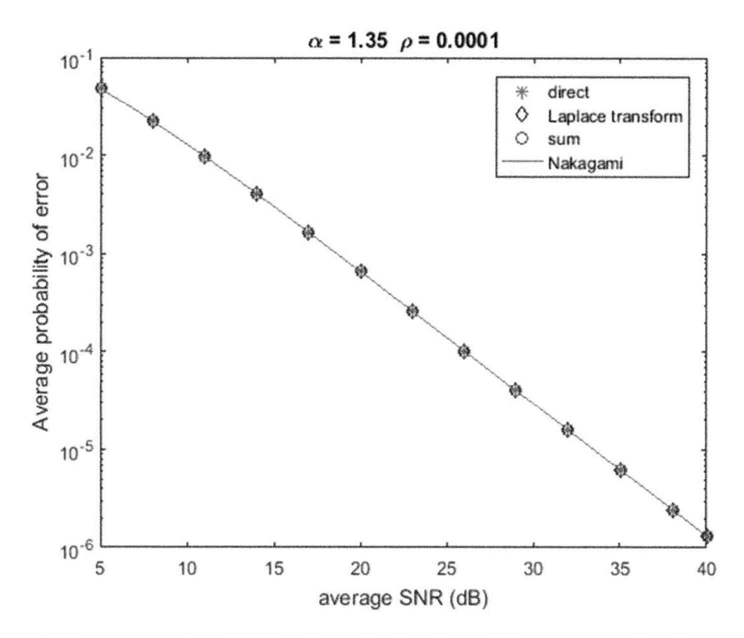

Fig. 4.102 Bit error rates in a McKay channel using three different approaches for a very low value of ρ. Bit error rate in a Nakagami channel is shown

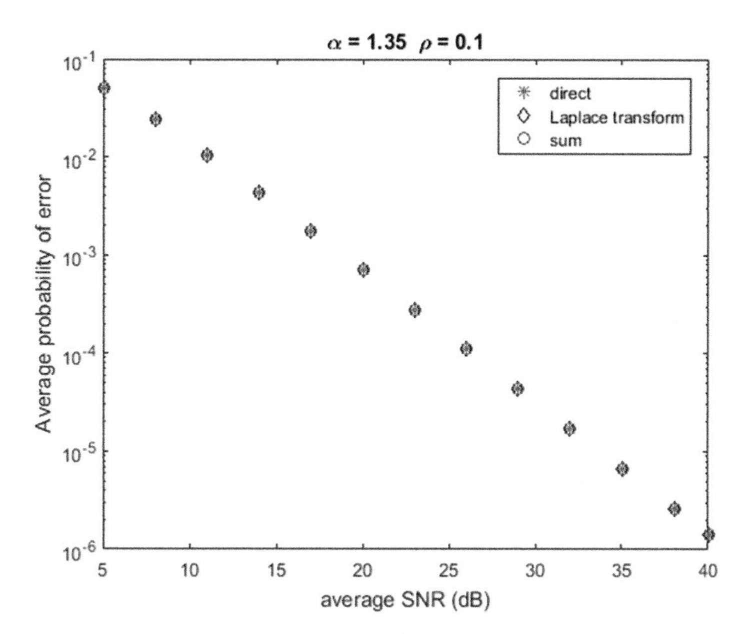

Fig. 4.103 Bit error rates in a McKay channel using three different approaches: $\rho = 0.1$

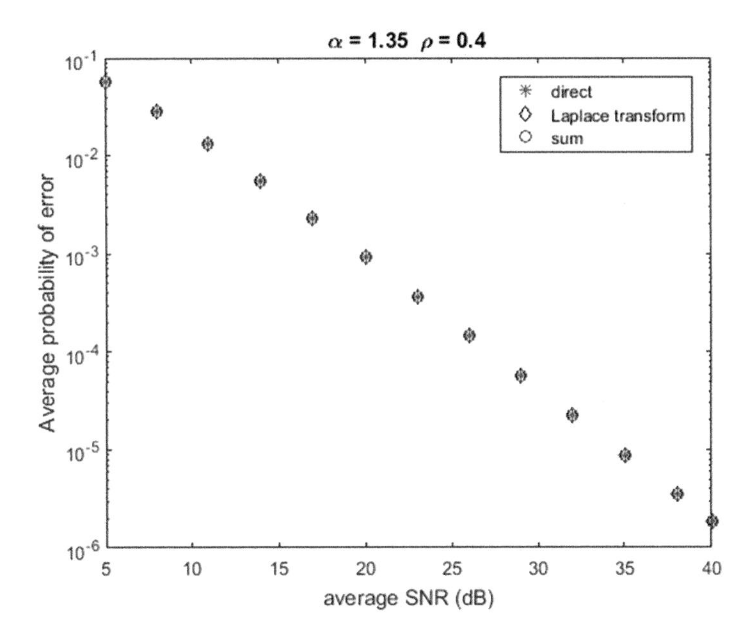

Fig. 4.104 Bit error rates in a McKay channel using three different approaches: $\rho = 0.4$

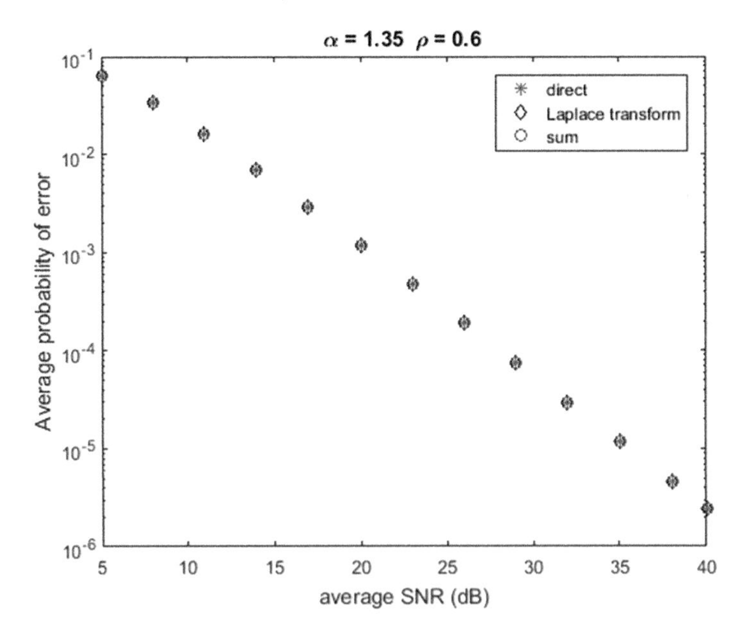

Fig. 4.105 Bit error rates in a McKay channel using three different approaches: $\rho = 0.6$

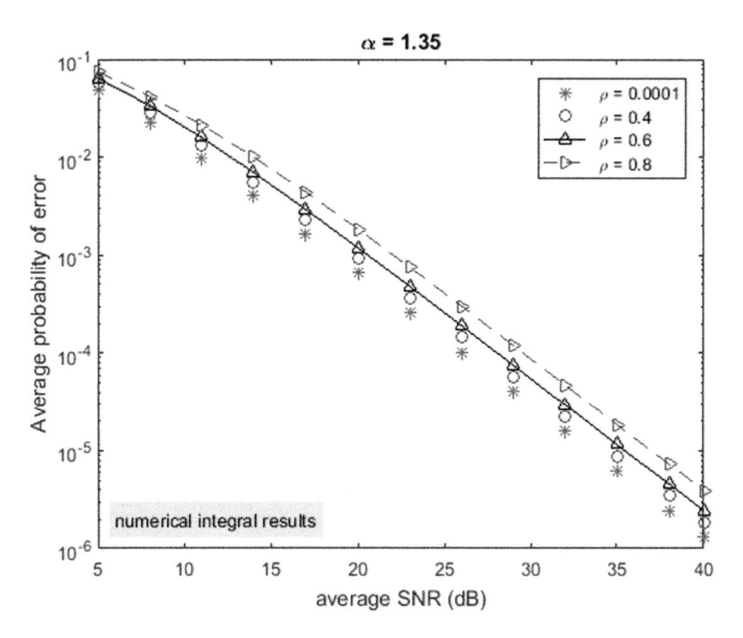

Fig. 4.106 Bit error rates in a McKay channel using numerical integration ($\alpha = 1.35$): multiple values of ρ

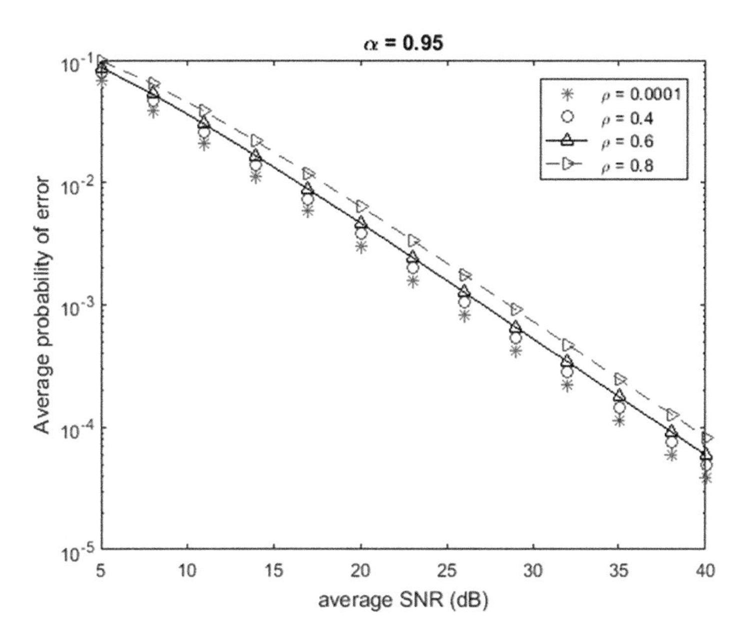

Fig. 4.107 Bit error rates in a McKay channel using numerical integration ($\alpha = 0.95$): multiple values of ρ

As seen from these results, the three approaches for the evaluation of the error rates lead to identical results. However, the approach based on Laplace transforms takes less time (not indicated here). The error rate simulation using random numbers will be discussed in Chap. 5 when fading mitigation using diversity is presented.

Outage Probability

Another measure of the performance in a fading channel is the outage probability. The system goes into outage when the instantaneous signal strength falls below a threshold value determined by the acceptable level of performance. As defined earlier, using the criterion of an error rate of 1e-4 for coherent BPSK, the threshold SNR (thr) will be 6.91. Outage probability is the CDF evaluated at the value of the threshold. Using the expression for the pdf in Eq. (4.281), the outage probability is

$$P_{out}(X) = \int_0^{thr} \frac{\sqrt{\pi}\; z^{\frac{\alpha-1}{2}}}{\left(2\frac{\alpha\sqrt{\rho}}{X(1-\rho)}\right)^{\frac{\alpha-1}{2}} \Gamma\left(\frac{\alpha}{2}\right)\left[\left(\frac{X}{\alpha}\right)^2(1-\rho)\right]^{\frac{\alpha}{2}}} \exp\left[-\frac{\alpha}{X(1-\rho)}z\right] I_{\frac{\alpha-1}{2}}\left(\frac{\alpha\sqrt{\rho}}{X(1-\rho)}z\right) dz \tag{4.315}$$

Using the CDF expression as a sum expressed in Eq. (4.302)

$$P_{\text{out}}(X) = \frac{\sqrt{\pi}\left(\sqrt{1-\rho}\right)^{\alpha}}{\Gamma\left(\frac{\alpha}{2}\right)} \sum_{k=0}^{\infty} \times \frac{\rho^k}{k!\Gamma\left(k+\frac{1}{2}+\frac{\alpha}{2}\right)2^{\alpha+2k-1}} \gamma\left(\alpha+2k, \frac{\alpha}{X(1-\rho)}\text{thr}\right). \tag{4.316}$$

Outage probabilities have been estimated using both equations, with the number of terms in the summation in Eq. (4.316) set to 10. The Matlab script and results are given below. The script appears only with the first set of results. The other sets are obtained with two other values of α. Results are shown in Figs. 4.108, 4.109, and 4.110.

```
function mckayoutageProbf
% Outage probabilities are calculated approximating the pdf as a sum. This
% leads to the CDF being the sum of incomplete gamma functions. Th enumber
% of terms in the sum is fixed at 7. The outage is also calculated by
% integrating the pdf. Threshold on power units corresponding to ber of
% 1e-4 for BPSK.
%
% symbolic toolbox is once again used. The script can updated by using a
% different value of alpha. One can also change the number of terms if
% needed.
% P M Shankar, July 2016
close all
thr=6.91;%threshold on power units corresponding to ber of 1e-4 for BPSK
Z0=[10:2:40];%average SNR
Z=10.^(Z0/10);
alpha=1.2;
rr=[0.0001,0.3,0.6,0.7];% rho values
LR=length(rr);
LZ=length(Z);
poutS=zeros(LR,LZ);
poutD=zeros(LR,LZ);
for k=1:LR
    rho=rr(k);
for kk=1:LZ
    outageS=double(mckaycdfsum(alpha,rho,Z(kk),7)); % outage using summation
    poutS(kk,k)=outageS; % Outage using the SUM
    macpdfn= mckpdf(alpha,rho,Z(kk));
    funcpdf=matlabFunction(macpdfn);% integrand for the CDF integral
    poutD(kk,k)=integral(funcpdf,0,thr);% Outage directly
end;
end;
semilogy(Z0,poutS(:,1),'-r',Z0,poutD(:,1),'r*',...
    Z0,poutS(:,2),'-k',Z0,poutD(:,2),'ko',...
    Z0,poutS(:,3),'-b',Z0,poutD(:,3),'bd',...
    Z0,poutS(:,4),'m--',Z0,poutD(:,4),'m^')
xlabel('Average SNR dB'),ylabel('Outage Probability')
legend(['sum \rho = ',num2str(rr(1))],['direct \rho = ',num2str(rr(1))],...
    ['sum \rho = ',num2str(rr(2))],['direct \rho = ',num2str(rr(2))],...
    ['sum \rho = ',num2str(rr(3))],['direct \rho = ',num2str(rr(3))],...
    ['sum \rho = ',num2str(rr(4))],['direct \rho = ',num2str(rr(4))])
title(['\alpha = ',num2str(alpha)])
end


function mckCDF=mckaycdfsum(alpha,rho,Z,MM)
% create the McKay CDF
thr=6.91;%%threshold on power units corresponding to ber of 1e-4 for BPSK
syms a r X M
syms k
xa=a/(X*(1-r));
a2=a/2;
%symbolic toolbox igamma is upper INC GAMMA it computes igamms differently
gamaprt=gamma(a+2*k)-igamma(a+2*k,xa*thr);
Nr=gamaprt*r^k;
Dr=factorial(k)*gamma(k+1/2+a2)*2^(a+2*k-1);
CDF=((sqrt(sym(pi))*(1-r)^(a2))/gamma(a2))*symsum(Nr/Dr,k,0,M);
mckCDF=subs(CDF,[a, X, r, M],[alpha,Z,rho,MM]); % alpha, X, rho, M
end

function macpdfn= mckpdf(alpha,rho,Z)
syms z a r X  positive
```

```
% get the expression for density directly from the notes
Nr=sqrt(sym(pi))*z^((1/2)*a-1/2)*exp(-a*z/(X*(1-r)))...
    *besseli((1/2)*a-1/2, a*sqrt(r)*z/(X*(1-r)));
Dr=gamma((1/2)*a)*(X^2*(1-r)/a^2)^((1/2)*a)...
    *(2*a*sqrt(r)/(X*(1-r)))^((1/2)*a-1/2);
pdf=Nr/Dr;
macpdfn=subs(pdf,[a,r,X],[alpha,rho,Z]);% this the pdf
end
```

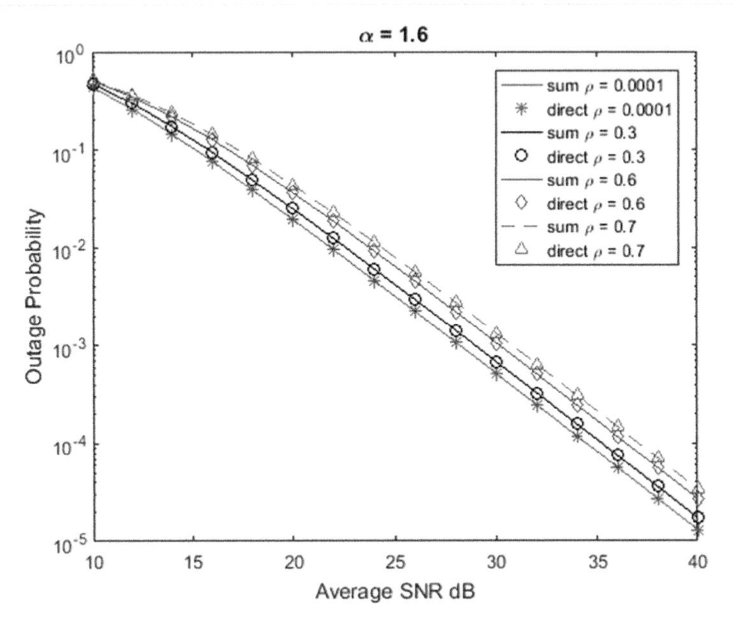

Fig. 4.108 Outage probabilities in a McKay channel ($\alpha = 1.2$): multiple values of ρ

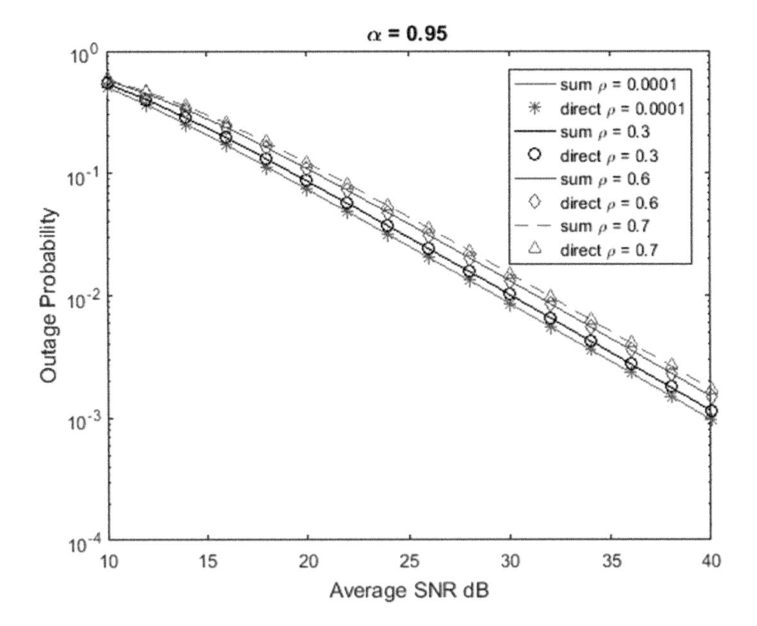

Fig. 4.109 Outage probabilities in a McKay channel ($\alpha = 0.95$): multiple values of ρ

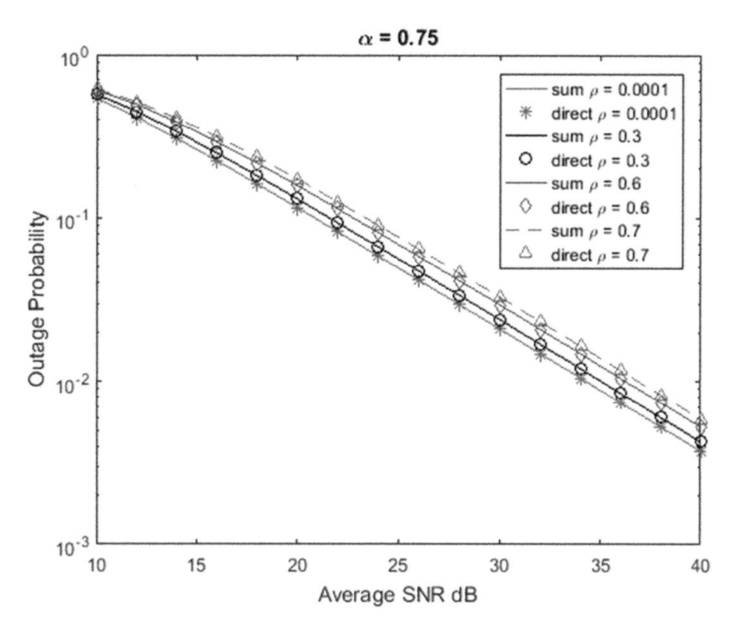

Fig. 4.110 Outage probabilities in a McKay channel ($\alpha = 0.75$): multiple values of ρ

Results on the McKay fading channel models clearly demonstrate its potential in modeling short-term fading channels ranging from the pre-Rayleigh fading including Nakagami–Hoyt to Nakagami–Rician and post-Rayleigh fading conditions. For $\rho = 0$, post-Rayleigh channel is represented by $\alpha > 1$. The three parameter distribution offers flexibility in modeling the signal strength fluctuations. Even though the presence of the Bessel function in the density poses some computational difficulties in undertaking the analysis, the form of the density, and CDF as a summation offers ways of mitigating these problems.

4.11.1.5 McKay Fading and Shadowing (Shadowed Fading Channels)

The McKay faded channel also undergoes shadowing concurrently. The effect of shadowing can be taken into account by treating the mean SNR X of the McKay density as a random variable with a lognormal density as it was shown earlier in Sect. 4.5. The density of the SNR in a shadowed fading channel becomes

$$f(z) = \int_0^\infty \frac{\sqrt{\pi}\; z^{\frac{\alpha-1}{2}}}{\left(2\frac{\alpha\sqrt{\rho}}{w(1-\rho)}\right)^{\frac{\alpha-1}{2}} \Gamma\left(\frac{\alpha}{2}\right)\left[\left(\frac{w}{\alpha}\right)^2 (1-\rho)\right]^{\frac{\alpha}{2}}} \exp\left[-\frac{\alpha}{w(1-\rho)} z\right] I_{\frac{\alpha-1}{2}}\left(\frac{\alpha\sqrt{\rho}}{w(1-\rho)} z\right) f(w) dw. \tag{4.317}$$

In Eq. (4.317), McKay pdf appears having X replaced by w. The lognormal density $f(w)$ earlier is

$$f(w) = \frac{K}{\sqrt{2\pi\sigma^2 w^2}} \exp\left(-\frac{[10\log_{10}(w) - \mu]^2}{2\sigma^2}\right). \tag{4.318}$$

In Eq. (4.318), K is the logarithmic conversion factor (same as A_0 in Eq. (4.68)),

$$K = \frac{10}{\log_e 10}. \tag{4.319}$$

The shadowing level is σ (dB) and the mean SNR is μ (dB). The relationship between the average in a lognormal channel and the average in McKay fading channel is related through

$$\mu = 10\log_{10}(X) - \frac{\sigma^2}{2K} = X_{dB} - \frac{\sigma^2}{2K}. \tag{4.320}$$

Equation (4.320) allows the comparison of the performance levels with and without shadowing since the plots are displayed with SNR in terms of X and not in terms of μ.

The error rate in a shadowed McKay fading channel is

$$p_e(X) = \int_0^\infty f(z)\frac{1}{2}erfc\left(\sqrt{z}\right)dz. \tag{4.321}$$

In Eq. (4.321), $f(z)$ is the density given in integral form in Eq. (4.317) resulting in the error rate becoming a double integral. Since no analytical expression exists for the error rate in Eq. (4.321), estimation of the error rates requires numerical (double) integration. The double integration can be avoided by using the McKay density as a sum. Using the density as a summation in Eq. (4.296) and the error rate in the McKay fading channel expressed in terms of hypergeometric functions in Eq. (4.308), the error rate becomes

$$\begin{aligned} p_e(X) = \int_0^\infty \sum_{k=0}^\infty \frac{\left(\frac{\alpha}{2w}\right)^{\alpha+2k}\rho^k}{k!\Gamma\left(\frac{\alpha}{2}\right)\Gamma\left(\frac{\alpha_k+1}{2}\right)} \frac{\Gamma\left(\alpha_k+\frac{1}{2}\right)}{\alpha_k(1-\rho)^{2k+\frac{\alpha}{2}}} {}_2F^2 1 \\ \times\left(\left[\alpha_k, \alpha_k+\frac{1}{2}\right], [1+\alpha_k], -\frac{\alpha}{(1-\rho)w}\right) f(w)dw. \end{aligned} \tag{4.322}$$

Note that X was replaced by w in Eq. (4.308) to get Eq. (4.322). It should be noted that Eq. (4.322) still requires numerical evaluation in addition to infinite summation.

The error probabilities are evaluated using both approaches numerically in Matlab. For the case of summation in Eq. (4.322), the number of terms was limited to six. Special care was paid to the limits to ensure that the results stayed finite. The Matlab script and results for one set of shadowing levels for the estimation of error rates are provided below showing that both approaches led to identical answers. Since the double integral was substantially time consuming, the remaining results are limited to one using the hypergeometric function in Eq. (4.322). Results are shown in Figs. 4.110, 4.111, and 4.112.

```
function ber_mckayshadowing_hypergeom_all
% bit error rate in McKay shadowing estimated by first obtaining the BER
% using the summation and then performing a single integral over the
% lognormal density. Summation uses hypergeometric functions and only
% requires a single integral
% results are verified through double integration (direct). Results for the
% case of no shadowing obtained using Laplace transforms is also plotted.
% input is alpha and rho. The number of terms in the summation is 6
% P M Shankar, October 2016

close all
global Z0 alpha rh
Z0=[10:2:30];
for krr=1:2
    if krr==1
alpha=0.65; rh=0.4;
    else
 alpha=1.65; rh=0.5;
    end;
sigv=[3,5,6];
LS=length(sigv);
LK=length(Z0);
KN=10/log(10);
MM=6;% number of terms in the summation
% create the integrand using symbolic variables
syms a x r M sigm MU K k
% the integrand
f1=gamma(a+1/2)*(a/x)^a*hypergeom([a, a+1/2], [a+1], -a/((1-r)*x))/...
    (gamma((1/2)*a)*a*gamma(1)*gamma(1/2+(1/2)*a)*2^a*(1-r)^((1/2)*a));
f2=symsum(gamma(a+2*k+1/2)*(a/x)^(a+2*k)*r^k*hypergeom([a+2*k, a+2*k+1/2],...
    [a+2*k+1], -a/((1-r)*x))/...
    (gamma((1/2)*a)*(a+2*k)*gamma(k)*gamma(k+1/2+(1/2)*a)*2^(a+2*k)*...
    (1-r)^(2*k+(1/2)*a)),k,1,M);
f=f1+f2;
ff=sym(K)*exp(-(10*log10(x)-MU)^2/(2*sigm^2))/sqrt((2*sym(pi)*x*x)*sigm^2);
fun=f*ff; % this is the integrand
pesh=zeros(LK,LS);
pedb=zeros(LK,LS);
for kr=1:length(sigv)
    sig=sigv(kr);% sigma value
    mu=Z0-sig^2/(2*KN); % average measured in dB
for k=1:LK
% now substitute the values
funer=subs(fun,[K,a,r,sigm,MU,M],[10/log(10),alpha,rh,sig,mu(k),MM]);
% create the in-line function
mfuner=matlabFunction(funer);
% perform the integration
pesh(k,kr)=integral(mfuner,0.,inf); % result using the summation
% get in-line function for the double integral
fdfuner = doubleintegralfun;
fdfer=subs(fdfuner,[K,a,r,MU,sigm],[10/log(10),alpha,rh,mu(k),sig]);
mfund=matlabFunction(fdfer);%order is dx dz necessary for putting limits
pedb(k,kr)=integral2(mfund,0.1,1e5,1e-4,0.995*pi/2);%second limit in tan(.)

end;

end;
[ peL] =laplace_ber;%error rate no shadowing
figure
semilogy(Z0,peL(:,1),'-',Z0,pesh(:,1),'r*',Z0,pedb(:,1),'--k',...
    Z0,pesh(:,2),'kd',Z0,pedb(:,2),'-.g',Z0,pesh(:,3),'bo',...
```

```
    Z0,pedb(:,3),':')
legend('No shadowing',['\sigma = ',num2str(sigv(1)),'dB (hypergeom)'],...
    ['\sigma = ',num2str(sigv(1)),'dB (dblintegral)'],...
   ['\sigma = ',num2str(sigv(2)),'dB (hypergeom)'],...
   ['\sigma = ',num2str(sigv(2)),'dB (dblintegral)'],...
   ['\sigma = ',num2str(sigv(3)),'dB (hypergeom)'],...
   ['\sigma = ',num2str(sigv(3)),'dB (dblintegral)'],'location','southwest')
 xlabel('Average SNR (dB)')
 ylabel('Average probability of error')
 title(['\alpha = ',num2str(alpha),'  \rho = ',num2str(rh)])

end;

end



function [ peL] =laplace_ber
% p m shankar, BER calculation using Laplace when no shadowing is present
global Z0 alpha rh
Z=10.^(Z0/10);
syms a X
syms s r y
fLzer=a^a/(a^2+s^2*X^2+2*s*X*a-s^2*X^2*r)^((1/2)*a);
LK=length(Z);
peL=zeros(LK);% pe using summation
for k=1:LK
     fun1L=subs(fLzer,[a, X, r],[alpha,Z(k),rh]);
        fun2L=subs(fun1L,s,1/(sin(y))^2);
         mckayf2=matlabFunction(fun2L);
         peL(k)=(1/pi)*integral(mckayf2,0,0.999975*pi/2);
end;


end


function ffun = doubleintegralfun
% created using symbolic toolbox
syms z a x MU K r sigm y
z=tan(y); % conver to tangent(.); variable representing the snr
Nr=sqrt(sym(pi))*z^((1/2)*a-1/2)*exp(-a*z/(x*(1-r)))...
    *besseli((1/2)*a-1/2, a*sqrt(r)*z/(x*(1-r)));
Dr=gamma((1/2)*a)*(x^2*(1-r)/a^2)^((1/2)*a)...
    *(2*a*sqrt(r)/(x*(1-r)))^((1/2)*a-1/2);
% (1+z)^2 needed for transformation from dz to dy, z=tan(y)
fun=Nr*(1+z^2)/Dr; % pdf McKay depends on X, alpha (a) and rho(r)
fun1=0.5*erfc(sqrt(z))*fun;% error rate integrand
ff=sym(K)*exp(-(10*log10(x)-MU)^2/(2*sigm^2))/sqrt((2*sym(pi)*x*x)*sigm^2);
ffun=fun1*ff;


end
```

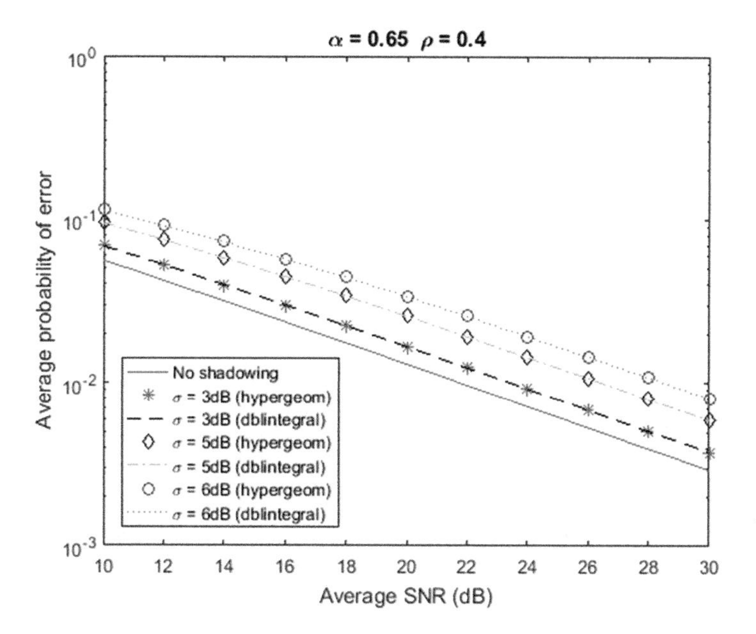

Fig. 4.111 Error rates in McKay-lognormal channel

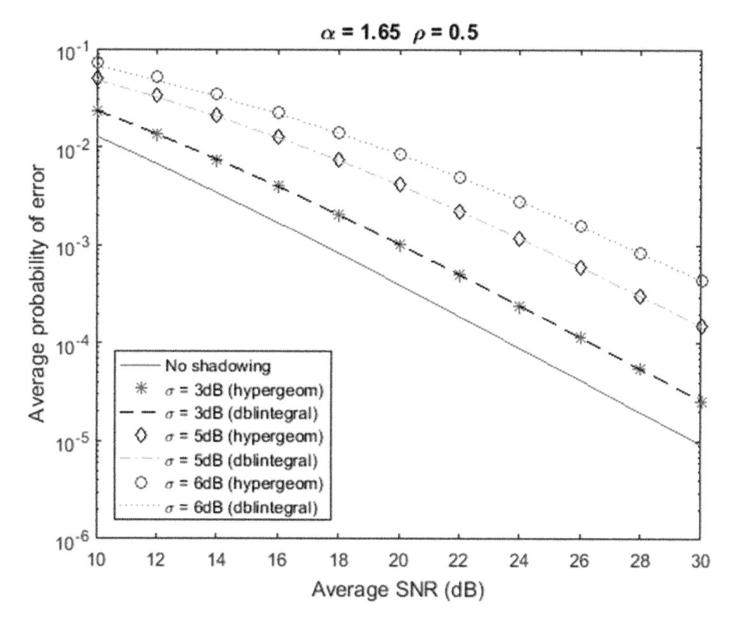

Fig. 4.112 Error rates in McKay-lognormal channel

It is possible to replace the lognormal density in Eq. (4.317) by a Meijer G-function representing the product of gamma variables. The Nakagami-N gamma model for fading was presented earlier. A similar procedure can be undertaken here as well. In this case, $f(w)$ in Eq. (4.317) will consist of Meijer G-function as shown in Eq. (4.121).

4.11.1.6 Generalized McKay Channel

The range of fading models encompassed in the McKay density can be expanded by obtaining a generalized form of the McKay density by defining a new random variable by power scaling the SNR Z to

$$P = Z^{\frac{1}{s}},\quad s > 0. \tag{4.323}$$

Using the transformation of random variables, the density function of P becomes

$$f_P(p) = \frac{s\sqrt{\pi}p^{\frac{s(\alpha+1)}{2}-1}}{\Gamma\left(\frac{\alpha}{2}\right)\left(\frac{2\alpha\sqrt{\rho}}{X(1-\rho)}\right)^{\frac{(\alpha-1)}{2}}\left[\left(\frac{X}{\alpha}\right)^2(1-\rho)\right]^{\frac{\alpha}{2}}} \exp\left[-\frac{\alpha}{X(1-\rho)}p^s\right] I_{\frac{(\alpha-1)}{2}}\left(\frac{\alpha\sqrt{\rho}}{X(1-\rho)}p^s\right). \tag{4.324}$$

Examining the generalized McKay density in Eq. (4.324), it can be seen that the density can morph into the density of the SNR in α-λ-μ, generalized gamma (Stacy's distribution), $\alpha - \mu$, $\eta - \mu$ along with the other models mentioned earlier. These relationships are tabulated in Table 4.5.

Table 4.5 Relationship of the generalized McKay channel to other fading channels

s	α	$0 \le \rho \le 1$	Fading channels
>0	>0	>0	$\alpha - \lambda - \mu$
>0	>0	0	Generalized gamma (Stacy's distribution)
1	>0	>0	McKay
1	≥1	>0	$\alpha - \mu$ or $\eta - \mu$ (special case of McKay)
1	1	>0	Nakagami–Hoyt (special case of McKay)
1	>0	0	Gamma (special case of McKay)
1	≥½	0	Nakagami (special case of McKay)
1	1	0	Rayleigh(special case of McKay)

4.11.2 Product Model for Shadowed Fading Channels

The short-term fading in wireless channels can be modeled as a cascaded process resulting in the density of the received SNR being described in terms of a Meijer G-function as

$$f_Z(z) = \frac{1}{z\Gamma^N(m)} G_{0,N}^{N,0}\left(\frac{m^N}{Z_0}z \middle| \begin{matrix} - \\ m,..,m \end{matrix}\right), \quad m \geq \frac{1}{2}, N = 1,2,3,\cdots. \tag{4.325}$$

As explained in Sect. 4.7, N being 1 represents the Nakagami channel and for $N = 1$, the density in Eq. (4.325) is the gamma density. As N increases, the level of fading increases and the performance of the channel goes down. In the presence of shadowing, the average SNR Z_0 becomes a random variable and the density function of the SNR in a shadowed fading channel becomes

$$f_Z(z|y) = \frac{1}{z\Gamma^N(m)} G_{0,N}^{N,0}\left(\frac{m^N}{y}z \middle| \begin{matrix} - \\ m,..,m \end{matrix}\right), \quad m \geq \frac{1}{2}, N = 1,2,3,\cdots. \tag{4.326}$$

In Eq. (4.326), y represents the shadowing component assumed to have a lognormal density.

While short-term fading is modeled as a cascading process, it was also shown that the shadowing is best explained in terms of a product-process (which results in a lognormal density for the shadowing component), which can also be described by another Meijer G-function. The probability density of the shadowing component in the cascaded model is

$$f_Y(y) = \frac{1}{y\Gamma^M(c)} G_{0,M}^{M,0}\left(\frac{c^M}{Z_0}y \middle| \begin{matrix} - \\ c,c,..,c \end{matrix}\right), \quad c > 0, \quad M = 1,2,\cdots. \tag{4.327}$$

Note that when $M = 1$, Eq. (4.327) becomes the gamma density. When $N = M = 1$, the shadowed fading channel becomes a generalized K fading channel. When N and M exceed unity, the density function of the SNR in a shadowed fading channel (Shankar 2012) becomes

$$\begin{aligned} f(z) = \int_0^\infty f(z|y)f(y)dy = \int_0^\infty \frac{1}{z\Gamma^N(m)} G_{0,N}^{N,0}\left(\frac{m^N}{Z_0}z \middle| \begin{matrix} - \\ m,..,m \end{matrix}\right) \\ \times \frac{1}{y\Gamma^M(c)} G_{0,M}^{M,0}\left(\frac{c^M}{Z_0}y \middle| \begin{matrix} - \\ c,c,..,c \end{matrix}\right) dz. \end{aligned} \tag{4.328}$$

Using the properties of Meijer G-functions, Eq. (4.328) simplifies to

$$f(z)=\frac{1}{z\Gamma^{N}(m)\Gamma^{M}(c)}G_{0,N+M}^{N+M,0}\left(\frac{c^{M}m^{N}}{Z_{0}}z\middle|\begin{matrix}-\\ m,..,m,c,..,c\end{matrix}\right). \tag{4.329}$$

Equation (4.329) represented the composite cascaded density for the shadowed fading channel. It is also viewed as a product model for shadowed fading since the density results from the products of NM independent gamma variables, N of one type and M of another type.

Using Eq. (4.329), the error rate (Coherent BPSK) in a shadowed fading channel modeled as a cascaded process becomes

$$\begin{aligned}p_{e}(Z_{0})&=\int_{0}^{\infty}\frac{1}{2}\,erfc\left(\sqrt{z}\right)\frac{1}{z\Gamma^{M}(c)\Gamma^{N}(m)}G_{0,N+M}^{N+M,0}\left(\frac{c^{M}m^{N}}{Z_{0}}z\middle|\begin{matrix}-\\ m,..,m,c,..,c\end{matrix}\right)dz.\\&=\frac{1}{2\sqrt{\pi}\Gamma^{M}(c)\Gamma^{N}(m)}G_{2,N+M+1}^{N+M,2}\left(\frac{c^{M}m^{N}}{Z_{0}}\middle|\begin{matrix}\frac{1}{2},1\\ m,..,m,c,..,c,0\end{matrix}\right)\end{aligned} \tag{4.330}$$

For a threshold SNR of Z_T, the outage probability becomes

$$\begin{aligned}P_{out}(Z_{0})&=\int_{0}^{Z_{T}}\frac{1}{z\Gamma^{M}(c)\Gamma^{N}(m)}G_{0,N+M}^{N+M,0}\left(\frac{c^{M}m^{N}}{Z_{0}}z\middle|\begin{matrix}-\\ m,..m,c,..,c\end{matrix}\right)dz.\\&=\frac{1}{\Gamma^{M}(c)\Gamma^{N}(m)}G_{1,N+M+1}^{N+M,1}\left(\frac{c^{M}m^{N}}{Z_{0}}Z_{T}\middle|\begin{matrix}1\\ m,..m,c,..,c,0\end{matrix}\right)\end{aligned} \tag{4.331}$$

Thus, the complete description of shadowed fading is possible through a multiple scattering process. It is possible to see the advantage of using a cascaded approach to describe the signal strength fluctuations in a shadowed fading channel. The probability density, cumulative distribution function, error rates, and outage probabilities can be expressed in closed forms in terms of Meijer G-functions. The only numerical computation needed is the estimation of c. It should be noted that c is related to the number of cascaded components M in shadowing and the severity of shadowing (shadowing level) quantified in terms of σ (dB). This was discussed in Sect. 4.7.

An example of error calculations is shown in Fig. 4.113. The Matlab script is also given.

```
        function  cascaded_shadowed_fading_channel
 %  example of cascaded fading (N) and cascaded shadowing (M)
 % m is the Nakagami parameter and c is the shadowing term obtained using
 % the relationship between shadowing level sigma for a specific value of M
 % in this example, values of m and c are chosen arbitrarily to illustrate
 % the use of Matlab in obtaining error rates and outage probabilities.
 % there is a need to ensure that (m-c) is not zero. To overcome this
 % problem, add 1e-5 to m and or 1e-6 to c
 % P M Shankar October 2016
 close all
 m=2.5;
 m1=m+1e-5;
 MN=[2,3,4]; % values of N
 c=6.511; M =4;
 c1=c+1e-6;
 Z0=5:30; % average SNR in dB
 Z=10.^(Z0/10);% average SNR in absolute units
 LZ=length(Z);
 ber1=zeros(3,LZ);
 for kk=1:3
     N=MN(kk);
      gm=(m1^N)*(c1^M);
 GM=sqrt(pi)*(gamma(m1)^N)*(gamma(c1)^M);
 for k=1:LZ
     Z1=Z(k);
     if N==2
x1=double(evalin(symengine,sprintf('Meijer G([[1/2], [1]], [[0,%e,%e,%e,%e,%e,%e], []],
%e)',m1,m1,c1,c1,c1,c1,gm/Z1)));
          elseif N==3
x1=double(evalin(symengine,sprintf('Meijer G([[1/2], [1]], [[0,%e,%e,%e,%e,%e,%e,%e], []],
%e)',m1,m1,m1,c1,c1,c1,c1,gm/Z1)));
     else
  x1=double(evalin(symengine,sprintf('Meijer G([[1/2], [1]], [[0,%e,%e,%e,%e,%e,%e,%e,%e], []],
%e)',m1,m1,m1,m1,c1,c1,c1,c1,gm/Z1)));
     end;
    ber1(kk,k)=(1/2)-(1/2)*x1/GM;
 end;
 end;

 semilogy(Z0,ber1(1,:),'r-',Z0,ber1(2,:),'k--',Z0,ber1(3,:),'m-.','linewidth',1.5)
 legend(['N = ',num2str(MN(1))],['N = ',num2str(MN(2))],['N = ',num2str(MN(3))])
 xlabel('Average SNR (dB)')
 ylabel('Average probability of error')
 title({'Cascaded Shadowed fading channel';['  m = ',num2str(m),'  , M = ',num2str(M),...
     ',  c = ',num2str(c)]})


end
```

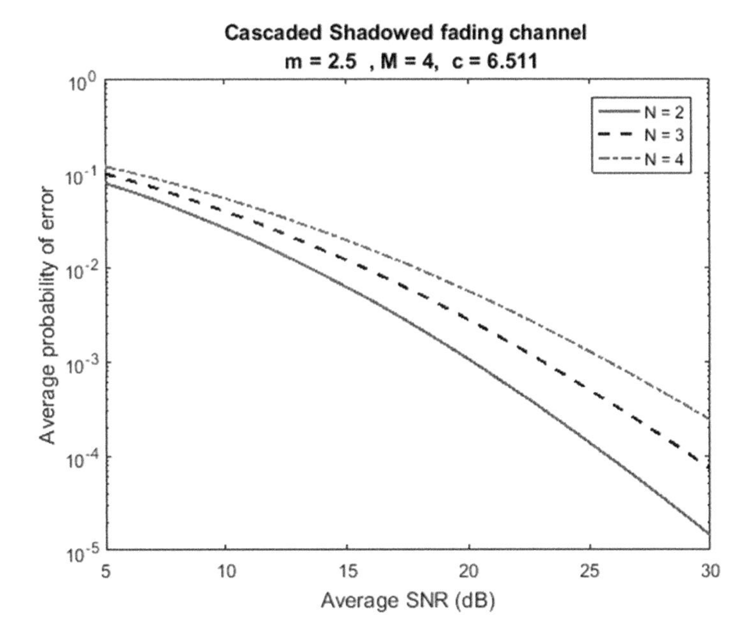

Fig. 4.113 Error rates in cascaded shadowed fading channel

4.11.3 Mixture Models for Fading, Shadowing, and Shadowed Fading

Another approach to modeling of the statistical fluctuations relies on the concept of fractional or weighted contribution to the statistics of the SNR from multiple sources. Contrary to the densities that rely on modified Bessel functions or Meijer G-functions (resulting from the sum or product models), the densities arising out of weighted contributions can be built from simple densities as shown in Chap. 2. For example, it is possible to have densities created by a weighted summation of either gamma, Gaussian, lognormal, or any similar (Shankar et al. 2003; Atapattu et al. 2011; Jung et al. 2014; Büyükçorak et al. 2015; Selim et al. 2015). This might allow for simpler analytical evaluation of the performance of the channel with or without diversity since error rates and outage probabilities can be expressed in known forms.

The mixture models have been used in image analysis and interpretation as described in Chap. 2. If one treats the images to be formed from contributions from several factors, it is possible to model the statistics of the image intensity to be made up of contributions from two or more sources. For example, if a certain region of the image is examined, the heterogeneous nature of the image can be interpreted as resulting from multiple statistical factors contributing to the intensity in any region of interest. Specifically, examining regions containing abnormal growths or tumors, one can attribute the intensity to have arisen from contributions from normal tissue regions, tissue regions containing abnormal cells, and tissue regions containing tiny blood vessels, each having its own statistical signature. Depending on the relative strengths of each form of these regions, the composite or mixture density can be formed. If $f(x)$ is the density of the pixels in the region of interest (Papoulis and Pillai 2002a, b)

$$f(x) = a_1 f_1(x) + a_2 f_2(x) + \cdots, \quad a_1 + a_2 + \cdots = 1, 0 < a_k < 1. \tag{4.332}$$

Note that in Eq. (4.332), a's are the weights and $f_1(.), f_2(x), \cdots$ are valid density functions. Such weighted densities can also be used to model boundaries in the region of interest.

While it is possible to attribute the existence of mixture densities in image analysis from a basic perspective of image formation, the efforts in wireless channel modeling seem to be based on making channels models more accurate (analytically simpler). It is also possible to argue for the existence of mixture models based on scattering by treating the channel as a mixture. The concept of weighted contributions leads directly to the density $f(z)$ of the received SNR Z as

$$f(z) = \sum_{k=1}^{n} p_k f_k(z). \tag{4.333}$$

In Eq. (4.333) $f_k(z)$ represent density functions that might be of the same type (gamma, Gaussian, lognormal, etc.) each having a set of unique parameters associated with it such as m_k and Ω_k for gamma, μ_k and σ_k for Gaussian, μ_k and σ_k (in decibel units) for lognormal, etc. These density functions might also be different. Taking note of the fact that $f(z)$ is a valid density function and $f_k(z)$ are also valid density functions,

$$\sum_{k=1}^{n} p_k = 1, \quad 0 \leq p_k \leq 1. \tag{4.334}$$

It can be seen that if $n = 1$, Eq. (4.333) becomes the density function of fading in a Nakagami channel if we treat $f_k(z)$ to be gamma density. As the value of n goes up, the density function is likely to be more accurate in predicting the statistical fluctuations of the SNR. While the accuracy goes up, the mathematical analysis of the performance will still remain simple as the density functions are of standard type. However, modeling of Eq. (4.333) itself becomes more complex since it might require evaluation of several parameters. As an example for $n = 2$, if $f_k(z)$ is considered to be a two parameter gamma density, one would require estimation of five parameters and if $n = 3$, estimation of 8 parameters and so on, imposing severe burden on parameter estimation. As expected, the higher the value of n, the better is the accuracy of the model. But any accuracy achieved is likely to come at a heavy cost of parameter estimation.

4.11.3.1 Gamma Mixture Models

The starting point of the discussion on the mixture densities is the gamma density associated with the Nakagami fading channel. When the number of variables in the mixture $n = 1$ in Eq. (4.333), the mixture density becomes the simple gamma density associated with the SNR in a Nakagami channel. As the fading in the channel moves away from the simple Nakagami fading conditions and becomes Nakagami–Hoyt, Nakagami–Rice, $\kappa - \mu$ or $\eta - \mu$, etc., these scenarios might be modeled by increasing the number of components (Papoulis and Pillai 2002a, b; Venturini et al. 2008). The existence of shadowing can also be modeled similarly by varying the number of components in the mixture. The gamma mixture density can be expressed as (Atapattu et al. 2011)

$$f(z) = \sum_{k=1}^{n} p_k \frac{1}{\Gamma(m_k)} \left(\frac{m_k}{\Omega_k}\right)^{m_k} z^{m_k - 1} e^{-\left(\frac{m_k}{\Omega_k}\right) z}. \tag{4.335}$$

In Eq. (4.335), the primary gamma density associated with a Nakagami fading channel is

$$f_k(z) = \frac{1}{\Gamma(m_k)} \left(\frac{m_k}{\Omega_k}\right)^{m_k} z^{m_k - 1} e^{-\left(\frac{m_k}{\Omega_k}\right) z}. \tag{4.336}$$

The cumulative distribution of the gamma mixture is

$$F(z) = \sum_{k=1}^{n} p_k \frac{1}{\Gamma(m_k)} \gamma\left(m_k, \frac{m_k}{\Omega_k} z\right). \tag{4.337}$$

In Eq. (4.337), $\gamma(.,.)$ is the lower incomplete gamma function. The moment generating function of the gamma mixture becomes

$$M(s) = \sum_{k=1}^{n} p_k \frac{1}{\left(1 - s\frac{\Omega_k}{m_k}\right)^{m_k}}. \tag{4.338}$$

The moments of the gamma mixture are given by

$$E(Z^r) = \sum_{k=1}^{n} p_k \frac{\Gamma(m_k + r)}{\Gamma(m_k)} \left(\frac{\Omega_k}{m_k}\right)^r. \tag{4.339}$$

The mean Z_0 is

$$E(Z) = \sum_{k=1}^{n} p_k \Omega_k. \tag{4.340}$$

The amount of fading (AF) in a fading channel modeled by the gamma mixture is

$$AF=\frac{\langle Z^2\rangle}{\langle Z\rangle^2}-1=\frac{\sum_{k=1}^{n}p_k\left(\frac{m_k+1}{m_k}\right)\Omega_k^2}{\left[\sum_{k=1}^{n}p_k\Omega_k\right]^2}-1. \tag{4.341}$$

The bit error of a coherent BPSK modem becomes

$$p_e(Z_0)=\sum_{k=1}^{n}p_k\left(\frac{1}{2\sqrt{\pi}}\right)\frac{\Gamma\left(m_k+\frac{1}{2}\right)}{m_k\Gamma(m_k)}\left(\frac{m_k}{\Omega_k}\right)^{m_k}{}_2F^21\left(\left[m_k,\frac{1}{2}+m_k\right],[1+m_k],-\frac{m_k}{\Omega_k}\right). \tag{4.342}$$

In Eq. (4.342), ${}_2F_1(.)$ is the hypergeometric function (Gradshteyn and Ryzhik 2007) and Z_0 is the average SNR given by

$$Z_0=\sum_{k=1}^{n}p_k\Omega_k. \tag{4.343}$$

The advantage offered by a mixture model is clearly seen from the use of well-established expressions for the error rates [Eq. (4.342)] and outage probabilities (in terms of CDF). The advantage of traditional models is that the associations among the values of the Nakagami parameter, average SNR, shadowing levels (in dB), and fading and shadowing conditions are known. The shortcoming of the mixture model is that the parameters of the gamma mixture models have to be evaluated for each value of the Nakagami parameter, shadowing levels, and the average SNR. The parameter estimation methods often are neither simple nor straightforward. For the efficient implementation of the gamma mixture models for fading, shadowing, and shadowed fading, one needs to investigate numerical methods for obtaining the parameters of the model, namely (3n-1) parameters for a gamma mixture with n components. The number of parameters is 3n-1 because of the total probability theorem in Eq. (4.334), regardless of the number of mixtures involved or the forms of the component densities involved. However, one also needs to determine the minimum number of components needed to maintain acceptable levels of compatibility and matching to the exact models being replaced. This would avoid use of higher n values to get finer matches to the actual densities without providing any extra accuracy in error rates and outage probabilities. Two methods are described to accomplish the goal of estimating the parameters including the number of components n and the criterion used for determining the optimum number of components.

Two primary methods employed to estimate the parameters of mixture densities are the maximum likelihood estimation (MLE) and the expectation maximization method (EM) (Davenport et al. 1988; Almhana et al. 2006). While these methods provide the estimate of the parameters for a chosen value of n, the mean square error (MSE) criterion is used to determine the minimum value of the number of components needed. These methods, namely MLE and EM, and the criterion MSE all can be implemented in Matlab as discussed in Chap. 2. The work reported here uses the MLE approach in Matlab. The optimum number of components can also be determined on the basis of another criterion, namely the Bayes Information Criterion (BIC). This is described in detail in Chap. 5 where mixture density is reintroduced in connection with diversity combining algorithms.

Some of the conventional models are now explored in terms of gamma mixtures.

Gamma Mixtures and Nakagami Fading

This is a trivial case since gamma density describes the statistics of SNR in a Nakagami fading channel. In other words, the SNR in a Nakagami fading channel is modeled with a mixture consisting of a single gamma density.

Gamma Mixtures and Nakagami-Lognormal Shadowed Fading

As presented earlier, the shadowed fading channel leads to density function expressed in integral form making it difficult to undertake the analysis of performance of wireless channels. Mixture densities offer a means to overcome this problem allowing the estimation of error rates shown in Eq. (4.342). The first step is the determination of the number of mixture components and the appropriate parameters of the mixture. The next step is to determine the optimum number of components required for a satisfactory fit to the gamma-lognormal density (of the SNR).

The detailed Matlab script written for the gamma mixture parameter estimation for the gamma-lognormal fit is given below. As described in Chap. 2, one of the critical needs is the choice of the initial or starting guesses for the parameters. For a chosen value of n, $k_{\mathrm{means}}(.)$ is to split the data into n groups and the appropriate parameters are determined by assuming each group to follow the gamma density. To simplify the analysis, the number of components n is reduced if the weights fall below 10%. In other words, if $p_k < 0.1$, the actual number of components used will be less than n. This step is only necessary to keep the computational time low.

The plots of the densities and characteristics of the mixture densities are shown in Figs. 4.114, 4.115, 4.116, and 4.117.

```
function gammamixture_shadowedfading_sim
% P M Shankar, July 2016
% shadowed fading channels are modeled using gamma mixtures. The MSE values
% are calculated and displayed on the pdf plots (histogram vs. the
% estimated pdf from mixture density).
% parameters of the gamma mixture are estimated using mle technique
% separately in another function (below)
close all
Z0=[5 20] ;%dB
Z=10.^(Z0/10);%power units
K=10/log(10);%conversion factor
Numb=1e5; % number of samples
m=1.7; % Nakagami parameter
sig=[2,5]; % shadowing level
n=5; % starting number of components for gamma mixture
for kk=1:length(Z0);
for ks=1:length(sig);
sigm=sig(ks);
 sigma2=sigm^2;
muZ=Z0(kk)-sigma2/(2*K);%convert Z0 (dB) to mu
g1=normrnd(muZ,sigm,1,Numb);%generate Gaussian
gg=10.^(g1/10);%lognormal random
hh=gamrnd(m,1/m,1,Numb).*gg;% cascaded gamma-lognormal set with mean of mu
[ff,xr]=ksdensity(hh);
% determine if any negative values of xr exist. discard them.
NK=[]; % eliminate negative values of xr
NK=find(xr<0);
if isempty(NK)==0
    NK1=max(NK)+1;% find the largest index and add 1 to start the non-zero values
    xr=xr(NK1:end);
    ff=ff(NK1:end);
else
end;
X=hh;

[nn,w,alpha,beta] = gammamixture_MLEf(X,n);
% nn is the actual number of components
L=length(xr);
fx=zeros(1,L);
nL=length(w);% to account for the reduced number of components
for k=1:nL
    fx=fx+w(k)*gampdf(xr,alpha(k),beta(k));
end;
figure, plot(xr,ff,'r-',xr,fx,'--k','linewidth',1.5)
legend('histogram','mixture fit','location','southeast')
xlabel('SNR'),ylabel('pdf')
MSE=(1/L)*sum(ff-fx).^2;
title({['  Average SNR = ',num2str(Z0(kk)),...
    ' dB,  \sigma = ',num2str(sig(ks)), ' dB,  m = ',num2str(m)];...
    ['No. of components =',num2str(nn),' , MSE=',num2str(MSE)]},'color','b')
text(0.5*max(xr),0.95*max(max(ff,fx)),...
    'p                    \alpha                    \beta')
val=[w;alpha;beta]';
text(0.45*max(xr),0.7*max(max(ff,fx)),num2str(val))
clear xr ff fx w alpha beta hh gg
end;

end;



end
```

```
function [nn,p,a,b] = gammamixture_MLEf(Y,n)
% P M Shankar, July 2016
% this one is seto to a maximum of 9 components
% the initial guesses of the parameters needed for MLE are obtained using
% clustering. If the clustering probability is less than 10%, the kmeans is
% repetade with fewer clusters. Therefore, acutual number of components
% will be between 2 and n.
%
[my,ny]=size(Y);
L=my*ny; % length of the data
Y=reshape(Y,[L,1]);
opt = statset('mlecustom');
opt = statset(opt,'FunValCheck','off','MaxIter',6000,'TolX',1e-4,'TolFun',1e-5,'TolBnd',1e-
5','MaxFunEvals', 6000);
% chose the initial guess from kmeans
index = kmeans(Y,n);
w = zeros(1,n); aa = zeros(1,n); bb= zeros(1,n);
for k=1:n
    w(k) = sum(index==k)/L;
    aa(k) = (mean(Y(index==k)))^2/var(Y(index==k));
    bb(k) = var(Y(index==k))/mean(Y(index==k));
end
disp('Initial weights'),disp(w),disp('set a 10% threshold and regroup if necessary')
M=sum(w<=.1); % number classes less than 10%
if M>0 && n>2 % nn must be at least 2
    nn=n-M;
    if nn==1 % there must be at least two classes
        nn=2;
    else
    end;
    index = kmeans(Y,nn);
    w = zeros(1,nn); aa = zeros(1,nn); bb= zeros(1,nn);
    for k=1:nn
        w(k) = sum(index==k)/L;
        aa(k) = (mean(Y(index==k)))^2/var(Y(index==k));
        bb(k) = var(Y(index==k))/mean(Y(index==k));
    end
 disp(['Number of mixtures=',num2str(nn)])
 disp('Initial weights after regrouping')
disp(w)
else
    disp('Number of mixtures remain unchanged')
end;
pq=w(2:end);% the first value is not needed
pqr=[aa,bb,pq]; % initial guesses
% these are
        % a's of the gamma pdf
        % b's of the gamma pedf
        % pq are the weights. Note that to reduce computation,
        % the number of pq values will be one less since the total
        % probability is unity and the remaining weight will be 1- sum of
        % the (nn-1) weights
pf =@(x,a,b)  gampdf(x,a,b); % the gamma pdf as an in-line function
% Use mle to fit the mixture.
% the output if mle will be 3*nn-1 values; the first nn values will be a's
% the next nn values will be b's and the remaining nn-1 values are the
% weights and weight of the first components will be 1-sum of the (nn-1)
% weights
```

```
if nn==2
    mypdf = @(x,a1,a2,b1,b2,pq) (1-pq)*pf(x,a1,b1) + pq*pf(x,a2,b2);
pp = mle(Y,'pdf',mypdf,'start',pqr, 'options',opt); % fit mypdf to the data xx
[my,ny]=size(Y);
L=my*ny; % length of the data
Y=reshape(Y,[L,1]);
opt = statset('mlecustom');
opt = statset(opt,'FunValCheck','off','MaxIter',6000,'TolX',1e-4,'TolFun',1e-5,'TolBnd',1e-
5','MaxFunEvals', 6000);
% chose the initial guess from kmeans
index = kmeans(Y,n);
w = zeros(1,n); aa = zeros(1,n); bb= zeros(1,n);
for k=1:n
    w(k) = sum(index==k)/L;
    aa(k) = (mean(Y(index==k)))^2/var(Y(index==k));
    bb(k) = var(Y(index==k))/mean(Y(index==k));
end
disp('Initial weights'),disp(w),disp('set a 10% threshold and regroup if necessary')
M=sum(w<=.1); % number classes less than 10%
if M>0 && n>2 % nn must be at least 2
    nn=n-M;
    if nn==1 % there must be at least two classes
        nn=2;
    else
    end;
    index = kmeans(Y,nn);
    w = zeros(1,nn); aa = zeros(1,nn); bb= zeros(1,nn);
    for k=1:nn
        w(k) = sum(index==k)/L;
        aa(k) = (mean(Y(index==k)))^2/var(Y(index==k));
        bb(k) = var(Y(index==k))/mean(Y(index==k));
    end
 disp(['Number of mixtures=',num2str(nn)])
 disp('Initial weights after regrouping')
disp(w)
else
    disp('Number of mixtures remain unchanged')
end;
pq=w(2:end);% the first value is not needed
pqr=[aa,bb,pq]; % initial guesses
% these are
        % a's of the gamma pdf
        % b's of the gamma pedf
        % pq are the weights. Note that to reduce computation,
        % the number of pq values will be one less since the total
        % probability is unity and the remaining weight will be 1- sum of
        % the (nn-1) weights
pf =@(x,a,b)  gampdf(x,a,b); % the gamma pdf as an in-line function
% Use mle to fit the mixture.
% the output if mle will be 3*nn-1 values; the first nn values will be a's
% the next nn values will be b's and the remaining nn-1 values are the
% weights and weight of the first components will be 1-sum of the (nn-1)
% weights
if nn==2
    mypdf = @(x,a1,a2,b1,b2,pq) (1-pq)*pf(x,a1,b1) + pq*pf(x,a2,b2);
pp = mle(Y,'pdf',mypdf,'start',pqr, 'options',opt); % fit mypdf to the data xx
a=[pp(1:2)];
b=[pp(3:4)];
p=[1-pp(5),pp(5)]; % all weights now
```

```
    elseif nn==3
mypdf = @(x,a1,a2,a3,b1,b2,b3,pq1,pq2) (1-pq1-pq2)*pf(x,a1,b1) + pq1*pf(x,a2,b2)+pq2*pf(x,a3,b3);
pp = mle(Y,'pdf',mypdf,'start',pqr, 'options',opt); % fit mypdf to the data xx
a=[pp(1:3)];
b=[pp(4:6)];
p=[1-pp(7)-pp(8),pp(7),pp(8)];
 elseif nn==4
mypdf = @(x,a1,a2,a3,a4,b1,b2,b3,b4,pq1,pq2,pq3) (1-pq1-pq2-pq3)*pf(x,a1,b1) +
pq1*pf(x,a2,b2)+pq2*pf(x,a3,b3)+...
pq3*pf(x,a4,b4);
pp = mle(Y,'pdf',mypdf,'start',pqr, 'options',opt); % fit mypdf to the data xx
a=[pp(1:4)];
b=[pp(5:8)];
p=[1-pp(9)-pp(10)-pp(11),pp(9),pp(10),pp(11)];
elseif nn==5
mypdf = @(x,a1,a2,a3,a4,a5,b1,b2,b3,b4,b5,pq1,pq2,pq3,pq4) (1-pq1-pq2-pq3)*pf(x,a1,b1) +
pq1*pf(x,a2,b2)+pq2*pf(x,a3,b3)+...
pq3*pf(x,a4,b4)+pq4*pf(x,a5,b5);
pp = mle(Y,'pdf',mypdf,'start',pqr, 'options',opt); % fit mypdf to the data xx
a=[pp(1:5)];
b=[pp(6:10)];
p=[1-pp(11)-pp(12)-pp(13)-pp(14),pp(11),pp(12),pp(13),pp(14)];
elseif nn==6
mypdf = @(x,a1,a2,a3,a4,a5,a6,b1,b2,b3,b4,b5,b6,pq1,pq2,pq3,pq4,pq5) (1-pq1-pq2-pq3)*pf(x,a1,b1)
+ pq1*pf(x,a2,b2)+pq2*pf(x,a3,b3)+...
pq3*pf(x,a4,b4)+pq4*pf(x,a5,b5)+pq5*pf(x,a6,b6);
pp = mle(Y,'pdf',mypdf,'start',pqr, 'options',opt); % fit mypdf to the data xx
a=[pp(1:6)];
b=[pp(7:12)];
p=[1-pp(13)-pp(14)-pp(15)-pp(16)-pp(17),pp(13) pp(14) pp(15) pp(16) pp(17)];
elseif nn==7
mypdf = @(x,a1,a2,a3,a4,a5,a6,a7,b1,b2,b3,b4,b5,b6,b7,pq1,pq2,pq3,pq4,pq5,pq6) (1-pq1-pq2-
pq3)*pf(x,a1,b1) + pq1*pf(x,a2,b2)+pq2*pf(x,a3,b3)+...
pq3*pf(x,a4,b4)+pq4*pf(x,a5,b5)+pq5*pf(x,a6,b6)+pq6*pf(x,a7,b7);
pp = mle(Y,'pdf',mypdf,'start',pqr, 'options',opt); % fit mypdf to the data xx
a=[pp(1:7)];
b=[pp(8:14)];
p=[1-pp(15)-pp(16)-pp(17)-pp(18)-pp(19)-pp(20),pp(15) pp(16) pp(17) pp(18) pp(19) pp(20)];
elseif nn==8
mypdf = @(x,a1,a2,a3,a4,a5,a6,a7,a8,b1,b2,b3,b4,b5,b6,b7,b8,pq1,pq2,pq3,pq4,pq5,pq6,pq7) (1-pq1-
pq2-pq3)*pf(x,a1,b1) + pq1*pf(x,a2,b2)+pq2*pf(x,a3,b3)+...
pq3*pf(x,a4,b4)+pq4*pf(x,a5,b5)+pq5*pf(x,a6,b6)+pq6*pf(x,a7,b7)+pq7*pf(x,a8,b8);
pp = mle(Y,'pdf',mypdf,'start',pqr, 'options',opt); % fit mypdf to the data xx
a=[pp(1:8)];
b=[pp(9:16)];
p=[1-pp(17)-pp(18)-pp(19)-pp(20)-pp(21)-pp(22)-pp(23),pp(17) pp(18) pp(19) pp(20) pp(21) pp(22)
pp(23)];
elseif nn==9
mypdf =
@(x,a1,a2,a3,a4,a5,a6,a7,a8,a9,b1,b2,b3,b4,b5,b6,b7,b8,b9,pq1,pq2,pq3,pq4,pq5,pq6,pq7,pq8) (1-
pq1-pq2-pq3)*pf(x,a1,b1) + pq1*pf(x,a2,b2)+pq2*pf(x,a3,b3)+...
pq3*pf(x,a4,b4)+pq4*pf(x,a5,b5)+pq5*pf(x,a6,b6)+pq6*pf(x,a7,b7)+pq7*pf(x,a8,b8)+pq8*pf(x,a9,b9);
pp = mle(Y,'pdf',mypdf,'start', pqr); % fit mypdf to the data xx
a=[pp(1:9)];
b=[pp(10:18)];
p=[1-pp(19)-pp(20)-pp(21)-pp(22)-pp(23)-pp(24)-pp(25)-pp(26),pp(19) pp(20) pp(21) pp(22) pp(23)
pp(24) pp(25) pp(26)];
else
end;

end
```

```
Initial weights
    0.0543    0.4746    0.0105    0.1526    0.3080

set a 10% threshold and regroup if necessary
Number of mixtures=3
Initial weights after regrouping
    0.6852    0.2670    0.0478

Initial weights
    0.1397    0.0005    0.0048    0.0282    0.8268

set a 10% threshold and regroup if necessary
Number of mixtures=2
Initial weights after regrouping
    0.9779    0.0221

Initial weights
    0.3107    0.4895    0.0078    0.0494    0.1425

set a 10% threshold and regroup if necessary
Number of mixtures=3
Initial weights after regrouping
    0.6871    0.0489    0.2640

Warning: Maximum likelihood estimation did not converge.  Function evaluation
limit exceeded.
Initial weights
    0.1492    0.8093    0.0009    0.0340    0.0066

set a 10% threshold and regroup if necessary
Number of mixtures=2
Initial weights after regrouping
    0.9698    0.0302
```

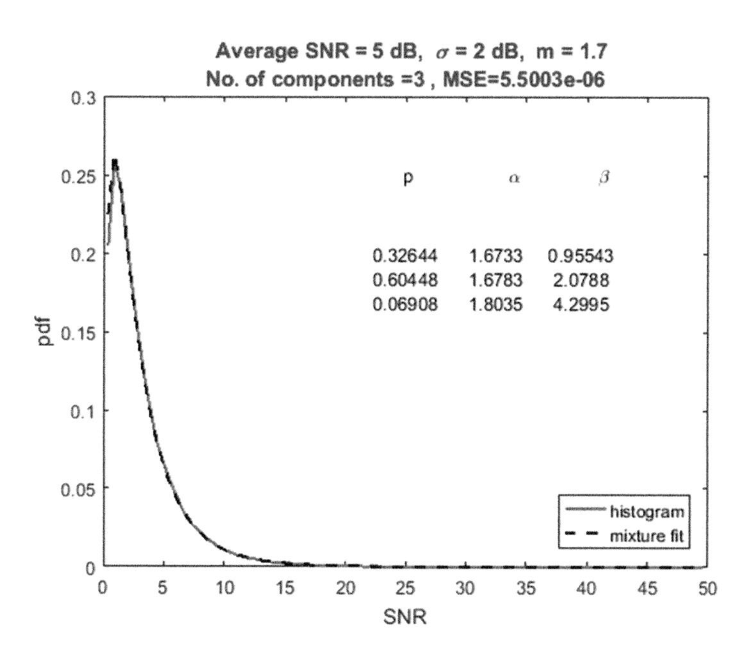

Fig. 4.114 Gamma mixture fit for Nakagami-lognormal fading channels (values shown in the title)

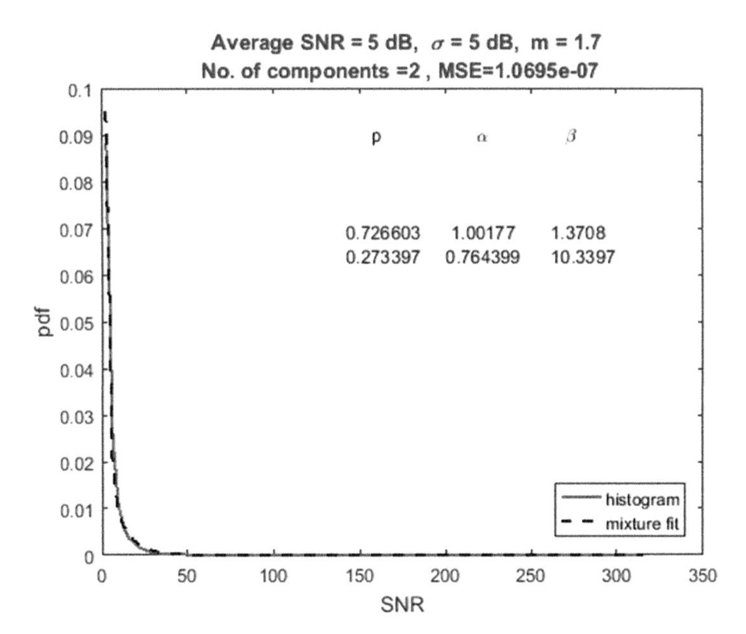

Fig. 4.115 Gamma mixture fit for Nakagami-lognormal fading channels (values shown in the title)

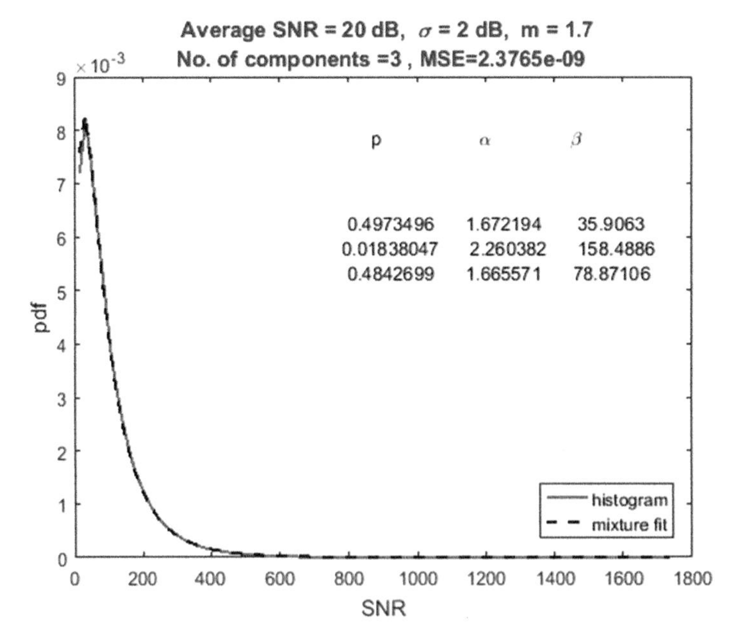

Fig. 4.116 Gamma mixture fit for Nakagami-lognormal fading channels (values shown in the title)

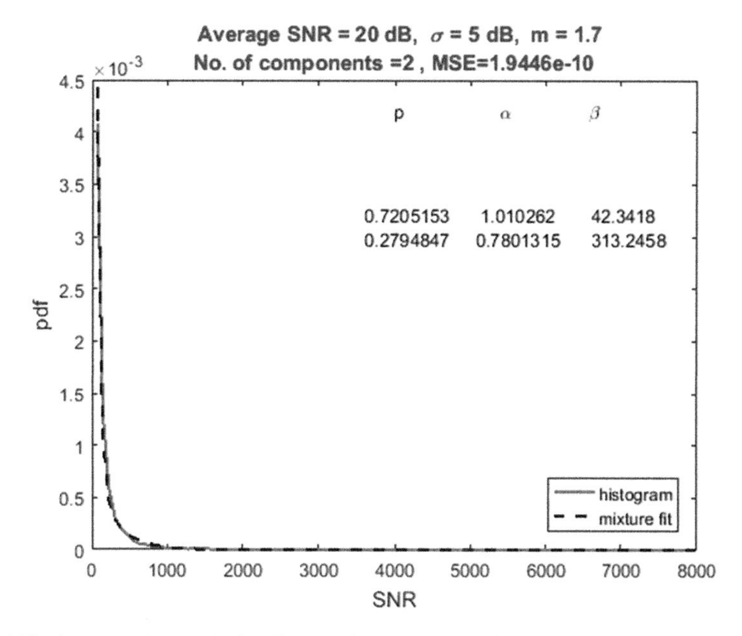

Fig. 4.117 Gamma mixture fit for Nakagami-lognormal fading channels (values shown in the title)

Gamma Mixtures and McKay Distribution

The McKay distribution can also be expressed using the mixture densities. Note that the McKay density is similar to the $\eta - \mu$ distribution. Results are shown in Figs. 4.118, 4.119, and 4.120.

```
function gammamixture_mckay
% P M Shankar, October 2016
% McKay density
close all
Numb=1e5; % number of samples
n=4;
for ks=1:3;
    if ks==1
        X=10;
        m=1.5;rh=0.4;X1=X*(1+sqrt(rh))/2;X2=X*(1-sqrt(rh))/2;
        m1=m/2;
        hh=gamrnd(m1,X1/m1,1,Numb)+gamrnd(m1,X2/m1,1,Numb);
    elseif ks==2
        X=15;
         m=0.75;rh=0.3;X1=X*(1+sqrt(rh))/2;X2=X*(1-sqrt(rh))/2;
        m1=m/2;
        hh=gamrnd(m1,X1/m1,1,Numb)+gamrnd(m1,X2/m1,1,Numb);
    else
        X=15;
        m=1.01;rh=0.6;X1=X*(1+sqrt(rh))/2;X2=X*(1-sqrt(rh))/2;
        m1=m/2;
        hh=gamrnd(m1,X1/m1,1,Numb)+gamrnd(m1,X2/m1,1,Numb);
    end;

[ff,xr]=ksdensity(hh);
% determine if any negative values of xr exist. discard them.
NK=[]; % eliminate negative values of xr
NK=find(xr<0);
if isempty(NK)==0
    NK1=max(NK)+1;% find the largest index and add 1 to start the non-zero values
    xr=xr(NK1:end);
    ff=ff(NK1:end);
else
end;
X=hh;

[nn,w,alpha,beta] = gammamixture_MLEf(X,n);
% nn is the actual number of components
L=length(xr);
fx=zeros(1,L);
nL=length(w);% to account for the reduced number of components
for k=1:nL
    fx=fx+w(k)*gampdf(xr,alpha(k),beta(k));
end;
figure, plot(xr,ff,'r-',xr,fx,'--k','linewidth',1.5)
legend('histogram','mixture fit','location','southeast')
xlabel('SNR'),ylabel('pdf')
MSE=(1/L)*sum(ff-fx).^2;
tit1={['\alpha = ',num2str(m),',  \rho = ',num2str(rh)];...
    ['No. of components =',num2str(nn),' , MSE=',num2str(MSE)]};
title(tit1,'color','b')
text(0.5*max(xr),0.95*max(max(ff,fx)),...
    'p                    \alpha                  \beta')
val=[w;alpha;beta]';
text(0.45*max(xr),0.7*max(max(ff,fx)),num2str(val))
clear xr ff fx w alpha beta hh gg
end;


end
```

```
function [nn,p,a,b] = gammamixture_MLEf(Y,n)
% P M Shankar, July 2016
% this one is seto to a maximum of 9 components
% the initial guesses of the parameters needed for MLE are obtained using
% clustering. If the clustering probability is less than 10%, the kmeans is
% repetade with fewer clusters. Therefore, acutual number of components
% will be between 2 and n.
%
[my,ny]=size(Y);
L=my*ny; % length of the data
Y=reshape(Y,[L,1]);
opt = statset('mlecustom');
opt = statset(opt,'FunValCheck','off','MaxIter',6000,'TolX',1e-4,'TolFun',1e-5,'TolBnd',1e-
5','MaxFunEvals', 6000);
% chose the initial guess from kmeans
index = kmeans(Y,n);
w = zeros(1,n); aa = zeros(1,n); bb= zeros(1,n);
for k=1:n
    w(k) = sum(index==k)/L;
    aa(k) = (mean(Y(index==k)))^2/var(Y(index==k));
    bb(k) = var(Y(index==k))/mean(Y(index==k));
end
disp('Initial weights'),disp(w),disp('set a 10% threshold and regroup if necessary')
M=sum(w<=.10) ;% number classes less with probability  than 10%
if M>0 && n>2 % nn must be at least 2
    nn=n-M;
    if nn==1 % there must be at least two classes
        nn=2;
    else
    end;
    index = kmeans(Y,nn);
    w = zeros(1,nn); aa = zeros(1,nn); bb= zeros(1,nn);
    for k=1:nn
        w(k) = sum(index==k)/L;
        aa(k) = (mean(Y(index==k)))^2/var(Y(index==k));
        bb(k) = var(Y(index==k))/mean(Y(index==k));
    end
 disp(['Number of mixtures=',num2str(nn)])
 disp('Initial weights after regrouping')
disp(w)
else
    disp('Number of mixtures remain unchanged')
end;
pq=w(2:end);% the first value is not needed
pqr=[aa,bb,pq]; % initial guesses
% these are
        % a's of the gamma pdf
        % b's of the gamma pedf
        % pq are the weights. Note that to reduce computation,
        % the number of pq values will be one less since the total
        % probability is unity and the remaining weight will be 1- sum of
        % the (nn-1) weights
pf =@(x,a,b)  gampdf(x,a,b); % the gamma pdf as an in-line function
% Use mle to fit the mixture.
% the output if mle will be 3*nn-1 values; the first nn values will be a's
% the next nn values will be b's and the remaining nn-1 values are the
% weights and weight of the first components will be 1-sum of the (nn-1)
% weights
```

```
if nn==2
    mypdf = @(x,a1,a2,b1,b2,pq) (1-pq)*pf(x,a1,b1) + pq*pf(x,a2,b2);
pp = mle(Y,'pdf',mypdf,'start',pqr, 'options',opt); % fit mypdf to the data xx
a=[pp(1:2)];
b=[pp(3:4)];
p=[1-pp(5),pp(5)]; % all weights now

    elseif nn==3
mypdf = @(x,a1,a2,a3,b1,b2,b3,pq1,pq2) (1-pq1-pq2)*pf(x,a1,b1) + pq1*pf(x,a2,b2)+pq2*pf(x,a3,b3);
pp = mle(Y,'pdf',mypdf,'start',pqr, 'options',opt); % fit mypdf to the data xx
a=[pp(1:3)];
b=[pp(4:6)];
p=[1-pp(7)-pp(8),pp(7),pp(8)];
 elseif nn==4
mypdf = @(x,a1,a2,a3,a4,b1,b2,b3,b4,pq1,pq2,pq3) (1-pq1-pq2-pq3)*pf(x,a1,b1) +
pq1*pf(x,a2,b2)+pq2*pf(x,a3,b3)+...
pq3*pf(x,a4,b4);
pp = mle(Y,'pdf',mypdf,'start',pqr, 'options',opt); % fit mypdf to the data xx
a=[pp(1:4)];
b=[pp(5:8)];
p=[1-pp(9)-pp(10)-pp(11),pp(9),pp(10),pp(11)];
elseif nn==5
mypdf = @(x,a1,a2,a3,a4,a5,b1,b2,b3,b4,b5,pq1,pq2,pq3,pq4) (1-pq1-pq2-pq3)*pf(x,a1,b1) +
pq1*pf(x,a2,b2)+pq2*pf(x,a3,b3)+...
pq3*pf(x,a4,b4)+pq4*pf(x,a5,b5);
pp = mle(Y,'pdf',mypdf,'start',pqr, 'options',opt); % fit mypdf to the data xx
a=[pp(1:5)];
b=[pp(6:10)];
p=[1-pp(11)-pp(12)-pp(13)-pp(14),pp(11),pp(12),pp(13),pp(14)];
elseif nn==6
mypdf = @(x,a1,a2,a3,a4,a5,a6,b1,b2,b3,b4,b5,b6,pq1,pq2,pq3,pq4,pq5) (1-pq1-pq2-pq3)*pf(x,a1,b1)
+ pq1*pf(x,a2,b2)+pq2*pf(x,a3,b3)+...
pq3*pf(x,a4,b4)+pq4*pf(x,a5,b5)+pq5*pf(x,a6,b6);
pp = mle(Y,'pdf',mypdf,'start',pqr, 'options',opt); % fit mypdf to the data xx
a=[pp(1:6)];
b=[pp(7:12)];
p=[1-pp(13)-pp(14)-pp(15)-pp(16)-pp(17),pp(13) pp(14) pp(15) pp(16) pp(17)];
elseif nn==7
mypdf = @(x,a1,a2,a3,a4,a5,a6,a7,b1,b2,b3,b4,b5,b6,b7,pq1,pq2,pq3,pq4,pq5,pq6) (1-pq1-pq2-
pq3)*pf(x,a1,b1) + pq1*pf(x,a2,b2)+pq2*pf(x,a3,b3)+...
pq3*pf(x,a4,b4)+pq4*pf(x,a5,b5)+pq5*pf(x,a6,b6)+pq6*pf(x,a7,b7);
pp = mle(Y,'pdf',mypdf,'start',pqr, 'options',opt); % fit mypdf to the data xx
a=[pp(1:7)];
b=[pp(8:14)];
p=[1-pp(15)-pp(16)-pp(17)-pp(18)-pp(19)-pp(20),pp(15) pp(16) pp(17) pp(18) pp(19) pp(20)];
elseif nn==8
mypdf = @(x,a1,a2,a3,a4,a5,a6,a7,a8,b1,b2,b3,b4,b5,b6,b7,b8,pq1,pq2,pq3,pq4,pq5,pq6,pq7) (1-pq1-
pq2-pq3)*pf(x,a1,b1) + pq1*pf(x,a2,b2)+pq2*pf(x,a3,b3)+...
pq3*pf(x,a4,b4)+pq4*pf(x,a5,b5)+pq5*pf(x,a6,b6)+pq6*pf(x,a7,b7)+pq7*pf(x,a8,b8);
pp = mle(Y,'pdf',mypdf,'start',pqr, 'options',opt); % fit mypdf to the data xx
a=[pp(1:8)];
b=[pp(9:16)];
p=[1-pp(17)-pp(18)-pp(19)-pp(20)-pp(21)-pp(22)-pp(23),pp(17) pp(18) pp(19) pp(20) pp(21) pp(22)
pp(23)];
elseif nn==9
mypdf =
@(x,a1,a2,a3,a4,a5,a6,a7,a8,a9,b1,b2,b3,b4,b5,b6,b7,b8,b9,pq1,pq2,pq3,pq4,pq5,pq6,pq7,pq8) (1-
pq1-pq2-pq3)*pf(x,a1,b1) + pq1*pf(x,a2,b2)+pq2*pf(x,a3,b3)+...
pq3*pf(x,a4,b4)+pq4*pf(x,a5,b5)+pq5*pf(x,a6,b6)+pq6*pf(x,a7,b7)+pq7*pf(x,a8,b8)+pq8*pf(x,a9,b9);
pp = mle(Y,'pdf',mypdf,'start', pqr); % fit mypdf to the data xx
```

```
a=[pp(1:9)];
b=[pp(10:18)];
p=[1-pp(19)-pp(20)-pp(21)-pp(22)-pp(23)-pp(24)-pp(25)-pp(26),pp(19) pp(20) pp(21) pp(22) pp(23)
pp(24) pp(25) pp(26)];
else
end;



end
```

```
Initial weights
    0.2995    0.0283    0.5492    0.1229

set a 10% threshold and regroup if necessary
Number of mixtures=3
Initial weights after regrouping
    0.2776    0.6558    0.0666

Initial weights
    0.6439    0.0196    0.0892    0.2473

set a 10% threshold and regroup if necessary
Number of mixtures=2
Initial weights after regrouping
    0.8705    0.1295

Initial weights
    0.0941    0.2423    0.0218    0.6417

set a 10% threshold and regroup if necessary
Number of mixtures=2
Initial weights after regrouping
    0.8598    0.1402
```

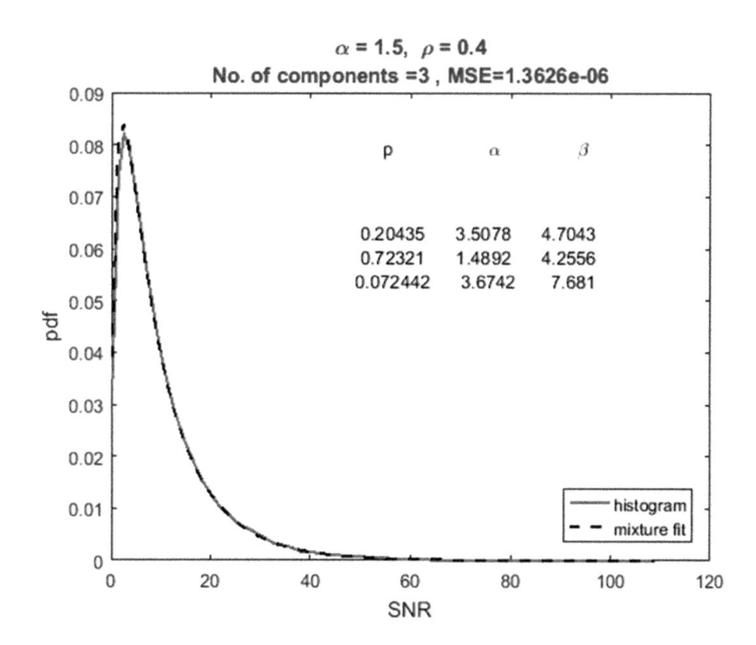

Fig. 4.118 Gamma mixture fit for McKay channels (values shown in the title)

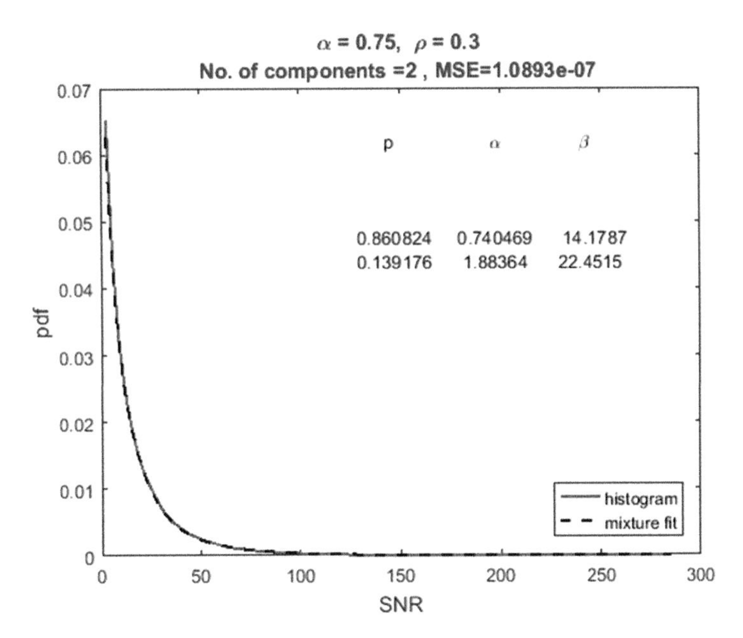

Fig. 4.119 Gamma mixture fit for McKay channels (values shown in the title)

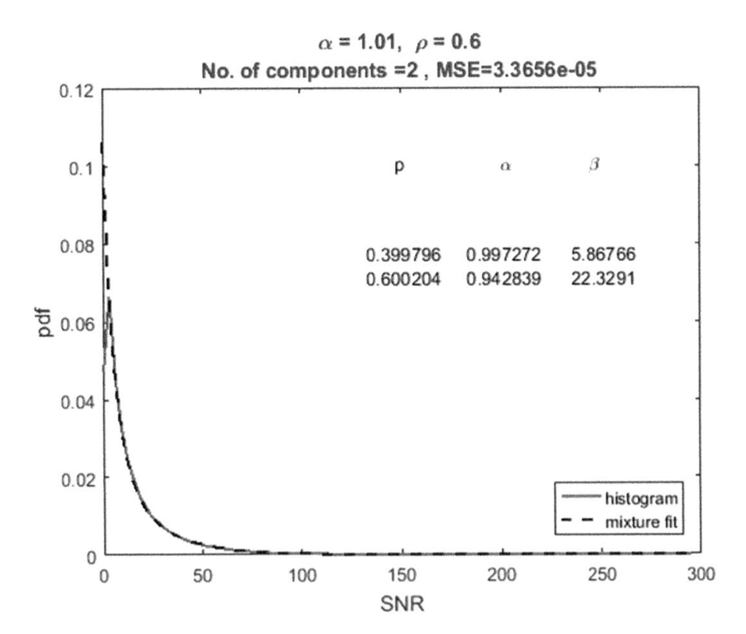

Fig. 4.120 Gamma mixture fit for McKay channels (values shown in the title)

Gamma Mixture and Rician Density

The Rician density can also be replaced and represented by a gamma mixture. Results are shown in Fig. 4.121, 4.122, and 4.123.

```
function gammamixture_rice
% P M Shankar, October  2016
% Rician
close all
Numb=1e5; % number of samples

n=4;
for ks=1:3;
    if ks==1
        x1=normrnd(0,2,1,Numb);
        x2=normrnd(1,2,1,Numb);
        hh=x1.^2+x2.^2;
        K=1/(2*2^2);
    elseif ks==2
        x1=normrnd(0,2,1,Numb);
        x2=normrnd(4,2,1,Numb);
        hh=x1.^2+x2.^2;
        K=16/(2*2^2);
    else
        x1=normrnd(0,2,1,Numb);
        x2=normrnd(8,2,1,Numb);
        hh=x1.^2+x2.^2;
        K=64/(2*2^2);
    end;

    [ff,xr]=ksdensity(hh);
    % determine if any negative values of xr exist. discard them.
    NK=[]; % eliminate negative values of xr
    NK=find(xr<0);
    if isempty(NK)==0
        NK1=max(NK)+1;% find the largest index and add 1 to start the non-zero values
        xr=xr(NK1:end);
        ff=ff(NK1:end);
    else
    end;
    X=hh;

    [nn,w,alpha,beta] = gammamixture_MLEf(X,n);
    % nn is the actual number of components
    L=length(xr);
    fx=zeros(1,L);
    nL=length(w);% to account for the reduced number of components
    for k=1:nL
        fx=fx+w(k)*gampdf(xr,alpha(k),beta(k));
    end;
    figure, plot(xr,ff,'r-',xr,fx,'--k','linewidth',1.5)
    legend('histogram','mixture fit','location','southeast')
    xlabel('SNR'),ylabel('pdf')
    MSE=(1/L)*sum(ff-fx).^2;
    tit1={['K = ',num2str(K)];...
        ['No. of components =',num2str(nn),' , MSE=',num2str(MSE)]};
    title(tit1,'color','b')
    text(0.5*max(xr),0.95*max(max(ff,fx)),...
        'p                    \alpha                    \beta')
    val=[w;alpha;beta]';
    text(0.45*max(xr),0.7*max(max(ff,fx)),num2str(val))
    clear xr ff fx w alpha beta hh gg
end;


end
```

```
Initial weights
    0.5221    0.2982    0.0403    0.1393

set a 10% threshold and regroup if necessary
Number of mixtures=3
Initial weights after regrouping
    0.6275    0.0796    0.2929

Initial weights
    0.3980    0.0613    0.1954    0.3453

set a 10% threshold and regroup if necessary
Number of mixtures=3
Initial weights after regrouping
    0.3647    0.5134    0.1218

Initial weights
    0.3653    0.0937    0.2827    0.2583

set a 10% threshold and regroup if necessary
Number of mixtures=3
Initial weights after regrouping
    0.4226    0.4078    0.1696
```

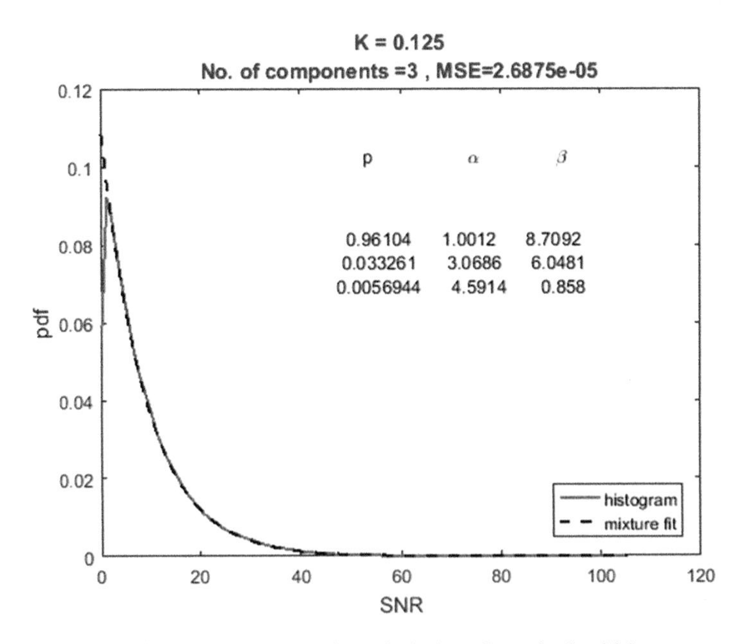

Fig. 4.121 Gamma mixture fit for Rician channels (values shown in the title)

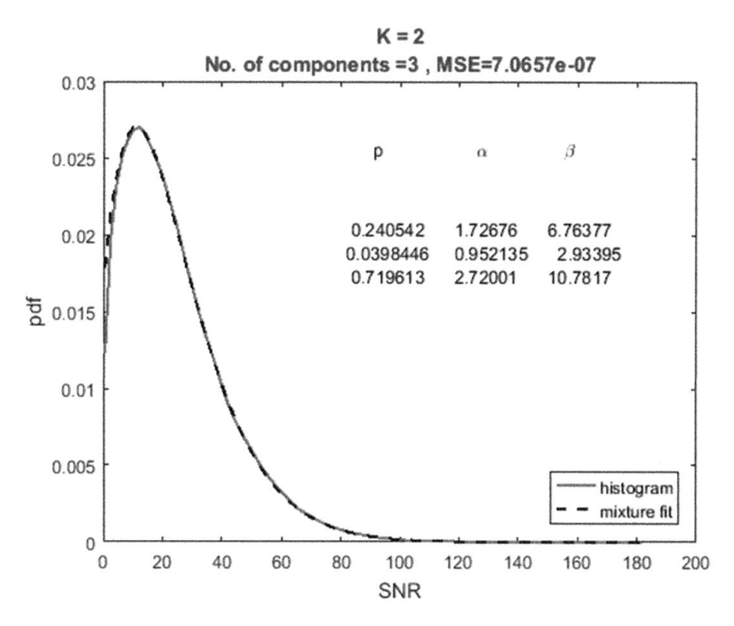

Fig. 4.122 Gamma mixture fit for Rician channels (values shown in the title)

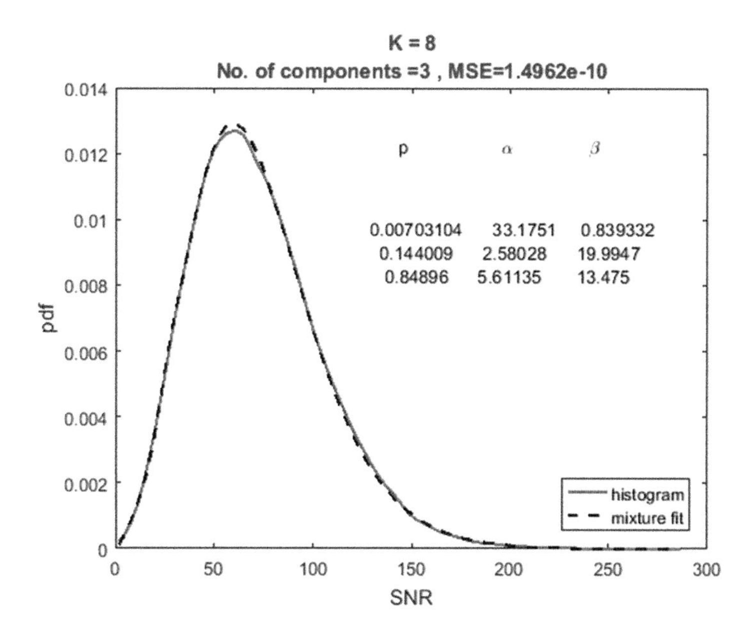

Fig. 4.123 Gamma mixture fit for Rician channels (values shown in the title)

Gamma Mixtures and Cascaded Nakagami Densities

The cascaded channels resulting in densities expressed in terms of the Meijer G-functions can be represented by gamma mixtures. Results are shown in Figs. 4.124, 4.125, and 4.126.

```
function gammamixture_Meijer G
% P M Shankar, October  2016
% gamma products
close all
Numb=1e4; % number of samples

n=7;
for ks=1:3;
    if ks==1
        m=1.5;
       x=gamrnd(m,2,3,Numb);
       N=3;
       hh=prod(x);
    elseif ks==2
        x=gamrnd(m,2,4,Numb);
       hh=prod(x);
       N=4;
    else
        x=gamrnd(m,2,5,Numb);
       hh=prod(x);
       N=5;
    end;

    [ff,xr]=ksdensity(hh);
    % determine if any negative values of xr exist. discard them.
    NK=[]; % eliminate negative values of xr
    NK=find(xr<0);
    if isempty(NK)==0
        NK1=max(NK)+1;% find the largest index and add 1 to start the non-zero values
        xr=xr(NK1:end);
        ff=ff(NK1:end);
    else
    end;
    X=hh;

    [nn,w,alpha,beta] = gammamixture_MLEf(X,n);
    % nn is the actual number of components
    L=length(xr);
    fx=zeros(1,L);
    nL=length(w);% to account for the reduced number of components
    for k=1:nL
        fx=fx+w(k)*gampdf(xr,alpha(k),beta(k));
    end;
    figure, plot(xr,ff,'r-',xr,fx,'--k','linewidth',1.5)
    legend('histogram','mixture fit','location','southeast')
    xlabel('SNR'),ylabel('pdf')
    MSE=(1/L)*sum(ff-fx).^2;
    tit1={['m = ', num2str(m),',  N = ',num2str(N)];...
        ['No. of components =',num2str(nn),' , MSE=',num2str(MSE)]};
    title(tit1,'color','b')
    text(0.5*max(xr),0.95*max(max(ff,fx)),...
        'p                  \alpha                 \beta')
    val=[w;alpha;beta]';
    text(0.45*max(xr),0.7*max(max(ff,fx)),num2str(val))
    clear xr ff fx w alpha beta hh gg
end;


end
```

```
Initial weights
    0.6777    0.2019    0.0034    0.0292    0.0115    0.0756    0.0007

set a 10% threshold and regroup if necessary
Number of mixtures=2
Initial weights after regrouping
    0.9468    0.0532

Initial weights
    0.8384    0.0315    0.1204    0.0011    0.0001    0.0080    0.0005

set a 10% threshold and regroup if necessary
Number of mixtures=2
Initial weights after regrouping
    0.9805    0.0195

Initial weights
    0.8284    0.0013    0.0002    0.0155    0.0382    0.1115    0.0049

set a 10% threshold and regroup if necessary
Number of mixtures=2
Initial weights after regrouping
    0.9997    0.0003
```

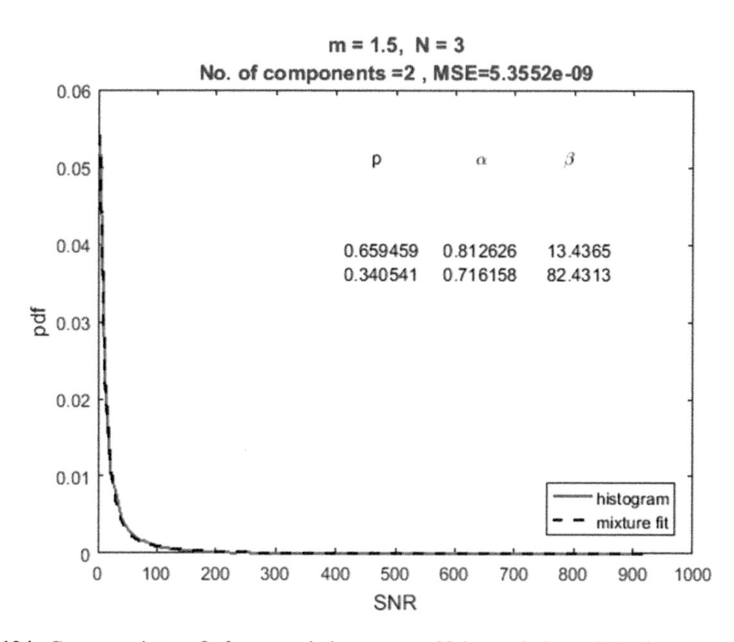

Fig. 4.124 Gamma mixture fit for cascaded gamma or Nakagami channels (values shown in the title)

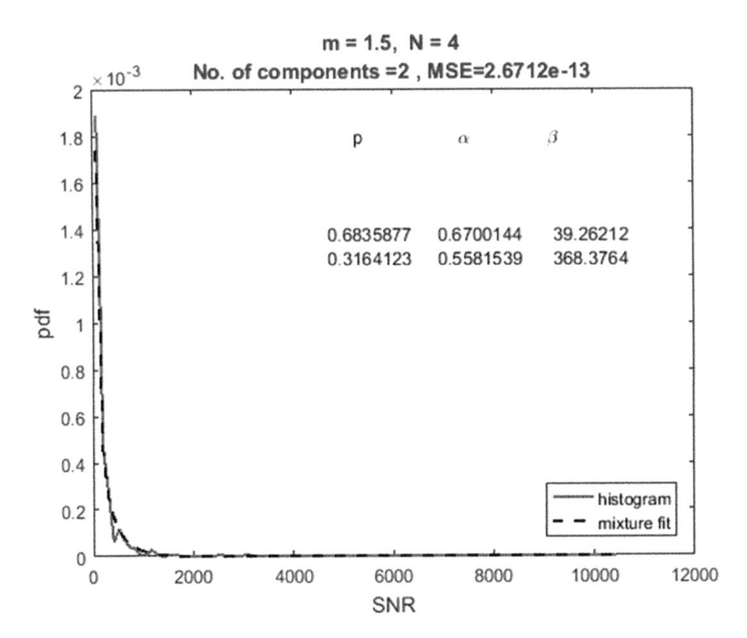

Fig. 4.125 Gamma mixture fit for cascaded gamma or Nakagami channels (values shown in the title)

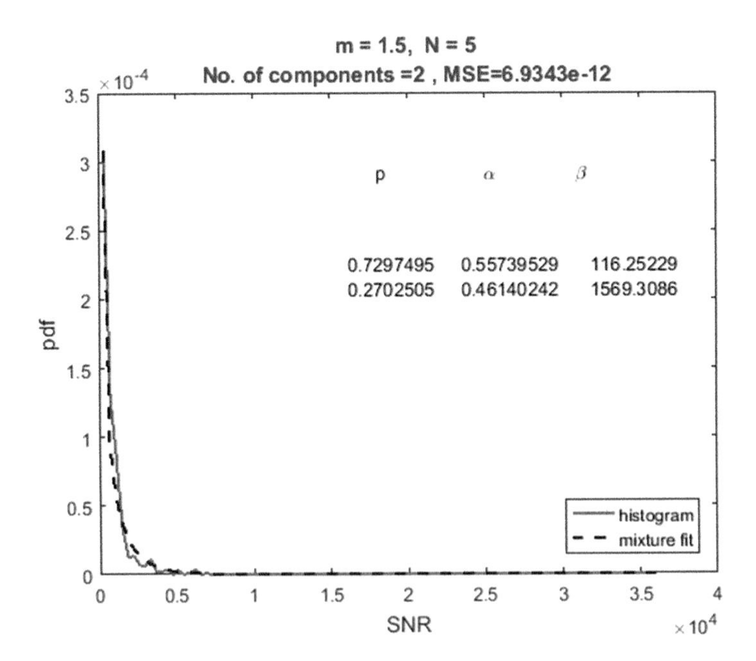

Fig. 4.126 Gamma mixture fit for cascaded gamma or Nakagami channels (values shown in the title)

4.11.3.2 Lognormal Mixtures

It has been assumed that a lognormal density is sufficient to model the shadowing in wireless channels. The concept of density mixtures can be extended to provide a better representation of a shadowing model based on a lognormal mixture (Liu et al. 2007; Büyükçorak et al. 2015). The lognormal density is the result of a large number of multiple scatterers being present in the scattering volume. However, in microcellular communication systems, there might be fewer scatterers to meet the requirements of the central limit theorem (CLT) for products. Note that the CLT is an essential requirement for the existence of the lognormal density. The insufficiency of the CLT for products can be mitigated through the use of a lognormal mixture for shadowing resulting in the density of the received signal strength (power) or SNR as

$$f(z) = \sum_{k=1}^{n} p_k \frac{1}{\sqrt{2\pi\sigma_k^2 z^2}} e^{-\frac{\left(10\log_{10}(z)-\mu_i\right)^2}{2\sigma_k^2}}. \tag{4.344}$$

Note that in Eq. (4.344), the parameters μ_k and σ_k are in decibel units. The parameters and the weights can be obtained using MLE approaches. However, the computation is not simple.

Taking advantage of the fact that a mixture of Gaussian densities is easier to handle, it was suggested that the lognormal mixture could be replaced by a Gaussian mixture. Gaussian mixtures have also been used differently in modeling the signal strength fluctuations as described below.

4.11.3.3 Gaussian Mixtures (Mixtures of Gaussian Distribution—MoG)

One of the basic densities is the Gaussian density and the fading channel amplitude can be expressed as (Selim et al. 2015)

$$f(x) = \sum_{k=1}^{n} p_k \frac{1}{\sqrt{2\pi\sigma_k^2}} e^{-\frac{(x-\mu_k)^2}{2\sigma_k^2}}. \tag{4.345}$$

Since wireless channels are characterized by the density of the SNR, the density function needs to be expressed in terms of the pdf of the SNR Z. The first step is the transformation of the random variable X to Z as

$$Z = X^2 Z_0. \tag{4.346}$$

In Eq. (4.346), Z_0 is the average SNR. The density function of the SNR in a channel modeled using GoM becomes

$$f(z)=\sum_{k=1}^{n}p_k\frac{1}{\sqrt{8\pi Z_0\sigma_k^2}}\left(\frac{1}{\sqrt{z}}\right)e^{-\frac{\left(\sqrt{\frac{z}{Z_0}}-\mu_k\right)^2}{2\sigma_k^2}}U(z). \tag{4.347}$$

In Eq. (4.347), the values of the parameters p_k , , μ_k, and σ_k are to be determined. The methods to evaluate these parameters will be described following the presentation of a few mixture models that have been reported in literature.

The *cumulative distribution function* of the SNR associated with the GoM is

$$F(z)=\int_0^z\sum_{k=1}^{n}p_k\frac{1}{\sqrt{8\pi Z_0\sigma_k^2}}\left(\frac{1}{\sqrt{w}}\right)e^{-\frac{\left(\sqrt{\frac{w}{Z_0}}-\mu_k\right)^2}{2\sigma_k^2}}dw. \tag{4.348}$$

The CDF can be expressed using Gaussian Q functions as

$$F(z)=\sum_{k=1}^{n}p_k\left[Q\left(-\frac{\mu_k}{\sigma_k}\right)-Q\left(-\frac{\sqrt{\frac{z}{Z_0}}-\mu_k}{\sigma_k}\right)\right]. \tag{4.349}$$

The *moments* of the MoG distribution are given by

$$E[Z^r]=\int_0^\infty\sum_{k=1}^{n}p_k\frac{z^r}{\sqrt{8\pi Z_0\sigma_k^2}}\left(\frac{1}{\sqrt{z}}\right)e^{-\frac{\left(\sqrt{\frac{z}{Z_0}}-\mu_k\right)^2}{2\sigma_k^2}}dz. \tag{4.350}$$

The moments of the MOG distribution can be expressed in terms of the moments of a Gaussian random variable

$$E\left[X^{2r}\right]=\int_{-\infty}^{\infty}x^{2r}\frac{1}{\sqrt{2\pi\sigma^2}}e^{-\frac{(x-\mu)^2}{2\sigma^2}}dx. \tag{4.351}$$

Using Eq. (4.351), the moments of the MoG distribution in Eq. (4.350) becomes

$$E[Z^r]=\sum_{k=1}^{n}p_kZ_0^rE\left[X_k^{2r}\right]. \tag{4.352}$$

Using the moments of the MoG distribution, the amount of fading can be expressed as

$$AF = \frac{E[Z^2]}{(E[Z])^2} - 1 = \frac{\sum_{k=1}^{n} p_k\left[\mu_k^4 + 6\mu_k^2\sigma_k^2 + 3\sigma_k^4\right]}{\left(\sum_{k=1}^{n} p_k\left[\mu_k^2 + \sigma_k^2\right]\right)^2} - 1. \tag{4.353}$$

4.12 Random Number Simulation

The results described in the previous sections contained examples of McKay and other densities examined using random number generation. In the first edition of the book, results on error rates in different fading channels were presented without providing the Matlab scripts and details on random number simulation. Before the McKay simulation is discussed, results on random number simulation of Nakagami, Nakagami-lognormal, generalized K, and cascaded channels are presented. In addition, estimation of error rates (coherent BPSK) is presented. In all cases, the simulated bit error rates are compared to the theoretical error rates obtained using analytical expressions for the error rates (either closed form expressions for the error rate or error rates obtained through numerical integration). The results based on analytical expressions were presented in previous sections.

We start with the case of a simple Nakagami fading channels. This is followed by the Nakagami-lognormal shadowed fading channels, generalized K shadowed fading channels, cascaded Nakagami channels, and finally, the McKay fading channels. Results of random number simulation are shown in Figs. 4.127–4.141.

4.12.1 Nakagami Channel

```
function wireless_simulation_NakagamiF
% Random number simulation Nakagami fading channel. Bit error rates
% compared to those obtained from theory.
% theoretical BER is obtained from the analytical expression involving the
% hypergeometric function. Created first in symbolic toolbox
% P M Shankar, October 2016
close all
global Z mm
m=[0.75,1.5,3,4];
ML=length(m);
Z0=5:30;
Z=10.^(Z0/10);
LK=length(Z);
bertN=zeros(ML,LK);
bersN=zeros(ML,LK);
for km=1:ML
    mm=m(km);
    for kk=1:LK
        SN=Z(kk);
        samp=gamrnd(mm,SN/mm,1,1e7); % generate gamma samples
        bersN(km,kk)=berest(samp);
    clear samp
    end;
      bertN(km,:)=bernakagami; % theoretical BER
end;
figure,semilogy(Z0,bertN(1,:),'-r',Z0,bersN(1,:),'ro',...
    Z0,bertN(2,:),'-k',Z0,bersN(2,:),'k*',...
    Z0,bertN(3,:),'-b',Z0,bersN(3,:),'bs',...
    Z0,bertN(4,:),'-m',Z0,bersN(4,:),'md')
ylim([1e-6,1]),xlabel('Average SNR (dB)')
ylabel('Average probability of error')
legend(['m = ',num2str(m(1)), '(th)'],['m = ',num2str(m(1)), '(sim)'],...
    ['m = ',num2str(m(2)), '(th)'],['m = ',num2str(m(2)), '(sim)'],...
     ['m = ',num2str(m(3)), '(th)'],['m = ',num2str(m(3)), '(sim)'],...
      ['m = ',num2str(m(4)), '(th)'],['m = ',num2str(m(4)), '(sim)'])

title('Nakagami fading channel')
end

function ber1= bernakagami % theoretical
% P M Shankar
% error rates in a Nakagami channel
global  Z mm
syms a X
F=((a/X)^a)*gamma(a+1/2)/(2*sqrt(sym(pi))*a*gamma(a));
er=F*hypergeom([a,a+1/2],[a+1],(-a/X));
ber1=zeros(1,length(Z));
for k=1:length(Z)
    ber1(k)=double(subs(er,[a,X],[mm,Z(k)]));
end;
end


function ber = berest(simdata) % get the error rate from using random samples
% P M Shankar
L=length(simdata);
```

```
nn=normrnd(0,1,1,L);%Gaussian noise of zero mean and unit variance
h=simdata;%data is in units of power
h=sqrt(h);%needs to be converted into envelope or amplitude units
%%input data is the BPSK bit sequence
ip = rand(1,L)>0.5;%generate 0's and 1's
indata=2*ip-1;%bipolar data
s=indata;
out= sqrt(2)*h.*s+nn;% this simulates sqrt(2E/T)*bk*a+n
x1=out>0;%detects 1s and zeros
x2=-1*(out<0);%detect-1's and zeros
outdata=x1+x2;%recreates the output bit stream in bipolar form
Diffdata=s-outdata;% will be either +2 or -2 when there is an error
DIF=abs(Diffdata);%will be 2 or zero
berN=sum(DIF>0);%counts how many times the abs difference exceeds zero.
ber=berN/L;

end
```

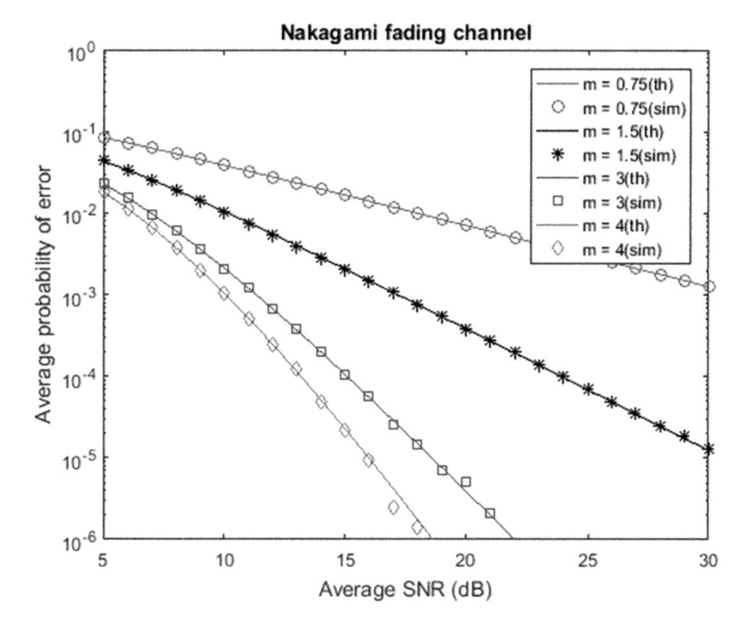

Fig. 4.127 Error rates in a Nakagami channel

4.12.2 Nakagami-Lognormal Shadowed Fading Channel

```
function wireless_simulation_shadowingF

% Random number simulation Nakagami-lognormal (NL) fading channel.
% Bit error ratescompared to those obtained from theory. Also compared
% to error rates a pure Nakagami fading channel (theory only)
%
% Theoretical BER in the NL channel estiamted using a double integral
%
% Nakagami theoretical BER is obtained from the analytical
% expression involving the hypergeometric function. Created first in
% symbolic toolbox
%
% Note that the average SNR Z0 needs to be converted to mu (dB) in
% shadowing. Whiel Z0 is the simply the value of expressed in dB, mu is the
% average when the power is actually measured in dB
% P M Shankar, October 2016
close all
global Z mm ss K MU
sigm=[2,4,6]; % shadowing levels dB
for ks=1:3
sig=sigm(ks); % shadowing level (dB)
ss=sig^2;
K=10/log(10);
m=[0.75,1.5,4];
ML=length(m);
Z0=5:30;
mu=Z0-ss/(2*K); % convert Z0 to mu
Z=10.^(Z0/10);
LK=length(Z);
bertNL=zeros(ML,LK);
bersNL=zeros(ML,LK);
bertN=zeros(ML,LK);
for km=1:ML
    mm=m(km);
    for kk=1:LK
        MU=mu(kk);
        samp1=gamrnd(mm,1/mm,1,1e7); % create a gamma variable of unit mean
        samp2=normrnd(MU,sig,1,1e7); % normal random variable of mean mu
        samp22=10.^(samp2/10); % convert into absolute power units
        samp=samp1.*samp22; % gamma-lognormal samples
        bersNL(km,kk)=berest(samp);
        bertNL(km,kk)=integral2(@nakalognorm,1e-4,inf,1e-4,inf);
    clear samp
    end;
      bertN(km,:)=bernakagami;
end;
figure,semilogy(Z0,bertNL(1,:),'-r',Z0,bersNL(1,:),'ro',...
    Z0,bertN(1,:),'^',Z0,bertNL(2,:),'-k',Z0,bersNL(2,:),'k*',...
    Z0,bertN(2,:),'v',Z0,bertNL(3,:),'-b',Z0,bersNL(3,:),'bs',...
    Z0,bertN(3,:),'p')
ylim([1e-6,1]),xlabel('Average SNR (dB)')
ylabel('Average probability of error')
legend(['m = ',num2str(m(1)), '(th-NL)'],['m = ',num2str(m(1)), '(sim-NL)'],...
    ['m = ',num2str(m(1)), '(th-N)'],...
    ['m = ',num2str(m(2)), '(th-NL)'],['m = ',num2str(m(2)), '(sim-NL)'],...
    ['m = ',num2str(m(2)), '(th-N)'],...
     ['m = ',num2str(m(3)), '(th-NL)'],['m = ',num2str(m(3)), '(sim-NL)'],...
     ['m = ',num2str(m(3)), '(th-N)'],  'location','southwest')
title(['Shadowed fading channel \sigma = ',num2str(sig),' dB'])
end;
end
```

```
function ber1= bernakagami % theoretical
% P M Shankar
% error rates in a Nakagami channel: expression from the book, first
% created in symbolic toolbox (Appendix Chapter 4)
global  Z mm
syms a X
F=((a/X)^a)*gamma(a+1/2)/(2*sqrt(sym(pi))*a*gamma(a));
er=F*hypergeom([a,a+1/2],[a+1],(-a/X));
ber1=zeros(1,length(Z));
for k=1:length(Z)
    ber1(k)=double(subs(er,[a,X],[mm,Z(k)]));
end;
end

function ber = berest(simdata) % get the error rate from using random samples
% P M Shankar
L=length(simdata);
nn=normrnd(0,1,1,L);%Gaussian noise of zero mean and unit variance
h=simdata;%data is in units of power
h=sqrt(h);%needs to be converted into envelope or amplitude units
%%input data is the BPSK bit sequence
ip = rand(1,L)>0.5;%generate 0's and 1's
indata=2*ip-1;%bipolar data
s=indata;
out= sqrt(2)*h.*s+nn;% this simulates sqrt(2E/T)*bk*a+n
x1=out>0;%detects 1s and zeros
x2=-1*(out<0);%detect-1's and zeros
outdata=x1+x2;%recreates the output bit stream in bipolar form
Diffdata=s-outdata;% will be either +2 or -2 when there is an error
DIF=abs(Diffdata);%will be 2 or zero
berN=sum(DIF>0);%counts how many times the abs difference exceeds zero.
ber=berN/L;

end


function zz = nakalognorm(x,xx) % function fo BER estimation in NL channel
% P M Shankar
global mm MU  ss K
fun1=gampdf(x,mm,xx/mm);
fun2=K*(exp(-(10*log10(xx)-MU).^2*1./(2*ss)))./sqrt(2*pi*ss*xx.^2);
zz=0.5*erfc(sqrt(x)).*fun1.*fun2;% error rate integrand
end
```

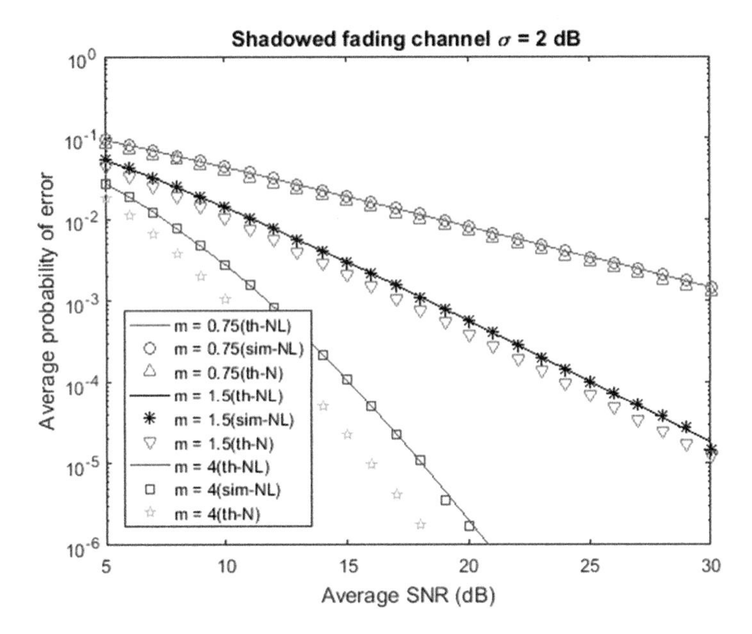

Fig. 4.128 Error rates in a Nakagami-lognormal shadowed fading channel (σ = 2 dB). Error rates for the Nakagami channel are also shown (theory)

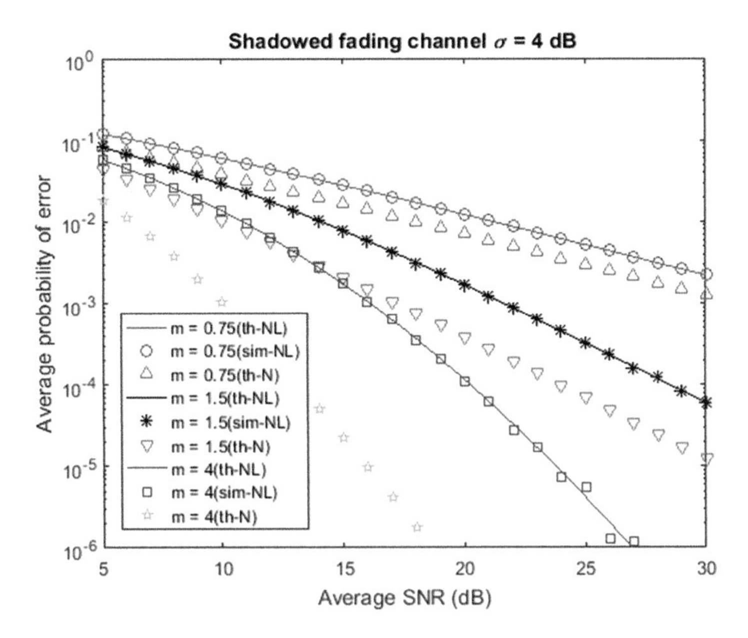

Fig. 4.129 Error rates in a Nakagami-lognormal shadowed fading channel (σ = 4 dB). Error rates for the Nakagami channel are also shown (theory)

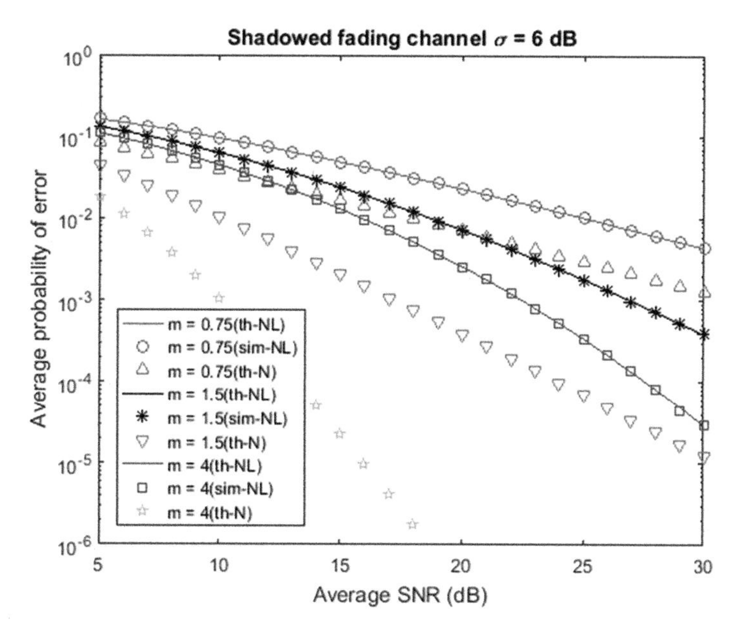

Fig. 4.130 Error rates in a Nakagami-lognormal shadowed fading channel (σ = 6 dB). Error rates for the Nakagami channel are also shown (theory)

4.12.3 GK Channels

```
function wireless_simulation_GKF
% Random number simulation GK channel. Bit error rates
% compared to those obtained from theory.
% the GK variable is the product of two independent gamma variables
% The theoretical BER is obtained in terms of hypergeometric functions
% Created first in symbolic toolbox
%
% GK variable is created from the product of two gamma variables of orders
% m1 and m2 . In the simulation while m2 is fixed, m1 is allowed to vary
% for the three plots.
%
% P M Shankar, October 2016
close all
global SN mm m2
mm2=[1.34, 2.47, 11.2];
for kq=1:3
  m2=mm2(kq);
m=[0.75,1.5,3]; % m1
ML=length(m);
Z0=5:30;
Z=10.^(Z0/10);
LK=length(Z);
bertN=zeros(ML,LK);
bersN=zeros(ML,LK);
```

```
for km=1:ML
    mm=m(km);
    for kk=1:LK
        SN=Z(kk);
        samp=gamrnd(m2,1/m2,1,1e7).*gamrnd(mm,SN/mm,1,1e7);% GK samples
        bersN(km,kk)=berest(samp);
       bertN(km,kk)=gkf_ber; % theoretical BER
    clear samp
    end;
end;
figure,semilogy(Z0,bertN(1,:),'-r',Z0,bersN(1,:),'ro',...
    Z0,bertN(2,:),'-k',Z0,bersN(2,:),'k*',...
    Z0,bertN(3,:),'-b',Z0,bersN(3,:),'bs')
ylim([1e-6,1]),xlabel('Average SNR (dB)'),ylabel('Average probability of error')
legend(['m_1 = ',num2str(m(1)), '(th)'],['m_1 = ',num2str(m(1)), '(sim)'],...
    ['m_1 = ',num2str(m(2)), '(th)'],['m_1 = ',num2str(m(2)), '(sim)'],...
     ['m_1 = ',num2str(m(3)), '(th)'],['m_1 = ',num2str(m(3)), '(sim)'],...
     'location','southwest')
title('GK fading channel (gamma product)')
text(6,5e-6,['m_2 = ',num2str(m2)],'backgroundcolor','y')
end;
end

function be = gkf_ber
global SN mm m2
% the analytical expression for the BER is taken from Chapter 4. It is
% created in symbolictoolbox and then converted for evaluation using
% symbolic substitution .
syms Z m1 c1
ZZ=m1*c1/Z;
p=sym(pi);
f1=1/2-(1/2)*p^2*csc(p*m1)*csc(p*c1)/...
    (gamma(1-m1)*gamma(1-c1)*gamma(m1)*gamma(c1));
f2=(1/(2*sqrt(p)))*hypergeom([c1,c1+1/2],[1+c1,1-m1+c1],ZZ)*...
    gamma(m1-c1)*gamma(c1+1/2)*ZZ^c1;
f22=f2/(gamma(m1)*gamma(c1+1));
f3=(1/(2*sqrt(p)))*hypergeom([m1,m1+1/2],[1+m1,1-c1+m1],ZZ)*...
    gamma(c1-m1)*gamma(m1+1/2)*ZZ^m1;
f33=f3/(gamma(c1)*gamma(m1+1));
f=f1+f22+f33;
be=double(subs(f,[m1,c1,Z],[mm+0.00121,m2+0.001231,SN]));
% add decimal places so that argument of the gamma(.) function is not a
% negative integer
end

function ber = berest(simdata) % get the error rate from using random samples
% P M Shankar
L=length(simdata);
nn=normrnd(0,1,1,L);%Gaussian noise of zero mean and unit variance
h=simdata;%data is in units of power
h=sqrt(h);%needs to be converted into envelope or amplitude units
%%input data is the BPSK bit sequence
ip = rand(1,L)>0.5;%generate 0's and 1's
indata=2*ip-1;%bipolar data
s=indata;
out= sqrt(2)*h.*s+nn;% this simulates sqrt(2E/T)*bk*a+n
x1=out>0;%detects 1s and zeros
x2=-1*(out<0);%detect-1's and zeros
outdata=x1+x2;%recreates the output bit stream in bipolar form
Diffdata=s-outdata;% will be either +2 or -2 when there is an error
DIF=abs(Diffdata);%will be 2 or zero
berN=sum(DIF>0);%counts how many times the abs difference exceeds zero.
ber=berN/L;

end
```

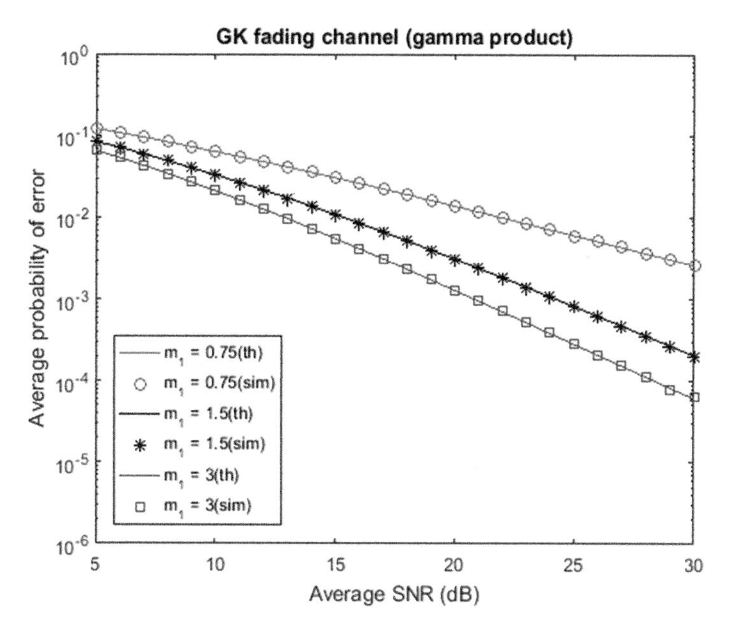

Fig. 4.131 Error rates in a GK channel ($m_2 = 1.34$)

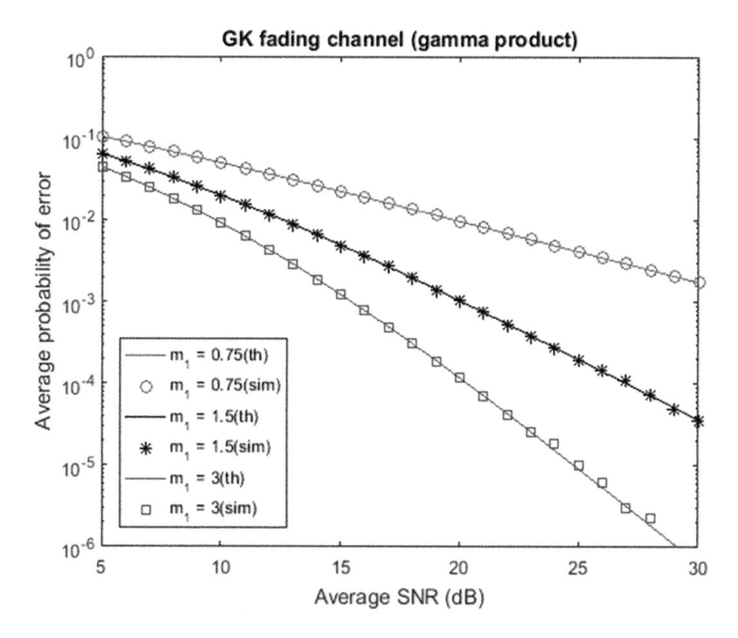

Fig. 4.132 Error rates in a GK channel ($m_2 = 2.47$)

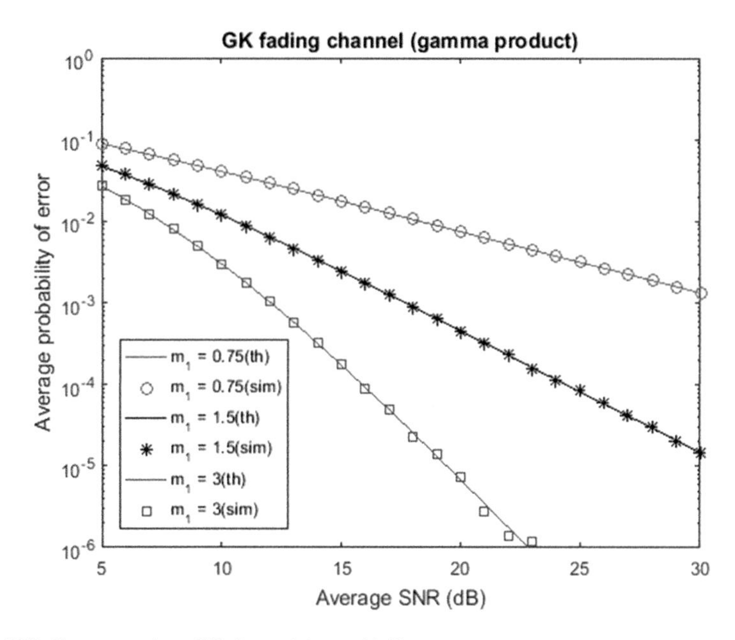

Fig. 4.133 Error rates in a GK channel ($m_2 = 11.2$)

4.12.4 Cascaded Nakagami Channels

```
function wireless_simulation_cascadedF
% Random number simulation cascaded channel Bit error rates
% compared to those obtained from theory.
%
% The theoretical BER is obtained in terms of Meijer G functions
% P M Shankar, October 2016
close all
for kn=1:5
N=kn;
m=[0.75,1.5,3]; % m1
ML=length(m);
Z0=5:30;
Z=10.^(Z0/10);
LK=length(Z);
bertN=zeros(ML,LK);
bersN=zeros(ML,LK);
for km=1:ML
    mm=m(km);
    for kk=1:LK
        SN=Z(kk);
       SNm=SN^(1/N); % the mean SNR will always be SNN regardless of N
       if N==1
            samp=gamrnd(mm,SNm/mm,1,1e7); % do not take the product
       else
        samp=prod(gamrnd(mm,SNm/mm,N,1e7));
       end;
        bersN(km,kk)=berest(samp);
       bertN(km,kk)=Meijer G_ber(N,mm,SN); % theoretical BER
    clear samp
    end;
end;
```

```
figure,semilogy(Z0,bertN(1,:),'-r',Z0,bersN(1,:),'ro',...
    Z0,bertN(2,:),'-k',Z0,bersN(2,:),'k*',...
    Z0,bertN(3,:),'-b',Z0,bersN(3,:),'bs')
ylim([1e-6,1]),xlabel('Average SNR (dB)'),ylabel('Average probability of error')
legend(['m = ',num2str(m(1)), '(th)'],['m = ',num2str(m(1)), '(sim)'],...
     ['m = ',num2str(m(2)), '(th)'],['m = ',num2str(m(2)), '(sim)'],...
      ['m = ',num2str(m(3)), '(th)'],['m = ',num2str(m(3)), '(sim)'],...
      'location','southwest')
 if N>1
title('Cascaded fading channel')
 text(15,5e-6,['N = ',num2str(N)],'backgroundcolor','y')
 else
     title('Nakagami fading channel')
 end;
end;

end

function [ber]=Meijer G_ber(N,m,ZZ)
% P M Shankar, October 2016
% ber is the error for coherent BPSK
% N number of cascades, m the parameter,  ZZ is the mean SNR (absolute units)
gm=gamma(m);
gM=gm^N;
if N==1
    ber=(1/2)-hypergeom([1/2, 1/2+m], [3/2], -ZZ/m)*gamma(1/2+m)/(sqrt(m/ZZ)*sqrt(pi)*gm);
elseif N==2
    ber=(1/2)-(1/2)*(1/sqrt(pi))*double(evalin(symengine,...
        sprintf('Meijer G([[1/2 ], [1 ]], [[0,%e,%e], []], %e)',m,m,(m^N)/ZZ)))/gM;
elseif N==3
     ber=(1/2)-(1/2)*(1/sqrt(pi))*double(evalin(symengine,...
         sprintf('Meijer G([[1/2 ], [1 ]], [[0,%e,%e,%e], []], %e)',m,m,m,(m^N)/ZZ)))/gM;
elseif N==4
       ber=(1/2)-(1/2)*(1/sqrt(pi))*double(evalin(symengine,...
           sprintf('Meijer G([[1/2 ], [1 ]], [[0,%e,%e,%e,%e], []], %e)',m,m,m,m,(m^N)/ZZ)))/gM;
elseif N==5
     ber=(1/2)-(1/2)*(1/sqrt(pi))*double(evalin(symengine,...
         sprintf('Meijer G([[1/2 ], [1 ]], [[0,%e,%e,%e,%e,%e], []],
%e)',m,m,m,m,m,(m^N)/ZZ)))/gM;
end;

end

function ber = berest(simdata) % get the error rate from using random samples
% P M Shankar
L=length(simdata);
nn=normrnd(0,1,1,L);%Gaussian noise of zero mean and unit variance
h=simdata;%data is in units of power
h=sqrt(h);%needs to be converted into envelope or amplitude units
%%input data is the BPSK bit sequence
ip = rand(1,L)>0.5;%generate 0's and 1's
indata=2*ip-1;%bipolar data
s=indata;
out= sqrt(2)*h.*s+nn;% this simulates sqrt(2E/T)*bk*a+n
x1=out>0;%detects 1s and zeros
x2=-1*(out<0);%detect-1's and zeros
outdata=x1+x2;%recreates the output bit stream in bipolar form
Diffdata=s-outdata;% will be either +2 or -2 when there is an error
DIF=abs(Diffdata);%will be 2 or zero
berN=sum(DIF>0);%counts how many times the abs difference exceeds zero.
ber=berN/L;

end
```

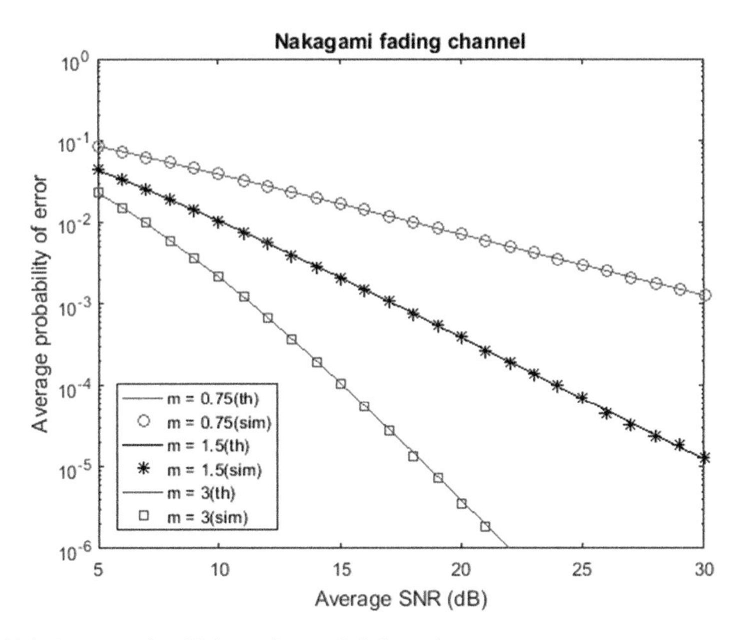

Fig. 4.134 Error rates in a Nakagami cascaded channel

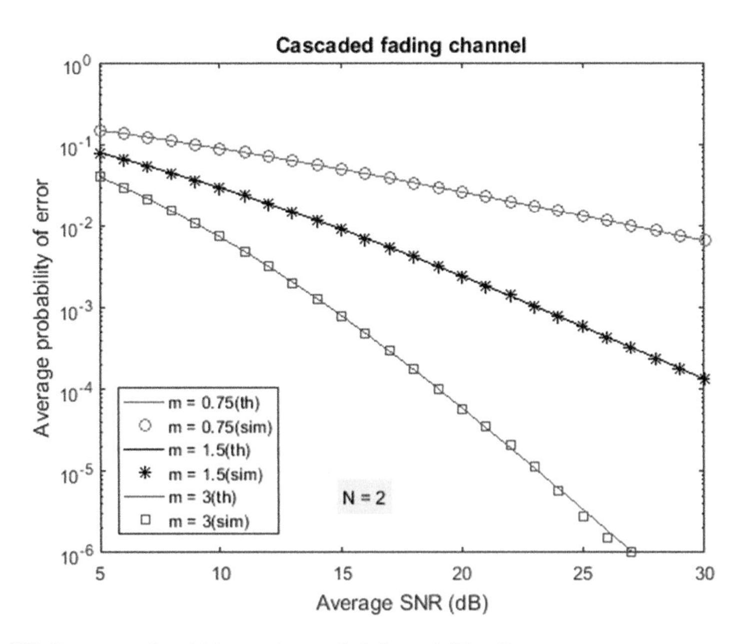

Fig. 4.135 Error rates in a Nakagami cascaded channel (N = 2)

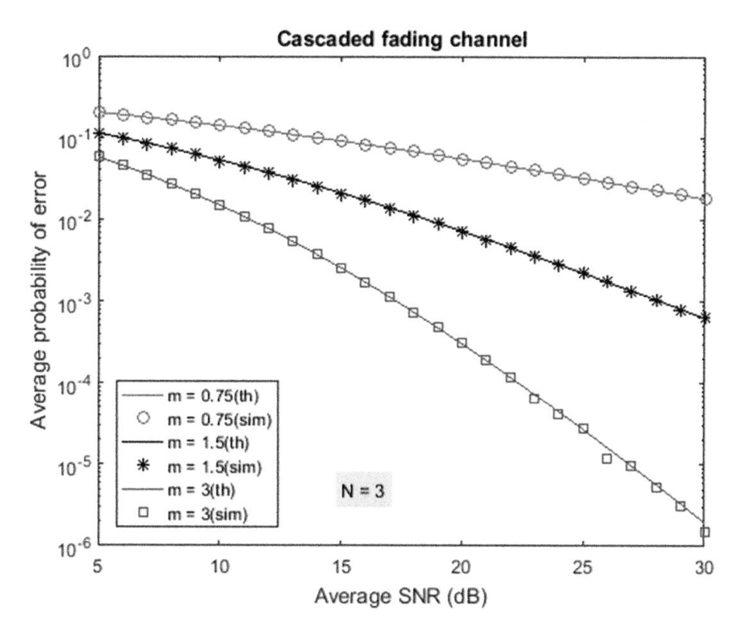

Fig. 4.136 Error rates in a Nakagami cascaded channel (N = 3)

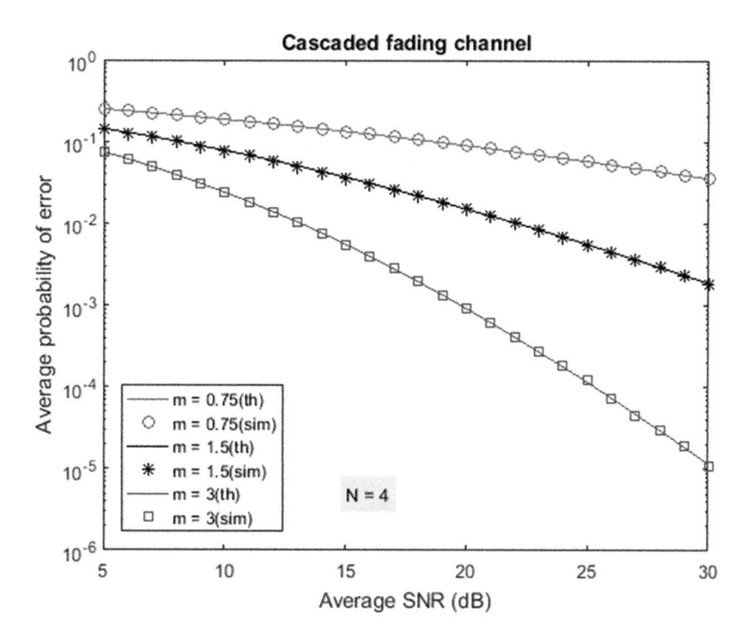

Fig. 4.137 Error rates in a Nakagami cascaded channel (N = 4)

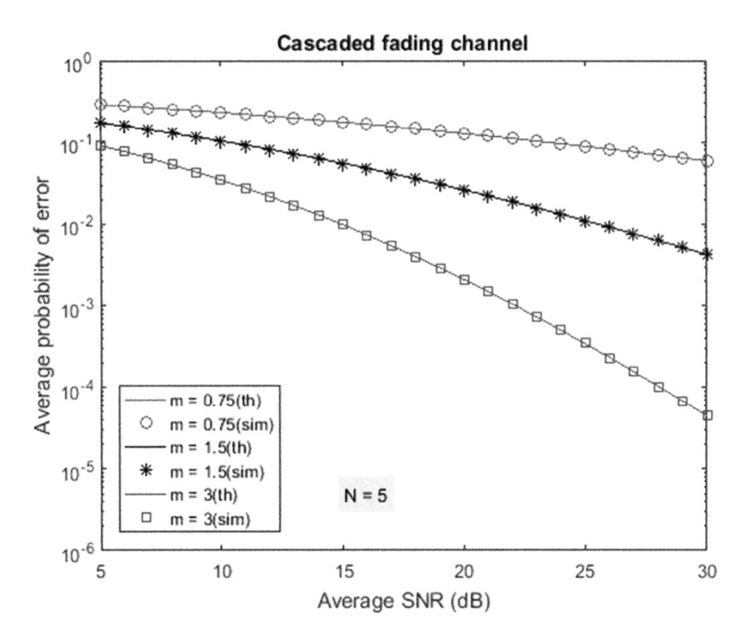

Fig. 4.138 Error rates in a Nakagami cascaded channel (N = 5)

4.12.5 *The McKay Channel*

Samples of McKay variables can be created as per the discussion earlier where it was shown that the McKay variable is the sum of two correlated identically distributed gamma variables or the sum of two independent but non-identical gamma variables of identical orders and different means. While simulation of correlated random numbers is possible, it is much easier to generate McKay variables using the latter concept of sum of two non-identical gamma variables. Results based on some of these approaches are reported in the earlier sections on McKay density. From the previous discussion, it is known that a McKay variable with parameters $\boldsymbol{\alpha}$, $\boldsymbol{X}$, and $\boldsymbol{\rho}$ is obtained from the sum of two gamma variables, each of order $\alpha/2$ and means of

$$X_1 = \frac{X}{2}\left(1 + \sqrt{\rho}\right) \tag{4.354}$$

$$X_2 = \frac{X}{2}\left(1 - \sqrt{\rho}\right). \tag{4.355}$$

The results of the simulation and appropriate match to the pdf and CDF of the McKay density are given below along with the Matlab script.

```
function  wireless_simulation_mckayF
% P M Shankar
% simulation of the error rates in a McKay channel. The simulated error
% rates compared to the theoretical ones obtained using Laplace transforms.
%
% McKay variable created from the summation of two independent gamma
% variables of unequal means. The means are related through the correlation
% recognizing that the sum of two correlated identical gamma variables lead
% to the same variable.
% October 2016
close all
Z0=5:3:30;
Z0=[Z0,30];
Z=10.^(Z0/10);LK=length(Z);
alpha=[0.5,0.8,1.5,2.5]; % values of alpha
LA=length(alpha);
rr=[0.1,0.4,0.7];% three values of the correlation
syms a r X  positive
% Laplace transform in symbolic form: creating the in-line function for
% integration
syms s y
fLzer=a^a/(a^2+s^2*X^2+2*s*X*a-s^2*X^2*r)^((1/2)*a);
%
for kkr=1:3;
    rh=rr(kkr);% correlation coefficient
pes=zeros(LK,LA); % pe simulation
peL=zeros(LK,LA);% pe using Laplace transforms or MGF

for kr=1:LA
    alp=alpha(kr);
    for k=1:LK
        fun1L=subs(fLzer,[a, X, r],[alp,Z(k),rh]); % make substitution
        fun2L=subs(fun1L,s,1/(sin(y))^2); % make substitution
        mckayf2=matlabFunction(fun2L); % in-line function for integration
        peL(k,kr)=(1/pi)*integral(mckayf2,0,0.999975*pi/2);
        % generate random numbers
        X1=Z(k)*(1+sqrt(rh))/2;       X2=Z(k)*(1-sqrt(rh))/2;
        a1=alp/2;
        x=gamrnd(a1,X1/a1,1,1e7)+gamrnd(a1,X2/a1,1,1e7); % McKay variable
        pes(k,kr)=berest(x);
        clear x
    end;
end;
figure,semilogy(Z0,peL(:,1),'r-',Z0,pes(:,1),'r*',...
    Z0,peL(:,2),'k--',Z0,pes(:,2),'kd',...
    Z0,peL(:,3),'-.b',Z0,pes(:,3),'bo',...
    Z0,peL(:,4),'-.m',Z0,pes(:,4),'mp')
xlim([5,30]),ylim([1e-6,1])
 legend(['\alpha =',num2str(alpha(1)),'  Th'],...
        ['\alpha =',num2str(alpha(1)),'  Sim'],...
        ['\alpha =',num2str(alpha(2)),'  Th'],...
        ['\alpha =',num2str(alpha(2)),'  Sim'],...
        ['\alpha =',num2str(alpha(3)),'  Th'],...
        ['\alpha =',num2str(alpha(3)),'  Sim'],...
         ['\alpha =',num2str(alpha(4)),'  th'],...
       ['\alpha =',num2str(alpha(4)),'  Sim'], 'location','southwest')
 xlabel('Average SNR (dB)')
 ylabel('Average probability of error')
 title(['McKay fading channel:  \rho = ',num2str(rh)])
end;
end
```

```
function ber = berest(simdata)
% P M Shankar, October 2016
L=length(simdata);
nn=normrnd(0,1,1,L);%Gaussian noise of zero mean and unit variance
h=simdata;%data is in units of power
h=sqrt(h);%needs to be converted into envelope or amplitude units
%%input data is the BPSK bit sequence
ip = rand(1,L)>0.5;%generate 0's and 1's
indata=2*ip-1;%bipolar data
s=indata;
out= sqrt(2)*h.*s+nn;% this simulates sqrt(2E/T)*bk*a+n
x1=out>0;%detects 1s and zeros
x2=-1*(out<0);%detect-1's and zeros
outdata=x1+x2;%recreates the output bit stream in bipolar form
Diffdata=s-outdata;% will be either +2 or -2 when there is an error
DIF=abs(Diffdata);%will be 2 or zero
berN=sum(DIF>0);%counts how many times the abs difference exceeds zero.
ber=berN/L;

end
```

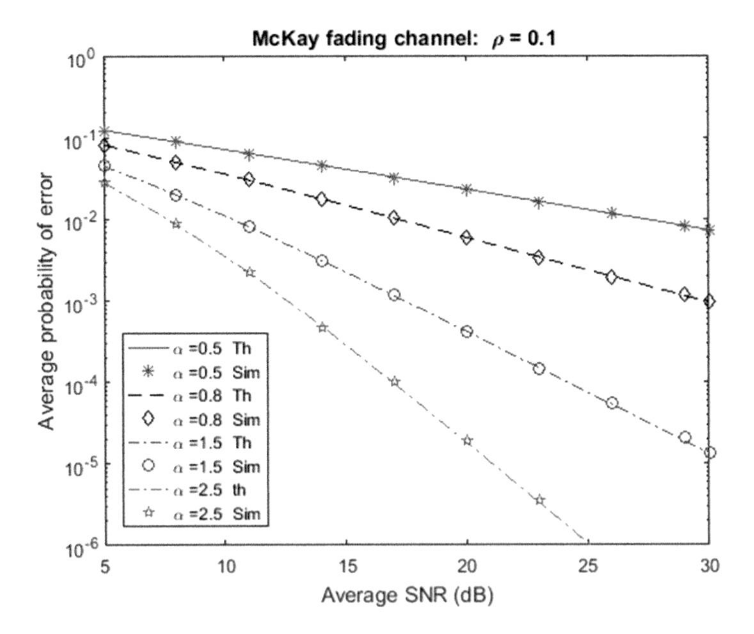

Fig. 4.139 Error rates in a McKay channel (ρ = 0.1)

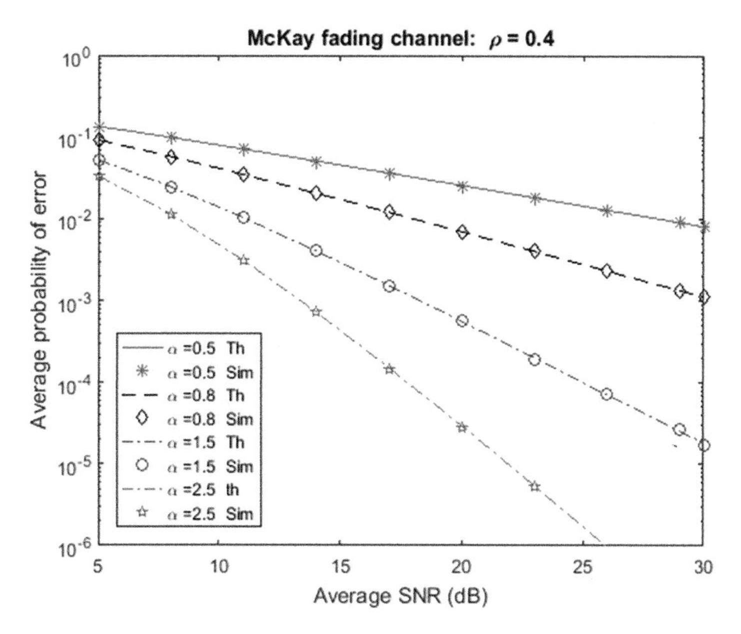

Fig. 4.140 Error rates in a McKay channel ($\rho = 0.4$)

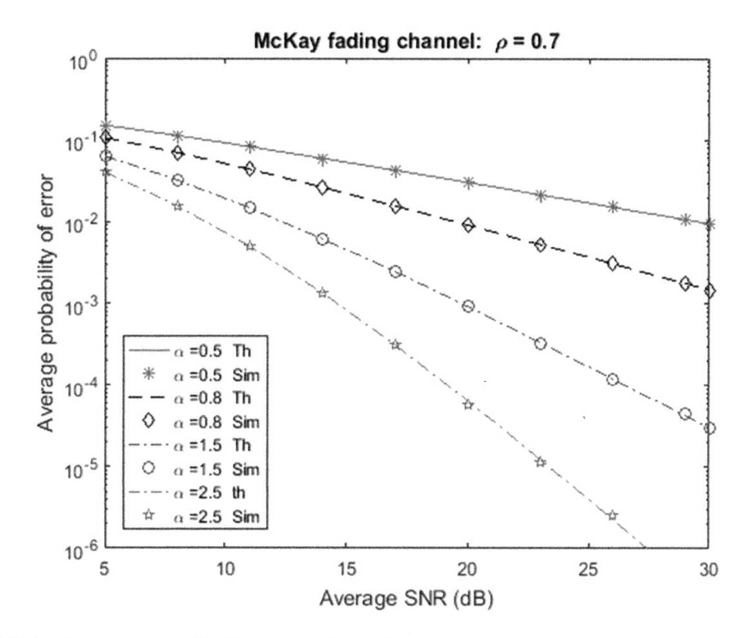

Fig. 4.141 Error rates in a McKay channel ($\rho = 0.7$)

4.13 Creation of Density Mixtures

While some examples of mixture densities are provided in previous sections, one also needs to explore ways of generating density mixtures in a general sense. The first approach uses a uniform random number set. Consider the case of a mixture of two densities

$$f(x) = pf_1(x) + (1-p)f_2(x). \tag{4.356}$$

If a set of uniform random numbers (N) is generated, it can be modified to create two sets of binary data sets, one with pN ones (with the rest remaining zeros) and the other one with $(1\text{-}p)N$ ones (with the rest remaining zeros). If these discrete sets of uniform random numbers are identified as U_1 and U_2, N samples of the mixture density are created as (https://www.mathworks.com/matlabcentral/answers/89318-random-numbers-from-custom-mixture-distribution?)

$$\begin{aligned} X &= U_1\text{randnumb\#1} + U_2\text{randnumb\#2} \\ &= U_1\text{randnumb\#1} + (1-U_1)\text{randnumb\#2}. \end{aligned} \tag{4.357}$$

In Eq. (4.357), *randnumb*#1 and *randnumb*#2 belong to the two densities in Eq. (4.356). This approach becomes cumbersome if the number of components in the mixture exceeds 2.

In a general case with multiple components ($n > 2$), the mixture can be generated directly by apportioning the total number of samples according to their weights. Both approaches are demonstrated in Matlab. The script and results (Fig. 4.142 and 4.143) are given below.

```
function mixture_creation_demo_shankar
% creation of mixtures of densities
% P M Shankar, October 2016 2016
% create a mixture of two densities using a Uniform random number
% generate a mixture of gamma and lognormal densities
close all
p=0.35; % weight; p and 1-p
N=1e6; % number of samples
a=1.5; b=4; % parameters of the gamma density
mu=2; sig=.5;% parameters of the lognorml density
% create the weights using a uniform random number
U = rand(1,N)<=p; % the number of 1's match p*N; remaining are 0's
% (1-U) is the number of 1's matching (1-p)*N
R = U.*gamrnd(a,b,1,N)+(1-U).*lognrnd(mu,sig,1,N);
 [fx,xi]=ksdensity(R);
 NK=find(xi<0); %eliminate negative values of xi generated by Matlab
```

```
if isempty(NK)==0
    NK1=max(NK)+1;% find largest index, add 1 to start non-zero values
    xi=xi(NK1:end);
    fx=fx(NK1:end);
else
end;
fth=p*gampdf(xi,a,b)+(1-p)*lognpdf(xi,mu,sig);
plot(xi,fth,'-r',xi,fx,'k*')
xlabel('value x'),ylabel('Estimated pdf')
legend('Theory','Simulation')
title({'Mixture of gamma & lognormal densities ';...
    ['pf_G(x)+(1-p)f_L(x),    p = ',num2str(p)]})
text(50,.02,{'use a uniform random number set'; 'split the samples'},...
    'backgroundcolor','y')
% create a gamma mixture with more than two components  (five)
clear NK p;
p=[0.1,.3,.2,.35,.05]; % weights
a=[1.5,2.5,3.5,4.5,6];% gamma parameter a
b=[5,4,2.3,1.2,1];% gamma parameter b
np=length(p); % number of mixtures = 5
N=1e6;
% split the samples into bins to match the weights
M=round(N*p(1:np-1));% apportion the total numbers into the first np-1 bins
M=[M,N-sum(M)] ;%remaining numbers for the last bin and create the bin set;
R=[]; % place holder
for k=1:np
    R=[R;gamrnd(a(k),b(k),M(k),1)];
end;
[fX,Xi]=ksdensity(R);
NK=find(xi<0);
if isempty(NK)==0
    NK1=max(NK)+1;% find largest index, add 1 to start non-zero values
    Xi=Xi(NK1:end);
    fX=fX(NK1:end);
else
end;
fthX=zeros(1,length(Xi));
for k=1:np
   fthX=fthX+p(k)*gampdf(Xi,a(k),b(k));
end
figure,plot(Xi,fthX,'-r',Xi,fX,'k*')
xlabel('value x'),ylabel('Estimated pdf')
legend('Theory','Simulation')
xlim([0,0.5*max(Xi)])
title(['Gamma mixture with ',num2str(np),' components'])
xlabel('value x'),ylabel('Estimated pdf')

end
```

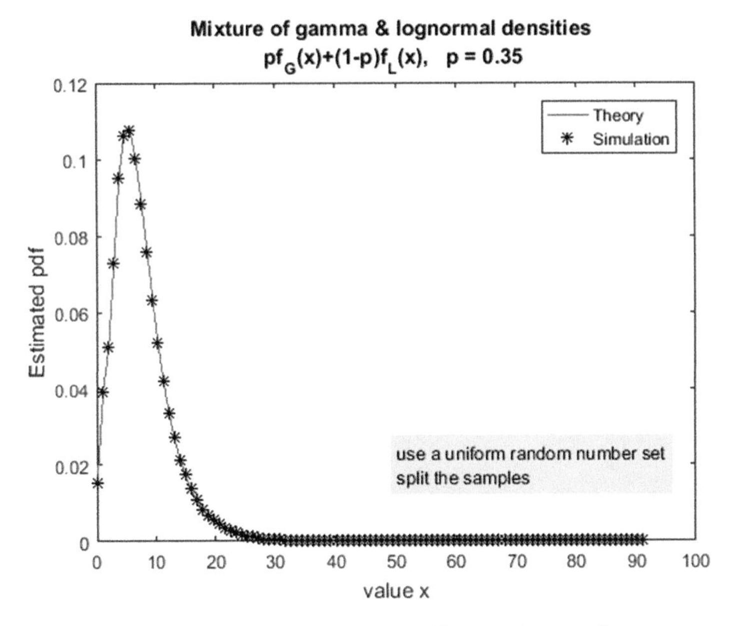

Fig. 4.142 Mixture of two densities created using a uniform random number set

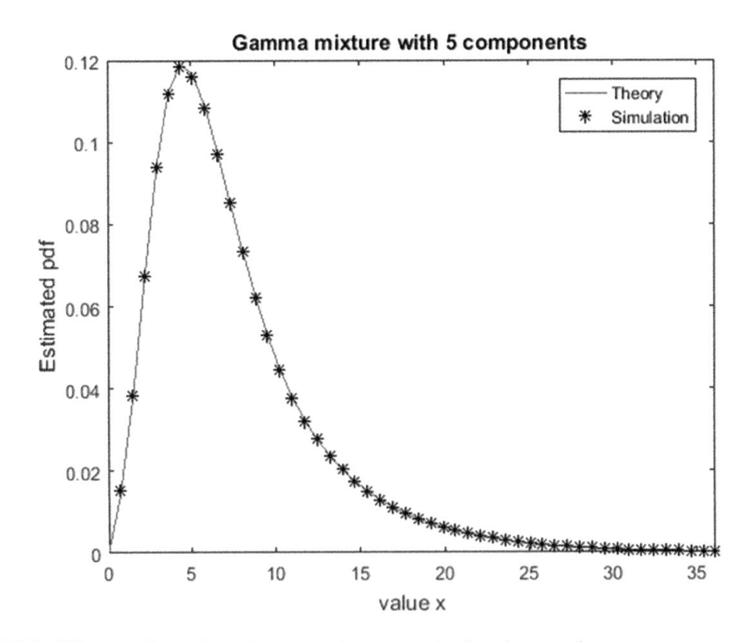

Fig. 4.143 Mixture of two densities created by apportioning the samples

4.14 Summary

We presented various models to describe the statistical fluctuations in wireless channels. The models ranged from the simple Rayleigh ones, to cascaded ones, and to complex models such as those based on $\kappa - \mu$ and $\eta - \mu$ distributions. The models were compared in terms of their density functions, distribution functions, and quantitative measures such as error rates and outage probabilities. The shadowing was examined using the traditional lognormal model and approaches based on similarities between lognormal pdf and other density functions. We looked at the simultaneous existence of short-term fading and shadowing using the Nakagami-lognormal density function and approximations to it using the GK model and the Nakagami-N-gamma model. To complete the study of these models for fading, shadowing, and shadowed fading channels, we examined some second order statistical properties for several models.

The updated sections provide detailed analysis of the McKay fading model and its properties. The model is also expanded to include the effects of shadowing. A general product model to model shadowed fading channel has also been presented. The mixture density introduced in Chap. 2 has now been expanded to see how it can be used to replace some of the existing statistical models of fading and shadowing. The error rates and outage probabilities have been evaluated. In all these cases, detailed Matlab scripts are provided (with complete annotation) to supplement the theoretical approach. A section on random number generation relevant to fading and shadowing modeling is also given to complete the pedagogic description of modeling.

Appendix

We have seen the flexibility offered through the use of the Meijer G-functions in expressing the density functions, distribution functions, error rates, and outage probabilities. They will also be used extensively in Chaps. 5 and 6 as well. Even though they were introduced in Chap. 2, their properties and functionalities which make them very versatile in the study of communication systems were not discussed. In this section, we will provide an overview of the definition, properties, and characteristics of these functions (Mathai and Saxena 1973; Gradshteyn and Ryzhik 2007; Wolfram 2011). We will also examine the relationship among functions commonly encountered in communications such as the incomplete gamma functions, hypergeometric functions, Bessel functions, error functions, complementary error functions, and the Meijer G-functions. Furthermore, closed form expressions for the error rates in cascaded Nakagami channels will be given in terms of Meijer G-functions.

Most of the software packages such as Maple (Maple 2011), Mathematica (Wolfram 2011), and Matlab (Matlab 2011) provide computations involving Meijer

G-functions. Mathematica also provides an excellent resource for understanding the properties of these functions. Additionally, Maple can provide a means to understand the relationships among Meijer G-functions and other functions.

The Meijer G-function $G(\mathrm{x})$ is defined as (Gradshteyn and Ryzhik 2007)

$$\begin{aligned} G_{p,q}^{m,n}\left(x\left|\begin{matrix} a_1,\ldots,a_n,\ldots,a_p \\ b_1,\ldots,b_m,\ldots,b_q \end{matrix}\right.\right) &= G_{p,q}^{m,n}\left(x\left|\begin{matrix} a_p \\ b_q \end{matrix}\right.\right) \\ &= G_{p,q}^{m,n}(x) = G(x) \\ &= \frac{1}{2\pi j}\int \frac{\prod_{i=1}^{m}\Gamma(b_i - s)\left|\prod_{i=1}^{m}\Gamma(1-a_i+s)\right.}{\prod_{i=m+1}^{q}\Gamma(1-b_i+s)\left|\prod_{i=n+1}^{p}\Gamma(a_i-s)\right.} x^s \mathrm{d}s. \end{aligned} \tag{4.358}$$

In most of the computational packages, the $G(\mathrm{x})$ function is expressed as (Maple 2011)

$$\begin{aligned} &G_{p,q}^{m,n}\left(x\left|\begin{matrix} a_1,\ldots,a_n,\ldots,a_p \\ b_1,\ldots,b_m,\ldots,b_q \end{matrix}\right.\right) = \\ &\mathrm{MeijerG}\left(\left[\underbrace{\left[\underbrace{a_1,\ a_2,\ \ldots,\ a_n}_{n\ \text{terms}}\right],\left[\underbrace{a_{n+1},\ a_{n+2},\ \ldots,\ a_p}_{(p-n)\ \text{terms}}\right]}_{p\ \text{terms}}\right]\right. \\ &\left.\left[\underbrace{\left[\underbrace{b_1,\ b_2,\ \ldots,\ b_m}_{m\ \text{terms}}\right],\left[\underbrace{a_{m+1},\ a_{m+2},\ \ldots,\ a_q}_{(q-m)\ \text{terms}}\right]}_{q\ \text{terms}}\right],x\right). \end{aligned} \tag{4.359}$$

A few properties of Meijer G-functions (Mathai and Saxena 1973; Gradshteyn and Ryzhik 2007; Wolfram 2011)

(a) Multiplication $G(\mathrm{w})$ with w^k:

$$w^k G_{pq}^{mn}\left(w\left|\begin{matrix} a_p \\ b_q \end{matrix}\right.\right) = G_{pq}^{mn}\left(w\left|\begin{matrix} a_p + k \\ b_q + k \end{matrix}\right.\right). \tag{4.360}$$

(b) Inversion of the argument

$$G_{pq}^{mn}\left(\frac{1}{w}\middle| \begin{matrix} a_p \\ b_q \end{matrix}\right) = G_{qp}^{mn}\left(w\middle| \begin{matrix} 1-b_q \\ 1-a_p \end{matrix}\right). \tag{4.361}$$

(c) Transformations

$$G_{pq}^{mn}\left(z\middle| \begin{matrix} a_1,\ldots,a_n,\ldots,a_p \\ b_1,b_1,\ldots,b_n,\ldots,b_{q-1},a_1 \end{matrix}\right) = G_{p-1,q-1}^{m,n-1} \times\left(z\middle| \begin{matrix} a_2,\ldots,a_n,\ldots,a_p \\ b_1,b_2,\ldots,b_n,\ldots,b_{q-1} \end{matrix}\right),\quad n,p,q \geq 1, \tag{4.362}$$

$$G_{p+1,q+1}^{m+1,n}\left(z\middle| \begin{matrix} a_1,\ldots,a_n,\ldots,a_p,1-r \\ 0,b_1,b_2,\ldots,b_n,\ldots,b_q \end{matrix}\right) = (-1)^r G_{p+1,q+1}^{m,n+1} \times\left(z\middle| \begin{matrix} 1-r,a_1,\ldots,a_n,\ldots,a_p \\ b_1,b_2,\ldots,b_n,\ldots,b_q,1 \end{matrix}\right),\quad r = 0,1,2,\ldots. \tag{4.363}$$

(d) Differentiation

$$\frac{\partial G_{p,q}^{m,n}\left(x\middle| \begin{matrix} a_p \\ b_q \end{matrix}\right)}{\partial x} = \frac{1}{x} G_{p+1,q+1}^{m,n+1}\left(x\middle| \begin{matrix} 0,a_1,a_2,\ldots,a_n,a_{n+1},\ldots,a_p \\ b_1,b_2,\ldots,b_m,1,b_{m+1},\ldots,b_q \end{matrix}\right)$$
$$= G_{p+1,q+1}^{m,n+1}\left(x\middle| \begin{matrix} -1,(a_1-1),(a_2-1),\ldots,(a_n-1),(a_{n+1}-1),\ldots,(a_p-1) \\ (b_1-1),(b_2-1),\ldots,(b_m-1),0,(b_{m+1}-1),\ldots,(b_q-1) \end{matrix}\right), \tag{4.364}$$

$$\frac{\partial G_{p,q}^{m,n}\left(x\middle| \begin{matrix} a_p \\ b_q \end{matrix}\right)}{\partial x} = (a_1-1) G_{p,q}^{m,n}\left(x\middle| \begin{matrix} a_1,a_2,\ \ldots,a_n,a_{n+1},\ \ldots,a_p \\ b_1,b_2,\ \ldots,b_m,b_{m+1},\ \ldots,b_q \end{matrix}\right) \tag{4.365}$$
$$= G_{p,q}^{m,n}\left(x\middle| \begin{matrix} (a_1-1),a_2,\ \ldots,a_n,a_{n+1},\ \ldots,a_p \\ b_1,b_2,\ \ldots,b_m,b_{m+1},\ \ldots,b_q \end{matrix}\right),$$

$$\frac{\partial\left\{x^{\alpha}G_{p,q}^{m,n}\left(x\middle|\begin{matrix}a_p\\b_q\end{matrix}\right)\right\}}{\partial x}=x^{\alpha-1}G_{p+1,q+1}^{m,n+1}\left(x\middle|\begin{matrix}-\alpha,a_1,a_2,\ldots,a_n,a_{n+1},\ldots,a_p\\b_1,b_2,\ldots,b_m,(1-\alpha),b_{m+1},\ldots,b_q\end{matrix}\right),\tag{4.366}$$

$$\frac{\partial^{v}G_{p,q}^{m,n}\left(x\middle|\begin{matrix}a_p\\b_q\end{matrix}\right)}{\partial x^{v}}=\frac{1}{z^{v}}G_{p+1,q+1}^{m,n+1}\left(x\middle|\begin{matrix}0,a_1,a_2,\ldots,a_n,a_{n+1},\ldots,a_p\\b_1,b_2,\ldots,b_m,v,b_{m+1},\ldots,b_q\end{matrix}\right).\tag{4.367}$$

(e) Integration

$$\int x^{\beta-1}G_{p,q}^{m,n}\left(ax\middle|\begin{matrix}a_p\\b_q\end{matrix}\right)\mathrm{d}x=x^{\beta}G_{p+1,q+1}^{m,n+1}\left(ax\middle|\begin{matrix}(1-\beta),a_p\\b_q,\beta\end{matrix}\right),\tag{4.368}$$

$$\int G_{p,q}^{m,n}\left(x\middle|\begin{matrix}a_p\\b_q\end{matrix}\right)\mathrm{d}x=G_{p+1,q+1}^{m,n+1}\left(x\middle|\begin{matrix}1,(a_p+1)\\(b_q+1),0\end{matrix}\right),\tag{4.369}$$

$$\int G_{p,q}^{m,n}\left(\lambda x\middle|\begin{matrix}a_1,\ldots,a_n,\ldots,a_p\\b_1,\ldots,b_m,\ldots,b_p\end{matrix}\right)G_{u,v}^{r,s}\left(\beta x\middle|\begin{matrix}c_1,\ldots,c_s,\ldots,c_u\\d_1,\ldots,d_r,\ldots,d_v\end{matrix}\right)\mathrm{d}x$$
$$\frac{1}{\lambda}G_{q+u,p+v}^{n+r,m+s}\left(\frac{\beta}{\lambda}\middle|\begin{matrix}-b_1,\ldots,-b_m,c_1,\ldots,c_s,\ldots,c_u,-b_{m+1},\ldots,-b_q\\-a_1,\ldots,-a_m,d_1,\ldots,d_r,\ldots,d_v,-a_{n+1},\ldots,-a_q\end{matrix}\right).\tag{4.370}$$

Relationships of Meijer G-functions to other functions.

Exponential Functions

$$\frac{1}{a}\exp\left(-\frac{x}{a}\right)=\frac{1}{a}G_{0,1}^{1,0}\left(\frac{x}{a}\middle|\begin{matrix}-\\0\end{matrix}\right)=\left(\frac{x}{a}\right)\frac{1}{x}G_{0,1}^{1,0}\left(\frac{x}{a}\middle|\begin{matrix}-\\0\end{matrix}\right)=\frac{1}{x}G_{0,1}^{1,0}\left(\frac{x}{a}\middle|\begin{matrix}-\\1\end{matrix}\right).\tag{4.371}$$

To arrive at the result on the far right-hand side of (4.371), we have made use of the identity in (4.360).

$$\exp\left(-\frac{x}{a}\right)=G_{0,1}^{1,0}\left(\frac{x}{a}\middle|\begin{matrix}-\\0\end{matrix}\right),\tag{4.372}$$

$$\exp\left(-\frac{x}{a}\right) = G_{0,1}^{1,0}\left(-\frac{x}{a}\middle| \begin{matrix} - \\ 0 \end{matrix}\right), \tag{4.373}$$

$$1 - \exp\left(-\frac{x}{a}\right) = 1 - G_{0,1}^{1,0}\left(\frac{x}{a}\middle| \begin{matrix} - \\ 0 \end{matrix}\right) = G_{1,2}^{1,1}\left(\frac{x}{a}\middle| \begin{matrix} - \\ 1,0 \end{matrix}\right). \tag{4.374}$$

Gaussian Function

$$\frac{1}{\sqrt{2\pi\sigma^2}}\mathrm{e}^{-\left(x^2/2\sigma^2\right)} = \frac{1}{\sqrt{2\pi\sigma^2}}G_{0,1}^{1,0}\left(\frac{x^2}{2\sigma^2}\middle| 0\right). \tag{4.375}$$

Gaussian Integrals

$$\int_{-\infty}^{y}\frac{1}{\sqrt{2\pi}}\exp\left(-\frac{x^2}{2}\right)\mathrm{d}x = \frac{1}{2} + \frac{1}{2\sqrt{\pi}}G_{1,2}^{1,1}\left(\frac{1}{2}y^2\middle| \begin{matrix} 1 \\ \frac{1}{2},0 \end{matrix}\right), \quad y > 0, \tag{4.376}$$

$$\int_{-\infty}^{y}\frac{1}{\sqrt{2\pi}}\exp\left(-\frac{x^2}{2}\right)\mathrm{d}x = \frac{1}{2} + \frac{1}{2\sqrt{\pi}}G_{1,2}^{1,1}\left(\frac{1}{2}y^2\middle| \begin{matrix} 1 \\ \frac{1}{2},0 \end{matrix}\right), \quad y < 0, \tag{4.377}$$

Complementary Error Functions (for Bit Error Rate Calculations)

$$\mathrm{erfc}\sqrt{x} = 1 = \frac{1}{\sqrt{\pi}}G_{1,2}^{1,1}\left(x^2\middle| \begin{matrix} 1 \\ \frac{1}{2},0 \end{matrix}\right) = \frac{1}{\sqrt{\pi}}G_{1,2}^{2,0}\left(x\middle| \begin{matrix} 1 \\ \frac{1}{2},0 \end{matrix}\right), \tag{4.378}$$

$$\frac{1}{2}\mathrm{erfc}\sqrt{x} = \frac{1}{2}\left\{1 - \frac{1}{\sqrt{\pi}}G_{1,2}^{1,1}\left(x^2\middle| \begin{matrix} 1 \\ \frac{1}{2},0 \end{matrix}\right)\right\}, \tag{4.379}$$

$$\begin{aligned} \frac{\mathrm{d}}{\mathrm{d}x}\left(\frac{1}{2}\mathrm{erfc}\sqrt{x}\right) &= -\frac{1}{2}\frac{1}{\sqrt{\pi x}}\mathrm{e}^{-x} = \frac{1}{2}\left(\frac{1}{\sqrt{\pi x}}\right)G_{0,1}^{1,0}(x|0) \\ &= -\left(\frac{1}{2\sqrt{\pi}}\right)G_{0,1}^{1,0}\left(x\middle| -\frac{1}{2}\right), \end{aligned} \tag{4.380}$$

$$Q\left(\sqrt{2\pi}\right) = \frac{1}{2}\mathrm{erfc}\left(\sqrt{z}\right) = \frac{1}{2}\left\{1 - \frac{1}{\sqrt{\pi}}G_{1,2}^{1,1}\left(x\left|\begin{matrix}1\\ \frac{1}{2},0\end{matrix}\right.\right)\right\}. \tag{4.381}$$

Natural Logarithm

$$\ln(x) = (x-1)G_{2,2}^{1,2}\left((x-1)\left|\begin{matrix}0,0\\ 0,-1\end{matrix}\right.\right) = G_{2,2}^{1,2}\left((x-1)\left|\begin{matrix}1,1\\ 1,0\end{matrix}\right.\right), \tag{4.382}$$

$$\ln(1+x) = xG_{2,2}^{1,2}\left(x\left|\begin{matrix}0,0\\ 0,-1\end{matrix}\right.\right) = G_{2,2}^{1,2}\left(x\left|\begin{matrix}1,1\\ 1,0\end{matrix}\right.\right). \tag{4.383}$$

Bessel Functions

$$K_{m-c}\left(n\frac{x}{Z}\right) = \frac{1}{2}G_{2,2}^{1,2}\left(\frac{n^2}{4}\left(\frac{x}{Z}\right)^2\left|\begin{matrix}-\\ \frac{1}{2}(m-c),\frac{1}{2}(c-m)\end{matrix}\right.\right), \tag{4.384}$$

$$K_{m-c}\left(n\sqrt{\frac{xmc}{Z}}\right) = \frac{1}{2}G_{0,2}^{2,0}\left(\frac{n^2xmc}{4\ Z}\left|\begin{matrix}-\\ \frac{1}{2}(m-c),\frac{1}{2}(c-m)\end{matrix}\right.\right), \tag{4.385}$$

$$J_n(ax) = G_{0,2}^{1,0}\left(\frac{1}{4}a^2x^2\left|\begin{matrix}-\\ \frac{n}{2},-\frac{n}{2}\end{matrix}\right.\right), \tag{4.386}$$

$$I_n(ax) = \frac{1}{(-1)^{n/2}}G_{0,2}^{1,0}\left(-\frac{1}{4}a^2x^2\left|\begin{matrix}-\\ \frac{n}{2},-\frac{n}{2}\end{matrix}\right.\right),\quad n \neq 0, \tag{4.387}$$

$$I_0(ax) = G_{0,2}^{1,0}\left(-\frac{1}{4}a^2x^2\left|\begin{matrix}-\\ 0,0\end{matrix}\right.\right). \tag{4.388}$$

Short-Term Fading Faded Channels

Probability density function of the amplitude (a) in a Nakagami faded channel is

$$f(a)=2\left(\frac{m}{Z_0}\right)^m\frac{a^{2m-1}}{\Gamma(m)}\exp\left(-\frac{m}{Z_0}a\right)=\frac{2}{a\Gamma(m)}G_{0,1}^{1,0}\left(\frac{m}{Z_0}a^2\middle|\begin{matrix}-\\m\end{matrix}\right). \tag{4.389}$$

Probability density function of the SNR in a Nakagami channel is

$$\begin{aligned}f(x)&=\left(\tfrac{m}{Z_0}\right)^m\frac{x^{m-1}}{\Gamma(m)}\exp\left(-m\frac{x}{Z_0}\right)=\left(\frac{m}{Z_0}\right)^m\frac{x^{m-1}}{\Gamma(m)}G_{0,1}^{1,0}\left(x\frac{x}{Z_0}\middle|\begin{matrix}-\\0\end{matrix}\right)\\&\frac{x^{-1}}{\Gamma(m)}\left(\frac{mx}{Z_0}\right)^m G_{0,1}^{1,0}\left(m\frac{x}{Z_0}\middle|\begin{matrix}-\\0\end{matrix}\right)=\frac{1}{\Gamma(m)x}G_{0,1}^{1,0}\left(m\frac{x}{Z_0}\middle|\begin{matrix}-\\m\end{matrix}\right).\end{aligned} \tag{4.390}$$

To arrive at the result on the far right-hand side of (4.390), we have made use of the identity in (4.360). The average SNR is Z_0.

The cumulative distribution function (CDF) of the SNR can take any one of the several forms expressed below. Note that $g(.,.)$ is the lower incomplete gamma function and $G(.,.)$ is the upper incomplete gamma function (Gradshteyn and Ryzhik 2007). The hypergeometric function is represented by ${}_pF_q(.)$.

$$\int_0^x\left(\frac{m}{Z_0}\right)^m\frac{y^{m-1}}{\Gamma(m)}\exp\left(-m\frac{y}{Z_0}\right)\mathrm{d}y=\begin{cases}\dfrac{\gamma(m,(mx/Z_0))}{\Gamma(m)}=\dfrac{\Gamma(m,(mx/Z_0))}{\Gamma(m)}.\\[2ex]\dfrac{(mx/Z_0)^m}{\Gamma(m)}G_{0,2}^{1,1}\left(\dfrac{mx}{Z_0}\middle|\begin{matrix}1-m\\0,\ -m\end{matrix}\right)\\[2ex].\qquad=\dfrac{1}{\Gamma(m)}G_{1,2}^{1,1}\left(\dfrac{mx}{Z_0}\middle|\begin{matrix}1\\m,0\end{matrix}\right).\\[2ex]1-\dfrac{1}{\Gamma(m)}G_{1,2}^{2,0}\left(\dfrac{m}{Z_0}x\middle|\begin{matrix}1\\0,m\end{matrix}\right)\\[2ex]\dfrac{1}{m\Gamma(m)}\left(\dfrac{mx}{Z_0}\right)^m{}_1F_1\left([m],[1+m],-\dfrac{mx}{Z_0}\right).\end{cases} \tag{4.391}$$

We can also express the density function of the SNR in a generalized gamma fading channel as

$$f(z)=\frac{\lambda z^{\lambda m-1}}{\Gamma(m)\beta^m}\exp\left(-\frac{z^\lambda}{\beta}\right)=\frac{\lambda}{\Gamma(m)z}G_{1,0}^{1,0}\left(\frac{z^\lambda}{\beta}\middle|\begin{matrix}-\\m\end{matrix}\right). \tag{4.392}$$

The density function of the SNR in a Weibull fading channel can be obtained from (4.392) by putting $m=1$. The probability density function of the SNR in a Rician faded channel (K_0 is the Rician factor defined earlier) can be expressed as the product of two Meijer G-functions as

$$f(z)=\frac{K_0+1}{Z_{\mathrm{Ri}}}\exp\left[-K_0-(K_0+1)\frac{z}{Z_{\mathrm{Ri}}}\right]I_0\left(2\sqrt{\frac{K_0(K_0+1)}{Z_{\mathrm{Ri}}}z}\right)$$
$$=\left(\frac{(K_0+1)}{Z_{\mathrm{Ri}}}\right)G_{0,1}^{1,0}\left(\left[K_0(K_0+1)\frac{z}{Z_{\mathrm{Ri}}}\right]\middle|\begin{matrix}-\\0\end{matrix}\right)G_{0,2}^{1,0}\left(\left[-K_0(K_0+1)\frac{z}{Z_{\mathrm{Ri}}}\right]\middle|\begin{matrix}-\\0,0\end{matrix}\right)\tag{4.393}$$

Shadowed Fading Channel: Gamma–Gamma PDF or Generalized **K** *PDF*

$$\begin{aligned}f(z)&=\int_0^\infty\left(\frac{m}{y}\right)^m\frac{z^{m-1}}{\Gamma(m)}\exp\left(-m\frac{z}{y}\right)\frac{y^{c-1}}{b^c\Gamma(c)}\exp\left(-\frac{y}{b}\right)\mathrm{d}y\\&=\int_0^\infty\left(\frac{m}{y}\right)^m\frac{z^{m-1}}{\Gamma(m)}\exp\left(-m\frac{z}{y}\right)\frac{y^{c-1}}{(Z_0/c)^c\Gamma(c)}\exp\left(-\frac{y}{(Z_0/c)}\right)\mathrm{d}y.\end{aligned}\tag{4.394}$$

In terms of Meijer G-functions, (4.394) becomes

$$f(z)=\int_0^\infty\frac{1}{z\Gamma(m)}G_{0,1}^{1,0}\left(\frac{mz}{y}\middle|\begin{matrix}-\\m\end{matrix}\right)\frac{1}{y\Gamma(c)}G_{0,1}^{1,0}\left(\frac{y}{(Z_0/c)}\middle|\begin{matrix}-\\c\end{matrix}\right)\mathrm{d}y.\tag{4.395}$$

The Meijer G-functions in (4.395) can be rewritten using the identities in (4.360) and (4.361) as

$$\int_0^\infty\frac{1}{z\Gamma(m)}G_{0,1}^{1,0}\left(\frac{mz}{y}\middle|\begin{matrix}-\\m\end{matrix}\right)=\frac{1}{z\Gamma(c)}G_{0,1}^{1,0}\left(\frac{y}{mx}\middle|\begin{matrix}1-m\\-\end{matrix}\right),\tag{4.396}$$

$$\frac{1}{y\Gamma(m)}G_{0,1}^{1,0}\left(\frac{cy}{Z_0}\middle|\begin{matrix}-\\c\end{matrix}\right)=\left(\frac{1}{\Gamma(c)}\right)\frac{1}{\Gamma(c)}G_{0,1}^{1,0}\left(\frac{cy}{Z_0}\middle|\begin{matrix}-\\c-1\end{matrix}\right).\tag{4.397}$$

Equation (4.395) now becomes

$$f(z)=\int_0^\infty\frac{1}{z\Gamma(m)}G_{0,1}^{1,0}\left(\frac{y}{mz}\middle|\begin{matrix}1-m\\-\end{matrix}\right)\left(\frac{c}{Z_0}\right)\frac{1}{\Gamma(c)}G_{0,1}^{1,0}\left(\frac{cy}{Z_0}\middle|\begin{matrix}-\\c-1\end{matrix}\right)\mathrm{d}y.\tag{4.398}$$

Using the integral of the product of Meijer G-function given in (4.370), (4.398) becomes

$$f(z) = \left(\frac{1}{z\Gamma(m)\Gamma(m)}\right)(mz)\left(\frac{c}{Z_0}\right)G_{0,2}^{2,0}\left(\frac{mc}{Z_0}\middle|\begin{matrix}-\\ m-1, c-1\end{matrix}\right). \tag{4.399}$$

Using the identity in (4.360), (4.399) becomes

$$f(z) = \frac{1}{z\Gamma(m)\Gamma(m)}G_{0,2}^{2,0}\left(\frac{mc}{Z_0}z\middle|\begin{matrix}-\\ m, c\end{matrix}\right). \tag{4.400}$$

Using the table of integrals (Gradshteyn and Ryzhik 2007), (4.394) becomes

$$f(z) = \frac{2}{\Gamma(m)\Gamma(c)}\left(\frac{mz}{Z_0}\right)^{(m+c)/2} z^{((m+c)/2)-1}K_{m-c}\left(2\sqrt{\frac{mcz}{Z_0}}\right). \tag{4.401}$$

The right-hand side of (4.401) can also be obtained from the conversion of Meijer G-function to Bessel functions in (4.385).

CDF of the SNR in a GK channel

$$\begin{aligned}\int_0^z &\frac{2}{\Gamma(m)\Gamma(c)}\left(\frac{mc}{Z_0}\right)^{(m+c)/2} y^{((m+c)/2)-1}K_{m-c}\left(2\sqrt{\frac{mcy}{Z_0}}\right)\mathrm{d}y\\ &= \frac{\Gamma(m-c)}{\Gamma(m)\Gamma(c+1)}\left(\frac{mcz}{Z_0}\right)^{c}{}_1F_2\left([c],[1+c,1-m+c],\frac{mcz}{Z_0}\right)\\ &+\frac{\Gamma(c-m)}{\Gamma(c)\Gamma(m+1)}\left(\frac{mcz}{Z_0}\right)^{m}{}_1F_2\left([m],[1+m,1-c+m],\frac{mcz}{Z_0}\right)\\ &= \frac{\Gamma(c-m)}{\Gamma(c)\Gamma(m+1)}\left(\frac{mcz}{Z_0}\middle|\begin{matrix}1\\ m,c,0\end{matrix}\right).\end{aligned} \tag{4.402}$$

Cascaded Channels

Cascaded gamma pdf (non-identical but independent) with

$$f(x_k) = \frac{1}{b_k^{m_k}\Gamma(m_k)}x_k^{m_k-1}\exp\left(-\frac{x_k}{b_k}\right),\quad k=1,2,\ldots,N, \tag{4.403}$$

$$Z = \prod_{k=1}^{N}X_k, \tag{4.404}$$

$$Z_0 = \prod_{k=1}^{N}b_k m_k = \prod_{k=1}^{N}m_k\prod_{k=1}^{N}b_k. \tag{4.405}$$

The pdf of the cascaded output (SNR) Z

$$f(x) = \frac{1}{z\prod_{k=1}^{N}\Gamma(m_k)} G_{0,N}^{N,0}\left(\frac{z\prod_{k=1}^{N} m_k}{Z_0}\middle|\begin{matrix} - \\ m_1, m_2, \ldots, m_N \end{matrix}\right). \tag{4.406}$$

CDF of the cascaded output

$$f(z) = \frac{1}{\prod_{k=1}^{N}\Gamma(m_k)} G_{1,N+1}^{N,1}\left(\frac{z\prod_{k=1}^{N} m_k}{Z_0}\middle|\begin{matrix} 1 \\ m_1, m_2, \ldots, m_N, 0 \end{matrix}\right). \tag{4.407}$$

For the case of N identical gamma channels,

$$f(z) = \frac{1}{z\Gamma(m)^N} G_{0,N}^{N,0}\left(\frac{m^N}{Z_0}\middle|\begin{matrix} - \\ m, m, \ldots, m \end{matrix}\right), \tag{4.408}$$

$$f(z) = \frac{1}{z\Gamma(m)^N} G_{1,N+1}^{N,1}\left(\frac{m^N}{Z_0}\middle|\begin{matrix} - \\ m, m, \ldots, m \end{matrix}\right). \tag{4.409}$$

Laplace Transforms

$$\begin{aligned} g(s) &= \int_0^\infty \frac{1}{z\Gamma(m)^N} G_{1,N+1}^{N,1}\left(\frac{m^N}{Z_0}z\middle|\begin{matrix} - \\ m, m, \ldots, m \end{matrix}\right)\exp(-sz)\mathrm{d}z \\ &= \left(\frac{sZ_0}{m^N}\right)\frac{1}{\Gamma(m)^N} G_{1,N}^{N,1}\left(\frac{m^N}{sZ_0}\middle|\begin{matrix} 2 \\ (m+1), (m+1), \ldots, (m+1) \end{matrix}\right). \end{aligned} \tag{4.410}$$

Using the multiplication property from (4.360), the Laplace transform becomes

$$\begin{aligned} g(s) &= \int_0^\infty \frac{1}{z\Gamma(m)^N} G_{0,N}^{N,0}\left(\frac{m^N}{Z_0}z\middle|\begin{matrix} - \\ m, m, \ldots, m \end{matrix}\right)\exp(-sz)\mathrm{d}z \\ &= \frac{1}{\Gamma(m)^N} G_{1,N}^{N,1}\left(\frac{m^N}{sZ_0}\middle|\begin{matrix} 1 \\ m, m, \ldots, m \end{matrix}\right). \end{aligned} \tag{4.411}$$

For the special case of $N = 1$ (gamma pdf or Nakagami channel), the Laplace transform simplifies to

$$g(s) = \int_0^\infty \frac{1}{z\Gamma(m)^N} G_{0,N}^{N,0}\left(\frac{m}{Z_0}z\middle|\begin{matrix} - \\ m \end{matrix}\right)\exp(-sz)\mathrm{d}z = \frac{1}{(1+(Z_0/m)s)^m}. \tag{4.412}$$

Bit Error Rates (Coherent BPSK) for Cascaded Channels

Using the PDF

$$\int_0^\infty \frac{1}{2}\operatorname{erfc}\left(\sqrt{z}\right)\frac{1}{\Gamma(m)^N z}G_{0,N}^{N,0}\left(\frac{m^N z}{Z_0}\middle|\begin{matrix} - \\ m,m,\ldots,m\end{matrix}\right)\mathrm{d}z$$
$$=\frac{1}{2}-\frac{1}{2\sqrt{\pi}\Gamma(m)^N}G_{2,N+1}^{N+1,1}\left(\frac{m^N}{Z_0}\middle|\begin{matrix}\frac{1}{2},1 \\ 0,m,m,\ldots,m\end{matrix}\right). \quad (4.413)$$

Using the CDF

$$\int_0^\infty \frac{1}{2\sqrt{\pi z}}\exp(-1)\frac{1}{\Gamma(m)^N}G_{1,N+1}^{N,1}\left(\frac{m^N z}{Z_0}\middle|\begin{matrix} 1 \\ m,m,\ldots,m\end{matrix}\right)\mathrm{d}z$$
$$=\frac{1}{2\sqrt{\pi}\Gamma(m)^N}G_{2,N+1}^{N,2}\left(\frac{m^N}{Z_0}\middle|\begin{matrix}\frac{1}{2},1 \\ 0,m,m,\ldots,m\end{matrix}\right). \quad (4.414)$$

It can be shown through computation that (4.413) and (4.414) lead to the same error rates. For $N = 1$ (Nakagami channel, SNR), the bit error rate for coherent BPSK, we have (using the CDF)

$$\int_0^\infty \frac{1}{2\sqrt{\pi z}}\exp(-z)\frac{1}{\Gamma(m)}G_{1,2}^{1,1}\left(\frac{mz}{Z_0}\middle|\begin{matrix} 1 \\ m,0\end{matrix}\right)\mathrm{d}z$$
$$=\left(\frac{1}{2\sqrt{\pi}}\right)\frac{\Gamma(m+(1/2))}{m\Gamma(m)}\left(\frac{m}{Z_0}\right)^m {}_2F_1\left(\left[m,\frac{1}{2}+m\right]1[+m],-\frac{m}{Z_0}\right) \quad (4.415)$$
$$=\left(\frac{1}{2\sqrt{\pi}}\right)\frac{1}{\Gamma(m)}G_{2,2}^{1,2}\left(\frac{m}{Z_0}\middle|\begin{matrix}1,\frac{1}{2} \\ m,0\end{matrix}\right)$$

References

Aalo, V., Piboongungon, T., et al. (2005). Bit-error rate of binary digital modulation schemes in generalized gamma fading channels. *IEEE Communications Letters, 9*(2), 139–141.

Abdi, A., & Kaveh, M. (1998). K distribution: An appropriate substitute for Rayleigh-lognormal distribution in fading-shadowing wireless channels. *Electronics Letters, 34*(9), 851–852.

Abdi, A., & Kaveh, M. (2000). Comparison of DPSK and MSK bit error rates for K and Rayleigh-lognormal fading distributions. *IEEE Communications Letters, 4*(4), 122–124.

Abdi, A., & Kaveh, M. (2002). Level crossing rate in terms of the characteristic function: A new approach for calculating the fading rate in diversity systems. *IEEE Transactions on Communications, 50*(9), 1397–1400.
Abdi, A., K. Wills, et al. (2000). Comparison of the level crossing rate and average fade duration of Rayleigh, Rice and Nakagami fading models with mobile channel data. In Vehicular Technology Conference, 2000. IEEE VTS-Fall VTC 2000 52nd (Vol. 4, pp. 1850–1857).
Abramowitz, M., & Segun, I. A. (Eds.). (1972). *Handbook of mathematical functions with formulas, graphs, and mathematical tables*. New York: Dover Publications.
Abu-Dayya, A. A., & Beaulieu, N. C. (1994). Outage probabilities in the presence of correlated lognormal interferers. *IEEE Transactions on Vehicular Technology, 43*(1), 164–173.
Adamchik, V. (1995). The evaluation of integrals of Bessel functions via G-function identities. *Journal of Computational and Applied Mathematics, 64*(3), 283–290.
Almhana, J., Liu, Z., Choulakian, V., & McGorman, R. (2006). A recursive algorithm for gamma mixture models. *Proceedings of IEEE International Conference on Communications, 1*, 197–202.
Alouini, M. S., & Simon, M. K. (1998). Generic form for average error probability of binary signals over fading channels. *Electronics Letters, 34*(10), 949–950.
Alouini, M. S., & Simon, M. K. (2006). Performance of generalized selection combining over Weibull fading channels. *Wireless Communications and Mobile Computing, 6*(8), 1077–1084.
Andersen, J. B. (2002a). Statistical distributions in mobile communications using multiple scattering. In Proceedings of General Assembly International Union of Radio Science, Maastricht, The Netherlands.
Andersen, J. B. (2002b). Statistical distributions in mobile communications using multiple scattering, presented at the 27th URSI General Assembly, Maastricht, Netherlands.
Annamalai, A., & Tellambura, C. (2001). Error rates for Nakagami-m fading multichannel reception of binary and M-ary signals. *IEEE Transactions on Communications, 49*(1), 58–68.
Annamalai, A., Tellambura, C., et al. (2001). Simple and accurate methods for outage analysis in cellular mobile radio systems-a unified approach. *IEEE Transactions on Communications, 49* (2), 303–316.
Annamalai, A., Tellambura, C., et al. (2005). A general method for calculating error probabilities over fading channels. *IEEE Transactions on Communications, 53*(5), 841–852.
Asplund, H., Molisch, A. F., Steinbauer, M. Mehta, N. B. (2002). Clustering of scatterers in mobile radio channels—Evaluation and modeling in the COST259 directional channel model. In Proceedings of IEEE ICC, New York (pp. 901–905).
Atapattu, S., Tellambura, C., & Jiang, H. (2011). A mixture gamma distribution to model the SNR of wireless channels. *IEEE Transactions on Wireless Communications, 10*, 4193–4203.
Aulin, T. (1979). A modified model for the fading signal at the mobile radio channel. *IEEE Transactions on Vehicular Technology, 28*(3), 182–203.
Bithas, P., Sagias, N., et al. (2006). On the performance analysis of digital communications over generalized-K fading channels. *IEEE Communications Letters, 10*(5), 353–355.
Braun, W. R., & Dersch, U. (1991). A physical mobile radio channel model. *IEEE Transactions on Vehicular Technology, 40*(2), 472–482.
Bullington, K. (1977). Radio propagation for vehicular communications. *IEEE Transactions on Vehicular Technology, 26*(4), 295–308.
Büyükçorak, S., Vural, M., & Kurt, G. K. (2015). Lognormal mixture shadowing. *IEEE Transactions on Vehicular Technology, 64*(10), 4386–4398.
Chatelain, F., Borgnat, P., Tourneret, J-Y Abry, P. (2008) Parameter estimation for sums of correlated gamma random variables: Application to anomaly detection in internet traffic. In IEEE International Conference on Acoustics, Speech and Signal Processing (ICASSP-2008) (pp. 3489–3492).
Chen, J. I. Z. (2007). A unified approach to average LCR and AFD performance of SC diversity in generalized fading environments. *Journal of the Franklin Institute-Engineering and Applied Mathematics, 344*(6), 889–911.

Chizhik, D., Foschini, G. J., Gans, M. J., & Valenzuela, R. A. (2002). Keyholes, correlations, and capacities of multielement transmit and receive antennas. *IEEE Transactions on Wireless Communications, 1*(2), 361–368.

Clark, R. H. (1968). A statistical description of mobile radio reception. *BSTJ, 47*, 957–1000.

Clark, J. R., & Karp, S. (1965). Approximations for lognormally fading optical signals. *Proceedings of the IEEE, 1970*(58), 1964–1965.

Cotton, S. L., & Scanlon, W. G. (2007). Higher order statistics for lognormal small-scale fading in mobile radio channels. *IEEE Antennas and Wireless Propagation Letters, 6*, 540–543.

Coulson, A. J., et al. (1998a). Improved fading distribution for mobile radio. *IEE Proceedings on Communications, 145*(3), 197–202.

Coulson, A. J., et al. (1998b). A statistical basis for lognormal shadowing effects in multipath fading channels. *IEEE Transactions on Communications, 46*(4), 494–502.

Da Costa, D. B., & Yacoub, M. D. (2008). Moment generating functions of generalized fading distributions and applications. *IEEE Communications Letters, 12*(2), 112–114.

Davenport, J. W., Bezdek, J. C., & Hathaway, R. J. (1988). Parameter estimation for finite mixture distributions. *Computers & Mathematcs with Applications, 15*(10), 819–828.

Di Renzo, M., Graziosi, F., & Santucci, F. (2010). Channel capacity over generalized fading channels: A novel MGF-based approach for performance analysis and Design of Wireless Communication Systems. *IEEE Transactions on Vehicular Technology, 59*(1), 127–149.

Dong, X., & Beaulieu, N. C. (2001). Average level crossing rate and average fade duration of selection diversity. *IEEE Communications Letters, 5*(10), 396–398.

Ermolova, N. Y. (2008). Moment generating functions of the generalized η-μ and κ-μ distributions and their applications to performance evaluation of communication systems. *IEEE Communications Letters, 12*(7), 502–504.

Ermolova, N. Y. (2009). Useful integrals for performance evaluation of communication systems in generalized −μ and −μ fading channels. *IET Communications, 3*(2), 303–308.

Filho, J. C. S. S., & Yacoub, M. D. (2005a). Highly accurate η-μ approximation to the sum of M independent nonidentical Hoyt variates. *IEEE Antennas and Wireless Propagation Letters, 4*, 436–438.

Filho, J. C. S., & Yacoub, M. D. (2005b). Highly accurate κ & m approximation to sum of M independent non-identical Ricean variates. *Electronics Letters, 41*(6), 338–339.

Fraile, R., Nasreddine, J., et al. (2007). Multiple diffraction shadowing simulation model. *IEEE Communications Letters, 11*(4), 319–321.

Galambos, J., & Simonelli, I. (2004). *Products of random variables*. New York: Marcel Dekker.

Gilbert, E. N. (1965). Energy reception for mobile radio. *BSTJ, 48*, 2473–2492.

Goodman, J. W. (1985). *Statistical optics*. New York: Wiley.

Gradshteyn, I. S., & Ryzhik, I. M. (2000). *Table of integrals, series, and products* (6th ed.). New York: Academic.

Gradshteyn, I. S., & Ryzhik, I. M. (2007). *Table of integrals, series and products*. Oxford: Academic.

Hajri, N., et al. (2009). A study on the statistical properties of double Hoyt fading channels. In Proceedings of the 6th international conference on Symposium on Wireless Communication Systems, Siena, Italy (pp. 201–205).

Hansen, F., & Meno, F. I. (1977). Mobile fading; Rayleigh and lognormal superimposed. *IEEE Transactions on Vehicular Technology, 26*(4), 332–335.

Haykin, S. S. (2001). *Digital communications*. New York: Wiley.

Holm, H. (2002). Adaptive coded modulation performance and channel estimation tools for flat fading channels. Ph. D. Thesis, Department of Telecommunications, Norwegian University of Science and Technology.

Holm, H., & Alouini, M.-S. (2004). Sum and difference of two squared correlated Nakagami variates in connection with the McKay distribution. *IEEE Transactions on Communications, 52*, 1367–1376.

Jakes, W. C. (1994). *Microwave mobile communications*. IEEE Press: Piscataway, NJ.

Jung, J., Lee, S. R., Park, H., Lee, S., & Lee, I. (2014). Capacity and error probability analysis of diversity reception schemes over generalized-fading channels using a mixture gamma distribution. IEEE.

Karadimas, P., & Kotsopoulos, S. A. (2008). A generalized modified Suzuki model with sectored and inhomogeneous diffuse scattering component. *Wireless Personal Communications, 47*(4), 449–469.

Karadimas, P., & Kotsopoulos, S. A. (2010). A modified loo model with partially blocked and three dimensional multipath scattering: Analysis, simulation and validation. *Wireless Personal Communications, 53*(4), 503–528.

Karagiannidis, G. K., Sagias, N. C. and T. A. Tsiftsis, T. A. (2006a) "Closed-form statistics for the sum of squared Nakagami-m variates and its applications." IEEE Transactions on Communications, 54:1353–1359.

Karagiannidis, G. K., Tsiftsis, T. A., & Mallik, R. K. (2006b). Bounds of multihop relayed communications in Nakagami-m fading. *IEEE Transactions on Communications, 54*, 18–22.

Karagiannidis, G. K., Sagias, N. C., et al. (2007). N*Nakagami: A novel stochastic model for cascaded fading channels. *IEEE Transactions on Communications, 55*(8), 1453–1458.

Karmeshu, J., & Agrawal, R. (2007). On efficacy of Rayleigh-inverse Gaussian distribution over K-distribution for wireless fading channels. *Wireless Communications and Mobile Computing, 7*(1), 1–7.

Kim, Y. Y., & Li, S. Q. (1999). Capturing important statistics of a fading shadowing channel for network performance analysis. *IEEE Journal on Selected Areas in Communications, 17*(5), 888–901.

Kostic, I. M. (2005). Analytical approach to performance analysis for channel subject to shadowing and fading. *IEE Proceedings Communications, 152*(6), 821–827.

Laourine, A., Stephenne, A., et al. (2007). Estimating the ergodic capacity of log-normal channels. *IEEE Communications Letters, 11*(7), 568–570.

Laourine, A., Alouini, M. S., et al. (2008). On the capacity of generalized-K fading channels. *IEEE Transactions onWireless Communications, 7*(7), 2441–2445.

Laourine, A., et al. (2009). On the performance analysis of composite multipath/shadowing channels using the G-distribution. *IEEE Transactions on Communications, 57*(4), 1162–1170.

Lee, W. C. Y. (1990). Estimate of channel capacity in Rayleigh fading environment. *IEEE Transactions on Vehicular Technology, 39*, 187–189.

Liu, Z., Almhana, J., Wang, F., & McGorman, R. (2007). Mixture lognormal approximations to lognormal sum distributions. *IEEE Communications Letters, 11*(9), 711–713.

Loo, C. (1985). A statistical model for a land mobile channel. *IEEE Transactions on Vehicular Technology, 34*(3), 122–127.

Malhotra, J., Sharma, A. K., et al. (2009). On the performance analysis of wireless receiver using generalized-gamma fading model. *Annales Des Telecommunications-Annals of Telecommunications, 64*(1–2), 147–153.

Maple (2011). http://www.maplesoft.com/

Mathai, A. M., & Haubold, H. J. (2008). *Special functions for applied scientists*. New York: Springer Science+Business Media.

Mathai, A. M., & Saxena, R. K. (1973). *Generalized hypergeometric functions with applications in statistics and physical sciences*. Berlin: Springer.

Matlab (2011). http://www.mathworks.com/

McKay, A. T. (1932). A Bessel distribution function. *Biometrika, 24*, 39–44.

Mendes, J. R., & Yacoub, M. D. (2007). A general bivariate Ricean model and its statistics. *IEEE Transactions on Vehicular Technology, 56*(2), 404–415.

Nadarajah, S., & Kotz, S. (2006a). Bivariate gamma distributions, sums and ratios. *Bulletin of the Brazilian Mathematical Society, New Series, 37*, 241–274.

Nadarajah, S., & Kotz, S. (2006b). On the bit-error rate for generalized gamma fading channels. *IEEE Communications Letters, 10*(9), 644–645.

Nadarajah, S., & Kotz, S. (2006c). On the product and ratio of gamma and Weibull random variables. *Econometric Theory, 22*(02), 338–344.

Nakagami, M. (1960). The m-distribution—A general formula of intensity distribution of rapid fading. In W. C. Hoffman (Ed.), *Statistical methods in radio wave propagation*. Elmsford, NY: Pergamon.

Ohta, M., & Koizumi, T. (1969). Intensity fluctuation of stationary random noise containing an arbitrary signal wave. *IEEE Proceedings, 1969*(57), 1231–1232.

Okui, S., Morinaga, N., & Namekawa, T. (1981). Probability distributions of ratio of fading signal envelopes and their generalizations. *Electronics and Communications in Japan, 64-B*, 1228–1235.

Pahlavan, K., & Levesque, A. (1995). *Wireless information networks*. New York: Wiley.

Pahlavan, K., & Levesque, A. H. (2005). *Wireless information networks*. Wiley: Hoboken, NJ.

Papazafeiropoulos, A. K., & Kotsopoulos, S. A. (2011a). The $\alpha-\lambda-\mu$ and $\alpha-\eta-\mu$ small-scale general fading distributions: A unified approach. *WIRE, 57*, 735–751.

Papazafeiropoulos, A. K., & Kotsopoulos, S. A. (2011b). The a-l-m and a-κ-m small scale general fading distributions: A unified approach. *Wireless Personal Communications, 57*, 735–751.

Papoulis, A., & Pillai, S. U. (2002a). *Probability, random variables and stochastic processes* (4th ed.). New York: McGraw-Hill.

Papoulis, A., & Pillai, S. U. (2002b). *Probability, random variables, and stochastic processes*. Boston: McGraw-Hill.

Paris, J. F. (2009a). Nakagami-q (Hoyt) distribution function with applications. *Electronics Letters, 45*(4), 210–211.

Paris, J. F. (2009b). Nakagami-q (Hoyt) distribution function with applications (erratum). *Electronics Letters, 45*(8), 210–211.

Patzold, M. (2002). *Mobile fading channels*. Chichester: Wiley.

Peppas, K., Lazarakis, F., Alexandridis, A., & Dankais, K. (2009). Error performance of digital modulation schemes with MRC diversity reception over η-μ fading channels. *IEEE Transactions on Wireless Communications, 8*(10), 4974–4980.

Podolski, H. (1972). The distribution of a product of n independent random variables with generalized gamma distribution. *Demonstratio Mathematica, 4*(2), 119–123.

Prabhu, G., & Shankar, P. M. (2002). Simulation of flat fading using MATLAB for classroom instruction. *IEEE Transactions on Education, 45*(1), 19–25.

Proakis, J. G. (2001). *Digital communications*. Boston: McGraw-Hill.

Radaydeh, R. M., & Matalgah, M. M. (2008). Compact formulas for the average error performance of noncoherent m-ary orthogonal signals over generalized Rician, Nakagami-m, and Nakagami-q fading channels with diversity reception. *IEEE Transactions on Communications, 56*, 32–38.

Rappaport, T. S. (2002). *Wireless communications: Principles and practice*. Prentice Hall PTR: Upper Saddle River, NJ.

Rohatgi, V. K., & Saleh, A. K. M. E. (2001). *An introduction to probability and statistics*. New York: Wiley.

Sagias, N. C., & Tombras, G. S. (2007). On the cascaded Weibull fading channel model. *Journal of the Franklin Institute, 344*(1), 1–11.

Sagias, N. C., et al. (2004). Channel capacity and second-order statistics in Weibull fading. *IEEE Communications Letters, 8*(6), 377–379.

Sagias, N. C., et al. (2005). New results for the Shannon channel capacity in generalized fading channels. *IEEE Communications Letters, 9*(2), 97–99.

Salo, J., El-Sallabi, H. M., et al. (2006a). The distribution of the product of independent Rayleigh random variables. *IEEE Transactions on Antennas and Propagation, 54*(2), 639–643.

Salo, J., El-Sallabi, H. M., et al. (2006b). Statistical analysis of the multiple scattering radio channel. *IEEE Transactions on Antennas and Propagation, 54*(11), 3114–3124.

Selim, B., Alhussein, O., Muhaidat, S., Karagiannidis, G. K., & Liang, J. (2015). Modeling and analysis of wireless channels via the mixture of Gaussian distribution. *IEEE Transactions on Vehicular Technology, 65*(10), 8309–8321.

Shankar, P. (2002a). *Introduction to wireless systems*. New York: Wiley.

Shankar, P. (2002b). Ultrasonic tissue characterization using a generalized Nakagami model. *IEEE Transactions on Ultrasonics, Ferroelectrics, and Frequency Control, 48*(6), 1716–1720.

Shankar, P. M. (2004). Error rates in generalized shadowed fading channels. *Wireless Personal Communications, 28*(3), 233–238.

Shankar, P. M. (2005). Outage probabilities in shadowed fading channels using a compound statistical model. *IEE Proceedings on Communications, 152*(6), 828–832.

Shankar, P. M. (2010a). Statistical models for fading and shadowed fading channels in wireless systems: A pedagogical perspective. *Wireless Personal Communications, 60*, 191–213. doi:10.1007/s11277-010-9938-2.

Shankar, P. M. (2010b). A Nakagami-N-gamma model for shadowed fading channels. *WIRE, 64*, 665–680. doi:10.1007/s11277-010-0211-5.

Shankar, P. M. (2011a). Maximal ratio combining in independent identically distributed N∗Nakagami fading channels. *IET Proceedings on Communications, 5*(3), 320–326.

Shankar, P. M. (2011b). Performance of N*Nakagami cascaded fading channels in dual selection combining diversity. In IWCMC 2011.

Shankar, P. M. (2012). A Nakagami-N-gamma model for shadowed fading channels. *WIRE, 64*, 665–680.

Shankar, P. M. (2013). A statistical model for the ultrasonic backscattered echo from tissue containing microcalcifications. *IEEE Transaction on UFFC, 60*, 932–942.

Shankar, P. M. (2015). A composite shadowed fading model based on the McKay distribution and Meijer G functions. *Wireless Personal Communications, 81*(3), 1017–1030.

Shankar, P. M. and C. Gentile (2010). Statistical analysis of short term fading and shadowing in ultra-wideband systems. In IEEE International Conference on Communications (ICC) (pp. 1–6).

Shankar, P. M., Forsberg, F., & Lown, L. (2003). Statistical modeling of atherosclerotic plaque in carotid B mode images a feasibility study. *Ultrasound in Medicine & Biology, 29*, 1305–1309.

Shepherd, N. H. (1977). Radio wave loss deviation and shadow loss at 900 MHZ. *IEEE Transactions on Vehicular Technology, 26*(4), 309–313.

Shin, H., & Lee, J. H. (2004). Performance analysis of space-time block codes over keyhole Nakagami-m fading channels. *IEEE Transactions on Vehicular Technology, 53*(2), 351–362.

Simon, M. K., & Alouini, M.-S. (2005). *Digital communication over fading channels*. Wiley: Hoboken, NJ.

Sklar, B. (2001). *Digital communications: Fundamentals and applications*. Prentice-Hall PTR: Upper Saddle River, NJ.

de Souza, R. & Yacoub, M. D. (2009). Bivariate Nakagami-q (Hoyt) Distribution. In IEEE International Conference on Communications, 2009. ICC '09 (pp. 1–5).

Springer, M., & Thompson, W. (1966). The distribution of products of independent random variables. *SIAM Journal on Applied Mathematics, 14*(3), 511–526.

Springer, M., & Thompson, W. (1970). The distribution of products of beta, gamma and Gaussian random variables. *SIAM Journal on Applied Mathematics, 18*(4), 721–737.

Steele, R. and L. Hanzó (1999). Mobile radio communications: Second and third generation cellular and WATM systems.; Chichester, Wiley

Stein, S. (1987). Fading channel issues in system engineering. *IEEE Journal on Selected Areas in Communications, 5*(2), 68–89.

Stuber, G. L. (2002). *Principles of mobile communication*. New York: Kluwer Academic.

Suzuki, H. (1977). A statistical model for urban radio propagation. *IEEE Transactions on Communications, 25*, 673–680.

Talha, B. and Patzold, M. (2007). On the statistical properties of double Rice channels. In Proceedings of 10th International Symposium on Wireless Personal Multimedia Communications, WPMC 2007, Jaipur, India (pp. 517–522).

Tellambura, C., & Annamalai, A. (1999). An unified numerical approach for computing the outage probability for mobile radio systems. *IEEE Communications Letters, 3*(4), 97–99.

Tjhung, T. T., & Chai, C. C. (1999). Fade statistics in Nakagami-lognormal channels. *IEEE Transactions on Communications, 47*(12), 1769–1772.

Trigui, I., Laourine, A., et al. (2009). On the performance of cascaded generalized κ fading channels. In Global telecommunications conference, 2009. GLOBECOM 2009 (pp. 1–5). IEEE.

Uysal, M. (2005). Maximum achievable diversity order for cascaded Rayleigh fading channels. *Electronics Letters, 41*(23), 1289–1290.

Vaughn, R. A., & Andersen, J. B. (2003). *Channels, propagation and antennas for mobile communications*. Herts: IEE.

Venturini, S., Dominici, F., & Parmigiani, G. (2008). Gamma shape mixtures for heavy-tailed distributions. *Annals of Applied Statistics, 2*, 756–776.

Wolfram (2011). http://functions.wolfram.com/, Wolfram Research, Inc.

Wongtrairat, W., & Supnithi, P. (2009). Performance of digital modulation in double Nakagami-m fading channels with MRC diversity. *IEICE Transactions on Communications, E92b*(2), 559–566.

Wongtrairat, W. et al. (2008). Performance of M-PSK modulation in double Rician fading channels with MRC diversity. In 5th international conference on electrical engineering/electronics, computer, telecommunications and information technology, 2008. ECTI-CON 2008 (Vol. 1, pp. 321–324).

Yacoub, M. D. (2000). Fading distributions and co-channel interference in wireless systems. *IEEE Magazine Antennas and Propagation, 42*(1), 150–159.

Yacoub, M. D. (2007a). The alpha-mu distribution: A physical fading model for the Stacy distribution. *IEEE Transactions on Vehicular Technology, 56*, 27–34.

Yacoub, M. D. (2007b). The κ-μ distribution and the η-μ distribution. *IEEE Antennas and Propagation Magazine, 49*(1), 68–81.

Yacoub, M. D. (2007c). The alpha-mu distribution: A physical fading model for the Stacy distribution. *IEEE Transactions on Vehicular Technology, 56*(1), 27–34.

Yacoub, M. D. (2007d). The kappa-mu distribution and the eta-mu distribution. *IEEE Magazine Antennas and Propagation, 49*(1), 68–81.

Yacoub, M. D., J. E. Vargas et al. (1998). On the statistics of the Nakagami-m signal. In Telecommunications symposium, 1998. ITS '98 proceedings (Vol. 2, pp. 377–382). SBT/IEEE International.

Yacoub, M. D., et al. (1999). On higher order statistics of the Nakagami-m distribution. *IEEE Transactions on Vehicular Technology, 48*(3), 790–794.

Yacoub, M. D., et al. (2001). Level crossing rate and average fade duration for pure selection and threshold selection diversity-combining systems. *International Journal of Communication Systems, 14*(10), 897–907.

Yilmaz, F. & Alouini, M. S. (2009). Product of the powers of generalized Nakagami-m variates and performance of cascaded fading channels. In Global telecommunications conference, 2009. GLOBECOM 2009 (pp. 1–5). IEEE.

Yilmaz, F., & Alouini, M.-S. (2012). A novel unified expression for the capacity and bit error probability of wireless communication systems over generalized fading channels. *IEEE Transactions on Communications, 60*, 1862–1876.

Yuan, L., Jian, Z., et al. (2008). Simulation of doubly-selective compound K fading channels for mobile-to-mobile communications. In Wireless communications and networking conference, 2008. WCNC 2008 (pp. 1–5). IEEE.

Zlatanov, N., et al. (2008). Level crossing rate and average fade duration of the double Nakakagmi-m random process and its applications to MIMO keyhole fading channels. *IEEE Communications Letters, 12*(11), 1–3.

Zogas, D., & Karagiannidis, G. K. (2005). Infinite-series representations associated with the bivariate Rician distribution and their applications. *IEEE Transactions on Communications, 53*(11), 1790–1794.

Chapter 5
Diversity Techniques

5.1 Introduction

In Chap. 4, we examined the statistical characteristics of signals in wireless channels. Those signals are subject to short-term fading arising from multipath effects and long-term fading or shadowing arising from multiple scattering. While short-term fading results in random fluctuations in power, long-term fading leads to randomness in the average power. Both these effects lead to a worsening performance in wireless channels in terms of higher values of error rates and outage probabilities. Using the basic concept that with the availability of multiple independent versions of the same signals, it is statistically unlikely that all of them would have low signal-to-noise ratio (SNR) in all the versions at any given time (or at any given location); strategies can be developed to exploit this aspect. Diversity techniques constitute the means to create such multiple versions of the signals (Hausman 1954; Brennan 1959; Parsons et al. 1975; Stein 1987; Jakes 1994; Schwartz et al. 1996; Steele and Hanzo 1999; Karagiannidis et al. 2006; Yilmaz and Alouini 2012). Various signal processing methods (algorithms) can then be developed and implemented to combine these multiple signals efficiently to lower the error rates and outage probabilities to enhance the channel performance. Depending on the physical locations where the diversity is implemented, diversity techniques can be broadly classified into two categories, microdiversity and macrodiversity (Bernhardt 1987; Vaughan and Andersen 1987; Jakes 1994).

Short-term fading is mitigated through microdiversity approaches using multiple antennas at the base station (receiver). Mitigation of shadowing requires macrodiversity approaches which rely on multiple base stations (Jakes 1994; Abu-Dayya and Beaulieu 1994a, b; Bdira and Mermelstein 1999; Jeong and Chung 2005; Shankar 2009). The improvement in performance obtained through diversity techniques depends on the modulation and coding techniques used for data transmission, the number of microdiversity channels, the number of base stations in the macrodiversity arrangement, correlation among the microdiversity

© Springer International Publishing AG 2017

P.M. Shankar, *Fading and Shadowing in Wireless Systems*,
DOI 10.1007/978-3-319-53198-4_5

channels, correlation among the base stations, and so on. Performance is also dependent on the algorithms employed to combine the signals generated through diversity (Van Wambeck and Ross 1951; Blanco and Zdunek 1979; Eng et al. 1996; Simon and Alouini 2005).

While one can speak of generic improvement in performance through diversity, performance improvement can be quantitatively expressed in terms of several measures. These include SNR enhancement, reduction in the amount of fading (AF), improvements in the sensitivity of the receiver measured through the reduction in threshold SNR required to maintain a specified bit error rate, and the reduction in outage probabilities. Even though all these measures are related (as we observed in Chap. 4), each of them provides a unique means to establish the enhancement brought on by diversity.

We will first provide a detailed description of the diversity approaches and the various signal processing approaches, i.e., algorithms, used to combine the signals. This will be followed by a review of the different quantitative measures for comparing the performance of the different diversity combining algorithms. Although Chap. 4 provided an analysis and discussion of several models for the statistical modeling of fading and shadowing, we will not be examining the implications of diversity for all those models. While we concentrate on the most commonly identified models for fading and shadowing, we will also look at some of the more complicated models, specifically those involving cascaded channels which describe the fading in terms of products of several random variables (as explored in Chap. 4). Thus, most of the analysis and discussion in this chapter is limited to Nakagami or gamma short-term fading models, Nakagami-lognormal or Nakagami-gamma shadowed fading models, and cascaded gamma or Nakagami models. We will also limit the discussion to bit error rates and outage probabilities. Thus, we will not cover the improvements in channel capacity and second order statistical characteristics, such as level crossings and fade duration.

In this updated edition, analysis of improvements in performance obtained through diversity implementation in some of the newer fading models presented in Chap. 4 is studied. The amount of fading, error rates, and outage probabilities are obtained using analytical means and random number simulation. All the Matlab scripts used are provided with special attention paid to simulation using random numbers.

5.2 Concept of Diversity

As stated earlier, an important consequence of the existence of short-term fading in wireless systems is to make the received signal random. This suggests that we can express the received signal as

$$x(t) = as(t) + n(t). \tag{5.1}$$

In (5.1), $x(t)$ is the received signal expressed as the sum of the white noise term $n(t)$ and $a.s(t)$ where a represents the randomness (short-term fading) introduced by the channel and $s(t)$ is considered to be the transmitted signal of unit power. We are assuming that besides having short-term fading that is flat (i.e., no frequency changes are introduced by the channel), the signal $s(t)$ is received without any distortions and signal power greater or smaller than unity is included in the term a. Because of the random nature of the channel, a is described only by its probability density function (pdf)

$$f_A(a) = \frac{2a}{P_0}\exp\left(-\frac{a^2}{P_0}\right)U(a), \tag{5.2}$$

where we have assumed that the short fading is Rayleigh (Rayleigh channel). In (5.2), P_0 is average power. We will discuss its relation to the average SNR shortly. We will look at other short-term fading models later. Let us remember that we had used the pdf of the envelope in a Rayleigh faded channel in Chap. 4.

The fading leads to increased outage and bit error rates (see Chap. 4). These detrimental effects can be mitigated through diversity techniques as mentioned in the Sect. 5.1. This would require generation of a set of "diverse" or "replicas" of the signals as the term "diversity" implies. Creation of such "multiple" signals is accomplished through spatial, frequency, angle, or, polarization diversity (Jakes 1994; Schwartz et al. 1996; Molisch 2005). While spatial diversity results from multiple transmitters or (receivers) located at a site, frequency diversity is realized by transmitting information over multiple frequency bands (White 1968; Jakes 1994; Diggavi 2001). Angular diversity corresponds to the use of directional antennas using nonoverlapping angular beams; polarization diversity results from transmitting the same information in two orthogonal polarizations (Giuli 1986; Perini and Holloway 1998; Vaughan 1990). Time diversity is attainable if the same information is transmitted repeatedly even though such a diversity approach is impractical in wireless systems. A practical form of time diversity is achieved in code division multiple access (CDMA) based systems when one uses a RAKE receiver (Efthymoglou and Aalo 1995; Win et al. 2000; Rappaport 2002; Molisch 2005). Regardless of how diversity is implemented in the creation of multiple signals, we assume that we have M such signals and that they are independent (cases of correlated signals will be considered later). This would mean that the diversity produces M outputs expressed as

$$x_k(t) = a_k s(t) + n(t), \quad k = 1, 2 \ldots M. \tag{5.3}$$

Note that (5.3) assumes that the additive white Gaussian noise in all the diversity branches is identical. The received signal power P_k is obtained as

$$P_k = \frac{1}{T_s}\int_0^{T_s} a_k^2 [s(t)]^2 \mathrm{d}t = a_k^2, \tag{5.4}$$

where T_s is the symbol duration. Since a' are Rayleigh distributed, as indicated in (5.2), the power is exponentially distributed and the pdf of the power is given by

$$f(p_k) = \frac{1}{P_0} \exp\left(-\frac{p_k}{P_0}\right) U(p_k). \tag{5.5}$$

If the noise power is N_0,

$$N_0 = \overline{n(t)^2}, \tag{5.6}$$

the instantaneous SNR in any one of the branches can be written as

$$Z_k = \frac{P_k}{N_0}. \tag{5.7}$$

Defining the average SNR Z_0 as

$$Z_0 = \frac{\langle P_k \rangle}{N_0} \tag{5.8}$$

we can write the pdf of the SNR as

$$f(z_k) = \frac{1}{Z_0} \exp\left(-\frac{z_k}{Z_0}\right) U(z_k). \tag{5.9}$$

Note that we have assumed that all the branches are identically distributed with equal average powers; (5.9) was also derived in Chap. 4. We will now take a detailed look at different ways of combining the signals from the diversity branches and we will compare the performances using several quantitative measures (Schwartz et al. 1996). As discussed in Chap. 4, comparison of (5.5) and (5.9) shows that the power and SNR have the same form of distribution. We can interchange the density functions if necessary without affecting the fundamental characteristics or nature of the pdf.

Before we examine the various approaches to combining the outputs from the diversity branches, we will briefly view diversity techniques (methods by which multiple versions of the signals are generated). Note that most of this chapter will be devoted to various ways of combining the outputs, statistical analysis of the outcome of the combining efforts, and analysis of the performance of wireless systems after the diversity is implemented. We will be devoting our attention to diversity at the receiver, even though it is possible to implement diversity at the transmitter as well as simultaneous implementation of diversity at the receiver and transmitter.

Based on the propagation mechanism and physical assemblage of the arrangement, we can broadly classify the diversity techniques (receiver) as

1. Space or spatial diversity
2. Frequency diversity
3. Polarization diversity
4. Time diversity (signal repetition)
5. Multipath or Rake diversity (a form of time diversity)

Note that in an ideal circumstance, the number of diversity branches can be unlimited except in the case of polarization diversity where there is a limit of two branches (Vaughan 1990; Schwartz et al. 1996). One can intuitively see that as the number of diversity branches increase, it might become more and more difficult to keep the branches uncorrelated. Use of uncorrelated branches will result in a reduction in processing gain in diversity. Because of practical issues there will be an upper limit on the number of diversity branches.

5.2.1 *Space Diversity*

Consider the case of a single transmitting antenna and two receiving antennas. Figure 5.1. shows a single transmitter and two receiving antennas separated by D, assumed to be much smaller than the separation between the transmitter and receiver(s). We treat the receiving antennas as having identical beamwidths (Schwartz et al. 1996; Perini and Holloway 1998). The concept behind the spatial diversity relies on the fact that the paths coming to the receivers are sufficiently separated and independent of each other. Thus, it is clear that if the spacing between the receivers is reduced, the paths arriving at the two receivers will be correlated and the signal components at the two (or if more than two receivers are present) or more receivers will no longer be uncorrelated. The independence of signals at these

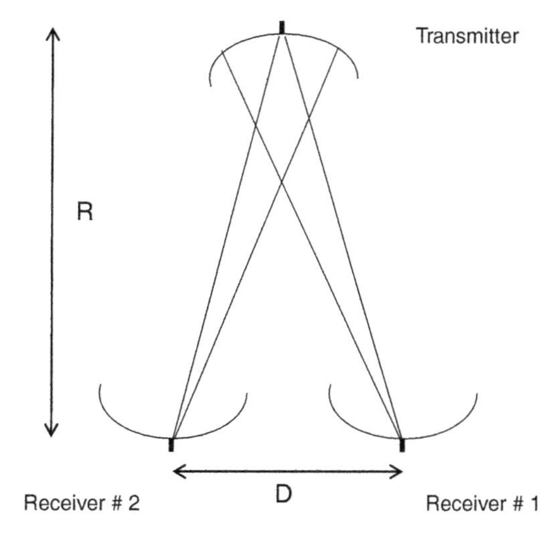

Fig. 5.1 Space diversity arrangement. The distance R between the transmitter and receivers is assumed to be much larger than the separation between receivers #1 and #2

receivers is essential for the improvement in performance expected from diversity. Treating the components of the complex values signal to be Gaussian, it can be argued that "uncorrelatedness" implies independence, thus providing independent signals from these receivers. Based on the beam patterns of the antennas, spacing of $\lambda/2$ (λ is the mean wavelength of the transmitted signal) between antennas is sufficient to produce uncorrelated signals from each of them (Vaughan and Andersen 1987; Schwartz et al. 1996). In practice, minimum spacing might be larger depending on the nature of scattering medium in which the propagation is taking place. Note also that the number of receiving antennas (M) will be limited due to the finite width of the transmitting beam.

5.2.2 *Frequency Diversity*

If signals transmitted over different carrier frequencies are sufficiently separated, we can treat these multipath signals (i.e., fading components) as uncorrelated. The separation between the carriers must be larger than the coherence bandwidth of the channel. This bandwidth is the minimum frequency separation needed so that two signals corresponding to the frequencies will be uncorrelated. Even though higher separation between the carrier frequencies assures uncorrelated signals, increasing separation will be at the cost of reduced bandwidth available for data transmission. Thus, there is a tradeoff between the maximum usable bandwidth and the number of frequency diversity branches available. The concept is illustrated in Fig. 5.2.

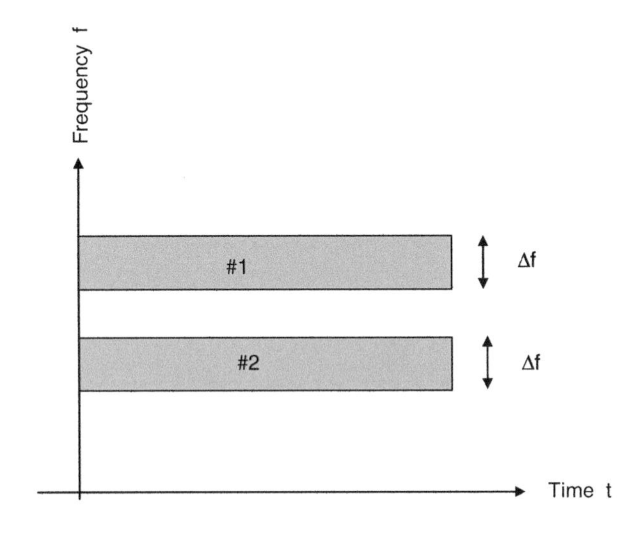

Fig. 5.2 Concept of frequency diversity. The same signal is transmitted over two frequency bands (#1 and # 2)

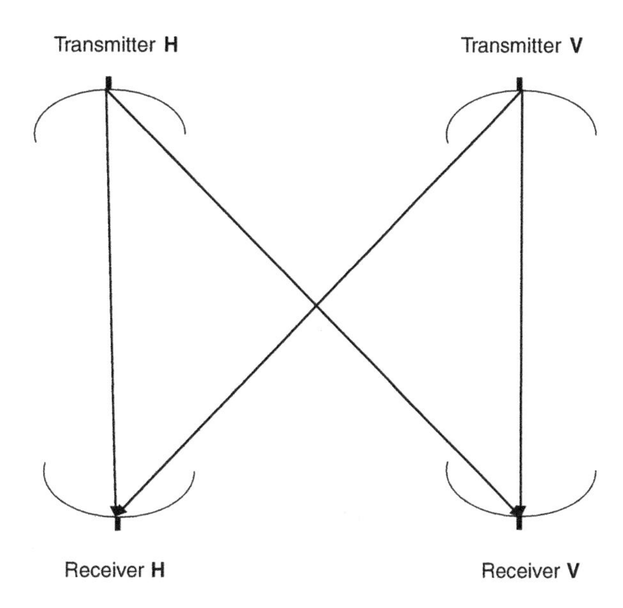

Fig. 5.3 Dual polarization transmitter/receiver pair V and H represents the vertical and horizontal polarizations

5.2.3 Polarization Diversity

Polarization diversity relies on the fact that signals transmitted or received in two orthogonal polarizations exhibit uncorrelated behavior and hence independence (Vaughan 1990). It further relies on the fact that even when the transmitter operates in a single (vertical or horizontal) polarization, the scattering takes place in the propagating channel which can couple some energy from one polarization to the other. Thus, if we have two antennas at the receiver, one for each polarization, we can achieve a diversity of order two ($M = 2$). Since the two polarizations are orthogonal, the spacing requirements associated with the space diversity do not play a role making the positioning and locating the antennas an easier task with polarization diversity. The polarization diversity concept can be extended further by using two transmitting antennas and two receiving antennas providing pairs of orthogonal polarization at both transmitter and receiver as shown in Fig. 5.3. It must also be noted that polarization diversity is a form of space diversity since multiple antennas are used.

5.2.4 Time Diversity

Time diversity is based on the premise that if information is transmitted sequentially through a random fading channel, the multiple replicas of the signal will be uncorrelated if the time or temporal separation among the samples is sufficiently large. Note the relative motion of the transmitter or receiver (i.e., if the mobile unit is moving) plays a crucial role in achieving "decorrelation" since coherence time is dependent on the speed of the transmitter/receiver. If there is no motion, time diversity ideally does not produce uncorrelated replicas. In practice and with time, even in the absence of any motion, one can treat the channel as dynamic from motions of the scatterers from wind, change in temperature, and so on. Time diversity also requires that information be stored at the transmitter and receiver which makes it rather inconvenient to use time diversity. Additionally, repeated transmission of the same information and subsequent processing produces time delay (latency effects). Because of this, time diversity is often used sparingly; space, frequency, and polarization diversity techniques find greater uses in wireless communications.

5.2.5 Multipath Diversity

In Chap. 3 we had seen that fading normally results in multipath signals arriving at the receiver such that these signals (or paths) are not resolvable. However, if the signal bandwidth is increased well beyond the channel bandwidth (for example in the case of CDMA schemes), the width of the transmitted temporal signal elements (known otherwise as "chips") is small enough that the multipath components arriving at the receiver become resolvable. Since the chips constituting the CDMA signals decorrelate with themselves if the delay is one chip period, these multipath resolvable components provide a form of time diversity with uncorrelated signals without suffering from the latency effects associated with conventional time diversity (Wang et al. 1999; Win and Kostic 1999; Win et al. 2000; Molisch 2005). A receiver that combines these resolvable components is referred to as the RAKE receiver. This concept of the multipath diversity is shown in Fig. 5.4 which shows three multipaths, each resulting in a replica of the transmitted signal. The signals are resolvable as seen in Fig. 5.4 and make it possible to apply diversity.

5.3 Diversity Combining Algorithms

We can now look at three different ways of combining the outputs of the M "branches" (Schwartz et al. 1996; Eng et al. 1996). The outline of the combining algorithm is shown in Fig. 5.5. Each output from the diversity branch is multiplied by a gain factor g_k. The three algorithms differ in the manner of operation of the gain factor on the multiple signal set. There are three primary combining algorithms, Selection Combining (SC), maximal ratio combining (MRC), and equal gain combining (EGC). We will now explore some of the details of these algorithms.

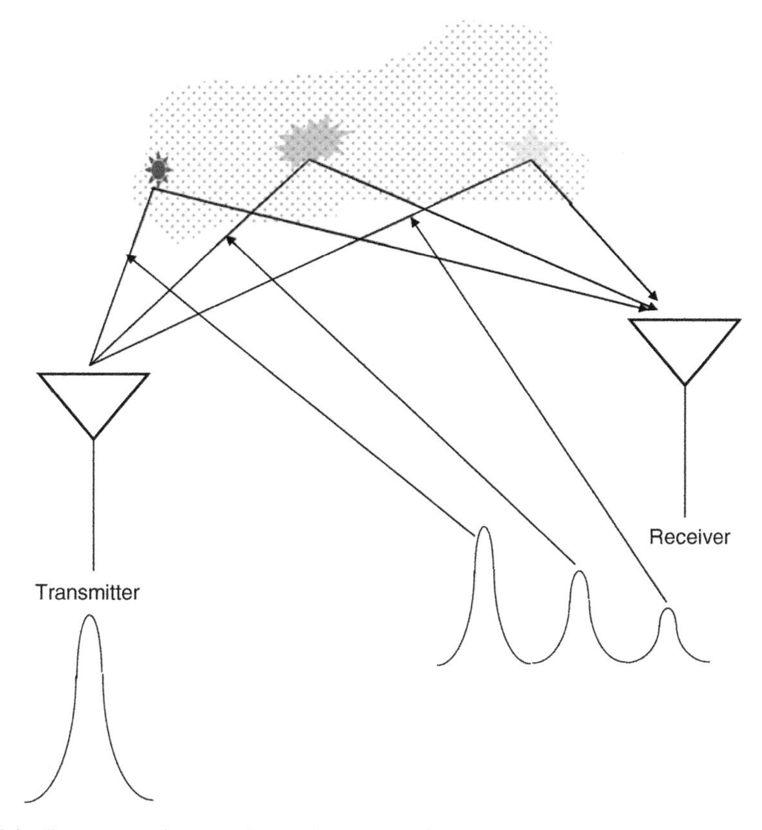

Fig. 5.4 The concept of multipath diversity. Three different paths are shown. The received pulses are resolvable

5.3.1 Selection Combining

The first approach for combining the signals is the selection combining (SC) algorithm. From the view point of practical implementation this is the simplest and easiest of the three algorithms. The algorithm can be described as the selection of the values of g_k such that

$$g_k = \begin{cases} 1 & z = \max\{Z_k\}_{k=1,2,\ldots M} \\ 0 & \text{otherwise.} \end{cases} \tag{5.10}$$

Equation (5.10) translates into picking the output or branch having the highest value of the SNR expressed as

$$Z_{SC} = \max\{Z_k\}_{k=1,2,\ldots M}. \tag{5.11}$$

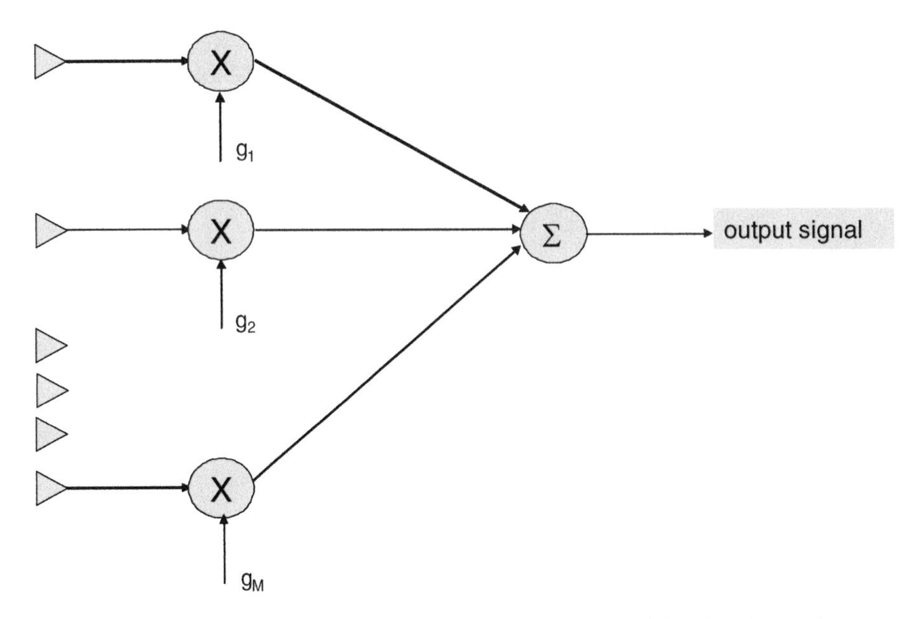

Fig. 5.5 Concept of diversity combining is shown. The outputs of the diversity receivers are combined. The weights for the branches are given by g_k

To understand the performance of the SC algorithm and to compare it to the other algorithms, it is necessary to determine the pdf of the output of the SC algorithm (Annamalai 1997; Simon and Alouini 2005). Let us evaluate the cumulative distribution function (CDF) of the output of the SC in (5.11) first. We can write the CDF as

$$F(z_{\mathrm{SC}}) = \mathrm{prob}\{Z_1 \le z_{\mathrm{SC}}, Z_2 \le z_{\mathrm{SC}}, \ldots, Z_M \le z_{\mathrm{SC}}\}. \tag{5.12}$$

Since we have assumed that all the branches are independent and identically distributed, the CDF becomes

$$F(z_{\mathrm{SC}}) = [\mathrm{prob}\{\mathrm{Z} \le \mathrm{z_{SC}}\}]^M = \left[1 - \exp\left(-\frac{\mathrm{z_{SC}}}{Z_0}\right)\right]^M. \tag{5.13}$$

The pdf of the output of the SC algorithm is obtained by differentiating (5.13) which results in

$$f_{\mathrm{SC}}(z) = \frac{M}{Z_0}\left[1 - \exp\left(-\frac{z}{Z_0}\right)\right]^{M-1} \exp\left(-\frac{z}{Z_0}\right) U(z). \tag{5.14}$$

In (5.14) we have dropped the subscript (SC) from the SNR, Z, and moved it as a subscript of the pdf, f. We will return to (5.14) after we have discussed the other algorithms for combining the signals. Note that details on the derivation of the pdf in (5.14) are given in Chap. 2.

5.3.2 *Maximal Ratio Combining*

Let us go back to the sketch of the processing algorithm in Fig. 5.5. We are now seeking a set of weights *g's* so that the output SNR is maximized (Al-Hussaini and Al-Bassiouni 1985; Aalo 1995; Schwartz et al. 1996; Annamalai et al. 1999). We will ignore the presence of $s(t)$ in (5.1) and express the output of the processing algorithm as

$$x(t) = \sum_{k=1}^{M} g_k a_k + n(t) \sum_{k=1}^{M} gk. \tag{5.15}$$

Note that (5.15) recognizes the fact that when the signals from the branch are scaled, the noise is scaled as well. The signal power P will be

$$P = \left[\sum_{k=1}^{M} g_k a_k \right]^2 \tag{5.16}$$

and the noise power N will be

$$N = N_0 \sum_{k=1}^{M} g_k^2 \tag{5.17}$$

Equations (5.16) and (5.17) are justified by our assumption that noise in every branch is the same, and noise and signal are uncorrelated. The instantaneous SNR becomes

$$Z_{\text{MRC}} = \frac{\left[\sum_{k=0}^{M} g_k a_k \right]^2}{N_0 \sum_{k=1}^{M} g_k^2}. \tag{5.18}$$

The MRC processing expects to maximize the SNR in (5.18). To perform this we invoke the *Chebyshev* inequality (Schwartz et al. 1996; Haykin 2001)

$$\left[\sum_{k=1}^{M} g_k a_k \right]^2 \leq \left(\sum_{k=1}^{M} g_k^2 \right) \left(\sum_{k=1}^{M} a_k^2 \right). \tag{5.19}$$

Equality is obtained when g and a are related through a scalar factor k such that

$$gk = ka_k. \tag{5.20}$$

Setting

$$k = \sqrt{N_0}, \tag{5.21}$$

Equation (5.18) becomes

$$Z_{\mathrm{MRC}} = \frac{\sum_{k=1}^{M} a_k^2}{N_0} = \sum_{k=1}^{M} \frac{a_k^2}{N_0} = \sum_{k=1}^{M} Z_k. \tag{5.22}$$

In other words, the out SNR of the MRC algorithm is the sum of the individual SNRs from the M branches.

We will now look at the third algorithm for combining the signals which is a special case of the MRC algorithm. This is described below.

5.3.3 *Equal Gain Combining*

If the gain factors are all equal, the output SNR of the EG combiner becomes (Beaulieu and Abu-Dayya 1991; Schwartz et al. 1996; Annamalai et al. 2000)

$$Z_{\mathrm{EGC}} = \frac{\left[\sum_{k=1}^{M} a_k\right]^2}{MN_0} = \frac{(1/M)\left[\sum_{k=1}^{M} a_k\right]^2}{N_0}. \tag{5.23}$$

In (5.23), the M in the denominator arises from the addition of noise powers from the M diversity branches in (5.17). Comparing (5.22) and (5.23), the differences and similarities between the output SNRs in MRC and EGC are clear,

$$Z_{\mathrm{MRC}} = \sum_{k=1}^{M} \frac{a_k^2}{N_0} = \sum_{k=1}^{M} Z_k, \tag{5.24}$$

$$Z_{\mathrm{EGC}} = \frac{1}{M}\left[\sum_{k=1}^{M} \frac{a_k}{\sqrt{N_0}}\right]^2 = \frac{1}{M}\left[\sum_{k=1}^{M} \sqrt{Z_k}\right]^2. \tag{5.25}$$

Note that a_k's represents the envelope values; consequently, they are always positive. In writing (5.24) and (5.25), we have assumed that the noise power in each branch is unity. If we now consider a simple case where a_k's are all equal and deterministic, we have

$$Z_{\mathrm{MRC}} = Z_{\mathrm{EGC}} \tag{5.26}$$

and if a_k's are not all equal,

$$Z_{\text{MRC}} \geq Z_{\text{EGC}} \tag{5.27}$$

establishing that MRC is better than EGC (at least in terms of SNR). We will now show that the same is true when fading exists by treating the short-term fading to be Nakagami-m distributed.

5.3.4 *Preliminary Comparison of the three Combining Algorithms*

We have seen the basic forms of the outputs of the three main combining algorithms. It is now time to look at these algorithms in greater detail so that we can start examining the differences among them. We will use the general form of the statistical model for short-term fading, namely the Nakagami-m distribution.

When short-term fading is described in terms of the Nakagami m-distribution, the pdf of the received signal power (or SNR) is given by

$$f(z) = \left(\frac{m}{Z_0}\right)^m \frac{z^{m-1}}{\Gamma(m)} \exp\left(-\frac{m}{Z_0}z\right) U(z), \quad m \geq \frac{1}{2}. \tag{5.28}$$

Note that if we remove the restriction on m to be always greater than 1/2, (5.28) can be identified as the gamma pdf as discussed earlier in Chap. 2. Note that Z_0 is average SNR. Thus, for all practical purposes, we can argue that the pdf of the signal power will have a gamma distribution when the short-term fading has the Nakagami-m pdf for the amplitude or envelope. Before we examine the case of MRC and EGC, let us go back and obtain the pdf of the output of the selection combining (SC). To derive the expression for the pdf of the output, we will need the expression for the CDF associated with the pdf in (5.28). The CDF can be written as

$$F(z_{\text{SC}}) = \gamma\left(m, \frac{m z_{\text{SC}}}{Z_0}\right) [\Gamma(m)]^{-1}, \tag{5.29}$$

where $\gamma(,)$ is the incomplete gamma function.

The CDF of the output of the SC algorithm given in (5.12) now becomes

$$F(z_{\text{SC}}) = \left[\gamma\left(m, \frac{m z_{\text{SC}}}{Z_0}\right) [\Gamma(m)]^{-1}\right]^M. \tag{5.30}$$

The pdf of the output of the SC algorithm is obtained by differentiating (5.30), leading to

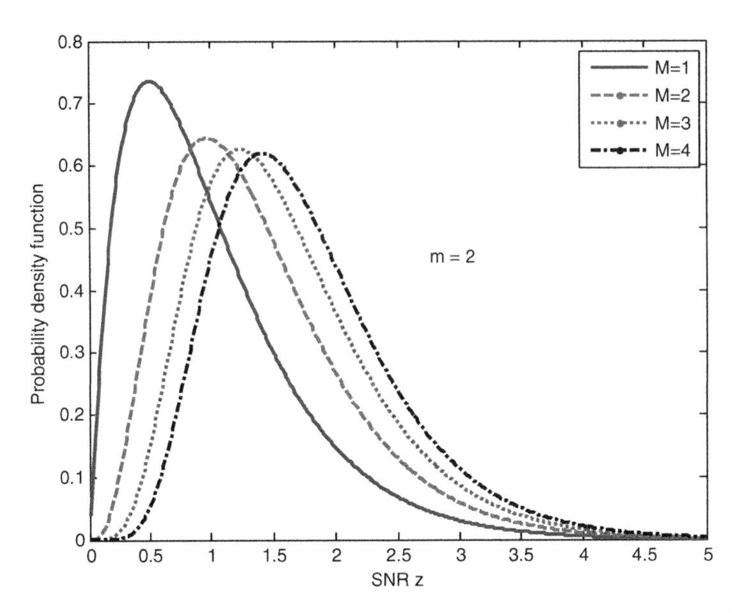

Fig. 5.6 The density functions of the SNR in Nakagami-m channels at the output of the SC algorithm (average SNR/branch Z0 = 1)

$$f_{\mathrm{SC}}(z) = M\left(\frac{m}{Z_0}\right)^m \frac{z^{m-1}}{\Gamma(m)} \exp\left(-\frac{m}{Z_0}z\right)\left[\gamma\left(m, \frac{mz}{Z_0}\right)[\Gamma(m)]^{-1}\right]^{M-1} U(z), \ m > \frac{1}{2}. \tag{5.31}$$

Note that the subscript (SC) attached to Z has been dropped. Equation (5.31) becomes (5.14) when $m = 1$ (Rayleigh pdf of the envelope). Figure 5.6 shows the density of the Selection Combiner.

Let us now return to the output of the MRC algorithm when we have a Nakagami faded channel. The pdf of the output SNR or power given in (5.24) is derived in the original work by Nakagami (1960). The pdf of the output of the MRC algorithm becomes

$$f_{\mathrm{MRC}}(z) = \left(\frac{m}{Z_0}\right)^{mM} \frac{z^{mM-1}}{\Gamma(mM)} \exp\left(-\frac{m}{Z_0}z\right) U(z), \quad m > \frac{1}{2}. \tag{5.32}$$

Simply put, the pdf of the MRC output is another gamma distributed random variable with the order of mM. The average SNR of the processed output is MZ_0. These results also follow from the properties of the sum of gamma random variables described in Chap. 2. The densities of the MRC outputs are shown in Fig. 5.7.

Comparison of the plots in Figs. 5.6 and 5.7 shows that the peaks of the densities with the MRC algorithm occur at higher SNR values than those for SC suggesting that MRC algorithm is likely to result in better performance in fading. Association between peak densities and peak performance was discussed in Chap. 4.

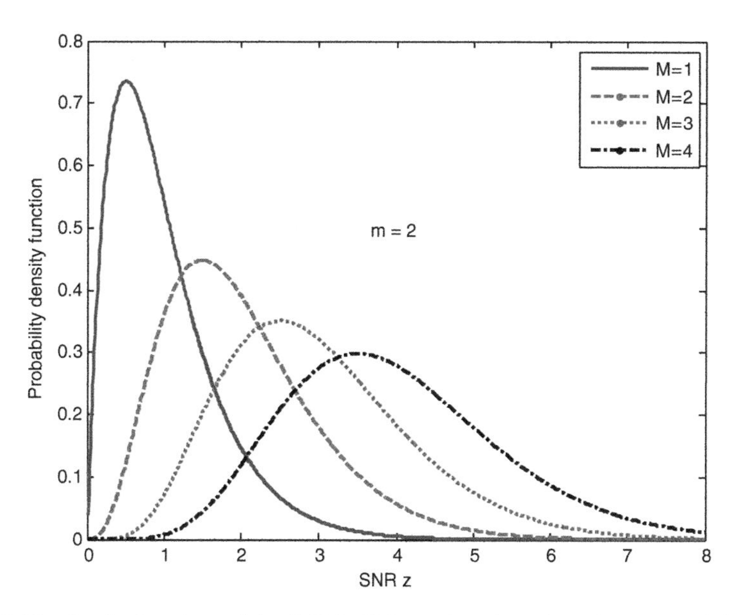

Fig. 5.7 The density functions of the SNR in Nakagami-m channels at the output of the MRC algorithm (average SNR/branch Z0 = 1)

We will now look at the output of the EGC algorithm. Several researchers have proposed ways of obtaining an analytical expression for the pdf of the output. All of them are approximations. It is also possible to get an approximate expression using the results from the original work by Nakagami (1960). If we express W as the sum of M independent identically distributed Nakagami variables,

$$W = \sum_{k=1}^{M} X_k, \tag{5.33}$$

w will be Nakagami distributed with a parameter εMm and average power M^2 $Z_0\delta$where

$$0.95 \leq \varepsilon \leq 1.1 \tag{5.34}$$

and

$$\delta = \left(1 - \frac{1}{5m}\right). \tag{5.35}$$

Equation (5.34) is valid for most practical values of m and $M < 8$ suggesting that we can assume that $\varepsilon \approx 1$. The pdf of the output of the EGC algorithm in (5.25) will be the result of the squaring of W followed by scaling by $(1/M)$. Thus, the pdf will be similar to the pdf of the MRC algorithm with a slightly lower value of the average SNR. The pdf of the SNR of EGC output can be expressed as

$$f_{\mathrm{EGC}}(z)=\left(\frac{m\varepsilon}{\delta Z_0}\right)^{m\varepsilon M}\frac{z^{m\varepsilon M-1}}{\Gamma(m\varepsilon M)}\exp\left(-\frac{m\varepsilon}{\delta Z_0}z\right)U(z),\quad m>\frac{1}{2}. \tag{5.36}$$

Equation (5.38) is obtained by noting that the EGC output is the scaled ($1/M$) power of W in (5.33).

We can now attempt to calculate the improvement in SNR following the diversity processing.

$$\langle Z_{\mathrm{MRC}}\rangle = MZ_0, \tag{5.37}$$

$$\langle Z_{\mathrm{ERC}}\rangle \approx M\left(1-\frac{1}{5m}\right)Z_0. \tag{5.38}$$

There is no simple analytical expression available for the case of SC except for the case of $m=1$ (Rayleigh fading).

$$\langle Z_{\mathrm{SC}}\rangle = Z_0\sum_{k=1}^{M}\frac{1}{k}. \tag{5.39}$$

For $M=2$, an analytical expression can be obtained for the average SNR of the SC output for all values of m as

$$\langle Z_{\mathrm{SC}}\rangle = Z_0\left[1+\frac{\Gamma(2m)}{2^{2m-1}\Gamma(m)\Gamma(m+1)}\right]. \tag{5.40}$$

Equation (5.39) points out an interesting aspect of the SC algorithm. The largest improvement from diversity is obtained from a two-branch receiver and any subsequent gain from diversity becomes less and less significant as M goes above two. In other words, selection combining leads to a waste of resources if the number of diversity branches goes up since the gains realized go down as M increases.

We can also compare the three major algorithms for diversity in terms of the density functions and CDFs following the diversity through random number simulations. This allows us an overview of three algorithms and how they are likely to impact the overall performance of data transmission in wireless channels. Gamma random numbers (Nakagami channels) can be generated using Matlab. The three different combining algorithms can then be implemented, and the pdfs and CDFs estimated using Matlab. Figure 5.8 shows the plots of the pdfss for $M=4$, 5, 6, and 7. We can see that peak of the density functions move toward increasing values of the SNR when diversity is implemented. Note that the shift of the peaks alone might not signify any improvement since peak shift even when average SNR changes. This shifting is the lowest for the selection combiner and highest for the MRC combiner, with EGC coming closer to the MRC case. One can also see that as M increases the peak of the density functions moves farther and farther to the right,

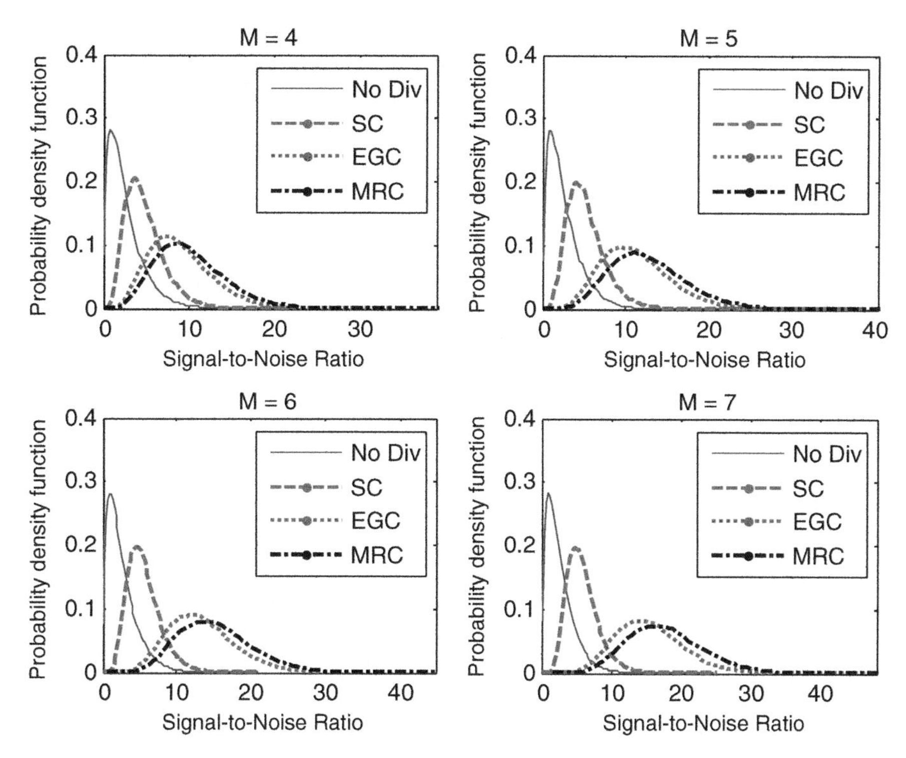

Fig. 5.8 Comparison of the density functions: no diversity, SC, EGC, MRC for $M = 4, 5, 6, 7$. The value of m used was 1.5 and the average SNR/channel was 4 dB

Table 5.1 SNR improvement following diversity. Results from random number simulation

M	SC	EGC	MRC
4	1.89	3.55	4
5	2.04	4.35	5
6	2.17	5.24	6
7	2.28	6.1	7

with the slowest movement seen with the case of SC. As discussed in Chap. 2, the shifting of the peaks to the right shows an improvement in performance, a simple measure of which is the increase in average SNR with diversity. For the simulation used in Fig. 5.7, the SNR improvement over the case of no diversity is tabulated in Table 5.1.

As we can observe, the gain in SNR for selection combining goes up at a very slow rate while the gains for EGC and MRC go up fast. The gains of EGC and MRC are also very close. The CDFs obtained from the simulation are shown in Fig. 5.9.

The CDFs for EGC and MRC are close to each other while the CDF for the selection combiner is close to the case of no diversity. The plots of the CDFs allow another means of visualizing the effect of diversity at the output. A slow rising CDF

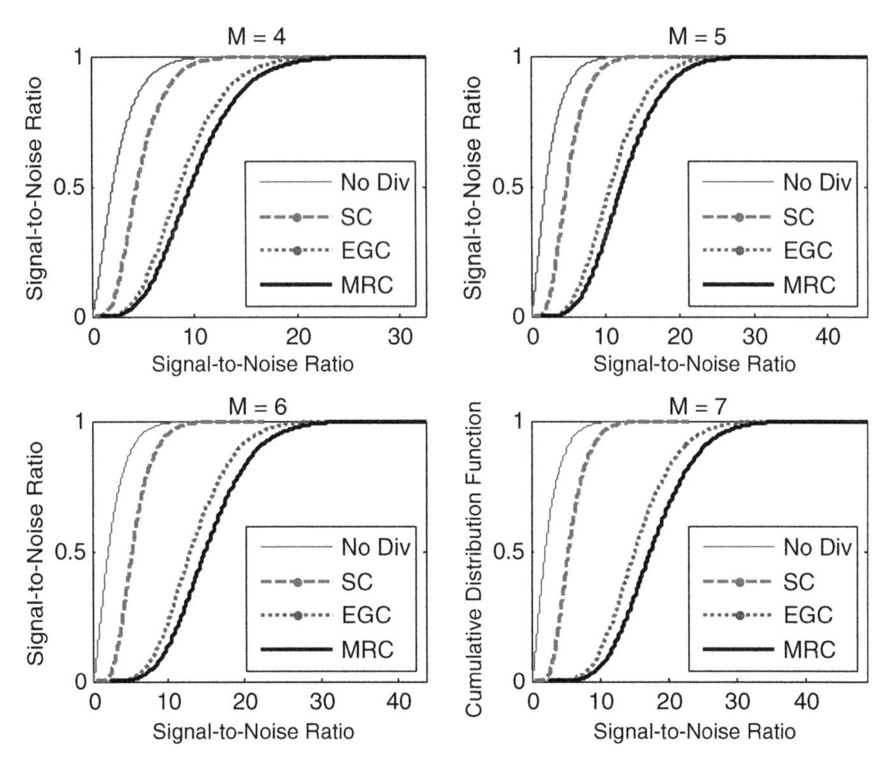

Fig. 5.9 Comparison of the cumulative distribution functions (CDFs): no diversity, SC, EGC, MRC for $M = 4, 5, 6, 7$. The value of m used was 1.5 and the average SNR/channel was 4 dB

signifies that the outage probability defined in terms of the probability that SNR fails to reach a set threshold will be small, while a fast rising CDF signifies the existence of higher values of outage probabilities when compared with these of slowly rising CDFs.

We will return later to a detailed comparison of the SNR improvement following diversity combining. First we will examine a special case of SC, namely, switched and stay combining (SSC).

5.3.5 Selection Combining and Switched and Stay Combining

Ideally in selection combining, one expects to pick the branch with the highest SNR. For the case of two-branch diversity, this would mean a continuous monitoring of the two branches and decisions on which one to choose. This is practically impossible since it puts a strain on the logic and switching circuits. So, a practical way is to set a predetermined threshold and choose the branch that meets that criteria and stay with that branch regardless of whether the other branch exceeds the threshold or not. Switching to the other branch will occur only when the current branch fails to meet the criteria. But, this also requires the choice of an appropriate threshold since choosing a high threshold would necessitate frequent switching. Choice of a lower threshold would reduce the need to switch. This will come at the cost of reduced gain from diversity since the other branch could have had a higher SNR. Such an approach of a slightly "intelligent" choice of the output results in the SSC. To understand this choice, let us examine the pdf of the output of this SSC algorithm (Abu-Dayya and Beaulieu 1994b; Young-Chai et al. 2000; Tellambura et al. 2001; Simon and Alouini 2005).

We assume that the two diversity branches are operating in identical Nakagami channels and they are independent. Let Z_1 and Z_2 be the respective values of the SNR in the two channels. First we determine the CDF of the output of the SSC algorithm. Let Z be the output and Z_T is the threshold. The expression for the CDF has been derived (Abu-Dayya and Beaulieu 1994a, b) and it can be written as

$$F_{\mathrm{SSC}}(z) = \mathrm{Prob}\{Z_T \le Z_1 \le z\} + \mathrm{Prob}\{Z_1 \le z\}\mathrm{Prob}\{Z_2 \le Z_T\}. \tag{5.41}$$

Simplifying, we have

$$F_{\mathrm{SSC}}(z) = \begin{cases} \mathrm{Prob}\{Z_1 \le z\}\,\mathrm{Prob}\{Z_2 \le Z_T\}, & z \le Z_T \\ \mathrm{Prob}\{Z_1 \le Z_T \le z\} + \mathrm{Prob}\{Z_1 \le z\}\mathrm{Prob}\{Z_2 \le Z_T\}, & z > Z_T. \end{cases} \tag{5.42}$$

Making use of the CDF of the SNR in (5.29) and (5.42) becomes

$$F_{\mathrm{SSC}}(z) = \begin{cases} \gamma\left(m,\frac{mZ_T}{Z_0}\right)\gamma\left(m,\frac{mz}{Z_0}\right)[\Gamma(m)]^{-2}, & z \le Z_T \\ \left(\gamma\left(m,\frac{mz}{Z_0}\right)-\gamma\left(m,\frac{mZ_T}{Z_0}\right)\right)[\Gamma(m)]^{-1}+\gamma\left(m,\frac{mZ_T}{Z_0}\right)\gamma\left(m,\frac{mz}{Z_0}\right)[\Gamma(m)]^{-2}, & z > Z_T. \end{cases} \tag{5.43}$$

The pdf of the output of the SSC algorithm is obtained by differentiating (5.43), resulting in

$$f_{\mathrm{SSC}}(z) = \begin{cases} \frac{\gamma(m,(mZ_T/Z_0))}{\Gamma(m)}\left(\frac{m}{Z_0}\right)^m \frac{z^{m-1}}{\Gamma(m)}\exp\left(-\frac{m}{Z_0}z\right), & z \le Z_T \\ \left(\frac{\gamma(m,(mZ_T/Z_0))}{\Gamma(m)}+1\right)\left(\frac{m}{Z_0}\right)^m \frac{z^{m-1}}{\Gamma(m)}\exp\left(-\frac{m}{Z_0}z\right), & z > Z_T. \end{cases} \tag{5.44}$$

The CDF of SSC is compared with the CDF of SC and a single branch receiver (no diversity) in Fig. 5.10 for an average SNR/branch of unity ($Z_0 = 0$ dB).

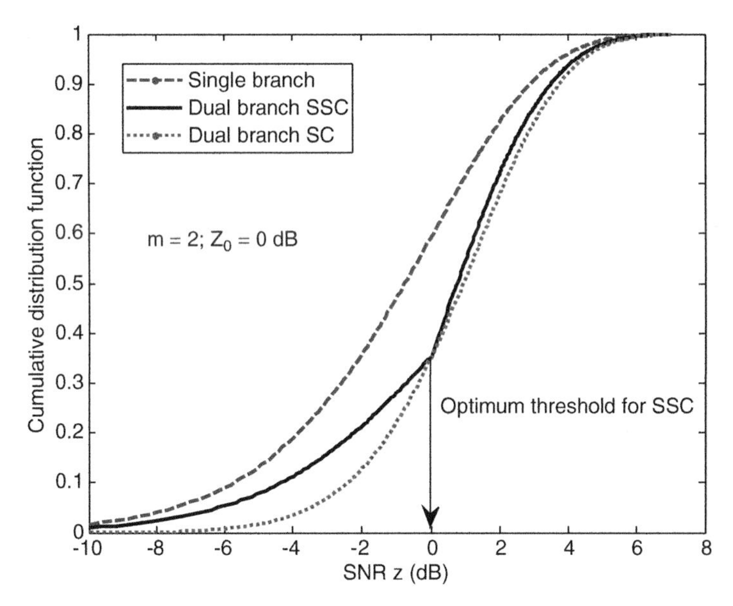

Fig. 5.10 CDF of SC vs. SSC dual branch (average SNR/branch of unity). Optimum threshold is equal to the average SNR

The density functions of the SC and SSC are compared in Fig. 5.11.

The average SNR of the output of the SSC algorithm can be obtained as

$$\langle Z_{\mathrm{SSC}} \rangle = Z_0 \left[1 + \left(\frac{m}{Z_0} Z_{\mathrm{T}} \right)^m \frac{\exp(-(m/Z_0) Z_{\mathrm{T}})}{\Gamma(m+1)} \right]. \tag{5.45}$$

Optimum value of the threshold is obtained by differentiating (5.45) and

$$Z_{\mathrm{T}} \Big|_{\mathrm{opt}} = Z_0. \tag{5.46}$$

Using (5.46), the optimum value of the average SNR becomes

$$\langle Z_{\mathrm{SSC}} \rangle = Z_0 \left[1 + \frac{m^{m-1} \exp(-m)}{\Gamma(m)} \right]. \tag{5.47}$$

Now that we have obtained expressions for the SNR following diversity, we can compare the performances of these combining algorithms. We will first look at the three basic algorithms, namely, SC, EGC, and MRC. In Fig. 5.9, we have already seen the results based on simulation.

The results on average SNR enhancement obtained from the analytical expressions for the pdf are shown in Fig. 5.12. The SNR enhancement is the ratio of the SNR following the diversity to the SNR before implementing the diversity. Note that the SNR of the SC algorithm as mentioned earlier does not lead to an analytical

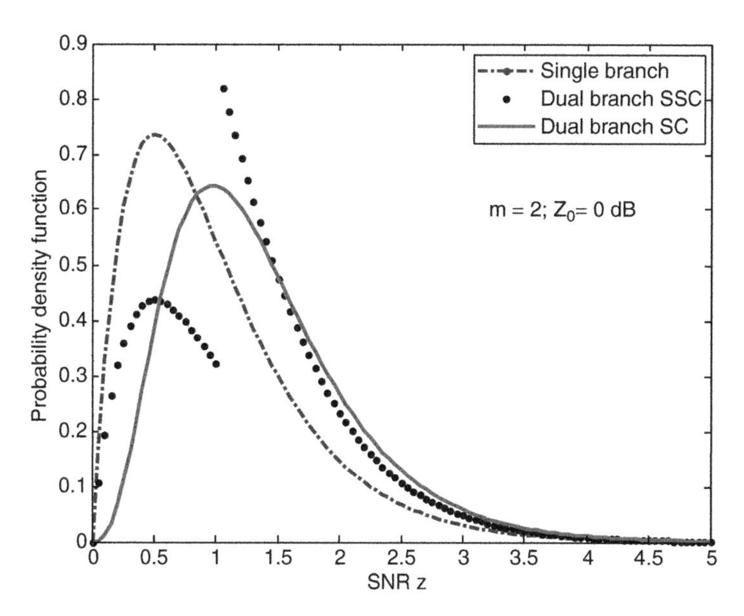

Fig. 5.11 The densities of SC vs. SSC dual branch (average SNR/branch of unity). Optimum threshold is equal to the average SNR

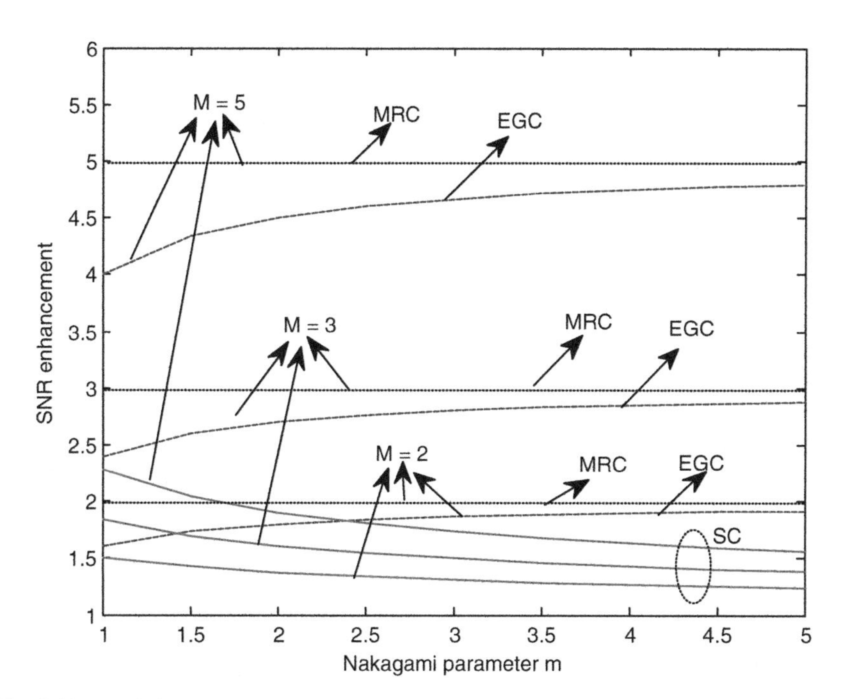

Fig. 5.12 The SNR improvement following diversity is plotted for the three combining algorithms, MRC, EGC, and SC for $M = 2$, 3, and 5. As the value of the Nakagami parameter m increases, the SNR improvement for EGC approaches that of MRC. On the other hand, the SNR improvement from SC declines as the Nakagami parameter increases. The rise in SNR improvement with M is at a very slow rate for the SC algorithm. For this reason, the SC algorithm is generally implemented with two or three branches ($M = 2$ or 3)

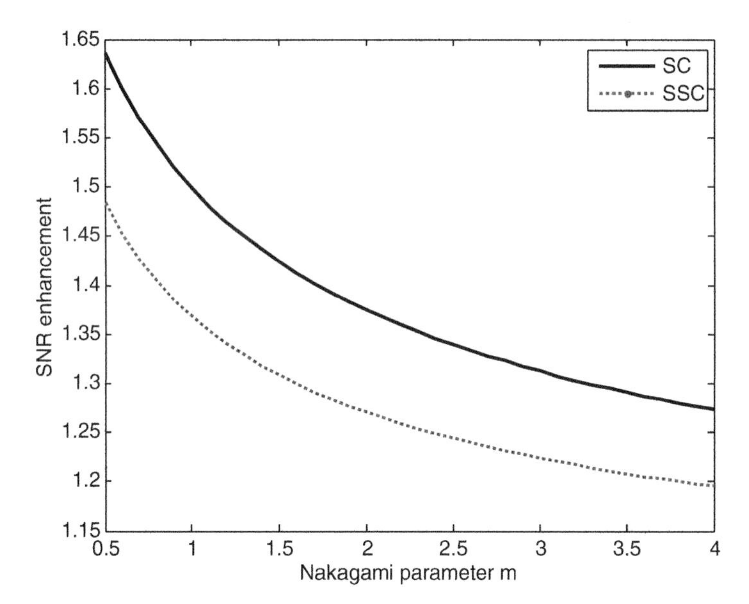

Fig. 5.13 SNR enhancement of SC vs. SSC (optimum threshold)

expression; the SNR was evaluated by performing the numerical integration. While the MRC algorithm results in SNR improvement of M regardless of the value of the Nakagami parameter m, both the EGC and SC cases lead to SNR improvements which depend on m.

As the value of the Nakagami parameter m increases, the SNR improvement for EGC approaches that of MRC. The SNR improvement from SC on the other hand declines as the Nakagami parameter increases. The rise in SNR improvement with M is at a very slow rate for the SC algorithm. Consequently, the SC algorithm is generally implemented with two or three branches ($M = 2$ or 3). It can also be seen that the SNR improvement following EGC is close to that of the MRC.

We will now compare the performance of the SC and SSC algorithms. The SNR improvements are plotted in Fig. 5.13. Only dual diversity ($M = 2$) is considered here. The input SNR of each branch Z_0 has been taken to be unity. Of the four algorithms, namely, MRC, EGC, SC. and SSC, it is easily seen that SSC performs the worst of the four algorithms. However, in terms of implementation, SSC is the simplest of all.

5.3.6 *Effects of Branch Correlation on Combining Algorithms*

In the previous analysis, it was assumed that the diversity branches are all independent. Often, correlation exists among the diversity branches. The improvement following diversity is adversely impacted by this correlation. Since the performance improvement is only slightly less with EGC compared with MRC, we will only consider MRC systems. This is in light of the availability of exact expression for the pdf of the diversity output (Aalo 1995; Simon and Alouini 2005). With regard to SC, we will only consider the case of a two-branch (dual) diversity; the enhancement obtained is only marginal from any further increase in the number of diversity branches.

When the diversity branches are correlated the pdf of the MRC output can be obtained by using the results available for correlated gamma random variables if the correlation is exponential (Kotz and Adams 1964; Aalo 1995; Simon and Alouini 2005). Having an exponential correlation instead of equal correlation among the branches is more realistic since the correlation with branches (for example in spatial diversity) located farther will be less. For the case of exponential correlation, results on gamma random variables are available in literature. It was shown that the pdf of the output of the MRC algorithm with correlated branches could be expressed as

$$f_{\mathrm{MRC}r}(z) = \left(\frac{mM}{rZ_0}\right)^{mM^2/r} \frac{z^{\left(mM^2/r\right)-1}}{\Gamma\left(mM^2/r\right)} \exp\left(-\frac{mM}{rZ_0}z\right) U(z), \tag{5.48}$$

where r is related to the correlation coefficient ρ through

$$r = M + \frac{2\rho}{1-\rho}\left(M - \frac{1-\rho^M}{1-\rho}\right). \tag{5.49}$$

It can be seen that (5.48) is similar to the pdf of the MRC output when the branches are uncorrelated; the effect of the correlation is to reduce the effective value of the order of the pdf and lower the value of the SNR. Equation (5.48) is a gamma pdf of order (mM^2/r). The effect of correlation on the MRC densities is seen in Fig. 5.14, where the pdfs are plotted for the case of $M = 3$ and $m = 2$. It has been assumed that the average SNR per channel Z_0 equals unity. It is seen that as the correlation declines, the density function moves to the right which suggests realizing the full potential of the diversity implementation. This improvement is also evident if one looks at the plots of the gamma densities of increasing values of m. Note that when the order of the gamma distribution increases, the gamma pdf moves closer and closer to a normal pdf and the channel becomes a pure Gaussian channel or an ideal channel. To illustrate this, pdfs of a few gamma random variables of increasing values of the order are shown in Fig. 5.15 (as seen in previous chapters).

Comparing (5.32) and (5.48), the ratio of the order of the pdfs of the correlated case to the uncorrelated case O_{MRC}

$$O_{\mathrm{MRC}} = \frac{\left(mM^2/r\right)}{mM} = \frac{M}{r} = \frac{1}{1 + (2\rho/1-\rho)[1 - (1 - \rho^M/M(1-\rho))]} < 1. \tag{5.50}$$

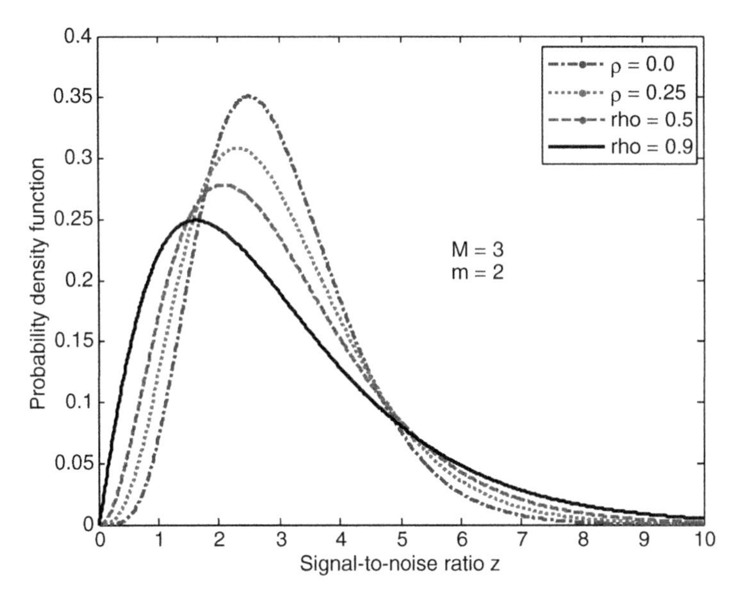

Fig. 5.14 Plots of the density functions of the MRC output for $M = 3$ and $m = 2$ for four values of the correlation coefficient. Average SNR/branch =1

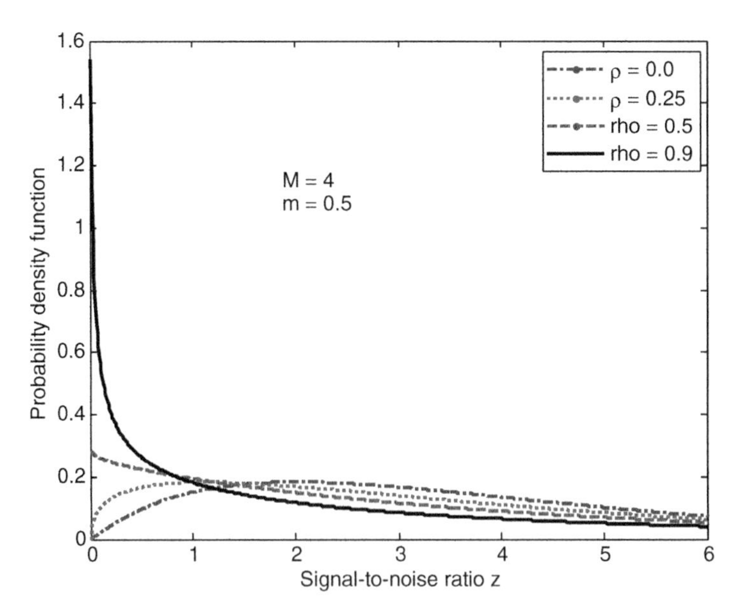

Fig. 5.15 Plots of the density functions of the MRC output for $M = 4$ and $m = 0.5$ for four values of the correlation coefficient. Average SNR/branch =1

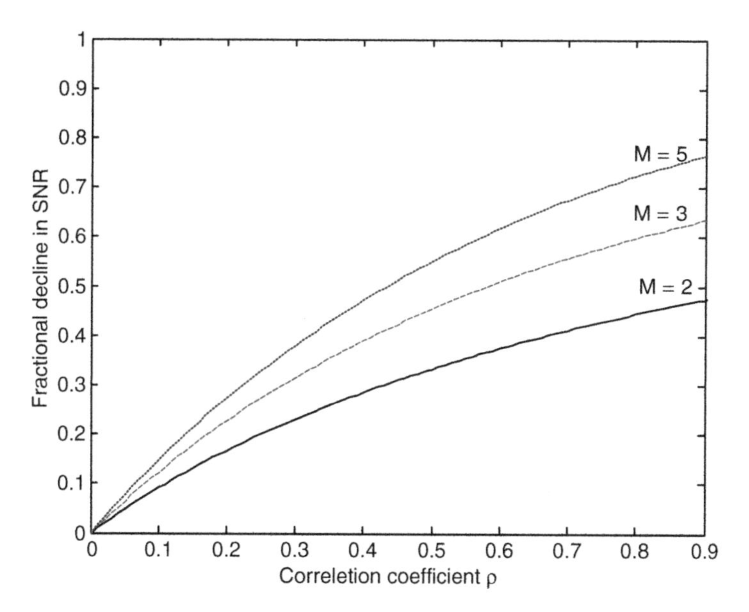

Fig. 5.16 Fractional decline in SNR as a function of the correlation coefficient for MRC

Since the SNR improvement in MRC is directly proportional to the order of the gamma pdf after diversity, O_{MRC} will be a measure of the decline in SNR enhancement when the branches are correlated. The decline in SNR enhancement in the presence of branch correlation is plotted in Fig. 5.16 as a function of the branch correlation. It can be seen that for low values of correlation, the decline in SNR is not significant.

When the number of branches is only two, specifically with the case of SC algorithm, one can use the results from the original work by Nakagami to obtain the expression for the pdf of the output of the SC algorithm. The joint pdf of the SNR outputs of the two branches can be expressed as (Kotz and Adams 1964; Simon and Alouini 2005)

$$\begin{aligned} f(z_1,z_2) = {} & \left(\frac{m}{Z_0}\right)^{m+1} \frac{(z_1 z_2/\rho)^{((m-1)/2)}}{\Gamma(m)(1-\rho)} \exp\left[-\left(\frac{m}{Z_0}\right)\frac{(z_1+z_2)}{(1-\rho)}\right] \\ & \times I_{m-1}\left[\left(\frac{m}{Z_0}\right)\frac{2\sqrt{\rho z_1 z_2}}{(1-\rho)}\right], \quad z_1>0, z_2>0, \end{aligned} \tag{5.51}$$

where $I_{\mathrm{m}-1}()$ is the modified Bessel function of the first kind of order $(m-1)$. The pdf of the output of the SC algorithm was discussed in Chap. 2 and it can be expressed as

$$fs_{Cr}(z) = 2\left(\frac{m}{Z_0}\right)^m \frac{z^{m-1}}{\Gamma(m)} \exp\left(-\frac{m}{Z_0}z\right) \times \left[1 - Q_m\left(\sqrt{\frac{2m\rho}{1-\rho}\left(\frac{z}{Z_0}\right)}, \sqrt{\frac{2m}{1-\rho}\left(\frac{z}{Z_0}\right)}\right)\right], \tag{5.52}$$

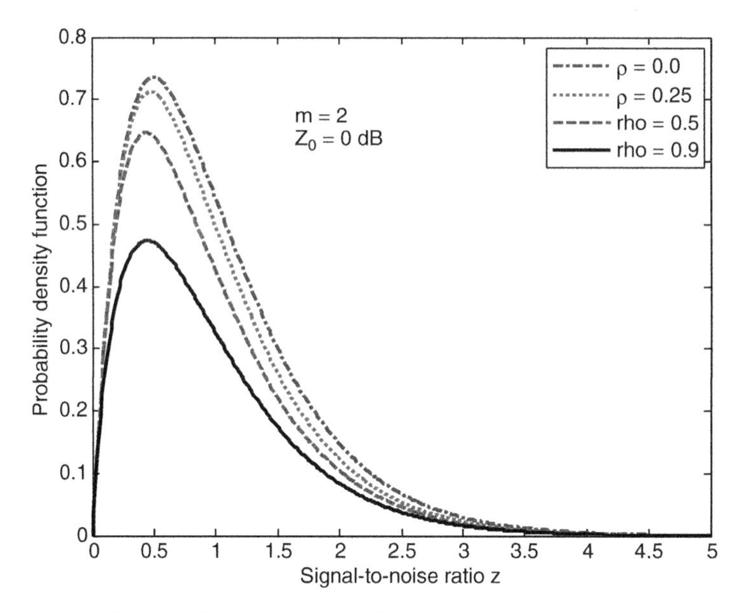

Fig. 5.17 Density function of the correlated dual SC output. The value of the Nakagami parameter m is 2 and the average SNR/branch Z_0 is unity

where $Q_m(\alpha,\beta)$ is the Marcum's Q function given by (Nuttall 1975; Simon and Alouini 2005)

$$Q_{\mathrm{m}}(\alpha,\beta)=\frac{1}{\alpha^{m-1}}\int_{\beta}^{\infty}x^{m}\exp\left[-\frac{x^{2}+\alpha^{2}}{2}\right]I_{m-1}(\alpha x)\,dx. \tag{5.53}$$

The existence of correlation between the two branches reduces the SNR of the output of the SC algorithm. The pdf of the output SNR in a dual diversity (correlated) case is shown in Fig. 5.17.

The SNR at the output of the two branch correlated SC receiver is (Ko et al. 2000)

$$\langle Z_{\mathrm{SC}r}\rangle=Z_{0}\left[1+\frac{\Gamma(2m)\sqrt{1-\rho}}{2^{2m-1}\Gamma(m)\Gamma(m+1)}\right]. \tag{5.54}$$

Similarly, the SNR at the output of the two branch correlated SSC receiver is (Abu-Dayya and Beaulieu 1994a, b; Simon and Alouini 2005)

$$\langle Z_{\mathrm{SSC}r}\rangle=_{0}\left[1+\frac{(1-\rho)m^{m-1}e^{-m}}{\Gamma(m)}\right]. \tag{5.55}$$

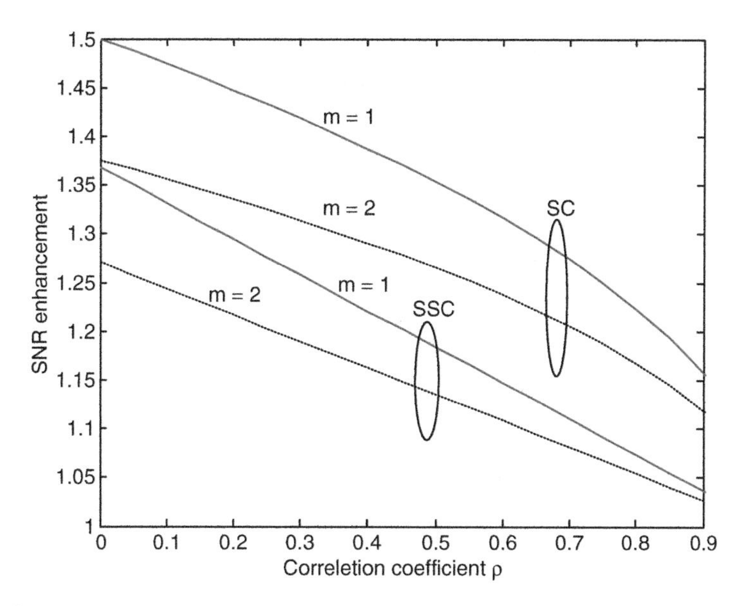

Fig. 5.18 SNR enhancement of SC and SSC as a function of the branch correlation for two values of the Nakagami parameter, $m = 1, 2$

The effect of branch correlation on the performance of a two-branch SC algorithm is compared with that of a similarly correlated SSC and shown in Fig. 5.18 for two values of the Nakagami parameter, $m = 1$ & 2.

A little more insight into (5.51) can be gained by expanding the modified Bessel function as (Gradshteyn and Ryzhik 2007; Shankar 2009)

$$I_n(x) = \sum_{l=0}^{\infty} \frac{(x/2)^{2l+n}}{\Gamma(l+1)\Gamma(n+l+1)}. \tag{5.56}$$

Equation (5.51) can now be rewritten using (5.56) as

$$f(z_1, z_2) = \sum_{l=0}^{\infty} D_l f_l(z_1) f_l(z_2), \tag{5.57}$$

where

$$D_l = \frac{\rho^l (1-\rho)^m \Gamma(m+l)}{\Gamma(m)\Gamma(l+1)} \tag{5.58}$$

and

$$f_l(z_1) = \left[\frac{1}{Z_0(1-\rho)}\right]^{m+l} \frac{z_1^{m+l-1}}{\Gamma(m+l)} \exp\left[-\frac{m z_1}{Z_0(1-\rho)}\right], \tag{5.59}$$

$$f_l(z_2) = \left[\frac{m}{Z_0(1-\rho)}\right]^{m+l} \frac{z_2^{m+l-1}}{\Gamma(m+l)} \exp\left[-\frac{mz_2}{Z_0(1-\rho)}\right]. \tag{5.60}$$

Note that when $\rho = 0$, $l = 0$ and (5.57) becomes the product of the marginal density functions of the SNR of the two branches, use of the series expansion for the modified Bessel functions allows us to write the joint pdf of the correlated outputs as the weighted sum of the products of marginal pdfs with the weights being decided by the correlation coefficient.

The joint CDF of (z_1, z_2) now becomes

$$\begin{aligned} F(z_1,z_2) = \sum_{l=0}^{\infty} D_l F_l(z_1) F_l(z_2) = \sum_{i=0}^{\infty} D_l \frac{\gamma(m+l,((m+l)z_1/Z_0(1-\rho)))}{[\Gamma(m+l)]} \\ \times \frac{\gamma(m+l,((m+l)z_2/Z_0(1-\rho)))}{[\Gamma(m+l)]}. \end{aligned} \tag{5.61}$$

Using the definition of the pdf of the maximum of two random variables, the pdf of the SC algorithm can be expressed as (Shankar 2009)

$$\begin{aligned} f_{\mathrm{SC}_r}(z) = 2\sum_{l=0}^{\infty} D_l \frac{\gamma(m+l,((m+l)z/Z_0(1-\rho)))}{[\Gamma(m+l)]} \\ \times \left[\frac{m}{Z_0(1-\rho}\right]^{m+l} \frac{z^{m+l-1}}{\Gamma(m+l)} \exp\left[-\frac{mz}{Z_0(1-\rho)}\right]. \end{aligned} \tag{5.62}$$

It is easy to see that when the two branches are independent ($\rho = 0$), (5.62) is identical to (5.31) for the case of $M = 2$. Equation (5.62) is much simpler for analytical purposes compared with (5.52) because of the absence of Marcum's Q function. We will use (5.62) later for the estimation of average probabilities of error and outage.

5.4 Shadowing Mitigation and Macrodiversity

The diversity techniques described so far are classified as microdiversity approaches, and diversity is implemented at a base station (or in rare instances at the mobile unit). Diversity can also be implemented so that the branches involved in the combining algorithm come from multiple base stations (Jakes 1994; Abu-Dayya and Beaulieu 1994a, b; Bdira and Mermelstein 1999; Shankar 2002, 2009). In other words, the branches are separated by much larger distances typically on the order of the cell size. Uses of multiple base stations constitute what are described as the macrodiversity techniques. While microdiversity techniques mitigate short-term fading commonly modeled using the Nakagami m distribution, macrodiversity techniques mitigate long-term fading or shadowing existing in wireless channels.

The most general model to describe the shadowing is the lognormal one. In this model, the pdf of the SNR expressed in decibel units is given by

$$f(z_{\mathrm{dB}})=\frac{1}{\sqrt{2\pi\sigma_{\mathrm{dB}}^{2}}}\exp\left[-\frac{(z_{\mathrm{dB}}-\mu)^{2}}{2\sigma_{\mathrm{dB}}^{2}}\right]. \tag{5.63}$$

In (5.63), m is the SNR (dB) and σ_{dB} is the standard deviation of shadowing (dB). Typical values of σ_{dB} fall in the range of 2–9 dB, with higher values corresponding to levels of severe shadowing. Note that the received SNR in absolute units (W/mW) can also be expressed as

$$f_{L}(z)=\frac{A}{z\sqrt{2\pi\sigma_{\mathrm{dB}}^{2}}}\exp\left[-\frac{(10\log_{10}z-\mu)^{2}}{2\sigma_{\mathrm{dB}}^{2}}\right]U(z). \tag{5.64}$$

The parameter A was defined in Chap. 4. Even though all the three diversity combining algorithms can be used in macrodiversity, the most commonly used algorithm is the selection combining because it is the simplest one to implement, i.e., the base station with the highest signal power or SNR is chosen to serve the mobile unit. Because of this, we will limit ourselves to the analysis of the SC algorithm only. While the microdiversity might involve several antennas, the number of base stations participating in the macrodiversity might be two or three, even though ideally there is no limit to the number of base stations involved. The use of two or three base stations is also reasonable since it is possible to imagine a case where the distances from the mobile unit to the base stations are almost the same (Jakes 1994; Alouini and Simon 2002; Piboongungon and Aalo 2004). Once we have looked at the case of macrodiversity, we will analyze the case of combining micro- and macro-diversity approaches to mitigate fading and shadowing simultaneously present in the channels. A typical three-base station scenario is shown in Fig. 5.19. We will assume that correlation exists among the base stations.

We will also treat the branches to be identical. Since we will be dealing with macrodiversity alone, it is very convenient to work with the normal pdf of the type in (5.63) instead of the lognormal pdf in (5.64). The CDF of the output of the SC algorithm has been derived and it is given by (Tellambura 2008; Skraparlis et al. 2009, 2010)

$$F(z_{\mathrm{dB}})=\frac{3}{2}\phi(z_{\mathrm{ndB}})-33T\left(z_{\mathrm{ndB}},\sqrt{\frac{1-\rho}{1+\rho}}\right)-6S\left(z_{\mathrm{ndB}},\sqrt{\frac{1-\rho}{1+\rho}},\sqrt{\frac{1}{1+2\rho}}\right), \tag{5.65}$$

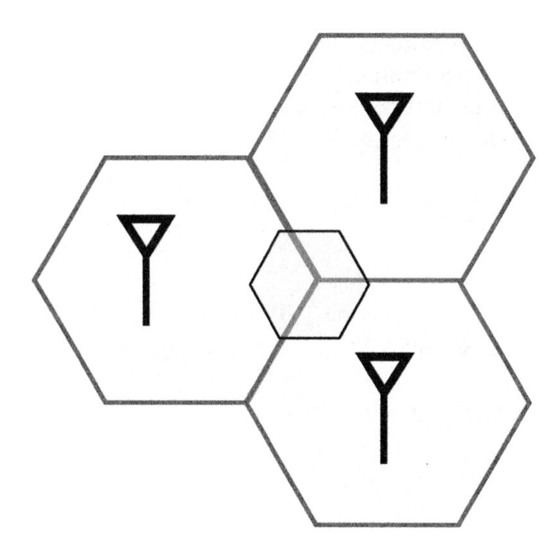

Fig. 5.19 A 3-base station macrodiversity arrangement

where

$$\phi(a) = \frac{1}{\sqrt{2\pi}} \int_{-\infty}^{a} \exp\left(-\frac{t^2}{2}\right) \mathrm{d}t, \tag{5.66}$$

$$T(a,b) = \frac{1}{2\pi} \int_{0}^{b} \frac{1}{(1+x^2)} \exp\left[-\frac{a^2}{2}(1-x^2)\right] \mathrm{d}x, \tag{5.67}$$

$$S(a,b,c) = \frac{c}{2\pi} \int_{0}^{1} \frac{\phi\left(a\sqrt{1+b^2+b^2c^2t^2}\right)}{(1+c^2t^2)\sqrt{1+b^2+b^2c^2t^2}} \mathrm{d}t. \tag{5.68}$$

Note that ρ is the power correlation coefficient. The normalized Gaussian variable Z_{ndB} is given by

$$z_{\mathrm{ndB}} = \frac{z_{\mathrm{dB}} - \mu}{\sigma_{\mathrm{dB}}}. \tag{5.69}$$

For the case of two branches (i.e., base stations), the CDF of the output of the SC algorithm is given by

$$F_{\mathrm{SC}}(z_{\mathrm{dB}}) = \frac{1}{2\pi\sqrt{1-\rho^2}} \int_{0}^{z_{\mathrm{ndB}}} \int_{0}^{z_{\mathrm{ndB}}} \exp\left[-\frac{(x^2 - 2\rho xy + y^2)}{2(1-\rho^2)}\right] \mathrm{d}x\mathrm{d}y. \tag{5.70}$$

Using (5.65), the amount of fading can be estimated following the implementation of SC. The amount of fading in triple diversity is (Tellambura 2008)

$$AF_{\text{SC3}} = \frac{\exp(\sigma_{\text{dB}}^2)}{3}\left[\frac{\phi\left(\sigma_{\text{dB}}\sqrt{2(1-\rho)}\right) - 2T\left(\sigma_{\text{dB}}\sqrt{2(1-\rho)}, \frac{1}{\sqrt{3}}\right)}{\phi\left(\sigma_{\text{dB}}\sqrt{\frac{(1-\rho)}{2}}\right) - 2T\left(\sigma_{\text{dB}}\sqrt{\frac{(1-\rho)}{2}}, \frac{1}{\sqrt{3}}\right)}\right] - 1. \quad (5.71)$$

For the case of a two-branch diversity, the AF becomes

$$AF_{\text{SC3}} = \frac{\exp(\sigma_{\text{dB}}^2)}{3}\left[\frac{\phi\left(\sigma_{\text{dB}}\sqrt{2(1-\rho)}\right)}{\phi\left(\sigma_{\text{dB}}\sqrt{\frac{(1-\rho)}{2}}\right)}\right] - 1. \quad (5.72)$$

Since most of the macrodiversity implementations involve two base stations, we will only look at the SNR improvement in dual diversity. The SNR following SC can be expressed as

$$Z_{\text{SCLN}r} = 2Z_{0\text{LN}}Q\left(-\frac{\sigma_{\text{dB}}}{\text{A}_0}\sqrt{\frac{1-\rho}{2}}\right), \quad (5.73)$$

where $Z_{0\text{LN}}$ is the average SNR/branch. Note that we are now using the average SNR expressed in absolute units instead of decibels. For the lognormal pdf [see Chap. 4 for the relationship between μ (dB) and the average of the SNR having the lognormal pdf and the definition of A_0], given by

$$Z_{0\text{LN}} = e^{\frac{\mu}{A_0}+\frac{\sigma_{\text{dB}}^2}{2A_0^2}} \quad (5.74)$$

and

$$Q(h) = \frac{1}{\sqrt{2\pi}}\int_h^{\infty} e^{-(x^2/2)}\text{d}x = 1 - \phi(h). \quad (5.75)$$

Thus, the SNR improvement after selection combining becomes

$$\text{SNRE}_{\text{SCLN}} = 2Q\left(-\frac{\sigma_{\text{dB}}}{A_0}\sqrt{\frac{1-\rho}{2}}\right) \quad (5.76)$$

Proceeding in a manner similar to the case of dual correlated Nakagami channels, the SNR enhancement after SSC in dual correlated lognormal channel becomes (Alouini and Simon 2002)

$$\text{SNRE}_{\text{SSCLN}} = 2Q\left[-\frac{\sigma_{\text{dB}}(1-\rho)}{2A_0}\right]. \quad (5.77)$$

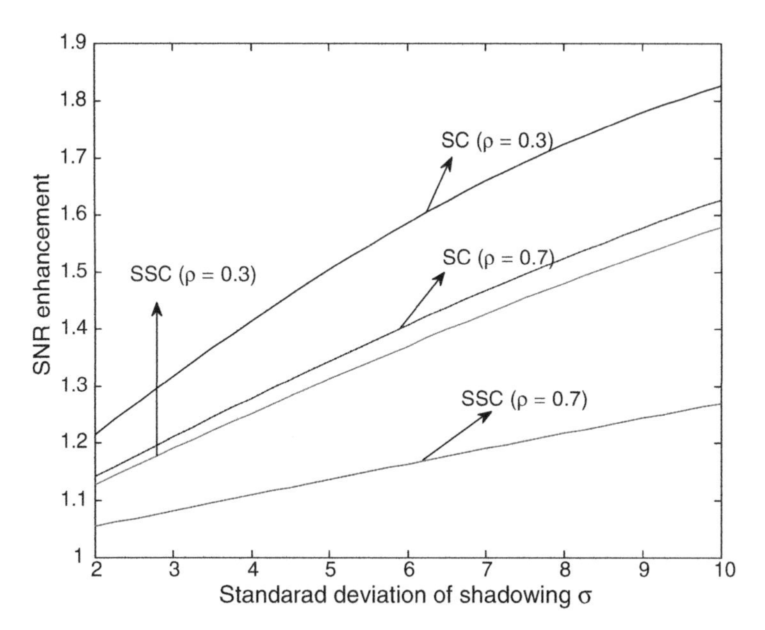

Fig. 5.20 SNR enhancement lognormal shadowing (two branch)

The SNR enhancement is plotted in Fig. 5.20 for two values of the correlation coefficient $\rho = 0.3, 0.7$. We see that the performance of the SSC is worse than that of dual channel SC, mirroring the performance in Nakagami faded channels. SNR enhancement also decreases with increasing branch correlation.

Even though macrodiversity can be implemented (as described above), we seldom deal with macrodiversity alone because fading occurs along with shadowing and microdiversity needs to be implemented simultaneously. To mitigate short-term fading and shadowing, microdiversity, hybrid schemes need to be examined. These involve the implementation of diversity at base stations and combining signals from multiple base stations. The need for this strategy was evident in the error and outage analysis carried out in Chap. 4 where it is seen that error rate and outage tend to be much higher when fading and shadowing are present instead of short-term fading only.

Such hybrid approaches, especially suited to improve performance in shadowed fading channels are described next.

5.5 Macro- and Microdiversity Systems (Hybrid Diversity)

While microdiversity techniques mitigate the effects of short-term fading and macrodiversity techniques mitigate the effects of long-term fading or shadowing, often wireless systems operate in channels which simultaneously suffer from fading and shadowing. Consequently, macrodiversity and microdiversity techniques have to be implemented in tandem (Abu-Dayya and Beaulieu 1994a; Mukherjee and Avidor 2003; Shankar 2008a, b, 2009). In shadowed fading channels which exhibit fading and shadowing simultaneously, the received SNR (assuming short fading to be Nakagami distributed) was described in Chap. 4 as

$$f(z|y) = \left(\frac{m}{y}\right)^m \frac{z^{m-1}}{\Gamma(m)} \exp\left(-\frac{m}{y}z\right). \tag{5.78}$$

In (5.78), the mean SNR is now a random variable Y which represents the effects of shadowing. The pdf of the received signal SNR in shadowed fading channels is reproduced below for the sake of immediate relevance as

$$\begin{aligned} f(z) &= \int_0^\infty f(z|y)f(y)\,dy \\ &= \int_0^\infty \left(\frac{m}{y}\right)^m \frac{z^{m-1}}{\Gamma(m)} \exp\left(-\frac{m}{y}z\right) \frac{A_0}{y\sqrt{2\pi\sigma_{\mathrm{dB}}^2}} \exp\left[-\frac{(10\log_{10}y-\mu)^2}{2\sigma_{\mathrm{dB}}^2}\right] \mathrm{d}y, \end{aligned} \tag{5.79}$$

where it is assumed that the shadowing follows lognormal statistics. Equation (5.81) has no closed form solution and, therefore, analytical explorations of diversity combining methods through (5.79) would be difficult. One of the ways in which this complexity can be resolved is through the use of alternate models for shadowing.

It was shown that a gamma pdf can approximate the lognormal pdf and, hence, it would be an ideal replacement for the lognormal pdf.

Using such an approach, we saw in Chap. 4 that a closed form solution to the pdf of the SNR in shadowed fading channels can be obtained as

$$f_{\mathrm{GK}}(z) = \frac{2}{\Gamma(m)\Gamma(c)} \left(\frac{b}{2}\right)^{c+m} z^{((c+m)/2)-1} K_{c-m}\left(b\sqrt{z}\right) U(z). \tag{5.80}$$

The pdf in (5.80) was earlier identified as the generalized K distribution (GK). The parameter b is related to the average SNR Z_0 through

$$Z_0 = \langle Z \rangle_{\mathrm{GK}} = \langle Z|Y \rangle_y = cy_0 = mc\left(\frac{2}{b}\right)^2. \tag{5.81}$$

One obtains the so called K distribution for the shadowed fading channels when the short-term fading is modeled as Rayleigh. The K distribution is obtained by putting $m = 1$ in (5.80). The pdf now becomes

$$f_K(z) = \frac{2}{\Gamma(c)} \left(\frac{b}{2}\right)^{c+1} z^{\left((c+1/2)^{-1}\right.} K_{c-1}\left(b\sqrt{z}\right) U(z). \tag{5.82}$$

Thus, while (5.80) is the pdf for the received SNR in Nakagami-lognormal shadowed fading channels, (5.82) is the pdf of the received SNR in Rayleigh lognormal shadowed fading channels. It was shown that the GK distribution is an excellent fit to model shadowed fading channels.

The availability of the analytical expression for the received SNR makes it convenient to undertake the study of hybrid diversity schemes which involve the use of microdiversity techniques to mitigate the effects of short-term fading and macrodiversity techniques to mitigate the effects of shadowing, thus improving the data transmission capabilities in wireless channels. We will consider the case of M microdiversity branches and L base stations participating in macrodiversity. We assume that the microdiversity branches are correlated (exponential correlation), but are otherwise identical. We will limit ourselves to the case of three base stations ($L = 3$), which are identical but correlated as described earlier. Since the simplest form of diversity that can be implemented at the macrolevel is the selection diversity, we will take the case of MRC–SC (MRC at the microlevel and SC at the macrolevel) and rewrite (5.32) to include the effects of shadowing as

$$f(z_l|y_l) = \left(\frac{m}{y_l}\right)^{mM} \frac{z_l^{mM-1}}{\Gamma(mM)} \exp\left(-\frac{m}{y_l}z_l\right), \quad l = 1, 2, 3. \tag{5.83}$$

In (5.83), $l = 1,2,3$ correspond to the three base stations of the macrodiversity. Joint pdf of the received SNRs from the three base stations will be

$$f(z_1, z_2, z_3) = \int_0^\infty \int_0^\infty \int_0^\infty f(z_1|y_1)f(z_2|y_2)f(z_3|y_3)f(y_1, y_2, y_3)\, \mathrm{d}y_1\, \mathrm{d}y_2\, \mathrm{d}y_3, \tag{5.84}$$

where $f(y1,y2,y3)$ is the joint pdf of the shadowing components. This joint pdf can be expressed using the L-dimensional gamma pdf given by (Warren 1992; Karagiannidis et al. 2003; Nomoto et al. 2004; Holm and Alouini 2004a, b; Shankar 2008a, b, 2009)

$$f(y_1, y_2, \dots y_L) = \frac{e^{-(1/y_0(1-\rho))\left[y_1 + y_L(1+\rho)\sum_{i=2}^{L-1} y_i\right]} \rho^{(-(L-1)(c-1))/2}}{y_0^{L+c-1}\Gamma(c)(1-\rho)^{L-1}} (y_1 y_L)^{((c-1)/2)} \prod_{l=1}^{L-1} I_{c-1}\left[\frac{2\sqrt{\rho y_l y_{l+1}}}{y_0(1-\rho)}\right]. \tag{5.85}$$

Using (5.85), the joint pdf $f(y1,y2,y3)$ becomes

$$f(y_1, y_2, y_3) = \sum_{k_1=0}^{\infty} \sum_{k_2=0}^{\infty} C_{k_1,k_2} f(y_1) f(y_2) f(y_3), \tag{5.86}$$

$$C_{k_1,k_2} = \frac{\rho^{k_1+k_2}(1-\rho)^c \Gamma(k_1+k_2+c)}{(1+\rho)^{k_1+k_2+c}\Gamma(k_1+1)\Gamma(k_2+1)\Gamma(c)}, \tag{5.87}$$

$$f(y_1) = \frac{y_1^{c+k_1} e^{-(y_1/\chi)}}{\Gamma(c+k_1)\chi^{c+k_1}}, \tag{5.88}$$

$$f(y_2) = \frac{y_2^{c+k_1+k_2-1} e^{-(((1+\rho)y_2)/\chi)}}{\Gamma(c+k_1+k_2)(\chi/(1+\rho))^{c+k_1+k_2}}, \tag{5.89}$$

$$f(y_3) = \frac{y_3^{c+k_2-1} e^{-(y_3/\chi)}}{\Gamma(c+k_2)\chi^{c+k_2}}, \tag{5.90}$$

$$\chi = y_0(1-\rho). \tag{5.91}$$

Even though correlation exists among the three base stations, (5.86) points out the fact that it is possible to write the joint pdf f(y1,y2,y3) as the sum of product marginal pdfs and making it possible to obtain an analytical expression for the pdf of the output SNR of the MRC–SC scheme. Equation (5.84) can now be simplified using (5.86), (5.87), (5.88), (5.89), (5.90), and (5.91) as

$$f(z_1,z_2,z_3) = \sum_{k_1=0}^{\infty}\sum_{k_2=0}^{\infty} C_{k_1,k_2} f_1(z) f_2(z) f_3(z), \tag{5.92}$$

where

$$f_1(z) = \frac{2}{\Gamma(mM)\Gamma(c+k_1)}\left(\frac{b_1}{2}\right)^{c+k_1+mM} z^{((c+k_1+mM)/2)-1} K_{c+k_1-mM}\left(b_1\sqrt{z}\right)U(z), \tag{5.93}$$

$$f_2(z) = \frac{2}{\Gamma(mM)\Gamma(c+k_1+k_2)}\left(\frac{b_2}{2}\right)^{c+k_1+k_2+mM} z^{((c+k_1+k_2+mM)/2)-1} K_{c+k_1+k_2-mM}\left(b_2\sqrt{z}\right)U(z), \tag{5.94}$$

$$f_3(z) = \frac{2}{\Gamma(mM)\Gamma(c+k_2)}\left(\frac{b_3}{2}\right)^{c+k_2+mM} z^{((c+k_2+mM)/2)-1} K_{c+k_2-mM}\left(b_3\sqrt{z}\right)U(z), \tag{5.95}$$

where

$$b_1 = 2\sqrt{\frac{m}{\chi}}, \tag{5.96}$$

$$b_2 = 2\sqrt{\frac{m(1+\rho)}{\chi}}, \tag{5.97}$$

$$b_3 = b_1. \tag{5.98}$$

Note that

$$Z_0 = cy_0, \tag{5.99}$$

the average SNR/branch of the microdiversity. The CDF of the output SNR of the MRC–SC algorithm becomes

$$F_{\mathrm{MRC-SC}}(z) = \int_0^z \int_0^z \int_0^z f(z_1, z_2, z_3)\,\mathrm{d}z_1\,\mathrm{d}z_2\,\mathrm{d}z_3. \tag{5.100}$$

Using (5.92), (5.100) becomes

$$F_{\mathrm{MRC-SC}}(z) = \sum_{k_1=0}^{\infty}\sum_{k_2=0}^{\infty} C_{k_1,k_2} F_1(z) F_2(z) F_3(z), \tag{5.101}$$

where $F_1(z)$–$F_3(z)$ are the CDFs corresponding to the pdfs $f_1(z)$–$f_3(z)$, respectively. They can easily be obtained as

$$F_1(z) = \frac{\Gamma(mM-c-k_1)\left(zb_1^2/4\right)^{c+k_1}}{\Gamma(mM)\Gamma(c+k_1+1)}\,{}_1F_2\left(c+k_1,[1-mM+c+k_1,1+c+k_1],\frac{zb_1^2}{4}\right) + \frac{\Gamma(c+k_1-mM)\left(zb_1^2/4\right)^{mM}}{\Gamma(mM+1)\Gamma(c+k_1)}\,{}_1F_2\left(mM,[1-c-k_1+mM,1+mM],\frac{zb_1^2}{4}\right), \tag{5.102}$$

$$F_2(z) = \frac{\Gamma(mM-c-k_1-k_2)\left(zb_2^2/4\right)^{c+k_1+k_2}}{\Gamma(mM)\Gamma(c+k_1+k_2+1)}\,{}_1F_2\left(c+k_1+k_2,[1-mM+c+k_1+k_2,1+c+k_1+k_2],\frac{zb_2^2}{4}\right) + \frac{\Gamma(c+k_1+k_2-mM)\left(zb_2^2/4\right)^{mM}}{\Gamma(mM+1)\Gamma(c+k_1+k_2)}\,{}_1F_2\left(mM,[1-c-k_1-k_2+mM,1+mM],\frac{zb_2^2}{4}\right), \tag{5.103}$$

$$F_3(z) = \frac{\Gamma(mM-c-k_2)\left(zb_3^2/4\right)^{c+k_2}}{\Gamma(mM)\Gamma(c+k_2+1)}\,{}_1F_2\left(c+k_2,[1-mM+c+k_2,1+c+k_2],\frac{zb_3^2}{4}\right) + \frac{\Gamma(c+k_2-mM)\left(zb_3^2/4\right)^{mM}}{\Gamma(mM+1)\Gamma(c+k_2)}\,{}_1F_2\left(mM,[1-c-k_2+mM,1+mM],\frac{zb_3^2}{4}\right), \tag{5.104}$$

where ${}_1F_2$ is the hypergeometric function (Gradshteyn and Ryzhik 2007). The pdf of the output SNR of the MRC–SC algorithm now becomes

$$f_{\mathrm{MRC-SC}}(z)=\sum_{k_1=0}^{\infty}\sum_{k_2=0}^{\infty}C_{k_1,k_2}[f_1(z)F_2(z)F_3(z)+f_2(z)F_1(z)F_3(z)+f_3(z)F_1(z)F_2(z)]. \tag{5.105}$$

The case of correlation at the microlevel can easily be incorporated since the correlation only at the microlevel still results in a gamma pdf for the output of the MRC algorithm, as seen in (5.48).

We can now consider a few special cases of the diversity implementation in shadowed fading channels.

If microdiversity alone is implemented in shadowed fading channels, the pdf of the output SNR of the MRC diversity becomes (zero correlation)

$$\begin{aligned} f_{\mathrm{MRC}}(z)&=\int_0^{\infty}f_{\mathrm{MRC}}(z|y)f(y)\,\mathrm{d}y=\int_0^{\infty}\left(\frac{m}{y}\right)^{mM}\frac{z^{mM-1}}{\Gamma(mM)}\exp\left(-\frac{m}{y}z\right)\frac{y^{c-1}}{y_0^c\Gamma(c)}\exp\left(-\frac{y}{y_0}\right)\mathrm{d}y\\ &=\frac{2}{\Gamma(mM)\Gamma(c)}\left(\frac{b}{2}\right)^{c+mM}z^{((c+mM)/2)-1}K_{mM-c}\left(b\sqrt{z}\right). \end{aligned} \tag{5.106}$$

In terms of the average SNR/branch of diversity Z_0, we have

$$Z_0=cy_0, \tag{5.107}$$

Equation (5.106) now becomes

$$f_{\mathrm{MRC}}(z)=\frac{2}{\Gamma(mM)\Gamma(c)}\left(\sqrt{\frac{mc}{Z_0}}\right)^{c+mM}z^{((c+mM)/2)-1}K_{mM-c}\left(2\sqrt{\frac{mc}{Z_0}z}\right). \tag{5.108}$$

The density function in (5.108) is plotted in Fig. 5.21.

Figure 5.22 shows the effect of diversity as M increases. As M increases the performance certainly improves as seen by the shift of the peaks to the right. The effect of correlated branches is shown in Fig. 5.23.

If selection combining is implemented at the microlevel, the pdf of the output SNR (zero correlation) can be written in terms of the conditional SNR $f(z/y)$ as

$$f_{\mathrm{SC}}(z|y)=M\left[\frac{\gamma(m,(mz/y))}{[\Gamma(m)]}\right]^{M-1}\left(\frac{m}{y}\right)^{m}\frac{z^{m-1}}{\Gamma(m)}\exp\left(-\frac{m}{y}z\right). \tag{5.109}$$

Substituting for the pdf of the shadowing, the pdf of the SNR in selection combining becomes

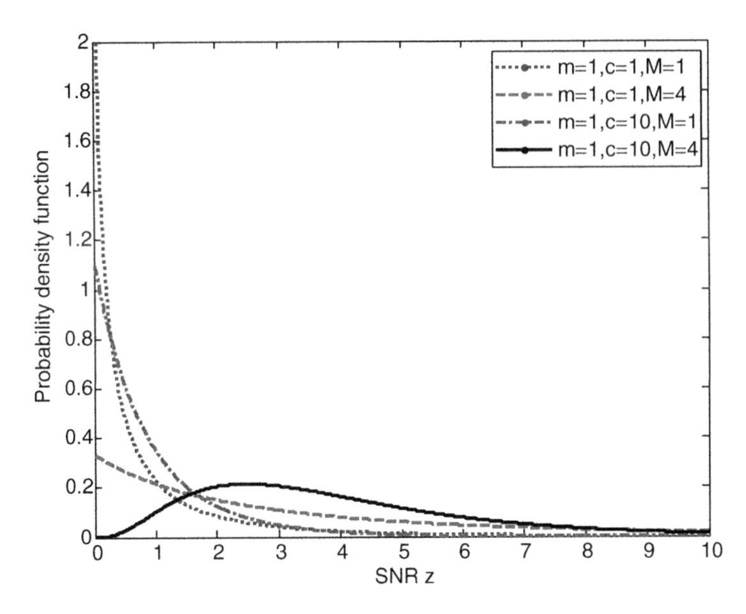

Fig. 5.21 Plot of the density functions showing the effects of microdiversity in shadowed fading channels (average SNR/branch of unity)

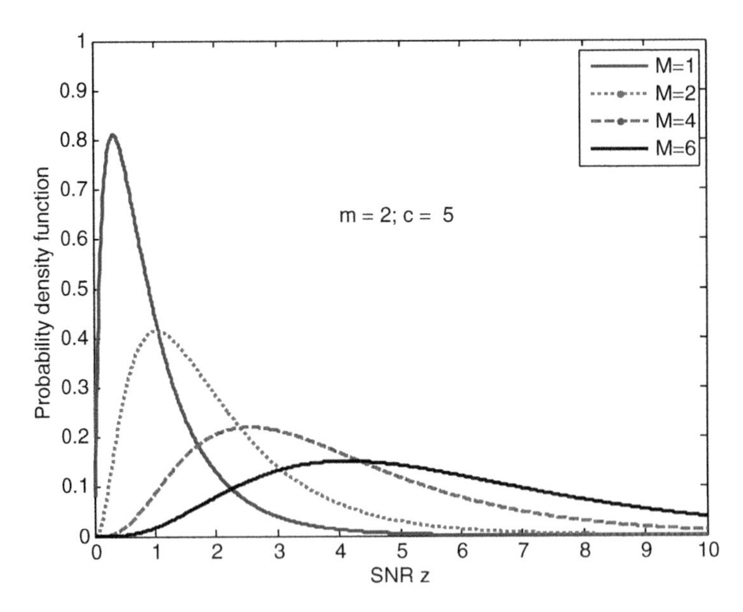

Fig. 5.22 The effect of increasing order of diversity in shadowed fading channels (average SNR/channel is unity)

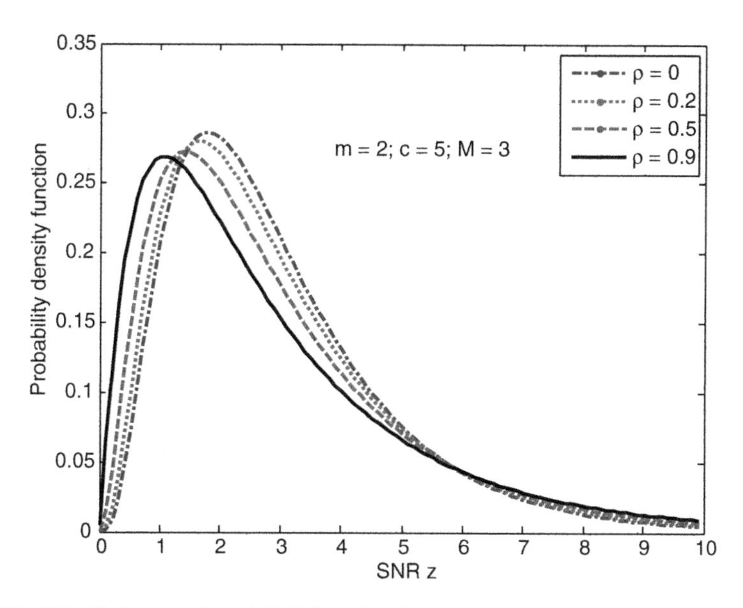

Fig. 5.23 GK pdf after correlated MRC diversity alone (average SNR of unity)

$$f_{SC}(z) = \int_0^\infty f_{SC}(z|y)f(y)\,\mathrm{d}y, \tag{5.110}$$

where

$$f_{SC}(z) = \int_0^\infty M\left[\frac{\gamma(m,(mz/y))}{[\Gamma(m)]}\right]^{M-1}\left(\frac{m}{y}\right)^m \frac{z^{m-1}}{\Gamma(m)}\exp\left(-\frac{m}{y}z\right)\left(\frac{c}{Z_0}\right)^c \frac{y^{c-1}}{\Gamma(c)}\exp\left(-\frac{c}{Z_0}y\right)\mathrm{d}y. \tag{5.111}$$

Once again, (5.111) has been expressed in terms of the average SNR/branch Z_0 using (5.107). The pdf is shown in Fig. 5.24 for $Z_0 = 1$.

There is another hybrid combining systems (also called a hybrid diversity system) which employs both selection combining and MRC at a given location. We will explore the attributes of hybrid diversity systems employed to mitigate fading and shadowing later in the next section. For the general case of MRC–SC, with microdiversity of order of M and macrodiversity of order L with independent and identical branches (microdiversity) and identical base stations (macrodiversity), the pdf of the received SNR will be

$$f_{MRC-SC}(z) = L[F_M(z)]^{L-1}\frac{2}{\Gamma(mM)\Gamma(c)}\left(\frac{b}{2}\right)^{c+mM} z^{(c+mM)/2-1}K_{c-mM}\left(b\sqrt{z}\right) \tag{5.112}$$

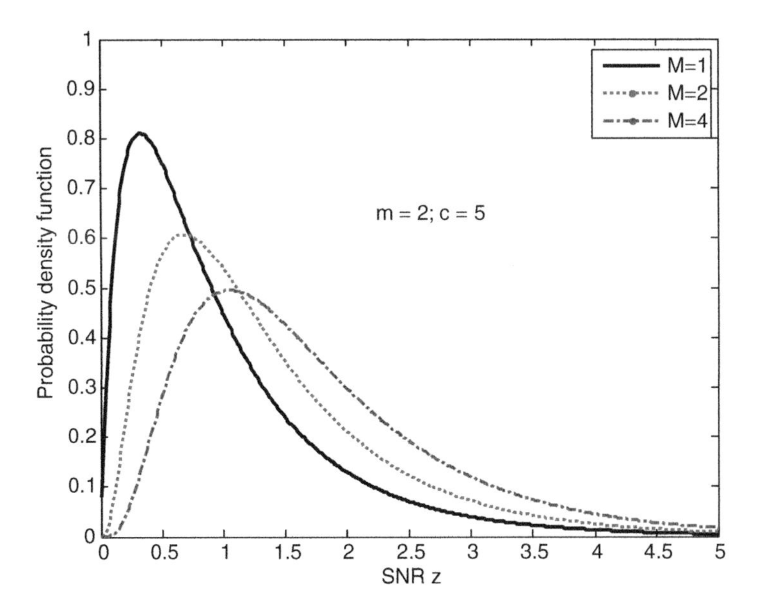

Fig. 5.24 Density function of SC only at the microlevel (average SNR/branch =1)

with

$$F_M(z) = \frac{\Gamma(mM-c)\left(\frac{zb^2}{4}\right)^c}{\Gamma(mM)\Gamma(c+1)}\,{}_1F_2\left(c,[1-mM+c,1+c],\frac{zb^2}{4}\right) + \frac{\Gamma(c-mM)\left(zb^2/4\right)^{mM}}{\Gamma(mM+1)\Gamma(c)}\,{}_1F_2\left(mM,[1-c+mM,1+mM],\frac{zb^2}{4}\right). \tag{5.113}$$

Using MeijerG functions, the density function can be expressed as

$$f_{\text{MRC-SC}}(z) = L\left[\frac{1}{\Gamma(c)\Gamma(mM)}G_{1,3}^{2,1}\left(\frac{mcz}{Z_0}\left[\begin{matrix}1\\ Mm,c,0\end{matrix}\right]\right)\right]^{L-1}\frac{1}{z\Gamma(c)(Mm)}G_{0,2}^{2,0}\left(\frac{mcz}{Z_0}\middle|\begin{matrix}-\\ Mm,c\end{matrix}\right). \tag{5.114}$$

The density functions are plotted in Fig. 5.25. We will return to these density functions when we look at the estimation of outage probabilities.

Another way to take a closer look at the effect of diversity in shadowed fading channels is to use random number generation. The shadowed fading channel can be simulated by generating random numbers (SNR) as

$$Z = XY. \tag{5.115}$$

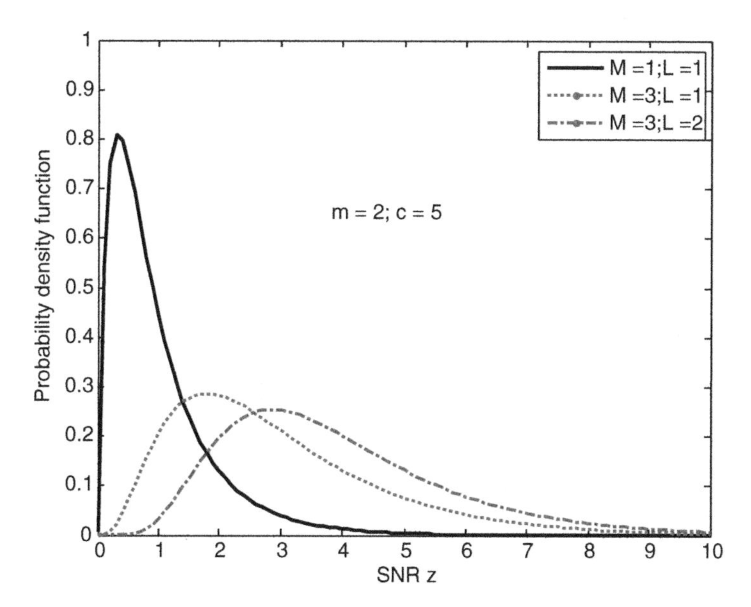

Fig. 5.25 Plots of the MRC–SC pdf (uncorrelated); average SNR/branch =1

As in the case of the GK model, both X and Y are gamma random variables with orders m and c. The diversity can then be implemented by generating several sets of random variables, one set for one base station and the other set for a second base station simulating macrodiversity of order two (dual). Results of such a simulation are shown in Figs. 5.26 and 5.27. Figure 5.26 shows the plots of the CDFs of the SNR obtained for the case of four-branch microdiversity and two-branch macrodiversity. The CDF shifts to the right indicating expected improvement in the channel conditions as one goes from no diversity to the case of hybrid diversity of MRC–SC. For comparison, the case of SC–SC is shown. It refers to selection combining at the microlevel followed by selection combining at the macrolevel.

Figure 5.27 shows the results on the comparison of the pdfs. Once again, the shift of the peak toward higher values of the SNR is seen, with diversity with the maximum shift occurring with MRC–SC. It is also seen that MRC alone at the microdiversity level is better than implementing selection combining at the microlevel followed by selection combining at the macrolevel.

We will return to the case of simultaneous implementation of micro- and macrodiversity later when we examine additional quantitative measures of improvement gained through diversity.

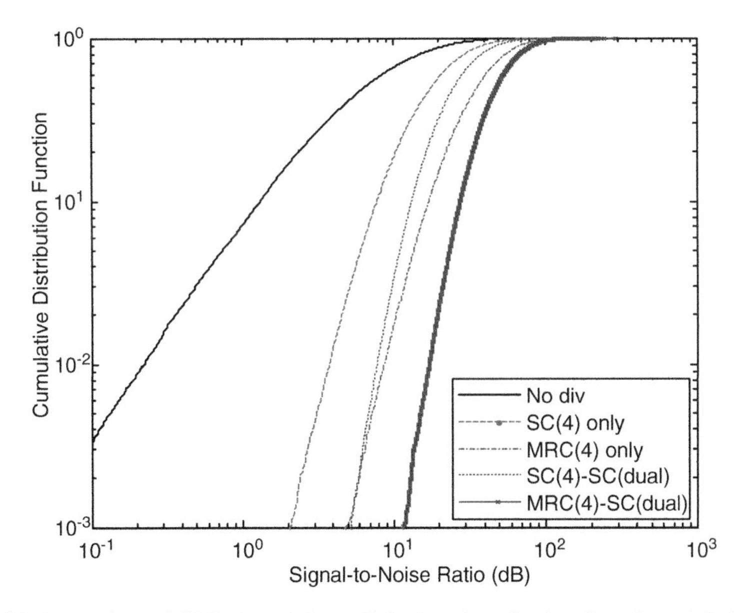

Fig. 5.26 Comparison of CDFs ($m = 1.5$; $c = 3$) for four-branch microdiversity and dual branch macrodiversity

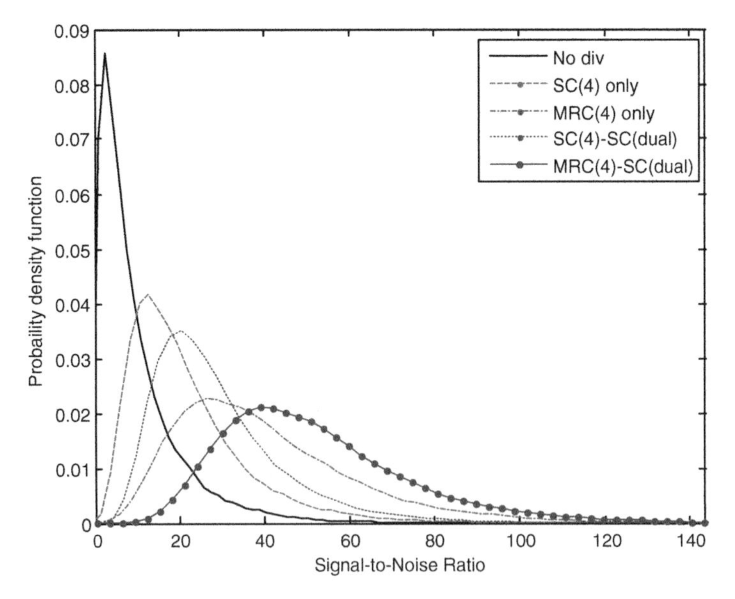

Fig. 5.27 Comparison of probability density functions (pdfs) ($m = 1.5$; $c = 3$) for four-branch microdiversity and dual branch macrodiversity

5.6 Generalized Selection Combining

We have so far explored the straightforward diversity combining algorithms, namely Selection Combining (including SSC as well), EGC, and MRC. These algorithms are implemented as a part of the microdiversity techniques. A hybrid diversity technique employing both microdiversity and macrodiversity combining techniques are examined for mitigating short-term fading and shadowing occurring concurrently. There is also yet another hybrid diversity combining algorithm used to mitigate short-term fading implying that it is a microdiversity approach implemented at a base station (or a mobile unit). This technique, known as the generalized selection combining (GSC) technique, uses both selection combining and MRC without using multiple base stations (Kong and Milstein 1999; Alouini and Simon 1999; Ma and Chai 2000; Ning and Milstein 2000; Mallik and Win 2002; Annamalai and Tellambura 2003; Theofilakos et al. 2008; Malhotra et al. 2009). Since selection combining only uses a single branch from M diversity branches, the SC algorithm wastes resources by not using the remaining $(M-1)$ branches. This underutilization of resources is a critical issue within the context of CDMA systems which rely on RAKE reception to mitigate short-term fading. On the other hand, with MRC using the complete set of M branches also poses some problems since some of the branches might be very weak and this can lead to problems in the estimation of channel characteristics crucial to the successful implementation of the MRC algorithm. Thus, there is merit in pursuing a hybrid strategy of using both selection combining of a few of the strongest branches and following this up with the MRC (Kong and Milstein 1999; Alouini and Goldsmith 1999). This would remove the weaker branches from consideration, eliminating the problems encountered during the channel estimation, and making use of a significant number of branches instead of a single branch mitigating the underutilization issue. The GSC algorithm relies on this concept.

The GSC algorithm uses a certain number of strongest branches (Select the strongest branches-SC) and then coherently combines these strongest ones in the MRC algorithm. Based on the two combining algorithms, this technique can be identified as SC/MRC algorithm which uses the strongest M_c branches of the M branches $(1 \leq M_c \leq M)$. Without any detailed examination it is obvious that if $M_c = 1$, we have the traditional SC algorithm, and if $M_c = M$, we have the MRC algorithm, and if $M_c < M$, we have the SC/MRC algorithm. It must be noted that we can also use the EGC instead of the MRC in the GSC approach, even though such an approach is not commonly pursued.

Before we examine the case of a Nakagami channel, we will look at the case of a Rayleigh channel (Nakagami parameter $m = 1$) since analytical expressions for the SNR output of the SC and MRC algorithms are readily available for any value of M while analytical expression for the SNR of the SC algorithm in a Nakagami channel is available only for the case of $M = 2$. A block diagram of the GSC algorithm is sketched in Fig. 5.28.

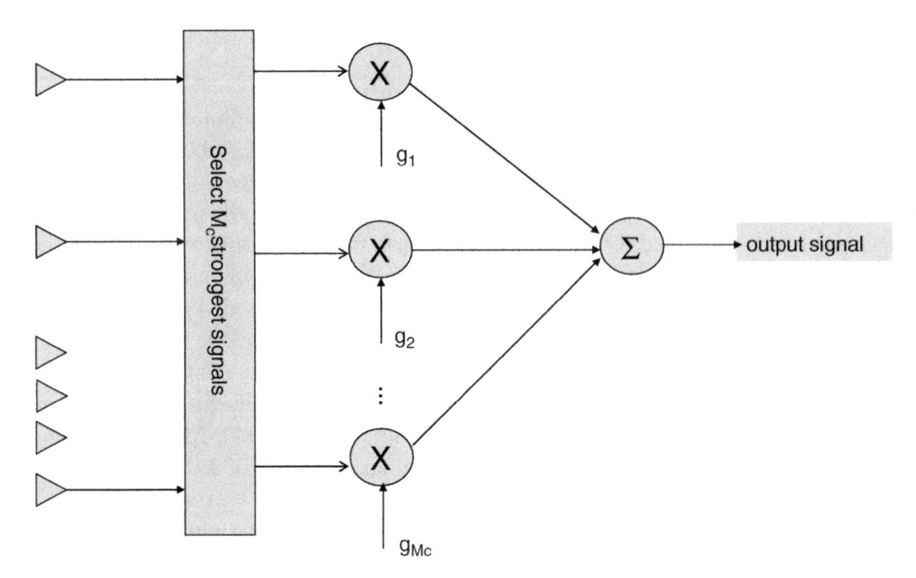

Fig. 5.28 Concept of generalized selection combining

The M_c signals with the strongest SNR are selected (SC) and weighted appropriately before combining them (MRC). The pdf of the SNR at the output of any one of the branches prior to MRC combining is given in (5.9), where Z_0 is the average SNR. The joint pdf of the M_c strongest signals of the M branches can be expressed as

$$f(z_1, z_2, \ldots, z_{M_c}) = M(M-1)\ldots(M-M_c+1)[F(z_{M_c})]^{M-M_c} \prod_{k=1}^{M_c} f(z_k), \quad z_1 \ldots \geq z_{M_c} \geq 0. \tag{5.116}$$

Thus, if $z_1; z_2;\ldots; z_{Mc}$ constitute the M_c strongest SNRs, the MRC combining will result in an output of

$$Z_{\mathrm{SC/MRC}} = Z_1 + Z_2 + \ldots Z_{M_c}, \tag{5.117}$$

the average output SNR of the GSC combing will be given by

$$\langle Z \rangle_{\mathrm{SC/MRC}} = \int_0^\infty \int_0^\infty \cdots \int_{z_{Mc}}^\infty (z_1 + z_2 + \cdots + z_{Mc}) \underbrace{\mathrm{d}z_1\,\mathrm{d}z_2 \cdots \mathrm{d}z_{Mc}}_{M_c-\text{fold}} \tag{5.118}$$

It has been shown that (5.118) becomes

$$\langle Z \rangle_{\mathrm{SC/MRC}} = Z_0 \left[M_c + \frac{M_c}{M_c+1} + \frac{M_c}{M_c+2} + \cdots + \frac{M_c}{M-1} + \frac{M_c}{M} \right], 1 \leq M_c \leq M. \tag{5.119}$$

Note that (5.119) contains $(M-M_c) + 1$ term. We can now look at the two special cases of (5.119) or the GSC combining algorithm. If $M_c = 1$, GSC becomes the traditional selection combining algorithm and (5.119) becomes

$$\langle Z\rangle_{SC/MRC,M_c=1} = Z_0\left[1+\frac{1}{2}+\frac{1}{3}+\cdots+\frac{1}{M-1}+\frac{1}{M}\right] = Z_0\sum_{k=1}^{M}\frac{1}{k}, \tag{5.120}$$

which was the same result obtained for selection combining. If $M_c = M$, we are using all the branches in the algorithm and, hence, this case is nothing but the MRC combining algorithm. Equation (5.119) now becomes

$$\langle Z\rangle_{SC/MRC,M_c=M} = MZ_0. \tag{5.121}$$

The SNR enhancements for SC, MRC, and SC/MRC are plotted in Fig. 5.29 as a function of the number M_c of the strongest branches selected for the case of $M = 8$. It can be seen that, as expected, the SNR enhancement of GSC falls between SC (minimum) and MRC (maximum).

We can now look at the case of a Nakagami channel, once again assuming that the branches are independent and identical. By selecting the strongest M_c branches out of the M branches, the joint pdf of the selected branch SNRs can be expressed using (5.116) as (Alouini and Simon 1999)

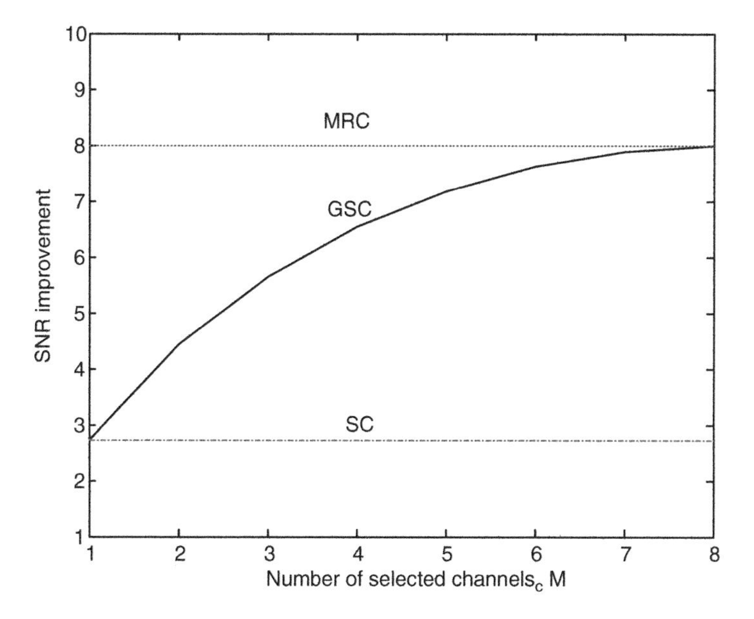

Fig. 5.29 SNR enhancement in Rayleigh channels (MRC, SC, and GSC)

$$f(z_1,\ldots,z_{M_c}) = M_c!\binom{M}{M_c}\left[\frac{\gamma(m,mz_{M_c}/Z_0)}{[\Gamma(m)]}\right]^{M-M_c}$$
$$\times\prod_{k=1}^{M_c}\left[\left(\frac{m}{Z_0}\right)^m\frac{z_{z_k}^{m-1}}{\Gamma(m)}\exp\left(-\frac{m}{Z_0}z_k\right)\right],\quad z_1\ldots\geq z_{M_c}. \quad (5.122)$$

Note that (5.122) is similar to (5.116) expect for the exact expressions for the pdf and CDF from (5.28) to (5.29), respectively. Just as in the case of selection combining, it is difficult to obtain an analytical expression for the SNR of the GSC algorithm. Because of this, we will consider a case of $M = 4$ and $M_c = 2$ and 3. For the case of MRC/SC with $M = 4$ and $M_c = 2$, the joint pdf in (5.122) becomes

$$f(z_1,z_2) = 12\left[\frac{\gamma(m,mz_2/Z_0}{[\Gamma(m)]}\right]^2\prod_{k=1}^{2}\left[\left(\frac{m}{Z_0}\right)^m\frac{z_{z_k}^{m-1}}{\Gamma(m)}\exp\left(-\frac{m}{Z_0}z_k\right)\right], z_1\geq z_2 \quad (5.123)$$

and for $M = 4$ and $M_c = 3$, the joint pdf becomes

$$f(z_1,z_2,z_3)=24\left[\frac{\gamma(m,mz_2/Z_0}{[\Gamma(m)]}\right]\prod_{k=1}^{3}\left[\left(\frac{m}{Z_0}\right)^m\frac{z_{z_k}^{m-1}}{\Gamma(m)}\exp\left(-\frac{m}{Z_0}z_k\right)\right], z_1\geq z_2\geq z_3\geq 0. \quad (5.124)$$

The average SNR after diversity combining becomes

$$\langle Z\rangle_{M_c=2,M=4} = \int_0^\infty\int_0^\infty 12z_{24}\left[\frac{\gamma(m,mz_2/Z_0}{[\Gamma(m)}\right]^2\left(\frac{m}{Z_0}\right)^{2m}\frac{(z_{z1}z_{z2})^{m-1}}{[\Gamma(m)]^2}\exp\left[-\frac{m}{Z_0}z_{24}\right]\mathrm{d}z_2\,\mathrm{d}z_1 \quad (5.125)$$

and

$$\langle Z\rangle M_{c=3,M=4} = \int_0^\infty\int_0^{z_1}\int_0^{z_2} 24z_{34}\left[\frac{\gamma(m,mz_3/Z_0}{[\Gamma(m)]}\right]$$
$$\times\left(\frac{m}{Z_0}\right)^{3m}\frac{(z_1z_2z_3)^{m-1}}{[\Gamma(m)]^3}\exp\left[-\frac{m}{Z_0}z_{34}\right]\mathrm{d}z_3\,\mathrm{d}z_2\,\mathrm{d}z_1, \quad (5.126)$$

where

$$z_{24} = z_1 + z_2 \quad (5.127)$$

and

$$z_{34} = z_1 + z_2 + z_3. \quad (5.128)$$

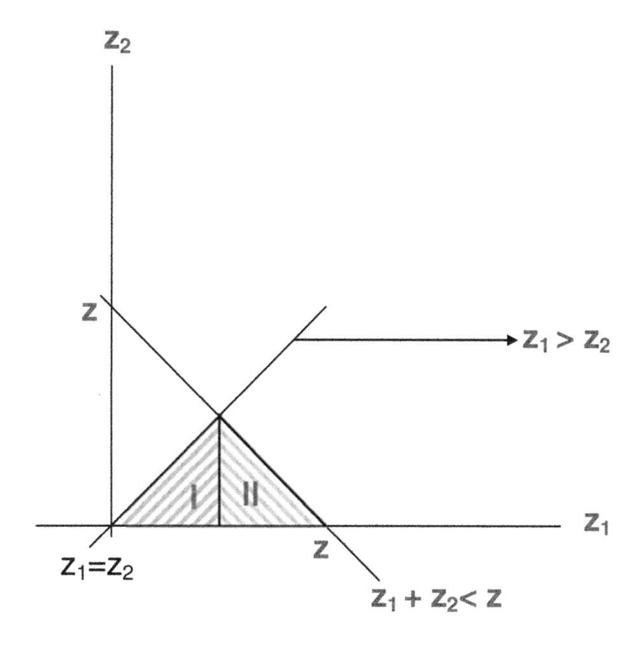

Fig. 5.30 Method to evaluate the CDF in GSC. The region below the line $z_1 = z_2$ corresponds to $z_1 > z_2$. Thus, the CDF is given by the volume contained in areas I and II as indicated

We will examine the CDFs of the GSC outputs for the case of a three-branch GSC. We will compare the CDF with the CDF of the SC as well as MRC. In this case, the joint pdf becomes

$$f(z_1, z_2) = 6f(z_1)f(z_2)F(z_2), z_2 < z_1. \tag{5.129}$$

In (5.129), the pdf $f(.)$ is the gamma pdf associated with the Nakagami fading channel and $F(.)$ is the corresponding CDF. The GSC output will be

$$Z = Z_1 + Z_2. \tag{5.130}$$

To obtain the CDF of the GSC output, we can use a graphical approach as shown in Fig. 5.30.

The CDF which is the volume contained within the shaded area can be expressed as

$$F_{\mathrm{GSC}}(z) = \left[F\left(\frac{z}{2}\right)\right]^3 + 3\int_{z/2}^{z} f(x)[F(Z-x)]^2\,\mathrm{d}x. \tag{5.131}$$

Note that the CDF of the pure selection combining algorithm with three branches will be

$$F_{\mathrm{SC}}(z) = [F(z)]^3 \tag{5.132}$$

In (5.131) and (5.132), *f*() and *F*(.) are

$$f(z) = \left(\frac{m}{Z_0}\right)^m \frac{z^{m-1}}{\Gamma(m)} \exp\left(-\frac{m}{Z_0}z\right), \quad (5.133)$$

$$F(z) = \frac{\gamma(m,(mz/Z_0))}{\Gamma(m)}. \quad (5.134)$$

For the case of a three-branch MRC, the CDF will be

$$F(z) = \frac{\gamma(3m,(mz/Z_0))}{\Gamma(3m)}. \quad (5.135)$$

The CDFs are plotted in Fig. 5.31.

The corresponding densities can be obtained by differentiating the CDFs. They are shown in Fig. 5.32.

As done before, we will compare the SNR enhancement achieved through the GSC approaches to those of the stand alone conventional SC and MRC algorithms. The SNR enhancement after the implementation of the algorithm is obtained by carrying out the numerical integration of the expressions in (5.125) and (5.126). The SNR enhancements for stand alone SC and MRC were obtained earlier. They can also be obtained by simplifying (5.125) or (5.126) by putting $M_c = 1$ to get the SC and $M_c = 4$ to get the MRC results. Figure 5.33 shows the SNR enhancements as a function of the Nakagami parameter *m*.

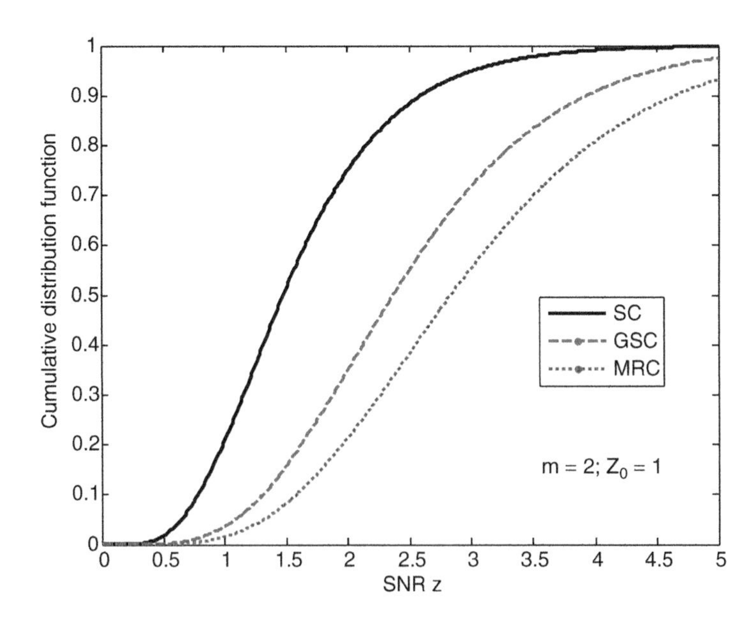

Fig. 5.31 CDFs of a three-branch diversity receiver SC, MRC, and GSC (3,2) are shown

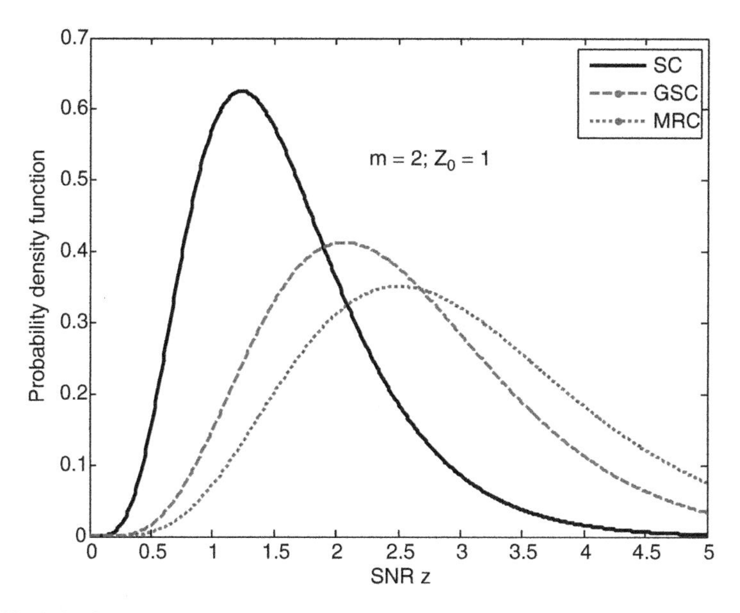

Fig. 5.32 Pdfs of a three-branch diversity receiver SC, MRC, and GSC (3,2) are shown

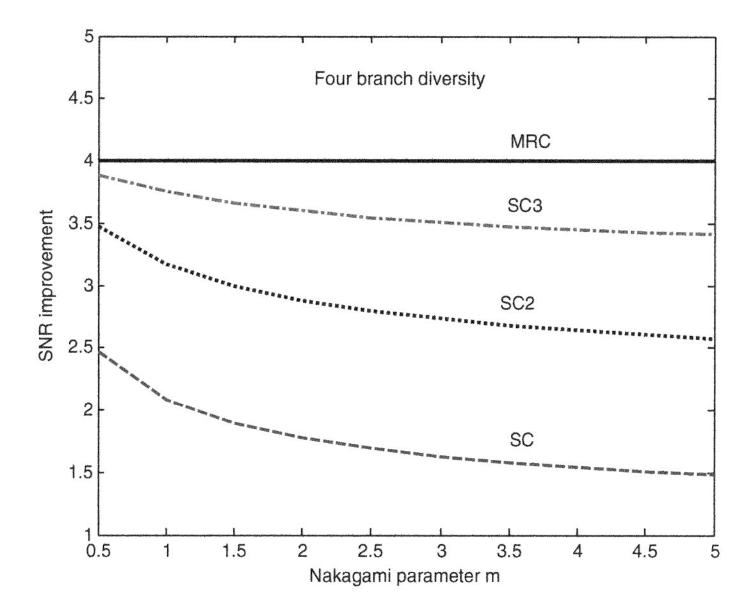

Fig. 5.33 SNR enhancement in Nakagami channels GSC vs. MRC and SC

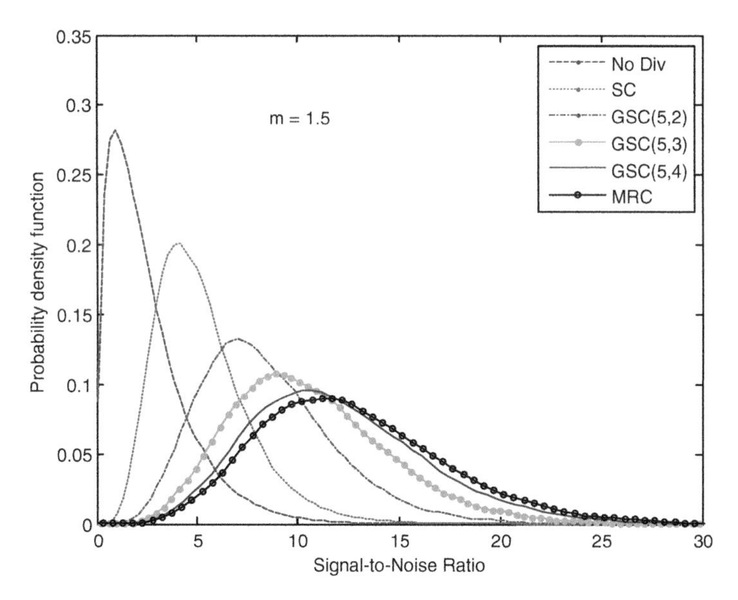

Fig. 5.34 Probability density functions of the various combining schemes in Nakagami-*m* faded channels

As we go from pure selection combining (SC) to pure MRC, it is seen that the SNR enhancement continues to up with M_c, the number of strongest selected branches out of the four total branches.

As we did earlier, we can explore the behavior of the density functions and distribution functions of the SNR following diversity including GSC through random number simulation. Sets of gamma random variables were generated and the diversity was implemented. The densities and distributions can be estimated using *ksdensity* and *ecdf* in Matlab. The results of this study for the case of a five-branch diversity are shown in Figs. 5.34 and 5.35.

Figure 5.34 shows the pdfs for the cases of No Diversity, selection combining, MRC, and GSC(5,2), GSC(5,3), and GSC(5,4). One can see that the peaks of the density functions lie between those of the density functions of no diversity and MRC. The case of GSC(5,4) is very close to that of the MRC, once again demonstrating that GSC becomes SC when only selection combining alone is carried out and GSC approaches MRC as the number of branches (M_c) increases.

Figure 5.35 shows the corresponding CDFs. Once again, CDF of the SNR in GSC is bounded by SC and MRC.

We will examine the effect of diversity in wireless channels modeled using a cascaded approach later.

We will now compare the performance of all the diversity combining algorithms using additional quantitative measures besides the SNR enhancement.

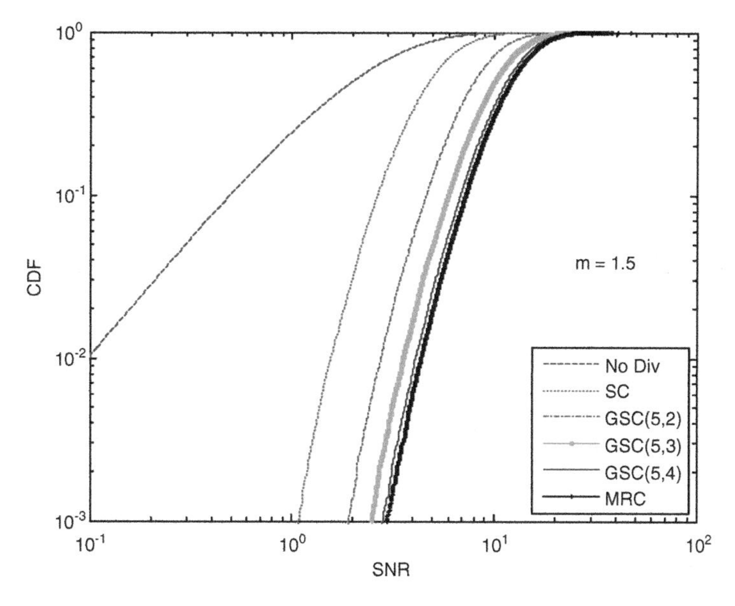

Fig. 5.35 CDFs of the various combining schemes in Nakagami-m faded channels

5.7 Quantitative Comparison of Diversity Combining Algorithms

The different diversity schemes employed to mitigate the effects of short term fading, shadowing as well as shadowing and fading simultaneously have been described. Even though the efforts were mainly devoted to the Nakagami-m short term faded channels, other fading channels such as Weibull and generalized gamma channels can be easily studied using the general expressions for the density functions for the Selection combining, MRC, and GSC. The main model explored for the case of stand alone shadowing was the lognormal one as it the primary model used. For the shadowed fading channel, the effort was centered on the Nakagami-m short term fading. Lognormal shadowing was replaced by the gamma shadowing.

While results on independent branches and independent base stations are much simpler, we also included the most general cases when the diversity branches are correlated in the case of selection combining and MRC. The case of the correlation among the base stations was also considered. There are several analytical ways to examine the performance improvement gained through the implementation of the diversity techniques. One of them is the straightforward estimation of the SNR before and after the diversity. This was undertaken earlier in this work along with the derivation and discussion of the pdfs above. However, the enhancement in SNR after diversity does not necessarily lead to a proportionate reduction in the probability of error. This is due to the fact that the relationship between the probability of error and SNR is not a linear one. Similarly, the enhancement in SNR after diversity will not result in a proportional decline in outage probability. Thus, we require additional means to explore the effects of diversity on wireless data transmission systems. Before we look at the probability of error and outage probability, we will examine a few additional parameters to quantify the improvement in wireless systems gained through the implementation of diversity techniques. These include the AF, error rates, and outage probabilities.

5.7.1 Amount of Fading

The AF in a wireless channel was discussed in Chap. 4 and it is given by (Charesh 1979; Nakagami 1960; Simon and Alouini 2005)

$$\mathrm{AF} = \frac{\langle Z^2 \rangle}{\langle Z \rangle^2} - 1 \tag{5.136}$$

For a short term faded Nakagami channel,

$$\mathrm{AF} = \frac{1}{m} \tag{5.137}$$

and for the case of MRC diversity,

$$\mathrm{AF}_{\mathrm{MRC}} = \frac{1}{mM}. \tag{5.138}$$

Equation (5.138) indicates that level fading goes down inversely with the number of branches of diversity, M. For a given site, the number of independent spatial or frequency diversity branches available is limited. Once M is increased beyond that value, the branches become correlated. The reduction in fading gained when correlation exists among the M branches is less than what is given in (5.138).

The AF for the SC can be evaluated numerically by computing the first and second moments of the pdf in (5.111). The AF for the case of independent identically distributed branches for MRC and SC are plotted in Fig. 5.36. These results once again show that MRC is better at mitigating fading than SC.

Similarly, the AF in pure shadowing which follows lognormal pdf can be expressed as (Stuber 2002; Simon and Alouini 2005)

$$\mathrm{AF} = e^{(\sigma_{\mathrm{dB}}/A_0)^2} - 1. \tag{5.139}$$

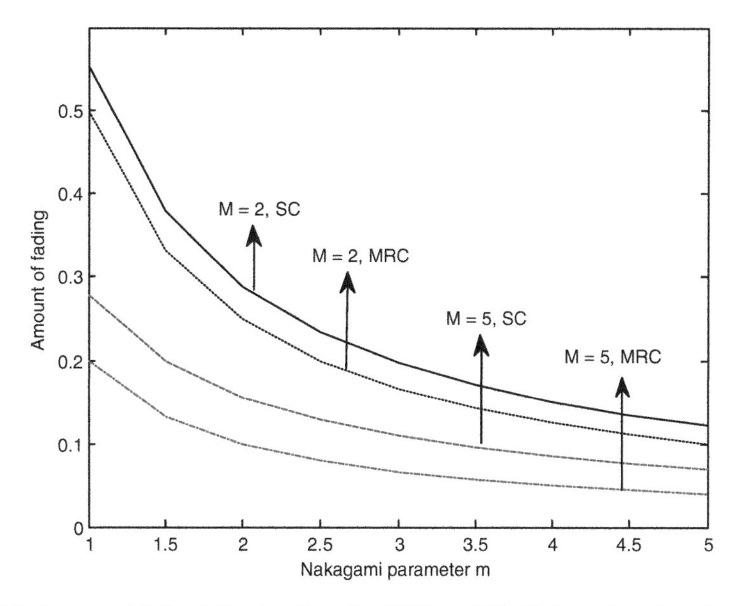

Fig. 5.36 Amount of fading following diversity: MRC vs. SC in Nakagami-*m* channels

For the case of a two channel selection combining diversity, AF becomes

$$\mathrm{AF_{SCLN}} = e^{(\sigma_{\mathrm{dB}}/\mathrm{A_0})^2} \frac{Q\left(-(\sigma_{\mathrm{dB}}/A_0)\sqrt{2(1-\rho)}\right)}{2\left[Q\left(-(\sigma_{\mathrm{dB}}/A_0\sqrt{2})\sqrt{1-\rho}\right)\right]^2} - 1. \tag{5.140}$$

In the case of switch and stay combining (SSC), the AF becomes (Corazza and Vatalaro 1994; Alouini and Simon 2002; Tellambura 2008)

$$\mathrm{AF_{SSCLN}} = e^{(\sigma/A)^2} \left[\frac{Q\left(\frac{10\log_{10}(Z_T)-\mu}{\sigma_{\mathrm{dB}}} - 2\frac{\sigma_{\mathrm{dB}}}{\mathrm{A_0}}\right) + Q\left(\frac{\sigma_{\mathrm{dB}}}{\mathrm{A_0}}2\rho - \frac{10\log_{10}(Z_T)-\mu}{\sigma_{\mathrm{dB}}}\right)}{\left(Q\left(\frac{10\log_{10}(Z_T)-\mu}{\sigma} - \frac{\sigma_{\mathrm{dB}}}{\mathrm{A_0}}\right) + Q\left(\frac{\sigma_{\mathrm{dB}}}{\mathrm{A_0}}\rho - \frac{10\log_{10}(Z_T)-\mu}{\sigma_{\mathrm{dB}}}\right)\right)^2} \right] - 1. \tag{5.141}$$

Note that in the absence of a simple analytical expression for the optimal threshold that minimizes the AF, as it was in the case with selection combining, the AF in a two-branch SSC algorithm in lognormal channels depends on the threshold SNR Z_T and the average power or SNR μ(dB). The AF for the case of lognormal shadowing is plotted in Fig. 5.37 for SC.

The AF for the case of lognormal shadowing in the case of SSC is plotted in Fig. 5.38 ($\sigma_{\mathrm{dB}} = 3$) and Fig. 5.39 ($\sigma_{\mathrm{dB}} = 7$), respectively.

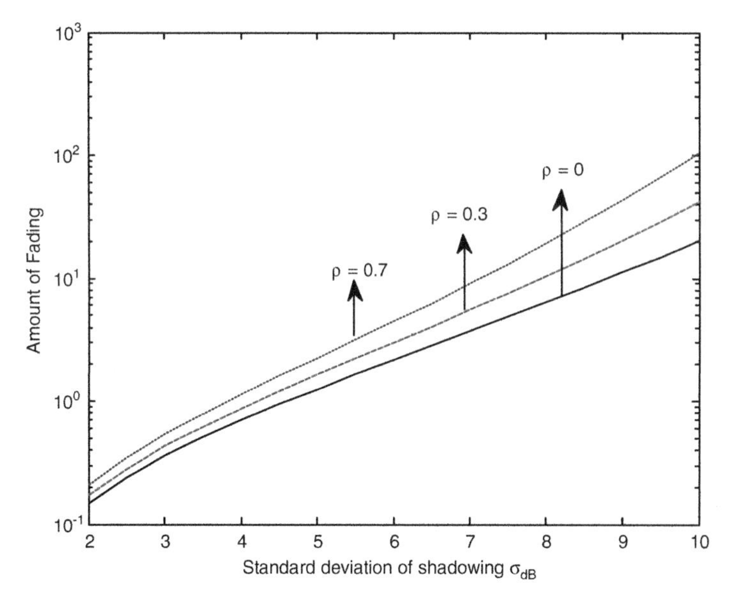

Fig. 5.37 Amount of fading in a lognormal shadowing channel following dual branch SC

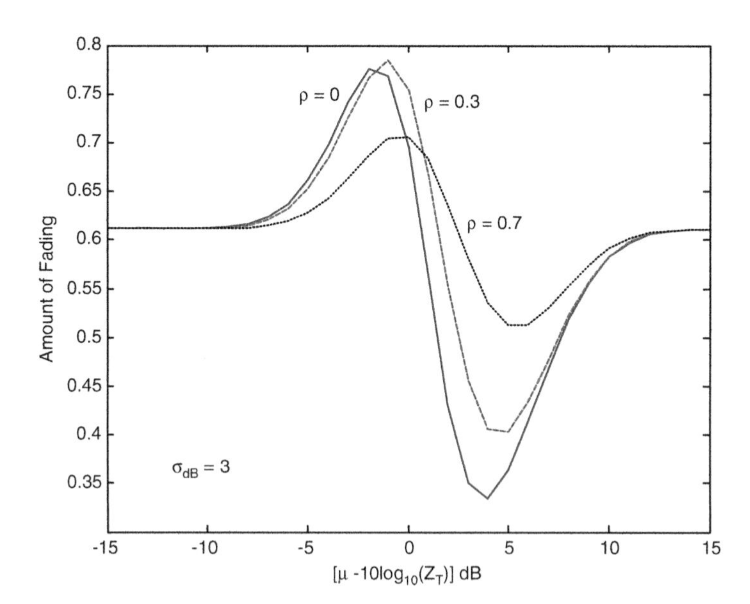

Fig. 5.38 Amount of fading in a lognormal shadowing channel in dual SSC ($\sigma_{\mathrm{dB}} = 3$)

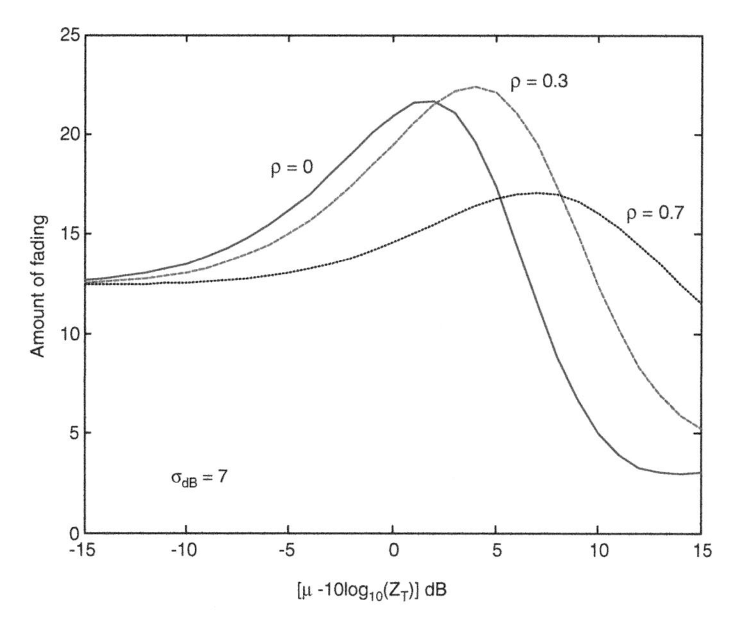

Fig. 5.39 Amount of fading in a lognormal shadowing channel in dual SSC ($\sigma_{\mathrm{dB}} = 7$)

We will now go back to GSC and estimate the AF after the implementation of diversity. Once again, we will use the example of four diversity branches and allow the number of strongest branches selected to vary from one to four. The second moment of the pdf can be evaluated similarly to the approach used in (5.125) and (5.126), where z_{24} and z_{34} are replaced by

$$z_{24} = (z_1 + z_2)^2 \tag{5.142}$$

and

$$z_{34} = (z_1 + z_2 + z_2)^2, \tag{5.143}$$

respectively. The AF as a function of the Nakagami parameter m is plotted in Fig. 5.40. The advantage of GSC over SC is clearly seen in the reduction of the fading levels (AF) as one moves from stand alone SC to stand alone MRC.

A few more observations on the AF are in order. We will use a Nakagami channel with a parameter m for this discussion. Comparing (5.137) and (5.138), it is obvious that as the number of independent branches M increase, the deleterious effect of fading will be less. Yet another way to see the benefits of diversity in reducing the effects of fading is to examine the pdfs gamma variables seen earlier. As the number of branches increases, the order of the gamma pdf increases from m to Mm, pushing the pdf further to the right. The pdf of the SNR at the output of the diversity approaches the Gaussian pdf making the Nakagami channel approximate to a Gaussian channel.

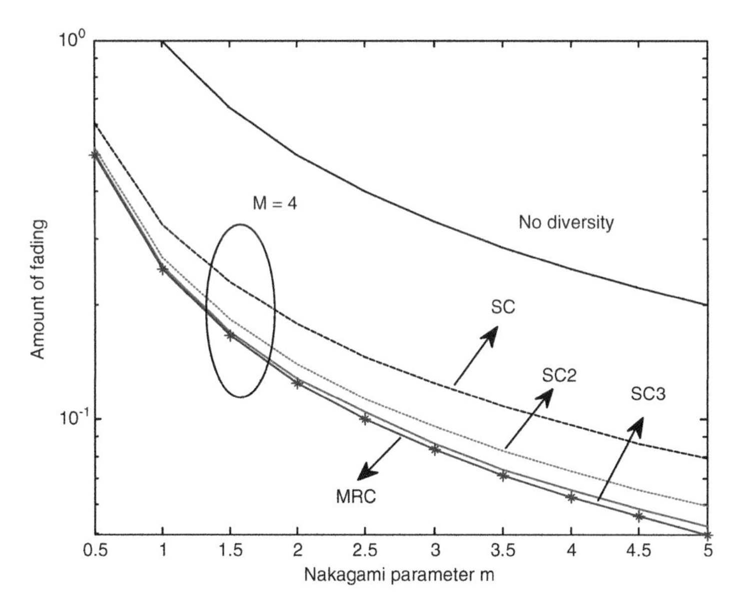

Fig. 5.40 Amount of fading following GSC

5.7.2 *Average Probability of Error*

There are two remaining important quantitative measures to compare the performance enhancement achieved through diversity and combining algorithms. One is to calculate the bit error rates before diversity (Chap. 4) and after diversity and determine the reduction in SNR achieved with diversity to maintain a specific bit error rate. The average error rate following diversity can be expressed as

$$p(e)_{\mathrm{av}} = \int_0^{\infty} \mathrm{ber}(z) f_{\mathrm{div}}(z)\,\mathrm{d}z. \tag{5.144}$$

In (5.144), ber(z) is the bit error rate (probability of error) in the absence of fading (ideal Gaussian channel) and $f_{\mathrm{div}}(z)$ is the pdf of the output SNR of the diversity combining algorithms. For the case of a coherent binary phase shift keying (CBPSK), (5.144) becomes

$$p(e)_{\mathrm{av}} = \int_0^{\infty} Q\left(\sqrt{2z}\right) f_{\mathrm{div}}(z)\,\mathrm{d}z, \tag{5.145}$$

where $Q(.)$ was defined earlier in (5.75) as well as in Chap. 3. Note that there are two key approaches of estimating the probability of error in fading channels (either prior to the diversity combining or after diversity combining). The first approach is to perform the integration in (5.145) analytically if possible. Otherwise, one needs to perform the integration in (5.145) analytically. If the pdf of the SNR is unavailable analytically, the use of (5.145) becomes computationally tedious since it is likely to involve multidimensional integration. If that is the case, we use the second approach using moment generating function (MGF) approaches (Alouini and Simon 2000; Annamalai et al. 2005; Goldsmith 2005). We will use the example of the MRC diversity in independent Nakagami channels to illustrate both techniques. Let us start with the case of the probability of error estimation using MGF first.

One of the points to be noted in this analysis is the fact that density functions of the SNR in fading before and after diversity is a nonnegative random variable. Thus, we can express the MGF of the SNR Z as (Papoulis and Pillai 2002; Annamalai et al. 2005; Goldsmith 2005)

$$M_z(s) = \int_0^\infty f(z)\exp(sz)\,\mathrm{d}z. \tag{5.146}$$

Note also that we can relate the MGF and the density function through the Laplace transform as

$$M_z(-s) = L[f_Z(z)]. \tag{5.147}$$

The Laplace transform of a Nakagami random variable is given by (Annamalai et al. 2005; Goldsmith 2005)

$$M_Z(s) = \left(1 - \frac{sZ_0}{m}\right)^{-m}. \tag{5.148}$$

Let us try to evaluate the average probability of error in (5.145) for the case of a Nakagami channel with no diversity. For this purpose, we will use an alternate form of the Q function instead of the one in (5.75).

$$Q(x) = \frac{1}{\pi}\int_0^{\pi/2} \exp\left[-\frac{x^2}{2\sin^2(\theta)}\right]\mathrm{d}\theta \tag{5.149}$$

and the expression for the probability of error for a coherent BPSK modem can be expressed as

$$\mathrm{der}(z) = \frac{1}{\pi}\int_0^{\pi/2} \exp\left[-\frac{z}{\sin^2(\theta)}\right]\mathrm{d}\theta. \tag{5.150}$$

The average probability of error in a Nakagami channel now becomes

$$p_e = \frac{1}{\pi}\int_0^\infty \int_0^{\pi/2} \exp\left[-\frac{z}{\sin^2(\theta)}\right]\mathrm{d}\theta f(z)\,\mathrm{d}z, \tag{5.151}$$

where $f(z)$ is the pdf of the SNR in (5.28). Using the Laplace transform of the pdf given in (5.148)

$$\begin{aligned} p_{\mathrm{e}} &= \frac{1}{\pi}\int_0^{\pi/2}\left[\int_0^{\infty}\exp\left[-\frac{z}{\sin^2(\theta)}\right]f_z(z)\,\mathrm{d}z\right]\mathrm{d}\theta \\ &= \frac{1}{\pi}\int_0^{\pi/2} M_Z\left(-\frac{1}{\sin^2(\theta)}\right)\mathrm{d}\theta, \end{aligned} \tag{5.152}$$

where $M_Z(.)$ is given in (5.148). Equation (5.152) now becomes

$$p_{\mathrm{e}} = \frac{1}{\pi}\int_0^{\pi/2}\left(1+\frac{Z_0}{m\sin^2(\theta)}\right)^{-m}\mathrm{d}\theta. \tag{5.153}$$

Equation (5.153) can be easily evaluated since it involves only ordinary trigonometric functions. We will now look at the case of MRC diversity in Nakagami channels.

Since the output of the MRC algorithm consists of the sum of the individual SNRs, the MGF of the output of MRC algorithm will be the product of the MGFs. The average probability of error at the output of the maximal ratio combiner will be

$$p_{\mathrm{eMRC}} = \int_0^{\infty}\int_0^{\infty}\cdots\int_0^{\infty} Q\left(\sqrt{2(z_1+z_2+\cdots+z_M)}\right)f(z_1)f(z_1)\ldots f(z_M) \tag{5.154}$$

which is an M-fold integral, and $f(z_1)$ through $f(z_M)$ are the marginal pdf's of the SNRS from each of the M branches, treating the branches as independent. Using the expression for the Q function in (5.149), (5.154) becomes

$$p_{\mathrm{eMRC}} = \frac{1}{\pi}\int_0^{\pi/2}\prod_{k=1}^{M} M_{Z_k}\left(-\frac{1}{\sin^2(\theta)}\right)\mathrm{d}\theta. \tag{5.155}$$

When all the branches are identically distributed, (5.155) becomes

$$p_{\mathrm{eMRC}} = \frac{1}{\pi}\int_0^{\pi/2}\left[M_z\left(-\frac{1}{\sin^2(\theta)}\right)\right]^M\mathrm{d}\theta = \frac{1}{\pi}\int_0^{\pi/2}\left[1+\frac{Z_0}{m\sin^2(\theta)}\right]^{-mM}\mathrm{d}\theta. \tag{5.156}$$

One can now write the expression for the average probability of error by directly using (5.156) as

$$p_{\mathrm{eMRC}} = \int_0^{\infty} Q\left(\sqrt{2z}\right)\left(\frac{m}{Z_0}\right)^{mM}\frac{z^{mM-1}}{\Gamma(mM)}\exp\left(-\frac{m}{Z_0}\right)\mathrm{d}z, \tag{5.157}$$

where we have used the pdf in (5.32) for the pdf of the output of the MRC algorithm. The analytical expression for the error rate is expressed in terms of hypergeometric functions ${}_2F_1(.)$ and the expression is obtained using the table of integrals (Gradshteyn and Ryzhik 2007; Wolfram 2011). Note that (5.156) can be evaluated without deriving the density function for the SNR following the diversity.

Equation (5.144) can also be modified slightly for use if the CDF is available in analytical form. Integrating by parts, (5.144) becomes

$$p(e)_{\text{av}} = -\int_0^{\infty} \frac{\mathrm{d}[ber(z)]}{\mathrm{d}z} F_{\text{div}}(z)\,\mathrm{d}v. \tag{5.158}$$

In (5.158), $F_{\text{div}}(.)$ is the CDF of the output of the diversity combiner. Using the derivative property of the Q function defined in (5.75), (5.158) becomes

$$p(e)_{\text{av}} = \int_0^{\infty} \frac{1}{2\sqrt{\pi}} \left[\frac{\exp(-z)}{\sqrt{z}}\right] F_{\text{div}}(z)\,\mathrm{d}z. \tag{5.159}$$

Equation (5.159) might be easier to integrate because the CDF always lies between 0 and 1 while the pdf can lie between 0 and ∞.

Equation (5.145) has been evaluated for the case of Nakagami short-term faded channel for a few diversity combining algorithms. In Chap. 4 we had seen the effect of increasing values of m on the error rates in fading channels. As the value of m increased, the excess power or SNR required to achieve a certain bit error rate went down. One can now view the diversity approaches as a means to reduce this power penalty as we will see now. The error rate in a Nakagami channel when MRC diversity is implemented is shown in Fig. 5.41. As M increases, the error rate

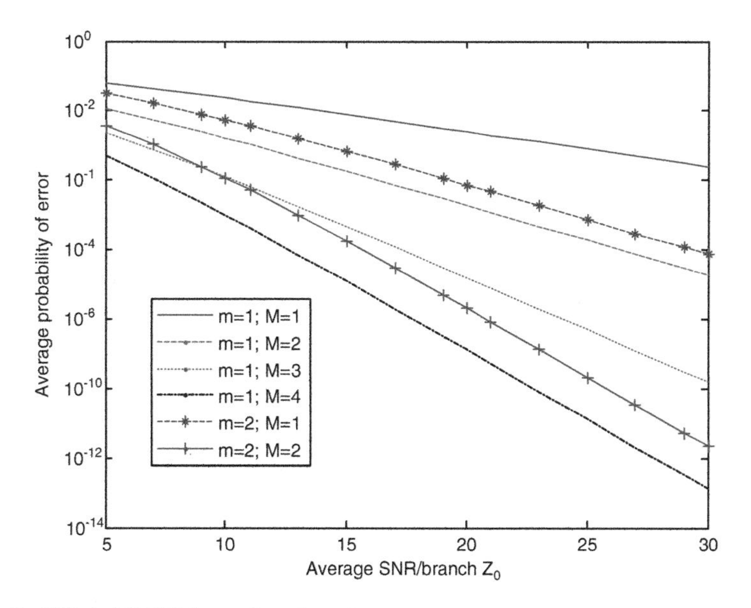

Fig. 5.41 BER in MRC Nakagami-m channel

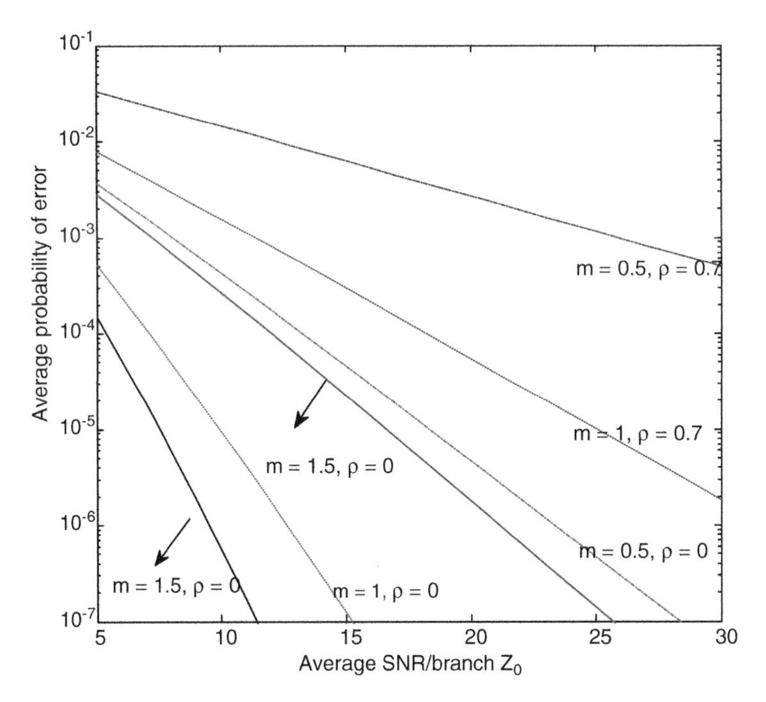

Fig. 5.42 BER for MRC for Nakagami channels ($M = 4$)

declines. This can be viewed in terms of increasing order of the gamma pdf as the order of the diversity M increases.

The effect of branch correlation in MRC is seen in the error rate plots shown in Fig. 5.42 for the case of a four-branch ($M = 4$) diversity. The error rates increase as the correlation goes up. A more detailed picture of the effects of correlation on error rates is shown in Fig. 5.43 for the case of $m = 2$ and $M = 3$.

The effect of selection combining on mitigating short-term fading is sketched in Fig. 5.42 for two values of the Nakagami parameter $m = 0.5$ and 1.5. As discussed earlier, arguing that most of the gains are realized with dual branch case for SC, we have only considered the case of a two-branch diversity where the branch correlation is taken into account. As the correlation decreases, the probability of error curves move toward the bottom left hand corner, indicating the reduction in fading penalty described in the previous paragraph. As the correlation increases, the probability of error plots moves toward the curves for no diversity for the respective values of the Nakagami parameter.

It was shown through an analysis of the SNR enhancement that MRC algorithm performs better than the SC algorithm in mitigating diversity. This aspect is shown in Figs. 5.44 and 5.45. Figure 5.44 shows the plots of the probability of error for the MRC algorithm for $M = 2$ and Fig. 5.45 shows the results for a correlated dual branch SC receiver. Comparing the curves in Figs. 5.44 and 5.45 it is seen that MRC algorithm leads to further reduction in power penalty over the SC algorithm.

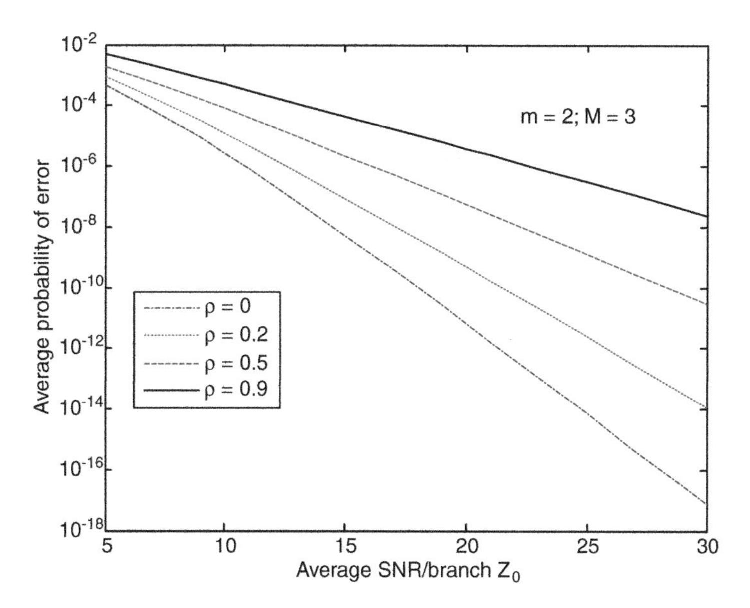

Fig. 5.43 MRC BER for Nakagami-m ($m = 2$) and $M = 3$

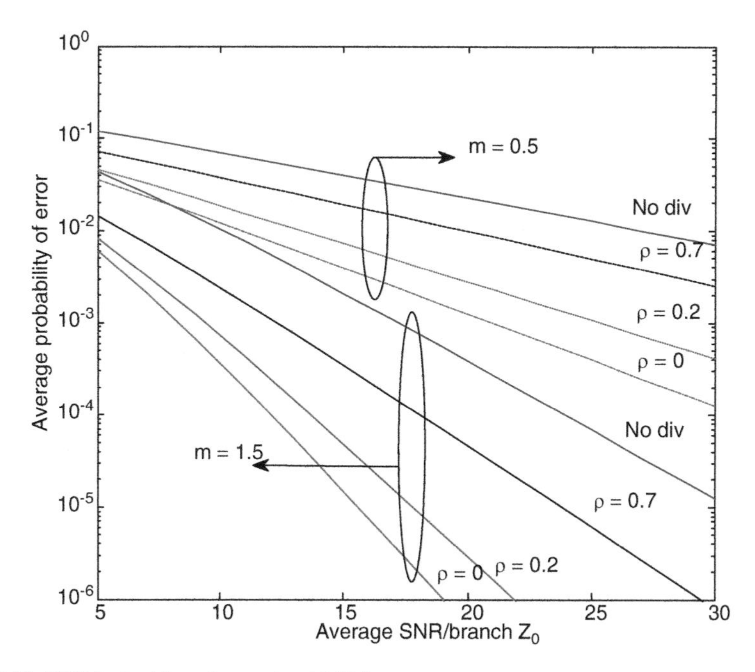

Fig. 5.44 BER in dual branch correlated MRC

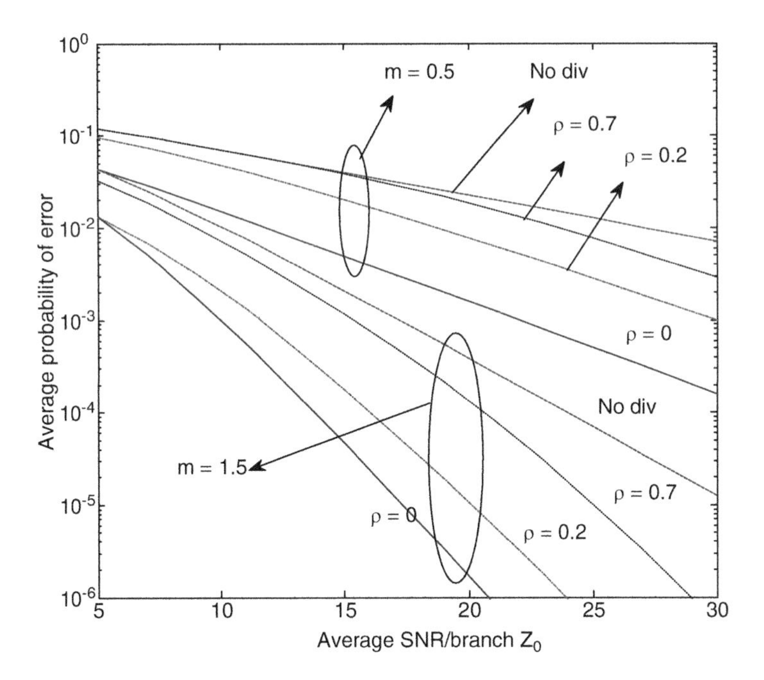

Fig. 5.45 BER in dual branch SC

We will explore the differences between MRC and SC algorithm and their relationship to the GSC algorithm in Fig. 5.46. A four-branch diversity ($M = 4$) is considered where it is assumed that the branches are independent. For the first case, we will also assume that the channel is Rayleigh. As discussed earlier, if all the four branches are selected, we have the conventional MRC algorithm. The curve identified as SC3 refers to the case where the three strongest branches are selected for the MRC algorithm. Similarly, SC2 refers to the case where the two strongest branches are selected for the MRC algorithm. One can see that probability of error values for SC3 is better than those for SC2. The error rates were calculated numerically by triple integration (SC3) and double integration (SC2) with the appropriate density functions given in (5.122). The performance (Fig. 5.46) is bounded by MRC at the better end and SC at the worse end, once again supporting the argument that MRC is the better of the two algorithms between MRC and SC, and SSC is a means to move the performance closer to that of MRC.

Figure 5.47 shows similar results for the case of a Nakagami channel with $m = 1.5$. Once again, the results shown here follow the trends seen in Fig. 5.24, reinforcing the SSC technique that allows the performance to approach that of the MRC algorithm in mitigating short-term fading.

When shadowing is present along with short-term fading, one needs to use diversity at the microlevel (at the base station or at the MU unit as the case may be) and macrodiversity at the macrolevel which involve the use of multiple base stations. Thus, if one only resorts to diversity implementation at the microlevel

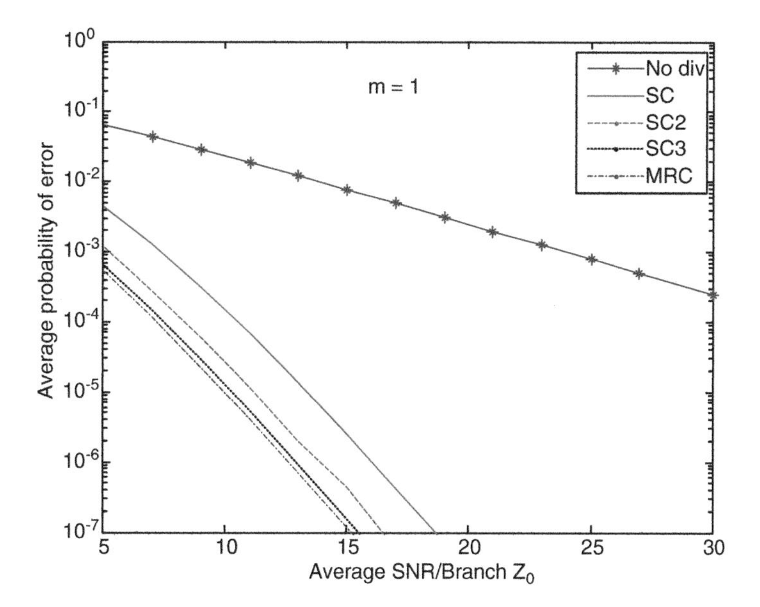

Fig. 5.46 GSC BER for the case of $m = 1$

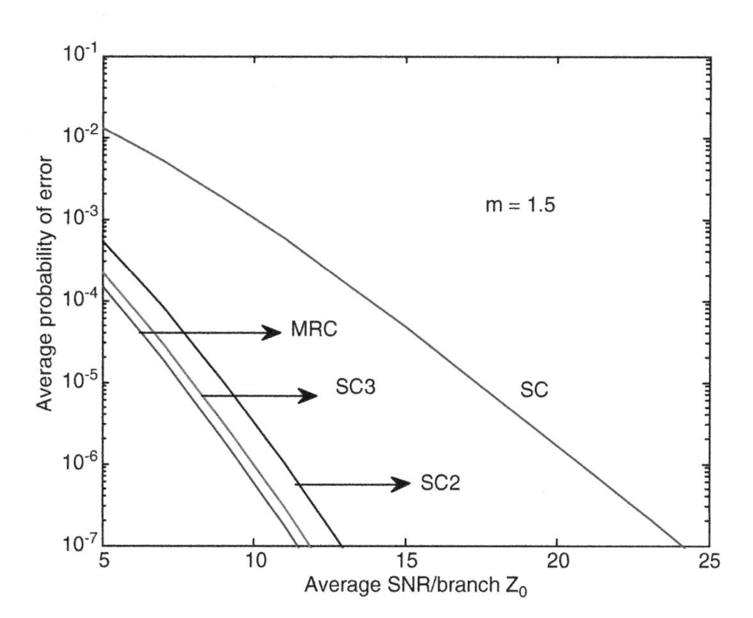

Fig. 5.47 GSC BER for $m = 1.5$

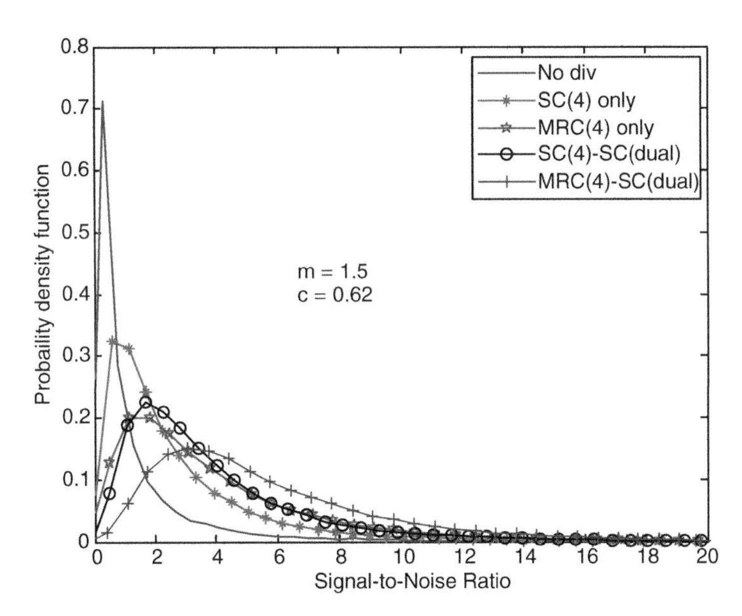

Fig. 5.48 Plots of the density functions following diversity $m = 1.5$ and $c = 0.62$ (shadowing level ~ 8 dB)

(for example MRC), the diversity gain might not be sufficient to overcome the performance degradation as discussed in connection with Fig. 5.27. To demonstrate this, we revisit the simulation of the densities of the SNR for different sets of values of m and c based on the GK model as an approximation to the Nakagami-lognormal model, discussed previously. Let us remember that low values of m correspond to severe shadowing and high values of c correspond to weak shadowing and shadowing vanishing when c approaches infinity. The Nakagami parameter m is taken to be unity and average SNR/channel is taken to be 0 dB. Figure 5.48 shows the densities for the case of $c = 0.62$ (shadowing level ~ 8 dB).

It can be seen that Selection Combining at the microlevel is likely to show only marginal improvement in performance. MRC at the microlevel shows that we can expect higher level of improvement. On the other hand, a hybrid diversity with both microdiversity and macrodiversity is likely to result in the best performance for the case of a four-branch microdiversity followed by a two-branch macrodiversity. Use of SC at both levels certainly provide more improvement than MRC or SC alone at the microlevel. But, it never matches the MRC–SC. Figure 5.49 shows the results for moderate shadowing at a value of $c = 1.13$ (shadowing level ~ 5 dB).

One can see that selection combining starts to produce some improvement as the peaks corresponding to the densities shift to higher SNR values. But, the performance of SC–SC is likely to be almost identical to MRC alone at the microlevel. As expected, MRC–SC is likely to provide the maximum enhancement. The decline in AF can be expected with diversity, with MRC–SC providing the maximum decline as seen in the shifts of the peaks of the densities in Fig. 5.49.

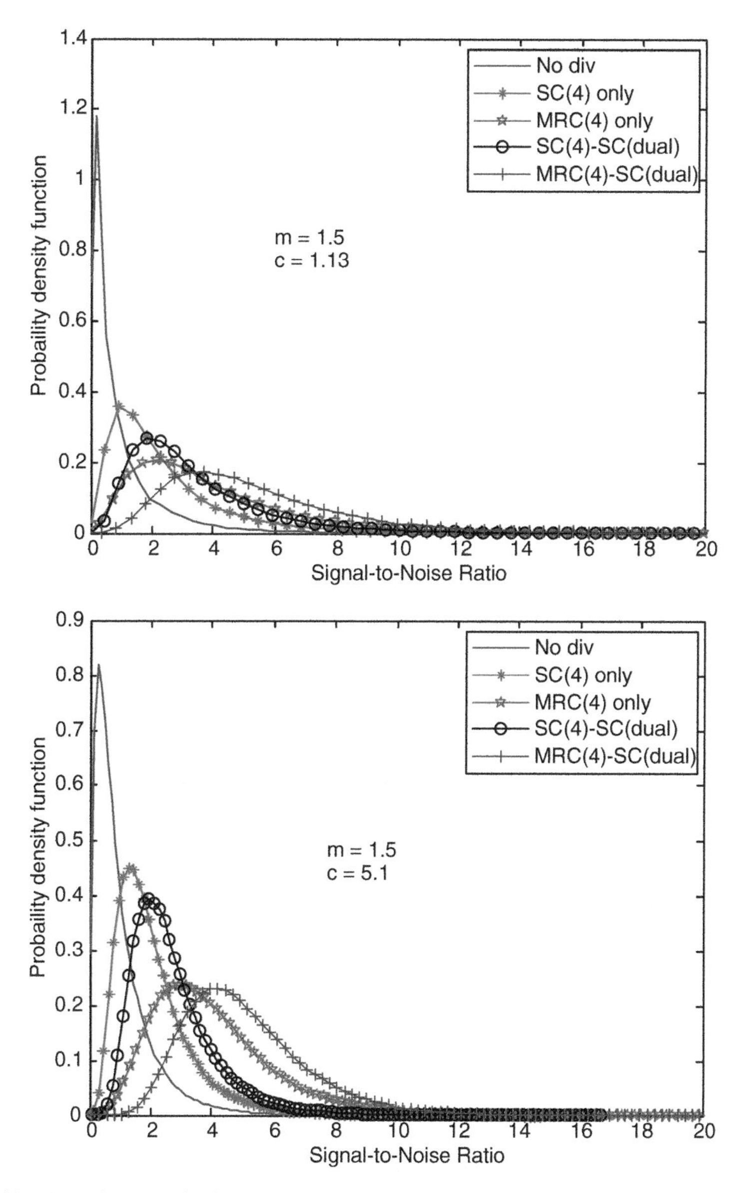

Fig. 5.49 Plots of the density functions following diversity (**a**) $m = 1.5$ and $c = 1.13$ (shadowing level ~ 5 dB), (**b**) $m = 1.5$ and $c = 5.1$ (shadowing level ~ 2 dB)

The density plots for the case of negligible shadowing realized with a value of $c = 5.1$ (shadowing level ~ 2 dB) are shown in Fig. 5.50.

The densities for MRC and MRC–SC are close. The densities for SC and SC–SC are also very close. In addition, we also see the typical performance expected from MRC and SC, with MRC likely to outperform SC significantly as the peak of the

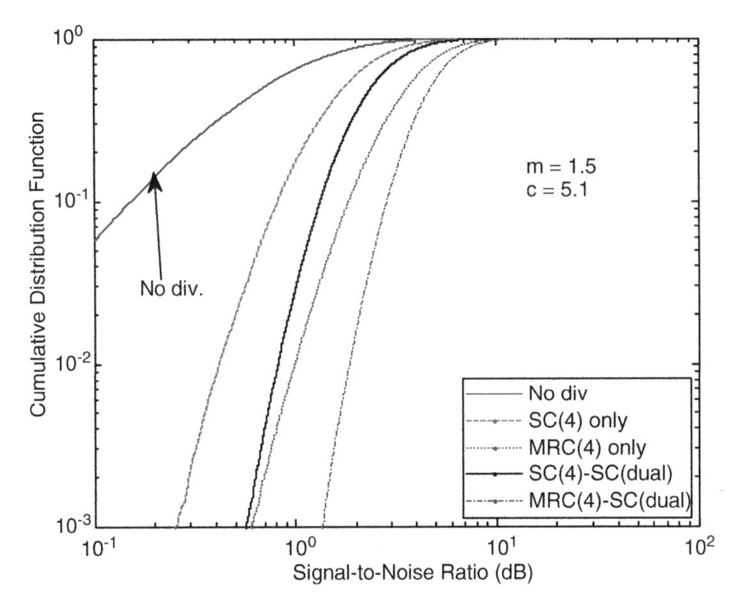

Fig. 5.50 Plots of the density functions following diversity $m = 1.5$ and $c = 5.1$ (shadowing level ~ 2 dB)

density associated with MRC is far removed to higher values of the SNR than the peak of the SC density. The behavior of the channels under these different diversity schemes can be observed slightly clearer with the plots of the CDFs. The CDFs associated with the pdfs in Fig. 5.50 are shown in Fig. 5.51.

We can now look at the implementation of diversity to mitigate the problems in a shadowed fading channel (Shankar 2006, 2008a, b; Bithas et al. 2007a, b). We will do this in two steps, first by looking at the effects of MRC at the base station (or the mobile unit in certain cases) and then using a hybrid diversity approach using an MRC at the base station and SC by using diversity at the macrolevel with the notion of involving multiple base stations. Figure 5.51 contains the plots of the average probability of error in shadowed fading channels when only MRC is implemented at the microlevel. Plots for the case of weak shadowing are also shown. It is clear that in a shadowed fading channel, diversity use at the microlevel alone is not sufficient. As explained earlier, one needs to resort to implementing the diversity at the macrolevel as well. For the results shown here, a gamma pdf was used to model the shadowing (discussed earlier), and shadowing levels were estimated using the equations from Chaps. 2 and 4 which relates the parameters of the gamma pdf and lognormal pdf. In other words, the analysis follows the GK model discussed in Chap. 4 and Sect. 5.5. All the results shown here were obtained numerically.

Figure 5.52 shows the case of a four-branch microdiversity.

The performance of SC and MRC at the microlevel without resorting to SC at the macrolevels can be observed in Fig. 5.53. Results are shown for the case of no diversity. As we had discussed, with the aid of the density functions, the

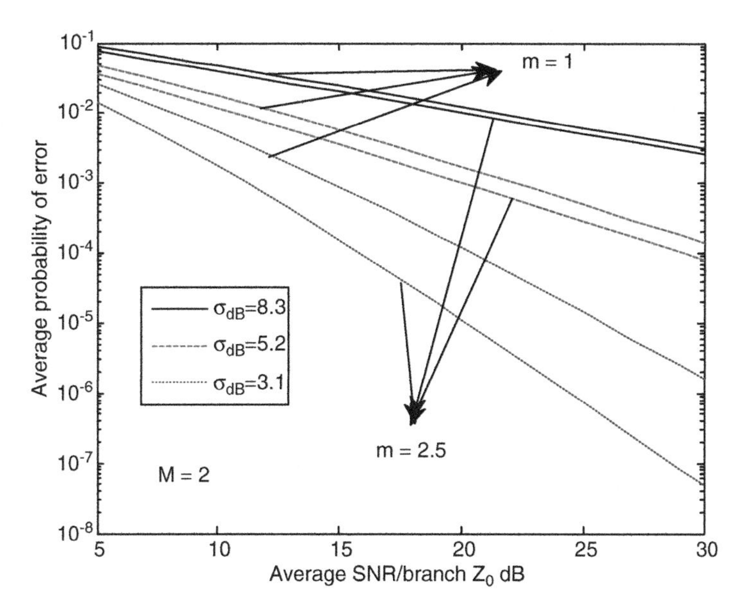

Fig. 5.51 Dual branch MRC in shadowed fading channels for two values of m

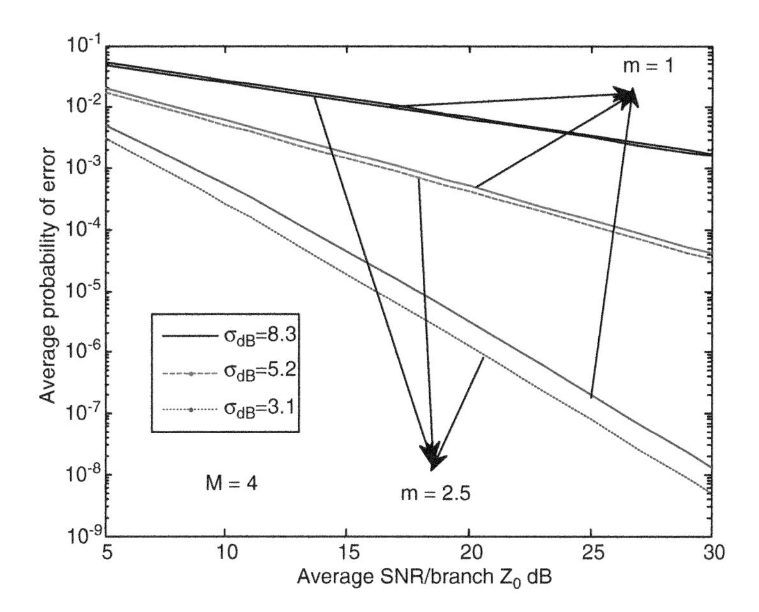

Fig. 5.52 Four channel MRC in shadowed fading channels for two values of m

performance of SC lags behind the performance of MRC and the error rates might still be high making it necessary to have higher power penalties unless macrodiversity is also explored. Figure 5.54 shows the microdiversity results for the case of $m = 2.5$.

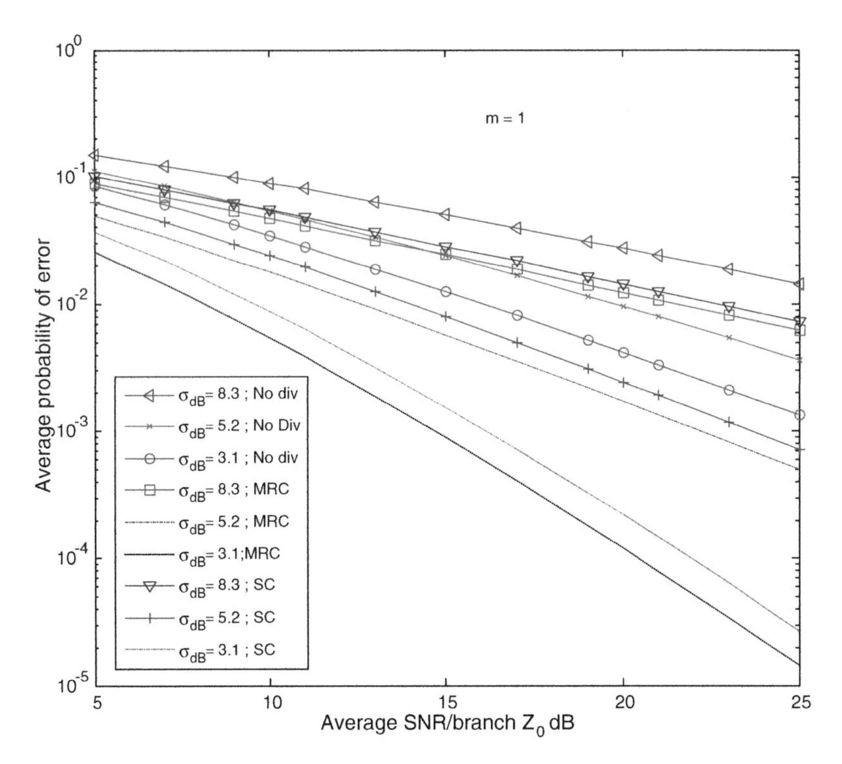

Fig. 5.53 Dual branch MRC is compared with dual branch SC in shadowed fading channels. The case of no diversity is also shown, $M = 1$

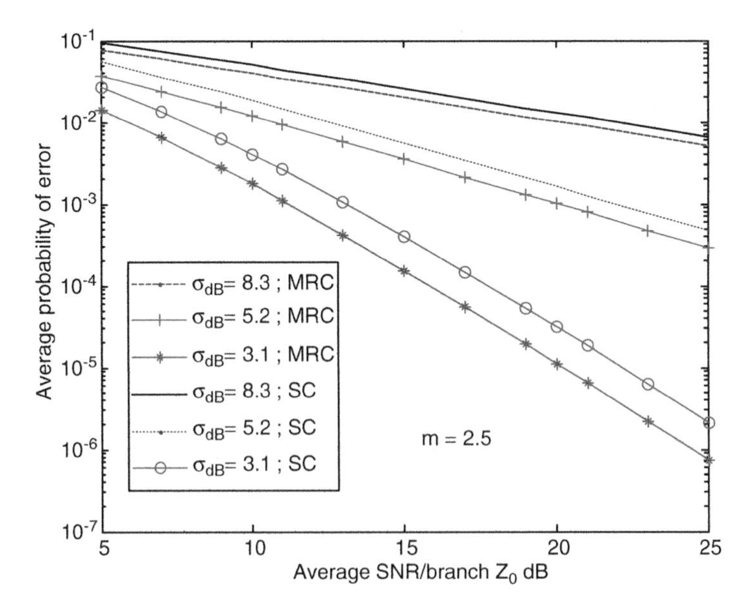

Fig. 5.54 Dual branch MRC is compared with dual branch SC in shadowed fading channels. The case of no diversity is also shown, $m = 2.5$

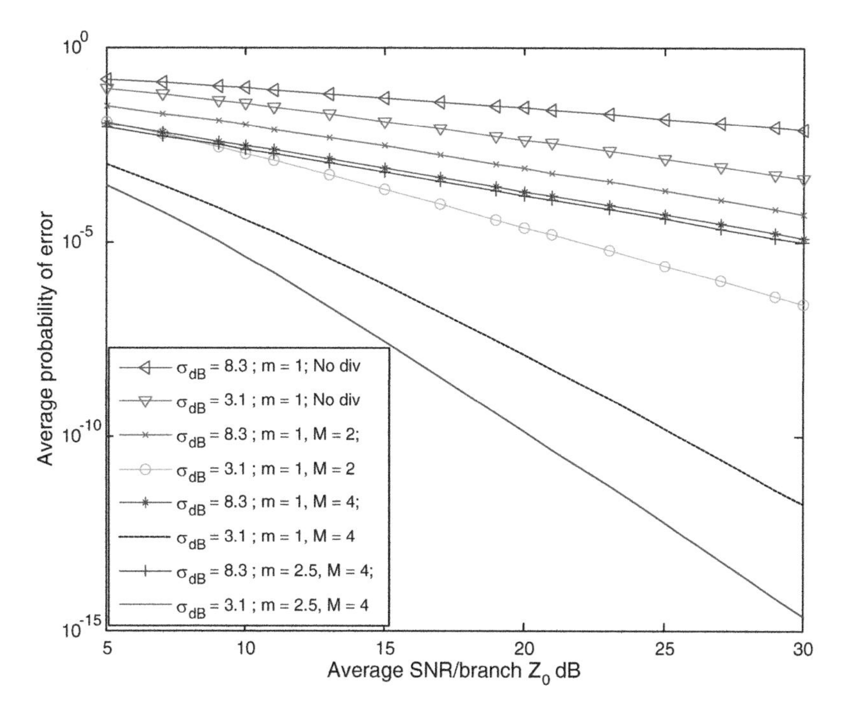

Fig. 5.55 MRC followed by dual SC in shadowed fading channels. The case of no diversity is also shown for $m = 1$. Two values of m are shown

Figure 5.55 shows the plots of the average probability of error when diversity is implemented at the microlevel as well as at the macrolevel. A hybrid diversity scheme is considered with MRC at the microlevel and a dual SC at the macrolevel. The benefits of mitigating the degradation in shadowed fading channels are clearly seen as the error rate curves come down.

Figure 5.56 shows the effects of shadowing levels on the error probabilities in the absence of any diversity at the macrolevel. Let us remember that the parameter c has an inverse relationship to the shadowing levels with small values of c corresponding to strong shadowing and higher values of c corresponding to weak shadowing levels. We can compare these results to those in Fig. 5.57, which shows the error rates as functions of shadowing levels when diversity is implemented at the microlevel (MRC) as well as macrolevel (SC). There is a significant reduction in error rates when hybrid diversity is implemented (Fig. 5.57).

5.7.3 Outage Probability

Another approach to quantify the improvement in performance gained using the diversity techniques involves the estimation of the outage probability before and after the implementation of diversity (Sowerby and Williamson 1992; Tellambura

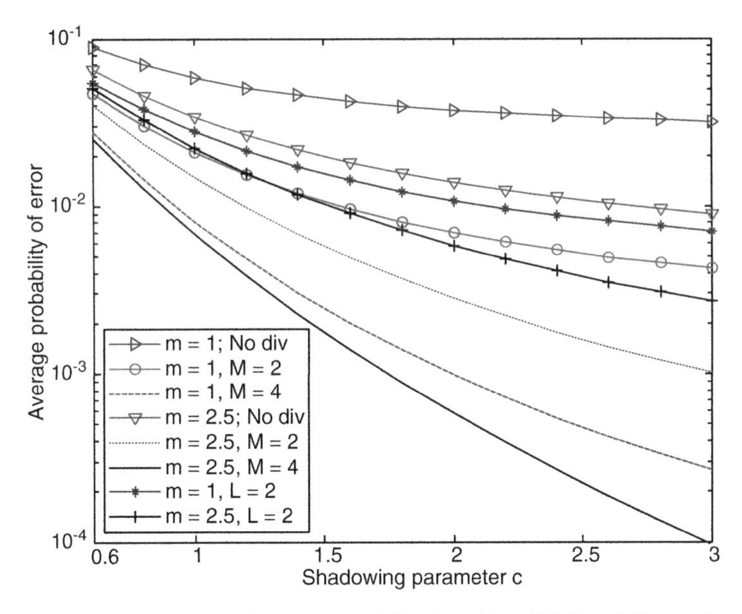

Fig. 5.56 BER in shadowed fading channels following either MRC or SC as a function of shadowing levels

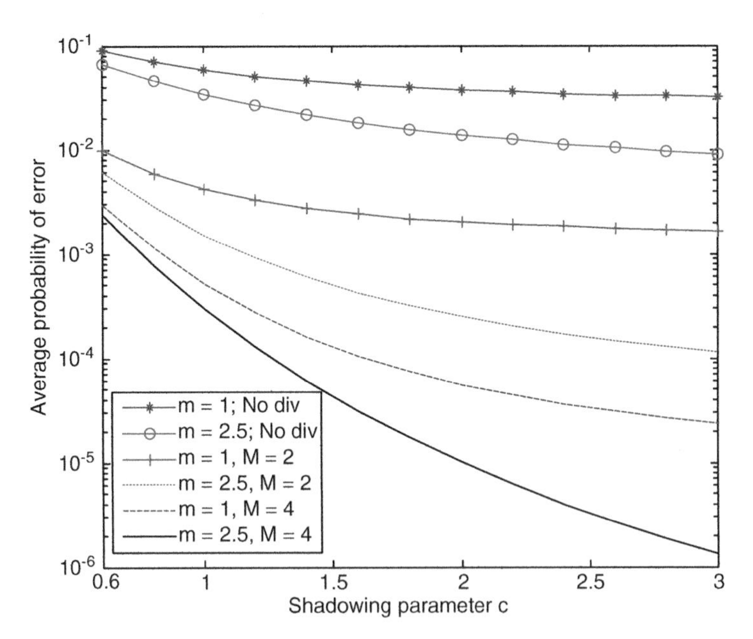

Fig. 5.57 BER in shadowed fading channels following MRC–SC as a function of shadowing levels—only dual SC considered

and Annamalai 1999; Annamalai et al. 2001; Kang et al. 2002; Simon and Alouini 2005). Based on the modem used and the minimum required bit error rate, the sensitivity (minimum SNR required to maintain an acceptable level of the average probability of error) can be calculated. If the minimum SNR needed to maintain acceptable levels of performance is Z_T, the outage probability is given by

$$P_{\text{out}} = \int_0^{Z_T} f(z)\,\mathrm{d}z. \tag{5.160}$$

In (5.160), $f(z)$ is the pdf of the SNR before or after the implementation of diversity. The expression for the outage probability in (5.160) is identical to the CDF evaluated at $z = Z_T$,

$$P_{\text{out}} = F(Z_T) \tag{5.161}$$

where $F(.)$ is the CDF of the SNR before or after the implementation of diversity. Since the choice of Z_T is dictated by the modem, minimum bit error rate tolerated, (5.161) also provides a reasonably complete picture of the effects of fading and shadowing and enhancements in performance obtained by implementing diversity techniques. In a purely short-term faded channel, the outage probability was derived in Chap. 4 and it is given by

$$P_{\text{out}} = \gamma\left(m, \frac{mZ_T}{Z_0}\right)[\Gamma(m)]^{-1}. \tag{5.162}$$

The outage probabilities have been evaluated in a Nakagami channel for the case of a BPSK modem with a threshold bit error rate of 1e-4 in Chap. 4 corresponding to a threshold SNR Z_T of 6.9. We use the same SNR that has been assumed in this chapter for the evaluation of outage probabilities following diversity.

Considering microdiversity first, for the case of the output of the MRC algorithm, the outage probability can be expressed as

$$P_{\text{outMRC}} = \gamma\left(mM, \frac{mZ_T}{Z_0}\right)[\Gamma(mM)]^{-1}, \tag{5.163}$$

where Z_0 is the average SNR/branch at the microlevel and M is the order of microdiversity. The outage probability for the case of the SC algorithm will be

$$P_{\text{outSC}} = \left[\gamma\left(m, \frac{mZ_T}{Z_0}\right)[\Gamma(m)]^{-1}\right]^M. \tag{5.164}$$

Figure 5.58 shows the plot of outage probabilities in a Nakagami faded channel for the case of MRC and SC diversity algorithms. The outage is estimated as a function of the average SNR/branch for two values of the Nakagami parameter

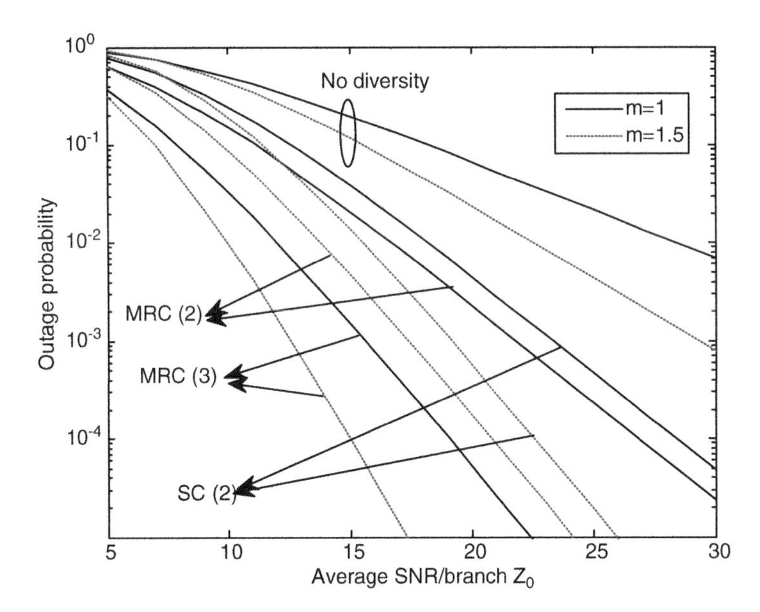

Fig. 5.58 Outage probabilities in Nakagami fading channels following diversity

$m = 1$ and 1.5. One can see the benefits of the diversity algorithms clearly in terms of a decline in outage probabilities with increase in Z_0 as well as M. The higher improvement obtained through MRC over SC is also seen. Comparing the outage probabilities after diversity to the outage with no diversity, the improvements brought on by the diversity in mitigating the effects of diversity are clear.

We will now look at the outage probabilities in shadowed fading channels. The outage probability in a shadowed fading channel modeled using the GK pdf when no diversity is implemented becomes

$$P_{\text{out1}} = \frac{\Gamma(m-c)}{\Gamma(m)\Gamma(c+1)}\, {}_1F_2\left(c, [1-m+c, 1+c], \frac{Z_{\text{T}} b^2}{4}\right)\left(\frac{Z_{\text{T}} b^2}{4}\right)^c + \frac{\Gamma(c-m)}{\Gamma(m+1)\Gamma(c)}\, {}_1F_2\left(m, [1-c+m, 1+m], \frac{Z_{\text{T}} b^2}{4}\right)\left(\frac{Z_{\text{T}} b^2}{4}\right)^m, \quad (5.165)$$

In (5.165), Z_{T} is once again the threshold SNR needed to maintain a specific BER and ${}_1F_2(.)$ is the hypergeometric function (Gradshteyn and Ryzhik 2007). If MRC is applied at the microlevel and no effort is taken to mitigate shadowing, the outage probability conditioned on the existence of shadowing becomes

$$P_{\text{out}}(Z_{\text{T}}|y) = \gamma\left(mM, \frac{mZ_{\text{T}}}{y}\right)[\Gamma(mM)]^{-1}. \quad (5.166)$$

The outage probability in a shadowed fading channel when only MRC is applied at the microlevel can now be written as

$$\begin{aligned}P_{\text{outMRC\_1}} &= \int_0^\infty P_{\text{out}}(Z_{\text{T}}|y) f(y)\,\text{d}y \\ &= \int_0^\infty \gamma\left(mM, \frac{mZ_{\text{T}}}{y}\right)[\Gamma(mM)]^{-1}\frac{y^{c-1}}{y_0^c\Gamma(c)}\exp\left(-\frac{y}{y_0}\right)\text{d}y.\end{aligned} \tag{5.167}$$

This equation can be simplified resulting in

$$\begin{aligned}P_{\text{outMRC\_1}} &= \frac{\Gamma(mM-c)}{\Gamma(mM)\Gamma(c+1)}\,{}_1F_2\left(c, [1-mM+c, 1+c], \frac{Z_{\text{T}}b^2}{4}\right)\left(\frac{Z_{\text{T}}b^2}{4}\right)^c \\ &+ \frac{\Gamma(c-mM)}{\Gamma(mM+1)\Gamma(c)}\,{}_1F_2\left(mM, [1-c+mM, 1+mM], \frac{Z_{\text{T}}b^2}{4}\right) \\ &\times \left(\frac{Z_{\text{T}}b^2}{4}\right)^{mM}.\end{aligned} \tag{5.168}$$

If only selection combining is applied at the microlevel and no mitigation of shadowing is undertaken, the conditional outage probability becomes

$$P_{\text{out}}(Z_{\text{T}}|y) = \left[\gamma\left(m, \frac{Z_{\text{T}}}{y}\right)[\Gamma(m)]^{-1}\right]^M. \tag{5.169}$$

As it was done for the case of MRC, the outage probability now becomes

$$P_{\text{out1\_SC}} = \int_0^\infty \left[\frac{\gamma(m, (mZ_{\text{T}}/y))}{[\Gamma(m)]}\right]^M \frac{y^{c-1}}{y_0^c\Gamma(c)}\exp\left(-\frac{y}{y_0}\right)\text{d}y. \tag{5.170}$$

When MRC is implemented at the microlevel (order M) and SC is implemented at the macrolevel (order L), the outage probability becomes

$$P_{\text{outMRC\_SC}} = (P_{\text{outMRC\_1}})^{\text{L}}. \tag{5.171}$$

For a different number of diversity branches and combining algorithms, the outage probabilities can be evaluated. The outage probabilities have been evaluated for several cases. For all these calculations, two values of the Nakagami parameter were considered, $m = 1$ and $m = 1.5$.

Figure 5.59 shows the outage probabilities in shadowed fading channels when only microdiversity is implemented with no effort being made to mitigate the shadowing. Two cases of shadowing is considered, moderate shadowing ($\sigma_{\text{dB}} = 5$) and heavy shadowing ($\sigma_{\text{dB}} = 9$). We can also probe the effects of shadowing levels on the outage probabilities. These results are shown in Fig. 5.60. One can clearly see the serious impact of shadowing in wireless systems in the high outage

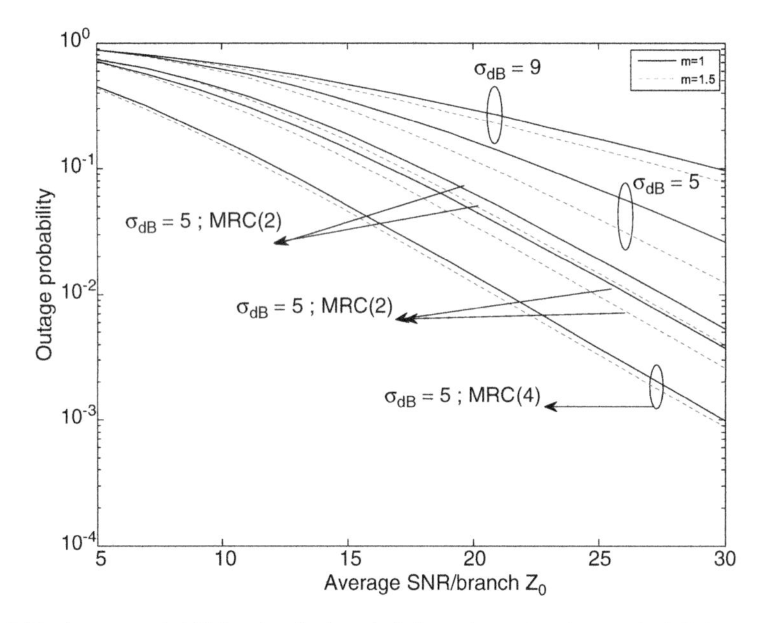

Fig. 5.59 Outage probabilities in shadowed fading channels when only MRC or SC is implemented

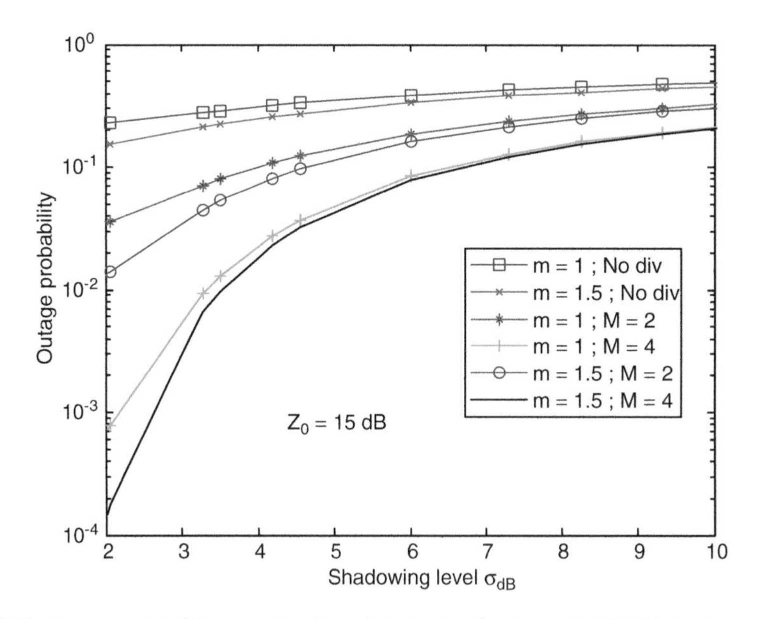

Fig. 5.60 Outage probabilities as a function of shadowing levels—only MRC is implemented

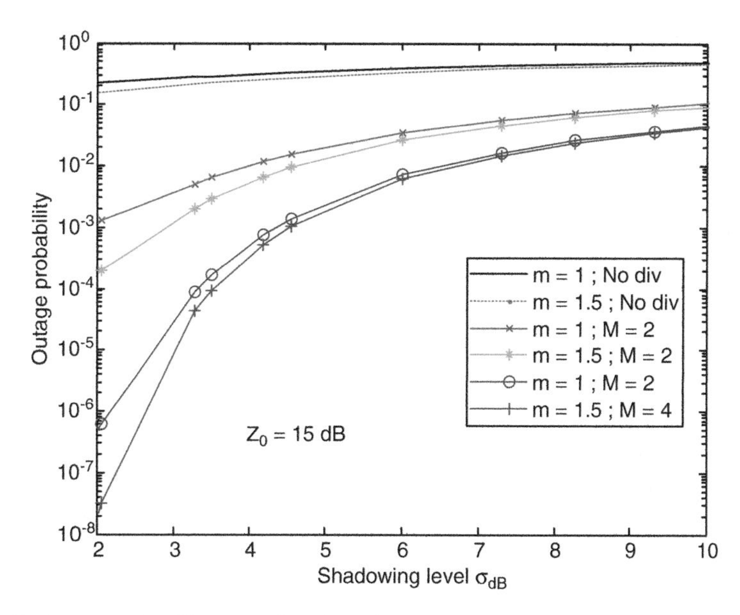

Fig. 5.61 Outage probabilities as a function of shadowing levels when only MRC-dual SC is implemented

probabilities even when diversity is implemented at the macrolevel. This also supports the need for additional fading mitigation through macrodiversity, as we had seen in connection with the error probability estimation earlier in this chapter.

The benefits of having a dual selection combining (macrolevel) along with microdiversity are seen in Fig. 5.61.

5.8 Diversity in Generalized Gamma and Weibull Channels

The presentation as well as discussion so far was limited primarily to short-term faded channels that could be modeled in terms of the Nakagami-*m* distribution and Rayleigh distributions which is a special case of the Nakagami-*m* distribution. However, other models do exist for describing the statistical behavior of fading. These include the generalized gamma (GG), Weibull, double Rayleigh, double Nakagami, and Rician channels. We will briefly look at the effects of diversity in those channels. We will assume that the branches are independent.

We will start with the generalized gamma pdf for the SNR in short-term faded channels. While the Nakagami pdf (or the resulting‘ gamma pdf for the SNR) is a two-parameter distribution, the generalized gamma pdf is a three-parameter distribution which allows more flexibility in modeling statistical changes in the channels. The pdf of the SNR is expressed as

$$f_{\mathrm{GG}}(z) = s\left(\frac{\beta}{Z_0}\right)^{ms}\frac{z^{ms-1}}{\Gamma(m)}\exp\left[-\left(\frac{\beta z}{Z_0}\right)^{s}\right]. \tag{5.172}$$

In (5.172), s is a scaling factor which can take positive values. The parameter β is given by

$$\beta = \frac{\Gamma(m+(1/s))}{\Gamma(m)}. \tag{5.173}$$

By letting s take different values, it is possible to get some of the fading conditions described earlier. For example, if $s = 1$, (5.172) reverts to the case of a Nakagami faded channel and if $m = s = 1$, we have Rayleigh channel. Equation (5.172) is also known as the Stacy's distribution and has been extensively used in radar system analysis for modeling clutter. Note also that (5.172) is one of the several forms of the generalized gamma pdf as mentioned in Chaps. 2 and 4.

We can observe the flexibility of the generalized gamma fading model by estimating the AF. Using the moments of the pdf in (5.172), the AF can be expressed as

$$\mathrm{AF}_{\mathrm{GG}} = \frac{\Gamma(m+(2/s))\Gamma(m)}{[\Gamma(m+(1/s))]^2} - 1. \tag{5.174}$$

By putting $s = 1$ in (5.174), the AF becomes $(1/m)$, the value in a Nakagami faded channel. Equation (5.174) is plotted in Fig. 4.14. It shows that the amount fading is substantially high for low values of s (0.25 and 0.5) and goes down as the value of s exceeds unity (Nakagami channel). The utility of having a three-parameter distribution to model short-term fading is obvious from Fig. 4.1.4.

One of the difficulties with the generalized gamma distribution is the analytical complexities involved in deriving the expressions for the density functions of the SNR at the output of the diversity combining algorithms. We can however explore the likely improvements in the wireless channels following diversity by exploring the densities and distribution functions through random number generators as it was done earlier. Since the GG variable is an exponentially scaled version of the gamma variable, it is easy to generate generalized gamma variables and implement the algorithms.

The probability functions of the output SNR of the diversity combining algorithms for the case of a generalized gamma channel are shown in Fig. 5.62. Note that values of s less than unity make the channel characteristics worse than a Nakagami channel with an identical value of m. This can be observed by comparing the pdfs in Fig. 5.62 to those of the pdfs of the output SNR in Nakagami channels shown in Fig. 5.34. The respective CDFs are shown in Fig. 5.63.

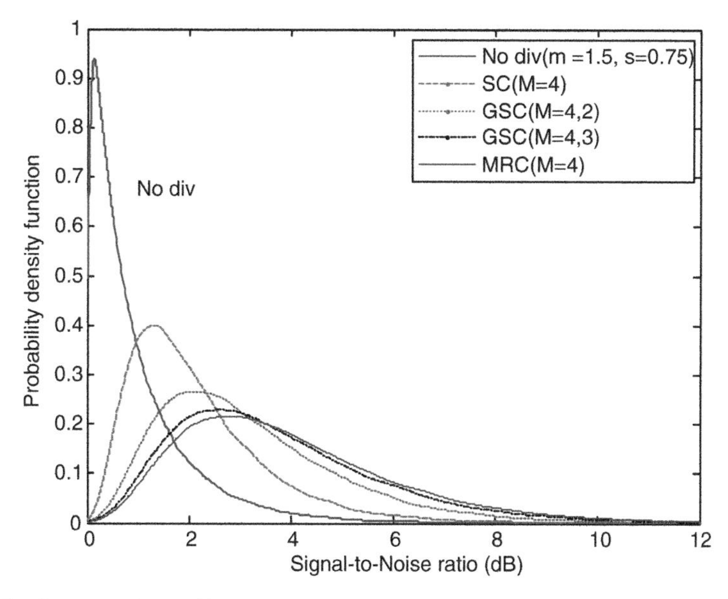

Fig. 5.62 The plots of the pdfs of the SNR for various diversity combining algorithms. Average SNR/branch is unity, $m = 1.5$; $s = 0.75$

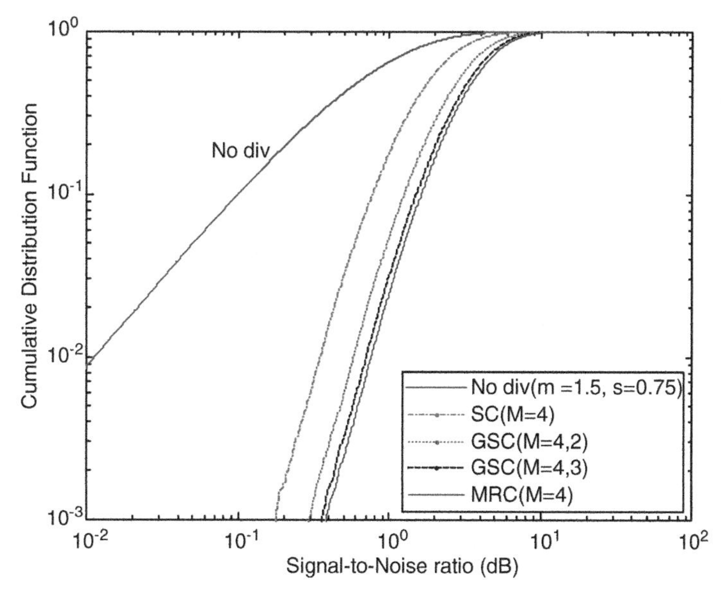

Fig. 5.63 The plots of the CDFs of the SNR for various diversity combining algorithms. Average SNR/branch is unity and $m = 1.5$; $s = 0.75$

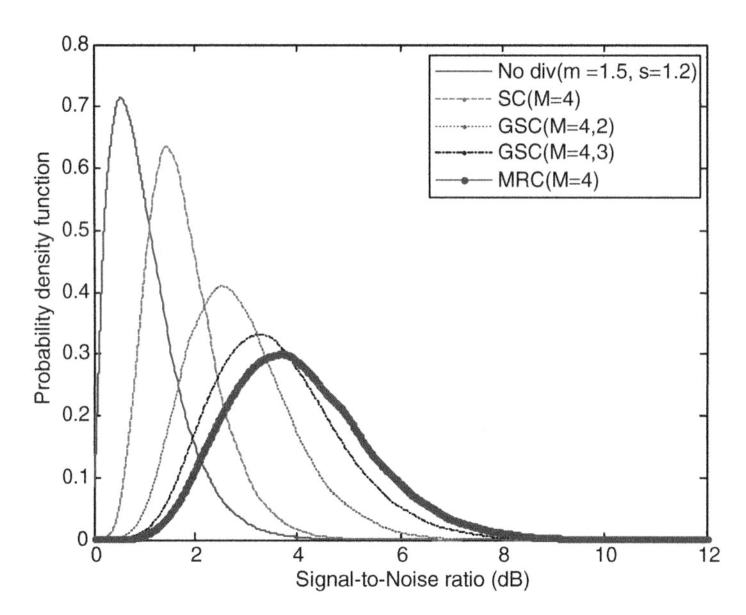

Fig. 5.64 The plots of the pdfs of the SNR for various diversity combining algorithms. Average SNR/branch is unity and $m = 1.5$; $s = 1.2$

One can now compare the CDFs in Fig. 5.63 to the pdfs of the output SNR in a Nakagami channel shown in Fig. 5.35. The contrast between the Nakagami channel and the generalized gamma channel can be further examined by looking at the values of the AF and peaks of the density functions. The corresponding values of the AF were [1.2; 0.42; 0.3; 0.38; 0.30]. These values are higher than the corresponding values for the Nakagami channel.

We will look at the case of a higher value of s. Figure 5.64 shows the plots of the densities for the case of $m = 1.5$ and $s = 1.2$. Comparing the densities in Figs. 5.62 and 5.64, it can be seen that the peaks of the densities have shifted to right indicating that fading mitigation will be better when $s > 1$ compared with the case when $s < 1$. This can also be seen from the values of the peaks of the densities in Fig. 5.64. The corresponding values of the AF were [0.47; 0.16; 0.13; 0.118; 0.116].

The CDFs for the densities in Fig. 5.64 are plotted in Fig. 5.65.

Before we look at the diversity combining techniques, we will examine the degradation in performance in terms of the bit error rates in a generalized gamma fading channel. This was undertaken in Chap. 4 through numerical integration. Here we will use a slightly different approach (Aalo et al. 2005).

Instead of using the case of a coherent BPSK modem, we will now use a more general expression for the probability of error (or the bit error rate) as a function of the SNR z,

$$\mathrm{ber}(z) = \frac{\Gamma(b, az)}{2\Gamma(b)}. \tag{5.175}$$

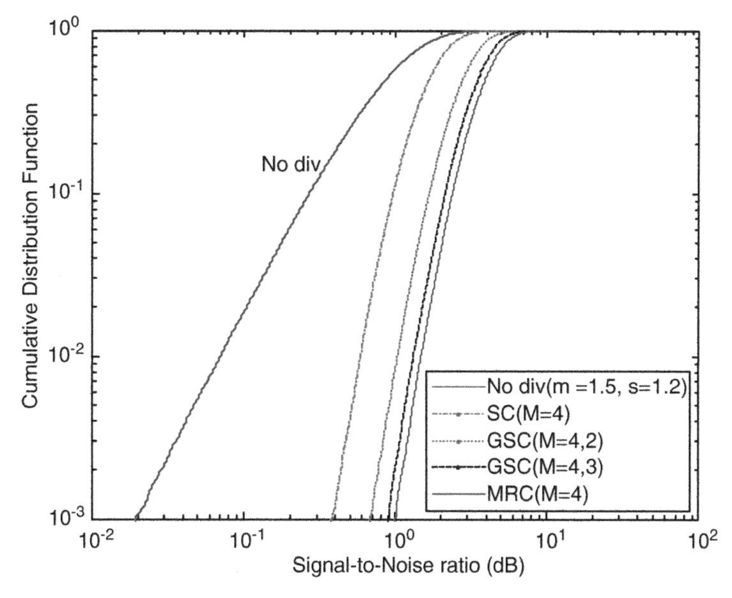

Fig. 5.65 The plots of the CDFs of the SNR for various diversity combining algorithms. Average SNR/branch is unity, $m = 1.5$; $s = 1.2$

By choosing differently specified sets of values of a and b, (5.175) gives the error rates in both coherent and noncoherent modems as mentioned in Chap. 3 (Wojnar 1986). Of these, a accounts for the modulation type (FSK or PSK)

$$a = \begin{cases} \frac{1}{2}, & \text{BFSK(orthogonal)} \\ 1, & \text{BPSK} \end{cases} \tag{5.176}$$

and b accounts for the detection scheme (coherent or noncoherent)

$$b = \begin{cases} \frac{1}{2}, & \text{Coherent,} \\ 1, & \text{Non-coherent.} \end{cases} \tag{5.177}$$

The function, $\Gamma(b,az)$ is the complimentary incomplete gamma function given by

$$\Gamma(b, az) = \int_{az}^{\infty} x^{b-1} \exp(-x)\, \mathrm{d}x, \tag{5.178}$$

The average probability of error in a generalized gamma-fading channel becomes

$$p_{\mathrm{eGG}} = \int_{az}^{\infty} \left[\frac{\Gamma(b, az)}{2\Gamma(b)}\right] s \left(\frac{\beta}{Z_0}\right)^{ms} \frac{z^{ms-1}}{\Gamma(m)} \exp\left[-\left(\frac{\beta_z}{Z_0}\right)^{s}\right] \mathrm{d}z. \tag{5.179}$$

Equation (5.179) can be expressed in closed form for the special case of

$$s = l/k \tag{5.180}$$

where both l and k are integers as (Aalo et al. 2005; Simon and Alouini 2005)

$$p_{\mathrm{eGG}} = A(k,l) G_{2l,k+l}^{k,2l}\left[\left(\frac{l\beta}{aZ_0 k^{k/l}}\right)^l \middle| \begin{matrix} c_1, \ldots, c_{2l} \\ d_1, \ldots c_{k+1} \end{matrix}\right]. \tag{5.181}$$

In (5.181),

$$\mathrm{A}(k,l) = \frac{l^{b-(1/2)} k^{m-(1/2)} (2\pi)^{(2-l-k)/2}}{2\Gamma(b)\Gamma(m)}, \tag{5.182}$$

$$c_n = \begin{cases} 1 - \dfrac{n+b-1}{l}, & n = 1, 2, \ldots l, \\ 1 - \dfrac{n-l-1}{l}, & n = l+1, l+2, \ldots, 2l, \end{cases} \tag{5.183}$$

$$d_n = \begin{cases} \dfrac{n+m-1}{k}, & n = 1, 2, \ldots, k, \\ 1 - \dfrac{n-k}{l}, & n = k+1, k+2, \ldots k+l, \end{cases} \tag{5.184}$$

and $G(.)$ is the MeijerG function (Gradshteyn and Ryzhik 2007). If (5.180) is not met, one needs to use numerical integration to evaluate the average probability of error as it was done earlier in Chap. 4. To demonstrate the versatility of the generalized gamma fading, the average error probability was evaluated for a few cases of m and s (BPSK, $a = 1$, $b = 1/2$). The results are shown in Fig. 5.66.

We can now explore the implementation of diversity in a generalized gamma fading channel. We will limit ourselves to the case of independent identically distributed diversity branches and selection combining and MRC algorithms. Using the approach described in the previous section, the pdf of the output SNR of an M branch selection combiner can be expressed as

$$f_{\mathrm{SCGG}}(z) = M[F(z)]^{M-1} f(z). \tag{5.185}$$

In (5.185), $F(.)$ is the cumulative distribution of the SNR with a pdf of $f(z)$ given in (5.172). It can be easily expressed as

$$F(z) = \frac{1}{\Gamma(m)} \gamma\left[m, \left(\frac{z\beta}{Z_0}\right)^s\right], \tag{5.186}$$

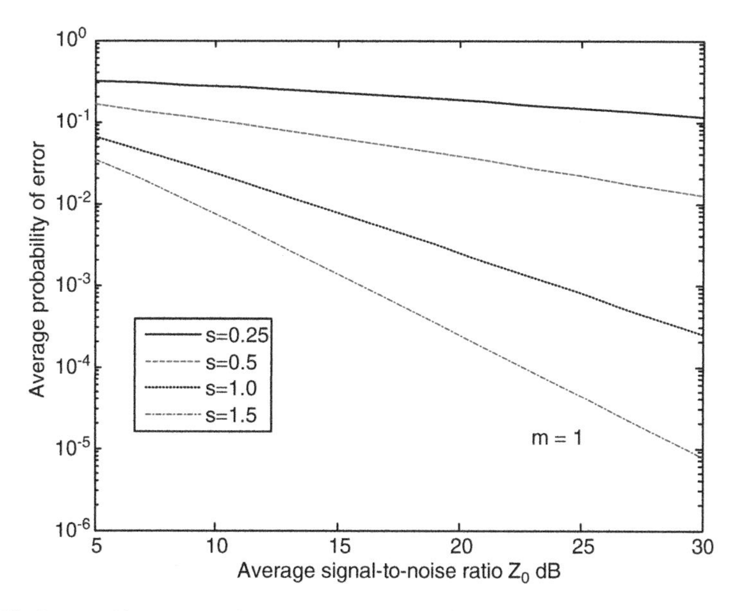

Fig. 5.66 Average bit error rates in a generalized gamma fading channel for $m = 1$. Four values of s are considered

where $\gamma(.,.)$ is the incomplete gamma function (Abramowitz and Segun 1972; Gradshteyn and Ryzhik 2007) Using (5.186), the pdf of the SNR at the output of a selection combiner in a generalized gamma-faded channel becomes

$$f_{\mathrm{SCGG}}(z) = M\left[\frac{1}{\Gamma(m)}\right]^{M-1}\gamma\left[m,\left(\frac{z\beta}{Z_0}\right)^{s}\right]^{M-1} s\left(\frac{\beta}{Z_0}\right)^{ms}\frac{z^{ms-1}}{\Gamma(m)}\exp\left[-\left(\frac{\beta_z}{Z_0}\right)^{s}\right]. \tag{5.187}$$

The average probability of error can be evaluated using (5.144). It is also possible to write the expression for the average probability of error by first estimating the MGF and evaluating the error probability as in (5.152). The MGF can be expressed in terms of MeijerG functions (Mathai 1993; Mathai and Saxena 1973). Note that one would have to resort to numerical integration regardless of the approach used. The results here were obtained using the direct evaluation done in Maple for the case of a dual branch selection combiner in generalized gamma channel.

We will also look at the case of a dual diversity when MRC algorithm is used. The pdf of the SNR at the output of the MRC combiner can easily be written in terms of the pdf (5.172), knowing that the pdf of the sum of two random variables is the convolution of the marginal pdfs.

The expression for the pdf becomes

$$f_{\mathrm{MRCGG}}(z) = \int_0^z \left(\frac{\beta}{Z_0}\right)^{2ms} \frac{[x(z-x)]^{ms-1}}{(\Gamma(m))^2} \exp\left[-\left(\frac{\beta}{Z_0}\right)^{ms} (x^s - (z-x)^s)\right] \mathrm{d}x. \tag{5.188}$$

Even though an analytical solution to (5.188) is unavailable, one can still evaluate the average probability of error for a dual MRC case using this equation. The average probability of error can be written as

$$p_{\mathrm{eMRCGG}} = \int_0^\infty \frac{1}{2}\mathrm{erfc}\sqrt{2}\left[\int_0^z s^2\left(\frac{\beta}{Z_0}\right)^{2ms} \frac{[x(z-x)]^{ms-1}}{(\Gamma(m))^2} \exp\left[-\left(\frac{\beta}{Z_0}\right)^{ms} (x^s - (z-x)^s)\right] \mathrm{d}x\right] \mathrm{d}z. \tag{5.189}$$

Equation (5.189) can be evaluated in MAPLE or MATLAB. Results obtained using MAPLE are shown in Fig. 5.67 for two values of m (1 and 1.5) and various values of s (0.25, 0.5, 1, 1.5). Use of (5.189) allows one to observe the effects of lower values of s on the worsening error probabilities since (5.189) can be evaluated for all values of (m,s). Results for the case of $m = 1.5$ are shown in Fig. 5.68.

As indicated, one can obtain an analytical expression for the MGF of the SNR of the MRC output (Aalo et al. 2005, 2007). The MGF for MRC diversity (M branch) is given by

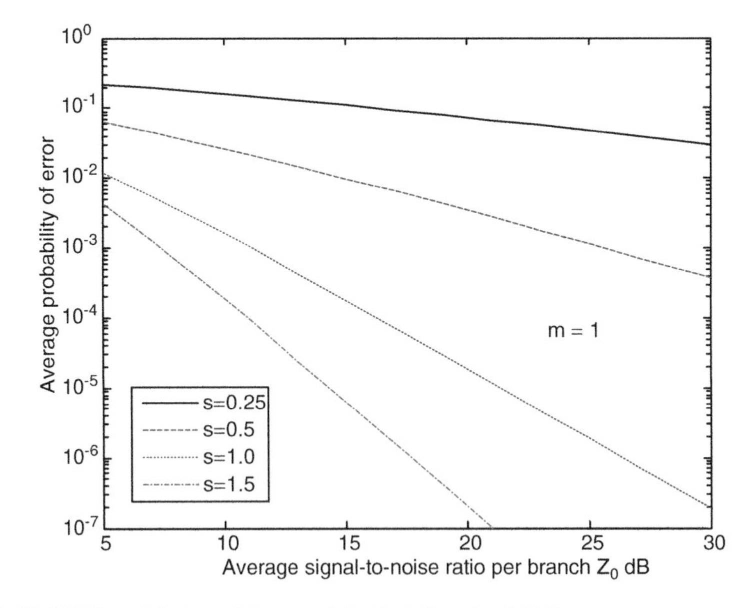

Fig. 5.67 BER in a GG channel for $m = 1$ for dual diversity (MRC)

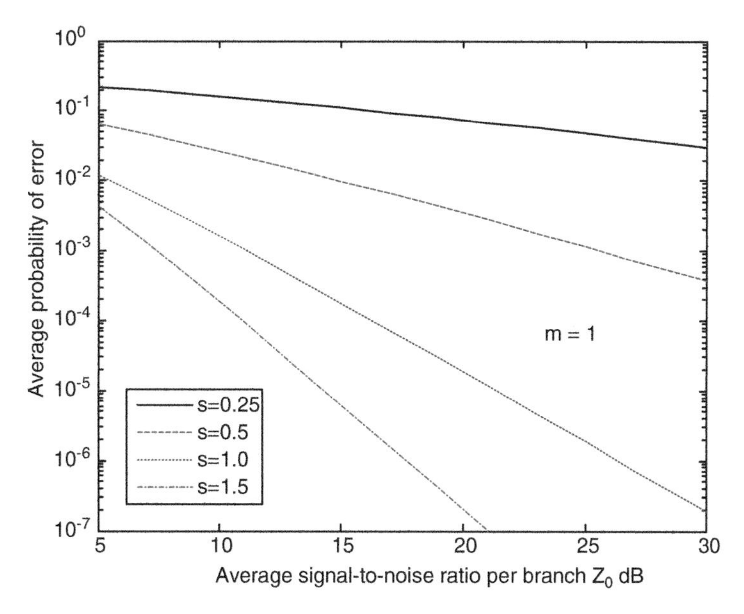

Fig. 5.68 BER in a GG channel for $m = 1.5$ for dual diversity (MRC)

$$M(s)=\prod_{j=1}^{M}\frac{k_j^m\sqrt{l_j/k_j}}{\Gamma(m_j)}(2\pi)^{\left(2-l_j-k_j\right)/2}G_{k_j,l_j}^{l_j,k_j}\left[\left(\frac{sZ_0}{\beta_j}\right)^{l_j}k_j^{k_j}\middle|\begin{matrix}1-\frac{m_j}{k_j},1-\frac{m_{j+1}}{k_j},\ldots,1-\frac{k_j+m_j+1}{k_j}\\0,\frac{1}{l_j},\ldots,\frac{l_j-1}{l_j}\end{matrix}\right].\tag{5.190}$$

In (5.190), $G[\ ]$ is the MeijerG function. Using (5.153) the average probability of error can be determined. One would still require the use of numerical integration techniques to estimate the average probability of error.

The effects of dual selection combining in a generalized gamma-fading channel are shown in Fig. 5.69 ($m = 1$) and Fig. 5.70 ($m = 1.5$).

The average probability of error in Weibull channels could be obtained from the results of the generalized gamma channel. A generalized gamma channel with $m = 1$ is a Weibull channel and, hence, the results shown in Fig. 5.69 for the case of $m = 1$ for different values of s (equal to β) provides the quantitative measures of a Weibull channel.

While generalized gamma (GG) and Weibull distributions provide a broad approach to a description of fading channels over the simple Nakagami channels, yet another form of modeling is coming into existence based on cascaded channels. In this approach, the received signal can be treated as a result of multiple bounces/ scattering such that it is a product of the individual responses. Such models also represent a more realistic condition of the present day wireless systems where low power transmitters are used. Use of low power transmitters would require the

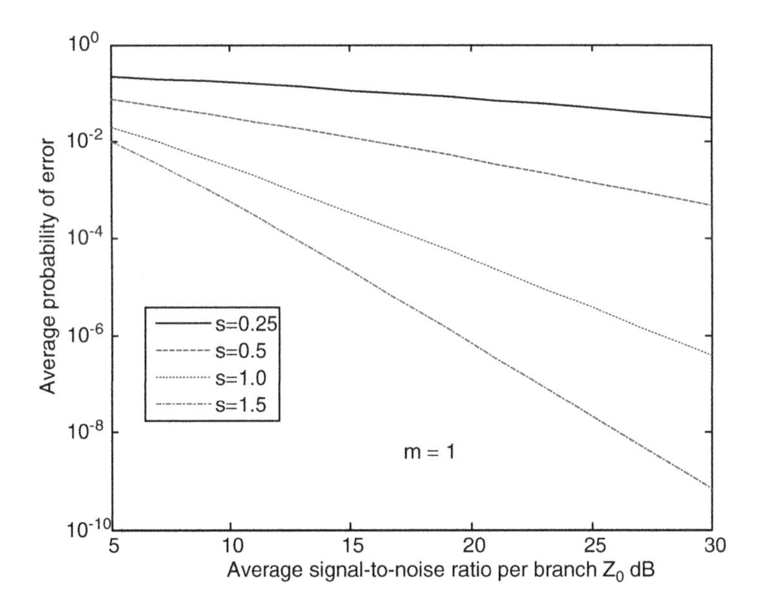

Fig. 5.69 BER in a GG channel for $m = 1$. Dual SC

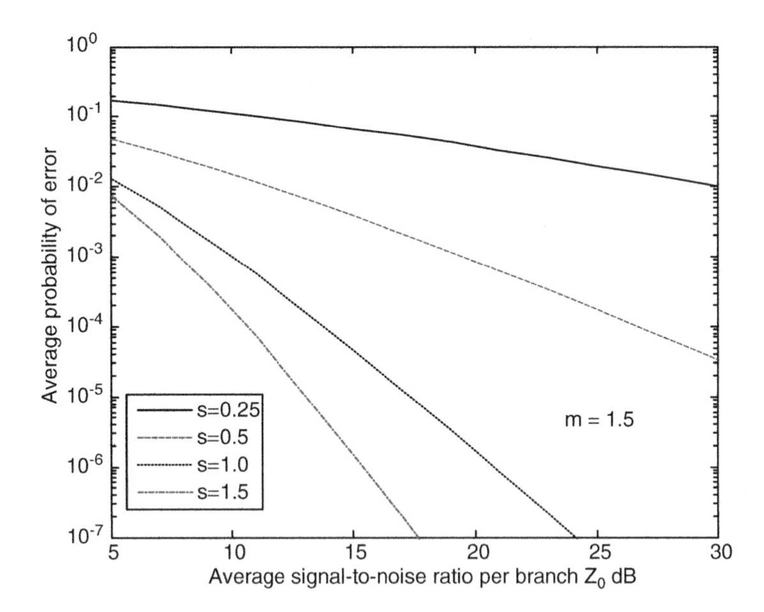

Fig. 5.70 BER in a GG channel for $m = 1.5$. Dual SC

stationing of several intermediate transponder stations (multihop relayed communication systems), thus, producing a multiplicative effect. Such effects are also sometimes referred to as "keyhole" scattering where the received signal is a result of passage through several keyholes, simulation multiplication conditions.

5.9 Diversity in Cascaded Nakagami Channels

We will now examine the implementation of diversity in cascaded N*Nakagami channels We will briefly restate a few equations from Chap. 4 for completeness. If the overall fading in the channel is the result of N multiple scattering components, the received signal power Z of the cascaded channel can be expressed as the product of N gamma distributed variables (Karagiannidis et al. 2009; Shankar 2011a, b)

$$Z = \prod_{k=1}^{N} Z_k. \tag{5.191}$$

In the absence of any cascading effects (i.e., $N = 1$), (5.191) suggests that conditions still exist for fading to occur; this will be the result of multipath fading described in Chap. 4. The density function of the SNR can now be rewritten in terms of the average SNR as

$$f_N(z) = \frac{1}{z\Gamma^N(m)} G_{0,N}^{N,0}\left(\frac{m^N}{Z_0} z \middle| \underbrace{m, m, \ldots, m}_{N-\text{terms}}.\right), \quad N = 1, 2, \ldots \tag{5.192}$$

An insight into the pdf in (5.192) and its effects on the wireless channel can be seen from plotting the pdf in (5.192) for a few values of N as it was done in Chap. 4. It is seen that as N increases, the pdf moves closer and closer to lower values of the SNR; this can be interpreted as leading to degradation in performance as N increases. This deterioration in the wireless channel performance can also be seen if we examine the AF in a cascaded channel. The AF in a cascaded channel can be written as

$$\mathrm{AF}_N \frac{\langle Z \rangle^2}{\langle Z \rangle^2} - 1 = \frac{(m^2 + m)^N}{m^2 N} - 1 = \left(1 + \frac{1}{m}\right)^N - 1. \tag{5.193}$$

It is observed from (5.193) that as N increases, the AF also goes up. As the peaks of the pdf shifts towards lower SNR values, the amount fading goes up, thus demonstrating the serious consequences of the existence of cascaded fading in the channel.

The CDF was obtained in Chap. 4 as

$$F_N(z) = \int_0^z f_N(\xi)\,\mathrm{d}\xi = \frac{1}{\Gamma^N(m)} G_{1,N+1}^{N,1}\left(\frac{m^N}{Z_0} z \,\middle|\, \begin{matrix} 1 \\ \underbrace{m, m, \ldots m, 0}_{(N+1)\text{terms}}. \end{matrix} \right), \quad N = 1, 2, \ldots. \tag{5.194}$$

Before we examine the analytical nature of the densities and distribution functions following diversity in a cascaded channel, it is worthwhile to have a better understanding of the expectations of mitigation that can be gained through diversity by undertaking computer simulations using random number generators as was done with the Nakagami-m channel. Also, the analytical expressions for the density functions and cumulative distributions might not be available in closed form for all the diversity combining algorithms and hence, a reasonable understanding could be gained through simulation. The steps involved in the simulation are similar to those described earlier. All three main diversity combining algorithms can be implemented in Matlab. A cascaded SNR variable can be created by multiplying N gamma variables. The study of such simulation is shown in a few figures next. The density functions of the SNR for a dual cascaded Nakagami channel ($m = 1.5$) are shown in Fig. 5.71 with the number of diversity branches $M = 4$. It has been assumed that average SNR/branch is unity. The specific example of SC, MRC, GSC (4,2), and GSC(4,3) are shown.

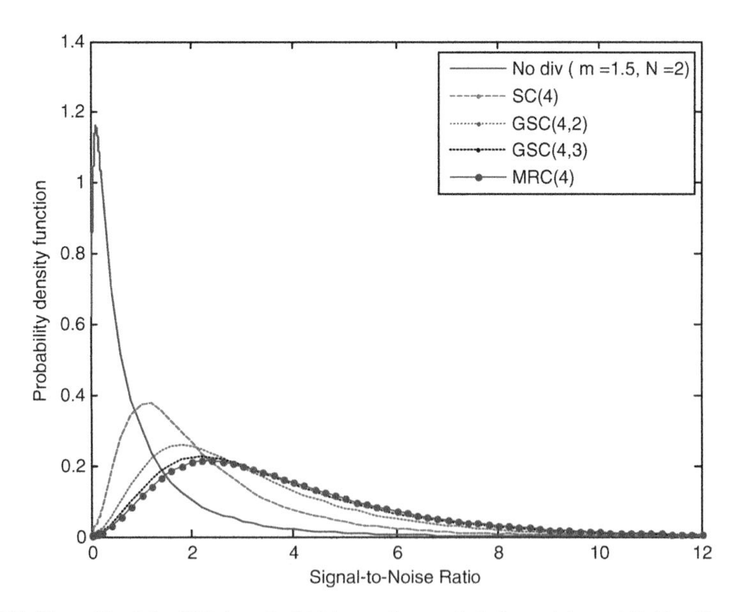

Fig. 5.71 The pdfs of the SNR in a dual Nakagami cascaded channel ($m = 1.5$; $N = 2$). Average SNR/branch is unity

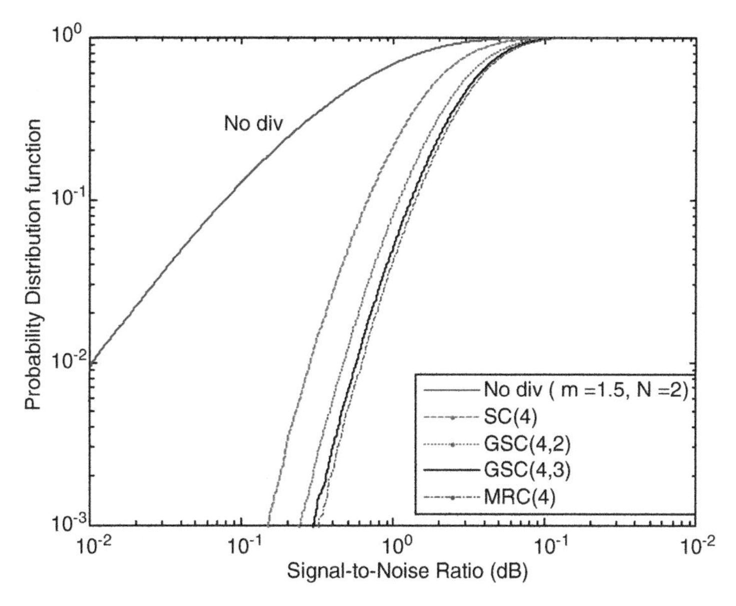

Fig. 5.72 The CDFs of the SNR in a dual Nakagami cascaded channel ($m = 1.5$; $N = 2$). Average SNR/branch is unity

One can see that the density functions of the SNR following diversity have their peaks shifting to the higher values of the SNR and demonstrating likely improvement in channel conditions as one goes from SC to MRC, with the GSC falling in between MRC and SC, with GSC(4,3) approaching the density function for the MRC.

Figure 5.72 shows the plots of the CDFs corresponding to the densities in Fig. 5.71. The proximity of the CDFs for GSC(4,3) and MRC follows the behavior seen in Fig. 5.71. The likelihood of lower outages following diversity can be surmised from the plots of the CDF.

The density functions for $N = 3$ are shown in Fig. 5.73. The corresponding CDFs are shown in Fig. 5.74.

The densities for $N = 4$ are shown in Fig. 5.75. The corresponding CDFs are shown in Fig. 5.76.

We can now consider the simple case of a two-branch selection combining diversity algorithm. We will look at the simple case where the branches are independent and identical, each with the pdf and CDF of the forms expressed in (5.192) and (5.194), respectively. Using the concepts of order statistics, the pdf of the SNR of the dual SC algorithm is given by the product of the CDF and pdf scaled by two. Thus, the pdf of the SNR of the output of the dual selection combining diversity can be expressed as (Papoulis and Pillai 2002; Shankar 2011a, b)

$$F_{\text{N\_SC}} = \left\{ \frac{1}{\Gamma^N(m)} G_{1,N+1}^{N,1} \left(\frac{m^N}{Z_0} z \middle| \begin{matrix} 1 \\ m, m, \ldots m, 0 \end{matrix} \right) \right\}^2, \quad N = 1, 2, \ldots, \tag{5.195}$$

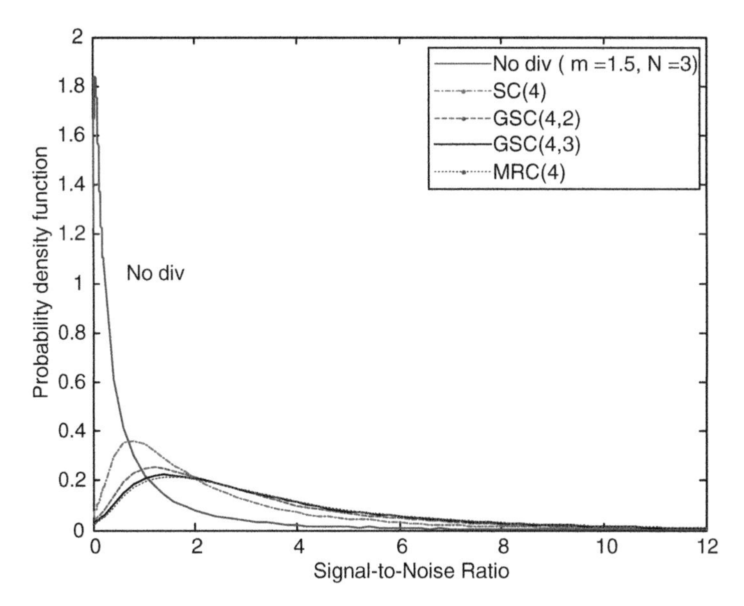

Fig. 5.73 The pdfs of the SNR in a triple Nakagami cascaded channel ($m = 1.5$; $N = 3$). Average SNR/branch is unity

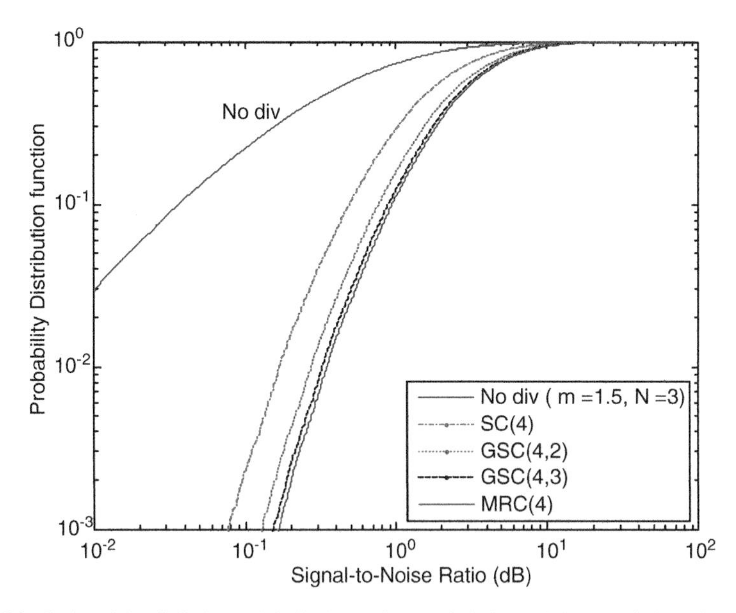

Fig. 5.74 CDFs of the SNR in a triple Nakagami cascaded channel ($m = 1.5$; $N = 3$). Average SNR/branch is unity

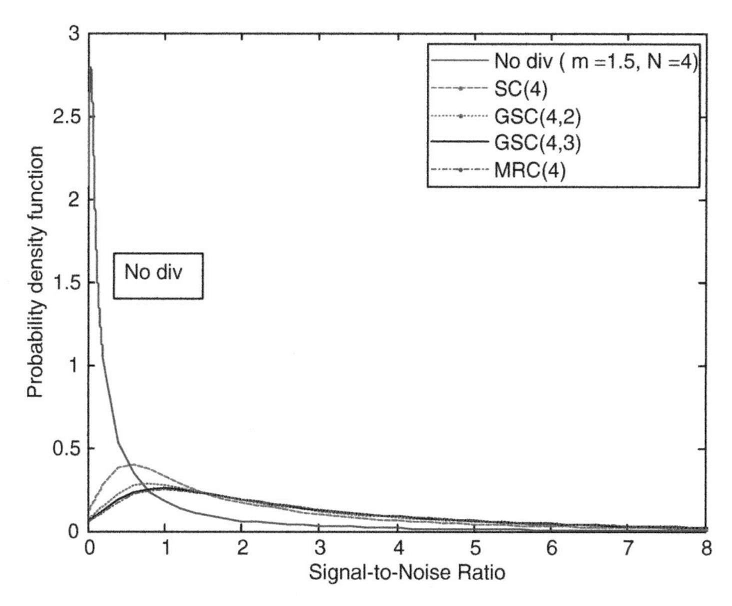

Fig. 5.75 The pdfs of the SNR in a quadruple Nakagami cascaded channel ($m = 1.5$; $N = 4$). Average SNR/branch is unity

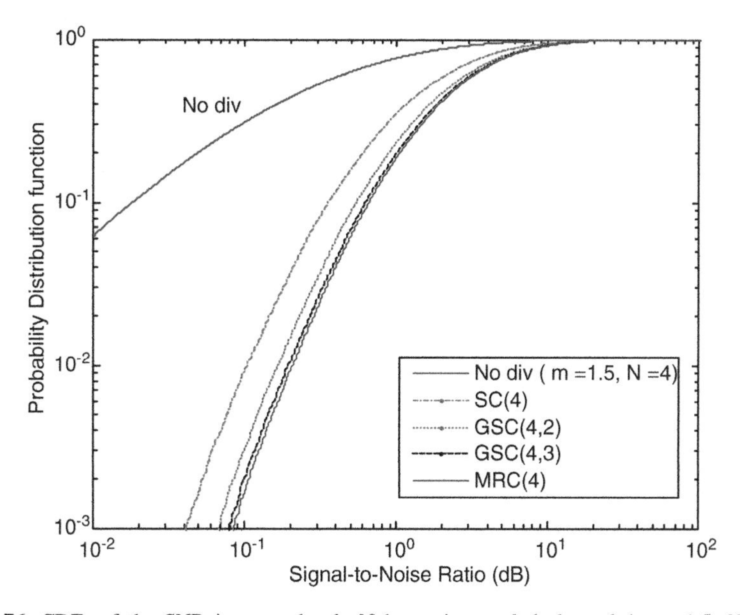

Fig. 5.76 CDFs of the SNR in a quadruple Nakagami cascaded channel ($m = 1.5$; $N = 4$). Average SNR/branch is unity

$$f_{\mathrm{N\_SC}} = \frac{2}{z\Gamma^N(m)} G_{0,N}^{N,0}\left(\frac{m^N}{Z_0}z\middle| m,m,\ldots,m\right)\frac{1}{\Gamma^N(m)} \times G_{1,N+1}^{N,1}\left(\frac{m^N}{Z_0}z\middle|\begin{matrix}1\\ m,m,\ldots,m,0\end{matrix}\right),\quad N=1,2,\ldots \tag{5.196}$$

With the availability of the pdf of the SNR at the output of the selection combiner, we can examine the performance of the channel before and after the implementation of the selection combining algorithm.

The AF can be estimated from moments. The result is shown in Fig. 5.77 where AF is plotted as a function of the Nakagami parameter and the number of cascading components. The reduction in the amount of fading following diversity implementation can be seen.

The CDFs obtained from (5.195) are plotted in Fig. 5.78 for $m = 1.5$. The improvement in performance expected to be gained from diversity can be seen from the change in the shape of the CDFs.

The pdfs are plotted in Fig. 5.79 for the case of dual SC with $m = 1.5$. One can compare these plots to those from Chap. 4. It can be observed that with diversity, the peaks of the density functions move to higher SNR values.

We can now study the error rate performances for cascaded systems with and without diversity.

Figure 5.80 shows the results on average probabilities of error ($N = 4$) when a dual selection combining is implemented. The benefits of diversity are clearly seen in terms of reduction in error rates.

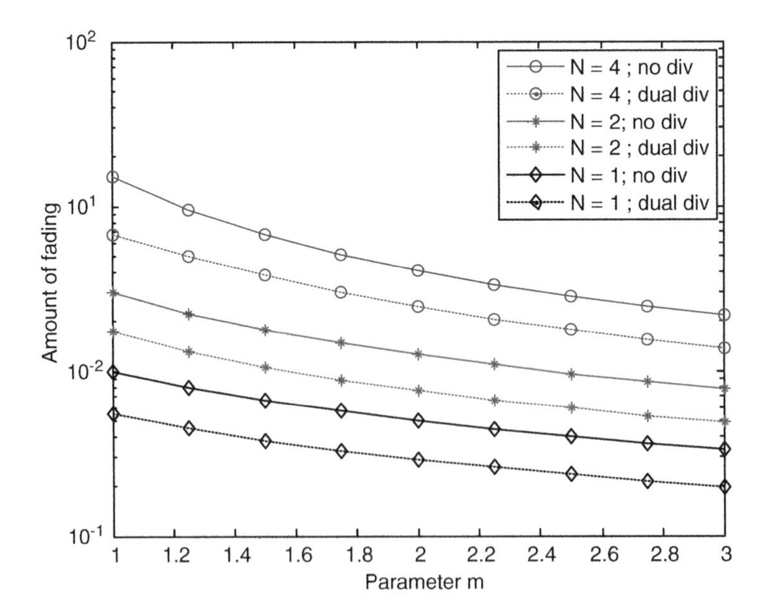

Fig. 5.77 Amount of fading is plotted as a function of the parameter m for $N = 1$, 3, and 4 with dual selection combining diversity and no diversity

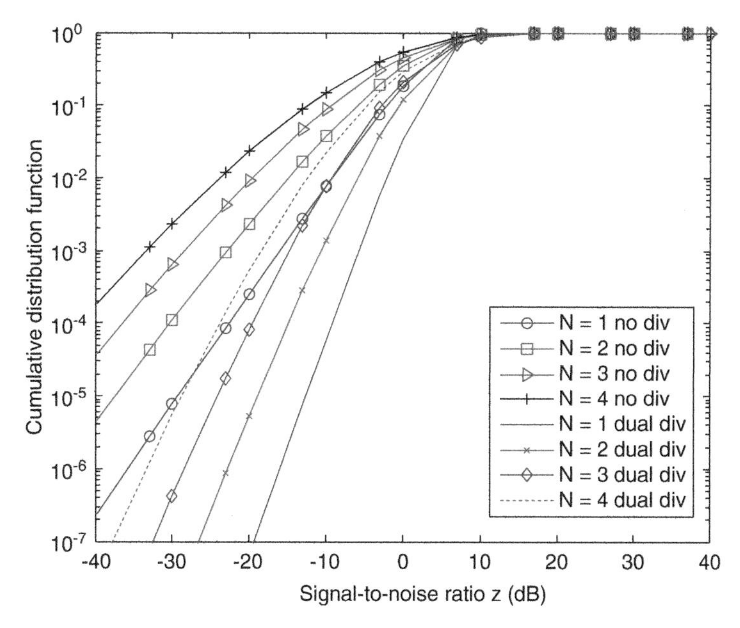

Fig. 5.78 The CDFs of the SNR for cascaded channels in the absence of diversity as well as for the case of dual selection combining diversity for $m = 1.5$ and $Z_0 = 5$ dB

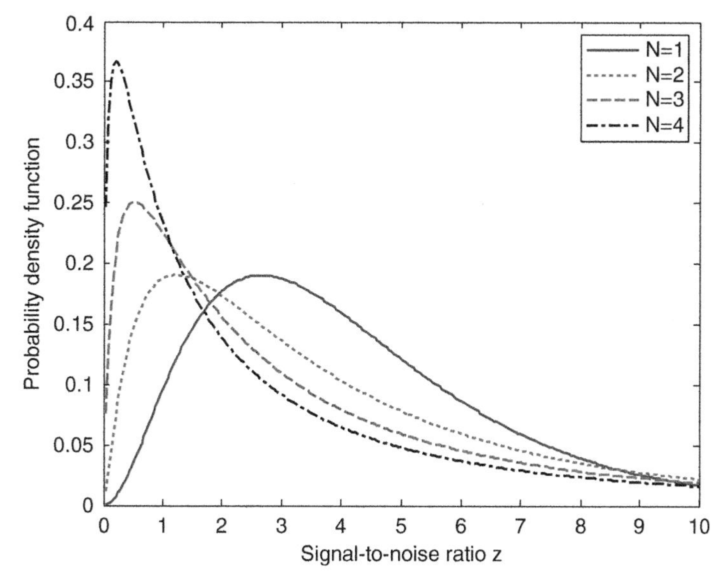

Fig. 5.79 The pdf of the SNR in cascaded channels for the case of $m = 1.5$ and $Z_0 = 5$ dB when dual selection combining diversity is implemented

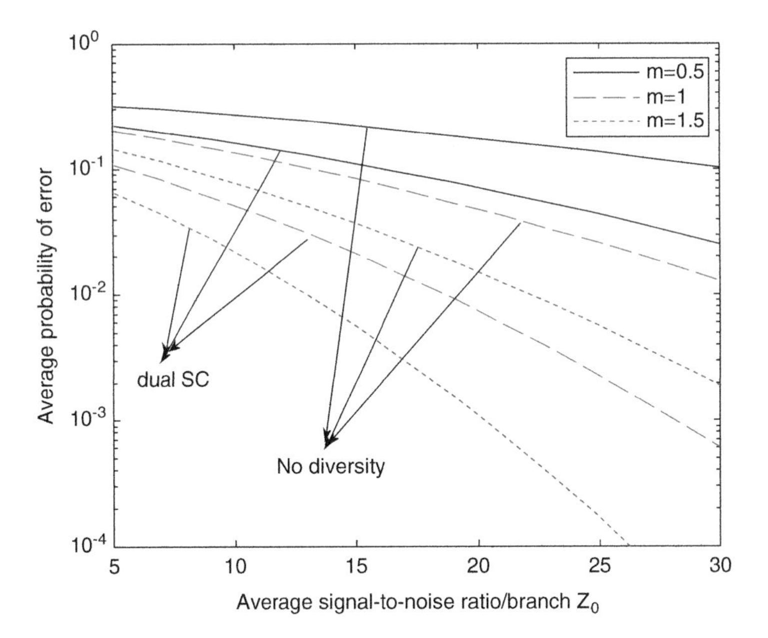

Fig. 5.80 BER in a cascaded channel ($N = 4$) dual SC

The outage probabilities following diversity can now be easily estimated to examine the level of improvement gained. For a fixed error rate (1e-4), the threshold SNR (Z_{T}) can be calculated in an ideal channel. Using the definition of outage probability, the expression for the outage probability in a cascaded channel becomes

$$P_{\mathrm{out}N} = [F_{\mathrm{N}}(Z_{\mathrm{T}})]^2 = \left\{ \frac{1}{\Gamma^N(m)} G^{N,1}_{1,N+1}\left(\frac{m^N}{Z_0} Z_{\mathrm{T}} \middle| \begin{matrix} 1 \\ m, m, \ldots, m, 0 \end{matrix} \right) \right\}^2. \tag{5.197}$$

Outage probabilities for triple and quadruple cascading are plotted in Fig. 5.81. Comparing these results to the outage probabilities obtained in Chap. 4 in the absence of any diversity.

The selection combining offers a simple means of improving the channel characteristics. However, as we had seen earlier, the enhancement will show further gains if we implement the maximal ratio combining algorithm.

We will now look at the case of MRC for fading mitigation in cascaded channels (Malhotra et al. 2008; Wongtraitrat and Supnithi 2009; Shankar 2011a). We will assume that there are M diversity branches and that they are independent and identically distributed. Therefore, the SNR Z at the output of the MRC combiner will be the sum of the individual SNRs, namely,

$$Z = \sum_{q=1}^{M} Z_{N_q}. \tag{5.198}$$

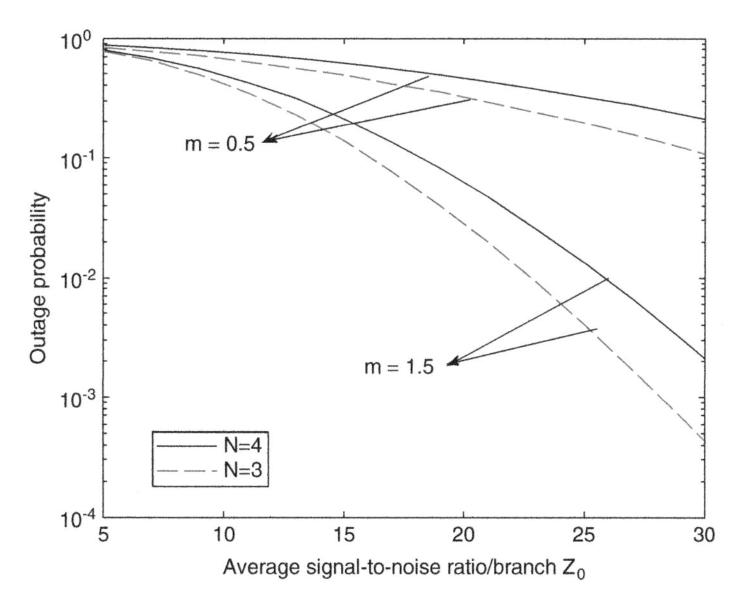

Fig. 5.81 Outage probabilities in dual SC in triple and quadruple cascaded channels

As we had seen in Chap. 2, we would require an M-fold convolution of expressions of the pdf of the form in (5.192) to obtain the pdf of Z. Even though we could estimate the pdf through random number simulation as shown earlier, obtaining analytical expressions for the pdf of the output of the MRC is not an easy task because of the M-fold convolution of MeijerG function involved. Instead, we can use the MGFs and characteristic functions (CHF) to obtain expressions for the pdf and CDF of Z as explained earlier. Before we look at ways of estimating the average probability of error and outage probability, we can look at the AF in a cascaded Nakagami channel and the reduction in the AF realized through diversity, since it can be estimated from the moments of the SNRs of the individual branches.

The AF following the implementation of MRC diversity can be obtained from the moments of the gamma distribution and it becomes

$$\mathrm{AF}_{\mathrm{MRC}} = \frac{1}{M}\left[\left(\frac{m+1}{m}\right)^{N} - 1\right], \tag{5.199}$$

demonstrating an M-fold decline in the level of fading in cascaded channels following the MRC algorithm.

Since we do not have an expression for the pdf of the SNR at the output of the maximal ratio combiner, we will use the MGFs to estimate the average bit error rate. We will consider the case of a coherent BPSK modem as an example.

MGF of the cascaded SNR has been derived, and the analytical expression for the MGF associated with the SNR Z_N in any one of the branches of diversity

($q = 1,2,\ldots M$) is given by (Mathai and Saxena 1973; Mathai 1993; Adamchik 1995; Karagiannidis et al. 2007)

$$\psi_{Z_N}(s) = \frac{1}{\Gamma^N(m)} G_{1,N}^{N,1}\left(-\frac{m^N}{sZ_0}\middle| \begin{matrix} 1 \\ m, m, \ldots, m \end{matrix}\right). \tag{5.200}$$

Noting that the MGF of the sum of the random variables is the product of the individual MGFs, the average error probability at the output of the maximal ratio combiner in a N*Nakagami channel is given by

$$p_e = \frac{1}{\pi}\int_0^{\pi/2}\left\{\psi_{Z_N}\left[-\frac{1}{\sin^2(x)}\right]\right\}^M dx. \tag{5.201}$$

Here we have made use of the relationship between MGF and average error probability in a fading channel, discussed earlier. Equation (5.201) can now be expressed explicitly using the MGF of each branch given in (5.200) as

$$p_e = \frac{1}{\pi}\int_0^{\pi/2} \frac{1}{\Gamma^{NM}(m)} G_{1,N}^{N,1}\left(\frac{m^N}{Z_0}\sin^2(x)\middle| \begin{matrix} 1 \\ m, m, \ldots, m \end{matrix}\right)^M dx \tag{5.202}$$

Equation (5.202) can be estimated numerically.

Estimation of the outage probability requires access to the cumulative distribution function (CDF) of the output SNR following the MRC algorithm. Once again, an analytical expression for the CDF of the output SNR following diversity is not available. However, it is possible CHF to estimate the outage probability through the CHF. Outage occurs when the instantaneous SNR fails to reach a threshold SNR required to maintain an acceptable bit error rate. If the threshold SNR is Z_T, the outage probability becomes

$$P_{out} = F(Z_T). \tag{5.203}$$

In (5.203), $F()$ is the CDF of the output in (5.198) of the maximal ratio combiner. Using the approach suggested by Gil-Pelaez, we can write an expression for the CDF as (Gil-Pelaez 1951; Nuttall 1969; Beaulieu 1990)

$$F_Z(z) \approx \frac{1}{2} - \frac{1}{\pi}\int_0^{\infty} \frac{1}{\omega}\Im\{\phi[-j\omega]\exp[-jz\omega]\}\, d\omega. \tag{5.204}$$

Note that in (5.204), $\Im\{\,\}$ denotes the imaginary component of the argument and $\phi()$ is the CHF of the output SNR. Note that the CHF is nothing but the MGF in (5.200) with s replaced by $(-j\omega)$. The CDF in (5.204) can be evaluated using trapezoidal rule (Abramowitz and Segun 1972), and the CDF becomes

$$F_Z(z) \approx \frac{1}{2} - \frac{2}{\pi \Gamma^{NM}(m)} \sum_{n=0}^{\infty} \frac{1}{u} \Im \left\{ G_M \exp\left[-ju\omega_0 \left(\frac{z}{Z_0} \right) \right] \right\}. \tag{5.205}$$

In (5.205), the parameter ω_0 is related to T which is the limiting value of the SNR beyond which the CDF is negligible (Beaulieu 1990; Tellambura et al. 2003)

$$\omega_0 = \frac{2\pi}{T}. \tag{5.206}$$

The other parameters in (5.205) are

$$u = 2n + 1, \tag{5.207}$$

$$G_M = G_{1,N}^{N,1} \left(\frac{jm^N}{\omega_0 u} \middle| \begin{matrix} 1 \\ m, m, \ldots, m \end{matrix} \right)^M. \tag{5.208}$$

As shown by several researchers, even though the evaluation of (5.205) appears to suggest the need for infinite summation, only a finite number of terms are needed. Note that a higher value of T would result in larger number of terms (Beaulieu 1990).

A typical CDF obtained through this approach is shown in Fig. 5.82. The slopes of the CDF plots clearly indicate that outage probabilities will be less with the implementation of the diversity algorithm.

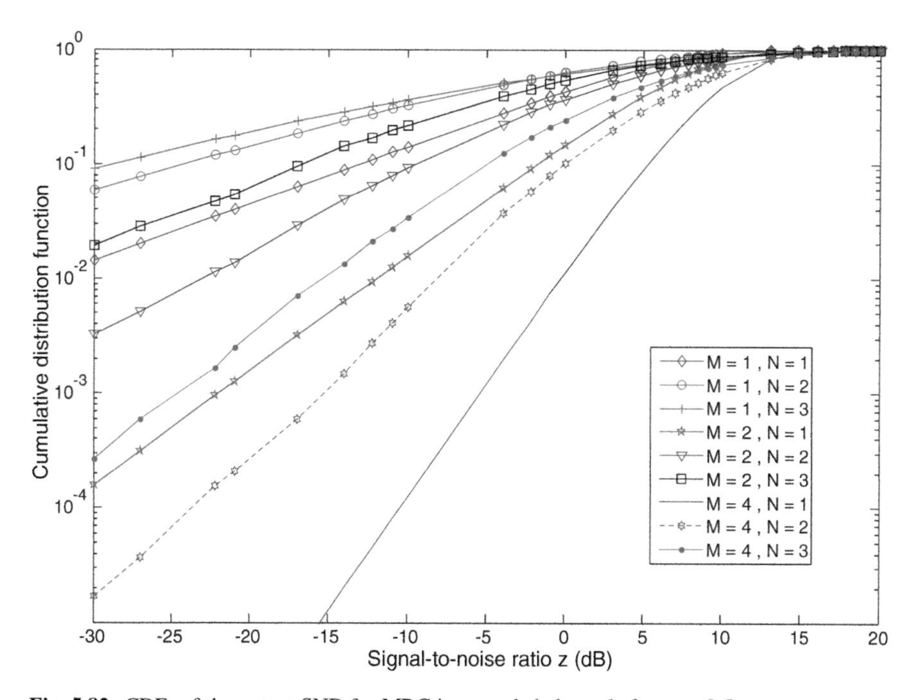

Fig. 5.82 CDFs of the output SNR for MRC in cascaded channels for $m = 0.5$

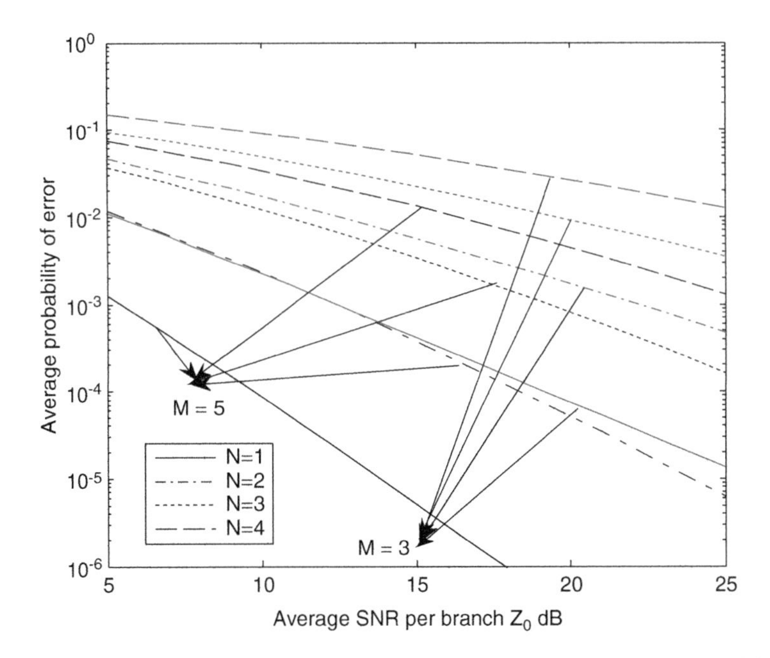

Fig. 5.83 Average probability of error values plotted as a function of the average SNR/branch Z_0 for three values of M (3 and 5) for the case of $m = 0.5$. The error probabilities decline substantially as M goes from 3 to 5

The performance of the cascaded N*Nakagami channel can now be explored in terms of three measures: AF, average probability of error, and outage probability. It is clear from (5.193) and (5.199) that the AF decreases as the number of diversity branches, M, increases.

The error rates can be obtained from (5.202) through numerical integration. One such result is shown in Fig. 5.83 which shows the improvement in bit error rate performance of the cascaded channel following the MRC algorithm ($m = 0.5$). As one would expect, error probabilities are higher with increasing values of N, clearly indicating worsening channel conditions with N (Shankar 2011a). As the number of diversity branches increase, the error rates come down.

The outage probabilities can be similarly evaluated for the case of a coherent BPSK receiver to have an acceptable error rate of 1e-4 which provided the values of the threshold SNR Z_T. To reduce the number of computations necessary, the CDF was normalized to the average SNR. The (5.203) can therefore be written explicitly in terms of the CDF as

$$P_{\text{out}} \approx \frac{1}{2} - \frac{2}{\pi \Gamma^{NM}(m)} \sum_{n=0}^{\infty} \frac{1}{u} \Im\left\{ G_M \exp\left[-ju\omega_0 \left(\frac{Z_T}{Z_0} \right) \right] \right\}, \tag{5.209}$$

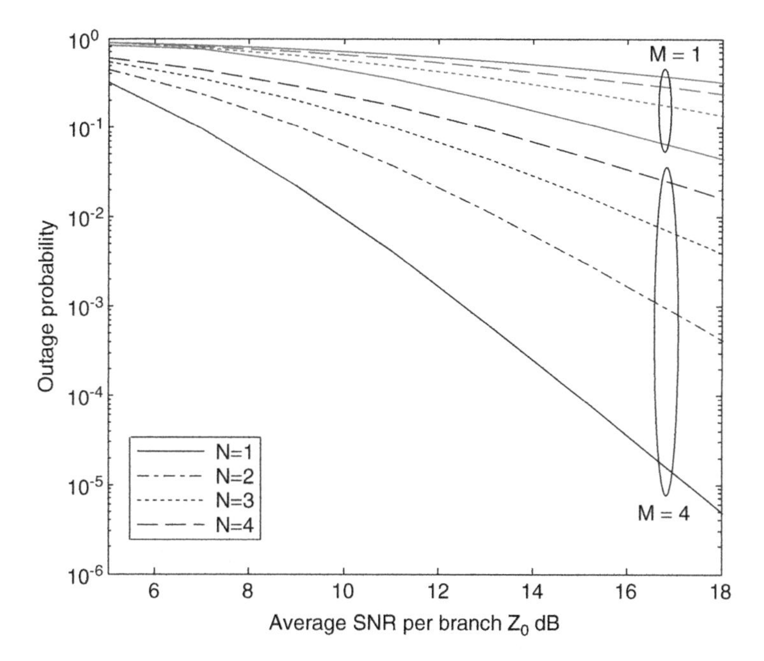

Fig. 5.84 Outage probabilities functions of the average SNR/branch Z_0 for $M = 1$ (no diversity) and $M = 4$ for the case of $m = 1.5$. Outage probabilities are higher with increasing values of N. The improvement in performance with diversity is seen in the lower values of the outage probabilities for $M = 4$

The outage probabilities are shown in Fig. 5.84 for the case of $m = 1.5$. The plot shows the reduction in outage probabilities when one goes from the case of no diversity ($M = 1$) to the case of a four-branch MRC.

5.10 Generalized Selection Combining

We will now look at the GSC algorithm for fading mitigation in cascaded channels. Since it was not possible to obtain an analytical expression for the density function of the SNR following MRC in cascaded channel, it must be clear that we will not be able to obtain analytical expressions for the density functions of the SNR following generalized selection combining. We can follow the path taken earlier in Sect. 5.7 in connection with the performance analysis of GSC in Nakagami channels. The numerical computation becomes relatively easy when we only use the two strongest branches out of the M diversity branches. In other words, we will look at performance represented by GSC(M,2). The density functions, distribution functions, error rates and outage probabilities can be estimated from appropriate functions given in (5.112).

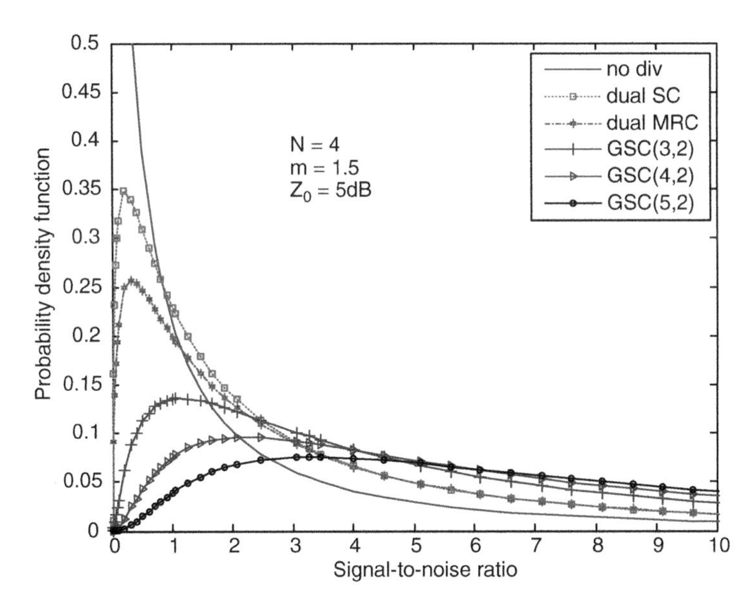

Fig. 5.85 Comparison of the pdfs in GSC. The pdf in the absence of any diversity is compared with the cases of dual SC, dual MRC, and GSC with 3, 4, and 5 branches, $N = 4$; $m = 1.5$. Average SNR/branch =5 dB

The density functions obtained for the output SNR of the GSC algorithm are shown in Fig. 5.85. The shifts in the peaks of the density functions to higher values of the SNR with diversity are clearly seen. The corresponding CDFs are plotted in Fig. 5.86. The slopes of the cumulative distributions show behavior consistent with improvements in performance achievable with generalized selection combining. A typical plot of the average probabilities of error is shown in Fig. 5.87. A typical plot of the outage probabilities is shown in Fig. 5.88. Comparing the results from Chap. 4, one can clearly see the benefits of diversity in reducing the effects of cascaded fading.

5.11 Diversity in McKay Fading Channels

The McKay fading channel was introduced in Chap. 4 (Shankar 2015). The improvement in performance obtained through diversity in the McKay channels can now be examined.

As seen in previous sections, the performance improvement in selection combining (SC) is the lowest and the improvement in maximal ratio combining (MRC) is the highest (Radaydeh and Matalgah 2008; Peppas et al. 2009). Because of this, the discussion is limited to these two combining algorithms and EGC is not considered initially. The generalized section combining (GSC) is also not treated

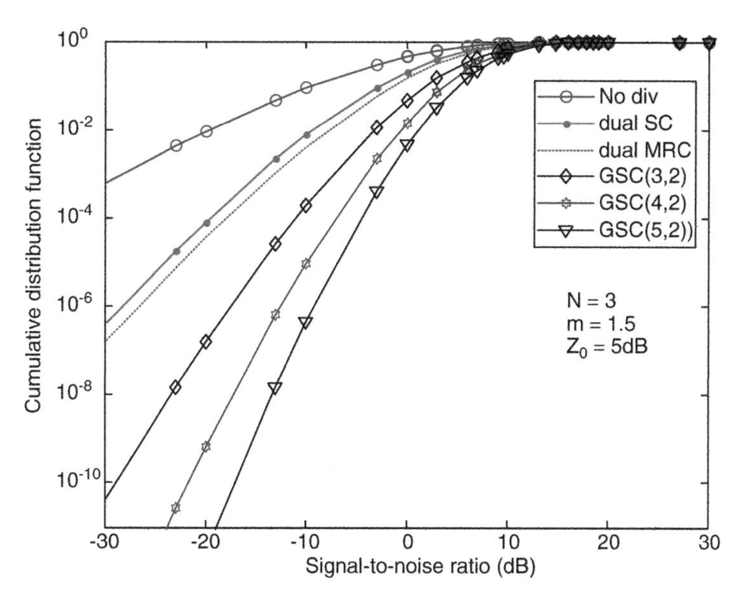

Fig. 5.86 Comparison of the CDFs in GSC. The CDF in the absence of any diversity is compared with the cases of dual SC, dual MRC, and GSC with 3, 4, and 5 branches, $N = 3$; $m = 1.5$. Average SNR/branch =5 dB

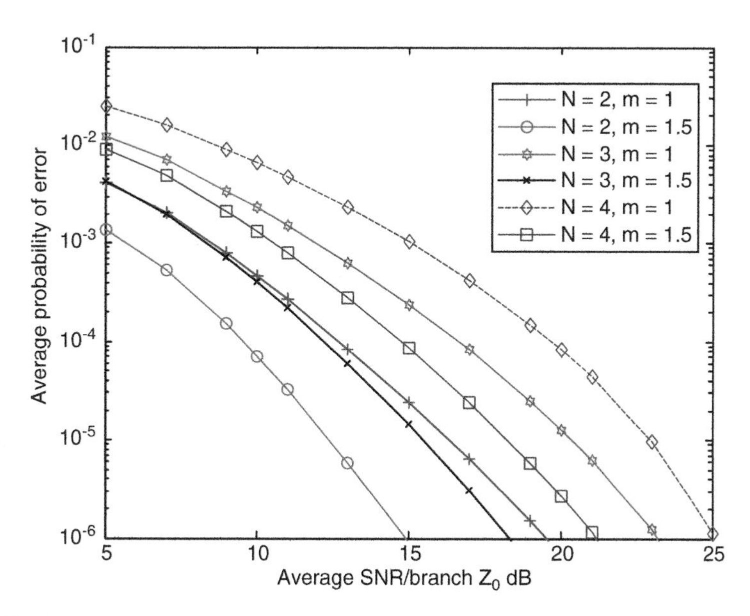

Fig. 5.87 Average error probabilities for the case of GSC(4,2) for two values of m (1 and 1.5) and for $N = 2, 3, 4$

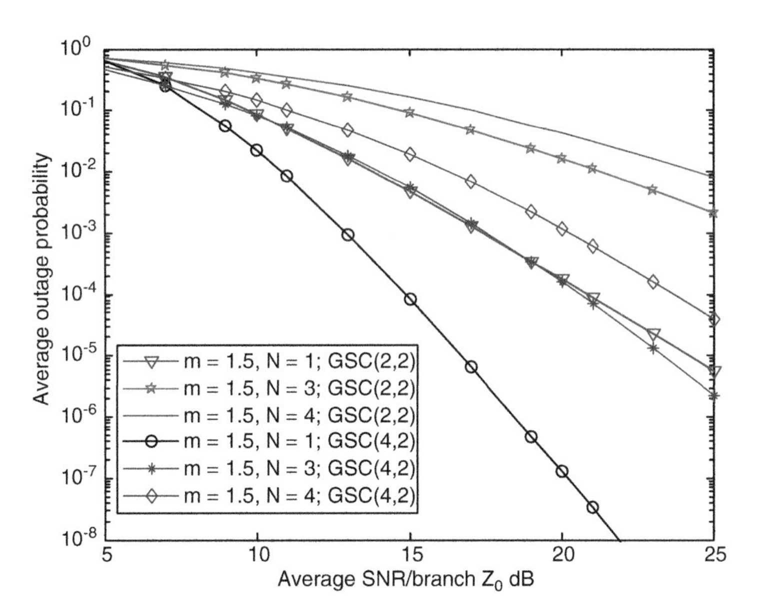

Fig. 5.88 Outage probabilities for dual MRC case are compared with those of GSC (4,2) for *m* 1.5. GSC(2,2) represents dual MRC

in this analysis. However, both EGC and GSC are discussed later when results based on random number simulation are presented.

5.11.1 Selection Combining

As seen earlier, the SC algorithm picks the strongest of a set of M diverse branches that are considered to be independent and identically distributed. The density function of the SNR in each of these ($i = 1,2,\ldots,M$) branches is

$$f(z_i) = \frac{\sqrt{\pi}}{\left(\frac{2\alpha\sqrt{\rho}}{X(1-\rho)}\right)^{\frac{(\alpha-1)}{2}}\Gamma\left(\frac{\alpha}{2}\right)\left[\left(\frac{X}{\alpha}\right)^2(1-\rho)\right]^{\frac{\alpha}{2}}}\exp\left[-\frac{\alpha}{X(1-\rho)}z_i\right]z_i^{\frac{(\alpha-1)}{2}}I_{\frac{(\alpha-1)}{2}}\left(\frac{\alpha\sqrt{\rho}}{X(1-\rho)}z_i\right) \tag{5.210}$$

The corresponding CDF in each of these branches is

$$F(z_i) = \frac{\sqrt{\pi}(1-\rho)^{\frac{\alpha}{2}}}{\Gamma\left(\frac{\alpha}{2}\right)}\sum_{k=0}^{\infty}\frac{\rho^k}{k!\Gamma\left(k+\frac{1}{2}+\frac{\alpha}{2}\right)2^{\alpha+2k-1}}\gamma\left(\alpha+2k,\frac{\alpha}{X(1-\rho)}z_i\right) \tag{5.211}$$

If Z_{sc} is the output of the SC algorithm, the CDF of the SNR following diversity is [Eq. (5.12)]

$$F(z_{SC}) = [F_{Z_i}(z_{sc})]^M \tag{5.212}$$

The pdf is

$$f_{Z_{SC}}(z_{SC}) = M[F_{Z_i}(z_{sc})]^{M-1} f_{Z_i}(z_{sc}) \tag{5.213}$$

In Eqs. (5.212) and (5.213), $F_{Z_i}(.)$ is given in Eq. (5.211) and $f_{Z_i}(.)$is given in Eq. (5.210).

The outage probability following diversity is

$$P_{outSC}(X) = F_{Z_{SC}}(z_{thr}). \tag{5.214}$$

In Eq. (5.214), z_{thr} is the threshold SNR required to maintain an acceptable bit error rate. While the outage probability estimation is simple and straightforward, error rate estimation requires numerical approaches. The numerical computation can be simplified if one uses the error rate calculations based on the CDF. For the case of a coherent BPSK modem, the error rate expressed in terms of the pdf as

$$p_e(X) = \int_0^\infty \frac{1}{2} erfc(\sqrt{z}) f(z) dz \tag{5.215}$$

and it can be expressed in terms of the CDF by performing integration by parts leading to

$$p_e(X) = \int_0^\infty \frac{1}{2} \frac{\exp(-z)}{\sqrt{\pi z}} F(z) dz. \tag{5.216}$$

Using Eq. (5.216), the error rate following SC diversity in a McKay fading channel is

$$p_{eSC}(X) = \int_0^\infty \frac{1}{2} \frac{\exp(-z)}{\sqrt{\pi z}} [F_Z(z)]^M dz. \tag{5.217}$$

In Eq. (5.217), $F(z)$ is the marginal CDF in Eq. (5.211).

5.11.2 Maximal Ratio Combining

If the MRC output represented by Z_{MRC} is given by

$$Z_{MRC} = \sum_{i=1}^{M} Z_i. \tag{5.218}$$

The pdf of the MRC output can be obtained easily from the Laplace transform of the McKay density given in Chap. 4. The Laplace transform of the McKay density is (Holm and Alouini 2004a, b; Filho and Yacoub 2005; Da Costa and Yacoub 2008; Ermolova 2008, Ermolova 2009; Peppas et al. 2009)

$$L_Z(s) = \frac{\alpha^{\alpha}}{\left[\alpha^2 + 2s\alpha X + X^2 s^2 (1-\rho)\right]^{\frac{\alpha}{2}}} \tag{5.219}$$

Assuming independence of the branches, the Laplace transform of the pdf of the MRC output is

$$L_{Z_{MRC}} = [L_Z(s)]^M = \left[\frac{\alpha^{\alpha}}{\left[\alpha^2 + 2s\alpha X + X^2 s^2 (1-\rho)\right]^{\frac{\alpha}{2}}}\right]^M. \tag{5.220}$$

Equation (5.220) can be interpreted by defining a new McKay random variable Y as having parameters $M\alpha$, MX and ρ. Using Eq. (5.219), the Laplace transform of Y will be

$$\begin{aligned} L_Y &= \frac{(M\alpha)^{M\alpha}}{\left[M^2\alpha^2 + 2sM^2\alpha X + M^2 X^2 s^2 (1-\rho)\right]^{\frac{M\alpha}{2}}} \\ &= \frac{M^{M\alpha}\alpha^{M\alpha}}{M^{M\alpha}\left[\alpha^2 + 2s\alpha X + X^2 s^2 (1-\rho)\right]^{\frac{M\alpha}{2}}} \end{aligned} \tag{5.221}$$

It can be seen that (5.221) and (5.220) are identical suggesting that the pdf of the sum of the M independent McKay variables is another McKay variable with parameters $M\alpha$, MX, and ρ. The pdf of the output of the MRC algorithm can be written as

$$f_{Z_{MRC}}(z_{MRC}) = A\exp\left[-\frac{\alpha}{X(1-\rho)} z_{MRC}\right] z_{MRC}^{\frac{(M\alpha-1)}{2}} I_{\frac{(M\alpha-1)}{2}}\left(\frac{\alpha\sqrt{\rho}}{X(1-\rho)} z_{MRC}\right). \tag{5.222}$$

In Eq. (5.222), the constant A is

$$A = \frac{\sqrt{\pi}}{\left(\frac{2\alpha\sqrt{\rho}}{X(1-\rho)}\right)^{\frac{(M\alpha-1)}{2}} \Gamma\left(\frac{M\alpha}{2}\right)\left[\left(\frac{X}{\alpha}\right)^{2}(1-\rho)\right]^{\frac{M\alpha}{2}}} \tag{5.223}$$

The CDF of the output of the MRC algorithm becomes

$$F_{Z_{MRC}}(z_{MRC}) = \sqrt{\pi}\frac{(1-\rho)^{\frac{M\alpha}{2}}}{\Gamma\left(\frac{M\alpha}{2}\right)} \sum_{k=0}^{\infty} \frac{\rho^{k}}{k!\Gamma\left(k+\frac{1}{2}+\frac{M\alpha}{2}\right)2^{M\alpha+2k-1}}\gamma\left(M\alpha+2k, z_{MRC}\frac{\alpha}{X(1-\rho)}\right). \tag{5.224}$$

The outage probability following MRC will be

$$P_{outMRC}(X) = F_{Z_{MRC}}(z_{thr}). \tag{5.225}$$

The error rate following the MRC can be obtained using the Laplace transforms as

$$p_{e}(X) = \frac{1}{\pi}\int_{0}^{\frac{\pi}{2}} L_{X}\left(\frac{1}{\sin^{2}(\theta)}\right)d\theta = = \frac{1}{\pi}\int_{0}^{\frac{\pi}{2}} \left[\frac{\alpha^{\alpha}}{\left[\alpha^{2}+2\frac{\alpha}{\sin^{2}(\theta)}X+X^{2}\frac{1}{\sin^{4}(\theta)}(1-\rho)\right]^{\frac{\alpha}{2}}}\right]^{M} d\theta \tag{5.226}$$

While outage probability and error rates provide powerful measures to compare the performance of the diversity combining algorithms, two simple measures of comparison are the enhancement in SNR and reduction in the amount of fading (AF). These two parameters can be obtained from the moments.

The improvement in performance in McKay channels from MRC and SC can now be studied. Along with evaluation based on analytical expressions, random number simulations have also been undertaken to support analytical approaches.

5.11.3 Density Functions, Distribution Functions, SNR, and AF

The probability density function and the CDF of the MRC output along with the Matlab script used for the generation are given below.

Figures 5.89, 5.90, and 5.91 show the densities and distribution functions, Figs. 5.92 and 5.93 show the amount of fading and SNR following the implementation of diversity (MRC and SC) in McKay fading channels. As expected, the performance following MRC is expected to be better than the performance following SC based on the peaks of the densities, AF and SNR seen in these plots.

```
function Mckaydiversity_pdf_cdf_comparison

% Obtains  pdf and CDF the output SNR of SC and MRC combining algorithm in
% a McKay fading channel. The random numbers are generated and the reults
% from simulations are compred to those obtained from the analytical
% approach. Note that the CDF is obtained in terms of a summation. AS much
% as possible, symbolictoolbox is used to generate expressions including
% the summartions
%
% P M Shankar, September 2016
close all
alpha=1.25;X=15;r=0.45; M=3;
pdf_CDF_MRC(alpha,X,r,M)
pdf_CDF_SC(alpha,X,r,M)
alpha=1.25;X=15;r=0.45; M=5;
pdf_CDF_MRC(alpha,X,r,M)
pdf_CDF_SC(alpha,X,r,M)
alpha=1.4;X=15;r=0.85; M=4;
pdf_CDF_MRC(alpha,X,r,M)
pdf_CDF_SC(alpha,X,r,M)
% now compare the diversity schemes
diversity_analysis_MRC_SC
end

function pdf_CDF_MRC(alpha,X,r,M)
a=alpha;
X1=X*(1+sqrt(r))/2;
X2=X*(1-sqrt(r))/2;
a1=a/2;
x=gamrnd(a1,X1/a1,M,5e6)+gamrnd(a1,X2/a1,M,5e6);
if M>1
    x=sum(x);
else
end;

[fzs,zi]=ksdensity(x);
NK=find(zi<0);
if isempty(NK)==0
    NK1=max(NK)+1;% find largest index, add 1 to start non-zero values
    zi=zi(NK1:end);
    fzs=fzs(NK1:end);
else
end;
mckpdf = mckaypdf(a*M,r,X*M);
fzm=mckpdf(zi);
figure,subplot(2,2,1),plot(zi,fzm,'-r',zi,fzs,'*')%
xlabel('SNR z'),ylabel(['pdf (MRC), M = ',num2str(M)])
legend('Theory','sim')
title(['\alpha = ',num2str(alpha),', \rho = ',num2str(r),...
    ',  X = ',num2str(X)])
%better way to get CDF with fewer points: use ksdensity instead of ecdf
Fzs=ksdensity(x,zi,'function','cdf');%obtain CDF at same points used in pdf
 mckCDF=mckaycdf(M*a,r,M*X,20); %at high rho values more summation terms needed
 Fzm=mckCDF(zi);
subplot(2,2,2),plot(zi,Fzm,'-r',zi,Fzs,'ko')
xlabel('SNR z'),ylabel(['CDF (MRC), M = ',num2str(M)])
% title(['\alpha = ',num2str(alpha),', \rho = ',num2str(r),...
%     ',  X = ',num2str(X),', M = ',num2str(M),'(MRC)'])
ylim([0,1.1]),legend('Theory','sim','location','southEast')
grid on
```

```
end

function pdf_CDF_SC(alpha,X,r,M)
a=alpha;
X1=X*(1+sqrt(r))/2;
X2=X*(1-sqrt(r))/2;
a1=a/2;
x=gamrnd(a1,X1/a1,M,5e6)+gamrnd(a1,X2/a1,M,5e6);
if M>1
    x=max(x);
else
end;

[fzs,zi]=ksdensity(x);
NK=find(zi<0);
if isempty(NK)==0
    NK1=max(NK)+1;% find largest index, add 1 to start non-zero values
    zi=zi(NK1:end);
    fzs=fzs(NK1:end);
else
end;
mckpdf1 = mckaypdf(a,r,X);
mckayCDF1=mckaycdf(a,r,X,20);
fzm=M*mckpdf1(zi).*mckayCDF1(zi).^(M-1);
subplot(2,2,3),plot(zi,fzm,'-r',zi,fzs,'*')%
xlabel('SNR z'),ylabel(['pdf (SC), M = ',num2str(M)])
legend('Theory','sim')
title(['\alpha = ',num2str(alpha),', \rho = ',num2str(r),...
    ',  X = ',num2str(X)])
%better way to get CDF with fewer points: use ksdensity instead of ecdf
Fzs=ksdensity(x,zi,'function','cdf');%obtain CDF at same points used in pdf
 Fzm=mckayCDF1(zi).^M;
subplot(2,2,4),plot(zi,Fzm,'-r',zi,Fzs,'ko')
xlabel('SNR z'),ylabel(['CDF (SC), M = ',num2str(M)])
% title(['\alpha = ',num2str(alpha),', \rho = ',num2str(r),...
%     ',  X = ',num2str(X),', M = ',num2str(M),'(SC)'])
ylim([0,1.1]),legend('Theory','sim','location','southEast')
grid on

end

function diversity_analysis_MRC_SC
% unit mean is assumed
alpha=0.9;r=0.8;
[M1,meMRC,AFMRC]=MRC_analysis(alpha,r);
[M2,meSC,AFSC]=SC_analysis(alpha,r);
figure,subplot(2,1,1),plot(M1,meMRC,'linewidth',2)
xlabel('Order of diversity M'),ylabel('SNR improvement'),legend('MRC')
subplot(2,1,2),plot(M1,AFMRC,'linewidth',2)
xlabel('Order of diversity M'),ylabel('Reduction in AF'),legend('MRC')
figure,subplot(2,1,1),plot(M2,meSC,'linewidth',2)
xlabel('Order of diversity M'),ylabel('SNR improvement'),legend('SC')
subplot(2,1,2),plot(M2,AFSC,'linewidth',2)
xlabel('Order of diversity M'),ylabel('Reduction in AF'),legend('SC')
figure,subplot(2,1,1),plot(M1,meMRC,'-',M2,meSC,'--','linewidth',2)
xlabel('Order of diversity M'),ylabel('SNR improvement'),legend('MRC','SC')
subplot(2,1,2),plot(M1,AFMRC,'-',M2,AFSC,'--','linewidth',2)

xlabel('Order of diversity M'),ylabel('Reduction in AF'),legend('MRC','SC')
```

```
end
function [MD,firstM,AFF]=MRC_analysis(alpha,r)
% deteremine mean, second moment and AF
X=1;
% deteremine mean, second moment and AF
a=alpha;
M=1:15;
a1=M*a;
X1=M*X;
firstM=X1;
secM=(X1.^2).*(1+a1+r)*1./a1;
AF=secM./firstM.^2-1;
AFF=AF/AF(1);
MD=M;
end

function [MD,firstM,AFF]=SC_analysis(alpha,r)
% deteremine mean, second moment and AF
X=1;% unit mean
a=alpha;
syms M1 z
mckpdf1 = mckaypdf(a,r,X);
mckayCDF1=mckaycdf(a,r,X,20);
fun1=z*M1*mckpdf1*mckayCDF1^(M1-1);
fun2=z^2*M1*mckpdf1*mckayCDF1^(M1-1);
MD=1:15;
firstM=zeros(1,length(MD));
secM=zeros(1,length(MD));
AF=zeros(1,length(MD));
for m=1:length(MD)
    M=MD(m);
funM=subs(fun1,M1,M);
funz1=matlabFunction(funM);
firstM(m)=integral(funz1,0,100*X);
funM=subs(fun2,M1,M);
funz2=matlabFunction(funM);
secM(m)=integral(funz2,0,100*X);
AF(m)=secM(m)/firstM(m)^2-1;
end;
AFF=AF/AF(1);

end


function mckpdf = mckaypdf(alpha,rho,Z)
syms z a r X   positive
% get the expression for density directly from the notes

Nr=sqrt(sym(pi))*z^((1/2)*a-1/2)*exp(-a*z/(X*(1-r)))...
    *besseli((1/2)*a-1/2, a*sqrt(r)*z/(X*(1-r)));
Dr=gamma((1/2)*a)*(X^2*(1-r)/a^2)^((1/2)*a)...
    *(2*a*sqrt(r)/(X*(1-r)))^((1/2)*a-1/2);
pdf=Nr/Dr;
% pdf McKay depends on X, alpha (a), rho(r) and z in order mcpdf
pdff=subs(pdf,[X,a,r],[Z,alpha,rho]);
mckpdf=matlabFunction(pdff); % function of z
end


function mckCDF=mckaycdf(alpha,rho,Z,KK)
```

```
% create the McKay CDF
% KK is the number of terms in the summation
syms a r X K z
syms k
xa=a/(X*(1-r));
a2=a/2;
gamaprt=gamma(a+2*k)-igamma(a+2*k,xa*z);
Nr=gamaprt*r^k;
Dr=factorial(k)*gamma(k+1/2+a2)*2^(a+2*k-1);
CDF=((sqrt(sym(pi))*(1-r)^(a2))/gamma(a2))*symsum(Nr/Dr,k,0,K);
%in order of variables @(K,X,a,r,z)
CDFF=subs(CDF,[X,a,r,K],[Z,alpha,rho,KK]);
mckCDF=matlabFunction(CDFF); % function of z only
end
```

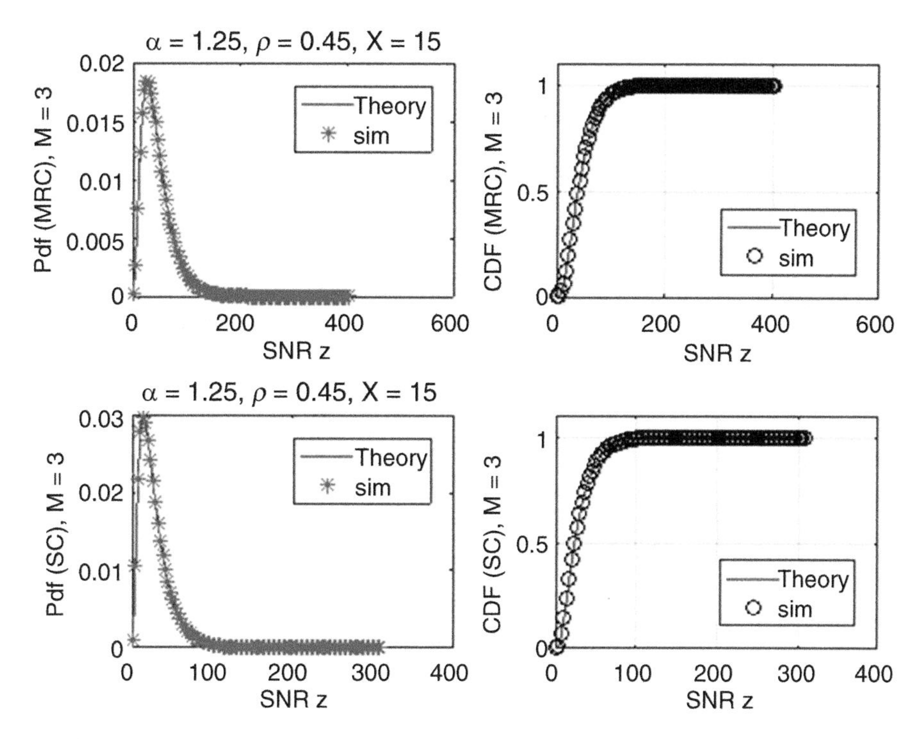

Fig. 5.89 Probability density functions and cumulative distributions functions $M = 3$

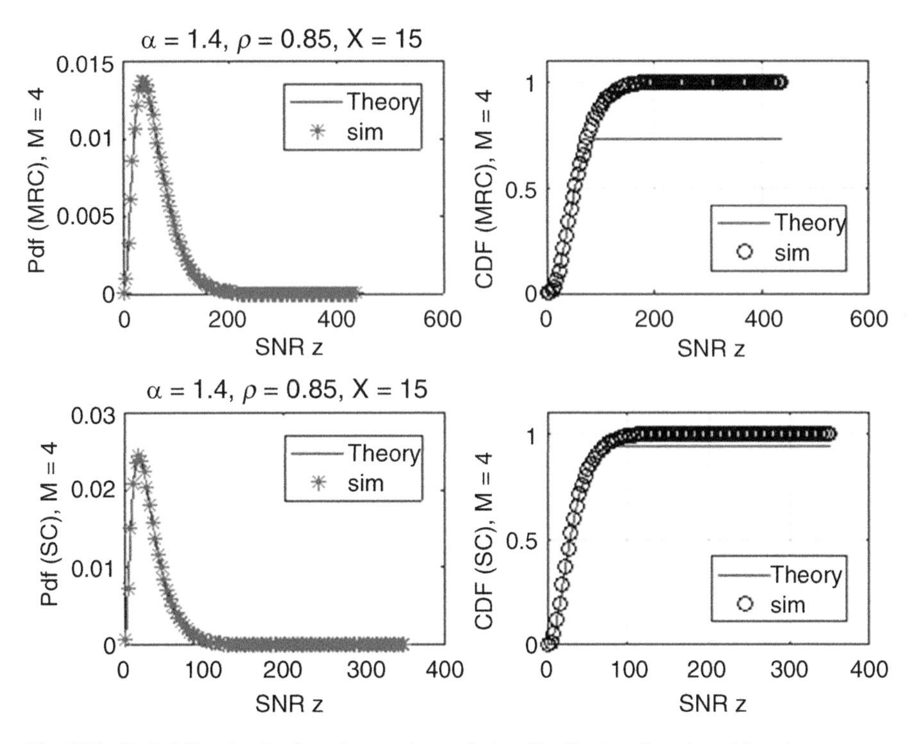

Fig. 5.90 Probability density functions and cumulative distributions functions $M = 4$

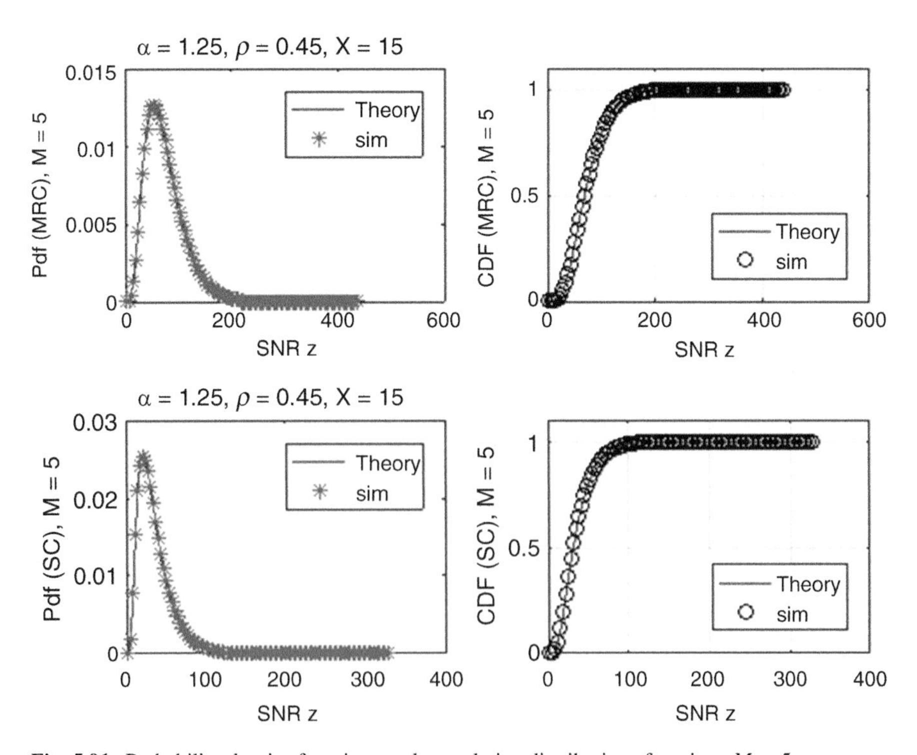

Fig. 5.91 Probability density functions and cumulative distributions functions $M = 5$

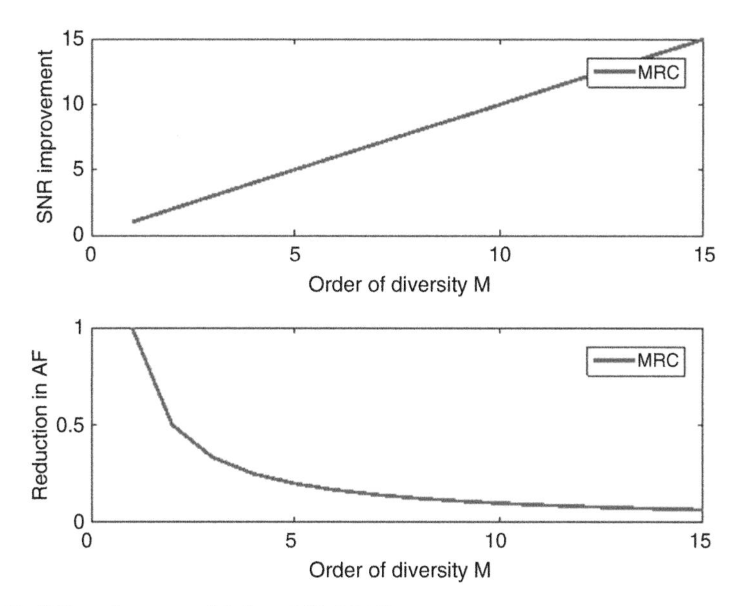

Fig. 5.92 SNR and amount of fading (AF): MRC

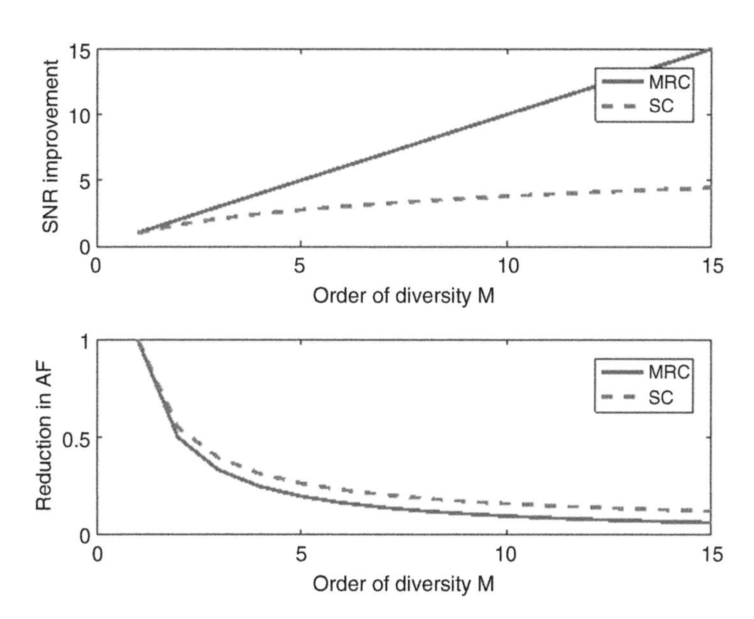

Fig. 5.93 SNR and amount of fading (AF): MRC and SC (comparison)

5.11.4 Outage Probability

Outage probabilities following SC can now be examined. The threshold SNR is obtained for maintaining the error rate of 1e-4. The number of terms in the summation in Eq. (5.211) is limited to 20. The Matlab script appears first followed by the results. Figures 5.94 and 5.95 display the outage probabilities following SC, Figs. 5.96 and 5.97 show the results for the MRC. Figures 5.98 and 5.99 compare the outage probabilities seen with MRC and SC in McKay fading channels. As expected, MRC shows larger improvement over SC.

```
function mckay_outage_diversitySCbook
% obtains the outage probability in SC diversity
% the threshold is 7 (not in dB) to get a BER of 1e-4;
close all
MM=20; % number of terms in the summation
Z0=10:2:30;
Z=10.^(Z0/10);
LK=length(Z);
alpha=1.2;
rho=0.85;
MN=[1,3,5];
MQ=length(MN);
pTSC=zeros(LK,MQ);
for km=1:MQ
    M=MN(km);
    for k=1:LK
          fn=mckaycdf(alpha,rho,Z(k),MM);
         pTSC(k,km)=fn(7)^M;
    end;
end;
figure
semilogy(Z0,pTSC(:,1),'r-',Z0,pTSC(:,2),'k--',Z0,pTSC(:,3),'b-.',...
    'linewidth',1);
     xlabel('average SNR/branch (dB)')
     ylabel('Outage Probability'),ylim([1e-6,1])
 legend('No-diversity',...
   ['M = ',num2str(MN(2))],['M = ',num2str(MN(3))])
    title(['\alpha = ',num2str(alpha),',   r = ',num2str(rho)])


end

function mckCDF=mckaycdf(alpha,rho,Z,KK)
% create the McKay CDF KK is the number of terms in the summation
syms a r X K z
syms k
xa=a/(X*(1-r));
a2=a/2;
gamaprt=gamma(a+2*k)-igamma(a+2*k,xa*z);
Nr=gamaprt*r^k;
Dr=factorial(k)*gamma(k+1/2+a2)*2^(a+2*k-1);
CDF=((sqrt(sym(pi))*(1-r)^(a2))/gamma(a2))*symsum(Nr/Dr,k,0,K);
%in order of variables @(K,X,a,r,z)
CDFF=subs(CDF,[X,a,r,K],[Z,alpha,rho,KK]);
mckCDF=matlabFunction(CDFF); % function of z only
end
```

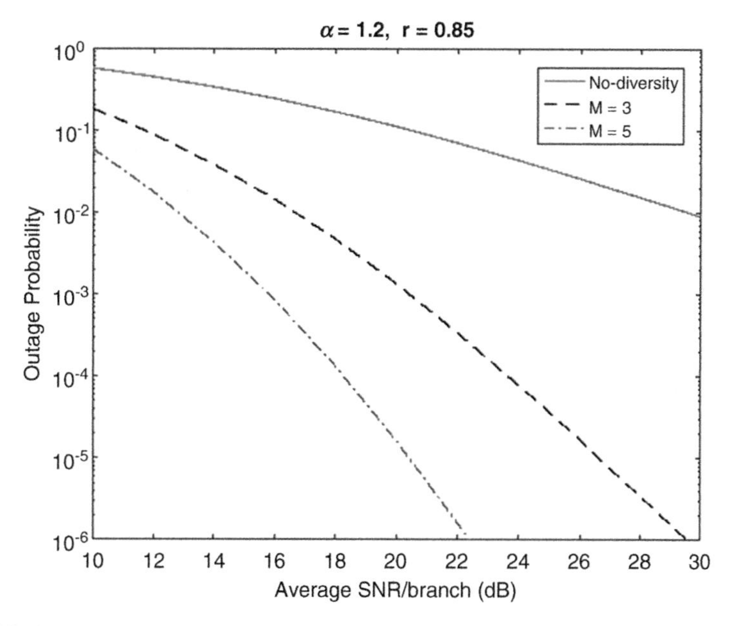

Fig. 5.94 Outage probability (SC): $\alpha = 1.2$, $\rho = 0.85$

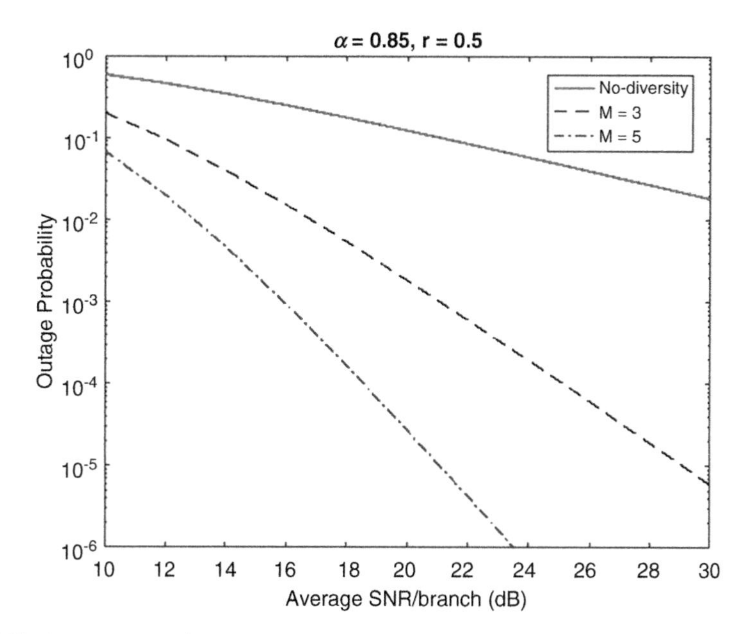

Fig. 5.95 Outage probability (SC): $\alpha = 0.85$, $\rho = 0.5$

```
function mckay_outage_diversityMRCbook
% obtains the outage probability in MRC diversity
% the threshold is 7 (not in dB) to get a BER of 1e-4;
close all
MM=20; % number of terms in the summation
Z0=10:2:30;
Z=10.^(Z0/10);
LK=length(Z);
alpha=1.2;
rho=0.85;
MN=[1,3,5];
MQ=length(MN);
pTSC=zeros(LK,MQ);
for km=1:MQ
    M=MN(km);
    for k=1:LK
          fn=mckaycdf(M*alpha,rho,M*Z(k),MM);
         pTSC(k,km)=fn(7);
    end;
end;
figure
semilogy(Z0,pTSC(:,1),'r-',Z0,pTSC(:,2),'k--',Z0,pTSC(:,3),'b-.',...
     'linewidth',1);
      xlabel('average SNR/branch (dB)')
      ylabel('Outage Probability'),ylim([1e-6,1])
 legend('No-diversity',...
   ['M = ',num2str(MN(2))],['M = ',num2str(MN(3))])
    title(['\alpha = ',num2str(alpha),',    r = ',num2str(rho)])


end

function mckCDF=mckaycdf(alpha,rho,Z,KK)
% create the McKay CDF KK is the number of terms in the summation
syms a r X K z
syms k
xa=a/(X*(1-r));
a2=a/2;
gamaprt=gamma(a+2*k)-igamma(a+2*k,xa*z);
Nr=gamaprt*r^k;
Dr=factorial(k)*gamma(k+1/2+a2)*2^(a+2*k-1);
CDF=((sqrt(sym(pi))*(1-r)^(a2))/gamma(a2))*symsum(Nr/Dr,k,0,K);
%in order of variables @(K,X,a,r,z)
CDFF=subs(CDF,[X,a,r,K],[Z,alpha,rho,KK]);
mckCDF=matlabFunction(CDFF); % function of z only
end
```

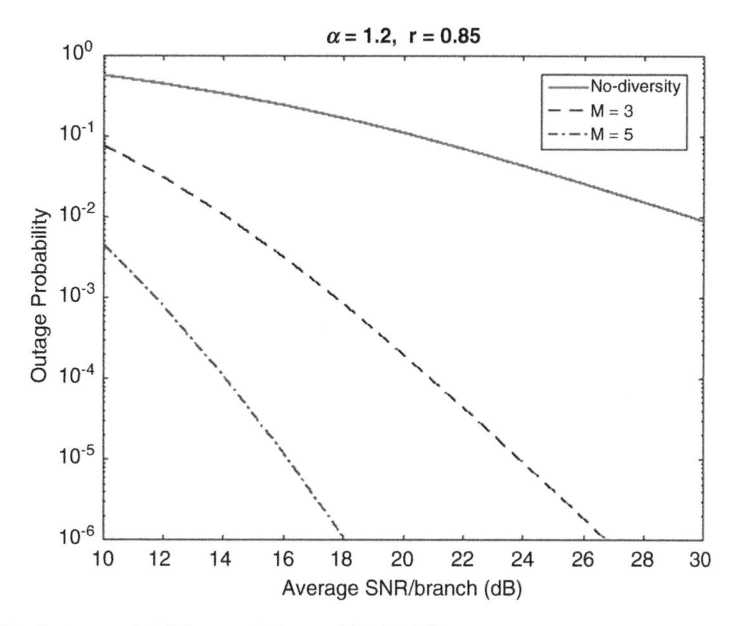

Fig. 5.96 Outage probability, $\alpha = 1.2$, $\rho = 0.85$ (MRC)

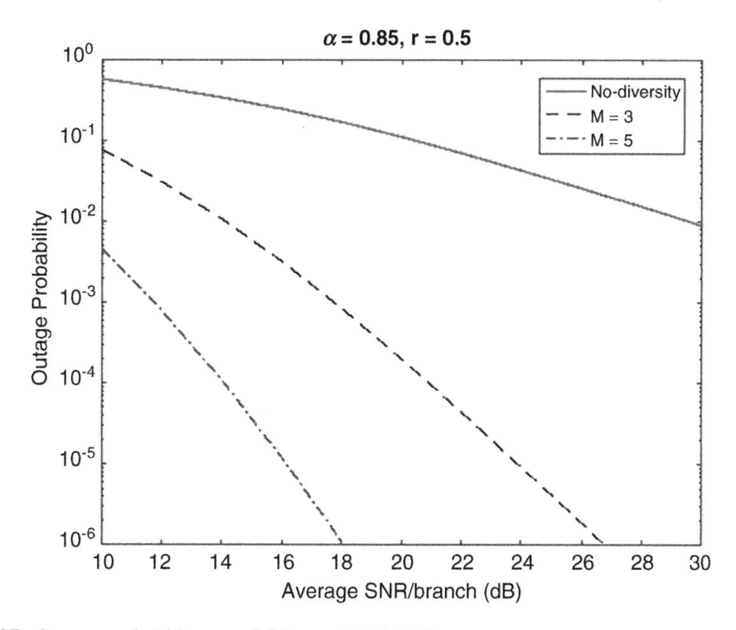

Fig. 5.97 Outage probability $\alpha = 0.85$, $\rho = 0.5$ (MRC)

```
function mckay_outage_diversityMRC_SCbook
% obtains the outage probability in both MRC and SC diversity
% the threshold is 7 (not in dB) to get a BER of 1e-4;
close all
MM=20; % number of terms in the summation
Z0=10:2:30;
Z=10.^(Z0/10);
LK=length(Z);
alpha=1.1;
rho=0.45;
MN=[2,6];
MQ=length(MN);
pTSC=zeros(LK,MQ);
pTMR=zeros(LK,MQ);
pout1=zeros(1,LK);
for k=1:LK
    ffn=mckaycdf(alpha,rho,Z(k),MM); % no diversity
    pout1(k)=ffn(7);
end;
for km=1:MQ
    M=MN(km);
    for k=1:LK
          fn=mckaycdf(M*alpha,rho,M*Z(k),MM);
         pTMR(k,km)=fn(7);
         fns=mckaycdf(alpha,rho,Z(k),MM);
          pTSC(k,km)=fns(7)^M;
    end;
end;
figure
semilogy(Z0,pout1,'r-',Z0,pTSC(:,1),'k--',Z0,pTMR(:,1),'b-.',...
    Z0,pTSC(:,2),'g-.',Z0,pTMR(:,2),'m-.','linewidth',1)
     xlabel('average SNR/branch (dB)')
     ylabel('Outage Probability'),ylim([1e-6,1])
 legend('No-diversity',...
   ['SC-M = ',num2str(MN(1))],['MRC-M = ',num2str(MN(1))],...
   ['SC-M = ',num2str(MN(2))],['MRC-M = ',num2str(MN(2))])
    title(['\alpha = ',num2str(alpha),',    r = ',num2str(rho)])


end

function mckCDF=mckaycdf(alpha,rho,Z,KK)
% create the McKay CDF KK is the number of terms in the summation
syms a r X K z
syms k
xa=a/(X*(1-r));
a2=a/2;
gamaprt=gamma(a+2*k)-igamma(a+2*k,xa*z);
Nr=gamaprt*r^k;
Dr=factorial(k)*gamma(k+1/2+a2)*2^(a+2*k-1);
CDF=((sqrt(sym(pi))*(1-r)^(a2))/gamma(a2))*symsum(Nr/Dr,k,0,K);
%in order of variables @(K,X,a,r,z)
CDFF=subs(CDF,[X,a,r,K],[Z,alpha,rho,KK]);
mckCDF=matlabFunction(CDFF); % function of z only
end
```

5.11.5 Bit Error Rates

Average error rates in a McKay channel following diversity are estimated and results are presented. Error rate expressed as a summation is used in the case of SC algorithm. Error rates following MRC are estimated using the Laplace based approach.

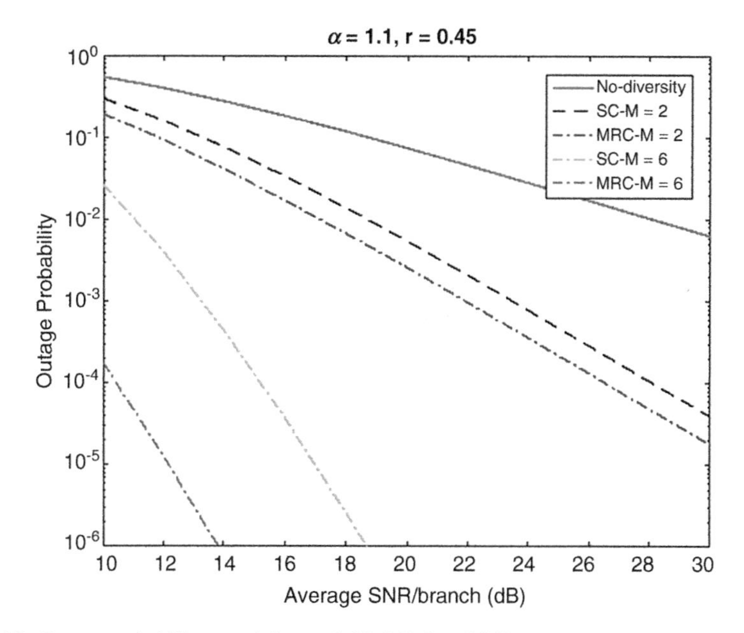

Fig. 5.98 Outage probability $\alpha = 1.1$, $\rho = 0.45$ (MRC and SC)

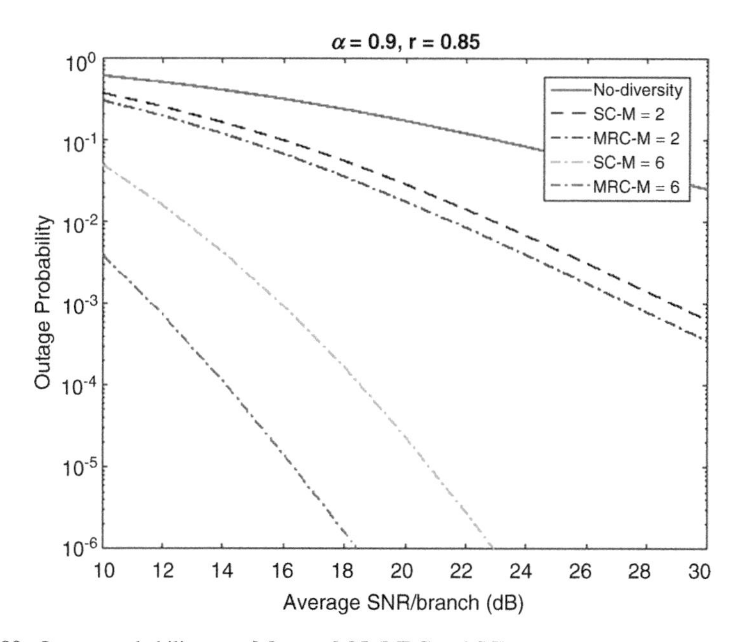

Fig. 5.99 Outage probability $\alpha = 0.9$, $\rho = 0.85$ (MRC and SC)

The use of the symbolic toolbox provides a simple and convenient way to obtain error rates. Figures 5.100 and 5.101 show the error rates following SC and Figs. 5.102 and 5.103 display the results for MRC. Figures 5.104 and 5.105 compare the error rates in SC and MRC. The improvements of MRC over SC are clearly seen.

```
function  ber_mckay_SC_diversity_book
% P M Shankar
% BER in SC diversity is calculated.. use the CDF
%July 2016
close all
global Z LK Z0 MM alpha rho MN
MM=20; % number of terms in the summation
 Z0=5:2:31;
Z=10.^(Z0/10);
LK=length(Z);
alpha=1.1;
rho=0.65;
MN=[1,3,4,6];
ber_mckay_M


end

function  ber_mckay_M
% P M Shankar
global LK Z0 MM alpha rho MN M Z
MQ=length(MN);
peT1=zeros(LK,MQ);
for km=1:MQ
    M=MN(km);
    for k=1:LK
        X=Z(k);
         peT1(k,km)=berMcKay(alpha,rho,X,MM);
    end;
end;
%
 figure,semilogy(Z0,peT1(:,1),'-r',Z0,peT1(:,2),'--k',...
  Z0,peT1(:,3),'-.b',Z0,peT1(:,4),'--.m','linewidth',1.5)
  xlabel('average SNR/branch (dB)'),xlim([5,30]),ylim([1e-6,.1])
  ylabel('Average probability of error')
  title(['\alpha = ',num2str(alpha),',  \rho = ',num2str(rho)])
 legend('No diversity',['M = ',num2str(MN(2))],['M  = ',num2str(MN(3))],...
     ['M  = ',num2str(MN(4))])
end


function peTT =berMcKay(aa,rr,ZZ,KK)
% create the McKay CDF % KK is the number of terms in the summation
global M
syms a r X K z
syms k
xa=a/(X*(1-r));
a2=a/2;

gamaprt=gamma(a+2*k)-igamma(a+2*k,xa*z);
Nr=gamaprt*r^k;
Dr=factorial(k)*gamma(k+1/2+a2)*2^(a+2*k-1);
CDF=((sqrt(sym(pi))*(1-r)^(a2))/gamma(a2))*symsum(Nr/Dr,k,0,K);
%in order of variables @(K,X,a,r,z)
CDFF=subs(CDF,[X,a,r,K],[ZZ,aa,rr,KK]);
integraN=(1/2)*1/sqrt(z*sym(pi))*exp(-z)*CDFF^M;
mckCDF=matlabFunction(integraN); % function of z only

   peTT=integral(mckCDF,1e-5,inf);
end
```

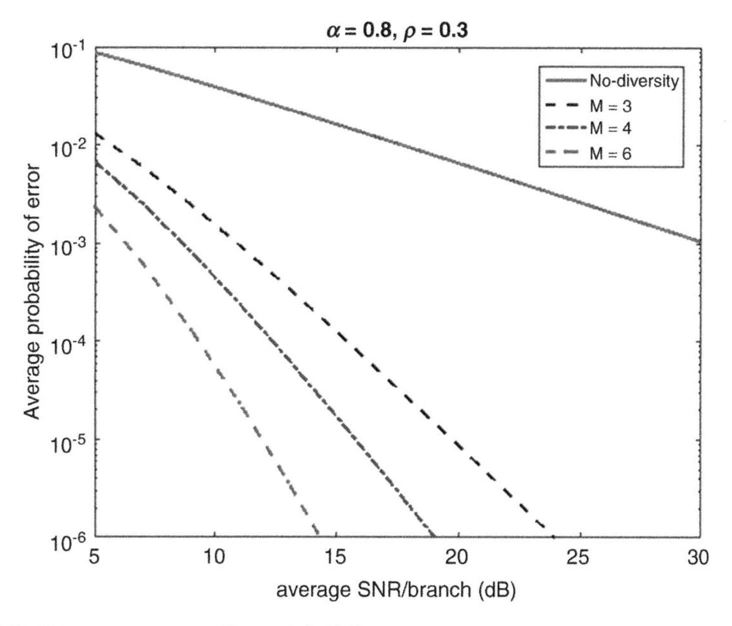

Fig. 5.100 Bit error rates, $\alpha = .8$, $\rho = 0.3$ (SC)

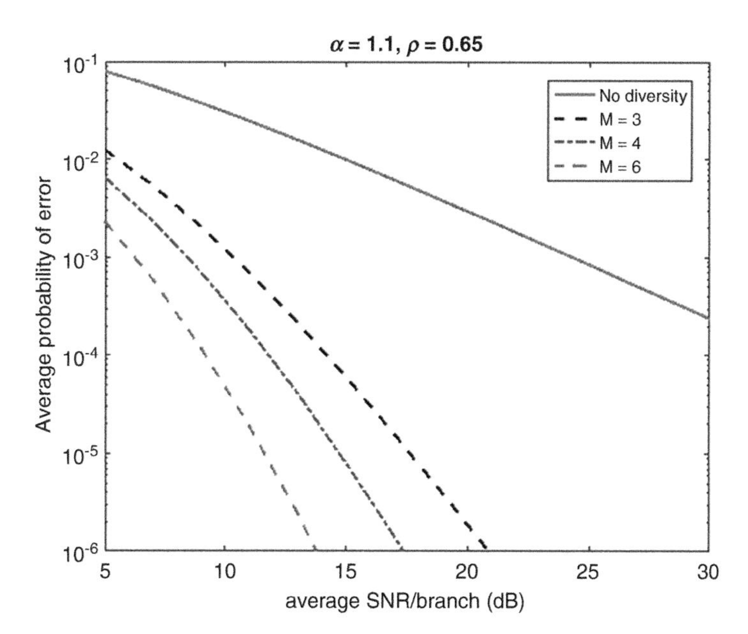

Fig. 5.101 Bit error rates, $\alpha = 1.1$, $\rho = 0.65$ (SC)

```
function  ber_mckay_MRC_diversity_book
% P M Shankar, Augsut 2016
% BER in MRC diversity is calculated. Laplace
close all
global Z LK Z0 alpha rho MN
 Z0=[5:2:30];
Z=10.^(Z0/10);
LK=length(Z);
alpha=1.1;
rho=0.65;
MN=[1,2,4,6];
ber_mckay_M

end

function  ber_mckay_M
% P M Shankar
global Z LK Z0 alpha rho MN
MQ=length(MN);
peL1=zeros(LK,MQ); % only Laplace transform is used
for km=1:MQ
    M=MN(km);
    for k=1:LK
        peL1(k,km)=laplace_ber_func(alpha,rho,Z(k),M);
    end;
end;
%
 figure,semilogy(Z0,peL1(:,1),'-r',Z0,peL1(:,2),'--k',Z0,peL1(:,3),'-.b',...
     Z0,peL1(:,4),'--.m','linewidth',1.5)
 xlabel('average SNR/branch (dB)'),xlim([5,25]),ylim([1e-6,.1])
 ylabel('Average probability of error')
 title(['\alpha = ',num2str(alpha),',  \rho = ',num2str(rho)])
 legend('No diversity',['M = ',num2str(MN(2))],['M = ',num2str(MN(3))],...
     ['M = ',num2str(MN(4))]);
end



function pe = laplace_ber_func(alpha,rho,Z,M) % Laplace based
syms s r a X y
fLr=a^a/(a^2+s^2*X^2+2*s*X*a-s^2*X^2*r)^((1/2)*a);
fLRf=subs(fLr,[a,X,r],[alpha,Z,rho]);
f1M=fLRf^M; % laplace transform MRC
fun2M=subs(f1M,s,1/(sin(y))^2);
         mckayf2=matlabFunction(fun2M);
        pe=(1/pi)*integral(mckayf2,0,0.999975*pi/2);
end
```

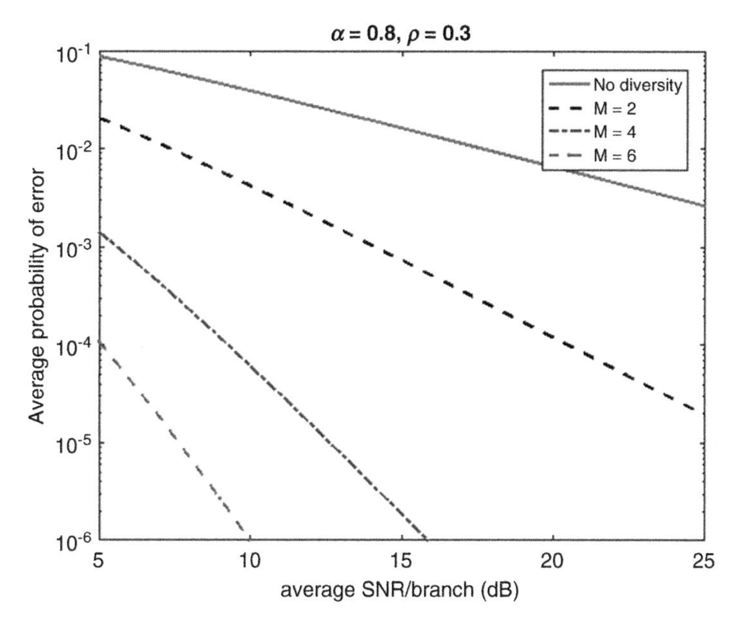

Fig. 5.102 Bit error rates, $\alpha = .8$, $\rho = 0.3$ (MRC)

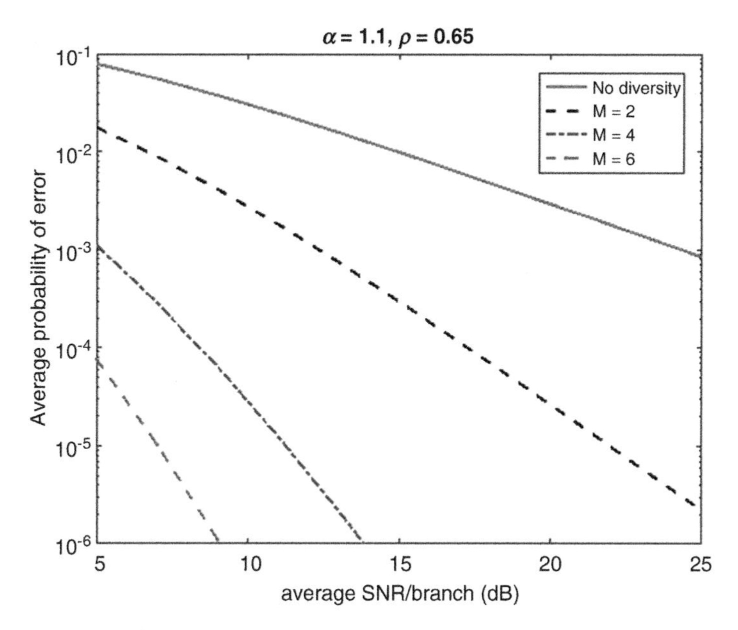

Fig. 5.103 Bit error rates $\alpha = 1.1$, $\rho = 0.65$ (MRC)

```
function  ber_mckay_SC_MRC_book
% P M Shankar
% BER in SC diversity is calculated.. use the CDF
%July 2016
close all
global Z LK Z0 MM alpha rho MN
MM=20; % number of terms in the summation
 Z0=5:2:31;
Z=10.^(Z0/10);
LK=length(Z);
alpha=0.85;
rho=0.6;
MN=[1,3,6];
ber_mckay
end

function  ber_mckay
% P M Shankar
global LK Z0 MM alpha rho MN M Z
MQ=length(MN);
peSC=zeros(LK,MQ);
peMR=zeros(LK,MQ); % only Laplace transform is used
for km=1:MQ
    M=MN(km);
    for k=1:LK
        X=Z(k);
         peSC(k,km)=berMcKaySC(alpha,rho,X,MM);
             peMR(k,km)=laplace_ber_func(alpha,rho,Z(k),M);
    end;
end;
%
 figure,semilogy(Z0,peMR(:,1),'r-*',Z0,peSC(:,2),'k-.+',...
   Z0,peMR(:,2),'k--s',Z0,peSC(:,3),'m-.o',Z0,peMR(:,3),'k--^',...
   'linewidth',1)
  xlabel('average SNR/branch (dB)'),xlim([5,30]),ylim([1e-6,.1])
  ylabel('Average probability of error')
  title(['\alpha = ',num2str(alpha),',  \rho = ',num2str(rho)])
 legend('No diversity',['SC-M = ',num2str(MN(2))],...
     ['MRC-M  = ',num2str(MN(2))],['SC-M = ',num2str(MN(3))],...
     ['MRC-M  = ',num2str(MN(3))])
end
```

```
function peTT =berMcKaySC(aa,rr,ZZ,KK)
% create the McKay CDF % KK is the number of terms in the summation
global M
syms a r X K z
syms k
xa=a/(X*(1-r));
a2=a/2;
gamaprt=gamma(a+2*k)-igamma(a+2*k,xa*z);
Nr=gamaprt*r^k;
Dr=factorial(k)*gamma(k+1/2+a2)*2^(a+2*k-1);
CDF=((sqrt(sym(pi))*(1-r)^(a2))/gamma(a2))*symsum(Nr/Dr,k,0,K);
%in order of variables @(K,X,a,r,z)
CDFF=subs(CDF,[X,a,r,K],[ZZ,aa,rr,KK]);
integraN=(1/2)*1/sqrt(z*sym(pi))*exp(-z)*CDFF^M;
mckCDF=matlabFunction(integraN); % function of z only
   peTT=integral(mckCDF,1e-5,inf);
end


function pe = laplace_ber_func(alpha,rho,Z,M) % Laplace based
syms s r a X y
fLr=a^a/(a^2+s^2*X^2+2*s*X*a-s^2*X^2*r)^((1/2)*a);
fLRf=subs(fLr,[a,X,r],[alpha,Z,rho]);
f1M=fLRf^M; % laplace transform MRC
fun2M=subs(f1M,s,1/(sin(y))^2);
         mckayf2=matlabFunction(fun2M);
        pe=(1/pi)*integral(mckayf2,0,0.999975*pi/2);
end
```

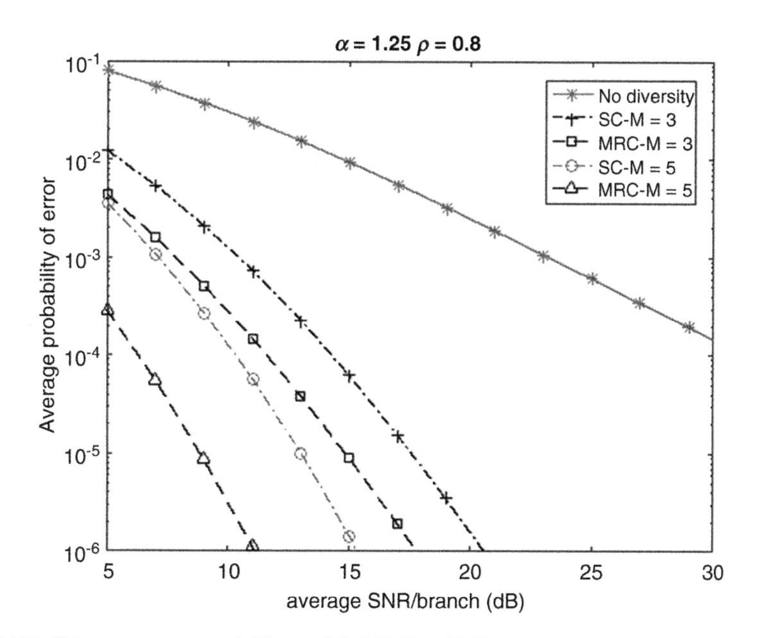

Fig. 5.104 Bit error rates, $\alpha = 1.25$, $\rho = 0.8$ (MRC and SC)

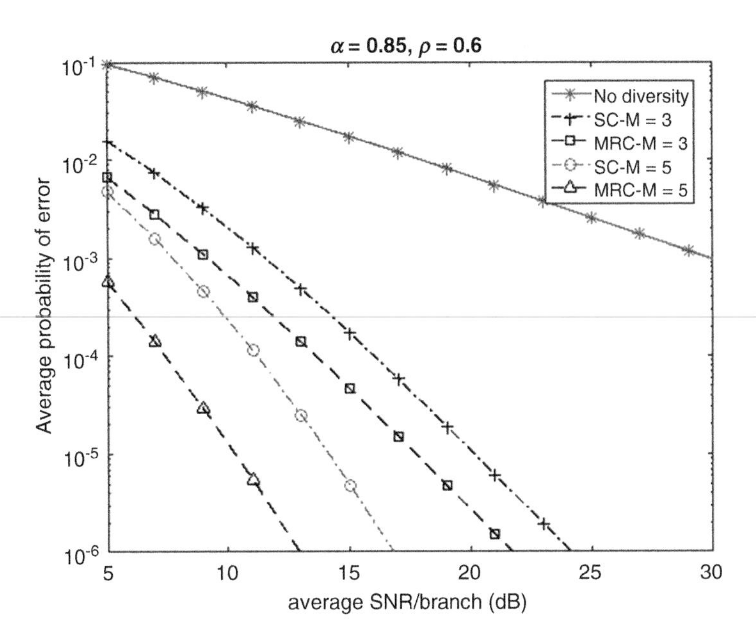

Fig. 5.105 Bit error rates, $\alpha = .85$, $\rho = 0.6$ (MRC and SC)

5.12 Diversity in McKay Faded Shadowed Channels

Since wireless channels undergo fading and shadowing simultaneously, it is also of interest to examine the performance of diversity combining algorithms when McKay fading and lognormal shadowing are present (Shankar 2012a, b; Shankar 2015). The probability density function of the SNR in a shadowed fading channel was given in Chap. 4 (Eq. 4.316) and it is given by

$$f(z) = \int_0^\infty \frac{\sqrt{\pi} \quad z^{\frac{\alpha-1}{2}}}{\left(2\frac{\alpha\sqrt{\rho}}{w(1-\rho)}\right)^{\frac{\alpha-1}{2}} \Gamma\left(\frac{\alpha}{2}\right) \left[\left(\frac{w}{\alpha}\right)^2 (1-\rho)\right]^{\frac{\alpha}{2}}} \times \exp\left[-\frac{\alpha}{w(1-\rho)} z\right] I_{\frac{\alpha-1}{2}}\left(\frac{\alpha\sqrt{\rho}}{w(1-\rho)} z\right) f(w)dw. \tag{5.227}$$

Assuming that the shadowing is same at all the diversity components (spatial or frequency diversity), Eq. (5.227) can be updated to reflect the diversity implementation by replacing the McKay density with the pdf in Eq. (5.222). Replacing X by w, Eq. (5.222) becomes

$$f_{Z_{MRC}}(z_{MRC}|w) = A(w)\exp\left[-\frac{\alpha}{w(1-\rho)} z_{MRC}\right] z_{MRC}^{\frac{(M\alpha-1)}{2}} I_{\frac{(M\alpha-1)}{2}}\left(\frac{\alpha\sqrt{\rho}}{w(1-\rho)} z_{MRC}\right). \tag{5.228}$$

In Eq. (5.228), the constant $A(w)$ can be written from Eq. (5.223) as

$$A(w) = \frac{\sqrt{\pi}}{\left(\frac{2\alpha\sqrt{\rho}}{w(1-\rho)}\right)^{\frac{(M\alpha-1)}{2}} \Gamma\left(\frac{M\alpha}{2}\right)\left[\left(\frac{w}{\alpha}\right)^2(1-\rho)\right]^{\frac{M\alpha}{2}}}. \tag{5.229}$$

Note that the density function can also be represented as an infinite sum as it was shown earlier. Using the summation, the error rate becomes

$$p_{eSh}(X) = \int_0^{\infty} \sum_{k=0}^{\infty} A_{k2}F^2 1\left(\left[\alpha+2k, \alpha+2k+\frac{1}{2}\right], [\alpha+2k+1], -\frac{\alpha}{w(1-r)}\right) f(w)dw. \tag{5.230}$$

In Eq. (5.230), the constant A_k is given by

$$A_k = \frac{\Gamma\left(\alpha+2k+\frac{1}{2}\right)\left(\frac{\alpha}{2w}\right)^{m+2k} r^k}{\Gamma\left(k+\frac{1}{2}+\frac{\alpha}{2}\right)\Gamma\left(\frac{\alpha}{2}\right)(\alpha+2k)(1-r)^{2k+\frac{\alpha}{2}}k!.} \tag{5.231}$$

The error rate can also be obtained from the Laplace transform. Replacing X with w in Eq. (5.226) and multiplying by $f(w)$ and integrating, the error rate following MRC diversity becomes

$$p_{esh}(X) = \frac{\alpha^{M\alpha}}{\pi} \int_0^{\infty} \int_0^{\frac{\pi}{2}} \left(\alpha^2 + \frac{w^2}{\sin^4(\theta)} + 2s\frac{w}{\sin^2(\theta)}\alpha - s^2\frac{w^2}{\sin^4(\theta)}r\right)^{-\frac{M\alpha}{2}} f(w)d\theta dw \tag{5.232}$$

In Eqs. (5.230) and (5.232), note that the relationship between X and μ has to be used,

$$\mu = 10\log_{10}(X) - \frac{\sigma^2}{2K}. \tag{5.233}$$

The error rate following SC diversity can be obtained using the cumulative distribution of the output of the SC algorithm. In terms of the error rate calculation using the CDF, the error rate following SC diversity in McKay faded shadowing channels will be

$$p_{eSh-SC}(X) = \int_0^{\infty}\int_0^{\infty} [F_Z(z_{sc}|w)]^M \frac{1}{2\sqrt{\pi z_{sc}}} \exp(-z_{sc}) f(w) dz_{sc} dw. \tag{5.234}$$

In Eq. (5.234),

$$F_Z(z_{sc}|w) = \sqrt{\pi}\frac{(1-r)^{\frac{\alpha}{2}}}{\Gamma\left(\frac{\alpha}{2}\right)} \sum_{k=0}^{\infty} \frac{r^k}{k!\Gamma\left(k+\frac{1}{2}+\frac{\alpha}{2}\right)2^{\alpha+2k-1}} \gamma\left(\alpha+2k, \frac{\alpha}{w(1-r)} z_{sc}\right) \tag{5.235}$$

The Matlab script used for the computation of error rates for MRC and SC appears below. The results for MRC are displayed in Figs. 5.106 and 5.107. Results for SC are displayed in Figs. 5.108 and 5.109.

```
function ber_mckayshadowing_MRC_book
% MRC to mitigate the effects of fading. (micro diversity). BER is first
% evaulated for MRC diversity keeping  SNR as a random variable. The error
% is further averaged over the lognormal density. The plots are obtained
% from a selected value of alpha, rho and sigma as a function of the order
% of diversity. Only requires a single integration.
% P M Shankar, October 2016
close all
Z0=[10:2:40];
alpha=.95;
rh=0.35;
sig=4;
MN=[1,3,5];
MQ=length(MN);
LK=length(Z0);
KN=10/log(10);
mu=Z0-sig^2/(2*KN); % average measured in dB
MM=10;% number of terms in the summation
% create the integrand using symbolic variables
syms a x r MS sigm MU K k MD
% the integrand
x1=x*MD; % scaled mean X
f1=gamma(a+1/2)*(a/x1)^a*hypergeom([a, a+1/2], [a+1], -a/((1-r)*x1))/...
    (gamma((1/2)*a)*a*gamma(1)*gamma(1/2+(1/2)*a)*2^a*(1-r)^((1/2)*a));
f2=symsum(gamma(a+2*k+1/2)*(a/x1)^(a+2*k)*r^k*hypergeom([a+2*k, a+2*k+1/2],...
    [a+2*k+1], -a/((1-r)*x1))/...
    (gamma((1/2)*a)*(a+2*k)*gamma(k)*gamma(k+1/2+(1/2)*a)*2^(a+2*k)*...
    (1-r)^(2*k+(1/2)*a)),k,1,MS);
f=f1+f2;
ff=sym(K)*exp(-(10*log10(x)-MU)^2/(2*sigm^2))/sqrt((2*sym(pi)*x*x)*sigm^2);
fun=f*ff; % this is the integrand
pesh=zeros(LK,MQ);
for kr=1:MQ
    M=MN(kr);% sigma value
    alphaM=alpha*M;
for k=1:LK
% substitute the values
funer=subs(fun,[K,a,r,sigm,MU,MS,MD],[KN,alphaM,rh,sig,mu(k),MM,M]);
% create the in-line function
mfuner=matlabFunction(funer);
% perform the integration
pesh(k,kr)=integral(mfuner,1e-1,inf);
% generate random numbers
clear X % clear symbolic X

end;

end;

% figure
 semilogy(Z0,pesh(:,1),'-r',Z0,pesh(:,2),'--k', Z0,pesh(:,3),'-.b',...
     'linewidth',1)
  xlabel('average SNR (dB)'),ylim([1e-6,1])
   legend(['M = ',num2str(MN(1))],...
   ['M = ',num2str(MN(2))],['M = ',num2str(MN(3))]);
   ylabel('Average probability of error')
  title(['\alpha = ',num2str(alpha),'  r = ',num2str(rh),...
      '  \sigma = ',num2str(sig),' dB'])


end
```

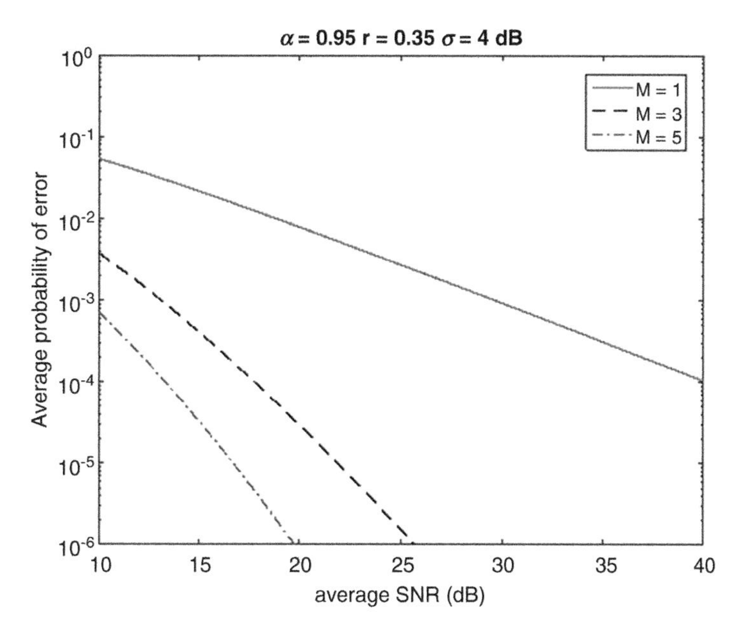

Fig. 5.106 Error rates $\alpha = .95$, $\rho = 0.35$, $\sigma = 4$ dB (MRC)

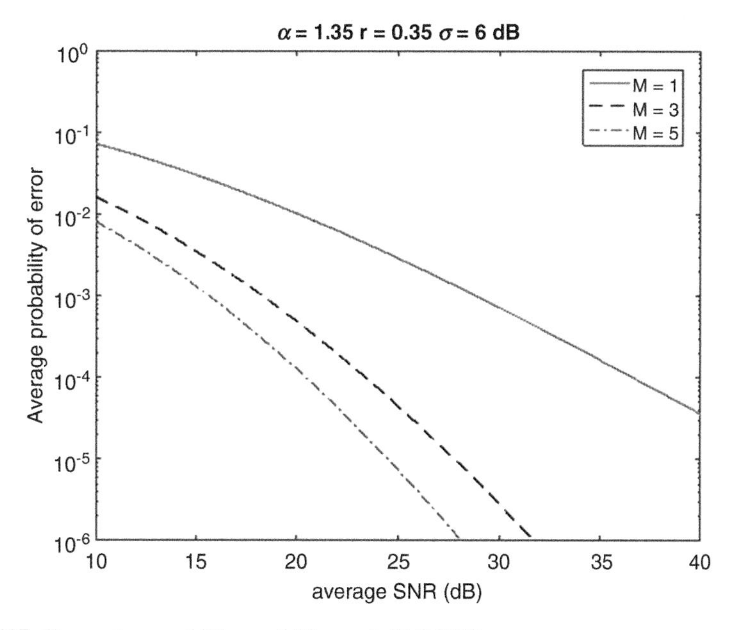

Fig. 5.107 Error rates $\alpha = 1.35$, $\rho = 0.35$, $\sigma = 6$ dB (MRC)

```
function ber_mckayshadowing_SC_book
% SC to mitigate the effects of fading. (micro diversity). The BER is first
% evaulated for SC diversity keeping SNR as a random variable. The error
% is further averaged over the lognormal density. The plots are obtained
% from a selected value of alpha, rho and sigma as a function of the order
% of diversity. Requires a double integral.
% Results are comprared to random number simulations.. with 1e7 set. the
% error rates are accurate only up to 1e-6.

% P M Shankar, July 2016
close all
Z0=[10:2:40];
alpha=1.15;
rh=0.45;
sig=4;
MN=[1,3,5];
MQ=length(MN);
LK=length(Z0);
KN=10/log(10);
mu=Z0-sig^2/(2*KN); % average measured in dB
MM=10;% number of terms in the summation
% create the integrand using symbolic variables
syms a x r sigm MU km MD z MS
% X is taken as 1 for shadowed fading
xa=a/(x*(1-r));
a2=a/2;
gamaprt=gamma(a+2*km)-igamma(a+2*km,xa*z);
Nr=gamaprt*r^km;
Dr=factorial(km)*gamma(km+1/2+a2)*2^(a+2*km-1);
CDF1=((sqrt(sym(pi))*(1-r)^(a2))/gamma(a2))*symsum(Nr/Dr,km,0,MS);
CDF=CDF1^MD;
fun=CDF*0.5*exp(-z)/sqrt(z*sym(pi));
ff=sym(KN)*exp(-(10*log10(x)-MU)^2/(2*sigm^2))/sqrt((2*sym(pi)*x*x)*sigm^2);
fun=fun*ff; % this is the integrand
pesh=zeros(LK,MQ);
%pesd=zeros(LK,MQ); % simulation
for kr=1:MQ
    M=MN(kr);% sigma value
for k=1:LK
% substitute the values
funer=subs(fun,[a,r,sigm,MU,MS,MD],[alpha,rh,sig,mu(k),MM,M]);
% create the in-line function
mfuner=matlabFunction(funer);
% perform the integration
pesh(k,kr)=integral2(mfuner,0.1,1e5,0.1,1e5);

 end;

end;
figure
 semilogy(Z0,pesh(:,1),'-r',Z0,pesh(:,2),'b-',...
     Z0,pesh(:,3),'--k','linewidth',1)
  xlabel('average SNR (dB)'),ylim([1e-6,1])
   legend(['M = ',num2str(MN(1))],['M = ',num2str(MN(2))],...
   ['M = ',num2str(MN(3))]);
   ylabel('Average probability of error')
  title(['\alpha = ',num2str(alpha),'  \rho = ',num2str(rh),...
      '  \sigma = ',num2str(sig),' dB'])

end
```

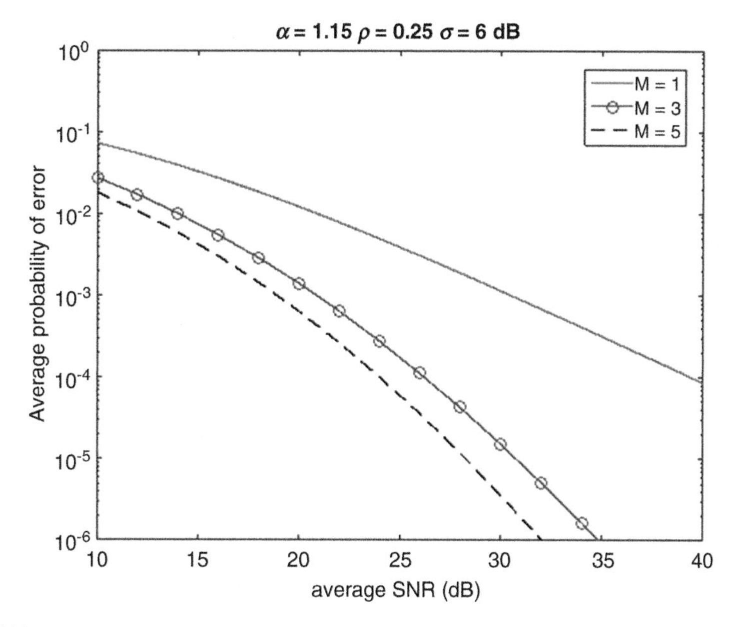

Fig. 5.108 Error rates $\alpha = 1.15$, $\rho = 0.25$, $\sigma = 6$ dB (SC)

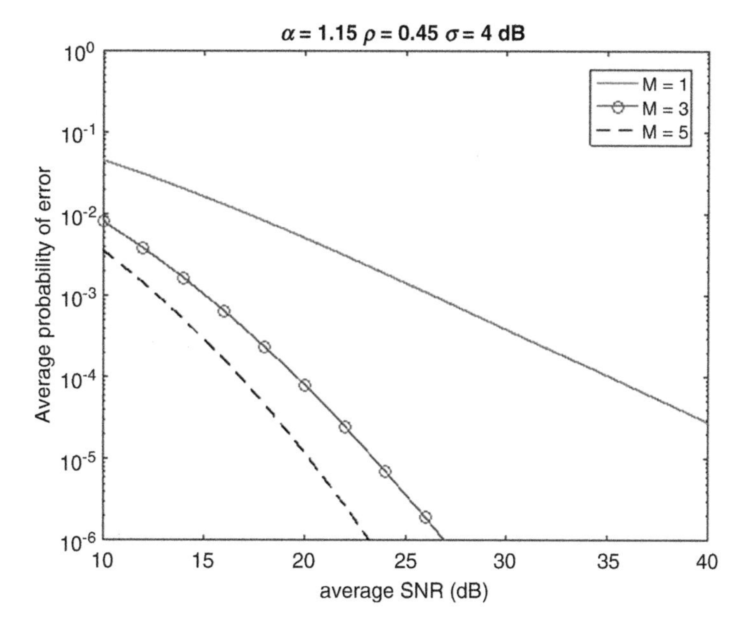

Fig. 5.109 Error rates $\alpha = 1.15$, $\rho = 0.45$, $\sigma = 4$ dB (SC)

5.13 Diversity in Channels Modeled as Gamma Mixtures

Gamma mixtures provide simple ways to overcome the lack of appropriate analytical expressions for the error rates and outage probabilities in fading channels (Atapattu et al. 2011). With equal gain combining and generalized selection combining algorithms, the analytical expressions for the pdf of the output of the algorithms are in integral forms making the estimation of error rates and outages computationally tedious. Following earlier discussions in Chaps. 2 and 4, the mixture density is expressed as (Chatelain et al. 2008; Venturini et al. 2008)

$$f_Z(z) = \sum_{i=1}^{n} p_i f_i(z). \tag{5.236}$$

In Eq. (5.236), the weights p's satisfy the total probability criterion

$$\sum_{i=1}^{n} p_i = 1 \tag{5.237}$$

The components of the gamma mixture are

$$f_i(z) = z^{\alpha_i - 1} \frac{e^{-\frac{z}{\beta_i}}}{\beta_i^{\alpha_i} \Gamma(\alpha_i)}, \quad i = 1, 2, \cdots, n \tag{5.238}$$

Note that n is the number of components in the mixture and density in Eq. (5.238) has a mean of $\alpha_i\beta_i$. In other words, Eq. (5.238) represents the density of the SNR in a Nakagami channel with a parameter $m = \alpha_i$ and average SNR$Z_0 = \alpha_i\beta_i$.

The success of the mixture model depends on the determination of the optimum number of components needed to have the density in Eq. (5.236) match the density of the output of the diversity combining algorithm. When the number of components n becomes large, chi-square testing becomes meaningless (typical number of bins such as 5 or 10 will be smaller than the number of parameters being estimated) making it necessary to look for other means of determining the optimum value of n (Papoulis and Pillai 2002). One of the simpler options was discussed in Chaps. 2 and 4 which uses the mean square error (MSE). Even though the fit improves with n, higher and higher values of n would require more computations as the number of parameters for an n-component gamma mixture is $(3n-1)$. Therefore, the interest is not only finding a better match, but having a match that is optimum in terms of computational complexity as well. The selection or the choice of the match will be more appropriate if a probability value can be associated with n, the most appropriate fit being the one with a probability nearest to unity. Such a probabilistic measure can be obtained using the concept of the Bayes Information Criterion (BIC). To understand BIC (Schwartz 1978; Atkinson 1981; Kim and Taylor 1995;

Kass and Raftery 1995; Mayrose et al. 2005; Almhana et al. 2006; Claeskens 2016), there is a need to go back to the log likelihood function defined in Chap. 2. If an optimum set of parameters is obtained for n, the log likelihood function LLF(n) can be written as

$$LLF(n)=\sum_{j=1}^{N}\log[f(z_j)]=\sum_{j=1}^{N}\left(\log\sum_{i=1}^{n}p_i f_i(z_j|\alpha_i,\beta_i)\right). \tag{5.239}$$

In Eq. (5.239), N is the number of samples of the data available. The Bayes Information Criterion takes into account the log likelihood function, the number of mixture components n, and the number of samples N. The relationship of BIC to the LLF, n, and N is given as (Schwartz 1978; Atkinson 1981; Neath and Cavanaugh 2012; Claeskens 2016)

$$BIC=-2\log(LLF)+[3n-1]\log(N) \tag{5.240}$$

It has been shown that BIC is a concave function of n and therefore, one chooses the value of n (number of mixture components) for which BIC is minimum (Claeskens 2016). Equation (5.240) clearly indicates that BIC values go up with increasing values of n, displaying the risk of overfitting with higher values of n. It also shows that the more samples are used, the greater is the value of BIC. But, the concavity of BIC will assure the optimum value of n.

While BIC offers a reasonable option, some more refining is needed since BIC values might fall in 1000s as seen from its dependence on N. When BIC values are very high and the differences among the BIC values are small, the decision of rejecting values of n is a difficult task. This problem can be overcome through the development of a posterior probability associated with each value of n. To understand this, one needs to find the minimum value of BIC and the differences between this minimum and the other values of BIC,

$$BIC_{\min}=\text{minimum}[BIC(1),BIC(2),\ldots BIC(n)] \tag{5.241}$$

$$\Delta_k=BIC(k)-BIC_{\min},\quad k=1,2,\ \ldots,n \tag{5.242}$$

It is clear that for the model to be selected ($n=$ k), $\Delta_k$will be zero and one needs to justify a basis for rejecting the rest of the values of n. A threshold value proposed to accomplish this is the following:

$$\begin{matrix}\Delta_k\le 2, & \text{do not reject the model with k component densities}\\ \Delta_k>10, & \text{reject the model with k component densities}\end{matrix}. \tag{5.243}$$

The step in Eq. (5.243) is only the first one in the quest for the best way to reject certain values of n. Since the values of BIC are quite large, the range between rejection and acceptance is rather broad. Therefore, there is a need to rank the models (corresponding to the values of n) by assigning a value of "1" for the best fit and "0" for the least likely fit. An appropriate quantity would be the a posteriori (posterior) probability identifying the model with a specific value of $\boldsymbol{n}$ being the most likely fit, given the data Z. Using Bayes' rule, the posterior probability $P(k|Z)$ can be expressed as (Neath and Cavanaugh 2012)

$$P(k|Z)P(Z) = P(Z|k)P(k), \quad k = 1, 2, \cdots n. \tag{5.244}$$

In Eq. (5.244) $P(k)$is the a priori probability of choosing a specific value of k from the set, 1, 2, . . ., n and

$$P(Z) = \sum_{k=1}^{n} P(Z|k)P(k). \tag{5.245}$$

Since the minimum value of BIC is associated with the best fit, the conditional probability $P(Z|k)$can be expressed as

$$P(Z|k) = e^{-\frac{1}{2}\Delta_k}. \tag{5.246}$$

The value of $\Delta_k$is given in Eq. (5.242). Using Eqs. (5.245) and (5.246), the posterior probability $P(k|Z)$becomes

$$P(k|Z) = \frac{P(Z|k)P(k)}{\sum_{k=1}^{n} P(Z|k)P(k)}. \tag{5.247}$$

Since one could have started exploring with any value of n to seek the fit, the a priori probability of choosing any value of $n = k$ is

$$P(k) = \frac{1}{n}, \quad k = 1, 2, \ldots, n. \tag{5.248}$$

Using Eq. (5.248), the posterior probability becomes

$$P(k|Z) = \frac{P(Z|k)}{\sum_{k=1}^{n} P(Z|k)}. \tag{5.249}$$

The posterior probability associated with kth model (k = 1, 2, . . ., n) is

$$P(k|Z) = \frac{e^{-\frac{1}{2}\Delta_k}}{\sum_{k=1}^{n} e^{-\frac{1}{2}\Delta_k}}. \tag{5.250}$$

It is now possible to observe the strength of BIC as a qualifier since

$$P(k|Z) = \frac{e^{-\frac{1}{2}\Delta_k}}{\sum_{k=1}^{n} e^{-\frac{1}{2}\Delta_k}} = \frac{1}{\sum_{k=1}^{n} e^{-\frac{1}{2}\Delta_k}}, \quad BIC(k) = BIC_{\min}, \quad 1 \le k \le n \tag{5.251}$$

Since the maximum value of the numerator in Eq. (5.250) is unity, the posterior probability forms the basis for the weight of evidence for the best fit for a specific value of $n = k$. Thus, Eq. (5.250) offers an opportunity to rank the models (defined in terms of the number of components n) using a measure that ranges from 1 to 0, with the model with $P(k|Z)=1$ offering an absolute certainty that none of the other values of n fit the data.

This concept is demonstrated in an example. A number of gamma variables are mixed and the data set analyzed using the technique described in Chaps. 2 and 4 to obtain the parameter sets for $n = 1,2,\ldots,6$. Results are summarized in several figures. Figure 5.110 shows the fit to a single gamma density and the parameters. The slight misfit to the histogram of the data can be seen. Neither the BIC value nor the MSE value conveys any specific information on the strength of the fit even though the MSE value is small. Figure 5.111 shows the fit with $n = 3$. One can see that BIC value has come down while MSE still very low. Figure 5.112 shows the fit $n = 5$. The BIC value has come down w.r.t the values for $n = 1$ and 3. Figure 5.113 shows the fit for $n = 6$. It can be seen that the BSC value has gone up. These observations have been summarized in Fig. 5.114 (generated automatically in

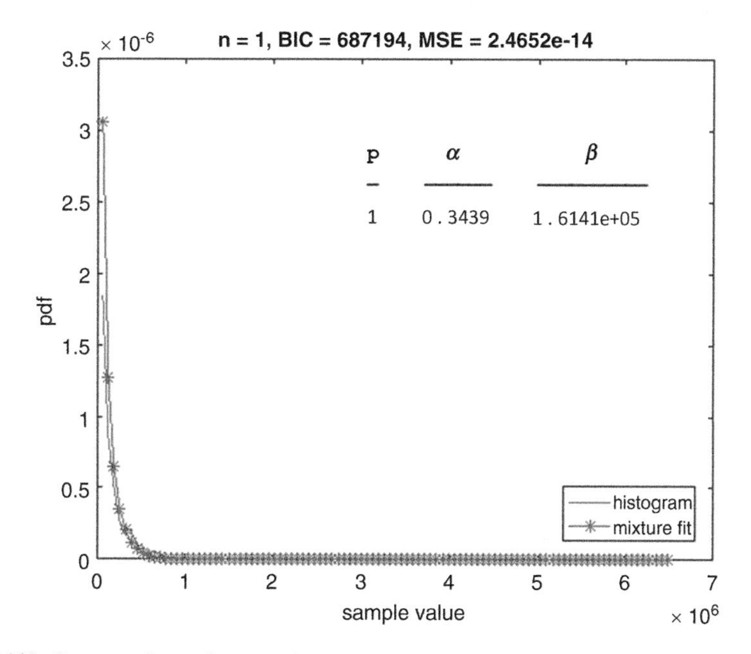

Fig. 5.110 Gamma mixture fit to $n = 1$

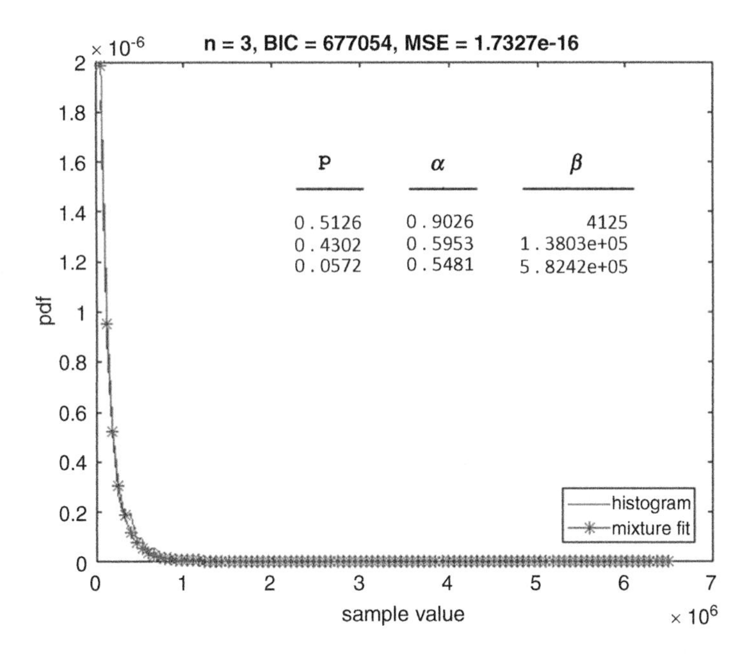

Fig. 5.111 Gamma mixture fit to $n = 3$

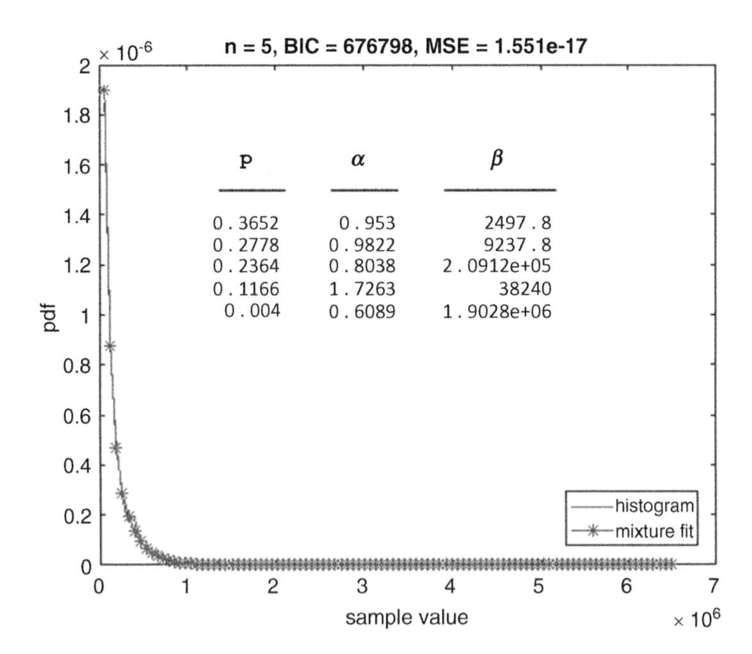

Fig. 5.112 Gamma mixture fit to $n = 5$

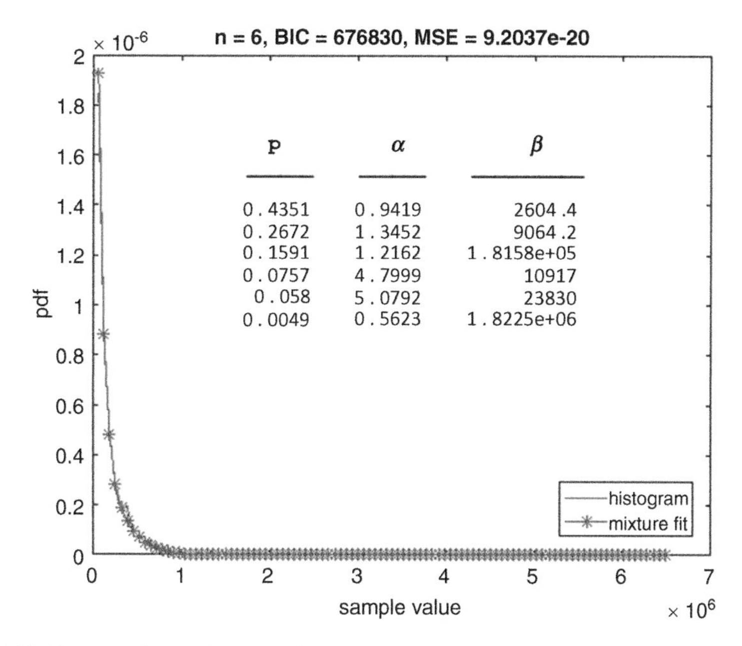

Fig. 5.113 Gamma mixture fit to $n = 6$

n	p	α	β	BIC	Post. Prob.	MSE	$\chi^2_T\,[\chi^2_t]$
1	1	0.34	161410	687194	0	2.4652e−14	>60 [14]
2	0.51 0.49	0.91 0.5	4323 216333	677575	0	1.8999e−16	>60 [9]
3	0.5126 0.4302 0.0572	0.9 0.6 0.55	4125 138027 582418	677054	0	1.7326e−16	>60 [4]
4	0.4153 0.3846 0.189 0.011	0.97 0.64 0.9 0.52	2884 181188 14334 1285622	676848	0	1.469e−16	[−−]
5	0.3587 0.2781 0.2204 0.1389 0.004	0.95 0.98 0.85 1.46 0.62	2458 8766 207713 43504 1910304	676796 ⇐ min	1	1.3319e−17	[−−]
6	0.4333 0.2691 0.1591 0.076 0.0575 0.0049	0.94 1.33 1.22 4.79 5.12 0.56	2600 9123 181562 10972 23719 1822407	676830	0	8.9846e−20	[−−]

Fig. 5.114 Summary fits showing BIC, posterior probability, and chi-square test stats

n	p	α	β	BIC		Post. Prob.	MSE	$\chi_T^2\,[\chi_t^2]$
1	1	1.97	46.77	54130		0.0006	5.575e-09	>60 [14]
2	0.8281 0.1719	2.25 3.19	34.17 51.82	54115	⇐ min	0.9994	4.552e-09	9 [9]
3	0.7923 0.1435 0.0642	2.39 4.3 14.85	29.65 44.01 9.15	54135		0	4.338e-09	10 [4]
4	0.668 0.2138 0.0619 0.0563	2.5 3.61 16.86 27.6	25.11 46.46 5.7 5.4	54159		0	4.2247e-09	[--]
5	0.5763 0.2402 0.0699 0.0583 0.0552	2.65 3.51 53.75 41.36 46.01	21 46.36 2.09 1.83 3.41	54176		0	4.1368e-09	[--]
6	0.6069 0.1834 0.0744 0.0624 0.0517 0.0213	2.65 4.21 60.38 44.22 60.69 61.35	21.41 42.88 1.84 1.72 2.44 2.96	54200		0	4.1426e-09	[--]

Fig. 5.115 Gamma mixture exploration of EGC (4) with $Z_0 = 15$ dB; $\alpha = 0.75$; $\rho = 0.7$

Matlab) showing all the parameter value sets, BIC, MSE, posterior probabilities, and the chi-square test statistic for $n = 1$, 2, and 3 for 10 bins. It is seen that posterior probability is unity for $n = 5$ suggesting that that is the best fit. Even though the MSE value goes down with $n = 6$, the results suggest that $n = 5$ is sufficient and $n > 6$ leads to higher levels of computations without offering any better fits.

This concept is utilized to obtain the fits for the outputs of EGC ($M = 4$ and $M = 2$) and GSC diversity ($M = 4$) in McKay fading channels. Results for EGC are shown in Figs. 5.115, 5.116, and 5.117. They show that $n = 2$ is a sufficient fit. A number of samples (5000) of McKay variables were created for the diversity analysis. In the first step, the gamma mixture concept was explored for n = 1, 2, . . ., 6 and the results are displayed in Fig. 5.115.

The analysis was repeated five times and the average values are summarized in Fig. 5.116.

A similar simulation was undertaken for $M = 2$ and the summary of the results is given in Fig. 5.117.

The GSC fit is shown in Fig. 5.118 with MRC applied to three strongest branches. Once again, the fit with $n = 2$ is appropriate.

Once the best fit parameters are obtained, the error rates and outage probabilities can be evaluated. While the technique seems direct in terms of having an analytical expression for the equivalent pdf of the SNR following diversity processing, the

n	p	α	β	BIC		Posterior Probability	$\chi_T^2\ [\chi_{thr}^2]$
1	1	1.9506	47.1806	54149		0.0003	30 [14]
2	0.70154 0.29846	2.3398 2.2294	33.7209 40.9598	54133	⇐ BIC_{min}	0.9996	6 [9]
3	0.7276 0.22152 0.0509	2.4238 5.3812 12.0544	27.7328 37.3338 14.5879	54155		0	3 [4]
4	0.67608 0.21192 0.06828 0.04372	2.5038 9.8446 15.204 31.497	25.8017 29.2947 16.2997 5.68204	54178		0	[– –]

Average of 5 simulations
Input: EGC McKay

Z_0 = 15 dB
α = 0.75
ρ = 0.7
EGC (4)

Fig. 5.116 Gamma mixture fit data for EGC ($M = 4$)

n	p	α	β	BIC		Posterior Probability	$\chi_T^2\ [\chi_{thr}^2]$
1	1	1.44	3.8213	26700		0	27 [14]
2	0.66302 0.33698	1.7552 2.2258	3.1457 2.1545	26672	⇐ BIC_{min}	0.9999	9 [9]
3	0.54544 0.29992 0.15466	1.9284 3.3098 8.3344	2.1068 2.6403 1.5772	26691		0.0001	5 [4]
4	0.62034 0.23178 0.09682 0.05106	1.934 8.4996 16.4094 15.9254	1.6501 2.6804 0.45906 1.1817	26712		0	[– –]

Average of 5 simulations
Input: EGC McKay

Z_0 = 5 dB
α = 1.2
ρ = 0.8
EGC (2)

Fig. 5.117 Gamma mixture fit data for EGC ($M = 2$)

computational issues related to the determination of the optimum parameter set are not simple.

As suggested in Chap. 4, other mixture densities including Gaussian mixtures have been proposed to obtain analytical expression for the densities.

n	p	α	β	BIC		Posterior Probability	$\chi_T^2\ [\chi_{thr}^2]$
1	1	1.8526	66.0612	57141		0	28 [14]
2	0.66138 0.33862	2.4626 2.7308	56.2876 31.7108	57106	⇐ BIC_{min}	1	9 [9]
3	0.63748 0.24998 0.11254	2.48 5.8456 9.7136	45.821 32.331 19.5986	57127		0	7 [4]
4	0.6762 0.16812 0.0989 0.0568	2.5298 7.709 18.4682 24.6878	31.3012 43.1126 15.259 9.04436	57152		0	[--]

Average of 5 simulations
Input: GSC(4,3) McKay

Z_0= 15 dB
α = 0.75
ρ = 0.7
GSC(4,3)

Fig. 5.118 Gamma mixture fit data for GSC(4,3)

5.14 Additional Examples of Random Number Simulation

It is possible to verify the results through random number simulation. The results on the error rates and outage probabilities reported earlier are based on analytical expressions.

5.14.1 Simulation of McKay Channels

A Matlab script that undertakes the comparison of error rates obtained using random number simulation and analytical methods appears below. The analytical approaches use Laplace transforms and summation using Eq. (5.224) for the CDF (See Figs. 5.119, 5.120, 5.121, 5.122, 5.123, 5.124, 5.125, and 5.126).

```
function  ber_mckay_MRC_diversity
% P M Shankar
% BER in MRC diversity is calculated
% The error rate is calculated using random number simulation. Theoretical
% error rates calculated using the CDF and Laplace transforms for
% comparison

%July 2016
close all
global Z LK Z0 MM alpha rho MN
MM=20; % number of terms in the summation
 Z0=[5:2:25];
Z=10.^(Z0/10);
LK=length(Z);
MN=[1,2,4];
for kq = 2:2
    if kq==1
alpha=1.5; rho=0.35;
    elseif kq==2
     alpha=1.5; rho=0.65;
    else
          alpha=2.5; rho=0.5;
    end;
ber_mckay_M
end;
end

function  ber_mckay_M
% P M Shankar
global Z LK Z0 MM alpha rho MN
MQ=length(MN);
peS1=zeros(LK,MQ);% pe using random number simulation
peT1=zeros(LK,MQ);
% using Laplace transforms
peL1=zeros(LK,MQ);
%
for km=1:MQ
    M=MN(km);
    alphaM=M*alpha;
    for k=1:LK
        X=Z(k);
        XM=M*X;
        X1=X*(1+sqrt(rho))/2;
X2=X*(1-sqrt(rho))/2;
a1=alpha/2;
x=gamrnd(a1,X1/a1,M,1e7)+gamrnd(a1,X2/a1,M,1e7);
if M>1
x=sum(x);
else
end;
         peS1(k,km)=berest(x);
         peT1(k,km)=berMcKay(alphaM,rho,XM,MM);
         peL1(k,km)=laplace_ber_func(alpha,rho,X,M);
         clear x
    end;
end;
%
```

```
figure,semilogy(Z0,peT1(:,1),'-r',Z0,peS1(:,1),'r*',Z0,peL1(:,1),'r^',...
 Z0,peT1(:,2),'--k',Z0,peS1(:,2),'ko',Z0,peL1(:,2),'ks',...
 Z0,peT1(:,3),'-.b',Z0,peS1(:,3),'bd',Z0,peL1(:,3),'bh')
xlabel('average SNR (dB)'),xlim([5,25]),ylim([1e-6,.1])
ylabel('Average probability of error')
title(['\alpha = ',num2str(alpha),',  \rho = ',num2str(rho)])
legend(['Th (M = ',num2str(MN(1)),')'],...
 ['S (M = ',num2str(MN(1)),')'],...
 ['L (M = ',num2str(MN(1)),')'],...
 ['Th (M  = ',num2str(MN(2)),')'],...
 ['S (M  = ',num2str(MN(2)),')'],...
 ['L (M = ',num2str(MN(2)),')'],...
 ['Th (M  = ',num2str(MN(3)),')'],...
 ['S (M  = ',num2str(MN(3)),')'],...
 ['L (M = ',num2str(MN(3)),')'])
end


function ber = berest(simdata)
L=length(simdata);
nn=normrnd(0,1,1,L);%Gaussian noise of zero mean and unit variance
h=simdata;%data is in units of power
h=sqrt(h);%needs to be converted into envelope or amplitude units
%%input data is the BPSK bit sequence
ip = rand(1,L)>0.5;%generate 0's and 1's
indata=2*ip-1;%bipolar data
s=indata;
out= sqrt(2)*h.*s+nn;% this simulates sqrt(2E/T)*bk*a+n
   x1=out>0;%detects 1s and zeros
   x2=-1*(out<0);%detect-1's and zeros
   outdata=x1+x2;%recreates the output bit stream in bipolar form
   Diffdata=s-outdata;% will be either +2 or -2 when there is an error
   DIF=abs(Diffdata);%will be 2 or zero
   berN=sum(DIF>0);%counts how many times the abs difference exceeds zero.
   ber=berN/L;

end

function peTT =berMcKay(aa,rr,ZZ,MM)
% p m shankar, BER calculation using the summation
% the function is obtained symbolically first
syms a X k M r
ak=a+2*k;
Xa=a/(X*(1-r));
X2=2*X;
Nr1=(a/X2)^ak*r^k*gamma(ak+1/2);
Dr1=factorial(k)*gamma(1/2+ak/2)*gamma(a/2)*ak*(1-r)^(2*k+a/2);
NDR=Nr1/Dr1;
pes=symsum(NDR*hypergeom([ak,ak+1/2],[ak+1],(-Xa)),k,0,M);
    pesk=subs(pes,[a,X,r,M],[aa,ZZ,rr,MM]);
   peTT=double(pesk); % this is the ber
end

function pe = laplace_ber_func(alpha,rho,Z,M) % Laplace based
syms s r a X y
fLr=a^a/(a^2+s^2*X^2+2*s*X*a-s^2*X^2*r)^((1/2)*a);
fLRf=subs(fLr,[a,X,r],[alpha,Z,rho]);
f1M=fLRf^M; % laplace transform MRC
fun2M=subs(f1M,s,1/(sin(y))^2);
         mckayf2=matlabFunction(fun2M);
        pe=(1/pi)*integral(mckayf2,0,0.999975*pi/2);
end
```

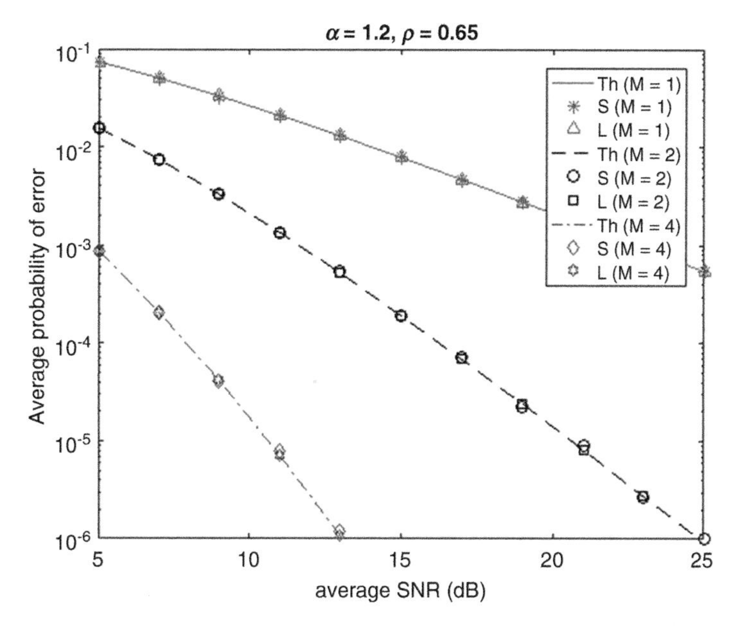

Fig. 5.119 Bit error rates (MRC), random number simulation (S), summation with 20 terms for CDF (Th), and Laplace (L) [$\alpha = 1.2$, $\rho = 0.65$]

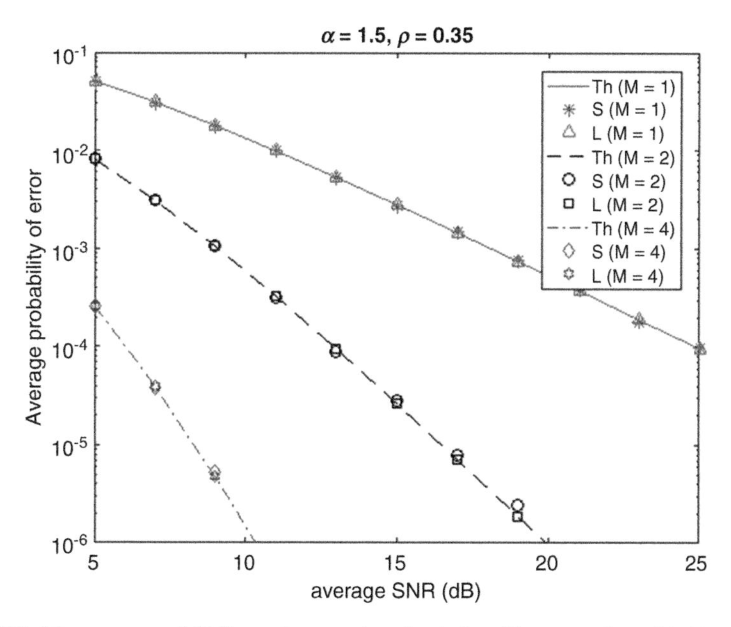

Fig. 5.120 Bit error rates (MRC), random number simulation (S), summation with 20 terms for CDF (Th), and Laplace (L) [$\alpha = 1.5$, $\rho = 0.35$]

```
function  ber_mckay_SC_diversity
% P M Shankar
% BER in SC diversity is calculated.. use the CDF
%July 2016
close all
global Z LK Z0 MM alpha rho MN
MM=20; % number of terms in the summation
 Z0=5:2:25;
Z=10.^(Z0/10);
LK=length(Z);
alpha=1.5;
rho=0.35;
MN=[1,3,4];
ber_mckay_M


end

function  ber_mckay_M
% P M Shankar
global Z LK Z0 MM alpha rho MN M
MQ=length(MN);
peS1=zeros(LK,MQ);% pe using random number simulation
peT1=zeros(LK,MQ);
for km=1:MQ
    M=MN(km);
    for k=1:LK
        X=Z(k);
        X1=X*(1+sqrt(rho))/2;
X2=X*(1-sqrt(rho))/2;
a1=alpha/2;
x=gamrnd(a1,X1/a1,M,1e7)+gamrnd(a1,X2/a1,M,1e7);
if M>1
x=max(x);
else
end;
         peS1(k,km)=berest(x);
         peT1(k,km)=berMcKay(alpha,rho,X,MM);
         clear x
    end;
end;
%
 figure,semilogy(Z0,peT1(:,1),'-r',Z0,peS1(:,1),'r*',...
  Z0,peT1(:,2),'--k',Z0,peS1(:,2),'ko',Z0,peT1(:,3),'-.b',Z0,peS1(:,3),'b^')
 xlabel('average SNR (dB)'),xlim([5,25]),ylim([1e-6,.1])
 ylabel('Average probability of error')
 title(['\alpha = ',num2str(alpha),',  \rho = ',num2str(rho)])
 legend(['Th (M = ',num2str(MN(1)),')'],...
  ['S (M = ',num2str(MN(1)),')'],...
  ['Th (M  = ',num2str(MN(2)),')'],...
  ['S (M  = ',num2str(MN(2)),')'],...
  ['Th (M  = ',num2str(MN(3)),')'],...
  ['S (M  = ',num2str(MN(3)),')'])
end


function ber = berest(simdata)
L=length(simdata);
nn=normrnd(0,1,1,L);%Gaussian noise of zero mean and unit variance
h=simdata;%data is in units of power
h=sqrt(h);%needs to be converted into envelope or amplitude units
%%input data is the BPSK bit sequence
ip = rand(1,L)>0.5;%generate 0's and 1's
indata=2*ip-1;%bipolar data
s=indata;
```

```
out= sqrt(2)*h.*s+nn;% this simulates sqrt(2E/T)*bk*a+n
   x1=out>0;%detects 1s and zeros
   x2=-1*(out<0);%detect-1's and zeros
   outdata=x1+x2;%recreates the output bit stream in bipolar form
   Diffdata=s-outdata;% will be either +2 or -2 when there is an error
   DIF=abs(Diffdata);%will be 2 or zero
   berN=sum(DIF>0);%counts how many times the abs difference exceeds zero.
   ber=berN/L;

end

function peTT =berMcKay(aa,rr,ZZ,KK)
% create the McKay CDF % KK is the number of terms in the summation
global M
syms a r X K z
syms k
xa=a/(X*(1-r));
a2=a/2;

gamaprt=gamma(a+2*k)-igamma(a+2*k,xa*z);
Nr=gamaprt*r^k;
Dr=factorial(k)*gamma(k+1/2+a2)*2^(a+2*k-1);
CDF=((sqrt(sym(pi))*(1-r)^(a2))/gamma(a2))*symsum(Nr/Dr,k,0,K);
%in order of variables @(K,X,a,r,z)
CDFF=subs(CDF,[X,a,r,K],[ZZ,aa,rr,KK]);
integraN=(1/2)*1/sqrt(z*sym(pi))*exp(-z)*CDFF^M;
mckCDF=matlabFunction(integraN); % function of z only

   peTT=integral(mckCDF,1e-5,inf);
end
```

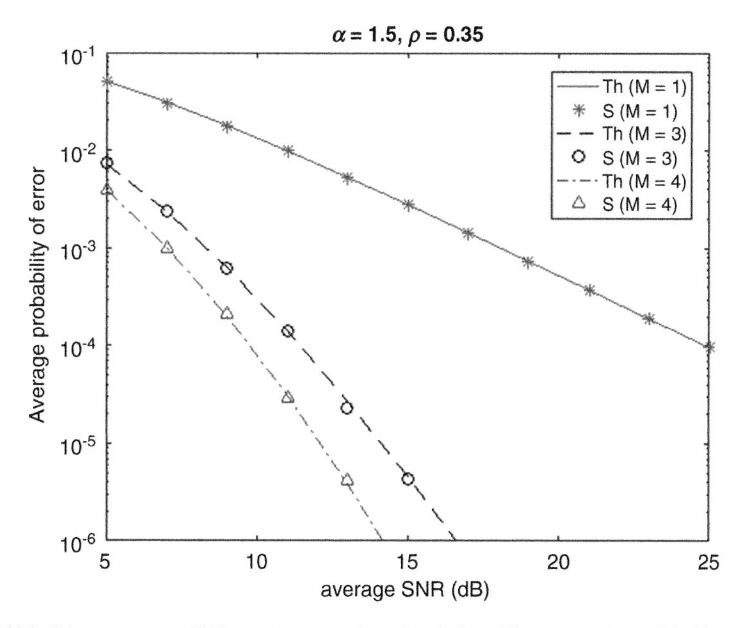

Fig. 5.121 Bit error rates (SC), random number simulation (S), summation with 20 terms for CDF (Th) [$\alpha = 1.5$, $\rho = 0.35$]

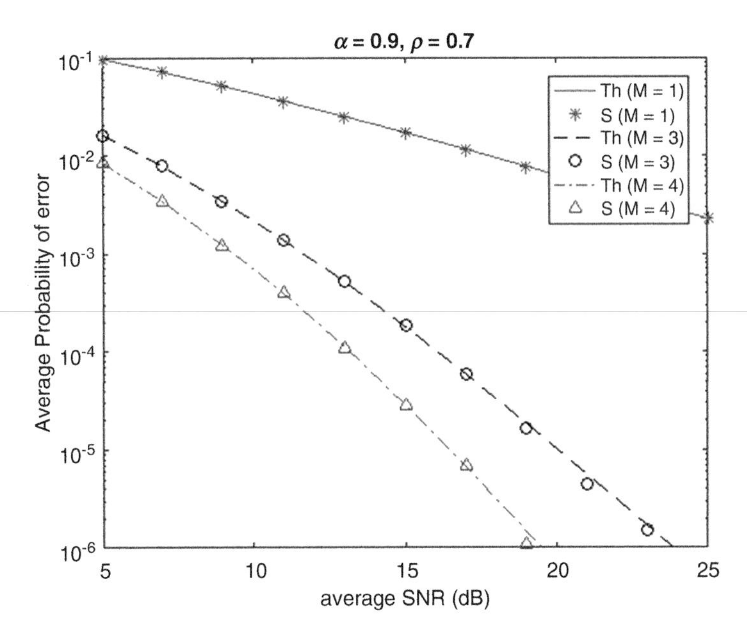

Fig. 5.122 Bit error rates (SC), random number simulation (S), summation with 20 terms for CDF (Th) [$\alpha = 0.9$, $\rho = 0.7$]

5.14.2 *Simulation of Diversity in Nakagami Fading Channels SC, EGC, GSC, and MRC*

```
function diversityALL_demo

% diversity demo for mckay fading. Random number simulation only
%SC, EGC, MRC if M=3 or more,  also GSC for two cases GSC(M,2) GSC(M,3)
% PDF's are obtained and error rates estimated. P M Shankar, October 2016
close all
numb=1e6;%numbers used for simulation
Z0=10;
m=1.5;
M=5;
Z=10^(Z0/10);
a=m; % alpha
r=0.7; % rho
X1=Z*(1+sqrt(r))/2;
X2=Z*(1-sqrt(r))/2;
a1=a/2;
x=gamrnd(a1,X1/a1,M,numb)+gamrnd(a1,X2/a1,M,numb);

xm=max(x);%SC
xs=sqrt(x);%for EGC
xem=(1/M)*sum(xs).^2;%EGC
xmr=sum(x);%MRC
```

```
ttl=strcat('(',num2str(M),')');
name1=strcat('SC',ttl);name2=strcat('EGC',ttl);name3=strcat('MRC',ttl);
tt=strcat([' \alpha = ',num2str(m),',  \rho = ',num2str(r),', X = ',num2str(Z0),' dB']);
[f1,x1]=ksdensity(x(:)); [f2]=ksdensity(xm(:),x1);
[f3]=ksdensity(xem(:),x1);[f4]=ksdensity(xmr(:),x1);
figure, plot(x1,f1,'k-',x1,f2,'m--',x1,f3,'r-.',x1,f4,'--+b','linewidth',1.2)
xlabel('Signal-to-Noise ratio '),ylabel('Probability density function')
xlim([0,0.6*max(x1)])
title(tt)
legend('No Div',name1,name2,name3),grid on

%generalized selection combining
if M>2
    y=sort(x,'descend');%sorts the columns in descending order
    ym2=sum(y(1:2,:));%GSC of 2 out of M
    ym3=sum(y(1:3,:));%GSC of 3 out of M
    [fg2]=ksdensity(ym2(:),x1);[fg3]=ksdensity(ym3(:),x1);
 name2g=strcat('GSC(',num2str(M),',2)'); name3g=strcat('GSC(',num2str(M),',3)');
    figure
    plot(x1,f1,'k-',x1,f2,'r--',x1,fg2,'m-.',x1,fg3,'b--+',x1,f4,'r:h')
    xlim([0,0.6*max(x1)]), title(tt)
    xlabel('Signal-to-Noise ratio'),ylabel('Probability density function')
    legend('No Div',name1,name2g,name3g,name3),grid on
else
end;
clear Z0 x xs xm xmr y ym2 ym3
a=1.2;
a1=a/2;
r=0.75;
Z0=5:.5:20; % average SNR
Z=10.^(Z0/10);
LZ=length(Z);
pe1=zeros(1,LZ);
pes=zeros(1,LZ);
per=zeros(1,LZ);
pee=zeros(1,LZ);
pey3=zeros(1,LZ);
for k=1:LZ
    X1=Z(k)*(1+sqrt(r))/2;
X2=Z(k)*(1-sqrt(r))/2;
    x=gamrnd(a1,X1/a1,M,4e7)+gamrnd(a1,X2/a1,M,4e7);
    xx=x(:);
    pe1(k)=berest(xx');% no diversity ber
    xm=max(x);%SC
    pes(k)=berest(xm);% SC ber
    xs=sqrt(x);%for EGC
xem=(1/M)*sum(xs).^2;%EGC
pee(k)=berest(xem);% EGC ber
xmr=sum(x);%MRC
per(k)=berest(xmr);% MRC ber
y=sort(x,'descend');%sorts the columns in descending order
    ym3=sum(y(1:3,:));%GSC of 3 out of M
     pey3(k)=berest(ym3);% %GSC of 3 out of M  ber
     clear x xs xm xmr y  ym3
end;
   figure,semilogy(Z0,pe1,'r-*',Z0,pes,'k-o',...
       Z0,pey3,'b--p',Z0,per,'-m^')
   ylim([1e-6,0.1])
legend('No diversity',name1,name3g,name3)
 figure,semilogy(Z0,pe1,'r-*',Z0,pes,'k-o',...
       Z0,pee,'b--p',Z0,per,'-m^')
   ylim([1e-6,0.1])
legend('No diversity',name1,name2,name3)

end
```

```
function ber = berest(simdata)
L=length(simdata);
nn=normrnd(0,1,1,L);%Gaussian noise of zero mean and unit variance
h=simdata;%data is in units of power
h=sqrt(h);%needs to be converted into envelope or amplitude units
%%input data is the BPSK bit sequence
ip = rand(1,L)>0.5;%generate 0's and 1's
indata=2*ip-1;%bipolar data
s=indata;
out= sqrt(2)*h.*s+nn;% this simulates sqrt(2E/T)*bk*a+n
   x1=out>0;%detects 1s and zeros
   x2=-1*(out<0);%detect-1's and zeros
   outdata=x1+x2;%recreates the output bit stream in bipolar form
   Diffdata=s-outdata;% will be either +2 or -2 when there is an error
   DIF=abs(Diffdata);%will be 2 or zero
   berN=sum(DIF>0);%counts how many times the abs difference exceeds zero.
   ber=berN/L;

end
```

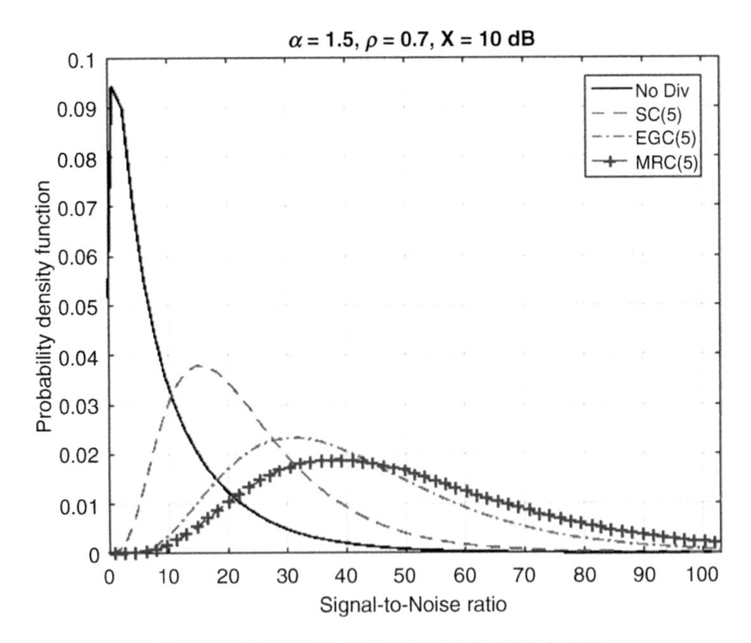

Fig. 5.123 Plots of pdf for $\alpha = 1.5$, $\rho = 0.7$, X = 10 dB (SC, EGC, MRC)

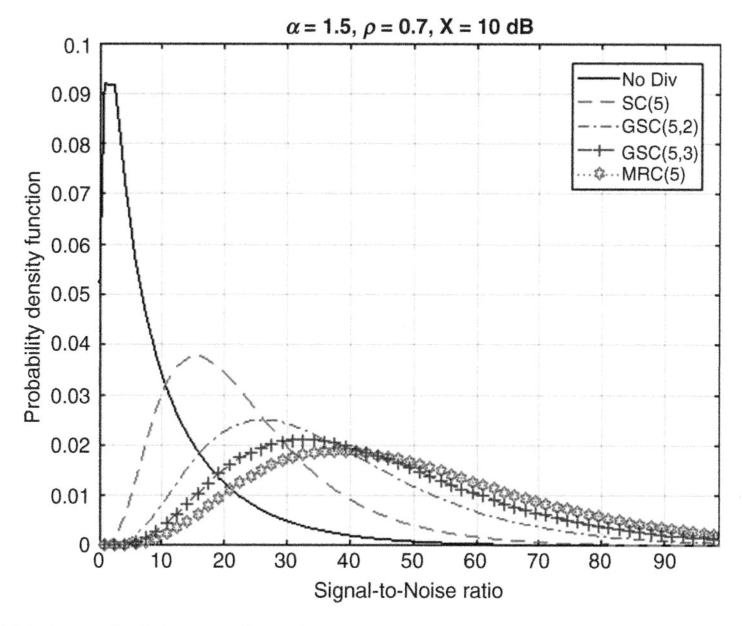

Fig. 5.124 Plots of pdf for $\alpha = 1.5$, $\rho = 0.7$, X = 10 dB (SC, GSC, MRC)

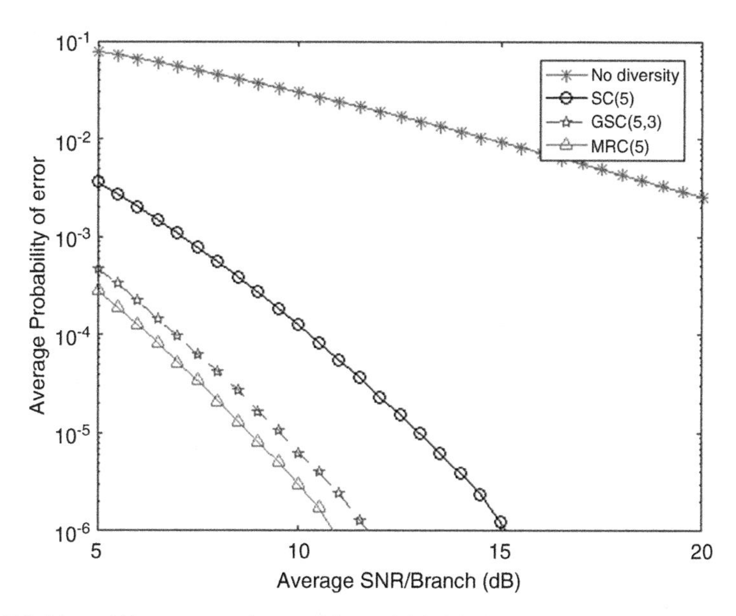

Fig. 5.125 Plots of bit error rates for $\alpha = 1.2$, $\rho = 0.75$ (SC, GSC, MRC)

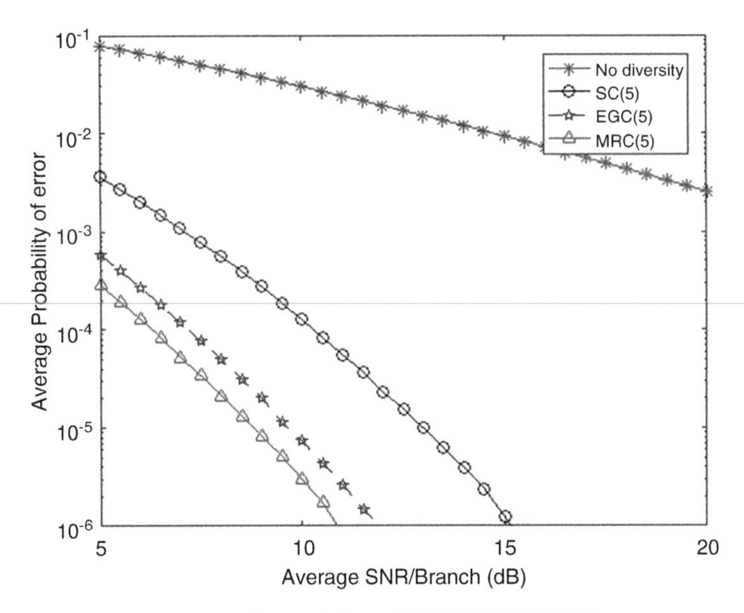

Fig. 5.126 Plots of bit error rates for $\alpha = 1.2$, $\rho = 0.75$ (SC, EGC, MRC)

5.15 Summary

In this chapter, we investigate the impact of diversity in wireless channels. Our study is limited to those fading and shadowing models based on the Nakagami or gamma densities. The benefits of diversity can be understood by carrying out the computations of the error rates and outage probabilities. The error rates and outage probabilities show reductions with diversity. Maximal ratio combining (MRC) diversity provides maximum improvement in performance.

We also explored the need to have diversity at the microlevel and macrolevel. This mitigates the channel impairments in shadowed fading channels. The hybrid diversity implemented at the microlevel through generalized selection combining offers a compromise between pure selection combining on one end and MRC on the other end. It must be understood that the trends in improvement with diversity will be similar regardless of the specific models used for fading or shadowing.

There is a large body of published work available in the literature of wireless which examines diversity in fading and shadowed fading channels modeled using other density functions. The improvements seen in Weibull channels through the implementation of the GSC algorithm have been reported by several researchers (Bithas et al. 2005; Alouini and Simon (2006) Wireless Comm. & Mobile Comp. 6:1077–1084). Results on other diversity algorithms in Weibull channels are additionally available (Sagias et al. (2003) Electronics Letters 39(20):1472–1474; Sagias et al. (2004) Communications, IEEE Transactions on 52(7):1063–1067; Karagiannidis et al. (2005) Wireless Communications, IEEE Transactions on 4

(3):841–846). While the diversity in generalized gamma channels was briefly discussed earlier on, several published results which undertook detailed studies of generalized gamma fading channels are available (Bithas et al. (2007a) Communications Letters, IEEE 11(12):964; Aalo et al. (2007) Communications, IET Proceedings 1(3):341–347; Samimi and Azmi (2008) Int. J. Electronics & Comm. (AEU) 62:496–605). The case of shadowed fading channels modeled using Rician-lognormal density has been studied by researchers (Wang and Stuber (1999) Vehicular Technology, IEEE Trans. on 48(2):429–436; Zhang and Aalo (2001) Communications, IEEE Transactions on 49(1):14–18). Diversity in Nakagami–Hoyt channels has also been studied (Iskander and Mathiopoulos (2005) Proceedings. IEEE 233–239; Zogas et al. (2005) Wireless Communications, IEEE Transactions on 4(2):374–379; Fraidenraich et al. (2008) Communications, IEEE Transactions on 56(2):183–188).

In the updated edition, diversity in McKay faded channels has been studied. The effect of shadowing on the improvement in performance from diversity is also discussed. Gamma mixtures are examined as replacements for the density of SNR following diversity. In all cases, Matlab scripts have been provided to complete the analysis of diversity in wireless systems.

References

Aalo, V. (1995). Performance of maximal-ratio diversity systems in a correlated Nakagami- fading environment. *IEEE Transactions on Communications, 43*(8), 2360–2369.

Aalo, V., Efthymoglou, G., Piboongungon, T., & Iskander, C. D. (2007). Performance of diversity receivers in generalised gamma fading channels. *IET Proceedings Communications, 1*(3), 341–347.

Aalo, V., Piboongungon, T., & Iskander, C. D. (2005). Bit-error rate of binary digital modulation schemes in generalized gamma fading channels. *IEEE Communications Letters, 9*(2), 139–141.

Abramowitz, M., & Segun, I. A. (1972). *Handbook of mathematical functions with formulas, graphs, and mathematical tables*. New York: Dover Publications.

Abu-Dayya, A., & Beaulieu, N. (1994a). Switched diversity on microcellular Ricean channels. *IEEE Transactions on Vehicular Technology, 43*(4), 970–976.

Abu-Dayya, A. A., & Beaulieu, N. C. (1994b). Analysis of switched diversity systems on generalized fading channels. *IEEE Transactions on Communications, 42*(11), 2959–2966.

Abu-Dayya, A. A., & Beaulieu, N. C. (1994c). Micro- and macrodiversity NCFSK (DPSK) on shadowed Nakagami-fading channels. *IEEE Transactions on Communications, 42*(9), 2693–2702.

Adamchik, V. (1995). The evaluation of integrals of Bessel functions via G-function identities. *Journal of Computational and Applied Mathematics, 64*(3), 283–290.

Al-Hussaini, E., & Al-Bassiouni, A. (1985). Performance of MRC diversity systems for the detection of signals with Nakagami fading. *IEEE Transactions on Communications, 33*(12), 1315–1319.

Almhana, J., Liu, Z., Choulakian, V., & McGorman, R. (2006). A recursive algorithm for gamma mixture models. *Proceedings of IEEE International Conference on Communications, 1*, 197–202.

Alouini, M. S., & Goldsmith, A. J. (1999). A unified approach for calculating error rates of linearly modulated signals over generalized fading channels. *IEEE Transactions on Communications, 47*(9), 1324–1334.

Alouini, M. S., & Simon, M. K. (1999). Performance of coherent receivers with hybrid SC/MRC over Nakagami-*m* fading channels. *IEEE Transactions on Vehicular Technology, 48*(4), 1155–1164.

Alouini, M. S., & Simon, M. K. (2000). An MGF-based performance analysis of generalized selection combining over Rayleigh fading channels. *IEEE Transactions on Communications, 48*(3), 401–415.

Alouini, M. S., & Simon, M. K. (2002). Dual diversity over correlated log-normal fading channels. *IEEE Transactions on Communications, 50*(12), 1946–1959.

Alouini, M.-S., & Simon, M. K. (2006). Performance of generalized selection combining in Weibull channels. *Wireless Communications and Mobile Computing, 6*, 1077–1084.

Annamalai, A. (1997). Analysis of selection diversity on Nakagami fading channels. *Electronics Letters, 33*(7), 548–549.

Annamalai, A., & Tellambura, C. (2001). A new approach to performance evaluation of generalized selection diversity receivers in wireless channels. *IEEE, 4*, 2309–2313.

Annamalai, A., & Tellambura, C. (2003). Performance evaluation of generalized selection diversity systems over Nakagami-*m* fading channels. *Wireless Communications and Mobile Computing, 3*(1), 99–116.

Annamalai, A., Tellambura, C., & Bhargava, V. K. (1999). Exact evaluation of maximal-ratio and equal-gain diversity receivers for *M*-ary QAM on Nakagami fading channels. *IEEE Transactions on Communications, 47*(9), 1335–1344.

Annamalai, A., Tellambura, C., & Bhargava, V. K. (2000). Equal-gain diversity receiver performance in wireless channels. *IEEE Transactions on Communications, 48*(10), 1732–1745.

Annamalai, A., Tellambura, C., & Bhargava, V. K. (2001). Simple and accurate methods for outage analysis in cellular mobile radio systems-a unified approach. *IEEE Transactions on Communications, 49*(2), 303–316.

Annamalai, A., Tellambura, C., & Bhargava, V. K. (2005). A general method for calculating error probabilities over fading channels. *IEEE Transactions on Communications, 53*(5), 841–852.

Atapattu, S., Tellambura, C., & Jiang, H. (2011). A mixture gamma distribution to model the SNR of wireless channels. *IEEE Transactions on Wireless Communications, 10*, 4193–4203.

Atkinson, A. C. (1981). Likelihood ratios, posterior odds and information criteria. *Journal of Econometrics, 16*, 15–20.

Bdira, E. B., & Mermelstein, P. (1999). Exploiting macrodiversity with distributed antennas in microcellular CDMA systems. *Wireless Personal Communications, 9*, 179–196.

Beaulieu, N. C. (1990). An infinite series for the computation of the complementary probability distribution function of a sum of independent random variables and its application to the sum of Rayleigh random variables. *IEEE Transactions on Communications, 38*(9), 1463–1474.

Beaulieu, N. C., & Abu-Dayya, A. A. (1991). Analysis of equal gain diversity on Nakagami fading channels. *IEEE Transactions on Communications, 39*(2), 225–234.

Bernhardt, R. (1987). Macroscopic diversity in frequency reuse radio systems. *IEEE Journal on Selected Areas in Communications, 5*(5), 862–870.

Bithas, P. S., Mathiopoulos, P. T., & Kotsopoulos, S. A. (2007a). Diversity reception over generalized-K (KG) fading channels. *IEEE Transactions on Wireless Communications, 6* (12), 4238–4243.

Bithas, P. S., Sagias, N. C., & Mathiopoulos, P. T. (2007b). GSC diversity receivers over generalized-gamma fading channels. *IEEE on Communications Letters, 11*(12), 964.

Bithas, P. S., Karagiannidis, G. K., Sagias, N. C., Mathiopoulos, P. T., Kotsopoulos, S. A., & Corazza, G. E. (2005). Performance analysis of a class of GSC receivers over nonidentical Weibull fading channels. *IEEE Transactions on Vehicular Technology, 54*(6), 1963–1970.

Blanco, M., & Zdunek, K. (1979). Performance and optimization of switched diversity systems for the detection of signals with Rayleigh fading. *IEEE Transactions on Communications, 27*(12), 1887–1895.

Brennan, D. G. (1959). Linear diversity combining techniques. *Proceedings of the IRE, 47*(1), 1075–1102.

Butterworth, K. S., Sowerby, K. W., & Williamson, A. G. (1997). Correlated shadowing in an in-building propagation environment. *Electronics Letters, 33*(5), 420–422.

Charesh, U. (1979). Reception through Nakagami fading multipath channels with random delays. *IEEE Transactions on Communications, 27*(4), 657–670.

Chatelain, F., Borgnat, P., Tourneret, J-Y & Abry, P. (2008). Parameter estimation for sums of correlated gamma random variables: Application to anomaly detection in internet traffic. *IEEE International Conference on Acoustics, Speech and Signal Processing (ICASSP-2008)*, 3489–3492.

Claeskens, G. (2016). Statistical model choice. *Annual Review of Statistics and its Application, 3*, 233–256.

Corazza, G. E., & Vatalaro, F. (1994). A statistical model for land mobile satellite. *IEEE Transactions on Vehicular Technology, 43*(2), 738–742.

Da Costa, D. B., & Yacoub, M. D. (2008). Moment generating functions of generalized fading distributions and applications. *IEEE Communications Letters, 12*(2), 112–114.

Di Salvo, F. (2008). A characterization of the distribution of a weighted sum of gamma variables through multiple hypergeometric functions. *Integral Transforms and Special Function, 19*, 563–575.

Diggavi, S. (2001). On achievable performance of spatial diversity fading channels. *IEEE Transactions on Information Theory, 47*(1), 308–325.

Efthymoglou, G., & Aalo, V. (1995). Performance of RAKE receivers in Nakagami fading channel with arbitrary fading parameters. *Electronics Letters, 31*(18), 1610–1612.

Eng, T., Kong, N., & Milstein, L. B. (1996). Comparison of diversity combining techniques for Rayleigh-fading channels. *IEEE Transactions on Communications, 44*(9), 1117–1129.

Ermolova, N. Y. (2008). Moment generating functions of the generalized η- μ and κ-μ distributions and their applications to performance evaluation of communication systems. *IEEE Communications Letters, 12*(7), 502–504.

Ermolova, N. Y. (2009). Useful integrals for performance evaluation of communication systems in generalized η-μ and κ-μ fading channels. *IET Communications, 3*(2), 303–308.

Filho, J. C. S. S., & Yacoub, M. D. (2005). Highly accurate η-μ approximation to the sum of *M* independent nonidentical Hoyt variates. *IEEE Anetannas and Wireless Propagation Letters, 4*, 436–438.

Fraidenraich, G., Yacoub, M. D., Mendes, J. R., & Santos Filho, J. C. S. (2008). Second-order statistics for diversity-combining of non-identical correlated Hoyt signals. *IEEE Transactions on Communications*, *56*(2). doi: 10.1109/TCOMM.2008.050024.

Gil-Pelaez, J. (1951). Note on the inversion theorem. *Biometrika, 38*, 481–482.

Giuli, D. (1986). Polarization diversity in radars. *IEEE Proceedings, 74*(2), 245–269.

Goldsmith, A. (2005). *Wireless communications*. New York: Cambridge University Press.

Gradshteyn, I. S., & Ryzhik, I. M. (2000). *Table of integrals, series, and products* (6th ed.). New York: Academic.

Gradshteyn, I. S., & Ryzhik, I. M. (2007). *Table of integrals, series and products*. Oxford: Academic.

Hausman, A. H. (1954). An analysis of dual diversity receiving systems. *Proceedings of the IRE, 42*(6), 944–947.

Haykin, S. S. (2001). *Digital communications*. New York: Wiley.

Holm, H., & Alouini, M. S. (2004a). Sum and difference of two squared correlated Nakagami variates in connection with the McKay distribution. *IEEE Transactions on Communications, 52*(8), 1367–1376.

Holm, H., & Alouini, M.-S. (2004b). Sum and difference of two squared correlated Nakagami variates in connection with the McKay distribution. *IEEE Transactions on Communications, 52*, 1367–1376.

Iskander, C.-D., & Mathiopoulos, P. T. (2005). Exact performance analysis of dual-branch coherent equal-gain combining in Nakagami-*m*, rice and Hoyt fading. In *Southeast Con, 2005. Proceedings. IEEE* (pp. 233–239).

Jakes, W. C. (1994). *Microwave mobile communications*. IEEE Press: Piscataway, NJ.
Jeong, W.-C., & Chung, J.-M. (2005). Analysis of macroscopic diversity combining of MIMO signals in mobile communications. *International Journal of Electronics & Communication (AEU), 59*, 454–462.
Karagiannidis, G. K., Sagias, N. C., & Mathiopoulos, P. T. (2007). N*Nakagami: A novel statistical model for cascaded fading channels. *IEEE Transactions on Communications, 55* (8), 1453–1458.
Karagiannidis, G. K., Sagias, N. C., & Tsiftsis, T. A. (2006). Closed-form statistics for the sum of squared Nakagami-*m* variates and its applications. *IEEE Transactions on Communications, 54*, 1353–1359.
Karagiannidis, G. K., Zogas, D. A., Sagias, N. C., Kotsopoulos, S. A., & Tombras, G. S. (2005). Equal-gain and maximal-ratio combining over nonidentical Weibull fading channels. *IEEE Transactions on Wireless Communications, 4*(3), 841–846.
Karagiannidis, G. K., Zogas, D. A., & Kotsopoulos, S. A. (2003). On the multivariate Nakagami-m distribution with exponential correlation. *IEEE Transactions on Communications, 51*(8), 1240–1244.
Karagiannidis, G. K., Sagias, N. C., & Mathiopoulos, P. T. (2007). *N**Nakagami: A novel statistical model for cascaded fading channels. *IEEE Transactions on Communications, 55* (8), 1453–1458.
Kang, M., Alouini, M. S., & Yang, L. (2002). Outage probability and spectrum efficiency of cellular mobile radio systems with smart antennas. *IEEE Transactions on Communications, 50* (12), 1871–1877.
Kass, R. E., & Raftery, A. E. (1995). Bayes factors. *Journal of the American Statistical Association, 90*, 773–795.
Kim, D. K., & Taylor, J. M. G. (1995). The restricted EM algorithm for maximum likelihood estimation under linear restrictions on the parameters. *Journal of the American Statistical Association, 90*, 708–716.
Ko, Y.-C., Alouini, M.-S., & Simon, M. K. (2000). Average SNR of dual selection combining over correlated Nakagami-m fading channels. *IEEE Communications Letters, 4*(1), 12–14.
Kong, N., & Milstein, L. B. (1999). Average SNR of a generalized diversity selection combining scheme. *IEEE Communications Letters, 3*(3), 57–59.
Kotz, S., & Adams, J. (1964). Distribution of sum of identically distributed exponentially correlated gamma-variables. *The Annals of Mathematical Statistics*, 277–283.
Ma, Y., & Chai, C. C. (2000). Unified error probability analysis for generalized selection combining in Nakagami fading channels. *IEEE Journal on Selected Areas in Communications, 18*(11), 2198–2210.
Malhotra, J., Sharma, A. K., & Kaler, R. S. (2008). On the performance of wireless receiver in cascaded fading channel. *African Journal of Information and Communication Technology, 4* (3), 65–72.
Malhotra, J., Sharma, A. K., & Kaler, R. S. (2009). On the performance analysis of wireless receiver using generalized-gamma fading model. *Annals of Telecommunications-Annales Des Télécommunications, 64*(1–2), 147–153.
Mallik, R. K., & Win, M. Z. (2002). Analysis of hybrid selection/maximal-ratio combining in correlated Nakagami fading. *IEEE Transactions on Communications, 50*(8), 1372–1383.
Mathai, A. M. (1993). *A handbook of generalized special functions for statistical and physical sciences*. Oxford: Oxford University Press.
Mathai, A. M., & Saxena, R. K. (1973). *Generalized hypergeometric functions with applications in statistics and physical sciences. Berlin*. New York, Springer.
Mayrose, I., Friedman, N., & Pupko, T. A. (2005). Gamma mixture model better accounts for among site rate heterogeneity. *Bioinformatics, 21*, ii151–ii158.
Molisch, A. F. (2005). *Wireless communications*. U. K. John Wiley & Sons: Chichester.
Mukherjee, S., & Avidor, D. (2003). Effect of microdiversity and correlated macrodiversity on outages in cellular systems. *IEEE Transactions on Wireless Communications, 2*(1), 50–58.

Nakagami, M. (1960). The m-distribution—A general formula of intensity distribution of rapid fading. In W. C. Hoffman (Ed.), *Statistical methods in radio wave propagation*. Elmsford, NY: Pergamon.

Neath, A. A., & Cavanaugh, J. E. (2012). The Bayesian information criterion: Background, derivation, and applications. *Wiley Interdisciplinary Reviews: Computational Statistics, 4*, 199–203.

Ning, K., & Milstein, L. B. (2000). SNR of generalized diversity selection combining with nonidentical Rayleigh fading statistics. *IEEE Transactions on Communications, 48*(8), 1266–1271.

Nomoto, S., Kishi, Y., & Nanba, S. (2004). Multivariate Gamma distributions and their numerical evaluations for M-branch selection diversity study. *Electronics and Communications in Japan (Part I: Communications), 87*(8), 1–12.

Nuttall, A. H. (1969). Numerical evaluation of cumulative probability distribution functions directly from characteristic functions. *IEEE Proceedings, 57*(11), 2071–2072.

Nuttall, A. H. (1975). Some integrals involving the Q_M function (Corresp.) *IEEE Transactions on Information Theory, 21*(1), 95–96.

Papoulis, A., & Pillai, S. U. (2002). *Probability, random variables, and stochastic processes*. Boston: McGraw-Hill.

Parsons, J. D., Henze, M. I. G. U. E. L., Ratliff, P. A., & Withers, M. J. (1975). Diversity techniques for mobile radio reception. *Radio and Electronic Engineer, 45*(7), 357–367.

Peppas, K., Lazarakis, F., Alexandridis, A., & Dankais, K. (2009). Error performance of digital modulation schemes with MRC diversity reception over η-μ fading channels. *IEEE Transactions on Wireless Communications, 8*(10), 4974–4980.

Perini, P. L., & Holloway, C. L. (1998). Angle and space diversity comparisons in different mobile radio environments. *IEEE Transactions on Antennas and Propagation, 46*(6), 764–775.

Piboongungon, T., & Aalo, V. A. (2004). Outage probability of L-branch selection combining in correlated lognormal fading channels. *Electronics Letters, 40*(14), 886–888.

Radaydeh, R. M., & Matalgah, M. M. (2008). Compact formulas for the average error performance of noncoherent m-ary orthogonal signals over generalized Rician, Nakagami-m, and Nakagami-q fading channels with diversity reception. *IEEE Transactions on Communications, 56*, 32–38.

Rappaport, T. S. (2002). *Wireless communications: Principles and practice*. Upper Saddle River, NJ: Prentice Hall PTR.

Sagias, N. C., Karagiannidis, G. K., Zogas, D. A., Mathiopoulos, P. T., & Tombras, G. S. (2004). Performance analysis of dual selection diversity in correlated Weibull fading channels. *IEEE Transactions on Communications, 52*(7), 1063–1067.

Sagias, N. C., Zogas, D. A., Karagiannidis, G. K., & Tombras, G. S. (2003). Performance analysis of switched diversity receivers in Weibull fading. *Electronics Letters, 39*(20), 1472–1474.

Samimi, H., & Azmi, P. (2008). Performance analysis of equal gain diversity receivers over generalized gamma fading channels. *International Journal on Electronics & Communication (AEU), 62*, 496–605.

Schwartz, G. (1978). Estimating the order of a model. *The Annals of Statistics, 2*, 461–464.

Schwartz, M., Bennett, W. R., & Stein, S. (1996). *Communication systems and techniques*. Piscataway, NJ: IEEE Press.

Shankar, P. (2010). Statistical models for fading and shadowed fading channels in wireless systems: A pedagogical perspective. *Wireless Personal Communications, 60*(2), 1–23.

Shankar, P. M. (2002). *Introduction to wireless systems*. New York: John Wiley.

Shankar, P. M. (2006). Performance analysis of diversity combining algorithms in shadowed fading channels. *Wireless Personal Communications, 37*(1), 61–72.

Shankar, P. M. (2008a). Analysis of microdiversity and dual channel macrodiversity in shadowed fading channels using a compound fading model. *AEU - International Journal of Electronics and Communications, 62*(6), 445–449.

Shankar, P. M. (2008b). Outage probabilities of a MIMO scheme in shadowed fading channels with micro- and macrodiversity reception. *IEEE Transactions on Wireless Communications, 7*(6), 2015–2019.

Shankar, P. M. (2009). Macrodiversity and microdiversity in correlated shadowed fading channels. *IEEE Transactions on Vehicular Technology, 58*(2), 727–732.

Shankar, P. M. (2011a). Maximal ratio combining in independent identically distributed n Nakagami fading channels. *IET Proceedings on Communications, 5*(3), 320–326.

Shankar, P. M. (2011b) "'Performance of N*Nakagami cascaded fading channels in dual selection combining diversity," IWCMC 2011, July 2011.

Shankar, P. M. (2012a). A Nakagami-N-gamma model for shadowed fading channels. *WIRE, 64*, 665–680.

Shankar, P. M. (2012b). A Nakagami-N-gamma model for shadowed fading channels. *Wireless Personal Communications, 64*, 1–16.

Shankar, P. M. (2015). A composite shadowed fading model based on the McKay distribution and Meijer G functions. *Wireless Personal Communications, 81*(3), 1017–1030.

Shankar, P. M. and C. Gentile (2010). Statistical analysis of short term fading and shadowing in ultra-wideband systems. Communications (ICC), 2010 I.E. International Conference on. 1–5.

Simon, M. K., & Alouini, M.-S. (2005). *Digital communication over fading channels* (–Wiley-Interscience). Hoboken, N.J.

Skraparlis, D., Sakarellos, V. K., Panagopoulos, A. D., & Kanellopoulos, J. D. (2009). Performance of N-branch receive diversity combining in correlated lognormal channels. *IEEE Communications Letters, 13*(7). doi: 10.1109/LCOMM.2009.090466.

Skraparlis, D., Sandell, M., Sakarellos, V. K., Panagopoulos, A. D., & Kanellopoulos, J. D. (2010). On the effect of correlation on the performance of dual diversity receivers in lognormal fading. *IEEE Communications Letters, 14*(11), 1038–1040.

Sowerby, K. W., & Williamson, A. G. (1992). Outage possibilities in mobile radio systems suffering cochannel interference. *IEEE Journal on Selected Areas in Communications, 10*(3), 516–522.

Steele, R., & Hanzo, L. (1999). *Mobile radio communications: Second and third generation cellular and WATM systems. Chichester, England.* New York: John Wiley & Sons, Inc..

Stein, S. (1987). Fading channel issues in system engineering. *IEEE Journal on Selected Areas in Communications, 5*(2), 68–89.

Stuber, G. L. (2002). *Principles of mobile communication*. New York: Kluwer Academic.

Tellambura, C. (2008). Bounds on the distribution of a sum of correlated lognormal random variables and their application. *IEEE Transactions on Communications, 56*(8), 1241–1248.

Tellambura, C., & Annamalai, A. (1999). An unified numerical approach for computing the outage probability for mobile radio systems. *IEEE Communications Letters, 3*(4), 97–99.

Tellambura, C., Annamalai, A., & Bhargava, V. K. (2001). Unified analysis of switched diversity systems in independent and correlated fading channels. *IEEE Transactions on Communications, 49*(11), 1955–1965.

Tellambura, C., Annamalai, A., & Bhargava, V. K. (2003). Closed form and infinite series solutions for the MGF of a dual- diversity selection combiner output in bivariate Nakagami fading. *IEEE Transactions on Communications, 51*(4), 539–542.

Theofilakos, P., Kanatas, A., & Efthymoglou, G. P. (2008). Performance of generalized selection combining receivers in K fading channels. *IEEE Communications Letters, 12*(11), 816–818.

Van Wambeck, S. H., & Ross, A. H. (1951). Performance of diversity receiving systems. *Proceedings of the IRE, 39*(3), 256–264.

Vaughan, R. G. (1990). Polarization diversity in mobile communications. *IEEE Transactions on Vehicular Technology, 39*(3), 177–186.

Vaughan, R. G., & Andersen, J. B. (1987). Antenna diversity in mobile communications. *IEEE Transactions on Vehicular Technology, 36*(4), 149–172.

Venturini, S., Dominici, F., & Parmigiani, G. (2008). Gamma shape mixtures for heavy-tailed distributions. *Ann. Appl. Stat., 2*, 756–776.

Wang, J. B., Zhao, M., Zhou, S. D., & Yao, Y. (1999). A novel multipath transmission diversity scheme in TDD-CDMA systems. *IEICE Transactions on EB Communications, 82*, 1706–1709.

Wang, L.-C., & Stuber, G. L. (1999). Effects of Rician fading and branch correlation on local mean based macrodiversity cellular systems. *IEEE Transactions on Vehicular Technology, 48*(2), 429–436.

Warren, D. (1992). A multivariate gamma distribution arising from a Markov model. *Stochastic Hydrology and Hydraulics, 6*, 183–190.

White, R. (1968). Space diversity on line-of-sight microwave systems. *IEEE Transactions on Communication Technology, 16*(1), 119–133.

Win, M. Z., & Kostic, Z. A. (1999). Virtual path analysis of selective rake receiver in dense multipath channels. *IEEE Communications Letters, 3*(11), 308–310.

Win, M. Z., Chrisikos, G., & Sollenberger, N. R. (2000). Performance of rake reception in dense multipath channels: Implications of spreading bandwidth and selection diversity order. *IEEE Journal on Selected Areas in Communications, 18*(8), 1516–1525.

Wojnar, A. H. (1986a). Unknown bounds on performance in Nakagami channels. *IEEE Transactions on Communications, 34*, 22–24.

Wojnar, A. H. (1986b). Unknown bounds on performance in Nakagami channels. *IEEE Transactions on Communications, 34*(1), 22–24.

Wolfram (2011). http://functions.wolfram.com/, Wolfram Research, Inc.

Wongtraitrat, W., & Supnithi, P. (2009). Performance of digital modulation in double Nakagami *m* fading channels with MRC diversity. *IEICE Transactions on Communications, E92B*(2), 559–566.

Yilmaz, F., & Alouini, M.-S. (2012). A novel unified expression for the capacity and bit error probability of wireless communication systems over generalized fading channels. *IEEE Transactions on Communications, 60*, 1862–1876.

Young-Chai, K., Alouini, M. S., & Simon, M. K. (2000). Analysis and optimization of switched diversity systems. *IEEE Transactions on Vehicular Technology, 49*(5), 1813–1831.

Zhang, J., & Aalo, V. (2001). Effect of macrodiversity on average-error probabilities in a Rician fading channel with correlated lognormal shadowing. *IEEE Transactions on Communications, 49*(1), 14–18.

Zogas, D. A., Karagiannidis, G. K., & Kotsopoulos, S. A. (2005). Equal gain combining over Nakagami-n (Rice) and Nakagami-q (Hoyt) generalized fading channels. *IEEE Transactions on Wireless Communications, 4*(2), 374–379.

Chapter 6
Interference in Wireless Channels

6.1 Introduction

The analysis and discussion so far in Chaps. 4 and 5 examined the effects of fading and shadowing in wireless systems while ignoring the possibility that the signal under consideration might be impacted by other signals, typically from other channels operating at the same frequency band. The contribution from other channels is often referred to as the cochannel interference or CCI. In simple terms, if the contribution from CCI exceeds a certain value, the ability to detect the signal of interest is adversely impacted leading to outage as well as further degradation in error rate performance (MacDonald 1979; Aalo and Zhang 1999; Muammar and Gupta 1982; Winters 1984; Abu-Dayya and Beaulieu 1991; Cardieri and Rappaport 2001; Yang and Alouini 2003). Often, there might be several cochannels instead of a single one. Even these cochannels might not be operating all the time, making it necessary to model the number of cochannels as a random variable. Thus, there are two factors that affect the performance of wireless systems in fading or shadowed fading channels, namely the noise and CCI. The analysis and studies carried out in Chaps. 4 and 5 can be treated as noise limited. If noise is neglected, we will be dealing with interference limited systems. In reality, wireless systems operating in fading should be analyzed when both the noise and interference cannot be ignored (Okui 1992; Yao and Sheikh 1992; Simon and Alouini 2005). The interference from cochannels affects both the outage probabilities and error rates. In fact, outage probabilities and error rates go up in the presence of CCI. The existence of CCI even leads to outage floors or error floors, suggesting that any further increase in the desired signal power or SNR will have no impact on the outages or error rates.

The effect of CCI will be analyzed in two steps. First, we examine the effect on outage probabilities. This will be followed by the examination of error rates.

© Springer International Publishing AG 2017

P.M. Shankar, *Fading and Shadowing in Wireless Systems*,
DOI 10.1007/978-3-319-53198-4_6

6.2 Outage Probabilities

We will start the discussion by looking at the simple case of Rayleigh channel before proceeding to other fading and shadowed fading channels. As we had done in Chap. 5, we will be examining fading and shadowing channels modeled using gamma densities including Rayleigh ones.

6.2.1 Rayleigh Channels

Let us consider the case of a single interfering channel first. Since the interfering channel is undesirable, the level of the interference will play a role in outage.

We assume that we have a Rayleigh fading environment. This means that the desired channel is undergoing Rayleigh fading and we have an interfering channel which is also undergoing Rayleigh fading. It is intuitive that we will have outage if the desired signal power falls below the interfering signal power. Let the desired signal power (average) be P_{s0} Watts. Let the interfering power (average) at the MU be P_{c0} Watts. Note that both the desired signal power (ps) and interfering power (pc) will be exponentially distributed as

$$f(p_s) = \frac{1}{P_{s0}} \exp\left(-\frac{p_s}{P_{s0}}\right), \tag{6.1}$$

$$f(p_c) = \frac{1}{P_{c0}} \exp\left(-\frac{p_c}{P_{c0}}\right). \tag{6.2}$$

Outage occurs when the interfering channel power exceeds the desired signal power (Sowerby and Williamson 1987; Abu-Dayya and Beaulieu 1991). The outage probability Pout is

$$P_{out} = \text{Pr}ob(P_c > P_s). \tag{6.3}$$

The outage probability is given by the shaded area in Fig. 6.1.

Treating the desired signal and interfering signal to be independent, the outage probability becomes

$$P_{out} = \int_0^\infty f(p_s) \left[\int_0^{p_s} f(p_c) dp_c \right] dp_s = \frac{P_{c0}}{P_{c0} + P_{s0}} = \frac{1}{1 + {}^{P_{s0}}/_{P_{c0}}}. \tag{6.4}$$

Equation (6.4) indicates that the outage probability depends only on the ratio of the average powers and not on the individual powers (signal or interference). The outage probability is plotted in Fig. 6.2.

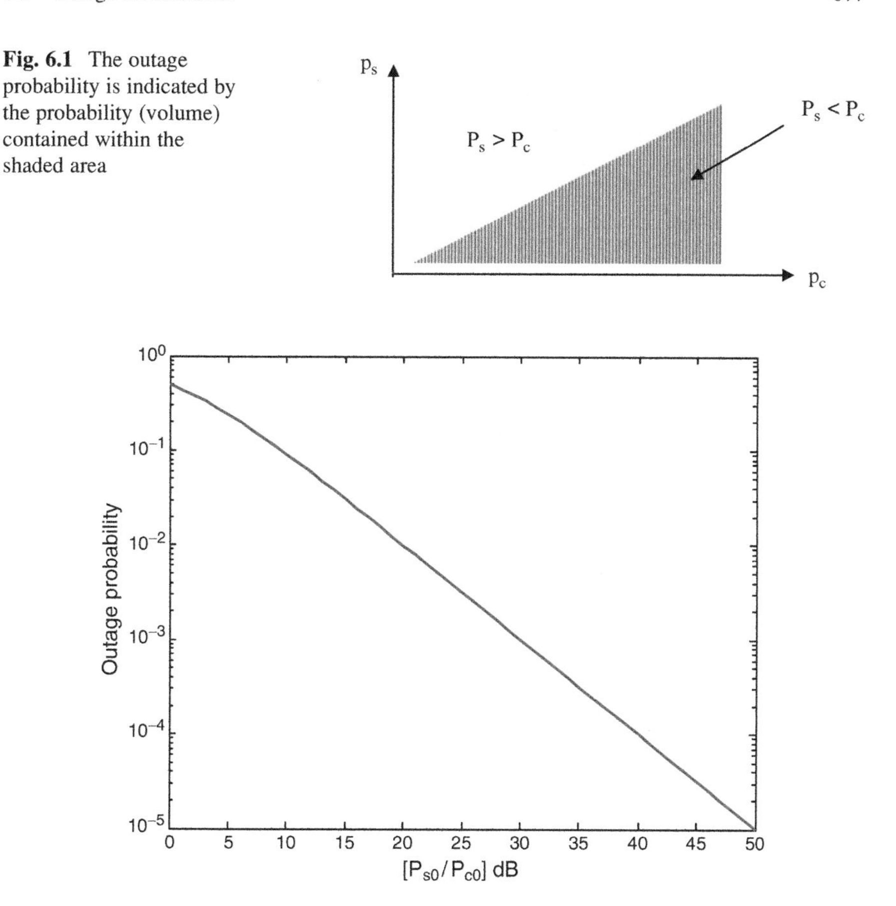

Fig. 6.1 The outage probability is indicated by the probability (volume) contained within the shaded area

Fig. 6.2 Outage probability as a function of the ratio of the average powers

The results in Fig. 6.2 demonstrate the existence of potential problems in wireless communication from interfering signals. We will now explore the effect of interfering signals in detail.

As we had done in the previous chapters, we will transition from power to signal-to-noise ratio since the density functions will be similar. Let us consider the case of N cochannels. Since the base stations from where they originate are sufficiently separated from one another, so that we will assume that they lack coherence, and the total contribution from the cochannels can be expressed as the sum of the signal-to-noise ratio terms from the N cochannels (Sowerby and Williamson 1988a, b; Linnartz 1992; Sowerby and Williamson 1992, 2002; Aalo and Chayawan 2000). Let Z represent the received SNR from the channel of interest. Let the total contribution to CCI from the N cochannels be Zc such that

$$Z_c = \sum_{k=1}^{N} Z_{ck}. \tag{6.5}$$

In (6.5), each term in the summation is the contribution from the cochannel. In a simple case, we will assume that all the cochannels are identical and the contributions are equal (i.e., are not weighted). We must also note that it is possible to have different forms of fading affecting the signal of interest and the CCI. Let us consider the simple case of Rayleigh fading taking place over the whole geographical region so that we can treat both the signal channel and the CCI channels as undergoing Rayleigh fading. This means that the density function of the signal-to-noise ratio signal of interest will be exponentially distributed as

$$f(z) = \frac{1}{Z_0} \exp\left(-\frac{z}{Z_0}\right). \tag{6.6}$$

Since all the cochannel powers add incoherently, the density function of the SNR from the cochannels in (6.5) will be pdf of the sum of N independent and identically distributed exponential random variables. This pdf can be written as (Moschopoulos 1985; Papoulis and Pillai 2002)

$$f(z_c) = \left(\frac{1}{Z_{0c}}\right)^N \frac{z_c^{N-1}}{\Gamma(N)} \exp\left(-\frac{z_c}{Z_{0c}}\right). \tag{6.7}$$

In obtaining (6.7), we have made use of the fact that an exponential pdf is a gamma pdf of order unity and that the pdf of the sum of independent and identically distributed gamma random variables is another gamma pdf or more specifically an Erlang pdf since N is an integer. In (6.6) and (6.7) Z_0 is the average SNR of the channel of interest and Z_{0c} is the average SNR in any one of the cochannels, respectively. On a channel to channel basis, we can define the signal-to-interference ratio (SIR) as

$$\text{SIR} = \frac{Z_0}{Z_{0c}}. \tag{6.8}$$

Note that using the definition in (6.8), SIR goes down with an increasing number of cochannels. It is also possible to define the signal-to-interference ratio on a composite basis as SIRN,

$$\text{SIR}_N = \frac{Z_0}{\sum_{k=1}^{N} Z_{0ck}}. \tag{6.9}$$

In (6.9), each cochannel is assumed to have a different SNR. For our discussion, we will use the definition in (6.8), which suggests that SIR goes down as the number of cochannels increases. To prevent outage from occurring, there is a need to maintain a minimum protection ratio q given as

$$q = \frac{Z}{Z_c}. \tag{6.10}$$

Typical values of q might be in the range of 10–15 dB depending on how much separation is expected between the desired signal of interest and the cochannels (Helstrom 1986; Abu-Dayya and Beaulieu 1991, 1992; Cui and Sheikh 1999; Mostafa et al. 2004; Shankar 2005). Note that outage will also occur if the SNR of the desired signal falls below a certain threshold (as discussed in Chaps. 4 and 5), with the value of the threshold being determined by the modulation and maximum acceptable error rates. Therefore, the outage is governed by two factors, the threshold and the protection factor. The outage probability can now be expressed as

$$\begin{aligned} P_{\text{out}} &= 1 - \text{Prob}\left\{ Z > Z_T, \frac{Z}{Z_c} > q \right\} \\ &= 1 - \int_{Z_T}^{\infty} f(z) \left[\int_{0}^{(z/q)} f(z_c) dz_c \right] dz. \end{aligned} \tag{6.11}$$

In (6.11) ZT is the threshold SNR required to maintain an acceptable bit error rate. In writing down the expression for the outage, we have assumed that the signal of interest (desired signal) and the CCI are independent. The outage probability can be simplified to

$$P_{out} = 1 - \int_{Z_T}^{\infty} \frac{1}{Z_0} \exp\left(-\frac{z}{Z_0}\right) \frac{\gamma[N, (z/qZ_{0c})]}{\Gamma(N)} dz, \tag{6.12}$$

where γ (,) is the incomplete gamma function defined in Chap. 2. In terms of SIR, we can rewrite Eq. (6.12) as

$$P_{\text{out}} = 1 - \int_{Z_T}^{\infty} \frac{1}{Z_0} \exp\left(-\frac{z}{Z_0}\right) \frac{\gamma[N, (Sz/Z_0)]}{\Gamma(N)} dz. \tag{6.13}$$

In (6.13), S is

$$S = \frac{\text{SIR}}{q}. \tag{6.14}$$

Equation (6.13) can be further simplified to

$$P_{\text{out}} = 1 - \int_{ZZ}^{\infty} \exp(-y) \frac{\gamma[N, Sy]}{\Gamma(N)} dy, \tag{6.15}$$

with

$$ZZ = \frac{Z_T}{Z_0}. \tag{6.16}$$

Before we examine the case of several cochannels, we will look at the case of $N = 1$. In this case, (6.15) becomes

$$P_{\text{out}} = 1 - \int_{ZZ}^{\infty} \exp(-y)[1 - \exp(-Sy)]dy. \tag{6.17}$$

Equation (6.17) reduces to

$$P_{\text{out}} = \frac{(1+S)[1-\exp(-ZZ)] + \exp[-ZZ(1+S)]}{1+S}. \tag{6.18}$$

Figure 6.3 shows the outage probability in a Rayleigh faded channel when there is only a single CCI channel expressed in (6.18). The most obvious outcome seen from Fig. 6.3 is the existence of "outage floors." The outage probabilities appear to reach certain values depending on the threshold, and often these values are significantly higher than what is expected in terms of outage levels.

Let us examine (6.18) to understand the outage performance in terms of the two limiting cases, one related to CCI and the other related to noise. If CCI is weak, i.e., we have a noise limited case, S is large and (6.18) becomes

$$P_{\text{out}} = \frac{(1+S)[1-\exp(-ZZ)] + \exp[-ZZ(1+S)]}{1+S} \approx 1 - \exp(-ZZ). \tag{6.19}$$

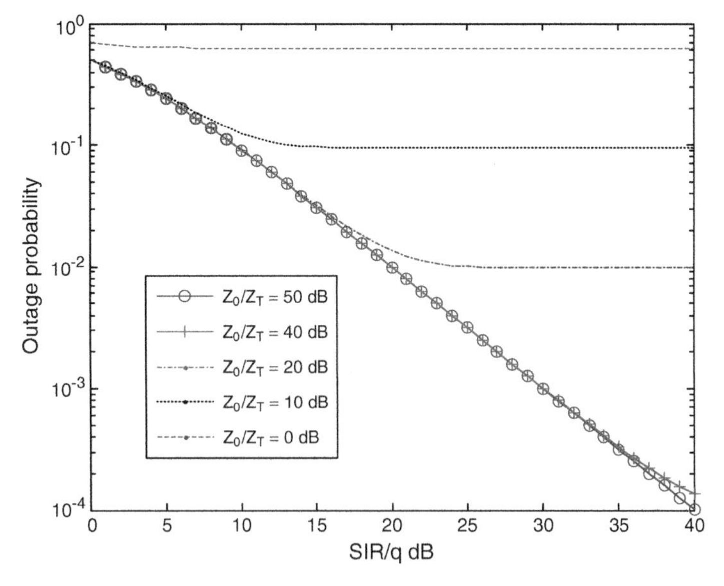

Fig. 6.3 Outage probability is plotted against SIR/q for various values of the ratio of the average SNR in the desired channel to the threshold SNR. Only one cochannel is present

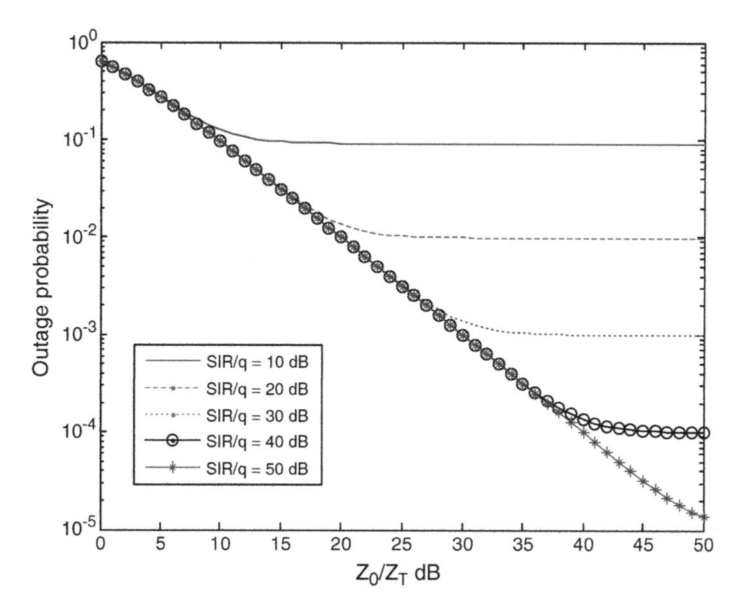

Fig. 6.4 Outage probability vs. Z_0/ZT for several values of SIR/q. Only a single cochannel is present

The outage probably depends only on the ZZ (note that this is also the outage probability we had obtained in Chap. 4, where no CCI was considered). This means that outage probability will continue to decrease as the average SNR Z_0 increases. The limiting value is determined by the average SNR of the desired signal.

If we have a CCI limited case, i.e., ZZ is small, (6.18) becomes

$$P_{\text{out}} = \frac{(1+S)[1-\exp(-ZZ)] + \exp[-ZZ(1+S)]}{1+S} \approx \frac{1}{(1+S)}. \tag{6.20}$$

The outage probability is inversely proportional to the signal-to-CCI ratio as seen in Fig. 6.3. The outage is plotted against Z_0/ZT for a few values of SIR/q to further illustrate this behavior (Fig. 6.4). The outage floors are clearly seen.

The result in (6.20) can also be obtained directly by considering the effect of CCI alone. If the channel has very little noise, outage occurs when the SNR in the desired channel falls below the threshold determined by the protection ratio. In other words, the outage probability is given by

$$P_{\text{out}} = \text{Prob}\left(\frac{Z}{Z_c} < q\right). \tag{6.21}$$

Using the density functions, the outage in (6.21) is given by

$$P_{\text{out}} = \int_0^{\infty} \frac{1}{Z_{0c}} \exp\left(-\frac{z_c}{Z_{0c}}\right) \left[\int_0^{q z_c} \frac{1}{Z_0} \exp\left(-\frac{z}{Z_0}\right) dz\right] dz_c = \frac{1}{1+S}. \tag{6.22}$$

The outage probability in (6.22) can also be interpreted as the outage probability in the absence of any constraints on the desired signal power. Note that the result in (6.22) is identical to the one obtained in (6.4).

Examining these figures, we see that CCI presents a serious problem in wireless systems: we see outage probabilities becoming almost a constant (floor). This value might be higher than what is tolerable and acceptable for data communication services which expect uninterrupted transmission. The only means to bring the outage probabilities lower is to operate at higher values of SIR (low CCI value) as well as average SNR in the desired signal (Z_0 high).

So far, we have only examined the existence of a single cochannel. We will now look at the effect of a fixed number of channels as well as a random number of cochannels. The results for the case of a fixed number of cochannels obtained using (6.15) are shown in Fig. 6.5.

To highlight the effect of no constraints on the desired signal power, the outage probabilities are plotted for the high value of Z_0/ZT of 50 dB for several values of N in Fig. 6.6. We clearly see that the outage probabilities go down and outage floor behavior is absent. Thus, the floor behavior is associated with the existence of CCI along with the requirement on the desired signal through the existence of the threshold ZT.

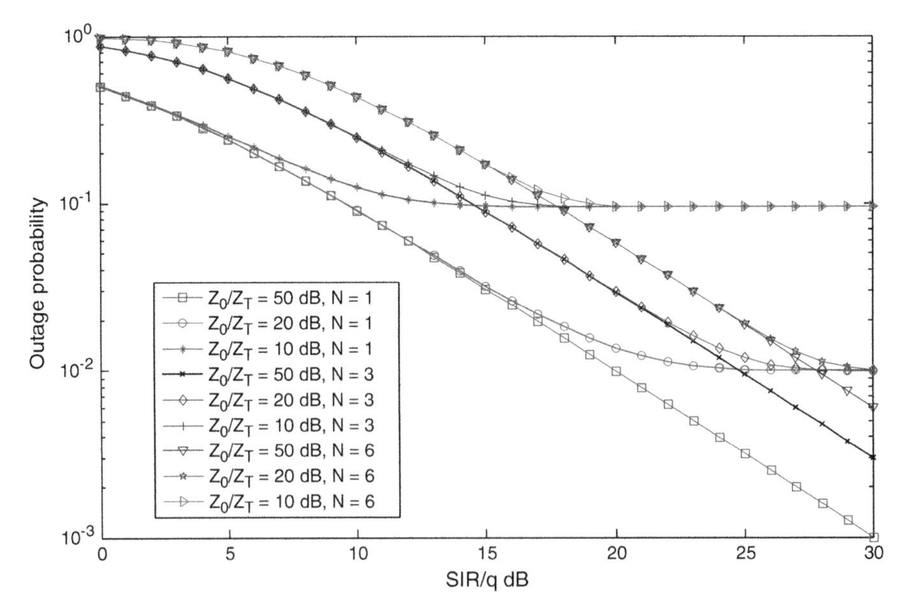

Fig. 6.5 Outage probabilities for multiple cochannels. The number of cochannels is fixed ($N = 1$, 3, and 6)

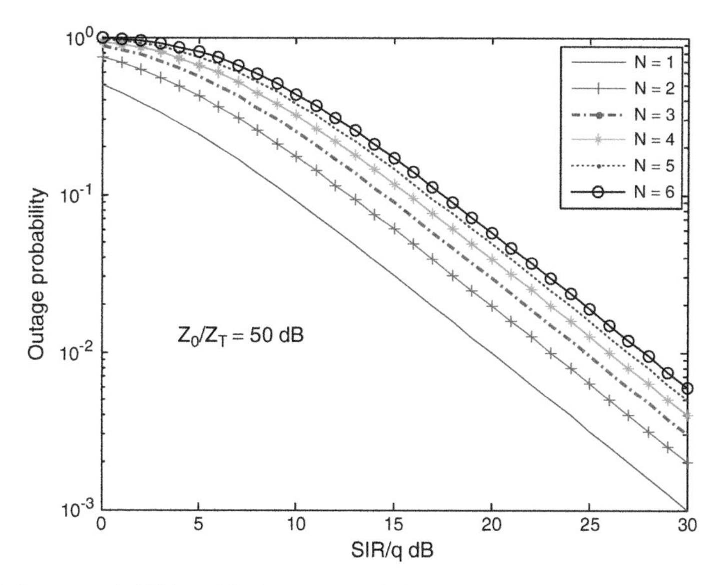

Fig. 6.6 Outage probabilities with almost no restrictions on the desired signal power or SNR

It is possible that all the cochannels may not be operating at the same time (Abu-Dayya and Beaulieu 1992; Yang and Petropulu 2003; Shankar 2005). This can be taken into consideration by treating the number of interfering channels as a Poisson random variable. The number of cochannels in use is described by

$$P(n = N) = \frac{[N_{av}]^N e^{-N_{av}}}{N!}, \tag{6.23}$$

where Nav is the average number of cochannels. The outage probability, PoutNr, in the presence of random number of cochannels can be evaluated by calculating the outage for various values of N from a Poisson distributed set of random numbers with an average of Nav followed by averaging over the Poisson probability. The expression for PoutNr becomes

$$P_{\text{outNr}} = e^{-N_{av}} \sum_{k=0}^{\infty} P_{outk} \frac{[N_{av}]^k}{k!}, \tag{6.24}$$

where Poutk is given in (6.15) for $k = N$. The outage probabilities are compared for the case of three cochannels in Fig. 6.7. It is seen that the effect of random number of cochannels is a slight reduction in outage probabilities.

Another way of exploring the existence of random number of cochannels is through the use of a binomial model. The average outage probability can now be expressed as

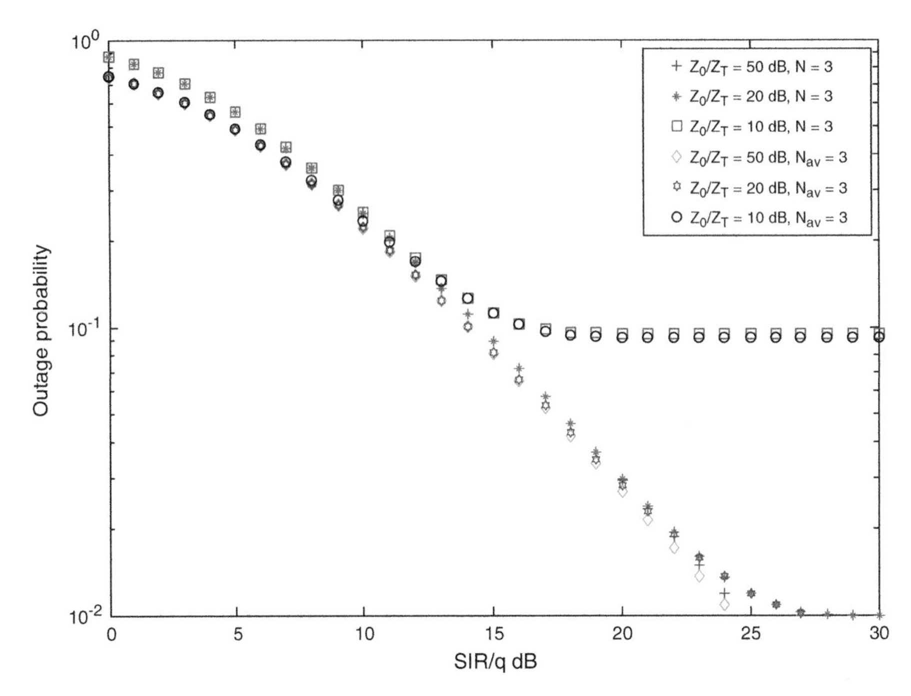

Fig. 6.7 Outage when the number of cochannel interference (CCI) channels is random (Nav = 3)

$$P_{\text{out}}^{av} = \sum_{N=0}^{L} P_{\text{out}N} * P(N). \tag{6.25}$$

where PoutN is the outage in equation for N = 0,1,2,... ,L, and P(N) is the probability that the number of interfering channels is N. This probability $P(N)$ is determined by the number of voice channels (Ns) and the blocking probability B. It can be expressed as (Abu-Dayya and Beaulieu 1991; Reig and Cardona 2000)

$$P(N) = \binom{L}{N} B^{N/N_s} \left(1 - B^{1/N_s}\right)^{L-N}. \tag{6.26}$$

Equation (6.25) can now be evaluated using (6.15) and (6.26) with the maximum value of N being 6. Outage probabilities in a random number of cochannels were evaluated for Ns = 10 and B = 0.02 that are shown in Fig. 6.8.

6.2.2 Nakagami Channels

We can now look at the general case of a desired signal in a Nakagami channel when the cochannels are also operating in Nakagami faded channels (Okui 1992; Abu-Dayya and Beaulieu 1991). The density function of the SNR in the desired channel becomes

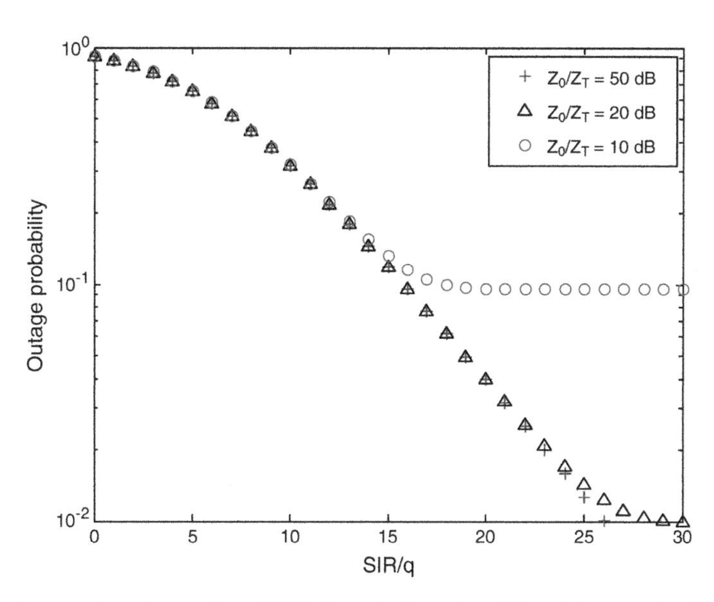

Fig. 6.8 Outage when the number of CCI channels is random. Binomial approximation. Maximum number of channels N is 6. The number of voice channels Ns is 10 and the blocking probability B is 0.02

$$f(z) = \left(\frac{m}{Z_0}\right)^m \frac{z^{m-1}}{\Gamma(m)} \exp\left(-\frac{m}{Z_0}z\right). \tag{6.27}$$

If we assume that all the cochannels are identical, each with a Nakagami parameter m_1, the density function of the SNR of the cochannels is (Aalo and Chayawan 2000; Simon and Alouini 2005)

$$f(z_c) = \left(\frac{m_1}{Z_{0c}}\right)^{m_1 N_I} \frac{z_0^{m_1 N-1}}{\Gamma(m_1 N)} \exp\left(-\frac{m_1}{Z_{0c}} z_c\right). \tag{6.28}$$

The outage probability in (6.11) now becomes

$$P_{\text{out}} = 1 - \int_{ZZ}^{\infty} m^m \frac{y^{m-1}}{\Gamma(m)} \exp(-my) \frac{\gamma[m_I N, mSy]}{\Gamma(m_I N)} dy. \tag{6.29}$$

The outage probability in a Nakagami channel in the presence of a single Nakagami CCI channel is plotted in Fig. 6.9 for three values of Z_0/ZT. The outage probability, plotted as multiple interferers (3 and 5), is compared with the case of a single interferer in Fig. 6.10.

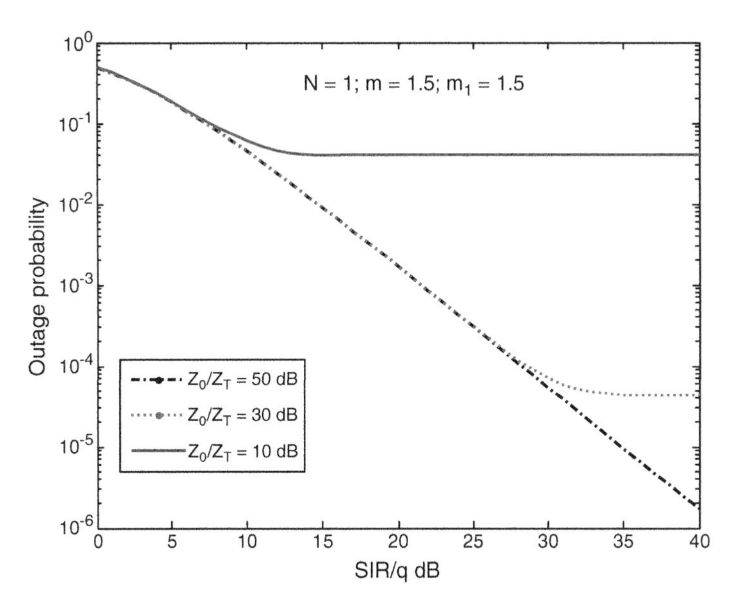

Fig. 6.9 Outage probability in Nakagami-m faded channel in the presence of a single Nakagami CCI channel

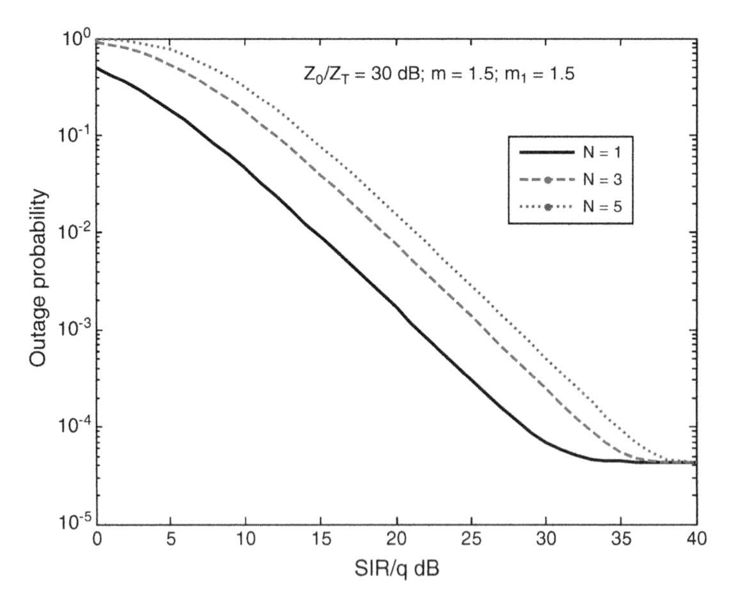

Fig. 6.10 Outage probability in Nakagami-m faded channel in the presence of CCI coming from $N = 1$, 3, and 5

6.2.3 Shadowed Fading Channels

We will now explore the effect of CCI in shadowed fading channels. Instead of using the Nakagami-lognormal model for the shadowed fading channels, we will use the equivalent Nakagami-gamma or the GK model for the shadowed fading channel (Shankar 2004, 2005). In this model, the pdf of the SNR in the desired channel can be written as

$$f(z) = \frac{2}{\Gamma(m)\Gamma(\nu)}\left(\sqrt{\frac{m\nu}{Z_0}}\right)^{\nu+m} z^{((\nu+m)/2)-1} K_{m-\nu}\left(2\sqrt{\frac{m\nu}{Z_0}z}\right). \tag{6.30}$$

Note that ν is the gamma parameter related to the shadowing level σ_{dB}, with low values ν corresponding to severe levels of shadowing (high values of σ_{dB}) and vice versa. First, we assume that CCI channels do not undergo shadowing and they are all independent and identically distributed Nakagami fading channels as discussed in the previous sections. This means that the density function of the SNR of the CCI component is gamma distributed as in (6.28), where m_1 is the Nakagami parameter and N one again is the number of CCI channels. Using (6.11) the outage probability becomes

$$\begin{aligned} P_{\text{out}} = {} & 1 - \int_{Z_T}^{\infty} \frac{2}{\Gamma(m)\Gamma(\nu)}\left(\sqrt{\frac{m\nu}{Z_0}}\right)^{\nu+m} z^{((\nu+m)/2)-1} K_{m-\nu}\left(2\sqrt{\frac{m\nu}{Z_0}z}\right) \\ & \times \int_0^{z/q} \left(\frac{m_1}{Z_{0c}}\right)^{m_1N_1} \frac{z_0^{m_1N-1}}{\Gamma(m_l N)} \exp\left(-\frac{m_1}{Z_{0c}}z_c\right) dz_c dz. \end{aligned} \tag{6.31}$$

Using (6.16) and (6.29), the outage probability becomes

$$P_{\text{out}} = 1 - \int_{ZZ}^{\infty} \frac{2}{\Gamma(m)\Gamma(\nu)} \left(\sqrt{m\nu}\right)^{\nu+m} y^{\left(\frac{\nu+m}{2}\right)-1} K_{m-\nu}\left(2\sqrt{m\nu y}\right) \frac{\gamma[m_1 N, mSy]}{\Gamma(m_1 N)} dy. \tag{6.32}$$

The outage probability in a shadowed fading channel in the presence of a single Nakagami-m distributed cochannel is shown in Fig. 6.11 for three different levels of shadowing. Figure 6.12 shows the results for $N = 1$ and 5.

Next, we consider the case when both the desired channel and the interfering cochannels undergo short-term fading and shadowing. We will once again use the equivalent gamma model for shadowing. We make the assumption that the cochannels are identical where the short-term fading components are Nakagami-m distributed and the shadowing can be treated the same in all the cochannels. This means that the density function of the SNR of the CCI can be expressed in the form of conditional density function as (Shankar 2005, 2007)

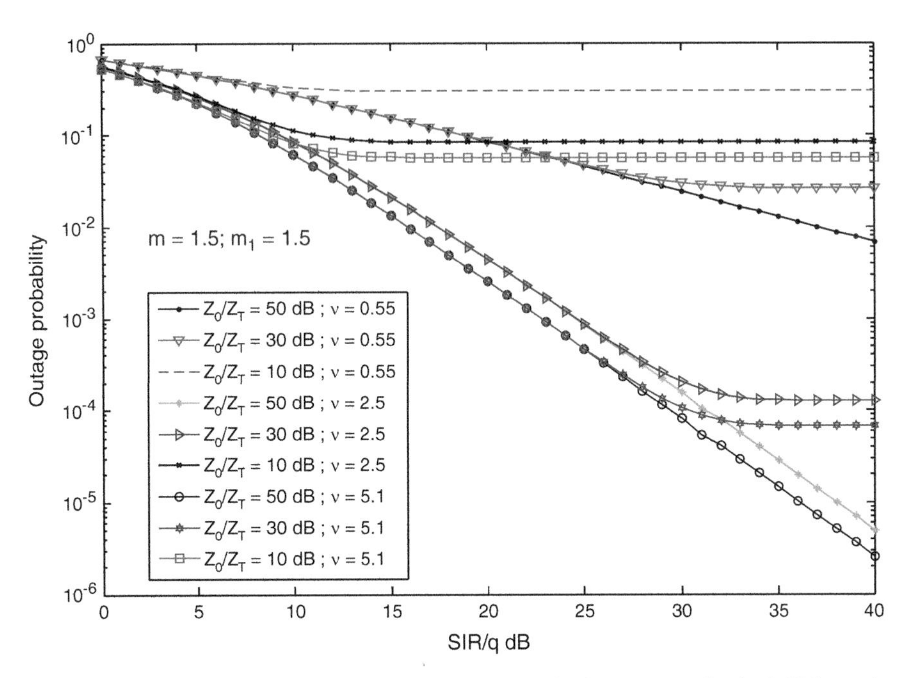

Fig. 6.11 Outage probability in a shadowed fading channel in the presence of a single Nakagami-m distributed interferer for three levels of shadowing $\nu = 0.55$ (*heavy shadowing*), $\nu = 2.5$ (*moderate shadowing*), and $\nu = 5.5$ (*light shadowing*)

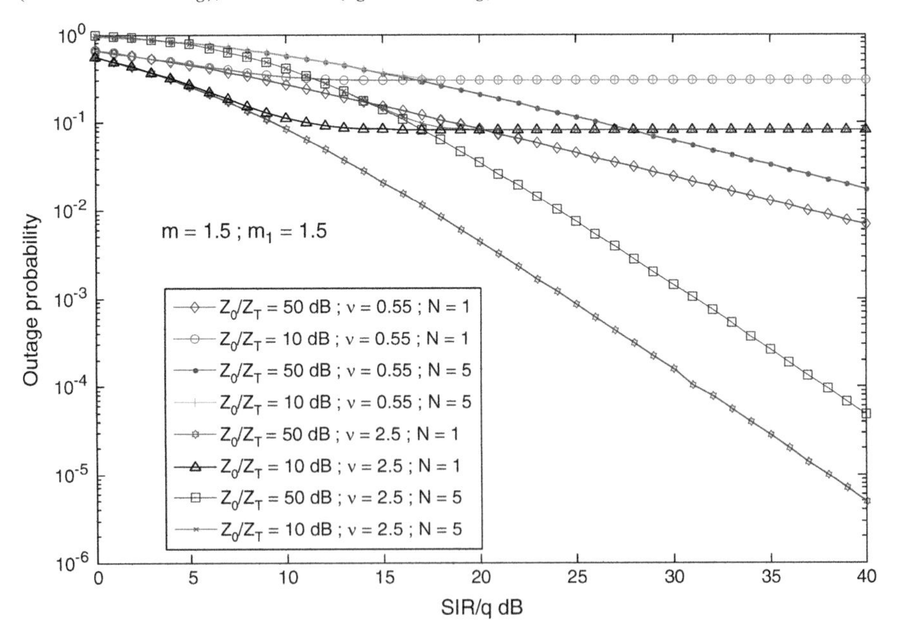

Fig. 6.12 Outage probability in a shadowed fading channel in the presence of Nakagami-m distributed interferers for two levels of shadowing $\nu = 0.55$ (*heavy shadowing*) and $\nu = 2.5$ (*moderate shadowing*) for $N = 1$ and $N = 5$

$$f(z_c|w) = \left(\frac{m_1}{w}\right)^{Nm_1} \frac{z_c^{Nm_1-1}}{\Gamma(Nm_1)} \exp\left(-\frac{m_1}{w} z_c\right). \tag{6.33}$$

The density function of the shadowing component is expressed as

$$f(w) = \left(\frac{\nu_1}{Z_{0c}}\right)^{\nu_1} \frac{w^{\nu_1-1}}{\Gamma(\nu_1)} \exp\left(-\frac{\nu_1}{Z_{0c}} w\right). \tag{6.34}$$

Note that c_1 is the gamma parameter which is related to the shadowing in the CCI channel. Using (6.34), the density function of the SNR of the interference becomes

$$f(z_c) = \frac{2}{\Gamma(Nm_1)\Gamma(\nu_1)} \left(\sqrt{\frac{Nm_1\nu_1}{Z_{0c}}}\right)^{\nu_1+Nm_1} z_c^{\left(\frac{\nu_1+Nm_1}{2}\right)-1} K_{Nm_1-\nu_1}\left(2\sqrt{\frac{Nm_1\nu_1}{Z_{0c}} z_c}\right). \tag{6.35}$$

Following the procedures discussed earlier, the outage probability becomes

$$\begin{aligned} P_{\text{out}} = \; & 1 - \int_{Z_T}^{\infty} \frac{2}{\Gamma(m)\Gamma(\nu)} \left(\sqrt{\frac{m\nu}{Z_0}}\right)^{\nu+m} z^{((\nu+m)/2)-1} K_{m-\nu}\left(2\sqrt{\frac{m\nu}{Z_0} z}\right) \\ & \times \left[\int_0^{z/q} \frac{2}{\Gamma(Nm_1)\Gamma(\nu_1)} \left(\sqrt{\frac{Nm_1\nu_1}{Z_{0c}}}\right)^{\nu_1+Nm_1} z_c^{((\nu_1+m)/2)-1} \right. \\ & \left. \times K_{Nm_1-\nu_1}\left(2\sqrt{\frac{Nm_1\nu_1}{Z_{0c}} z_c}\right) dz_c \right] dz. \end{aligned} \tag{6.36}$$

Equation (6.36) can be rewritten as

$$\begin{aligned} P_{\text{out}} = \; & 1 - \int_{ZZ}^{\infty} \frac{2}{\Gamma(m)\Gamma(\nu)} \left(\sqrt{m\nu}\right)^{\nu+m} y^{((\nu+m)/2)-1} K_{m-\nu}\left(2\sqrt{m\nu y}\right) \\ & \times \left[\int_0^{Sy} \frac{2}{\Gamma(Nm_1)\Gamma(\nu_1)} \left(\sqrt{Nm_1\nu_1}\right)^{\nu_1+Nm_1} w^{((\nu_1+Nm_1)/2)-1} \right. \\ & \left. \times K_{Nm_1-\nu_1}\left(2\sqrt{Nm_1\nu_1 w}\right) dw \right] dy. \end{aligned} \tag{6.37}$$

The inner integral becomes (Gradshteyn and Ryzhik 2007)

$$\int_0^{Sy} \frac{2}{\Gamma(Nm_1)\Gamma(\nu_1)}(Nm_1\nu_1)^{\nu_1+Nm_1} w^{\left(\frac{\nu_1+Nm_1}{2}\right)-1} K_{Nm_1-\nu_1}\left(2\sqrt{Nm_1\nu_1 w}\right)dz_c = F(y)$$
$$= \frac{\Gamma(Nm_1-\nu_1)}{\Gamma(Nm_1)\Gamma(1+\nu_1)}[m_1N\nu_1Sy]^{\nu_1}{}_1F^12([\nu_1],[\nu_1+1,\nu_1+1-m_1N],m_1N\nu_1Sy)$$
$$+\frac{\Gamma(\nu_1-Nm_1)}{\Gamma(Nm_1+1)\Gamma(\nu_1)}[m_1N\nu_1Sy]^{Nm_1}{}_1F^12([Nm_1],[Nm_1+1,Nm_1+1-\nu_1],m_1N\nu_1Sy). \tag{6.38}$$

The outage probability becomes

$$P_{\text{out}} = 1 - \int_{ZZ}^{\infty} \frac{2}{\Gamma(m)\Gamma(\nu)}\left(\sqrt{m\nu}\right)^{\nu+m} y^{\left(\frac{\nu+m}{2}\right)-1} K_{m-\nu}\left(2\sqrt{m\nu y}\right)F(y)dy. \tag{6.39}$$

where $F(y)$ is given in (6.38). The outage probability in a single interfering channel (both desired and cochannel components are shadowed fading ones) is shown in Fig. 6.13.

The case of multiple interfering shadowed fading channels is shown in Fig. 6.14. These results can now be extended to any combination of models of fading and shadowing such as Rician, Weibull, and generalized gamma. We expect that trends in behavior will be similar to the ones we have seen with Nakagami-m channels with or without shadowing being present in the wireless environment as the case may be.

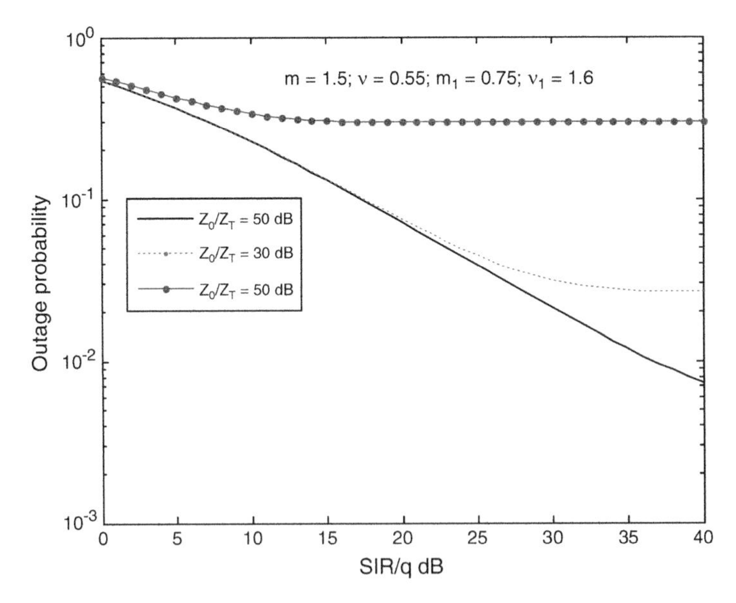

Fig. 6.13 Outage probabilities in a shadowed fading channel in the presence of a single interferer

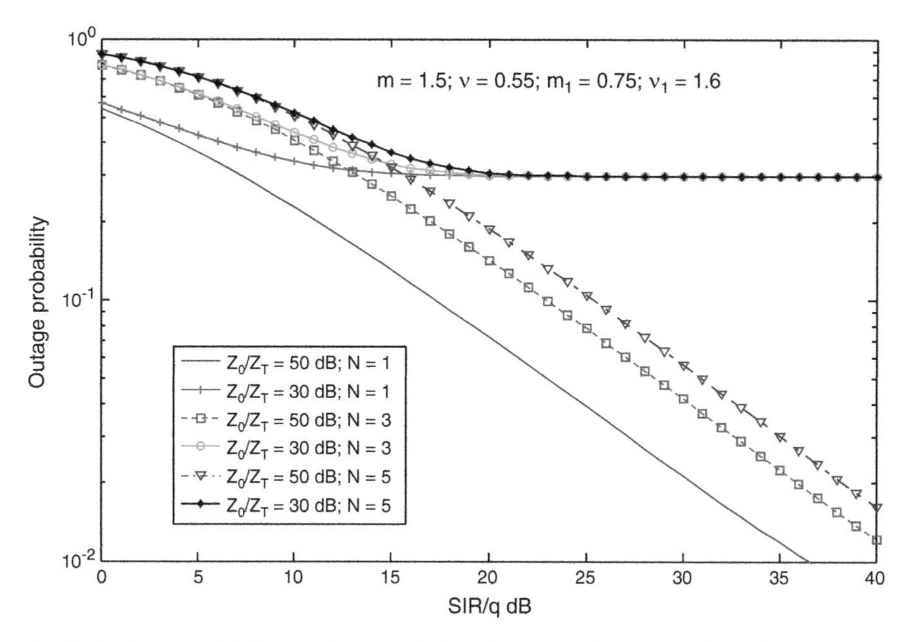

Fig. 6.14 Outage probabilities in shadowed fading channels in the presence of multiple shadowed fading interferers

6.3 Average Probability of Error

Even though outage probabilities are affected by the existence of interference from cochannels leading to outage "floors," the effect of the existence of CCI on bit error rates must also be examined to fully understand the degrading effects of CCI. The analysis of the effects of CCI on error rates is far more complicated than the calculations involved with the outages since we need to obtain explicit expressions for the density function of the received SNR when the CCI is present. Treating CCI as unwanted random component, the SNR of the desired signal can be written as (Winters 1984; Aalo and Zhang 1999; Beaulieu and Cheng 2004).

$$U = \frac{P}{N_s + P_{\text{CCI}}}. \tag{6.40}$$

In (6.40), P is the signal power which can be deterministic or random (due to fading, shadowing, or both) and $PCCI$ is the power from the CCI, similarly considered to be deterministic or random (due to fading, shadowing, or both). The noise power is given by Ns. Eq. (6.40) can be rewritten as

$$U = \frac{\left(\frac{P}{N_s}\right)}{1 + \left(\frac{P_{CCI}}{N_s}\right)} = \frac{Z}{1 + Z_c}. \tag{6.41}$$

In (6.41), Z is the SNR of the desired signal and Zc is the SNR of the CCI component as defined earlier in this chapter. The average probability of error when CCI exists can now be expressed as

$$p_{av}(e) = \int_0^\infty f(u)\frac{1}{2}erfc(\sqrt{u})du. \tag{6.42}$$

In (6.42) we have considered, as before, the case of a coherent BPSK modem. To evaluate (6.42), there is a need to obtain the probability density function (pdf) of the SNR U in (6.41).

We will obtain general expressions for the pdf and CDF of the SNR U. Using the results from Chap. 2 on the pdf of the ratio of two random variables, the density function of U in (6.41) becomes (Rohatgi and Saleh 2001; Papoulis and Pillai 2002)

$$f_U(u) = \int_0^\infty (1 + z_c) f_z[u(1 + z_c)] f(z_c) dz_c. \tag{6.43}$$

In arriving at (6.43) we have assumed, as we did earlier in this chapter, that the SNR of the desired channel Z and the CCI Zc are independent random variable. As before, we will start with the case of Rayleigh channels before we look into other fading and shadowed fading channels. This will also be done in two steps: obtaining the density function in (6.43) for each case and then estimating the error rates.

6.3.1 Probability Density Function (Rayleigh Channels)

For the case of the Rayleigh (desired) channel in the presence of a single CCI channel (Rayleigh), the density function of the overall SNR U in (6.43) can be obtained as

$$f(u) = \frac{Z_0 + uZ_{0c} + Z_0 Z_{0c}}{(Z_0 + uZ_{0c})^2} \exp\left(-\frac{u}{Z_0}\right). \tag{6.44}$$

In expressing (6.44), we have used the density function of the SNR of the desired signal and the SNR of the CCI component in (6.6) and (6.7), respectively, ($N = 1$). Note that (6.44) becomes (6.6) when the average SNR of the CCI component Z_{0c} is

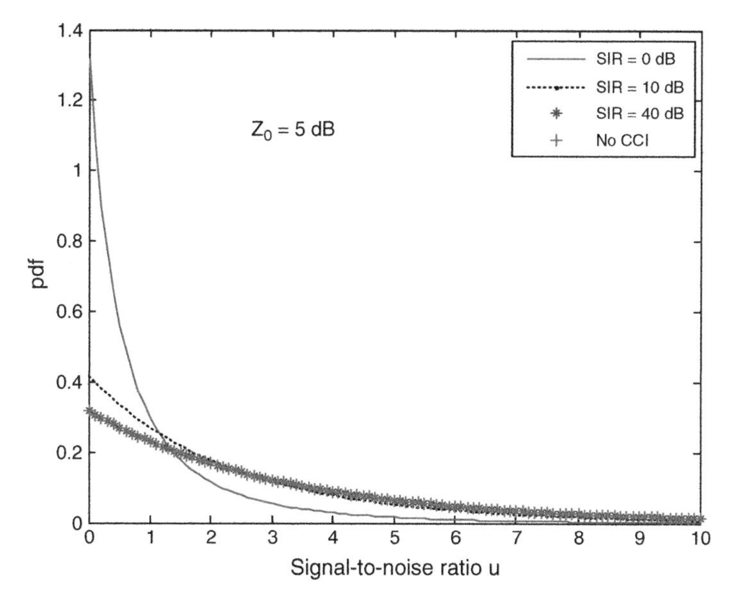

Fig. 6.15 Probability density function of the SNR in the presence of CCI (single channel)

much smaller than the average SNR of the desired signal, Z_0. In terms of the signal-to-interference ratio SIR defined in (6.8), the pdf in (6.44) becomes

$$f(u) = \frac{\left(1 + \frac{u}{SIR} + \frac{Z_0}{SIR}\right)}{Z_0\left(1 + \frac{u}{SIR}\right)^2} \exp\left(-\frac{u}{Z_0}\right). \tag{6.45}$$

The density function is plotted in Fig. 6.15.

Similarly, we can obtain an expression for the pdf of the SNR when there are N cochannels, and the desired cochannels are all Rayleigh faded. Using (6.6), (6.7), and (6.43), we get the pdf of the SNR as

$$f_U(u) = \left[\frac{Z_0^{N-1}}{(Z_0 + uZ_{0c})^N} + \frac{NZ_0^N Z_{0c}}{(Z_0 + uZ_{0c})^{N+1}}\right] \exp\left(-\frac{u}{Z_0}\right). \tag{6.46}$$

Using the definition of SIR in (6.8), the density function becomes

$$f_U(u) = \left[\frac{1 + \left(\frac{u}{SIR}\right) + N\left(\frac{Z_0}{SIR}\right)}{Z_0\left(1 + \frac{u}{SIR}\right)^{N+1}}\right] \exp\left(-\frac{u}{Z_0}\right). \tag{6.47}$$

Note that (6.47) becomes (6.45) when $N = 1$ (single cochannel). The density function is plotted in Fig. 6.16.

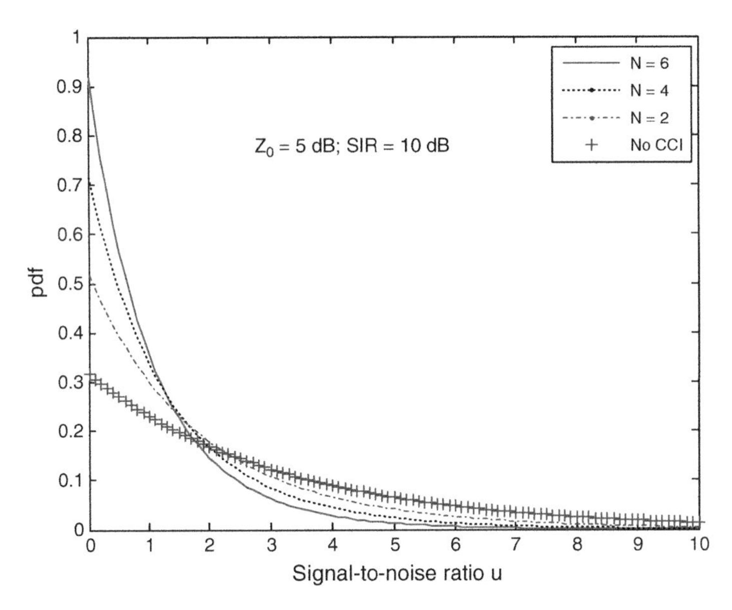

Fig. 6.16 Density function of the SNR in the presence of N-cochannels. All channels are Rayleigh channels

6.3.2 Probability Density Function (Nakagami Channels)

We will now attempt to get the density function of the SNR when both the desired channel and the CCI channel are described in terms of the Nakagami density functions as in (6.27) and (6.28). Substituting in (6.43), we have

$$f_U(u) = \int_0^\infty (1+z_c)\frac{m^m}{\Gamma(m)Z_0^m}[u(1+z_c)]^{m-1}\exp\left[-\frac{mu(1+z_c)}{Z_0}\right] \times \left(\frac{m_1}{Z_{0c}}\right)^{Nm_1}\frac{z_c^{Nm_1-1}}{\Gamma(Nm_1)}\exp\left(-\frac{m_1}{Z_{0c}}z_c\right)dz_c. \tag{6.48}$$

The pdf of the SNR is plotted in Fig. 6.17 for $m = 2.2$, $m_1 = 1.5$, $N = 4$, and three values of SIR. The pdf of the SNR in the absence of CCI is also plotted alongside.

The density functions are plotted in Fig. 6.18 for $N = 1, 2, 4$, and 6 as a function of the SNR in a channel with CCI.

6.3.3 Probability Density Function (Shadowed Fading Channels)

We can now include the effects of shadowing. First, we consider the desired channel to be a shadowed fading channel while CCI channels are considered to

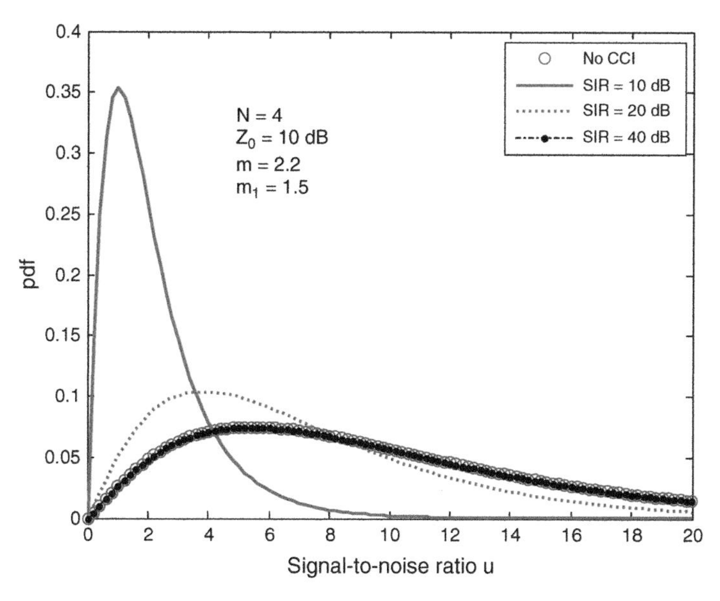

Fig. 6.17 The probability density function of the SNR for $m = 2.2$, $m_1 = 1.5$, $N = 4$, and $Z_0 = 10$ dB (Nakagami channels)

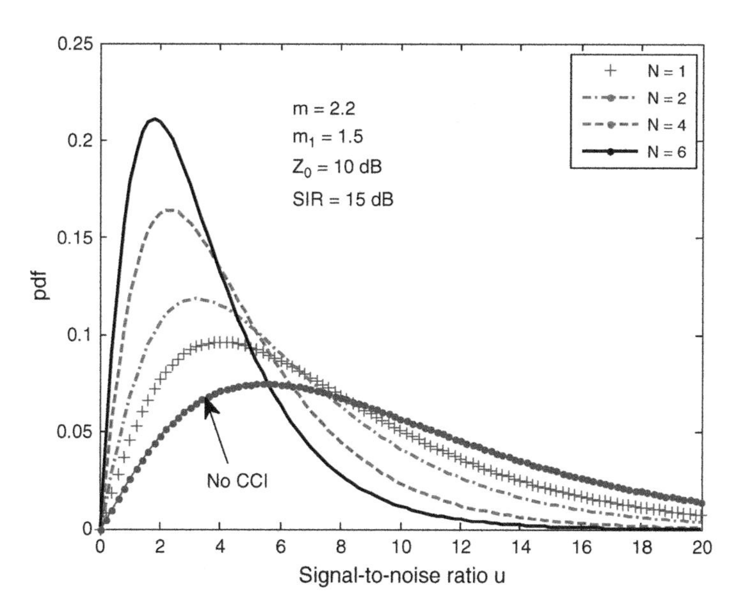

Fig. 6.18 Density functions for four different values of N compared with the case of no CCI (Nakagami channels)

be independent and identically distributed Nakagami channels. In terms of (6.43), the density function of the SNR in the presence of CCI now becomes

$$\begin{aligned} f_U(u) = \int_0^\infty (1+z_c)\frac{2}{\Gamma(m)\Gamma(\nu)}\left(\sqrt{\frac{m\nu}{Z_0}}\right)^{\nu+m} [u(1+z_c)]^{\left(\frac{\nu+m}{2}\right)-1} K_{m-\nu} \\ \times \left(2\sqrt{\frac{m\nu}{Z_0}u(1+z_c)}\right)\left(\frac{m_1}{Z_{0c}}\right)^{Nm_1}\frac{z_c^{Nm_1-1}}{\Gamma(Nm_1)}\exp\left(-\frac{m_1}{Z_{0c}}z_c\right)dz_c. \end{aligned} \quad (6.49)$$

The density function in (6.49) can be obtained through numerical integration. The density function for $N = 1$, 3, and 6 is plotted along with the case of no CCI in Fig. 6.19 for the case of $m = 1.5$, $n = 2.5$, $m_1 = 1.2$, and SIR of 10 dB. The problems associated with the increasing values of N are seen from the shifting of the peaks of the densities to the left as N increases.

Figure 6.20 shows the density functions for three values of the signal-to-CCI ratio for the case of four interfering channels along with the density function in the absence of any CCI. Once again, as expected, the density function of the signal-to-noise ratio approaches the pdf of the SNR in the absence of CCI as signal-to-CCI ratio increases.

We can now examine the last case considered in connection with the outage probabilities, namely the presence of interferers which also undergo shadowing and fading when the desired channel is undergoing shadowing and fading.

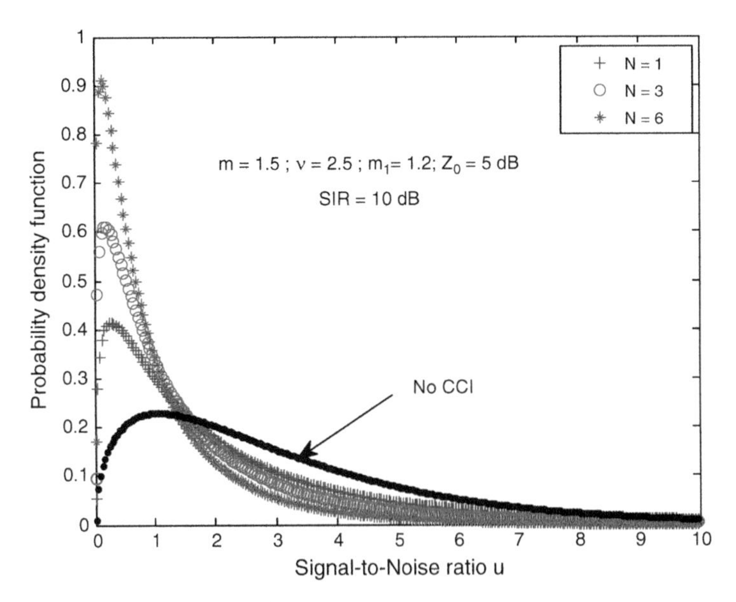

Fig. 6.19 Probability density functions of the signal-to-noise ratio in the presence of CCI (shadowed fading channels; Nakagami CCI). $m = 1.5$, $n = 2.5$, $m_1 = 1.2$, SIR = 10 dB. Average SNR of the desired channel is 5 dB

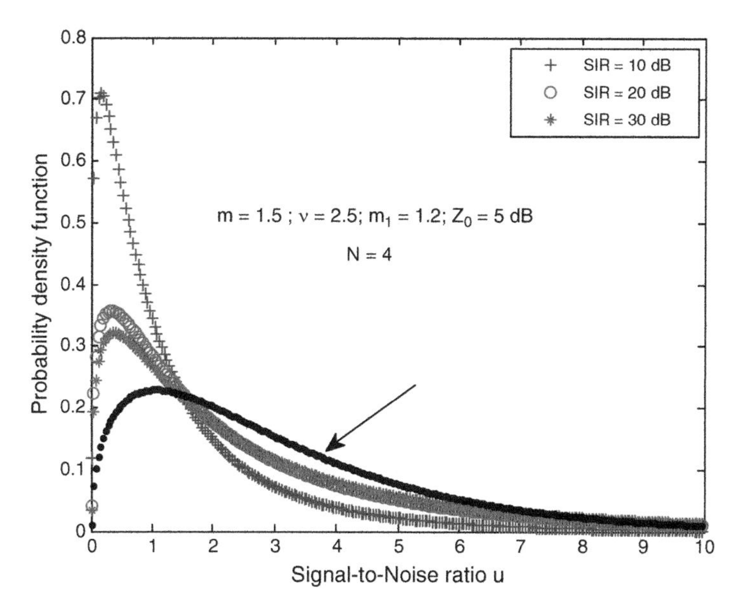

Fig. 6.20 Probability density functions of the signal-to-noise ratio in the presence of CCI (shadowed fading channels; Nakagami CCI). $m = 1.5$, $n = 2.5$, $m_1 = 1.2$, $N = 4$. Average SNR of the desired channel is 5 dB

The difficulty in obtaining an expression for the pdf of the SNR in the presence of CCI lies in the fact that the density function of the sum of a number of independent GK distributed variables (treating the Nakagami-lognormal shadowed fading channel to be equivalent to a GK channel) is not readily available analytically. One of the ways of overcoming this hurdle is to use the approximation to the sum of GK variables by another GK variable as suggested by Al-Ahmadi and Yanikomeroglu (2010) and others (Sriv et al. 2005; Chatzidiamantis et al. 2009; Al-Ahmadi and Yanikomeroglu 2010). Considering N interfering channels, each with parameters m_1 and ν_1 such that the density functions of any one of the channels can be expressed as

$$f(w_i) = \frac{2}{\Gamma(m)\Gamma(\nu)}\left(\sqrt{\frac{m_1\nu_1}{Z_{0c}}}\right)^{\nu_1+m_1}[w_i]^{\left(\frac{\nu_1+m_1}{2}\right)-1}K_{m_1-\nu_1}\left(2\sqrt{\frac{m_1\nu_1}{Z_{0c}}w_i}\right), \quad i = 1,2,..,N. \tag{6.50}$$

$$Z_c = \sum_{i=1}^{N} W_i. \tag{6.51}$$

the density function of the CCI component can be written as

$$f(z_c) = \frac{2}{\Gamma(m_n)\Gamma(\nu_n)} \left(\sqrt{\frac{m_n \nu_n}{N Z_{0c}}} \right)^{\nu_n + m_n} [z_c]^{\left(\frac{\nu_n + m_n}{2}\right) - 1} K_{m_n - \nu_n} \left(2 \sqrt{\frac{m_n \nu_n}{N Z_{0c}} z_c} \right). \tag{6.52}$$

In (6.52),

$$m_n = \frac{(1+a) + \sqrt{(1+a)^2 + (4/N\nu_1^2)k_1}}{2(a + (1/\nu_1) + 1)} m_1 N. \tag{6.53}$$

$$\nu_n = \frac{m_n}{a}. \tag{6.54}$$

The parameters a and k_1 are

$$a = \frac{m_1}{\nu_1}, \tag{6.55}$$

$$k_1 = 1 + m_1 + \nu_1. \tag{6.56}$$

Using (6.52), the density function in (6.43) of the signal-to-noise ratio in the presence of CCI becomes

$$\begin{aligned} f_U(u) = & \int_0^{\infty} (1+z_c) \frac{2}{\Gamma(m)\Gamma(\nu)} \left(\sqrt{\frac{m\nu}{Z_0}} \right)^{\nu+m} [u(1+z_c)]^{\left(\frac{\nu+m}{2}\right)-1} K_{m-\nu} \left(2\sqrt{\frac{m\nu}{Z_0} u(1+z_c)} \right) \\ & \times \frac{2}{\Gamma(m_n)\Gamma(\nu_n)} \left(\sqrt{\frac{m_n \nu_n}{N Z_{0c}}} \right)^{\nu_n + m_n} [z_c]^{\left(\frac{\nu_n + m_n}{2}\right)-1} K_{m_n - \nu_n} \left(2\sqrt{\frac{m_n \nu_n}{N Z_{0c}} z_c} \right) dz_c. \end{aligned} \tag{6.57}$$

The density function is plotted in Fig. 6.21 for three values of N and for the case of absence of any CCI.

The density functions for two values of the SIR are shown in Fig. 6.22.

Now that we have seen the density functions of the SNR in the presence of CCI, we can estimate the error rates and understand how the presence of CCI will impact the error rates in wireless channels.

6.3.4 Error Rates (Rayleigh Channels)

Having seen the density functions, we can estimate the error rates (Winters 1984; Beaulieu and Cheng 2004; Sivanesan and Beaulieu 2004; Ismail and Matalgah 2007). The bit error rate in a Rayleigh channel in the presence of a single Rayleigh interferer can be expressed in integral form using (6.42) and (6.45)

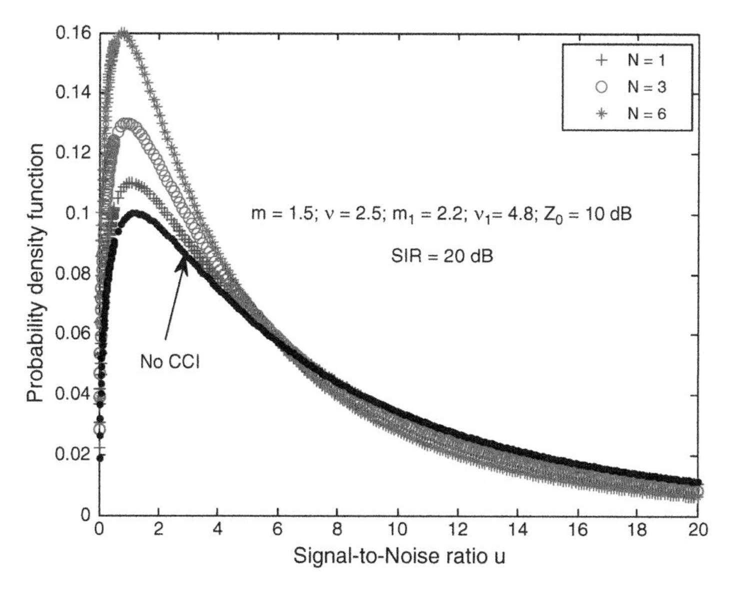

Fig. 6.21 Probability density functions for $N = 1, 3, 6$ and for the case of no CCI. The average SNR of the desired channel is 10 dB and the SIR is 20 dB

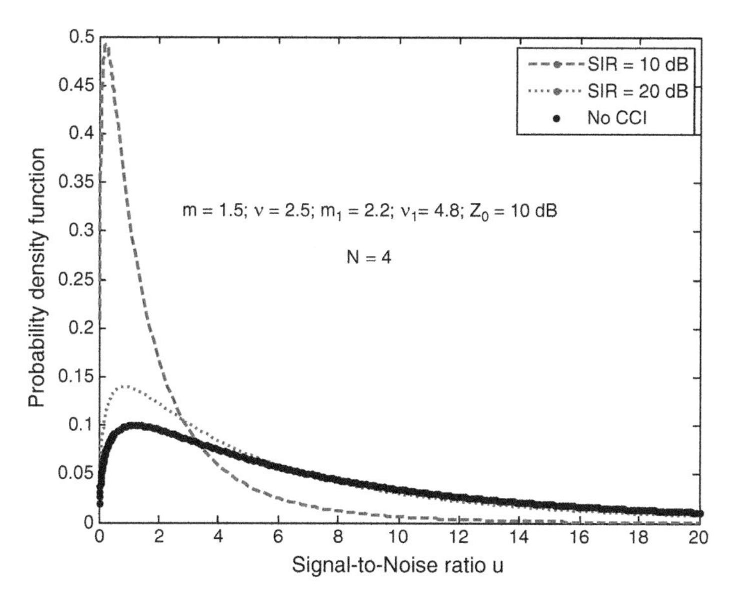

Fig. 6.22 Probability density functions for SIR =10 and 20 dB and for the case of no CCI. The average SNR of the desired channel is 10 dB and the number of interfering channels is four

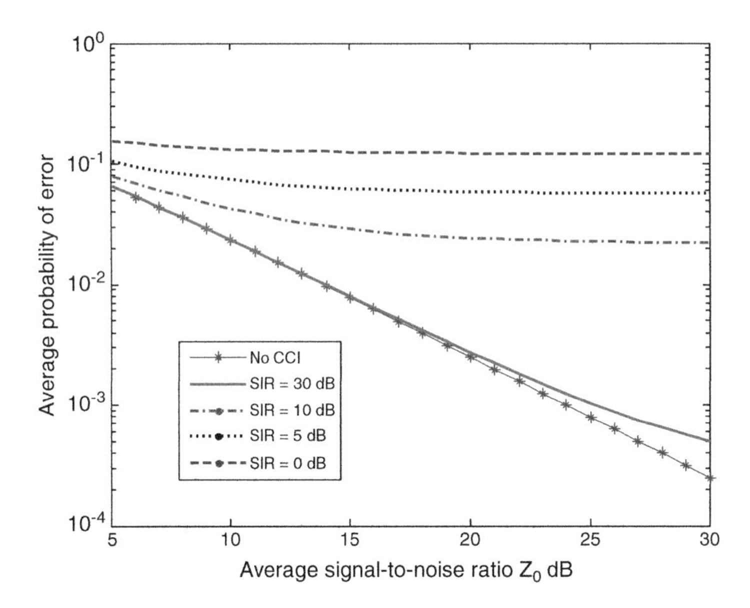

Fig. 6.23 Average probability of error in a Rayleigh channel with a single interferer from a Rayleigh cochannel

$$p_{av}(e) = \int_0^\infty \frac{1}{2} erfc\sqrt{u} \frac{(1 + (u/\mathrm{SIR}) + (Z_0/\mathrm{SIR}))}{Z_0(1 + (u/\mathrm{SIR}))^2} \exp\left(-\frac{u}{Z_0}\right) du. \tag{6.58}$$

Equation (6.58) can be integrated numerically. The results are shown in Fig. 6.23 as a function of the SNR of the desired signal and SIR. Error rates when multiple interferers are present can be evaluated using (6.47). The average error rate in the presence of multiple Rayleigh interferers will be

$$p_{av}(e) = \int_0^\infty \frac{1}{2} erfc\sqrt{u} \left[\frac{1 + (u/\mathrm{SIR}) + N(Z_0/\mathrm{SIR})}{Z_0(1 + (u/\mathrm{SIR}))^{N+1}}\right] \exp\left(-\frac{u}{Z_0}\right) du. \tag{6.59}$$

The results are shown in Fig. 6.24.

6.3.5 Error Rates (Nakagami Channels)

The error probabilities in Nakagami-m faded channels can now be evaluated (Aalo and Zhang 1999; Beaulieu and Cheng 2004). We will make use of the expression for the pdf of the SNR in (6.48). The expression for the average error probability becomes

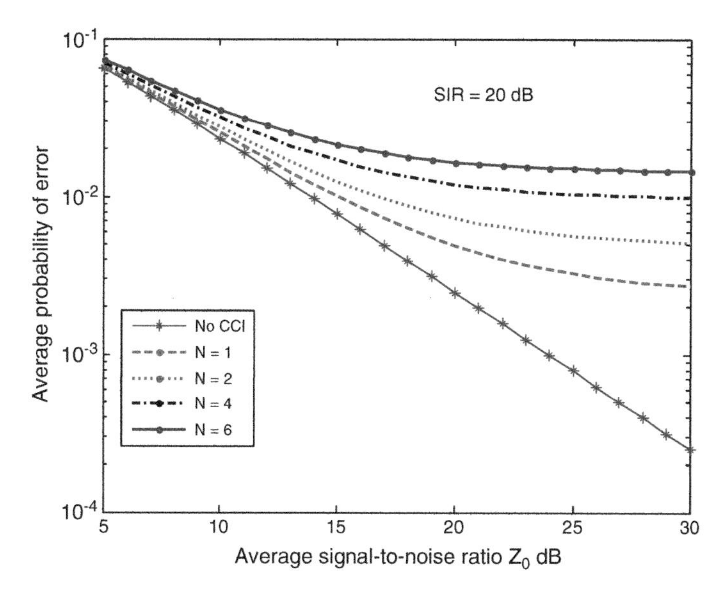

Fig. 6.24 Average BER in Rayleigh fading channel in the presence of N Rayleigh distributed cochannels for several values of N (SIR =20 dB)

$$p_{av}(e) = \int_0^\infty \frac{1}{2} erfc\sqrt{u} \int_0^\infty (1+z_c)\frac{m^m}{\Gamma(m)Z_0^m}[u(1+z_c)]^{m-1}\exp\left[-\frac{mu(1+z_c)}{Z_0}\right] \times \left(\frac{m_1}{Z_{0c}}\right)^{Nm_1} \frac{z_c^{Nm_1-1}}{\Gamma(Nm_1)}\exp\left(-\frac{m_1}{Z_{0c}}z_c\right)dz_c du. \tag{6.60}$$

Defining

$$Z_u = \frac{Z_0}{(1+z_c)}, \tag{6.61}$$

we can rewrite (6.60) as

$$p_{av}(e) = \int_0^\infty \int_0^\infty \frac{1}{2} erfc(\sqrt{u}) \left(\frac{m}{Z_u}\right)^m \frac{u^{m-1}}{\Gamma(m)}\exp\left(-\frac{mu}{Z_u}\right)du \times \left(\frac{m_1}{Z_{0c}}\right)^{Nm_1} \frac{z_c^{Nm_1-1}}{\Gamma(Nm_1)}\exp\left(-\frac{m_1}{Z_{0c}}z_c\right)dz_c. \tag{6.62}$$

The double integral in (6.62) can be converted to a single integral by performing the integral over the variable u. Then we have

$$\begin{aligned} I &= \int_0^\infty \frac{1}{2} erfc\left(\sqrt{u}\right) \left(\frac{m}{Z_u}\right)^m \frac{u^{m-1}}{\Gamma(m)} \exp\left(-\frac{mu}{Z_u}\right) du \\ &= \int_0^\infty \frac{\exp(-u)}{2\sqrt{\pi u}} \frac{1}{\Gamma(m)} G_{1,2}^{1,1}\left(\frac{mu}{Z_u}\middle| \begin{matrix} 1 \\ m, 0 \end{matrix}\right) du. \end{aligned} \tag{6.63}$$

In writing down (6.63), we have made use of the fact that we can express the error rates using the CDF as discussed in Chaps. 4 and 5. We also expressed the gamma pdf as a Meijer G-function which has as its CDF, another Meijer G-function as follows (Mathai 1993; Mathai and Haubold 2008). We have the density function of the gamma pdf as

$$f(u) = \left(\frac{m}{Z_u}\right)^m \frac{u^{m-1}}{\Gamma(m)} \exp\left(-\frac{mu}{Z_u}\right) = \frac{1}{u\Gamma(m)} G_{0,1}^{1,0}\left(\frac{mu}{Z_u}\middle| \begin{matrix} - \\ m \end{matrix}\right), \tag{6.64}$$

and the corresponding CDF as

$$F(u) = \frac{1}{\Gamma(m)} G_{1,2}^{1,1}\left(\frac{mu}{Z_u}\middle| \begin{matrix} 1 \\ m, 0 \end{matrix}\right). \tag{6.65}$$

The integral I in (6.63) becomes

$$I = \frac{1}{2} \frac{\Gamma\left(m+\frac{1}{2}\right)\left[\frac{m(1+z_c)}{Z_0}\right]^m}{\Gamma(m) m \sqrt{\pi}} {}_2F_1\left(\left[m, m+\frac{1}{2}\right], [1+m], -\frac{m(1+z_c)}{Z_0}\right). \tag{6.66}$$

Note that ${}_2F_1([.,.],[.],.)$ is the hypergeometric function (Mathai 1993; Abramowitz and Segun 1972; Gradshteyn and Ryzhik 2007).

$$\begin{aligned} p_{av}(e) = \int_0^\infty & \frac{1}{2} \frac{\Gamma\left(m+\frac{1}{2}\right)\left[\frac{m(1+z_c)}{Z_0}\right]^m}{\Gamma(m) m \sqrt{\pi}} {}_2F_1\left(\left[m, m+\frac{1}{2}\right], [1+m], -\frac{m(1+z_c)}{Z_0}\right) \\ & \times \left(\frac{m_1}{Z_{0c}}\right)^{Nm_1} \frac{z_c^{Nm_1-1}}{\Gamma(Nm_1)} \exp\left(-\frac{m_1}{Z_{0c}} z_c\right) dz_c. \end{aligned} \tag{6.67}$$

Equation (6.67) can be integrated numerically.

The results for the case of three values of N are compared with the case of the absence of CCI for an SIR of 20 dB, shown in Fig. 6.25. The value of m is 1.5 and m_1 is 0.75.

The effect of the variation in SIR values is shown in Fig. 6.26, where the error rates are plotted for the case of six interfering channels ($m = 1.5$ and $m_1 = 0.75$).

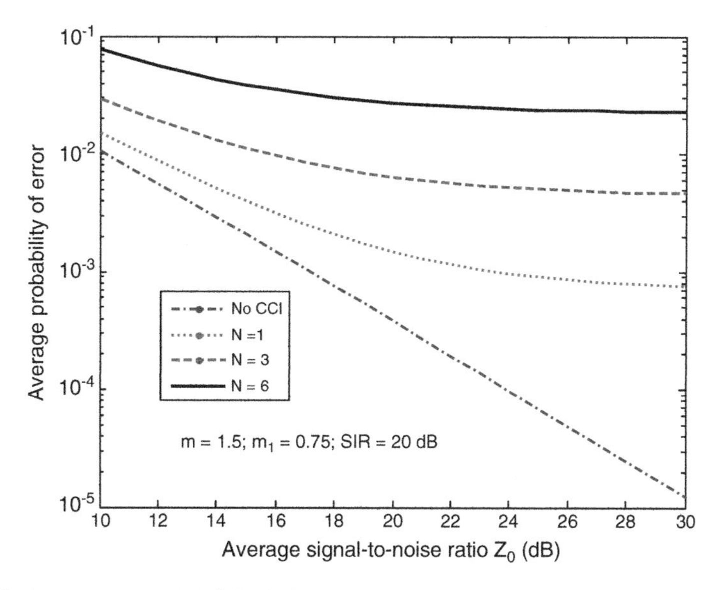

Fig. 6.25 Average error probabilities in Nakagami faded channels for the case of Nakagami faded cochannels ($N = 1$, 3, and 6; $m = 1.5$, $m_1 = 0.75$, and SIR = 20 dB)

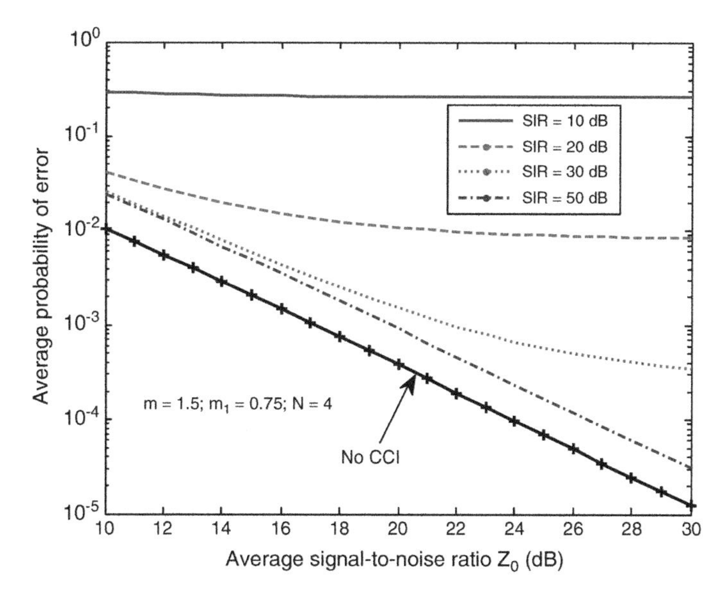

Fig. 6.26 Average error probabilities in Nakagami faded channels for the case of four Nakagami faded cochannels ($m = 1.5$, $m_1 = 0.75$, and SIR = 10, 20, 30, and 50 dB)

6.3.6 Error Rates (Shadowed Fading Channels)

We can now study the error rates in a shadowed fading channel when the interferers operate in a Nakagami faded environment. The error rate in a shadowed fading channel in the presence of N Nakagami interferers can be expressed as a double integral as

$$p_{av}(e) = \int_0^\infty \frac{1}{2} erfc\sqrt{u} \int_0^\infty (1+z_c) \times \frac{2}{\Gamma(m)\Gamma(\nu)} \left(\sqrt{\frac{m\nu}{Z_0}}\right)^{\nu+m} [u(1+z_c)]^{((\nu+m)/2)-1} \times K_{m-\nu}\left(2\sqrt{\frac{m\nu}{Z_0}u(1+z_c)}\right) \left(\frac{m_1}{Z_{0c}}\right)^{Nm_1} \frac{z_c^{Nm_1-1}}{\Gamma(Nm_1)} \exp\left(-\frac{m_1}{Z_{0c}}z_c\right) dz_c du. \tag{6.68}$$

In arriving at (6.68), we have made use of the expression for the error rate in (6.43) and the density function of the SNR in (6.49). As we had done before, the double integral in (6.68) can be converted to a single integral using

$$I_1 = \int_0^\infty \frac{1}{2} erfc(\sqrt{u})(1+z_c) \frac{2}{\Gamma(m)\Gamma(\nu)} \left(\sqrt{\frac{m\nu}{Z_0}}\right)^{\nu+m} [u(1+z_c)]^{((\nu+m)/2)-1} \times K_{m-\nu}\left(2\sqrt{\frac{m\nu}{Z_0}u(1+z_c)}\right) du. \tag{6.69}$$

Using the table of integrals (Gradshteyn and Ryzhik 2007; Wolfram 2011), (6.69) becomes

$$I_1 = \frac{1}{2} - \frac{1}{2}\left\{\pi^2 \frac{\csc(\pi m)\csc(\pi\nu)}{\Gamma(1-m)\Gamma(1-\nu)\Gamma(m)\Gamma(\nu)} - W\right\}. \tag{6.70}$$

The parameter W in (6.70) is

$$W = \left[\tfrac{m\nu}{Z_0}(1+z_c)\right]^m \frac{{}_2F_2\left(\left[m, m+\frac{1}{2}\right], [1+m, 1+m-\nu], \left(\frac{m\nu}{Z_0}\right)(1+z_c)\right)}{\sqrt{\pi}\Gamma(m+1)\Gamma(\nu)} \times \Gamma\left(m+\frac{1}{2}\right)\Gamma(-m+\nu) + \left[\tfrac{m\nu}{Z_0}(1+z_c)\right]^\nu \times \frac{{}_2F_2\left(\left[\nu, \nu+\frac{1}{2}\right], [1+\nu, 1+\nu-m], \left(\frac{m\nu}{Z_0}\right)(1+z_c)\right)}{\sqrt{\pi}\Gamma(\nu+1)\Gamma(m)} \Gamma\left(\nu+\frac{1}{2}\right)\Gamma(-\nu+m). \tag{6.71}$$

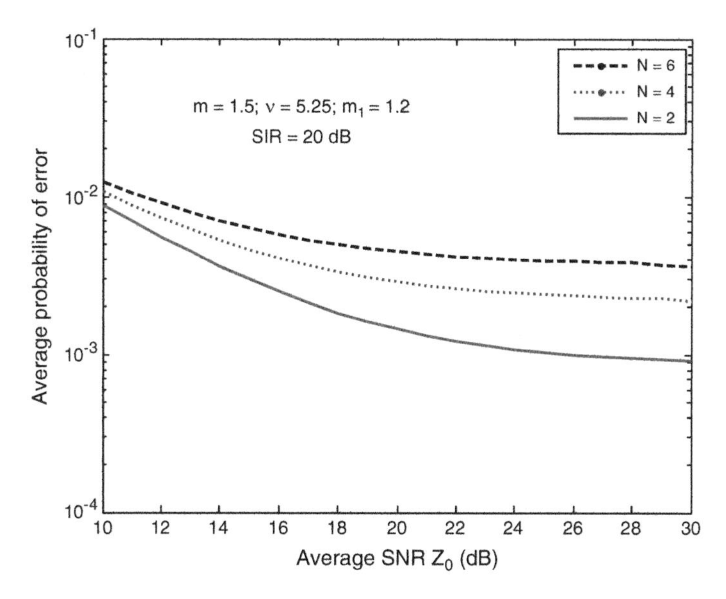

Fig. 6.27 Average error rates in shadowed fading channels in the presence of Nakagami faded cochannels ($m = 1.5$, $\nu = 5.25$, $m_1 = 1.2$, and SIR = 20 dB)

The error rate in (6.68) now becomes

$$p_{av}(e) = \int_0^\infty I_1 \left(\frac{m_1}{Z_{0c}}\right)^{Nm_1} \frac{z_c^{Nm_1-1}}{\Gamma(Nm_1)} \exp\left(-\frac{m_1}{Z_{0c}} z_c\right) dz_c. \tag{6.72}$$

The effect of CCI on the bit error rates in shadowed fading channels can be obtained from (6.72), which can be integrated numerically. The error rates for three values of N are shown in Fig. 6.27.

Figure 6.28 shows the effects of the shadowing on the error rates. Error rates are plotted for three levels of shadowing corresponding to $\nu = 5.25$, 2.2, and 1.1.

Error rates in shadowed fading channels, while the cochannels are also undergoing fading and shadowing, can be obtained using (6.57) and (6.43). The error rate now becomes

$$\begin{aligned} p_{av}(e) = &\int_0^\infty \frac{1}{2}\operatorname{erfc}\sqrt{u} \int_0^\infty (1+z_c) \frac{2}{\Gamma(m)\Gamma(\nu)} \left(\sqrt{\frac{m\nu}{Z_0}}\right)^{\nu+m} [u(1+z_c)]^{\left(\frac{\nu+m}{2}\right)-1} \\ &\times K_{m-\nu}\left(2\sqrt{\frac{m\nu}{Z_0}u(1+z_c)}\right) \\ &\times \frac{2}{\Gamma(m_n)\Gamma(\nu_n)} \left(\sqrt{\frac{m_n\nu_n}{NZ_{0c}}}\right)^{\nu_n+m_n} [z_c]^{\left(\frac{\nu_n+m_n}{2}\right)-1} K_{m_n-\nu_n}\left(2\sqrt{\frac{m_n\nu_n}{NZ_{0c}} z_c}\right) dz_c du. \end{aligned} \tag{6.73}$$

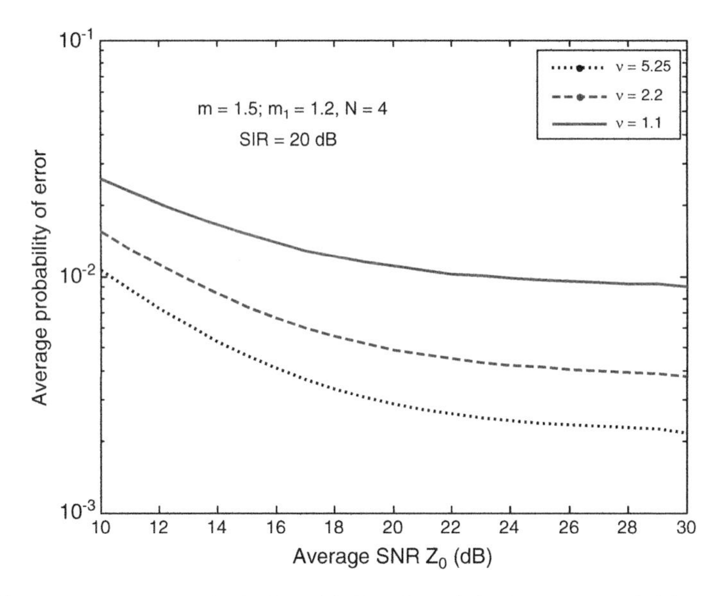

Fig. 6.28 Average error rates in shadowed fading channels in the presence of Nakagami faded cochannels ($m = 1.5$, $m_1 = 1.2$, $N = 4$, and SIR = 20 dB, $\nu = 5.25, 2.2, 1.1$)

Note what we have made of the approximation for the pdf of the sum of N i.i.d GK variables. Using (6.69–6.71), the error rate in a shadowed fading channel in the presence of N shadowed fading cochannels is

$$p_{av}(e) = \int_0^\infty I_1 \frac{2}{\Gamma(m_n)\Gamma(\nu_n)} \left(\sqrt{\frac{m_n \nu_n}{N Z_{0c}}}\right)^{\nu_n + m_n} [z_c]^{\left(\frac{\nu_n + m_n}{2}\right) - 1} K_{m_n - \nu_n}\left(2\sqrt{\frac{m_n \nu_n}{N Z_{0c}} z_c}\right) dz_c. \tag{6.74}$$

The error rates in shadowed fading channels in the presence of cochannels which also undergo shadowed fading are plotted in Fig. 6.29.

6.3.7 Error Rates Following Diversity

We have examined the effect of CCI in wireless systems in the absence of any diversity to mitigate the fading (Shah and Haimovich 1998, 2000; Cui et al. 1997; Aalo and Zhang 1999; Aalo and Chayawan 2000). We will now look at the error rates when diversity is implemented for the mitigation of short-term fading modeled in terms of the Nakagami-m distribution. As an example, a maximal ratio combiner (MRC) is considered. While the diversity is implemented for the desired channel, the cochannels are treated as if no mitigation is applied to the cochannels. The probability density function of the SNR in the presence of CCI can now be expressed using (6.48) as

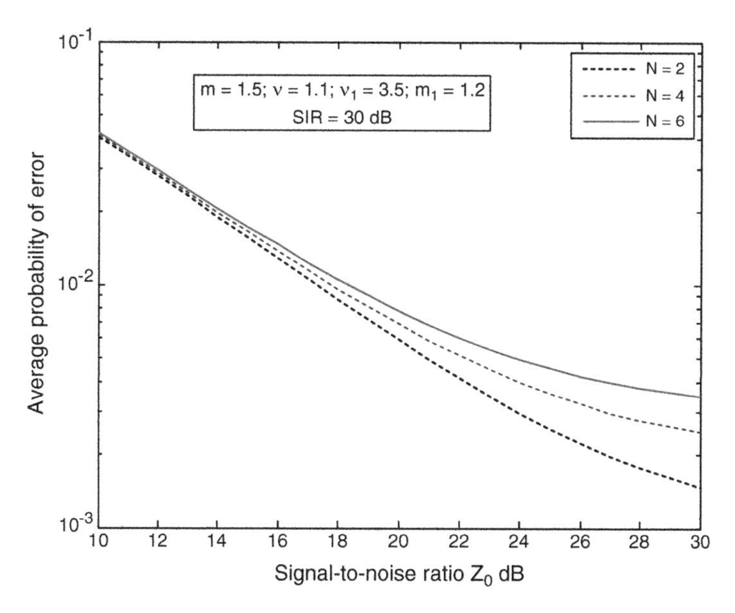

Fig. 6.29 Average probability of error in shadowed fading channels when the cochannels also undergo fading and shadowing ($m = 1.5$; $\nu = 1.1$; $\nu 1 = 3.5$; $m_1 = 1.2$; SIR = 30 dB, and $N = 2, 4, 6$)

$$f_U(u) = \int_0^\infty (1+z_c)\frac{m^{mM}}{\Gamma(mM)Z_0^{mM}}[u(1+z_c)]^{mM-1}\exp\left[-\frac{mu(1+z_c)}{Z_0}\right] \times \left(\frac{m_1}{Z_{0c}}\right)^{Nm_1}\frac{z_c^{Nm_1-1}}{\Gamma(Nm_1)}\exp\left(-\frac{m_1}{Z_{0c}}z_c\right)dz_c. \tag{6.75}$$

The probability density functions are plotted in Fig. 6.30 for $M = 1$ (no diversity), $M = 2$, and 4 for $N = 4$ and SIR of 15 dB. The average SNR is 10 dB.

In (6.75), M is the order of diversity. The error rate following diversity can be written using (6.67) as

$$p_{av}(e) = \frac{1}{2}\frac{\Gamma\left(mM+\frac{1}{2}\right)\left[\frac{m(1+z_c)}{Z_0}\right]^{mM}}{\Gamma(mM)mM\sqrt{\pi}} {}_2F_1\left(\left[mM, mM+\frac{1}{2}\right], [1+mM], -\frac{m(1+z_c)}{Z_0}\right) \times \left(\frac{m_1}{Z_{0c}}\right)^{Nm_1}\frac{z_c^{Nm_1-1}}{\Gamma(Nm_1)}\exp\left(-\frac{m_1}{Z_{0c}}z_c\right)dz_c. \tag{6.76}$$

The average probability of error is plotted in Fig. 6.31 for $M = 2$ and compared with the case of no diversity.

Figure 6.32 shows the BER for $N = 3$ for the case of $M = 2$ and 3.

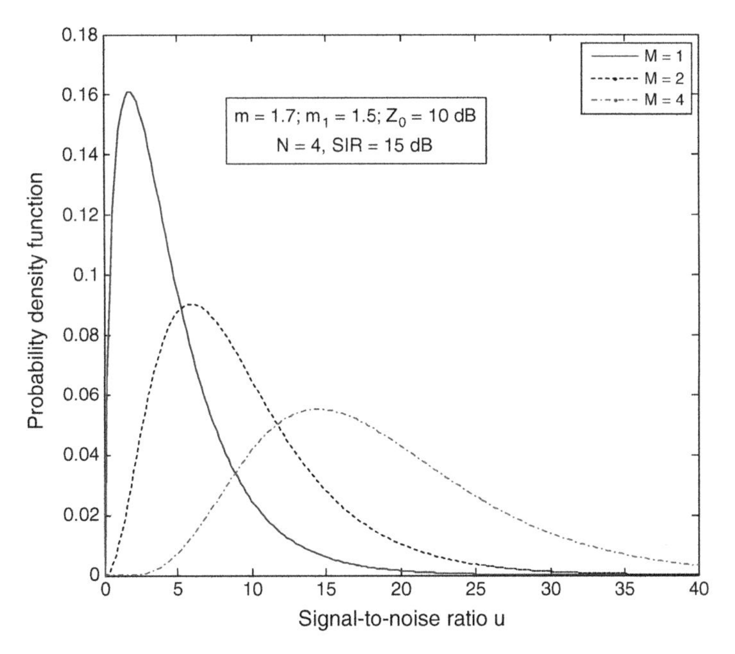

Fig. 6.30 The probability density functions are plotted in Fig. 6.31 for $M = 1$ (no diversity), $M = 2$, and $M = 4$ for $N = 4$ and SIR of 15 dB. The average SNR is 10 dB

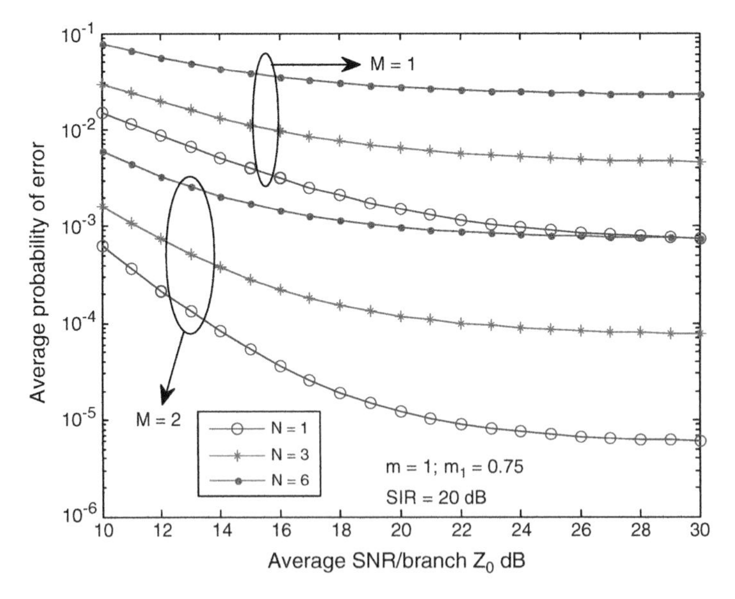

Fig. 6.31 BER in maximal ratio combiner (MRC) diversity in the presence of CCI for $N = 1$, 3, and 6 (SIR = 20 dB, $m = 1.5$, $m_1 = 0.75$). The case of no diversity is also shown

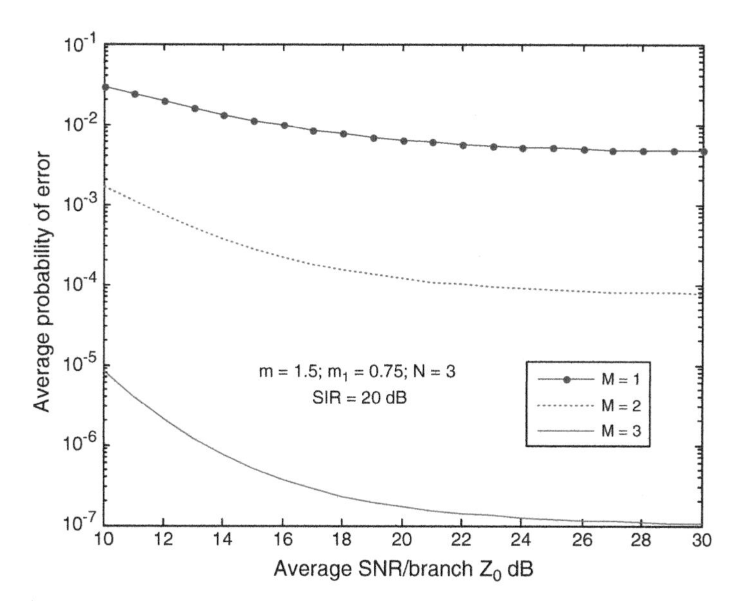

Fig. 6.32 BER in MRC diversity in the presence of CCI for $N = 3$, SIR $= 20$ dB, $m = 1.5$, $m_1 = 0.75$ ($M = 1$ (no diversity), 2, and 3)

6.4 Summary

We presented the effects of the existence of cochannels. Starting with the simple notion of outage when the interfering signal strength exceeds the strength of the desired signal, outage probabilities in the presence of multiple cochannels were derived. The density functions of the signal-to-noise ratio (by taking noise as well as CCI) were obtained for the case of pure short-term fading and shadowed fading. The error probabilities were then estimated to illustrate the degrading effects of CCI. Using the example of maximal ratio combining algorithm we demonstrated the improvement gained through diversity.

References

Aalo, V., & Chayawan, C. (2000). Outage probability of cellular radio systems using maximal ratio combining in Rayleigh fading channel with multiple interferers. *Electronics Letters, 36* (15), 1314–1315.

Aalo, V., & Zhang, J. (1999). On the effect of cochannel interference on average error rates in Nakagami-fading channels. *IEEE Communications Letters, 3*(5), 136–138.

Abu-Dayya, A. A., & Beaulieu, N. C. (1991). Outage probabilities of cellular mobile radio systems with multiple Nakagami interferers. *IEEE Transactions on Vehicular Technology, 40*(4), 757–768.

Abu-Dayya, A. A., & Beaulieu, N. C. (1992). Outage probabilities of diversity cellular systems with cochannel interference in Nakagami fading. *IEEE Transactions on Vehicular Technology, 41*(4), 343–355.
Abu-Dayya, A. A., & Beaulieu, N. C. (1999). Diversity MPSK receivers in cochannel interference. *IEEE Transactions on Vehicular Technology, 48*(6), 1959–1965.
Abramowitz, M., & Segun, I. A. (Eds.). (1972). *Handbook of mathematical functions with formulas, graphs, and tables*. New York, NY: Dover Publications.
Al-Ahmadi, S., & Yanikomeroglu, H. (2010). On the approximation of the PDF of the sum of independent generalized-K RVs by another generalized-K PDF with applications to distributed antenna systems. In Wireless communications and networking conference (WCNC) (pp. 1–5). IEEE.
Beaulieu, N. C., & Cheng, J. (2004). Precise error-rate analysis of bandwidth-efficient BPSK in Nakagami fading and cochannel interference. *IEEE Transactions on Communications, 52*(1), 149–158.
Cardieri, P., & Rappaport, T. (2001). Statistical analysis of co-channel interference in wireless communications systems. *Wireless Communications and Mobile Computing, 1*(1), 111–121.
Chatzidiamantis, N. D., Karagiannidis, G. K., et al. (2009). On the distribution of the sum of gamma-gamma variates and application in MIMO optical wireless systems. In Global telecommunications conference, GLOBECOM 2009 (pp. 1–6). IEEE.
Cui, J., Falconer, D., et al. (1997). Performance evaluation of optimum combining and maximal ratio combining in the presence of co channel interference and channel correlation for wireless communication systems. *Mobile Networks and Applications, 2*(4), 315–324.
Cui, J., & Sheikh, A. U. H. (1999). Outage probability of cellular radio systems using maximal ratio combining in the presence of multiple interferers. *IEEE Transactions on Communications, 47*(8), 1121–1124.
Gradshteyn, I. S., & Ryzhik, I. M. (2007). *Table of integrals, series and products*. Oxford: Academic.
Helstrom, C. (1986). Calculating error probabilities for Intersymbol and cochannel interference. *IEEE Transactions on Communications, 34*(5), 430–435.
Ismail, M., & Matalgah, M. (2007). Exact and approximate error-rate analysis of BPSK in Weibull fading with cochannel interference. *IEEE Transactions on Communications, 1*(2), 203–208.
Linnartz, J.-P. (1992). Exact analysis of the outage probability in multiple-user mobile radio. *IEEE Transactions on Communications, 40*(1), 20–23.
MacDonald, V. (1979). The cellular concept. *Bell System Technical Journal, 58*(1), 15–41.
Mathai, A. M. (1993). *A handbook of generalized special functions for statistical and physical sciences*. Oxford: Oxford University Press.
Mathai, A. M., & Haubold, H. J. (2008). *Special functions for applied scientists*. New York, NY: Springer Science+Business Media.
Moschopoulos, P. (1985). The distribution of the sum of independent gamma random variables. *Annals of the Institute of Statistical Mathematics, 37*(1), 541–544.
Mostafa, R., Annamalai, A., et al. (2004). Performance evaluation of cellular mobile radio systems with interference of dominant interferers. *IEEE Transactions on Communications, 52*(2), 326–335.
Muammar, R., & Gupta, S. (1982). Cochannel interference in high-capacity mobile radio systems. *IEEE Transactions on Communications, 30*(8), 1973–1978.
Okui, S. (1992). Probability of co-channel interference for selection diversity reception in the Nakagami m-fading channel. *IEEE Proceedings on Communications, Speech and Vision, 139* (1), 91–94.
Papoulis, A., & Pillai, S. U. (2002). *Probability, random variables, and stochastic processes*. Boston, MA: McGraw-Hill.
Reig, J., & Cardona, N. (2000). Approximation of outage probability on Nakagami fading channels with multiple interferes. *Electronics Letters, 36*(19), 1649–1650.

Rohatgi, V. K., & Saleh, A. K. M. E. (2001). *An introduction to probability and statistics*. New York, NY: Wiley.
Shah, A., & Haimovich, A. M. (1998). Performance analysis of optimum combining in wireless communications with Rayleigh fading and cochannel interference. *IEEE Transactions on Communications, 46*(4), 473–479.
Shah, A., & Haimovich, A. M. (2000). Performance analysis of maximal ratio combining and comparison with optimum combining for mobile radio communications with cochannel interference. *IEEE Transactions on Vehicular Technology, 49*(4), 1454–1463.
Shankar, P. M. (2004). Error rates in generalized shadowed fading channels. *Wireless Personal Communications, 28*(3), 233–238.
Shankar, P. M. (2005). Outage probabilities in shadowed fading channels using a compound statistical model. *IEEE Proceedings on Communications, 152*(6), 828–832.
Shankar, P. (2007). Outage analysis in wireless channels with multiple interferers subject to shadowing and fading using a compound pdf model. *International Journal of Electronics and Communications (AEU), 61*(4), 255–261.
Sivanesan, K., & Beaulieu, N. C. (2004). Exact BER analyses of Nakagami/Nakagami CCI BPSK and Nakagami/Rayleigh CCI QPSK systems in slow fading. *IEEE Communications Letters, 8* (1), 45–47.
Simon, M. K., & Alouini, M.-S. (2005). *Digital communication over fading channels*. Wiley-Interscience: Hoboken, NJ.
Sriv, T., Chayawan, C., et al. (2005). An analytical model of maximal ratio combining systems in K-distribution fading channel and multiple co-channel interferers. Networks. In *Jointly held with the 2005 7th and 13th IEEE Malaysia international conference on communication* (pp. 1011–1016). IEEE.
Sowerby, K., & Williamson, A. (1987). Outage probability calculations for a mobile radio system having two log-normal interferers. *Electronics Letters, 23*(25), 1345–1346.
Sowerby, K., & Williamson, A. (1988a). Outage probability calculations for mobile radio systems with multiple interferers. *Electronics Letters, 24*(17), 1073–1075.
Sowerby, K. W., & Williamson, A. (1988b). Outage probability calculations for multiple cochannel interferers in cellular mobile radio systems. *IEE Proceedings F Communications, Radar and Signal Processing, 135*(3), 208–215.
Sowerby, K. W., & Williamson, A. G. (1992). Outage possibilities in mobile radio systems suffering cochannel interference. *IEEE Journal on Selected Areas in Communications, 10* (3), 516–522.
Sowerby, K., & Williamson, A. (2002). Outage possibilities in mobile radio systems suffering cochannel interference. *IEEE Journal on Selected Areas in Communications, 10*(3), 516–522.
Winters, J. H. (1984). Optimum combining in digital mobile radio with cochannel interference. *IEEE Transactions on Vehicular Technology, 33*(3), 144–155.
Wolfram (2011), Wolfram Research, Inc., http://functions.wolfram.com/.
Yang, H., & Alouini, M. (2003). Outage probability of dual-branch diversity systems in presence of co-channel interference. *IEEE Transactions on Wireless Communications, 2*(2), 310–319.
Yang, X., & Petropulu, A. P. (2003). Co-channel interference modeling and analysis in a Poisson field of interferers in wireless communications. *IEEE Transactions on Signal Processing, 51* (1), 64–76.
Yao, Y. D., & Sheikh, A. U. H. (1992). Investigations into cochannel interference in microcellular mobile radio systems. *IEEE Transactions on Vehicular Technology, 41*(2), 114–123.

Chapter 7
Cognitive Radio

7.1 Introduction

Spectral band available for use in wireless communications is a limited commodity regardless of whether it is used by the traditional wireless subscribers or by digital television. It is also known that these limited resources often remain underutilized raising the possibility that the unused channels may be utilized during the downtime of the authorized users. This opportunistic access to the wireless spectrum (of licensed bands) by unlicensed users when the licensed users are not online constitutes the concept of the cognitive radio (Urkowitz 1967; Haykin et al. 2009). This means that such unlicensed users, generally referred to as secondary users (as opposed to primary users who own the license), must rely on a mechanism to detect the absence of any activity in the spectral band by the licensed (primary) user. The secondary users are allowed to use the licensed spectrum only when such a use does not cause any interference to the primary users. This means that adequate sensing mechanism needs to be placed to detect the presence of the primary users and the monitoring needs to take place continuously. In other words, techniques must exist to undertake what is termed as "spectrum sensing" (Haykin et al. 2009). Spectrum sensing might be accomplished through the use of simple energy detection, pilot based coherent detection, or any other such techniques that are able to detect the presence of signals effectively. Since energy detection is the simplest form of detection, it will be used as the primary mechanism in this discussion.

7.2 Energy Detection in Ideal Channels

Energy detection is a noncoherent form of signal processing which involves estimation of energy during a certain period of observation (Urkowitz 1967; Digham et al. 2007; Alam et al. 2012; Hossain et al. 2012; Yu et al. 2012;

© Springer International Publishing AG 2017

P.M. Shankar, *Fading and Shadowing in Wireless Systems*,
DOI 10.1007/978-3-319-53198-4_7

Umar et al. 2014). Energy is estimated by taking samples of the received signals. The collection of samples is followed by a comparison of the estimated energy to a set threshold to conclude whether the signal from the primary user is present or absent. The simplicity of the energy detection schemes arises from the fact that it can be implemented using low cost devices.

The fundamental process of the energy detection scheme can be best described using the concept of hypothesis testing introduced in Chap. 2. The two hypotheses are H_0 and H_1, the former implying the absence of primary users and the latter implying the presence of the primary or the licensed user. These can be stated as (Digham et al. 2007; Atapattu et al. 2010; Sithamparanathan and Giorgetti 2012)

$$\begin{array}{ll} x(n) = w[n], \quad n = 1, 2, \cdots, N_s & H_0 \\ x(n) = hs[n] + w[n], \quad n = 1, 2, \cdots, N_s & H_1 \end{array}. \tag{7.1}$$

In Eq. (7.1), samples of the received signals are represented by $x[n]$. The samples of the signal from the primary user are represented by $s(n)$ while $w[n]$ are the noise samples. The channel gain is represented by h. Noise samples $w[n]$ come from zero mean additive white Gaussian noise with power spectral density of N_0. The number of independent samples is given by N_s.

Using the appropriate representation of noise as a bandpass process with inphase and quadrature components, Eq. (7.1) can be transformed into units of energy or signal-to-noise ratio as

$$Z = \sum_{n=1}^{N_s} |x[n]|^2 \tag{7.2}$$

If T is observation time and W is the bandwidth, the time bandwidth product u is given by (Urkowitz 1967)

$$u = TW. \tag{7.3}$$

The number of samples $N_s = u$. Since the noise is zero mean Gaussian, the energy detected will be

$$Z = \begin{cases} \chi_{2u}^2 & H_0 \\ \chi_{2u}^2(2\gamma) & H_1 \end{cases} \tag{7.4}$$

In Eq. (7.4), γ is the instantaneous signal-to-noise ratio of the primary user defined as

$$\gamma = h^2 \frac{E_s}{N_0}. \tag{7.5}$$

While Z is a chi-square random variable, χ_{2u}^2, of order u under the hypothesis H_0, Z is a non-central chi-square variable, $\chi_{2u}^2(2\gamma)$, with mean of $2u$ and 2γ under the hypothesis H_1. The probability density functions of these random variables are (Papoulis and Pillai 2002; Digham et al. 2003)

$$f_Z(z) = \begin{cases} \dfrac{1}{2^u \Gamma(u)} z^{u-1} e^{-\frac{z}{2}} & H_0 \\ \dfrac{1}{2}\left(\dfrac{z}{2\gamma}\right)^{\frac{u-1}{2}} e^{-\frac{2\gamma+z}{2}} I_{u-1}\left(\sqrt{2\gamma z}\right) & H_1 \end{cases}. \tag{7.6}$$

In Eq. (7.6), $I_{u-1}(.)$ is the modified Bessel function of order $(u - 1)$ of the first kind (Gradshteyn and Ryzhik 2007). From Chap. 2, one can now formulate a hypothesis testing setup to determine whether the primary user is present or absent.

First, let us examine the nature of these two densities. In an ideal channel, the average energy or SNR γ is constant while in a fading or shadowed fading channel γ will be a random variable. For an ideal channel (only noise is present and fading is absent), the two densities in Eq. (7.6) are shown in Fig. 7.1 for the case of $u = 3$, 5 and $\gamma = 5$ dB.

```
% densities cognitiveRadio
% pdfplots  August 2016
clear ;clc; close all
u=3;
gdB=5;
g=10^(gdB/10);
x=0:.5:80;
ncx21 = ncx2pdf(x,2*u,2*g); % non-central chi square pdf
chi21 = chi2pdf(x,2*u); % Central chi square pdf
subplot(2,1,1),plot(x,chi21,'r--',x,ncx21,'k-','linewidth',2)
xlabel('snr '),ylabel('pdf')
legend('f(z|H_0)','f(z|H_1)')
title(['u = ',num2str(u),', \gamma = ',num2str(gdB),'dB'])
xlim([0,40])
hold on
xx=[10,10];
yy=[0,0.5*max(max(chi21,ncx21))];
plot(xx,yy,'linewidth',1.5)
text(9.9,1.4*yy(2),'\lambda','fontweight','bold','color','b')
u=5;
ncx22 = ncx2pdf(x,2*u,2*g);
chi22 = chi2pdf(x,2*u);
subplot(2,1,2),plot(x,chi22,'r--',x,ncx22,'k-','linewidth',2)
xlabel('snr '),ylabel('pdf')
legend('f(z|H_0)','f(z|H_1)')
title(['u = ',num2str(u),', \gamma = ',num2str(gdB),'dB'])
hold on
xx=[16,16];
yy=[0,0.5*max(max(chi21,ncx21))];
plot(xx,yy,'linewidth',1.5)
text(15.9,1.2*yy(2),'\lambda','fontweight','bold','color','b')
xlim([0,60])
```

If an arbitrary threshold λ is chosen, the probability of false alarm (P_f) and the probability of detection (P_d) are given by (van Trees 1968)

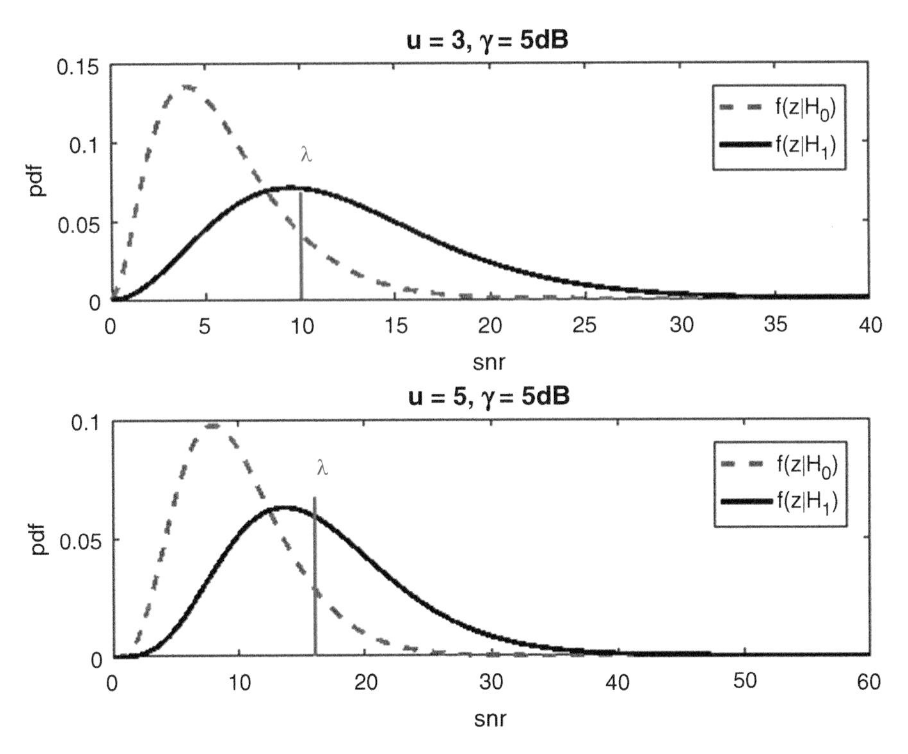

Fig. 7.1 The probability density function of the energy (SNR) for two values of u. Threshold is at λ

$$\begin{aligned} P_f &= \int_\lambda^\infty f(z|H_0)dz \\ P_d &= \int_\lambda^\infty f(z|H_1)dz\cdot \end{aligned} \tag{7.7}$$

Substituting the appropriate density functions, the probabilities can be expressed in compact form as

$$P_f = P_f(u, \lambda) = \frac{\Gamma\left(u, \frac{\lambda}{2}\right)}{\Gamma(u)} \tag{7.8}$$

$$P_d = P_d(u, \gamma, \lambda) = Q_u\left(\sqrt{2\gamma}, \sqrt{\lambda}\right) \tag{7.9}$$

In Eq. (7.8), $\Gamma(.,.)$ is the upper incomplete gamma function given by (Gradshteyn and Ryzhik 2007)

$$\Gamma(a, b) = \int_b^\infty x^{a-1}\exp(-x)dx. \tag{7.10}$$

In Eq. (7.9), $Q_u(.,.)$ is the generalized Marcum Q function expressed in integral form as (Simon and Alouini 2005)

$$Q_u(\alpha,\beta) = \frac{1}{\alpha^{u-1}} \int_{\beta}^{\infty} x^u \exp\left(-\frac{x^2+\alpha^2}{2}\right) I_{u-1}(\alpha x) dx \tag{7.11}$$

For integer values of u, the probability of detection can be obtained from Matlab using the command *marcumq*(.). Once the probabilities of false alarm and detection are available, it is possible to create the ROC plots. The Matlab script used for the creation of the ROC curves and estimation of A_z is given below along with the results. While A_z is used in medical and clinical research to represent the area under the ROC curve, often the term AUC is used in cognitive radio. There is a need to pay close attention to the use of the polyarea(.) command to ensure that the area is completely accounted for in the estimation. This aspect is explained in the Matlab script.

```
function ROC_analysis_idealChannel
% ROC curves in an indeal channel. The area under the ROC curve is obtained
% using polyarea(.) command. Since time bandwidth product is an integer,
% the Marcum Q function is obtained using the command marcumq(.)
% two plots are obtained. First one shows the ROC curves for a fixed value
% of the SNR for different value of u while the second one is a plot of the
% AUC versus u for a few values of the SNR
% P M Shankar, September 2016
close all
PF=valuesofpf; % get the values of PF
LF=length(PF); % count the PF
uu=1:20;
KU=length(uu); % values of u and their count
PD=zeros(LF,KU);
lam=mylamb(uu); % get values of lambda for this set of u
% no fading
Z=10^(5/10); % 5 dB
for k=1:KU
    u=uu(k);
    PD(:,k)=marcumq(sqrt(2*Z),sqrt(lam(:,k)),u);
    if k==1
    plot(PF,PD(:,k),'-r*')
    elseif k==KU;
        plot(PF,PD(:,k),'--ko')
    else
       plot(PF,PD(:,k))
    end;
    hold on
end;
% create an arrow
xx=[0.4 0.2];yy=[0.5 0.8];
annotation('textarrow',xx,yy,'String',...
    ['u = ',num2str(max(uu)),' to  u = ',num2str(min(uu))],...
    'color','r','fontweight','bold')
title(['Ideal Channel : \gamma = ',num2str(10*log10(Z)), ' dB'])
xlabel('P_f')
ylabel('P_d')
text(0.2,0.9,'u = 1')
text(0.46,0.7,[' u = ',num2str(max(uu))])
hold off
```

```
zdB=[0,5,8,10];
LZ=length(zdB);
Azz=zeros(LZ,KU);
for k1=1:LZ
z=10^(zdB(k1)/10);
for k=1:KU
    u=uu(k);
    PD(:,k)=marcumq(sqrt(2*z),sqrt(lam(:,k)),u);
    PF1=[0,PF,1];
    PD1=[0;PD(:,k);1];
    % padding is needed so that the ends are not missed. otherwise,
    % it will give a lower area
    Azz(k1,k)=polyarea(PF1',PD1)+0.5;%0.5 accounts for the area below the
    % diagonal. Polyarea only gives area between the curve & diagonal
end;
end;
% get the plot of the AUC
figure,plot(uu,Azz(1,:),'r-*',uu,Azz(2,:),'k-o',uu,Azz(3,:),'b-s',...
    uu,Azz(4,:),'m-d'),title('Ideal Channel')
xlabel('time bandwidth product u'),ylabel('Area under the ROC curve A_z')
legend(['\gamma = ',num2str(zdB(1)),' dB'],...
    ['\gamma = ',num2str(zdB(2)),' dB'],...
    ['\gamma = ',num2str(zdB(3)),' dB'],...
    ['\gamma = ',num2str(zdB(4)),' dB'])
ylim([0.5,1.1])
end

function PF1=valuesofpf
PF1=[1e-12 5e-12 1e-11 5e-11 1e-10 1e-9,...
    1e-8 1e-7 1e-6 1e-5 1e-4...
    .001 .01 0.03 0.05 0.08 .1 0.14 0.16 .2 0.25 .3 0.35 .4 ...
    .5 0.55 .6 0.65 .7 .75 .8 .85 .9 0.91 .92 0.93 .94 0.95 ...
    .96 0.965 0.968 0.97 0.975 0.978 0.98 0.985 .99 0.995 ...
    0.99999 0.999999 0.9999999];
end

function lambda=mylamb(U) % get values of lambda
PF=valuesofpf; %values of PF
LF=length(PF);  KU=length(U);
lambda=zeros(LF,KU);
for ku=1:KU
    lambda(:,ku)=2*gammaincinv(PF,U(ku),'upper');%invert PF & get threshold
end;

end
```

Several plots of the receiver operating characteristics are shown in Fig. 7.2. As the number of samples goes up, the ROC curves move closer and closer to the diagonal indicating that the performance worsens. As described in Chap. 2, the best measure of the performance of the energy detection scheme is the area under the receiver operating characteristics (ROC) curve (AUC), generally identified as A_z (Metz 1978; Hanley and McNeil 1982). From the ROC curves, the AUC can be evaluated directly in Matlab using the command *polyarea*(.). The AUC values are plotted in Fig. 7.3.

One can see that as the value of u increases, the area under the ROC curves (AUC) comes down. This seems to be an apparent contradiction since one expects to have performance levels increase (higher values of AUC) as the dimensionality

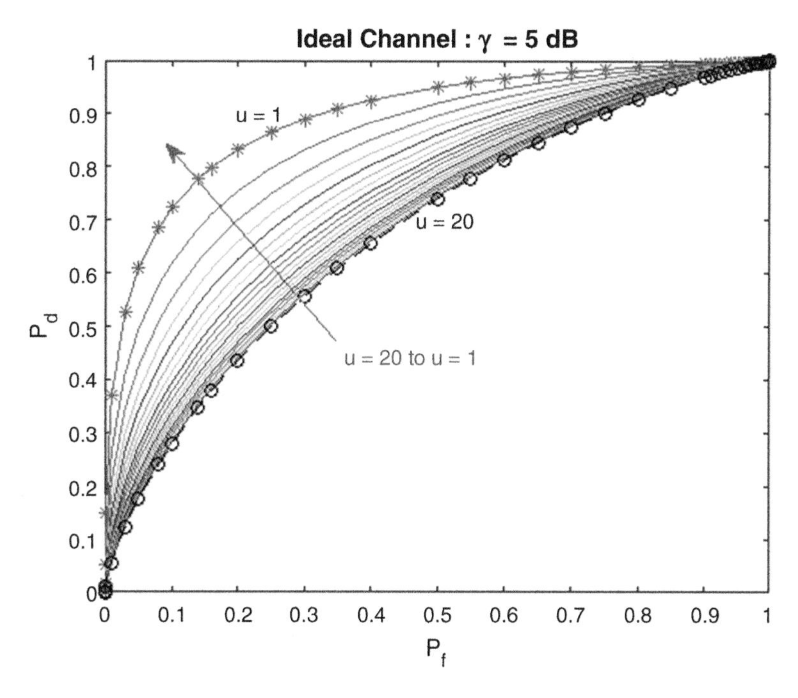

Fig. 7.2 ROC curves for SNR = 5 dB for several values of u from $u = 1$ to $u = 20$

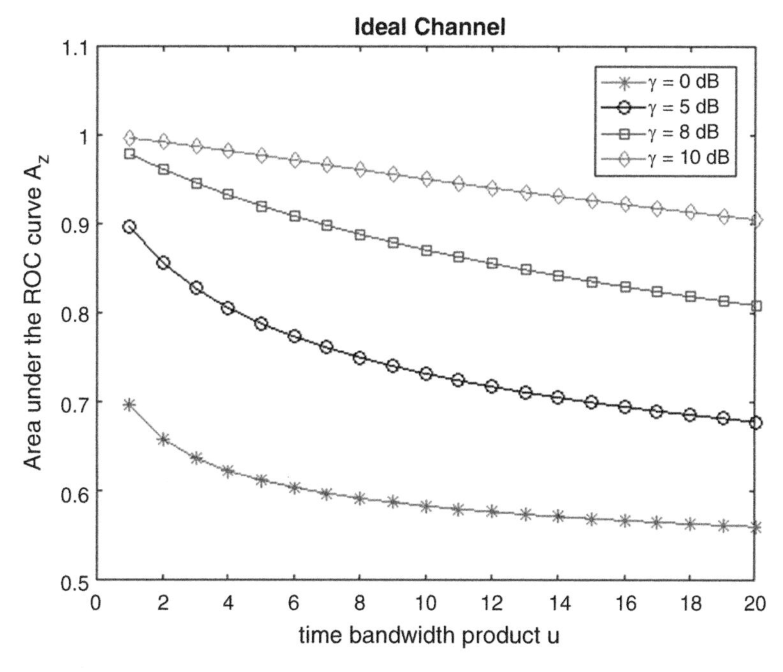

Fig. 7.3 AUC SNR as a function of the time bandwidth product

defined by the time bandwidth product goes up (Urkowitz 1967; Atapattu et al. 2010; Banjade et al. 2012). Indeed, the best performance occurs when the time bandwidth product is unity and it goes down as u increases to 20. As expected, for a fixed value of the time bandwidth product, the performance improves with SNR. The behavior of the AUC can be explored further through the concept of a performance index.

7.2.1 Performance Index and AUC

The unexpected trend in the decline in performance with increasing values of u can be explained using the concept of performance index. It can be defined in terms of the moments of the central chi-square and non-central chi-square variables having densities in Eq. (7.6). The mean and variance are (Evans et al. 2000; Papoulis and Pillai 2002)

$$\left.\begin{aligned}\mu &= 2u,\\ \sigma^2 &= 2(2u)\end{aligned}\right\}H_0$$
$$\left.\begin{aligned}\mu &= 2u+2\gamma,\\ \sigma^2 &= 2(2u+4\gamma)\end{aligned}\right\}H_1 \tag{7.12}$$

Using the concepts used in image analysis where the performance of a detector is quantified in terms of a detection index or contrast, the performance index $\boldsymbol{\eta}$ in cognitive radio is defined as (Patterson and Foster 1983; Evans and Nixon 1995)

$$\eta = \frac{|(\mu|H_1) - (\mu|H_0)|}{\sqrt{\frac{1}{2}(\sigma^2|H_1) + (\sigma^2|H_0)}} \tag{7.13}$$

Using the moments in Eq. (7.12), the performance index $\boldsymbol{\eta}$ becomes

$$\eta = \eta(u,\gamma) = \frac{\gamma}{\sqrt{u+\gamma}} \tag{7.14}$$

Equation (7.14) shows that the performance index goes down when u increases. In other words, it is expected that the performance of an energy detector characterized in terms of AUC is expected to go down when the time bandwidth product u goes up. Thus, the performance index provides a quantitative explanation of the decline in AUC with increasing values of u.

The plots of the performance index and AUC are explored next. The Matlab script used is given next along with the results.

```
function ROC_analysis_idealChannel_performance
% ROC curves in an indeal channel. The area under the ROC curve is obtained
% using polyarea(.) command. Since time bandwidth product is an integer,
% the Marcum Q function is obtained using the command marcumq(.)
% two plots are obtained. First one shows the ROC curves for a fixed value
% of the SNR for different value of u while the second one is a plot of the
% AUC versus u for a few values of the SNR
% P M Shankar, September 2016
close all
PF=valuesofpf; % get the values of PF
LF=length(PF); % count the PF
uu=1:20;
KU=length(uu); % values of u and their count
PD=zeros(LF,KU);
lam=mylamb(uu); % get values of lambda for this set of u

zdB=[3,5,7];
LZ=length(zdB);
Azz=zeros(LZ,KU);
perf=zeros(LZ,KU);
for k1=1:LZ
z=10^(zdB(k1)/10);
for k=1:KU
    u=uu(k);
    PD(:,k)=marcumq(sqrt(2*z),sqrt(lam(:,k)),u);
    PF1=[0,PF,1];
    PD1=[0;PD(:,k);1];
    % padding is needed so that the ends are not missed. otherwise,
    % it will give a lower area
    Azz(k1,k)=polyarea(PF1',PD1)+0.5;%0.5 accounts for the area below the
    % diagonal. Polyarea only gives area between the curve & diagonal
    perf(k1,k)=z/sqrt(u+z);
end;
end;
% get the plot of the AUC
figure,plot(uu,Azz(1,:),'r-+',uu,Azz(2,:),'k-o',uu,Azz(3,:),'b-s',...
    uu,perf(1,:),'m--d',uu,perf(2,:),'m--^',uu,perf(3,:),'m--p' )
title('Ideal Channel')
xlabel('time bandwidth product u'),ylabel(' A_z  or    \eta')
legend([' A_z, \gamma = ',num2str(zdB(1)),' dB'],...
    [' A_z, \gamma = ',num2str(zdB(2)),' dB'],...
    [' A_z, \gamma = ',num2str(zdB(3)),' dB'],...
    [' \eta, \gamma = ',num2str(zdB(1)),' dB'],...
    [' \eta, \gamma = ',num2str(zdB(2)),' dB'],...
    [' \eta, \gamma = ',num2str(zdB(3)),' dB'])
end

function PF1=valuesofpf
PF1=[1e-12 5e-12 1e-11 5e-11 1e-10 1e-9,...
    1e-8 1e-7 1e-6 1e-5 1e-4...
    .001 .01 0.03 0.05 0.08 .1 0.14 0.16 .2 0.25 .3 0.35 .4 ...
    .5 0.55 .6 0.65 .7 .75 .8 .85 .9 0.91 .92 0.93 .94 0.95 ...
    .96 0.965 0.968 0.97 0.975 0.978 0.98 0.985 .99 0.995 ...
    0.99999 0.999999 0.9999999];
end

function lambda=mylamb(U) % get values of lambda
PF=valuesofpf; %values of PF
LF=length(PF);  KU=length(U);
lambda=zeros(LF,KU);
for ku=1:KU
    lambda(:,ku)=2*gammaincinv(PF,U(ku),'upper');%invert PF & get threshold
end;

end
```

Figure 7.4 shows the AUC and the performance index as a function of u. As the time bandwidth product u increases, the AUC goes down. The downward trajectory is consistent with the decrease in performance index η with increasing values of the time bandwidth product demonstrating the usefulness of the performance index as a reliable indicator of the decline in AUC (Urkowitz 1967; Atapattu et al. 2010; Banjade et al. 2012).

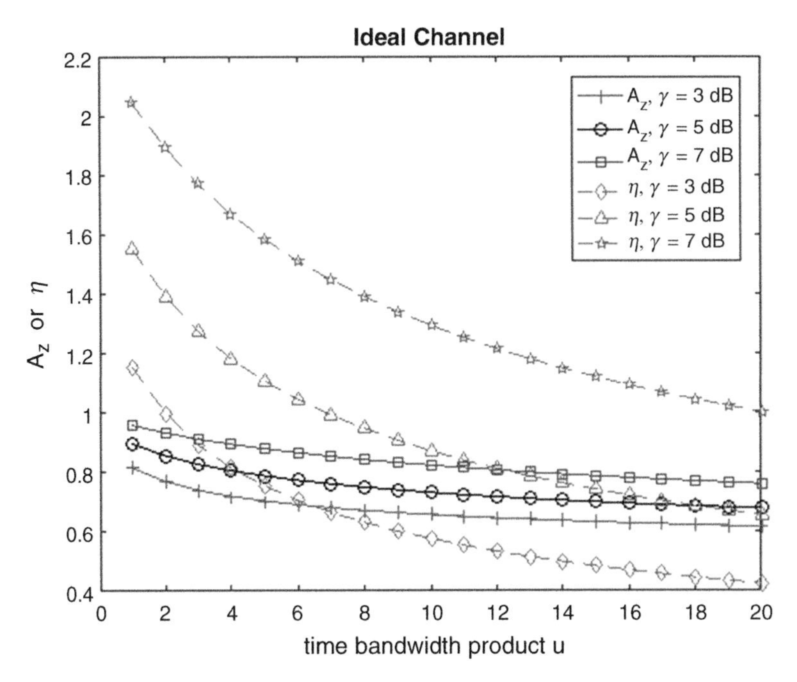

Fig. 7.4 AUC and performance index η

7.2.2 *Direct Evaluation of AUC*

While AUC can be obtained from the probability of false alarm versus probability of detection plots, often it is necessary to examine the behavior of AUC without using these ROC plots so that the effect of fading on the performance in cognitive radio can be interpreted and understood directly. It is possible to evaluate AUC without generating the ROC curves by extending the concept of the area to an integral. The area under the ROC curve (AUC) can be mathematically expressed as (Metz 1978; Papoulis and Pillai 2002; Atapattu et al. 2010)

$$A_z = \int_0^1 P_d(\lambda) d\big[P_f(\lambda)\big]. \tag{7.15}$$

As the threshold λ goes from $0 \to \infty$, the probabilities of false alarm and detection go from $1 \to 0$. Taking note of this, the expression for A_z in Eq. (7.15) becomes

$$A_z = A_z(u,\gamma) = -\int_0^{\infty} P_d(\gamma, u, \lambda) \frac{\partial P_f}{\partial \lambda}(u,\lambda) d\lambda \tag{7.16}$$

Using the definition of the gamma density, Eq. (7.16) becomes

$$A_z = 1 - \int_0^{\infty} f_Z(\lambda|H_0) F_Z(\lambda|H_1) d\lambda = \int_0^{\infty} \frac{1}{2^u \Gamma(u)} \lambda^{u-1} e^{-\frac{\lambda}{2}} Q_u\left(\sqrt{2\gamma}, \sqrt{\lambda}\right) d\lambda \quad (7.17)$$

One can see the benefits of estimating the AUC using the integral since there is no need to determine the threshold from the probability of false alarm. This aspect is explored next. Note that an analytical solution to Eq. (7.17) does not exist and one needs to use numerical integration. The Matlab script appears next. The results of the numerical integration are compared to those obtained using the *polyarea*(.) command directly.

```
function ROC_analysis_compareAUC
% ROC curves in an indeal channel. The area under the ROC curve is obtained
% using polyarea(.) command. Since time bandwidth product is an integer,
% the Marcum Q function is obtained using the command marcumq(.)
% ARea is obtained directly using the integral
% instead of finding the probability of detection. The results are compared
% to those obtained using the polyarea.
% P M Shankar, September 2016
close all
global u z
PF=valuesofpf; % get the values of PF
LF=length(PF); % count the PF
uu=1:20;
KU=length(uu); % values of u and their count
PD=zeros(LF,KU);
lam=mylamb(uu); % get values of lambda for this set of u

zdB=[3,5,7];
LZ=length(zdB);
Azz=zeros(LZ,KU);
Azint=zeros(LZ,KU);
Azsum=zeros(LZ,KU);
for k1=1:LZ
z=10^(zdB(k1)/10);
for k=1:KU
    u=uu(k);
    PD(:,k)=marcumq(sqrt(2*z),sqrt(lam(:,k)),u);
    PF1=[0,PF,1];
    PD1=[0;PD(:,k);1];
    % padding is needed so that the ends are not missed. otherwise,
    % it will give a lower area
    Azz(k1,k)=polyarea(PF1',PD1)+0.5;%0.5 accounts for the area below the
    % diagonal. Polyarea only gives area between the curve & diagonal
    Azint(k1,k)=integral(@aucintf,0,inf);
end;
end;
% get the plot of the AUC
figure,plot(uu,Azz(1,:),'r-', uu,Azint(1,:),'rd',...
uu,Azz(2,:),'k--',uu,Azint(2,:),'k--^', uu,Azz(3,:),'b-.',...
   uu,Azint(3,:),'bp' )
title('Ideal Channel')
xlabel('time bandwidth product u'),ylabel(' A_z ')
legend([' A_z, \gamma = ',num2str(zdB(1)),' dB (polyarea)'],...
    [' A_z, \gamma = ',num2str(zdB(1)),' dB (integral)'],...
    [' A_z, \gamma = ',num2str(zdB(2)),' dB (polyarea)'],...
    [' \eta, \gamma = ',num2str(zdB(2)),' dB (integral)'],...
    [' \eta, \gamma = ',num2str(zdB(3)),' dB (polyarea)'],...
    [' \eta, \gamma = ',num2str(zdB(3)),' dB (integral)'])


end

function PF1=valuesofpf
PF1=[1e-12 5e-12 1e-11 5e-11 1e-10 1e-9,...
    1e-8 1e-7 1e-6 1e-5 1e-4...
    .001 .01 0.03 0.05 0.08 .1 0.14 0.16 .2 0.25 .3 0.35 .4 ...
    .5 0.55 .6 0.65 .7 .75 .8 .85 .9 0.91 .92 0.93 .94 0.95 ...
    .96 0.965 0.968 0.97 0.975 0.978 0.98 0.985 .99 0.995 ...
    0.99999 0.999999 0.9999999];
end
```

```
function lambda=mylamb(U) % get values of lambda
PF=valuesofpf; %values of PF
LF=length(PF);  KU=length(U);
lambda=zeros(LF,KU);
for ku=1:KU
    lambda(:,ku)=2*gammaincinv(PF,U(ku),'upper');%invert PF & get threshold
end;

end

function F=aucintf(x)
% function to evaluate AUC directly instead of using PF, PD and polyarea
global u z
f1=x.^(u-1);
f2=exp(-x/2);
f3=marcumq(sqrt(2*z),sqrt(x),u);
F=f1.*f2.*f3/(2^u*gamma(u));

end
```

Figure 7.5 shows that the results of the numerical integration and polyarea(.) match. It is also possible to eliminate the need for numerical integration by expressing the modified Bessel function in Eq. (7.11) in series form allowing the Marcum Q function to be expressed as (Simon and Alouini 2005; Atapattu et al. 2010; Olabiyi and Annamalai 2012; Alam et al. 2012)

$$Q_u\left(\sqrt{2\gamma},\sqrt{\lambda}\right)=\sum_{k=0}^{\infty}\gamma^k e^{-\gamma}\frac{\Gamma\left(u+k,\frac{\lambda}{2}\right)}{\Gamma(u+k)\Gamma(k+1)}. \tag{7.18}$$

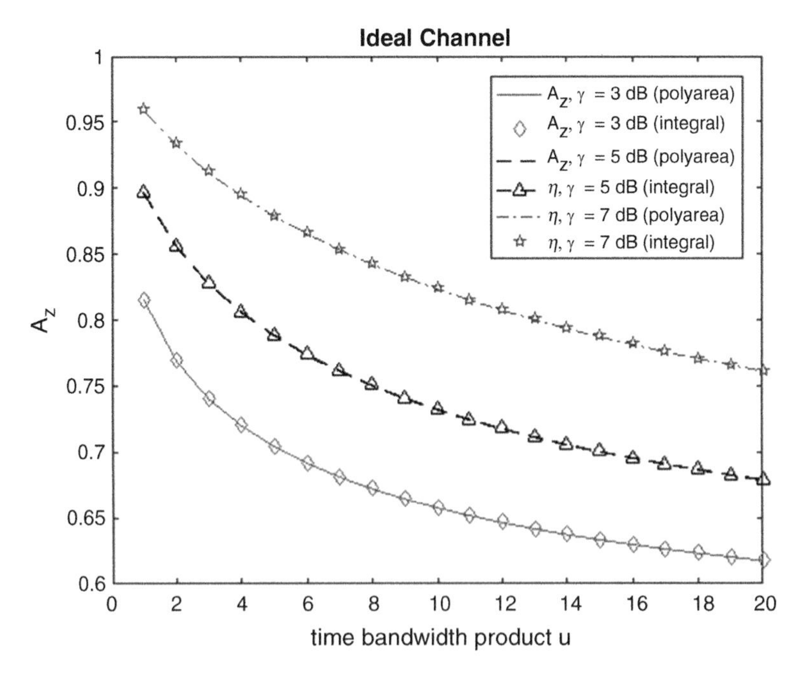

Fig. 7.5 The results of numerical integration compared to those obtained using polyarea(.) for three values of SNR

Using the series expansion, the expression for the AUC becomes

$$
\begin{aligned}
A_z &= \sum_{k=0}^{\infty} \gamma^k e^{-\gamma} \int_0^{\infty} \frac{\Gamma\left(u+k, \frac{\lambda}{2}\right) \lambda^{u-1} e^{-\frac{\lambda}{2}}}{2^u \Gamma(u) \Gamma(u+k) \Gamma(k+1)} d\lambda \\
&= \sum_{k=0}^{\infty} \gamma^k e^{-\gamma} \int_0^{\infty} \frac{\Gamma(u+k, y) y^{u-1} e^{-y}}{\Gamma(u) \Gamma(u+k) \Gamma(k+1)} dy
\end{aligned}
\tag{7.19}
$$

Using the table of integrals [Ref. Eq. (#6-45)] from Gradshteyn and Ryzhik (2007), Eq. (7.19) becomes

$$
A_z = \sum_{k=0}^{\infty} \gamma^k e^{-\gamma} \frac{\Gamma(2u+k)}{u\Gamma(u)\Gamma(u+k)\Gamma(k+1)2^{2u+k}} {}_2F_1\left([1, 2u+k]; [1+u]; \frac{1}{2}\right) \tag{7.20}
$$

In Eq. (7.20), ${}_2F_1(.)$ is the hypergeometric function (Gradshteyn and Ryzhik 2007). Simplifying further, the expression for AUC becomes

$$
A_z = \sum_{k=0}^{\infty} \gamma^k e^{-\gamma} \frac{\Gamma(2u+k)}{\Gamma(u+1)\Gamma(u+k)\Gamma(k+1)2^{2u+k}} {}_2F_1\left([1, 2u+k]; [1+u]; \frac{1}{2}\right). \tag{7.21}
$$

One of the unknowns here is the actual number of terms necessary to get satisfactory results. Equation (7.21) can be expressed as a finite sum of K as

$$
A_z = \sum_{k=0}^{K} \gamma^k e^{-\gamma} \frac{\Gamma(2u+k)}{\Gamma(u+1)\Gamma(u+k)\Gamma(k+1)2^{2u+k}} {}_2F_1\left([1, 2u+k]; [1+u]; \frac{1}{2}\right). \tag{7.22}
$$

It is possible to obtain an estimate of K by varying it and comparing the results to AUC obtained using other methods such as those reported earlier using *polyarea*(.) command. The Matlab script appears next and results are displayed in Figs. 7.6 and 7.7. Two values of $K(=10, 20)$ are used.

```
function ROC_analysis_compareAUC_sum
% ROC curves in an indeal channel. The area under the ROC curve is obtained
% using polyarea(.) command. Since time bandwidth product is an integer,
% the Marcum Q function is obtained using the command marcumq(.)
% two plots are obtained. ARea is obtained using the summation.
% The results are comparedto those obtained using the polyarea.
% P M Shankar, September 2016
close all
global u z
PF=valuesofpf; % get the values of PF
LF=length(PF); % count the PF
```

```
uu=1:20;
KU=length(uu); % values of u and their count
PD=zeros(LF,KU);
lam=mylamb(uu); % get values of lambda for this set of u
ZdB=[3,5,7];
LZ=length(ZdB);
Azz=zeros(LZ,KU);
Azsum10=zeros(LZ,KU);
Azsum20=zeros(LZ,KU);
for k1=1:LZ
Z=10^(ZdB(k1)/10);
for k=1:KU
    u=uu(k);
    PD(:,k)=marcumq(sqrt(2*Z),sqrt(lam(:,k)),u);
    PF1=[0,PF,1];
    PD1=[0;PD(:,k);1];
    % padding is needed so that the ends are not missed. otherwise,
    % it will give a lower area
    Azz(k1,k)=polyarea(PF1',PD1)+0.5;%0.5 accounts for the area below the
    % diagonal. Polyarea only gives area between the curve & diagonal
     K=10;% number of terms in the summation
    Azsum10(k1,k)=aucsum(K);
     K=20;% number of terms in the summation
    Azsum20(k1,k)=aucsum(K);
end;
end;

figure
plot(uu,Azsum10(1,:),'-rd',uu,Azsum20(1,:),'-rs',...
uu,Azsum10(2,:),'k--^',uu,Azsum20(2,:),'k--+',...
uu,Azsum10(3,:),'b-.*',uu,Azsum20(3,:),'b--o')
title('Ideal Channel')
xlabel('time bandwidth product u'),ylabel(' A_z ')
legend([' A_z, \gamma = ',num2str(ZdB(1)),' dB (K = 10)'],...
    [' A_z, \gamma = ',num2str(ZdB(1)),' dB (K = 20)'],...
    [' A_z, \gamma = ',num2str(ZdB(2)),' dB (K = 10)'],...
    [' A_z, \gamma = ',num2str(ZdB(2)),' dB (K = 20)'],...
    [' A_z, \gamma = ',num2str(ZdB(3)),' dB (K = 10)'],...
    [' A_z, \gamma = ',num2str(ZdB(3)),' dB (K = 20)'])

figure,plot(uu,Azz(1,:),'r-', uu,Azsum20(1,:),'rd',...
uu,Azz(2,:),'k--',uu,Azsum20(2,:),'k--^', uu,Azz(3,:),'b-.',...
   uu,Azsum20(3,:),'bp' )
title('Ideal Channel')
xlabel('time bandwidth product u'),ylabel(' A_z ')
legend([' A_z, \gamma = ',num2str(ZdB(1)),' dB (polyarea)'],...
    [' A_z, \gamma = ',num2str(ZdB(1)),' dB (K = 20)'],...
    [' A_z, \gamma = ',num2str(ZdB(2)),' dB (polyarea)'],...
    [' \eta, \gamma = ',num2str(ZdB(2)),' dB (K = 20)'],...
    [' \eta, \gamma = ',num2str(ZdB(3)),' dB (polyarea)'],...
    [' \eta, \gamma = ',num2str(ZdB(3)),' dB (K = 20)'])
end

function PF1=valuesofpf
PF1=[1e-12 5e-12 1e-11 5e-11 1e-10 1e-9,...
    1e-8 1e-7 1e-6 1e-5 1e-4...
    .001 .01 0.03 0.05 0.08 .1 0.14 0.16 .2 0.25 .3 0.35 .4 ...
    .5 0.55 .6 0.65 .7 .75 .8 .85 .9 0.91 .92 0.93 .94 0.95 ...
    .96 0.965 0.968 0.97 0.975 0.978 0.98 0.985 .99 0.995 ...
    0.99999 0.999999 0.9999999];
end

function lambda=mylamb(U) % get values of lambda
PF=valuesofpf; %values of PF
LF=length(PF);  KU=length(U);
lambda=zeros(LF,KU);
for ku=1:KU
    lambda(:,ku)=2*gammaincinv(PF,U(ku),'upper');%invert PF & get threshold
end;

end

function F=aucsum(K)
% function to evaluate AUC as a sum in place of integral or polyarea
% symbolic calculations provide greater stability with hypergeom (.)
global u Z
syms k u1
u1=sym(u);
f1=sym(Z)^k*gamma(2*u1+k)*hypergeom([1,k+2*u1],1+u1,1/2);
f2=factorial(k)*gamma(u1+k)*2^(2*u1+k);
F1=exp(-sym(Z))*symsum(f1/f2,k,0,sym(K));
FF=F1/gamma(u1+1);
F=double(FF); % convert to double precision from symbolic
end
```

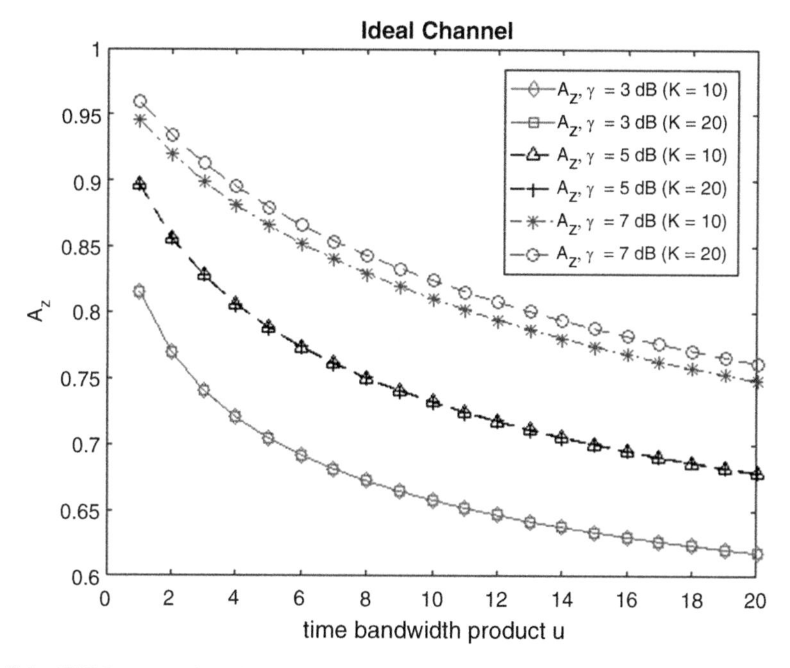

Fig. 7.6 AUC for two value of $K = 10, 20$ shown

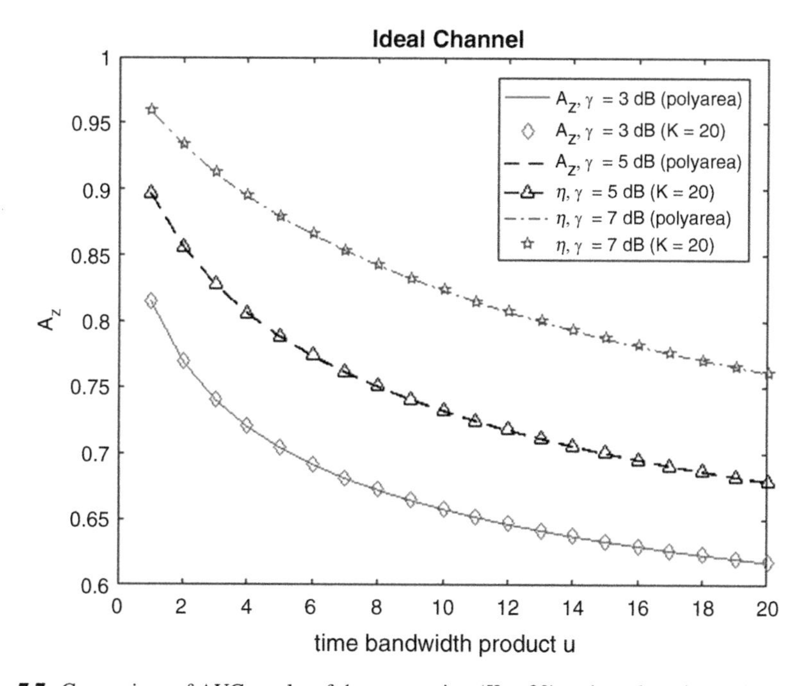

Fig. 7.7 Comparison of AUC results of the summation ($K = 20$) and results using *polyarea*(.)

Figure 7.6 shows that at high values of the average SNR, the AUC values do not match indicating the likelihood that $K = 10$ is insufficient. To test whether $K = 20$ is sufficient, the results of the summation are compared to those obtained using the *polyarea*(.) command. Figure 7.7 shows that for $K = 20$, the results of the summation and those of *polyarea*(.) match. Another reason for the use of the *polyarea* command is that the time taken by summation is far greater than the computational time needed to perform the *polyarea*(.) command.

7.3 Energy Detection in Fading and Shadowed Fading Channels

The analysis so far has been limited to an ideal channel. As described in Chap. 4, wireless channels undergo fading, shadowing, and often both concurrently. Signal strength fluctuations arising from fading necessitate modification of the model of the energy detector in cognitive radio. The performance is likely to worsen when the channel conditions are not ideal (similar to the degradation of error rates and outage probabilities seen in Chap. 4). The effects of fading and shadowed fading channels on the energy detector are explored next (Ghasemi and Sousa 2007; Herath et al. 2009; Herath et al. 2011; Kakkar et al. 2014).

7.3.1 Nakagami Fading Channel

In a Nakagami fading channel, the SNR associated with the primary user becomes a random variable, with the SNR being described as a gamma random variable. The density function of the SNR is

$$f(\gamma) = \frac{1}{\Gamma(m)} \left(\frac{m}{\gamma_0}\right)^m \gamma^{m-1} e^{-\frac{m}{\gamma_0}\gamma} U(\gamma). \tag{7.23}$$

In Eq. (7.23), m is the Nakagami parameter and γ_0 is the average SNR. While the signal strength fluctuations from fading have no impact on the probability density of the energy in the absence of the primary user in Eq. (7.6), the detected energy from the primary user becomes a random variable. This means that the probability of false alarm remains unchanged in a fading channel. But, the probability of detection and consequently, the area under the ROC curve become random variables and this aspect needs to be addressed. Since the performance index also depends on the SNR, the fading effects the performance index defined earlier. The performance index in a Nakagami fading channel becomes

$$\eta = \eta(u, \gamma_0) = \int_0^\infty \frac{\gamma}{\sqrt{u+\gamma}} f(\gamma) d\gamma = \int_0^\infty \frac{\gamma}{\sqrt{u+\gamma}} \frac{1}{\Gamma(m)} \left(\frac{m}{\gamma_0}\right)^m \gamma^{m-1} e^{-\frac{m}{\gamma_0}\gamma} d\gamma. \tag{7.24}$$

The performance indices are calculated and the Matlab script appears below.

```
function performance_Nakagamichannel
% performance index in a Nakagami channel
% P M Shankar, September 2016
close all
global m Z u
for kk=1:4
m=0.5*kk;
uu=1:20;
KU=length(uu); % values of u and their count
ZdB=[3,5,7];
LZ=length(ZdB);
perf=zeros(LZ,KU);
perfN=zeros(LZ,KU);
for k1=1:LZ
Z=10^(ZdB(k1)/10);
for k=1:KU
    u=uu(k);
    perf(k1,k)=Z/sqrt(u+Z);
    perfN(k1,k)=integral(@perfun,0,inf);
end;
end;
% get the plot of the AUC
figure,plot(uu,perf(1,:),'r-',uu,perfN(1,:),'r-*',...
uu,perf(2,:),'k--',uu,perfN(2,:),'k--o',...
    uu,perf(3,:),'b-.',uu,perfN(3,:),'b-.s' )
title(['Nakagami Channel, m = ',num2str(m)])
xlabel('time bandwidth product u'),ylabel('Performance index \eta')
legend(['\gamma_0 = ',num2str(ZdB(1)),' dB'],...
    ['\gamma_0 = ',num2str(ZdB(1)),' dB, m = ',num2str(m)],...
    ['\gamma_0 = ',num2str(ZdB(2)),' dB'],...
    ['\gamma_0 = ',num2str(ZdB(2)),' dB, m = ',num2str(m)],...
    ['\gamma_0 = ',num2str(ZdB(3)),' dB'],...
    ['\gamma_0 = ',num2str(ZdB(3)),' dB, m = ',num2str(m)])
end;
end

function F=perfun(x) % external function for integration
global m Z u
f1=(m/Z)^m*(x.^(m-1)).*exp(-m*x/Z)/gamma(m);
f2=x./sqrt(x+u);
F=f1.*f2;
end
```

Figures 7.8, 7.9, 7.10, and 7.11 show the results for the different values of the Nakagami parameter.

The performance index declines in the presence of fading and it starts to regain as the fading level decreases (higher values of the Nakagami parameter). The trend with respect to the time bandwidth product is identical to the ideal case with performance index declining with increase in the time bandwidth product.

To determine the degradation in AUC, the probability of detection in the presence of fading needs to be calculated. The probability of detection in a Nakagami fading channel becomes (Digham et al. 2007; Altrad and Muhaidat 2013)

$$\begin{aligned} P_d = P_d(u, m, \gamma_0, \lambda) &= \int_0^\infty Q_u\left(\sqrt{2\gamma}, \sqrt{\lambda}\right) f(\gamma) d\gamma \\ &= \int_0^\infty Q_u\left(\sqrt{2\gamma}, \sqrt{\lambda}\right) \frac{1}{\Gamma(m)} \left(\frac{m}{\gamma_0}\right)^m \gamma^{m-1} e^{-\frac{m}{\gamma_0}\gamma} d\gamma \end{aligned} \tag{7.25}$$

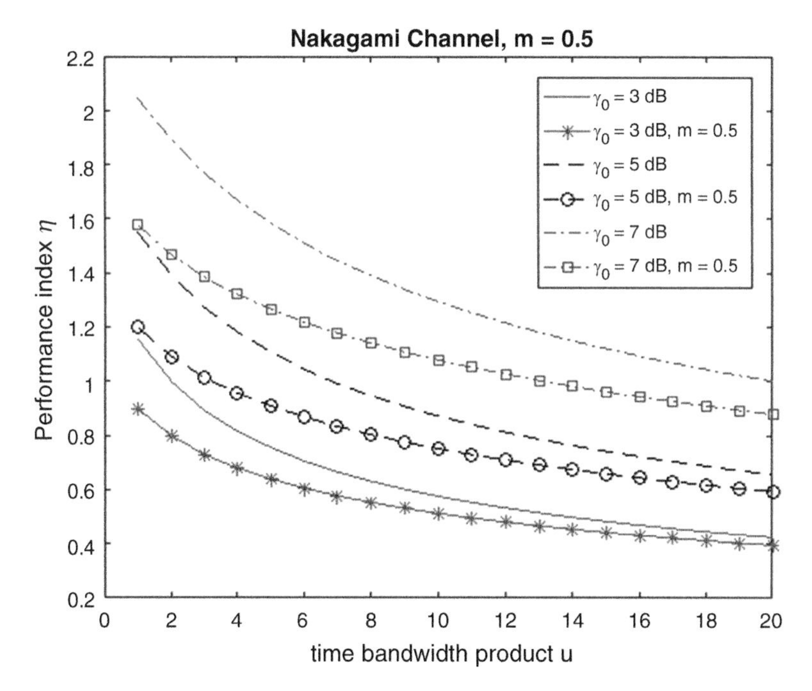

Fig. 7.8 Performance index for $m = 0.5$ for a few values of the average SNR. The performance index in the absence of fading is also shown

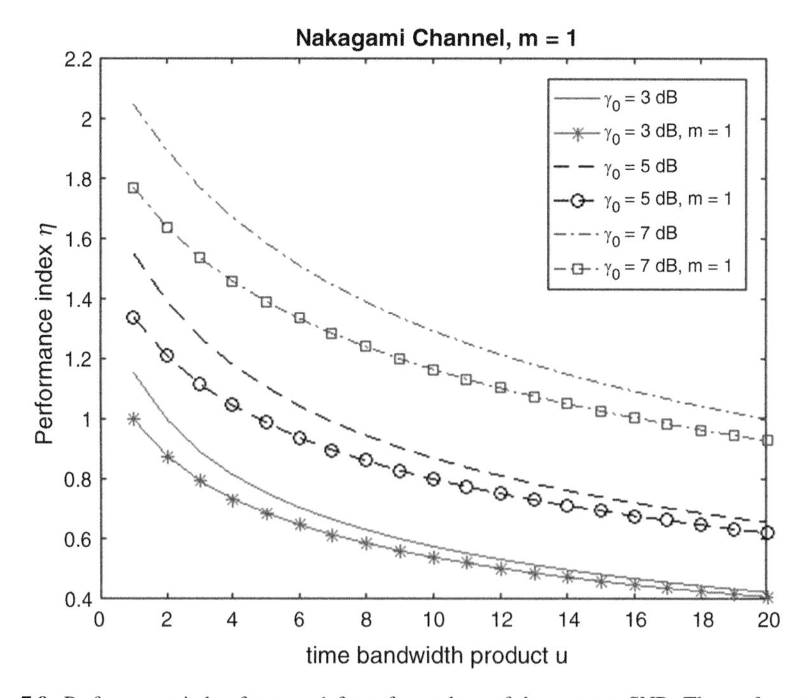

Fig. 7.9 Performance index for $m = 1$ for a few values of the average SNR. The performance index in the absence of fading is also shown

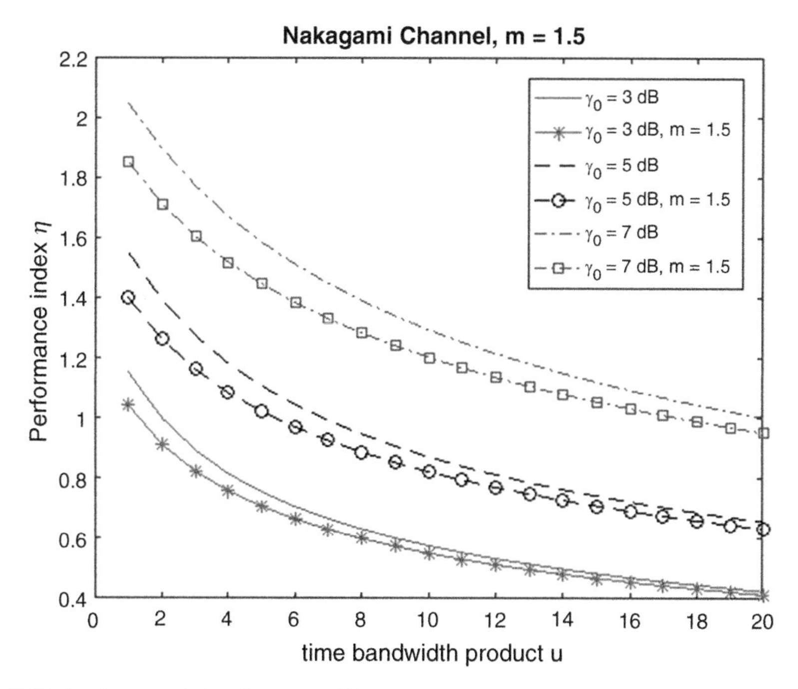

Fig. 7.10 Performance index for $m = 1.5$ for a few values of the average SNR. The performance index in the absence of fading is also shown

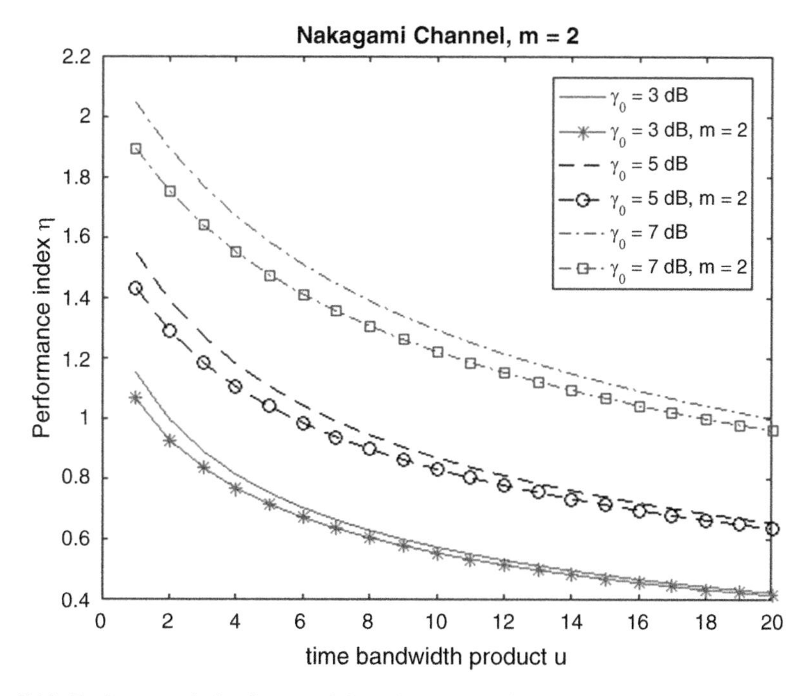

Fig. 7.11 Performance index for $m = 2$ for a few values of the average SNR. The performance index in the absence of fading is also shown

The area under the ROC curve becomes

$$A_z = \int_0^{\infty}\int_0^{\infty} \frac{1}{2^u\Gamma(u)}\lambda^{u-1}e^{-\frac{\lambda}{2}}Q_u\left(\sqrt{2\gamma},\sqrt{\lambda}\right)\frac{1}{\Gamma(m)}\left(\frac{m}{\gamma_0}\right)^m \gamma^{m-1}e^{-\frac{m}{\gamma_0}\gamma}d\lambda d\gamma. \quad (7.26)$$

As it was done earlier, the AUC can be obtained using the *polyarea*(.) command by plotting the ROC curve, with the probability of detection in Eq. (7.25) avoiding the double integral in Eq. (7.26). First, Eq. (7.25) needs to be simplified. To accomplish this step, the Marcum Q function is written as a series using Eq. (7.18) as

$$P_d(u,m,\gamma_0,\lambda) = \int_0^{\infty}\sum_{k=0}^{\infty}\gamma^k e^{-\gamma}\frac{\Gamma\left(u+k,\frac{\lambda}{2}\right)}{\Gamma(u+k)\Gamma(k+1)}\frac{1}{\Gamma(m)}\left(\frac{m}{\gamma_0}\right)^m\gamma^{m-1}e^{-\frac{m}{\gamma_0}\gamma}d\gamma \quad (7.27)$$

Simplifying further,

$$P_d(u,m,\gamma_0,\lambda) = \sum_{k=0}^{\infty}\frac{\Gamma\left(u+k,\frac{\lambda}{2}\right)}{\Gamma(u+k)\Gamma(k+1)}\int_0^{\infty}\frac{e^{-\gamma}\gamma^k}{\Gamma(m)}\left(\frac{m}{\gamma_0}\right)^m\gamma^{m-1}e^{-\frac{m}{\gamma_0}\gamma}d\gamma \quad (7.28)$$

After completing the integration, the probability of detection becomes

$$P_d(u,m,\gamma_0,\lambda) = \sum_{k=0}^{\infty}\frac{\Gamma\left(u+k,\frac{\lambda}{2}\right)}{\Gamma(u+k)\Gamma(k+1)}\frac{\Gamma(m+k)m^m\gamma_0^k}{\Gamma(m)(m+\gamma_0)^{k+m}}. \quad (7.29)$$

The first set of results is generated by performing the integration in Eq. (7.25) and estimating the AUC using the *polyarea*(.) command in Matlab. The Matlab script appears below.

```
function ROC_Nakagami_book
% ROC in a Nakagami channel. The probability of detection is evaluated
% using numerical integration and AUC estimated using polyarea(.)
% P M Shankar, Sept. 2016
close all
global m lamb Z u
PF=valuesofpf; % get the values of PF
LF=length(PF); % count the PF
uu=1:20;  KU=length(uu); % values of u and their count
PD=zeros(LF,KU);
lam=mylamb(uu); % get values of lambda for this set of u
% no fading
Z=10^(5/10);
for k=1:KU
    u=uu(k);
    PD(:,k)=marcumq(sqrt(2*Z),sqrt(lam(:,k)),u);
    PF1=[0,PF,1]; % padding is needed so that the ends are not missed.
    % otherwise, it will give a lower area
    Az(k)=polyarea(PF1',[0;PD(:,k);1])+0.5; % area with no fading
end;
```

```
% study of Az as a function of the Nakagami parameter
Mk=[0.5,1,1.5,2,3];
ZdB=5;
Z=10^(ZdB/10);
for km=1:length(Mk);
    m=Mk(km);
    for k=1:KU
        u=uu(k);
        for kk=1:LF
            lamb=lam(kk,k);
            pdF(kk,k)=integral(@myfun,0,inf);
        end;
        Azm(km,k)=polyarea(PF1',[0;pdF(:,k);1])+0.5;
    end
end;
figure,plot(uu,Az,'-k',uu,Azm(1,:),'-*',uu,Azm(2,:),...
    uu,Azm(3,:),'--o',uu,Azm(4,:),'-d',uu,Azm(5,:),'--^')
legend('No fading',['m = ',num2str(Mk(1))],['m = ',num2str(Mk(2))],...
    ['m = ',num2str(Mk(3))],['m = ',num2str(Mk(4))],...
    ['m = ',num2str(Mk(5))])
xlim([min(uu),max(uu)])
ylim([0.6,.9])
title(['Nakagami fading channel: Average SNR = ',num2str(ZdB),'dB'])
xlabel('Time bandwidth product u')
ylabel('Area under the ROC curve A_z')
% study a few values of SNR as a function of u
mm=1.5;
m=mm;
Z1dB=[0,5,10,15];
ZZ=10.^(Z1dB/10);
for km=1:length(ZZ);
    Z=ZZ(km);
    for k=1:KU
        u=uu(k);
        for kk=1:LF
            lamb=lam(kk,k); % this is the original set
            pdF(kk,k)=integral(@myfun,0,inf);
        end;
        Azz(km,k)=polyarea(PF1',[0;pdF(:,k);1])+0.5;
    end
end;
figure,plot(uu,Azz(1,:),'-*',uu,Azz(2,:),'-s',...
    uu,Azz(3,:),'--o',uu,Azz(4,:),'-d')
legend(['SNR = ',num2str(Z1dB(1)),'dB'],...
    ['SNR = ',num2str(Z1dB(2)),'dB'],...
    ['SNR = ',num2str(Z1dB(3)),'dB'],['SNR = ',num2str(Z1dB(4)),'dB'])
xlim([min(uu),max(uu)])
ylim([0.5,1])
title(['Nakagami fading channel: m = ',num2str(mm)])
xlabel('Time bandwidth product u')
ylabel('Area under the ROC curve A_z')
% a few values of u as a function of SNR
mm=1.5;
m=mm;
Z1dB=[0:20];
ZZ=10.^(Z1dB/10);
UU=[1,3,5,7];
lamm=mylamb(UU); % new set of lambda values for this U and PF
for km=1:length(ZZ);
    Z=ZZ(km);
    for k=1:4
        u=UU(k);
        for kk=1:LF
            lamb=lamm(kk,k);
            pdF(kk,k)=integral(@myfun,0,inf);
        end;
```

```
        Azu(km,k)=polyarea(PF1',[0;pdF(:,k);1])+0.5;
    end
end;
figure,plot(Z1dB,Azu(:,1),'-*',Z1dB,Azu(:,2),'-s',...
    Z1dB,Azu(:,3),'--o',Z1dB,Azu(:,4),'-d')
legend(['u = ',num2str(UU(1))],['u = ',num2str(UU(2))],...
    ['u = ',num2str(UU(3))],['u = ',num2str(UU(4))],'location','best')
ylim([0.6,1])
title(['Nakagami fading channel: m = ',num2str(mm)])
xlabel('Average SNR (dB)')
ylabel('Area under the ROC curve A_z')
% area for a few value of m
u=4;
lamU=mylamb(u); % value of lambda for the set pf PF
mm=[1,2,3,4];
for km=1:length (mm)
   m=mm(km);
  for k=1:length(ZZ)
      Z=ZZ(k);
      for kk=1:LF
            lamb=lamU(kk);
            pdF(kk,k)=integral(@myfun,0,inf);
        end;
        Az1(km,k)=polyarea(PF1',[0;pdF(:,k);1])+0.5;
  end;
end;
figure,plot(Z1dB,Az1(1,:),'-*',Z1dB,Az1(2,:),'-s',...
    Z1dB,Az1(3,:),'--o',Z1dB,Az1(4,:),'-d')
legend(['m = ',num2str(mm(1))],['m = ',num2str(mm(2))],...
    ['m = ',num2str(mm(3))],['m = ',num2str(mm(4))],'location','best')
ylim([0.5,1])
title({'Nakagami fading channel';[' u = ',num2str(u)]})
xlabel('Average SNR (dB)')
ylabel('Area under the ROC curve A_z')

end

function PF1=valuesofpf
PF1=[1e-12 5e-12 1e-11 5e-11 1e-10 1e-9,...
    1e-8 1e-7 1e-6 1e-5 1e-4...
    .001 .01 0.03 0.05 0.08 .1 0.14 0.16 .2 0.25 .3 0.35 .4 ...
    .5 0.55 .6 0.65 .7 .75 .8 .85 .9 0.91 .92 0.93 .94 0.95 ...
    .96 0.965 0.968 0.97 0.975 0.978 0.98 0.985 .99 0.995 ...
    0.99999 0.999999 0.9999999];
end


function Fg=myfun(x)
% Nakagami channel
global m lamb Z u
Fg=gampdf(x,m,Z/m).*marcumq(sqrt(2*x),sqrt(lamb),u);
end


function lambda=mylamb(U) % get values of lambda
PF=valuesofpf; %values of PF
LF=length(PF);  KU=length(U);
lambda=zeros(LF,KU);
for ku=1:KU
    lambda(:,ku)=2*gammaincinv(PF,U(ku),'upper');%invert PF & get threshold
end;

end
```

Area under the ROC curve for a few values of the Nakagami parameter is shown in Fig. 7.12 as a function of the time bandwidth product u. As m increases, the AUC starts approaching the AUC of an ideal channel. Figure 7.13 shows the results for a fixed value of m for a set of average SNR values. Figure 7.14 shows the AUC as a function of the average SNR for a few values of the time bandwidth product. As expected, the performance worsens with increasing values of u. Figure 7.15 shows the results for a few value of the Nakagami parameter for a fixed value of the time bandwidth product.

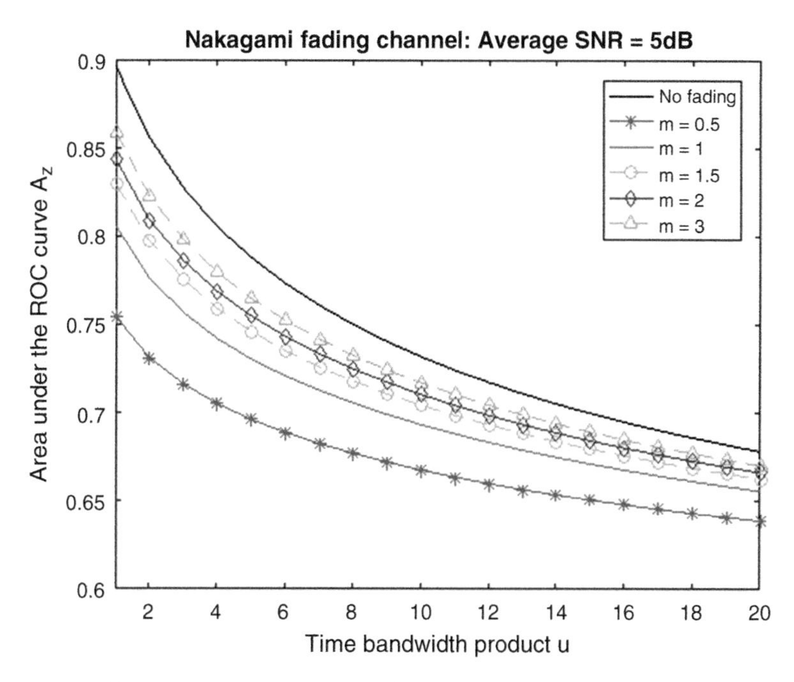

Fig. 7.12 AUC in a Nakagami fading channel obtained by numerically integrating the expression for the probability of detection followed by polyarea(.)

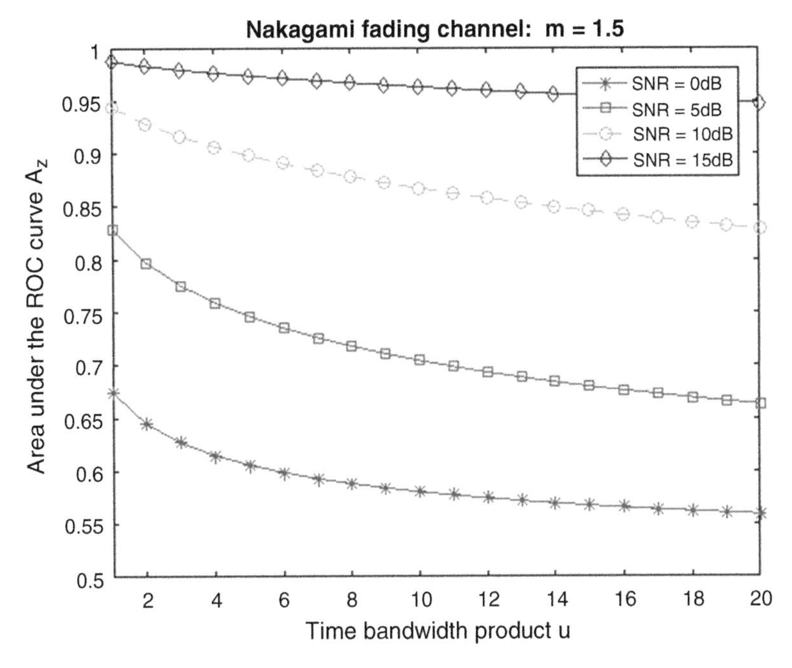

Fig. 7.13 AUC in a Nakagami fading channel. The performance improves with increasing values of the average SNR

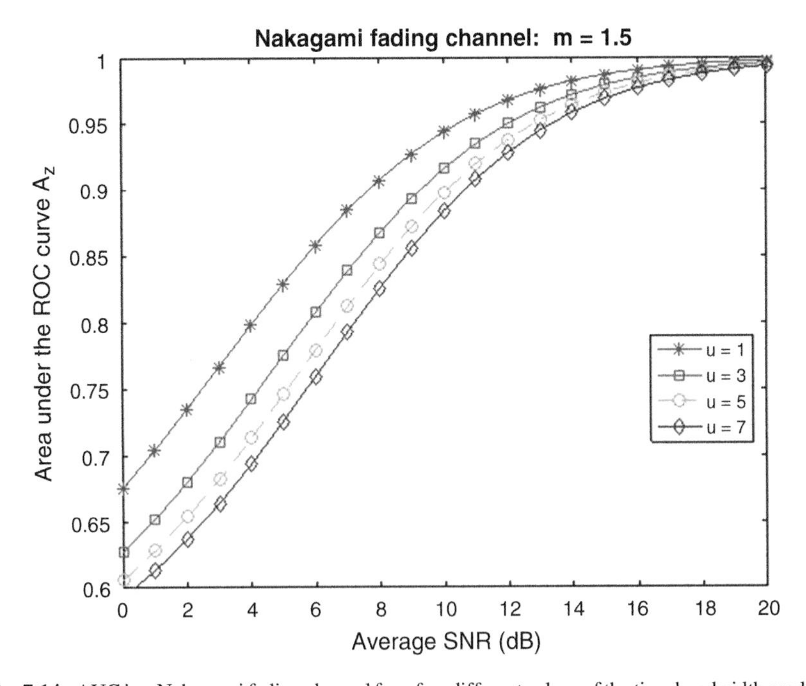

Fig. 7.14 AUC in a Nakagami fading channel for a few different values of the time bandwidth product

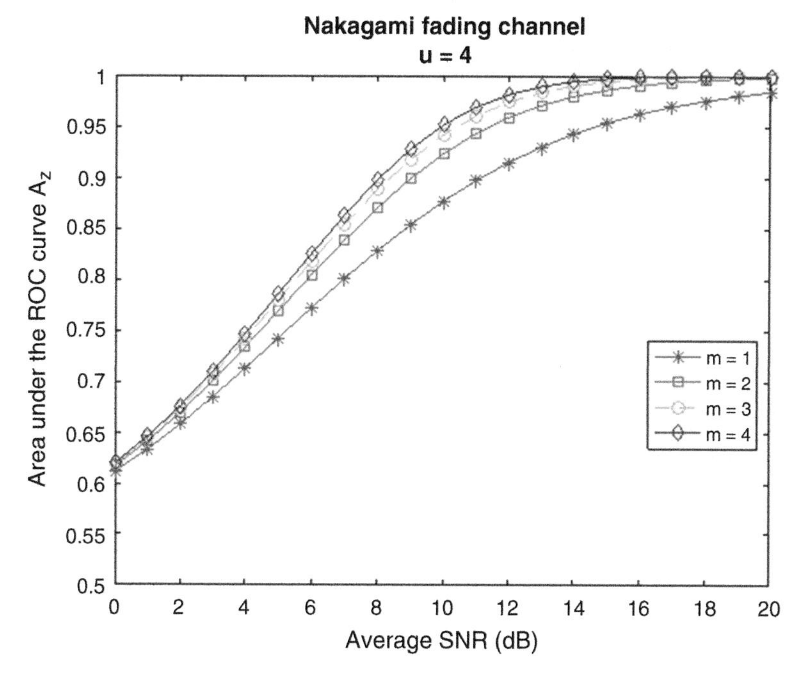

Fig. 7.15 AUC as a function of the Nakagami parameter for $u = 4$

The results of the numerical integration are compared to those obtained using the summation in Eq. (7.29). Computations were done with the number of terms limited to 35. The Matlab script appears below.

```
function ROC_Nakagami_book_pdsum
% ROC in a Nakagami channel. The probability of detection is evaluated
% as a sum and AUC estimated using polyarea(.)
% P M Shankar, Sept. 2016
close all
global m lamb Z u
K=35; % number of terms in the summation
PF=valuesofpf; % get the values of PF
LF=length(PF); % count the PF
uu=1:20;  KU=length(uu); % values of u and their count
PD=zeros(LF,KU);
lam=mylamb(uu); % get values of lambda for this set of u
% no fading
Z=10^(5/10);
for k=1:KU
    u=uu(k);
    PD(:,k)=marcumq(sqrt(2*Z),sqrt(lam(:,k)),u);
    PF1=[0,PF,1]; % padding is needed so that the ends are not missed.
    % otherwise, it will give a lower area
    Az(k)=polyarea(PF1',[0;PD(:,k);1])+0.5; % area with no fading
end;
```

```
% study of Az as a function of the Nakagami parameter
Mk=[0.5,1.5,3];
ZdB=5;
Z=10^(ZdB/10);
for km=1:length(Mk);
    m=Mk(km);
    for k=1:KU
        u=uu(k);
        for kk=1:LF
            lamb=lam(kk,k);
            pdFS(kk,k)=pdfsum(K);
                pdF(kk,k)=integral(@myfun,0,inf);
        end;
         AzI(km,k)=polyarea(PF1',[0;pdF(:,k);1])+0.5;
        AzS(km,k)=polyarea(PF1',[0;pdFS(:,k);1])+0.5;
    end
end;
figure,plot(uu,Az,'-k',uu,AzI(1,:),'r-*', uu,AzS(1,:),'r-s',...
    uu,AzI(2,:),'b--o',uu,AzS(2,:),'b-->',...
    uu,AzI(3,:),'m-.^', uu,AzS(3,:),'m-.p')
legend('No fading',['m = ',num2str(Mk(1)),'(integral)'],...
  ['m = ',num2str(Mk(2)),'(integral)'],['m = ',num2str(Mk(2)),'(sum)'],...
      ['m = ',num2str(Mk(3)),'(integral)'],['m = ',num2str(Mk(3)),'(sum)'])
xlim([min(uu),max(uu)])
ylim([0.6,.9])
title(['Nakagami fading channel: Average SNR = ',num2str(ZdB),'dB'])
xlabel('Time bandwidth product u')
ylabel('Area under the ROC curve A_z')
% study a few values of SNR as a function of u
mm=1.5;
m=mm;
Z1dB=[0,3,7];
ZZ=10.^(Z1dB/10);
for km=1:length(ZZ);
    Z=ZZ(km);
    for k=1:KU
        u=uu(k);
        for kk=1:LF
            lamb=lam(kk,k); % this is the original set
            pdFS(kk,k)=pdfsum(K);
                pdF(kk,k)=integral(@myfun,0,inf);
        end;
            AzI(km,k)=polyarea(PF1',[0;pdF(:,k);1])+0.5;
        AzS(km,k)=polyarea(PF1',[0;pdFS(:,k);1])+0.5;
    end
end;
figure,plot(uu,AzI(1,:),'r-*', uu,AzS(1,:),'r-s',...
    uu,AzI(2,:),'b--o',uu,AzS(2,:),'b-->',...
    uu,AzI(3,:),'k-.^', uu,AzS(3,:),'k-.p')
legend(['SNR = ',num2str(Z1dB(1)),'dB (integral)'],...
    ['SNR = ',num2str(Z1dB(1)),'dB (sum)'],...
    ['SNR = ',num2str(Z1dB(2)),'dB (integral)'],...
    ['SNR = ',num2str(Z1dB(2)),'dB (sum)'],...
    ['SNR = ',num2str(Z1dB(3)),'dB (integral)'],...
    ['SNR = ',num2str(Z1dB(3)),'dB (sum)'])
xlim([min(uu),max(uu)]),ylim([0.5,1])
title(['Nakagami fading channel: m = ',num2str(mm)])
xlabel('Time bandwidth product u'),ylabel('Area under the ROC curve A_z')
end
```

```
function PF1=valuesofpf
PF1=[1e-12 5e-12 1e-11 5e-11 1e-10 1e-9,...
    1e-8 1e-7 1e-6 1e-5 1e-4...
    .001 .01 0.03 0.05 0.08 .1 0.14 0.16 .2 0.25 .3 0.35 .4 ...
    .5 0.55 .6 0.65 .7 .75 .8 .85 .9 0.91 .92 0.93 .94 0.95 ...
    .96 0.965 0.968 0.97 0.975 0.978 0.98 0.985 .99 0.995 ...
    0.99999 0.999999 0.9999999];
end
function pd=pdfsum(K)
% Nakagami channel prob detection as a sum
global m lamb Z u
syms ss k
u1=sym(u);
m1=sym(m);
Z1=sym(Z);
    f1=gamma(m1+k)*Z1^k;
    f2=gamma(u1+k)*gamma(k+1)*(m1+Z1)^(k+m1);
    f3=igamma(u1+k,sym(lamb)/2);
    ss=(m1^m1)*(1/gamma(m1))*symsum(f3*f1/f2,k,0,sym(K));
pd=double(ss);
end

function lambda=mylamb(U) % get values of lambda
PF=valuesofpf; %values of PF
LF=length(PF);  KU=length(U);
lambda=zeros(LF,KU);
for ku=1:KU
    lambda(:,ku)=2*gammaincinv(PF,U(ku),'upper');%invert PF & get threshold
end;
end

function Fg=myfun(x)
% Nakagami channel
global m lamb Z u
Fg=gampdf(x,m,Z/m).*marcumq(sqrt(2*x),sqrt(lamb),u);
end
```

The results of the comparison of the numerical integration and summation are shown in Figs. 7.16 and 7.17. It can be seen that one can use either approach to obtain the AUC.

Equation (7.26) for the estimation of AUC as an integral can also be simplified. Note that the area under the ROC curve obtained for the ideal channel expressed as a sum in Eq. (7.22) can be used to replace Eq. (7.26) as

$$A_z = \int_0^\infty \sum_{k=0}^{K} \gamma^k e^{-\gamma} \left(\frac{m}{\gamma_0}\right)^m \frac{\gamma^{m-1}}{\Gamma(m)} e^{-\frac{m}{\gamma_0}\gamma} \frac{\Gamma(2u+k)}{\Gamma(u+1)\Gamma(u+k)\Gamma(k+1)2^{2u+k}} \times {}_2F_1\left([1, 2u+k]; [1+u]; \frac{1}{2}\right) \times d\gamma \tag{7.30}$$

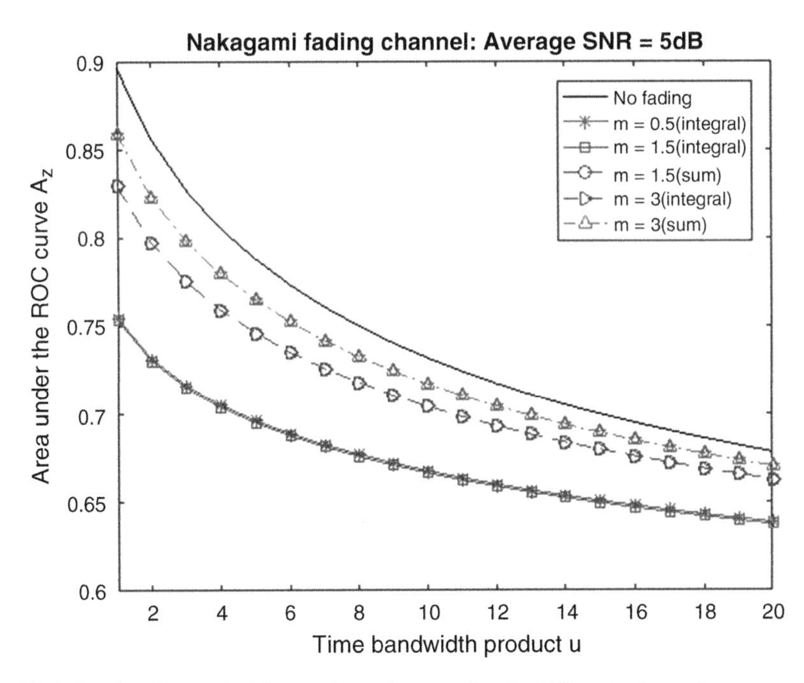

Fig. 7.16 Results of numerical integration and summation for different values of m

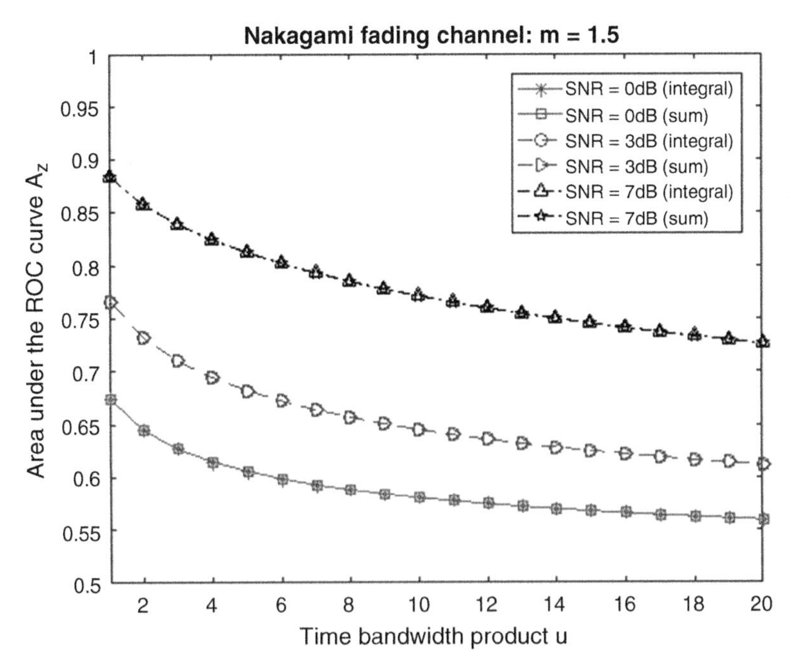

Fig. 7.17 Results of numerical integration and summation for different values of average SNR

Equation (7.30) can be further simplified by performing the integration w. r. t. γ resulting in the expression for A_z as

$$A_z = \frac{m^m}{\Gamma(u+1)\Gamma(m)} \sum_{k=0}^{\infty} \frac{\Gamma(m+k)\Gamma(2u+k)\gamma_0^k}{\Gamma(u+k)\Gamma(k+1)(m+\gamma_0)^{k+m}2^{2u+k}} {}_2F_1\left([1,2u+k];[1+u];\frac{1}{2}\right) \tag{7.31}$$

Results on directly estimating AUC using the summation in Eq. (7.31) are obtained next. The Matlab script appears below. Figures 7.18 and 7.19 compare the results of Eq. (7.31) to those obtained using the integration. Note that the number of terms in the summation was limited to 10.

```
function ROC_Nakagami_AUCsum
% ROC in a Nakagami channel. AUC is directly deteremined as a sum without
% using polyarea and compared to the result from polyarea
% P M Shankar, Sept. 2016
close all
global m lamb Z u
K=10; % number of terms in the summation
PF=valuesofpf; % get the values of PF
LF=length(PF); % count the PF
uu=1:20;  KU=length(uu); % values of u and their count
PD=zeros(LF,KU);
lam=mylamb(uu); % get values of lambda for this set of u
% no fading
Z=10^(5/10);
for k=1:KU
    u=uu(k);
    PD(:,k)=marcumq(sqrt(2*Z),sqrt(lam(:,k)),u);
    PF1=[0,PF,1]; % padding is needed so that the ends are not missed.
    % otherwise, it will give a lower area
    Az(k)=polyarea(PF1',[0;PD(:,k);1])+0.5; % area with no fading
end;
% study of Az as a function of the Nakagami parameter
Mk=[0.5,1.5,3];
ZdB=5;     Z=10^(ZdB/10);
for km=1:length(Mk);
    m=Mk(km);
    for k=1:KU
        u=uu(k);
        for kk=1:LF
            lamb=lam(kk,k);
                pdF(kk,k)=integral(@myfun,0,inf);
        end;
         AzI(km,k)=polyarea(PF1',[0;pdF(:,k);1])+0.5;
        AzS(km,k)=aucsum(K);
    end
end;
figure,plot(uu,Az,'-k',uu,AzI(1,:),'r-*', uu,AzS(1,:),'r-s',...
    uu,AzI(2,:),'b--o',uu,AzS(2,:),'b-->',...
    uu,AzI(3,:),'m-.^', uu,AzS(3,:),'m-.p')
legend('No fading',['m = ',num2str(Mk(1)),'(integral)'],...
  ['m = ',num2str(Mk(2)),'(integral)'],['m = ',num2str(Mk(2)),'(sum)'],...
      ['m = ',num2str(Mk(3)),'(integral)'],['m = ',num2str(Mk(3)),'(sum)'])
xlim([min(uu),max(uu)])
ylim([0.6,.9])
title(['Nakagami fading channel: Average SNR = ',num2str(ZdB),'dB'])
xlabel('Time bandwidth product u')
ylabel('Area under the ROC curve A_z')
```

```
% study a few values of SNR as a function of u
mm=1.5;   m=mm;
Z1dB=[0,3,7];
ZZ=10.^(Z1dB/10);
for km=1:length(ZZ);
    Z=ZZ(km);
    for k=1:KU
        u=uu(k);
        for kk=1:LF
            lamb=lam(kk,k); % this is the original set
                pdF(kk,k)=integral(@myfun,0,inf);
        end;
             AzI(km,k)=polyarea(PF1',[0;pdF(:,k);1])+0.5;
        AzS(km,k)=aucsum(K);
    end
end;
figure,plot(uu,AzI(1,:),'r-*', uu,AzS(1,:),'r-s',...
    uu,AzI(2,:),'b--o',uu,AzS(2,:),'b-->',...
    uu,AzI(3,:),'k-.^', uu,AzS(3,:),'k-.p')
legend(['SNR = ',num2str(Z1dB(1)),'dB (integral)'],...
    ['SNR = ',num2str(Z1dB(1)),'dB (sum)'],...
    ['SNR = ',num2str(Z1dB(2)),'dB (integral)'],...
    ['SNR = ',num2str(Z1dB(2)),'dB (sum)'],...
    ['SNR = ',num2str(Z1dB(3)),'dB (integral)'],...
    ['SNR = ',num2str(Z1dB(3)),'dB (sum)'])
xlim([min(uu),max(uu)]),ylim([0.5,1])
title(['Nakagami fading channel: m = ',num2str(mm)])
xlabel('Time bandwidth product u'),ylabel('Area under the ROC curve A_z')
end

function PF1=valuesofpf
PF1=[1e-12 5e-12 1e-11 5e-11 1e-10 1e-9,...
    1e-8 1e-7 1e-6 1e-5 1e-4...
    .001 .01 0.03 0.05 0.08 .1 0.14 0.16 .2 0.25 .3 0.35 .4 ...
    .5 0.55 .6 0.65 .7 .75 .8 .85 .9 0.91 .92 0.93 .94 0.95 ...
    .96 0.965 0.968 0.97 0.975 0.978 0.98 0.985 .99 0.995 ...
    0.99999 0.999999 0.9999999];
end

function auc=aucsum(K)
% AUC as a sum directly
global m  Z u
syms ss k
u1=sym(u);  m1=sym(m);  Z1=sym(Z);
    f1=gamma(m1+k)*gamma(2*u1+k)*Z1^k;
    f2=gamma(u1+k)*gamma(k+1)*(m1+Z1)^(k+m1);
    f3=hypergeom([1,2*u1+k],[1+u1],1/2);
    ss=(m1^m1)*(1/(gamma(m1)*gamma(u1+1)))*symsum(f1*f3/f2,k,0,sym(K));
auc=double(ss);
end

function lambda=mylamb(U) % get values of lambda
PF=valuesofpf; %values of PF
LF=length(PF);  KU=length(U);
lambda=zeros(LF,KU);
for ku=1:KU
    lambda(:,ku)=2*gammaincinv(PF,U(ku),'upper');%invert PF & get threshold
end;
end

function Fg=myfun(x)
% Nakagami channel; computation of the probability of detection
global m lamb Z u
Fg=gampdf(x,m,Z/m).*marcumq(sqrt(2*x),sqrt(lamb),u);
end
```

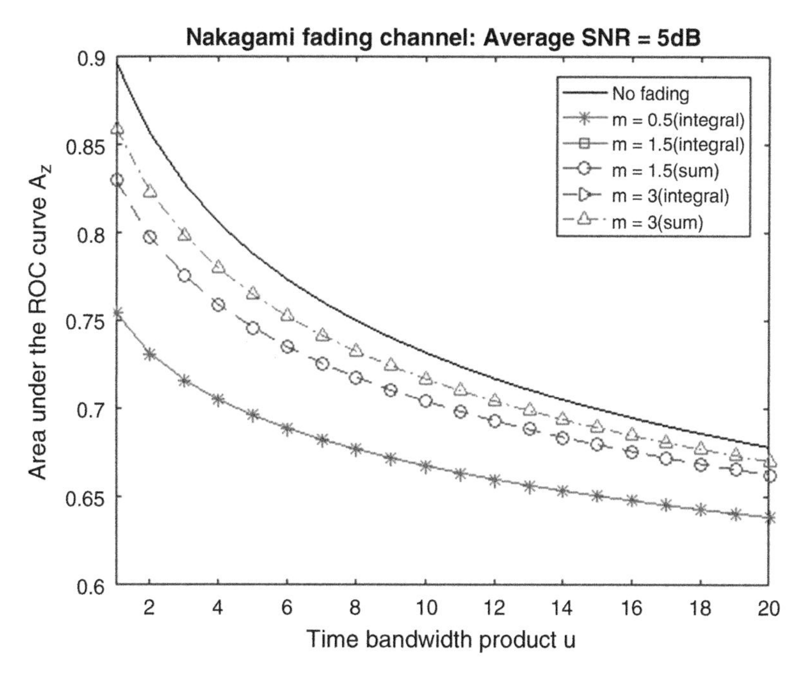

Fig. 7.18 Comparison of AC from Eq. (7.31) and numerical integration (fixed value of the average SNR)

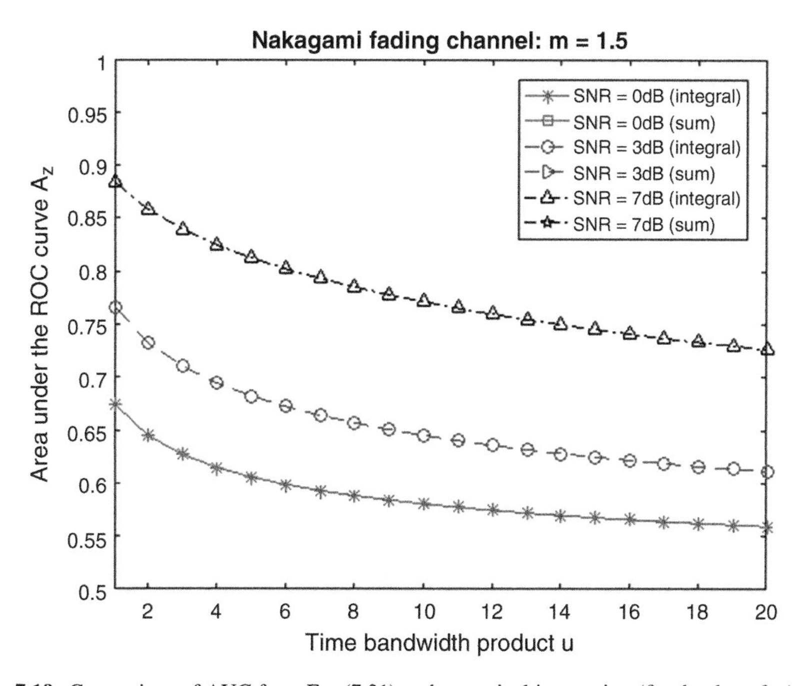

Fig. 7.19 Comparison of AUC from Eq. (7.31) and numerical integration (fixed value of m)

7.3.2 Shadowed Fading Channels

One of the models that can describe the statistical fluctuations in wireless channels when both fading and shadowing are present simultaneously is the generalized K model (Chap. 4). The probability density function of the SNR in a GK model is

$$f(\gamma) = 2\left(\sqrt{\frac{mc}{\gamma_0}\gamma}\right)^{m+c} K_{m-c}\left(2\sqrt{\frac{mc}{\gamma_0}\gamma}\right)U(\gamma) \tag{7.32}$$

In Eq. (7.32), m is the Nakagami parameter and c describes the level of shadowing in the channel while K_{m-c} (.) is the modified Bessel function of the second kind of order $(m-c)$. Lower values of c correspond to higher levels of shadowing and higher values of c indicate low levels of shadowing. As described in Chap. 4, the GK model provides a simple analytical description of the effects of shadowed fading in place of the exact representation offered by the gamma-lognormal model.

The performance index in a shadowed fading channel can be expressed as

$$\begin{aligned} \eta = \eta(u, \gamma_0, m, c) &= \int_0^{\infty} \frac{\gamma}{\sqrt{u+\gamma}} f(\gamma) d\gamma \\ &= \int_0^{\infty} \frac{\gamma}{\sqrt{u+\gamma}} 2\left(\sqrt{\frac{mc}{\gamma_0}\gamma}\right)^{m+c} K_{m-c}\left(2\sqrt{\frac{mc}{\gamma_0}\gamma}\right) d\gamma \end{aligned} \tag{7.33}$$

The performance index declines with increasing levels of the amount of fading given by

$$\mathrm{AF} = \frac{1}{m} + \frac{1}{c} + \frac{1}{mc}. \tag{7.34}$$

The performance index in a GK channel is estimated for a few set of different values of m and c and the results are displayed below along with the Matlab script.

The Matlab script appears below. The performance index calculations in a GK channel are shown in Fig. 7.20, 7.21, 7.22, and 7.23. In each case, results are compared to the case of an ideal channel (no fading/shadowing).

```
function performance_GKchannel
% performance index in a GK channel
% P M Shankar, September 2016
close all
global m Z u c
cc=[2.1, 8.5];
mm=[1, 2.5];
for k1=1:2
    c=cc(k1);
for k2=1:2
    m=mm(k2);
uu=1:20;
KU=length(uu); % values of u and their count
ZdB=[3,5,7];
LZ=length(ZdB);
perf=zeros(LZ,KU);
perfN=zeros(LZ,KU);
for k1=1:LZ
Z=10^(ZdB(k1)/10);
for k=1:KU
    u=uu(k);
    perf(k1,k)=Z/sqrt(u+Z);
    perfN(k1,k)=integral(@perfun,0,inf);
end;
end;
% get the plot of the AUC
figure,plot(uu,perf(1,:),'r-',uu,perfN(1,:),'r-*',...
uu,perf(2,:),'k--',uu,perfN(2,:),'k--o',...
    uu,perf(3,:),'b-.',uu,perfN(3,:),'b-.s' )
title(['GK Channel, m = ',num2str(m), ', c = ',num2str(c)])
xlabel('time bandwidth product u'),ylabel('Performance index \eta')
legend(['\gamma_0 = ',num2str(ZdB(1)),' dB'],...
    ['\gamma_0 = ',num2str(ZdB(1)),' dB, GK channel'],...
    ['\gamma_0 = ',num2str(ZdB(2)),' dB'],...
    ['\gamma_0 = ',num2str(ZdB(2)),' dB, GK channel'],...
    ['\gamma_0 = ',num2str(ZdB(3)),' dB'],...
    ['\gamma_0 = ',num2str(ZdB(3)),' dB, GK channel'])
ylim([0.2,2.2])
grid on
end;
end;
end

function F=perfun(x) % external function for integration
global m Z u c
f1=2*((sqrt(c/Z)*sqrt(m*x)).^(c+m)).*...
    besselk(c-m, 2*sqrt(c/Z)*sqrt(m*x))*1./(x*gamma(m)*gamma(c));
f2=x./sqrt(x+u);
F=f1.*f2;
end
```

The properties of cognitive radio in a GK channel can be studied similar to the one undertaken for the case of a Nakagami fading channel. Results are obtained using the simple approach of obtaining the probability of detection in a GK channel first followed by the estimation of AUC using the *polyarea*(.) command in Matlab. The probability of detection in a GK channel becomes (Alhennawi et al. 2014)

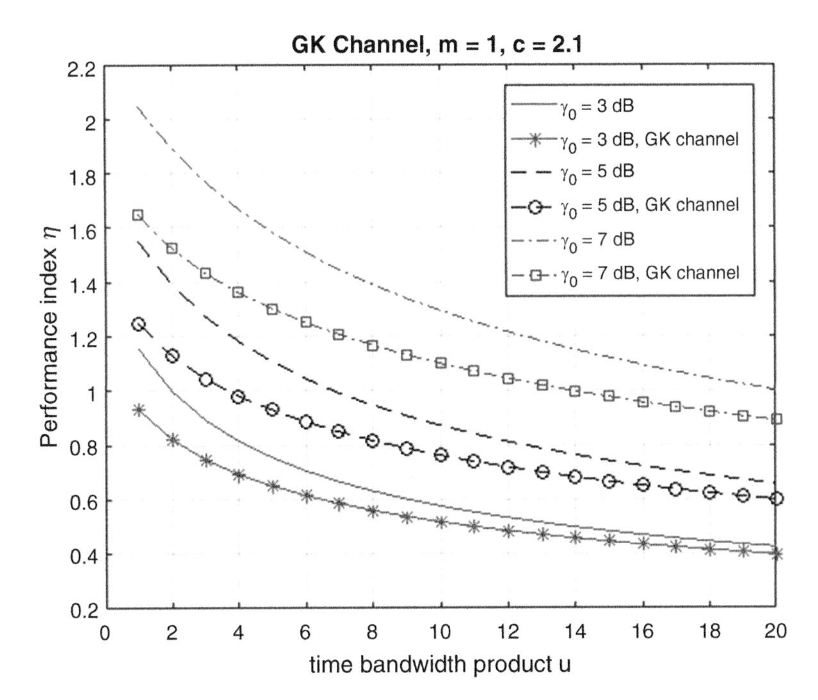

Fig. 7.20 Performance index for $m = 1$ and $c = 2.1$ for three values of the average SNR. The performance index in ideal channels (for identical values of the average SNR) is also shown

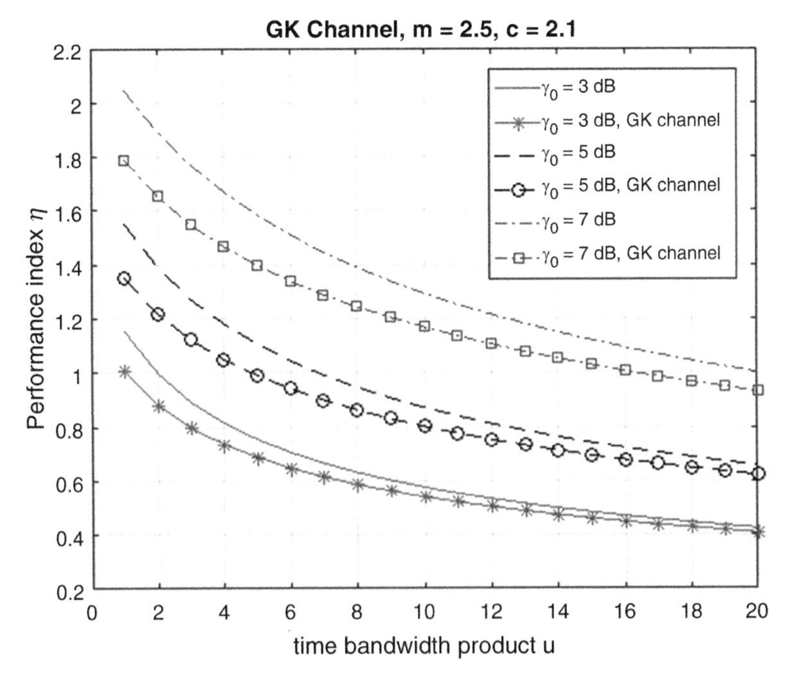

Fig. 7.21 Performance index for $m = 2.5$ and $c = 2.1$ for three values of the average SNR. The performance index in ideal channels (for identical values of the average SNR) is also shown

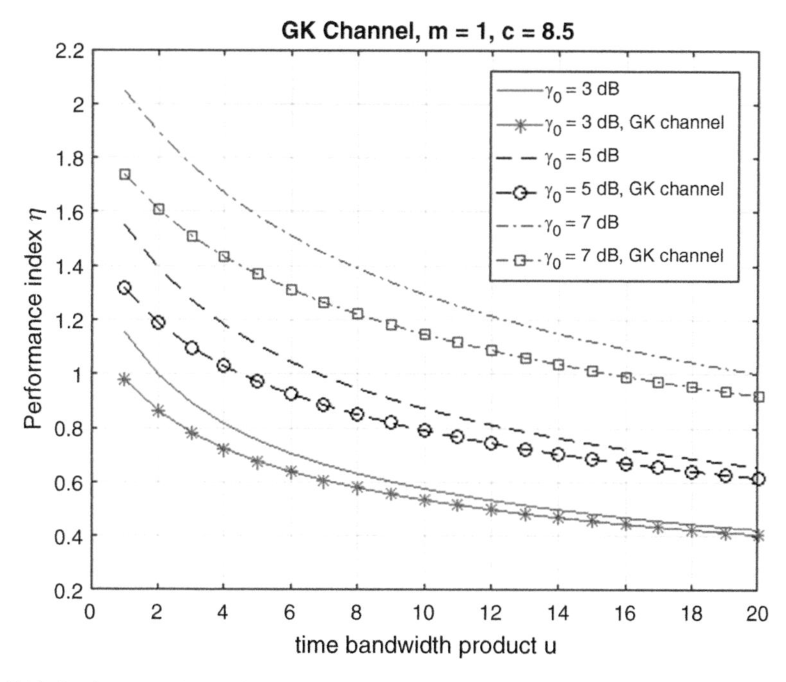

Fig. 7.22 Performance index for $m = 1$ and $c = 8.5$ for three values of the average SNR. The performance index in ideal channels (for identical values of the average SNR) is also shown

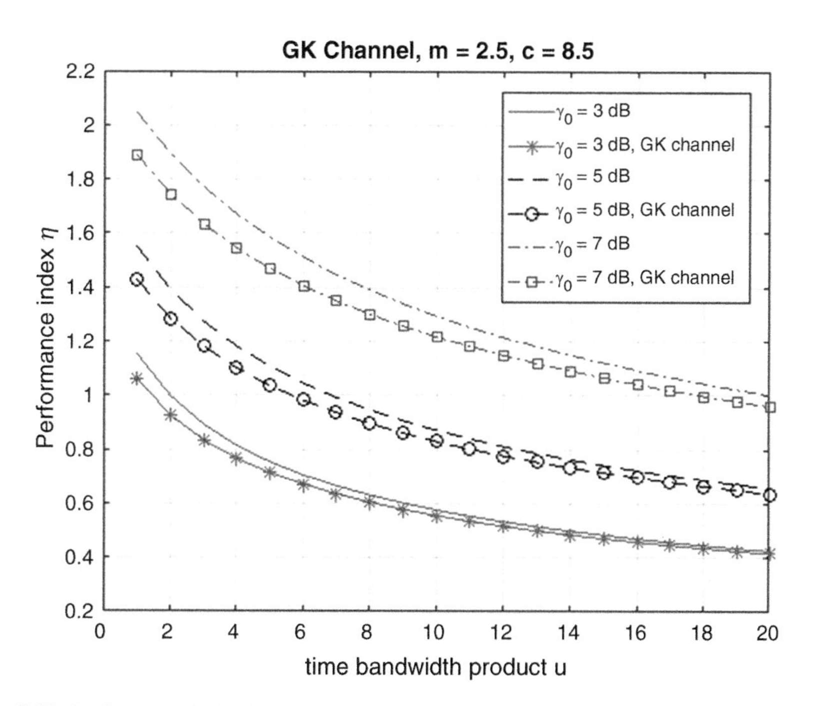

Fig. 7.23 Performance index for $m = 2.5$ and $c = 8.5$ for three values of the average SNR. The performance index in ideal channels (for identical values of the average SNR) is also shown

$$P_d(u, \gamma_0, m, c, \lambda) = \int_0^{\infty} Q_u\left(\sqrt{2\gamma}, \sqrt{\lambda}\right) f(\gamma) d\gamma$$
$$= \int_0^{\infty} Q_u\left(\sqrt{2\gamma}, \sqrt{\lambda}\right) 2\left(\sqrt{\frac{mc}{\gamma_0}\gamma}\right)^{m+c} K_{m-c}\left(2\sqrt{\frac{mc}{\gamma_0}\gamma}\right) d\gamma \quad (7.35)$$

The Matlab script used is given below. The results are displayed in Figs. 7.24, 7.25, 7.26, 7.27, and 7.28.

```
function ROC_analysis_fading_GK
% complete analysis of ROC. Examines the area under the ROC as a function
% of TBW product, ave. SNR, Nakagami parameter, shadowing parameter etc.
% The probability of detection is evaluated numerically.
% shadowed fading channel.   P M Shankar, September 2016
close all
global m lamb Z u c
PF=valuesofpf; % get the values of PF
LF=length(PF); % count the PF
uu=1:20;  KU=length(uu); % values of u and their count
lam=mylamb(uu); % get values of lambda for this set of u
   PF1=[0,PF,1];
% no fading
Z=10^(5/10);
for k=1:KU
    u=uu(k);
    PD(:,k)=marcumq(sqrt(2*Z),sqrt(lam(:,k)),u);
    PF1=[0,PF,1]; % padding is needed so that the ends are not missed.
    % otherwise, it will give a lower area
    Az(k)=polyarea(PF1',[0;PD(:,k);1])+0.5; % area with no fading
end;
% study of Az as a function of the Nakagami parameter and u; SNR & c fixed
c=5.1;
Mk=[0.5,1,2,4];
ZdB=5;
Z=10^(ZdB/10);
for km=1:length(Mk);
    m=Mk(km);
    for k=1:KU
        u=uu(k);
        for kk=1:LF
            lamb=lam(kk,k);
            pdF(kk,k)=integral(@myfun,0,inf);
        end;
        Azm(km,k)=polyarea(PF1',[0;pdF(:,k);1])+0.5;
    end

end;
figure,plot(uu,Az,'-k',uu,Azm(1,:),'r-.*',uu,Azm(2,:),'g-^',...
    uu,Azm(3,:),'b--o',uu,Azm(4,:),'m-d','linewidth',1.3)
legend('No fading',['m = ',num2str(Mk(1))],['m = ',num2str(Mk(2))],...
    ['m = ',num2str(Mk(3))],['m = ',num2str(Mk(4))])
xlim([min(uu),max(uu)])
ylim([0.5,1])
title({'Shadowed fading channel';
    ['\gamma_0 = ',num2str(ZdB),'dB; c = ',num2str(c)]})
xlabel('Time bandwidth product u')
ylabel('Area under the ROC curve A_z')
```

```
% study a few values of SNR as a function of u; m & c fixed
mm=1.5;
m=mm;
Z1dB=[0,5,10,15];
ZZ=10.^(Z1dB/10);
for km=1:length(ZZ);
    Z=ZZ(km);
    for k=1:KU
        u=uu(k);
        for kk=1:LF
            lamb=lam(kk,k); % this is the original set
            pdF(kk,k)=integral(@myfun,0,inf);
        end;
        Azz(km,k)=polyarea(PF1',[0;pdF(:,k);1])+0.5;
    end
end;
figure,plot(uu,Azz(1,:),'k-*',uu,Azz(2,:),'r-s',...
    uu,Azz(3,:),'b--o',uu,Azz(4,:),'m-d')
legend(['\gamma_0 = ',num2str(Z1dB(1)),'dB'],...
    ['\gamma_0 = ',num2str(Z1dB(2)),'dB'],...
    ['\gamma_0 = ',num2str(Z1dB(3)),'dB'],['\gamma = ',num2str(Z1dB(4)),'dB'])
xlim([min(uu),max(uu)])
ylim([0.5,1])
title({'Shadowed fading channel';['m = ',num2str(mm),'; c = ',num2str(c)]})
xlabel('Time bandwidth product u'),ylabel('Area under the ROC curve A_z')
% a few values of u as a function of SNR: m and c fixed
mm=1.5;
m=mm;
Z1dB=[0:20];
ZZ=10.^(Z1dB/10);
UU=[1,3,5,7];
lamm=mylamb(UU); % new set of lambda values for this U and PF
for km=1:length(ZZ);
    Z=ZZ(km);
    for k=1:4
        u=UU(k);
        for kk=1:LF
            lamb=lamm(kk,k);
            pdF(kk,k)=integral(@myfun,0,inf);
        end;
        Azu(km,k)=polyarea(PF1',[0;pdF(:,k);1])+0.5;
    end
end;
figure,plot(Z1dB,Azu(:,1),'r-*',Z1dB,Azu(:,2),'b-s',...
    Z1dB,Azu(:,3),'k--o',Z1dB,Azu(:,4),'m-d')
legend(['u = ',num2str(UU(1))],['u = ',num2str(UU(2))],...
    ['u = ',num2str(UU(3))],['u = ',num2str(UU(4))])
ylim([0.5,1])
title({'Shadowed fading channel';
    ['m = ',num2str(mm),'; c = ',num2str(c)]})
xlabel('Average SNR \gamma_0 (dB)'),ylabel('Area under the ROC curve A_z')
% area for a few value of m. u & c fixed
u=4;
lamU=mylamb(u); % value of lambda for the set pf PF
mm=[1,2,3,4];
for km=1:length (mm)
   m=mm(km);
  for k=1:length(ZZ)
      Z=ZZ(k);
      for kk=1:LF
            lamb=lamU(kk);
            pdF(kk,k)=integral(@myfun,0,inf);
        end;
        Az1(km,k)=polyarea(PF1',[0;pdF(:,k);1])+0.5;
  end;
end;
```

```
figure,plot(Z1dB,Az1(1,:),'r-*',Z1dB,Az1(2,:),'b-s',...
    Z1dB,Az1(3,:),'k--o',Z1dB,Az1(4,:),'m-d')
legend(['m = ',num2str(mm(1))],['m = ',num2str(mm(2))],...
    ['m = ',num2str(mm(3))],['m = ',num2str(mm(4))])
ylim([0.5,1])
title({['Shadowed fading channel ; c = ',num2str(c)];...
    [' u = ',num2str(u)]})
xlabel('Average SNR \gamma_0 (dB)'),ylabel('Area under the ROC curve A_z')

% study of Az as a function of the shadowing parameter c and u; SNR fixed
m=1.8;
Ck=[1.2,3.5,12.2];
ZdB=5;
Z=10^(ZdB/10);
for km=1:length(Ck);
    c=Ck(km);
    for k=1:KU
        u=uu(k);
        for kk=1:LF
            lamb=lam(kk,k);
            pdF(kk,k)=integral(@myfun,0,inf);
        end;
        Azm(km,k)=polyarea(PF1',[0;pdF(:,k);1])+0.5;
    end

end;
figure,plot(uu,Az,'-k',uu,Azm(1,:),'r-*',uu,Azm(2,:),'b-s',...
    uu,Azm(3,:),'m--o')
legend('No fading',['c = ',num2str(Ck(1))],['c = ',num2str(Ck(2))],...
    ['c = ',num2str(Ck(3))])
xlim([min(uu),max(uu)])
ylim([0.5,1])
title({'Shadowed fading channel';
    ['\gamma_0 = ',num2str(ZdB),'dB; m = ',num2str(m)]})
xlabel('Time bandwidth product u')
ylabel('Area under the ROC curve A_z')
end

function PF1=valuesofpf
PF1=[1e-12 5e-12 1e-11 5e-11 1e-10 1e-9,1e-8 1e-7 1e-6 1e-5 1e-4...
    .001 .01 0.03 0.05 0.08 .1 0.14 0.16 .2 0.25 .3 0.35 .4 ...
    .5 0.55 .6 0.65 .7 .75 .8 .85 .9 0.91 .92 0.93 .94 0.95 ...
    .96 0.965 0.968 0.97 0.975 0.978 0.98 0.985 .99 0.995 ...
    0.99999 0.999999 0.9999999];
end

function Fg=myfun(x) % external function for the integral
% GK channel
global m lamb Z u c
pd=2*((sqrt(c/Z)*sqrt(m*x)).^(c+m)).*...
    besselk(c-m, 2*sqrt(c/Z)*sqrt(m*x))*1./(x*gamma(m)*gamma(c));
Fg=pd.*marcumq(sqrt(2*x),sqrt(lamb),u);
end


function lambda=mylamb(U) % get values of lambda
PF=valuesofpf; %values of PF
LF=length(PF);  KU=length(U);
lambda=zeros(LF,KU);
for ku=1:KU
    lambda(:,ku)=2*gammaincinv(PF,U(ku),'upper');%invert PF & get threshold
end;

end
```

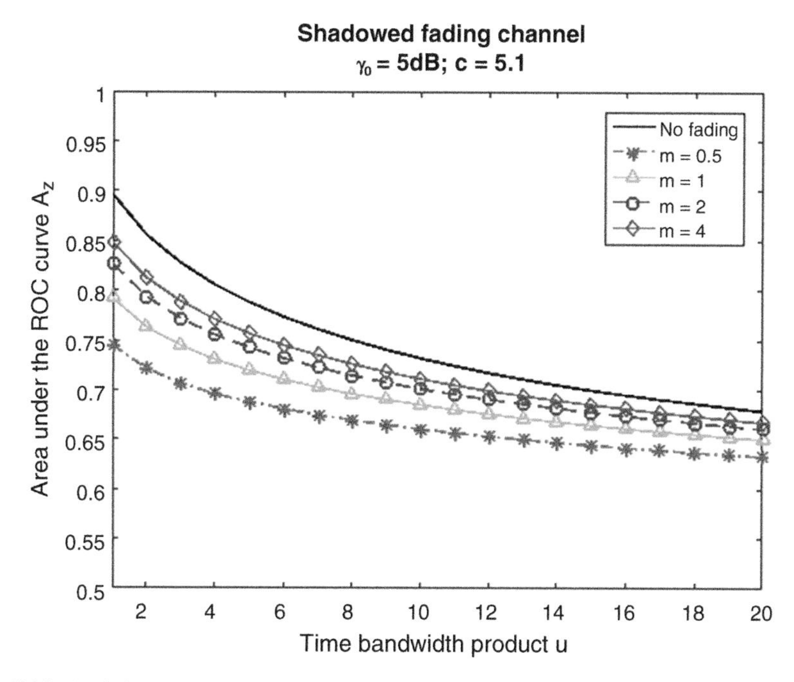

Fig. 7.24 AUC for a few values of the Nakagami parameter (fixed value of average SNR and shadowing level). The performance in an ideal channel is also shown

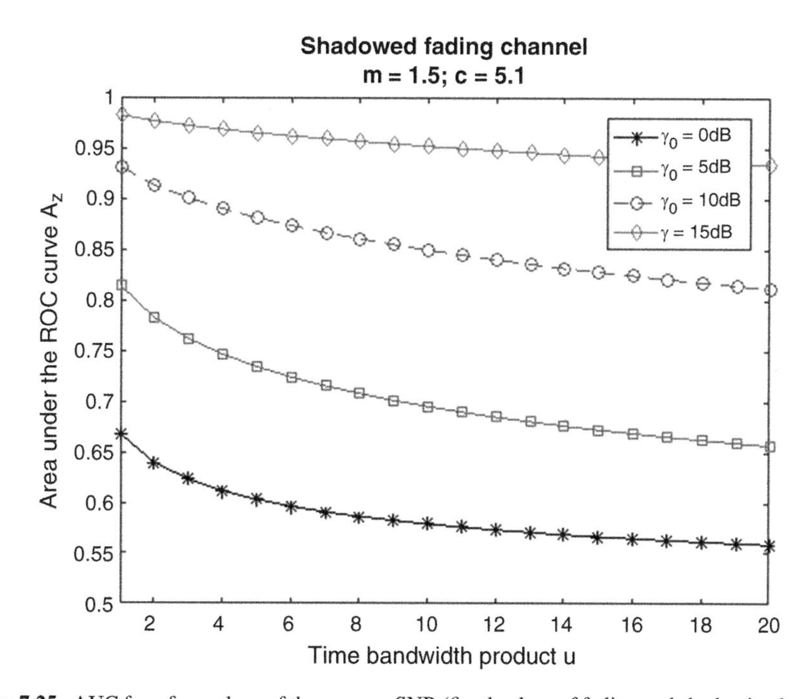

Fig. 7.25 AUC for a few values of the average SNR (fixed values of fading and shadowing levels)

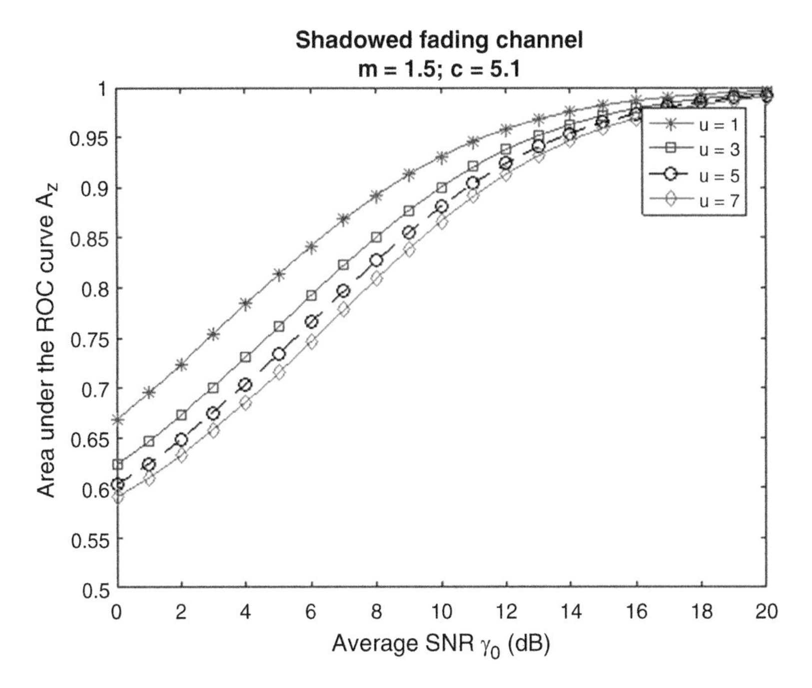

Fig. 7.26 AUC for a few values of the time bandwidth product

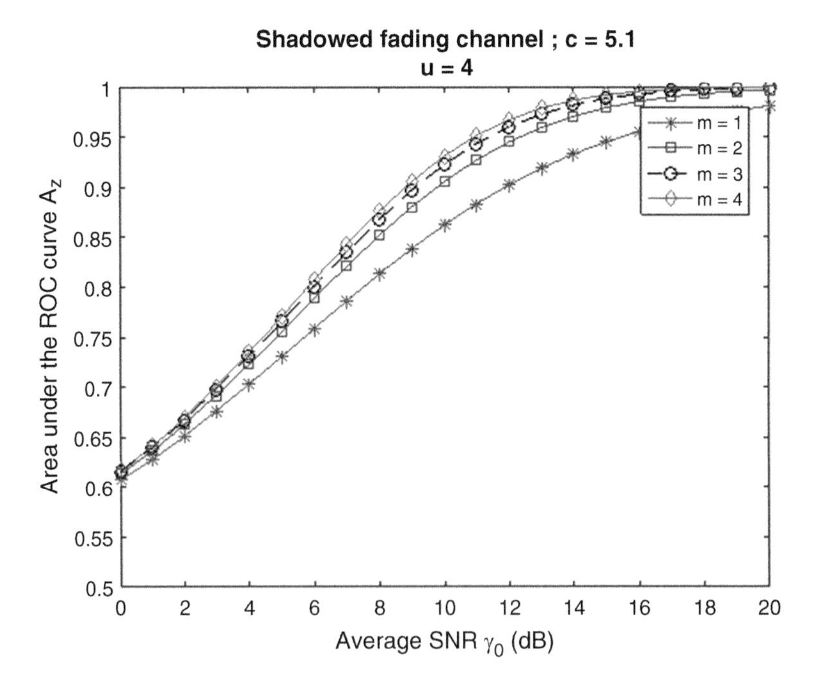

Fig. 7.27 AUC for different values of m

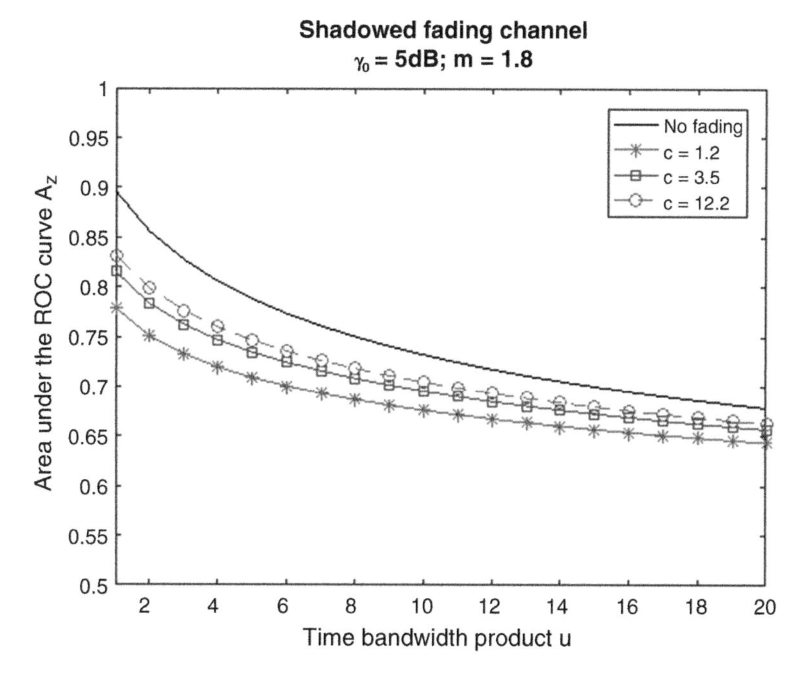

Fig. 7.28 AUC for different values of shadowing levels

7.3.3 Partial AUC

While the area under the ROC curve provides very useful information, it becomes difficult to compare the performances of two or more channels or detectors when they have almost identical AUC values and one observes that the ROC curves actually crossover. An example is shown in here where one sees that the ROC curves cross. The Matlab script used here is given below. Results are shown in Fig. 7.29.

```
function partial_AUC_concept
% Concept of partial AUC.
%
% Two ROC curves are generated showing a cross-over point eventhough the
% AUC values are almost equal.
% P M Shankar, September 2016
close all
global m lamb Z u c
PF=[1e-12 5e-12 1e-11 5e-11 1e-10 1e-9,1e-8 1e-7 1e-6 1e-5 1e-4...
    .001 .01 0.03 0.05 0.08 .1 0.14 0.16 .2 0.25 .3 0.35 .4 ...
    .5 0.55 .6 0.65 .7 .75 .8 .85 .9 0.91 .92 0.93 .94 0.95 ...
    .96 0.965 0.968 0.97 0.975 0.978 0.98 0.985 .99 0.995 ...
    0.99999 0.999999 0.9999999];
LF=length(PF); %count the PF values
u=2;
  lam=2*gammaincinv(PF,u,'upper');%invert PF & get threshold
        for kk=1:LF
            lamb=lam(kk);
            c=2.8;m=4.5; ZdB=5.5;Z=10^(ZdB/10);
            pdfGK(kk)=integral(@myfunGK,0,inf); % prob detection GK
           m=3.1;ZdB=5;Z=10^(ZdB/10);
              pdfN(kk)=integral(@myfunN,0,inf); % prob detection gamma
        end;
   loglog(PF,pdfGK,'r-',PF,pdfN,'--k','linewidth',1.2)
   xlabel('P_f'),ylabel('P_d'), xlim([1e-2,1])
    AUCGK=polyarea([0,PF,1],[0,pdfGK,1])+.5;%fill edges to complete shape
    AUCN=polyarea([0,PF,1],[0,pdfN,1])+.5;%fill edges to complete shape
   legend(['A_z = ',num2str(AUCGK)],['A_z = ',num2str(AUCN)],...
       'location','best')
   title(['Two ROC curves; u = ',num2str(u)]),grid on
   text(0.11,0.59,'\Leftarrow Crossover point','color','b',...
       'fontweight','bold')

end

function Fg=myfunGK(x) % external function for the integral
% GK channel
global m lamb Z u c
pd=2*((sqrt(c/Z)*sqrt(m*x)).^(c+m)).*...
    besselk(c-m, 2*sqrt(c/Z)*sqrt(m*x))*1./(x*gamma(m)*gamma(c));
Fg=pd.*marcumq(sqrt(2*x),sqrt(lamb),u);
end

function F=myfunN(x) % external function for the Nakagami channel
global m Z u lamb
pd=gampdf(x,m,Z/m);
F=pd.*marcumq(sqrt(2*x),sqrt(lamb),u);
end
```

The ROC curves in Fig. 7.29 suggest that one needs to examine closely the ROC in the region of interest characterized by a specific value of the probability of false alarm required. To understand this issue, two false alarm probability values are indicated in Fig. 7.30 and a region of interest is drawn.

It appears from Fig. 7.30 that the ROC curves # 1 and # 2 might end up with identical AUC values. It is possible to imagine a scenario where both probabilities of detection and false alarm are high. Such a case is not an ideal one since one must have a high probability of detection and very low probability of false alarm (McClish 1989; Obuchowski 2003; Dodd and Pepe 2003; Li and Liao 2008). This

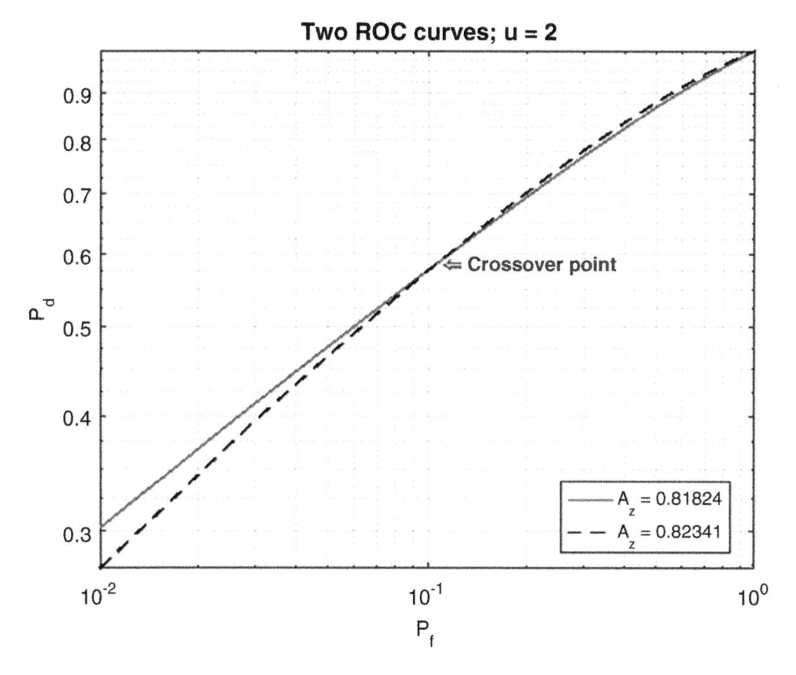

Fig. 7.29 Concept of the partial area under the ROC curves. One of the plots represents the ROC in a GK channel while the other one is in a Nakagami channel

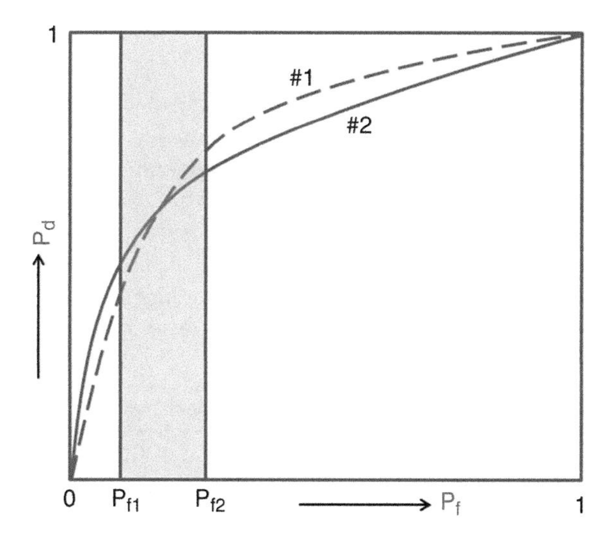

Fig. 7.30 Two ROC curves (#1 and #2) with a crossover point. Two values of the probability of false alarm are marked to indicate the region of interest

would require that greater examination of the AUC plot at the low end of the probability of false alarm.

If λ_1 and λ_2 are the threshold values of the energy corresponding to the two values of the probabilities of false alarm P_{f1} and P_{f2}, it is possible to define a partial AUC as (Adebola et al. 2014; Shankar 2016)

$$pA_z = -\int_0^\infty \int_{\lambda=\lambda_1}^{\lambda_2} \frac{1}{2^u\Gamma(u)} \lambda^{u-1} e^{-\frac{\lambda}{2}} Q_u\left(\sqrt{2\gamma}, \sqrt{\lambda}\right) f(\gamma) d\lambda d\gamma \tag{7.36}$$

Notice the presence of the negative sign reflecting the trend of the probability of false alarm. Using the series expansion for the Marcum Q function and inverting the limits of the threshold,

$$pA_z = \int_0^\infty \int_{\lambda=\lambda_2}^{\lambda_1} \frac{1}{2^u\Gamma(u)} \lambda^{u-1} e^{-\frac{\lambda}{2}} \sum_{k=0}^{\infty} \gamma^k e^{-\gamma} \frac{\Gamma\left(u+k, \frac{\lambda}{2}\right)}{\Gamma(u+k)\Gamma(k+1)} f(\gamma) d\lambda d\gamma \tag{7.37}$$

In Eqs. (7.36) and (7.37), $f(\gamma)$ is the density function of the SNR in a fading channel.

While partial AUC is conceptually useful, it has practical limitations. A close examination of the rectangle in Fig. 7.30 suggests clearly that the value of partial AUC depends on the interval of the probabilities of false alarm chosen. The interval is

$$\Delta P_f = P_{f_2} - P_{f_1}. \tag{7.38}$$

Note that the typical values of the range lie between 0.1 and 0.3, a better way to utilize the partial AUC is to define a partial AUC index as (Shankar 2016)

$$pAI = \frac{pA_z}{\Delta P_f}. \tag{7.39}$$

The partial AUC index will lie between 0 and 1 providing a normalized parameter for comparison. For an ideal detector, the partial AUC index is the area of the rectangle divided by the difference in the values of the probabilities of false alarm.

To demonstrate the characteristics of the partial AUC index, consider the case of two fading distributions. The first one is the density function of a selection combining algorithm in a Nakagami channel of order 3 resulting in a pdf (Chap. 4)

$$f(\gamma) = f(\gamma_1) = 3\left[1 - \frac{\Gamma\left(m, \frac{m\gamma_1}{\gamma_{01}}\right)}{\Gamma(m)}\right]^2 \left(\frac{m}{\gamma_{01}}\right)^m \frac{\gamma_1^{m-1}}{\Gamma(m)} e^{-\frac{m}{\gamma_{01}}\gamma_1}. \tag{7.40}$$

The second density is the pdf associated with the maximal ratio combining algorithm (order 3) resulting in a density of (Chap. 4)

$$f(\gamma) = f(\gamma_2) = \left(\frac{m}{\gamma_{02}}\right)^{3m} \frac{\gamma_2^{3m-1}}{\Gamma(3m)} e^{-\frac{m}{\gamma_{02}}\gamma_2} \tag{7.41}$$

For this analysis, it has been assumed that

$$\gamma_{01} = 2\gamma_{02} \tag{7.42}$$

Instead of using the summation in Eq. (7.37), it is much easier to use the double integral in Eq. (7.36) since one set of the limits (thresholds) will be of finite extent and numerical double integration can be easily carried out. For the two sets of density functions, $f(\gamma)$ in Eq. (7.36) is replaced by the two densities in Eqs. (7.40) and (7.41).

The concept of partial AUC is demonstrated by considering the case of the energy detection in a Nakagami fading channel. The two cases for the ROC are constituted by the probability densities in Eqs. (7.40) and (7.41). Results are displayed in Figs. 7.31 and 7.32. The Matlab script appears below.

```
function partial_AUC_calculations
% Concept of partial AUC.
%
% Two densities are used to examine the partial AUC index. Numerical double
% integration is used to estimate the partial AUC index.
% P M Shankar, September 2016
close all
global m  Z u
u=4;
ZZ=0:20; % SNR in dB
LZ=length(ZZ);
m=1.5;
for kk=1:2 % two sets of PF values
    if kk==1
PF1=0.001;PF2=0.005; DP=PF2-PF1;
lam1=2*gammaincinv(PF1,u,'upper');%invert PF & get threshold
  lam2=2*gammaincinv(PF2,u,'upper');%invert PF & get threshold
    else
   PF1=0.0001;PF2=0.001; DP=PF2-PF1;
lam1=2*gammaincinv(PF1,u,'upper');%invert PF & get threshold
  lam2=2*gammaincinv(PF2,u,'upper');%invert PF & get threshold
    end;
  for k=1:LZ
      Z=10.^(ZZ(k)/10);
  p11=integral2(@myfun1,lam2,lam1,0,inf);%note  change in order of limits
  p22=integral2(@myfun2,lam2,lam1,0,inf);%note  change in order of limits
    p1(k)=p11/DP;
    p2(k)=p22/DP;
  end;
 figure, plot(ZZ,p1,'k-o',ZZ,p2,'--r*')
  legend('f(\gamma_1)','f(\gamma_2)')
  xlabel('average SNR \gamma_{02} dB')
  ylabel('Partial AUC index')
  title(['P_f_1 = ',num2str(PF1),',  P_f_2 = ',num2str(PF2),...
      ', m = ',num2str(m),', u = ',num2str(u)])
end;
end
```

```
function F1=myfun1(x,y) % external function for the integral
global u m Z
Z1=2*Z;
f1=gampdf(x,u,2).*marcumq(sqrt(2*y),sqrt(x),u);
f2=3*gampdf(y,m,Z1/m).*(gamcdf(y,m,Z1/m)).^2;
F1=f1.*f2;
end

function F2=myfun2(x,y) % external function for the Nakagami channel
global m u Z
M=3;
f1=gampdf(x,u,2).*marcumq(sqrt(2*y),sqrt(x),u);
f2=gampdf(y,M*m,M*Z/(m*M));
F2=f1.*f2;
end
```

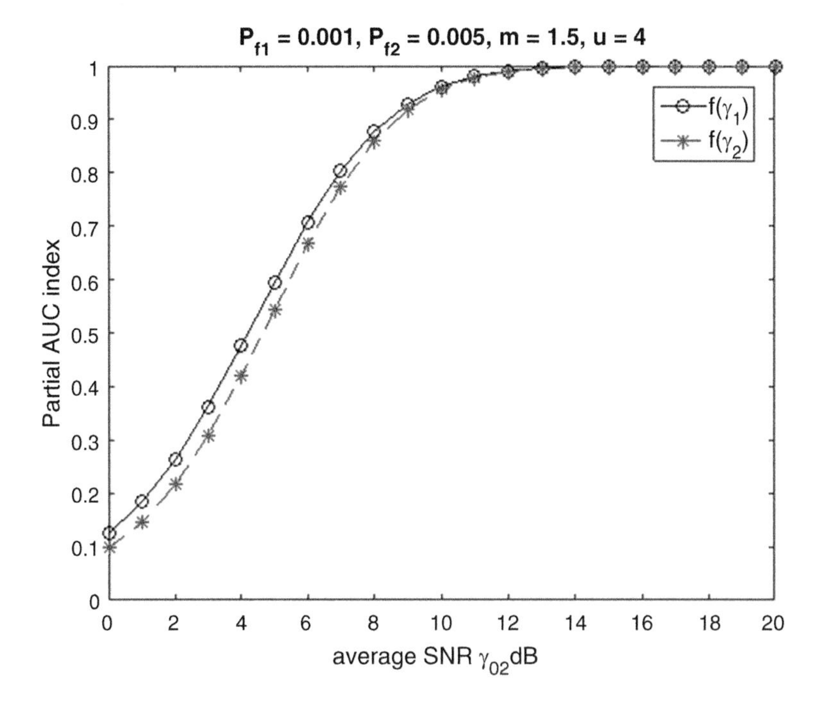

Fig. 7.31 Partial AUC index in a Nakagami fading channel for a set of values of P_f (0.001 and 0.005). The first density $f(\gamma_1)$ corresponds to a Nakagami fading channel and the second density $f(\gamma_2)$ corresponds to a channel where selection combining has been implemented

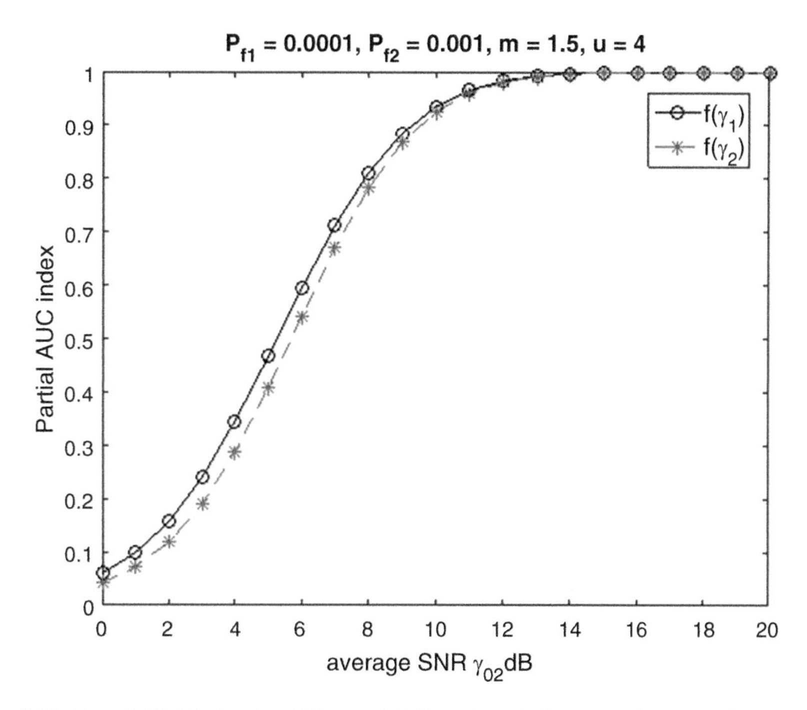

Fig. 7.32 Partial AUC index in a Nakagami fading channel for a set of values of P_f (0.0001 and 0.0001). The first density $f(\gamma_1)$ corresponds to a Nakagami fading channel and the second density $f(\gamma_2)$ corresponds to a channel where selection combining has been implemented

7.3.4 N*Nakagami Channels

The range of fading that can be modeled using the Nakagami model is limited. It is possible to expand the range of fading from very severe through mild using the N*Nakagami models as it was shown in Chap. 4 (Karagiannidis et al. 2007; Shankar 2013; Ilhan 2014; Shankar 2016). In this case, the fading is treated as a cascading process resulting in the SNR becoming the product of a number of independent and identically distributed gamma variables. If the number of "cascades" is N, the density of the received SNR was shown to be

$$f(\gamma) = \frac{1}{\gamma \Gamma^N(m)} G_{0,N}^{N,0}\left(\frac{m^N}{\gamma_0}\gamma \middle| \begin{matrix} - \\ m,..,m \end{matrix}\right), \quad N = 1, 2, 3, \ldots \tag{7.43}$$

In Eq. (7.43), $G_{0,N}^{N,0}\left(\frac{m^N}{\gamma_0}\gamma \middle| \begin{matrix} - \\ m,..,m \end{matrix}\right)$ is the Meijer G-function and γ_0 is the average SNR (see Appendix, Chap. 4). For $N = 1$, the N*Nakagami channel becomes a Nakagami channel with m taking values larger than 0.5. In the most general case, the only requirement on the value of m is that it is positive. As N increases, the fading becomes more and more severe. Using the density of the SNR in Eq. (7.43), the probability of detection in cognitive radio in Eq. (7.25) becomes

$$P_d = \int_0^{\infty} Q_u\left(\sqrt{2\gamma}, \sqrt{\lambda}\right) \frac{1}{\gamma \Gamma^N(m)} G_{0,N}^{N,0}\left(\frac{m^N}{\gamma_0}\gamma \middle| \begin{matrix} - \\ m, .., m \end{matrix}\right) d\gamma. \tag{7.44}$$

The area under the ROC curve becomes

$$A_z = \int_0^{\infty} \frac{1}{\gamma \Gamma^N(m)} G_{0,N}^{N,0}\left(\frac{m^N}{\gamma_0}\gamma \middle| \begin{matrix} - \\ m, .., m \end{matrix}\right) \int_0^{\infty} \frac{1}{2^u \Gamma(u)} \lambda^{u-1} e^{-\frac{\lambda}{2}} Q_u\left(\sqrt{2\gamma}, \sqrt{\lambda}\right) d\lambda d\gamma \tag{7.45}$$

Using the series expansion for the Marcum Q function, the probability of detection in Eq. (7.44) becomes

$$P_d = \int_0^{\infty} \sum_{k=0}^{\infty} \gamma^k e^{-\gamma} \frac{\Gamma\left(u+k, \frac{\lambda}{2}\right)}{\Gamma(u+k)\Gamma(k+1)} \frac{1}{\gamma \Gamma^N(m)} G_{0,N}^{N,0}\left(\frac{m^N}{\gamma_0}\gamma \middle| \begin{matrix} - \\ m, .., m \end{matrix}\right) d\gamma \tag{7.46}$$

Using the table of integrals [Gradshteyn and Ryzhik 2007, Eq. 7-813], the probability of detection in Eq. (7.46) becomes

$$P_d = \sum_{k=0}^{\infty} \frac{\Gamma\left(u+k, \frac{\lambda}{2}\right)}{\Gamma(u+k)\Gamma(k+1)} \frac{1}{\Gamma^N(m)} G_{1,N}^{N,1}\left(\frac{m^N}{\gamma_0} \middle| \begin{matrix} 1-k \\ m, .., m \end{matrix}\right) \tag{7.47}$$

Using the table of integrals [Gradshteyn and Ryzhik 2007, Eqs. 7-813 and 6.455], the expression for AUC in Eq. (7.45) becomes

$$A_z = \frac{1}{u 2^{2u} \Gamma(u) \Gamma^N(m)} \sum_{k=0}^{\infty} G_{1,N}^{N,1}\left(\frac{m^N}{\gamma_0} \middle| \begin{matrix} 1-k \\ m, .., m \end{matrix}\right) \frac{\Gamma(2u+k)}{2^k \Gamma(u+k) k!} {}_2F_1\left([1, 2u+k]; u+1; \frac{1}{2}\right) \tag{7.48}$$

The partial AUC becomes

$$pA_z = \int_0^{\infty} \frac{1}{\gamma \Gamma^N(m)} G_{0,N}^{N,0}\left(\frac{m^N}{\gamma_0}\gamma \middle| \begin{matrix} - \\ m, .., m \end{matrix}\right) \int_{\lambda_2}^{\lambda_1} \frac{1}{2^u \Gamma(u)} \lambda^{u-1} e^{-\frac{\lambda}{2}} \sum_{k=0}^{\infty} \gamma^k e^{-\gamma} \frac{\Gamma\left(u+k, \frac{\lambda}{2}\right)}{\Gamma(u+k)\Gamma(k+1)} d\lambda d\gamma \tag{7.49}$$

Using the table of integrals [Gradshteyn and Ryzhik 2007, Eq. 7–813], the expression for partial AUC becomes

$$pA_z = \sum_{k=0}^{\infty} \frac{G_{1,N}^{N,1}\left(\frac{m^N}{\gamma_0} \middle| \begin{matrix} 1-k \\ m, .., m \end{matrix}\right)}{2^u \Gamma^N(m) \Gamma(u) \Gamma(u+k) \Gamma(k+1)} \int_{\lambda_2}^{\lambda_1} x^{u-1} e^{-\frac{x}{2}} \Gamma\left(u+k, \frac{x}{2}\right) dx. \tag{7.50}$$

Results on the performance in cascaded channels are shown in Figs. 7.33, 7.34, 7.35, 7.36, 7.37, and 7.38. The Matlab script appears below. Results on partial AUC are available elsewhere (Shankar 2016).

```
function cascaded_AUC_book
% cognitive radio performance on a cascaded N*Nakagami channel. Numerical
% integration is used to estimate the probability of detection and polyarea
% command is used to obtain AUC
% Meijer G function is separately generated for N=1,2,3, and 4 and the
% concept can be extended to other values of N
% P M Shankar
close all
global  m lamb Z u N
PF=valuesofpf; % get the values of PF
LF=length(PF); % count the PF
uu=1:2:21;  KU=length(uu); % values of u and their count
pdF=zeros(LF,KU);
lam=mylamb(uu); % get values of lambda for this set of u
mm=[1.5,2.5,3.5];
ZdB=[5,10];
for kz=1:2
    Z=10^(ZdB(kz)/10);
for kkm=1:length(mm)
m=mm(kkm);
NN=[1:4];
for kn=1:length(NN);
    N=NN(kn);
for k=1:KU
    u=uu(k);
    for kk=1:LF
        lamb=lam(kk,k);
        pdF(kk,k)=integral(@cascad_pdF,0.01,0.9995*pi/2);
    end;
    PF1=[0,PF,1];
    % padding is needed so that the ends are not missed. otherwise, it
    % will give a incorrect lower area
    Azn(kn,k)=polyarea(PF1',[0;pdF(:,k);1])+0.5;
end
end;
figure, plot(uu,Azn(1,:),'r-*',uu,Azn(2,:),'k-s',uu,Azn(3,:),'b-o',...
    uu,Azn(4,:),'m--^')
legend('N = 1','N = 2','N = 3', 'N= 4')
xlabel('u'),ylabel('AUC A_z')
title(['Cascaded N*Nakagami channel:  m = ',num2str(m),',  \gamma_0 = ',...
    num2str(10*log10(Z))])
xlim([min(uu),max(uu)])
end;
end;
end

function y = cascad_pdF(xx)
global N Z m lamb u
yy=tan(xx);
for k=1:length(yy);
    x=yy(k);
    pdf=Meijer Gpdf(N,m,x,Z);% gets the pdf
    y(k)=pdf*marcumq(sqrt(2*x),sqrt(lamb),u)*(1+x^2);
end
end

function pd=Meijer Gpdf(N,m,x,Z)
% N number of cascades, m the parameter, x is the variable, Z is the mean SNR
gm=gamma(m);
xZ=x/Z;
if N==1
    pd=gampdf(x,m,Z/m);
elseif N==2
    pd=(1/x)*double(evalin(symengine,sprintf('Meijer G([[ ], [ ]], [[%e,%e], []],
%e)',m,m,(m^N)*xZ)))/gm^N;
elseif N==3
```

```
    pd=(1/x)*double(evalin(symengine,sprintf('Meijer G([[ ], [ ]], [[%e,%e,%e], []],
%e)',m,m,m,(m^N)*xZ)))/gm^N;
elseif N==4
    pd=(1/x)*double(evalin(symengine,sprintf('Meijer G([[ ], [ ]], [[%e,%e,%e,%e], []],
%e)',m,m,m,m,(m^N)*xZ)))/gm^N;
elseif N==5
    pd=(1/x)*double(evalin(symengine,sprintf('Meijer G([[ ], [ ]], [[%e,%e,%e,%e,%e], []],
%e)',m,m,m,m,m,(m^N)*xZ)))/gm^N;
end;
end

function PF1=valuesofpf
PF1=[1e-12 5e-12 1e-11 5e-11 1e-10 1e-9,1e-8 1e-7 1e-6 1e-5 1e-4...
    .001 .01 0.03 0.05 0.08 .1 0.14 0.16 .2 0.25 .3 0.35 .4 ...
    .5 0.55 .6 0.65 .7 .75 .8 .85 .9 0.91 .92 0.93 .94 0.95 ...
    .96 0.965 0.968 0.97 0.975 0.978 0.98 0.985 .99 0.995 ...
    0.99999 0.999999 0.9999999];
end


function lambda=mylamb(U) % get values of lambda
PF=valuesofpf; %values of PF
LF=length(PF);  KU=length(U);
lambda=zeros(LF,KU);
for ku=1:KU
    lambda(:,ku)=2*gammaincinv(PF,U(ku),'upper');%invert PF & get threshold
end;

end
```

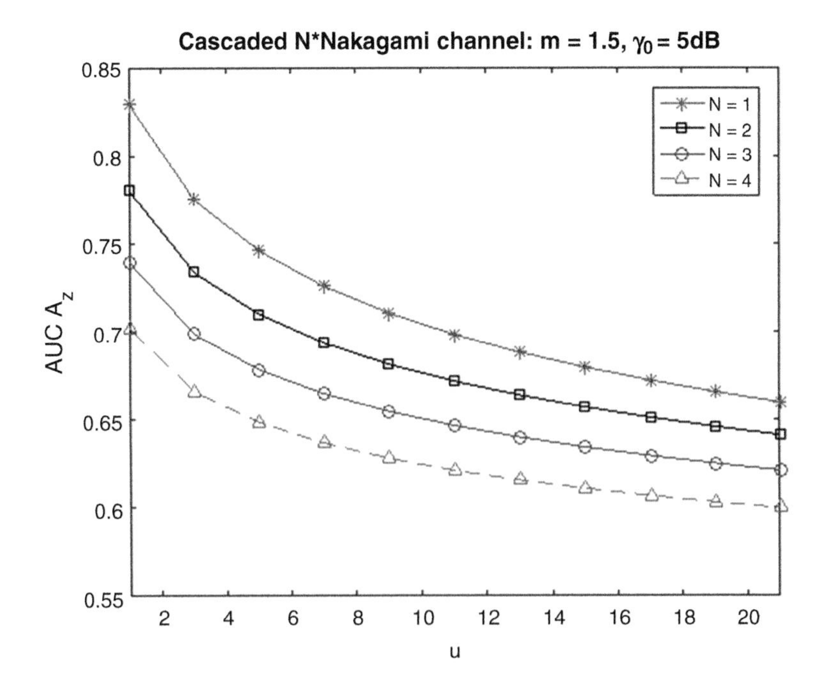

Fig. 7.33 AUC for $N = 1,2,3,4$ ($m = 1.5$, $\gamma_0 = 5$ dB)

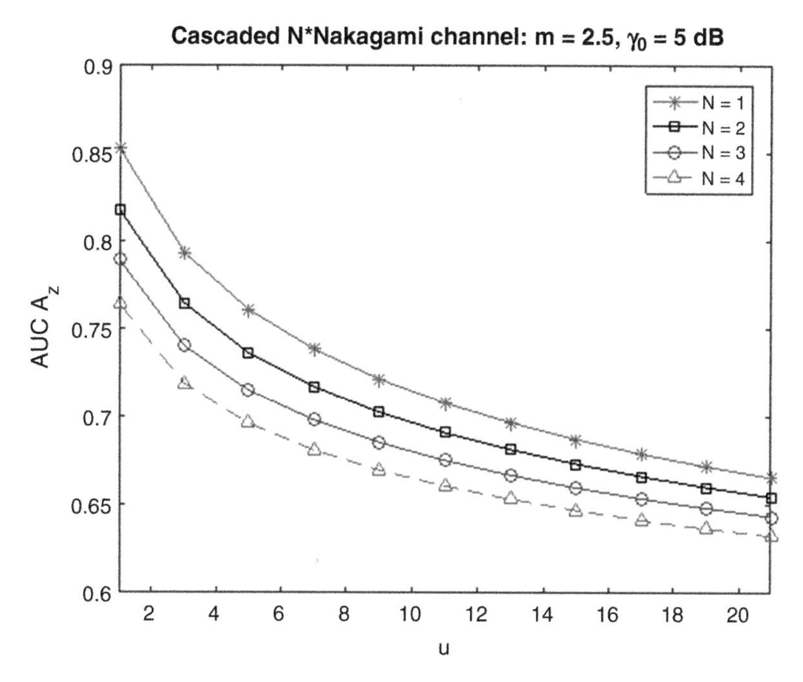

Fig. 7.34 AUC for $N = 1,2,3,4$ ($m = 2.5$, $\gamma_0 = 5$ dB)

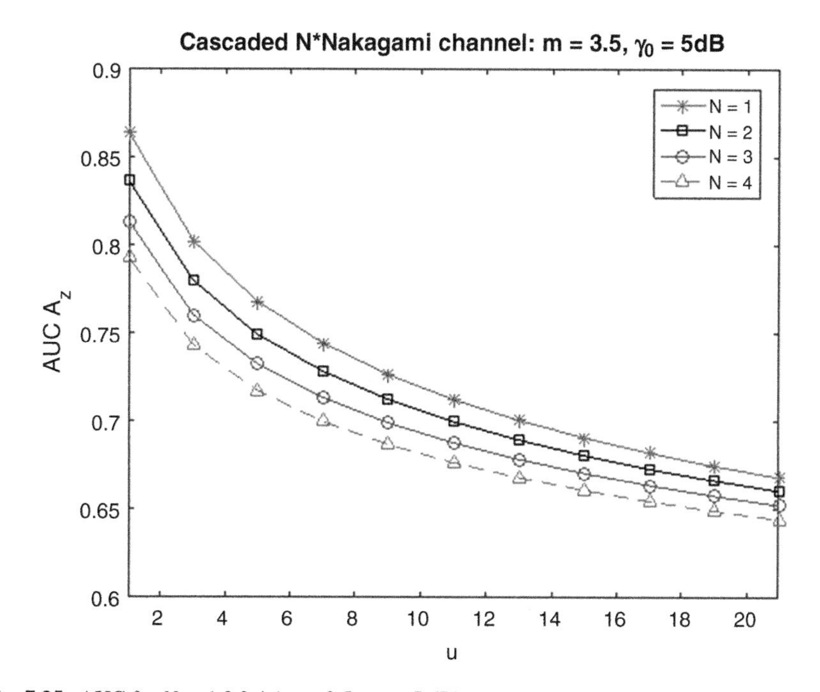

Fig. 7.35 AUC for $N = 1,2,3,4$ ($m = 3.5$, $\gamma_0 = 5$ dB)

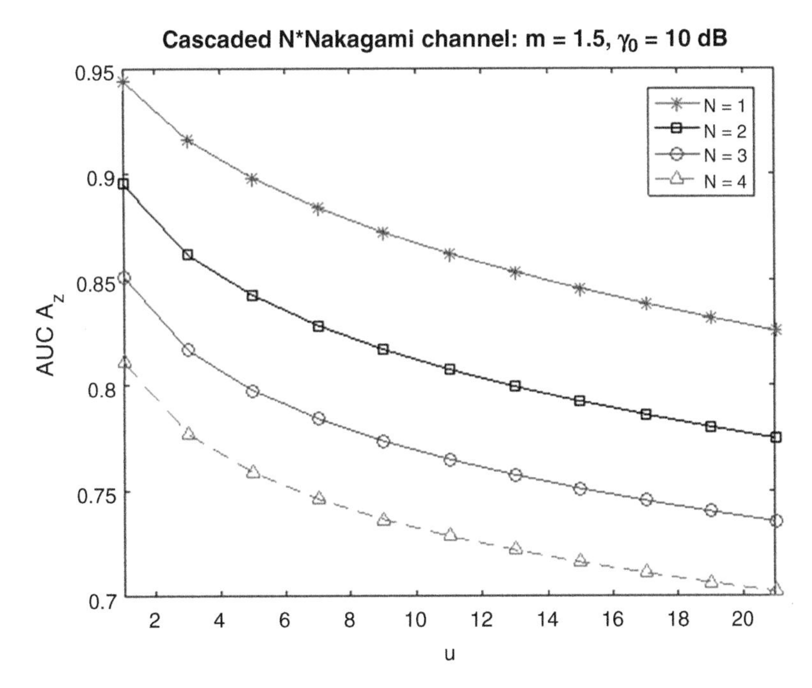

Fig. 7.36 AUC for $N = 1,2,3,4$ ($m = 1.5$, $\gamma_0 = 10$ dB)

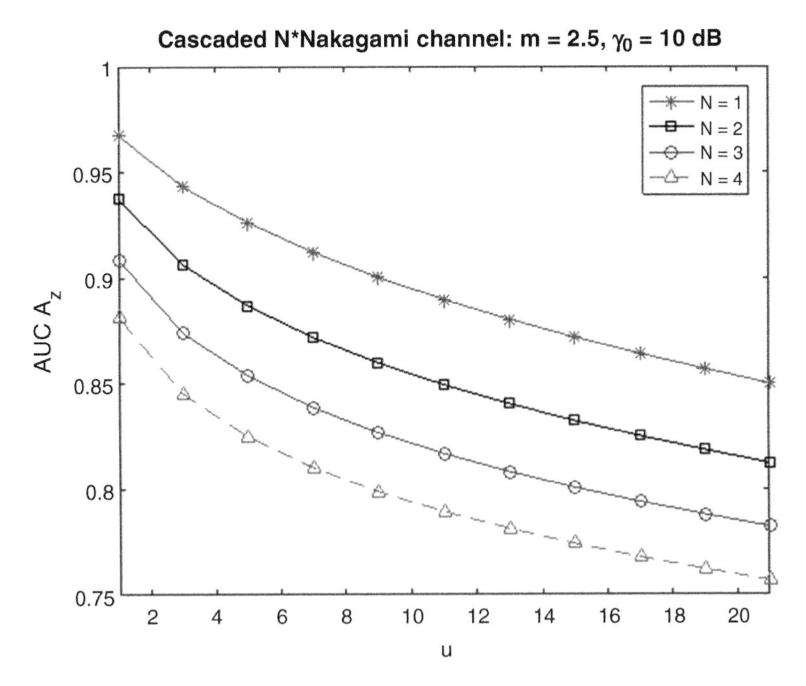

Fig. 7.37 AUC for $N = 1,2,3,4$ ($m = 2.5$, $\gamma_0 = 10$ dB)

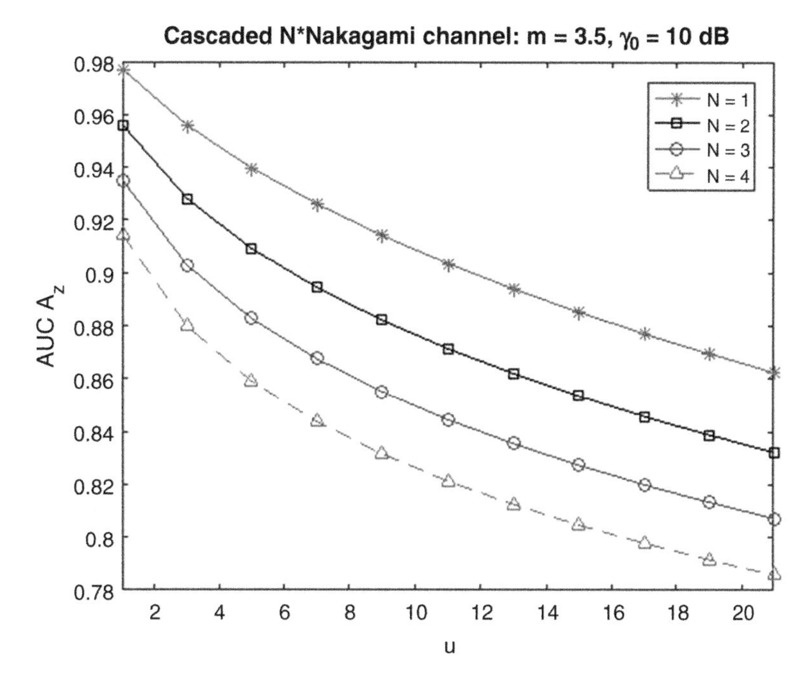

Fig. 7.38 AUC for $N = 1,2,3,4$ ($m = 3.5$, $\gamma_0 = 10$ dB)

7.3.5 Diversity in Cognitive Radio

As seen in Chap. 5, diversity techniques are implemented to mitigate the effects of fading. The diversity techniques also can be employed in cognitive radio to improve the probability of detection and consequently the area under the ROC curve. Since the probability of false alarm does not depend on the signal strength it will not be impacted by the presence of diversity. Going back to Eq. (7.25) and rewriting it, the probability of detection becomes (Digham et al. 2007; Atapattu et al. 2011; Sun et al. 2011)

$$P_d = \int_0^{\infty} Q_u\left(\sqrt{2\gamma_{\text{div}}}, \sqrt{\lambda}\right) f(\gamma_{\text{div}}) d\gamma_{\text{div}}. \tag{7.51}$$

In Eq. (7.51), γ_{div} is the output of the diversity combing algorithm and $f(\gamma_{\text{div}})$ is its probability density function. Consider the case of a Nakagami fading channel with an average SNR of γ_0 and Nakagami parameter m. The branches of diversity are treated as independent and identically distributed.

If the order of diversity is M, the probability density function of the output of the selection combining algorithm will be (Chap. 5)

$$f(\gamma_{SC}) = M\left[1-\frac{\Gamma\left(m,\frac{m\gamma_{SC}}{\gamma_0}\right)}{\Gamma(m)}\right]^{M-1}\left(\frac{m}{\gamma_0}\right)^m\frac{\gamma_{SC}^{m-1}}{\Gamma(m)}e^{-\frac{m}{\gamma_0}\gamma_{SC}}. \tag{7.52}$$

In the case of the MRC diversity, the probability density function of the output of the MRC algorithm will be (Chap. 5)

$$f(\gamma_{MRC}) = \left(\frac{m}{\gamma_0}\right)^{Mm}\frac{\gamma_{MRC}^{Mm-1}}{\Gamma(Mm)}e^{-\frac{m}{\gamma_0}\gamma_{MRC}}. \tag{7.53}$$

Results of the diversity analysis are shown in Figs. 7.39, 7.40, 7.41, 7.42, 7.43, 7.44, 7.45, 7.46, and 7.47. The Matlab script used appears below. The improvement in performance gained through diversity is clearly seen with MRC performing better than SC as expected.

```
function ROC_Nakagami_diversity
% ROC in a Nakagami channel. The probability of detection is evaluated
% using numerical integration and AUC estimated using polyarea(.)
% Both MRC and SC are implemented and results are ompared to the case of no
% P M Shankar, Sept. 2016
close all
global m lamb Z u M
PF=valuesofpf; % get the values of PF
LF=length(PF); % count the PF
uu=1:20;  KU=length(uu); % values of u and their count
lam=mylamb(uu); % get values of lambda for this set of u
 PF1=[0,PF,1]; % padding is needed so that the ends are not missed.
% study of Az as a function of the Nakagami parameter
for kz=1:2
    if kz==1
ZdB=3;
    else
        ZdB=5;
    end;
Z=10^(ZdB/10);
for km=1:4
    if km==1
      m=1.5; M=2;
    elseif km==2
        m=1.5; M=4;
    elseif km==3
        m=2; M=2;
    else
        m=2; M=5;
    end;

    for k=1:KU
        u=uu(k);
        for kk=1:LF
            lamb=lam(kk,k);
            pdF(kk,k)=integral(@myfun1,0,inf);
              pdFSC(kk,k)=integral(@myfunSC,0,inf);
                pdFMR(kk,k)=integral(@myfunMRC,0,inf);
        end;
```

```
        Az(k)=polyarea(PF1',[0;pdF(:,k);1])+0.5;
          AzS(k)=polyarea(PF1',[0;pdFSC(:,k);1])+0.5;
            AzM(k)=polyarea(PF1',[0;pdFMR(:,k);1])+0.5;
    end

figure,plot(uu,Az,'r-*',uu,AzS,'k-s',uu,AzM,'m-o')
  legend('No Div.','SC','MRC')
title(['Nakagami fading channel: m = ', num2str(m),', Average SNR = ',...
    num2str(ZdB),'dB, M = ',num2str(M)])
xlabel('Time bandwidth product u')
ylabel('Area under the ROC curve A_z')
end;
end;
end

function PF1=valuesofpf
PF1=[1e-12 5e-12 1e-11 5e-11 1e-10 1e-9,...
    1e-8 1e-7 1e-6 1e-5 1e-4...
    .001 .01 0.03 0.05 0.08 .1 0.14 0.16 .2 0.25 .3 0.35 .4 ...
    .5 0.55 .6 0.65 .7 .75 .8 .85 .9 0.91 .92 0.93 .94 0.95 ...
    .96 0.965 0.968 0.97 0.975 0.978 0.98 0.985 .99 0.995 ...
    0.99999 0.999999 0.9999999];
end

function Fg=myfun1(x) % Nakagami channel No diversity
global m lamb Z u
pd=1-ncx2cdf(lamb,2*u,2*x);% use the CDF
Fg=gampdf(x,m,Z/m).*pd;
end


function Fg=myfunMRC(x) % Nakagami channel MRC diversity
global m lamb Z u M
pd=1-ncx2cdf(lamb,2*u,2*x);% use the CDF
Fg=gampdf(x,m*M,Z/m).*pd; % note that M*Z/(M*m) is same as Z/m
end

function Fg=myfunSC(x) % Nakagami channel SCC diversity
global m lamb Z u M
pd=1-ncx2cdf(lamb,2*u,2*x);% use the CDF
pdfSC=M*gampdf(x,m,Z/m).*(gamcdf(x,m,Z/m)).^(M-1);
Fg=pdfSC.*pd; % note that M*Z/(M*m) is same as Z/m
end


function lambda=mylamb(U) % get values of lambda
PF=valuesofpf; %values of PF
LF=length(PF);  KU=length(U);
lambda=zeros(LF,KU);
for ku=1:KU
    lambda(:,ku)=2*gammaincinv(PF,U(ku),'upper');%invert PF & get threshold
end;

end
```

A discussion of diversity in cognitive radio will be incomplete without introducing another form of diversity typically used with noncoherent systems, namely the square law combining (SLC) diversity. This is similar to MRC without the need for channel estimation [3, 4]. This leads to M signals being added regardless of

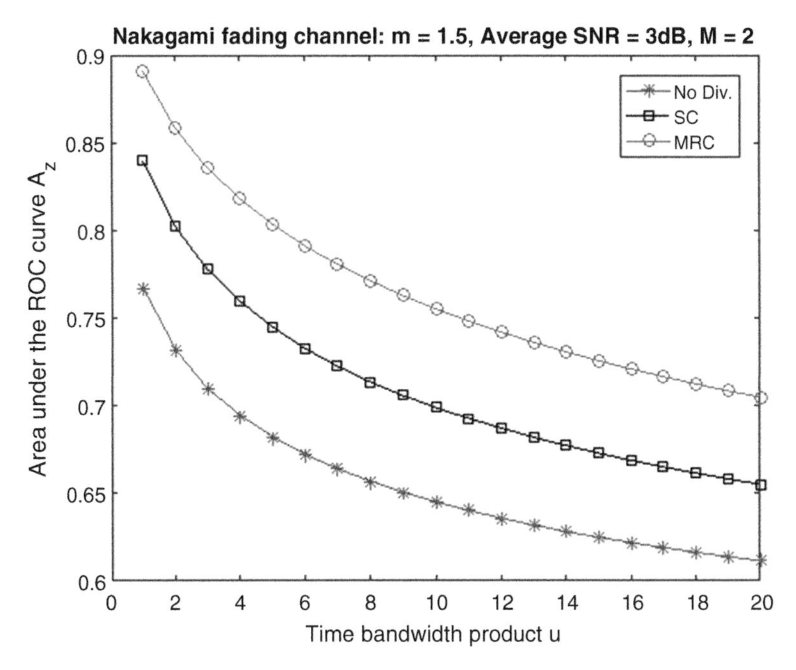

Fig. 7.39 AUC in a Nakagami fading channel: Improvement through diversity (m = 1.5, Average SNR = 3 dB, M = 2)

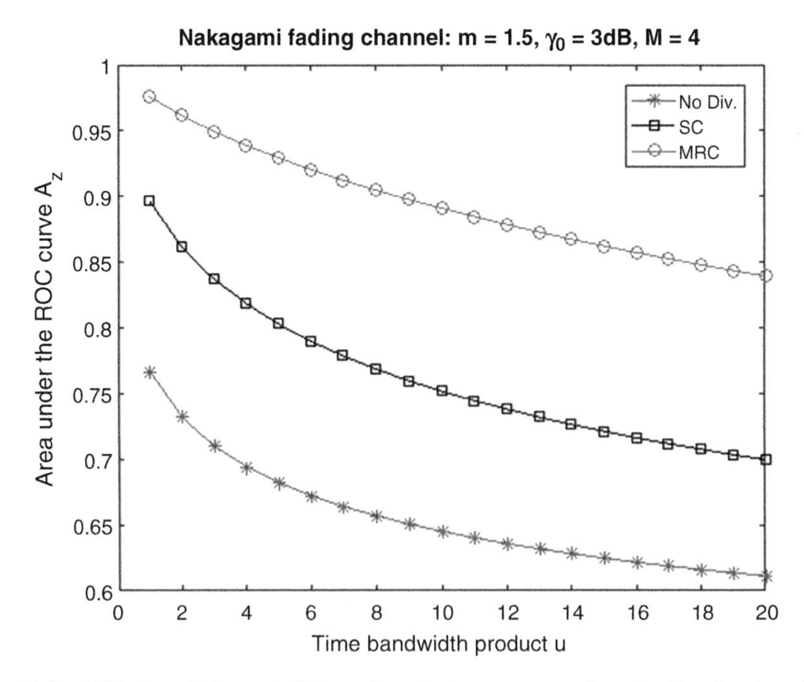

Fig. 7.40 AUC in a Nakagami fading channel: Improvement through diversity (m = 1.5, γ_0 = 3 dB, M = 4)

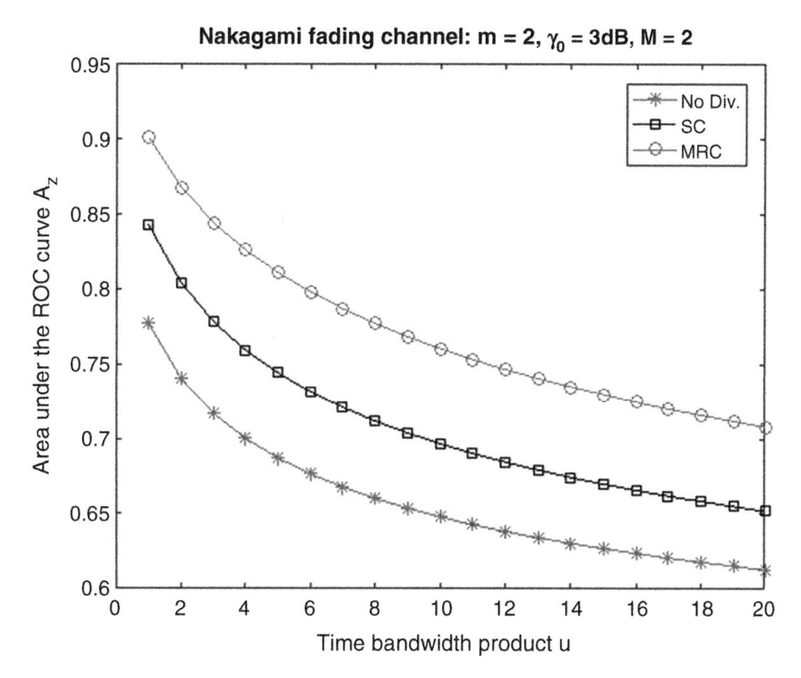

Fig. 7.41 AUC in a Nakagami fading channel: Improvement through diversity (m = 2, $\gamma_0 = 3$ dB, M = 2)

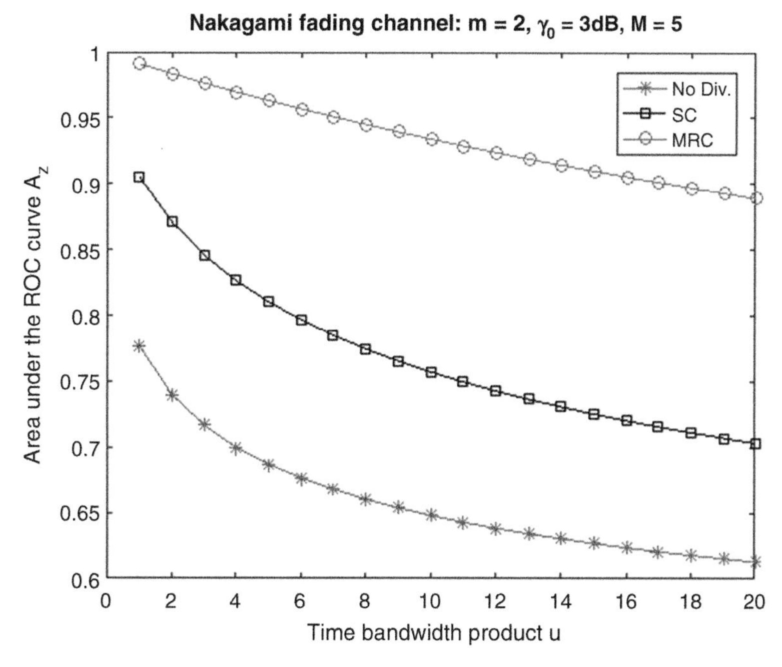

Fig. 7.42 AUC in a Nakagami fading channel: Improvement through diversity (m = 2, $\gamma_0 = 3$ dB, M = 5)

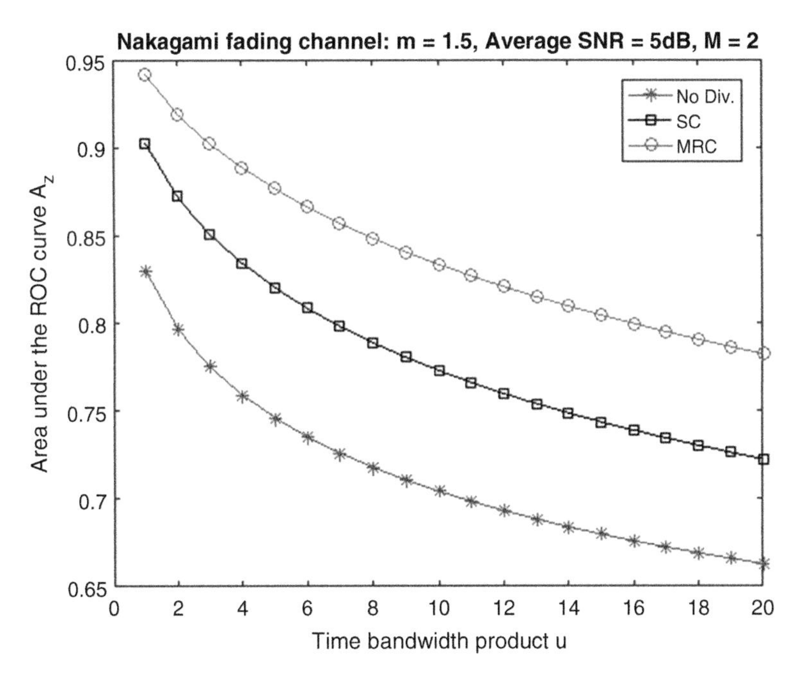

Fig. 7.43 AUC in a Nakagami fading channel: Improvement through diversity (m = 1.5, Average SNR = 5 dB, M = 2)

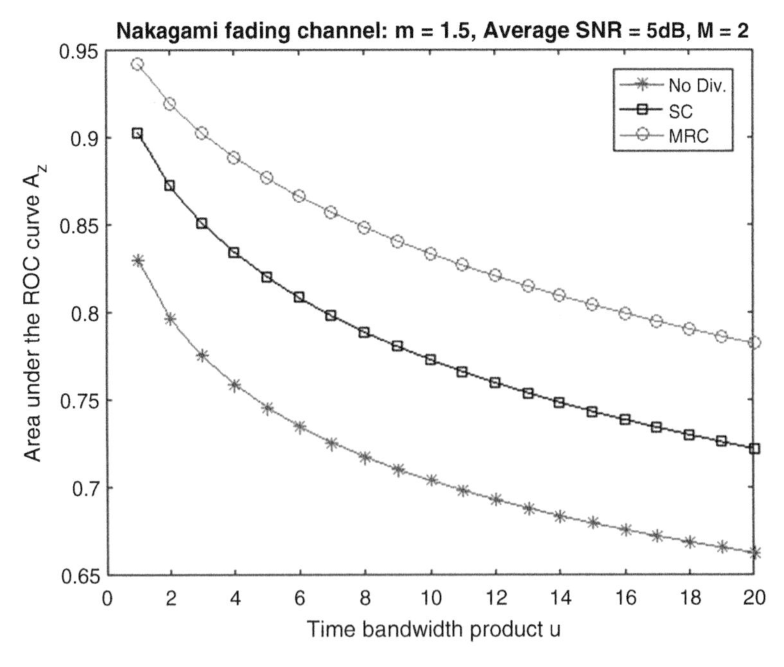

Fig. 7.44 AUC in a Nakagami fading channel: Improvement through diversity (m = 1.5, Average SNR = 5 dB, M = 2)

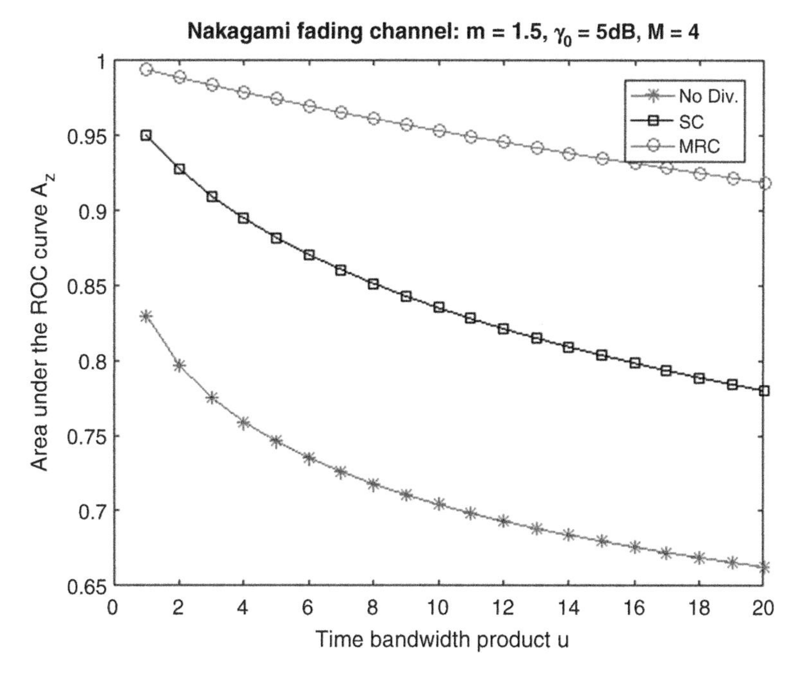

Fig. 7.45 AUC in a Nakagami fading channel: Improvement through diversity (m = 1.5, $\gamma_0 = 5$ dB, M = 4)

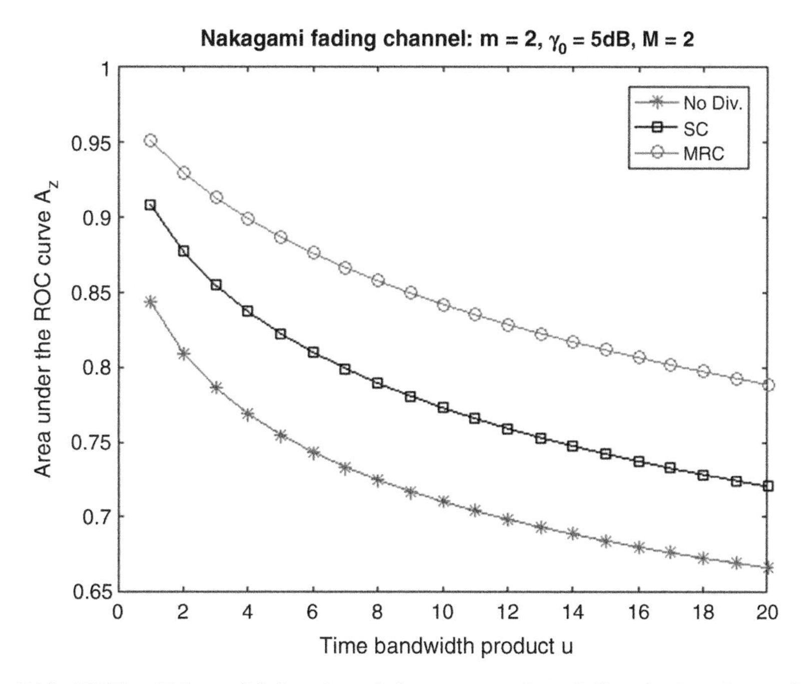

Fig. 7.46 AUC in a Nakagami fading channel: Improvement through diversity (m = 2, $\gamma_0 = 5$ dB, M = 2)

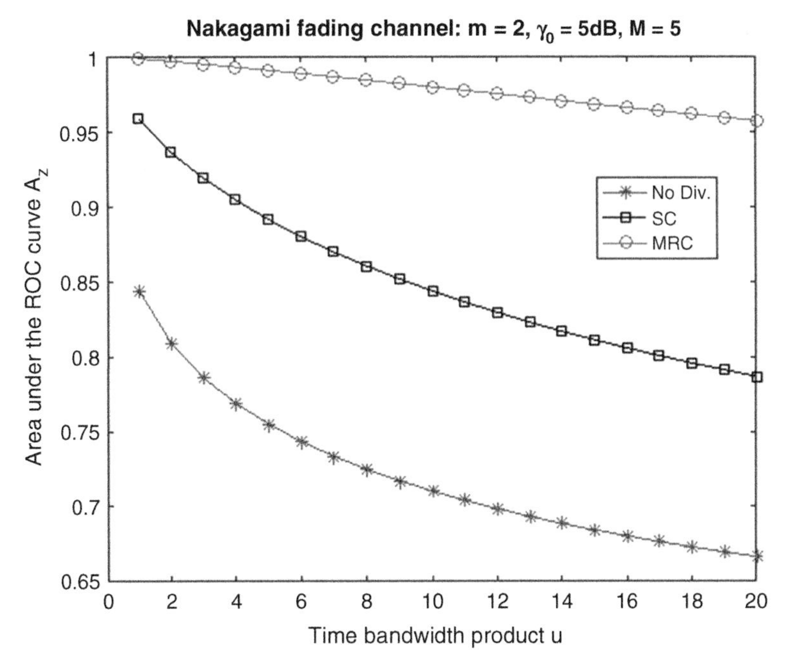

Fig. 7.47 AUC in a Nakagami fading channel: Improvement through diversity (m = 2, $\gamma_0 = 5$ dB, M = 5)

whether primary user is present or not. Therefore, the energy received under the two hypotheses now becomes

$$Z = \begin{cases} \chi^2_{2\text{Mu}} & H_0 \\ \chi^2_{2\text{Mu}}(2\gamma_{\text{SLC}}) & H_1 \end{cases}. \tag{7.54}$$

It can be seen that the chi-square variable is now of order Mu instead of u as it was with the case of MRC and SC. The probability of false alarm now becomes

$$P_f = \frac{\Gamma\left(\text{Mu}, \frac{\lambda}{2}\right)}{\Gamma(\text{Mu})}. \tag{7.55}$$

The probability of detection becomes

$$P_{d|\gamma_{\text{SLC}}} = Q_{\text{Mu}}\left(\sqrt{2\gamma_{\text{SLC}}}, \sqrt{\lambda}\right) \tag{7.56}$$

The density function of the SNR is similar to the pdf of the SNR at the output of the MRC algorithm and it is given by

$$f(\gamma_{\mathrm{SLC}}) = \frac{1}{\Gamma(mM)}\left(\frac{m}{\gamma_0}\right)^{mM} \gamma_{\mathrm{SLC}}^{mM-1} e^{-\frac{m}{\gamma_0}\gamma_{\mathrm{SLC}}}, \quad M = 1, 2, 3, .. \quad \text{(SLC Diversity)} \tag{7.57}$$

Since Eq. (7.56) is conditioned on the output of the SLC algorithm, the probability of detection becomes

$$\begin{aligned} P_d &= \int_0^\infty P_{d|\gamma_{\mathrm{SLC}}} f(\gamma_{\mathrm{SLC}}) d\gamma_{\mathrm{SLC}} \\ &= \int_0^\infty Q_{\mathrm{Mu}}\left(\sqrt{2\gamma_{\mathrm{SLC}}}, \sqrt{\lambda}\right) \frac{1}{\Gamma(mM)}\left(\frac{m}{\gamma_0}\right)^{mM} \gamma_{\mathrm{SLC}}^{mM-1} e^{-\frac{m}{\gamma_0}\gamma_{\mathrm{SLC}}} d\gamma_{\mathrm{SLC}}. \end{aligned} \tag{7.58}$$

The Matlab script used for the comparison is given below.

```
function ROC_Nakagami_diversity_comparison
% ROC in a Nakagami channel.
% MRC, SC and SLC are compared. NO diversity is also considered.
% P M Shankar, Sept. 2016
close all
global m lamb Z u M
PF=valuesofpf; % get the values of PF
LF=length(PF); % count the PF
uu=1:10;  KU=length(uu); % values of u and their count
lam=mylamb(uu); % get values of lambda for this set of u
PF1=[0,PF,1]; % padding is needed so that the ends are not missed.
% study of Az as a function of the Nakagami parameter
for kz=1:2
    if kz==1
        ZdB=3;
    else
        ZdB=5;
    end;
    Z=10^(ZdB/10);
            m=1.5; M=3;
        for k=1:KU
            u=uu(k);
            for kk=1:LF
                lamb=lam(kk,k);
                pdF(kk,k)=integral(@myfun1,0,inf);
                pdFSC(kk,k)=integral(@myfunSC,0,inf);
                pdFMR(kk,k)=integral(@myfunMRC,0,inf);
                 PFmt(kk,k)=1-chi2cdf(lamb,2*u*M); % theory SLC
                 pdFMt(kk,k)=integral(@myfun2,0,inf);
            end;

            Az(k)=polyarea(PF1',[0;pdF(:,k);1])+0.5;
            AzS(k)=polyarea(PF1',[0;pdFSC(:,k);1])+0.5;
            AzM(k)=polyarea(PF1',[0;pdFMR(:,k);1])+0.5;
            AzML(k)=polyarea([0;PFmt(:,k);1],[0;pdFMt(:,k);1])+0.5;
        end
```

```
        figure,plot(uu,Az,'r-*',uu,AzS,'k-s',uu,AzM,'m-o',uu,AzML,'b-d')
        legend('No Div.','SC','MRC','SLC')
        title(['Nakagami fading channel: m = ', num2str(m),', \gamma_0 = ',...
            num2str(ZdB),'dB, M = ',num2str(M)])
        xlabel('Time bandwidth product u')
        ylabel('Area under the ROC curve A_z')

end;

end

function PF1=valuesofpf
PF1=[1e-12 5e-12 1e-11 5e-11 1e-10 1e-9,...
    1e-8 1e-7 1e-6 1e-5 1e-4...
    .001 .01 0.03 0.05 0.08 .1 0.14 0.16 .2 0.25 .3 0.35 .4 ...
    .5 0.55 .6 0.65 .7 .75 .8 .85 .9 0.91 .92 0.93 .94 0.95 ...
    .96 0.965 0.968 0.97 0.975 0.978 0.98 0.985 .99 0.995 ...
    0.99999 0.999999 0.9999999];
end

function Fg=myfun1(x) % Nakagami channel No diversity
global m lamb Z u
pd=1-ncx2cdf(lamb,2*u,2*x);% use the CDF
Fg=gampdf(x,m,Z/m).*pd;
end

% the integral for PD is same for SLC and MRC
% only PF is different
function Fg=myfunMRC(x) % Nakagami channel MRC diversity
global m lamb Z u M
pd=1-ncx2cdf(lamb,2*u,2*x);% use the CDF
Fg=gampdf(x,m*M,Z/m).*pd; % note that M*Z/(M*m) is same as Z/m
end


function Fg=myfun2(x)  % SLC: integral function
global m lamb Z u M
pd=1-ncx2cdf(lamb,2*u*M,2*x);%1- non-central chiSquare CDF instead of MarcumQ
Fg=gampdf(x,M*m,Z/m).*pd;
end


function Fg=myfunSC(x) % Nakagami channel SCC diversity
global m lamb Z u M
pd=1-ncx2cdf(lamb,2*u,2*x);% use the CDF
pdfSC=M*gampdf(x,m,Z/m).*(gamcdf(x,m,Z/m)).^(M-1);
Fg=pdfSC.*pd; % note that M*Z/(M*m) is same as Z/m
end


function lambda=mylamb(U) % get values of lambda
PF=valuesofpf; %values of PF
LF=length(PF);  KU=length(U);
lambda=zeros(LF,KU);
for ku=1:KU
    lambda(:,ku)=2*gammaincinv(PF,U(ku),'upper');%invert PF & get threshold
end;

end
```

Figures 7.48 and 7.49 compare the AUC in Nakagami channel for SC, MRC, and SLC diversity. The performance of SLC lies between SC and MRC.

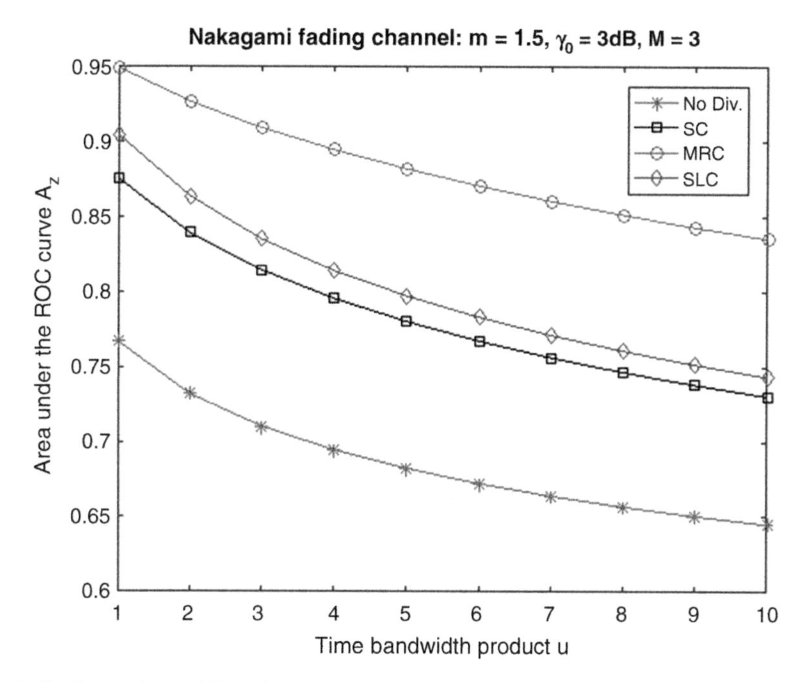

Fig. 7.48 Comparison of diversity schemes in cognitive radio ($m = 1.5$, $\gamma_0 = 3$ dB, $M = 3$)

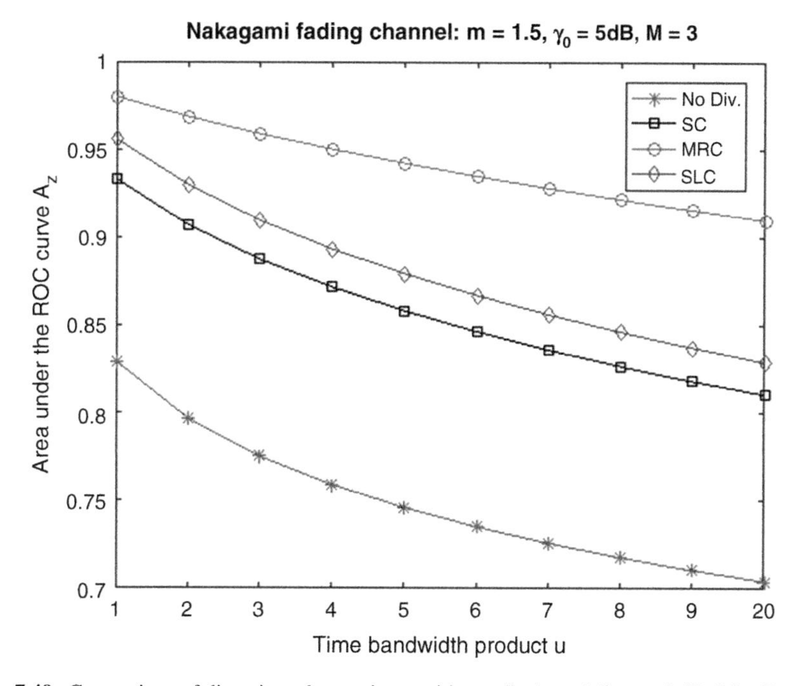

Fig. 7.49 Comparison of diversity schemes in cognitive radio ($m = 1.5$, $\gamma_0 = 5$ dB, $M = 3$)

7.4 Random Number Simulation

Having seen the results obtained analytically, we can now examine how random number simulation of cognitive radio matches those results. Matlab scripts and results are provided spanning the performance in ideal and fading channels described in previous sections. The density functions corresponding to the two hypotheses in Eq. (7.4) in an ideal Gaussian channel can be expressed in terms of Matlab commands as

$$f(z) = \begin{cases} cx2pdf(z, 2u) & H_0 \\ ncx2pdf(z, 2u, 2\gamma) & H_1 \end{cases}. \tag{7.59}$$

The probabilities of false alarm and detection can be expressed as

$$\begin{aligned} P_f &= 1 - cx2\mathrm{cdf}(\lambda, 2u) \\ P_d &= 1 - ncx2cdf(\lambda, 2u, 2\gamma). \end{aligned} \tag{7.60}$$

It is important to mention that Eq. (7.60) for the probability of detection avoids the need for Marcum Q function. For random number simulation, the number (N) of samples used was fixed at 1e6. While the generation of samples in an ideal channel is relatively easy and straightforward, generation of samples in fading and shadowed fading channels required some extra manipulations. Table 7.1 provides the Matlab commands for the generation of the various samples. For the case of Nakagami-lognormal shadowed fading channels, we must be aware of the fact that the average SNR in a showed fading channel μ is in decibel units while SNR

Table 7.1 Random number simulation of ideal, fading, and shadowed fading channels

Channel description	H_0	H_1
Ideal (no fading)	chi2rand(2*u,1,N)	ncx2rnd(2*u,2*γ_0,1,N)
Nakagami (m,γ_0)	chi2rand(2*u,1,N)	h = (gamrnd(m,γ_0/m ,1,N))
		ncx2rnd(2*u,2*h,1,N)
Nakagami (m,γ_0), MRC(M)	chi2rand(2*u,1,N)	h = sum(gamrnd(m,γ_0/m ,M,N))
$M = 2, 3, \ldots$		ncx2rnd(2*u,2*h,1,N)
Nakagami (m,γ_0), SLC(M)	chi2rand(2*u*M,1,N)	h = sum(gamrnd(m,γ_0/m ,M,N))
$M = 2, 3, \ldots$		ncx2rnd(2*u*M,2*h,1,N)
Nakagami (m,γ_0) , SC(M)	chi2rand(2*u,1,N)	h = max(gamrnd(m,γ_0/m ,M,N))
$M = 2, 3, \ldots$		ncx2rnd(2*u,2*h,1,N)
Shadowed fading (m,μ,σ)	chi2rand(2*u,1,N)	h = normrnd(m, σ,1,N)
		hh = 10.^(h/10)
		hhh = gamrnd(m,1/m,1,N).*hh
		ncx2rnd(2*u,2*hhh,1,N)
Generalized K (m,c,γ_0)	chi2rand(2*u,1,N)	$r1$ = gamrnd(m,1/m,1,N)
		r = $r1$.*gamrnd(c,γ_0/c,1,N)
		ncx2rnd(2*u,2*r,1,N)

N is the number of samples

γ_0 in other channels is in absolute units (even though the plots show the values in dB units). For the case of MRC diversity, results were obtained using both the infinite series (terms limited to 50) and numerical integration with Eq. (7.53) to obtain the probability of detection. For the case of selection diversity and GK channel, *integral*(.) was used to perform the numerical integration. For the case of the shadowed fading channel, *integral2*(.) was used to perform double integration. Additionally, for the shadowed fading channels, the limits were converted to 0 and $\pi/2$ by transforming the variable to *tan*(.). The integrals were not used for finding the area under the ROC curve (AUC). Only the probabilities of detection and false alarm were estimated and the command *polyarea*(.) was used instead to get the AUC.

7.4.1 Ideal Channel and Nakagami Channel (No Diversity)

The performance of the cognitive radio in a Nakagami channel is simulated using random numbers. Figure 7.50 shows the ROC for an ideal channel. Figures 7.51 and 7.52 show the results the cognitive radio operating in a Nakagami channel.

The decline in performance with increasing values of *u* is seen. The results of the simulation match the analytical results.

```
function cognitive_radio_SIM_Nakagami_fading
% Computer simulation of the ROC in cognitive radio operating in a Nakagami
% channel. ROC for an ideal channel and Nakagami channel obtained using
% theory and random number simulation. Instead of MarcumQ function, the
% non-central chi square pdf and CDF available in the Statistis and Machine
% Learning Tool box is used.
% probability of detection is estimated using integral(numerical
% integration) and summation
% P M Shankar September 2016
close all
Numb=1e6;
global m lamb g u
% no fading; ideal channel
Z=5;% dB
g=10.^(Z/10);
uu=[1,5,9];LU=length(uu);
% create labels for the legend
uu1=['u = ',num2str(uu(1))];uu2=['u = ',num2str(uu(2))];
uu3=['u = ',num2str(uu(3))];
L=100; % number of steps for PF
LT=L+1; % actual number of threshold values will be (L+1)
% place holders
PFs=zeros(LU,LT); PDs=zeros(LU,LT); % simulation
PFt=zeros(LU,LT); PDt=zeros(LU,LT); % theory
Az0s=zeros(1,LT);
Az0t=zeros(1,LT);
for kk=1:LU
    u=uu(kk);
    x1=chi2rnd(2*u,1,Numb);% H0
    x2=ncx2rnd(2*u,2*g,1,Numb); % H1
    xx=[x1,x2];%to determine maximum and minimum of the set for threshold
    min1=min(xx);max1=max(xx);
    step1=(max1-min1)/L;
    thr=min1:step1:max1; % these will be L+1 steps
    for k=1:LT;
```

```
        PFs(kk,k)=sum(x1>thr(k))/Numb; % False alarm sim
        PDs(kk,k)=sum(x2>thr(k))/Numb;
        PFt(kk,k)=1-chi2cdf(thr(k),2*u); % False alarm theory
        PDt(kk,k)=1-ncx2cdf(thr(k),2*u,2*g); % use the CDF
        %this one as well as mrcumQ works
        %    PDt(kk,k)=marcumq(sqrt(2*g),sqrt(thr(k)),u); % theory
    end;
    clear x1 x2 xx
    Az0s(kk)=0.5+polyarea([0,PFs(kk,:),1],[0,PDs(kk,:),1]);
    Az0t(kk)=0.5+polyarea([0,PFt(kk,:),1],[0,PDt(kk,:),1]);
end;
figure, plot(PFs(1,:),PDs(1,:),'r*',PFt(1,:),PDt(1,:),'r-',...
    PFs(2,:),PDs(2,:),'bo',PFt(2,:),PDt(2,:),'b-.',...
    PFs(3,:),PDs(3,:),'ks',PFt(3,:),PDt(3,:),'k--')
legend(['A_z = ',num2str(Az0s(1)),', ', uu1,' (sim)'],...
    ['A_z = ',num2str(Az0t(1)),', ', uu1,' (theory)'],...
    ['A_z = ',num2str(Az0s(2)),', ', uu2,' (sim)'],...
    ['A_z = ',num2str(Az0t(2)),', ', uu2,' (theory)'],...
    ['A_z = ',num2str(Az0s(3)),', ', uu3,' (sim)'],...
    ['A_z = ',num2str(Az0t(3)),', ', uu3,' (theory)'],...
    'location','best')
xlabel('P_f'),ylabel('P_d'),grid on
title(['Ideal Channel: \gamma = ',num2str(Z),' dB'])
hold on
plot([0,1],[0,1],':m')
% simulation of the Nakagami channel
clear Z
mm=[1.3,2.8,4.1];
ZZ=[3,5,8];% average SNR dB
for mk=1:3
    m=mm(mk);

    g=10^(ZZ(mk)/10);
    LU=length(uu);
    % place holders
    PFms=zeros(LU,LT); PDms=zeros(LU,LT); % simulation
    PFmt=zeros(LU,LT);PDmt=zeros(LU,LT); % theory
    PDSmt=zeros(LU,LT);% prob. detection theory as a sum
    Azs=zeros(1,LT);
    Azt=zeros(1,LT);
    AzSt=zeros(1,LT);
    for kk=1:3
        u=uu(kk);
        x1=chi2rnd(2*u,1,Numb); % hypothesis H0
        %hypothesis H1: non-centrality parameter a sample of a gamma variable
        % generate a non-central chi square sample set with a non-centrality
        % parameter = 2* gamma random variable of order m and mean g and it needs
        % to match the size
        x2=ncx2rnd(2*u,2*gamrnd(m,g/m,1,Numb),1,Numb); % H1
        xx=[x1,x2];%to determine the maximum and minimum of the set for threshold
        min1=min(xx);max1=max(xx);
        step1=(max1-min1)/L;
        thr=min1:step1:max1; % these will be L+1 steps
        for k=1:LT;
            lamb=thr(k); % lambda
            PFms(kk,k)=sum(x1>lamb)/Numb; % sim
            PDms(kk,k)=sum(x2>lamb)/Numb; % sim
            PDmt(kk,k)=integral(@myfun,0,inf);% theory
            PFmt(kk,k)=1-chi2cdf(thr(k),2*u); % theory
            PDSmt(kk,k)=probdetnakagami(50); %Pd as a summation in a Nakagami channel
        end;
```

```
        Azs(kk)=0.5+polyarea([0,PFms(kk,:),1],[0,PDms(kk,:),1]);
        Azt(kk)=0.5+polyarea([0,PFmt(kk,:),1],[0,PDmt(kk,:),1]);
        AzSt(kk)=0.5+polyarea([0,PFmt(kk,:),1],[0,PDSmt(kk,:),1]);
    end;
    figure,plot(PFms(1,:),PDms(1,:),'r*',PFmt(1,:),PDmt(1,:),'r-',...
        PFmt(1,:),PDSmt(1,:),'r+',...
        PFms(2,:),PDms(2,:),'mo',PFmt(2,:),PDmt(2,:),'m--',...
        PFmt(2,:),PDSmt(2,:),'m>',...
        PFms(3,:),PDms(3,:),'ks',PFmt(3,:),PDmt(3,:),'k-.',...
        PFmt(3,:),PDSmt(3,:),'kd')
    legend(['A_z = ',num2str(Azs(1)),', ', uu1,' (sim)'],...
        ['A_z = ',num2str(Azt(1)),', ', uu1,' (theory)'],...
        ['A_z = ',num2str(AzSt(1)),', ', uu1,' (theory-SUM)'],...
        ['A_z = ',num2str(Azs(2)),', ', uu2,' (sim)'],...
        ['A_z = ',num2str(Azt(2)),', ', uu2,' (theory)'],...
        ['A_z = ',num2str(AzSt(2)),', ', uu2,' (theory-SUM)'],...
        ['A_z = ',num2str(Azs(3)),', ', uu3,' (sim)'],...
        ['A_z = ',num2str(Azt(3)),', ', uu3,' (theory)'],...
        ['A_z = ',num2str(AzSt(3)),', ', uu3,' (theory-SUM)'],...
        'location','best')
    xlabel('P_f'),ylabel('P_d'),grid on
    title(['Nakagami Channel: m = ',num2str(m),', \gamma_0 = ',...
        num2str(ZZ(mk)),' dB'])
    hold on
    plot([0,1],[0,1],':m')
    clear x1 x2 xx
end;
end


function Fg=myfun(x)  % Nakagami channel: integral function
global m lamb g u
pd=1-ncx2cdf(lamb,2*u,2*x);%1- non-central chiSquare CDF instead of MarcumQ
%Fg=gampdf(x,m,g/m).*marcumq(sqrt(2*x),sqrt(lamb),u);
Fg=gampdf(x,m,g/m).*pd;
end
function Pdet=probdetnakagami(KN) %Pd as a summation in a Nakagami channel
global m  u lamb g
pdd=0;
for kks=1:KN
    ks=kks-1;
    pd1=gammainc(lamb/2,u+ks,'upper')/factorial(ks);
    % Matlab description of gammainc(.) includes the division by
    % gamma(u+ks)
    pd2=gamma(ks+m)*(g^ks)/(g+m)^(ks+m);
    pdd=pdd+pd1*pd2;
end;
Pdet=pdd*(m^m)/gamma(m);
end
```

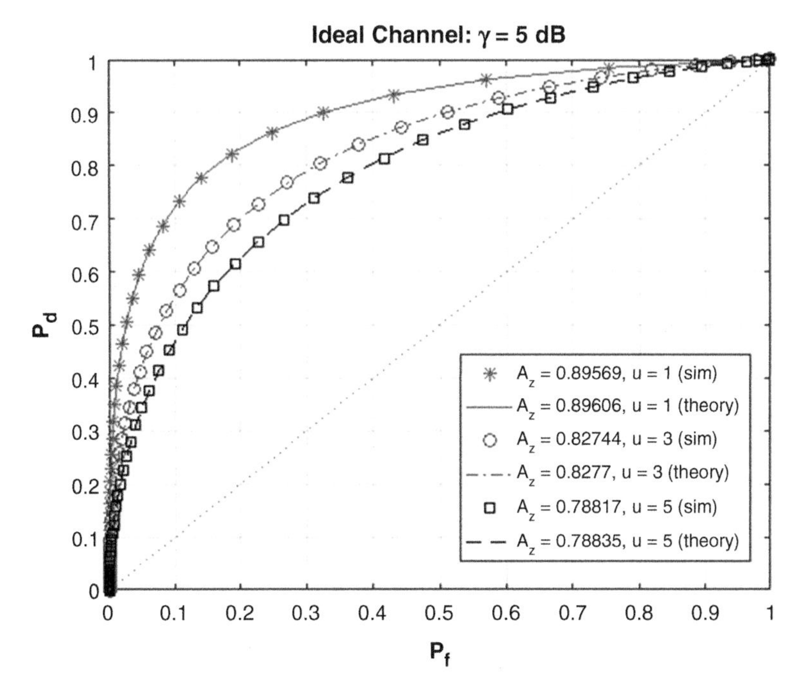

Fig. 7.50 AUC in an ideal channel ($\gamma = 5$ dB)

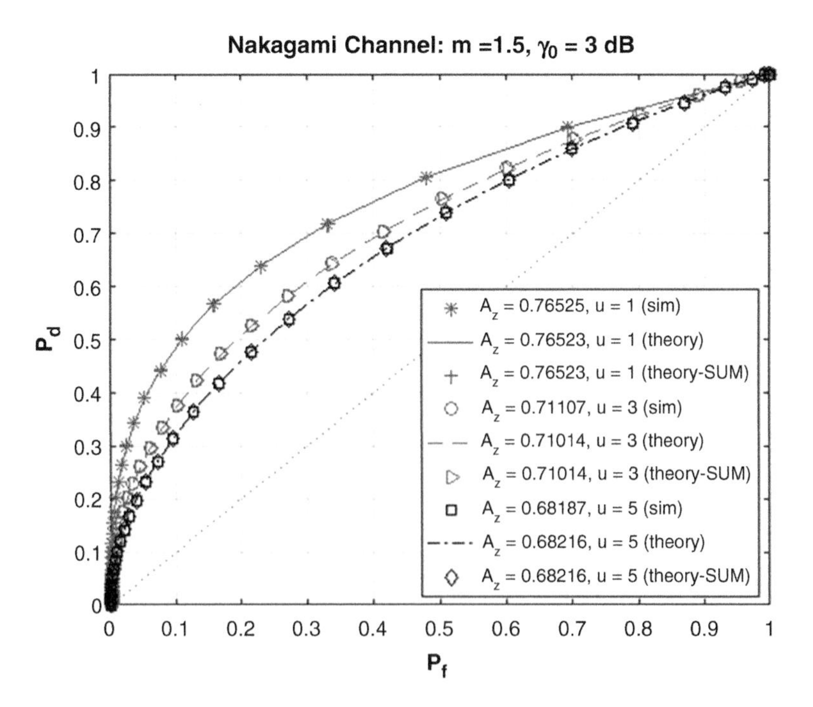

Fig. 7.51 AUC in a Nakagami channel ($m = 1.5$, $\gamma_0 = 3$ dB)

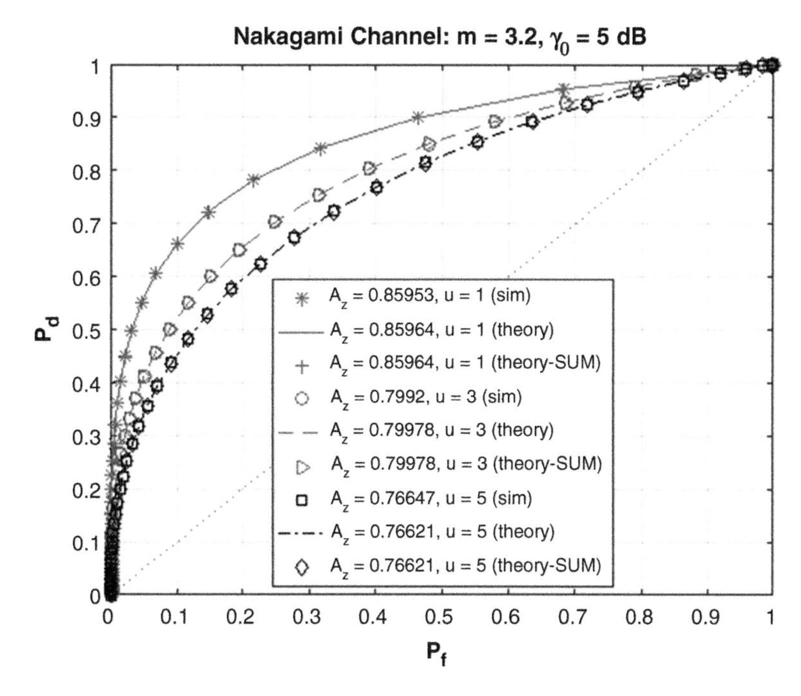

Fig. 7.52 AUC in a Nakagami channel (m = 3.2, $\gamma_0 = 5$ dB)

7.4.2 *Generalized* K *Channel (Shadowed Fading)*

Next, the performance is simulated in a generalized *K* channel. Note that in this case, higher values of *c* correspond to lower levels of shadowing. The Matlab script appears below.

```
function cognitive_radio_SIM_GKchannel
% Computer simulation of the ROC in cognitive radio operating in a GK
% channel.
% ROC for an ideal channel and GK channel obtained using
% theory and random number simulation.
% Instead of MarcumQ function, the
% non-central chi square pdf and CDF available in the Statistis and Machine
% Learning Tool box is used.
% P M Shankar November 2016
close all
Numb=1e6; % number of samples
global m lamb g u c
uu=[1,5,9];LU=length(uu);
% create labels for the legend
uu1=['u = ',num2str(uu(1))];uu2=['u = ',num2str(uu(2))];
uu3=['u = ',num2str(uu(3))];
L=150; % number of steps for PF
LT=L+1; % actual number of threshold values will be (L+1)
% simulate the ideal channel
```

```
Z=5 ;% dB
g=10^(Z/10);
% place holders
PFs=zeros(LU,LT); PDs=zeros(LU,LT); % simulation
PFt=zeros(LU,LT); PDt=zeros(LU,LT); % theory
Az0s=zeros(1,LT);
Az0t=zeros(1,LT);
for kk=1:LU
    u=uu(kk);
    x1=chi2rnd(2*u,1,Numb);% H0
    x2=ncx2rnd(2*u,2*g,1,Numb); % H1
    xx=[x1,x2];%to determine maximum and minimum of the set for threshold
    min1=min(xx);max1=max(xx);
    step1=(max1-min1)/L;
    thr=min1:step1:max1; % these will be L+1 steps
    for k=1:LT;
        PFs(kk,k)=sum(x1>thr(k))/Numb; % False alarm sim
        PDs(kk,k)=sum(x2>thr(k))/Numb;
        PFt(kk,k)=1-chi2cdf(thr(k),2*u); % False alarm theory
        PDt(kk,k)=1-ncx2cdf(thr(k),2*u,2*g); % use the CDF
         %this one as well as mrcumQ works
    %    PDt(kk,k)=marcumq(sqrt(2*g),sqrt(thr(k)),u); % theory
    end;
    clear x1 x2 xx
     Az0s(kk)=0.5+polyarea([0,PFs(kk,:),1],[0,PDs(kk,:),1]);
    Az0t(kk)=0.5+polyarea([0,PFt(kk,:),1],[0,PDt(kk,:),1]);
end;
figure, plot(PFs(1,:),PDs(1,:),'r*',PFt(1,:),PDt(1,:),'b-',...
    PFs(2,:),PDs(2,:),'bo',PFt(2,:),PDt(2,:),'b-.',...
    PFs(3,:),PDs(3,:),'ks',PFt(3,:),PDt(3,:),'k--')

legend(['A_z = ',num2str(Az0s(1)),', ', uu1,' (sim)'],...
    ['A_z = ',num2str(Az0t(1)),', ', uu1,' (theory)'],...
    ['A_z = ',num2str(Az0s(2)),', ', uu2,' (sim)'],...
    ['A_z = ',num2str(Az0t(2)),', ', uu2,' (theory)'],...
  ['A_z = ',num2str(Az0s(3)),', ', uu3,' (sim)'],...
    ['A_z = ',num2str(Az0t(3)),', ', uu3,' (theory)'],...
    'location','best')
xlabel('P_f'),ylabel('P_d'),grid on
title(['Ideal Channel: \gamma = ',num2str(Z),' dB'])
hold on
plot([0,1],[0,1],':m')

% simulation of the GK channel
for kQ=1:6
    if kQ==1
m=1.5;c=2.2; Z=0;
    elseif kQ==2
        m=1.5;c=8.2;Z=5;
    elseif kQ==3
        m=2;c=5.2;Z=8;
    elseif kQ==4
         m=2.5;c=5.2;Z=3;
         elseif kQ==5
         m=2.5;c=15.2;Z=8;
    else
        m=3.1;c=12.2;Z=10;
    end;
```

```
  g=10^(Z/10);
% place holders
PFms=zeros(LU,LT); PDms=zeros(LU,LT); % simulation
PFmt=zeros(LU,LT);PDmt=zeros(LU,LT); % theory
Azs=zeros(1,LT);
Azt=zeros(1,LT);
for kk=1:3
    u=uu(kk);
        clear x1 x2 xx
    x1=chi2rnd(2*u,1,Numb); % hypothesis H0
    %hypothesis H1: non-centrality parameter a sample of a gamma variable
% generate a non-central chi square sample set with a non-centrality
% parameter = 2* gamma random variable of order m and mean g and it needs
% to match the size
x2=ncx2rnd(2*u,2*gamrnd(m,1/m,1,Numb).*gamrnd(c,g/c,1,Numb),1,Numb); % H1
    xx=[x1,x2];%to determine the maximum and minimum of the set for threshold
    min1=min(xx);max1=max(xx);
    step1=(max1-min1)/L;
    thr=min1:step1:max1; % these will be L+1 steps
    for k=1:LT;
        lamb=thr(k); % lambda
        PFms(kk,k)=sum(x1>lamb)/Numb; % sim
        PDms(kk,k)=sum(x2>lamb)/Numb; % sim
        PDmt(kk,k)=integral(@myfunGK,0,inf);% theory
        PFmt(kk,k)=1-chi2cdf(thr(k),2*u); % theory
    end;
    Azs(kk)=0.5+polyarea([0,PFms(kk,:),1],[0,PDms(kk,:),1]);
    Azt(kk)=0.5+polyarea([0,PFmt(kk,:),1],[0,PDmt(kk,:),1]);
end;
figure,plot(PFms(1,:),PDms(1,:),'r*',PFmt(1,:),PDmt(1,:),'r-',...
    PFms(2,:),PDms(2,:),'ko',PFmt(2,:),PDmt(2,:),'k--',...
    PFms(3,:),PDms(3,:),'bs',PFmt(3,:),PDmt(3,:),'b-.')
legend(['A_z = ',num2str(Azs(1)),', ', uu1,' (sim)'],...
    ['A_z = ',num2str(Azt(1)),', ', uu1,' (theory)'],...
    ['A_z = ',num2str(Azs(2)),', ', uu2,' (sim)'],...
    ['A_z = ',num2str(Azt(2)),', ', uu2,' (theory)'],...
  ['A_z = ',num2str(Azs(3)),', ', uu3,' (sim)'],...
    ['A_z = ',num2str(Azt(3)),', ', uu3,' (theory)'],...
    'location','best')
xlabel('P_f'),ylabel('P_d'),grid on
title(['GK Channel: m = ',num2str(m),', c = ',num2str(c),...
    ',  \gamma_0 = ',num2str(Z),' dB'])
hold on
plot([0,1],[0,1],':m')
end;

end
function Fg=myfunGK(x) % external function for the integral
% GK channel
global m lamb g u c
pd=2*((sqrt(c/g)*sqrt(m*x)).^(c+m)).*...
    besselk(c-m, 2*sqrt(c/g)*sqrt(m*x))*1./(x*gamma(m)*gamma(c));
%1- non-central chiSquare CDF instead of MarcumQ
Fg=pd.*(1-ncx2cdf(lamb,2*u,2*x));
end
```

Results are displayed in Figs. 7.53, 7.54, and 7.55.

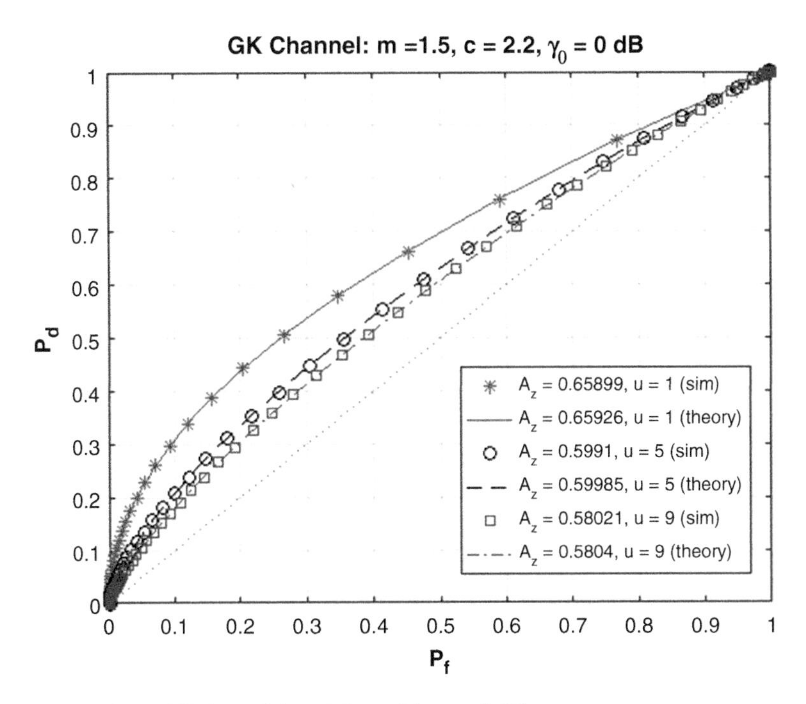

Fig. 7.53 AUC in a GK channel (m = 1.5, c = 2.2, $\gamma_0 = 0$ dB)

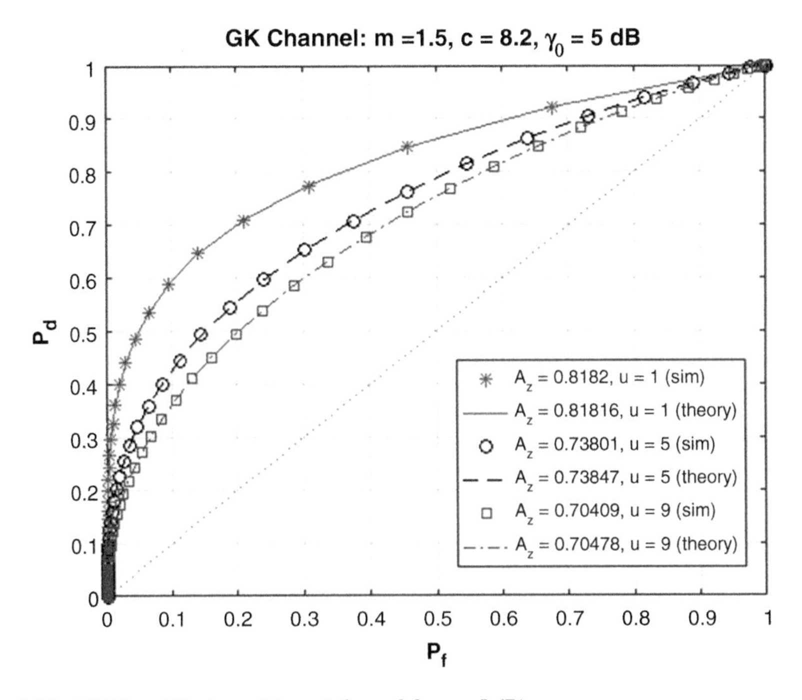

Fig. 7.54 AUC in a GK channel (m = 1.5, c = 8.2, $\gamma_0 = 5$ dB)

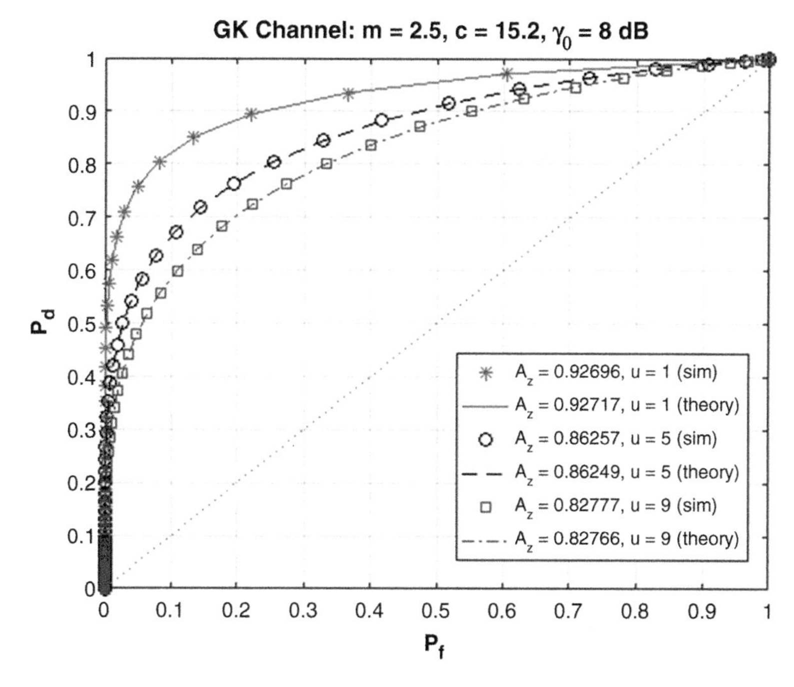

Fig. 7.55 AUC in a GK channel (m = 2.5, c = 15.2, γ_0 = 8 dB)

7.4.3 Nakagami-Lognormal Channel (Shadowed Fading)

The performance in a Nakagami-lognormal channel is simulated. The Matlab script appears below.

```
function cognitive_radio_SIM_shadowedfading
% Computer simulation of the ROC in cognitive radio operating in a Nakagami
% lognormalchannel. ROC for an ideal and shadowed fading channels obtained
% using theory and random number simulation. Instead of MarcumQ function,
% the non-central chi square pdf and CDF available in the Statistis and
% Machine Learning Tool box is used.
% P M Shankar September 2016
close all
Numb=1e6;
global m lamb g u sigm
% no fading; ideal channel
Z=5;% dB
g=10.^(Z/10);
uu=[1,5,9];
% create labels for the legend
uu1=['u = ',num2str(uu(1))];uu2=['u = ',num2str(uu(2))];
uu3=['u = ',num2str(uu(3))];
PF=[1e-12 5e-12 1e-11 5e-11 1e-10 1e-9,...
    1e-8 1e-7 1e-6 1e-5 1e-4 0.001 0.004 0.008...
    .009 .01 0.02 0.03 0.04 0.05 0.08 .1 0.14 0.16 .2 0.25 .3 0.35 .4 ...
    .5 0.55 .6 0.65 .7 .75 .8 .85 .9 0.91 .92 0.93 .94 0.95 ...
    .96 0.965 0.968 0.97 0.975 0.978 0.98 0.985 .99 0.995 ...
    0.99999 0.999999 0.9999999];

LT=length(PF);
LU=length(uu);
% place holders
PFs=zeros(LU,LT); PDs=zeros(LU,LT); % simulation
PFt=zeros(LU,LT); PDt=zeros(LU,LT); % theory
Az0s=zeros(1,LT);Az0t=zeros(1,LT);
for kk=1:LU
    u=uu(kk);
    x1=chi2rnd(2*u,1,Numb);% H0
    x2=ncx2rnd(2*u,2*g,1,Numb); % H1
    thr=2*gammaincinv(PF,u,'upper'); % get the threshold values
    for k=1:LT;
        PFs(kk,k)=sum(x1>thr(k))/Numb; % False alarm sim
        PDs(kk,k)=sum(x2>thr(k))/Numb;
        PFt(kk,k)=1-chi2cdf(thr(k),2*u); % False alarm theory
        PDt(kk,k)=1-ncx2cdf(thr(k),2*u,2*g); % Prob. det: use the CDF
    end;
    clear x1 x2 xx
    Az0s(kk)=0.5+polyarea([0,PFs(kk,:),1],[0,PDs(kk,:),1]);
    Az0t(kk)=0.5+polyarea([0,PFt(kk,:),1],[0,PDt(kk,:),1]);
end;
figure, plot(PFs(1,:),PDs(1,:),'r*',PFt(1,:),PDt(1,:),'b-',...
    PFs(2,:),PDs(2,:),'bo',PFt(2,:),PDt(2,:),'b-.',...
    PFs(3,:),PDs(3,:),'ks',PFt(3,:),PDt(3,:),'k--')
legend(['A_z = ',num2str(Az0s(1)),', ', uu1,' (sim)'],...
    ['A_z = ',num2str(Az0t(1)),', ', uu1,' (theory)'],...
    ['A_z = ',num2str(Az0s(2)),', ', uu2,' (sim)'],...
    ['A_z = ',num2str(Az0t(2)),', ', uu2,' (theory)'],...
    ['A_z = ',num2str(Az0s(3)),', ', uu3,' (sim)'],...
    ['A_z = ',num2str(Az0t(3)),', ', uu3,' (theory)'],...
    'location','best')
xlabel('P_f'),ylabel('P_d'),grid on
title(['Ideal Channel: \gamma = ',num2str(Z),' dB'])
hold on
plot([0,1],[0,1],':m')
% simulation of the Nakagami -lognormal channel
clear Z
mm=[1.1,2,3.5];
ZZ=[3,5,8];% average SNR dB
sd=[2,4,6];% shadowing levels in dB
for kks=1:3
```

```
    sigm=sd(kks); % shadowing sigma
    for mk=1:3
        m=mm(mk);
        g=ZZ(mk); % note that g is in dB here
        LU=length(uu);
        % place holders
        PFms=zeros(LU,LT); PDms=zeros(LU,LT); % simulation
        PFmt=zeros(LU,LT);PDmt=zeros(LU,LT); % theory
        Azs=zeros(1,LT); Azt=zeros(1,LT);
        for kk=1:3
            u=uu(kk);
            x1=chi2rnd(2*u,1,Numb); % hypothesis H0
    %hypothesis H1: non-centrality parameter a sample of a gamma variable
    %generate a non-central chi square sample set with a non-centrality
    %parameter = 2*gammalognormal random number
            thr=2*gammaincinv(PF,u,'upper'); % get  threshold values
            g1=normrnd(g,sigm,1,Numb);%generate Gaussian
            gg=10.^(g1/10);%lognormal random
            hh=gamrnd(m,1/m,1,Numb).*gg;% gamma-lognormal
            x2=ncx2rnd(2*u,2*hh,1,Numb); % H1
            for k=1:LT;
                lamb=thr(k); % lambda
                PFms(kk,k)=sum(x1>lamb)/Numb; % sim
                PDms(kk,k)=sum(x2>lamb)/Numb; % sim
                PDmt(kk,k)=integral2(@myfunL,0,pi/2,0,pi/2);%theory
                PFmt(kk,k)=1-chi2cdf(thr(k),2*u); % theory
            end;
            Azs(kk)=0.5+polyarea([0,PFms(kk,:),1],[0,PDms(kk,:),1]);
            Azt(kk)=0.5+polyarea([0,PFmt(kk,:),1],[0,PDmt(kk,:),1]);
        end;
        figure,plot(PFms(1,:),PDms(1,:),'r*',PFmt(1,:),PDmt(1,:),'r-',...
            PFms(2,:),PDms(2,:),'ko',PFmt(2,:),PDmt(2,:),'k--',...
            PFms(3,:),PDms(3,:),'bs',PFmt(3,:),PDmt(3,:),'b-.')
        legend(['A_z = ',num2str(Azs(1)),', ', uu1,' (sim)'],...
            ['A_z = ',num2str(Azt(1)),', ', uu1,' (theory)'],...
            ['A_z = ',num2str(Azs(2)),', ', uu2,' (sim)'],...
            ['A_z = ',num2str(Azt(2)),', ', uu2,' (theory)'],...
            ['A_z = ',num2str(Azs(3)),', ', uu3,' (sim)'],...
            ['A_z = ',num2str(Azt(3)),', ', uu3,' (theory)'],...
            'location','best')
        xlabel('P_f'),ylabel('P_d'),grid on
        title(['shadowed fading: m = ',num2str(m),', \sigma = ',...
            num2str(sigm),' dB, \gamma_0 = ',num2str(ZZ(mk)),' dB'])
        xlim([0,1]),ylim([0,1])
        hold on
        plot([0,1],[0,1],':m')
        clear x1 x2 xx
    end;
end;
end

function Fg=myfunL(xx,yy)  % Nakagami lognormal channel: integral function
global m lamb g u sigm
x=tan(xx);
y=tan(yy);
pd=1-ncx2cdf(lamb,2*u,2*x);%1- non-central chiSquare CDF instead of MarcumQ
K=10/log(10);
f1=K*1./sqrt(2*pi*y.*y*sigm^2);
f2=exp(-(10*log10(y)-g).^2/(2*sigm^2));
fL=f1.*f2;
pdfL=gampdf(x,m,y/m).*fL;
Fg=pdfL.*pd.*(1+x.^2).*(1+y.^2);
end
```

The results are shown in Figs. 7.56, 7.57, and 7.58.

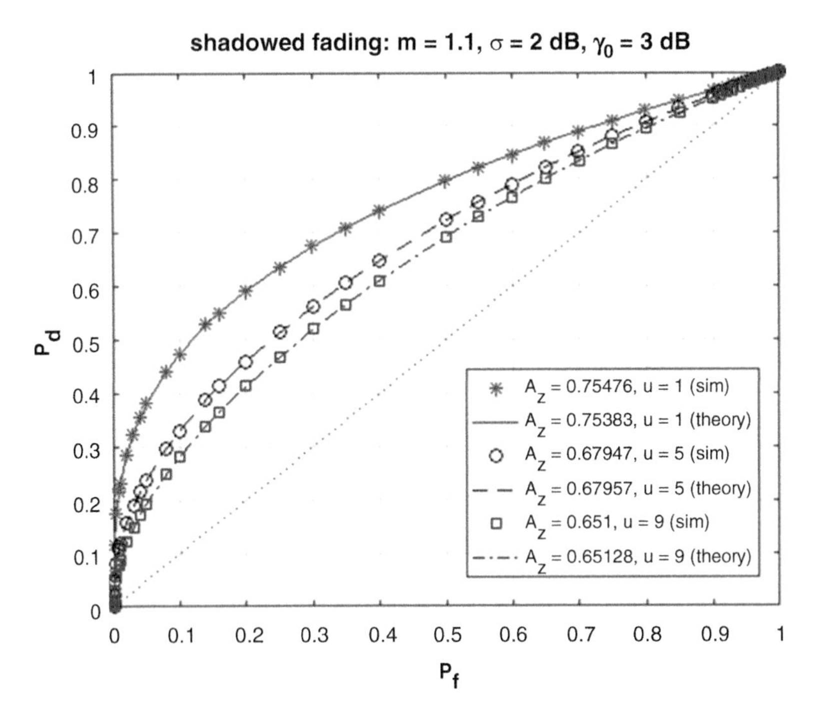

Fig. 7.56 AUC in a Nakagami-lognormal channel (m = 1.1, $\sigma = 2$ dB, $\gamma_0 = 3$ dB)

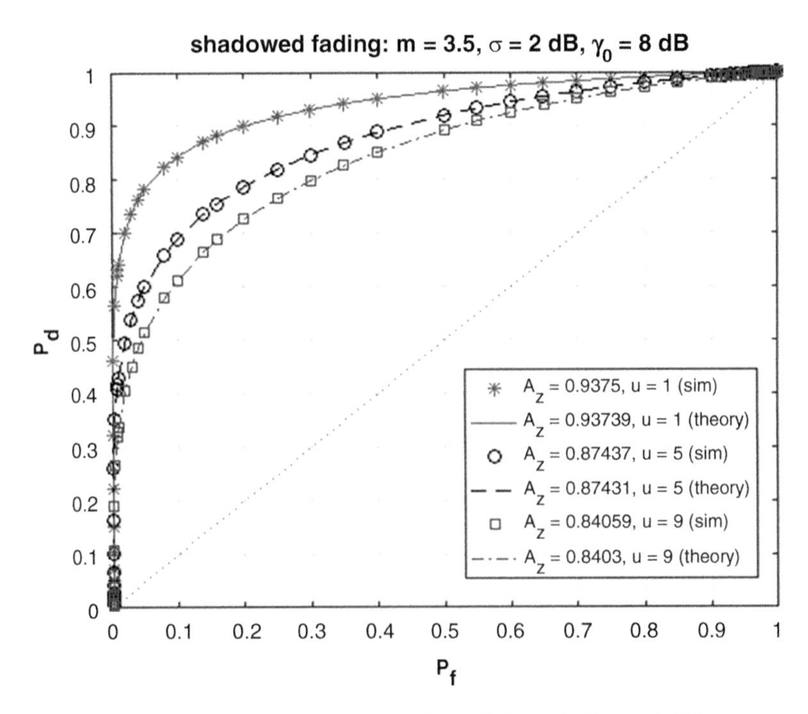

Fig. 7.57 AUC in a Nakagami-lognormal channel (m = 3.5, $\sigma = 2$ dB, $\gamma_0 = 8$ dB)

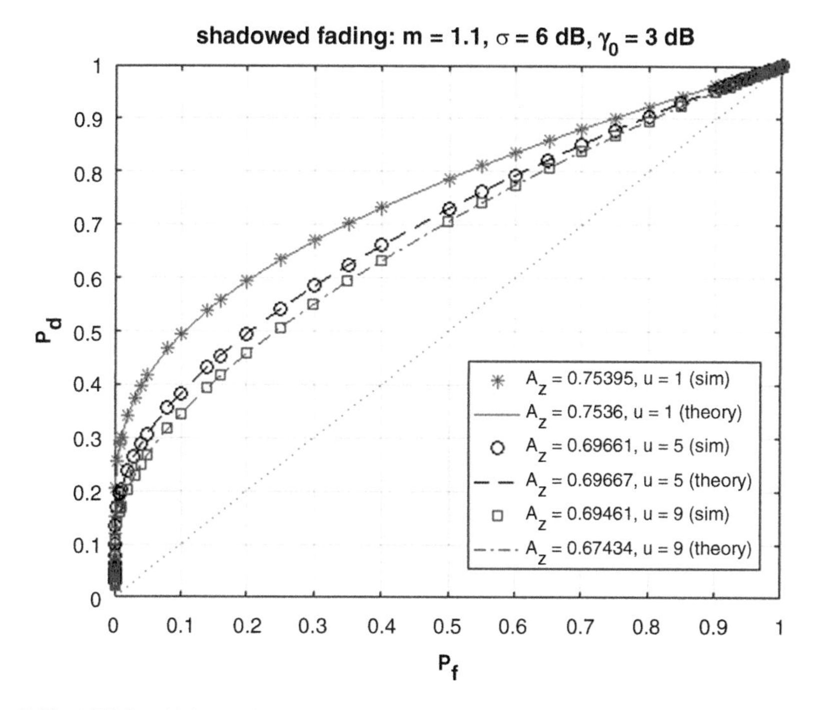

Fig. 7.58 AUC in a Nakagami-lognormal channel ($m = 3.5$, $\sigma = 6$ dB, $\gamma_0 = 3$ dB)

7.4.4 Nakagami Channel (MRC and SC Diversity)

The simulation of cognitive radio when MRC and SC diversity algorithms are implemented is undertaken next. The Matlab script appears below.

```
function cognitive_radio_SIM_Nakagami_channel_div
% Computer simulation of the ROC in cognitive radio operating in a Nakagami
% channel. ROC for an ideal channel and Nakagami channel obtained using
% theory and random number simulation. Instead of MarcumQ function, the
% non-central chi square pdf and CDF available in the Statistis and Machine
% Learning Tool box is used.
% In this simulation, MRC and SC implemented. Note that M =1 corresponds to
% a Nakagami channel (no diversity)
% P M Shankar September 2016
close all
Numb=1e6;
global m lamb g u M
uu=[1,5,9];
L=100; % number of steps for PF
LT=L+1; % actual number of threshold values will be (L+1)
LU=length(uu);
% generate three sets of chi square random numbers: H0
xr=zeros(LU,Numb);
for kk=1:LU
    xr(kk,:)=chi2rnd(2*uu(kk),1,Numb);
end;
% create labels for the legend
uu1=['u = ',num2str(uu(1))];uu2=['u = ',num2str(uu(2))];
uu3=['u = ',num2str(uu(3))];
clear Z
% simulation of the Nakagami channel MRC including M=1 (no diversity)
MM=[1,2,4];% order of diversity
mm=[0.5,1,1.5];
ZZ=[0,3,5];% average SNR dB
for k1=1:3
    M=MM(k1);
    for mk=1:3
        m=mm(mk);
        g=10^(ZZ(mk)/10);
        % place holders
PFms=zeros(LU,LT); PDms=zeros(LU,LT); % MRC simulation
PFmt=zeros(LU,LT);PDmt=zeros(LU,LT); % MRC theory
Azs=zeros(1,LT); Azt=zeros(1,LT);
for kk=1:LU
    u=uu(kk);
    x1=xr(kk,:);% hypothesis H0
    %   x1=chi2rnd(2*u,1,Numb); % hypothesis H0
    %hypothesis H1: non-centrality parameter a sample of a gamma variable
    % generate a non-central chi square sample set with a non-centrality
    % parameter = 2* gamma random variable of order m and mean g and it needs
    % to match the size
    % generate an MRC sample
    if M==1 % summation not needed
        x2m=ncx2rnd(2*u,2*gamrnd(m,g/m,1,Numb),1,Numb); % H1
    else
        x2m=ncx2rnd(2*u,2*sum(gamrnd(m,g/m,M,Numb)),1,Numb); % H1
    end;
    xx=[x1,x2m];%to determine the maximum and minimum of the set for threshold
    min1=min(xx);max1=max(xx);
    step1=(max1-min1)/L;
    thr=min1:step1:max1; % these will be L+1 steps
    for k=1:LT;
        lamb=thr(k); % lambda
        PFms(kk,k)=sum(x1>lamb)/Numb; % sim
        PDms(kk,k)=sum(x2m>lamb)/Numb; % sim
        PDmt(kk,k)=integral(@myfunmrc,0,inf);% theory
        PFmt(kk,k)=1-chi2cdf(thr(k),2*u); % theory
    end;
    Azs(kk)=0.5+polyarea([0,PFms(kk,:),1],[0,PDms(kk,:),1]);
    Azt(kk)=0.5+polyarea([0,PFmt(kk,:),1],[0,PDmt(kk,:),1]);
end;
```

```
        figure,plot(PFms(1,:),PDms(1,:),'r*',PFmt(1,:),PDmt(1,:),'r-',...
            PFms(2,:),PDms(2,:),'ko',PFmt(2,:),PDmt(2,:),'k--',...
            PFms(3,:),PDms(3,:),'bs',PFmt(3,:),PDmt(3,:),'b-.')
        legend(['A_z = ',num2str(Azs(1)),', ', uu1,' (sim)'],...
            ['A_z = ',num2str(Azt(1)),', ', uu1,' (theory)'],...
            ['A_z = ',num2str(Azs(2)),', ', uu2,' (sim)'],...
            ['A_z = ',num2str(Azt(2)),', ', uu2,' (theory)'],...
            ['A_z = ',num2str(Azs(3)),', ', uu3,' (sim)'],...
            ['A_z = ',num2str(Azt(3)),', ', uu3,' (theory)'],...
            'location','best')
        xlabel('P_f'),ylabel('P_d'),grid on
        if M==1
            title(['m = ',num2str(m),', \gamma_0 = ',...
                num2str(ZZ(mk)),' dB: No Diversity'])
        else
            title(['m = ',num2str(m),', \gamma_0 = ',...
                num2str(ZZ(mk)),' dB:  MRC (M = ', num2str(M),')'])
        end;
        hold on
        plot([0,1],[0,1],':m')
        clear x1 x2m xx
    end;
end;
% now SC
for k1=2:3 % M=1 is not needed
    M=MM(k1);
    for mk=1:3
        m=mm(mk);
        g=10^(ZZ(mk)/10);
        % place holders
        PFmsS=zeros(LU,LT); PDmsS=zeros(LU,LT); % simulation
        PFmtS=zeros(LU,LT);PDmtS=zeros(LU,LT); % theory
        AzsS=zeros(1,LT); AztS=zeros(1,LT);
        for kk=1:LU
            u=uu(kk);
      x1=xr(kk,:);% hypothesis H0
      %   x1=chi2rnd(2*u,1,Numb); % hypothesis H0
      %hypothesis H1: non-centrality parameter a sample of a gamma variable
      % generate a non-central chi square sample set with a non-centrality
      % parameter = 2* gamma random variable of order m and mean g and it needs
      % to match the size
      % generate  SC sample
      x2s=ncx2rnd(2*u,2*max(gamrnd(m,g/m,M,Numb)),1,Numb); % H1
      xx=[x1,x2s];%to determine the maximum and minimum of the set for threshold
      min1=min(xx);max1=max(xx);
      step1=(max1-min1)/L;
      thr=min1:step1:max1; % these will be L+1 steps
      for k=1:LT;
          lamb=thr(k); % lambda
          PFmsS(kk,k)=sum(x1>lamb)/Numb; % sim
          PDmsS(kk,k)=sum(x2s>lamb)/Numb; % sim
          PDmtS(kk,k)=integral(@myfunsc,0,inf);% theory
          PFmtS(kk,k)=1-chi2cdf(thr(k),2*u); % theory
      end;
      AzsS(kk)=0.5+polyarea([0,PFmsS(kk,:),1],[0,PDmsS(kk,:),1]);
      AztS(kk)=0.5+polyarea([0,PFmtS(kk,:),1],[0,PDmtS(kk,:),1]);
  end;
figure,plot(PFmsS(1,:),PDmsS(1,:),'r*',PFmtS(1,:),PDmtS(1,:),'r-',...
      PFmsS(2,:),PDmsS(2,:),'ko',PFmtS(2,:),PDmtS(2,:),'k--',...
      PFmsS(3,:),PDmsS(3,:),'bs',PFmtS(3,:),PDmtS(3,:),'b-.')
  legend(['A_z = ',num2str(AzsS(1)),', ', uu1,' (sim)'],...
      ['A_z = ',num2str(AztS(1)),', ', uu1,' (theory)'],...
      ['A_z = ',num2str(AzsS(2)),', ', uu2,' (sim)'],...
      ['A_z = ',num2str(AztS(2)),', ', uu2,' (theory)'],...
      ['A_z = ',num2str(AzsS(3)),', ', uu3,' (sim)'],...
      ['A_z = ',num2str(AztS(3)),', ', uu3,' (theory)'],...
      'location','best')
```

```
        xlabel('P_f'),ylabel('P_d'),grid on
        title(['m = ',num2str(m),', \gamma_0 = ',...
            num2str(ZZ(mk)),' dB:  SC (M = ', num2str(M),')'])
        hold on
        plot([0,1],[0,1],':m')
        clear x1 x2s xx
    end;
end;

end

function Fg=myfunmrc(x)  % MRC: integral function
global m lamb g u M
pd=1-ncx2cdf(lamb,2*u,2*x);%1- non-central chiSquare CDF instead of MarcumQ
Fg=gampdf(x,M*m,g/m).*pd;
end

function Fg=myfunsc(x)  % SC: integral function
global m lamb g u M
pd=1-ncx2cdf(lamb,2*u,2*x);%1- non-central chiSquare CDF instead of MarcumQ
Fg=M*(gamcdf(x,m,g/m).^(M-1)).*gampdf(x,m,g/m).*pd;
end

function F=probdetnakagami
global m KN u lamb g
 pdd=0;
    for kks=1:KN
        ks=kks-1;
        pd1=gammainc(lamb/2,u+ks,'upper')/factorial(ks);
        pd2=gamma(ks+m)*(g^ks)/(g+m)^(ks+m);
        pdd=pdd+pd1*pd2;
    end;
F=pdd*(m^m)/gamma(m);
end
```

The results are displayed in Figs. 7.59, 7.60, 7.61, 7.62, 7.63, 7.64, and 7.65.

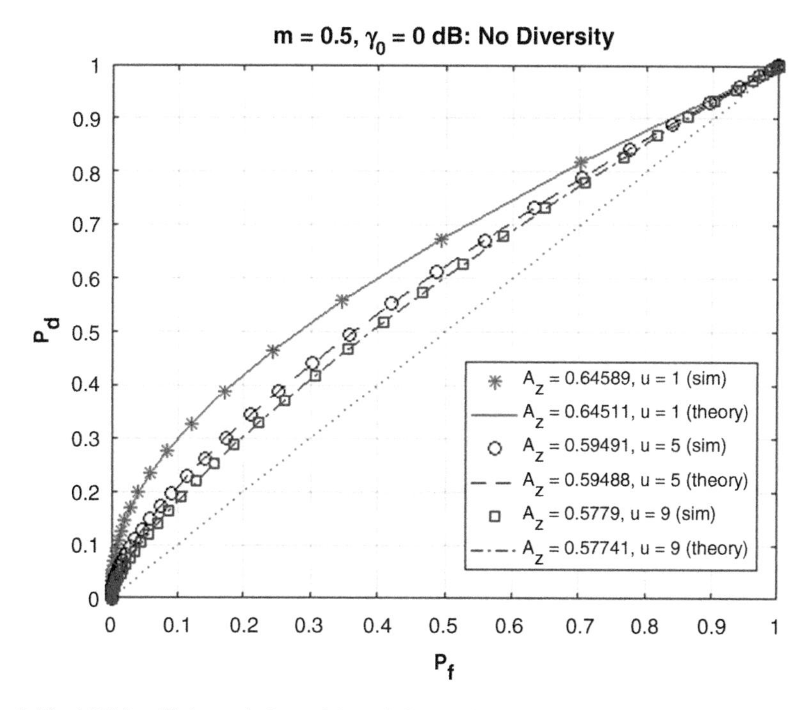

Fig. 7.59 AUC in a Nakagami channel (m = 0.5, γ_0 = 0 dB: No diversity)

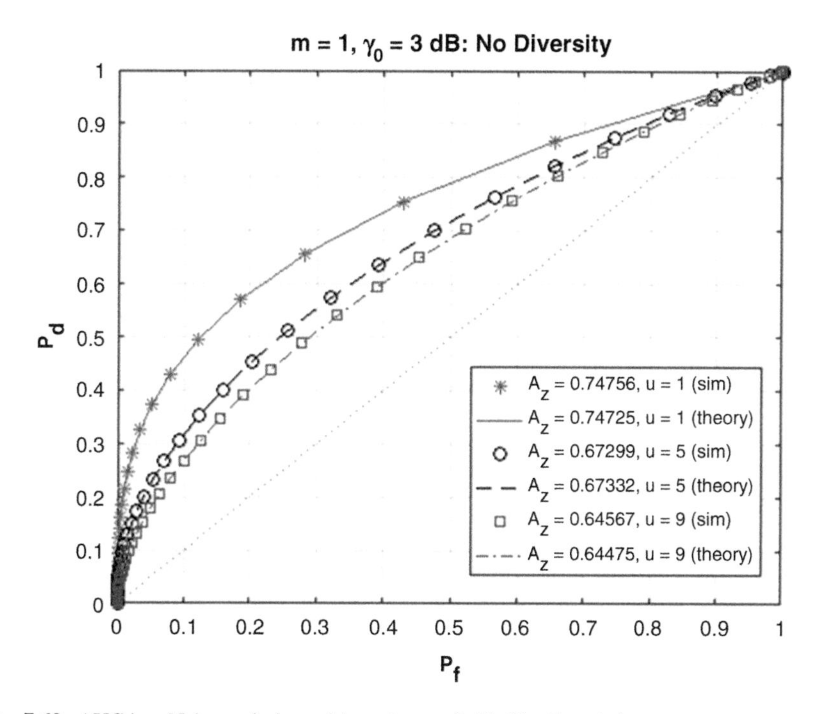

Fig. 7.60 AUC in a Nakagami channel (m = 1, γ_0 = 3 dB: No diversity)

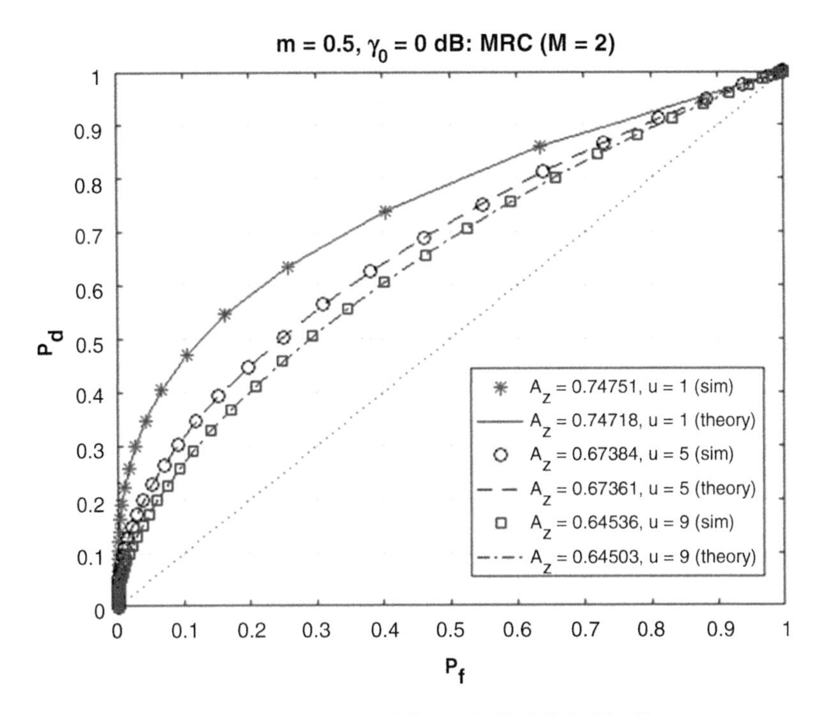

Fig. 7.61 AUC in a Nakagami channel: $m = 0.5$, $\gamma_0 = 0$ dB: MRC ($M = 2$)

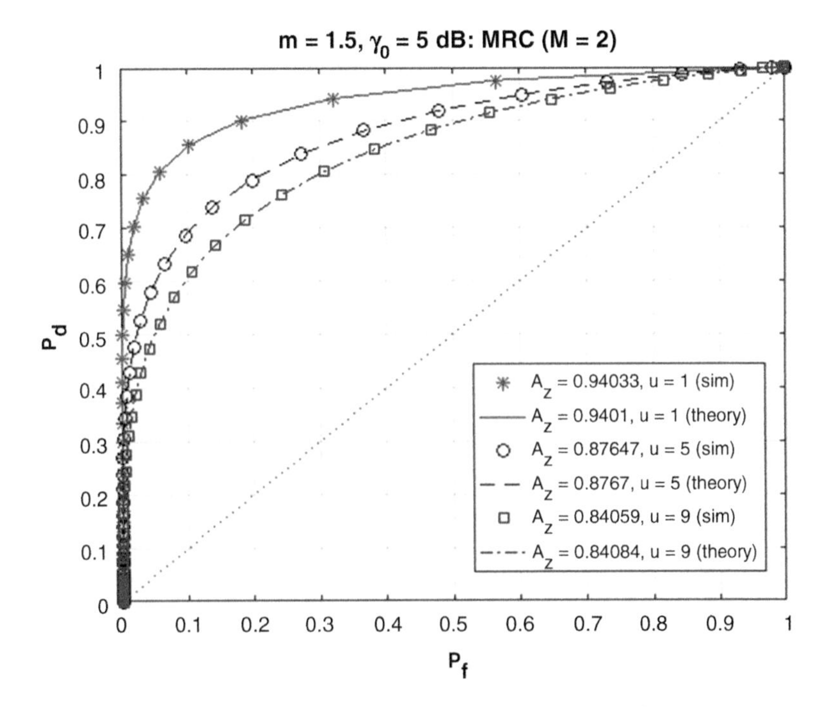

Fig. 7.62 AUC in a Nakagami channel: $m = 1.5$, $\gamma_0 = 5$ dB: MRC ($M = 2$)

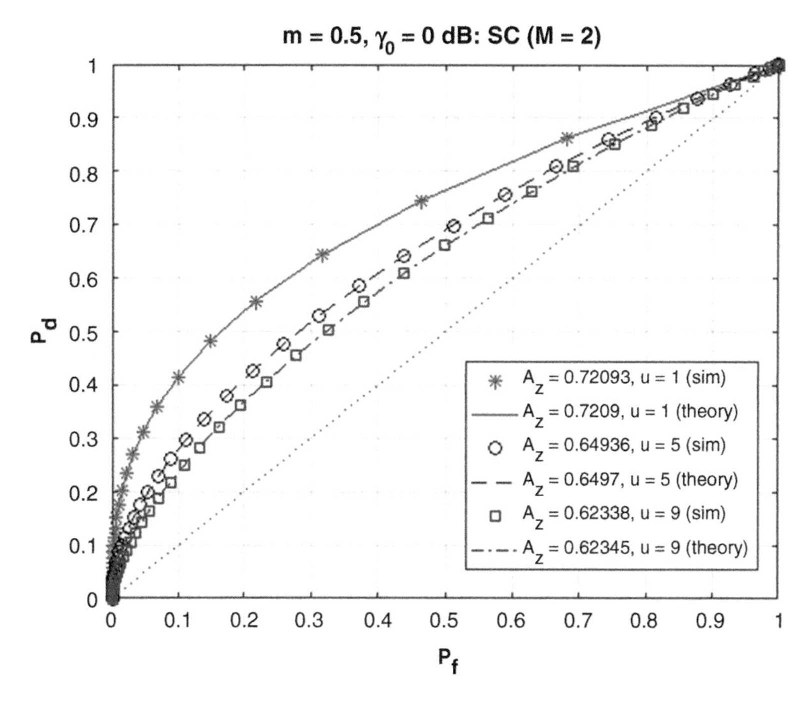

Fig. 7.63 AUC in a Nakagami channel: m = 0.5, $\gamma_0 = 0$ dB: SC (M = 2)

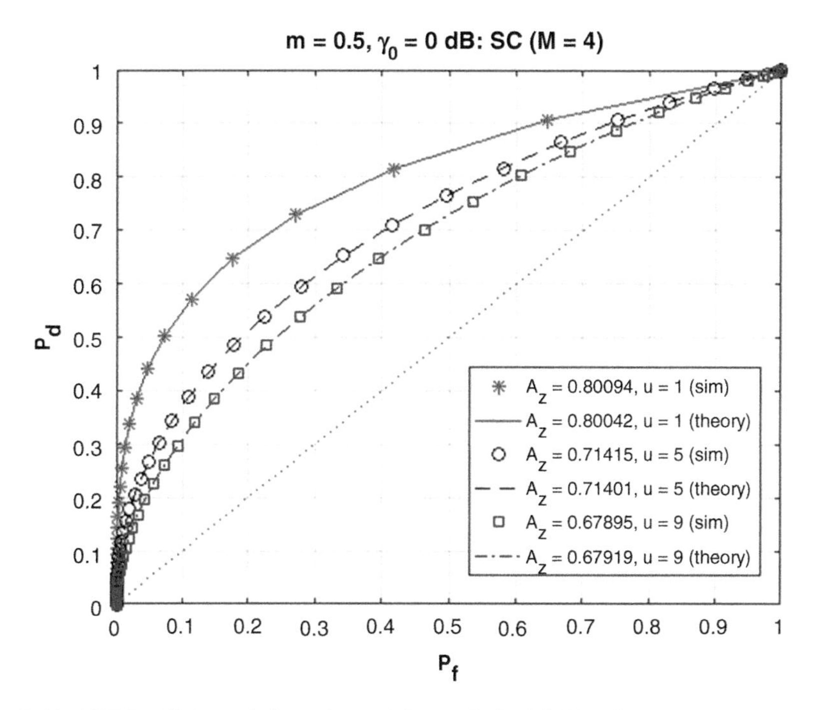

Fig. 7.64 AUC in a Nakagami channel: m = 0.5, $\gamma_0 = 0$ dB: SC (M = 4)

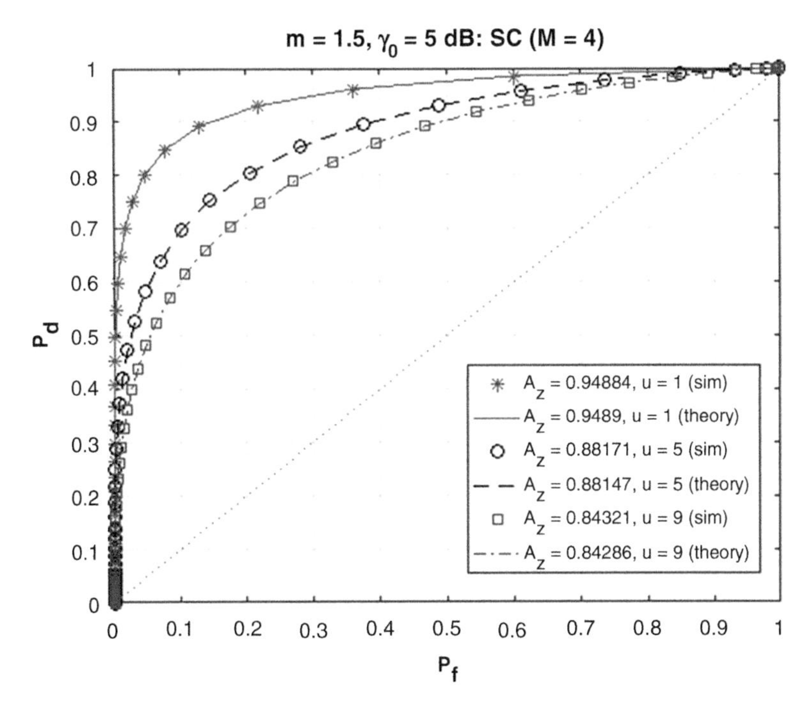

Fig. 7.65 AUC in a Nakagami channel: m = 1.5, $\gamma_0 = 5$ dB: SC (M = 4)

7.4.5 Nakagami Channel (Square Law Combining Diversity)

The next simulation undertaken pertains to the square law combining diversity. The Matlab script appears below.

```
function cognitive_radio_SIM_Nakagami_fading_sqlawdiv
% Computer simulation of the ROC in cognitive radio operating in a Nakagami
% channel. ROC for an ideal channel and Nakagami channel obtained using
% theory and random number simulation. Instead of MarcumQ function, the
% non-central chi square pdf and CDF available in the Statistis and Machine
% Learning Tool box is used. Instead of the MarcumQ function, theretical
% results are also obtained using the summation (Pd expressed as a series.
% In this simulation, SQUARE LAW COMBINING implemnted
% P M Shankar September 2016
close all
Numb=1e6;
global m lamb g u M
uu=[1,5,11];
% create labels for the legend
uu1=['u = ',num2str(uu(1))];uu2=['u = ',num2str(uu(2))];
uu3=['u = ',num2str(uu(3))];
L=100; % number of steps for PF
LT=L+1; % actual number of threshold values will be (L+1)
LU=length(uu);
clear Z
```

```
% simulation of the Nakagami channel MRC
MM=[1,2,4];% order of diversity
mm=[0.75,1,1.5];
zz=[0,3,5];% average SNR dB
for k1=1:3
    M=MM(k1);
for mk=1:3
  m=mm(mk);
  g=10^(zz(mk)/10);
% place holders
PFms=zeros(LU,LT); PDms=zeros(LU,LT); % simulation
PFmt=zeros(LU,LT);
PDmt1=zeros(LU,LT); % theory
PDmt2=zeros(LU,LT); % theory
Azs=zeros(1,LT); Azt1=zeros(1,LT);Azt2=zeros(1,LT);
for kk=1:3
    u=uu(kk);
    x1=chi2rnd(2*u*M,1,Numb); % hypothesis H0
    %hypothesis H1: non-centrality parameter a sample of a gamma variable
% generate a non-central chi square sample set with a non-centrality
% parameter = 2* gamma random variable of order m and mean g and it needs
% to match the size
% generate an MRC sample
if M==1 % summation not needed
    x2=ncx2rnd(2*u*M,2*gamrnd(m,g/m,1,Numb),1,Numb); % H1
else
x2=ncx2rnd(2*u*M,2*sum(gamrnd(m,g/m,M,Numb)),1,Numb); % H1
end;
    xx=[x1,x2];%to determine the maximum and minimum of the set for threshold
    min1=min(xx);max1=max(xx);
    step1=(max1-min1)/L;
    thr=min1:step1:max1; % these will be L+1 steps
    KN=40;
    for k=1:LT;
        lamb=thr(k); % lambda
        PFms(kk,k)=sum(x1>lamb)/Numb; % sim
        PDms(kk,k)=sum(x2>lamb)/Numb; % sim
       PDmt1(kk,k)=integral(@myfun,0,inf);% theory
       PDmt2(kk,k)= probdetnakagami(KN); % theory; summation
        PFmt(kk,k)=1-chi2cdf(lamb,2*u*M); % theory
    end;
    Azs(kk)=0.5+polyarea([0,PFms(kk,:),1],[0,PDms(kk,:),1]);
    Azt1(kk)=0.5+polyarea([0,PFmt(kk,:),1],[0,PDmt1(kk,:),1]);
       Azt2(kk)=0.5+polyarea([0,PFmt(kk,:),1],[0,PDmt2(kk,:),1]);
end;

figure,plot(PFms(1,:),PDms(1,:),'r-',...
    PFmt(1,:),PDmt1(1,:),'r*',...
    PFmt(1,:),PDmt2(1,:),'r^',...
    PFms(2,:),PDms(2,:),'k--',...
    PFmt(2,:),PDmt1(2,:),'ko',...
    PFmt(2,:),PDmt2(2,:),'kd',...
    PFms(3,:),PDms(3,:),'b-.',...
    PFmt(3,:),PDmt1(3,:),'bs',...
    PFmt(3,:),PDmt2(3,:),'b>')
legend(['A_z = ',num2str(Azs(1)),', ', uu1,' (sim)'],...
    ['A_z = ',num2str(Azt1(1)),', ', uu1,' (theory)'],...
      ['A_z = ',num2str(Azt2(1)),', ', uu1,' (theory-sum)'],...
    ['A_z = ',num2str(Azs(2)),', ', uu2,' (sim)'],...
    ['A_z = ',num2str(Azt1(2)),', ', uu2,' (theory)'],...
     ['A_z = ',num2str(Azt2(2)),', ', uu2,' (theory-sum)'],...
  ['A_z = ',num2str(Azs(3)),', ', uu3,' (sim)'],...
    ['A_z = ',num2str(Azt1(3)),', ', uu3,' (theory)'],...
    ['A_z = ',num2str(Azt2(3)),', ', uu3,' (theory-sum)'],...
    'location','best')
```

```
xlabel('P_f'),ylabel('P_d'),grid on
if M==1
title(['m = ',num2str(m),', \gamma_0 = ',...
    num2str(ZZ(mk)),' dB: No Diversity'])
else
    title(['m = ',num2str(m),', \gamma_0 = ',...
    num2str(ZZ(mk)),' dB:  SLC (M = ', num2str(M),')'])
end;
hold on
plot([0,1],[0,1],':m','linewidth',1.2)
clear x1 x2 xx
end;

end;
end


function Fg=myfun(x)  % SLC: integral function
global m lamb g u M
pd=1-ncx2cdf(lamb,2*u*M,2*x);%1- non-central chiSquare CDF instead of MarcumQ
Fg=gampdf(x,M*m,g/m).*pd;
end

function Pdet=probdetnakagami(KN) %Pd sum in a Nakagami channel (SLC)
global m  u lamb g M
Mm=M*m;
pdd=0;
for kks=1:KN
    ks=kks-1;
    pd1=gammainc(lamb/2,u*M+ks,'upper')/factorial(ks);
    % Matlab description of gammainc(.) includes the division by
    % gamma(u+ks)
    pd2=gamma(ks+Mm)*(g^ks)/(g+m)^(ks+Mm);
    pdd=pdd+pd1*pd2;
end;
Pdet=pdd*(m^Mm)/gamma(Mm);
end
```

The results are displayed in Figs. 7.66, 7.67, and 7.68.

7.4.6 Performance Comparison

In the next simulation (no analytical results), performances of the cognitive radio in all diversity combining algorithms are compared. The Matlab script appears below. The results are displayed in Fig. 7.69, 7.70, and 7.71 for three different values of M.

The performance of cognitive radio under MRC, SC, and SLC ($M = 2$) is better than the performance in an ideal Gaussian channel. This requires additional explanation. For the case of MRC and SC, it should be noted that the probabilities of false alarm matched those of the ideal channel while probabilities of detection improved substantially. The case of SLC can be explained in terms of the higher values of the order of the chi-square distribution and SNR. The probability of false alarm goes up with M as seen in Eq. (7.55) while the probability of detection goes up faster since the average SNR goes up by M resulting in improved performance associated with SLC diversity. To verify these conclusions, an extra simulation was undertaken for the ideal Gaussian channel with an SNR $= M\gamma_0$ matching the SNR in MRC

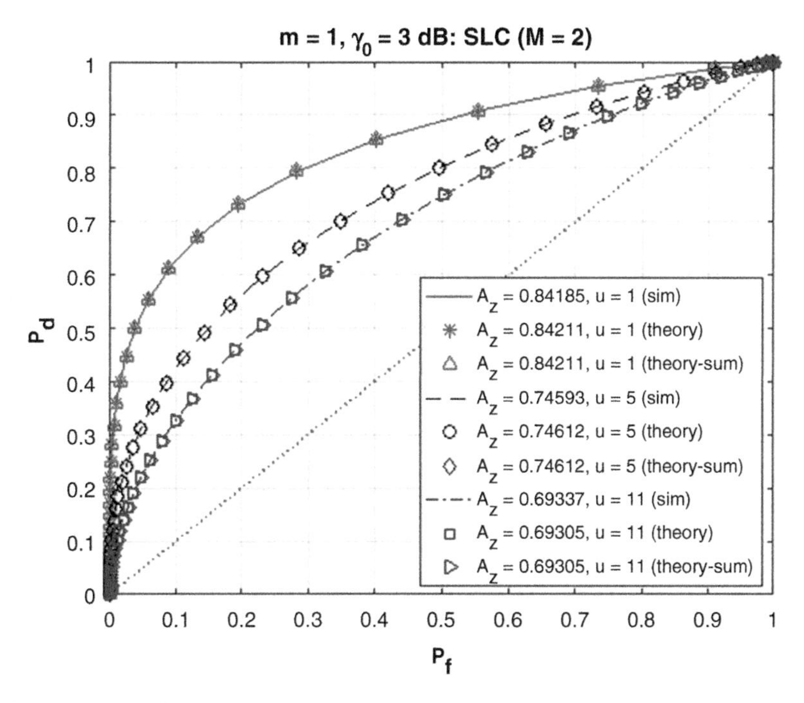

Fig. 7.66 AUC in a Nakagami channel: m = 1, $\gamma_0 = 3$ dB: SLC (M = 2)

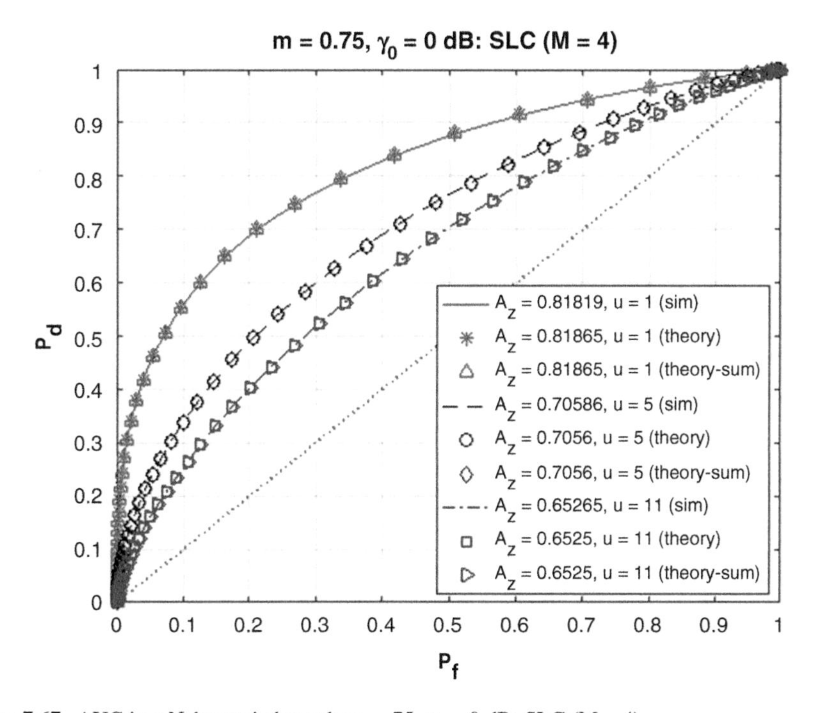

Fig. 7.67 AUC in a Nakagami channel: m = .75, $\gamma_0 = 0$ dB: SLC (M = 4)

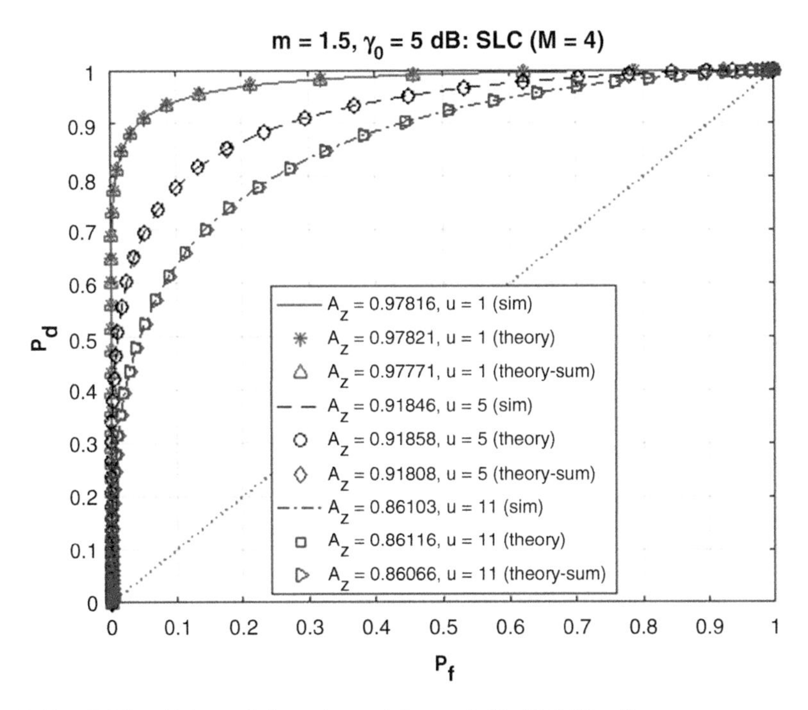

Fig. 7.68 AUC in a Nakagami channel: m = 1.5, $\gamma_0 = 5$ dB: SLC (M = 4)

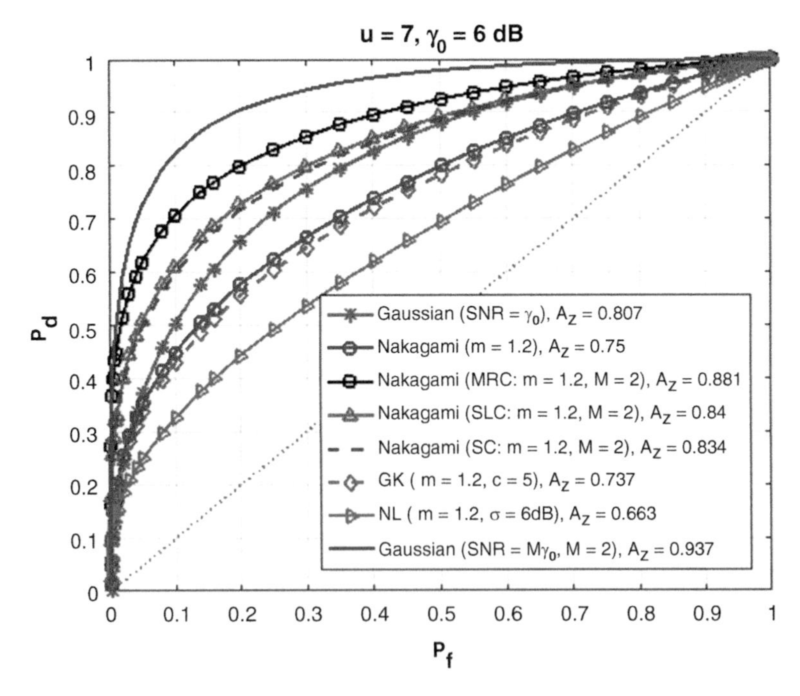

Fig. 7.69 Comparison of AUC (all forms of diversity): u = 7, $\gamma_0 = 6$ dB (different channels/ diversity)

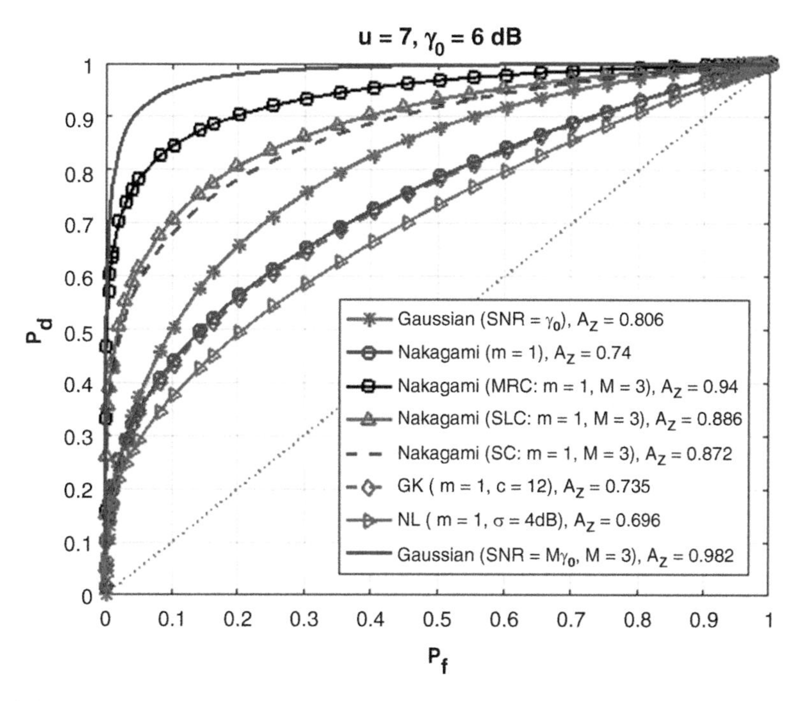

Fig. 7.70 Comparison of AUC (all forms of diversity): u = 7, $\gamma_0 = 6$ dB (different channels/diversity)

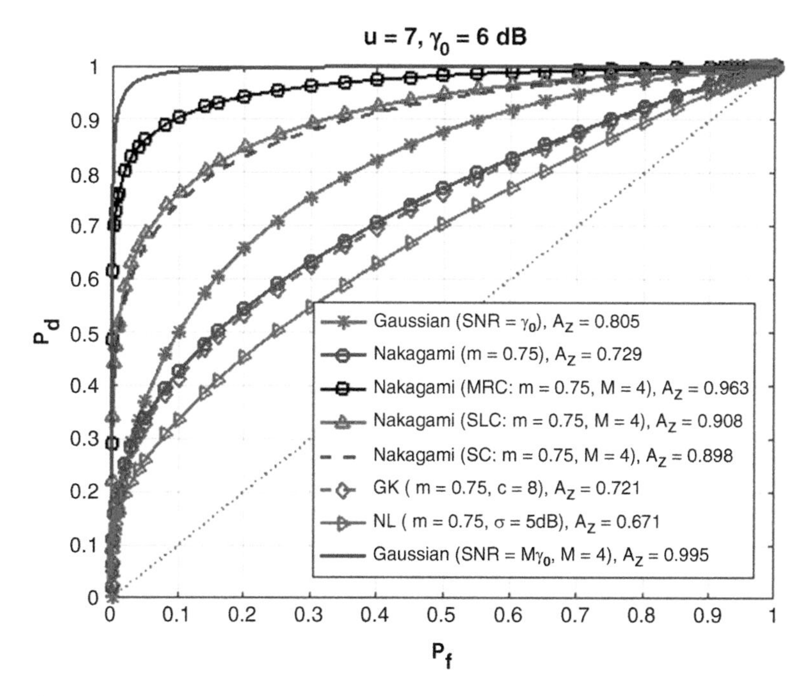

Fig. 7.71 Comparison of AUC (all forms of diversity): u = 7, $\gamma_0 = 6$ dB (different channels/diversity)

diversity. The results clearly show that the ideal channel performs the best when the SNR is scaled to match the SNR in MRC.

```
function CR_allchannels_simulationONLY
% Cognitive radio simulation: ideal channel, Nakagami, Nakagami (MRC),
% Nakagami (SC), GK, shadowed fading                                 his
% notice that PF remains the same after diversity: This is due to the fact
% that there is only noise in the channel regardless of diversity. T
% leads to PF being same even when diversity is present.
% with MRC, the average SNR goes up by M and this leads to better
% performance.
% square law combining requires PF separately
% P M Shankar SEptember 2016
close all                                                              ...
Numb=1e5; % number of samples                                        ...
PF=[1e-6 1e-4 0.001 0.004 0.006 .009 .01 0.02 0.03 0.04 0.05 0.08  ...
     .1 0.14 0.16 .2 0.25 .3 0.35 .4 0.45 .5 0.55 .6 0.65 .7
     .75 .8 .85 .9 .92 0.93 .94 0.95 0.96 0.965 0.968 0.97 0.975
    0.978 0.98 0.985 .99 0.995 0.99999 0.999999]; % values of PF
LL=length(PF);
for kp=1:3
    if kp==1
        m=1.2;% Nakagami parameter
         c=5;% for GK channel
        sigm=6; % shadowing level (dB)
        M=2;
    elseif kp==2
         m=1;% Nakagami parameter
         c=12;% for GK channel
        sigm=4; % shadowing level (dB)
        M=3;

    else
          m=0.75;% Nakagami parameter
         c=8;% for GK channel
        sigm=5; % shadowing level (dB)
        M=4;
    end;
u=7; % time bandwidth product
g0=6;% average SNR (dB)
g=10^(g0/10); % average SNR (absolute units)
%convert average SNR in dB in fading to mu (dB) in shadowing
mu=g0-sigm^2/(2*10/log(10));
thr=2*gammaincinv(PF,u,'upper'); % get the threshold values
thrL=2*gammaincinv(PF,u*M,'upper'); % get the threshold values for SLC
%
x1=chi2rnd(2*u,1,Numb);% H0
xL=chi2rnd(2*u*M,1,Numb);% H0 for SQLW combining
% now H1
y1=ncx2rnd(2*u,2*g,1,Numb); % no fading
y11=ncx2rnd(2*u,2*g*M,1,Numb); % no fading; SNR scaled to match SNR in MRC
y2=ncx2rnd(2*u,2*gamrnd(m,g/m,1,Numb),1,Numb); % Nakagami
y3=ncx2rnd(2*u,2*sum(gamrnd(m,g/m,M,Numb)),1,Numb); %Nakagami MRC
yL=ncx2rnd(2*u*M,2*sum(gamrnd(m,g/m,M,Numb)),1,Numb); %Nakagami SLC
y4=ncx2rnd(2*u,2*max(gamrnd(m,g/m,M,Numb)),1,Numb); %Nakagami SC
y5=ncx2rnd(2*u,2*gamrnd(m,1/m,1,Numb).*gamrnd(c,g/c,1,Numb),1,Numb);% GK
            g1=normrnd(mu,sigm,1,Numb);%generate Gaussian
            gg=10.^(g1/10);%lognormal random
            hh=gamrnd(m,1/m,1,Numb).*gg;% gamma-lognormal
y6=ncx2rnd(2*u,2*hh,1,Numb); %shadowed fading
% place holders for prob. false and detection
pFs=zeros(1,LL);pDy1=zeros(1,LL);pDy2=zeros(1,LL);pDy3=zeros(1,LL);
pDy4=zeros(1,LL);pDy5=zeros(1,LL);pDy6=zeros(1,LL);
pFsL=zeros(1,LL);pDyL=zeros(1,LL);pDy11=zeros(1,LL);
% get PF and PD by counting the numbers exceeding the threshold
 for k=1:LL;
```

```
pFs(k)=sum(x1>thr(k))/Numb;
pFsL(k)=sum(xL>thrL(k))/Numb;
pDy1(k)=sum(y1>thr(k))/Numb;
pDy11(k)=sum(y11>thr(k))/Numb;
pDy2(k)=sum(y2>thr(k))/Numb;
pDy3(k)=sum(y3>thr(k))/Numb;
pDy4(k)=sum(y4>thr(k))/Numb;
pDy5(k)=sum(y5>thr(k))/Numb;
pDy6(k)=sum(y6>thr(k))/Numb;
pDyL(k)=sum(yL>thrL(k))/Numb;
 end;
 AUC1=0.5+round(polyarea([0,pFs,1],[0,pDy1,1])*1000)/1000; % three digits
 AUC2=0.5+round(polyarea([0,pFs,1],[0,pDy2,1])*1000)/1000;
 AUC3=0.5+round(polyarea([0,pFs,1],[0,pDy3,1])*1000)/1000;
 AUC4=0.5+round(polyarea([0,pFs,1],[0,pDy4,1])*1000)/1000;
 AUC5=0.5+round(polyarea([0,pFs,1],[0,pDy5,1])*1000)/1000;
 AUC6=0.5+round(polyarea([0,pFs,1],[0,pDy6,1])*1000)/1000;
 AUC7=0.5+round(polyarea([0,pFsL,1],[0,pDyL,1])*1000)/1000;
 AUC11=0.5+round(polyarea([0,pFs,1],[0,pDy11,1])*1000)/1000; % three digits
figure
plot(pFs,pDy1,'-r*',pFs,pDy2,'b-o',pFs,pDy3,'k-s',pFsL,pDyL,'r-^',...
    pFs,pDy4,'b--',pFs,pDy5,'m--d',pFs,pDy6,'->',pFs,pDy11,'-b',...
    'linewidth',1.5)
xlabel('P_f'),ylabel('P_d'),grid on
xlim([0,1]),ylim([0,1]),hold on, plot([0,1],[0,1],':m','linewidth',1.2)
legend(['Gaussian (SNR = \gamma_0), A_z = ',num2str(AUC1)],...
    ['Nakagami (m = ',num2str(m),'), A_z = ',num2str(AUC2)],...
     ['Nakagami (MRC: m = ',num2str(m),...
     ', M = ',num2str(M),'), A_z = ',num2str(AUC3)],...
      ['Nakagami (SLC: m = ',num2str(m),...
     ', M = ',num2str(M),'), A_z = ',num2str(AUC7)],...
       ['Nakagami (SC: m = ',num2str(m),...
     ', M = ',num2str(M),'), A_z = ',num2str(AUC4)],...
        ['GK ( m = ',num2str(m),...
     ', c = ',num2str(c),'), A_z = ',num2str(AUC5)],...
      ['NL ( m = ',num2str(m),...
     ', \sigma = ',num2str(sigm),'dB), A_z = ',num2str(AUC6)],...
 ['Gaussian (SNR = M\gamma_0, M = ',num2str(M),') A_z = ',...
 num2str(AUC11)],'location','southeast')
  title([' u = ',num2str(u),', \gamma_0 = ',num2str(g0), ' dB'])
end;
end
```

7.5 Summary

In this chapter, the performance of cognitive radio has been studied. Quantitative measures such as the performance index, area under the ROC curve, partial AUC index have been used to compare the performance of energy detectors commonly used in cognitive radio. The adverse effects of fading and shadowing have been examined along with improvements obtained through diversity. Along with analytical and numerical methods for evaluating the performance, steps to simulate the cognitive radio in ideal and faded channels have been provided along with complete description of Matlab scripts for analytical evaluation and Monte Carlo simulation using random numbers.

References

Adebola, E., & Annamalai A (2015). On the performance of energy detectors in generalized fading environments. In *Proceedings Military Communications Conference, MILCOM-2015* (pp. 1536–1541).
Adebola, E., Olaluwe, A., & Annamalai, A. (2014). Partial area under the receiver operating characteristics curves of diversity-enabled energy detectors in generalised fading channels. *IET Communications, 8*, 1637–1647.
Alam, S., Olabiyi, O., Odejide, O., & Annamalai, A. (2012). Simplified performance analysis of energy detectors over myriad fading channels: Area under the ROC curve approach. *International Journal of Wireless and Mobile Networks, 4*, 33–52.
Alhennawi, H. R., Ismail, M. H., & Mourad, H. A. (2014). Performance evaluation of energy detection over extended generalized-K composite fading channels. *Electronics Letters, 22*, 1643–1645.
Altrad, O., & Muhaidat, S. (2013). A new mathematical analysis of the probability of detection in cognitive radio over fading channels. *EURASIP Journal on Wireless Communications and Networking, 1*, 1–11.
Atapattu, S., Tellambura, C., & Jiang, H. (2010). Analysis of area under the ROC curve of energy detection. *IEEE Transactions on Wireless Communications, 9*, 1216–1225.
Atapattu, S., Tellambura, C., & Jiang, H. (2011). MGF based analysis of area under the ROC curve in energy detection. *IEEE Communications Letters, 15*, 1301–1303.
Banjade, V. R. S., Rajatheva, N., & Tellambura, C. (2012). Performance analysis of energy detection with multiple correlated antenna cognitive radio in Nakagami-m fading. *IEEE Communications Letters, 16*, 502–505.
Digham, F., Alouini, M.-S., & Simon, M. K. (2007). On the energy detection of unknown signals over fading channels. *IEEE Transactions on Communications, 55*, 21–24.
Digham, F., Alouini, M.S., & Simon, M.K. (2003). On the energy detection of unknown signals over fading channels. In *IEEE International Conference on Communications, ICC'03* (Vol. 5, pp. 3575–3579).
Dodd, L. E., & Pepe, M. S. (2003). Partial AUC estimation and regression. *Biometrics, 59*, 614–623.
Evans, A. N., & Nixon, M. S. (1995). Mode filtering to reduce ultrasound speckle for feature extraction. *IEE Proceedings, Vision, Image and Signal Process, 142*, 87–94.
Evans, M., Hastings, N., & Peacock, B. (2000). *Statistical distributions*. New York, NY: Wiley.
Ghasemi, A., & Sousa, E. S. (2007). Opportunistic spectrum access in fading channels through collaborative sensing. *Journal of Communication, 2*, 71–82.
Gradshteyn, I., & Ryzhik, I. (2007). *Table of integrals, series and products*. San Diego, CA: Academic.
Hanley, J. A., & McNeil, B. (1982). The meaning and use of the area under a receiver operating characteristic (ROC) curve. *Radiology, 143*, 29–36.
Haykin, S., Thomson, D., & Reed, J. (2009). Spectrum sensing for cognitive radio. *Proceedings of the IEEE, 97*, 849–877.
Herath, S.P., Rajatheva, N., & Tellambura, C. (2009). On the energy detection of unknown deterministic signal over Nakagami channels with selection combining. In *Canadian Conference on Electrical and Computer Engineering, IEEE, CCECE'09* (pp. 745–749).
Herath, S. P., Rajatheva, N., & Tellambura, C. (2011). Energy detection of unknown signals in fading and diversity reception. *IEEE Transactions on Communications, 59*, 2443–2453.
Hossain, M. S., Abdullah, M. I., & Hossain, M. A. (2012). Energy detection performance of spectrum sensing in cognitive radio. *International Journal of Information Technology and Computer Science, 4*, 11–17.
Ilhan, H. (2014). Analysis of energy detection over cascaded Nakagami-m fading channels. In *International Conference on Multidisciplinary Trends in Academic Research*, Bangkok, Thailand.

Kakkar, D., Khosla, A., & Uddin, M. (2014). Sensing performance evaluation over multihop system with composite fading channel. *International Journal of Advances in Science and Technology, 62*, 19–30.

Karagiannidis, G. K., Sagias, N. C., & Mathiopoulos, P. T. (2007). N*Nakagami: A novel stochastic model for cascaded fading channels. *IEEE Transactions on Communications, 55*, 1453–1458.

Li, C-R, Liao, C-T & Liu, J. P. (2008). A noninferiority test for diagnostic accuracy based on the paired partial areas under the ROC curves. Statistics in Medicine 27: 1762–1776

Li, G., Fang, J., Tan, H., & Li, J. (2010). The impact of time-bandwidth product on the energy detection in the cognitive radio. In *Proceedings of the 2010 I.E. International Conference on In Broadband Network and Multimedia Technology (IC-BNMT)* (pp. 634–638).

McClish, D. K. (1989). Analyzing a portion of the ROC curve. *Medical Decision Making, 9*, 190–195.

Metz, C. E. (1978). Basic principles of ROC analysis. *Seminars in Nuclear Medicine, 8*, 283–298.

Obuchowski, N. A. (2003). Receiver operating characteristic curves and their use in radiology. *Radiology, 229*, 3–8.

Olabiyi, O. & Annamalai, A. (2012). New series representations for generalized Nuttall Q-function with applications. In *Proceedings of the IEEE Consumer Communications and Networking Conference (CCNC-2012)* (pp. 782–786).

Papoulis, A., & Pillai, S. U. (2002). *Probability, random variables and stochastic processes* (4th ed.). Boston, U S A: McGraw-Hill.

Patterson, M. S., & Foster, F. S. (1983). The improvement and quantitative assessment of B-mode image produced an annular array/cone hybrid. *Ultrasonic Imaging, 5*, 195–213.

Shankar, P. M. (2013). Diversity in cascaded N*Nakagami channels. *Annals of Telecommunications, 5*, 320–326.

Shankar, P. M. (2016). Performance of cognitive radio in N*Nakagami cascaded channels. *Wireless Personal Communications, 88*(3), 657–667.

Shankar, P. M. (2017). A pedagogical perspective of cognitive radio in fading channels. Under review.

Simon, M. K., & Alouini, M.-S. (2005). *Digital communication over fading channels.* Wiley-Interscience: Hoboken, N.J.

Sithamparanathan, K., & Giorgetti, A. (2012). *Cognitive radio techniques: Spectrum sensing, interference mitigation, and localization.* Boston, MA: Artech House.

Sun, H., Nallanathan, A., Jiang, J. & Wang, C.X. (2011). Cooperative spectrum sensing with diversity reception in cognitive radios. In *International ICST Conference on in Communications and Networking in China (CHINACOM-2011)* (pp. 216–220).

Umar, R., Sheikh, A. U., & Deriche, M. (2014). Unveiling the hidden assumptions of energy detector based spectrum sensing for cognitive radios. *IEEE Communication Surveys and Tutorials, 16*, 713–728.

Urkowitz, H. (1967). Energy detection of unknown deterministic signals. *Proceedings of the IEEE, 55*, 523–531.

van Trees, H. (1968) Detection, Estimation, and Modulation Theory, Part 1, Wiley, NY.

Yu, G., Long, C., Xiang, M., & Xi, W. (2012). A novel energy detection scheme based on dynamic threshold in cognitive radio systems. *The Journal of Computer Information Systems, 8*, 2245–2252.

Index

© Springer International Publishing AG 2017
P.M. Shankar, *Fading and Shadowing in Wireless Systems*,
DOI 10.1007/978-3-319-53198-4

FSC
www.fsc.org
MIX
Papier aus verantwortungsvollen Quellen
Paper from responsible sources
FSC® C141904

Druck:
Customized Business Services GmbH
im Auftrag der KNV-Gruppe
Ferdinand-Jühlke-Str. 7
99095 Erfurt